Exploring Genetic Mechanisms

Exploring Genetic Mechanisms

Edited by

MAXINE SINGER

PRESIDENT, CARNEGIE INSTITUTION OF WASHINGTON

SCIENTIST EMERITUS, NATIONAL INSTITUTES OF HEALTH

PAUL BERG

ROBERT W. AND VIVIAN K. CAHILL PROFESSOR
IN CANCER RESEARCH AND BIOCHEMISTRY

DIRECTOR, BECKMAN CENTER
FOR MOLECULAR AND GENETIC MEDICINE
STANFORD UNIVERSITY MEDICAL CENTER

University Science Books
Sausalito, California

University Science Books
55D Gate Five Road
Sausalito, CA 94965
Fax: (415) 332-5393

Production manager: *Eugene R. Bailey*
Copy editor: *Sylvia Stein Wright*
Designer: *Robert Ishi*
Illustrators: *Georg Klatt, Audre Newman, John and Judy Waller*
Compositor: *Compset Inc.*
Printer & Binder: *Edwards Brothers*

This book is printed on acid-free paper.

Library of Congress Cataloging-in-Publication Data

Exploring genetic mechanisms / Maxine Singer and Paul Berg.
p. cm.
Includes bibliographical references and index.
ISBN 0-935702-70-9 (cloth)
1. Gene expression. 2. Genetic regulation. 3. Genetic engineering. I. Singer, Maxine. II. Berg, Paul, 1926-
[DNLM: 1. Genetics, Biochemical. 2. Genetic Engineering. QH 430 E96 1997]
QH450.E948 1997
575.1—dc20
DNLM/DLC
for Library of Congress 96-28916
CIP

Printed in the United States of America
10 9 8 7 6 5 4 3 2 1

To the mentors who inspired our devotion to research:
Harland G. Wood, Arthur Kornberg,
Joseph S. Fruton, and Leon A. Heppel.

Abbreviated Contents

List of Contributors

Numbers in parentheses indicate the pages on which the authors' contributions begin.

Frederick W. Alt (337), Childrens Hospital, Enders 861, 300 Longwood Ave., Boston, MA 02115

Stylianos Antonarakis (301), University of Geneva Medical School, 9 avenue de Champel, Geneva, Switzerland.

Paul Berg (1), (405), Beckman Center, B-062, Stanford University Medical Center, Stanford, CA 94305

Jianzhu Chen (337), Center for Cancer Research, Massachusetts Institute of Technology, E17-128C, 77 Massachusetts Ave., Cambridge, MA 02139

Stephen P. Goff (133), Higgins Professor of Biochemistry, Howard Hughes Medical Institute, Columbia University College of Physicians and Surgeons, 701 W. 168th St., New York, NY 10032.

Tony Hunter (203), Molecular Biology and Virology Laboratory, The Salk Institute, 10010 N. Torrey Pines Rd., La Jolla, CA 92037

Haig H. Kazazian, Jr. (301), Chair, Department of Genetics, University of Pennsylvania, School of Medicine, 475 CRB, 415 Curie Blvd., Philadelphia, PA 19104-6145.

Cynthia N. Kiser (515), Department of Chemistry, California Institute of Technology, Pasadena, CA 91125

Thomas B. Kornberg (459), Department of Biochemistry, University of California at San Francisco, Box 0054, San Francisco, CA 94143

John H. Richards (515), Department of Chemistry, California Institute of Technology, Pasadena, CA 91125

Matthew P. Scott (459), Howard Hughes Medical Institute, Department of Developmental Biology, Stanford University, Stanford, CA 94305.

Maxine Singer (1), (405), Carnegie Institution of Washington, 1530 P St., NW, Washington, DC 20005

Tim Stewart (565), Genentech, 490 Point San Bruno, S. San Francisco, CA 94080

Ellen G. Strauss (71), Division of Biology, California Institute of Technology, Pasadena, CA 91125-0001

James H. Strauss (71), Division of Biology, California Institute of Technology, Pasadena, CA 91125-0001

Ray White (271), Huntsman Cancer Institute, University of Utah, Bldg. 533 Suite 7410, Salt Lake City, UT 84112

Patricia Zambryski (597), Dept. of Molecular Plant Biology, Hilgard Hall, University of California at Berkeley, Berkeley, CA 94720

Contents

CHAPTER 3

Retroviruses: Infectious Genetic Elements 133

STEPHEN P. GOFF

CHAPTER 4

Oncogenes, Growth Suppressor Genes, and Cancer 203

TONY HUNTER

CHAPTER 5

Mapping Markers and Genes in the Human Genome 271

RAY WHITE

CHAPTER 6

Molecular Genetics of the Hemoglobin Genes 301

HAIG H. KAZAZIAN, JR., and STYLIANOS ANTONARAKIS

CHAPTER 7

Generation of Antigen Receptor Diversity 337

JIANZHU CHEN and FREDERICK W. ALT

CHAPTER 8

Biosynthesis of Intercellular Messenger Peptides 405

MAXINE SINGER and PAUL BERG

CHAPTER 9

Regulation of *Drosophila* Development by Transcription Factors and Cell-Cell Signaling 459

THOMAS B. KORNBERG and MATTHEW P. SCOTT

CHAPTER 10

Manipulating Protein Structure 515

CYNTHIA N. KISER and JOHN H. RICHARDS

CHAPTER 11

Genetic Modification of Animals 565

TIM STEWART

CHAPTER 12

Genetic Manipulation of Plants 597

PATRICIA ZAMBRYSKI

Preface

Our earlier textbook, *Genes and Genomes*, used well-studied examples to show how modern experimental techniques provide rich molecular detail about genetics. That book and other textbooks with similar intentions stress the unity of genetic processes among different organisms through a broad overview of the structure, organization, and expression of genes and genomes. However, such overviews tend to mask the enormous variety of biological mechanisms used to convert genotypes into successful organisms. Molecular analysis has only begun to reveal the varied genetic tactics that account for the diversity of organismal form, habitat, behavior, and function. In fact, the number of genetic strategies living organisms employ seems to be limitless. The principal generalization is that anything that can work is likely to occur somewhere in nature. This book aims to introduce to students and scientists how such complexity is being analyzed.

Classical biochemistry and genetics epitomize what is called the **reductionist approach** to biology; single biological elements (e.g., a protein, an enzyme, or a gene) are examined in great detail through a series of narrowly defined questions and experimental protocols. This approach can illuminate many phenotypic characteristics of complex systems, including those of whole organisms, if those phenotypes are the direct consequence of a single gene's function. An example is the common red-green color blindness phenotype of human males. Thus, normal males possess a single gene for the red-sensitive pigment and varying numbers (one to three) of genes for the green-sensitive pigment, arrayed in tandem on the long arm of the X chromosome. After cloning the human genes encoding the protein moieties of the pigments for red and green perception, scientists established that aberrant red-green color perception is often associated with mutations caused by recombination such as unequal crossing-over or gene conversion among these highly homologous and closely spaced genes.

Unlike color blindness, the vast majority of organisms' phenotypic characteristics arises from complex interactions between the products of many genes and the environment. To understand these phenotypes, we must study the physiological properties of cells, organelles, whole tissues, and populations of organisms and their interactions with their environments. Now it is clear that the expanding resources of molecular genetics are applicable to the analysis of such complex multigenic systems. Thus, like single gene phenotypes, complex phenotypes are being described in molecular terms, and mastering the intricacies of isolated genetic elements must be viewed as a first step in solving the higher order complexities of living organisms. We can anticipate a continuum of biological understanding starting with the fundamental properties of genes and genomes and extending to the complex, hierarchical interactions fundamental to living things. The tasks will be challenging because many different organisms and diverse kinds of systems and levels of organization will need to be considered. In time, the debate about the relative merits of the holistic and reductionist approaches to understanding complex biological phenomena will become irrelevant.

The subjects covered in this book are not intended to give a comprehensive picture of the many ways molecular genetics is being applied to the analysis of complex systems. The choice of what to include was challenging, and, regrettably, important topics have been omitted. Thus, we elected to emphasize advances that reveal fundamental features of complex genetic systems and to omit developments that are more related to commercial and medical applications—for example, the discovery and production of therapeutic proteins, the emergence of novel diagnostic strategies, and the fledgling gene therapies. Nevertheless, it is evident even from the coverage in these twelve chapters that many long recognized biological mysteries are giving way to experimental analysis.

For example, the life cycles of viruses (Chapters 1, 2, and 3) provide a relatively simple model for how the regulation of gene expression can establish an orderly progression in time of events that lead to multiplication: a useful model for the developmental processes of higher organisms. Indeed, the goal of understanding the relation between a complete genome and the consequent phenotype has, to date, been most closely approached with certain viruses. The viral systems also illustrate the way alternative outcomes can result from a single set of viral genes, depending on the cellular environment. Moreover, studies of the events following infection of mammalian cells by DNA viruses have revealed important aspects of cell cycle control and of the unpredictable role of recombination in cellular events. Thus, for these viruses to replicate their genomes, they must perturb the systems that regulate the passage of cells from the quiescent G1 phase into S phase. And chance recombinations that incorporate viral genes into the infected cell's genome can result in a heritable alteration of the cell's phenotype—often an oncogenic transformation.

Viruses whose genomes consist of RNA use distinctive mechanisms to replicate their genomes. In this instance, RNA acts as the template to direct the assembly of progeny RNA genomes. Furthermore, although these virus's own genetic program accounts for the production of their mRNAs, their expression depends on the host cell's machinery for translating them into proteins. As will be evident in Chapter 2, the analysis of their varied and complex life cycles was greatly facilitated and expanded by the ability to clone and analyze their genes and genomes as DNA.

Retroviruses reveal still another solution for virus multiplication: the conversion of a single-strand RNA genome into a double-strand DNA and its multiplication as an integral part of the infected cell's own chromosomal complement. Here, too, the vagaries of recombination between virus and cell genomic segments led to the recognition of protooncogenes—cellular genes whose incorporation into viral genomes renders them oncogenic. Many known protooncogenes were first identified as components of retroviral genomes. There has also been a direct, if unpredictable, track from fundamental research on retroviruses that cause tumors in birds and nonhuman mammals to the urgent clinical problems associated with the acquired immune deficiency syndrome, AIDS. AIDS is caused by a retrovirus called HIV-1 that brings about a devastating destruction of the immune response. Because of the prior intensive study of vertebrate retroviruses and the availability of recombinant DNA techniques, the structure of HIV and its biological properties were rapidly elucidated. This permitted the cloning of viral genes, their expression in *E. coli*, and diagnostic tests for the virus that culminated in the exclusion of contaminated blood from use in transfusions, an important element in controlling the disease's spread. Another important consequence of the study of retroviruses has been their adaptation as recombinant vectors. These vectors have proven to be particularly useful for introducing new genes into early mammalian embryos and thus into essentially all the cells, including germ line cells, of such experimental animals.

The discovery of retroviral oncogenes and the recognition that they are altered forms of normal cellular genes radically altered the prevailing paradigm of cancer research (Chapter 4). Soon after such protooncogenes were recognized, it was clear that mutations in these cellular genes can create an oncogenic phenotype. More than 100 protooncogenes and their oncogenic counterparts have now been identified and their normal and aberrant biological activities in regulating cell growth and multiplication studied. More recently, our understanding of cancer has broadened to include tumor suppressor genes, whose activities relate to the mechanisms retarding cell multiplication.

Underlying the fruitful analysis of viral genetic systems are correlations between physical and genetic maps, detailed information about genome structure, and an interplay between structural, genetic, and biochemical data. Similar information is critical for the eventual undertanding of more complex genetic systems such as those of mammals. The development of techniques for mapping very large, complex genomes like the human genome is a great challenge (Chapter 5). But the early problems stemming from both the large genomic size and the paucity of genetic markers can now be circumvented. Consequently, the human genetic map is no longer sparsely populated, and the new methods are being applied to other organisms. For example, maize chromosomes are being mapped in this way, an endeavor of importance to both biology and agriculture.

A fertilized egg yields, through successive cell divisions, hundreds of kinds of cells and many highly differentiated tissues. The manifold shapes and functions of differentiated cells are phenotypes caused by the selective expression of specific genes in particular cell types or tissues at precise and limited times during development. Other genes appear to be expressed in most if not all cell types. The positional and temporal regulation of cell- and tissue-specific gene expression depends primarily on controlling the initiation of transcription. Moreover, differential gene expression in specific cells and tissues involves not just single genes, but sets of genes whose expression is turned on and off in a coordinated manner. One approach to understanding differential gene expression during development is to analyze already differentiated systems, for example, the expression of different α- and β-globin genes in vertebrate red blood cell precursors at different developmental stages (Chapter 6). Similarly, the ontogeny of lymphocytes is closely correlated with specific gene rearrangements at different stages of B and T cell development, respectively (Chapter 7). Such rearrangements have three functional consequences. They lead to the formation of complete immunoglobulin and T cell receptor genes by joining DNA segments that are not contiguous in the germ line genome. They allow efficient transcription initiation by bringing promoter and enhancer into proximity. They generate diversity in the segments encoding immune protein variable regions and thereby provide for an enormous repertoire of antigenic responses.

The flow of gene products is also regulated by posttranslational events that modify nascent polypeptides to ensure that they are localized to specific intracellular sites or are secreted into the extracellular medium. Such modifications are typical of the myriad polypeptides that serve as intercellular messengers in multicellular organisms (Chapter 8). Typically, the genes that encode the various intercellular peptide messengers encode polypeptides that are several hundred amino acids long. The messenger peptide itself, the product of posttranslational processing, may be as small as a pentapeptide, as are, for example, the two enkephalins that act as natural opiates in the brain and regulate peristaltic contractions in the gut. Typical of all proteins destined to be secreted, primary translation products of these genes are modified by a series of steps as they pass through the complex series of cytoplasmic membrane systems—the endoplasmic reticulum and the Golgi complex.

The earliest practical goal of recombinant DNA research was to produce medically and economically important proteins such as vaccines and intercellular messenger peptides (e.g., insulin, growth hormone, and oxytocin). This proved to be more difficult than expected, but the initial problems have been overcome, and important polypeptides are now readily produced by recombinant DNA methods. Quite apart from its application to the synthesis of therapeutic polypeptides, the ability to synthesize substantial quantities of specific polypeptides in bacteria has impressive implications for the study of protein structure and function (Chapter 10). The three-dimensional structure and consequently the biological activity of each protein depend largely on its unique amino acid sequence. Chemical studies have shown that modification of individual amino acid side chains or groups of side chains strikingly alters a protein's ability to renature and form a fully active secondary, tertiary, or quaternary structure. Analysis of proteins produced by genes containing mutations in coding regions tells the same story. However, rather than depending on random mutations, the amino acid sequence of a protein can now be systematically changed by site-specific mutagenesis of its cloned gene or cDNA. Thus, the ramifications of the recombinant DNA technique have been extended to chemistry and foster an understanding of proteins at a previously impossible level of sophistication.

One phenomenon that has captured biologists' imaginations and attention for centuries is the development of a complex multicellular organism from a zygote, the process of differentiating primordial cells into the diverse cell types and tissues found in adult organisms. Molecular genetics has opened new windows on this marvelous process. Many interesting experiments are being performed on plants, invertebrates, and vertebrates, including mammals. One aim of a molecular approach to understanding determination and differentiation is to correlate spatial and temporal changes in embryonic cells with differential gene expression. This approach has been particularly successful with *Drosophila melanogaster*, largely because mutants with defects in morphological development are available (Chapter 9). Thus, genes regulating various stages of embryonic development in flies have been identified, cloned, and characterized. The picture that emerges is that specific transcription factors are deposited at defined locations in the egg during oogenesis and that, with the onset of embryogenesis, an increasingly diverse but topographically constrained set of transcription factors is produced. These in turn regulate different sets of genes that ultimately define the body plan of the mature fly. It is a continuous wonder that genes that control various stages of embryogenesis in the fly have their structural and functional counterparts in mammals.

Among mammals, mice are the most promising animals for experimental work (Chapter 11). Because extensive genetic information exists, many mouse mutations are known, and cells from early mouse developmental stages are accessible, scientists have developed transgenic mice—mice into which a gene is introduced or ablated in the germ line. For some types of experiments, the transgene is allowed to integrate at random sites in the recipient germ line genome. But for many purposes, it is most useful or even imperative to introduce new genetic information at specific chromosomal sites or to modify existing genes in directed ways. This can readily be accomplished with yeast and *Drosophila*, but considerable progress along these lines has now been made with mammalian cells by designing special vectors that recombine with their homologous sequences in chromosomes. Using these methods and special techniques for preferentially recovering the appropriately modified cells and introducing them into early mouse embryos, scientists can produce mice with specific mutations in selected genes, eliminate mutations from defective genes, or eliminate certain genes entirely.

The modification of plant genotypes shares with the animal experiments the goal of understanding fundamental biological processes. Here, too, specialized means for modifying plant genomes have revolutionized the kind of experiments that are possible. This is illustrated in Chapter 12 by the level at which the genetic control of flowering is understood. This line of research is also motivated by a desire to improve the quality of agriculturally important crops. Thus, genes that increase crop resistance to herbicides and insect pests are a major research focus.

The concepts and techniques of molecular genetics will eventually be used to modify biological molecules and systems in many, presently unknown ways. Currently, although protein chemists utilize *in vitro* mutagenesis to produce altered pure proteins in bulk, biologists are more interested in studying the effect of altered protein structure, regulatory regions, or genomic architecture on the phenotypes of whole cells and organisms in order to understand normal and abnormal states. To achieve this aim, altered genes are introduced into cells and whole organisms. Being able to accomplish such alterations accounts for the most profound implications of the recombinant DNA techniques. Here biology changes from its traditional pursuit of understanding how living things are constructed and function to a constructive science that can make permanent, heritable changes in organisms. Moreover, here the reductionism inherent in molecular genetics is being transformed into studies of whole cells and organisms, thereby opening the way to analyze the pleiotropic effects of single genes on physiological, morphological, and developmental systems.

Naturally, the rapidly developing ability to modify multicellular organisms genetically has focused attention on the possibility of applying this technology to altering somatic cells of humans. The experimental approaches currently involve transforming cells *ex vivo* with a recombinant vector containing the gene of interest (or the corresponding cDNA) and reintroducing the cells into the individual from which they were obtained. This is the basic protocol presently being tested for therapeutic intervention in some human genetic diseases. Some experiments utilize bone marrow, which includes the stem cells that are the progenitors of circulating blood cells. In other approaches, liver cells are the targets of modification. The principal means for such somatic modifications are viruses, either appropriately modified retroviruses or adenoviruses, although other viruses are being considered. The premise is that viruses are the most efficient means for introducing functional mammalian genes of interest. We do not yet know if the clinical trials warrant hope that they can be used to treat debilitating human genetic diseases.

Somatic cell alteration of the previously outlined type does not introduce heritable changes into multicellular organisms. To do so requires modifying germ line cells. Experimental animals with specific germ line modifications provide extraordinary opportunities for studying the regulation of gene expression during development and differentiation. Among the questions that can be answered are the following. Is the gene expressed in all cells or only in the cell in which the normal gene would be expressed? What regulatory DNA segments must accompany the newly introduced gene if expression is to mimic that of the normal gene both temporally and as to cell type? Is the chromosomal position of the newly introduced gene important for its appropriate regulation during development and differentiation? Although germ line modification is a powerful experimental paradigm, there is considerable concern that such capabilities may be applied to humans, either for benevolent or malevolent purposes. But germ line modification as a means for gene therapy is problematic, uncertain, and remains to be justified.

Molecular genetics has made a beginning. A wealth of detail about many biological systems is already available. Still, the successes do not amount to a complete or even a very profound understanding. On the contrary, current ignorance is vaster than current knowledge. Nothing in the humanmade world rivals the complexity and diversity of living things. No information system we have devised approaches, in content, the amount of data encoded in genomes or, in complexity, the intricate regulatory networks that control gene expression. There remain to be discovered mechanisms and concepts that no one has yet even imagined. In some instances, we have learned enough at least to identify important areas of ignorance. Certain of these concern long-standing questions, like development and differentiation, or the molecular basis of mind. Others are new questions, raised by the very achievements of molecular genetics itself. And, of course, we should be wary; some things that we now think we know may become less clear or even prove wrong in the years to come.

The fast pace of modern biology is driven by technology; by major challenges in health, agriculture, and industry; and, ultimately, by curiosity. Because we and our offspring are the products of functional genetic systems, our intellectual curiosity is buttressed by personal involvement. This same combination of intellectual curiosity and personal relevance is responsible for the intense public interest in the recombinant DNA revolution and its product, genetic engineering. And it is why these achievements will identify twentieth-century biology as a great historical landmark. Thus far, the revolution has been a positive endeavor. Our understanding of ourselves and other organisms has deepened. Important products (such as hormones, vaccines, and enzymes for research and commerce) are being developed and produced. Harmless microorganisms are being manipulated to make them beneficial environmental agents. Still, there are disquieting aspects to the revolution. Care is required to assure that no unexpected detrimental properties accompany the beneficial characteristics introduced into altered microorganisms. Deep thought and cautious analysis are required if we are to avoid unwise use of sophisticated diagnostic techniques afforded by the new methods or of somatic cell gene therapy. The challenges are great, particularly because the course of future research is inherently unpredictable.

Almost nothing recorded in this book was known when we completed our formal educations about 40 years ago. Our chief regret is that we will not likely see the achievements 40 years hence. We can be certain of only one thing: from future research will emerge major new concepts, which will be as unexpected in their time as were introns and movable elements to the present generation of biologists. A changing perspective is the history of science, and molecular genetics will not be immune to that imperative.

Acknowledgments

Our original, and unrealistic, plan for *Genes and Genomes* (University Science Books, 1991) envisioned the inclusion of a series of "case studies," descriptions of particular biological systems, selected to demonstrate how the general principles of molecular genetics work in complex genetic systems. As we worked on *Genes and Genomes,* we realized that, besides the problem of space, the emerging information about specific genetic systems was of extraordinary breadth and depth and that presenting that material in a coherent fashion required the knowledge of experts in the various fields. Consequently, we elected to put the case studies in a separate volume and to recruit additional authors to contribute chapters for this book. We are enormously indebted to these active scientists for their willingness to undertake this task and for the superb manner in which they executed it. We especially appreciate their patience, particularly those who submitted their manuscripts early, only to wait and then have to revise and update because the project lagged.

As in the preparation of *Genes and Genomes,* the facilities and magnificent settings of Stanford University's Hopkins Marine Station at Pacific Grove, California, and the Institute for Advanced Study at Princeton, New Jersey, enabled us to work without interruption. We are especially indebted to the director of the institute, Dr. Phillip Griffiths, and his staff for their efforts on our behalf and to Dr. Franz Moehn for his enthusiastic welcomes and gracious hospitality. The

Arnold and Mabel Beckman Center of the National Academies of Science and Engineering at Irvine, California, provided another quiet retreat.

We appreciate Bruce Armbruster's willingness to publish this follow-on to *Genes and Genomes,* and we are grateful to Sylvia Stein Wright, Jane Ellis, and Eugene Bailey for their contributions to transforming manuscripts into the printed versions in this book. As before, it was a privilege and pleasure working with Audre Newman and Georg Klatt, who converted sketches and vague ideas into the illustrations that enhance the texts.

Dot Potter at Stanford and Sharon Bassin at the Carnegie Institution of Washington have been patient and steady assistants. Dot Potter had the especially onerous task of typing and retyping marked-up manuscripts, and for that we are especially grateful.

Our families deserve our heartfelt gratitude for tolerating the nine years spent completing *Genes and Genomes* and the additional five years needed to finish *Dealing with Genes* and the present book.

Maxine Singer and Paul Berg

CHAPTER 1

Mammalian DNA Viruses: Papovaviruses as Models of Cellular Genetic Function and Oncogenesis

PAUL BERG AND MAXINE SINGER

1.1 Introduction

Even a cursory reading of the history of prokaryotic molecular biology reveals the seminal contributions and insights provided by bacteriophage studies. Although the motivation for most of the early studies on mammalian DNA viruses centered on their pathogenicity, and particularly their carcinogenicity, these viruses also proved to be extremely fruitful systems for analysis. Indeed, some of the most significant discoveries concerning mammalian genome function—the existence and properties of circular DNAs, the widespread occurrence of introns and splicing, the recognition of enhancers and the complexity of eukaryotic promoters, the enzymatic machinery and mechanisms for mammalian DNA replication—were made in studies of mammalian DNA viruses. We have learned that viral gene expression involves an interacting network of regulatory signals, coding sequences, and gene products in which individual functions are subject to multiple coordinating influences. The importance of timing and the effects of subtle environmental and genetic modulations on the outcome of infections and the ability to manipulate these parameters experimentally have made these viruses attractive models for the infinitely more complex problems of development and differentiation of eukaryotes.

Several important attributes contributed to making viruses so useful in molecular biological studies. Even before the advent of molecular cloning, many viral DNA genomes were obtainable in pure form and virtually unlimited quantities. Moreover, mammalian virus genomes are considerably smaller and simpler than those of their hosts, and the complexity of the genetic functions is, accordingly, much reduced. In one sense, viruses provide a cloned set of genes, organized in a physiologically meaningful array on a single DNA molecule. Besides contributing substrates for studies of genomic expression, function, and replication, the ready availability of pure viral DNA provides probes for elucidating these processes. Study of the progress and outcome of viral infections is greatly simplified by the availability of established lines of cultured mammalian cells that serve admirably as hosts for infection or as recipients for specifically altered viral genomes, allowing for a variety of carefully designed experimental approaches. The relatively synchronous course of virus infections also simplifies analysis of the temporal patterns of virus gene expression and host response.

An unanticipated, but extremely fruitful, bonus from studies of mammalian DNA viruses is their utility as vectors for transducing discrete nonviral DNA segments into appropriate hosts. Indeed, the growing knowledge of how various parts of viral DNAs work makes it possible to assemble novel minichromosomes as specialized, transducing vectors for mammalian cells (see *Genes and Genomes*, section 1.3i and chapter 5). Viral vectors have made human gene therapy experimentation a reality.

This chapter begins with an overview of the structural features common to many types of DNA viruses and a general discussion of the outcomes following infections of various host cells (Section 1.2). Following this, we focus on the papovavirus group, dealing in detail with **Simian virus 40 (SV40)** (Section 1.3), then with some of its close relatives, **BK virus (BKV)** and **JC virus (JCV)** (Section 1.4a) and with **polyoma** (Section 1.4b). The principles derived from the studies of the SV40-polyoma group of papovaviruses are relevant to the more complex papovaviruses, the **papilloma viruses** (Section 1.4c). Section 1.5 describes how DNA viruses other than papovaviruses utilize similar and alternative strategies.

1.2 General Features of Mammalian DNA Viruses

a. Virus-Host Specificity

The mammalian DNA viruses generally have restricted host ranges, that is, they infect only cells of certain species and often only certain tissues. Host specificity is frequently associated with special structural features of the virion capsid structure that mediate adsorption to specific receptors on the cell surface and subsequent penetration to the inside of the cell. In other cases, incompatibility is due to a lack of, or to a difference in, a cellular function that is essential for expression of a critical viral gene or for DNA replication. For example, both polyoma and SV40 can penetrate rodent cells and express their early genes, but only polyoma replicates in rodent cells; SV40 DNA replicates in primate cells, but polyoma DNA does not. The mouse cellular DNA replication machinery apparently lacks one or more components required for SV40 replication. Transcription of essential viral genes may be limited by the availability or levels of certain transcription factors in some hosts and not in others. These are some but by no means all of the reasons that could account for the specificity of virus-host interactions. Occasionally, the host restriction existing in animals is abrogated with cell cultures.

b. Virion Morphology and Constituents

Mammalian DNA viruses are a diverse lot, varying widely in their size, the amount of DNA that constitues their genome, and the nature of the coat that encloses the genome's nucleoprotein (Table 1.1). Most of the viruses are roughly spherical in appearance and consist of the genome enclosed in a shell of one or more layers (Figure 1.1). The shell or **capsid** surrounding the DNA is constructed of proteins encoded by the viral genes. The SV40 and polyoma virions are the simplest, consisting of the genome associated with a capsid consisting of three polypeptides; the adenovirus capsid is made of about 12 discrete polypeptide chains, 2 of which constitute 12 identical protein spikes embedded in and protruding from the capsid. The herpesviruses and hepatitis B virions also contain a capsid consisting of virus-encoded polypeptides, but the capsid is surrounded by a lipid membrane whose composition resembles the host cell's plasma membrane; in both cases, the virion's lipid bilayer is studded with virus-encoded glycoproteins.

The spherical appearance of most capsids is a consequence of the organization of the structural subunit (**capsomers**) into particles with icosohedral symmetry (**icosohedrons** are structures with 20 triangular faces and 12 apices). The size and composition of the capsomers depend on the precise number and type of viral-encoded coat proteins and varies among the different viruses. Irrespective of capsid size, the organization of capsomers into icosohedronlike capsids is analogous among known viruses.

The genomes of the mammalian DNA viruses vary in size from about 3×10^3 to about 2.8×10^5 base pairs (Table 1.1), reflecting the presence of different numbers of genes. Very little of the DNA is nonfunctional, and no pseudogenes have been detected. Except in the very large genomes, like those of the herpesviruses, DNA sequence repetition is rare. However, parvoviruses and adenoviruses contain at their termini short, repeated DNA sequences that are important for their DNA replication strategies. Thus, circular DNA genomes lack repeated sequences, but linear DNA genomes contain terminal repeats. Herpesviruses contain extensive repeated domains that include coding regions. Some viral genomes (such as SV40, polyoma, and papilloma) occur as supercoiled circular duplexes; others are double-stranded linear DNAs (adeno- and

Table 1.1 ***Physical Characteristics of Mammalian DNA Viruses***

Virus	Virion (diameter, nm)	DNA (size, kbp)
Parvovirus	Naked (18–26)	ss linear (5)
Papovavirus		
Polyoma/SV40	Naked (45)	ds circular (5)
Papilloma	Naked (55)	ds circular (8)
Hepadnavirus		
Hepatitis B	Enveloped (42)	Gapped ds circular (3)
Adenovirus	Naked (70–90)	ds linear (36)
Herpesvirus	Enveloped (150–200)	ds linear (120–240)
Poxvirus	Enveloped (375x200)	ds linear* (130–280)

*Covalently closed, "hairpin" ends.

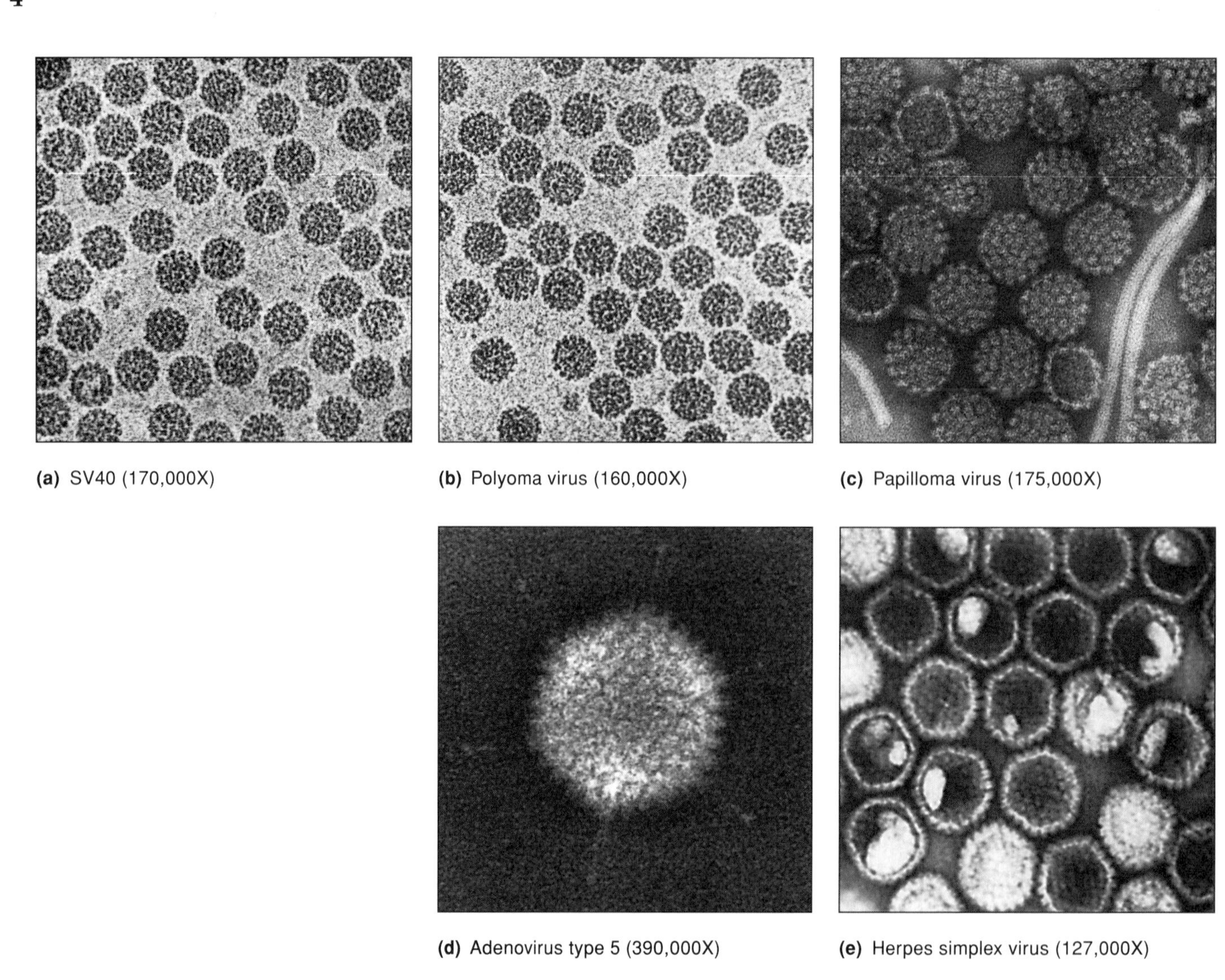

(a) SV40 (170,000X)

(b) Polyoma virus (160,000X)

(c) Papilloma virus (175,000X)

(d) Adenovirus type 5 (390,000X)

(e) Herpes simplex virus (127,000X)

Figure 1.1
Electron micrographs of several DNA viruses (the magnification for each image is indicated). (a) Frozen hydrated SV40 virions; (b), (c), and (d) negatively stained polyoma, papilloma, and adenovirus, respectively; (e) herpes simplex virus capsids (the lipid envelope is absent). (Part a provided by Timothy S. Baker; parts b, c, d provided by Harold W. Fisher.)

herpesviruses); one (hepatitis B) even has a gapped circular duplex. One group of DNA viruses, parvoviruses, contains a single-stranded linear genome.

Each of these forms implies a corresponding special strategy for both viral gene expression and replication in infected cells. The DNA in virions invariably exists as a nucleoprotein. In the papovaviruses, the nucleoprotein resembles cellular chromatin, employing the four histones, H2A, H2B, H3, and H4, to condense the DNA. The other viruses encode their own basic proteins to provide the means for DNA compaction.

c. Virus Life Cycles

Infections by mammalian DNA viruses have both common and distinctive features. Each must initiate infection by adsorbing to and penetrating the cell surface. The virus binds to a specific cell membrane–associated structure, often a

Figure 1.2
Plaques produced by SV40 infection of a monkey kidney cell monolayer. The clear (white) areas reveal the sites at which infections were initiated and infected cells were killed; the uninfected cell monolayer appears black, but in reality it is usually red due to staining of the live cells with a red dye.

glycoprotein. Lipid-enveloped viruses then enter the cell by membrane fusion; viruses lacking an envelope penetrate the membrane by an active endocytic process. After endocytosis, these viruses are enclosed in endosomes, and their capsid proteins enable the virions to escape through the endosomal membrane into the cytoplasm. Disassembly of the virion begins as soon as it enters the cell. Except for poxviruses, which remain and multiply in the cytoplasm, the viral nucleoproteins of DNA viruses reach the nucleus; the participants in and mechanism of the transport to the nucleus is presently murky. The viral chromosome is then expressed and replicated in the nucleus.

The outcome of infections by the mammalian DNA viruses depends on the kind of infected cell and a variety of less well defined conditions. In so-called **permissive cells**, the viral genome is fully expressed and replicated; multiple progeny virions are produced and the infected cells die. This so-called **cytolytic** or **lytic response** provides a convenient assay for virus titers. The assay involves infecting appropriate cell monolayers in petri dishes with various amounts of virus. After a suitable period for the infection to begin, the cells are covered with a layer of agar, which prevents the diffusion of progeny virus from the infection sites. After multiple rounds of infection have occurred, the agar is removed, and the cell monolayer is stained with a dye that stains living but not dead cells. Clear areas appear where a virus initiates an infection and kills the cells surrounding the initially infected cell (Figure 1.2). The virus titer can be calculated from the number of plaques, assuming that each plaque results from infection by a single virion. This number is usually lower than the number of virion particles in the innoculum because of the low probability that any one particle will initiate an infection.

If the infected cells are **nonpermissive**, that is, unable to support the lytic response, no new virions are produced; and, depending on the virus and cell, some or most cells survive with no apparent aftereffects. However, cells infected with some viruses (e.g., SV40 and polyoma) can, with low frequency, acquire new and heritable phenotypic properties; such cells are referred to as **transformed**. The ability to transform cells stably after infection is a property of the viral genome because transfection of such cells with purified viral DNA also leads to transformation. It is now clear that the transforming activity of this group of viruses is a consequence of the stable acquisition and continued expression of one or a few viral genes.

The distinction between permissive and nonpermissive cells is not absolute. For any particular virus, some cell types are semipermissive. The virus may replicate very inefficiently, and the cells may survive and grow but continually shed low levels of progeny virions; such carrier states are more often found *in vivo* than *in vitro*. Other semipermissive states depend more on the culture conditions or the physiological state of the cells at the time of infection. In some cases, intracellular virus is maintained in a nonreplicating or latent state; cell growth is apparently normal except the virus in a small fraction of the cells recurrently escapes the latent state and replicates to produce infectious progeny.

d. Lytic Infections

Expression of the viral genes follows the dismantling of the viral capsid and establishment of the viral genome in the infected cell nucleus. Early expression of papovaviruses is primarily from a specific subset of genes whose protein products are essential for initiation of viral DNA replication. The adeno- and herpesviruses have a somewhat more complex early transcriptional program: two sets of genes are expressed sequentially to produce the needed proteins for DNA replication. In each case, however, a previously silent set of genes begins to be transcribed concurrent with the onset of DNA replication; their mRNAs encode the virus's structural proteins. Thus, as a general matter, the lytic cycle is divided into stages: early (and in some cases, immediate early, early, and delayed early) preceding viral DNA replication, followed by a late stage to produce the protein constituents of virions. As the infection proceeds, the virus competes with its host for energy and the macromolecular machinery. Ultimately, the virus wins out and the cell dies. Progeny virions are assembled in the nucleus from newly replicated viral genomes and virion structural proteins and are released by dissolution of the killed cells. The herpes and hepatitis B virions acquire their lipid-glycoprotein envelopes when the capsids containing the viral DNA bud off from the infected cell membrane; here, too, the infected cells die.

The growth status of infected cells can dramatically influence viral replication. Many enzymatic activities needed for viral growth are limiting in resting cells, so one or more viral gene products generally promote entry of the cells into S phase (Section 1.5b). This involves overcoming the regulatory blocks to entry into S phase and transcriptional activation of cellular genes encoding enzymes needed for efficient DNA replication (e.g., DNA polymerases, topoisomerases, helicases, and enzymes involved in deoxynucleotide synthesis). Once papovavirus infections are initiated with the aid of the viral-encoded early proteins, the transcription, translation, and replication machinery needed for virus gene expression and replication are provided by the infected cell. However, the extent to which DNA viruses depend on host machinery varies. For example, the relatively complex adenoviruses and herpesviruses encode their own DNA polymerases but the simpler papovaviruses depend entirely on host cell DNA polymerases. Poxviruses rely very little on the cell's transcription and replication machinery as they encode the proteins needed for these processes. Most DNA virus genomes encode regulatory proteins that influence transcription of viral and cellular genes by the host's RNA polymerase. Hepatitis B virions contain a reverse transcriptase that converts RNA transcripts of the infecting DNA back into DNA.

e. Genetic Transformation of Infected Cells

Virus infections of nonpermissive cells do not lead to virus production. Instead of viral DNA replication, viral protein synthesis, and cell death, some of the surviving cells acquire new properties. Such cellular transformations are heritable because they result either from stable integration of all or part of the viral genome into the cell's chromosomes or, as with bovine papillomavirus, from maintenance of the viral genome as an autonomously replicating extra chromosomal DNA molecule. In most cases, the acquisition and continuous expression of one or more viral genes and the attendant genetic and phenotypic alterations lead to an oncogenic state (i.e., the ability to form malignant tumors). In many instances, transformation is accompanied (and perhaps enhanced) by chromosomal aberrations such as rearrangements, deletions, and amplifications.

Regardless of the genomic changes involved, cell transformation is generally first detected as a phenotypic alteration. Because only a small fraction of cells becomes transformed following virus infection, most transformation assays rely on changes that can be readily observed or selected for with great sensitivity. De-

Table 1.2 *General Features of Normal and Virus-Transformed Cells**

Growth Phenotype	Normal	Transformed
Life span	Finite in cell culture	Immortal
Serum requirement for growth in culture	High levels	Low levels
Growth pattern	Monolayers of elongated and parallel array of fibroblasts or sheetlike array of epithelial cells; requires solid surface for growth	Many layers of randomly oriented cells; form foci or piled-up colonies; can grow in some solid or liquid media because of anchorage-independent growth
Cell cycle control	Arrests normally at G1-S phase boundary	Little or no arrest at G1-S phase boundary
Tumorigenesis	Not tumorigenic in recipient animals	Two types: one shows most or all properties mentioned above but fails to cause tumors in animals; a second shows most properties mentioned above and is highly tumorigenic in animals

*See also Table 1.4.

pending on the virus, the target cell, and the conditions, a variable number of infected nonpermissive cells undergoes transient transformation. Such cells show one or more of the phenotypic changes characteristic of stably transformed cells, but these persist only briefly. The cells eventually die or revert to their normal phenotype. This implies two stages of cell transformation: first, the transient alteration of the normal state and then additional changes that result in maintenance of the altered state over many cell generations.

Stable transformation can be assayed in a variety of ways (Table 1.2). Cells judged to be transformed by one criterion need not show all the phenotypic alterations listed in Table 1.2. Transformed cells often have only a subset of the characteristic phenotypes. The particular subset varies with the nature of the virus, the species and tissue of origin of the cell, and the infection conditions. Even when a cloned population of recipient cells is used, individual isolates of transformed cells derived in a single experiment can have different phenotypic characteristics. Occasionally, viral transformation displays phenotypes not listed in Table 1.2—for example, altered cell morphology (e.g., round instead of flat), increased agglutinability of the cells by plant lectins, loss of the cells' organized cytoskeleton structure, and increased secretion of plasminogen activator, which promotes proteolysis of extracellular proteins.

The growth phenotypes of cultured cells can be ordered in a continuous spectrum from normal to highly tumorigenic. Generally, virus-transformed cells are grouped into one of two broad classes. One, commonly referred to as **immortalized**, occurs frequently when the infected cells are freshly obtained from animals. Such freshly obtained cells, referred to as **primary cells**, normally have a finite life in culture, being able to go through only a limited number of cell divisions before they lose their ability to divide. When such primary cells are transformed by virus infection, they become immortalized, that is, they grow and divide indefinitely. Immortalization is often accompanied by a change in the cells' nutritional requirements. For example, there may be a decreased requirement for factors present in the mammalian serum that is normally included in

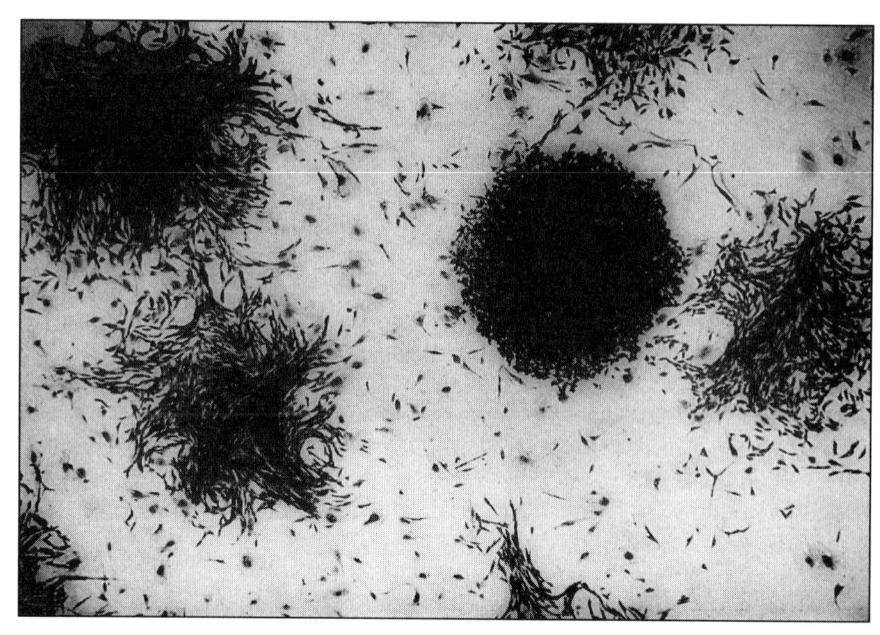

Figure 1.3
The colonial morphology of normal and polyoma transformed baby hamster kidney cells in soft agar medium. The single compact colony to the right of center illustrates the distinctive multilayered and randomly oriented arrangement of the cells comprising a colony of transformed cells. The remaining colonies illustrate the colonial morphology of untransformed cells. (Courtesy of Michael G. P. Stoker.)

nutrient media used to support cell growth. However, in spite of their indefinite life span and altered nutritional requirements, immortalized cells are contact inhibited. Thus, normal cells, immortalized cells stop dividing when they contact neighboring cells.

Immortalized cells generally do not produce tumors when innoculated into appropriate experimental animals; therefore, immortalized cells are not tumorigenic. However, transformation is often manifested by a second class of phenotypic change termed **tumorigenicity**. Tumorigenic cells generally do not cease growth when they contact neighboring cells; they acquire the ability to grow in multilayered, dense colonies (Figure 1.3), which are relatively easy to detect. In such colonies, termed **foci**, the cells are often oriented randomly with respect to one another, whereas untransformed cells grow as monolayers in which the cells assume parallel arrangements. Another characteristic feature of tumorigenic cells in culture is their **anchorage independence**. Untransformed cells usually require attachment to a solid surface to grow and divide. In contrast, transformed cells can form colonies even in a semisolid medium like soft agar and occasionally can grow in suspension in liquid cultures. Cells that form foci and grow in soft agar are not necessarily tumorigenic in animals, but most tumorigenic cells form foci and grow in soft agar.

f. Viral Transformation as a Model for Carcinogenesis

In addition to their ability to transform cultured cells *in vitro*, some DNA viruses are oncogenic when injected in certain animals and may be associated with tumor formation in nature. Cells from such tumors can be removed and propagated as cell lines in culture dishes while retaining their potential to proliferate into tumors. Consequently, the neoplastic properties of cells transformed *in vitro* or *in vivo* provide an important model for studying natural tumorigenesis.

How well does the model fit what is known about carcinogenesis? A great deal of biological and epidemiological data indicates that carcinogenesis is a multistep process in which successive genetic alterations influence the regulation of cell growth. Similarly, transformation of normal cells in culture by DNA viruses, retroviruses (Chapter 3), and oncogenes (Chapter 4) proceeds through stages that result in immortalization and acquisition of tumorigenic properties. These properties probably result from several genetic changes that alter critical cellular processes.

Virus-induced tumors in animals generally continue to express viral-encoded proteins, some of which elicit specific antibodies in the tumor-bearing animals; these are termed tumor (or T) antigens. These antibodies also detect the same or related T antigens in both lytically infected and viral-transformed cells in culture. In most instances, these proteins are found in the nucleus, although their antigenic determinants (epitopes) have also been detected on the surface of the transformed or tumor cells. Their appearance is another permanent phenotypic change associated with transformation. The early proteins encoded by several DNA viruses alter the regulatory mechanisms that control the pace of cell division. Whether this perturbation of the cell's division cycle is sufficient for oncogenesis to ensue or whether additional cellular mutations are needed is unresolved.

1.3 SV40: The Best Understood Papovavirus

More than any other group of DNA viruses, the papovaviruses (for **pa**pilloma, **po**lyoma, and **va**cuolating) have been one of the most influential models for understanding the molecular and genetic characteristics of eukaryotes. Papillomaviruses have been known for more than 50 years, first as the causative agent of wartlike tumors in rabbits (Shope papillomavirus) and subsequently in human and bovine warts. Initially, the bovine papillomaviruses (BPV) were the most intensively studied, but the indictment of human papillomaviruses (HPV) as causative agents of cervical and other cancers has shifted attention to the molecular anatomy and physiology of the human viruses. About 30 years ago, polyoma, a naturally occurring infectious agent in feral mice, was found to produce tumors in other rodents. More important, because its DNA alone could transform normal cells to tumorlike cells in culture, intensive molecular biological investigations sought to identify its **oncogene**. Shortly thereafter, analogues of polyoma were discovered in monkeys (**Simian virus 40** or **SV40**) and humans (**BKV** and **JCV**). For technical reasons and because of the early characterization of its genome, SV40 became the preferred experimental model of the papova group. Thus, we examine SV40 first and follow with a description of the distinctive features of the BK-, JC-, polyoma-, and papillomaviruses.

SV40 was discovered in the course of the development of a poliomyelitis vaccine. At that time, poliovirus was being propagated in rhesus monkey (*Macaca mulatta*) kidney cells, and there was concern that the cells and culture fluids from which the poliovirus was harvested might contain infectious agents that were indigenous to monkeys. As a precaution, the cell culture fluids were routinely screened for the presence of infectious material by adding them to cultured cells of other species. Such screens revealed several simian viruses, one of them being SV40. Interestingly, rhesus monkey cells are relatively resistant to SV40 infection (indeed, SV40 is endemic in wild rhesus monkeys without untoward effects), but African green monkey (*Cercopithecus aethiops*) cells are rapidly killed. Almost immediately after its discovery, SV40 was found to produce tumors upon innoculation into newborn hamsters. Shortly thereafter, normal rodent cells were shown to undergo tumorigenic transformation after infection with SV40 *in vitro*. Fortunately, although several early batches of poliovirus vaccine contained viable SV40, and the virus multiplied in the innoculated human hosts, there has been no increase in tumor frequency among innoculated individuals in the ensuing 40 years.

a. Virion Structure

By direct electron microscopy (Figure 1.1a), SV40 virions appear as nearly spherical particles with a diameter of 45 nm, but electron diffraction analysis shows that they have icosahedral symmetry. The particles consist of 88 percent protein and 12 percent DNA. About 75 percent of the virion protein is contained in the outer shell or virion capsid, which consists of three virus-coded proteins: VP1 (45 kDa) is 86 percent of the total, and VP2 (42 kDa) and VP3 (30 kDa) make up the rest. The capsid consists of 360 copies of VP1 assembled as 72 pentamers, 12 of which occupy positions of fivefold symmetry, and the remaining 60 pentamers are surrounded by 6 other pentamers, suggesting that the specificity of bonding among the protein subunits is not conserved (Figure 1.4). The pentamers are joined by extended polypeptide arms (the C termini of VP1) rather than by contacts between complementary surfaces. Thus, five C terminal arms are inserted into the folded structures of their neighbors. Each pentamer appears to be associated through an amino terminal sequence of VP1 with a carboxy terminal domain of VP2 and VP3. The structural association between the pentameric capsid and the minichromosome within is unclear.

The entire SV40 genome is contained in the single DNA molecule enclosed in the capsid. The 5243 bp DNA, a covalently closed, circular, double-stranded DNA, is organized in a chromatinlike structure called a **minichromosome** (Figure 1.5). Electron micrographs and biochemical experiments reveal that the minichromosomes contain 24 ± 2 nucleosomes, each composed of two molecules each of host histones H2a, H2b, H3, and H4. Although present in nonvirion nuclear minichromosomes, histone H1 is absent from the virion minichromosomes.

Except for one discrete region of the viral DNA, nucleosomes are evenly distributed on the minichromosome. Some minichromosomes isolated from the nuclei of infected cells have a nucleosome-free region encompassing 350–400 bp;

Figure 1.4 (*facing page*)
Structural features of the SV40 virion. The structure of the SV40 virion has been solved to 3.8Å resolution using X-ray crystallography with synchroton radiation. (a) Overall view of the SV40 virion. The shell is composed of 360 copies of VP1 organized into 72 pentamers. There are twelve 5-coordinated pentamers (pentamer B) and sixty 6-coordinated pentamers (the central pentamer surrounded by pentamers A-F). The α, α′, and α″ subunits of the 6-coordinated pentamer form a 3-fold interaction, the β and β′ subunits form one kind of 2-fold interaction, and the α subunits form another kind of 2-fold interaction. (b) Illustration of how carboxy-terminal extensions from each VP1 subunit interact with adjacent pentamers; the extension is designated "linking arm" in (b), (c), and (d). The principal interpentamer contacts are made by interactions of the linking arm from one subunit with a subunit of an adjacent pentamer. Thus, each pentamer receives five invading arms, one from each of the adjacent pentamers, and donates five arms to surrounding pentamers. (c) A single pentamer, showing how the linking arm extends from each VP1 subunit. (d) Richardson representation of a VP1 subunit. α-helices are represented as barrels and β-sheets as flat arrows; narrow tubes connect the two kinds of secondary structure elements. The α-helices (αC′) and β-sheet (J′) are part of the carboxy-terminal arm from an adjacent VP1 subunit; note that the invading J′ arm from the adjacent subunit is clamped in place by interaction with this subunit's amino-terminal A sheet. The β-sheet segment labeled A″ and β-sheet segment labeled G1″ indicate how neighboring VP1 subunits in the same pentamer interact. The segment shown as J‴ is a carboxy-terminal fragment from one subunit interacting with the N-terminal A sheet (A″) of a neighboring subunit. (Figure provided by S. C. Harrison.)

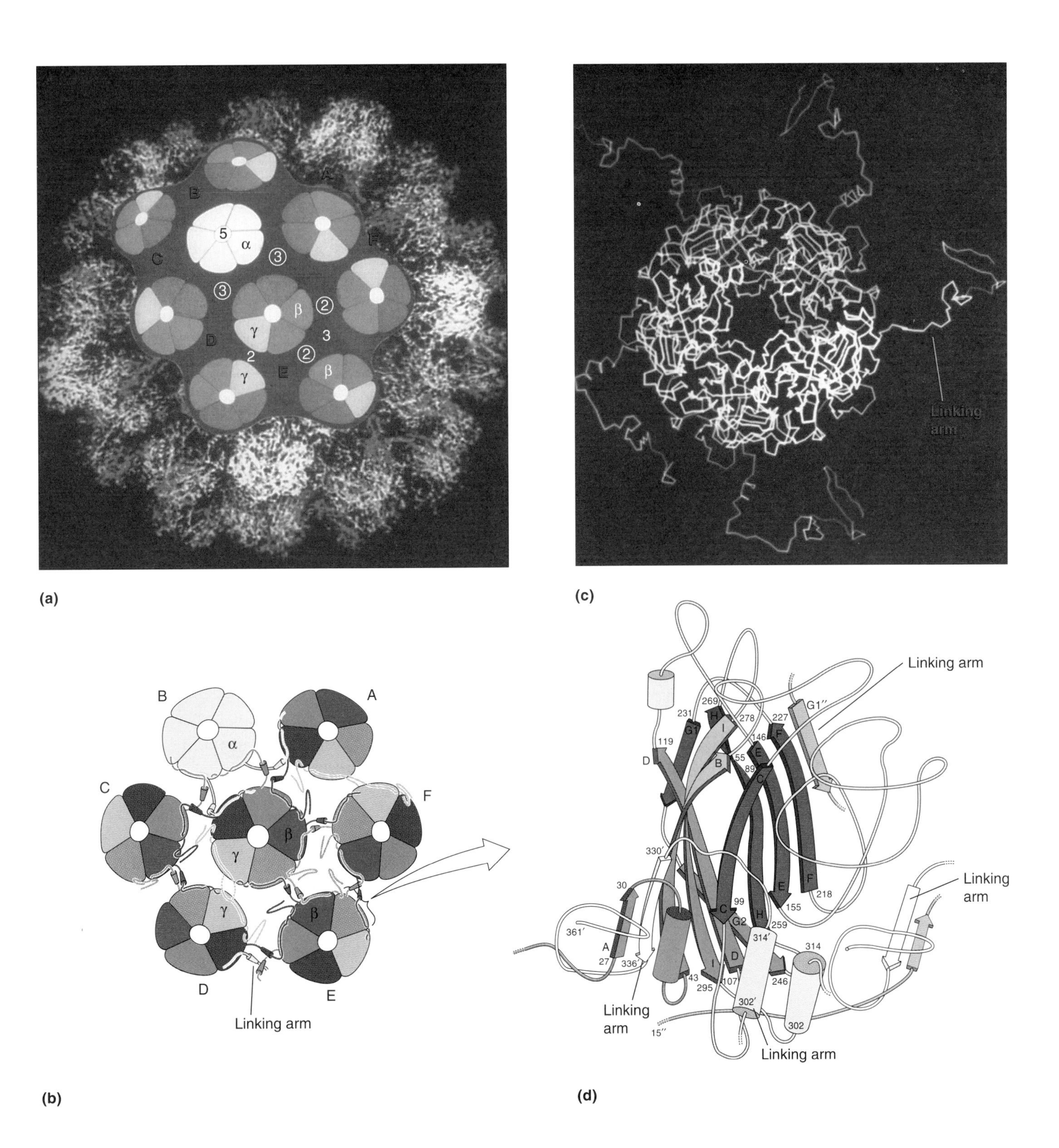
5
α
③
③
β
②
γ
3
2
②
γ
β
A
B
C
D
E
F
Linking arm
(a)
(c)
B
A
α
C
F
β
γ
γ
β
D
E
Linking arm
(b)
Linking arm
231
269
278
227
G1″
119
G1
H
I
146
F
D
B
55
E
89
C
330′
30
99
E
F
218
155
C
G2
H
259
361′
A
27
336′
314′
314
43
I
D
295
107
246
Linking arm
302′
302
15″
Linking arm
Linking arm
(d)

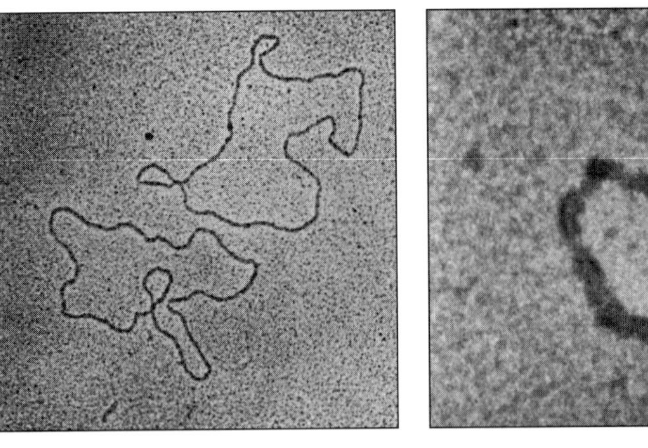
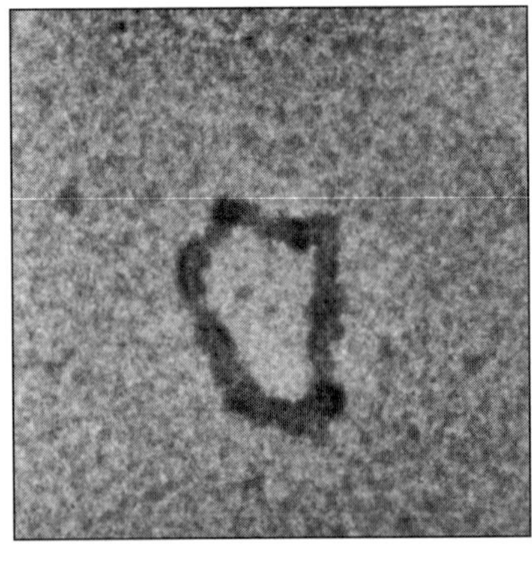
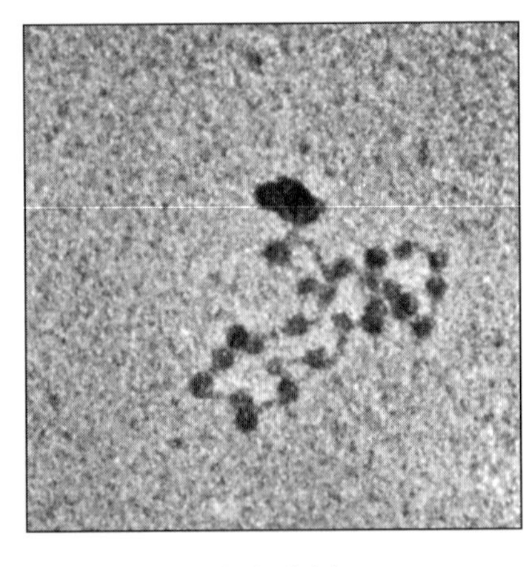

(a) SV40 DNA **(b)** SV40 condensed minichromosome **(c)** SV40 extended minichromosome

Figure 1.5
Various states of SV40 DNA. (a) Electron micrograph of free SV40 DNA; (b) the same SV40 DNA associated with the five histones (H1, H2a, H2b, H3, and H4) to form a condensed chromatin-like structure often referred to as a **minichromosome;** (c) minichromosomes lacking H1 are extended; the nucleosomes, consisting of pairs of the other histones, appear as small spheres spaced along the DNA.

this region contains the sequences that regulate transcription of the early and late viral genes as well as the origin of DNA replication (Sections 1.3d and e).

The minichromosomes appear in the electron microscope as extended rings, but the DNA isolated free of protein from the minichromosomes is a **superhelical** circular molecule (Figure 1.6). This superhelicity or **supercoiling** stems from the fact that the duplex DNA in the nucleosomes is underwound due to its association with the nucleosomes, and when the histones are removed, the number of right-handed helical turns characteristic of the B-DNA helix is restored. Because the duplex DNA circle is covalently closed, and therefore topologically constrained, restoration of the correct number of right-handed helical turns must be compensated for by twisting the duplex DNA an equal number of times in the opposite (negative) sense. Thus, the number of negative superhelical turns in isolated viral DNA (24 $\pm$ 2) is directly related to the number of nucleosomes with which it is associated in the minichromosomes. This equivalence, between the number of superhelical turns and the number of nucleosomes per circular DNA molecule, has been validated experimentally by reconstructing minichromosomes from circular DNA and the four histones *in vitro.*

b. Outcomes of Infections

Permissive cells Infection of primary African green monkey cell cultures or established cell lines derived from such cultures (e.g., CV-1, Vero, BSC-1) results in cell death and dissolution and the production of progeny virus. Infections are readily initiated by adding a suspension of virus to a dish containing a confluent layer of cells, generally at a high enough multiplicity to ensure infection of all the cells (five to ten infectious virions per cell). How virions enter the cell and migrate to the nucleus within 30 minutes is unknown. Disassociation of the capsid and release of the nucleoprotein start an orderly progression of events that culminates two to four days later in the production of thousands of new virions and detachment of the dead cells from the surface of the dish (Figure 1.7). The first detectable consequence of the infection (6–10 hours) is the appearance of

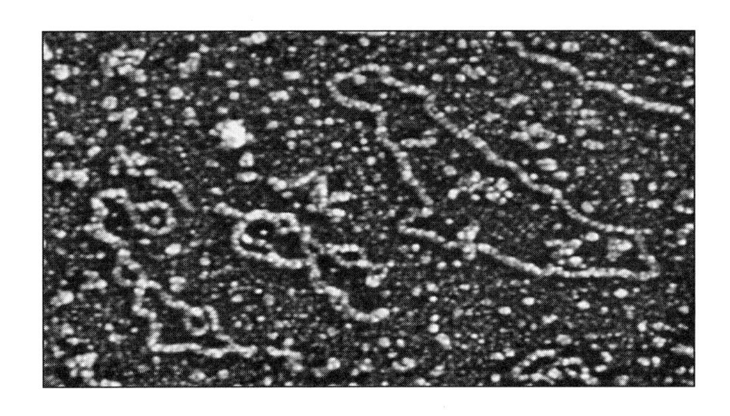

Figure 1.6
Electron micrograph of superhelical (supercoiled) and relaxed SV40 DNA. An extended, relaxed circular DNA is at the center right; two tightly supercoiled circular DNAs show in the lower left portion of the micrograph. (Courtesy of J. Griffith.)

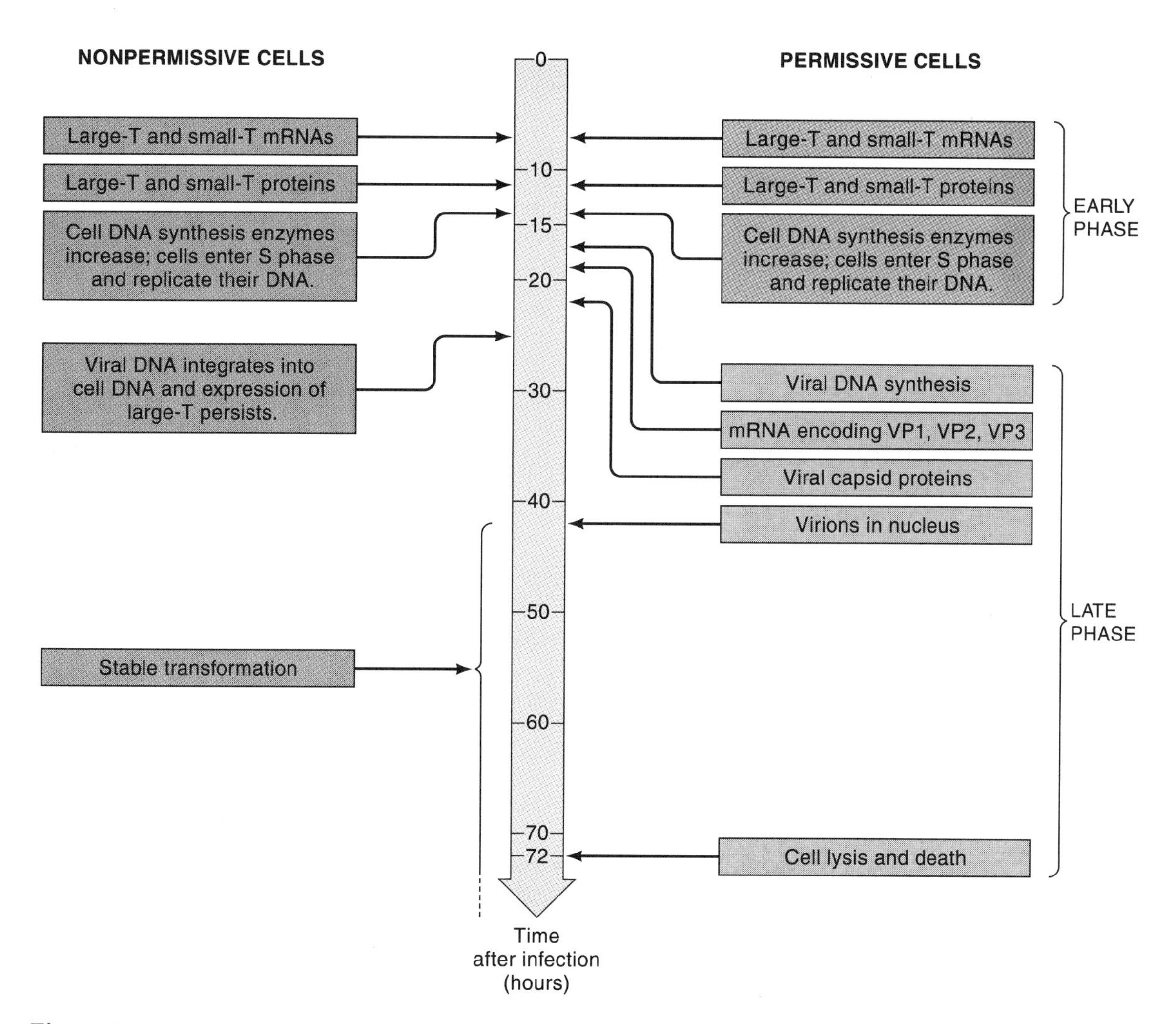

Figure 1.7
Course of SV40 infection of permissive and nonpermissive cells. The vertical arrow indicates the time (hr) following infection (0 hr) at which various events occur. Until about 14 hours (the early phase), both permissive and nonpermissive cells respond similarly to the infection. However, the ensuing events—viral DNA synthesis, late mRNA, viral capsid protein synthesis, the formation of virions and death of the cell—occur only in permissive cells. In nonpermissive cells, viral DNA integrates into the cellular DNA; transformation occurs if large-T expression persists.

mRNAs that encode the SV40 early proteins, large-T and small-T; by about 10–15 hours after infection, the two proteins are evident in the cell nucleus.

The small-T protein appears to be dispensable for most conditions of lytic infection of cultured cells, although this is less true with primary cells. However, none of the subsequent events in the virus growth cycle takes place in the absence of large-T. In addition to regulating early viral transcription and playing a key role in initiating viral DNA replication, large-T interferes with the function of cellular growth regulatory proteins and induces resting cells to enter the S phase of the cell cycle. This leads to the accumulation of elevated levels of host-specified enzymes related to DNA synthesis. Among these are DNA polymerases, DNA ligase, and a battery of enzymes involved in the synthesis of deoxynucleotides (e.g., thymidine kinase, dihydrofolate reductase, ribonucleotide reductase, and deoxycytidycate deaminase). Shortly after the infected cells enter S phase and begin to synthesize cellular DNA and histones (about 15 hours after infection), viral DNA synthesis begins and continues for days. Throughout this phase, the viral DNA associates with newly made histones, including histone H1, to form the minichromosome precursor of mature virions.

The onset of viral DNA synthesis initiates the events referred to as the late phase of the lytic infection. Thus, as the rate of viral DNA synthesis accelerates, the production of early mRNAs declines, and the so-called late mRNAs encoding the capsid proteins VP1, VP2, and VP3 become the predominant transcripts. At the height of the late phase, these three proteins represent almost 20 to 25 percent of the protein being made in the cell, most of which is VP1, the major capsid protein. With the accumulation of the minichromosomes and capsid proteins (pentamers plus VP2/VP3), virions are assembled in the nucleus, first into intermediate previrions and then into mature, infectious virions. Sometime during this process, histone H1 is dissociated from the minichromosome. Cell death, lysis, and release of the virion particles follow. Depending on the infection conditions, as many as 10^5 virions are produced per infected cell. Although the precise cause of cell death is not known, it may result from the accumulation of minichromosomes in the nucleus and the ensuing disruption of the cellular DNA replication machinery; cells infected with SV40 mutants that are unable to express large-T and therefore are unable to synthesize viral DNA are not killed.

Nonpermissive cells With rodent or other nonpermissive cells, the outcome of SV40 infection is quite different: the cells survive, most of them exhibit no permanent change in phenotype, but some become stably transformed to a new and often oncogenic phenotype. The early events in the infection of nonpermissive cells by SV40 are virtually the same as those that occur with permissive hosts: the same viral early mRNAs and proteins are made, and the infected cells enter into S phase as a consequence of the induction of the cellular DNA replication machinery by large-T protein. At this point, however, the courses of the infections of permissive and nonpermissive cells differ. Neither viral DNA replication nor late viral gene expression occurs in nonpermissive cells. Most of the infected cells proceed through several cell division cycles, then revert to their previous quiescent state. A small fraction of the infected cells (10^{-3}–10^{-4}) acquires new properties such as those mentioned in Table 1.1 and often the ability to produce tumors when innoculated into susceptible animals.

The capacity of SV40 to induce both transient cell divisions and stable transformation requires the formation of functional large-T protein. Thus, blocking early gene expression with interferon, for example, prevents both transient and stable transformations. Furthermore, stable transformation is prevented if the infected cells are unable to proceed through S phase, as happens in the presence of inhibitors of DNA synthesis.

The fundamental alteration in cells transformed by SV40 is integration of the viral DNA sequence into the cellular genome. Thereafter, the viral DNA replicates along with the rest of the genome and is passed to progeny cells during each cell division. However, the integration of viral DNA sequences into the cellular genome is not sufficient to assure the formation of stably transformed cells. The early genes must be transcribed and translated to form large-T protein in the cell nucleus.

There is at present no satisfactory explanation for the difference between permissive and nonpermissive cells. Two alternatives can be considered: nonpermissive cells might contain a factor that inhibits viral DNA replication, or they might lack an essential factor for viral DNA replication. An inhibitory factor in nonpermissive cells is unlikely because viral DNA can be released from the rodent genome and initiate a lytic infectious cycle when permissive African green monkey kidney cells are fused with SV40-transformed rodent cells. In such fusion hybrids, viral DNA is replicated, late genes are expressed, the DNA and VP1, VP2, and VP3 are assembled into virions, and virus is released. Thus, the presence of rodent cell constituents does not prevent the hybrid cells from carrying out a complete lytic cycle. It seems more likely that nonpermissive cells lack one or more factors that permissive cells contain that are required at least for viral DNA replication, if not for late gene expression. Moreover, the distinction between permissive and nonpermissive is not necessarily sharp. Many cell types, including human cells, are semipermissive for SV40 infection. When such cells are infected, some of them produce new virions, but others may become transformed. Perhaps the putative permissivity factor is present, but at too low a concentration to ensure the orderly progression of the lytic process. *In vitro* studies of SV40 DNA replication provide an insight into the identity of one of the permissive host factors (Section 1.3f).

c. Genetic Analysis of SV40 Functions

The earliest approach to identifying the genes responsible for SV40's multiplying and transforming abilities relied on the use of thermosensitive (ts) mutants. Such conditional mutants were selected after chemical mutagenesis (e.g., nitrosoguanidine, hydroxylamine, and 5-bromodeoxyuridine) on the basis of their ability to produce plaques at relatively low (32°) but not at high (39°–40°) temperatures. Analysis of these mutants followed two lines: viral infections were carried out either at the high temperature throughout or at a low temperature initially, followed by shifts to the high temperature at various times after infection. All the mutants could be classified in two major groups (Table 1.3). One group failed to initiate any of the sequelae characteristic of SV40 infection when the infection was initiated and maintained at the elevated temperature. How-

Table 1.3 ***Thermosensitive Mutants of SV40****

Class	Mutant	Defective Functions
I	*ts*A	Viral and cell DNA synthesis due to defective large-T
II	*ts*B *ts*C *ts*BC	Capsid formation due to defective VP1 protein
	*ts*D	Infectious virion formation due to defective VP3
	*dl*E**	Block in virion disassembly following infection due to defective VP2 protein

*Thermosensitive mutants are viable at 32° but not at 39°.
***dl*E contains a small deletion in the VP2 coding sequence.

ever, when permissive cells were infected with this group at a low temperature, a normal lytic cycle characteristic of that temperature ensued. Most important, this class of mutants could carry out a complete lytic cycle if infected cells were shifted from the low temperature to the high temperature once viral DNA replication began. This group, it was concluded, was defective in a function needed to initiate viral DNA synthesis but could proceed normally once this process had been initiated. This group of viruses was also impaired in its ability to induce cellular DNA synthesis or to transform nonpermissive cells at the high temperature, although both properties were normal at the low temperature. More strikingly, however, once this group of mutant cells had stably transformed at a low temperature, the transformed phenotype and tumorigenicity were lost when the cells were shifted to the high temperature. Viruses in this group of mutants were subsequently found to produce either an unstable or nonfunctional large-T protein. This provided the first evidence for the critical role of large-T protein in the initiation of both cellular and viral DNA synthesis, for transformation, and for the dependence of late gene expression on the initiation of viral DNA replication.

With one exception (a mutant that is unable to uncoat virion particles at the elevated temperature), mutants in the second group were able to initiate all the early events of an infection at a high temperature. But they were either unable to make functional capsid proteins or virions, or the virions they assembled were unstable or noninfectious at the high temperature. Thus, this class of mutants was defective in carrying out late functions. More refined analysis of the ts mutants of this latter class established that they were altered in the VP1, VP2, or VP3 genes.

All the mutants affected in a late function are fully able to transform cells in culture at the high temperature and to produce tumors in susceptible animals. This confirms the deduction that cellular transformation by SV40 depends exclusively on the expression of the early region, specifically, functional large-T protein.

d. Molecular Analysis of SV40 Genes

Mapping the location of the early and late genes on the viral DNA proved to be more formidable than obtaining the mutants and characterizing their phenotypes, that is, until restriction enzymes were found to digest SV40 DNA into discrete, separable fragments (Figure 1.8a). Assembling such a collection of fragments into a consistent physical map of the viral DNA was relatively straightforward (Figure 1.8a and b). Now the cleavage patterns for SV40 DNA after digestion with virtually every available restriction endonuclease are known,

---

Figure 1.8 (*facing page*)
Cleavage patterns of SV40 DNA by various restriction endonucleases. (a) The pattern of fragments was obtained by agarose gel electrophoresis after digestion of SV40 DNA with a mixture of *Hin*dII and *Hin*dIII restriction endonucleases. The fragments were stained with ethidium bromide to enable visualization after exposure to UV light. Smaller fragments migrate faster from the top than large fragments. (b) A mixture of *Hin*dII and *Hin*dIII restriction nucleases cleave SV40 DNA at 11 major sites (lines across the circle correspond to cleavage sites) to yield the 11 separated fragments shown in (a). The arrangement relative to one another was ascertained by determining which fragments were joined to one another in large fragments produced by partial digestions. Each fragment could be assigned to the region expressed early or late after infection by its ability to hybridize with the mRNA produced before viral DNA synthesis (early) and after the

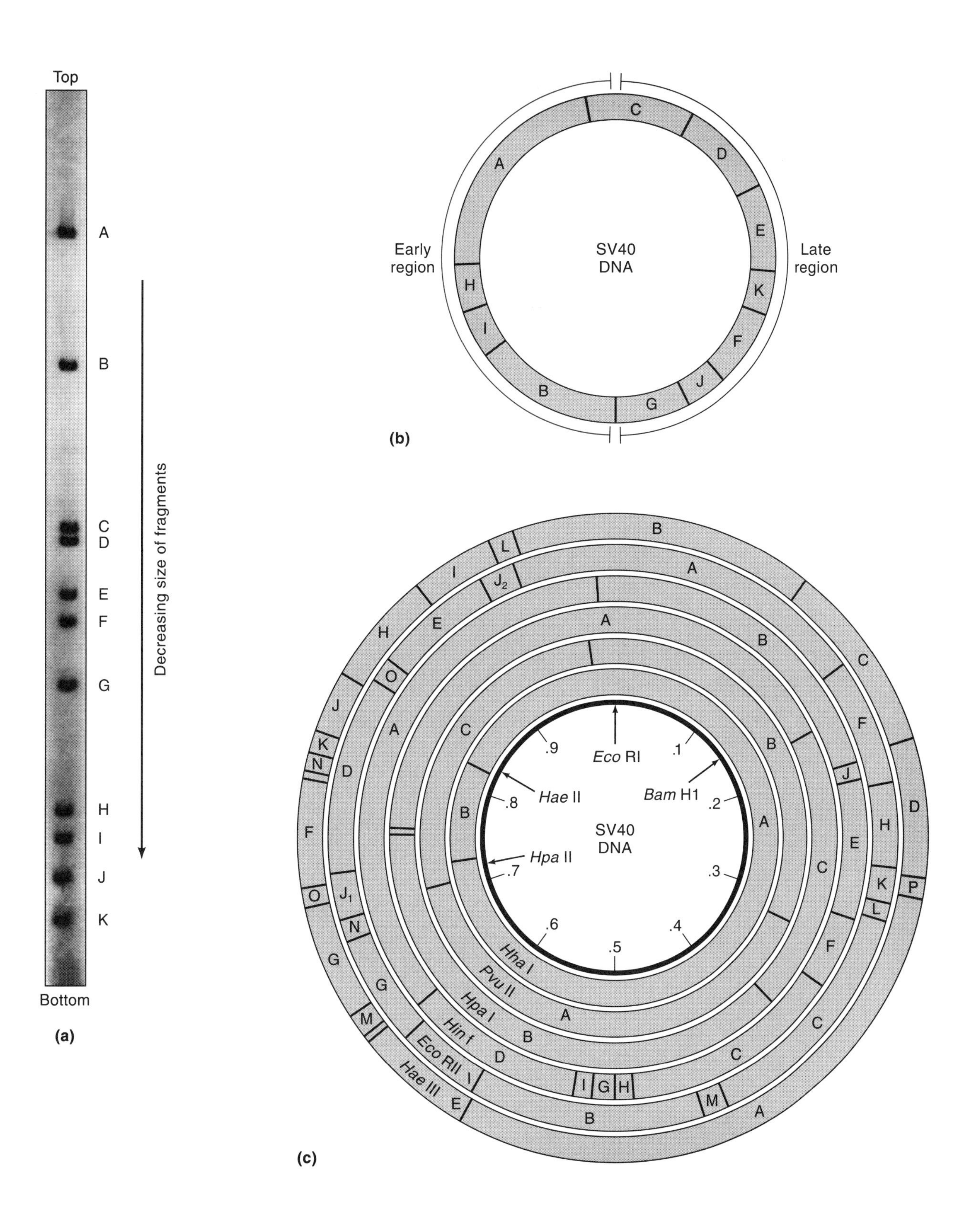

appearance of viral DNA and viral capsid mRNAs (late). The relative arrangement was confirmed by the order in which they were replicated from the origin of replication located in fragment C. (c) Restriction maps of SV40 DNA produced by digestion with a variety of restriction endonucleases. The arrows shown inside the inner circle represent single cleavage sites for the indicated restriction enzymes. The fractional numbers shown inside the inner circle represent fractional lengths (map units) of SV40 DNA relative to the position of the *Eco*R1 cleavage site. Based on a total length of 5243 bp, each 0.1 map unit is 524 bp. Thus, the *Bam*H1 cleavage site is at approximately 0.15 map units (786 bp), and the *Hpa* II cleavage site at about 0.72 (3775 bp), clockwise from the *Eco*RI cleavage site (5243/0 bp). (Part a courtesy of Dan Nathans.)

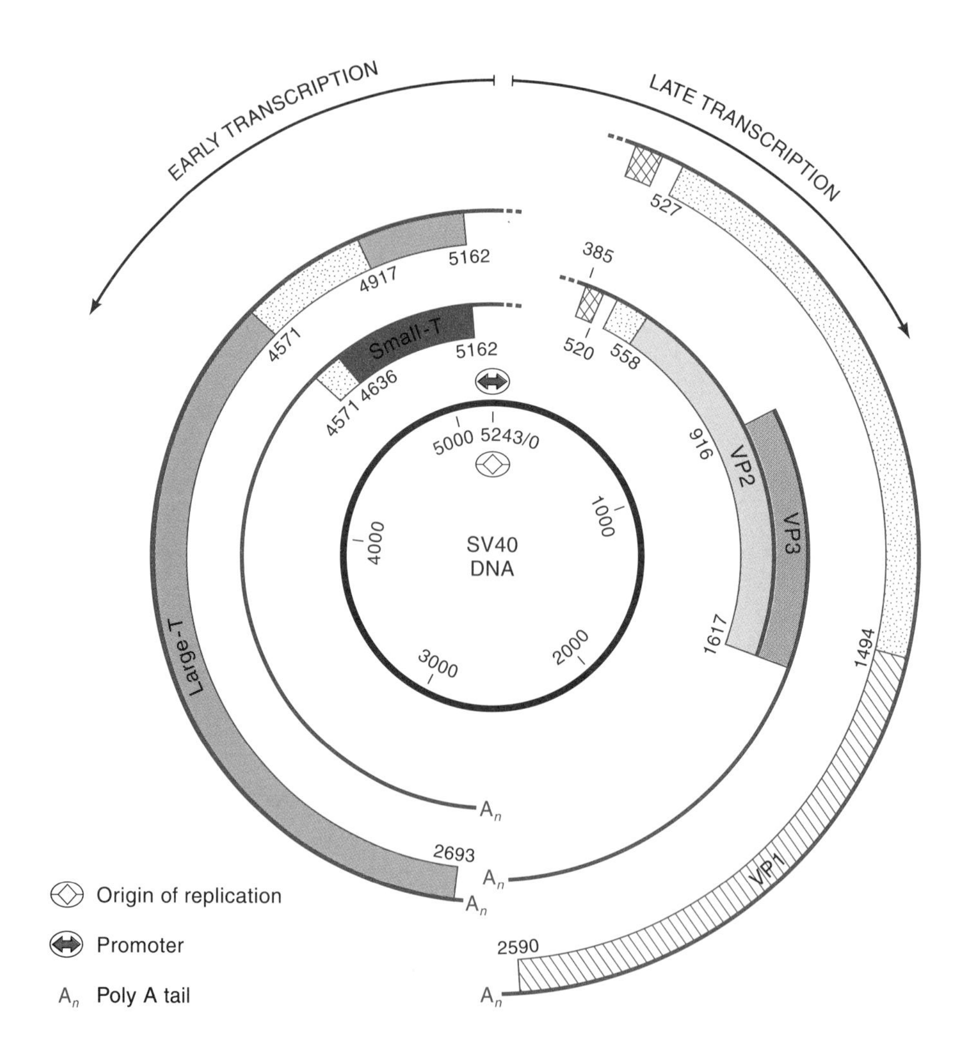

Figure 1.9
Transcription map of SV40. The physical map of SV40 is represented by the black inner circle with the bidirectional replication origin and early and late promoters at the top straddling bp 5243/0. Transcription of the early region is represented by the counterclockwise arrow and yields two mRNAs of nearly equal length, one encoding the large-T protein and the other, the small-T protein. Transcription of the late region is represented by the clockwise arrow, leading to one mRNA encoding both VP2 and VP3 and another encoding VP1. Each mRNA is terminated at its 3′ end by untranslated regions ending in poly A tails (A_n). Solid lines represent untranslated regions of the respective mRNAs and stippled regions indicate the segments that are spliced out to produce the respective mRNAs. The coding segments for large-T, small-T, VP2, and VP3 are each shown in a different shade of color. The coding segment of VP1 is shown hatched; the agnoprotein is crosshatched; the introns are stippled.

some of which are shown in Figure 1.8c. When used as hybridization probes, the isolated DNA segments corresponding to defined regions of the map made it possible to assign the templates for the early and late mRNAs to specific genomic regions. Subsequently, a large variety of deletions, insertions, and base changes was introduced at known locations in the viral DNA. The locations of each of these physical alterations were readily determined by altered restriction patterns or other means (see *Genes and Genomes*, section 7.2). By correlating the phenotypic changes resulting from each kind of mutational alteration, researchers were able to construct the first composite physical and genetic map of a viral genome (Figure 1.9).

A curious anomaly emerged when it was found that deletions at a particular location in the early region believed to encode the large-T protein had little or no effect on the structure of the large T-protein or the virus's ability to multiply or transform nonpermissive cells. This anomaly was resolved with the discovery that the SV40 early region contains an intron that is spliced in either of two ways to produce two early mRNAs, each about the same size, one coding for the large-T protein and the other for the small-T protein. The "anomalous" deletions were located in the large-T gene's intron, which contains some of the coding sequence of small-T protein, thereby explaining the failure to produce small-T by these mutants.

The most detailed and precise assignment of the positions of each SV40 coding region emerged when the entire 5243 bp sequence of SV40 DNA was completed (Figure 1.10). Then, and only then, could the precise start and stop codons of each of the viral proteins be assigned, the intron-exon organization be deduced, and the location of the putative regulatory regions be identified. A notable feature of the sequence is that the early genes are encoded on one DNA strand in the counterclockwise direction (Figure 1.10a, bp 5163 to bp 2694) and the late genes on the other strand in the clockwise direction (Figure 1.10b, bp 562 to bp 2590). Consequently, there is a paucity of noncoding sequence, most of which (about 300 bp) is located between bp 5235 and bp 300. This region contains the principal regulatory signals that govern where DNA replication begins, where early and late transcriptions start, and, most important, how the two transcription units are regulated temporally and quantitatively.

Figure 1.10 (*following pages*)
The sequence of SV40 DNA. (a) The early region coding strand; (b) the late region coding strand. The small-T and large-T protein coding sequences are indicated by the colors shown in the key at the bottom of the figure. The large-T (bp 4917–4571) and small-T (bp 4636–4571) introns are shown by stippling. The colors used to show the coding sequences for the late proteins—agnoprotein, VP1, VP2, and VP3—are indicated in the key at the bottom of the figure. Color shades, hatching, and crosshatching represent the same elements as in Figure 1.9. The intron for VP2-VP3 mRNA (bp 527–558) and for VP1 mRNA (bp 527–1494) are most easily seen in Figure 1.9.

```
      5233       5223       5213       5203       5193       5183       5173       5163       5153
CGCCTCGGCC TCTGAGCTAT TCCAGAAGTA GTGAGGAGGC TTTTTTGGAG GCCTAGGCTT TTGCAAAAAG CTTTGCAAAG ATGGATAAAG

      5143       5133       5123       5113       5103       5093       5083       5073       5063
TTTTAAACAG AGAGGAATCT TTGCAGCTAA TGGACCTTCT AGGTCTTGAA AGGAGTGCCT GGGGGAATAT TCCTCTGATG AGAAAGGCAT

      5053       5043       5033       5023       5013       5003       4993       4983       4973
ATTTAAAAAA ATGCAAGGAG TTTCATCCTG ATAAAGGAGG AGATGAAGAA AAAATGAAGA AAATGAATAC TCTGTACAAG AAAATGGAAG

      4963       4953       4943       4933       4923       4913       4903       4893       4883
ATGGAGTAAA ATATGCTCAT CAACCTGACT TTGGAGGCTT CTGGGATGCA ACTGAGGTAT TTGCTTCTTC CTTAAATCCT GGTGTTGATG

      4873       4863       4853       4843       4833       4823       4813       4803       4793
CAATGTACTG CAAACAATGG CCTGAGTGTG CAAAGAAAAT GTCTGCTAAC TGCATATGCT TGCTGTGCTT ACTGAGGATG AAGCATGAAA

      4783       4773       4763       4753       4743       4733       4723       4713       4703
ATAGAAAATT ATACAGGAAA GATCCACTTG TGTGGGTTGA TTGCTACTGC TTCGATTGCT TTAGAATGTG GTTTGGACTT GATCTTTGTG

      4693       4683       4673       4663       4653       4643       4633       4623       4613
AAGGAACCTT ACTTCTGTGG TGTGACATAA TTGGACAAAC TACCTACAGA GATTTAAAGC TCTAAGGTAA ATATAAAATT TTTAAGTGTA

      4603       4593       4583       4573       4563       4553       4543       4533       4523
TAATGTGTTA AACTACTGAT TCTAATTGTT TGTGTATTTT AGATTCCAAC CTATGGAACT GATGAATGGG AGCAGTGGTG GAATGCCTTT

      4513       4503       4493       4483       4473       4463       4453       4443       4433
AATGAGGAAA ACCTGTTTTG CTCAGAAGAA ATGCCATCTA GTGATGATGA GGCTACTGCT GACTCTCAAC ATTCTACTCC TCCAAAAAAG

      4423       4413       4403       4393       4383       4373       4363       4353       4343
AAGAGAAAGG TAGAAGACCC CAAGGACTTT CCTTCAGAAT TGCTAAGTTT TTTGAGTCAT GCTGTGTTTA GTAATAGAAC TCTTGCTTGC

      4333       4323       4313       4303       4293       4283       4273       4263       4253
TTTGCTATTT ACACCACAAA GGAAAAAGCT GCACTGCTAT ACAAGAAAAT TATGGAAAAA TATTCTGTAA CCTTTATAAG TAGGCATAAC

      4243       4233       4223       4213       4203       4193       4183       4173       4163
AGTTATAATC ATAACATACT GTTTTTTCTT ACTCCACACA GGCATAGAGT GTCTGCTATT AATAACTATG CTCAAAAATT GTGTACCTTT

      4153       4143       4133       4123       4113       4103       4093       4083       4073
AGCTTTTTAA TTTGTAAAGG GGTTAATAAG GAATATTTGA TGTATAGTGC CTTGACTAGA GATCCATTTT CTGTTATTGA GGAAAGTTTG

      4063       4053       4043       4033       4023       4013       4003       3993       3983
CCAGGTGGGT TAAAGGAGCA TGATTTTAAT CCAGAAGAAG CAGAGGAAAC TAAACAAGTG TCCTGGAAGC TTGTAACAGA GTATGCAATG

      3973       3963       3953       3943       3933       3923       3913       3903       3893
GAAACAAAAT GTGATGATGT GTTGTTATTG CTTGGGATGT ACTTGGAATT TCAGTACAGT TTTGAAATGT GTTTAAAATG TATTAAAAAA

      3883       3873       3863       3853       3843       3833       3823       3813       3803
GAACAGCCCA GCCACTATAA GTACCATGAA AAGCATTATG CAAATGCTGC TATATTTGCG GACAGCAAAA ACCAAAAAAC CATATGCCAA

      3793       3783       3773       3763       3753       3743       3733       3723       3713
CAGGCTGTTG ATACTGTTTT AGCTAAAAAG CGGGTTGATA GCCTACAATT AACTAGAGAA CAAATGTTAA CAAACAGATT TAATGATCTT

      3703       3693       3683       3673       3663       3653       3643       3633       3623
TTGGATAGGA TGGATATAAT GTTTGGTTCT ACAGGCTCTG CTGACATAGA AGAATGGATG GCTGGAGTTG CTTGGCTACA CTGTTTGTTG

      3613       3603       3593       3583       3573       3563       3553       3543       3533
CCCAAAATGG ATTCAGTGGT GTATGACTTT TTAAAATGCA TGGTGTACAA CATTCCTAAA AAAAGATACT GGCTGTTTAA AGGACCAATT

      3523       3513       3503       3493       3483       3473       3463       3453       3443
GATAGTGGTA AAACTACATT AGCAGCTGCT TTGCTTGAAT TATGTGGGGG GAAAGCTTTA AATGTTAATT TGCCCTTGGA CAGGCTGAAC

      3433       3423       3413       3403       3393       3383       3373       3363       3353
TTTGAGCTAG GAGTAGCTAT TGACCAGTTT TTAGTAGTTT TTGAGGATGT AAAGGGCACT GGAGGGGAGT CCAGAGATTT GCCTTCAGGT

      3343       3333       3323       3313       3303       3293       3283       3273       3263
CAGGGAATTA ATAACCTGGA CAATTTAAGG GATTATTTGG ATGGCAGTGT TAAGGTAAAC TTAGAAAAGA AACACCTAAA TAAAAGAACT

      3253       3243       3233       3223       3213       3203       3193       3183       3173
CAAATATTTC CCCCTGGAAT AGTCACCATG AATGAGTACA GTGTGCCTAA AACACTGCAG GCCAGATTTG TAAAACAAAT AGATTTTAGG

      3163       3153       3143       3133       3123       3113       3103       3093       3083
CCCAAAGATT ATTTAAAGCA TTGCCTGGAA CGCAGTGAGT TTTTGTTAGA AAAGAGAATA ATTCAAAGTG GCATTGCTTT GCTTCTTATG

      3073       3063       3053       3043       3033       3023       3013       3003       2993
TTAATTTGGT ACAGACCTGT GGCTGAGTTT GCTCAAAGTA TTCAGAGCAG AATTGTGGAG TGGAAAGAGA GATTGGACAA AGAGTTTAGT

      2983       2973       2963       2953       2943       2933       2923       2913       2903
TTGTCAGTGT ATCAAAAAAT GAAGTTTAAT GTGGCTATGG GAATTGGAGT TTTAGATTGG CTAAGAAACA GTGATGATGA TGATGAAGAC

      2893       2883       2873       2863       2853       2843       2833       2823       2813
AGCCAGGAAA ATGCTGATAA AAATGAAGAT GGTGGGGAGA AGAACATGGA AGACTCAGGG CATGAAACAG GCATTGATTC ACAGTCCCAA

      2803       2793       2783       2773       2763       2753       2743       2733       2723
GGCTCATTTC AGGCCCCTCA GTCCTCACAG TCTGTTCATG ATCATAATCA GCCATACCAC ATTTGTAGAG GTTTTACTTG CTTTAAAAAA

      2713       2703       2693       2683       2673       2663       2653       2643       2633
CCTCCCACAC CTCCCCCTGA ACCTGAAACA TAAAATGAAT GCAATTGTTG TTGTTAACTT GTTTATTGCA GCTTATAATG GTTACAAATA

      2623       2613
AAGCAATAGC ATCACAAATT TCAC
```

Intron | Large-T | Small-T

(a) Early region genes

GCCTCGGCCT CTGCATAAAT AAAAAAAATT AGTCAGCCAT GGGGCGGAGA ATGGGCGGAA CTGGGCGGAG TTAGGGGCGG GATGGGCGGA

GTTAGGGGCG GGACTATGGT TGCTGACTAA TTGAGATGCA TGCTTTGCAT ACTTCTGCCT GCTGGGGAGC CTGGGGACTT TCCACACCTG

GTTGCTGACT AATTGAGATG CATGCTTTGC ATACTTCTGC CTGCTGGGGA GCCTGGGGAC TTTCCACACC CTAACTGACA CACATTCCAC

AGCTGGTTCT TTCCGCCTCA GAAGGTACCT AACCAAGTTC CTCTTTCAGA GGTTATTTCA GGCCATGGTG CTGCGCCGGC TGTCACGCCA

GGCCTCCGTT AAGGTTCGTA GGTC**ATG**GAC TGAAAGTAAA AAAACAGCTC AACGCCTTTT TGTGTTTGTT TTAGAGCTTT TGCTGCAATT

TTGTGAAGGG GAAGATACTG TTGAGGGGAA ACGCAAAAAA CCAGAAAGGT TAACTGAAAA ACCAGAAAGT TAACTGGTAA GTTTAGTCTT

TTTGTCTTTT ATTTCAGGTC C**ATG**GGTGCT GCTTTAACAC TGTTGGGGGA CCTAATTGCT ACTGTGTCTG AAGCTGCTGC TGCTACTGGA

TTTTCAGTAG CTGAAATTGC TGCTGGAGAG GCCGCTGCTG CAATTGAAGT GCAACTTGCA TCTGTTGCTA CTGTTGAAGG CCTAACAACC

TCTGAGGCAA TTGCTGCTAT AGGCCTCACT CCACAGGCCT ATGCTGTGAT ATCTGGGGCT CCTGCTGCTA TAGCTGGATT TGCAGCTTTA

CTGCAAACTG TGACTGGTGT GAGCGCTGTT GCTCAAGTGG GGTATAGATT TTTTAGTGAC TGGGATCACA AAGTTTCTAC TGTTGGTTTA

TATCAACAAC CAGGA**ATG**GC TGTAGATTTG TATAGGCCAG ATGATTACTA TGATATTTTA TTTCCTGGAG TACAAACCTT TGTTCACAGT

GTTCAGTATC TTGACCCCAG ACATTGGGGT CCAACACTTT TTAATGCCAT TTCTCAAGCT TTTTGGCGTG TAATACAAAA TGACATTCCT

AGGCTCACCT CACAGGAGCT TGAAAGAAGA ACCCAAAGAT ATTTAAGGGA CAGTTTGGCA AGGTTTTTAG AGGAAACTAC TTGGACAGTA

ATTAATGCTC CTGTTAATTG GTATAACTCT TTACAAGATT ACTACTCTAC TTTGTCTCCC ATTAGGCCTA CAATGGTGAG ACAAGTAGCC

AACAGGGAAG GGTTGCAAAT ATCATTTGGG CACACCTATG ATAATATTGA TGAAGCAGAC AGTATTCAGC AAGTAACTGA GAGGTGGGAA

GCTCAAAGCC AAAGTCCTAA TGTGCAGTCA GGTGAATTTA TTGAAAAATT TGAGGCTCCT GGTGGTGCAA ATCAAAGAAC TGCTCCTCAG

TGGATGTTGC CTTTACTTCT AGGCCTGTAC GGAAGTGTTA CTTCTGCTCT AAAAGCTT**AT G**AAGATGGCC CCAACAAAAA GAAAAGGAAG

TTGTCCAGGG GCAGCTCCCA AAAAACCAAA GGAACCAGTG CAAGTGCCAA AGCTCGTCAT AAAAGGAGGA ATAGAAGTTC TAGGAGTTAA

AACTGGAGTA GACAGCTTCA CTGAGGTGGA GTGCTTTTTA AATCCTCAAA TGGGCAATCC TGATGAACAT CAAAAAGGCT TAAGTAAAAG

CTTAGCAGCT GAAAAACAGT TTACAGATGA CTCTCCAGAC AAAGAACAAC TGCCTTGCTA CAGTGTGGCT AGAATTCCTT TGCCTAATTT

AAATGAGGAC TTAACCTGTG GAAATATTTT GATGTGGGAA GCTGTTACTG TTAAAACTGA GGTTATTGGG GTAACTGCTA TGTTAAACTT

GCATTCAGGG ACACAAAAAA CTCATGAAAA TGGTGCTGGA AAACCCATTC AAGGGTCAAA TTTTCATTTT TTTGCTGTTG GTGGGGAACC

TTTGGAGCTG CAGGGTGTGT TAGCAAACTA CAGGACCAAA TATCCTGCTC AAACTGTAAC CCCAAAAAAT GCTACAGTTG ACAGTCAGCA

GATGAACACT GACCACAAGG CTGTTTTGGA TAAGGATAAT GCTTATCCAG TGGAGTGCTG GGTTCCTGAT CCAAGTAAAA ATGAAAACAC

TAGATATTTT GGAACCTACA CAGGTGGGGA AAATGTGCCT CCTGTTTTGC ACATTACTAA CACAGCAACC ACAGTGCTTC TTGATGAGCA

GGGTGTTGGG CCCTTGTGCA AAGCTGACAG CTTGTATGTT TCTGCTGTTG ACATTTGTGG GCTGTTTACC AACACTTCTG GAACACAGCA

GTGGAAGGGA CTTCCCAGAT ATTTTAAAAT TACCCTTAGA AAGCGGTCTG TGAAAAACCC CTACCCAATT TCCTTTTTGT TAAGTGACCT

AATTAACAGG AGGACACAGA GGGTGGATGG GCAGCCTATG ATTGGAATGT CCTCTCAAGT AGAGGAGGTT AGGGTTTATG AGGACACAGA

GGAGCTTCCT GGGGATCCAG ACATGATAAG ATACATTGAT GAGTTTGGAC AAACCACAAC TAGAATGCAG TGAAAAAAAT GCTTTATTTG

Intron   Agno protein   VP2   VP3   VP1

(b) Late region genes

This region is central to a discussion of SV40 functions. Figure 1.11 shows that the sequence between bp 5230 and bp 250 contains a 27 bp palindromic sequence within which are four repeats of 5′-GAGGC-3′ arranged in palindromic pairs, a 17 bp block of only A's and T's (AT-block), three tandemly repeated 21 bp sequences, each repeat containing two 5′-GGGCGG-3′ sequences (GC-boxes), and two identical direct repeats of 72 bp. The 5′-GAGGC-3′ sequence repeated in the 27 bp palindrome is also repeated three times in the same orientation between bp 5210 and bp 5185.

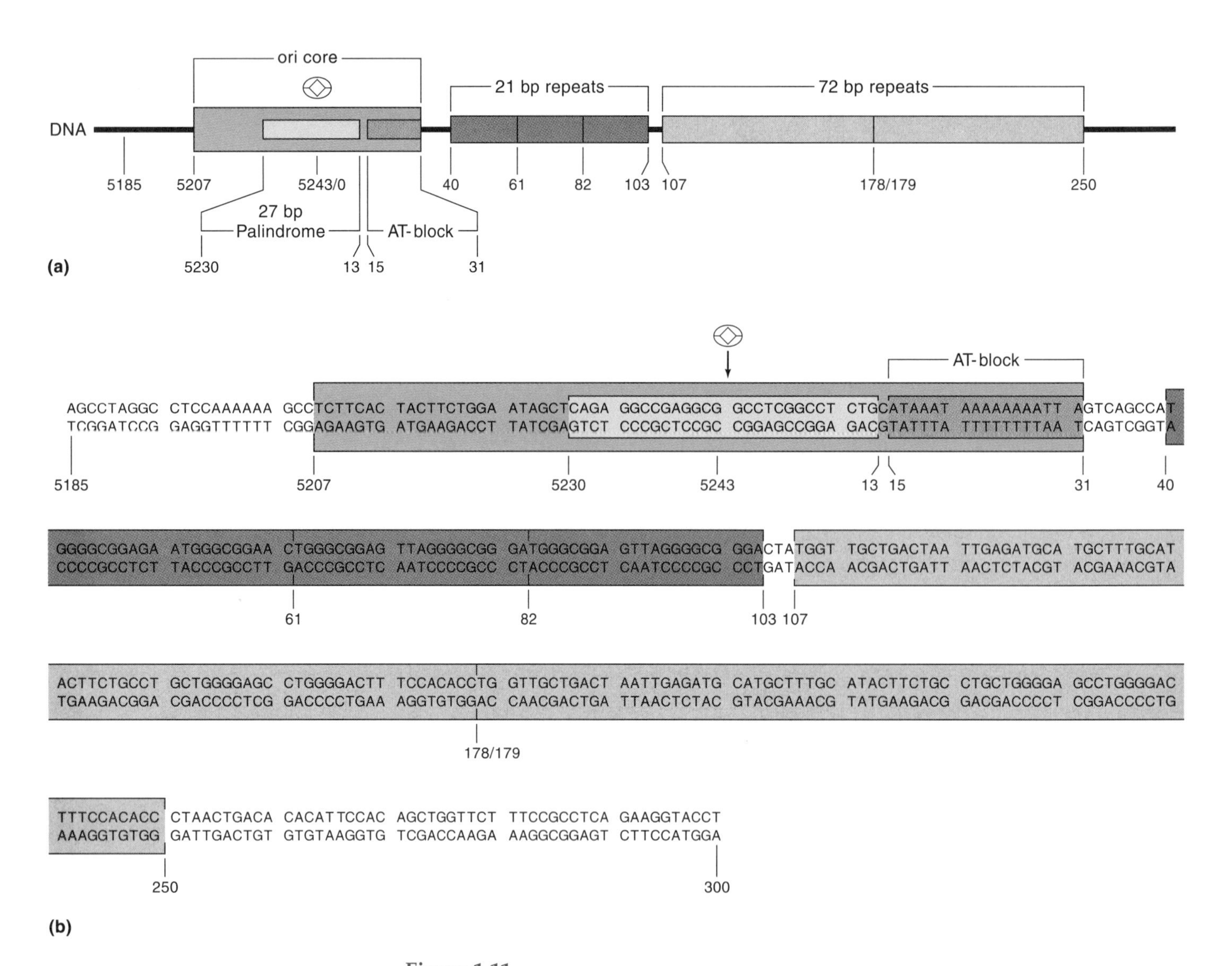

Figure 1.11
The regulatory region governing DNA replication and early and late gene expression. (a) The origin of DNA replication is symbolized by a diamond within an ellipse; it occurs within bp 5207 and 31, the most important elements of which are a 27-bp palindrome and an adjacent sequence of 17 AT bp (the AT-block). Transcription of the early and late regions is regulated by the bidirectional promoter that comprises the AT-block, the 21-bp repeats, and sequences within each identical 72-bp segment. (b) The bp sequence with colored highlighting of the region described in (a). (Color shades represent the same elements as in Figure 1.9.)

e. Transcriptional Regulation

Transcription of SV40's minichromosome is mediated by the host cell's RNA polymerase II in conjunction with a panoply of cellular accessory proteins needed to form functional transcription complexes (Figure 1.12). The GC-boxes and 72 bp

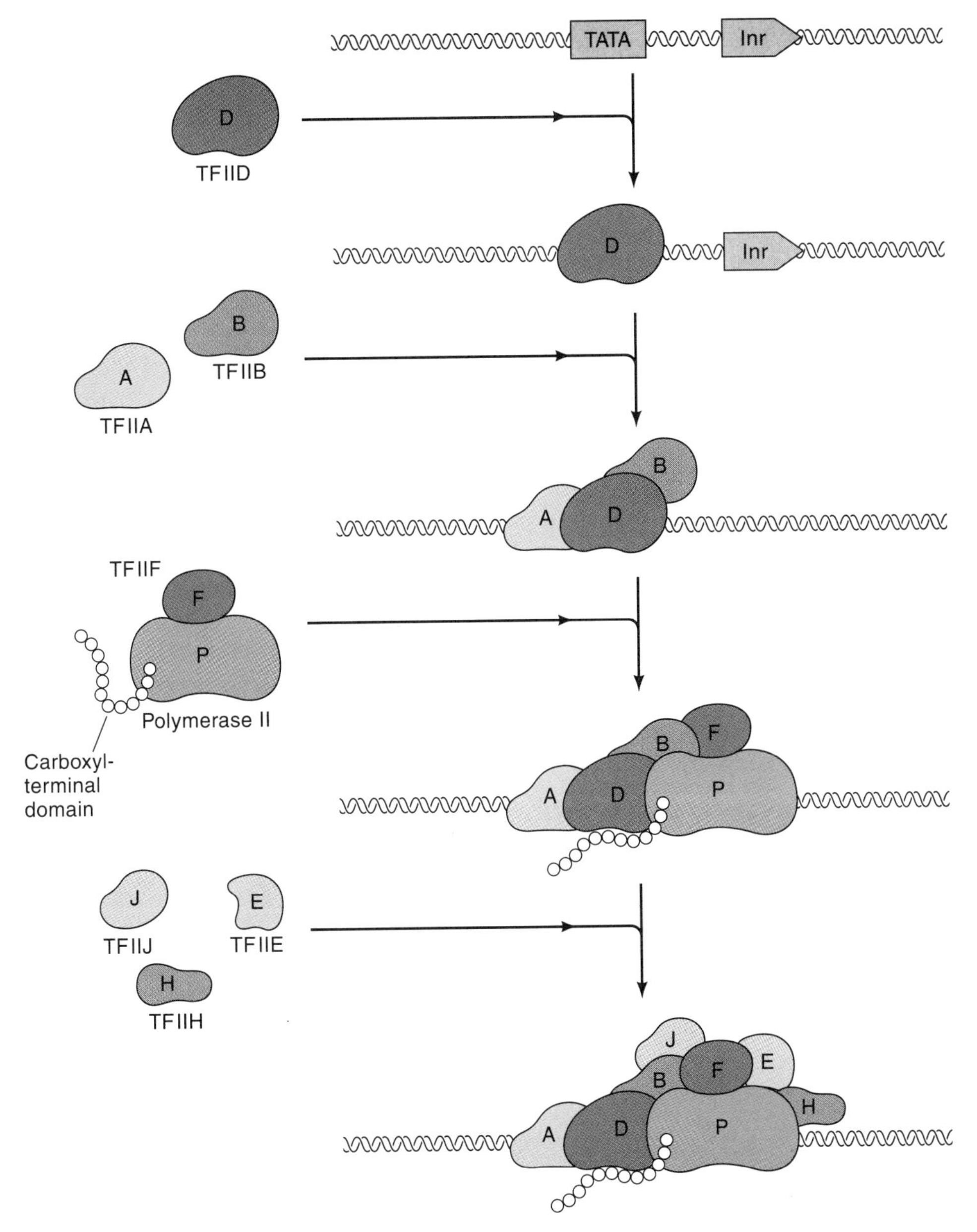

Figure 1.12
The stepwise assembly of the basal transcription complex at a class II promoter. The basal transcription complex consists of multiple protein complexes; these are designated TFIIA to TFIIJ and polymerase II. They are shown being assembled into a functional transcription complex in a quasi-stepwise order in the vicinity of the transcription start site. The protein complex, TFIID, is believed to bind first to the TATA sequence; additional protein complexes associate with TFIID, polymerase II, and with each other to form a "super" complex that initiates transcription at a site between TATA and Inv (a sequence that facilitates assembly of the complex in the absence of a TATA sequence). Some evidence suggests that such complexes and possibly ones containing other proteins (ones implicated in DNA repair) preexist and bind to the indicated region.

repeats constitute a set of specific sequences at which a variety of proteins (transcription factors) binds and modulates the rate of transcription initiation in complex ways (see *Genes and Genomes,* section 8.3). Although transcription initiation appears to be the principal mode of regulating the level of early and late mRNAs, differential splicing and polyadenylation also contribute to the overall regulation of mRNA types and quantities.

Shortly after the viral genome enters the nucleus of either permissive or nonpermissive cells, early transcription initiates at multiple sites closely spaced around bp 5230, continues counterclockwise (Figures 1.9 and 1.10a), and terminates somewhere beyond bp 2620; creation of the 5′ cap and 3′ poly A modifications occurs as described in *Genes and Genomes,* section 8.2. Further modification of the transcript involves splicing two overlapping introns (Figure 1.13) to produce two mRNAs, one encoding large-T protein and the other small-T. Removal of the larger intron removes an in-frame translation stop codon, enabling the mRNA to be translated into the 708 amino acid long large-T protein. Splicing of the small intron leaves the stop codon in place, permitting the production of only the 174 amino acid small-T protein. The large-T and small-T proteins share 82 amino acids at the N terminus, but their protein sequences differ beyond that. There are some indications that the relative amounts of the two mRNAs, and their corresponding proteins, differ in different cells and at different times during the lytic infection.

Deletion and insertion mutagenesis of the putative regulatory region define the elements comprising the early and late region promoters and the ori (*Genes and Genomes,* section 8.3). The AT-box is dispensable for early and late transcription, although it serves as a TATA-like motif and influences the position of the start sites for early region transcripts, as evidenced by the finding that, in its absence, early region transcripts have heterogeneous 5′ ends. The absence of a TATA-like sequence for late region transcription probably accounts for the many

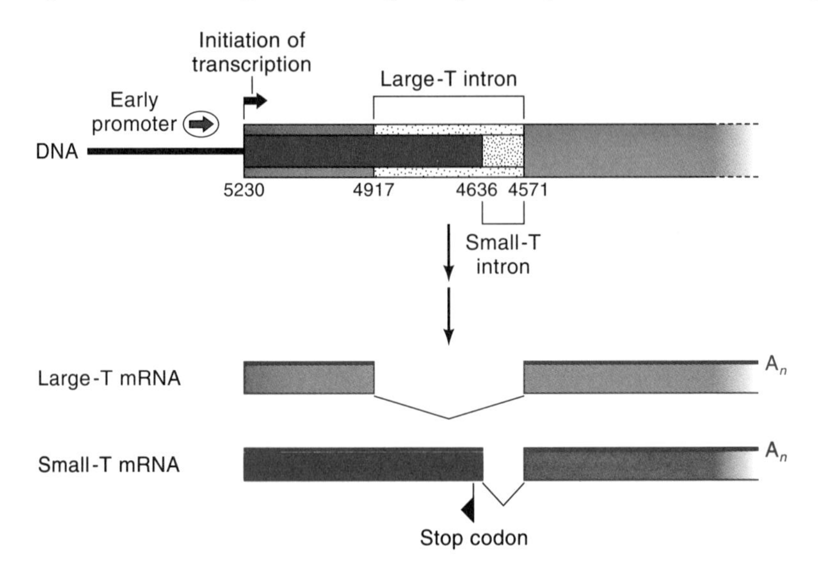

Figure 1.13
The organization of introns and the alternative splicing that yield large-T and small-T mRNAs. Early region transcription initiates at the black arrow pointing rightward. The large-T intron is shown by the lightly stippled region and the small-T intron as the more heavily stippled segment. A single transcript is spliced to remove either the small-T or the large-T intron, which yields two different kinds of mRNAs. A translation terminator codon located just before the small-T intron stops protein assembly at that point, thus yielding small-T protein. Splicing of the large-T intron removes the termination codon, which allows translation to proceed to the termination codon near the end of the mRNA, thus producing large-T protein. (Color shades conform to those used in Figure 1.9.)

different 5′ ends in late mRNAs. These sites are likely directed by a series of transcriptional initiator elements. The 21 bp repeats containing altogether six 5′-GGGCGG-3′ repeats are necessary for maximal rates of both early and late transcription, although some early transcription occurs with only two of the 5′-GGGCGG-3′ segments. Furthermore, very few or no early or late transcripts are produced if the 72 bp repeats are deleted. Regulation of early and late region transcription is mediated by sequence-specific DNA-binding proteins that bind to the GC-rich sequence motifs in the three 21 bp repeats and to the panoply of enhancer motifs throughout the 72 bp repeats (Figure 1.14). Although the precise mechanisms by which the various DNA-bound transcription factors accelerate transcription initiation are still unclear, current evidence indicates they facilitate the assembly of transcription complexes at the transcription start sites, either directly or indirectly, by interactions with one or more of the proteins of the transcription complex shown in Figure 1.12.

In early studies of SV40 lytic infections, it was noted that the initial high rate of early mRNA is not sustained and that the accumulation of functional large-T protein represses early region transcription; thus, with many ts mutants of large-T, early mRNA continues to be made at a high rate at the high temperature but is rapidly repressed at the low temperature. This repression is accounted for by the binding of large-T protein to three nonoverlapping sites between bp 5191 and bp 61 (Figure 1.15). The most avid binding occurs at site I (bp 5191 to bp 5213), which spans the sequence at which early region transcription begins. Repression of early region transcription by large-T binding to this site results from the interference with the assembly or progression of the transcription complex. The essential feature of site I is the sequence 5′-GAGGC-3′, two copies of which are arranged in tandem separated by an essential spacer of seven A-T base pairs. At somewhat higher levels of large-T, the protein binds to site II (bp 5230 to bp 11), inhibiting early region transcription still further. Binding of large-T to site III occurs at only very high levels of large-T and may not be physiologically relevant. These findings show that the early region promoter is regulated autologously—that is, the level of the gene's protein product regulates the production of its own mRNA.

Perhaps the most significant outcome of binding large-T to sites I and II is the initiation of the late phase of the lytic cycle. Thus, large-T protein bound at site II promotes the initiation of DNA replication at the ori sequence in site II and the initiation of late region transcription. It is not known whether the binding of large-T to site II contributes directly to the activation of late transcription or whether it functions indirectly by activating DNA replication, which in turn permits transcription of the late region to begin. Regardless of its role at site II, large-T helps activate late transcription at least in part through mechanisms that don't require DNA binding. Large-T may bind to certain cellular transcription factors, thereby altering their ability to stimulate late transcription. Alternatively, recent evidence indicates that late region transcription from the infecting unreplicated genomes is repressed by proteins that bind at or just downstream of the major late region mRNA start sites. When the number of potential transcription templates rises as a consequence of replication, repressor-free templates accumulate and serve for late region transcription. As DNA replication proceeds, the rate of late region transcription increases until it is 10 to 20 times greater than transcription of the early region.

The virus-encoded large-T protein is a key player in modulating many of the events of virus multiplication. Large-T protein has many vital roles in both the viral lytic and transforming "life-styles": (1) it functions as a transcriptional repressor in down-regulating its own expression; (2) it is a critical participant in initiating multiple rounds of viral DNA replication; (3) it provides the switch for shifting transcription from the early to the late region; (4) it overrides cell cycle control by disrupting the normal mechanisms for keeping cells in G1 or preventing them from entering S phase; and (5) it acts indirectly to activate transcription

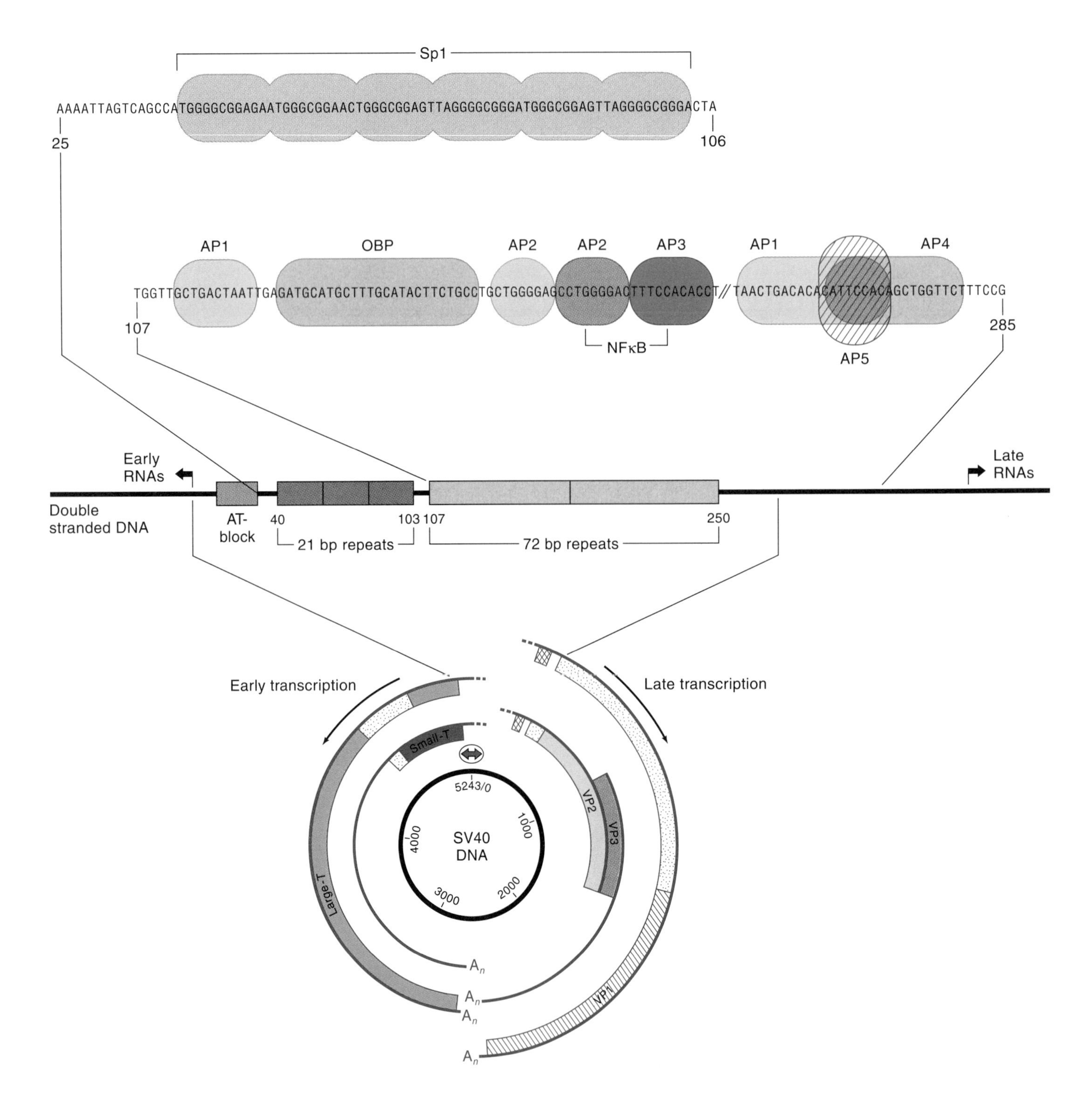

Figure 1.14
The sequences comprising the SV40 early and late promoter and their enhancer sequences. The lower portion of the figure repeats the information in Figure 1.9; the middle (linear area) summarizes the general features and organization of the early and late promoters and their enhancers. The transcription start sites for early and late mRNA are indicated by black arrows pointing in opposite directions. The upper portion illustrates the DNA sequence of the important promoter and enhancer elements between bp 25–235 and indicates by various shapes and coloring the transcription factors that bind to specific motifs within that sequence. (Color shades and hatching represent the same elements as in Figure 1.9.)

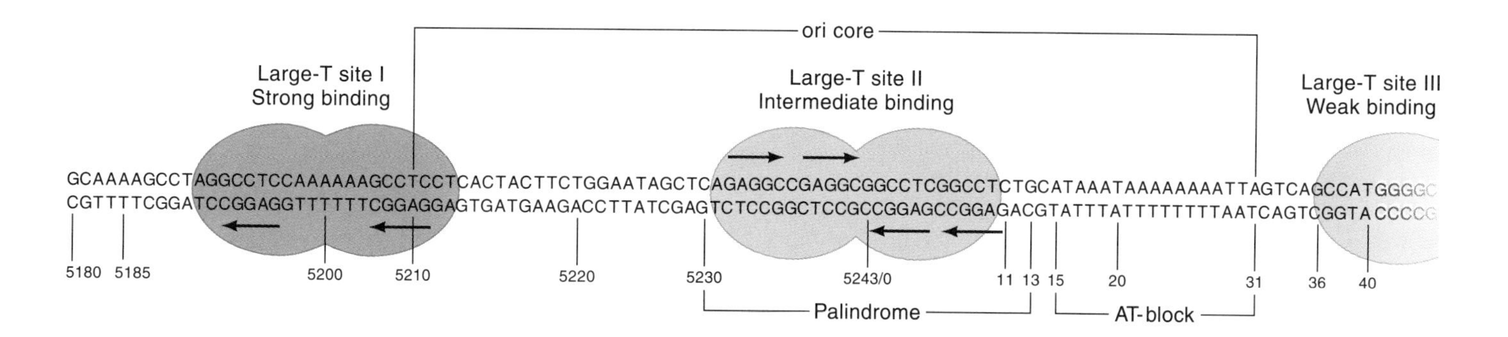

Figure 1.15
The SV40 origin of DNA replication and large-T binding sites. The SV40 origin is contained within bp 5210-31 (ori core) and contains a binding site for large-T (site II); the sequence that defines that binding site spans the central palindrome and AT-block. The paired palindromic sequences of GAGGC, shown as paired arrows oriented in different directions on the two strands, are the principal determinant for binding large-T. Another large-T binding site (site I) exists on the early region side of the ori between bp 5191–5213; this site contains two GAGGC sequences marked by two arrows on the same strand. Large-T's binding affinity for site I is greater than for site II. The binding of large-T to site I represses transcription of the early region. The existence of a third low-affinity binding site for large-T on the late region side of ori is problematic.

of a host of cellular genes. We discuss the structural basis for how large-T performs these myriad functions in detail in Section 1.3h.

Although the late mRNAs have a variety of 5′ ends, they all share a short common open reading frame that could encode a highly basic protein of 61 amino acids. This protein, referred to as the **agnoprotein**, is not incorporated into virions. However, because mutations that disrupt or even eliminate this open reading frame have only slight effects on the course of lytic infection of cultured cells, its function is unclear. There is suggestive evidence that the agnoprotein may have a subsidiary role in influencing the splicing pattern of late transcripts or in virion assembly or in their release from the dying cell.

f. Replication

Three approaches have contributed greatly to our current understanding of the participants in and mechanisms of SV40 DNA replication. The first is the study of the *trans-* and *cis-*elements encoded by the virus and required for replication and the structural features of the observed replicative intermediates *in vivo.* The second focuses on the cellular proteins and mechanisms involved in the replicative process as deduced from studies *in vitro.* The third concerns the structure and biochemical activities of large-T protein. The information generated by the two latter approaches revealed features and participants relevant to cellular DNA replication and how the large-T protein interacts with that machinery and subverts the normal controls for entry of cells into the S phase. Thus, one of the principal incentives for understanding the molecular details of SV40 DNA replication is the conviction that the identification of the requisite cellular replication proteins and their mechanism of action will help unravel the complexities of cellular chromosome replication.

SV40 DNA synthesis Initiation is the key step in SV40 DNA replication and the only SV40-specific event in the process. Replication begins when the large-T protein binds to ori and the cellular replication machinery consequently assembles

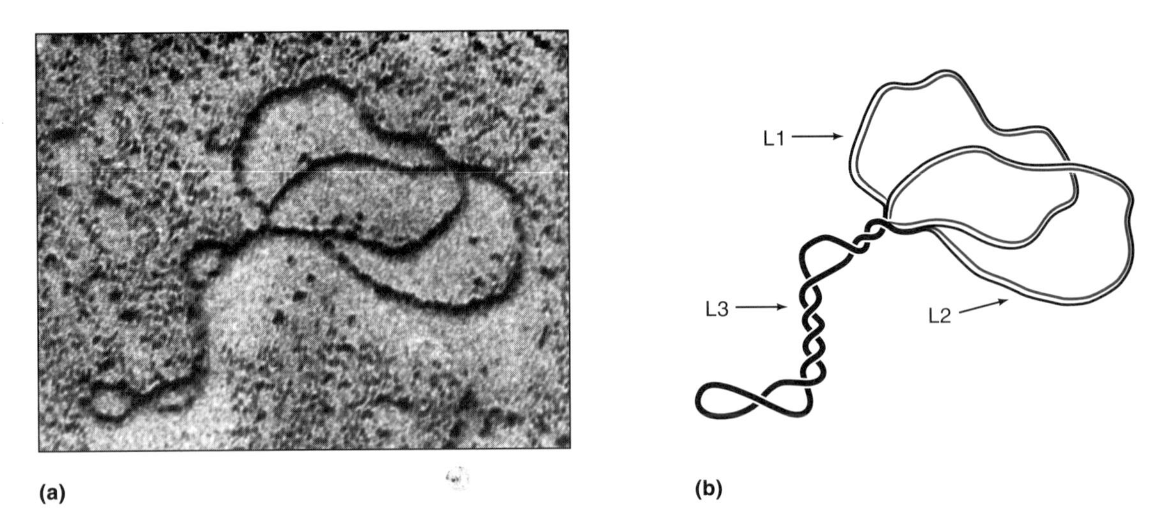

Figure 1.16
Replicating SV40 DNA. (a) An electron micrograph of partially replicated SV40 DNA; (b) diagrammatic representation of the electron micrographic image. L1 and L2 represent the newly replicated regions and are approximately equal in length; L3 represents the supercoiled region still to be replicated. The two oppositely moving replication forks are at the junction where L1 and L2 join L3. (Micrograph courtesy of M. L. DePamphilis.)

at that site. Once initiated, two replication forks are formed, and replication proceeds bidirectionally at nearly equal rates, creating molecules whose two replicated segments can be seen by electron microscopy (Figure 1.16). As DNA is unwound at the replication forks, the torsional strain is most probably relieved by the nicking activity of a topoisomerase I. As replication progresses, the newly replicated DNA is relaxed while the unreplicated region remains supercoiled (Figure 1.17). When the two replication forks meet about halfway around the circular DNA, replication stops, and most of the parental DNA between the forks is unwound to release two circular monomers each containing a gap in the nascent DNA strand. The gap is filled, and covalently closed circular DNA is produced, presumably by the sequential actions of a DNA polymerase and a DNA ligase. Occasionally, the two forks meet, running into each other, and form intertwined circular dimers referred to as catenanes, which can be separated by the action of topoisomerase II.

Replication at each fork occurs continuously on one template strand (the "leading" strand) and discontinuously on the other (the "lagging" strand) (Figure 1.18). This follows from the ability of all DNA polymerases to synthesize chains in only the 5′-to-3′ direction and not 3′-to-5′. Consequently, once initiated with a short RNA primer (about ten nucleotides) at ori, the leading strand is continuously extended to the completion point by DNA polymerase δ in association with several proteins, particularly one that ensures a high degree of processivity

---

Figure 1.17 (*facing page*)
SV40 DNA replication. Large-T and the host cell replication proteins bind at ori and replication proceeds bidirectionally, thereby causing more DNA to relax as DNA copying proceeds. Two kinds of end products result: catenated dimers and gapped monomeric circles. Catenated dimers are converted to monomeric-sized viral DNA by concerted cleavages and rejoining by topoisomerase II; gapped circles are completed by gap filling and ligation.

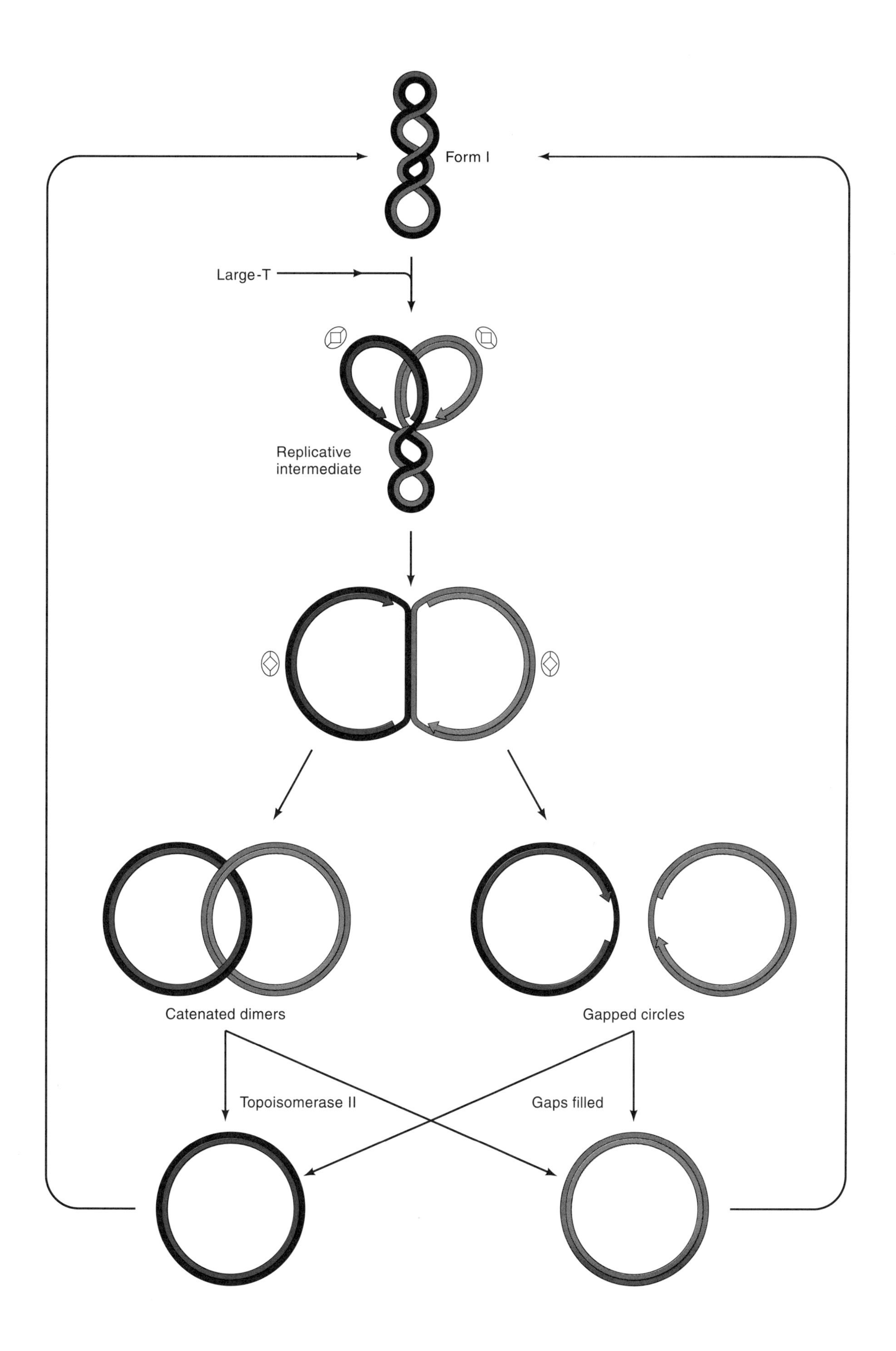
Form I
Large-T
Replicative
intermediate
Catenated dimers
Gapped circles
Topoisomerase II
Gaps filled

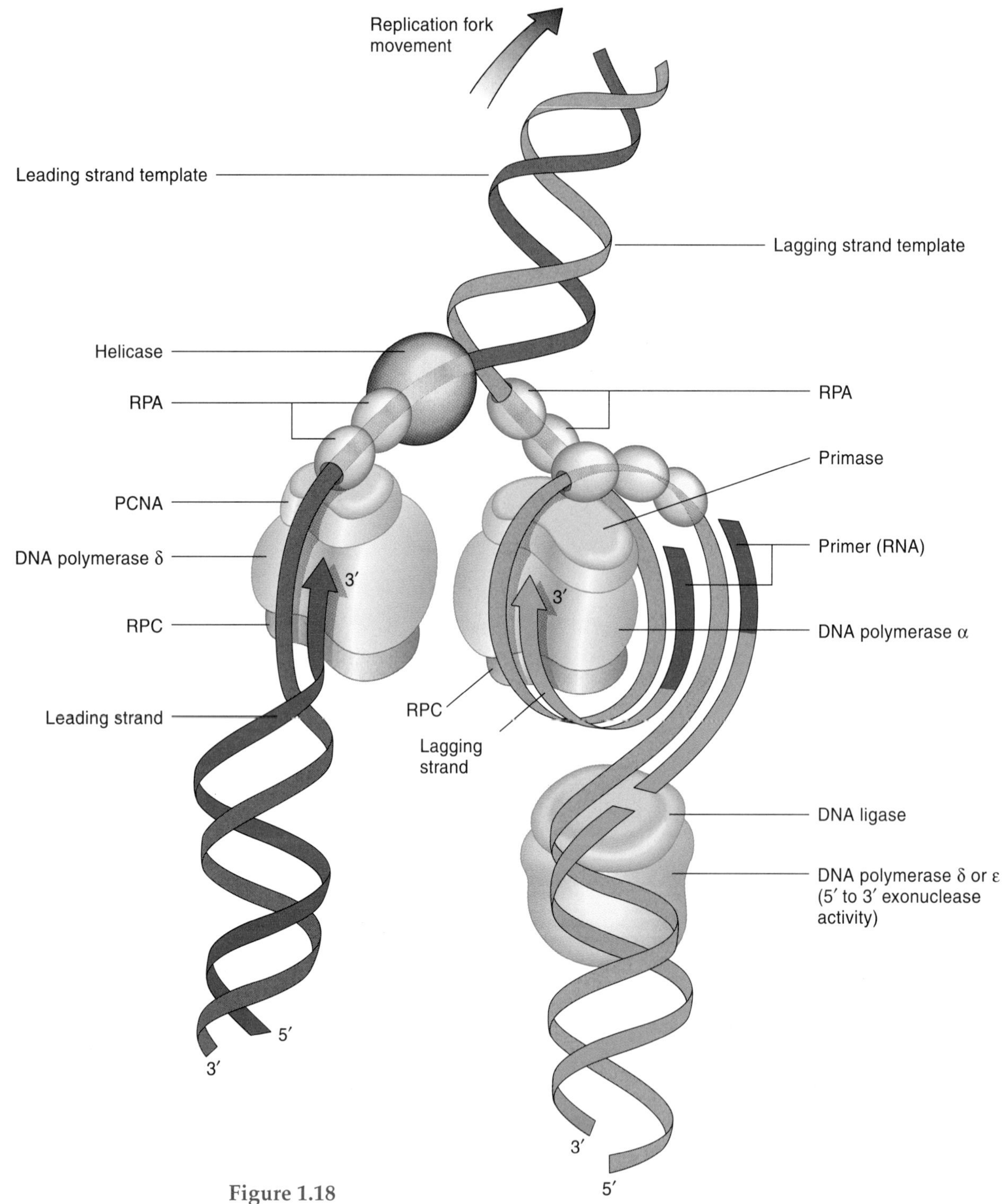

Figure 1.18
Model of DNA replication at a fork. The parental DNA is shown as two helical ribbons in light gray and dark gray tones. A helicase unwinds the duplex in advance of the replication proteins and the single strands become associated with RPA. The leading strand is synthesized continuously in the 5′ to 3′ direction by DNA polymerase δ in conjunction with RPC and PCNA. The lagging strand is synthesized in segments; each segment is initiated by the formation of a short RNA by a polymerase α-primase complex, and these RNAs serve as primers for elongation by DNA polymerase δ or ε. The growing end of a newly initiated lagging-strand fragment ultimately reaches the previously synthesized segment, the RNA end is removed, the gap is filled, and the ends are joined. Thus, although each segment is made in the 5′ to 3′ direction, the lagging strand grows in the 3′ to 5′ direction.

(e.g., proliferating cell nuclear antigen, PCNA). Discontinuous synthesis of the lagging strand occurs by frequent initiations to produce multiple DNA segments that are ultimately joined to the growing lagging strand. The frequent initiations needed for lagging strand assembly are made by the DNA polymerase α–primase complex, the first step being the synthesis of an RNA primer and its elongation by DNA polymerase α to produce segments about 100 to 200 nucleotides in length (Okazaki fragments). Thereafter, the RNA segments at the 5′ ends of the Okazaki fragments are eliminated by the combined action of RNase H and the 5′-to-3′ exonuclease activity of DNA polymerase α. After the gaps between Okazaki fragments are filled by either DNA polymerase δ or ϵ, possibly in conjunction with other proteins that facilitate gap filling, the fragments are joined to the end of the growing continuous lagging strand by a DNA ligase.

Replication initiation by large-T binding to ori The sequence that is both necessary and sufficient to permit the initiation of SV40 DNA replication, the **ori core,** lies between bp 5210 and bp 31 (Figure 1.15). This stretch includes the 27 bp palindromic, repeat straddling bp 5243/0, a 20 bp palindromic sequence flanking it on the early region side (EP), and the adjacent sequence of 17 A-T base pairs on the late region side. Modifications of sequences within the ori core, particularly in the 27 bp palindrome, eliminate ori's activity. Alterations in the flanking regions have variable but occasionally drastic consequences. Although not essential, the GC-rich repeat sequences up to about bp 100 facilitate the initiation process *in vivo,* possibly by their enhancement of early region transcription.

The principal determinants of ori function in replication occur in large-T binding site II (Figure 1.15); these are the four 5′-GAGGC-3′ sequences arranged in pairs in each arm of the 27 bp central palindrome. Binding of large-T to these sites positions the protein at the proper location and orientation for interaction with the cellular replication proteins.

Binding of large-T to the ori core is markedly stimulated by ATP and even nonhydrolyzable ATP analogues, indicating that hydrolysis of the pyrophosphate bond is not required. Upon binding ATP (or the analogues), large-T protein undergoes a conformational change permitting it to cover the entire ori sequence in close apposition to all faces of the DNA. Scanning transmission electron microscopy studies have shown that, in the presence of ATP, large-T binds to the ori core as a two-lobed structure, each lobe consisting of a hexamer of large-T (Figure 1.19). There is evidence that at low levels of large-T, the formation of the bilobed structure is facilitated by DNA polymerase α. In binding the bilobed structure, the DNA undergoes structural distortions at the short palindrome on the early region side of the central 27 bp palindrome and at the AT-block on the late region side; the end result is a local untwisting of the ori core DNA. Up to this point, no hydrolysis of ATP need occur.

In the next step, which is accompanied by ATP hydrolysis, the helix undergoes a limited bidirectional unwinding. This unwinding can be extended if topoisomerase I is available to relieve the torsional strain produced by the localized unwinding and if mammalian single-strand binding protein (mSSB or RPA) is present to stabilize the single strands that are produced. Thus, large-T protein is a virus-encoded DNA helicase that utilizes the energy of ATP hydrolysis to translocate on DNA strands in the 3′-to-5′ direction, melting any duplex regions it encounters. The SV40 large-T protein's ability to promote SV40 DNA synthesis therefore depends on its ability to bind specifically at the SV40 ori sequence and then, in an ATP-dependent manner, to create a regional bubble sufficient to permit initiation by the cellular replication machinery. This kind of bidirectional replication is characteristic of cellular chromosome replication.

In vitro replication Many of the components and mechanisms involved in the complete replication of SV40 DNA were identified by biochemical fractionation studies. Besides large-T protein, these include seven purified fractions: DNA

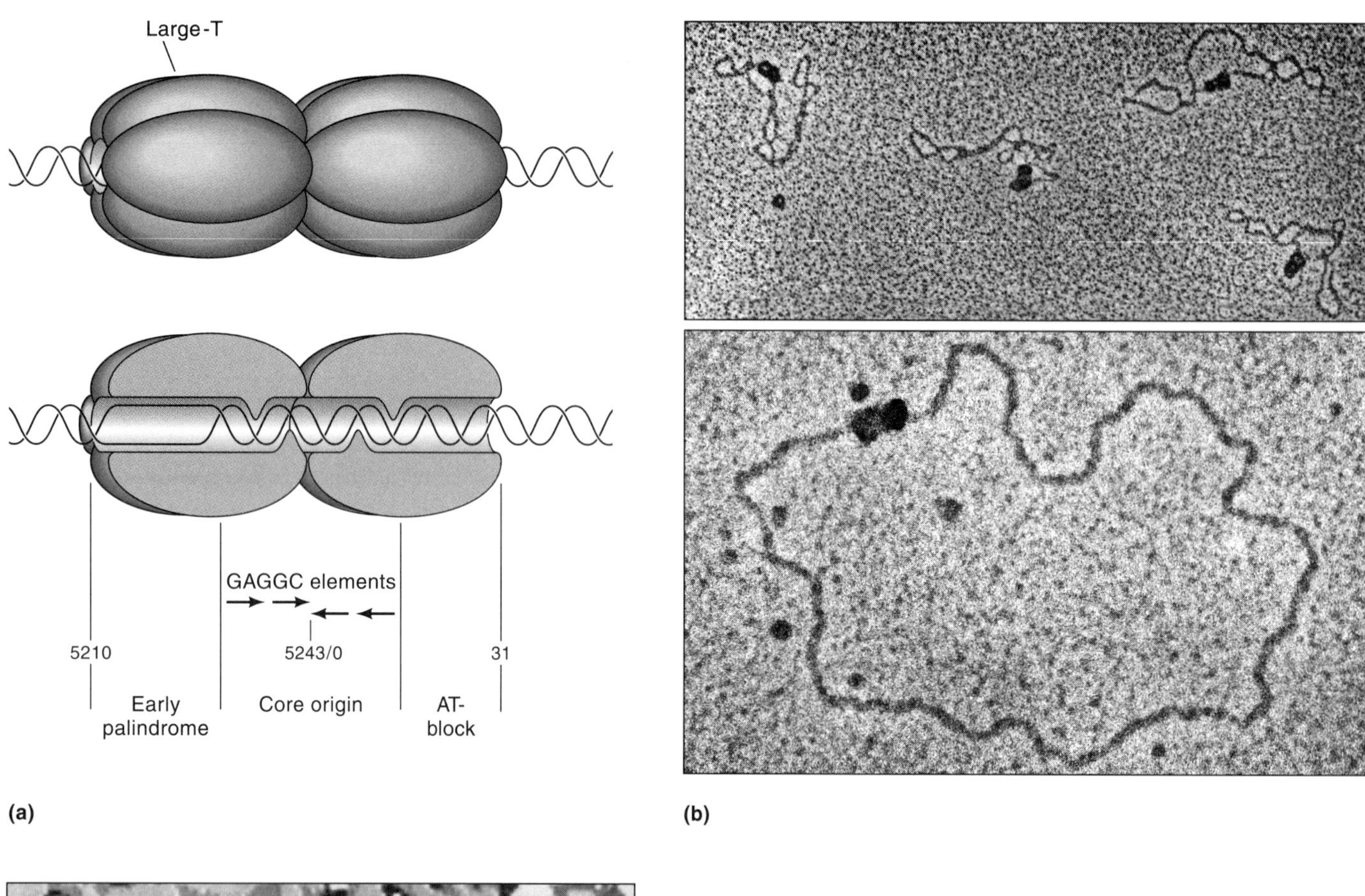

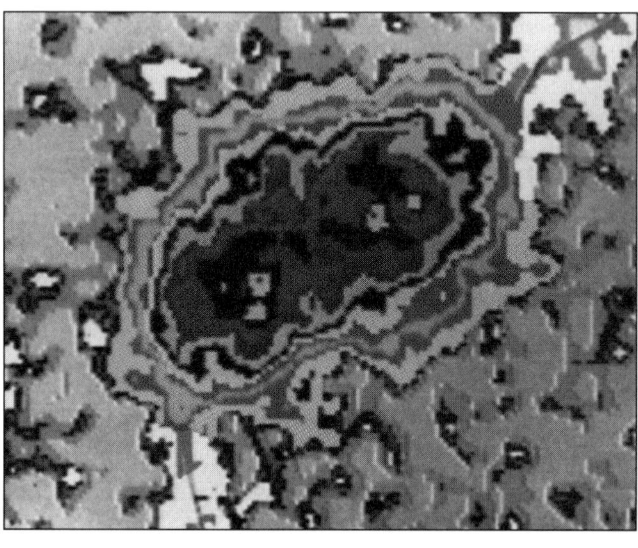

Figure 1.19
(a) Large-T binds to SV40 ori as paired hexamers. The drawing shows that the bilobed paired hexamer binds across the region containing the four GAGGC elements (shown as oppositely pointing short arrows) and the flanking palindrome and AT-block on the early and late region sides, respectively. (b) Electron micrographs showing large-T bound at ori of supercoiled DNA (top micrograph) and relaxed SV40 DNA (bottom micrograph). (c) STEM (scanning transmission electron microscopy) image of paired hexamers of large-T bound to SV40 ori. DNA strands can be seen emanating from the complex at the top right and bottom of the picture. The molecular mass estimated from STEM is consistent with a bilobed structure composed of paired hexamers of large-T. (The micrograph showing large-T bound to supercoiled DNA and the STEM micrograph were provided by J. Hurwitz; the micrograph of large-T bound to relaxed circular DNA was provided by J. Griffith.)

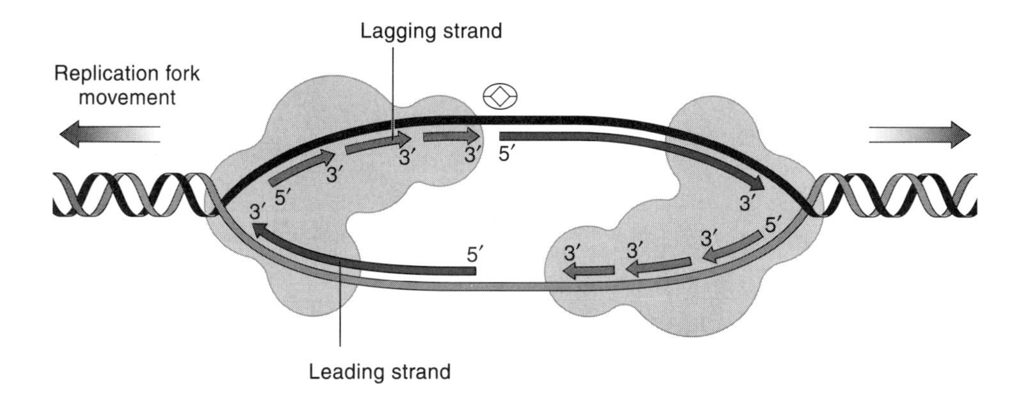

Figure 1.20
Bidirectional replication from SV40 origin. Binding of large-T, RPA, the DNA polymerase α-primase complex, and auxiliary factors at ori leads to the formation of a "bubble" within which replication begins at or near the center of the palindrome. Leading-strand synthesis is continuous, but lagging-strand production is discontinuous and is shown without the RNA segment at the 5′ ends and before ligation.

polymerase α–primase; DNA polymerase δ; the mSSB/RPA, which consists of 70 kDa, 34 kDa, and 13 kDa subunits; PCNA; replication factor C (RF-C); and topoisomerase I and II.

The sites and directions of RNA primer synthesis and extension of the primers as DNA are fixed through specific interactions between the large-T protein and DNA polymerase α–primase, both of which are bound at each end of the bubble (Figure 1.20). The initial extension of the leading strand is mediated by DNA polymerase α, probably assisted by RF-C. Subsequent elongation of the leading strand is catalyzed by DNA polymerase δ in association with RF-C and PCNA, which ensure high processivity. After the replication bubble is expanded somewhat, DNA polymerase α–primase initiates short RNA primers on the lagging strand templates within ori and continues the discontinuous mode of replication described earlier.

Although most studies of SV40 DNA replication *in vitro* have utilized free DNA as the template, minichromosomes containing a full array of nucleosomes can also serve as the template. Studies with purified proteins have shown that replication of the minichromosomes occurs without the displacement of the histone octamers (Figure 1.21). Moreover, the original nucleosomes are segregated equally between the leading and lagging strands at the replication fork; newly synthesized histones provide for the new octamers that must be associated with the newly formed daughter duplexes. This use of parental histone octamers may be important for maintaining the same differentiated chromatin state in the daughter duplexes as existed in the parental DNA before replication.

The *in vitro* studies of SV40 DNA replication have revealed a high degree of specificity in the interaction of large-T protein with parts of the cellular replication machinery. Thus, although the ATP-stimulated binding and untwisting of the ori core DNA is facilitated by any of several single-strand DNA binding proteins (e.g., *E. coli*, T4 phage, or mammalian), the subsequent ATP-dependent helicase action proceeds only with mSSB. Also, the 34 kDa subunit of the mSSB must be phosphorylated for it to function in DNA replication; this occurs by the action of a protein kinase that is activated by a cyclin made at the G1-S checkpoint in the cell cycle. Large-T must also be phosphorylated at a unique threonine residue by the same protein kinase and dephosphorylated at multiple serine sites before it can bind and untwist the DNA at ori.

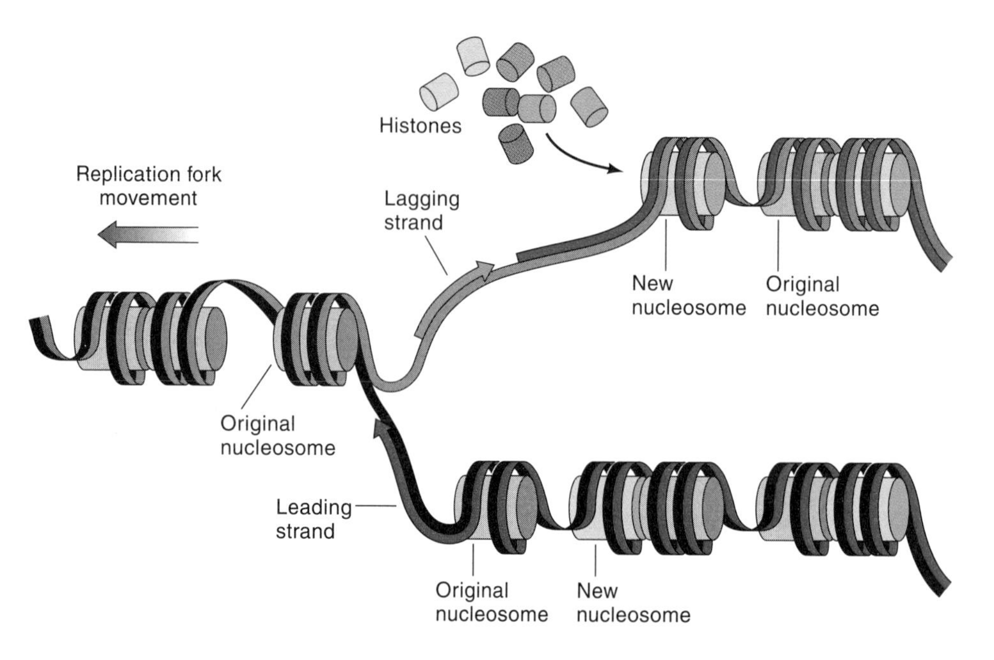

Figure 1.21
Fate of nucleosomes during DNA replication. A replication fork showing unreplicated DNA wrapped around nucleosomes at the left and newly replicated DNA associated with both old and new nucleosomes. The new nucleosomes are assembled as octamers from the four different kinds of histones (H2a, H2b, H3, and H4) shown in various shades of color.

Now we understand one of the reasons why SV40 replicates in primate cells and not in rodent cells. SV40 large-T protein interacts efficiently with the primate DNA polymerase α–primase complex but poorly with the corresponding mouse or hamster enzymes. By contrast, polyoma large-T interacts with the mouse DNA polymerase α more efficiently than with the corresponding primate enzyme.

g. Transformation

Section 1.3b describes SV40's capacity to induce a variety of new phenotypic properties, including oncogenicity, following virus infection or even viral DNA transfection of a wide variety of rodent cells in culture (Table 1.4). Such transformation rarely follows infection or transfection of permissive primate cells (monkey and human) but is a consistent outcome (frequencies 10^{-3} to 10^{-5} per infected cell) when nonpermissive cells (e.g., mouse, rat, and hamster) are exposed to the virus or its DNA. Neither tumors nor stable transformed cell lines induced by SV40 infection produce virus particles or express any of the virus late genes.

Large-T's ability to transform The principal virus protein needed for oncogenic transformation is large-T. Deletion and other nonconditional mutants that fail to express functional large-T and are thus unable to replicate their DNA following infection of permissive cells are also unable to effect cellular transformation of nonpermissive cells. SV40 mutants whose large-T protein is thermosensitive are defective for transformation if the infected cells are maintained at the nonpermissive temperature (usually 39°). Cells transformed with such ts mutants retain the transformed phenotype if maintained at a low temperature but revert to more normal morphology and growth characteristics when propagated at the elevated temperature. Thus, functional large-T is required to produce and main-

Table 1.4 *Properties of Cells Transformed by SV40*

Growth
High or indefinite saturation density*
Different, usually reduced, serum requirement*
Growth in agar or Methocel suspension—anchorage independence*
Tumor formation upon injection into susceptible animals
Not susceptible to contact inhibition of movement
Growth of cells in a less-oriented manner*
Growth on monolayers of normal cells*
Surface
Increased agglutinability by plant lectins*
Changes in composition of glycoproteins and glycolipids
Tight junctions missing
Fetal antigens revealed
Virus-specific transplantation antigen
Different staining properties
Increased transport rate of nutrients
Increased secretion of proteases or protease activators*
Intracellular
Disruption of the cytoskeleton
Altered amounts of cyclic nucleotides
Evidence of virus
Virus-specific antigenic proteins detectable
Viral DNA sequences detected
Viral mRNA present
Virus can be rescued in some cases

Note: Transformed cells show many, if not all, of these properties, which are not shared by untransformed parental cells.
*Several of these properties have formed the basis of selection procedures for isolating transformants.

tain the transformed properties. That large-T is also sufficient to cause cellular transformation stems from the finding that the injection of DNA containing only the SV40 early region into the nucleus of cultured cells leads to transformation of those cells that integrate and continue to express the large-T gene. Even the injection of purified large-T protein into susceptible cells transiently induces cellular changes characteristic of stably transformed cells. In some instances (e.g., with primary rodent cells and under certain conditions), and particularly when large-T is limiting, the expression of small-T influences the transformation efficiency and the properties of the transformed cells. Besides large-T, tumorigenesis in animals may require additional and as yet unidentified cellular events. Thus, mice carrying a large-T transgene, and expressing the large-T protein in many of their tissues, nevertheless develop only certain types of brain tumors. And when large-T is expressed in only certain cell types, only occasional and local tumors arise in those cells.

More than 95 percent of the large-T protein synthesized in the cytoplasm enters the nucleus, where it accumulates; small amounts of variously modified large-T protein are found in the cytoplasm and plasma membrane. The oncogenicity of large-T protein depends on its presence in the nucleus, where it interacts with proteins that control the cell division cycle. The function of the extranuclear forms of large-T are unknown, although plasma membrane–associated large-T-like protein appears to be the target for the immunological rejection of cells expressing large-T.

Integrated viral DNA in transformants Generally, integration of viral DNA at chromosomal sites that permit expression of large-T is the limiting determinant of the transformation frequency. DNA blotting of a variety of restriction endonuclease digests of transformed cell DNA demonstrate that the viral and cellular chromosomal sequences are joined covalently in the integrated DNA. Even more persuasive evidence for the state of the viral DNA is based on analysis of cloned DNA segments that includes the integrated viral DNA and its joints with the immediately flanking cellular DNA. These studies established that the viral DNA and cellular DNA are joined at many different, probably random, sites.

There is no apparent specificity for either the viral or cellular DNA sequences at the integration junctions. Double-strand breaks at random locations in the viral and cellular DNA, and nonhomologous, most often nonconservative, joining of viral and cellular DNA ends are believed to account for the diverse integration patterns. Most frequently, deletions and rearrangements of the cellular and viral DNA occur at the integration sites. Integration establishes the viral DNA as a permanent feature of the cell's genome that is passed on to the progeny at all subsequent cell divisions. Transformed cells often contain multiple copies of integrated viral DNA. At least one of the integrated copies must retain an intact and expressible early region, thereby permitting the continued production of large-T and small-T. Subsequent rearrangements leading to loss or disruption of the integrated sequence responsible for large-T production occasionally occur and lead to the disappearance of large-T protein and a loss of the transformed phenotype.

Mitogenic cellular responses induced by large-T expression Large-T protein is a multifunctional protein with distinctive domains, several of which regulate the expression of early and late viral genes and initiate viral DNA replication. Some of these, as well as other domains, are particularly relevant to large-T's transforming capability.

Among the earliest discoveries regarding transformation was the finding that virus infection of nondividing rodent cells induces enzymes involved in the synthesis of DNA precursors (i.e., nucleotides). Even more striking was the elevation in the level of enzymes devoted to the synthesis of DNA precursors and of DNA polymerase followed by a round of cellular DNA synthesis and mitosis (**mitogenesis**). The changes in enzyme levels and the onset of DNA synthesis occur in most of the infected cells. Generally, these changes are only transitory, a phenomenon called **abortive transformation**. Only those rare cells that become stably transformed maintain the elevated levels of the metabolic and synthetic enzymes and replicate their DNA without the long delay associated with a normal G1 phase. SV40 mutants that are unable to replicate their DNA because of an inactive large-T protein are defective for abortive as well as permanent transformation. A variety of experiments support the conclusion that large-T is responsible for the induction of enzyme synthesis and unscheduled cellular DNA replication. But how?

Large-T binding to cellular proteins A partial answer emerged from the finding that large-T binds to several cellular proteins, some of which regulate the critical transition from the G1 to the S phase of the cell cycle. In the absence of these proteins, whether by inactivation or deletion of their genes, certain normal constraints on cell division are relaxed or lost.

These cellular proteins were first recognized when extracts of SV40-infected or transformed cells were treated with antibodies to large-T protein. Among the precipitated proteins were several new entities. One (the p53 protein) has a molecular weight of 53 kDa, another (p105), of 105 kDa, and one (p107) of 107 kDa. Subsequently, the genes encoding these proteins were cloned, sequenced, and characterized. The gene encoding the p53 protein was ultimately discovered to be a tumor suppressor gene, that is, a gene whose inactivation or loss of both alleles predisposes individuals to a variety of tumors. The **retinoblastoma (Rb) gene**, which specifies the Rb protein, p105, is another tumor suppressor gene, identified independently in studies of individuals who, because of the loss of function of both Rb alleles, have an increased incidence of retinoblastomas as well as osteo- and other sarcomas. The gene encoding p107 has also been cloned and sequenced and shown to be a member of the Rb-like proteins; it binds several of the same cellular proteins as Rb.

Therefore, a plausible model is that the tumorigenicity of large-T protein is a consequence of the sequestration and functional inactivation of the tumor suppressor proteins Rb and p53 and perhaps other growth regulatory proteins. But how do such proteins regulate normal cell growth and division, and how does association with large-T protein disrupt those functions? To explore these questions, we need to consider briefly the roles of Rb and p53 proteins in regulating the G1-S transition in the cell division cycle.

Rb and p107 proteins Rb is a nuclear phosphoprotein whose extent of phosphorylation varies throughout the cell cycle. It is in a hypophosphorylated state during most of G1 and in nondividing and differentiated cells but becomes hyperphosphorylated in proliferating cells late in G1 and remains so during S, G2, and M (Figure 1.22). Thus, the change from the hypophosphorylated to hyperphosphorylated state and back again can serve as a switch for passage from G1 to S and from M to G1. But what does the switch activate? Hypophosphorylated Rb binds strongly to a number of proteins, at least one of which, E2F, is

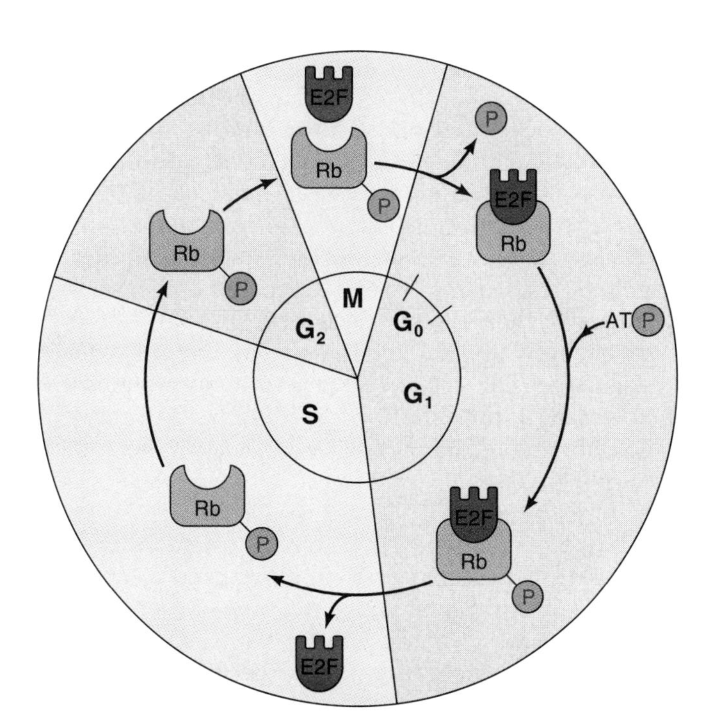

Figure 1.22
The role of phosphorylation and dephosphorylation in regulating the association and dissociation of Rb and E2F during the cell cycle. The transcription factor E2F is bound to hypophosphorylated Rb (shown without a P) as cells progress from the mitotic (M) phase of the cell cycle into G_1. Late in G_1, Rb becomes hyperphosphorylated (shown by the addition of a single P) by a G_1-specific cyclin-dependent protein kinase. E2F dissociates from hyperphosphorylated Rb and can activate transcription of genes whose products are needed for high rates of DNA synthesis. Conversion of Rb to a hypophosphorylated state during M phase (shown by the removal of P) promotes reassociation of E2F.

a cellular transcription factor that appears to activate transcription of many genes, several of which encode enzymes involved in DNA replication. When bound to a specific site in the Rb protein (the "Rb pocket"), E2F can still engage its target sites in DNA, but it does not activate transcription. Thus, in early to mid G1, the hypophosphorylated Rb protein is presumed to sequester and prevent the action of at least one transcriptional regulatory protein needed to mobilize the DNA replication system. Late in G1, when a constitutively expressed cyclin-dependent protein kinase (CDK) is activated by the appearance of a G1-S phase–specific cyclin (cyclin D or cyclin E), Rb protein becomes hyperphosphorylated on serines and threonines. Upon hyperphosphorylation, E2F dissociates from the Rb binding pocket and, in conjunction with other factors, activates transcription of target genes, enabling cells to assemble the enzymes and precursors needed for DNA synthesis during the S phase. Following mitosis, Rb protein returns to its hypophosphorylated state because of the rapid decay of the G1-S–specific cyclin and the attending inactivation of the CDK and possibly the activation of specific or general protein phosphatases. In this state, Rb can once again bind E2F and delay the cell's entry into the next S phase.

Large-T also binds to the hypophosphorylated Rb protein pocket and, in doing so, excludes E2F. Thus, the binding of large-T may override the normal Rb protein–mediated G1-S checkpoint by mimicking the effects of hyperphosphorylation. There is evidence that other proteins can bind the hypophosphorylated Rb protein pocket, and these too may be involved in regulating the G1 to S transition. These proteins are also displaced by the binding of large-T to Rb. Thus, the immortalizing, transforming, and oncogenic properties of large-T may be explained, at least in part, by its ability to inactivate those properties of Rb that help regulate the switch controlling passage from the quiescent G1 state to one of active chromosomal replication prior to cell division. Several observations support this interpretation of large-T activity. Mutations that prevent large-T's binding to the Rb pocket also inactivate its transforming ability, and mutational alterations in the Rb pocket that prevent the binding of large-T and E2F also inactivate the Rb protein's ability to control growth. Moreover, other viral oncoproteins—polyoma large-T, adenovirus E1A, and papilloma E7—share with SV40 large-T the ability to bind to the Rb pocket with similar outcomes.

The p107 Rb-like protein also has an E2F binding pocket, although it is somewhat larger and of a different design. At the onset of S phase, the p107-E2F complex associates with the S phase–specific cyclin A-CDK complex. This complex is able to phosphorylate histone H1 and possibly other DNA-bound proteins; E2F may facilitate CDK's action on DNA-bound proteins through its own DNA binding potential. In binding to p107, SV40 large-T alters the interaction with E2F, thereby abrogating the events mediated by the p107-E2F-cyclin A-CDK complex during S phase.

p53 protein Initially, p53 was believed to possess transforming activity because transfections with vectors expressing p53 immortalized primary cells in culture and cooperated with other oncogenes (e.g., with *ras*) to enhance the transformation of normal cells to tumorigenicity. Later, it was discovered that only certain mutated forms of p53 were oncogenes. It soon became clear that cotransfection of the wild-type p53 gene with a variety of oncogenes suppresses rather than helps the latter's tumorigenicity. Indeed, transfection of wild-type p53 gene into transformed and a variety of cancer cells, particularly ones lacking both p53 alleles, blocks cell proliferation by arresting cells late in G1. Now we know that mutant forms of p53 are tumorigenic in at least two ways: one is by a gain of function, the nature of which is unknown, but seems to be required for immortalization and/or transformation; the second way stems from the ability of mutant p53 to form complexes with the endogenous normal p53, thereby inhibiting p53's natural growth suppressor function. Thus, wild-type p53 behaves not as an oncogene but as a tumor suppressor. This view is further supported by the finding that loss or inactivation of p53 function occurs in more than 50 percent of a wide variety of human tumors. Furthermore, individuals with Li-Fraumeni syndrome who inherit only a single normal p53 allele develop a variety of tumors in which the normal p53 allele is lost or inactivated. A plausible presumption is that, by binding to p53 protein, large-T blocks p53's growth-suppressing property. To understand how, we need to consider the functions of p53.

p53 has been implicated in several different kinds of activities, each of which contributes to its putative growth suppressor function. First, wild-type p53, but not mutant forms, binds to a large number of different sites in DNA, many of which contain the consensus sequence 5′-PuPuPuC(A/T)(T/A)GPyPyPy-3′; additional binding motifs that have been noted are repeats of 5′-TGGCT-3′. Interestingly, some of the sequences to which p53 binds have been implicated as sites where cellular DNA replication initiates. This relates to the finding that sequences to which wild-type p53 binds have been mapped at or near replication origins. One of the sequences to which p53 binds is at the 5′-GGGCGG-3′ repeats in the three GC-boxes of the SV40 early promoter that is adjacent to the SV40 ori. Binding at that site inhibits replication, in part by inhibiting large-T's helicase activity and ability to interact with DNA polymerase α. Sequestration of p53 by large-T may therefore facilitate the replication of SV40 DNA. Perhaps p53's varying availability has a role in regulating the initiation of cellular DNA replication or processes involving DNA synthesis.

Second, there is considerable evidence that p53, by virtue of its ability to bind to specific promoter proximal sequences, can activate or inhibit transcription from the neighboring promoters. p53 has a demonstrable transcriptional activation domain at its amino terminus and a putative DNA-binding domain at its carboxy terminus. Moreover, when the amino terminal domain is fused to the GAL4 DNA binding domain, minimal promoters flanked by a GAL4 DNA binding site are activated (see *Genes and Genomes,* section 8.3f). Presumably, promoters with a p53 response element in their transcriptional control region could respond similarly. One such gene, *WAF1/Cip1,* which encodes a 21 kDa protein, has been identified. Its promoter contains a p53 binding site and is transcriptionally activated by the binding of p53. The WAF1/Cip1 protein binds to G1 cyclin–CDK complexes, preventing the phosphorylation of Rb and entry into S phase. Thus, p53, but not mutant p53, blocks the G1 to S phase transition indirectly by regulating the expression of *WAF1/Cip1,* which blocks the critical phosphorylation of Rb and probably other proteins as well. Thus, the actions of *WAF1/Cip1* and SV40 large-T proteins are completely antagonistic in the ways they regulate cell cycle progression.

Transcriptional repression by wild-type but not mutant p53 has been observed with a wide variety of promoters that lack p53 response elements, including several that are activated during the transition from G1 to S induced by serum growth factors: promoters regulating the expression of *Fos, Jun,* and PCNA, for example. There is clear indication that wild-type p53 can interact with the TATA-binding protein directly and interfere with the initiation of transcription. Thus, wild-type p53 can apparently transactivate genes that contain a p53-specific DNA binding site and, in a more general way, suppress the transcription of genes that lack a p53 binding site. Quite likely, p53 exerts its tumor suppressor activity by down-regulating genes governing proliferation and up-regulating genes involved in preventing growth.

Transcriptional activation by p53 can be blocked by binding to another protein, MDM-2, which itself has oncogenic properties when overexpressed; apparently, MDM-2's oncogenicity resides in its ability to modulate p53's transcriptional activities. Assuming that these findings are physiologically relevant, p53 may both activate and repress different sets of genes needed to maintain cells in G1 until the proper signals for progression into S phase are received.

A third and intriguing role for p53 purports its role in arresting cell growth in response to DNA damage. Cells exposed to UV light, ionizing radiation, or

DNA-damaging chemicals respond with an increased rate of p53 synthesis. The elevated level of p53 functions to delay transit from G1 to S phase and the onset of mitosis. This reduces the likelihood of incorporating the mutagenic effects of the DNA damage in the next generation. With excessive damage or a failure to effect sufficient repair before the onset of S phase, cells undergo a p53-assisted process of cell death known as **apoptosis**, a specialized form of programmed cell death accompanied by a characteristic pattern of chromosomal DNA fragmentation and cellular dissolution. Thus, virally transformed or spontaneously arising tumor cells lacking functional p53 protein are unable to arrest proliferation following exposure to DNA-damaging agents and to induce the apoptotic response. As a consequence, p53-deficient cells are particularly prone to sustaining additional mutations and to genetic instability such as chromosomal rearrangements and gene amplification in response to minimal amounts of DNA damage. Not surprisingly, transfection of p53-deficient cells with a normal p53 gene leads to a more normal growth arrest response and a decreased propensity for chromosomal anomalies and regional gene amplifications.

How p53 senses and transmits the existence of DNA damage is unclear. Nor is it known under what conditions and how p53 triggers the apoptotic response. p53 itself may serve that purpose or act in conjunction with other proteins. For example, damage to DNA might induce the expression of p53, and the elevated level of p53 may block the cell's entry into S phase until the damage is repaired. Alternatively, the elevated level of p53 could influence other gene products that are more directly implicated in controlling the cell cycle and the responses to its arrest. Support for the latter mechanism derives from p53's ability to activate transcription of the *WAF1/Cip1* gene, the product of which blocks progression of the cell cycle at the G1-S transition. However, it may also activate one or several genes that respond to DNA damage by arresting growth (e.g., GADD45, which contains a p53 response element in its intron). In this regard, cells from patients with the inherited autosomal recessive disease ataxia telangiectasia are sensitive to irradiation and predisposed to tumors; interestingly, the increase in p53 is delayed and GADD45 fails to be induced after such cells suffer DNA damage. The precise way in which p53 regulates the response to DNA damage remains unclear.

Thus, p53 can act as a brake on progression from G1 to S in at least three possible ways: activation of genes that negatively regulate cell growth, especially ones that respond to DNA damage; repression of genes whose products are needed for cell cycle progression; and inhibition of critical steps in the initiation of chromosomal DNA replication. Each of these activities could be impaired by the binding of p53 to large-T or to the adenovirus-transforming protein E1B (Section 1.5b). As discussed in Section 1.4c, one of the transforming proteins, E6, of highly oncogenic human papillomaviruses functions by complexing p53 and promoting its degradation by the ubiquitin-protease system.

h. Structure-Function of Large-T Protein

Previous sections described numerous activities of SV40 large-T protein. These include the regulation of the timing and level of its own expression, the initiation of viral DNA replication, and the activation of viral late gene expression. By its ability to bind to the Rb and p53 proteins, large-T attenuates the control these proteins exercise on the orderly transition from the G1 to the S phase of the cell cycle. These multiple and diverse activities of large-T raise the question of how the single polypeptide chain carries out so many functions and in an integrated and programmed fashion.

Biochemical and structural characteristics The large-T protein contains 708 amino acids with a calculated molecular weight of 82.5 kDa. Various modifications of

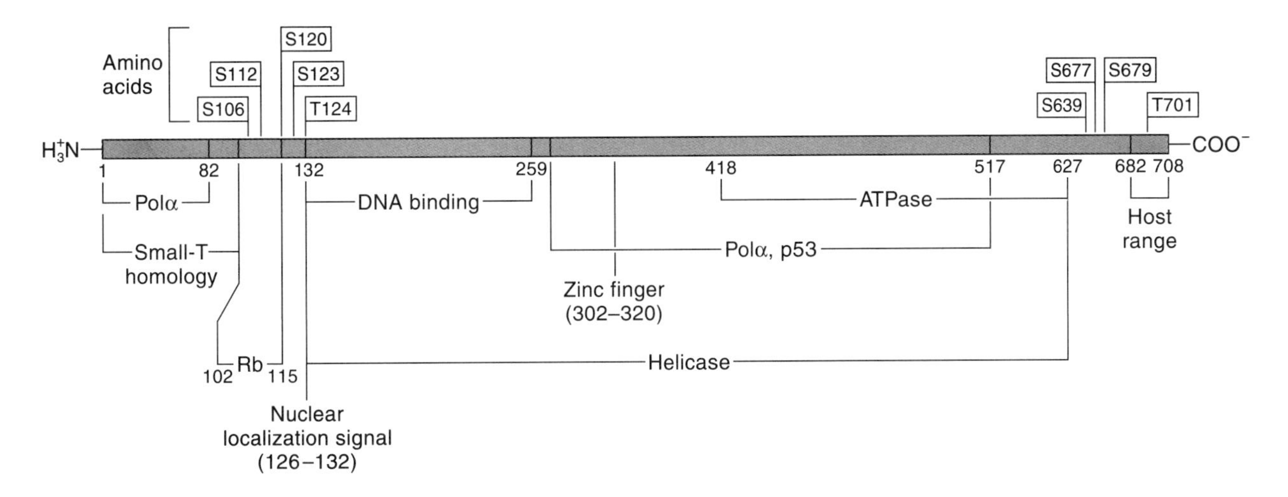

Figure 1.23
The functional domains of SV40 large-T. The solid bar represents the protein's length from amino to carboxy termini. Notable features and functional domains are indicated by brackets or by the amino acid residues that define a domain. Specific phosphorylated serine (S) and threonine (T) residues are indicated in rectangular boxes at their respective positions.

large-T have been detected (e.g., N terminal acetylation, O glycosylation, acylation, adenylylation, and poly ADP ribosylation), but the only modification with known functional significance is phosphorylation. As is commonly done for analysis of structure-function relationships of proteins (Chapter 10), a large variety of sequence modifications (e.g., deletions, truncations, and single or multiple base pair changes) was made in the gene encoding large-T to produce a commensurately wide range of altered large-T proteins. These altered proteins were then tested for the various known biological functions. Not surprisingly, different functions were ascribed to particular regions of the polypeptide chain, only several of which may constitute discrete three dimensional domains (Figure 1.23). Knowledge of the precise nature of the three-dimensional organization of the various regions awaits a high-resolution solution of the protein's structure.

Analysis of large-T proteins altered by deletion and expressed in *E. coli* established that amino acids 131 to about 250 are sufficient for sequence-specific binding to sites I and II on SV40 DNA, although this capacity is also influenced somewhat by other amino acids outside this region. The quite basic sequence from residue 126 to 132 is sufficient to localize large-T to the nucleus; indeed, if that heptapeptide is attached to a normally cytoplasmic protein, that modified protein is targeted to the nucleus. The region responsible for binding to the Rb pocket encompasses only about 13 amino acid residues (103 to 115), and, as expected, amino acid changes in this region impair complex formation. Two separate regions appear to interact with the host DNA polymerase α–primase: amino acids 1–83 and 259–517. The region assigned the helicase function covers most of the protein (residues 131–627); this region has also been shown to be responsible for ATP binding and ATPase activity, for association with DNA polymerase α, and for complexing with p53 (residues 259–517).

Regulatory consequences of large-T phosphorylation Large-T protein isolated from infected or transformed cells is phosphorylated at serine and threonine residues clustered within two small regions near the amino and carboxy termini (Figure 1.24). Several observations focused attention on the relevance of these phosphorylated residues. The fully phosphorylated large-T is inefficient at initiating SV40 DNA replication *in vitro*. However, if all the phosphates are removed by treat-

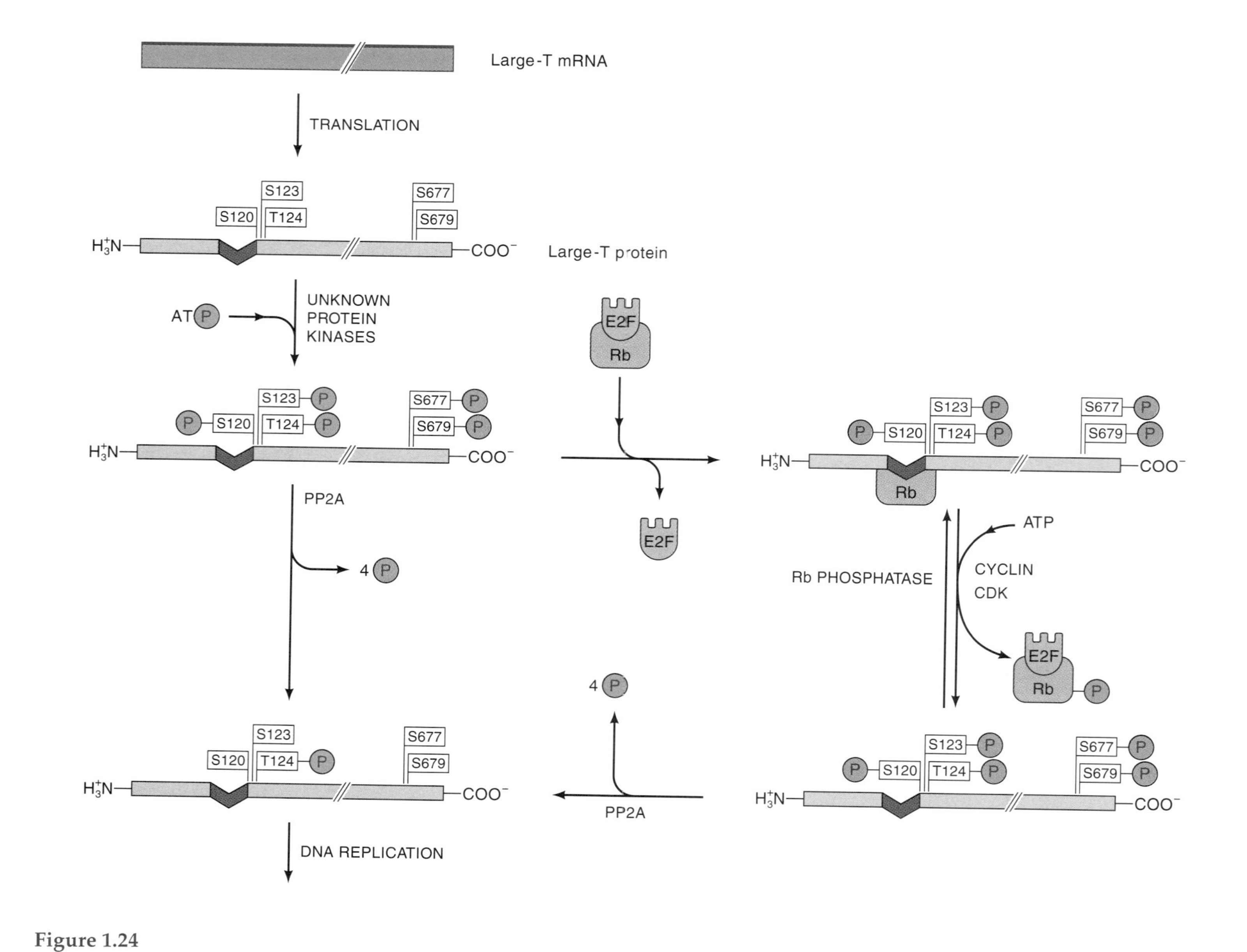

Figure 1.24
Phosphorylation and dephosphorylation of large-T and Rb control large-T's availability and activity for viral DNA replication. Translation of large-T mRNA yields a protein which is presumably rapidly phosphorylated at the indicated serine (S) and threonine (T) residues. This phosphorylated form cannot initiate viral DNA replication, but it can bind to hypophosphorylated Rb, which in turn induces cells in G_1 to enter S phase. Subsequently, phosphorylation of Rb by a cyclin-dependent protein kinase (Cdk) induces the release of hyperphosphorylated large-T. Removal of the serine phosphates in a PP2A-mediated reaction yields large-T that is able to bind to ori and initiate viral DNA synthesis. PP2A can also act directly on unbound hyperphosphorylated large-T to permit viral DNA replication.

ment with phosphatases, the protein is unable to bind to ori or support DNA replication; large-T synthesized in *E. coli* is similarly inactive. The ATPase and helicase activities of the two dephosphorylated large-T preparations are, in contrast, unaffected. If, rather than removing all the phosphates from the large-T isolated from animal cells, the serine-linked phosphates are removed selectively with alkaline phosphatase, the DNA binding and replication activities increase. Moreover, the same activation can be achieved by selectively removing the phosphates from only serine-120 and serine-123. These experiments identify serine-120 and serine-123 as sites for down-regulation of DNA binding and replicative activity by phosphorylation.

In contrast to the effects of serine phosphorylation and dephosphorylation, treatments that remove threonine-linked phosphates from large-T protein result in a complete loss of the protein's DNA binding and replicative activities. The absolute necessity for phosphorylation at threonine-124 for ori binding and replication was established in two ways. First, modification of the large-T coding sequence leading to substitution of alanine for threonine-124 produces an inactive large-T, but a change resulting in a similar substitution of threonine-701 has no effect on binding to ori and support of replication. More significantly, phosphorylation of a completely dephosphorylated large-T protein at only threonine-124 results in the reconstitution of full activity for DNA binding and initiation of replication.

Fully phosphorylated large-T is inefficient at initiating SV40 DNA replication; however, it is the preferred substrate for binding to Rb. If replication of viral DNA requires that the infected cells be activated into S phase and that this occurs only after large-T binding to Rb, then large-T would seem to become fully phosphorylated soon after it is made. Recall that the preferred substrate for large-T binding is hypophosphorylated Rb, the predominant state of Rb during G1. As cells enter S phase, Rb becomes phosphorylated, and large-T is dissociated, at which time it becomes differentially dephosphorylated and available for initiating viral DNA replication. How the proper phosphorylation state of large-T for DNA replication is attained and regulated is not understood.

Serine-120 and serine-123 must be dephosphorylated and threonine-124 must remain phosphorylated before large-T can promote SV40 DNA replication. Serine and threonine protein kinases are known, but enzymes specific for phosphorylating the large-T residues have not been identified. Threonine-124 is readily phosphorylated *in vitro* by a cyclin-dependent protein kinase, but whether this enzyme or an analogous protein in primate cells is responsible for this essential phosphorylation is also unclear. Lastly, protein phosphatase 2A (PP2A) can specifically dephosphorylate serine-120 and serine-123; interestingly, this enzyme's activity is also regulated by phosphorylation and dephosphorylation. Moreover, SV40 small-T, the protein encoded by the second early gene, binds to and inhibits the phosphatase activity of PP2A. How this influences the early events of cell activation is unclear.

SV40 large-T protein is clearly designed for a multiplicity of purposes. In primate cells, besides its direct involvement in amplifying the infecting genome and ensuring the requisite expression of the SV40 proteins necessary for propagation of the virus, it plays a key role in preparing cells to support those processes. Having to rely on the infected cell's machinery for DNA replication and other macromolecular syntheses, the virus evolved, in the large-T protein, the capacity to abrogate the signals that otherwise keep cells quiescent and to induce changes that enable primate cells to carry out the reactions needed for viral multiplication. In rodent cells, large-T's ability to abrogate the signals that maintain cells in the quiescent state lead to tumorigenicity rather then cell death. This depends on the cell's ability to incorporate the viral DNA into its own genome and to express large-T continuously; as a consequence, the cell is locked into a continuous proliferative mode. Progression from an initially immortalized cell to a highly tumorigenic one may be the consequence of additional genetic changes resulting from the impairment of a cell's capacity to monitor or respond to normally induced DNA damage.

i. SV40 as a Transducing Vector

As knowledge about SV40's genome organization, sequence, and expression grew and the recombinant DNA technology for dissecting and reassembling DNA molecules advanced, it was inevitable that SV40 DNA would be viewed as a potential vector for delivering foreign DNA segments into mammalian cells. The logic is straightforward: replace discrete segments of viral DNA whose functions are known with segments of DNA to be investigated, and transduce the reconstructed genomes into appropriate cells. The newly introduced segments may have entirely autonomous capabilities (e.g., a complete gene or an intact promoter), or they may be segments that rely on viral sequences for their expression or function (e.g., cDNAs, enhancers, oris, introns, etc.).

Section 5.7b in *Genes and Genomes* describes a variety of SV40-based vectors, but we review a few basic considerations here. It is advantageous to use virions carrying recombinant genomes as vectors for transducing new DNA into mammalian cells rather than the naked DNA because infection is much more efficient than transfection. Consequently, the number of cells expressing or utilizing the introduced DNA is substantially increased, facilitating analysis. But there is a considerable price for this increased efficiency. First, the size of the transduced DNA is limited by the amount of DNA that can be accommodated in mature virions. This is estimated to be about 5500 bp for SV40, but in practice, the transduced segment cannot exceed half that size in order to retain enough of the viral DNA to ensure the genome's propagation and maintenance as a virus. For example, replacement of part or all of the nearly 2 kbp late region segment encoding VP1, VP2, and VP3 is necessary to accommodate a segment of approximately the same size containing cDNAs or other protein-coding segments (Figure 1.25). In this example, the transduced segments are expressed because the necessary *cis-* and *trans-*acting transcription and processing elements are retained in the recombinant genome. These include the early and late region promoters, ori, splicing signals, poly A sites, and the ability to express the large-T protein for replication. In such cases, the recombinant virions are defective and can be propagated only in coinfections with so-called helper viruses. Such helper viruses supply the capsid proteins but rely on the coinfecting recombinant viruses to supply the early region functions needed for their replication. Analogous constructions in which the nearly 2 kbp segment encoding the T proteins is replaced with the desired segment permit the transduced segments to be expressed under the control of the early region promoter and poly A site. Such recombinants can be propagated only if a source of large-T is provided by an appropriate helper virus or special primate cells (COS) that have an expressible SV40 large-T gene incorporated in their genomic DNA.

Recognizing the robust activity of the SV40 early region promoter in a wide variety of vertebrate cells and understanding which of the SV40 DNA elements are needed to constitute an expressible transcription and translation unit led to the creation of plasmid-transducing vectors. Two examples, pSV2 and pcD, contain the SV40 ori and early region promoter, a spliceable intron, and a poly A site (Figure 1.26). Sequences inserted between the promoter and the intron-poly A segment (pSV2) or into a polylinker located between the promoter-intron segment and the poly A signal (pcD) are expressed transiently following transfections into a variety of cells or continuously in cells that integrate the transfected DNA into their DNA. Comparable transfections into COS cells result in considerable amplification of the vector and its insert DNA and therefore even higher levels of expression of the insert's coding sequences.

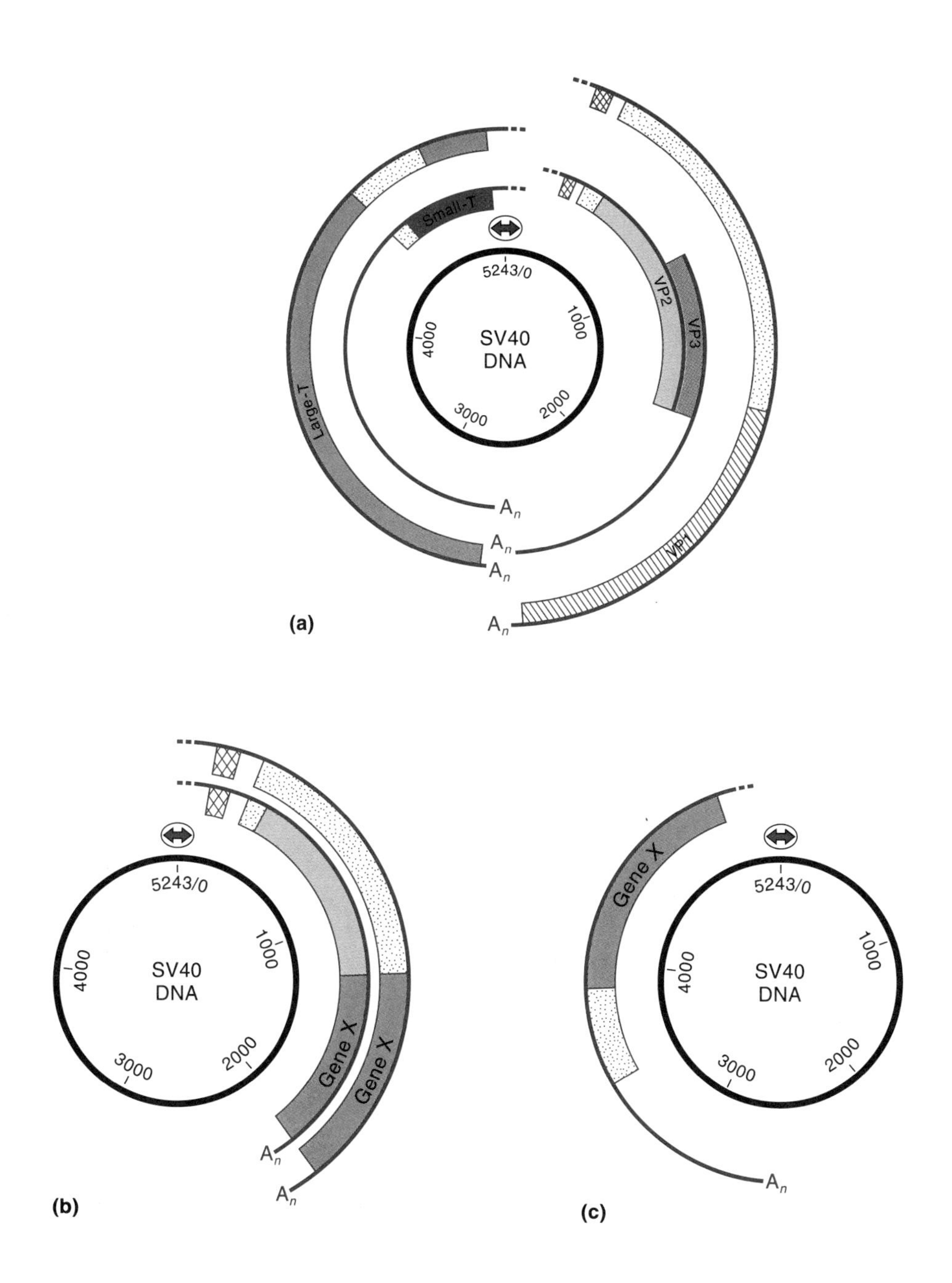

Figure 1.25
SV40 DNA-based transducing vectors. (a) Representation of the early and late transcripts shown in relation to the SV40 DNA map (see Figure 1.9 for a larger version and additional detail). (b) Replacement of the VP1 coding sequence with gene X permits the formation of late-region transcripts containing gene X; one of these corresponds to the spliced VP1 mRNA and the other to the VP2-VP3 RNA. Protein X is translated from the spliced VP1 mRNA only. (c) Placement of gene X in the early region in place of the large- and small-T coding region permits its expression from the early-region promoter. The small-T intron located beyond gene X's coding sequence permits splicing and maturation of that mRNA.

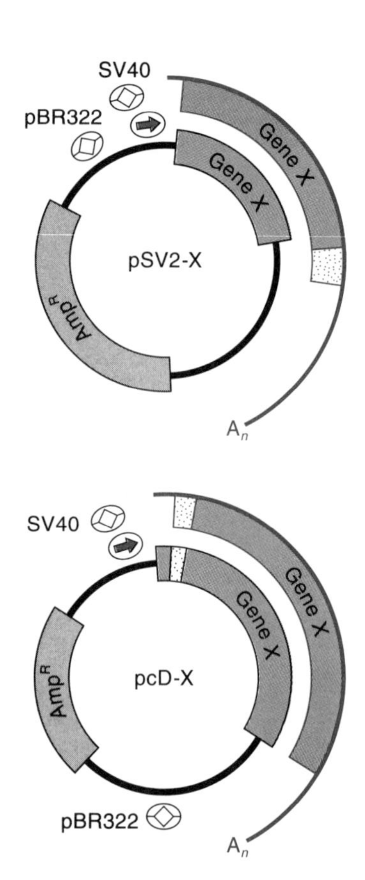

Figure 1.26
Plasmid transducing vectors. The basic backbone of the plasmids pSV2-X and pcD-X are pBR322 plasmid sequences containing the antibiotic resistance marker, Amp^R, and an origin of replication (diamond symbol) that enables pBR322-derived plasmids to be propagated in an *E. coli* host. In pSV2-X, the desired coding sequence, X, is flanked at its 5′ end by the SV40 early region promoter (colored arrow) oriented so that transcription proceeds toward gene X. The SV40 ori is present to permit amplification of the plasmid in cells expressing large-T (e.g., Cos cells). To permit splicing and polyadenylation of the gene X mRNA, the small-T intron and the SV40 early region poly A site follow the coding sequence. pcD-X relies on the same promoter and SV40 origin but contains a modified intron (derived from the VP1 intron) between the promoter and gene X (see Figure 1.9).

Such transducing vectors have proven to be extremely useful in conjunction with genes encoding proteins that confer a strong selective advantage for vertebrate cells harboring the vector. Thus, the expression of bacterial coding sequences that confer resistance to kanamycin in bacteria protects vertebrate cells against killing by a related aminoglycoside, G418. Vectors with genes that encode resistance to mycophenolic acid, hygromycin, puromycin, and histidinol are now widely used to select cells that have acquired the transfected DNA and any genes it contains. Cotransfection with vector DNAs carrying dominant selectable markers has also proven invaluable for introducing genes that do not themselves confer a selectable phenotype.

Altogether, the detailed knowledge of SV40's molecular and genetic structure and the ability to manipulate DNA segments and sequences in almost endless ways have provided new and powerful tools for analyzing vertebrate genome function.

1.4 Other Papovaviruses

a. JC and BK Viruses

BK virus (BKV) was first isolated from the urine of a renal allograft recipient undergoing immunosuppression, and JC virus (JCV) was identified in brain tissue from an immunosuppressed patient with the fatal disease progressive multifocal leukoencephalopathy (PML). The potential pathogenicity and the fact that they were the first human viruses that resembled SV40 encouraged a molecular biological characterization of their infectious properties.

As it turned out, both BKV and JCV are ubiquitous in human populations throughout the world; indeed, most children become infected by age ten, and virtually all adults have circulating antibodies against both viruses. There is no evidence for the existence of JCV in any other vertebrate. As far as is known, no disease symptoms can be attributed to primary infections by either of these viruses. However, both remain latent, and their multiplication appears to be activated by immunosuppressive measures.

BKV BKV appears to replicate primarily in the kidney *in vivo* and most efficiently in human embryonic kidney cell cultures *in vitro.* The virions are virtually indistinguishable morphologically from SV40, and their capsids contain three proteins analogous to SV40 VP1, VP2, and VP3. BKV's circular DNA genome contains about 5000 bp (4936 bp in one isolate and 5153 bp in another), and the complete sequences of several isolates are known. The BKV DNA sequence is 70 percent homologous to SV40, and the deduced protein sequences are 73 percent homologous. The organization of the BKV genes and the temporal patterns of their expression are virtually identical to those of SV40. Thus, two alternatively spliced mRNAs are transcribed from the early region to produce large-T and small-T proteins that are very homologous to those of SV40. Two differentially spliced late mRNAs encode the three capsid proteins and a protein analogous to the SV40 agnoprotein. Replication requires a functional large-T protein and initiates at an ori site that contains four repeats of the large-T binding motif 5′-GAGGC-3′. BKV's early promoter lacks sequences resembling the three tandem 21 bp repeats or their 5′-GGGCGG-3′ motifs. Instead, there are one or more repeats of a 68 bp sequence that appears to serve as both the promoter and enhancer for early region transcription. A striking fact is that the putative BKV enhancer-promoter (the 68 bp repeat) can provide enhancer function to an SV40 early region promoter lacking the 72 bp repeat. Furthermore, SV40 large-T can support BKV replication in the absence of a functional BKV large-T protein.

BKV is capable of transforming human cells, but the transformants have the characteristics of immortalized cells (Table 1.2) but they are not oncogenic in nude mice, nor do they grow in soft agar or in suspension. Transformed cells contain integrated BKV DNA and express large-T and small-T proteins. In spite of considerable screening of human tumors, none has been found to contain BKV DNA sequences or proteins.

JCV The molecular biology of JCV is much less advanced than that of BKV or SV40 because it multiplies poorly and only in certain cells of the human nervous system. Nevertheless, studies of lytic multiplication have yielded virion particles of about the same size and capsid construction as SV40. Similarly, the DNA extracted from these virions is a supercoiled, covalently circular duplex of 5130 bp with overall homology of 75 percent and 69 percent, respectively, to BKV and SV40. The sequence predicts six proteins, two early proteins closely resembling large-T and small-T proteins, and four late proteins corresponding to VP1, VP2, VP3, and the agnoprotein. Although there are many mismatches in the amino acid sequences of the JCV, BKV, and SV40 large-T proteins, the types of amino acids along the polypeptide chain suggest a similar domain structure.

The noncoding region between the early and late transcription units contains an ori and characteristic large-T protein binding motifs. Although sequences resembling the 5′-GGGCGG-3′ motif are lacking, JCV possesses a 98 bp tandem repeat quite different in sequence from the SV40 72 bp repeat or the BKV 68 bp repeat but still functions as an enhancer. In contrast to the interchangeability of the SV40 and BKV enhancers, the JCV enhancer functions only in neuronal cells, suggesting that it accounts for the virus's tissue tropism. Interestingly, independent isolates of JCV from both non-PML and PML patients show considerable sequence variability in the transcriptional regulatory elements. Deletions, insertions, and base changes in motifs thought to regulate transcription and replication abound, but their functional significance is unclear.

JCV is highly oncogenic following intracerebral innoculation of its DNA into hamsters and owl monkeys. The tumors are very variable in type and highly malignant. All tumors examined contain JCV DNA integrated at variable sites in the cellular DNA. So far as is known, JCV DNA has not been found in any human brain tumors, nor has the virus been found in any extraneural tissues in individuals with PML.

The distinctly different tissue specificities and outcomes of infections by BKV, JCV, and SV40 emphasize how critical the nature of the regulatory sequences is, and how even subtle changes in their sequences and arrangement can produce profound consequences in their biology.

b. Polyoma

There are, besides SV40 and its close relatives, BKV and JCV, a number of other papovaviruses whose molecular biology and transforming properties are epitomized, in most respects, by murine polyoma. For that reason, we focus on that virus here.

Polyomavirus is endemic and without pathological consequences in feral mice. However, it produces tumors at a wide variety of tissue sites (hence the name polyoma) when inoculated into newborn hamsters, rats, mice, and immunocompromised adult rodents. The virus proliferates rapidly in primary mouse embryo cell cultures and moderately efficiently in established mouse cell lines. In either case, the infected cells are killed, thereby providing the basis for a quantitative plaque assay similar to that described for SV40 (Figure 1.2). Although transfection of mouse cells with viral DNA is less efficient than virus infection, the outcome is the same: transfected cells die and infectious virus particles are produced. Hamster and rat cells fail to produce viruses following either virus infection or DNA transfection. Instead, such cells become transformed and tumorigenic.

Much of the molecular biology of polyoma's infectious cycle closely resembles SV40's in its general features, although there are some differences in detail. It is in the mechanism of cellular transformation and oncogenesis that polyoma differs strikingly from SV40. Our principal focus is on those features that distinguish polyoma and SV40.

Polyoma's capsid and DNA Polyoma virions are indistinguishable physically from SV40. The 72 pentamers are organized to yield 12 units with fivefold symmetry and 60 units with sixfold symmetry, just as occurs in the SV40 capsid. Indeed, VP1 alone can self-assemble into nearly perfect icosohedral capsids, but VP2 and VP3 are presumably associated with VP1 in the capsid. Phosphorylation of VP1 may be required to form a stable capsid. The capsid encloses a minichromosome consisting of a single double-stranded, covalently closed, circular DNA molecule of about 5333 bp; the approximately 24 nucleosomes associated with the DNA contain the same histone proteins as those in SV40 minichromosomes. As is the case with SV40 DNA, the nucleotides are unmodified, and the base composition does not deviate appreciably from that of its respective host cell's DNA.

In much the same way as was accomplished for SV40, a genetic and physical map of the DNA was derived by analyzing the phenotypes of genetically and physically altered mutants, initially with conditional mutants and plaque morphology variants, but subsequently by employing molecular techniques to introduce directed modifications into the DNA. Ultimately, the entire base pair sequence of polyoma DNA was determined, and the location of the different coding regions, the *cis*-acting regulatory sequences governing replication, transcriptional regulation, and RNA processing were identified (Figure 1.27).

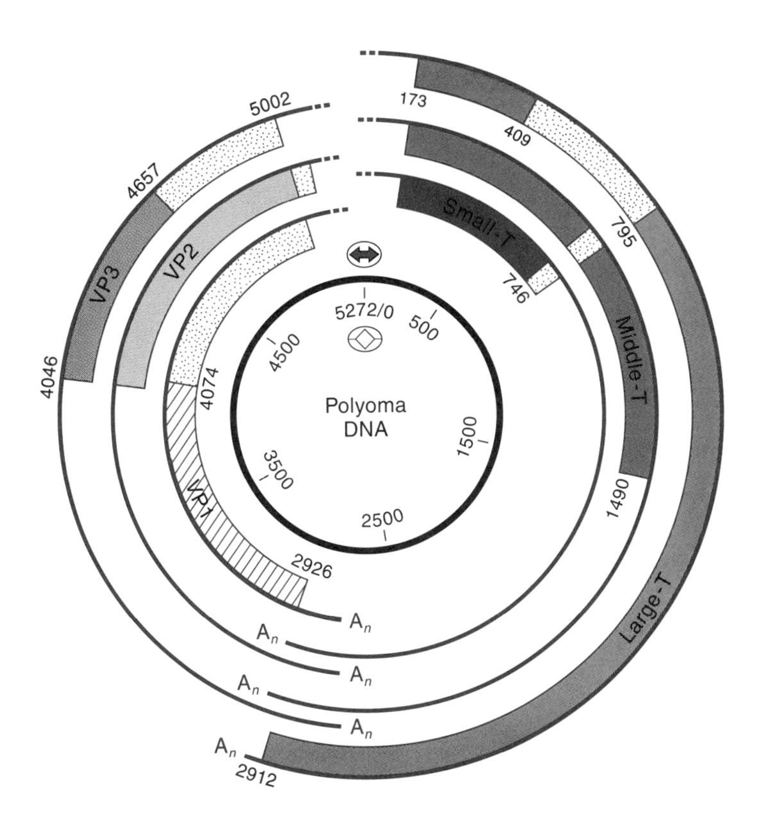

Figure 1.27
A physical, genetic, and transcription map of polyoma DNA. Polyoma DNA is represented as the black inner circle with its origin of replication and bidirectional promoter (the early region is clockwise and the late region is counterclockwise). The numbers within the circle represent the number of base pairs from the core ori. Three early mRNAs (large-T, middle-T, and small-T) are derived from the early region transcript by different patterns of splicing of the introns (stippled regions) (see Figure 1.28 for more detail); the coding region for the three early proteins are indicated by the colored portions; the untranslated region is shown as a line followed by the poly A sequence. Three late region mRNAs encoding VP_1, VP_2, and VP_3 also arise by differential splicing of late region transcripts. The coding regions for the late mRNAs are in color, the introns are stippled, the untranslated segments are solid lines, and the poly A sequences are indicated.

Alternative outcomes of infection As intimated earlier, polyoma, like SV40, initiates a cytolytic response in its permissive host (mouse cells) and either an abortive or a stable transformation in nonpermissive hosts (other rodent cells). Polyoma does not infect primate cells, probably because it does not enter or initiate the events needed to express its genome.

Following infection of either permissive or nonpermissive cells, the virions are transported to the nucleus, disassembled, and half of the DNA (the early region) is transcribed along one strand to produce a single RNA transcript, which is subsequently spliced and polyadenylated. Three alternative splicing patterns produce three distinctive early mRNAs encoding three related early proteins: large-T, middle-T, and small-T (Figure 1.28). As a consequence of this elaborate splicing pattern, three related but distinctive proteins are formed (Figure 1.29). All three proteins share the same 80 residue amino terminus; the small-T and middle-T also share an adjacent sequence of 129 amino acids that is lacking in large-T. As a result of the frame shift, middle-T and large-T have unique carboxy termini of 229 and 700 amino acids, respectively. The production of both large-T

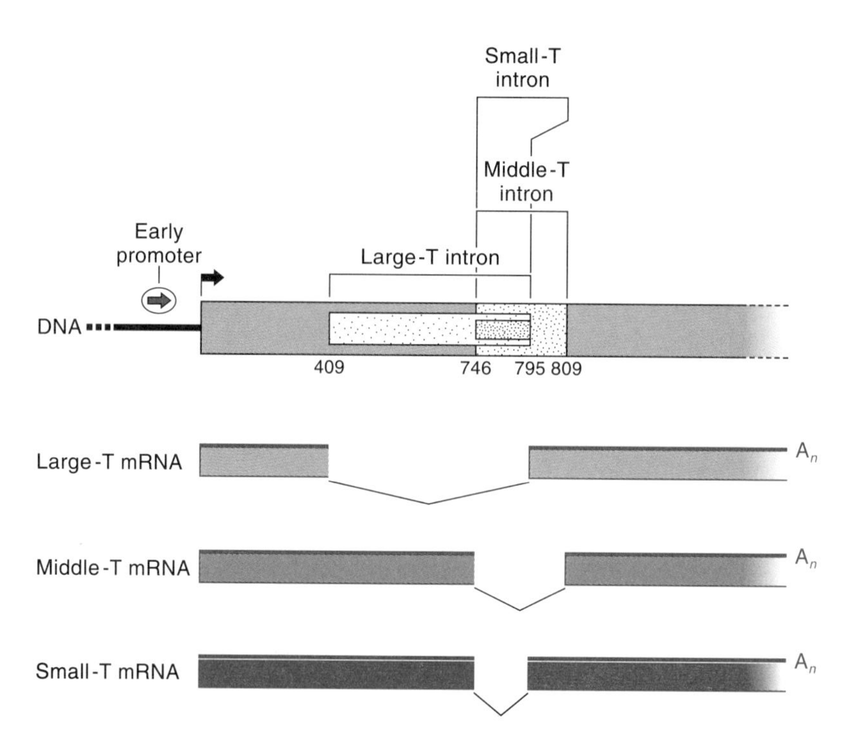

Figure 1.28
Splicing pattern responsible for the formation of polyoma large-T, middle-T, and small-T mRNAs. The polyoma early region DNA is represented at the top. The colored and black arrows represent the early region promoter and transcription start site, respectively. The three introns are shown in three different stippling intensities and the boundaries of the respective introns are indicated by the labeled brackets. The numbers refer to base pairs as indicated in Figure 1.27.

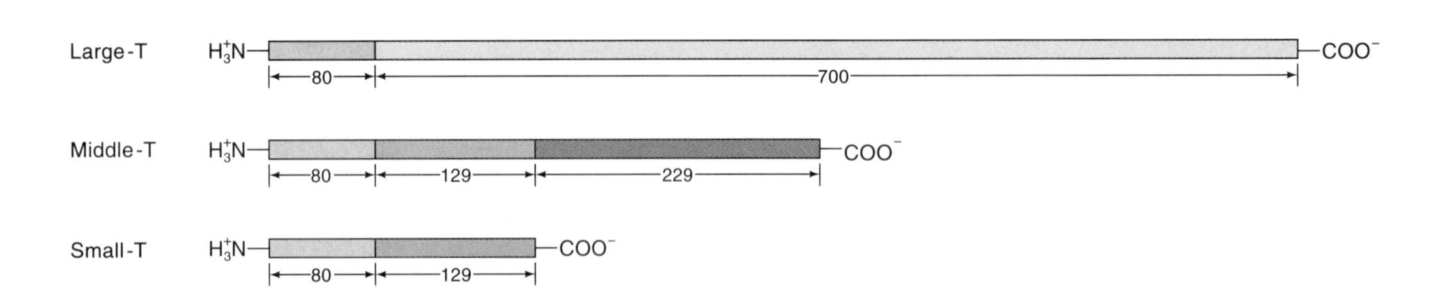

Figure 1.29
The shared and unique domains of polyoma large-T, middle-T, and small-T proteins. The length of each protein is indicated by the overall length of each bar; the number of amino acids is the sum of the numbers below each bar. The three proteins share a common sequence of 80 amino acids at their amino termini. Middle-T and small-T share an adjacent additional sequence of 129 amino acids. The 700 amino acids adjacent to the common amino terminal 80 residue stretch of large-T is unique, as is the 229 residue carboxy terminal sequence of middle-T.

and middle-T adds an additional dimension to the mechanism of polyoma's oncogenicity. Thus, although SV40 uses a single protein, large-T, for both immortalization and oncogenesis, polyoma employs two proteins for those purposes, large-T for immortalization and middle-T for oncogenesis.

In permissive hosts, the appearance of large-T initiates bidirectional DNA replication from a single ori located within the transcriptional regulatory region and, almost concomitantly, late region transcription. Unlike SV40, polyoma's late region transcript is spliced to produce three different mRNAs, each able to be translated into one of the three capsid polypeptides (Figure 1.27). Virion assembly occurs in the nucleus from accumulated capsid proteins and progeny minichromosomes, resulting in dissolution of the cell.

Ori and transcriptional regulatory elements Although the general organization and functional logic of the DNA elements needed to regulate polyoma and SV40 replication and transcription are analogous, they do differ in detail. A distinctive difference between the polyoma and SV40 ori is that the former requires the promoter enhancer sequences on the late gene side of ori, but the SV40 ori is autonomous and largely independent of the promoters. Four elements, α, β, core, and auxiliary sequences, have been identified as contributing to the function of polyoma ori (Figure 1.30a). The core and either α or β are essential for replication, with replication initiating in the core element. The roles of α and β in ori function follow from their transcriptional enhancer activity; thus α- or β-elements can activate the core for replication even if they are inverted or placed some distance from the core. Unlike SV40's ori promoter region, where early region transcription initiates within ori, polyoma early RNA is initiated well away from where replication begins (Figure 1.30b).

Polyoma also contains three large-T binding sequences, two of which, sites B and C, are outside of ori and are needed for autologous repression of early region transcription (Figure 1.30a). The other large-T binding site, site A, overlaps the ori core and auxiliary sequence. Curiously, large-T binds only weakly to these sites *in vitro*; specific modifications of the protein (phosphorylation?) or of the DNA may be needed for proper strong binding to occur *in vivo.* Conceivably, transcription factor(s) binding to the enhancer elements *in vivo* might account for stronger binding of large-T to the ori-associated binding site.

The α- and β-segments lie within regions that constitute the promoter enhancer for polyoma early and late region transcription. These contain binding sites for a host of cellular transcription factors. The region containing the α-element can function as a self-contained promoter. Thus, linking that segment to a reporter gene containing the polyoma TATA and transcription start site permits high-level transcription of the reporter sequence following transfection into mouse cells.

Interestingly, the polyoma promoter enhancer does not function in primate cells, suggesting the need for species-specific transcription factors. Also, although the polyoma promoter enhancer functions well in differentiated mouse cells, it is inactive in embryonic mouse cells. Polyoma variants competent for propagation in embryonic mouse cells have been obtained; some have alterations in the sequence organization of the α-element, and others have base pair changes in the β-element. Similarly, whereas the wild-type polyoma virus cannot infect neuroblastoma cells or Friend erythroleukemia cells, rearrangements and alterations in the enhancer segments produce viruses that propagate well in these cells.

Studies of polyoma multiplication in its natural host—the newborn mouse—has turned up rather revealing results. Infection of neonatal mice with wild-type polyoma leads to preferential multiplication in the kidney and more limited replication in salivary glands, bone marrow, and lung; but these organs' ability to support virus multiplication falls strikingly with age. Significantly, this characteristic tissue-specific pattern and duration of virus multiplication depends on

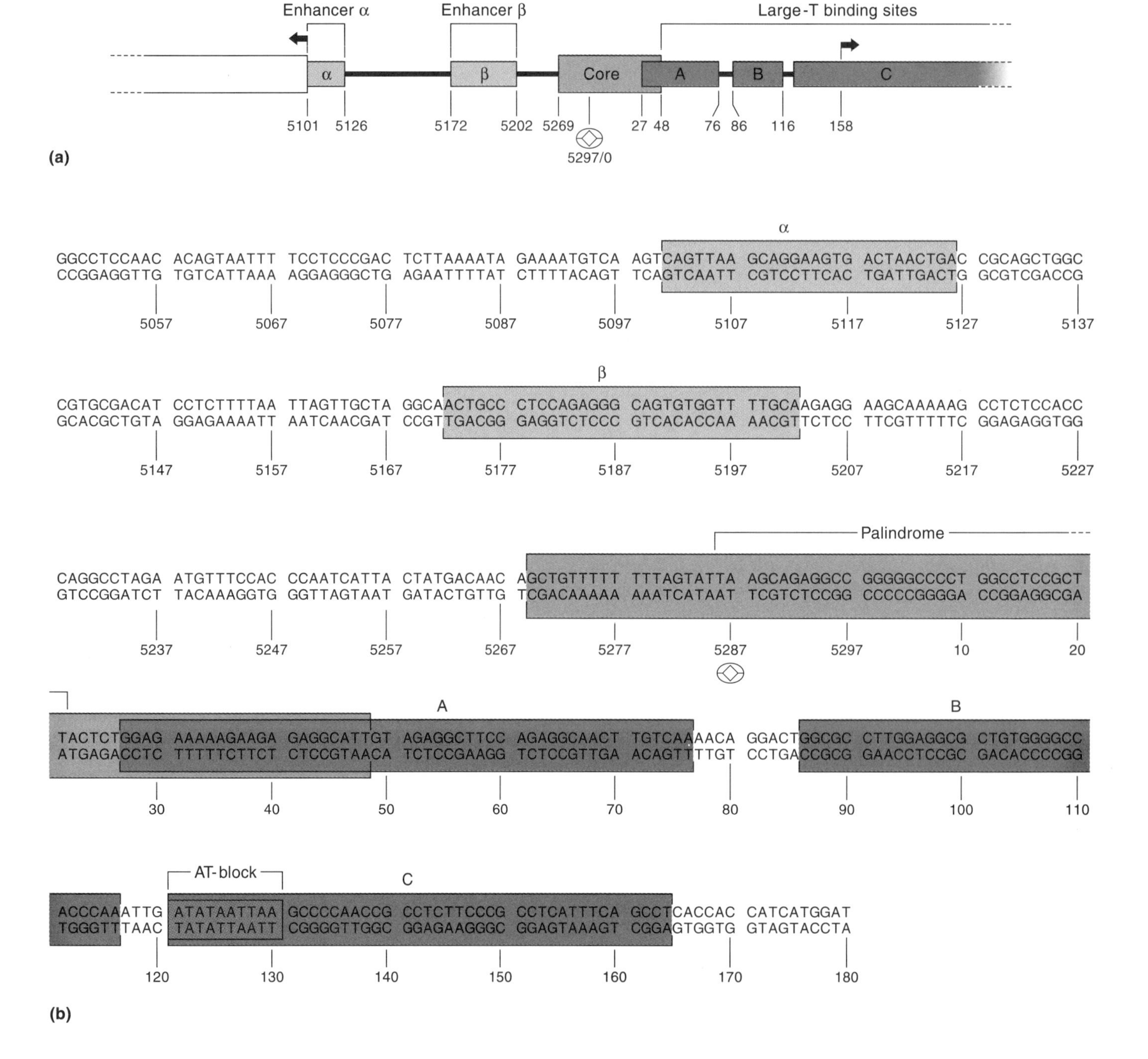

Figure 1.30
Sequence of polyoma's transcription control region and origin of DNA replication. (a) The overall arrangement of the core origin, the replication and transcription α and β enhancers, and the large-T binding sites, A, B, and C. Replication begins at the origin symbol within the core element. Transcription of the early and late regions starts at the rightward and leftward black arrows, respectively. (b) The base pair sequence of these elements. Note that the AT-block within large-T binding site C is some distance from the ori core but is between 25–30 base pairs upstream from the early region's transcription initiation site.

the enhancer structure. For example, certain enhancer variants possess different capacities to support latent infection of the kidney and the associated pathology. Other mutant enhancers enable the virus to multiply preferentially and to high levels in the pancreas during the acute phase, but, unlike the wild-type virus, they are unable to support persistent infections in that tissue. These differences probably reflect tissue- and age-related differences in either the availabiliy of enhancer-binding proteins or in their functional ability.

Transformation and oncogenesis Genetic studies have shown that large-T protein is essential early in transformation, although by itself it is unable to effect stable transformation or oncogenesis. Thus, certain mutations that render large-T protein thermosensitive permit transformation at the permissive temperature but not at the nonpermissive one. Nevertheless, once transformed at the permissive temperature, the transformed phenotype is stable even if the transformants' temperature is raised to the nonpermissive temperature. Moreover, many stable transformed cell lines and polyoma-induced tumors fail to express large-T. Recall that, in contrast, maintenance of the SV40-induced transformed or tumor phenotype requires continued expression of functional large-T.

Notably, nonpermissive cells infected with a polyoma mutant that expresses only large-T and small-T, but not middle-T, fail to become tranformed. In permissive cells, such middle-T mutants grow poorly and fail to induce even the transient transformation seen with wild-type virus. These observations indicate that middle-T provides function(s) necessary for inducing and maintaining the transformed state. The outcome of these middle-T functions appears to be comparable to what is achieved by SV40 large-T. However, whereas SV40 large-T influences cell cycle regulation by association with the Rb and p53 proteins, polyoma middle-T alters cell cycle control by another route.

Polyoma middle-T is an integral membrane protein of about 45 kDa. It is associated with the plasma membrane and the endoplasmic reticulum in productively infected or stably transformed cells. Association of middle-T with these membranes is believed to be accomplished posttranslationally by the insertion of its 20 amino acid carboxy terminal hydrophobic domain through the lipid bilayer.

Expression of middle-T profoundly alters cells. When the middle-T gene is introduced with a regulatable promoter, actin cables and cell adhesion, two characteristics of normal cells in culture, are lost when the promoter is activated. At high levels of middle-T expression, cells are able to form multilayered colonies, and, at still higher levels, the cells become anchorage independent and acquire the ability to cause tumors in animals (Table 1.1). The multifaceted actions of middle-T stem from its interactions with a plasma membrane–bound protein tyrosine kinase, c-Src, or related members of the Src family, and phosphatidylinositol 3-kinase (Figure 1.31). The key event is the formation of a complex between membrane-bound middle-T and c-Src, resulting in a marked increase (10- to 50-fold) in c-Src's protein tyrosine kinase activity. Normally, c-Src tyrosine kinase is inactive because of the phosphorylation of its tyrosine-527; it is activated by removal of that phosphate (or by deletion or substitution of tyrosine-527). Apparently, association of middle-T with c-Src overcomes the negative effects of the phosphate on tyrosine-527. There are indications that middle-T's ability to associate stably with c-Src is facilitated by or depends on phosphorylation of one or more of its serines.

One outcome of c-Src activation is the phosphorylation of middle-T tyrosine residues, one of which is tyrosine-315. This enables phosphorylated middle-T to associate with several proteins. One of these, phosphatidylinositol-3-kinase (PI3K), binds to middle-T's phosphotyrosine-315 through PI3K's SH2 domain. Proteins with SH2 domains recognize and bind to specific protein phosphotyrosine residues. Formation of the middle-T:c-Src:PI3K tripartite complex is es-

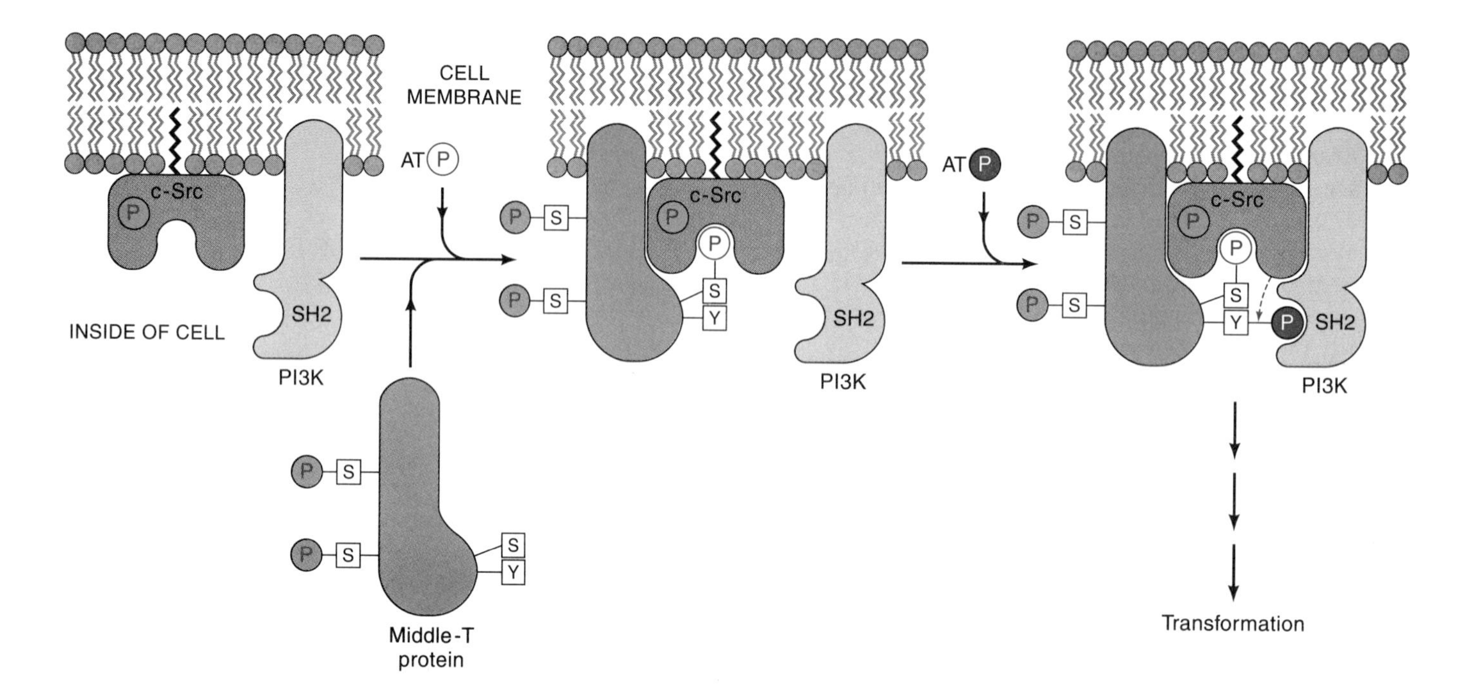

Figure 1.31

Model of the interaction of polyoma middle-T with membrane proteins to initiate cellular transformation. Phosphatidylinositol-3-kinase (PI3K) and c-Src are localized in the plasma membrane, the former through hydrophobic interactions and the latter through a myristoyl substituant at the protein's amino terminus. Following its synthesis, middle-T is phosphorylated (P) at specific serine (S) residues and is localized in the plasma membrane where it associates with c-Src. This association activates c-Src's protein tyrosine kinase activity, which causes phosphorylation of a middle-T tyrosine (Y). Phosphorylation of middle-T's tyrosine promotes association of middle-T with PI3K's SH2 domain, thereby triggering a signaling cascade of Raf kinase through the Ras protein and ultimately causing cellular transformation.

sential for efficient transformation because mutants whose middle-T lacks tyrosine-315, and therefore cannot be phosphorylated by c-Src, fail to bind PI3K; as a consequence, such mutants are severely impaired in their ability to transform cultured cells and to cause tumors *in vivo.*

How the formation of the middle-T:c-Src:PI3K ternary complex triggers the events leading to transformation is still being investigated. Complex formation appears to result in the activation of a serine/threonine kinase (Raf kinase) via the Ras protein signaling pathway, leading ultimately to the increased expression of the transcription factors Jun and Fos. Thus, the oncogenic function of polyoma middle-T relies on its ability to associate with membrane-bound and cytoplasmic cellular proteins and trigger a series of signals, possibly mediated by protein phosphorylations. These phosphorylations appear to govern the activity of transcription factors controlling genes involved in cell proliferation.

c. Papilloma

The Shope papillomavirus, discovered nearly 60 years ago in rabbit papillomas (warts), is the largest of the papovaviruses. The virions consist of an icosohedral capsid (Figure 1.1), made up of two proteins (L1 and L2), enclosing a single supercoiled, covalently circular DNA of approximately 8 kbp associated with the same four cellular histones found in the other papova minichromosomes. Besides the rabbit papillomaviruses, there are related viruses associated with warts of other vertebrates, the most intensively studied ones being the bovine (BPV) and human (HPV) papillomaviruses. Although the warts produced by most types of BPV and HPV are benign, both viral DNAs are capable of transforming cultured rodent cells. Considerable interest in their transforming activity has been provoked by the implication of several HPVs, notably HPV-16 and HPV-18, in the genesis of human cervical carcinoma.

The molecular biology of papillomavirus infections has been very difficult to study for lack of a cell culture system that allows them to multiply. Productive infections have been achieved only in fully differentiated squamous epithelial cells, the keratinized layer giving rise to the papilloma. Undifferentiated basal cells are either refractory to infection or, if infected, are able to maintain only low levels of the viral genome in a latent state. This barrier to a molecular analysis of papillomavirus infections was partially breached with the advent of DNA cloning. Small amounts of DNA could be recovered from warts, cloned and amplified, and ultimately completely sequenced. Besides revealing the organization and locations of the viral genes, this breakthrough also provided discrete viral DNA segments for use as hybridization probes to follow the genome's transcription and replication.

Genome organization Ten open reading frames (ORFs), presumably all corresponding to translated proteins, can be deduced from the complete BPV-1 sequence; their arrangement on the 7946 bp circular map and on a linear representation of the map (the ends being defined by a single *Hpa*I cleavage site) is shown in Figure 1.32. There are ten ORFs expressed during productive infection in papillomas. The position, size, and function of most of these ORFs, eight of which are involved in transcriptional regulation and replication and two of which encode the virion capsid proteins, are conserved among various papillomaviruses. A 1 kbp segment between bp 7093 and bp 89 lacks an ORF but contains an array of *cis*-acting elements that are implicated in the regulation of transcription and replication; it is generally referred to as the long control region (LCR).

Transcription and its regulation BPV-1 is the most thoroughly studied of the papillomaviruses because it transforms rodent cells, C127 mouse cells being the preferred host. In that host, BPV-1 replicates as an extrachromosomal plasmid and expresses all of the early genes needed to maintain the plasmid state and the cell's transformed state; the two late genes, L1 and L2, that encode the capsid proteins, are not expressed.

Only one of the viral strands serves as the transcription template; thus, all the promoters and mRNAs are oriented in the same direction (Figure 1.32). Because of multiple promoters (seven have been identified) and complex splicing patterns, there are many more mRNAs than ORFs. Four promoters—P_{89} P_{890} P_{2443} and P_{3080}—account for most of the early mRNAs responsible for plasmid replication and for maintenance of the transformed state; their transcripts are polyadenylated at a common site (A_E) just upstream of L2. Late mRNAs originate from P_L and after splicing are polyadenylated at A_L.

BPV-1 encodes three related proteins that regulate transcription by binding to enhancers located in the LCR. Full-length E2 (410 amino acids), which is encoded in an mRNA transcribed from the P_{2443} promoter, serves as a transactivator for transcription of the early genes by virtue of its binding at multiple 5′-ACCN$_6$GGT-3′ motifs in the two E2-responsive elements (E2RE1 and E2RE2). An E2-related protein, E2TR (248 amino acids), originates from an mRNA that is transcribed from the P_{3080} promoter and translated from an internal, inframe, ini-

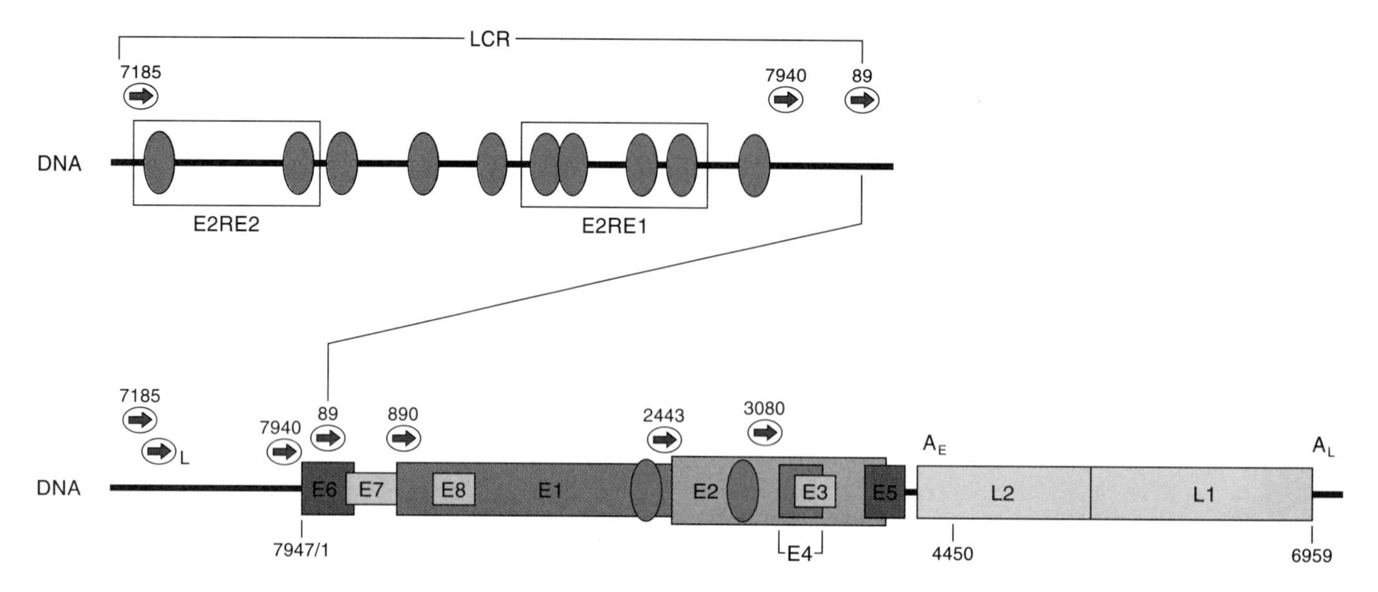

Figure 1.32
Linear representation of the circular bovine papilloma virus genome. The linear arrangement shown results from opening the circular BPV DNA at a point between the late region polyadenylation sequence (A_L) and the early region promoter (colored arrows) at position 7185. There are 10 open reading frames, 8 of which encode proteins that are involved in transcriptional regulation and replication (E1–E8), the remaining 2 encoding the virion capsid proteins (L1 and L2). The polyadenylation site for the early gene transcripts lies within L1. Note that there is considerable overlap in the early gene reading frames. The region at the left, from approximately 7100 to 89, is referred to as the long control region (LCR). It contains the replication origin and a number of *cis*-acting elements (colored ovals) that serve as binding sites for the transcriptional and replication activator protein E2 and possibly of modified forms of E2 (E2PEs). Colored arrows (in the top portion of figure) indicate the positions of multiple early region promoters and the single late region promoter.

tiator codon (Figure 1.33). E2TR functions as a repressor of E2 expression. A third but less abundant E2-related protein—a fusion between sequences specified by the E8 and E2 genes—also functions as a repressor of E2-activated transcription. Thus, in BPV-1, the relative abundance of the positive- and negative-acting E2 proteins determines the level of early gene expression. Note that the three E2-related proteins share a common 3′-proximal domain (the DNA binding and dimerization domain) but that only the complete E2 protein contains the transactivation domain. Therefore, repression by the E2TR or E8/E2 proteins could occur by competition for binding at the E2REs or by forming transcriptionally inactive heterodimers.

In HPV-16 and HPV-18 (Figure 1.34), the full-length E2 protein, but not the truncated or fusion E2-like proteins, is expressed, with transcription of its mRNA initiating from promoter P_{97}. E2 acts as transactivator at the E2REs in the LCR and as a repressor at the promoter responsible for the expression of E6, E7, and E1.

Replication and its regulation During infection of cultured rodent cells by BPV-1, the viral DNA is maintained as a multicopy nuclear plasmid. Both *trans*- and *cis*-acting viral functions are required for plasmid replication and maintenance. The viral E1 and E2 proteins were identified as the essential *trans*-acting factors by

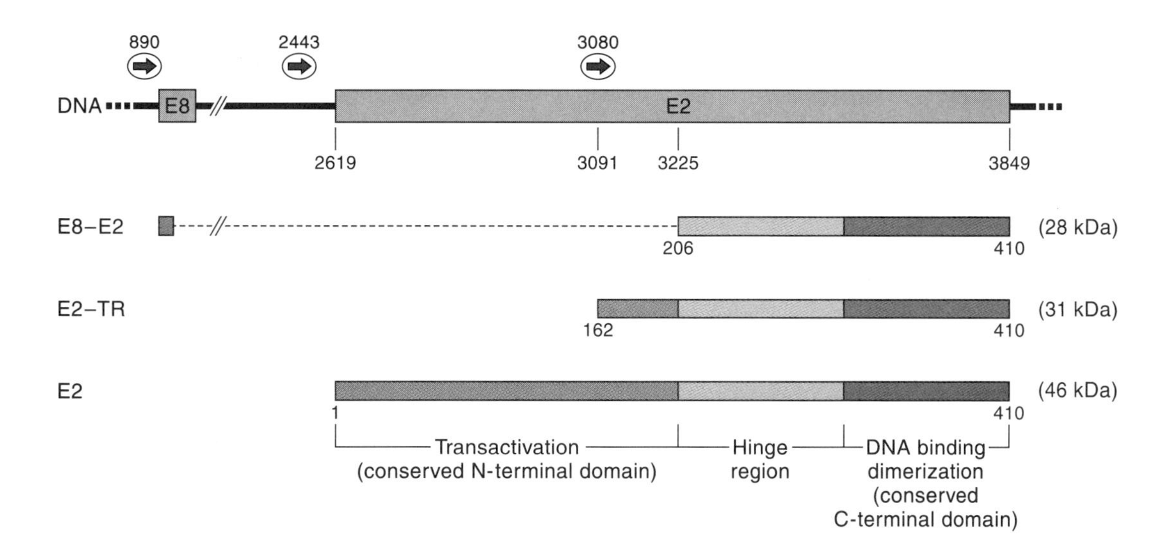

Figure 1.33
Functional domains of E2 and its various modified forms. BPV's genomic arrangement of the promoters and open reading frames responsible for expressing E2 and its modified forms is shown at the top. Full-length E2 (410 amino acids) is produced from an mRNA transcribed from the promoter at bp 2443. E2's transcription activator domain occupies the amino terminal half and the DNA binding and dimerization domains are located in the carboxy terminal quarter of the protein; a flexible hinge region separates the two functional domains. The truncated protein, E2-TR, which lacks the transcription activation domain but contains the DNA binding and dimerziation domains, derives from an mRNA initiated from a promoter at bp 3080 and translation from an internal start codon. The E8-E2 fusion protein, whose action is less clear but may have similar properties as E2-TR, is translated from an mRNA transcribed from the promoter at bp 890 but which is spliced in a way that fuses internal codons of E8 and E2.

experiments showing that plasmids containing a BPV-1 ori region can replicate in rodent cells if they are cotransfected with plasmids that express E1 and E2, but not if either E1 or E2 is expressed alone. The *cis*-element(s) defining ori have not been localized precisely because different experimental approaches give slightly different locations. The most widely accepted view is that ori resides within or very close to the LCR, where there are several repeats of the sequence 5′-ACCN$_6$GTT-3′, the preferred binding sites for the E1-E2 complex.

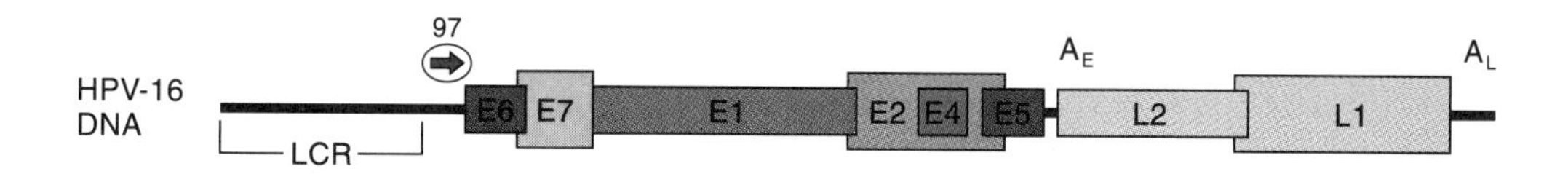

Figure 1.34
The genomic arrangement of HPV-16. HPV-16 (and HPV-18) has only 9 open reading frames, 7 early protein coding sequences and 2 for late proteins; a single polyadenylation sequence (A_E) serves for all the early mRNAs and a separate one (A_L) for the two late mRNAs. The long control region (LCR) is between A_L and the promoter (P_{97}) just upstream from the E6 coding sequence.

E1 is a short-lived nuclear phosphoprotein of 68 to 72 kDa. It shares with SV40 large-T protein regions that promote nuclear localization, amino acid phosphorylation, ATP binding, ATPase, and DNA helicase activities. Moreover, mutations of E1 in these shared regions render it replication defective, just as they do for large-T. Unlike large-T, BPV-1 E1 protein alone binds only weakly to the putative ori sequences mentioned previously; however, binding is markedly enhanced if E1 is associated with E2 protein.

Thus, the complex of E1 and E2, which binds with the same sequence specificity as E2 but more avidly and possesses the DNA helicase and other properties of E1, contributes the same initiation functions provided by the large-T proteins for initiating SV40 and polyoma DNA replication. Indeed, biochemical evidence from *in vitro* studies of BPV DNA replication shows that the E1-E2 complex promotes an ATP-dependent unwinding of the DNA surrounding E2 binding sites. There is some indication that E2TR protein functions as a repressor of replication, perhaps in the same way it acts as a repressor of E2-dependent transcription.

Genes responsible for transformation Both immortalization and oncogenic transformation are the result of the expression of papillomavirus early genes. Continuous expression of these genes appears to be mandatory for the maintenance of the neoplastic state in most cases. Generally, transformation by various papillomaviruses requires the cooperation of two genes. In BPV-1, they are E5 and E6; in BPV-4, E7 and possibly E8, and in HPV-16 and HPV-18, E6 and E7 are the oncogenes.

BPV-1 E5 appears to accentuate the activity of several growth factor receptors (e.g., EGF, PDGF), possibly by retarding the destruction of the receptors, resulting in elevated levels of the basal signaling of these receptors above the threshold for triggering a mitogenic response. The E6 protein is expressed by all papillomaviruses, although the protein's transforming activity might differ among the various virus genera and in different cells. The BPV-1 E6 protein is a zinc-containing protein (zinc is bound via Cys-X-X-Cys repeats) and behaves as a transcriptional activator; its transforming activity may be a consequence of enhanced expression of certain cellular genes or blocking the action of a cellular transcriptional repressor.

The roles of HPV-16 and HPV-18, E6 and E7 proteins in cellular transformation, and particularly in the etiology of cervical carcinomas are clearer and resemble mechanisms used by the large-T of SV40 (Section 1.3a) and the E1A and E1B proteins of the oncogenic adenoviruses (Section 1.5).

The expression of HPV-16 E7 protein alone is sufficient for complete oncogenic transformation of rodent fibroblasts, but transformation of primary human epithelial and fibroblastic cells requires coexpression of HPV-16 E6 protein. Moreover, in the context of the entire HPV-16 genome, disruption of the E1 and E2 genes increases the transformation frequency of primary human keratinocytes. How do the activities of these genes account for the high transforming and oncogenic potential of HPV-16 (and -18)?

The HPV-16 E7 gene encodes a 98 amino acid nuclear phosphoprotein that binds zinc at a pair of Cys-X-X-Cys motifs (Table 1.5). It binds to the hypophospharylated form of the Rb protein in much the same way as SV40 large-T does. However, except for the striking resemblance between the N terminal 38 amino acids of E7 and the sequences in the SV40 large-T domain responsible for binding to Rb (Section 1.3h), the two proteins differ outside this region. Notably, in both cases, the conserved regions are essential for cellular transformation. Alterations in the E7 protein's Rb-binding domain—the one with shared homology to large-T—eliminate the binding of E7 to Rb and E7's ability to transform even ro-

Table 1.5 ***Properties of the HPV-16 E7 and E6 Protein***

E7 Protein
Biochemical properties
98 amino acid, zinc binding, acidic nuclear phosphoprotein
Phosphorylated at serine residue(s) by casein kinase II
Carboxy terminal region contains two Cys-X-X-Cys sequence motifs
Predicted molecular weight: 11 kDa
Complexes with Rb protein
Amino terminal region (amino acid 1-329) structurally related to the Rb-binding regions of SV40 large-T and adenovirus E1A

Biological properties
Sufficient for transformation of established rodent fibroblasts (e.g., NIH 3T3 cells)
Transactivates adenovirus E2 promoter
Cooperates with *ras* to transform primary rodent cells
Necessary together with E6 for the efficient immortalization of primary human squamous epithelial cells

E6 Protein
Biochemical properties
151 amino acid, zinc binding, basic protein
Four copies of Cys-X-X-Cys sequence motif, which may serve as ligands for zinc binding
Binds to double-stranded DNA with high affinity
Can form a complex with the p53 tumor suppressor protein and promote its degradation

Biological properties
Cooperates with E7 for the efficient immortalization of primary human squamous epithelial cells
May have transcriptional regulatory activities

dent fibroblasts in culture. Moreover, the E7 proteins from poorly oncogenic HPVs bind less well to Rb than does E7 from the more highly oncogenic viruses HPV-16 and HPV-18. These findings suggest that the HPV-16 E7 and SV40 large-T protein's shared ability to bind Rb contributes to cellular transformation. Recall that hypophosphorylated Rb is presumed to block entry of cells into S phase by binding to E2F, the cellular transcription factor needed to activate genes whose proteins participate in DNA replication. Binding of E7 to Rb also causes the release of E2F, thereby contributing to the transition from the G1 to S phase.

The 151 amino acid long E6 protein binds zinc at four repeats of a Cys-X-X-Cys sequence (Table 1.5). The E6 protein's oncogenicity probably stems from its ability to bind the tumor suppressor protein p53. The consequence of binding p53 to E6 is the rapid destruction of p53 protein via a ubiquitin-dependent pathway. Not surprisingly, the sequence domains in E6 responsible for binding p53 are different from those in SV40 large-T. Whereas SV40 uses a single protein, large-T, to bind both Rb and p53, HPV-16 uses two proteins, E6 and E7. The end result in each case is the functional neutralization of Rb and p53, both of which are negative regulators of progression from the G1 to the S phase of the cell cycle.

Recall the earlier comment that mutations in E1 or E2 enhance the transforming activity of BPV-1 DNA. Related to this is the finding that, in cervical carcinomas containing integrated HPV DNA, the viral DNA is most often interrupted in the E1 and/or E2 genes. Why should inactivation of HPV-16's E1 and E2 genes enhance the tumorigenicity of the viral DNA for primary human epithelial cells? One explanation is that there is a greater probability that the viral DNA will be stably integrated into the infected cell's genome if the absence of E1 and/or E2 proteins prevents the ability to replicate and maintain the DNA as an autonomously replicating plasmid. Also, because E2 may be a repressor of E6 and E7 gene transcription, its absence would result in higher levels of E6 and E7. These higher levels may be necessary to disrupt the mechanisms controlling normal cell division and permit the selection of variants with enhanced tumorigenic potential.

1.5 Comparative Strategies for Cellular Activation by Papova- and Adenoviruses

a. Cellular Controls of the G1 to S and G2 to M Phase Transitions

Based largely on the genetic analysis of the cell cycle in fission and budding yeast and the conserved nature of cell cycle components, the major control points in the eukaryotic cell cycle have been defined as the G1 to S transition, at which cells become committed to DNA synthesis, and the G2/M boundary, where cells enter mitosis. In yeast, both decisions are controlled by the activity of a single protein-serine kinase, Cdc2, in conjunction with one of two different but related classes of Cdc2-activating proteins called cyclins. The G1 cyclins regulate the G1 to S transition. The mitotic cyclins, first discovered as proteins whose abundance oscillates during the cell cycles of early invertebrate embryonic development, regulate the G2 to M transition.

In higher eukaryotes, this picture of the decision points in the cell cycle is more complex. There are at least ten distinct cyclin genes and at least six known cyclin-dependent kinase (Cdk) partners for these cyclins encoded in the mammalian genome; however, only certain cyclin/Cdk complexes have been found to exist. The multiplicity of known cyclin/Cdk complexes implies that, in addition to the G1/S and G2/M cell cycle transitions, there are probably many regulated transitions in the vertebrate somatic cell cycle. The key substrates for these cyclin/Cdk complexes are not yet well defined, although some have been identified; for example, the phosphorylation of lamin is essential for nuclear lamina breakdown. The retinoblastoma (RB) growth suppressor gene product is a key cyclin/Cdk substrate in late G1 and must be phosphorylated for cells to progress into S phase. Cyclin/Cdk-mediated phosphorylation of RB releases transcription factors and other nuclear proteins that are specifically bound to the unphosphorylated form of RB (and RB-related proteins); these transcription factors mediate the induction of genes needed for progression into S phase. As mentioned earlier, one of these transcription factors, E2F, is a transcriptional regulator for expression of genes that are induced in late G1. Other likely targets for cyclin/Cdk complexes in G1 and S phases are p107, an RB-related protein, p53, replication origin-binding proteins, and the single-strand binding protein (RPA).

As pointed out previously, p53 also regulates the G1 to S transition, ostensibly by its ability to function both as a transcriptional activator of genes whose products prevent or delay the onset of S phase and as a repressor of genes whose expression facilitates that switch. p53 also appears to function as a monitor of chromosomal integrity by delaying the G1 to S switch and the G2 to M transition if chromosomal replication is incomplete or if DNA damage persists. This property is important in considering the phenomenon of tumor progression.

b. Abrogation of Cellular Control by Viral Genes

Because of the relatively limited genetic capacity of their genomes, the papovaviruses and adenoviruses must utilize the infected cell's DNA replication machinery for their multiplication. But because the targets of infections by these viruses are generally quiescent cells, they must have a way to overcome the normal cellular mechanisms that regulate the G1 to S phase transition.

Each of the papovaviruses employs one or more of its gene products to overcome that block. SV40 accomplishes this with a single protein that abrogates the control functions of both Rb and p53 (Figure 1.35). Binding of large-T to hypophosphorylated Rb mimics the effects of the phosphorylation associated with natural mitogenic signals. Thus, binding of large-T to Rb results in the release of E2F, allowing it to activate the expression of cellular genes needed for chromosome replication. Large-T also facilitates the entry of cells into S phase by binding to p53 and blocking its ability to function as a transcriptional activator and repressor.

Although polyoma large-T also appears to bind Rb, possibly with the same outcome, the principal mitogenic signal following polyomavirus infection results from interactions of a related viral protein—middle-T—with a membrane-associated cellular Src-family protein. This interaction initiates a protein phosphorylation cascade that causes, among other things, the activation of the cell cycle machinery for cell division. Because of the alternative splicing of the polyoma early region transcript, middle-T protein contains sequences that allow it to localize to the plasma membrane, where it can interact with c-Src and become phosphorylated at tyrosine-315 by c-Src.

The human papillomaviruses also encode two proteins that promote the entry of infected cells into S phase (Figure 1.35). One of these, E7, binds to hypophosphorylated Rb, causing the release of E2F, the result being the same as

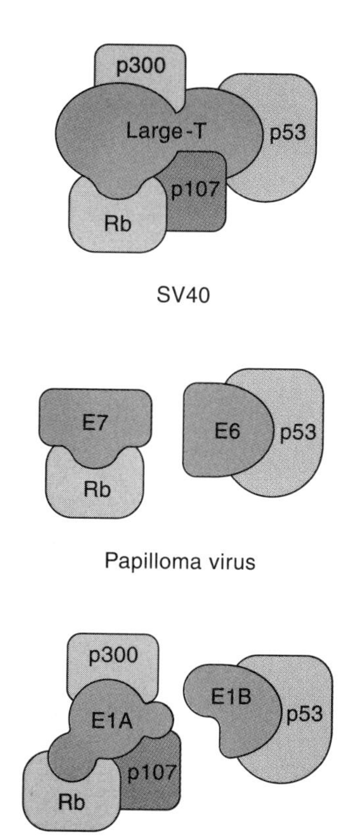

Figure 1.35
Known interactions between the oncogenic proteins of SV40, papilloma, and adenovirus with cellular proteins that regulate passage through the cell cycle. SV40 large-T interacts with the Rb family of proteins, which negates their ability to sequester transcription factors and other cell cycle regulatory proteins. Large-T also forms complexes with p53, which inhibits its transcriptional activation function, and thereby influences the cell's capacity to regulate entry into G1 and G2 in response to DNA damage. The human papilloma virus proteins, E7 and E6, accomplish the same end by their ability to bind Rb and p53, respectively; binding of E6 to p53 promotes p53's destruction. Adenoviruses rely on E1A's ability to bind the Rb family of proteins in ways similar to those of SV40 large-T and papilloma E7, thereby promoting entry into S phase. Adenovirus E1B interacts with p53 to neutralize the antiproliferative activities mentioned above. Another form of E1B (19 kDa) also binds to p53, which blocks its apoptotic function.

that caused by phosphorylation of Rb and binding of large-T. The second, HPV protein, E6, binds to p53 and activates its rapid destruction. Thus, the E6-facilitated elimination of p53 illustrates another mechanism for overcoming the regulatory block in the G1 to S transition.

Adenovirus genomes are larger than those of the papovaviruses (about 36 kbp of linear double-stranded DNA), and they encode proteins specialized for their rather unusual mode of genome replication (see *Genes and Genomes*, section 2.1g and figure 2.34). Nevertheless, they too rely on cellular proteins for the production of the nucleotide precursors and for their participation in the replication process per se. And because the cellular targets of adenovirus infection are generally quiescent cells, these viruses must also have ways to induce the transition from G1 into S phase. This is accomplished by the action of two adenovirus proteins: E1A and E1B.

Transcription of the adenovirus E1A gene yields multiple, alternately spliced mRNAs, the two most important being the 12S and 13S species that encode proteins of 243 and 289 amino acids, respectively. These proteins differ only in the 46 amino acid domain unique to the protein encoded in the 13S mRNA. Both proteins share two regions that are highly conserved among many adenoviral serotypes: CR1 and CR2. The larger protein contains a third conserved region, CR3, which permits the protein to function as a transcriptional activator for a variety of viral and cellular promoters, although the precise mechanism of its transcriptional activity is unsettled. Its transcriptional activity clearly depends on the integrity of the CR3 domain.

The 12S-encoded E1A protein is by itself able to induce DNA synthesis in growth-arrested mouse fibroblasts, to immortalize baby rat kidney cells, and, in conjunction with the Ha-*ras* oncogene, to effect oncogenic transformation of normal cells. Each of these activities is differentially dependent on the integrity of CR1 and CR2 and somewhat less so on the amino terminal 36 amino acids. The growth-promoting and transforming functions of E1A depend on interactions between the amino terminus, CR1 and CR2 domains with a variety of cellular proteins, the principal ones being approximately 300 kDa (p300), 130 kDa (p130), 107 kDa (p107), 60 kDa (p60), and 30 kDa (p30). The p300 protein appears to be a transcriptional activator, perhaps functions in conjunction with E2F; p130 and p107 are closely related in structure and possibly in function to p105, the Rb protein; p60 and p30 have been identified as human cyclin A and the cell cycle–dependent protein kinase Cdk2. Binding of these proteins to E1A appears either to abrogate or modify their normal functions in regulating the cell cycle.

The conserved E1A domains involved in the interactions with these growth-controlling proteins have been identified by mutational analysis: binding of p300 and p130 requires interactions with N terminal amino acids and with the C terminal half of CR1, and binding of the Rb-related proteins involves mostly CR2 and part of CR1. Thus, CR1 is involved in binding both members of the Rb family and p300, which probably accounts for the fact that loss of CR1 is accompanied by complete loss of cell cycle–stimulating activity.

The binding of E1A to the Rb family of proteins causes the dissociation of E2F protein, but it is still unclear to what extent the E1A-mediated stimulation of DNA synthesis is associated with the release of E2F. It is also likely that interactions between E1A and Rb, p107, and p130 influence the activity of other proteins, G1, or S phase cyclins. Thus, dissociation of G1 cyclins bound to hypophosphorylated Rb may represent another mechanism by which E1A activates the cell cycle. Inasmuch as p300 is a part of E2F transcription complexes, and p300 and the Rb-like proteins bind independently to E1A, E1A's transcriptional activity may be mediated indirectly by promoting association between p300 and E2F. Thus, E1A regulates both viral and cellular gene transcription, the latter contributing to the induction of the cell machinery for DNA replication and activation of the cell's proliferative capacity.

Adenoviruses are also able to overcome the antiproliferative activities of p53 with a second virus-encoded protein (Figure 1.35). That protein, E1B, is a 55 kDa nuclear protein expressed at about the same time as E1A. It binds p53, blocking its ability to influence transcription of, for example, the *WAF1/Cip1* gene, thereby promoting the cell's entry into S phase. As is the case with SV40 large-T, E1B binding occurs within the 300 amino acid domain near the center of the p53 sequence, the same region in which mutations block p53's transcriptional control properties. Another form of E1B (19 kDa form) also binds p53 and may account for prevention of apoptosis of adenovirus-infected cells.

Thus, adenoviruses, like the papovaviruses, encode proteins whose principal functions are to overcome the regulatory mechanisms for maintaining cells in a G1 quiescent phase in the absence of mitogenic stimuli and to prevent an ensuing apoptotic response. SV40 uses a single protein, large-T, to overcome the growth-suppressing abilities of hypophosphorylated Rb and of p53; by contrast, HPVs and adenoviruses each use two proteins to achieve the same end, although the inactivation of p53 by HPV's E6 protein and adenovirus E1B is achieved by quite different mechanisms. The polyoma protein devoted to activating cells for virus multiplication works in another way. In this case, middle-T, a modified form of the viral replication protein large-T , acts on the activation pathway initiated by phosphorylation of c-Src, one which is normally triggered by growth factor binding to receptors.

Of interest is the fact that SV40 large-T, HPV E7, and adenovirus E1A proteins share closely similar sequence domains for binding hypophosphorylated Rb; alterations of corresponding amino acid residues in their shared domains eliminate Rb binding activity. Not surprisingly, mutations in the binding pocket of Rb block the binding of the three proteins to about the same extent. The SV40 large-T and adenovirus E1B domains for binding p53 are likewise similar and through their binding block the same p53 functions. Large-T binding occurs within a 300 amino acid domain near the center of the p53 sequence, the same region in which mutations block p53's transcriptional control properties.

c. Cellular Transformation and Oncogenesis

Considering the necessity for papovaviruses and adenoviruses to activate the cellular machinery for their own multiplication, we may view transformation and oncogenesis as anomalous manifestations of this normal activation process. Thus, integration of viral DNA into the cellular genome, whether inadvertently during multiplication or more frequently when the virus fails to initiate its replication, often results in continued expression of the viral proteins that disrupt normal cell cycle regulation. Although integration of the SV40, polyoma, and adenovirus genomes is responsible for the continued misregulation of the cell cycle, continued disruption of cell cycle control in the case of papillomaviruses can result from either integration or maintenance of the viral genome as an autonomously replicating episome. Loss, or inactivation, of the viral genes encoding these proteins often leads to re-establishment of the normal regulatory mechanisms and reversion from the transformed state. In the case of HPV-induced cancers, stable integration of the HPV genome into the host cell's genome causes elevated and continued expression of E6 and E7 and establishment of a stable oncogenic state.

The disruption of cell cycle regulation, although sufficient to cause continued cellular proliferation and very often the altered cell morphology and nutritional requirements characteristic of transformed cells, only rarely confers enhanced tumorigenicity or metastatic potential *in vivo*. However, transformed cells that are initially nononcogenic often become tumorigenic after continued propagation *in vivo*, a phenomenon referred to as **tumor progression**.

A widely held view is that progression results from the accumulation of mutations that enable already altered cells to override the usual constraints for establishment and invasive spread under the conditions prevailing *in vivo.* Some of these mutations (e.g., *ras, c-myc,* and a variety of genes encoding growth factor receptor and signal transduction proteins) are dominant; others (e.g., NF-1—the neurofibromatosis gene—and p53) inactivate tumor suppressor functions. (For an in-depth discussion of dominant oncogenes and tumor suppressor genes, see Chapter 4.)

Nearly two-thirds of human tumors lack a functional p53 gene product, although it is present in normal levels in the surrounding unaffected tissues. Moreover, individuals who inherit only one functional p53 gene (Li-Fraumeni patients) develop a variety of malignant tumors with high frequency during their lifetimes; generally, cells in the tumors lack the originally normal p53 allele and are therefore devoid of any 53. Thus, the p53 gene product probably has an important role in both preventing the initial transforming event and, more significantly, preventing the progression of potentially cancer-causing cells into malignant tumors.

The mechanism underlying many of p53's functions, particularly its role in regulating normal cell proliferation, remains to be elucidated. The ability of SV40 large-T and adenovirus E1A to bind p53 and the enhanced degradation of p53 by the HPV E6 protein impairs some or all of p53's activities. Thus, inactivation of p53's ability to regulate entry into M phase in response to aberrant genome structures (i.e., incompletely replicated or damaged chromosomes), could account for the enhanced rate of mutational alterations that promote progression to a more tumorigenic potential.

1.6 Conclusion

Perhaps the most significant outcome of the molecular genetic and biochemical analyses of papovavirus genome structure, replication, and expression is the many valuable insights into how these viruses multiply and render normal mammalian cells tumorigenic. These studies, as well as comparable ones with adenoviruses, have profoundly influenced many aspects of the contemporary molecular biological paradigms of mammalian cells. For example, the discovery of introns in SV40 and adenovirus genes, and subsequently in cellular genes as well, transformed our views of the structure of eukaryotic genes and their origin. Furthermore, the necessity of removing introns from primary transcripts and the recognition that alternative splicing of viral and many mammalian genes can produce multiple and distinctive mRNAs from the same transcript required a revision of the long held one gene–one polypeptide model for gene expression. Analysis of the SV40 early region promoter also revealed the existence of novel DNA sequences (enhancers) that regulate transcription initiation at distances relatively far removed from and in either orientation relative to the sites of transcription initiation. Subsequently, the effects of viral and cellular enhancers were found to be mediated by a plethora of sequence-specific DNA-binding proteins that influence the activity of the transcription machinery. And, as discussed in Section 1.3f, studies of SV40 DNA replication *in vitro* revealed novel interactions between the SV40 replication initiator protein, large-T, and multiple components of the cellular DNA replication machinery: the DNA α-polymerase-primase, the single-strand DNA-binding protein (RPA), and proteins that ensure proper formation of the copying complex and processive DNA synthesis. Contrasting the three papovaviruses with another group of DNA viruses, the adenoviruses, with respect to multiplication in and transformation of their mammalian hosts has been both informative and interesting. Chapter 2 examines some of the same issues with the RNA viruses, and Chapter 3 discusses the replicating and transforming properties of the retroviruses.

Further Reading

1.1 and 1.2

J. Tooze (ed.). 1981. DNA Tumor Viruses: Molecular Biology of Tumor Viruses, 2nd ed. Cold Spring Harbor Laboratory, Cold Spring Harbor, New York.

B. S. Schaffhausen. 1982. Transforming Genes and Gene Products of Polyoma and SV40. *CRC Crit. Rev. Biochem.* 13 215–269.

L. Gross. 1983. Oncogenic Viruses, 3rd ed. Pergamon Press, Oxford.

N. P. Salzman (ed.). 1986. The Papovaviridae, vol. 1: The Polyoma Viruses, and vol 2: The Papillomaviruses. Plenum Press, New York.

E. P. Reddy, A. M. Skalka, and T. Curran (eds.). 1988. The Oncogene Handbook. Elsevier, New York.

R. Weinberg and M. Wigler (eds.). 1989. The Oncogenes. Cold Spring Harbor Laboratory, Cold Spring Harbor, New York.

B. N. Fields and D. M. Knipe (eds.). 1990. Virology, 2nd ed. Raven Press, New York.

B. N. Fields, D. M. Knipe, and P. M. Howley (eds.). 1995. Virology, 3rd ed. Raven Press, New York.

1.3a

D. L. D. Caspar and A. Klug. 1962. Physical Principles in the Construction of Regular Viruses. *Cold Spring Harbor Symp. Quant. Biol.* 27 1–22.

L. V. Crawford and P. N. Black. 1965. The Nucleic Acid of Simian Virus 40. *Virology* 24 388–392.

J. Vinograd and J. Lebowitz. 1966. Physical and Topological Properties of Circular DNA. *J. Gen. Physiol.* 49 103–125.

J. T. Finch and L. V. Crawford. 1975. Structure of Small DNA Containing Animal Viruses. In Comprehensive Virology, vol. 5, pp. 119–154. Plenum Press, New York.

J. E. Germond, B. Hirt, P. Oudet, M. Gross-Bellard, and P. Chambon. 1975. Folding of the DNA Double Helix in Chromatin-Like Structures from Simian Virus 40. *Proc. Natl. Acad. Sci. USA* 72 1843–1847.

M. Shure and J. Vinograd. 1976. The Number of Superhelical Turns in Native Virion SV40 DNA Determined by the Band Counting Method. *Cell* 8 215–226.

T. J. Kelly Jr. and D. Nathans. 1977. The Genome of Simian Virus 40. *Adv. Virus Res.* 21 85–173.

B. A. J. Ponder and L. V. Crawford. 1977. The Arrangement of Nucleosomes in Nucleoprotein Complexes from Polyoma and SV40. *Cell* 11 35–49.

S. Saragosti, G. Moyne, and M. Yaniv. 1980. Absence of Nucleosomes in a Fraction of SV40 Chromatin Between the Origin of Replication and the Region Coding for the Late Leader RNA. *Cell* 20 65–73.

R. C. Liddington, Y. Yan, J. Moulai, R. Sahli, T. L. Benjamin, and S. C. Harrison. 1991. Structure of Simian Virus 40 at 3.8Å Resolution. *Nature* 354 278–284.

J. Clever, D. A. Dean, and H. Kasamatsu. 1993. Identification of a DNA Binding Domain in Simian Virus 40 Capsid Proteins: VP2 and VP3. *J. Biol. Chem.* 268 20877–20883.

T. Stehle, Y. Yan, T. L. Benjamin, and S. C. Harrison. 1994. Structure of Murine Polyomavirus Complexed with an Oligosaccharide Receptor Fragment. *Nature* 369 160–163.

S. C. Harrison, J. J. Skehel, and D. C. Wiley. 1995. Principles of Virus Structure. In B. N. Fields, D. M. Knipe, and P. M. Howley (eds.), Virology, 3rd ed., chap. 3, pp. 59–99. Lippincott-Raven Publishers, Philadelphia.

1.3b

J. Tooze. 1982. Lytic Cycle of SV40 and Polyoma Virus, pp. 125–204, Transformation by Simian Virus 40 and Polyoma Virus, pp. 205–296, Human Papovaviruses, pp. 339–382, Lytic Infections by Adenoviruses, pp. 443–546, and Cell Transformation Induced by Adenoviruses, pp. 547–576. In DNA Tumor Viruses: Molecular Biology of Tumor Viruses, 2nd ed. Cold Spring Harbor Laboratory, Cold Spring Harbor, New York.

1.3c

C. J. Lai and D. Nathans. 1974. Deletion Mutants of Simian Virus 40 Generated by Enzymatic Excision of DNA Segments from the Viral Genome. *J. Mol. Biol.* 89 179–193.

J. E. Mertz and P. Berg. 1974. Viable Deletion Mutants of Simian Virus 40: Selective Isolation by Means of a Restriction Endonuclease from *Haemophilus parainfluenzae. Proc. Natl. Acad. Sci. USA* 71 4879–4883.

C. J. Lai and D. Nathans. 1975. A Map of Temperature Sensitive Mutants of Simian Virus 40. *Virology* 5 66–70.

C. N. Cole, T. Landers, S. P. Goff, S. Manteuil-Brutlag, and P. Berg. 1977. Physical and Genetic Characterization of Deletion Mutants of Simian Virus 40 Constructed *in Vitro. J. Virol.* 24 277–294.

J. M. Pippas. 1985. Mutations Near the Carboxyl Terminus of the Simian Virus 40 Large T Antigen Alter Viral Host Range. *J. Virol.* 54 569–575.

T. E. Shenk, J. Carbon, and P. Berg. 1976. Construction and Analysis of Mutants of Simian Virus 40. *J. Virol.* 18 664–671.

1.3d

W. Fiers, R. Contreras, G. Haegeman, R. Rogiers, A. van de Voorde, H. Van Heuverswyn, J. Herrewegh, G. Volckaert, and M. Yseboert. 1978. Complete Nucleotide Sequence of SV40 DNA. *Nature* 273 113–120.

V. B. Reddy, R. Thimmappaya, R. Dhar, K. N. Subramanian, B. S. Zain, J. Pan, P. K. Ghosh, M. L. Celma, and S. M. Weissman. 1978. The Genome of Simian Virus 40. *Science* 200 494–502.

A. R. Buchman, L. Burnett, and P. Berg. 1981. The SV40 Nucleotide Sequence. In J. Tooze (ed.), DNA Tumor Viruses, pp. 799–841. Cold Spring Harbor Laboratory, Cold Spring Harbor, New York.

1.3e

A. Graessman, M. Graessman, and C. Mueller. 1981. Regulation of SV40 Gene Expression. *Adv. Cancer Res.* 35 111–149.

R. Tjian. 1981. Regulation of Viral Transcription and DNA Replication by the SV40 Large-T Antigen. *Curr. Top. Microbiol. Immunol.* 93 5–24.

N. P. Salzman, V. Natarajan, and G. B. Selzer. 1986. Transcription of SV40 and Polyoma Virus and Its Regulation. In N. P. Salzman (ed.), The Papovaviridae, vol. 1, pp. 27–98. Plenum Press, New York.

1.3f

K. J. Danna and D. Nathans. 1972. Bidirectional Replication of Simian Virus 40 DNA. *Proc. Natl. Acad. Sci. USA* 69 3097–3100.

M. L. DePamphilis and P. M. Wassarman. 1980. Replication of Eukaryotic Chromosomes: A Closeup of the Replication Fork. *Annu. Rev. Biochem.* 49 627–666.

M. Seidman and N. P. Salzman. 1983. DNA Replication of Papovaviruses: *In Vivo* Studies. In Y. Becker (ed.), Replication of Viral and Cellular Genomes, pp. 29–52. Nijoff, The Hague.

J. Li and T. J. Kelly. 1984. Simian Virus 40 DNA Replication *in Vitro*. *Proc. Natl. Acad. Sci. USA* 81 6973–6977.

J. Li and T. J. Kelly. 1985. Simian Virus 40 DNA Replication *in Vitro*: Specificity of Initiation and Evidence for Bidirectional Replication. *Mol. Cell. Biol.* 5 1238–1246.

M. L. DePamphilis and M. Bradley. 1986. Replication of SV40 and Polyoma Virus Chromosomes. In N. P. Salzman (ed.), The Papovaviridae, vol. 1, pp. 99–246. Plenum Press, New York.

T. J. Kelly. 1988. SV40 DNA Replication. *J. Biol. Chem.* 263 17889–17892.

M. D. Challberg and T. J. Kelly. 1989. Animal Virus DNA Replication. *Annu. Rev. Biochem.* 58 671–717.

B. Stillman. 1989. Initiation of Eukaryotic DNA Replication *in Vitro*. *Annu. Rev. Cell Biol.* 5 197–245.

J. A. Borowiec, F. B. Dean, P. A. Bullock, and J. Hurwitz. 1990. Binding and Unwinding—How T Antigen Engages the SV40 Origin of DNA Replication. *Cell* 60 181–184.

R. Parsons, M. E. Anderson, and P. Tegtmeyer. 1990. Three Domains in the Simian Virus 40 Core Origin Orchestrate the Binding, Melting, and DNA Helicase Activities of T Antigen. *J. Virol.* 64 509–518.

C. Prives. 1990. The Replication Functions of SV40 T Antigen are Regulated by Phosphorylation. *Cell* 61 735–738.

L. F. Erdile, K. L. Collins, A. Russo, P. Simancek, D. Small, C. Umbricht, D. Virshup, L. Cheng, S. Randall, D. Weinberg, I. Moarefi, E. Fanning, and T. Kelly. 1991. Initiation of SV40 DNA Replication: Mechanism and Control. *Cold Spring Harbor Symp. Quant. Biol.* 56 303–313.

T. J. Kelly. 1991. DNA Replication in Mammalian Cells: Insights from the SV40 Model System. In The Harvey Lectures, ser. 85, pp. 173–188. Wiley-Liss, New York.

F. B. Dean, J. A. Borowiec, T. Eki, and J. Hurwitz. 1992. The Simian Virus 40 T Antigen Double Hexamer Assembles Around the DNA at the Replication Origin. *J. Biol. Chem.* 267 14129–14137.

S. K. Randall and T. J. Kelly. 1992. The Fate of Parental Nucleosomes During SV40 DNA Replication. *J. Biol. Chem.* 267 14259–14265.

K. Sugasawa, Y. Ishimi, T. Eki, J. Hurwitz, A. Kikuchi, and F. Hanaoka. 1992. Nonconservative Segregation of Parental Nucleosomes During Simian Virus 40 Chromosome Replication *in Vitro*. *Proc. Natl. Acad. Sci. USA* 89 1055–1059.

I. F. Moartefi, D. Small, I. Gilbert, M. Höpfner, S. K. Randall, C. Schneider, A. A. R. Russo, U. Ramsperger, A. K. Arthur, H. Stahl, T. J. Kelly, and E. Fanning. 1993. Mutation of the Cyclin-Dependent Kinase Phosphorylation Site in Simian Virus 40 (SV40) Large T Antigen Specifically Blocks SV40 Origin DNA Unwinding. *J. Virol.* 67 4992–5002.

Y. Murakami and J. Hurwitz. 1993. DNA Polymerase α Stimulates the ATP-Dependent Binding of Simian Virus Tumor T Antigen to the SV40 Origin of Replication. *J. Biol. Chem.* 268 11018–11027.

1.3g

M. Vogt and R. Dulbecco. 1960. Virus-Cell Interaction with a Tumor-Producing Virus. *Proc. Natl. Acad. Sci. USA* 46 365–370.

D. P. Lane and L. V. Crawford. 1979. T Antigen Is Bound to a Host Protein in SV40 Transformed Cells. *Nature* 278 261–263.

M. Botchan, W. C. Topp, and J. Sambrook. 1976. The Arrangement of Simian Virus 40 Sequences in the DNA of Transformed Cells. *Cell* 9 269–287.

R. G. Martin. 1981. The Transformation of Cell Growth and Transmogrification of DNA Synthesis by Simian Virus 40. *Adv. Cancer Res.* 34 1–68.

G. H. Sack Jr. 1981. Human Cell Transformation by Simian Virus 40—A Review. *In Vitro* 17 1–19.

B. Schaffhausen. 1983. Transforming Genes and Gene Products of Polyoma and SV40. *CRC Crit. Rev. Biochem.* 13 215–286.

A. E. Smith. 1984. Oncogenes: Growth Regulation and the Papovaviruses Polyoma and SV40. *J. Cell. Biochem.* 22 89–93.

R. Monier. 1986. Transformation by SV40 and Polyoma. In M. Salzman (ed.), The Papovaviridae, vol. 1, pp. 247–294. Plenum Press, New York.

T. Benjamin and P. K. Vogt. 1990. Cell Transformation by Viruses. In B. N. Fields, D. M. Knipe, and P. M. Howley (eds.), Virology, 2nd ed., chap. 14, pp. 317–367. Raven Press, New York.

K. Buchkovich, L. A. Duffy, and E. Harlow. 1989. The Retinoblastoma Protein Is Phosphorylated During Specific Phases of the Cell Cycle. *Cell* 58 1097–1105.

D. Malkin, F. O. P. Li, L. C. Strong, J. F. Fraumeni Jr., C. E. Nelson, D. Kin, J. Kassel, M. A. Gryka, F. Z. Bischoff, M. A. Fainsky, and S. H. Friend. 1990. Germ Line p53 Mutations in a Familial Syndrome of Breast Cancer, Sarcomas and Other Neoplasms. *Science* 250 1233–1238.

L. Raycroft, H. Wu, and G. Lozano. 1990. Transcriptional Activation by Wild-Type But Not Transforming Mutants of the p53 Anti-Oncogene. *Science* 249 1049–1051.

S. E. Kern, K. W. Kinzler, A. Bruskin, D. Jarosz, P. Friedman, C. Prives, and B. Vogelstein. 1991. Identification of p53 as a Sequence-Specific DNA-Binding Protein. *Science* 252 1708–1711.

A. J. Levine, J. Momand, and C. A. Finlay. 1991. The p53 Tumour Suppressor Gene. *Nature* 351 453–456.

L. Cao, B. Faha, M. Dembski, L.-H. Tsai, E. Harlow, and N. Dyson. 1992. Independent Binding of the Retinoblastoma Protein and p107 to the Transcription Factor E2F. *Nature* 355 176–179.

D. Cobrinik, S. F. Dowdy, P. W. Hinds, S. Mittnacht, and R. A. Weinberg. 1992. The Retinoblastoma Protein and the Regulation of Cell Cycling. *TIBS* 17 312–315.

S. W. Hiebert, S. P. Chellappan, J. M. Horowitz, and J. R. Nevins. 1992. *Genes and Development* 6 177–185.

D. P. Lane. 1992. Worrying About p53. *Current Biology* 2 581–583.

A. J. Levine. 1992. The p53 Tumor Suppressor Gene and Product. In A. J. Levine (ed.), Cancer Surveys: Tumor Suppressor Genes, the Cell Cycle and Cancer, pp. 59–79. Cold Spring Harbor Laboratory Press, Cold Spring Harbor, New York.

D. M. Livingston. 1992. Functional Analysis of the Retinoblastoma Gene Product and of Rb-SV40 T Antigen Complexes. In A. J. Levine (ed.), Cancer Surveys: Tumor Suppressor Genes, the Cell Cycle and Cancer, pp. 153–160. Cold Spring Harbor Laboratory Press, Cold Spring Harbor, New York.

J. Momand, G. P. Zambetti, D. C. Olson, D. George, and A. J. Levine. 1992. The *mdm*-2 Oncogene Product Forms a Complex with the p53 Protein and Inhibits p53-Mediated Transactivation. *Cell* 69 1237–1245.

E. Seto, A. Usheva, G. P. Zambetti, J. Momand, N. Horikoshi, R. Weinmann, A. J. Levine, and T. Shenk. 1992. Wild-Type p53 Binds to the TATA Binding Protein and Interferes with Transorption. *Proc. Natl. Acad. Sci. USA* 89: 12028–12032.

S. J. Ullrich, C. W. Anderson, W. E. Mercer, and E. Appella. 1992. The p53 Tumor Suppressor Protein, a Modulator of Cell Proliferation. *J. Biol. Chem.* 267 15259–15262.

R. A. Weinberg. 1992. The Retinoblastoma Gene and Gene Product. In A. J. Levine (ed.), Cancer Surveys: Tumor Suppressor Genes, the Cell Cycle and Cancer, pp. 43–57. Cold Spring Harbor Laboratory Press, Cold Spring Harbor, New York.

Y. Yin, M. A. Tainsky, F. Z. Bischoff, L. C. Strong, and G. M. Wahl. 1992. Wild-Type p53 Restores Cell Cycle Control and Inhibits Gene Amplification in Cells with Mutant p53 Alleles. *Cell* 70 937–948.

W. S. El-Deiry, T. Tokino, V. E. Velculescu, D. B. Levy, R. Parsons, J. M. Trent, D. Lin, W. E. Mercer, K. W. Kinzler, and B. Vogelstein. 1993. *WAF1*, a Potential Mediator of p53 Tumor Suppression. *Cell* 75 817–825.

C. A. Finlay. 1993. The mdm-2 Oncogene Can Overcome Wild-Type p53 Suppression of Transformed Cell Growth. *Mol. Cell. Biol.* 13 301–306.

J. W. Harper, G. R. Adami, N. Wei, K. Keyomarsi, and S. J. Elledge. 1993. The p21 Cdk-Interacting Protein Cip1 Is a Potent Inhibitor of G1 Cyclin-Dependent Kinases. *Cell* 75 805–816.

T. Hunter. 1993. Braking the Cycle. *Cell* 75 839–841.

X. Lu and D. P. Lane. 1993. Differential Induction of Transcriptionally Active p53 Following UV or Ionizing Radiation: Defects in Chromosome Instability Syndromes. *Cell* 75 765–778.

J. W. Ludlow. 1993. Interactions Between SV40 Large-Tumor Antigen and the Growth Suppressor Proteins pRB and p53. *FASEB J.* 7 866–871.

E. Moran. 1993. DNA Tumor Virus Transforming Proteins and the Cell Cycle. *Curr. Opin. Genet. Dev.* 3 63–70.

M. E. Perry and A. J. Levine. 1993. Tumor Suppressor p53 and the Cell Cycle. *Curr. Opin. Genet. Dev.* 3 50–54.

C. Prives. 1993. Doing the Right Thing: Feedback Control and p53. *Curr. Opin. Cell Biol.* 5 214–218.

K. G. Wiman. 1993. The Retinoblastoma Gene: Role in Cell Cycle Control and Cell Differentiation. *FASEB J.* 7 841–845.

G. P. Zambetti and A. J. Levine. 1993. A Comparison of the Biological Activities of Wild-Type and Mutant p53. *FASEB J.* 7 855–865.

M. E. Ewen. 1994. The Cell Cycle and the Retinoblastoma Protein Family. *Cancer Metastasis Rev.* 13 45–66.

J. R. Nevins. 1994. Cell Cycle Targets of the DNA Tumor Viruses. *Curr. Opin. Genet. Rev.* 4 130–134.

J. R. Nevins and P. K. Vogt. 1995. Cell Transformation by Viruses. In B. M. Fields, D. M. Knipe, and P. M. Howley (eds.), Virology, 3rd ed., chap. 11. Raven Press, New York.

1.3h

E. Fanning. 1992. Simian Virus 40 Large T Antigen: The Puzzle, the Pieces and the Emerging Picture. *J. Virol.* 66 1289–1293.

E. Fanning and R. Knippers. 1992. Structure and Function of Simian Virus 40 Large Tumor Antigen. *Annu. Rev. Biochem.* 61 55–85.

J. M. Pipas. 1992. Common and Unique Features of T Antigens Encoded by the Polyomavirus Group. *J. Virol.* 66 3979–3985.

E. Sontag, S. Fedorov, C. Kamibayashi, D. Robbins, M. Cobb, and M. Mumby. 1993. The Interaction of SV40 Small Tumor Antigen with Protein Phosphatase 2A Stimulates the Map Kinase Pathway and Induces Cell Proliferation. *Cell* 75 887–897.

1.3i

D. A. Jackson, R. H. Symons, and P. Berg. 1972. Biochemical Method for Inserting New Genetic Information into DNA of Simian Virus 40: Circular SV40 DNA Molecules Containing λ Phage Genes and the Galactose Operon of *E. coli*. *Proc. Natl. Acad. Sci. USA* 69 2904–2909.

S. P. Goff and P. Berg. 1976. Construction of Hybrid Viruses Containing SV40 and λ Phage DNA Segments and Their Propagation in Cultured Monkey Cells. *Cell* 9 695–705.

R. M. Mulligan and P. Berg. 1979. Synthesis of Rabbit β-Globin Recombinant Genome. *Nature* 277 108–114.

P. Berg. 1980. Dissections and Reconstructions of Genes and Genomes. In Les Prix Nobel, pp. 97–114. Almquist and Wiksell International, Stockholm, Sweden.

D. H. Hamer. 1980. DNA Cloning in Mammalian Cells with SV40 Vectors. In J. K. Setlow and A. Hollaender (eds.), Genetic Engineering, vol. 2, pp. 83–101. Plenum Press, New York.

R. C. Mulligan and P. Berg. 1980. Simian Virus 40 Mediated Expression of a Bacterial Gene in Mammalian Cells. *Science* 209 1422–1427.

1.4a

I. Seif, G. Khoury, and R. Dahn. 1979. The Genome of Human Papovavirus BKV. *Cell* 18 963–977.

R. J. Frisque, G. L. Bream, and M. T. Cannella. 1984. Human Polyomavirus JC Virus Genome. *J. Virol.* 51 458–469.

D. L. Walker and R. J. Frisque. 1986. The Biology and Molecular Biology of JC Virus. In M. P. Salzman (ed.), The Papovaviridae, pp. 327–377. Plenum Press, New York.

K. Yoshiike and K. K. Takemoto. 1986. Studies with BK Virus and Monkey Lymphotropic Papovavirus. In N. P. Salzman (ed.), The Papovaviridae, pp. 295–326. Plenum Press, New York.

K. V. Shah. 1995. Human Polyomaviruses. In B. N. Fields, D. M. Knipe, and P. M. Howley (eds.), Virology, 3rd ed., chap. 65, pp. 2027–2043. Lippincott-Raven Publishers, Philadelphia.

1.4b

R. Dulbecco and M. Vogt. 1963. Evidence for a Ring Structure of Polyoma Virus DNA. *Proc. Natl. Acad. Sci. USA* 50 236–243.

R. L. Mackey and R. A. Consigli. 1976. Early Events in Polyoma Virus Infection: Attachment, Penetration and Nuclear Entry. *J. Virol.* 19 620–636.

M. Fried and B. Griffin. 1977. Organization of the Genomes of Polyoma Virus and SV40. *Adv. Cancer Res.* 24 67–113.

R. J. Staneloni and T. L. Benjamin. 1977. Host Range Selection of Transformation Defective hr-t Mutants of Polyoma Virus. *Virology* 77 598–609.

M. Katinka, M. Yaniv, M. Vasseur, and D. Blangy. 1980. Expression of Polyoma Early Functions in Mouse Embryonal Carcinoma Cells Depends on Sequence Rearrangements in the Beginning of the Late Region. *Cell* 20 393–399.

W. Eckhart. 1981. Polyoma T-Antigens. *Adv. Cancer Res.* 35 1–25.

B. E. Griffin, E. Soeda, B. G. Barrell, and R. Staden. 1981. Sequence and Analysis of Polyoma Virus DNA. In J. Tooze (ed.), DNA Tumor Viruses, 2nd ed., pp. 843–913. Cold Spring Harbor Laboratory, Cold Spring Harbor, New York.

A. J. Levine. 1982. The Nature of the Host Range Restriction of SV40 and Polyoma Viruses in Embryonal Carcinoma Cells. *Curr. Top. Microbiol. and Immunol.* 10 1–30.

S. A. Courteneidge and A. E. Smith. 1983. Polyoma Virus Transforming Protein Associates with the Product of the c-src Cellular Gene. *Nature* 303 435–439.

B. E. Griffin and S. M. Dilworth. 1983. Polyomavirus: An Overview of Its Unique Properties. *Adv. Cancer Res.* 39 183–268.

G. M. Veldman, S. Lupton, and R. Kamen. 1985. Polyomavirus Enhancer Contains Multiple Redundant Sequence Elements That Activate Both DNA Replication and Gene Expression. *Mol. Cell. Biol.* 5 649–658.

D. M. Salunke, D. L. D. Caspar, and R. L. Garcea. 1986. Self Assembly of Purified Polyoma Virus Capsid Protein VP1. *Cell* 46 895–904.

W. Eckhart. 1990. Polyomavirinae and Their Replication. In B. N. Fields and D. M. Knipe (eds.), Virology, 2nd ed., pp. 1593–1607. Raven Press, New York.

R. Rochford, B. A. Campbell, and L. P. Villarreal. 1990. Genetic Analysis of the Enhancer Requirements for Polyomavirus DNA Replication in Mice. *J. Virol.* 64 476–485.

J. J. Wirth, A. Amalfitano, R. Gross, M. B. A. Oldstone, and M. M. Fluck. 1992. Organ- and Age-Specific Replication of Polyomavirus in Mice. *J. Virol.* 66 3278–3286.

D. H. Barouch and S. C. Harrison. 1994. Interactions Among the Major and Minor Coat Proteins of Polyomavirus. *J. Virol.* 68 3982–3989.

S. M. Dilworth, C. E. P. Brewster, M. D. Jones, L. Lanfrancone, G. Pellici, and P. G. Pellici. 1994. Transformation by Polyoma Virus Middle T-Antigen Involves the Binding and Phosphorylation of Shc. *Nature* 367 87–90.

C. N. Cole. 1995. Polyomavirinae: The Viruses and Their Replication. In B. N. Fields, D. M. Knipe, and P. M. Howley (eds.), Virology, 3rd ed., chap. 63. Lippincott-Raven Publishers, Philadelphia.

1.4c

H. Weiher and M. R. Botchan. 1984. An Enhancer Sequence from Bovine Papilloma Virus DNA Consists of Two Essential Regions. *Nucleic Acids Res.* 12 2901–2916.

H. zur Hausen and A. Schneider. 1987. The Role of Papillomaviruses in Human Anogenital Cancer. In P. N. Howley and N. P. Salzman (eds.), The Papovaviridae, vol. 2, pp. 245–263. Plenum Press, New York.

P. F. Lambert, C. C. Baker, and P. M. Howley. 1988. The Genetics of Bovine Papillomavirus Type 1. *Annu. Rev. Genet.* 22 235–258.

N. Dyson, P. M. Howley, K. Münger, and E. Harlow. 1989. The Human Papillomavirus E7 Oncoproten Is Able to Bind the Retinoblastoma Gene Product. *Science* 243 934–937.

W. C. Phelps and P. M. Howley. 1989. Regulation of Papillomavirus Gene Expression. In N. H. Colburn (ed.), Genes and Signal Transduction in Multistage Carcinogenesis, pp. 231–259. Marcel Dekker, New York.

C. C. Baker and L. M. Cousert. 1990. The Genomes of the Papillomaviruses: Genetic Maps. Locus of Complex Genomes. In S. J. O'Brien (ed.), Genetic Maps, Book 1, 5th ed., pp. 134–146. Cold Spring Harbor Laboratory Press, Cold Spring Harbor, New York.

B. A. Werness, A. J. Levine, and P. M. Howley. 1990. Association of Human Papillomavirus Types 16 and 18 E6 Proteins with p53. *Science* 248 76–79.

B. A. Werness, K. Münger, and P. M. Howley. 1991. Role of the Human Papillomavirus Oncoproteins in Transformation and Carcinogenic Progression. In V. T. DeVita Jr., S. Hellman, and S. A. Rosenberg (eds.), Important Advances in Oncology, pp. 3–18. J. B. Lippincott Press, Philadelphia.

P. M. Howley. 1991. Role of the Human Papillomaviruses in Human Cancer. *Cancer Res.* (Suppl.) 51 5019s–5022s.

P. F. Lambert. 1991. Papillomavirus DNA Replication. *J. Virol.* 65 3417–3420.

A. A. McBride, H. Romanczuk, and P. M. Howley. 1991. The Papillomavirus E2 Regulatory Proteins. *J. Biol. Chem.* 266 18411–18414.

L. Yang, R. Li, I. J. Mohr, R. Clark, and M. R. Botchan. 1991. Activation of BPV-1 Replication *in Vitro* by the Transcription Factor E2. *Nature* 353 628–632.

L. Yang, I. Mohr, R. Li, T. Nottoli, S. Sun, and M. Botchan. 1991. Transcription Factor E2 Regulates BPV-1 DNA Replication in Vitro by Direct Protein-Protein Interaction. *Cold Spring Harbor Symp. Quant. Biol.* 56 335–346.

M. S. Campo. 1992. Cell Transformation by Animal Papillomaviruses. *J. General Virol.* 73 217–222.

L. Gissmann. 1992. Papillomaviruses and Human Oncogenesis. *Curr. Opin. Gen. & Devel.* 2 97–102.

K. Münger, M. Scheffner, J. M. Huibregtse, and P. M. Howley. 1992. Interactions of HPV E6 and E7 Oncoproteins with Tumor Suppressor Gene Products. In Cancer Surveys: Tumor Suppressor Genes, the Cell Cycle and Cancer, pp. 197–217. Cold Spring Harbor Laboratory Press, Cold Spring Harbor, New York.

Y.-S. Seo, F. Müller, M. Lusky, E. Gibbs, H.-Y. Kim, B. Phillips, and J. Hurwitz. 1993. Bovine Papilloma Virus (BPV)-Encoded E2 Protein Enhances Binding of E1 Protein to the BPV Replication Origin. *Proc. Natl. Acad. Sci. USA* 90 2865–2869.

Y.-S. Seo, F. Müller, M. Lusky, and J. Hurwitz. 1993. Bovine Papilloma Virus (BPV)-Encoded E1 Protein Contains Multiple Activities Required for BPV DNA Replication. *Proc. Natl. Acad. Sci. USA* 90 702–706.

K. Vousden. 1993. Interactions of Human Papillomavirus Transforming Proteins with the Products of Tumor Suppressor Genes. *FASEB J.* 7 872–879.

M. Scheffner, H. Romanszuk, K. Münger, J. M. Huibregtse, J. A. Mietz, and P. M. Howley. 1994. Functions of Human Papillomavirus Proteins. *Current Topics in Microbiol. and Immunol.* 186 83–99.

P. M. Howley. 1995. Papillomavirinae. In B. N. Fields, D. M. Knipe, and P. M. Howley (eds.), Virology, 3rd ed., chap. 65. Raven Press, New York.

1.5a

P. Whyle, N. M. Williamson, and E. Harlow. 1989. Cellular Targets for Transformation by the Adenovirus E1A Proteins. *Cell* 56 67–75.

M. S. Barbosa, C. Edmonds, C. Fisher, J. T. Schiller, D. R. Lowy, and K. H. Vousden. 1990. The Region of HPV E7 Oncoprotein Homologous to E1A and SV40 Large-T Antigen Contains Separate Domains for Rb Binding and Casein Kinase II Phosphorylation. *EMBO J.* 9 153–160.

A. W. Braithwaite, C. C. Nelson, and A. J. D. Bellett. 1991. E1a Revisited: The Case for Multiple Cooperative Trans-Activation Domains. *The New Biol.* 3 18–26.

J. Flint and T. Shenk. 1991. Transcriptional and Transforming Activities of the Adenovirus E1A Protein. *Adv. Cancer Res.* 57 47–85.

N. Horikoshi, K. Maguire, A. Kralli, E. Maldonado, D. Reinberg, and R. Weinmann. 1991. Direct Interaction Between Adenovirus E1A Protein and the TATA Box Binding Transcription Factor IID. *Proc. Natl. Acad. Sci. USA* 88 5124–5128.

W. S. Lee, C. C. Kao, G. O. Bryant, X. Liu, and A. J. Berk. 1991. Adenovirus E1A Activation Domain Binds the Basic Repeat in the TATA Box Transcription Factor. *Cell* 67 365–376.

Y. Shi, E. Seto, L.-S. Chang, and T. Shenk. 1991. Transcriptional Repression by YY1, a Human GLI-Krüppel-Related Protein, and Relief of Repression by Adenovirus E1A Protein. *Cell* 67 377–388.

D. Barbeau, R. C. Marcellus, S. Bacchetti, S. T. Bayley, and P. E. Branton. 1992. Quantitative Analysis of Regions of Adenovirus E1A Products Involved in Interactions with Cellular Proteins. *Biochem. and Cell Biol.* 70 1123–1134.

N. Dyson and E. Harlow. 1992. Adenovirus E1 Targets Key Regulators of Cell Proliferation. In Cancer Surveys: Tumor Suppressor Genes, the Cell Cycle and Cancer, pp. 161–195. Cold Spring Harbor Laboratory Press, Cold Spring Harbor, New York.

S. Shirodkar, M. Ewen, J. A. DeCaprio, J. Morgan, D. M. Livingston, and T. Chittenden. 1992. The Transcription Factor E2F Interacts with the Retinoblastoma Product and a p107-Cyclin A Complex in a Cell Cycle–Regulated Manner. *Cell* 68 157–166.

E. Moran. 1993. DNA Tumor Virus Transforming Proteins and the Cell Cycle. *Curr. Opin. Genet. Devel.* 3 63–70.

E. Moran. 1993. Interaction of Adenoviral Proteins with pRB and p53. *FASEB J.* 7 880–885.

J. R. Nevins. 1994. Cell Cycle Targets of the DNA Tumor Viruses. *Curr. Opin. Genet. Dev.* 4 130–134.

J. R. Nevins and P. K. Vogt. 1995. Cell Transformation by Viruses. In B. H. Fields, D. M. Knipe, and P. M. Howley (eds.), Virology, 3rd ed., chap. 11. Raven Press, New York.

CHAPTER 2

Eukaryotic RNA Viruses: A Variant Genetic System

ELLEN G. STRAUSS and JAMES H. STRAUSS

2.1 Replication Strategies of Eukaryotic RNA Viruses

We define RNA viruses as those having RNA genomes and replicating without forming DNA intermediates. Retroviruses also have RNA genomes, but they replicate by means of DNA copies (Chapter 3). Before the advent of cloning and sequencing, relatively little was known about the structure, organization, and transcription of RNA virus genomes, the characteristics and amino acid sequences of the virus-specific polypeptides translated from the RNAs, or the nature and scope of their untranslated regions. Although the RNA genome of bacteriophage MS2 was sequenced in its entirety at the RNA level, direct RNA sequencing is difficult, inefficient, and tedious. Since the discovery of reverse transcriptase, the development of new sequencing strategies, and the advent of cDNA cloning, many RNA virus genomes have been sequenced in their entirety, and new sequences appear monthly. With the development of cloning methods and the ability to amplify the cloned segments as bacterial plasmids, it has become far easier to determine the nucleotide sequence of a gene encoding a particular protein than to purify the polypeptide and determine its amino acid sequence directly. This is especially true for those viral gene products that perform catalytic functions and therefore are synthesized in minute amounts in infected cells. More recently, the polymerase chain reaction (PCR) has been used in conjunction with reverse transcriptase to amplify rare RNA molecules as cDNA.

RNA viruses are a diverse group that infect prokaryotic as well as a wide variety of eukaryotic hosts, both plants and animals. Most RNA viruses have a single-stranded genome; the exceptions include members of the *Reoviridae* and three families of fungal viruses and virus-like particles: *Totiviridae*, *Partitiviridae* (Section 2.4). Unlike DNA, which has evolved as a stable repository of genetic information, cellular RNAs are primarily effector molecules. In particular, mRNAs are transient intermediates designed to convey information from the nucleus to the translational machinery. Cells therefore contain a variety of enzymes that degrade mRNA, a situation that makes isolating intact RNA molecules difficult.

RNA viruses can be considered a particular class of complex mRNAs that have evolved strategies both to replicate and to maintain themselves in the cytoplasm of cells and to pass undamaged from one host cell to another. To do this, they must encode a number of functions not usually required of mRNAs. These include proteins to form protective shells for the extracellular transit from one cell to another, proteins or other components that subvert the cellular machinery away from the production of cellular macromolecules toward the synthesis of viral products, and proteins with RNA-dependent RNA replicase/transcriptase enzymatic activities. The various RNA virus families have evolved different ways to meet these requirements.

RNA viruses come in a variety of sizes and shapes, a number of which are illustrated in Figure 2.1. The simplest consist of a single RNA molecule encapsidated with multiple copies of a single species of capsid protein. The geometry of the resulting virions is usually either icosahedral or helical, two structural solutions that allow identical protein subunits to protect a nucleic acid molecule completely. Most plant viruses (Figure 2.1b) are simple icosahedra or helices (rods). Some animal viruses are also simple icosahedra whose protein shells are composed of more than one kind of polypeptide chain. Several animal viruses consist of a nucleocapsid, either helical or icosahedral, which is further protected by either an outer concentric protein shell or an outer envelope (Figure 2.1a, "enveloped"). Viral envelopes consist of a lipid bilayer containing a number of virus-encoded integral transmembrane glycoproteins. In some families, the inner surface of the bilayer is coated with a hydrophobic protein known as the M or matrix protein, which also interacts with the nucleocapsid during morphogenesis.

RNA genomes consist of either a single strand of RNA or several discrete RNA pieces or segments. For animal viruses, the genomic segments are all encapsidated within a single virion. However, in plant viruses, different genomic segments may be individually encapsidated into separate particles. Thus, for these latter viruses, more than one particle is required to initiate the infectious cycle.

With the exception of the segmented minus-strand RNA viruses, replication of RNA viruses is wholly cytoplasmic and therefore can utilize only cytoplasmic or virus-encoded enzymes; nuclear functions such as RNA splicing are unavailable. Because their replication is cytoplasmic, RNA viruses are in general insensitive to transcription inhibitors, such as actinomycin D, which inhibits DNA-dependent RNA polymerase, or α-amanitin, which at very low concentrations inhibits RNA polymerase II (*Genes and Genomes,* section 3.9). Some RNA viruses can in fact replicate in enucleated cells. For others, enucleation results in

Table 2.1 *Human and Veterinary Pathogens Among RNA Viruses*

Virus Family	Some Pathogenic Members
Togaviridae	Rubella (German measles) Eastern equine encephalitis Western equine encephalitis Venezuelan equine encephalitis
Flaviviridae	Yellow fever Dengue St. Louis encephalitis Japanese encephalitis
Picornaviridae	Poliovirus Foot and mouth disease Rhinovirus (common cold)
Coronaviridae	Human coronavirus (common cold) Avian infectious bronchitis
Rhabdoviridae	Rabies
Paramyxoviridae	Measles Mumps Respiratory syncytial
Orthomyxoviridae	Influenza
Arenaviridae	Lassa fever Tacaribe complex
Bunyaviridae	La Crosse Rift Valley fever Korean hemorrhagic fever
Filoviridae	Ebola Marburg
Reoviridae	Human rotavirus (diarrheal infections) Bluetongue of sheep

depressed virus replication, presumably because most enveloped RNA viruses require glycoprotein modifications that occur in the Golgi, and enucleation removes some or all of the Golgi apparatus.

Medically, the RNA viruses are extremely important, and all groups of animal RNA viruses contain significant human or veterinary pathogens (Table 2.1). Their significance for the economic and social well-being of the world's population is one reason they have been so thoroughly studied; another is that they represent fascinating systems for studies in molecular biology and evolution. Today several laboratories are using recombinant DNA techniques to develop new vaccines for a large number of these pathogens in order to prevent future widespread epidemics of dengue fever, influenza, and other debilitating diseases. In addition, many RNA viruses are important plant pathogens that are being intensively studied because of their impact upon agriculture.

a. The Life Cycles of Three Types of RNA Viruses

Plus-strand RNA viruses **Plus-strand RNA** viruses are so named because their genomic RNA is of the messenger (plus) sense, and consequently the naked deproteinized RNA is infectious. They may be icosahedral or helical, with or without envelopes (Figure 2.1). Upon entry into a susceptible cell the input genomic

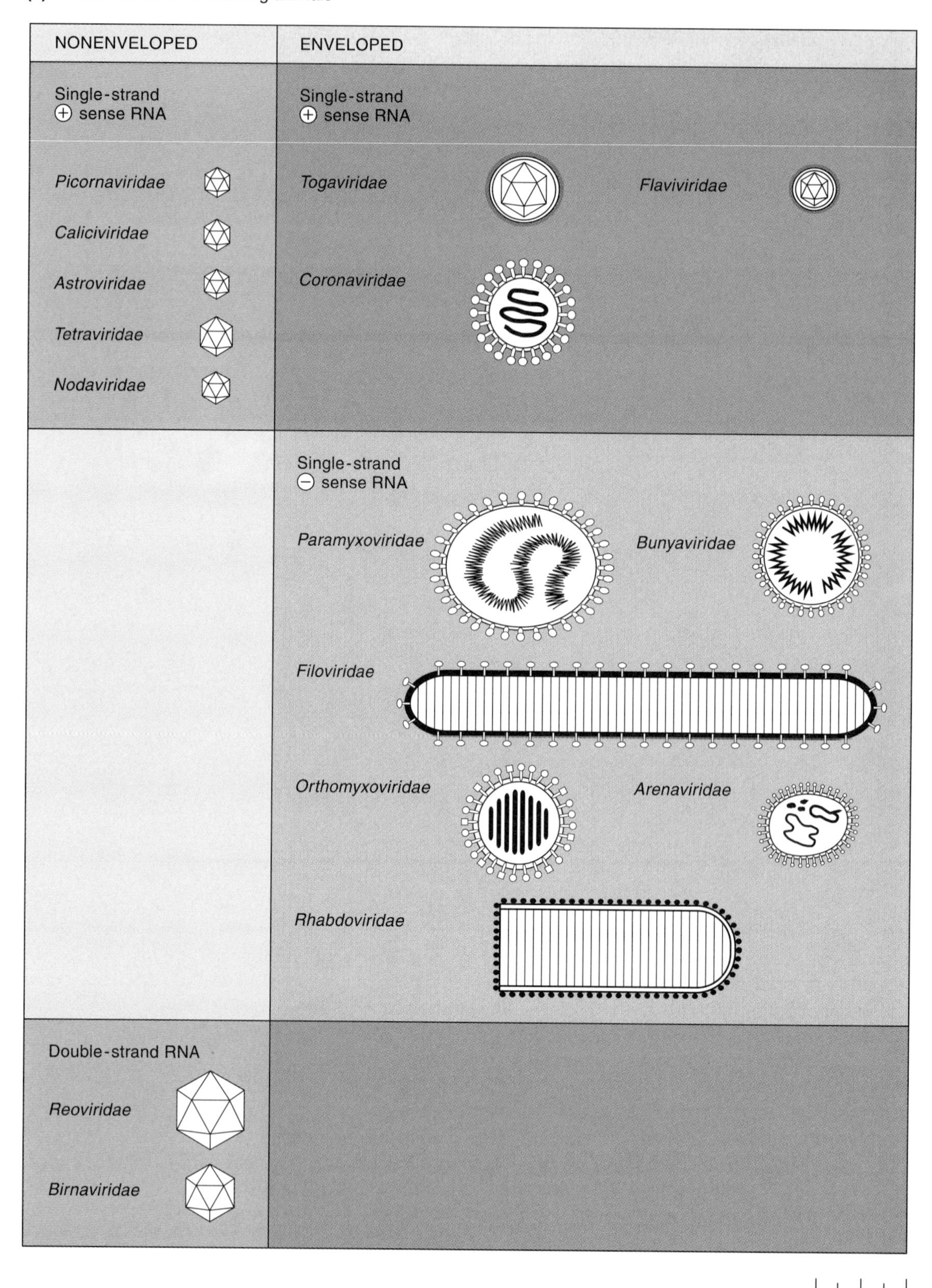

Figure 2.1
Morphology of RNA viruses. Representatives of families of viruses infecting (a) animals (both vertebrates and invertebrates) and (b) plants are shown drawn to scale. The families are grouped by the polarity of their RNA genomes and the presence or absence of

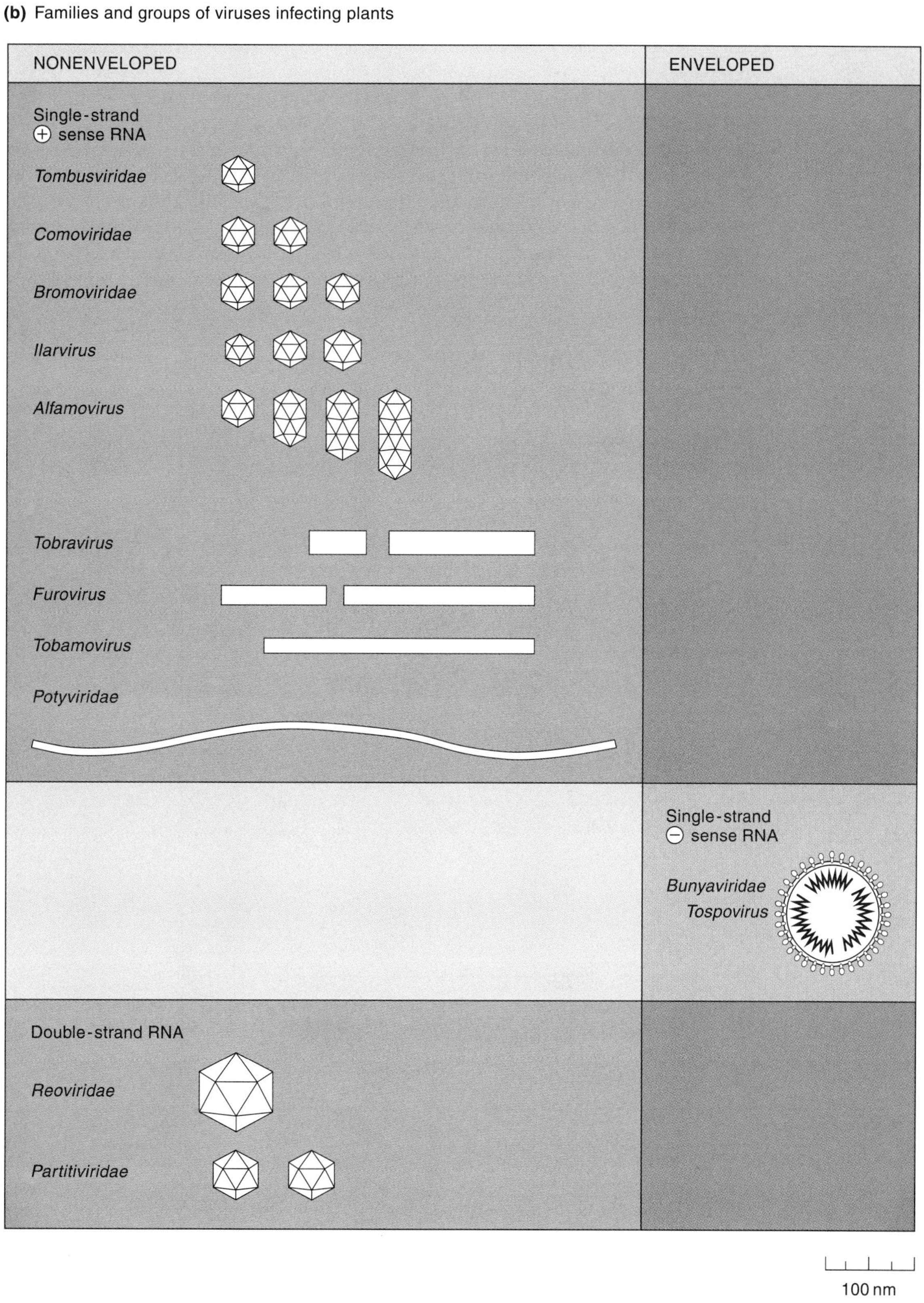

(b) Families and groups of viruses infecting plants

NONENVELOPED

ENVELOPED

Single-strand
⊕ sense RNA

Tombusviridae

Comoviridae

Bromoviridae

Ilarvirus

Alfamovirus

Tobravirus

Furovirus

Tobamovirus

Potyviridae

Single-strand
⊖ sense RNA

Bunyaviridae
Tospovirus

Double-strand RNA

Reoviridae

Partitiviridae

100 nm

a lipid envelope. Plus-strand RNAs can serve directly as mRNAs within a host cell. (Adapted from F. A. Murphy et al., eds. 1995. Viral Taxonomy: Sixth Report of the International Committee on Taxonomy of Viruses, pp. 19–23, Springer-Verlag, New York.)

strand is immediately translated to form the virus-specific RNA replicase (Figure 2.2). The replicase makes an exact minus-strand copy (called **vcRNA**, for virus complementary RNA) of the plus-strand, and this copy serves as a template from which full-length genomic RNAs are synthesized. For some viruses (picornaviruses, flaviviruses, bacteriophages, and some of the plant viruses), the genome-length RNA is the only messenger, and all virus-encoded proteins are translated from it. For other viruses, one (alphaviruses and some plant viruses) or more (caliciviruses, coronaviruses, and many plant viruses) mRNAs containing only part of the genome (subgenomic mRNAs) are also transcribed from the minus strand and translated into viral proteins. These subgenomic mRNAs en-

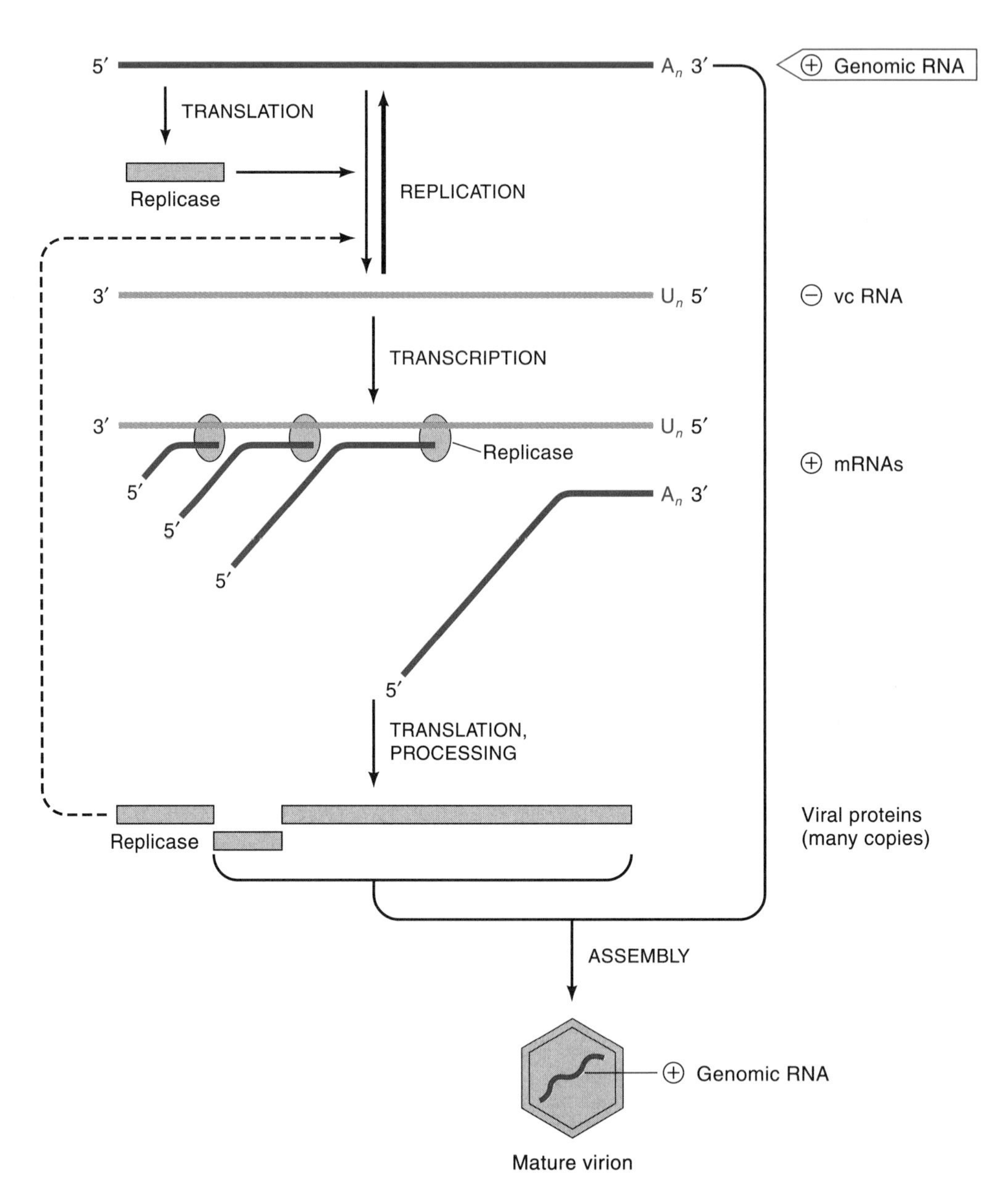

Figure 2.2
Replication of a typical plus-strand virus. Plus-strand RNAs are shown in dark color; minus-strand RNAs are shown in light color. The genomic RNA is highlighted by the colored arrow outline. *Replication* refers to synthesis of genomic RNA molecules and full-length RNAs complementary to the genomes (vcRNA) which serve as templates for genomic RNA synthesis. The term *transcription* is used exclusively to describe the production of mRNAs, which in general are not full-length (i.e., they are subgenomic).

code both the structural proteins, which self-assemble into virions containing genomic length plus strands, as well as other viral proteins that are not components of the virion.

Minus-strand RNA viruses In the life cycle of **minus-strand RNA** viruses, the genomic RNA, which may be a single molecule or segmented into several molecules, is of the antisense polarity and cannot serve as mRNA (Figure 2.3). For this reason, the virion must carry with it a viral polymerase that is responsible for

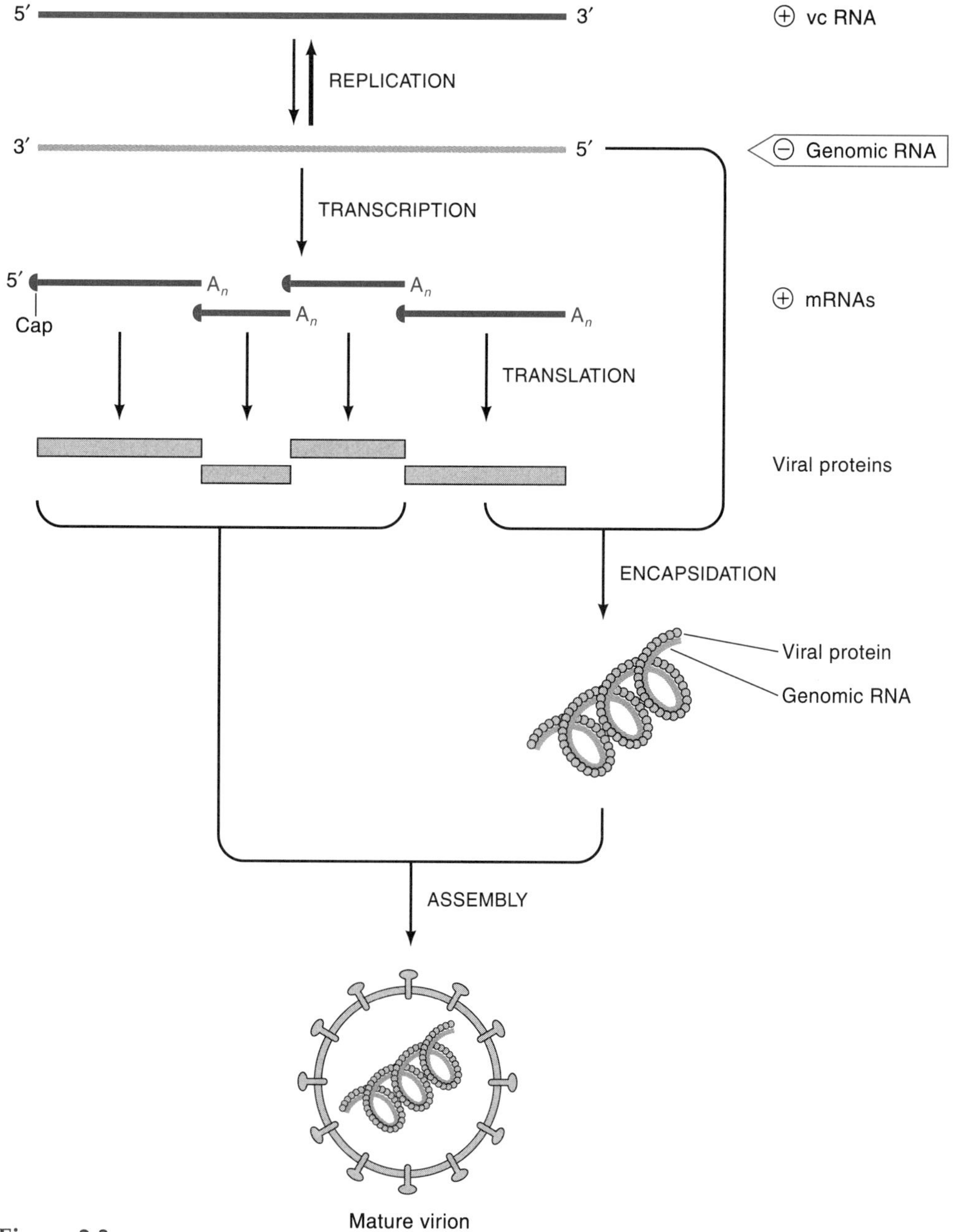

Figure 2.3
Replication of a typical minus-strand virus. Full-length plus-strand virus-complementary RNA (vcRNA) is produced by replication; subgenomic mRNAs are produced by transcription as shown. Plus-strand RNAs are dark colored; minus-strand RNAs are light colored. Genomic RNA is shown with colored arrow outline.

primary transcription. Although the envelopes of minus-strand viruses are removed in the early stages of cell penetration, the naked RNA is never present in the cell cytoplasm. The replication complex consists of the nucleocapsid, which contains a large number of copies of a nucleocapsid protein and minor amounts of other polypeptides, including the replicase/transcriptase enzymes. Primary transcription produces short RNAs transcribed from the 3′ terminus of the genomic RNA (leader or "le" RNAs) plus a series of plus-strand mRNAs that are not identical to the vcRNA used in replication (see Section 2.3). These primary transcripts are translated into protein products that are, in turn, responsible for setting up a secondary transcription cycle. During replication, a full-length plus-strand copy of the genome (vcRNA) is made that serves as a template for genomic full-length minus strands. The vcRNA as well as the genomic minus strand and various le RNAs are all encapsidated as soon as they are synthesized.

All the minus-strand viruses we know about are enveloped and, in addition to the nucleocapsid protein and the transcription enzymes, encode a number of glycoproteins. These glycoproteins are modified during passage through subcellular organelles (Chapter 8) and transported to appropriate cellular membranes, from which they are incorporated into mature virions by budding.

Double-strand RNA viruses In some ways, the double-strand RNA viruses may be considered a special case of the plus-strand viruses (Figure 2.4). The virion is only partially uncoated upon entering the cell, and the subviral core particle (containing transcriptase activity) is the site of synthesis of plus strands from all the RNA segments (ten in the case of reoviruses). Synthesis is temporally controlled, yielding early and late plus-strand transcripts, which serve as mRNAs for the viral proteins. Late in infection, these plus strands are encapsidated into a nascent core by the structural proteins. The RNAs within the nascent core are made into double strands by the synthesis of the minus-strand RNA as a final step in the virus morphogenesis.

b. Genetic Interactions Between Viral RNA Genomes

One characteristic that sets the RNA viruses apart as a "variant genetic system" is their mode of exchanging genetic information and generating diversity. Unlike the DNA genomes of their prokaryotic or eukaryotic hosts, or even the DNA viruses, most RNA viruses lack the ability to recombine efficiently during mixed infections (i.e., to produce progeny virus whose genomes contain covalently linked portions of the chromosomes of the two parental genomes) because they do not encode enzymes to break and rejoin RNA molecules. This means that, for most families of RNA viruses, genetic maps based upon frequency of recombination between markers do not exist. However, in the laboratory, complementation permits functional analysis. Thus, for most RNA virus families, catalogs of conditional lethal (usually temperature-sensitive) mutants exist and have been classified into complementation groups. The number of these groups indicates the number of essential functions that the virus has to perform.

Although in some cases the sequential arrangement of virus genes on the genome has been deduced from the order of translation of products from polycistronic messages or transcription of genes from the genomic RNA, for the most part, determining the genetic map of RNA virus genomes has depended on sequencing their genomes. From the sequence, the genome organization and translation strategy are usually clear, and similarities among strategies have been useful in deducing evolutionary relationships among viruses (for examples, see Sections 2.2a and 2.2b).

At least one member of most RNA virus families has been sequenced in its entirety, and it is now possible to generalize about their genomic organization.

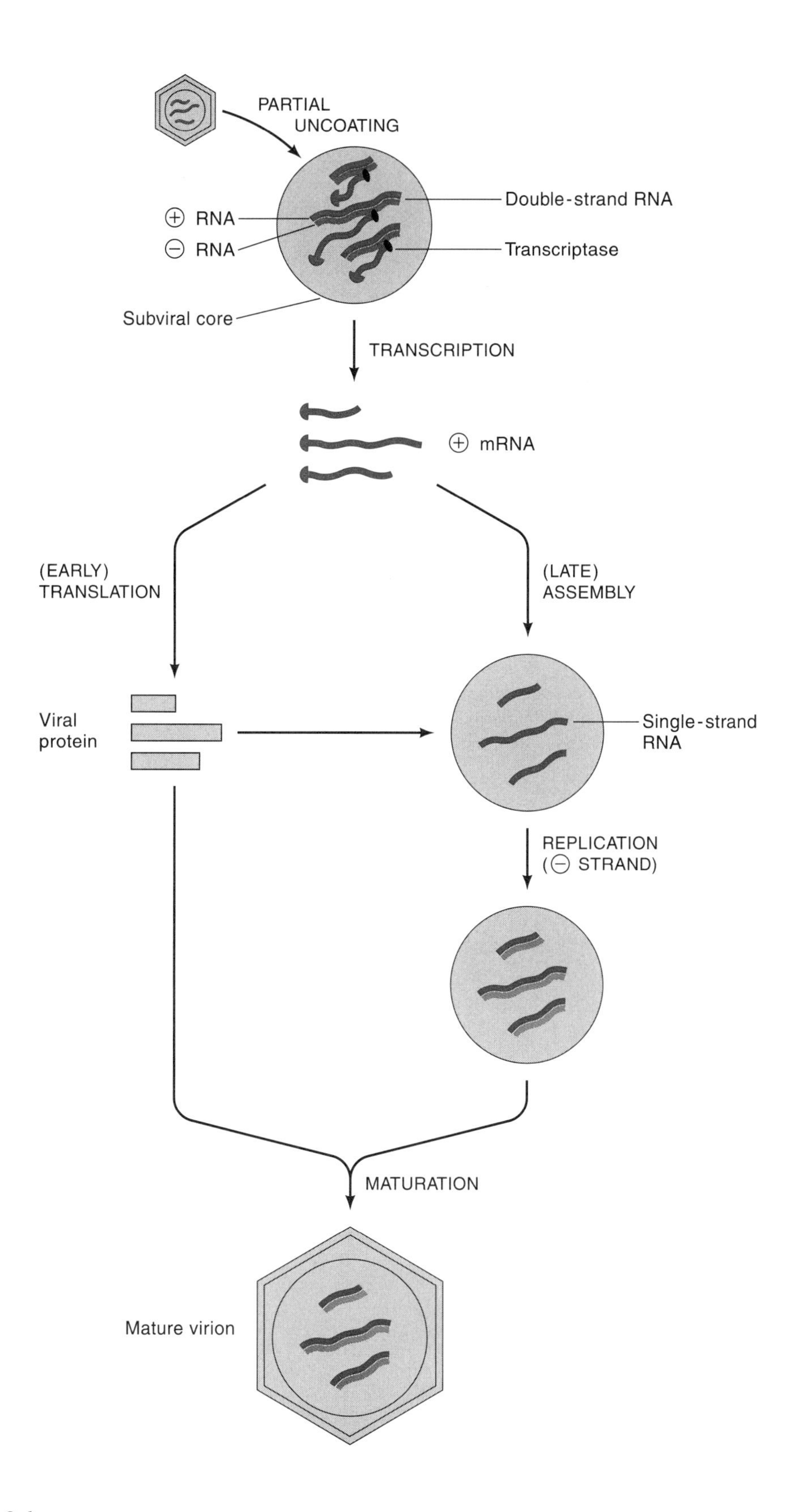

Figure 2.4
Replication of double-strand RNA viruses. The virion consists of a number of segments of double-strand RNA encapsidated in two concentric protein shells. The subviral core (virion minus its outer protein shell) is the transcription complex. Primary plus-strand transcripts (shown by dark color) are translated and encapsidated. Minus-strand replication (light color) occurs just before virion maturation.

First, there is very little untranslated sequence in their genomes. The **3′ untranslated regions (3′ UTR)** are usually short, of the order of 100 to 500 nucleotides, and **5′ untranslated regions (5′ UTR)** are also for the most part short. Second, many viruses, especially minus-strand viruses, employ a variety of strategies to increase the information content per nucleotide.

One significant question is how, without efficient recombination, do these RNA virus groups generate diversity, that is, how do they evolve to survive changing external conditions, extend their host range or evade the immunological defenses of their current hosts? Three independent mechanisms employed by viruses to generate diversity are discussed in the following paragraphs.

First, RNA genomes are much more mutable than DNA genomes because the replication enzymes lack the proofreading and correctional mechanisms of DNA polymerases (*Genes and Genomes,* chapter 2). RNA replicase enzymes have a mistake frequency on the order of 1 nucleotide in 10,000. Thus, for an RNA genome of 31,000 nucleotides (the largest RNA genome known), the RNA molecules produced contain on average of one or more errors. In fact, it has been suggested that this replicase error frequency sets an upper limit for the size of RNA genomes. A population of genomic RNA molecules for any RNA virus contains members with different mistakes, but the genome constantly reverts to the average sequence by selection. However, if the selection pressures change, the population can rapidly evolve new dominant variants. A familiar example of this is antigenic drift among influenza viruses.

A second mechanism by which these viruses can generate diversity is **reassortment**. Orthomyxoviruses (influenza viruses), arenaviruses, bunyaviruses, and reoviruses contain segmented genomes made up of a number of pieces of RNA. When a cell is infected by more than one strain of a given virus or by two closely related viruses, these pieces assort randomly among the newly formed virions, yielding all possible reassortants (rather than recombinants) with high frequency. Reassortment is an important mechanism for maintaining and promoting diversity in those viruses where it is possible. For example, reassortment between human and animal influenza viruses is probably responsible for the large antigenic shifts that occur every ten years or so in the major circulating strain of influenza and result in pandemics. In contrast, human reoviruses of Types 1, 2, and 3 do not appear to reassort in nature, although they do under certain laboratory conditions (Section 2.4).

A third mechanism for generating diversity among RNA viruses is the rare recombinational event that apparently occurs when the replicase switches from copying one RNA to copying another during replication (**copy-choice** mechanism). For coronaviruses and picornaviruses, such recombination is frequently observed in the laboratory, suggesting that copy-choice could affect the genetic diversity of these viruses in the native populations. For most of the other viruses, recombination is difficult to demonstrate in the laboratory, but even an exceedingly rare event would lead to genetic exchange on an evolutionary time scale. For example, in an alphavirus, Western equine encephalitis virus (WEE), the glycoproteins are very similar to those of Sindbis virus (SIN) and serologically cross-react, but the remainder of the genome (including the nucleocapsid protein) and the disease caused by the virus are very similar to those of Eastern equine encephalitis virus, suggesting that WEE is a natural recombinant. Recombination has recently been observed in a plant virus after experimental deletion of a region of the genome (Section 2.6b), and sequence analysis of numerous natural plant virus RNAs shows evidence of recombination. Furthermore, the presence of certain elements in virus genomes suggests that, at some time in their evolutionary history, many of them may have undergone recombination with host mRNAs to acquire host genes that were subsequently modified for their own use.

c. Cloning RNA Virus Genomes

Many of the earliest determinations of RNA sequence of RNA virus genomes were started before the regulatory agencies granted permission to obtain cDNA clones. Today, however, cloning has become routine, and the methods are found in many manuals. Viral RNA is isolated from either infected cells or virion particles, and a cDNA copy of it is made with reverse transcriptase using either specific or random primers. This is directly analogous to obtaining a cDNA copy of an mRNA using oligo dT as a primer. The resulting single-strand DNA is then made double-stranded by any of a variety of methods, inserted into a properly prepared plasmid or bacteriophage, and the nucleotide sequence of the cloned insert determined using standard methods (*Genes and Genomes,* chapters 6 and 7). Inserts from independent clones containing overlapping sequences can be digested with restriction endonucleases and ligated together until the entire genome sequence is present in a single plasmid.

Extensive sequence analysis has yielded a wealth of information about the genome organization and characteristics of the deduced proteins of particular viruses and about the evolutionary relationships among various RNA virus families. Cloned virus-specifc proteins can also be expressed either alone or as fusion proteins in sufficient quantities to serve as antigens and the specific antibodies generated used in many ways. In addition, from the known genomic sequences, oligonucleotide probes can be designed either for screening cDNA libraries or for use as PCR primers for amplification of virus-specific sequences for diagnosis.

Furthermore, cloned viral genomes can be altered at the DNA level to study the effects of introduced mutations. Several methods for introducing mutations into cloned segments are described in *Genes and Genomes,* chapter 6. Such mutations can be studied by transcription of RNA directly from the plasmids and translation *in vitro* or by expression of the mutated genes in bacterial or eukaryotic cells. Finally, from mutagenized, full-length "infectious clones," variant viruses can be rescued after transfection of susceptible cells with RNA transcribed *in vitro* (Section 2.6). The properties of these rescued variants can then be assessed in either tissue culture or animal models.

2.2 Plus-Strand RNA Virus Genomes

a. The Sindbis Virus Superfamily

As the genome organization of a number of plus-strand RNA viruses became known, it was obvious that alphaviruses, enveloped animal viruses belonging to the family *Togaviridae,* contain features in common with a number of plant viruses. These similarities include overall genome organization, readthrough of termination codons, common sequence motifs within the replicase components, and long stretches of low but significant sequence identity in the regions encoding the nonstructural proteins. Many treatments now consider these diverse groups together as the "Sindbis virus superfamily."

Alphaviruses The complete sequence of several alphavirus genomes, including that of the type virus, Sindbis virus, have been determined. A map of the Sindbis genome (11,703 nucleotides exclusive of the 3′ poly A tail) is shown in Figure 2.5. The genomic 49S RNA is of plus sense and serves as an mRNA in infected cells. It is translated into a polyprotein (P1234) whose sequence contains the nonstructural polypeptides of the virus (nsP1, nsP2, nsP3, and nsP4), which form the viral replicase/transcriptase. The four polypeptides are produced by cleavage

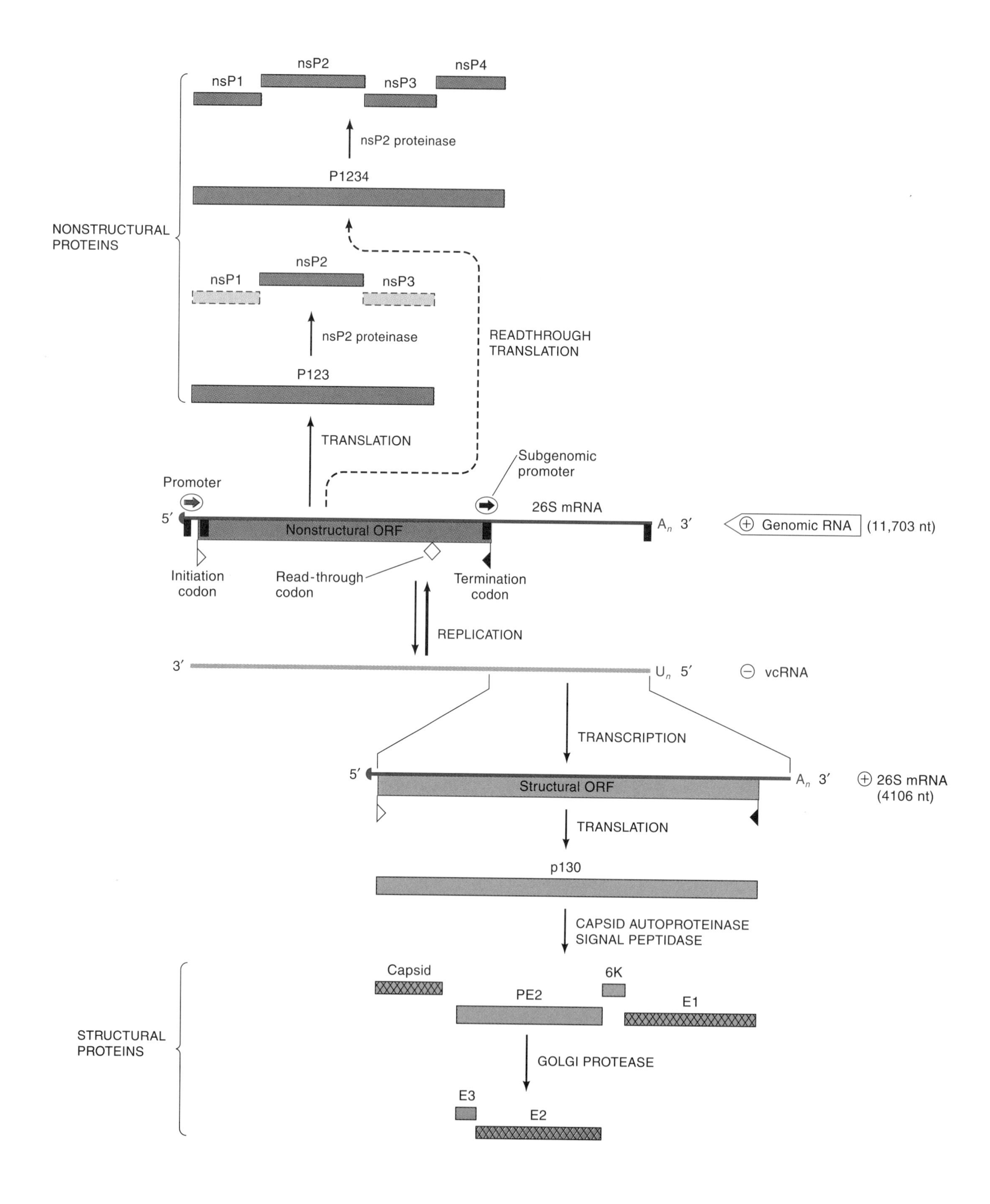

nsP1
nsP2
nsP3
nsP4
nsP2 proteinase
P1234
NONSTRUCTURAL PROTEINS
nsP1
nsP2
nsP3
nsP2 proteinase
READTHROUGH TRANSLATION
P123
TRANSLATION
Subgenomic promoter
Promoter
5′
26S mRNA
Nonstructural ORF
An 3′
Genomic RNA
(11,703 nt)
Initiation codon
Read-through codon
Termination codon
REPLICATION
3′
Un 5′
vcRNA
TRANSCRIPTION
5′
Structural ORF
An 3′
26S mRNA
(4106 nt)
TRANSLATION
p130
CAPSID AUTOPROTEINASE
SIGNAL PEPTIDASE
Capsid
6K
PE2
E1
STRUCTURAL PROTEINS
GOLGI PROTEASE
E3
E2

Figure 2.5 (*facing page*)
Replication strategy of Sindbis virus. Untranslated regions of the genomic RNA are shown as colored lines; the nonstructural open reading frame (ORF) is indicated by a shaded rectangle. The subgenomic RNA region is expanded below using the same conventions. Four vertical black boxes on the genome RNA are domains of conserved nucleotide sequence among alphaviruses; the two which function as transcriptional promoters are indicated with arrows. All translation products are shown. The final proteins found in the virion (in this case, E1, E2, and the capsid protein) are indicated by cross-hatched boxes. Note that precursor P123 serves mainly as a source of the nsP2 proteinase; the remaining components of the replicase complex, nsP1, nsP3, and nsP4, are derived primarily from the readthrough product P1234.

---

catalyzed by a proteinase encoded in the carboxy terminal half of nsP2. The polyprotein precursor of the viral structural proteins, C, E2, and E1, is translated exclusively from a subgenomic 26S mRNA, which represents the 3′ one-third of the genome; posttranslational cleavage is by an autoproteinase encoded near the carboxy terminus of the capsid protein as well as by two cellular proteases.

The genomic RNA sequence reveals a short 5′ UTR that varies in length from 43 to 84 nucleotides in different alphaviruses, followed by a long **open reading frame (ORF)**. In Sindbis virus, this ORF extends from nucleotide 60 to nucleotide 7598 and encodes the P1234 precursor of 2513 amino acids; it is followed by a short sequence (about 50 nucleotides) between the end of the nsP4 coding sequence and the first AUG codon in the subgenomic 26S RNA. The genome is so compact that the first nucleotide of the subgenomic transcript is found within the final codon of the ORF for P1234.

The structural proteins are encoded in 26S RNA by a second long ORF that begins with an AUG codon at position 7647 and that extends for 1357 amino acids (in Sindbis). The 3′ UTR, which for Sindbis is 322 nucleotides long, varies in length among alphaviruses from 112 to more than 500 nucleotides. This region also contains a number of repeated sequences, and the variability in length may reflect duplication events that are not selected against, as they might be in a coding region. The first AUG codon from the 5′ end of the RNA is used for translation initiation of both the 49S and 26S RNAs in Sindbis virus, but not in all alphaviruses.

An interesting feature of the Sindbis virus genome is the occurrence of a UGA codon at nucleotides 5748 to 5750 punctuating the P1234 ORF between the coding sequences for nsP3 and nsP4. Translation termination normally occurs at this codon to produce the precursor P123, but occasional readthrough leads to the translation of the nsP4 protein, albeit in small quantities. In both Middelburg and Ross River viruses, two Old World alphaviruses distantly related to Sindbis virus, as well as in three New World equine encephalitis viruses, a stop codon is present at this position. However, in two other alphaviruses, Semliki Forest virus (which is closely related to Middelburg and Ross River) and O'Nyong-nyong virus, there is no stop codon between nsP3 and nsP4; and in both cases, the UGA is replaced by CGA, an arginine codon (Figure 2.6). Because modulation of nsP4 production appears to be important for the virus life cycle, this implies that, in the latter two viruses, modulation of nsP4 must occur in some other way, perhaps by translation attenuation, rapid turnover, or differences in the processing pathway. The presence of this stop codon illustrates that clues to replication strategy may be found as unexpected bonuses from a nucleic acid sequence analysis.

We present the cloning and sequencing strategies used to study Sindbis virus genes briefly here because they illustrate methods employed for many other virus groups and emphasize some general problems and solutions unique

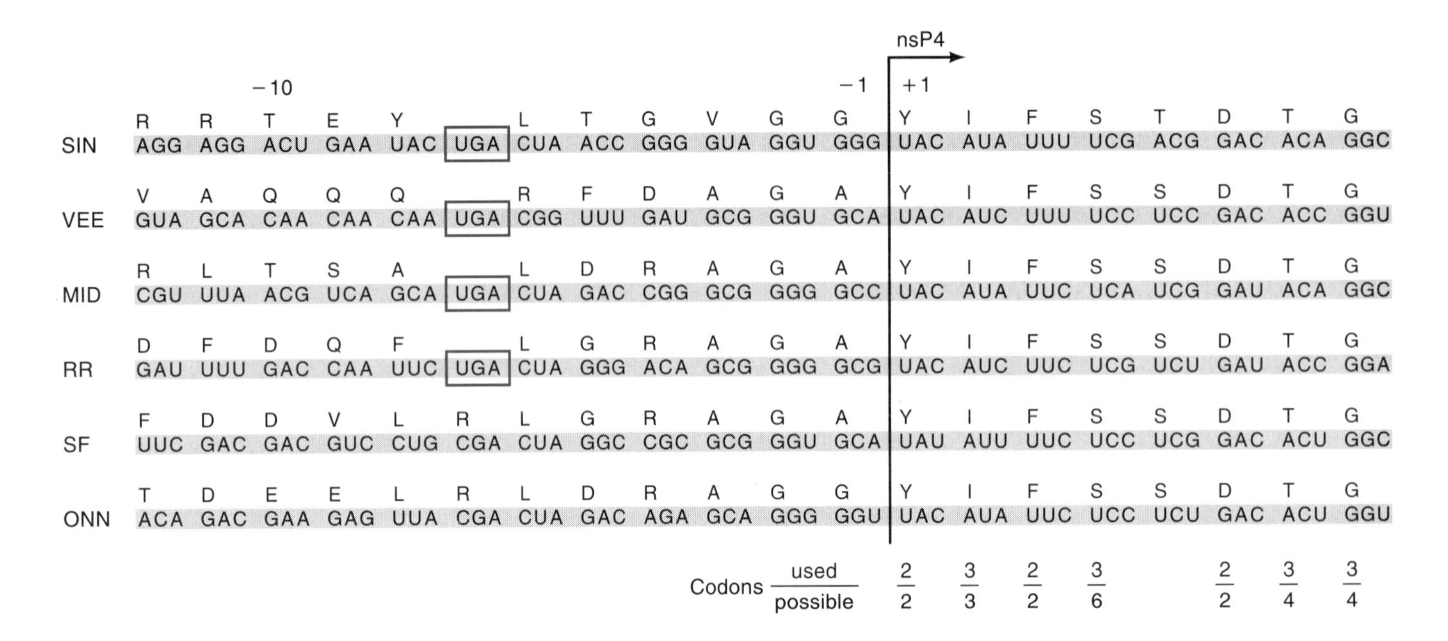

Figure 2.6
Codon usage in alphaviruses. The translated sequences at the end of nsP3 and beginning of nsP4 are shown for six alphaviruses: SIN = Sindbis; VEE = Venezuelan equine encephalitis; MID = Middelburg; RR = Ross River; SF = Semliki Forest; and ONN = O'Nyong-nyong. The start of nsP4 was determined for SF by direct amino-terminal sequencing of the protein. Opal codons are boxed. The number of codons used for each of seven conserved residues at the beginning of nsP4 is indicated, as well as the number of possible codons in the genetic code.

to the cloning and sequencing of RNA viruses. In general, success in cloning RNA genomes has depended upon the integrity of the initial genomic or subgenomic RNA preparations used: the more intact the RNA, the better the cDNA library obtained. Purified genomic 49S RNA is prepared from virions, and purified 26S mRNA is isolated from infected cell cytoplasm. First strand cDNA synthesis using reverse transcriptase is generally primed with a combination of oligo dT and randomly sheared calf thymus DNA. Second strand synthesis is primed either by nicked template RNA produced by RNase H digestion, by fold back of the first cDNA strand, or by oligo dG primers that anneal to poly dC tails that have been added at the 3′ end of the first strand. The cDNAs are inserted into suitable plasmids using appropriate linkers and cloned in *E. coli.* Approaches to sequencing the inserts are described in *Genes and Genomes,* chapter 7. Modifications of these general procedures have been used for a large number of viruses. When viral material is limited, the first strand cDNA can be converted to double-stranded cDNA and amplified by PCR, after which the amplified cDNA can be cloned and sequenced.

If the starting viral RNA is of good quality and the double-strand cDNA is size selected before the linkers are added, the library will contain predominately clones with viral cDNA inserts 1 kb to greater than 5 kb in length. Using convenient restriction sites, these can be ligated together to construct a single plasmid containing most of the alphavirus genome. However, the alphavirus cDNA libraries thus obtained generally lack short sequences from the exact 5′ terminus of the genome. This region usually has to be obtained in some other way, for example, by reverse transcribing the genomic RNA using a minus-strand oligonucleotide as a primer and converting that cDNA to duplex DNA as indicated previously. Clones containing the full-length genome can be obtained by linking the cDNA corresponding to the 5′ end to the rest of the cloned genome.

To obtain partial sequences of large numbers of closely related viruses (e.g., temperature-sensitive mutants, antibody escape variants, geographical isolates of a particular virus, attenuated candidate vaccine strains, etc.), it is usually unnecessary to obtain cDNA clones. Interesting regions may be reverse transcribed into cDNA and then amplified by PCR and the sequence of the PCR products determined directly (**r**everse **t**ranscription [RT]-PCR sequencing).

In general, more information can be gained by comparing the sequences of a number of related viruses than by studying any given sequence alone. Among the alphaviruses, the amino acid sequences of a particular protein may be 25 to 90 percent identical, depending upon the protein. This degree of similarity can be used to align clones of a new alphavirus with the complete sequence of the Sindbis genome. When the nucleotide sequence of short regions at the ends of new clones is determined and their deduced amino acid sequences in all six possible reading frames are compared to the translated proteins of the Sindbis 49S genome, it is possible to align the cloned segments of a new virus isolate unambiguously with a particular region of the Sindbis genome. This type of mapping to align sequences from related viruses is more informative than restriction mapping. Furthermore, such an alignment can be used to ascertain whether the entire genome of a new isolate is represented in its cDNA library.

One particularly striking feature of the RNA virus genomes is the extent of nucleotide sequence divergence among viruses, even in regions of high amino acid identity. Figure 2.6 shows that, among the indicated viruses, the number of codons used for a particular conserved amino acid is close to the maximum number permitted by the degeneracy of the amino acid code. Consequently, nucleic acid probes corresponding to one member of a family will not hybridize readily with other members, and unless they are highly degenerate, such probes cannot be used to identify regions of the genome. The divergent codon usage also implies that a long evolutionary history separates these viruses and indicates that there is particular significance to those domains in which the nucleic acid sequence (as opposed to amino acid sequence) has been conserved. Four such regions of nucleotide sequence conservation have been found in alphaviruses; their locations are shown as black boxes in Figure 2.5. One of these is the 19 nucleotide sequence directly adjacent to the poly A tail, which probably serves as a replicase recognition element for initiation of minus-strand synthesis. A second highly conserved region of 24 nucleotides occurs in the junction region just preceding the beginning of 26S RNA; this domain is the promoter for initiation of 26S subgenomic mRNA transcription. A slightly longer conserved sequence of 51 nucleotides begins about 150 nucleotides from the 5′ end of the RNA and is capable of forming a double hairpin structure whose function is presently unknown. The 5′ end of the genome also demonstrates limited sequence conservation. All the 5′ terminal sequences of alphaviruses, however, can be drawn as very similar stem-and-loop structures. Similar stem-and-loop structures are present at the termini of many plant virus genomes and are thought to be involved in RNA replication.

In addition to revealing details of the genome structure and translation strategy of the alphaviruses, the complete nucleotide sequence also defines the amino acid composition and sequence of each of the proteins encoded by the virus. The amino terminal sequence of each protein has been determined directly by amino acid sequencing and shown to correspond to the amino acid sequence deduced from the nucleotide sequence, but most of what we know about the composition and sequence of viral proteins comes from their coding sequences. In the nonstructural proteins, several amino acid sequence motifs related to replicase/transcriptase activities have been identified; these include the Gly-Asp-Asp (GDD) motif in elongation enzymes, the Asp-Glu-Ala-Asp

(DEAD) motifs in helicases, and a motif characteristic of a nucleoside triphosphate binding site. The deduced amino acid sequences of the virion glycoproteins have revealed hydrophobic signal sequences at the amino termini, hydrophobic domains that act as membrane anchors (which are carboxy terminal in alphavirus glycoproteins but may be either amino terminal or carboxy terminal in the glycoproteins of other viruses), and other landmarks of virus architecture. Glycosylation sites of the form asparagine-X-threonine/serine have been identified and the nature of the oligosaccharide chains attached to these asparagines determined. In addition, the origin and specificities of the proteinases effecting posttranslational processing have been deduced from the nature of the cleavage sites in the polyprotein precursors. Studies with these viruses have revealed that there are few, if any, proteinases free in the cytoplasm of a eukaryotic cell and that the cleavages of the polyproteins that occur in the cytosol are catalyzed by virus-encoded proteinases. Organelle-bound cellular proteases, such as signal peptidase or furin (the protease found within the Golgi apparatus that cleaves after double basic residues), also act on the viral structural glycoproteins as they traverse these cellular compartments (Chapter 8).

"Sindbis-like" plant viruses Progress in determining the replication strategy and genome organization of the plant viruses has been both promoted and hampered by the particular characteristics of these virus-host systems. On the one hand, large quantities of purified virus and viral RNA can be obtained from infected leaves, making studies on virion structure possible. On the other hand, biochemical studies have lagged behind structural studies because the host cells' cellulose walls make it difficult to infect cells in a controlled fashion and to label virus-specific products metabolically with radioactive precursors. Investigations of plant virus replication are often restricted to examining the translation of their RNAs in either rabbit reticulocyte or wheat germ *in vitro* systems, although the recent development of plant protoplasts has made biochemical studies of virus-infected cells more feasible.

For these reasons, until recently, plant viruses were classified primarily on the basis of their morphology (Figure 2.1). Now, however, this classification has been refined on the basis of the nucleotide sequences of their plus-strand RNA genomes. In this section, we discuss the genome organization of three of the viruses belonging to the Sindbis virus superfamily: tobacco mosaic virus (TMV), a helical rod; brome mosaic virus (BMV), an icosahedral tripartite virus; and carnation mottle virus (CarMV), an isometric monopartite virus. Although these three viruses differ remarkably from one another and from the alphaviruses in their morphology, the amino acid sequence similarities in their nonstructural proteins (i.e., their replicases) show that these groups are clearly related to one another and may have all descended from a common ancestor.

TMV is a monopartite plant virus whose genomic RNA is 6395 nucleotides long (Figure 2.7). The genome is capped and contains a short 5′ UTR of 68 nucleotides followed by a long ORF of 4848 nucleotides, which is interrupted by a UAG termination codon at nucleotide 3417. Two translation products are obtained from the genomic RNA both *in vitro* and *in vivo.* Both initiate at the methionine codon at nucleotide 69. The first product has a molecular weight of 126 kDa and terminates at the UAG; a second polypeptide of 183 kDa is produced by readthrough of this codon. At least two subgenomic mRNAs with a common 3′ terminus are also produced after TMV infection. The larger of these overlaps by some 21 nucleotides the ORF for the replicase and encodes a nonstructural protein of 27 kDa. The ORF for the 27 kDa protein in turn is overlapped by a small mRNA encoding the 17 kDa coat protein. The origin for encapsidation (i.e., the binding site for the first multimer of TMV coat protein) is found between nucleotides 5420 and 5545. The 3′ terminus of TMV RNA folds into a tRNA-like structure that is recognized by an amino acyl transferase.

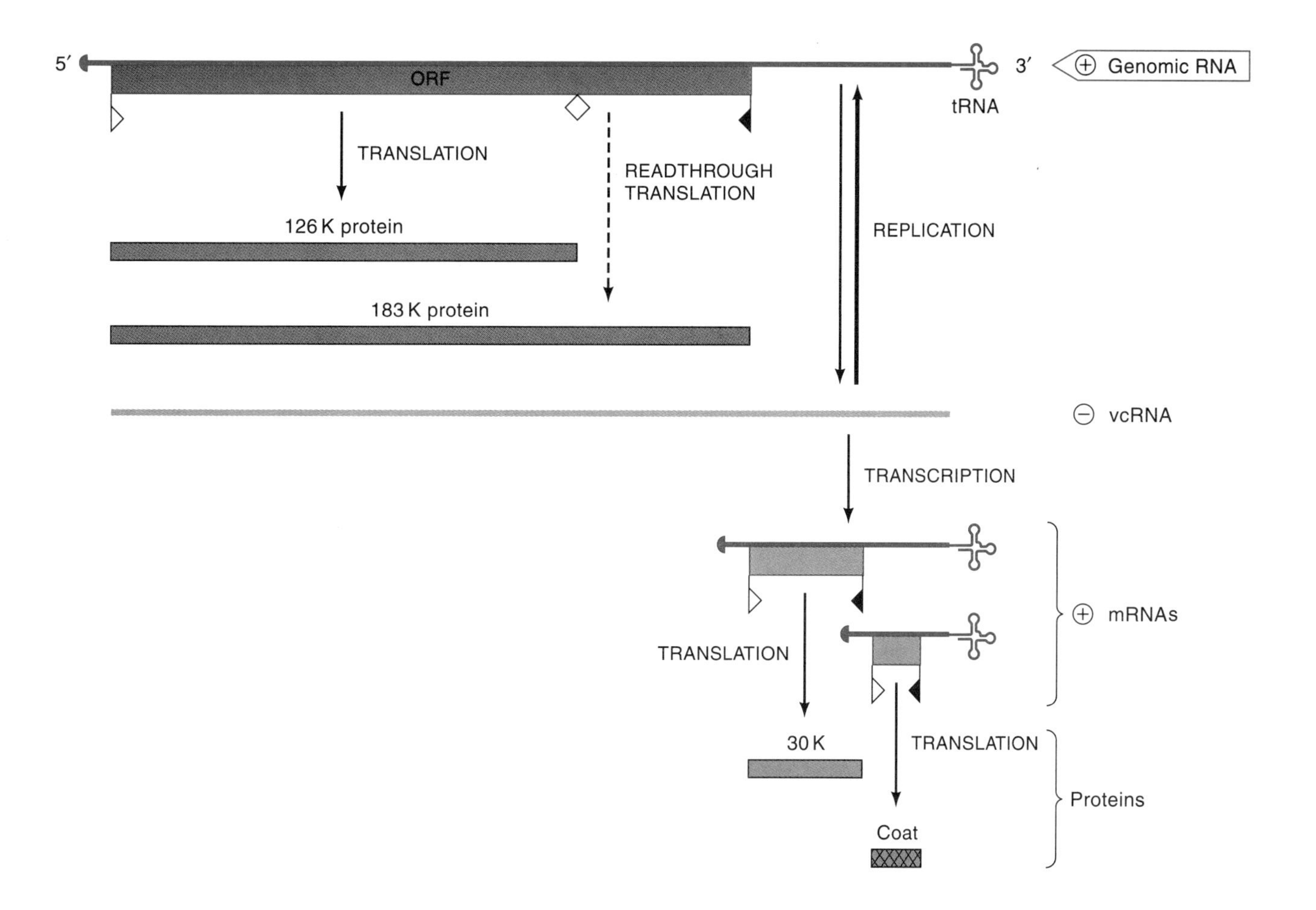

Figure 2.7
The transcription and translation strategy of tobacco mosaic virus. The top line shows the ORFs and translation products of the genomic RNA. The 183-K protein is produced by readthrough of a termination codon. The translation of two subgenomic RNAs is shown below. The cloverleaf at the 3′ terminus represents a t-RNA-like structure similar to that shown in Figure 2.8, which can be amino-acylated.

The BMV genome consists of three pieces of RNA encapsidated in three different capsids that can be separated on the basis of their density. The largest RNA (RNA1) is 3234 nucleotides long and is found in the heaviest particles. The next largest RNA (RNA2) is 2865 nucleotides long and found in the lightest particles. RNA3 (2114 nucleotides) and RNA4 (876 nucleotides), a subgenomic RNA derived from RNA3, are found encapsidated together in the medium density particles. Because both the RNAs and the particles are separable on the basis of their physical properties, the properties of each RNA species could be determined. All three genomic RNAs are required for infectivity, and RNAs 1 and 2 can replicate in the absence of RNA3. Although RNA3 contains an ORF for the virus coat protein, no coat protein is translated from RNA3; the coat protein is translated exclusively from the subgenomic message, RNA4.

Each of the RNAs is capped at the 5′ end, and the 3′ terminal sequences of all three RNA segments of BMV are highly conserved and presumably form a relatively complicated stem-and-loop structure (Figure 2.8), which is difficult to show in a two-dimensional representation. This structure, like the 3′ terminus of TMV RNA, is also recognized by an amino acyl transferase. This structure is also

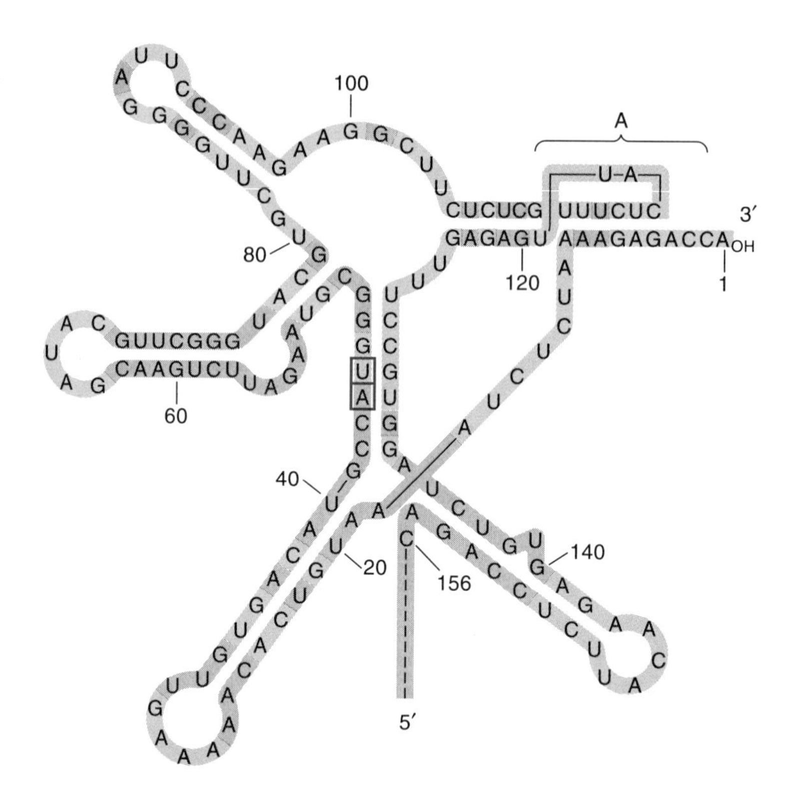

Figure 2.8
Stem-loop and pseudoknot structure found at the 3′ terminus of brome mosaic virus (BMV) RNAs. The sequence of the last 156 nucleotides of RNA 3 is shown, numbered from the 3′ terminus. Within this domain, the sequence is identical for BMV RNA 1 and RNA 2 except at positions 43 and 44 (color-outlined boxes). The pseudoknot is formed at "A" by hydrogen bonding of nucleotides 5 to 10 with the loop formed by nucleotides 111 to 116. Nucleotides identical to those in another bromovirus RNA, cowpea chlorotic mottle virus RNA 1, are shaded in color.

a recognition site for the BMV-specific, RNA-dependent RNA polymerase, and mutations introduced in several loops affect RNA replication. The 3′ termini of related tripartite viruses such as cowpea chlorotic mottle virus (CCMV), cucumber mosaic virus, and broad bean mottle virus, despite sequence divergence from BMV, form remarkably similar structures.

The nucleotide sequence of BMV is known, and its genome organization is shown in Figure 2.9. After their caps, both RNA1 and RNA2 have a short 5′ UTR, a single long open reading frame, and a short 3′ UTR that ends in the 3′ terminal structure described previously. RNA3 is somewhat more complicated: its capped 5′ end is followed by a 5′ UTR of 91 nucleotides and an ORF of 909 nucleotides. However, between the end of this ORF and the start of RNA4, there is an internal polyadenylate stretch that varies in length from 16 to 22 nucleotides. RNA4, the coat protein mRNA, is transcribed from a full-length, complementary minus-strand template by internal initiation, reminiscent of the production of Sindbis 26S mRNA; the entire template is not necessary, and a minus strand containing the complement of RNA4 and approximately 20 nucleotides upstream can be transcribed into RNA4. RNA transcripts of cDNA clones of the three RNAs of BMV have been found to be infectious (Section 2.6b).

CarMV is an isometric virus containing one of the smallest known plant virus genomes, 4003 nucleotides. In addition to the genomic RNA, two subgenomic mRNAs are also encapsidated into virions. Both mRNAs are colinear with the 3′ terminus of the genomic RNA, and the smaller (approximately 1600 nucle-

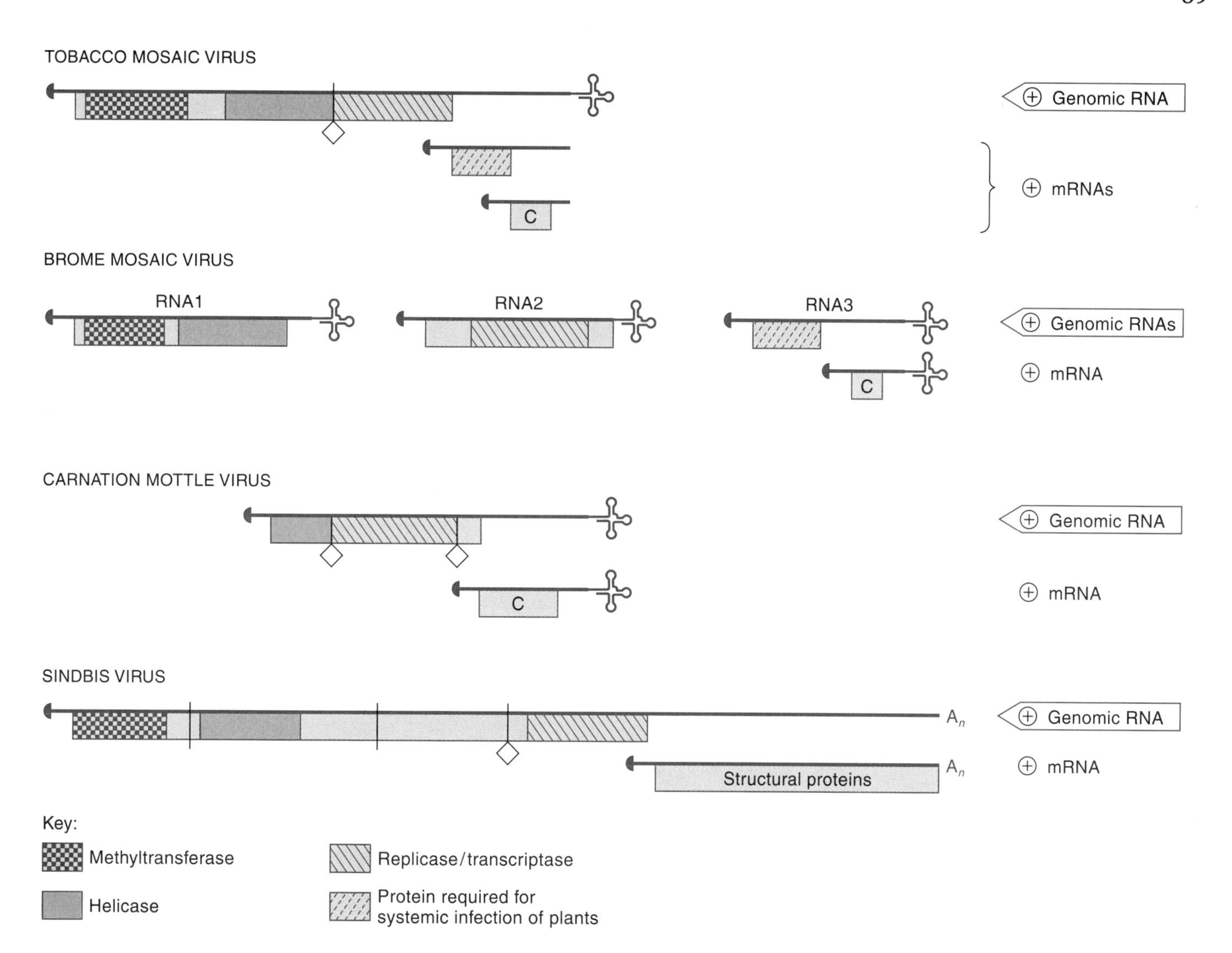

Figure 2.9
Homologies between the nonstructural proteins encoded by three plant viruses and Sindbis (SIN) virus. The translated regions of the genomes of tobacco mosaic virus (TMV), brome mosaic virus, and carnation mottle virus are shown diagramatically and compared with that of Sindbis virus. (More detailed translation strategies for SIN and TMV are shown in Figures 2.5 and 2.7.) Three domains of significant sequence similarity in the nonstructural proteins within the helicase, methyl transferase, and replicase genes, respectively, are shaded differentially (see key).

otides) encodes the coat protein. The genomic RNA contains, in addition to the ORF for the coat protein, a long ORF (nucleotides 70–2679) that contains termination codons at nucleotides 805–807 and 2359–2361. Two translation products of the genomic RNA are formed *in vitro*; one is a 27 kDa protein encoded between the 5′ terminal AUG codon and the first UAG codon; the second is a readthrough product of 86 kDa that terminates at the second UAG codon. The full-length translation product, which would require readthrough of both termination codons, has not been identified *in vitro* or *in vivo*.

The similarities between the genome organizations of Sindbis virus, TMV, BMV, amd CarMV are relevant when considering their replication strategies (Figure 2.9). The nonstructural protein precursor (P123) of Sindbis virus is formally equivalent to the 126 kDa protein of TMV, and Sindbis nsP4 (which requires readthrough of a UGA termination codon) is equivalent to the domain downstream of the UAG stop codon in the TMV 183 kDa protein. As is the case for the alphaviruses, the structural proteins of these plant viruses are translated from a 3′ coterminal subgenomic mRNA. Furthermore, the 126 kDa product of TMV corresponds to the protein translated from BMV RNA1, and the readthrough portion of the TMV 183 kDa polypeptide is analogous to the protein translated from RNA2. Also, the products translated from the two TMV subgenomic mRNAs are related to the proteins encoded by BMV RNA3 and RNA4. The genome organization of CarMV resembles that of TMV except that it lacks the corresponding 5′ end.

Comparison of the amino acid sequences of the proteins encoded by the four groups of viruses have revealed significant amino acid sequence similarity in three domains of the nonstructural proteins indicated by checkerboard, dark gray stippling, and diagonal lines in Figure 2.9. Within these three regions, the amino acid identities between TMV and BMV are 18.6 percent, 19.5 percent, and 21.4 percent, respectively. Surprisingly, the amino acid identities between TMV and Sindbis in the comparable domains are 12.9 percent, 18.2 percent, and 20.0 percent. Although the degree of similarity is not extreme, it is statistically significant, and these proteins exhibit similarity over many hundreds of amino acids, not just short motifs, suggesting that these three plant viruses and Sindbis virus have evolved from a common ancestral virus. The functions of the nonstructural proteins of the plant viruses are most likely analogous to those of the corresponding Sindbis virus proteins. This suggests that the checkerboard domains contain a methyl transferase or capping function, the dark gray domains, a helicase, and the diagonally striped domain, an RNA replicase/transcriptase. Note that the carboxy terminal half of Sindbis nsP2, which is known to be the nonstructural proteinase (Section 2.6c), has no counterpart among the plant viruses. No amino acid similarity is detectable among the structural proteins of these viruses.

b. The Poliovirus Superfamily

The *Picornaviridae* and the plant *Comoviridae* make up another superfamily, based on genome organization and amino acid sequence similarity. These groups may also be derived from a common ancestor, perhaps an insect virus, although none of the known extant representatives uses an insect vector.

Picornaviridae The family *Picornaviridae* is comprised of five genera (Enterovirus, Rhinovirus, Aphthovirus, Hepatovirus and Cardiovirus) as well as a number of currently unclassified picornaviruses. Poliovirus is the type virus of the genus Enterovirus. Picornavirus genomes consist of one molecule of single-stranded RNA of 7200 to 8400 nucleotides, depending on the virus. The RNA is contained within an icosahedral capsid consisting of 60 identical subunits each containing four different proteins. The 5′ end of the genomic RNA is covalently linked to a small viral protein called VPg in place of the 5′ methylated cap found in most other RNA viruses. However, the RNA has the characteristic polyadenylated 3′ end. Many members of this group are significant human and veterinary pathogens (Table 2.1) that have been studied intensively for years. However, only with the determination of the nucleotide sequence, and thus the deduced amino acid sequences of the proteins, were the translation strategy and the processing of the polyprotein products understood.

The nucleotide sequence of the poliovirus genome, first obtained before cDNA cloning of animal viruses was feasible, was the first sequence determined for an RNA-containing animal virus. The RNA was reverse transcribed into cDNA, and the sequence of the cDNA was determined by a modification of the chain termination protocol (*Genes and Genomes,* section 7.2c) using *E. coli* DNA polymerase I. The cDNA was the template, and the primers were oligoribonucleotides obtained by digestion of poliovirus genomic RNA with ribonuclease T1.

Shortly after the publication of this sequence, the sequence was determined again using a cloned cDNA. The first strand was synthesized with reverse transcriptase using an oligo dT primer annealed to the 3′ poly A tail. Second-strand cDNA synthesis was accomplished using the loop back method (*Genes and Genomes,* figure 4.20); and after S1 nuclease digestion, oligo dC tails were added, and the DNA was inserted into a vector with oligo dG tails. The exact 5′ terminus of the genomic RNA was obtained by primer extension using the parental RNA as template and a fragment from a downstream cDNA clone as primer. The sequence was determined from three overlapping clones that were subsequently joined into a single infectious cloned DNA containing the entire genome (Section 2.6a).

The genome organization of poliovirus is illustrated in Figure 2.10. The genomic RNA contains 7433 nucleotides exclusive of the 3′ poly A tail. The 5′ UTR

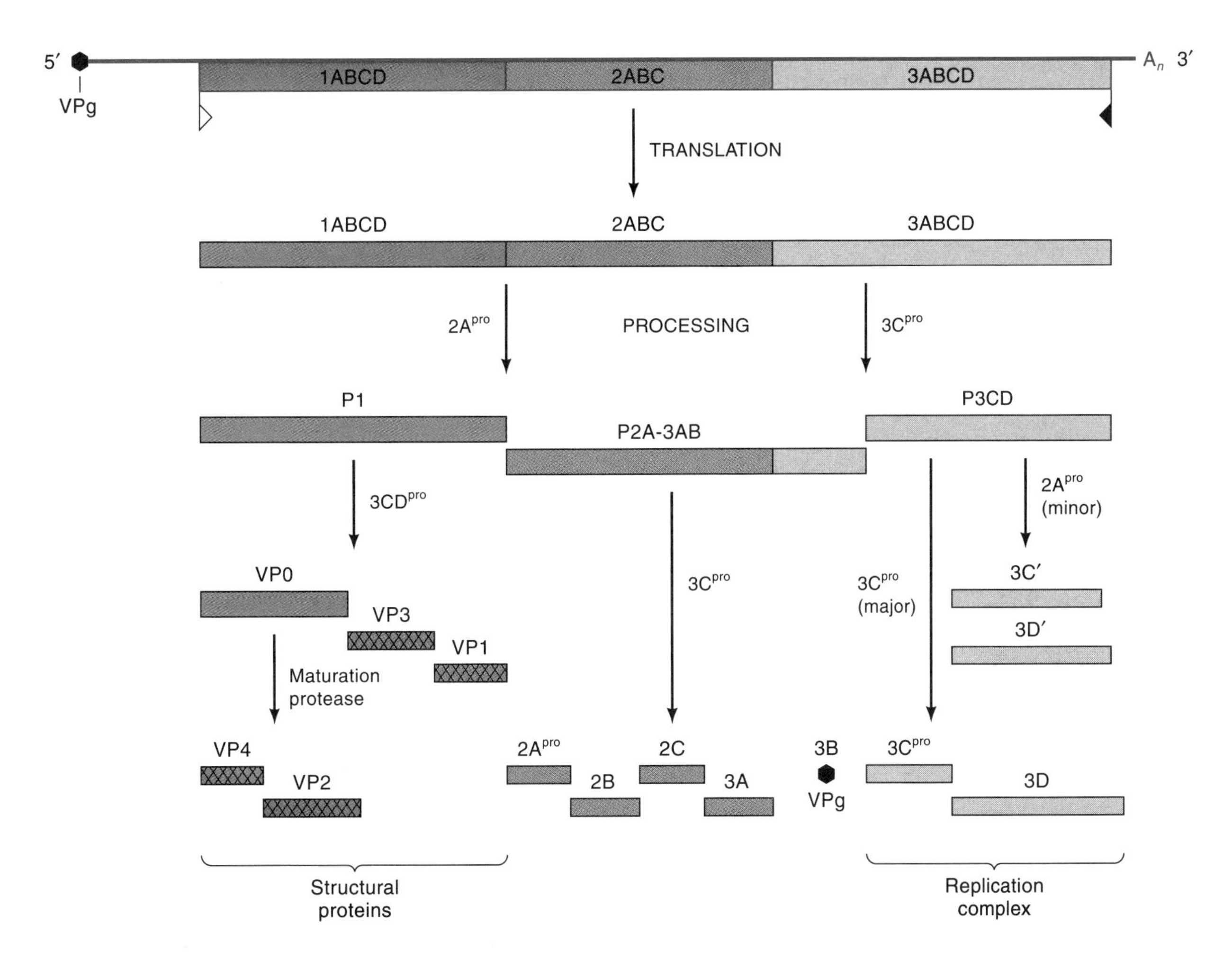

Figure 2.10
Translation of the picornavirus genome and processing of the poliovirus polyprotein precursor. Untranslated regions are shown as single lines; the three main domains of the ORF are differentially shaded in color. The VPg is shown as a solid hexagon, both in its location at the 5′ terminus of the genome and in the region in which it is encoded. Final virion capsid proteins are shown as crosshatched boxes. Three primary precursors are shown and the virus-specific proteases responsible for particular cleavages are indicated. Note that the exact identity of the virus-encoded maturation protease is not known.

is relatively long (742 nucleotides; the length varies between 500 and 700 nucleotides in other picornaviruses). Although there are several potential initiation codons (AUG) in this leader segment, these apparently are not used for initiation of translation; instead, translation begins at an **i**nternal **r**ibsome **e**ntry **s**ite, or IRES. The genome contains a single long ORF of 6621 nucleotides (2207 amino acids) and terminates with two contiguous in-phase termination codons. The polyprotein translated from this long ORF is processed in a number of discrete steps into the final products. Very rapid cleavage events separate the still nascent polyprotein into three primary precursors: P1, the precursor to the capsid proteins; P2A-3AB, the precursor to the middle region of unknown function; and P3CD, the replicase precursor (Figure 2.10).

A total of ten cleavage events is needed to process the poliovirus polyprotein into the mature products. Of these, the eight that take place between a glutamine and a glycine are all catalyzed by a cysteine proteinase encoded in the P3CD region of the viral genome (labelled $3C^{pro}$ in Figure 2.10). This proteinase acts first as an autoproteinase to cleave itself from the polyprotein precursor and then as a diffusible proteinase to cleave other similar bonds. Two other polio proteinases are also believed to be necessary for processing the polyprotein, both as autoproteinases; one ($2A^{pro}$) acts while the polyprotein is nascent and separates the P1 domain from P2A-3AB at a tyrosine-glycine bond, and the other cleaves one of the capsid protein precursors at an asparagine-serine bond during virion maturation.

The organization of the picornavirus genomes, with the structural proteins at the 5′ end of the RNA and the replicase proteins at the 3′ end, contrasts with the organization of alphavirus RNA (replicase proteins being encoded at the 5′ end and structural proteins at the 3′ end). However, with both groups, the genomic RNA sequences at the 3′ and 5′ termini are not complementary, implying that the signals for replicase recognition on the plus and minus strands must be distinct.

Among the picornaviruses, host protein synthesis is shut off after infection by inactivation of the cap-binding complex, which binds necessarily to the 5′ end of capped mRNAs to initiate translation. In enteroviruses and rhinoviruses, the inactivation of the largest component of the cap-binding complex, p220, is mediated by the viral-encoded polypeptide 2A. Picornavirus proteins are translated from the genomic RNA by cap-independent translation, which is enhanced by the IRES present in the long 5′ UTR. The 5′ UTR of encephalomyocarditis virus (EMC) is especially active for translation and has been used to improve the efficiency of numerous experimental expression systems that generate uncapped transcripts (Section 2.6). The sequence similarity at the nucleotide level among picornaviruses in this region is much higher than it is in translated regions, suggesting that the exact sequence of this domain is indispensable for replication, presumably because it forms a particular tertiary configuration.

Comoviridae As for alphaviruses and certain plant viruses (Section 2.2a), the genome structure and replication strategy of the *Picornaviridae* resemble that found among the *Comoviridae*, a family of plant viruses which includes the genus Comovirus (named from the type virus, **co**wpea **mo**saic virus), the genus Nepovirus (named for their mode of transmission and morphology, i.e., **ne**matode **po**lyhedral), and the genus Fabavirus. Fabaviruses appear to be similar to comoviruses, but have been little studied and will not be discussed further here. Like the picornaviruses, comoviruses, and nepoviruses, but not fabaviruses, contain small, plus-strand RNA genomes with VPgs covalently attached to the 5′ termini, an unusual structure for plant viruses. The VPgs are not required for infection by comoviruses but are required for nepovirus infectivity. In both groups, the genome consists of two RNAs separately encapsidated into independent virion particles; either both RNAs or both types of particles are required for productive infection. Although separate encapsidation of genome segments is unknown

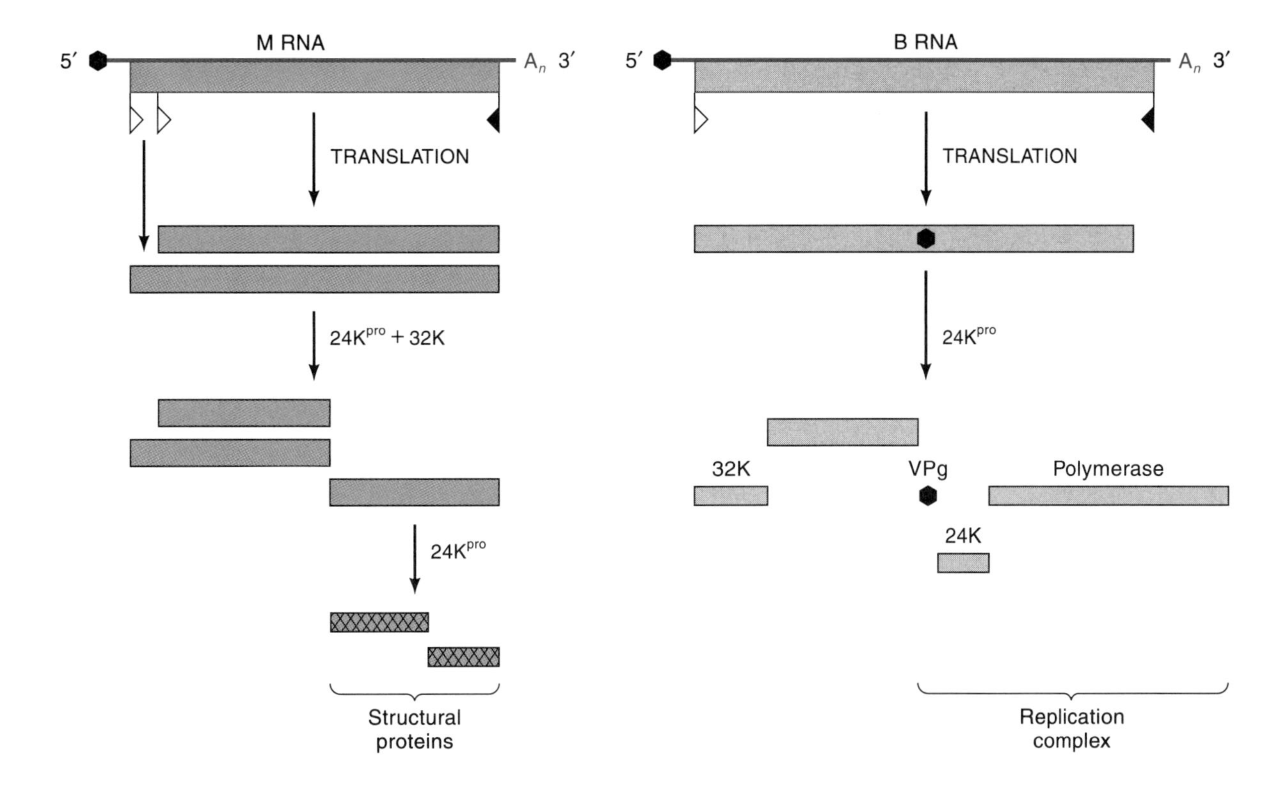

Figure 2.11
Translation of comovirus RNAs. The translation of the M RNA and B RNA of the cowpea mosaic virus are drawn to scale using the conventions shown in Figure 2.10. Proteins are named by their approximate molecular weights in kilodaltons.

among the animal viruses, it is relatively common among the plant viruses. Most plant viruses are transmitted mechanically, and a requirement for entry of more than one particle to initiate infection of a given cell may not be disadvantageous. Each of the RNAs is polyadenylated and translated into polyproteins that are processed posttranslationally, also unusual features among plant viruses (Figure 2.11). In comoviruses, the M segment of 3481 nucleotides corresponds to the region of poliovirus that encodes the structural proteins, and the B segment of 5889 nucleotides encodes the replication and polyprotein processing functions.

When the nucleotide sequence of a comovirus was obtained, the deduced amino acid sequences showed that there are clear sequence similarities in the 2C region and in the domains encoding VPg, $3C^{pro}$, and the replicase between polio and these two groups of plant viruses (Figure 2.12). Within these domains, the average amino acid identities were 20 percent to 30 percent over stretches of hundreds of amino acids.

c. Flaviviruses

Flavivirus is a genus containing about 70 viruses, many of which are of worldwide importance as human pathogens, including yellow fever virus (the type virus), dengue virus, St. Louis encephalitis virus, and Japanese encephalitis virus. Flaviviruses contain plus-strand RNA, are similar in structure to alphaviruses, and were originally grouped into the same family, the *Togaviridae.* Fundamental differences in replication strategy made clear by sequence analyses have now led to their reclassification as a genus in the family *Flaviviridae.* The

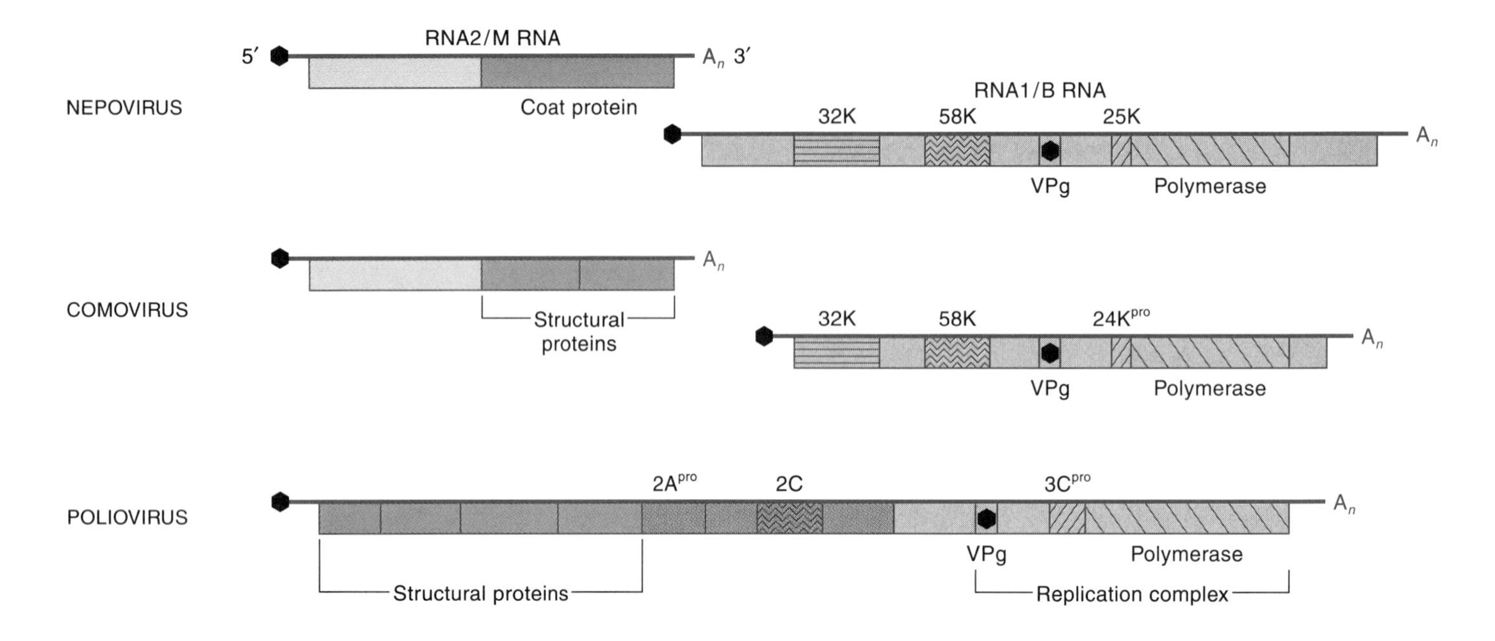

Figure 2.12
Similarities between comoviruses, nepoviruses, and picornaviruses. A simplified map of the genome of poliovirus is shown and compared to those of a comovirus and a nepovirus. Four regions in which significant amino acid sequence similarity has been found among the three viruses are indicated with different patterns as follows: a domain in the 32K protein with horizontal lines; the 2C protein and the 58K protein with wavy lines; the 3C protease and its homologs with narrow diagonal lines; and the polymerase domains with wide hatching.

first flavivirus for which the complete genome sequence was obtained was yellow fever virus, and today the sequences of a number of flaviviruses have been published. We outline the methods used for cloning and sequencing yellow fever RNA because they illustrate a number of technical challenges.

Most of the flaviviruses grow poorly in culture, the yield of virus released into the extracellular medium is low, and the virions are relatively unstable. Therefore, only small amounts of purified virus and extracted RNA can be obtained. Under these conditions, preparing intact RNA is difficult because, at low concentrations, most of it is digested by contaminating nucleases. The genomic RNA is about 11 kb in length and is capped at the 5′ end but lacks a 3′ poly A tail. The first stage was to generate a library of clones that contained about 90 percent of the RNA sequence in cDNAs obtained with random oligonucleotide primers. These clones were aligned by restriction mapping and sequenced by chemical methods. Each nucleotide position was determined on at least two different clones so as to obtain an average sequence and to eliminate possible cloning artifacts.

To obtain cDNA clones containing the ends of the RNA, researchers had to use different methods. A cDNA containing the sequence at the 5′ end was obtained by primer extension with a protocol analogous to that used for the 5′ end of alphavirus RNA. For the 3′ end, poly A polymerase was used to add a poly A tail onto the RNA followed by priming with oligo dT for first strand synthesis. Because secondary structures at the RNA's 3′ end made polyadenylation very inefficient, there was a chance that the clones obtained might arise from other regions of the genome or from contaminants in the RNA. However, a *Bgl*I restriction site had been identified near the 3′ terminus of the known sequence. Most *Bgl*I sites are unique because the recognition sequence contains five indetermi-

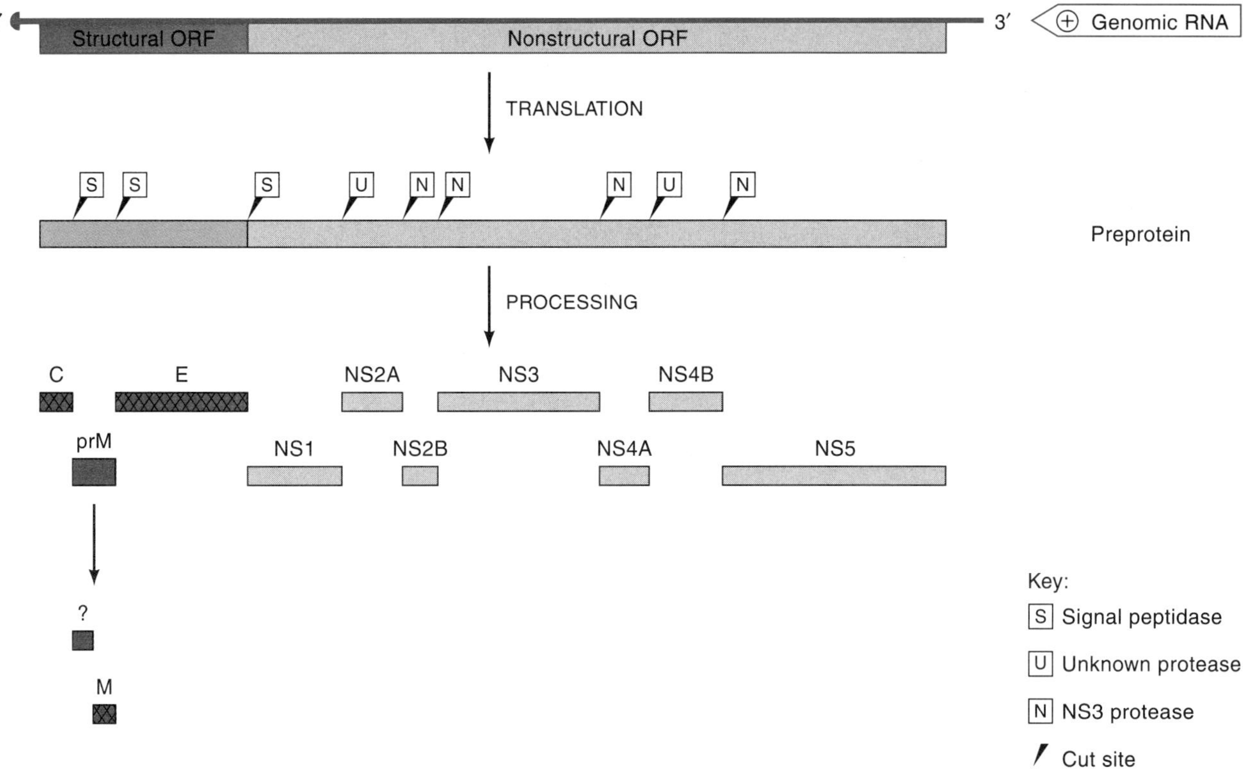

Figure 2.13
Map of the yellow fever virus genome. Conventions for translated regions are the same as those in Figure 2.10. The three crosshatched boxes represent polypeptides found in mature virions. The proteinases responsible for each cleavage are indicated with symbols above the site (see key).

nant nucleotides. Thus, the double-stranded cDNA synthesized with an oligo dT primer and the poorly polyadenylated yellow fever RNA as template was cleaved with *Bgl*I and ligated to the *Bgl*I site in a clone from the yellow fever library containing the sequence near the 3′ terminus of the genome. Only the authentic yellow fever cDNA fragment with the perfectly matched *Bgl*I end could be ligated, thus producing a clone with the proper 3′ end of yellow fever RNA.

The yellow fever virus RNA genome is 10,862 nucleotides long and contains a single long ORF (encoding 3411 amino acids) beginning at a methionine codon at nucleotides 119 to 121 and ending at nucleotide 10,351 with three termination codons that are in phase but not contiguous (Figure 2.13). There are no other significant open reading frames in either the genomic RNA or its complement, and the entire genome is translated as a single polyprotein precursor. Like other RNA viruses, flaviviruses make efficient use of their genomic RNA. Of the 10,862 nucleotides in yellow fever RNA, only 118 at the 5′ end and 511 at the 3′ end are not translated.

The yellow fever virion proteins have been positioned within the deduced sequence of the polyprotein precursor by direct amino terminal amino acid sequencing. Many of these proteins have been expressed as fusion proteins and used to produce monospecific antisera to each viral protein. Using these reagents, researchers have been able to determine the order of proteolytic processing and the enzymes responsible for most of the cleavages. The genome organization and translation strategy of yellow fever (Figure 2.13) is quite different from that of the alphaviruses and resembles in some ways that of picornaviruses. The genes for the structural proteins C, M, and E are present at the 5′ end of the open reading frame, and all the nonstructural proteins are encoded at the 3′ end of the genome. Because all the virus-specific polypeptides are translated from the genomic RNA, transcriptional controls cannot regulate the relative amounts of various polypepides synthesized. However, production of nonstructural proteins (encoded "downstream") may be somewhat attenuated by premature termination of translation.

The different proteinases that account for the processing of the polyprotein are indicated in Figure 2.13. Signal peptidase in the lumen of the endoplasmic reticulum is responsible for cleavages of the structural proteins, and many of the cleavages of the nonstructural proteins, which occur in the cytosol, are effected by a viral proteinase encoded in the amino terminal half of NS3. The cleavage of prM occurs late and may be due to an enzyme in the Golgi.

Several features at the ends of the RNA may be important for RNA replication. First, there are short, complementary sequences at the 3′ and 5′ ends of the genome. Such sequences are commonly found among the minus-strand viruses but not in plus-strand viruses. These terminal sequences may be sites at which the replicase binds, implying that the same sequence element is used for initiation of both minus- and plus-strand synthesis. Second, repeated sequences have been found in the 3′ untranslated region of flaviviruses, but their significance is currently unknown. Third, the sequence of the 3′ end of the yellow fever genome (and in other flaviviruses) can be drawn as a very stable secondary structure (Figure 2.14). With a stability of greater than 40 Kcal per mole, this structure would be expected to exist in solution under physiological conditions. This may explain the difficulties observed in chemically modifying the 3′ terminus of flavivirus RNA (e.g., polyadenylation) and the self-priming activity observed during cDNA synthesis. Because the structure is conserved among flaviviruses, it is presumably important for some aspect of viral replication.

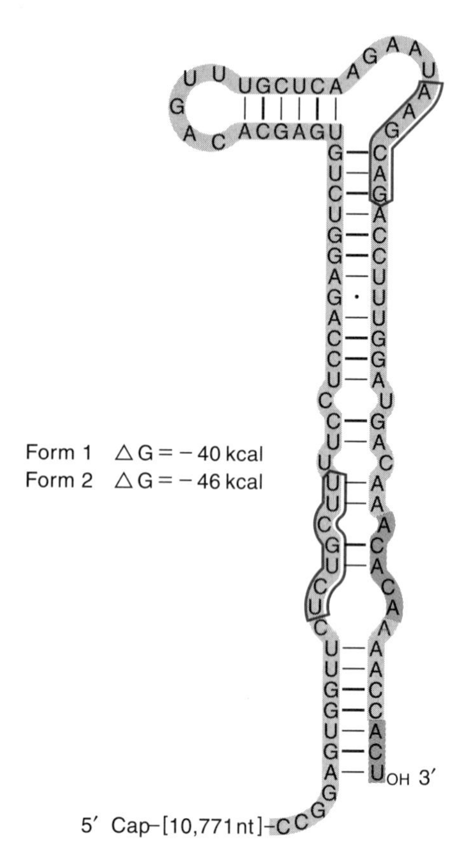

Figure 2.14
Structure at the 3′ end of yellow fever virus RNA. A proposed secondary structure for the 3′ terminal 87 nucleotides of yellow fever virus (strain 17D) is shown. Nucleotides shaded in color are shared with the 3′ terminus of yellow fever vcRNA. Form 1 is shown. Form 2 can be formed if the bases outlined in color are base paired to one another. (From C. M. Rice, et al. 1985. Nucleotide sequence of yellow fever virus: Implications for flavivirus gene expression and evolution. *Science* 229 726–733.)

The flaviviruses are closely related to one another, and all demonstrate considerable serological cross-reaction. However, the overall amino acid sequence identity is modest (e.g., 45 percent between yellow fever and West Nile viruses). A plot of amino acid similarity between these two viruses is shown in Figure 2.15. The similarity varies from below 10 percent to more than 90 percent in different domains of the various proteins. Shown below the similarity plot are two hydrophobicity plots indicating the characteristics of the individual virus polypeptides. It is easy to see that even in regions such as NS2A and NS2B, where these two flaviviruses have very little amino acid sequence similarity, their hydrophobicity profiles are virtually superimposable. This implies that although numerous individual amino acid substitutions have occurred during evolution, the size, shape, and many other characteristics of individual flavivirus proteins are largely conserved.

Similar comparisons of hydrophobicity profiles and genome organization have also led to the recent inclusion of the pestiviruses and hepatitis C virus as two additional genera within the family *Flaviviridae.* The pestiviruses, a group of important pathogens of domestic animals, include classical swine fever virus, bovine viral diarrhea virus, and border disease of sheep. The genome (12,700 nucleotides) contains a single long ORF and is translated into a polyprotein that is processed proteolytically by both cellular and viral enzymes. Hepatitis C virus is an extremely important human pathogen responsible for many cases of liver disease every year. The genome (approximately 10,000 nucleotides) is slightly smaller than most flaviviruses and again has a single ORF. Although no detectable amino acid sequence similarity (other than short motifs common to many viruses) has been found among flaviviruses, pestiviruses, and hepatitis C virus, functionally similar polypeptides are encoded in the same order on their genomes, and comparative hydrophobicity plots are suggestively similar.

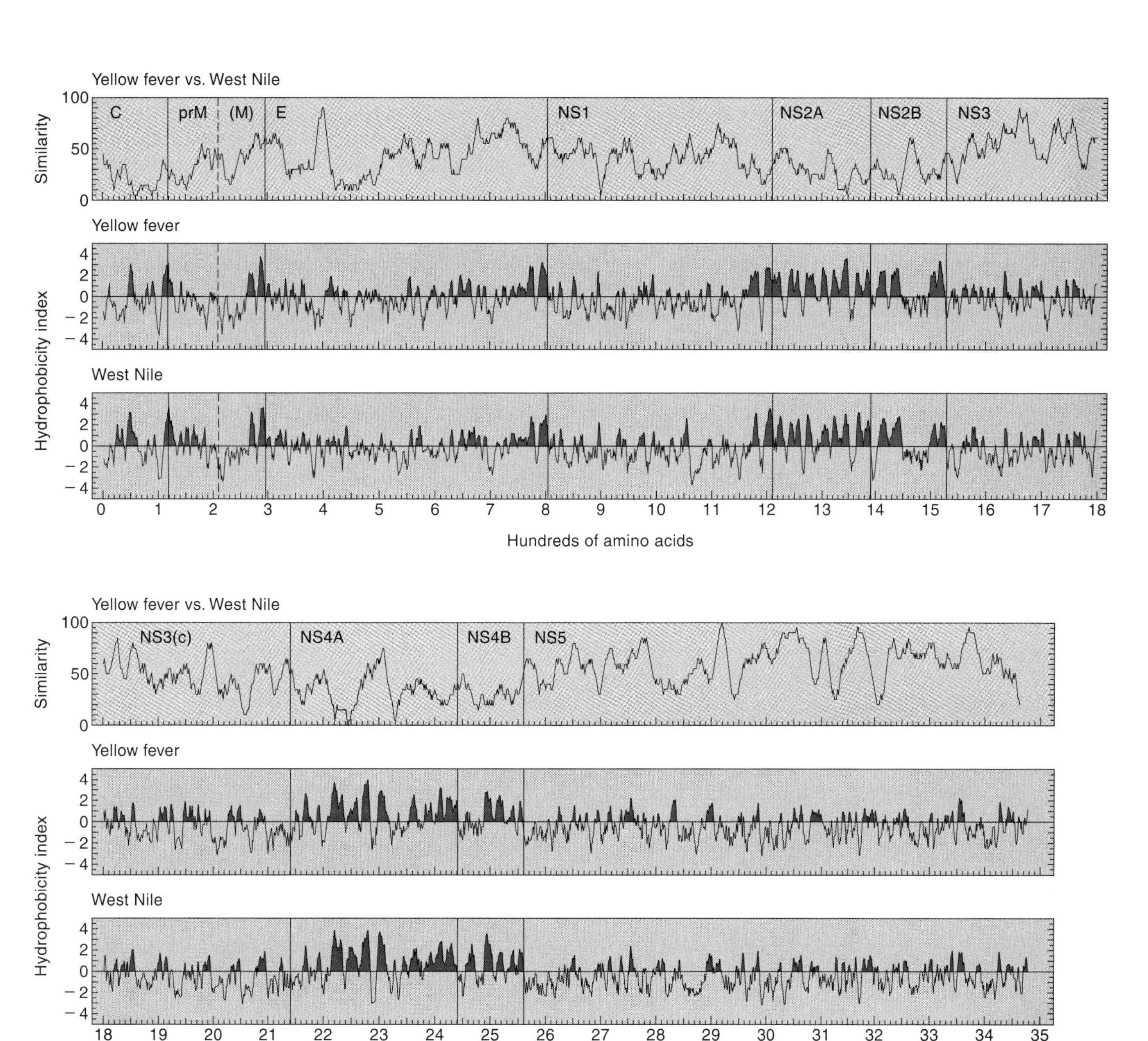

Figure 2.15
A similarity comparison and hydrophobicity plots for the deduced amino acid sequences of yellow fever and West Nile virus proteins. Deduced amino acid sequences of the ORFs of the two flaviviruses were aligned with as few gaps as possible. The percentage of amino acid similarity between the two aligned sequences is shown as a moving average by using a window length of 20. The sites of cleavage between the various protein products are indicated by vertical lines. Below the similarity profile are shown two individual hydrophobicity plots for yellow fever and West Nile, respectively, using a window of 7. Hydrophobic sequences (dark color-filled areas of the plot) are shown above the plot midline; hydrophilic sequences are shown below the midline.

d. Coronaviruses

Coronavirus genomes consist of a single-strand RNA from 27,000 to 31,000 nucleotides long, depending on the virus. These enveloped viruses appear in the electron microscope to have a crown or corona of spikes surrounding the lipid bilayer, hence their name. They contain a nucleocapsid of helical symmetry whose morphology is not as well defined as that of the nucleocapsids of minus-strand viruses. One of the two envelope transmembrane glycoproteins, E1 (or the matrix protein), has most of its mass on the cytoplasmic side or within the membrane and only a small region on the external side of the membrane, the **ectodomain**. The second, E2 (or S for spike protein), is a more typical viral glycoprotein with a large glycosylated ectodomain, a transmembrane sequence, and a short cytoplasmic tail. The S protein is translated as a precursor protein cleaved to the mature form late during morphogenesis.

The replication strategy of coronaviruses is particularly interesting (Figure 2.16). The virus produces a full-length minus-strand vcRNA and a nested set of mRNAs (with the virion RNA, RNA1, being a member of the set), which are 3′

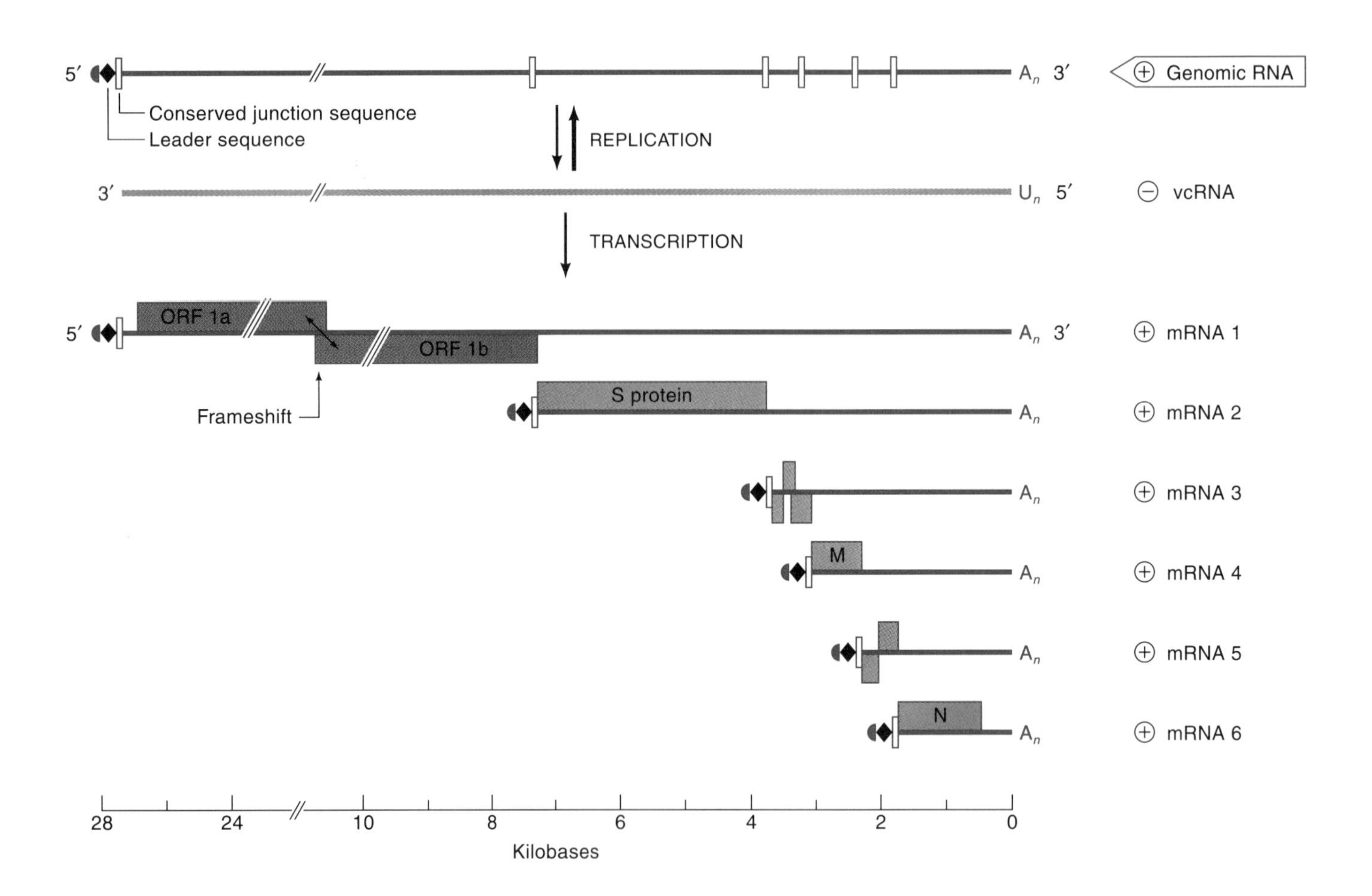

Figure 2.16
Transcription and translation strategy of coronaviruses. The genomic RNA, vcRNA, and subgenomic mRNAs for mouse hepatitis virus are shown. The dark-colored boxes are ORFs 1a and 1b, which encode the nonstructural (replicase) proteins; the light-colored boxes indicate the virion proteins M, N, and S. The leader sequence (solid black diamond) encoded at the 5′ end of the genome RNA segment 1 is attached directly to the 5′ terminus of the ORFs of the other segments at the conserved junction sequences indicated by the open vertical bars. ORF1a and ORF1b are thought to be translated as a single polyprotein precursor by a ribosomal frameshift at the arrow.

coterminal and polyadenylated. Only the most 5′ ORF of any given mRNA is translated, as is the case for the alphaviruses and for the plant viruses previously discussed. Each message begins with a leader sequence derived from the 5′ terminus of the genome. After transcription of the leader (by copying the 3′ end of the vcRNA), the polymerase is thought to "jump" to one of five conserved intergenic domains and continue transcription from there to the 5′ end of the minus-strand template (a mechanism termed "leader-primed transcription"). Recently, five subgenomic minus-strand RNAs have been isolated from coronavirus-infected cells, but their role in transcription is obscure. However, because coronavirus mRNAs are synthesized in vastly different amounts, the amount of each mRNA may be determined in part by differential rates of transcription of these minus-strand templates.

The genome of infectious bronchitis virus, an avian coronavirus, is 26,700 nucleotides long, and that of mouse hepatitis virus is greater than 31,000 nucleotides; these are the longest known RNA virus genomes. The two 5′ ORFs of infectious bronchitis virus, ORF1a and ORF1b, which encode the replicase/transcriptase functions, contain 21,798 nucleotides, more than the entire genome of most RNA viruses. These two ORFs overlap for 42 nucleotides, but because of ribosomal frame shifting in the overlap region (see Section 3.2a), a single protein with a molecular weight of greater than 800 kDa is produced. Both ORFs contain a number of motifs that are shared with other plus-strand viruses. ORF1a encodes two copies of a putative papainlike proteinase and a cysteine proteinase similar to $3C^{pro}$ of poliovirus; ORF1b has a helicase domain, polymerase motifs, and zinc finger domains. The presence of the proteinase domains suggests that the 800 kDa precursor is proteolytically processed posttranslationally, but the products have yet to be identified in infected cells.

2.3 Minus-Strand RNA Virus Genomes

The five families of minus-strand viruses (*Rhabdoviridae, Paramyxoviridae, Orthomyxoviridae, Arenaviridae,* and *Bunyaviridae*) are closely related to one another, suggesting that they have recently diverged from a common ancester.

A sixth family of minus-strand viruses, *Filoviridae,* includes Ebola and Marburg viruses, which are enveloped bacilliform particles with a single piece of RNA as their genome. In part because of their extreme pathogenicity, with up to 88 percent mortality in human outbreaks, they have not been as well studied as the other families, and we do not discuss them.

All negative-strand viruses are morphologically similar. Their helical nucleocapsids, which are surrounded by a lipid bilayer containing two or three different virally encoded glycoproteins, are composed mainly of a single type of protein (the N protein) and small amounts of two to four other polypeptides that compose the RNA replicase/transcriptase complex. In paramyxo-, rhabdo-, and orthomyxoviruses, a hydrophobic polypeptide called the matrix protein, M, lines the inner surface of the bilayer and interacts with the nucleocapsid during virion assembly.

The genome organizations of viruses in all five groups are similar even though rhabdovirus and paramyxovirus genomes contain a single RNA molecule, the influenza virus genome consists of eight different RNA segments, and the arenavirus and bunyavirus genomes contain two and three segments, respectively. To clone the nonsegmented virus genomes, researchers made cDNA copies of cytoplasmic mRNAs found in infected cells. This strategy results in clones from separated portions of the genome, which are then aligned and ordered by reference to spanning mRNAs that occur naturally at low frequency. For the segmented viruses, the virion RNAs have been the preferred templates for cDNA synthesis and subsequent cloning, in part because the mRNAs are incomplete copies of the virion RNAs. For orthomyxoviruses, bunyaviruses, and arenaviruses, all segments have a common sequence at the 3′ end and its complement at the 5′ end. Therefore, once the end of one segment is known, a single synthetic oligonucleotide can be used to prime both first and second strand cDNA synthesis for all segments. Maps of the genomes of representatives of all five families are shown in schematic form in Figure 2.17, with the genes aligned from 3′ to 5′ to illustrate the similarities in genome organization.

The replication cycles of all negative-strand viruses have five common features. (1) The 5′ and 3′ termini of their genomes, or of their genome segments, have short complementary sequences. This means that the sequence found at the 3′ end of the negative strand is the same as the sequence found at the 3′ end of the positive strand. If these small elements form part of the replicase-binding sites, then the transcription promoters for plus- and minus-strand RNA synthesis must be closely related; this contrasts with the situation discussed previously for most of the plus-strand viruses. These terminal elements are also encapsidation signals for both genomic RNA and virion complementary RNA (vcRNA); the absence of the terminal elements from the mRNAs assures that they are not encapsidated. (2) The vcRNAs—full-length, plus-strand template(s) for synthesizing the minus-strand genomic RNA—are not identical to the mRNAs found within the cell, although both are of the same polarity. This is obvious for the unsegmented viruses, which make five to eight monocistronic mRNAs from a single genomic RNA, but it is also true for the segmented viruses (Section 2.3b). (3) All viral RNA synthesis, plus and minus strand, occurs within a nucleocapsid structure. Viral RNA is never free in the infected cell cytoplasm. (4) Most of these viruses utilize one or more strategies to expand the coding potential of their genomes. On some mRNAs, translation is initiated at AUG codons that define two different overlapping ORFs. In some transcripts, one or more extra nucleotides are incorporated during transcription, thereby shifting translation to another ORF downstream of the insertion. In addition, some primary transcripts are spliced differentially to produce mRNAs that encode different products. (5) The viral polymerases "switch" from the almost exclusive transcription of mRNAs early in infection to replication of genomic RNA late in infection.

a. Nonsegmented Viruses

Rhabdoviruses One of the best studied families of minus-strand viruses is the *Rhabdoviridae,* which is composed of two genera: Vesiculovirus, of which vesicular stomatitis virus (VSV) is the best known, and Lyssavirus, whose most studied member is rabies virus. The RNA genome of VSV is 11,162 nucleotides long and consists of five genes as well as short leader sequences at both ends. These five genes, listed from 3′ to 5′ in the order of mRNA transcription, are N, P, M, G, and L; they encode the **n**ucleocapsid protein, a nonstructural **p**hosphoprotein, the **m**atrix protein, the **g**lycoprotein, and the replicase/transcriptase (or **l**arge protein), respectively (Figure 2.17). The mRNAs are transcribed sequentially and appear to decrease in amount in their order of transcription (i.e., the N mRNA is transcribed first and is the most abundant, and the L mRNA is made last and is the least abundant). Each mRNA consists of a 5′ UTR of 10 to 41 nucleotides that starts with a common initiating sequence (AACAG) followed, in order, by the open reading frame encoding the protein in question, a common termination sequence, and polyadenylation signals. Each poly A tail is apparently formed by "stuttering" of the enzyme at a template sequence of seven uridines at the end of the gene. Transcription occurs in the cytoplasm, and the mRNAs are capped by the viral transcriptase. In the virion RNA, each gene is separated from the next by two intergenic nucleotides that do not appear in mRNA.

The genome organization of the lyssaviruses is almost identical to that of VSV (Figure 2.17). Originally M1 and M2 were thought to encode two forms of

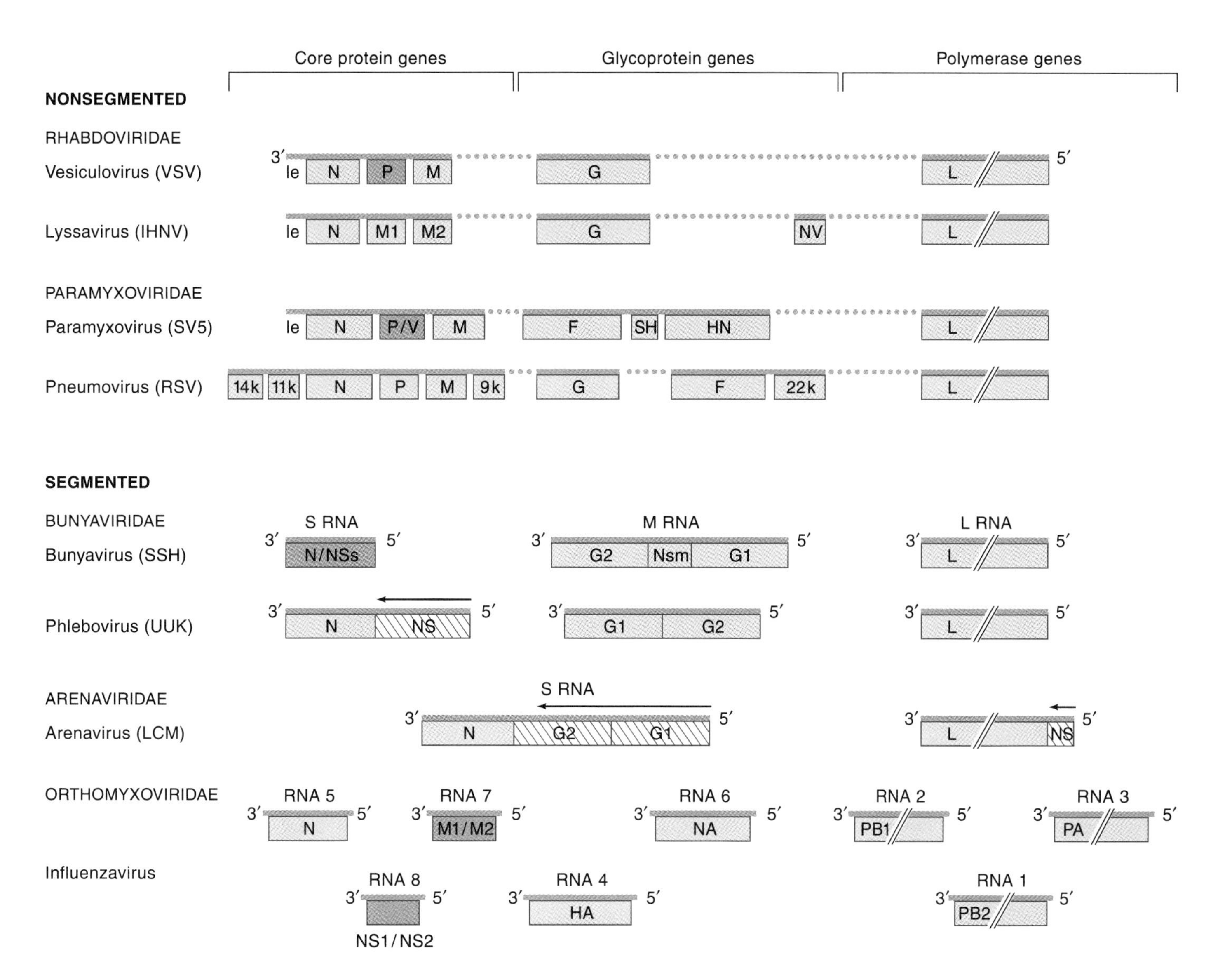

Figure 2.17
Comparative genome organizations of minus-strand viruses. The maps are only roughly to scale. Genes (boxes) are aligned to indicate similarities of genome organization. Genes shaded in darker color encode more than one protein in overlapping ORFs; genes with diagonal hatching are transcribed from vcRNA (ambisense). For vesiculovirus (VSV) the proteins encoded are N, the nucleocapsid protein, P, the phosphoprotein, M, the matrix protein, G, the glycoprotein, and L, the replicase (the large protein). The leaders (le) are short sequences that transcribed but not translated. For lyssavirus (IHNV), M1 is a phosphoprotein corresponding to P, M2 corresponds to M, and there is a small nonvirion protein called NV. Among the Paramyxoviridae, N is the nucleocapsid protein, P the phosphoprotein, M the matrix protein, F the fusion protein (a glycoprotein), and L the replicase. Paramyxovirus (SV5), contains HN, the hemagglutinin-neuraminidase protein (a glycoprotein) and a small, hydrophobic nonstructural protein (SH) located between F and HN. For pneumovirus (RSV) several additional genes are shown, including glycoprotein G. The genomes of the segmented viruses have been aligned such that segments are shown below their functional counterparts in the nonsegmented viruses. Bunyaviruses and arenaviruses encode the nucleocapsid protein N, two envelope glycoproteins G1 and G2, and the replicase L. Some also encode various NS or nonstructural proteins. The proteins encoded in the eight segments of influenza virus are the nucleoprotein N, the phosphorylated proteins NS1 (and NS2 and M2), the matrix protein M1, the hemagglutinin HA, the neuraminidase NA, and the three replicase components, PB1, PB2, and PA.

the matrix protein, but it is now clear that M1 corresponds to the P protein of the vesiculoviruses, whereas M2 is a true matrix protein. However, lyssaviruses have, between G and L, an additional gene that encodes a **n**on**v**irion protein (NV).

Paramyxoviruses The *Paramyxoviridae*, the other family of nonsegmented negative-strand viruses, contains three genera: Paramyxovirus, Morbillivirus, and Pneumovirus. These genera were originally distinguished from one another by the presence or absence of neuraminidase and hemagglutination activities in their major glycoprotein. The genomes of the *Paramyxoviridae* are 16,000 to 18,000 nucleotides long, with a genome organization strikingly similar to that of the *Rhabdoviridae* (Figure 2.17).

The genome of paramyxovirus simian virus 5 (SV5) contains the following seven genes, listed from 3′ to 5′: N, P, M, F, SH, HN, and L. These encode the **nu**cleocapsid protein, a **p**hosphoprotein, the **m**atrix protein, the **f**usion protein, a **s**mall **h**ydrophobic polypeptide, **h**emagglutinin-**n**euraminidase, and the **l**arge protein, which is the replicase. The HN protein is the counterpart of the VSV G protein, but the F protein and SH proteins have no counterparts in rhabdoviruses. The genome of pneumovirus respiratory syncytial virus (RSV), which is even more complicated, contains ten genes. It lacks the SH gene, but there are two genes encoding the 14 kDa and 11 kDa proteins upstream of the nucleocapsid protein; a gene for a 9.5 kDa protein that has no counterpart in other paramyxoviruses lies between the M and G genes; and a gene encoding a 22 kDa protein occurs between the F and L genes. The fusion proteins of respiratory syncytial virus and the parainfluenza virus SV5 are similar in structure and function and were probably derived from a single ancestral gene. The fact that G and F genes are rearranged in the RSV genome relative to the F and HN genes of SV5 indicates that gene order is not critical within this region.

b. Segmented Viruses

Orthomyxoviruses The family *Orthomyxoviridae* contains two genera: Influenza virus A and B and Influenza virus C. The morphology of the influenza viruses is very similar to that of the paramyxoviruses described in the preceding section. The RNA is present in helical nucleocapsids containing a single species of nucleocapsid protein and one of the eight segments of the genomic RNA. Small amounts of the replicase proteins PB2, PB1, and PA are also present in each nucleocapsid. The lipid envelope surrounding the nucleocapsid contains the matrix protein lining the interior of the lipid bilayer and two glycoproteins, NA (the neuraminidase) and HA (the hemagglutinin), embedded in the lipid membrane whose ectodomains form characteristic spikes visible in the electron microscope.

The influenza virus genome consists of eight segments of single-strand RNA of negative polarity that vary in length from 800 to 2341 nucleotides. The eight genomic segments can be separated by polyacrylamide gel electrophoresis under the proper conditions, and different strains of influenza viruses can be distinguished on the basis of the electrophoretic profile of their genome segments. In mixed infection, the genomic segments reassort randomly into progeny virions, and such recombinant (reassorted) progeny were used initially to correlate certain genome segments with particular encoded proteins. For example, if the HA proteins of two types of influenza can be distinguished antigenically, and all recombinants with RNA segment 4 from Type 1 have Type 1 HA, segment 4 encodes HA. These assignments have subsequently been confirmed by cloning and sequencing all eight segments and expression of their gene products.

The three largest RNA segments encode the components of the viral replicase/transcriptase: two basic polypeptides, PB1 and PB2, and the acidic protein PA. PB1 and PB2 are involved in the transcription of mRNAs, whereas PA appears to be important for virion RNA replication. Segment 4 encodes the HA, one of the external glycoproteins, segment 5 encodes the nucleocapsid protein (NP), and segment 6 encodes NA, the second external glycoprotein. Segment 7 encodes the M1 or matrix protein as well as a second (nonstructural) protein called M2. Segment 8 encodes two nonstructural proteins, NS1 and NS2. We discuss transcription and translation strategy of these two segments in greater detail in Section 2.2c.

In Figure 2.17, the eight genome segments are aligned beneath their functional counterparts in the unsegmented genomes of the *Paramyxoviridae* and the *Rhabdoviridae.* This comparison makes clear the correspondence of the gene products and the similarities among these virus groups. Although functionally equivalent, no sequence homology is detectable between, for example, the VSV G protein, the SV5 F protein, and influenza HA; but the F proteins of paramyxoviruses and the HA of influenza are similar in their organization and posttranslational processing, as are the HN protein of paramyxoviruses and the NA of influenza. Finally, limited but significant amino acid sequence similarity has been found between the matrix proteins of influenza and VSV. All these facts suggest that the paramyxoviruses, rhabdoviruses, and orthomyxoviruses have descended from a common ancestor and in this process the genome of the ancestral virus became fragmented in the branch that led to influenza.

Bunyaviruses and arenaviruses Bunyaviruses and arenaviruses are somewhat different in structure from the preceding three families of negative-strand viruses in that they lack a matrix protein. Like other minus-strand viruses, they contain helical nucleocapsids consisting of one species of nucleocapsid protein and the minus-strand genome segments, together with a few molecules of a replicase protein. The nucleocapsid is surrounded by a lipid bilayer that contains two species of integral membrane glycoproteins.

There are five genera of the *Bunyaviridae*: Bunyavirus, Phlebovirus, Nairovirus, Tospovirus, and Hantavirus. Tospoviruses are of special interest for evolutionary considerations because they are exclusively plant viruses. However, they are unusual for plant viruses in having a lipid envelope and minus-strand genomes. *Bunyaviridae* genomes consist of three segments of minus-strand RNA: the small (S), middle (M), and large (L) RNAs. Because these RNAs contain 50 to 100 nucleotide long self-complementary stretches at their two ends, they form largely circular, hydrogen-bonded, panhandle structures. In addition, the three nucleocapsids are circular as isolated from the virion.

Reassortant bunyaviruses containing one or more genome segments from each of two different parents can arise when cells are simultaneously infected with two viruses from the same genus but not with members of different genera. Members of each genus contain conserved sequences 11 to 20 nucleotides in length at the 3′ ends of all three RNA segments; these sequences differ between genera. The 3′ ends of the vcRNAs have a similar conserved sequence. Presumably, these sequences serve as promoters and encapsidation signals, and because they interact with the viral replicase/transcriptase in a genus-specific manner, they limit the genetic exchanges permissible within the family.

In all genera, bunyavirus replicase/transcriptase is encoded by a single long ORF in the virion L RNA. The M segment encodes the two virus glycoproteins, and the S segment encodes the nucleocapsid protein and often a nonstructural protein as well. The organization of genes on the S and M segments of bunyavirus RNA varies from genus to genus. Both the S and M RNAs of the hanta- and nairoviruses are transcribed into mRNAs with a single ORF. However, the S segments of bunyaviruses encode two proteins, each initiated at AUG codons in different reading frames; one ORF initiates at the first AUG (nucleotides 82 to 84) and is translated into the N protein of 235 amino acids, and the second ORF initiates at the AUG at nucleotides 101 to 103 to yield a nonstructural protein (NS_S) of 92 amino acids. Phlebovirus S RNAs encode two proteins: the N protein (243

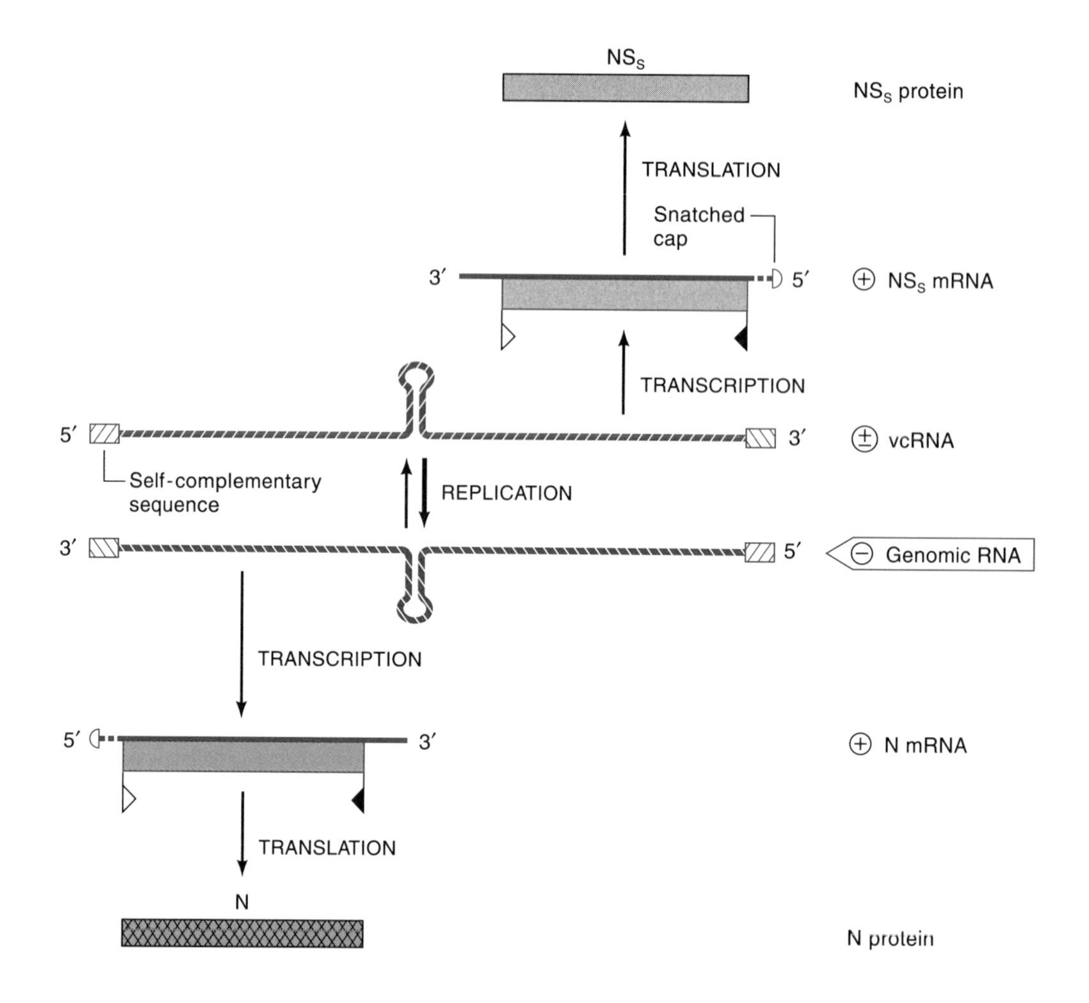

Figure 2.18
Ambisense transcription and translation strategy of the S RNA of a phlebovirus. The genomic RNA and the vcRNA are neither plus- nor minus-strand and therefore are shown as diagonally hatched dark-color lines. The plus-strand mRNAs are shown as solid dark-color lines. The 5′ caps (open symbols) and a few nucleotides at the 5′ ends (dashed lines) are derived from cellular mRNAs. The N mRNA that encodes the N protein (cross-hatched because it is a virion protein) is transcribed from the virion RNA itself. The NS$_S$ mRNA is transcribed from the vcRNA. Both transcripts terminate at a hairpin structure between the two ORFs. The diagonally hatched boxes at the ends of the genomic and vcRNA are self-complementary sequences not present in mRNAs.

amino acids long) is translated from the mRNA transcribed from the 3′ half of the genomic S RNA, and the NS$_S$ protein (250 amino acids) is translated from an mRNA transcribed from the 3′ end of the vcRNA (Figure 2.18). The nonstructural NS$_S$ protein is made in far smaller quantities than the N protein. This use of a genomic minus-strand sequence to encode two proteins, one ORF from the 3′ end and the other from the complement of the 5′ end of the sequence has been termed **ambisense**. The ambisense strategy may be a way of differentially regulating the transcription and translation of these two protein products. Both mRNAs terminate in the middle of the segment in a region where both the virion RNA and the vcRNA can form a very stable hairpin structure, which may be responsible for terminating transcription. Both S and M segments are ambisense in the plant tospoviruses.

There is only a single genus of the arenaviruses, although they are split into two subgroups, the Old World and the New World arenaviruses. The arenaviruses have two segments of viral RNA, termed the L and the S segments, which are separately encapsidated but enveloped together in a single virion particle. These viruses get their family name from the word "arena," meaning "sand," because they commonly envelop a number of host ribosomes during budding, causing a "sandy" appearance in the electron microscope. Like the genome segments of the bunyaviruses, the arenavirus L and S segments contain common sequences at their 3′ termini and complementary sequences at the 5′ termini, such that the RNAs can cyclize in solution.

Both arenavirus genome segments are expressed employing the ambisense strategy. The L segment encodes the replicase/transcriptase in a long ORF transcribed from genomic RNA and a short NS protein transcribed from the vcRNA (Figure 2.17). For the S segment, the mRNA for the N protein is transcribed from genomic RNA, and the mRNA for the two glycoproteins is transcribed from vcRNA, similar to the expression of the phlebovirus S segment (Figure 2.18). Note that the arenavirus S segment resembles the product of joining the N gene of a bunyavirus at its 5′end to the 3′end of a bunyavirus G mRNA.

Transcription The mRNAs transcribed from the segments of orthomyxoviruses, arenaviruses, and bunyaviruses are not exact complementary copies of the genome. Unlike other RNA viruses, segmented minus-strand viruses transcribe their mRNAs in the nucleus. Because the transcriptase encoded by these viruses lacks a capping activity, the formation of capped mRNAs occurs by "cap snatching" (i.e., transcription is primed with capped oligonucleotides 10 to 13 nucleotides long, which are cleaved from cellular, capped mRNAs by an endonucleolytic activity in the viral transcriptase). For orthomyxoviruses, transcription terminates 17 to 22 nucleotides from the 5′ terminus of the template at a series of several U residues. The poly A tail on these mRNAs is added by transcriptase "stuttering" at this point (Figure 2.19).

For bunyavirus segments like the L segment, transcription terminates 60 to 120 nucleotides from the 5′end of the genomic RNA at a signal that is not well understood; no poly A is added. Thus, each mRNA begins with 10 to 13 nucleotides that are not present at the 3′end of the genome and lacks between 17 and 120 nucleotides at the 5′end. Because the mRNAs lack the long terminal, self-complementary sequences, they cannot form the panhandlelike structures

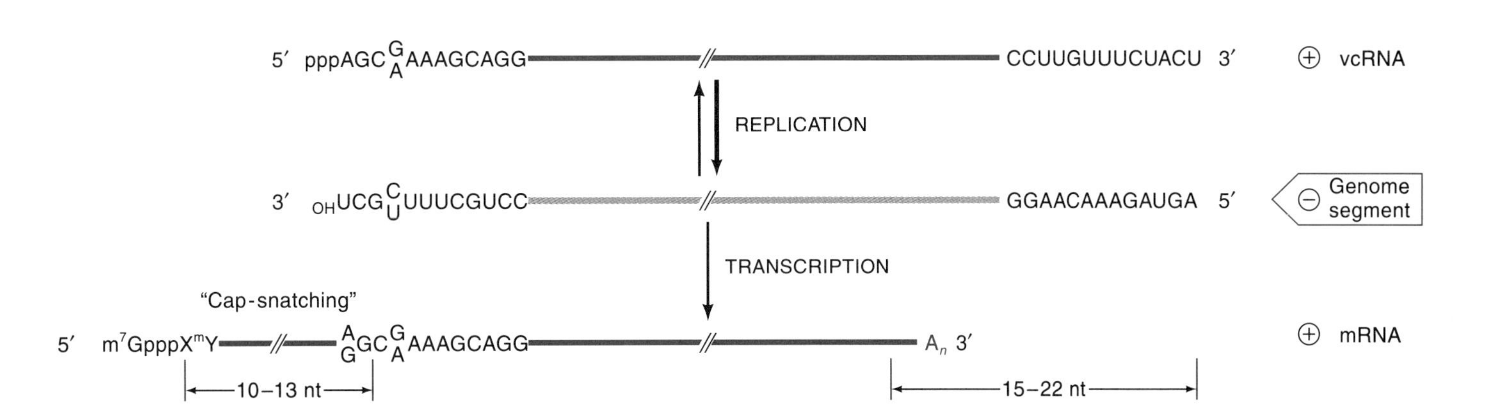

Figure 2.19
Relationship between genome RNAs, mRNAs, and vcRNA of orthomyxoviruses (influenza). The figure illustrates the nonidentity between the mRNA and the vcRNA for one representative genome segment. Plus-strand mRNAs have caps and leader sequences derived from cellular mRNAs (shown in Figure 2.18) and terminate with a poly A tail. Plus-strand templates (vcRNAs) are exact complements of the genomic minus-strand RNAs.

formed by the genome RNAs. For arenavirus RNAs and bunyavirus segments with an ambisense strategy, transcription terminates at the secondary structure that separates the two ORFs.

c. Translation Strategies of Minus-Strand RNA Viruses

Translation of most mRNAs of the nonsegmented viruses (the genes shown as open boxes in Figure 2.17) is illustrated in Figure 2.20, using the VSV N protein and G protein as models for nonglycosylated polypeptides and glycoproteins, respectively. In each case, translation initiates with the first available AUG codon (open triangle) in the ORF and ends with one of three termination codons (solid triangle).

The translation, modification, and processing of the G protein, the important features of which are described in Chapter 8, have been a model for glycoprotein synthesis and transport for many years. This same general pathway accounts for the maturation of other glycoproteins (F and HN of paramyxoviruses, NA and HA of influenza), as well as those of positive-strand viruses (e.g., E1 and E2 of alphaviruses, E and NS1 of flaviviruses).

The translation strategy for the individual influenza genome segments is similar to that shown for VSV mRNAs in Figure 2.20. The mRNAs transcribed from segments 1 to 6 consist of a short, untranslated 5′ region, an ORF encoding

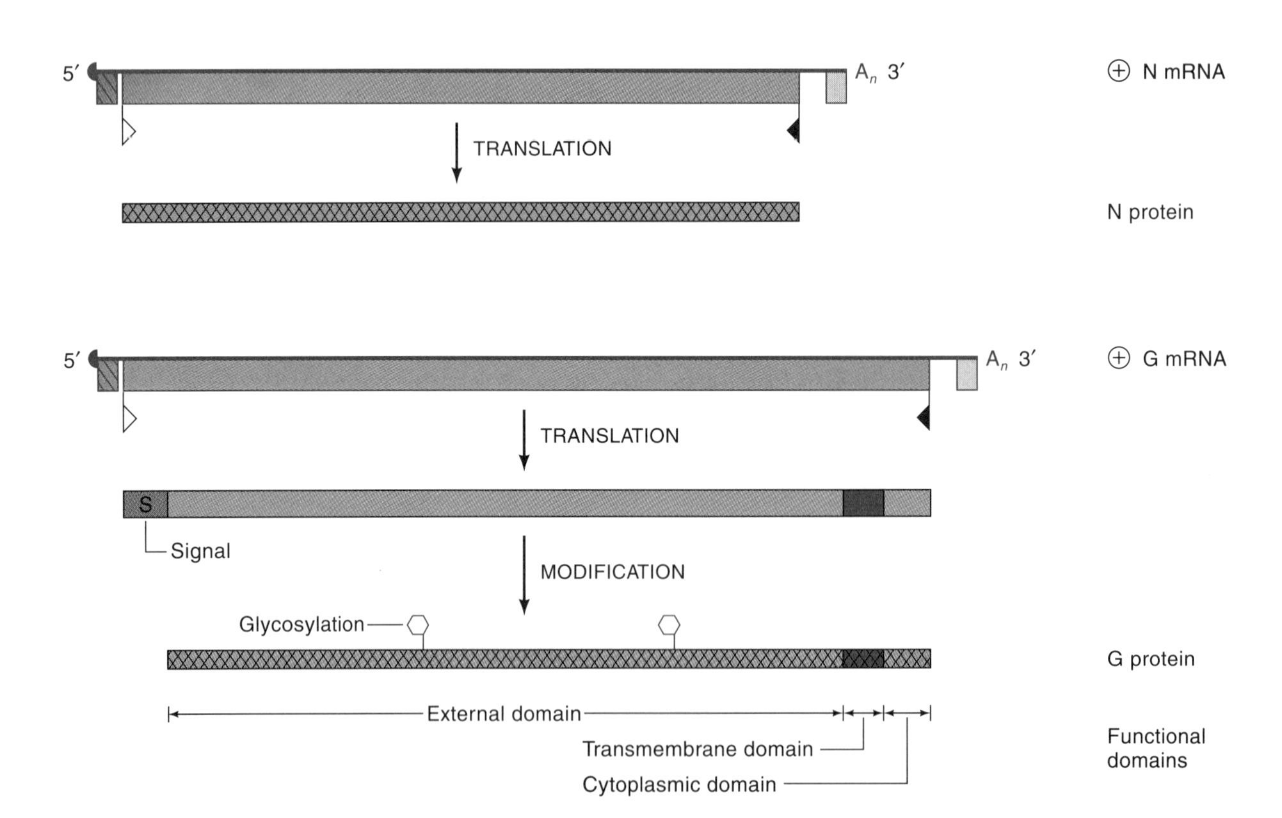

Figure 2.20
Generic translation strategy of monocistronic mRNAs. Two mRNAs from VSV and their translation products are shown. N is the nucleocapsid protein and G is the glycoprotein. The diagonally hatched box at the 5′ end is a short conserved nucleotide motif; the shaded box at the 3′ end is a polyadenylation signal. Functional domains of the protein are labelled; the open hexagons represent carbohydrate chains which are added post-translationally.

a single polypeptide, and a very short, untranslated 3′ region preceding the poly A tail. Both the NA and HA proteins are glycosylated, but only the HA polypeptide is processed posttranslationally by proteolytic cleavage.

The shaded boxes in Figure 2.17 are those for which the coding capacity has been expanded. The P mRNA of VSV, for example, can be translated into the P protein starting at the first AUG codon, or it can be translated into shorter products of 55 or 65 amino acids by internal initiation in a second ORF. Unlike the P protein, which is found in virions as part of the replicase/transcriptase complex, the small, highly basic products translated from the alternative ORF are nonstructural proteins.

The translation of segments 7 and 8 of influenza is somewhat different; segment 7 is shown in Figure 2.21. Two species of mRNAs are formed from segment 7; one is transcribed by the general mechanism shown in Figure 2.19 for the other segments and is translated into a major product, M1. The other is a spliced message in which a short region containing the capped primer and several additional nucleotides at the 5′ end of the mRNA is spliced to a downstream region in another ORF and produces a totally different translation product, M2, that has only a few amino terminal amino acids in common with M1. Segment 8 is transcribed and translated in a similar fashion to produce two nonstructural proteins, NS1 and NS2. All four products of segments 7 and 8 have been identified *in vivo*. Once again, the use of overlapping reading frames maximizes the coding capacity of the viral genome.

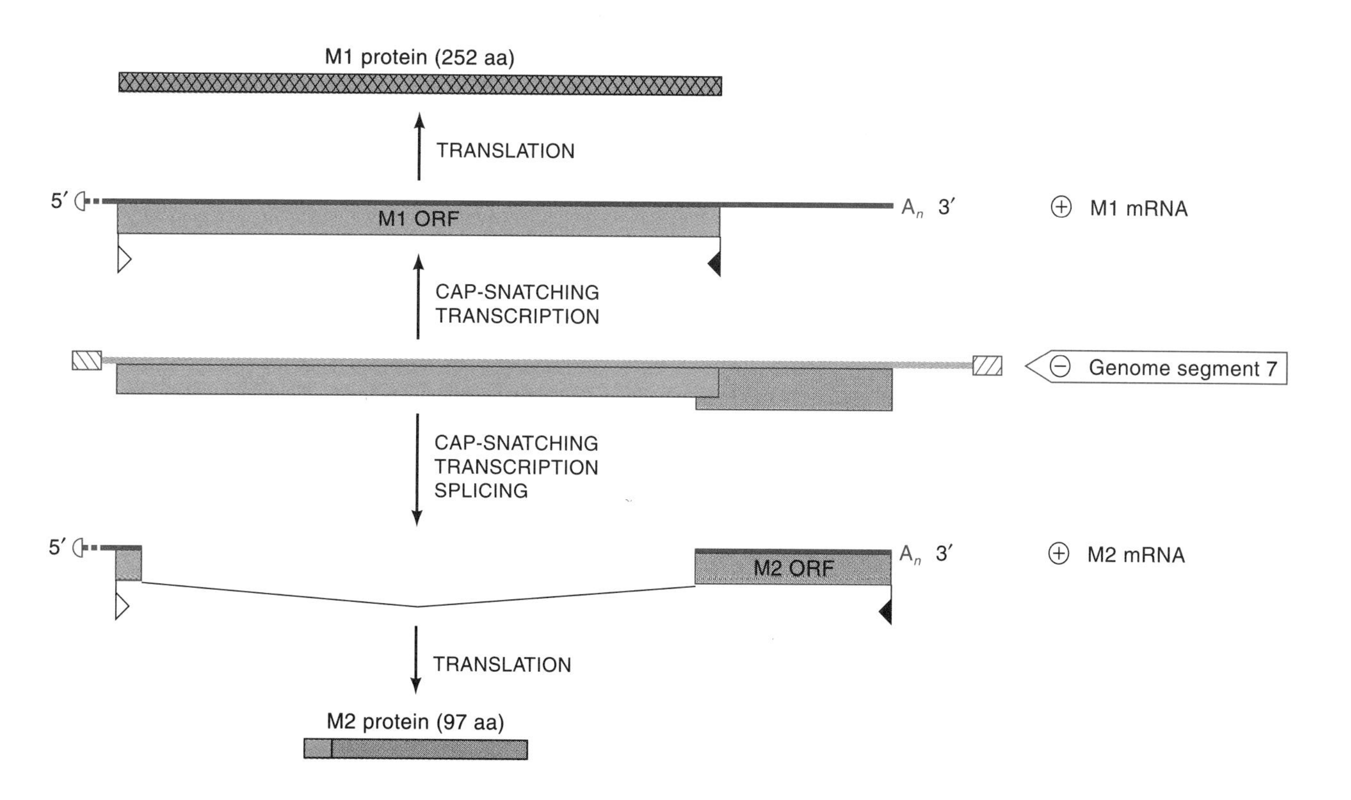

Figure 2.21
Transcription and translation strategies for influenza virus segment 7. The M1 mRNA is synthesized as shown in Figure 2.19. The M2 mRNA is made by splicing a short sequence from the 5′ end of the M1 ORF (indicated throughout this figure by colored boxes) to sequences containing a second ORF. Both mRNAs have caps (open symbols) derived from cellular capped messages ("cap-snatching"). Diagonally patterned boxes at the termini of the genome segment indicate short self-complementary sequences that are absent from the mRNAs.

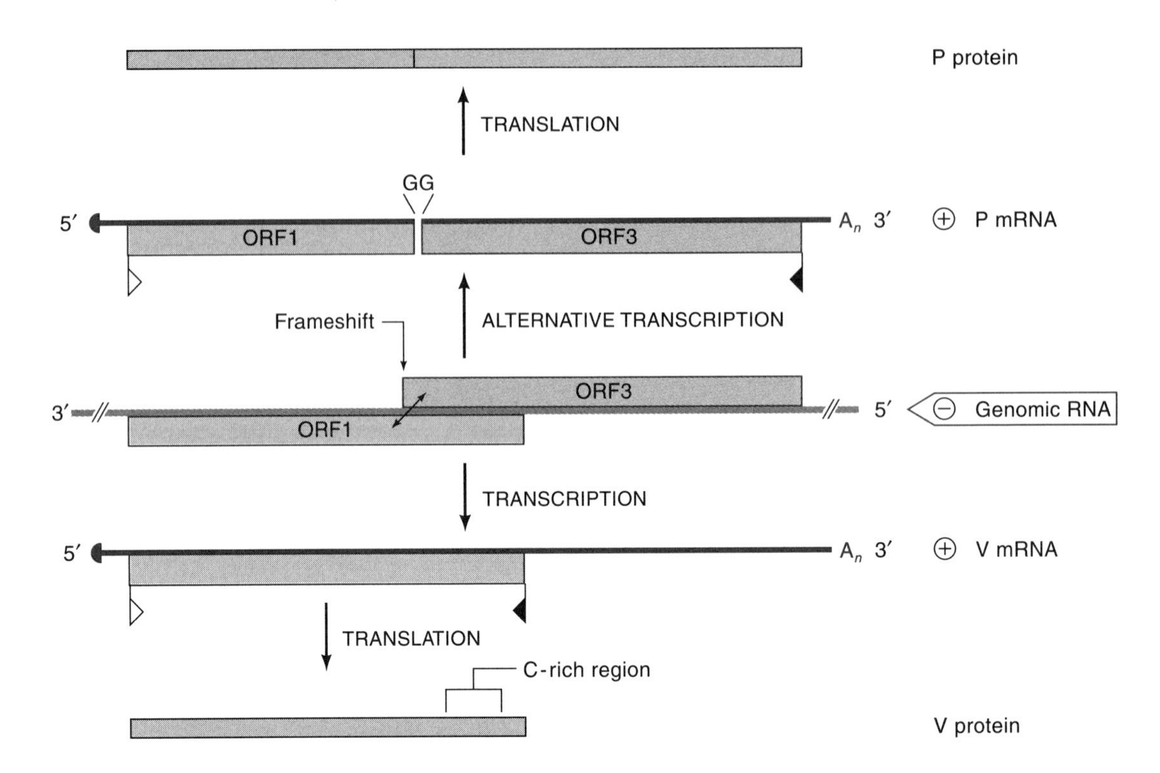

Figure 2.22
Nontemplated nucleotides in paramyxovirus mRNAs. Transcription of the P gene of SV5 to produce the P and V proteins is shown. Transcription of V mRNA begins by internal initiation on the paramyxovirus genome, and translation from the first AUG to the first stop codon (ORF1) to produce the V protein. Alternatively, two nontemplated G residues are added at a defined position during transcription to produce the P mRNA, which is translated into the P protein. Although the amino termini of the two proteins share an amino acid sequence, the characteristics of the proteins downstream of the insertion are drastically different because the downstream portion of the P protein is translated from ORF3.

Expressing the paramyxovirus and morbillivirus P genes illustrates another way of expanding coding capacity. In these instances, extra G residues are introduced at a specific point in the sequence by a process referred to as **RNA editing** (by which nontemplated nucleotides are introduced into mRNAs during transcription). These added nucleotides shift the translation frame from that point on so that the carboxy terminal half of the protein is very different from that translated from the mRNA without the insertion (Figure 2.22). In certain paramyxoviruses, transcripts that are exact complements of the genomic RNA, transcripts with one G inserted (shifting to ORF2), and transcripts with 2 Gs inserted (shifting to ORF3), respectively, are translated into three distinct proteins. Translation of a given nucleotide sequence in all three reading frames is the ultimate in coding efficiency!

2.4 Double-Strand RNA Virus Genomes

The family *Reoviridae* contains eight genera: Orthoreovirus, human viruses causing mild respiratory illness falling into three serotypes; Orbivirus, which includes a number of significant veterinary pathogens; Rotavirus, containing diarrheal viruses of many species but of special importance for human infants;

Coltivirus, containing numerous strains of **Col**orado **ti**ck fever virus; Aquareovirus, containing viruses of fish and shellfish; Cypovirus, containing **cy**toplasmic **po**lyhedrosis viruses of insects; and two genera of plant reoviruses, Phytoreovirus and Fijivirus.

Mature virions consist of two concentric protein shells within which are enclosed the 10 to 12 segments of double-strand RNA making up the viral genome. As shown by electron microscopy, the virion core consists of the inner protein shell within which the RNA is immobilized in a nearly paracrystalline array. Upon entry into a cell, the outer layer of protein is lost, and proteins in the inner shell undergo rearrangements to form uncoated, "activated" cores. These changes also appear to free the immobilized RNA; activated cores can transcribe mRNAs, whereas mature virus particles cannot.

The RNA segments can be separated electrophoretically on polyacrylamide gels according to their size; they fall into three major size groups. For members of the genus Orthoreovirus, there are three L or large segments (L1, L2, L3), three M or middle segments (M1, M2, M3), and four S or small segments (S1, S2, S3, S4) (Table 2.2). Most segments encode a single major polypeptide product. These polypeptides are named to correspond to their coding segments; thus, the λ-proteins are encoded by the L segments, the μ-proteins by the M segments, and the σ-proteins by the S segments. The outer protein shell of the virion consists largely of the μ1c and σ3 proteins, with small amounts of the σ1 protein as well; σ1 is the hemagglutinin and also contains the major neutralizing epitope. The inner protein shell and the RNA segments constitute a core particle that functions as a transcription and replication complex. The core contains three major protein components, λ2, λ1, and σ2, as well as small amounts of three other protein components that probably represent the enzymatic transcriptase/replicase complex.

The 10 genomic segments in the case of reovirus and 11 or 12 segments typical of the other genera are dsRNA with no poly A on either strand. One strand of each segment is transcribed into mRNA within the core particle, and the completed mRNAs, which are capped but not polyadenylated, are extruded from

Table 2.2 *Genome Organization and Translation Strategy of the Ten RNA Segments of the Human Reovirus Type 3 Genome*

Genome Segment	Protein Product	Length (nt)	5′ UT (nt)	ORF (aa)	3′ UT (nt)	Function of Protein	Abundance/Location
L1	λ3	3854	18	1267	53	Poly C polymerase, GDD motif	Minor core component
L2	λ2	3916	13	1289	36	Guanyl transferase	"Chimneys" on virion
L3	λ1	3896	13	1233	181*	Nucleotide binding motif, zinc finger	Core
M1	μ2	2304	13	736	83	Unknown	Very rare, core
M2	μ1->μ1C	2203	29	708	50	Major structural protein	Outer shell
M3	μNS(μNSC)	2235	18	719	60	Morphogenesis	Abundant, nonstructural
S1	σ1	1416	12	455	36	HA, neut Ag, attachment to cells, host DNA inhibition	24–36 copies in virion
	σ1NS			120			Nonstructural
S2	σ2	1331	18	418	57	Structural component	Core
S3	σNS	1198	27	366	73	Poly C polymerase, binds ss RNA, zinc finger	Nonstructural
S4	σ3	1196	32	365	69	Zinc finger	Outer capsid shell most abundant

*3′ UT contains a second ORF downstream of the first termination codon.

the core. These mRNAs are translated into the virion-specific polypeptides and then assemble into pseudocore or precore structures. Once within these core structures, each mRNA is transcribed once to form a double-stranded molecule. These double-stranded RNA segments in turn generate mRNAs during secondary transcription, leading to amplification of virus-encoded products. Late in infection, presumably triggered by the buildup of certain viral proteins that are added to the cores, the formation of new mRNAs ceases, and the particles complete their morphogenesis by acquiring an outer shell and becoming mature virions. Thus, many details of the replication of the dsRNA viruses are fundamentally different from those of the previous groups.

Cloning and sequencing of the reovirus RNAs have been very important for the elucidation of the molecular mechanisms discussed previously. The most detailed knowledge comes from work with the human reoviruses that replicate well in tissue culture systems, thereby permitting both molecular biological and biochemical studies. Rotaviruses, especially human rotaviruses, are more difficult to grow in tissue culture; therefore, much of what we know about them has been deduced from sequencing cDNAs of their genomic segments.

The cloning strategy used for the reovirus genomes made use of their unusual double-stranded nature. Individual segments were isolated after acrylamide gel electrophoresis, these were denatured into single strands, poly A tails were added to the 3′ ends with *E. coli* poly A polymerase, cDNAs of each strand were made as already described for other polyadenylated RNAs, the complementary cDNAs were annealed together after the RNA was removed, and the ds cDNAs were cloned into appropriate plasmid vectors.

Three serotypes of human reovirus have been distinguished. All ten genomic RNA segments from the Type 3 reovirus have been cloned and sequenced. Characteristics of the ten segments are shown in Table 2.2. The ten RNAs have a common sequence of four nucleotides (GCUA) at the 5′ terminus of the plus strand (i.e., the strand of the same polarity as the mRNAs), and a five nucleotide common sequence at the 3′ end (UCAUC). Each segment (except S1) contains a single long ORF. S1 has two nested ORFs, ORF1 (nt 14–1424 in Type 3) encoding protein σ1 and ORF2 (nucleotides 71–431)) encoding σ1NS. The L segments, encoding components of the RNA polymerase, are the most highly conserved of the three serotypes, the M segments are intermediate, and the S segments are the most divergent in sequence.

Viruses containing segments from more than one serotype have been constructed in the laboratory, but this probably does not occur often in nature. All the M and L segments of Type 1 are more closely related to their counterparts in Type 3 than Type 2, indicating that they are evolving as a unit. There is a single exception: the S1 RNA of Type 1 is most closely related to the S1 of Type 2.

Although the products of the ten segments vary a great deal in abundance in infected cells, no clues as to the source of this regulation are evident in the sequences themselves. Even changing the context of the two initiation codons in S1 fails to affect the ratio of σ1 to σ1NS produced. However, if the initiation codon for ORF1 is removed, more ORF2 product (σ1NS) is made.

2.5 Manipulation of Cloned Viral Genomes

Characteristics of many virus-encoded proteins have been determined from amino acid sequences deduced from the respective nucleotide sequences. With ever-increasing numbers of sequences available, certain motifs are becoming recognized as indicators of protein function, as already described for alphaviruses (Section 2.2a). In many cases, the function of a particular viral polypeptide can be deduced before it is ever detected in an infected cell.

Overproduction of a viral protein in one of the expression systems described later in this section can produce sufficient quantities of material for direct study or for use as an immunogen. But more important than just recognizing the function of viral proteins is the ability to test these functions directly by deletion or mutation of the genes encoding them. Once an RNA viral genome has been cloned as cDNA, any particular viral gene can be expressed in the same way as any cloned DNA gene (*Genes and Genomes,* chapters 5 and 7). Furthermore, cloned mutated viral genes can be incorporated into an infectious virion genome so that the effect of the mutations on pathogenesis in a natural host can be studied (Section 2.6).

a. Expression Systems

Expression in bacteria The use of recombinant DNA vectors able to express cloned genes in bacterial cultures is described in *Genes and Genomes,* chapters 5 and 7. There are a number of reasons for expressing genes from eukaryotic viruses in bacterial systems. These include the production of sufficient protein for physical or enzymatic studies, of protein antigens to elicit specific antibodies, and for possible use as vaccines. In some cases, sufficient protein can be obtained from small batches of cells, but bacterial cultures can readily be scaled up to produce much larger quantities.

One important use of expressed proteins is for the production of monospecific antibodies. High levels of expression are obtained when the viral gene is inserted within a gene that is efficiently expressed in bacteria. For example, the coding sequence for a viral protein can be inserted into the bacterial β-galactosidase gene so that a fusion protein can be expressed in *E. coli.* Such fusion proteins are readily purified from bacterial extracts by precipitation with commercially available antibodies to β-galactosidase. The purified fusion protein serves as an immunogen to produce antibodies to the viral protein. Some bacterial plasmids permit the viral gene product to be fused to an insoluble bacterial protein, which can be purified easily from the bacterial extract by differential centrifugation.

Expression in eukaryotic cells Many viral polypeptides undergo extensive post-translational modification. These modifications often do not take place in bacteria, necessitating expression in eukaryotic systems. Hybrid shuttle vector/expression plasmids have been designed with both bacterial origins of replication and eukaryotic transcription promoters. These plasmids replicate efficiently in bacteria and produce large amounts of DNA that can then be used to study the expression of cloned viral genes and the activities of their encoded products in a system that mimics the virus's natural host. Currently, there are several eukaryotic expression systems in common use, and they are constantly being refined and improved. Such systems also permit the expression of the viral genes alone or in various combinations to determine how each expressed product affects the cellular machinery or how combinations of products behave.

Several vector systems are based on DNA-containing animal viruses because they have very strong and largely unregulated promoters. These include the SV40 system, the vaccinia system, and the baculovirus system. In many cases, the new genetic material can be introduced into the cell of interest by infection with the engineered virus vector. Heterologous genes can also be expressed under the control of inducible cellular promoters so that production of the desired protein can be regulated. Transfected vectors may result in only transient expression of the desired gene, but certain vectors have been developed to permit stable integration into the host genome to be detected and allow for continuous expression of the transduced gene.

SV40 expression systems One means of achieving high levels of transient expression is to put the desired cloned insert into a vector in which expression is under

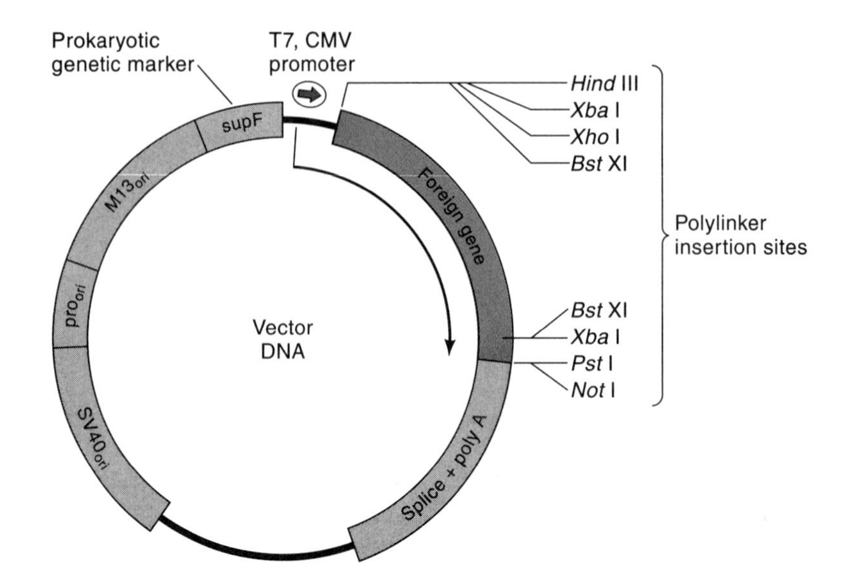

Figure 2.23
Map of a shuttle vector/COS cell expression vector. This vector is capable of being replicated in bacteria and expressed in COS cells. It contains an SV40 origin of replication ($SV40_{ori}$), eukaryotic transcription regulatory elements (Splice and poly A), a prokaryotic origin of replication (pro_{ori}), an M13 origin of replication ($M13_{ori}$), a T7 RNA polymerase promoter, and a polylinker insertion site. The inserted foreign gene (color) has its 5′ end of the message sense strand next to the T7 promoter and is transcribed and translated in the direction of the arrow.

the control of one of the SV40 promoters (*Genes and Genomes,* sections 5.7b and this volume, section 1.3c). Another approach to high-level expression, especially of surface glycoproteins, involves the use of COS cells. COS cells are derived from a line of African green monkey kidney cells stably transformed with an origin-defective SV40 virus. These cells produce the SV40 large-T protein, thereby permitting replication of any DNA that has an SV40 origin of replication (ori). Vectors for use with COS cells have bacterial oris (so that large amounts of DNA can be grown in bacteria), as well as the SV40 ori. Such vectors generally contain eukaryotic promoters, selectable markers, and an appropriate polylinker with unique cloning sites for insertion of cloned viral genes (Figure 2.23). The plasmid is replicated, transcribed, and translated efficiently. Transfected COS cells transiently express the viral product for up to a week.

Baculovirus expression systems When very large amounts of a gene product are desired (e.g., for crystallographic studies), *Autographa californica* multiply enveloped nuclear polyhedrosis virus (Ac MNPV or baculovirus) is often the vector of choice. Baculoviruses are insect viruses that are produced in two forms. One is embedded in large inclusions of proteinaceous material known as the polyhedrin protein. Polyhedrin is produced in prodigious amounts from an extremely active promoter but is not essential for replication of the virus. When a foreign (viral) gene is inserted under the control of the polyhedrin promoter, up to half the cellular protein late in infection can be the product of the introduced gene. Moreover, some but not all types of typical eukaryotic posttranslational processing steps occur in insect cells. This approach requires the generation of recombinant baculoviruses by homologous recombination between a wild-type baculovirus and a special plasmid, in which polyhedrin sequences flank both sides of the viral gene (Figure 2.24). Introduction of a foreign gene into the poly-

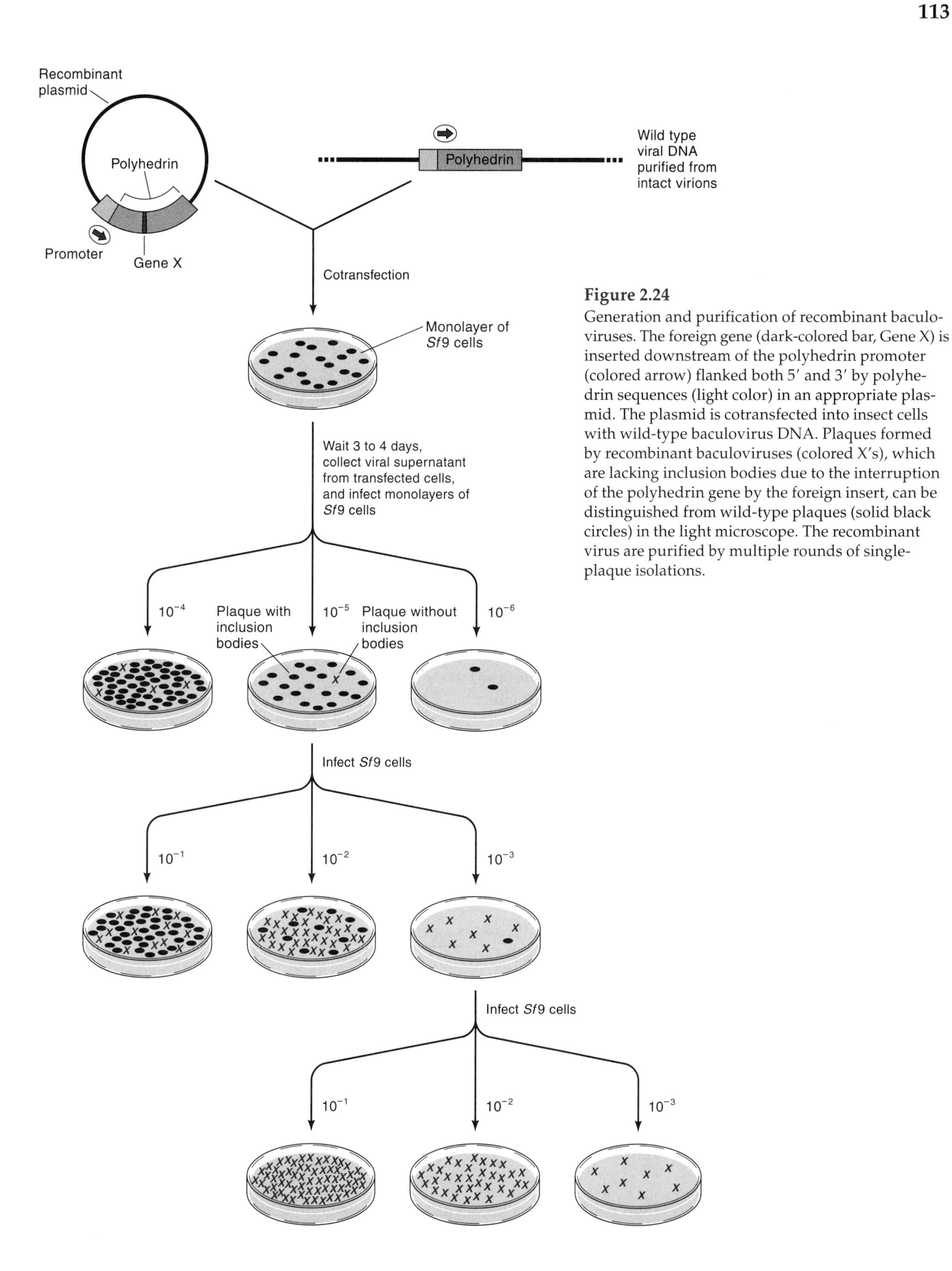

Figure 2.24
Generation and purification of recombinant baculoviruses. The foreign gene (dark-colored bar, Gene X) is inserted downstream of the polyhedrin promoter (colored arrow) flanked both 5′ and 3′ by polyhedrin sequences (light color) in an appropriate plasmid. The plasmid is cotransfected into insect cells with wild-type baculovirus DNA. Plaques formed by recombinant baculoviruses (colored X's), which are lacking inclusion bodies due to the interruption of the polyhedrin gene by the foreign insert, can be distinguished from wild-type plaques (solid black circles) in the light microscope. The recombinant virus are purified by multiple rounds of single-plaque isolations.

hedrin gene prevents polyhedrin formation, and hence no inclusion bodies are formed. Because plaques with and without inclusion bodies can be distinguished on a monolayer of insect cells under the light microscope, recombinants can be detected even though the level of homologous recombination is low (1 percent to 5 percent). Moreover, baculovirus expression is often used for genes that have been found, for whatever reason, to be toxic in bacteria.

Expression using recombinant vaccinia virus COS cells and other systems requiring transcription of the cloned gene in the nucleus are sometimes ineffective for the expression of RNA virus genes. Most RNA virus genomes replicate in the cytoplasm, and if their RNAs possess cryptic splice sites, their normal replication and transcription are unlikely to be affected because the splicing machinery is in the nucleus. However, when viral mRNAs are transcribed in the nucleus, splicing at such cryptic splice sites can cause deletions in the viral gene and lead to inactive proteins.

One way to avoid problems arising from cryptic splice sites is to use a virus vector that replicates in the cytoplasm; vaccinia virus is an excellent vector for this purpose. This virus also has other advantages as a vector for expressing foreign genes. Its genome is large (186 kb) and can incorporate significant amounts of foreign genetic material without losing viability. Moreover, several vaccinia promoters have been characterized, and it is possible to put different viral genes behind different promoters. Lastly, the virus titer can be readily measured by a plaque assay, the virus has a wide host range, and it grows well in tissue culture. Because the very large virus DNA is intractable, vaccinia recombinants are constructed largely by means of homologous recombination (Figure 2.25). A bacterial plasmid containing the vaccinia virus thymidine kinase (TK) gene sequence is constructed. The gene or cDNA under the transcriptional control of a vaccinia promoter is inserted into the middle of the TK gene, and the plasmid DNA is transfected into host cells, which are simultaneously infected with wild-type vaccinia virus. The vaccinia virus DNA undergoes homologous recombination with the plasmid DNA, leading to recombinant vaccinia viruses that are TK^-. TK^- recombinants can be selected by their ability to grow in 5-bromodeoxyuridine (BUdR); BUdR is ordinarily lethal for vaccinia, but if the virus lacks a functional TK gene, it cannot phosphorylate BUdR, thereby preventing its incorporation into DNA. Many different viral proteins have been cloned into vaccinia using this approach.

Vaccinia virus has also been used in a novel transient expression system. Eukaryotic cells infected with a vaccinia strain containing a cloned gene for the bacteriophage T7 RNA polymerase under the transcriptional control of a very active vaccinia promoter are then transfected with a bacterial plasmid in which the viral gene of interest resides next to a T7 promoter (Figure 2.26). In these cells, the plasmid-borne viral gene is transcribed by the T7 polymerase, but because these T7 transcripts are uncapped, they are translated inefficiently by eukaryotic host cells. To overcome this low translation efficiency, the long 5′ UTR of encephalomyocarditis virus (a picornavirus whose RNA is uncapped) is placed upstream of the viral gene start codon in the plasmid. This UTR contains an internal ribosome entry sequence (IRES) that permits cap-independent translation.

Because of the vaccinia genome's large capacity and the decades of experience in inoculating humans with vaccinia virus to protect against smallpox, vaccinia has been considered a potential vehicle for immunization against other viruses. Toward this aim, many viral glycoproteins, including influenza HA, the herpesvirus glycoprotein D, the rabies virus G protein, VSV G protein, yellow fever virus E and NS1 proteins, alphavirus E1 and E2, and the HIV glycoprotein, have been expressed in vaccinia. The prospects for creating vaccinia-based vaccines for other viral antigens are promising. Infection of animals with recombinant vaccinia that express surface antigens of several veterinary pathogens leads

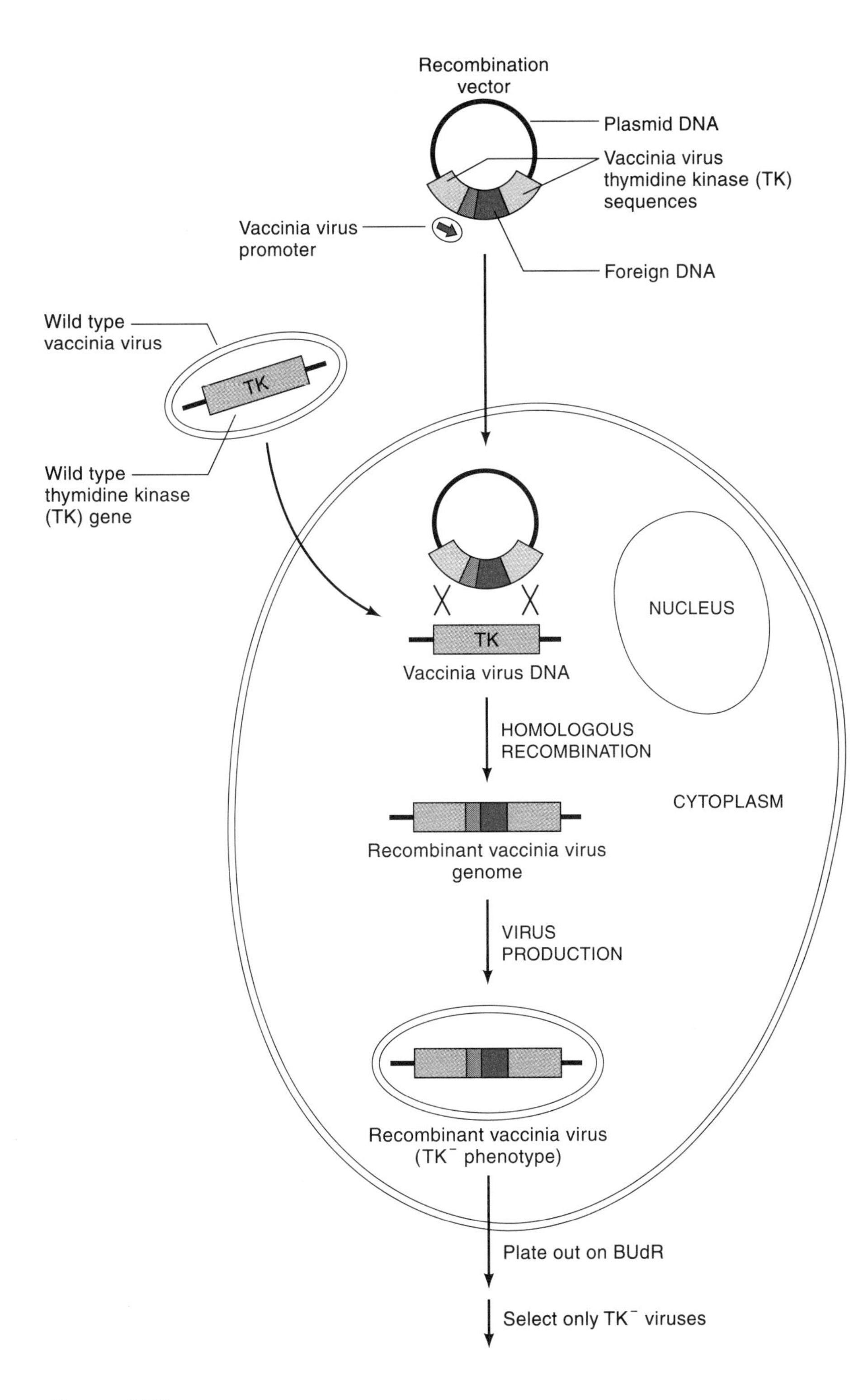

Figure 2.25
Construction of recombinant vaccinia for expression. A foreign gene (dark color) is inserted into a bacterial plasmid adjacent to a vaccinia promoter that has in turn been inserted into the vaccinia TK gene. Plasmid DNA is transfected into cells infected with wild-type (TK^+) vaccinia virus. Recombinant viral progeny (TK^-) produced by homologous recombination between vaccinia and the plasmid are all TK^-, due to the interruption of the TK gene by the foreign gene. These TK^- progeny are selected by growth in bromodeoxyuridine and screened by plaque hybridization for appropriate inserts. In subsequent cycles of infection, vaccinia express the foreign genes under the control of the vaccinia promoter. (From Brown, F., Schild, G. C., Ada, G. L. 1986. Recombinant vaccinia viruses as vaccines. *Nature* 319, p. 549–550.)

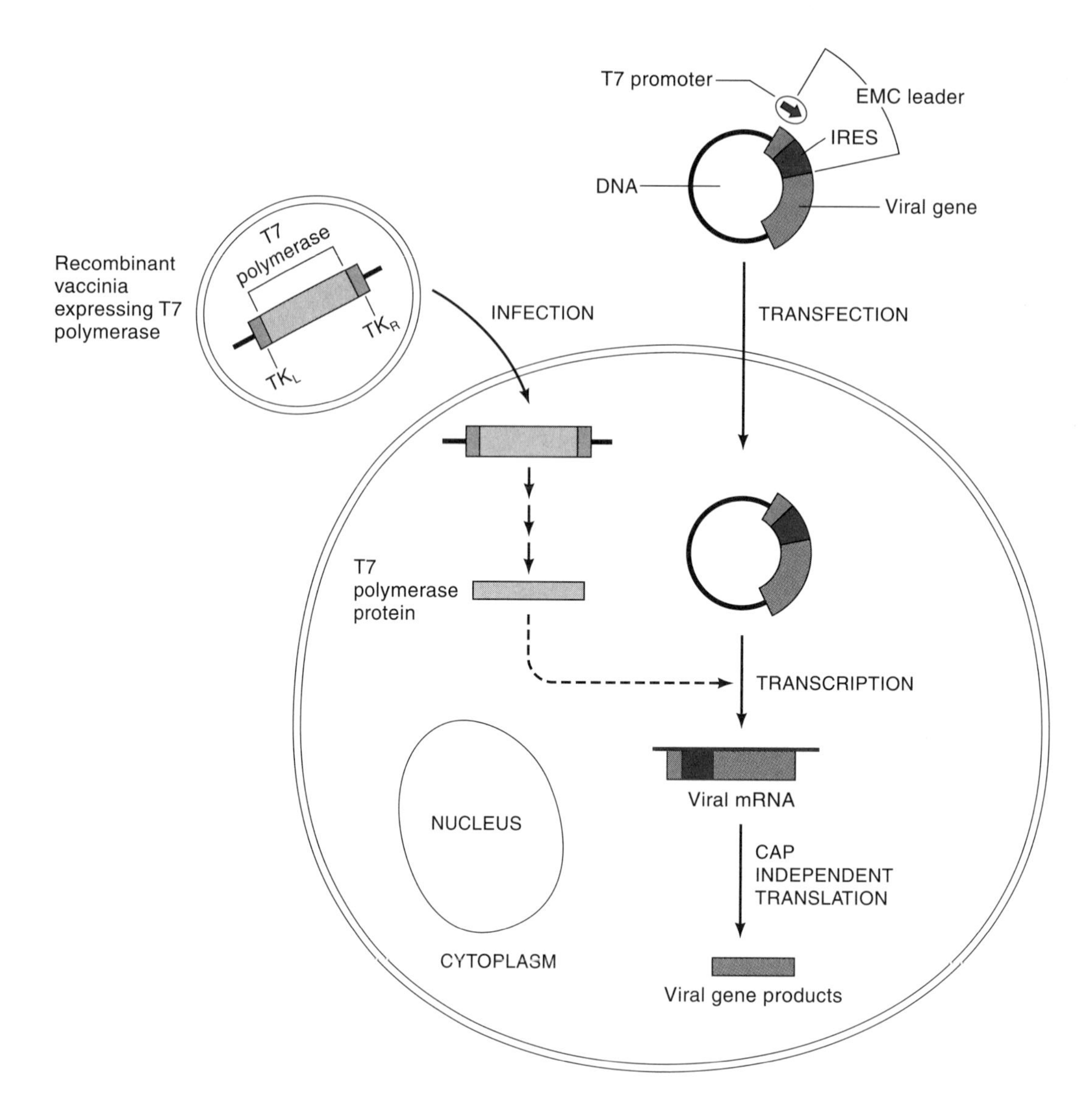

Figure 2.26
Expression of viral genes from T7 transcripts in eukaryotic cells. Eukaryotic cells are infected with recombinant vaccinia virus (see Figure 2.25) expressing T7 RNA polymerase under the control of a strong vaccinia promoter. These infected cells are transfected with plasmid DNA containing the foreign (viral) gene of interest under the control of a T7 promoter. T7 polymerase is expressed inside the cell and transcribes the viral gene on the plasmid. Inclusion of the EMC leader sequence containing the IRES (internal ribosome entry site) improves the efficiency of cap-independent translation of the uncapped T7 transcripts.

to protection of the vaccinated animal against subsequent challenge with virulent virus. Preliminary studies of human vaccination with vaccinia virus expressing HIV gp120 are also encouraging. Given the vaccinia genome's size, more than one viral gene could be inserted, and this may eventually lead to a vaccine that could be effective against multiple virus targets. But because a fraction of individuals inoculated with vaccinia virus develops undesirable side effects, further research is needed to reduce the virulence of vaccinia virus.

Transgenic plants resistant to viruses Plant virologists have known for a long time that plants infected with attenuated strains of certain plant viruses sometimes acquire resistance to the disease caused by more virulent strains of closely related viruses. However, the viral functions responsible for the reduction in subsequent disease symptomology during challenge were unknown. Now we know that constitutive expression of a plant virus coat protein by a transgenic plant can confer resistance to infection.

This resistance was first demonstrated by integrating the coat protein gene for tobacco mosaic virus into a tobacco plant's genome. In this case, the cloned coat protein gene was inserted into a vector containing a selectable marker downstream of the 35S promoter from cauliflower mosaic virus. After transfection of the vector into *Agrobacteria* containing a defective Ti insertion plasmid (Chapter 12), the resulting recombinant bacteria were inoculated onto leaf disks from which transgenic plants were generated. These plants expressed both the coat mRNA and coat protein and were less severely affected after infection with TMV. The resistance requires the production of the coat protein, and protection is restricted to the same or very closely related viruses, usually those whose coat proteins share 60 percent or more amino acid identity with the transgenic coat protein. For example, a transgenic plant expressing the TMV coat protein is not resistant to alfalfa mosaic virus. Because infection with deproteinized TMV RNA circumvents the resistance, the expressed coat protein may interfere in some way with uncoating the virus. Although the molecular mechanism of resistance is not yet understood, this approach is promising for protecting crop plants of economic importance against common viral pests.

Other expression systems Viral mRNAs made *in vitro* from a cDNA clone can be delivered directly to a cell cytoplasm. Many commercially available plasmids have an SP6 promoter and a T7 promoter flanking a polylinker so that transcripts of a viral DNA sequence can be made *in vitro* from one strand with SP6 polymerase and from the other with T7 polymerase (*Genes and Genomes*, section 6.4b). The amount and quality of such transcripts is constrained only by the amount and purity of DNA template available. Either the "sense" or "antisense" RNA can then be delivered to appropriate cells by microinjection, transfection using DEAE dextran, electroporation, or enclosure in liposomes that are then fused to the cell surface (lipofection). Introduction of *in vitro* transcripts of "infectious clones" are a special case of this procedure (Section 2.6c).

b. Site-Specific Mutagenesis

Many aspects of viral replication have now been examined using site-specific mutagenesis, deletion analysis, and formation of chimeric genes. The following examples are not intended to be comprehensive.

The G glycoprotein of VSV The assembly of enveloped viruses has been widely used as a model for membrane biogenesis. Similarly, the synthesis, modification, folding, and transport of viral glycoproteins have been studied as models for comparable processes involving cell surface glycoproteins.

The mRNAs of the rhabdovirus, VSV, are monocistronic, and the expression of a clone containing only the G protein sequence is unaffected by the absence of other viral components. Expression of the G protein cDNA was studied both by transient expression (using an SV40 vector in COS cells) and in permanently transformed cells. The cDNA encodes a G protein precursor with a number of different functional domains (Figure 2.20): an amino terminal signal sequence of 16 amino acids, which is removed by signal peptidase (S); the 446 amino acid long ectodomain, which is glycosylated at positions 179 and 336 (stalked symbols); a downstream transmembrane domain of 20 hydrophobic amino acids; and a carboxy terminal 29 amino acid long cytoplasmic domain, which is hydrophilic and basic. The effects of mutations in each of these domains on glycosylation, acylation, transport, and cellular localization of the glycoprotein have been studied.

Alterations in the cytoplasmic domain that decreased its hydrophilicity blocked transport of G out of the endoplasmic reticulum. However, replacement of the sequence encoding this cytoplasmic domain with sequences encoding comparable cytoplasmic domains from other virus glycoproteins yielded chimeric proteins that were transported normally to the cell surface.

Two types of studies have been performed on the membrane-spanning domain. Mutation of the single cysteine in this domain to a serine yielded a protein that was not acylated with fatty acids but was still transported normally to the cell surface. Presumably, cysteine acylation increases the hydrophobicity of the membrane-spanning region, but the exact function of acylation is unknown, and it is clearly not required for transport. In other experiments, the hydrophobic amino acids in the transmembrane domain were deleted in stepwise fashion. As long as more than 12 amino acids remained in this domain, the protein was transported to the surface of the cell, albeit less efficiently than when all 20 residues were present. If fewer than 12 transmembrane amino acids remained, the protein was stuck in the endoplasmic reticulum and did not progress further along the glycoprotein maturation pathway. If all transmembrane amino acids were deleted, the protein was no longer bound to the membrane and was secreted.

Additional experiments demonstrate that glycosylation of the external domain of the G protein is essential for transport to the cell membrane. Thus, the Asn-X-Thr/Ser glycosylation sites were selectively eliminated by site-specific mutagenesis. If either the site at 179 or the site at 336 is retained, G protein is efficiently transported, but if both sites are destroyed, the nonglycosylated polypeptide is not transported to the cell surface. These studies do not by themselves clarify why glycosylation is important for proper transport from the endoplasmic reticulum to the cell surface. However, these and other studies imply a major role for glycosylation in maintaining proper protein folding and suggest that the correct three-dimensional configuration of a membrane protein is essential for each step in the transport pathway.

Site-specific mutagenesis of a Sindbis virus proteinase The nonstructural proteins of Sindbis virus are translated as two polyproteins that are subsequently processed by a proteinase encoded in the carboxy terminal half of protein nsP2 (Section 2.2a). Virtually everything we know about the structure and function of this proteinase and its importance in regulation of RNA replication comes from manipulation of cDNA copies of the nonstructural protein genes.

RNA transcribed with SP6 polymerase from a full-length cDNA clone of Sindbis virus (Section 2.6c) can be translated in a rabbit reticulocyte system. Using polyclonal monospecific antibodies made against fusion proteins of each of the four Sindbis nonstructural proteins, we can immunoprecipitate the translated products, nsP1, nsP2, nsP3, and nsP4, as well as polyprotein intermediates containing these products. In the reticulocyte system, proteolytic processing of the Sindbis nonstructural polyprotein is virtually identical to that seen in the cytoplasm of infected cells. Deletions in Sindbis cDNA clones within the regions encoding nsP1, nsP3, or nsP4 have little effect on the processing of the corresponding polyprotein (Figure 2.27). In contrast, deletions in the carboxy terminal half of nsP2 abolish processing, and deletions in the amino terminal portion of nsP2 cause aberrant processing, possibly because of incorrect folding of the nsP2 polypeptide. Within the domain delimited by the deletion experiments, two cysteines and five histidines are conserved among the known alphavirus sequences. Each of these residues was individually altered by site-specific mutagenesis and the RNA transcripts translated *in vitro*. Any change in Cys481 or His558 of nsP2 abolishes proteolytic processing completely. Comparison of

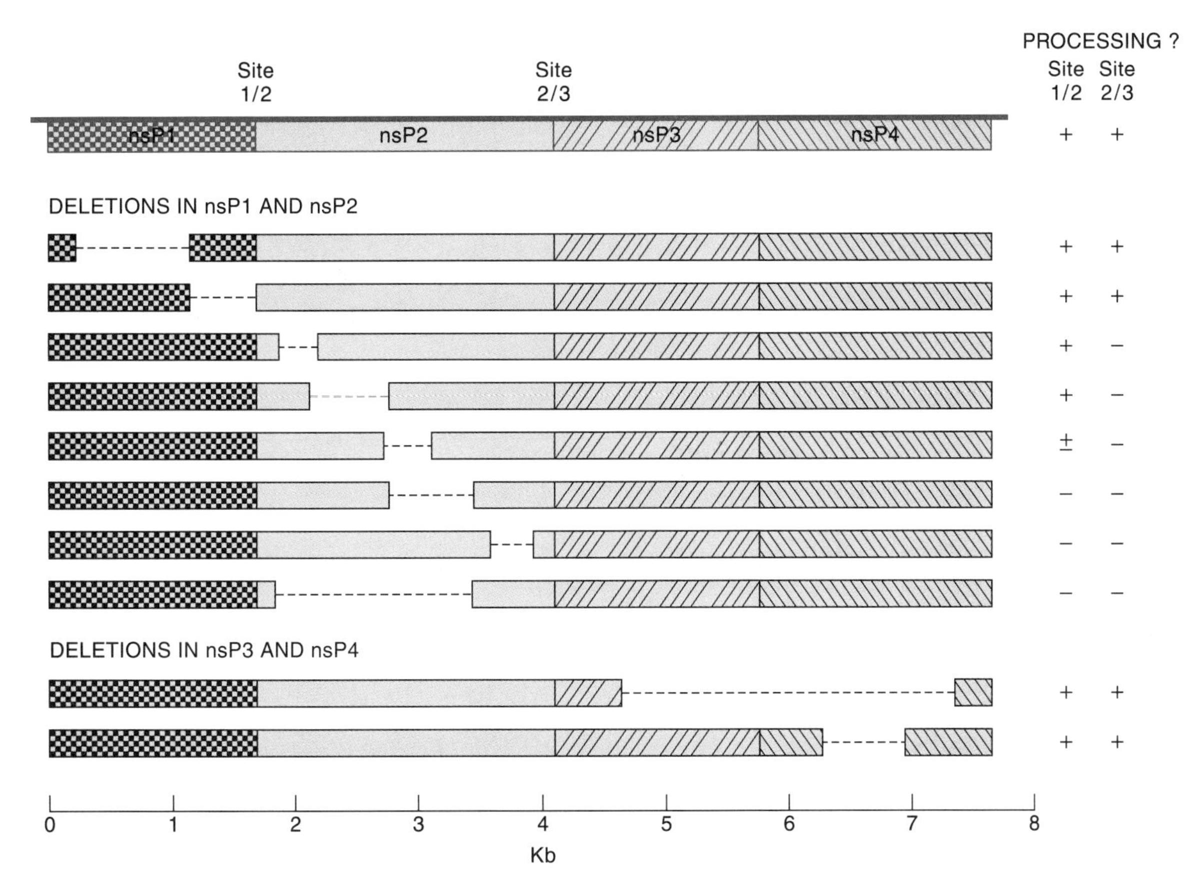

Figure 2.27
Deletion mapping of the nsP2 proteinase. Deletions in the full-length cDNA clone of Sindbis virus were made in all four nonstructural proteins by using convenient restriction sites. RNA transcripts were synthesized *in vitro* with SP6 polymerase and translated in reticulocyte lysates. Products were characterized by immunoprecipitation and acrylamide gel electrophoresis and scored for processing as shown in the two columns on the right. These results localized the nonstructural proteinase to the carboxy terminus of the nsP2 protein.

the sequences around these residues with sequences near the active site residues of known proteases suggests that the nsP2 proteinase is similar in structure to the papainlike proteases (Figure 2.28).

In alphaviruses, the penultimate amino acid (position P2 of each cleavage site) of each nonstructural protein is glycine. When these glycines are mutated to valine, the sites cannot be cleaved. Using this information, we could construct cDNA clones that expressed uncleavable polyproteins containing the nsP2 proteinase (such as P12, P23, P123, etc.) as well as clones expressing nsP2 alone in order to determine the specificities and cleavage site preferences of these intermediates. These reactions could be studied by using a "universal substrate" (i.e., the polyprotein containing cleavable sites but encoding an inactive proteinase, due to the mutation of Cys481 to Gly). When a reticulocyte lysate translating an uncleavable polyprotein is mixed with one translating the "universal substrate," efficient processing occurs in *trans.* Moreover, the uncleavable polyproteins cleave the three sites with very different kinetics.

All four nonstructural proteins are also components of the Sindbis replicase/transcriptase. Although no plus-strand RNA viral replicase that maintains its native properties has yet been purified, a great deal has been learned about

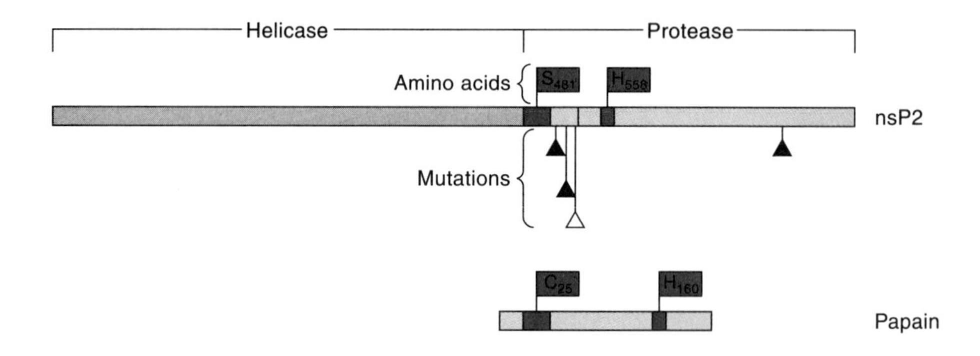

Original amino acid	Mutant amino acid	Processing
Cys-481	Ser Arg Gly	none none none
Cys-525	Ser Arg	normal none
Ser-535	Thr	normal
His-558	Ala Gln Tyr	none none none
His-709	Ala Tyr Arg	normal normal none
His-701	Ala	normal
His-702	Ala	normal
His-619	Ala	normal

Figure 2.28

Diagram of mutations in nsP2 and their effects on proteolytic processing by the nsP2 proteinase. Conserved cysteine, serine, and histidine residues in the carboxy-terminal half of nsP2 were altered by site-specific mutagenesis of the full-length clone of Sindbis virus and tested for proteolytic activity by transcription *in vitro* with SP6 polymerase and translation in reticulocyte lysates. Cysteine-481 and Histidine-558 were identified in this way as essential components of the catalytic site. The location of these two residues is mapped on a scale representation of nsP2, along with the locations of mutation in several temperature-sensitive RNA^- mutants (solid black and open triangles). The solid triangles are mutations that affect proteolytic processing; the open triangle is a mutation that does not affect processing. The location of the comparable residues in papain are shown for comparison

the activities of the various components by expressing them alone or in combinations as uncleavable polyproteins or mutant forms and testing their effects on RNA replication.

Studies of individual components and uncleavable polyproteins have been performed by transfection with *in vitro* transcribed RNA or with transient expression using the T7-vaccinia system (Section 2.5a). Other experiments have used infectious Sindbis variants containing the cleavage site mutants that had been rescued by the methods described in Section 2.6. Together, these two approaches have led to a model for temporal regulation of RNA synthesis during alphavirus replication. Although minus-strand (vcRNA) synthesis, which takes place only in the first four hours or so after infection, absolutely requires nsP4 and the uncleaved precursor, P123, efficient plus-strand synthesis (both subgenomic 26S RNA and genomic 49S RNA) requires cleaved replicase components. The relative efficiencies and specificities for particular cleavage sites of the nsP2 proteinase, alone and in the various intermediate polyproteins, produces a processing cascade that results in the switch from minus-strand synthesis early in infection to plus-strand synthesis late.

2.6 "Infectious Clones" and the Genetic Revolution

Following the determination of the nucleotide sequence and genome organization of the various RNA viruses, the single most important development has been the construction of infectious clones (i.e., cDNA copies of entire viral genomes in bacterial plasmids in which either the DNA itself or RNA transcribed from the cDNA *in vitro* is infectious). To date, it has been possible to construct infectious clones only of plus-strand RNA viruses. Because the deproteinized RNA of plus-strand viruses is infectious, whether isolated from virions or transcribed from cDNA clones, replication to produce infectious virus occurs without the addition of any other viral components. In contrast, infection by minus-strand viruses requires introduction of both ribonucleoprotein complexes and polymerase components into a cell; therefore, the expression of these viral genomes from cDNA is difficult.

The availability of a system in which infectious virus can be produced from cDNA clones has created boundless opportunities for a reverse genetic approach to the analysis of RNA virus gene structure and function. Today, viral genomes synthesized in bacteria can be manipulated as DNA in many ways. RNA can be transcribed from these modified viral genomes *in vitro* and introduced by a variety of means into appropriate cells to produce RNA-containing infectious progeny virions. The biological activities of such modified viruses can be studied in cell culture systems or in animal model systems.

Infectious clones now exist for alphaviruses, flaviviruses, enteroviruses, and a number of plant viruses. For RNA viruses whose genomes change rapidly during passage, such infectious DNA clones are a means to maintain stable virus seed stocks because, for the most part, cDNAs are replicated in bacteria with utmost fidelity. Infectious cDNA clones have been widely used for mapping preexisting conditional lethal mutations as well as second site revertants, for localizing determinants of neurovirulence, for both site-directed and random mutagenesis, for construction of chimeric viruses, and for development of new expression systems (Section 2.7).

a. The Poliovirus Infectious Clone

To date, the only cDNA clones for which the DNA is infectious are those of bacteriophage Qβ and two picornaviruses, poliovirus and coxsackie B virus. In order to sequence the genome of poliovirus, several cDNA inserts were cloned into plasmid pBR322 and subsequently assembled into a single clone containing the entire genome. After amplification in *E. coli*, transfection of the purified plasmid DNA into mammalian cells resulted in the production of infectious poliovirus; these were the first animal RNA viruses derived from a cloned cDNA. Although the initial finding was extremely exciting, the efficiency in terms of plaque-forming units per microgram of input DNA was disappointingly low. Subsequent modifications in which SV40 transcription and replication signals were added upstream of the insert and transfection was into COS cells (Section 2.5a) improved the efficiency markedly.

Such infectious clones have been used to delineate regions of the poliovirus genome that are important for virulence. There are three serotypes of poliovirus, Types 1, 2, and 3, and the Sabin live virus vaccine contains attenuated strains of all three. There are only ten nucleotide changes, resulting in three altered amino acids, between the Type 3 parental strain and the vaccine strain. The contribution of each nucleotide change to attenuation was determined by constructing recombinant clones, rescuing the virus after transfection, and testing the virus for virulence. Researchers thus determined that changing the C at position 472 to U (in the 5′ UTR) and C to U at 2034 (resulting in Ser91 to Phe in VP3) are sufficient for attenuation of the vaccine strain. The change at 2034 appears to be stable, and it is now possible to predict the neurovirulence of a particular lot of Sabin vaccine by measuring the content of U at nucleotide 472. The small number of changes between the pathogenic and avirulent strains may explain the high reversion frequency to virulence of this particular component of the Sabin vaccine.

The Sabin Type 1 vaccine strain has also been compared with its virulent parent, the Mahoney strain; there are 56 nucleotide differences between these two, resulting in 21 amino acid changes. After examining infectious recombinants between Sabin Type 1 and the virulent Mahoney parent, virologists proposed that the residues determining attenuation are scattered throughout the genome but that a change at nucleotide 480 in the 5′ UTR is especially important.

Sabin Type 2 was derived from a naturally attenuated strain of poliovirus, so this type of direct comparison is not possible. However, a virulent strain of Type 2 isolated from a case of vaccine-derived disease differs at 23 nucleotides. Using infectious recombinants the scientists could ascribe neurovirulence to two changes, a G to A at 480 in the 5′ UTR and Val143 to Ile in VP1. This type of information should be very useful in designing new vaccines in which attenuation is induced by deletion rather than single nucleotide substitutions, thereby preventing same site reversions.

b. Infectious Clones of Brome Mosaic Virus

The second type of infectious clone is the indirectly infectious clone, in which viral RNA is produced *in vitro* from an appropriate promoter inserted just upstream of the virus cDNA by an RNA polymerase. The RNA polymerases used have been from *E. coli* (with the phage λ repressor promoter) and from bacteriophages SP6, T7, and T3 (with their respective promoters). Such indirectly infectious cDNA clones are available for a number of plant viruses, a few insect viruses, several alphaviruses, and members of all the genera of *Picornaviridae,* including poliovirus. Although most of these infectious clones were generated by ligation of several shorter clones, a frequent difficulty of this approach is that the resultant clones are not infectious, presumably due to either point mutations or deletions that occur during the cloning steps or to rearrangements when long clones are replicated in bacteria.

The brome mosaic virus (BMV) genome consists of three RNA segments (Section 2.2a). Separate cDNA clones representing each one of the three different RNAs were not directly infectious in plant cells or plant protoplasts, either alone or in combination. Transcripts made from these cDNA clones *in vitro* were then tested for their infectivity in both plants and protoplasts. These transcripts were made with *E. coli* RNA polymerase, using the λ repressor promoter engineered just upstream of the transcription start site. Before transcription, the cDNA clones were cleaved at a restriction endonuclease site just downstream of the end of the BMV sequence. Thus, although the transcripts began precisely with the 5′ nucleotide of each BMV RNA (each was capped by cap analogues present in the transcription mixture), they had a few nonviral nucleotides at the 3′ end. After

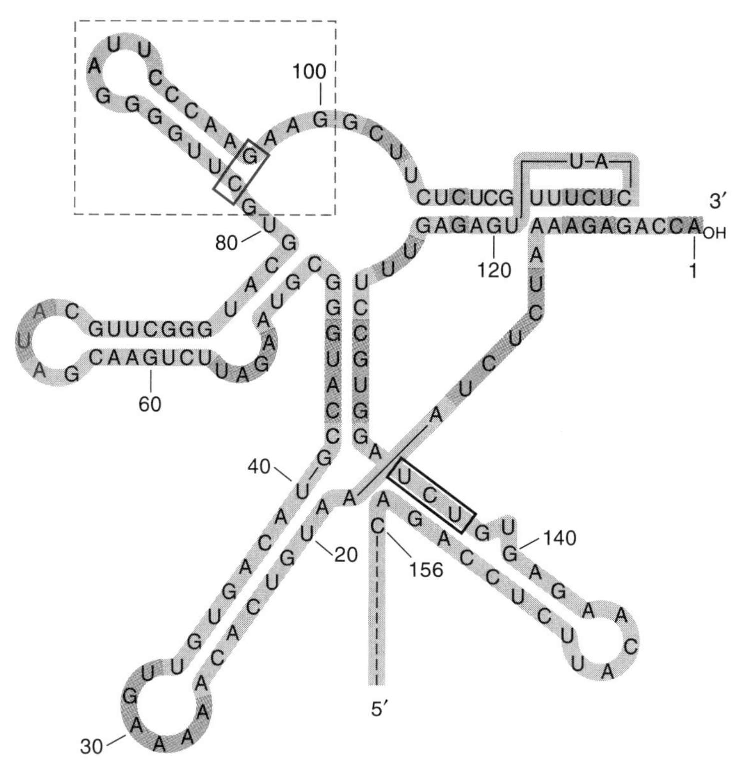

Figure 2.29
The 3′ terminus of RNA 3 of BMV. The 3′ terminal 156 nucleotides of RNA 3 of BMV are shown in a two-dimensional representation of the stem-loop and pseudoknot configuration. Various functional domains are indicated (see key). Nucleotides in the anticodon loop which had been mutated by site-specific mutagenesis were repaired *in vivo* by both recombination and same-site reversion.

transfection of the three RNAs into plant protoplasts, the newly synthesized virus RNAs terminated precisely at the 3′ terminal nucleotides of each BMV RNA, and the extra nucleotides had been removed. Evidently, the viral minus-strand replicase can ignore extra nucleotides and begin copying the 3′ end at the proper place.

By infecting cells with combinations of the BMV RNAs, using only RNA transcribed *in vitro* or combinations of RNA transcribed *in vitro* with RNA purified from virions, plant virologists showed unambiguously that all three genome segments are required for infectivity and production of progeny BMV. RNA segments 1 and 2 alone replicate, but no virus is produced.

Recombination between the RNA segments of BMV during productive infection has been demonstrated using genetically engineered cDNA. The three genomic RNAs contain similar but not identical structures at their 3′ termini. A deletion of 20 base pairs, (dashed colored-outline box in Figure 2.29), was made in this terminal sequence in the cDNA corresponding to RNA3. Plants were infected with RNA transcribed from cDNA clones of wild-type BMV RNA1 and RNA2 together with transcripts from the deleted RNA3 clone. The RNA3 containing the deletion was replicated less efficiently than the wild-type RNA3. Upon prolonged incubation for 15 days postinoculation, a new RNA3 emerged in which a wild-type 3′ terminus derived from either RNA1 or RNA2 had replaced the deleted sequence. Five recombinants of this type were isolated and

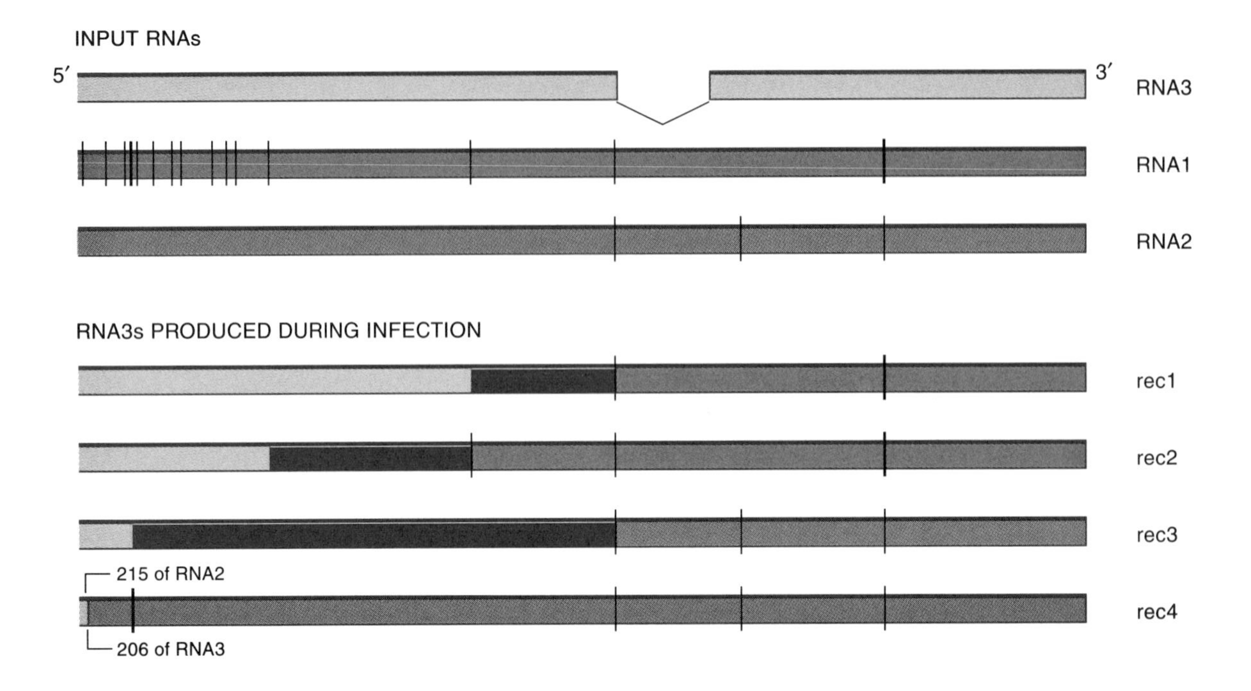

Figure 2.30
Structure of new BMV RNA 3′s generated by recombination. Whole plants were infected with a mixture of BMV RNA 1, RNA 2, and a deleted form of RNA 3. The 3′ terminal 215 nucleotides of the three input RNAs are shown diagramatically; vertical lines indicate nucleotides that are different from the RNA3 sequence. BMV RNA was isolated from individual lesions on the leaves and the sequences analyzed. Four types of repaired RNA 3 were found. In each case, the 3′ terminus of RNA1 or RNA2 was found covalently attached to upstream sequences from RNA 3. In recombinants 1, 2, and 3 the exact point of the cross-over could not be determined due to the small number of differences in sequence between RNA 3 and the other two RNAs, but in each case it must have taken place within the domain shown in dark color.

characterized by nucleotide sequencing, of which four are shown in Fig. 2.30. These studies were the first direct demonstration of recombination in RNA plant viruses containing single-stranded genomes.

c. The Sindbis Infectious Clone

Full-length cDNA clones from which infectious RNA can be transcribed *in vitro* exist for a number of alphaviruses, including Sindbis, Semliki Forest, Venezuelan equine encephalitis (VEE), and Ross River (RR) viruses. Most of the recent research on this group of viruses would not have been feasible without these clones. Infectious clones are particularly useful for characterizing phenotypic variants (e.g., temperature-sensitive mutants). The ability to rescue virus is also essential for the study of mutations affecting replication and pathogenicity in animals. Furthermore, infectivity implies, but does not guarantee, that the cDNA sequence is correct and that deletions and frameshift mutations have not occurred during the cloning. Indeed, a candidate VEE clone lacked 102 nucleotides in the nonconserved carboxy terminus of nsP3 but was infectious.

The first infectious alphavirus cDNA clone contained the entire Sindbis virus genome downstream of an SP6 RNA polymerase promoter in a vector derived from pBR322 (Figure 2.31). Capped RNA transcripts could be synthesized

Figure 2.31

Rescue of infectious virus from a cDNA clone of Sindbis virus. The plasmid pToto51 was linearized at a unique *Xho*I restriction endonuclease site downstream of the poly A tail, and RNA transcribed *in vitro* using SP6 polymerase. RNA can be introduced into susceptible cells by a variety of techniques including transfection after treatment with DEAE dextran, lipofection, and electroporation. After transfection, the cells were overlayered with solid medium and incubated until visible plaques appeared. Individual plaques were picked and used to infect new cells under liquid medium to make virus stocks.

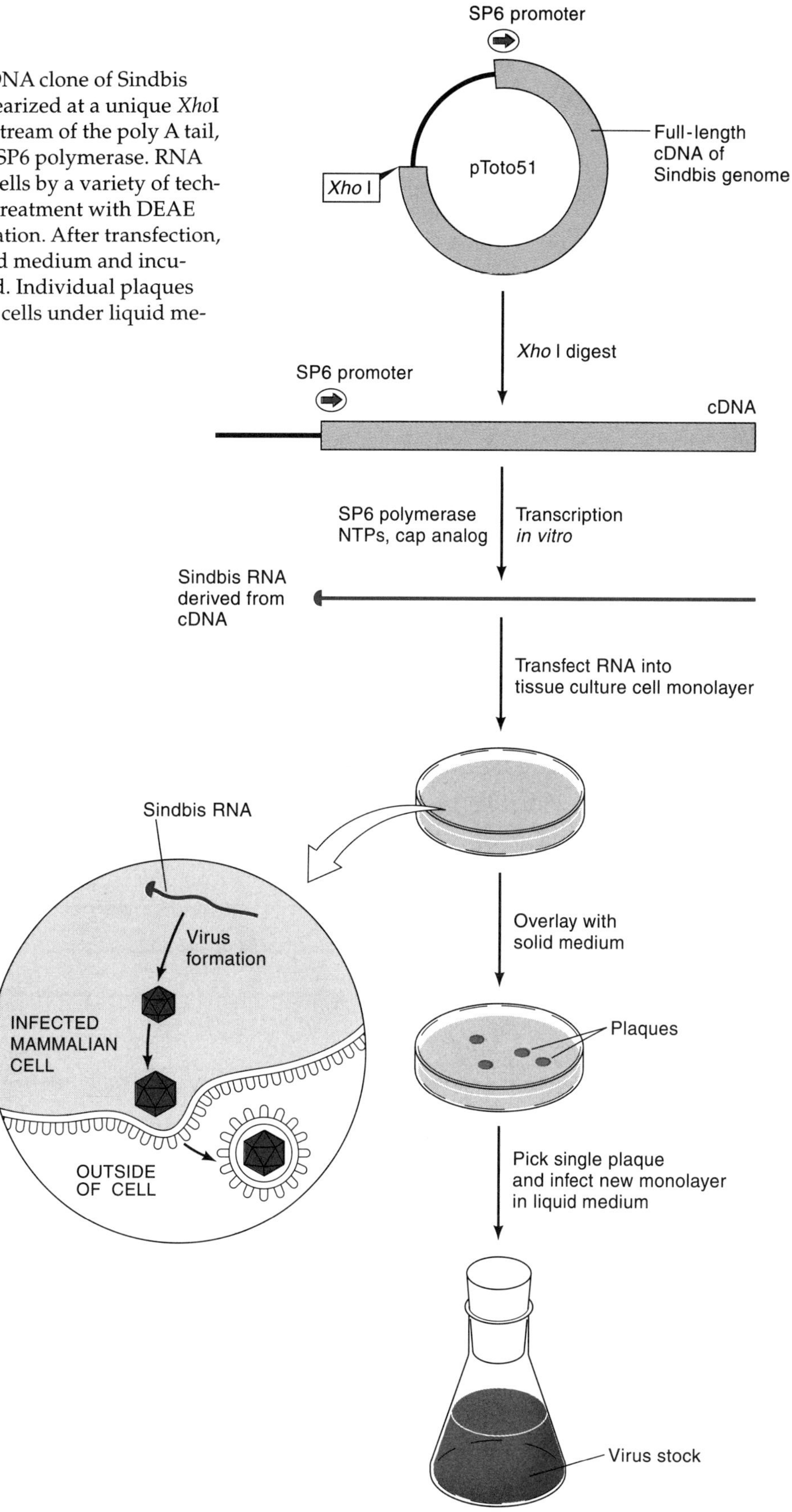

in high yield from this plasmid, and 10 percent or more of the transcripts were full length. The specific infectivity of the (full-length) transcripts was low, about 10^{-8} infections per RNA molecule or about 10 percent that of RNA extracted from virions and transfected under the same conditions with DEAE dextran, due perhaps to the presence of an additional G residue at the 5′ end of the genome sequence. Improved methods of delivering RNA to cells (lipofection and electroporation) have yielded higher specific infectivities for both transcripts and virion RNA. A family of related clones engineered to have useful properties, such as strategically placed unique restriction endonuclease sites, now exists. In addition, many mutant clones, containing insertions, deletions, or point mutations in the 5′ UTR and 3′ UTR and clones containing mutations in all the viral polypeptides (like those in the nonstructural proteinase, Section 2.5b), now exist. Variant genomes with additional subgenomic promoters to accommodate foreign gene sequences (Section 2.7b) and chimeric genomes (discussed later in this section) have been constructed.

Mapping temperature-sensitive mutants and revertants Several temperature-sensitive mutants of Sindbis virus have been isolated and their phenotypes described in detail. In order to determine which polypeptides are responsible for particular virus-encoded properties, researchers mapped these mutations to sequences encoding particular polypeptides using the infectious clone of Sindbis virus.

To map the mutations in the nonstructural region, segments of viral cDNAs believed to contain a mutation of interest were substituted for the corresponding region of a wild-type infectious clone, and the phenotype of the resulting virions was determined after transfection (Figure 2.32). To start, large substitutions con-

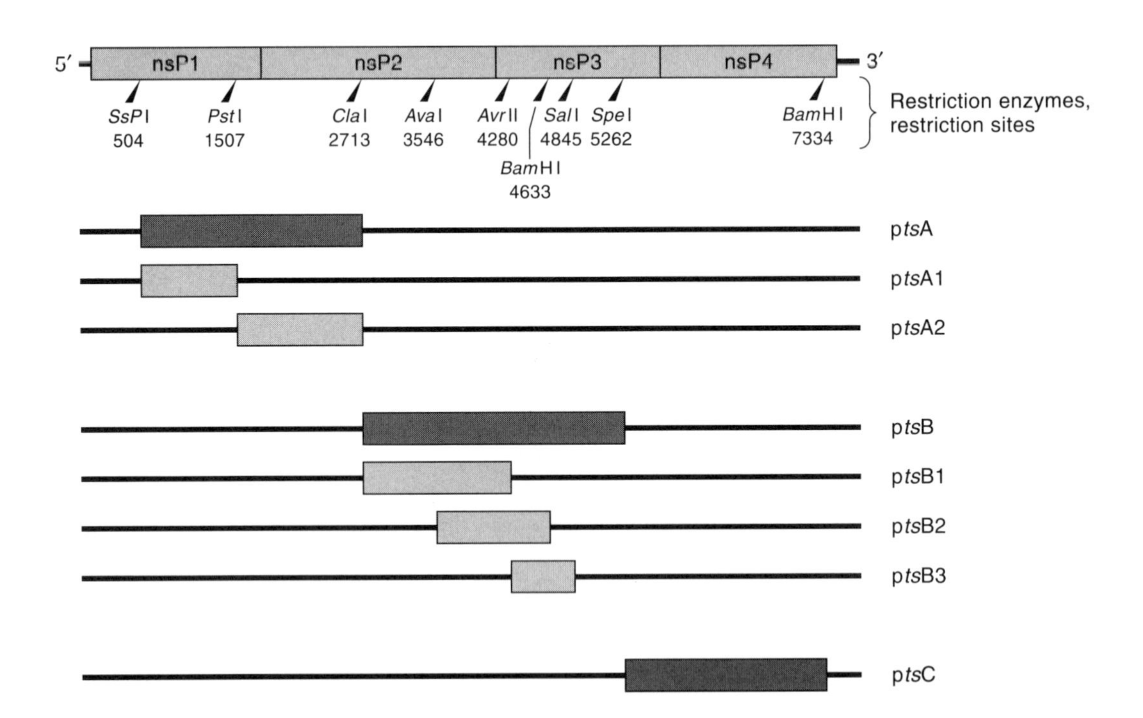

Figure 2.32
Mapping of temperature-sensitive mutants by the construction of hybrid genomes. The top line is a diagram of the region of the Sindbis virus infectious clone that encodes the nonstructural proteins. The diagram shows the boundaries of the nonstructural proteins and the locations of a number of restriction sites used to make the recombinants. To map a mutant, large fragments of the mutant or revertant genome (dark gray) are swapped into the infectious clone, virus is recovered, and its phenotype determined. When it has been determined that the mutation is located within a particular domain (e.g., A) shorter fragments (light gray boxes, p*ts*A1, p*ts*A2, etc.) are replaced for finer mapping.

taining up to one-third of the entire nonstructural region were tested (dark grey blocks). If one of the constructs conferred temperature sensitivity for virus propagation, smaller domains of this region were substituted in order to localize the mutation to a smaller region. When mutations had been localized to a fragment of no more than a few hundred nucleotides, they could be identified by sequence analysis. The phenotypes of rescued virus containing only a single change in a common genetic background were then compared to those of the original mutants, which often contained numerous exogenous nucleotide changes.

Localization of second site revertants is an extension of this procedure. In this case, the full-length clone containing the mutation in a common background was used; restriction fragments generated from cloned revertants were sequentially substituted into this clone, and the recovered virus was examined for the revertant phenotype. Alternatively, to avoid cloning large numbers of revertants, cDNA was synthesized from the revertant virion RNA with reverse transcriptase, amplified by PCR, cleaved into discrete segments with appropriate endonucleases, and each inserted directly into a properly prepared mutant full-length clone. Characterization of revertants can be more difficult than localization of mutations because, although many temperature-sensitive mutations result from only a single nucleotide change, revertants may arise from several suppressor mutations, each of which alone produces only partial reversion or none at all.

Mapping RNA regulatory regions As noted in Section 2.2a, there are four regions in the Sindbis genome in which the nucleotide and amino acid sequences are conserved among alphaviruses. All four have been altered by site-specific mutagenesis and deletion mapping, and the phenotypes of the resulting mutants have been examined in different cell culture lines as well as in animals. Most alterations in the 5′ UTR and the 3′ UTR produced viable mutants, which was surprising, considering the strong conservation in these regions. However, the effects of many of these mutations depended on the host cell. This implies that the conserved sequences present in natural isolates reflect an optimal compromise between the most efficient promoters for vertebrate and invertebrate cells because in nature these viruses undergo an obligate alternation between replication in mammalian or avian hosts and replication in mosquito vectors. The fact that mutations in these regions have host-dependent effects suggests that host components interact specifically with these elements during replication. Some of the mutations in the UTRs caused reduced multiplication in murine cells and concomitant attenuation of disease in mice, suggesting that systematic alteration of these domains could be used to produce attenuated, stable variants suitable for vaccines.

Chimeric viruses Using infectious clones, molecular biologists have been able to construct chimeric viruses in order to study the interactions between various parts of the genome. Viable chimeric viruses have been constructed from infectious clones of alphaviruses (Sindbis virus/Ross River virus) between picornaviruses (poliovirus/coxsackie B virus) and between plant bromoviruses (brome mosaic virus/cowpea chlorotic mottle virus). The exchanges in the plant virus chimeras involved the sequences encoding the replicase component in RNA2; the picornavirus chimeras involved the introduction of the long 5′ UTR of coxsackie B virus into the poliovirus RNA.

Alphavirus chimeras were constructed from infectious clones of Ross River (RR) and Sindbis (SIN) viruses. One such set consisted of viruses in which the 5′ UTR of Ross River was replaced with that of Sindbis virus (5′ SIN/RR) and vice versa (Figure 2.33). Other constructs replaced the 3′ UTR of Ross River with that of Sindbis (RR/3′ SIN) and vice versa. Both UTRs are thought to function as promoter elements for minus- and plus-strand synthesis, respectively, and thus to interact with the virus-encoded replicase components. Although there are con-

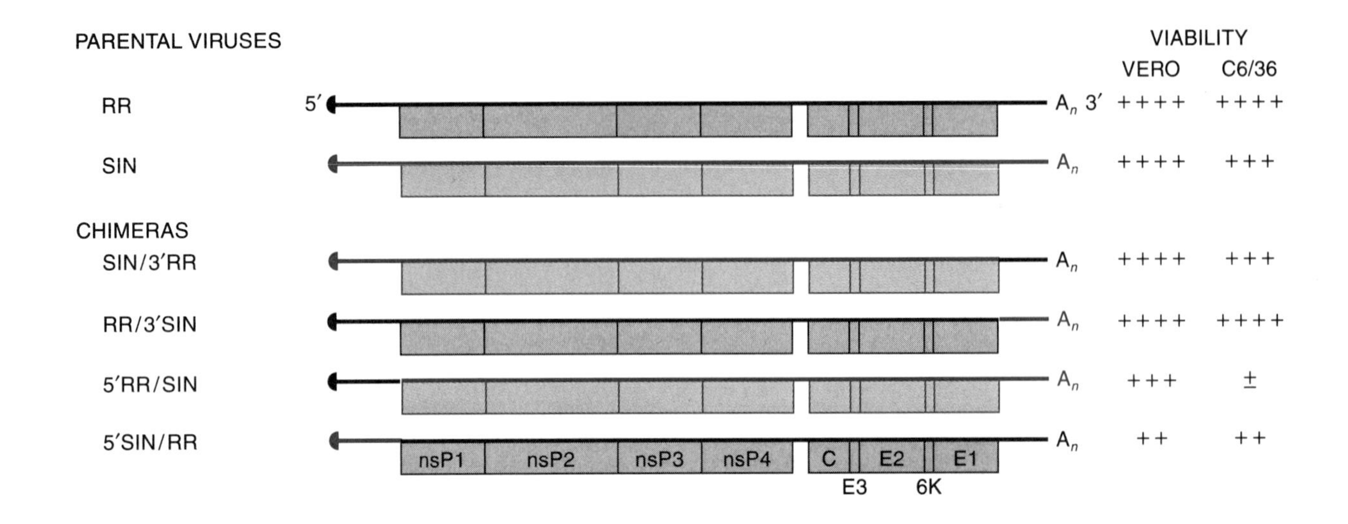

Figure 2.33
Chimeric viruses in which the 5′ and 3′ NTRs of Sindbis and Ross River viruses have been exchanged. The SIN 5′ UTR is 59 nucleotides long and the 3′ UTR is 322 nucleotides long; the RR 5′ UTR is 78 nucleotides long and the 3′ UTR is 524 nucleotides long. The parental viruses are shown, with RR in black and SIN in color. Chimeras were made in which the entire 5′ UTR of SIN was substituted in RR for its own 5′ UTR and vice versa, as shown below. The relative viability of each chimera in mammalian cells (VERO) and in mosquito cells (C6/36) is indicated by the number of plus signs in the right-hand columns.

served elements in these domains, they differ in both length and sequence between the two viruses. Surprisingly, all four chimeras yielded infectious virus. Exchange of the 3′ UTRs gave virus that replicated identically to its parent virus in both vertebrate and invertebrate cells (compare RR and RR/3′ SIN in Fig. 2.33). Exchange of the 5′ UTRs gave chimeras that replicated less efficiently than either parental virus, and this effect was more pronounced in mosquito cells, where, for example, 5′ RR/SIN grew 10^{-5}-fold as well as SIN. These results indicate that the 5′ and 3′ UTRs do interact with the encoded replicase proteins and with host cell factors.

2.7 RNA Viruses as Vectors

RNA viruses, particularly those for which naked RNA is infectious, are currently being developed as expression vectors in eukaryotic cells. Retroviruses have already been widely used as vectors (Chapter 3). In most RNA virus vector systems, RNA transcripts of appropriately substituted cDNA clones are synthesized *in vitro,* and then the RNA is introduced into cells. With the alphavirus vectors discussed in this section, molecular biologists are attempting to encapsidate the RNAs expressing heterologous genes to form "infectious" particles.

One advantage of plus-strand RNA viruses is that their replication is wholly cytoplasmic and thus aptly suited to the expression of gene products from cDNA copies of mRNAs, in which no splicing functions are required or desired. However, due to the inherent instability of RNA viruses (but not their cDNA counterparts), these viruses are generally useful only for transient expression.

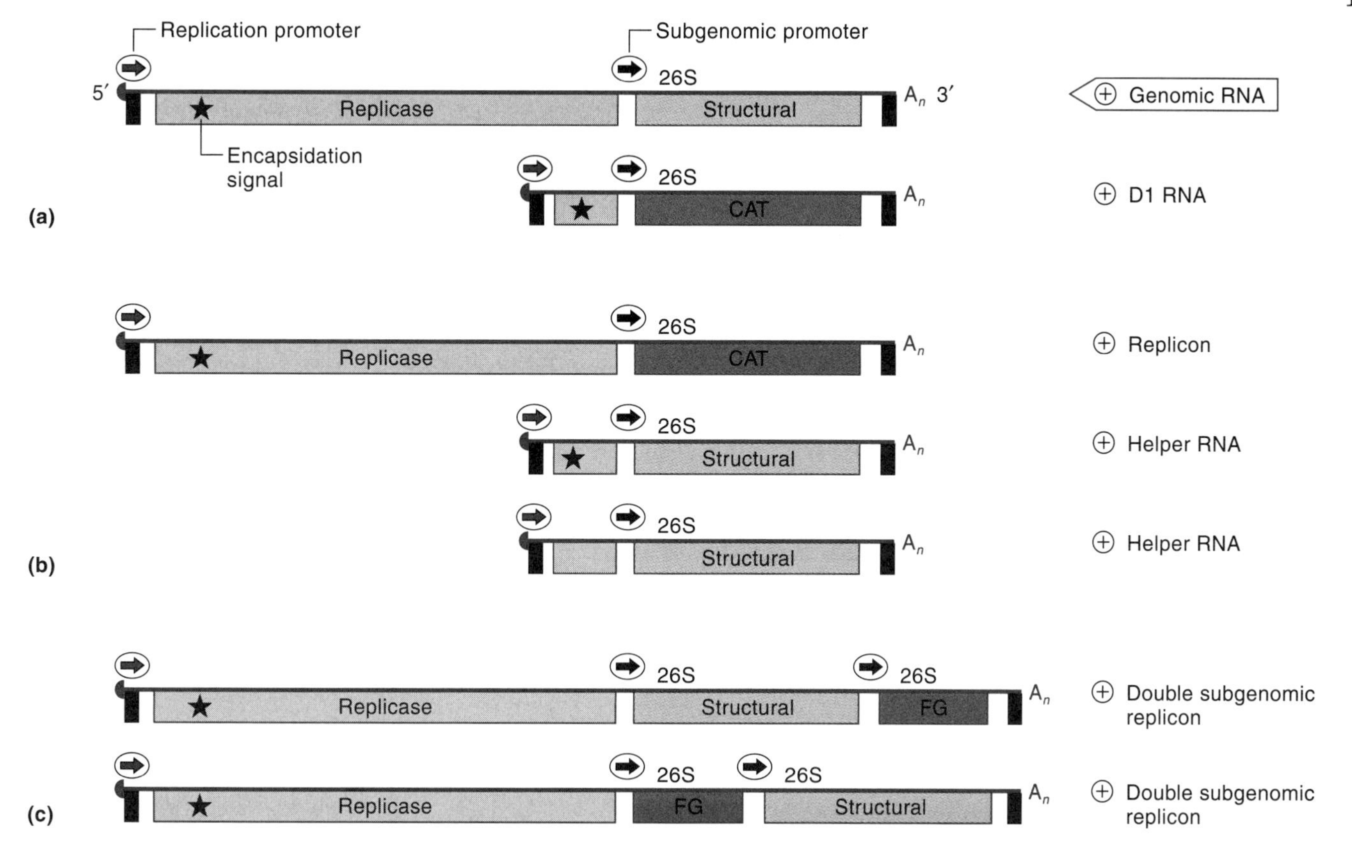

Figure 2.34
Alphavirus expression vectors. (a) The simplest expression system, in which wild-type alphavirus (labelled genomic RNA) and a DI expressing a foreign gene (CAT) under the control of the SIN subgenomic promoter are infected simultaneously into the same cell. (b) A replicon system in which the foreign gene has been inserted in place of the structural proteins in the SIN clone. The replicon can be used alone for transient expression, or can be packaged with either of the two helper RNAs shown. If the helper also contains the encapsidation signal (star) the virus yield will contain particles with a bipartite genome. If the helper lacks the encapsidation signal, theoretically the yield of packaged replicon will be helper-free. (c) Double subgenomic replicons, which contain the entire genome of Sindbis virus and a foreign gene (FG) inserted either 3′ or 5′ of the structural genes, under the control of a second copy of the 26S subgenomic promoter.

a. Prospects for Plant Virus Vectors

Most plant viruses are RNA viruses; it is therefore possible to find an RNA virus with almost any host range desired for expression of a heterologous gene. Plant viruses make large amounts of RNA replicases and RNA after infection. Moreover, studies of infectious transcripts of plant viruses have identified useful promoter elements. Many *cis*-acting elements have been found by studying the replication of the BMV RNA3 in protoplasts simultaneously transfected with RNA1 and RNA2 that express the replicase.

Although many genes in RNA virus vectors have been expressed in plant cell protoplasts, there has been less success with whole plants. Recent discoveries on how viral polypeptides facilitate cell-to-cell spread within a plant by altering the physical structure of plasmodesmata might alleviate this problem. Viral vectors might be restricted to systemic infection of single plant species if the gene responsible for systemic infection by a particular virus vector is deleted from the virus but is expressed as a transgene in the intended host.

Plant protoplasts transfected with transcripts of cDNA clones of BMV-RNA1 and BMV-RNA2 will support the replication of BMV-RNA3; they will also transcribe and express the subgenomic BMV-RNA4. A limited number of host cell components is necessary for this replication, but carrying this analysis further has been difficult due to the lack of molecular markers on plant chromosomes. *Cis*-acting elements necessary for transcription of RNA3 (which encodes the movement protein) and subgenomic RNA4 (which encodes the coat protein) have been identified (Figure 2.9). When foreign genes were inserted into BMV-RNA3 or BMV-RNA4 in place of the movement protein or the coat protein, respectively, the foreign protein was expressed in protoplasts. Moreover, yeast cells stably transformed with yeast plasmids expressing the proteins encoded in BMV-RNA1 and BMV-RNA2 synthesize chloroamphenicol acetyl-transferase (CAT) when transfected with BMV-RNA3 containing the CAT gene. Yeast cells clearly provide all the necessary host factors for BMV replication, and these factors can be identified by yeast genetics.

b. Alphavirus Vectors

A transient expression system based upon an alphavirus must contain the following elements: (1) the *cis*-acting elements necessary for efficient replication of the RNA, (2) a source of virus-specific replicase (the nonstructural proteins of the virus), and (3) the heterologous gene under the control of the *cis*-acting Sindbis transcriptional promoter. To encapsidate the expression vector into "infectious" particles requires these elements plus an encapsidation signal and a source of the structural proteins. How these elements are joined determines the characteristics of the system. Because alphaviruses in nature have such an extraordinarily wide host range, an infectious (encapsidated) vector based upon them could potentially be used for expression in almost all tissue culture cells. However, there appear to be limits on the size of RNAs that can be encapsidated, so the size of the inserted foreign gene may be limited.

The first alphavirus-based expression systems used a **defective interfering** (**DI**) genomic RNA into which the CAT gene had been inserted under the control of the subgenomic (26S) promoter. DI RNAs are functionally defective and can replicate only in the presence of helper or "standard" virus. Most of the coding sequence in DI RNAs has been deleted, but they contain such efficient *cis*-acting replication signals that they interfere with the replication of the wild-type helper virus and ultimately become the predominant RNA in infected cells. They also contain an encapsidation signal and can be encapsidated by structural proteins provided by the helper virus. Thus, DI RNAs can be introduced into cells by infection rather than transfection. However, the yield from infection will always be a mixture of DI and helper virus (Figure 2.34a).

Modifications of this arrangement include putting the CAT gene in place of the genes encoding the structural proteins in replication-competent helper virus; this necessitates supplying the structural proteins on a second replicon (Figure 2.34b). If both components contain a packaging signal, the virus stocks produced will have a bipartite genome. If the helper virus encoding the structural proteins does not contain a packaging signal, then "helper-free" progeny viruses are produced. This is then a suicide vector, which can infect and replicate in the next host and be expressed at high levels but cannot be transmitted because it cannot be packaged. However, the yield must be tested for infectious virus arising by recombination in the first transfection. Another approach is to construct nondefective virus vectors with two subgenomic promoters, one controlling structural protein synthesis and the other controlling the heterologous gene (Figure 2.34c).

For these, the size limitation appears to be significant because double subgenomic promoter constructs with heterologous inserts of more than 2000 nucleotides have proven to be unstable. Many of the problems with these animal virus vectors are primarily technical, and these vectors hold promise for particular kinds of uses.

2.8 Conclusion

Recombinant DNA technology clearly has been enormously important for the study and understanding of animal and plant RNA viruses. It is now possible to sequence readily the genomes of these viruses and, from these nucleotide sequences and the deduced amino acid sequences, to understand their genome organization and learn a great deal about their encoded proteins. Comparisons of the sequences, at both the nucleotide and amino acid levels, have given insights into the evolutionary relationships among the virus families. cDNA clones have been especially useful in the study of viruses; for example, cloning of the hepatitis C virus genome has permitted studies of its molecular biology, even though the virus cannot yet be grown in cell culture. Once a genomic RNA has been reverse transcribed into cDNA, it can be amplified either by PCR or in bacteria and put under the control of powerful promoters. It can be mutated or deleted in a number of ways. Then transcription and translation of the new RNA transcripts can be studied in wholly *in vitro* systems. With infectious clones, many changes can be introduced into the DNA and tested for their effects on viral replication in cell culture and pathogenesis in animals.

New technologies are constantly being developed to detect the presence of viruses with greater sensitivity and specificity. PCR, in particular, is becoming a valuable diagnostic tool for detecting the presence of viral nucleic acids in patient samples. Similarly, amplification of relatively short sequences of dozens of isolates of a given virus using PCR and sequence analysis of the products is very valuable for studying viral variation and dispersion in nature. With this technique, one can characterize sequential isolates during the course of an epidemic, compare geographical isolates to trace the origins of newly introduced strains, or follow sequence changes occurring with passage or even over the time course of an illness in a single individual patient.

Technologies to produce a new generation of vaccines and antiviral therapies are being developed. Avirulent viruses expressing antigens from related or unrelated pathogens are being explored. With greater understanding of the domains, motifs, and, in some cases, particular residues responsible for virulence, modifying virulent viruses specifically to produce avirulent but highly immunogenic vaccines will be possible. The more we understand the replication of a virus on a molecular level, the more we will be able to abort viral infection and reduce viral disease by selective interference with particular steps in viral replication.

Further Reading

2.1

E. G. Strauss and J. H. Strauss. 1983. Replication Strategies of the Single Stranded RNA Viruses of Eukaryotes. *Curr. Topics Micro. Immun.* 105 1–98.

W. K. Joklik. 1985. Recent Progress in Reovirus Research. *Ann. Rev. Genet* 19 537–575.

R. W. Goldbach. 1986. Molecular Evolution of Plant RNA Viruses. *Ann. Rev. Phytopathol.* 24 289–310.

2.2

T. J. Chambers, C. S. Hahn, R. Galler, and C. M. Rice. 1990. Flavivirus Genome Organization, Expression, and Replication. *Ann. Rev. Microbiol.* 44 649–699.

M. M. C. Lai. 1990. Coronavirus: Organization, Replication and Expression of Genome. *Ann. Rev. Microbiol.* 44 303–333.

H.-J. Lee, C.-K. Shieh, A. E. Gorbalenya, E. V. Koonin, N. La Monica, J. Tuler, A. Bagdzhadzhyan, and M. M. C. Lai. 1991. The Complete Sequence (22 kilobases) of Murine Coronavirus Gene 1 Encoding the Putative Proteases and RNA Polymerase. *Virology* 180 567–582.

P. Ahlquist. 1992. Bromovirus RNA Replication and Transcription. *Curr. Opin. in Genet. Dev.* 2 71–76.

J. H. Strauss and E. G. Strauss. 1994. The Alphaviruses: Gene Expression, Replication and Evolution. *Microbiol. Revs.* 58 491–562.

2.3

N. Tordo, P. De Haan, R. Goldbach, and O. Poch. 1992. Evolution of Negative-Stranded RNA Genomes. *Sem. in Virology* 3 341–358.

2.5

P. P. Abel, R. S. Nelson, B. De, N. Hoffmann, S. G. Rogers, R. T. Fraley, and R. N. Beachy. 1986. Delay of Disease Development in Transgenic Plants That Express the Tobacco Mosaic Virus Coat Protein Gene. *Science* 232 738–743.

T. R. Fuerst, E. G. Niles, F. W. Studier, and B. Moss. 1986. Eukaryotic Transient-Expression System Based on Recombinant Vaccinia Virus That Synthesizes Bacteriophage T7 RNA Polymerase. *Proc. Natl. Acad. Sci. USA* 83 8122–8126.

A. H. Rosenberg, B. N. Lade, D.-S. Chui, S.-W. Lin, J. L. Dunn, and F. W. Studier. 1987. Vectors for Selective Expression of Cloned DNAs by T7 RNA Polymerase. *Gene* 56 125–135.

2.6

V. R. Racaniello and D. Baltimore. 1981. Cloned Poliovirus Complementary DNA Is Infectious in Mammalian Cells. *Science* 214 916–919.

W. O. Dawson, D. L. Beck, D. A. Knorr, and G. L. Grantham. 1986. cDNA Cloning of the Complete Genome of Tobacco Mosaic Virus and Production of Infectious Transcripts. *Proc. Natl. Acad. Sci. USA* 83 1832–1836.

C. M. Rice, R. Levis, J. H. Strauss, and H. V. Huang. 1987. Production of Infectious RNA Transcripts from Sindbis Virus cDNA Clones: Mapping of Lethal Mutations, Rescue of a Temperature Sensitive Marker, and in Vitro Mutagenesis to Generate Defined Mutants. *J. Virol.* 61 3809–3819.

J. J. Bujarski and W. A. Miller. 1992. Use of *in Vitro* Transcription to Study Gene Expression and Replication of Spherical, Positive Sense RNA Plant Viruses. In Genetic Engineering with Plant Viruses, pp. 115–147. CRC Press, Boca Raton, Florida.

2.7

P. J. Bredenbeek and C. M. Rice. 1992. Animal RNA Virus Expression Systems. *Sem. in Virology* 3 297–310.

P. Janda and P. Ahlquist. 1993. RNA-Dependent Replication, Transcription, and Persistence of Brome Mosaic Virus in *S. cerevisiae*. *Cell* 72 961–970.

CHAPTER 3

Retroviruses: Infectious Genetic Elements

STEPHEN P. GOFF

3.1 Introduction

Retroviruses, originally termed RNA tumor viruses, consist of a large class of viruses whose genomes are RNA and replicate through a DNA intermediate. These viruses were discussed briefly in *Genes and Genomes* with respect to their utility as vectors and their relationship to transposable elements. This chapter provides a more comprehensive overview of the molecular biology of these viruses. Beyond their inherent interest to virologists, who have been intrigued by their bizarre life-style, retroviruses have broader significance. As mentioned in several chapters in *Genes and Genomes,* they are examples of genetic elements undergoing complex, regulated gene expression (chapter 8), elements contributing to genomic rearrangements (chapter 10), and of important tools for gene transfer (chapter 5). These viruses are also important as the etiologic agents of the disease now called AIDS, which is in epidemic growth worldwide. The study of retroviruses has become a vivid example of how basic research ultimately leads to important applications in biotechnology and medicine.

a. Discovery and History

The first retrovirus to be studied intensively was a tumor virus of birds, isolated in 1911 by Peyton Rous. He was able to demonstrate that cell-free filtrates of extracts from a spontaneous chicken sarcoma could induce in inoculated birds the growth of tumors similar to the original isolate. The virus responsible, ultimately termed the Rous sarcoma virus (RSV), was much later found to be able to induce the formation of morphologically transformed cells in chick fibroblast cell cultures, permitting quantitative measurements of the virus titer present in a preparation. In the 1950s, the entities exhibiting these biological properties were first visualized and characterized as spherical particles, 70–80 nm in diameter, surrounded by a lipid-containing membrane. Infected cells were shown to assemble the particles at the cell surface, budding off the progeny virions into the extracellular space without inducing cell lysis. The cells did not die after infection, but simply became producers of virus while continuing to grow and divide.

The discovery of reverse transcription Analysis of the virus particles revealed that the genetic material within them was RNA. The natural assumption, based on the precedents established by the study of other viruses, was that these viruses replicated their RNA by RNA polymerases in the infected cell. Researchers did not expect inhibitors of DNA synthesis, or of DNA-dependent RNA synthesis, to inhibit virus replication. But experiments in the early 1960s, testing the effects of various inhibitors on RSV replication, led to unexpected results. The addition of bromodeoxyuridine, an inhibitor of DNA synthesis, soon after infection blocked transformation of the cells and replication of the virus; similarly, addition of actinomycin D, an inhibitor of transcription, immediately blocked production of virus in infected cells. These results led to the proposal that the RNA in the infecting particle was converted to DNA, which then served as template for the formation of new viral RNA. The intracellular DNA form of the viral genome, termed the provirus, was proposed to integrate into one or more of the host chromosomes, by analogy to the behavior of phage λ. A crude version of the "provirus hypothesis," as the idea was named, is shown in Figure 3.1.

This unconventional life cycle required some mechanism by which RNA could be copied into DNA and thus flew in the face of the "central dogma" of molecular biology, which specified that genetic information flows from DNA to RNA to protein as it is expressed in the cell. The proposed reverse flow of information, from RNA to DNA, was widely greeted with skepticism until the presence of reverse transcriptase, an RNA-dependent DNA polymerase activity, in virions of RSV and other RNA tumor viruses, was discovered. As appreciation of the fundamental importance of the "backwards" or retrograde steps of the life cycle of the viruses increased, the new term **retroviruses** was coined to replace the name RNA tumor viruses. The term retroviruses is now applied to all viruses that use this kind of replication cycle, whether or not the viruses cause tumors.

Isolation of many different retroviruses During the 1950s, several mammalian viruses were isolated as filterable agents capable of inducing leukemias in mice. These viruses, collectively known as murine leukemia viruses, were named after

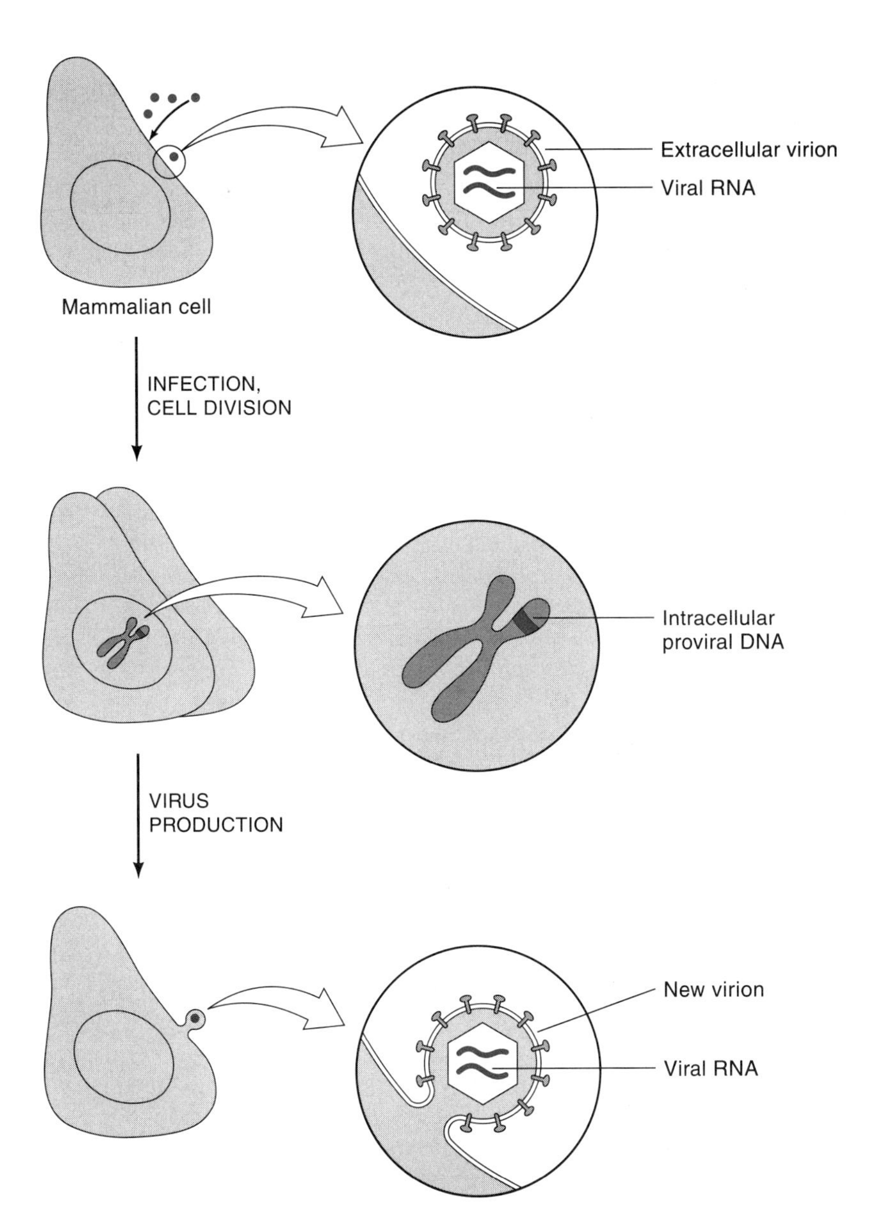

Figure 3.1
The provirus hypothesis in retrovirus replication. The hypothesis proposes that extracellular virions containing their genomic information in the form of RNA enter the cell (top), and then generate DNA copies that are retained in the infected cell and transmitted like a host gene to daughter cells during cell division (middle). The infected cell expresses the DNA to form viral RNAs and proteins that assemble into progeny virus (bottom). The complete infectious cycle thus requires conversion of the viral genetic information from RNA to DNA and from DNA back to RNA.

their discoverers; examples include the Gross virus, Friend virus, Moloney virus, and Rauscher virus. All these were able to induce the formation of leukemias when injected into newborn mice after latency periods of several months. In the 1960s, as these viruses were passed through mice and rats, new isolates were obtained that induced distinctive tumor types, in many cases sarcomas, with much shorter latency periods than those exhibited by the parental viruses. The mammalian viruses had one important advantage over the avian viruses: immortal cell lines that were susceptible to infection and to morphological transformation were readily obtained and cultured. Even more important, these cells could be plated at high dilution and clones of single cells obtained, permitting the biological cloning of the viruses that infected them. (To this day, well-behaved avian cell lines are rare.) As the search for tumor viruses widened, similar viruses were found in a variety of mammals, especially cats. The study of feline leukemia viruses has provided a large number of new isolates of transforming viruses. Since these early days, retroviruses have been isolated from many vertebrates, including reptiles, fish, birds, and many mammals, including primates and, not surprisingly, humans.

b. Taxonomy: Various Classification Systems

As the number of isolates of distinct viruses grew, various attempts were made to impose order and to classify these agents into groups. Some efforts were made to group them by the morphology of the virion particles as visualized under the electron microscope (a-type, b-type, c-type, d-type), by their antigenic cross-reactivity, by the type of tumor they induced, and by the species of host they could infect. Some of these classifications have been historically useful, but many are now considered obsolete or at least no longer useful in establishing functional or evolutionary relationships among the viruses. Two very fundamental groupings of the viruses have passed the test of time as meaningful aids to understanding their biological properties and will be used here.

Host range A key classification of retroviruses is based on their host range, their ability to infect particular cells. There are three levels of restriction of virus infection: the host species, the distribution of the cell surface receptors, and the phenomenon known as superinfection resistance.

The retroviruses are variously limited in the range of species they can infect. For example, the avian viruses can generally infect a wide array of avian species, such as chicken, quail, turkey, or duck, but not more distantly related species of animals. Similarly, the mammalian retroviruses are generally able to infect only mammals, and in many cases only those mammals close to the species of origin. Thus, grouping viruses by the species of origin is a meaningful way to organize them. A listing of some viruses grouped by host range is given in Table 3.1.

Table 3.1 ***Examples of Retroviruses Grouped by Host Range***

Virus Groups	Hosts
Avian retroviruses	
Subgroups A–E	Chicken, quail, duck
Subgroups F, G	Pheasant
Reticuloendothelial viruses	Chicken, other birds, some mammals
Murine leukemia viruses	
Ecotropic	Mouse, rat
Xenotropic	Mink, human, other mammals
Amphotropic	Most mammals
Mouse mammary tumor virus	Mouse
Feline leukemia viruses	Cat, human
HIV-1, -2	Human, chimpanzee

Note: Retroviruses can be grouped in many ways, but one classification system groups the viruses by their host range (i.e., the species they are able to infect). The host range depends on many specific interactions between viral and host gene products, including binding of the virus to the surface receptors on the host cell.

Subdivisions can also be discerned among the viruses isolated from a given species. Further analysis of the avian viruses, for example, revealed that chicken cells of different genetic backgrounds are differentially sensitive to different strains of a particular virus and that the susceptibility of a cell to infection by a given virus depends on the presence of a particular receptor for that virus. Studies of these systems allowed the definition of at least five different receptors on avian cells. Somewhat similar observations were made for the mammalian viruses. Some of the viruses isolated from mice are narrowly restricted to infecting cells of mouse, rat, and a few other rodents; these viruses are classified as **ecotropic** (eco-, for same) because they infect the same species as that of their origin. Other viruses isolated from mice are best able to infect nonmurine species such as mink or human cells. These viruses are termed **xenotropic** (xeno-, for other). Still other viruses can infect a very broad range of mammalian cells; these are termed **amphotropic** (ampho-, for all or ambivalent). These last viruses of very broad host range are very significant today for their planned use in gene transfer and gene therapy procedures in humans. The human retroviruses isolated to date are quite restricted in the species they infect. The human immunodeficiency viruses (HIVs), for example, infect only humans and very closely related primates. This restriction, like those of other retroviruses, is almost certainly at the level of the cell receptor for the HIVs, the CD4 surface molecule. The human T cell lymphotropic viruses (HTLVs) are also narrowly restricted to humans.

Resistance to superinfection Perhaps the sharpest subdivision of the retroviruses is provided by the phenomenon of **superinfection resistance**. As it became apparent in the early 1960s that preparations of transforming viruses were a mixture of helper viruses and replication-defective transforming viruses, researchers also realized that some cells previously exposed to virus but not successfully transformed were often resistant to subsequent infection (superinfection) by fresh virus. Ultimately, it became clear that infection of a cell by helper virus rendered the cell resistant to further infection, but only by viruses closely related to the initial one; infection by other viruses was not blocked. The pattern of superinfection resistance allowed for the definition of interference groups, that is, groups of viruses that all cross-interfered with each other. The steps in the life cycle blocked by superinfection are absorption and penetration of virions. The site of inhibition is the cell surface receptor, which is blocked by viral proteins synthesized within the cell. Thus, successful infection by a virus blocked infection by other viruses using the same receptor but not by viruses using a different receptor. Interference groups therefore define groups of viruses utilizing the same receptor. All the viruses in an interference group thus constitute a subdivision of the viruses that can infect a given host species.

Replication-competent vs. defective viruses Many of the acutely **transforming** viruses consist of mixtures of two kinds of viruses: a **replication-defective** virus capable of transforming cells but unable to induce the formation of progeny virus on its own and a **replication-competent** virus acting as a **helper** for the transmission of the defective partner. We now appreciate this pattern as extremely common. In fact, with the single exception of RSV, all the acute transforming viruses are replication-defective and depend on the presence of helper viruses for their replication. Only when a cell is simultaneously infected with both a transforming virus and a helper virus are progeny transforming viruses produced. Examples of replication-competent and replication-defective transforming retroviruses are given in Table 3.2.

This important concept of transmission of a defective virus is diagrammed in Figure 3.2. A cell infected with both a transforming and a helper virus will acquire each of the proviral DNAs in its chromosomes. Such a cell is capable of re-

Table 3.2 *Examples of Retroviruses Grouped by Their Ability to Replicate*

Replication-Competent Virus	Related Defective Viruses
Moloney murine leukemia virus (M-MuLV)	Moloney sarcoma virus (MSV)
	Kirsten murine sarcoma virus (KiSV)
	Harvey murine sarcoma virus (HaSV)
	Abelson murine leukemia virus (A-MuLV)
Feline leukemia virus (FeLV)	Gardner-Arnstein feline sarcoma virus (GA-FeSV)
	Gardner-Rasheed feline sarcoma virus (GR-FeSV)
	Hardy-Zuckerman 2 feline sarcoma virus (HZ2 FeSV)
Avian leukosis virus (ALV)	Avian erythroleukemia virus (AEV)
	Avian myeloblasosis virus (AMV)
	Avian myelocytomatosis virus (MC29)

Note: Many retroviruses, including the replication-competent leukemia viruses, can carry out all the steps of replication and thus can spread from cell to cell (entries on left). Other viruses, including the sarcoma viruses, are replication-defective (entries on right). The defective viruses depend on a replication-competent virus to help their transmission from cell to cell. In all cases, the defective viruses were derived from a related replication-competent virus by loss of one or more viral replication genes and substitution by host genetic sequences.

leasing virions because all the essential virion proteins are encoded by the helper virus genome. Both proviral DNAs are transcribed into RNA, and both RNAs are packaged as genomic information into virion particles; therefore, the virus preparation harvested from the cell contains a mixture of two RNA species. Both viruses can infect new cells and can be separated from one another if the infection occurs at a low multiplicity. Cells infected by the helper alone will simply produce more helper virus. Cells infected by the defective transforming virus alone will become transformed without producing any new virus. And if the infection is carried out at a high multiplicity, cells may be infected with both viruses. In this case, the cell becomes transformed and also becomes a virus-producing cell; it re-creates the doubly infected cell from which virus was collected originally. Only through these doubly infected cells can the replication-defective genome be replicated.

Figure 3.2 (*facing page*)
Cell-to-cell transfer of a replication-defective virus by a replication competent helper virus. A cell containing integrated proviral copies of both a replicaton-defective transforming virus and a replication-competent helper virus (top) expresses RNA and proteins from both genomes. The cell is morphologically transformed by expression of the transforming virus, and releases virus particles by expression of the helper virus. The virus particles contain genomic RNAs of both viruses. Upon infection, cells may receive the helper alone, becoming producers of nontransforming virus (left); or transforming virus alone, becoming transformed nonproducers (middle); or both viruses, becoming transformed producers similar to the starting cells (right). In this way the replication-defective virus can spread from cell to cell in concert with the helper.

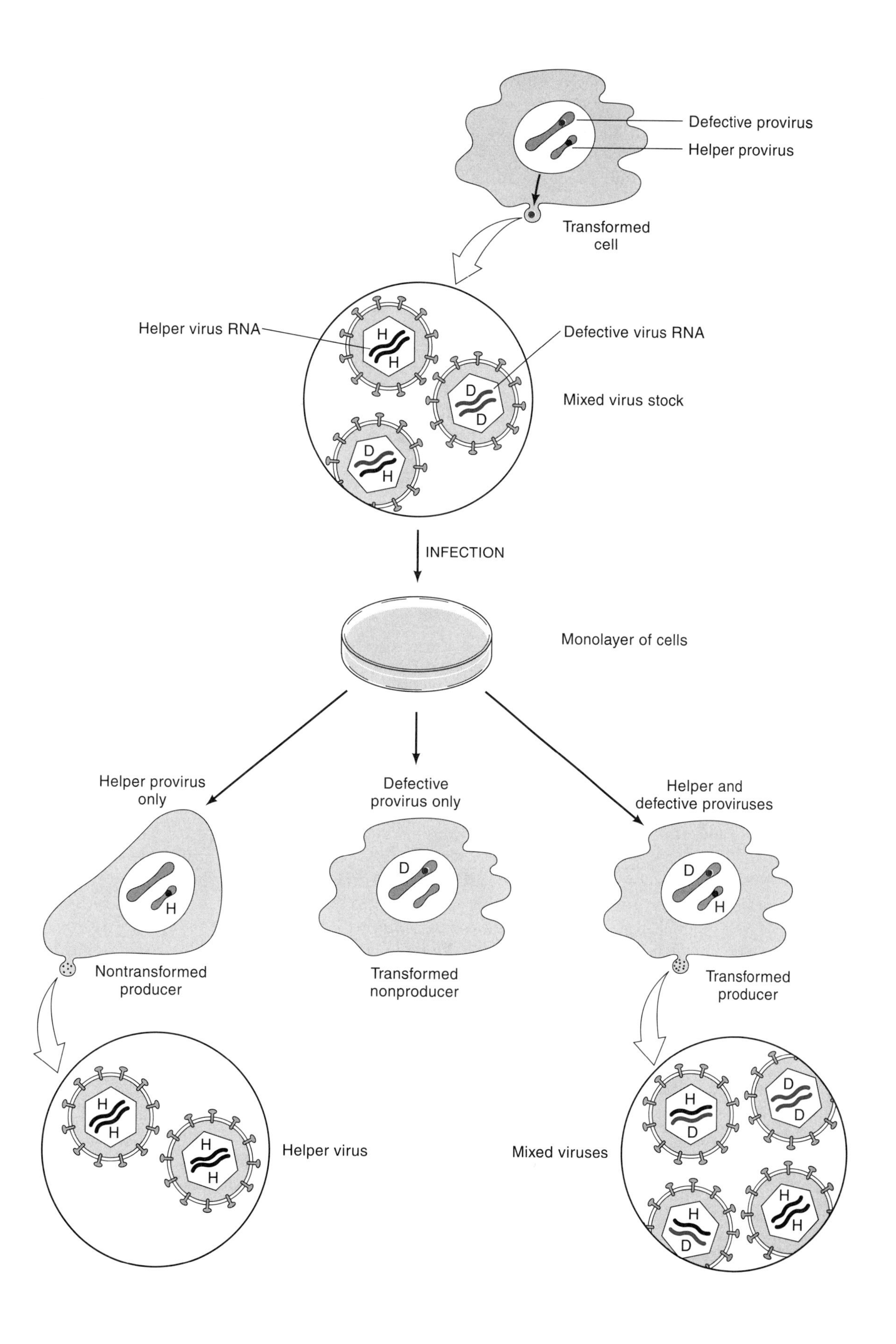
Defective provirus
Helper provirus
Transformed cell
Helper virus RNA
Defective virus RNA
Mixed virus stock
INFECTION
Monolayer of cells
Helper provirus only
Defective provirus only
Helper and defective proviruses
Nontransformed producer
Transformed nonproducer
Transformed producer
Helper virus
Mixed viruses

The distinction between replication-defective transforming virus and replication-competent helper virus is important because it dictates much of the biology of the resulting virus. The transforming viruses are generated from the helper viruses through a complex series of genomic rearrangements. They arise when sequences from the host genome are acquired by a virus; the new sequences almost always replace genes essential for replication. The loss of these essential genes accounts for the defectiveness of the resulting virus.

3.2 The Replication-Competent Retroviruses: Practically Perfect Parasites

Most of what we know about the replication of retroviruses we have obtained from the study of two classes of prototypical replication-competent viruses: the murine leukemia viruses and the avian leukosis viruses. These are such useful experimental subjects because they are relatively benign and the infection of cells in culture leads to no obvious effect on the cell. Moreover, their proviral DNA is expressed at high levels so that high titers of virus are released, an experimental convenience; a variety of probes for nucleic acid and protein structure is available; and the complete nucleic acid sequences of the genomes have been known for almost a decade.

a. Overview of the Life Cycle

The retroviral life cycle is divided into two phases: an early phase, in which the virus enters the cell and establishes the proviral state in which the DNA is integrated in the host genome, and a chronic late phase, in which the integrated DNA is expressed to give rise to progeny. The early phase is diagrammed in Figure 3.3.

Binding and entry Retroviruses absorb to the cell surface and enter the cytoplasm by means of cell surface receptors, probably using mechanisms in common with many other viruses. In most cases, the identity and nature of the receptor are not known. There are exceptions: the receptor for HIV is known to be the CD4 surface protein of human T lymphocytes; the receptor for the ecotropic murine leukemia viruses is known to be a membrane transporter for basic amino acids; and the gene encoding the receptor for one of the avian retroviral subgroups has recently been cloned. Each of these host proteins is likely to play a significant role in the life of the cell, and the viruses may have evolved to utilize these proteins as receptors without regard for those roles. These various proteins seem to have little in common, and there may be no general rules about their structure or function. In all cases, however, the virions bind to the receptors through the envelope protein, a virus-encoded glycoprotein on the virion surface.

After binding, the envelope (env) protein mediates the fusion of the virion and cell membranes, either at the cell surface or after internalization. For some retroviruses, notably HIV-1, the fusion event can occur at neutral pH at the cell surface; for others, probably Moloney MuLV, the fusion may be activated by the acidic pH found in intracellular endosomes. In both cases, the env protein is likely an active catalyst in the fusion process. The fusion event may require the recognition of other host proteins by the env protein, but little is known about the requirements for fusion. During the process, the virion and host membranes fuse, and the virion membrane turns inside out or everts; the membrane is left behind at the cell surface, and a membrane-free core particle is released into the cytoplasm. The complete list of the virion proteins retained on the intracellular

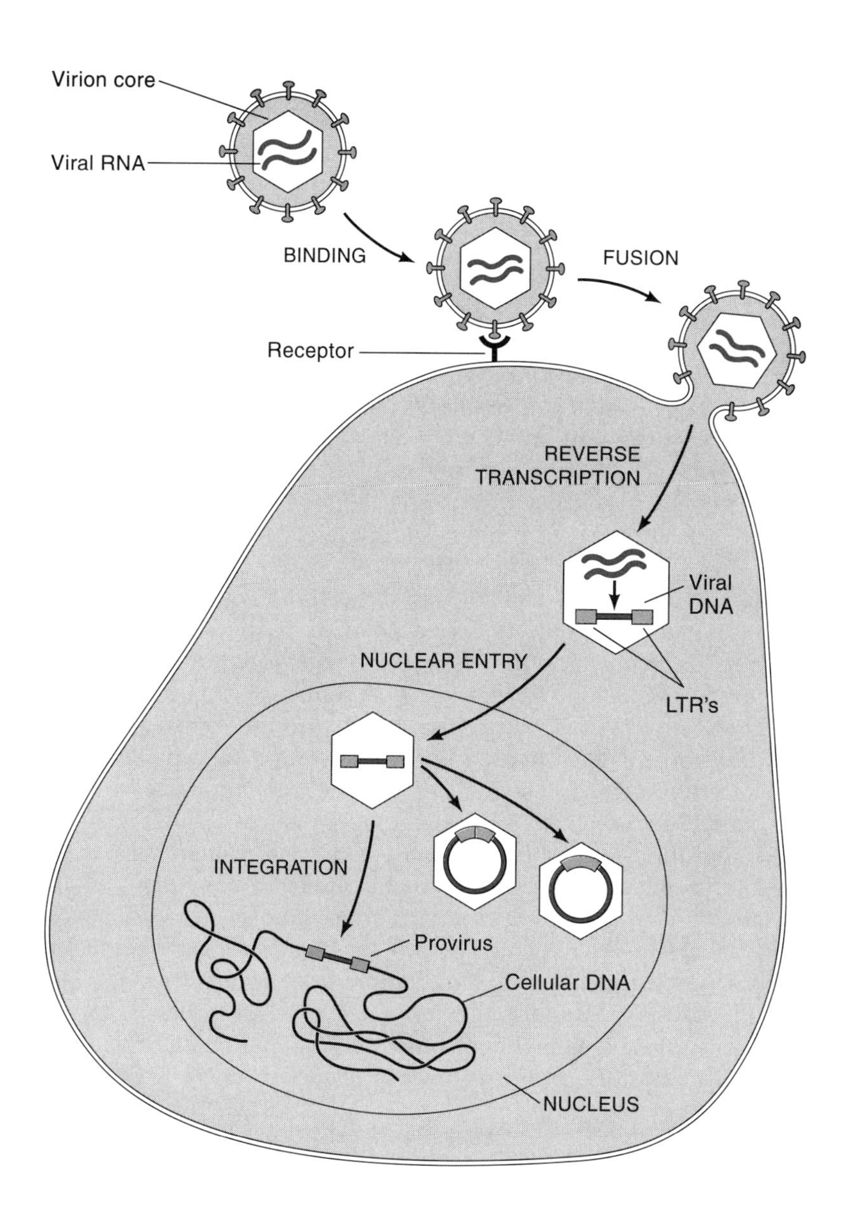

Figure 3.3
The early phase of the viral life cycle. Virion particles containing genomic viral RNA bind to receptor molecules on the cell surface, are internalized into the cytoplasm, and direct the reverse transcription of the viral RNA into linear DNA. The DNA enters the nucleus, and the linear DNA is stably integrated into the host genome, which forms the provirus.

particle is not known, but at a minimum, it is thought to include one or more structural proteins, reverse transcriptase, and the enzyme required for integrating the viral DNA.

Reverse transcription: from RNA to DNA Once in the cytoplasm, the virion core mediates the crucial reverse transcription reaction: the synthesis of viral DNA from the viral RNA. By convention, the virion RNA is defined as the (+) strand sequence; the reaction thus consists of the formation of both a (−) strand DNA, complementary to the viral RNA, and a (+) strand DNA, complementary to the (−) strand DNA and therefore of the same sequence as the viral RNA. The process is complex and involves many discrete intermediates.

The viral DNA synthesized early in infection is not a simple copy of the RNA brought into the cell. During the reverse transcription process, important sequence blocks termed long terminal repeats, or LTRs, are assembled from sequences dispersed in the virion RNA template. At the end of the reaction, these LTRs are arranged in the same or direct orientation and flank the open reading frames of the viral genome. Virtually all the key sequences for reverse transcription, integration of the viral DNA, and expression of the viral RNA are contained within the LTRs.

Integration of the viral DNA In the next critical step in the life cycle, the viral DNA is integrated into the host genome. This reaction is catalyzed by a virus-encoded protein, termed integrase, that enters the cell within the virion core and remains with the genome throughout the early phase of the life cycle. Integrase acts directly on the linear duplex viral DNA genome and inserts it into the host DNA to establish the integrated provirus. The choice of target site in the host genome is to a first approximation random; any sequence seems to be able to act as a recipient site, and targets on all chromosomes are certainly utilized. Recently, "hot spots"—sites of frequent insertion—have been identified for the avian retroviruses, but the combined use of all hot spots constitutes only a minority of all insertion events. Insertions are essentially permanent; there is no mechanism by which a provirus is normally excised. The provirus DNA becomes a permanent part of the infected host's genome for the rest of the life of the cell and of its progeny. The retroviruses therefore differ in a signficant way from the phages that integrate their DNA into their bacterial host's genome: such lysogenic phages can efficiently excise their DNA for vegetative replication.

Insertion is an extremely efficient and orderly process, not at all like the integration of DNAs mediated by DNA transformation procedures. Retroviral integration is so efficient that it is common for every cell in an infected population to undergo a clean insertion event. The development of methods for cloning genomic DNA, mapping DNAs, and determining DNA sequences has allowed the recovery and structural analysis of many integrated proviral DNAs. The inserted DNAs are always arranged in an identical way; the proviral sequences are in the same order, and the same viral sequences, namely, the outer edges of the LTRs, are always joined to the host. No host sequences are lost or grossly rearranged at the integration site. The insertion event is associated with very specific sequence alterations at the joints; a small number of base pairs, characteristic of each retrovirus and ranging from 4 to 6 bp, that were present only once at the preintegration site, is duplicated during the reaction so as to flank the ends of the viral DNA. The duplication of a small number of base pairs at the target site of retroviral insertions is remarkably similar to the duplication of target sites for transposition of mobile elements in other settings (*Genes and Genomes*, chapter 10). The integration of the provirus marks the end of the early phase of the retroviral life cycle.

Viral gene expression: gag, pol, *and* env Once inserted into the host chromosome, the provirus can be expressed by conventional mechanisms of transcription and translation to yield the viral gene products (Figure 3.4). For many retroviruses, that expression is simple. Transcription is initiated by a strong promoter for RNA polymerase II and results in the formation of a continuous RNA copy of the viral genome. With some important exceptions, the retroviral promoter is constitutively active and results in the unregulated expression of viral RNA at high levels. The RNA, like most host mRNAs, is capped at the 5′end and polyadenylated at the 3′end; all signals specifying these structures are similar to signals on host genes.

Some of the primary transcript is subsequently spliced to give rise to one or more subgenomic mRNAs, and the various mRNA species are exported to the cytoplasm and translated into viral proteins. The details of these processes differ for the different viruses, but certain features are invariant. All replication-competent retroviruses direct the synthesis of three major groups of proteins, encoded in three separate open reading frames termed gag, pol, and env. The *gag* gene, named for historical reasons for **g**roup-specific **a**nti**g**en, encodes the major proteins in the virion capsid or core; the *pol* gene, for **pol**ymerase, encodes the enzymes needed for replication of the viral genome in the virion; and the *env* gene, for **env**elope, encodes the transmembrane glycoprotein found in the virion envelope. These three open reading frames are arranged along the genome from 5′ to 3′ with respect to the direction of transcription (Figure 3.5). The full-length, unspliced transcript performs an additional role beyond that of acting as an mRNA for the production of proteins: this transcript serves as the genome of the virus and is packaged into the virion particle.

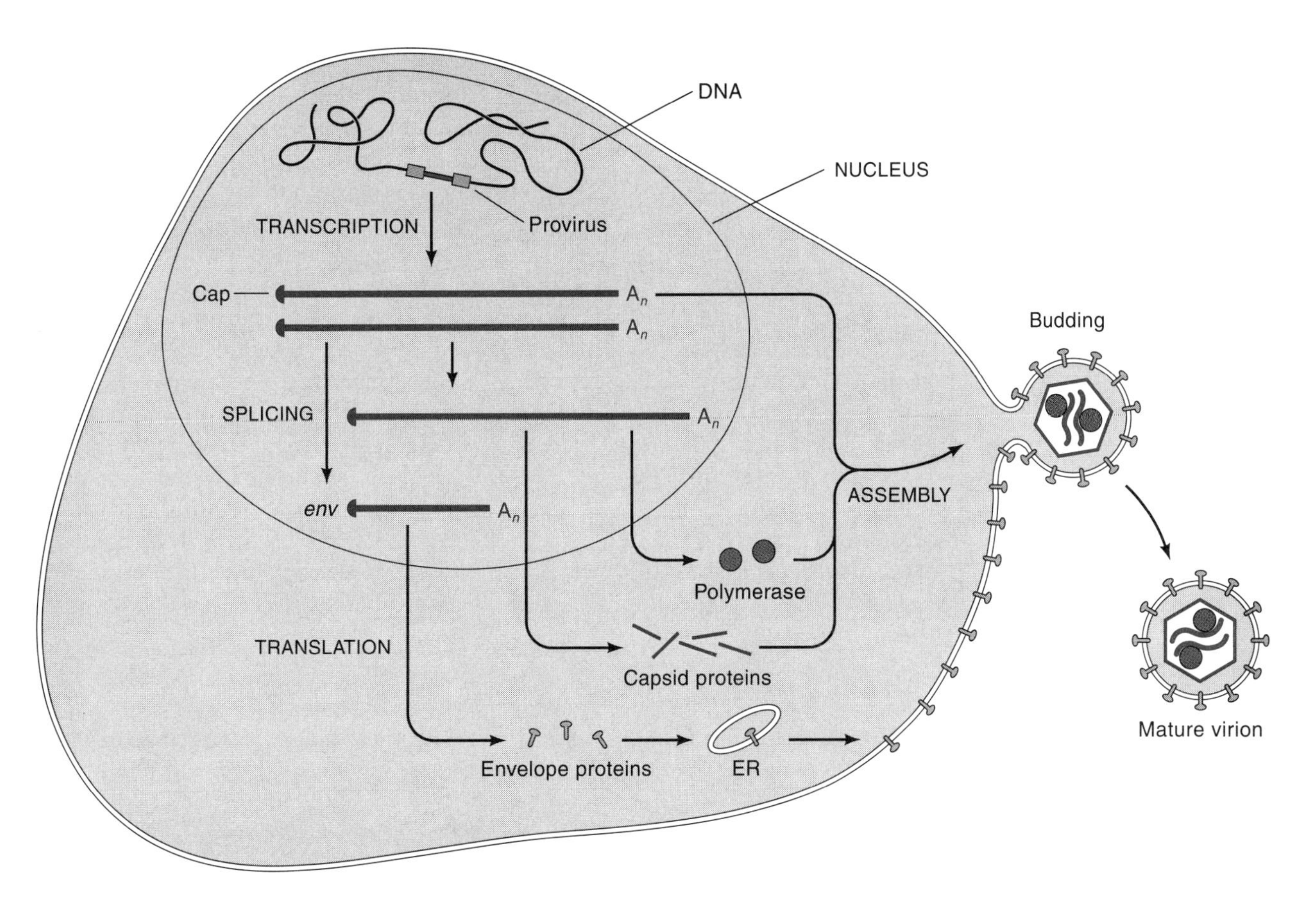

Figure 3.4
The late phase of the viral life cycle. The proviral DNA is transcribed by the host RNA polymerase II to form a variety of viral mRNAs, and the genomic RNA. The mRNAs are translated to form viral proteins. The proteins assemble, usually at the cell surface; RNA is packaged into the nascent particles and particles are released by budding.

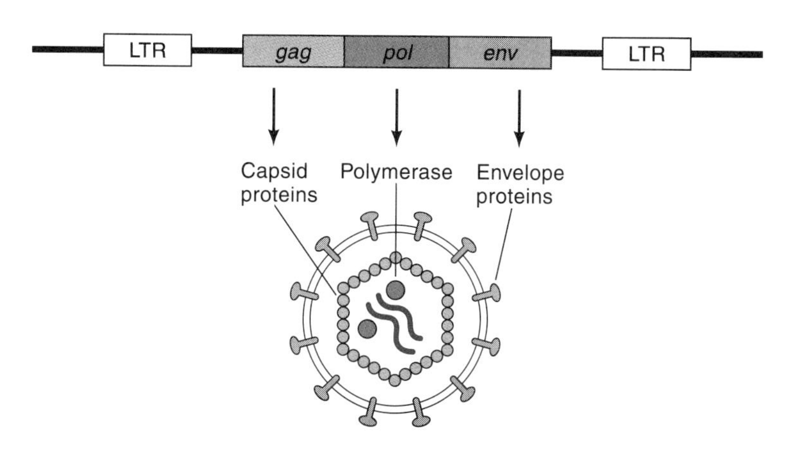

Figure 3.5
Retroviral genes. All retroviruses contain, at a minimum, three genes essential for replication: *gag, pol,* and *env.* The *gag* gene encodes proteins making up the virion core; the *pol* gene encodes a number of viral replication enzymes that are contained inside the virion; the *env* gene encodes a transmembrane glycoprotein located on the virion surface.

Assembly and release of progeny virions A chronically infected cell expresses the provirus to form the various virion proteins and then assembles the progeny virus from these protein products. The details of the assembly process are not well established, but some aspects are understood. First, the *gag* gene product is the engine driving the assembly process; the expression of this protein alone is sufficient to induce the formation and release of particles from the cell surface. For most retroviruses (the so-called c-type viruses), the particle forms at the plasma membrane in a structure known as a bud; for a few unusual viruses (the so-called d-type viruses), the virion cores preassemble in the cytoplasm and then migrate to the membrane. The genomic RNA is specifically encapsidated by the gag protein, bound into the particle, and annealed to another copy of the RNA genome to form a highly condensed, dimeric RNA structure. The *pol* gene products are incorporated into the core; the env proteins, positioned in the cell membrane, are attracted to the site of budding and there associate with the particle. As assembly proceeds, the gag, pol, and sometimes env protein precursors are proteolytically processed into their mature forms. These cleavages are associated with major rearrangements of the proteins in the virion and with activation of the virus infectivity.

b. Early Phase of the Life Cycle: Establishment of the Integrated Provirus

Before the availability of cloned viral DNAs, the two major reactions mediated by retroviral enzymes during the early phase of the life cycle—reverse transcription and DNA integration—were only vaguely understood. But with clones of all the major forms of the viral genomes in hand, researchers could discover the details of the process. Perhaps the determinations of the nucleotide sequence of the genome, and of the precise boundaries of the various sequence blocks, were the most significant aids. But other cloning technologies were also crucial. The abilities to label DNA to high specific activity to generate hybridization probes, to perform hybridization to DNA and RNA on filter supports with cloned DNA fragments, to use synthetic oligonucleotides as probes for individual sequences, to use DNA polymerase to elongate primers *in vitro* for mapping the positions of 5′ ends, and to use nucleases to determine whether DNAs are single or double stranded were all invaluable in obtaining a picture of the early steps in replication. Much of the story was advanced by the development of methods that allowed carrying out reverse transcription and integration in cell-free extracts. These processes are now understood in considerable depth.

Reverse transcription: formation of the LTRs The complex process of reverse transcription normally begins soon after infection in the cytoplasm. The reaction is thought to occur inside the virion core, through the action of the enzymes within that core; no host cell proteins seem to be required. Purified virion particles are in fact able to carry out the entire reaction, after the simple addition of low levels of detergent to disrupt the membrane and of deoxynucleoside triphosphates as precursors. It is not a rapid reaction either *in vitro* or *in vivo*; in both cases, several hours are needed to detect full-length, double-stranded DNA.

The overall reaction is diagrammed in Figure 3.6. Several sequence blocks

---

Figure 3.6 *(facing page)*
The process of reverse transcription of the retroviral genome. The complete formation of a double-strand DNA from the single-strand RNA requires several complex steps. See text for detailed explanation. (Adapted from J. Darnell et al. 1990. Molecular Cell Biology, 2nd ed., p. 972. Scientific American Books, New York.)

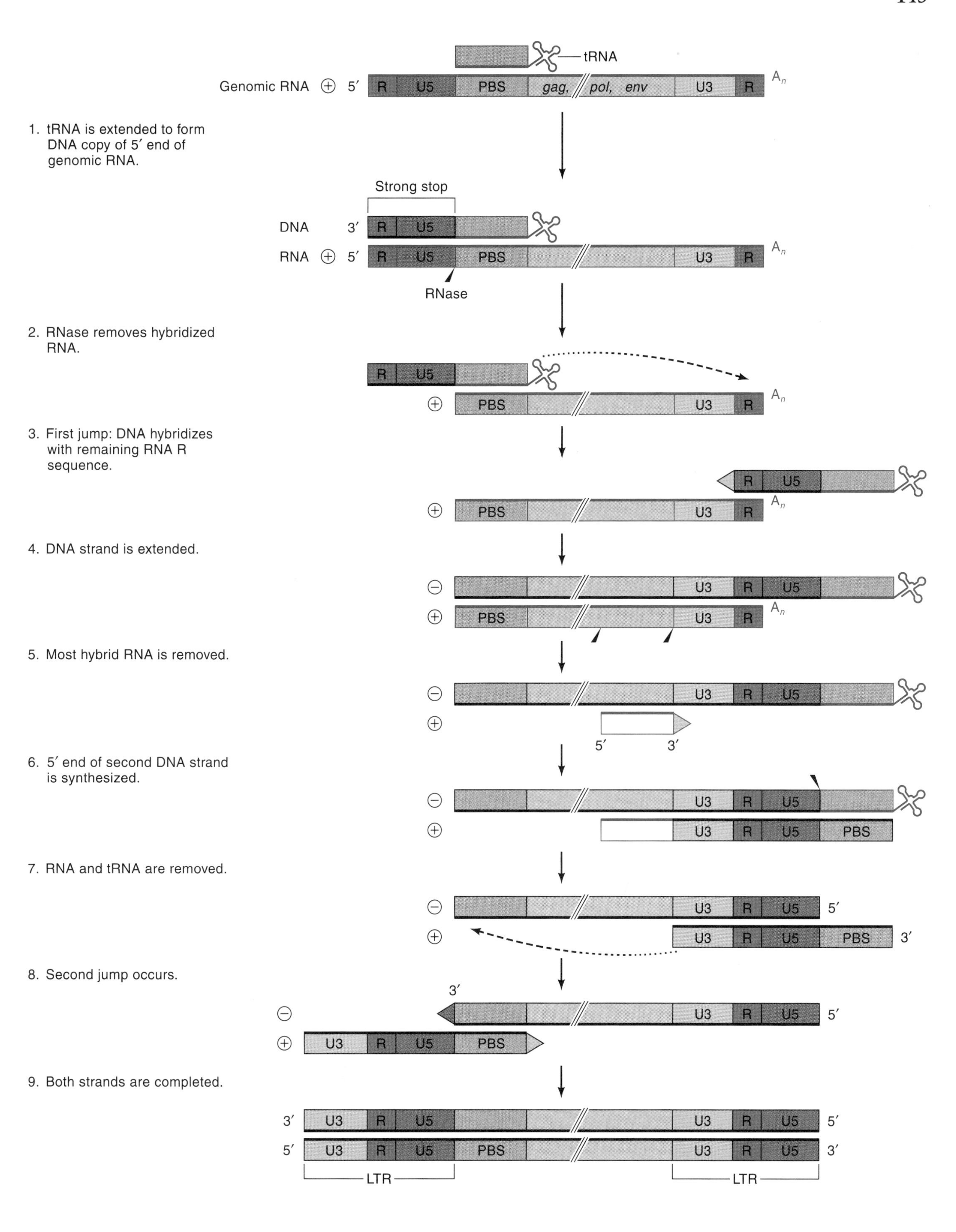
tRNA
Genomic RNA
5′
R
U5
PBS
gag,
pol,
env
U3
R
A_n
1. tRNA is extended to form DNA copy of 5′ end of genomic RNA.
Strong stop
DNA
3′
RNA
5′
RNase
2. RNase removes hybridized RNA.
3. First jump: DNA hybridizes with remaining RNA R sequence.
4. DNA strand is extended.
5. Most hybrid RNA is removed.
6. 5′ end of second DNA strand is synthesized.
7. RNA and tRNA are removed.
8. Second jump occurs.
9. Both strands are completed.
LTR
LTR

present in the RNA genome that are important in mediating various steps must be defined before the reaction can be understood. First, there is a short sequence redundancy at the 5′ and 3′ ends of the RNA genome, termed the "R" (for repeated) regions. Adjacent to the 5′ R is a block termed U5 (for unique 5′ sequence), and adjacent to the 3′ R is a block termed U3 (for unique 3′ sequence). These blocks are duplicated and rearranged during the course of the reaction. The virion core contains two copies of the RNA genome, both of which probably participate in the formation of provirus DNA.

DNA synthesis begins with the formation of minus-strand DNAs by elongation of a tRNA primer partially annealed to the genomic RNA. The tRNA is present in the virus particle before it enters the cell; it is packaged into the virion and positioned on the genome during the assembly process. For most viruses, 18 nucleotides at the 3′ end of the tRNA are hybridized to a complementary sequence near the 5′ end of the genomic RNA, termed the primer binding site (pbs). The initial product of DNA synthesis, extending from the primer tRNA to the 5′ end of the genomic RNA, is an intermediate termed **minus-strand strong stop** DNA.

The next step in synthesis consists of the translocation of the (−) strand strong stop DNA from the 5′ end of the genome to the 3′ end. That event requires the action of RNase H to degrade the genomic RNA annealed to the DNA, exposing at least a portion of that DNA in single-stranded form; the DNA can then anneal to the R sequence at the 3′ end of the RNA. The translocation thus utilizes the homology provided by the repeated R sequences. Recent evidence indicates that, in some instances, the strong stop DNA translocates to the R sequence at the 3′ end of the second RNA genome within the virion core.

After annealing, the 3′ end of the minus-strand strong stop DNA is suitably positioned to act as primer and is elongated to generate long minus-strand DNAs. These minus strands, when completed, are nearly complete copies of the genome and extend through the primer binding site. The U5 and R regions at the 5′ end of the genome are probably not available for copying, having been degraded by RNase H action. As synthesis proceeds, the genomic RNA transiently enters RNA:DNA hybrid form and is degraded by the RNase H. There is some evidence that these processes—DNA synthesis and RNA degradation—are coupled and that one RT molecule may simultaneously mediate both reactions (Figure 3.7). In such a process, only a short stretch of the newly synthesized DNA remains base paired to the RNA.

As degradation of the genomic RNA occurs, a special sequence near the 3′ end of the RNA genome is recognized by RNase H and is used in a special way. A sequence rich in purines, termed the **polypurine tract**, remains relatively resistant to RNase H action and serves the special role of priming DNA synthesis for the plus strand. Specific cleavages create an RNA oligonucleotide able to initiate synthesis; as soon as synthesis begins, the primer is removed by RNase H activ-

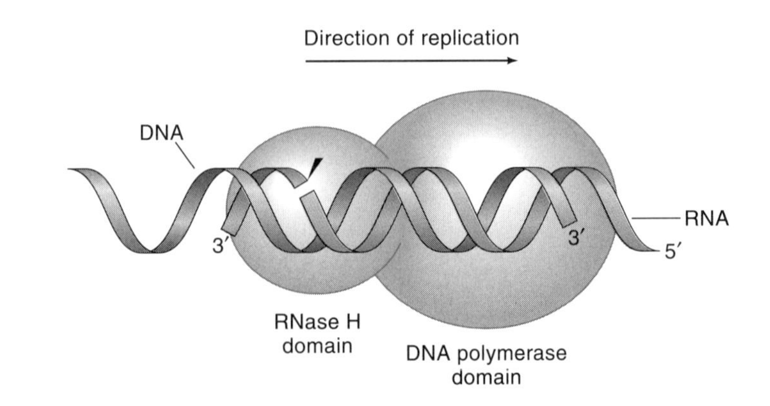

Figure 3.7
A model for the concerted action of the DNA polymerase and RNase H activities of reverse transcriptase in the course of DNA synthesis on an RNA template. As the DNA strand is elongated by extension of the growing 3′ end, the RNA template enters RNA:DNA duplex form and becomes susceptible to cleavage by RNase H. The RNase H activity of the reverse transcriptase is thought to cleave the RNA approximately 17 bp behind the 3′ end of the DNA.

ity. DNA is synthesized, using the minus-strand DNA as template, until the 5′ end of the minus strand is reached, forming the intermediate termed **plus-strand strong stop** DNA. This DNA in fact extends slightly beyond the 5′ end of the minus strand, copying bases of the tRNA that are still attached there (Figure 3.5).

Further synthesis of the plus strand requires another translocation. When the plus-strand strong stop is formed, a portion of the primer tRNA on the minus strand is placed into RNA:DNA duplex form. This renders the tRNA susceptible to degradation by RNase H, and the tRNA is thus removed from the minus-strand DNA. The removal of the tRNA exposes the last few nucleotides of the plus-strand strong stop DNA in single-strand form. These bases now permit the translocation to the other end of the genome. In this translocation, a circular intermediate is formed by the annealing of the 3′ end of the plus-strand strong stop DNA to the 3′ end of the long minus strand (Figure 3.8).

Both minus and plus strands are now in a position to be elongated to form the complete, full-length DNA genome. Further minus-strand synthesis must be associated with strand displacement. This displacement synthesis actually causes the translocation of the plus-strand strong stop from one end of its template to the other; as the minus strand is elongated, it "peels" its own 5′ end away from the plus-strand strong stop and transfers the DNA to its 3′ end (Figure 3.8). When the displacement is complete, the circle opens up into a linear form again. The final completion of the plus strand to its full length finishes the reverse transcription process.

Several features of the product DNA bear emphasis. First, it is actually longer than the starting RNA from which it is derived; aligning the body of the sequences shows that sequences in the DNA protrude beyond the limits of the RNA at both ends. The extra sequences are duplications of U3 and U5, generated by the two translocation events. Because of those duplications, the DNA is flanked by much longer redundancies than are present at the ends of the RNA. Whereas the RNA contains repeats of only the R region, the DNA contains repeats of U3-R-U5 at each end. These three blocks together constitute the LTR. The final DNA is blunt ended and probably remains associated with at least some viral proteins, including some of the virion gag proteins. The mechanism of the ultimate transport of the DNA to the nucleus, presumably essential for its integration into the host genome, remains mysterious. Experiments in which the cell cycle is arrested with various drugs have recently shown that host DNA synthesis is required for productive infection and that entry into the nucleus occurs only when the nuclear envelope is disassembled during mitosis. Recent work suggests that HIV-1 may be able to enter the nucleus of non-dividing cells by a distinct mechanism.

Besides the full-length linear DNA, two additional DNA forms, both circular, are found in the nucleus of the infected cell. One of these circular DNAs contains a single LTR and has an overall structure consistent with its formation from the linear DNA by homologous recombination between the two LTRs. The other contains two tandem LTRs and almost certainly forms by the end-to-end joining of the blunt termini of the linear DNA. Sequence analysis of the LTR-LTR junction region shows that some molecules retain the complete inverted repeats, forming a palindromic sequence; others have lost a few base pairs from either of the LTRs. Much attention was lavished on these molecules as potential precursors for formation of the integrated provirus, but more recent experiments suggest that these circular forms are rarely, if ever, utilized for integration. The current consensus is that these forms are probably dead-end products that do not give rise to progeny virus.

Integration of the proviral DNA The pre-integrative viral DNA, as noted previously, is a blunt, duplex, linear DNA containing the two LTRs in direct orientation. Each LTR, however, contains short, inverted repeats at its termini; thus, the

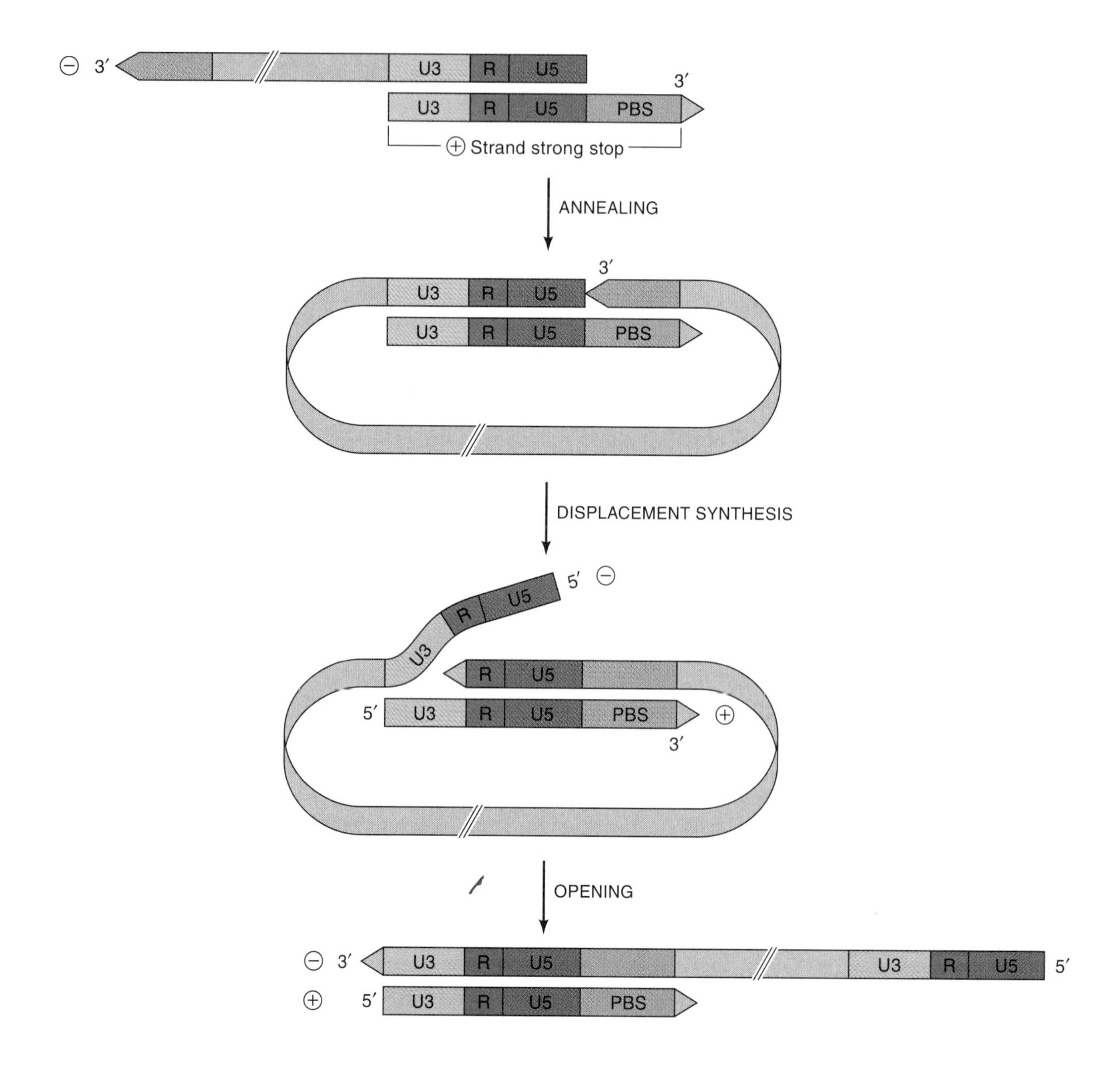

Figure 3.8
Detail of the translocation of the plus-strand strong stop DNA. The translocation of the plus-strand strong stop DNA is thought to proceed through a circular intermediate formed by annealing of complementary sequences at the 3′ ends of the minus and plus strands. Elongation of the minus strand coupled to displacement of the 5′ end of the minus strand from the plus-strand template results in the opening of the circle into a linear DNA. The net effect is the "jumping" of the plus-strand strong stop from the 5′ end to the 3′ end of the minus strand.

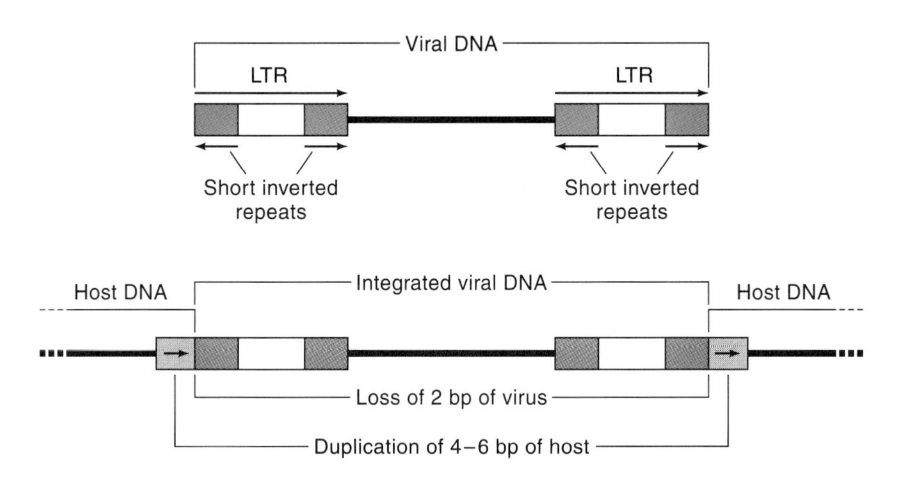

Figure 3.9
Structures of unintegrated and integrated viral DNAs. The full-length linear viral DNA contains sequence blocks repeated in direct orientation at the two termini—the Long Terminal Repeats (LTRs). The edges of the LTRs consist of short inverted repeats (small arrows). The arrangement of the LTRs implies that the viral DNA overall contains short inverted repeats at the termini. The integrated proviral DNA is collinear with the unintegrated DNA and is always joined to the host DNA at the outermost inverted repeat sequences. For most viruses, the two terminal bp are lost during the integration reaction. The provirus is flanked by short direct repeats (4–6 bp) derived by duplication of sequences present only once in the original target sequence.

linear DNA as a whole contains short, inverted repeats at its termini. The integrated DNA is essentially colinear with the product of the reverse transcription reaction, with only minor changes at the termini. During integration, a few base pairs at the very ends of the viral DNA—usually exactly 2 bp—are lost, and a few base pairs at the target site are duplicated so as to flank the inserted provirus (Figure 3.9).

The mechanism of the reaction that accounts for these features has only recently been determined through the advent of *in vitro* systems that recapitulate the process. The product of the reverse transcription reaction, the full-length, double-stranded, linear DNA, is the immediate precursor and substrate for the integration reaction. The process is divisible into two steps: a preliminary **trimming** or processing of the viral DNA, mediated by the endonucleolytic activity of the integrase (IN) protein, and a climactic integration step in which a strand transfer joins one viral strand at each end to one strand of host DNA (Figure 3.10).

The sequences at the two termini of the linear DNA are identical or very similar to each other, marked by the presence of short, imperfect, inverted repeats present at the two edges of the LTR. In the murine leukemia viruses, these sequences are 13 bp long, perfect repeats; in other viruses, these sequences are shorter and interspersed with imperfect nucleotide matches. Near the 3′ends, a conserved CA dinucleotide is always present within these inverted repeats, and usually the CA is two bases from the very end. In the first step of integration, the blunt termini are trimmed by an endonucleolytic cleavage of the DNA immediately 3′ to the CA dinucleotide; the cleavage releases the very terminal nucleotides at the 3′ ends of the molecule (Figure 3.10). For most viruses, a TT dinucleotide is released. The cleavage reaction creates a new, recessed 3′ end, which will be the site of joining between the virus and the host DNA.

Mutational studies, in which alterations were introduced into the terminal sequences, have shown that the cleavage *in vivo* requires the presence of specific

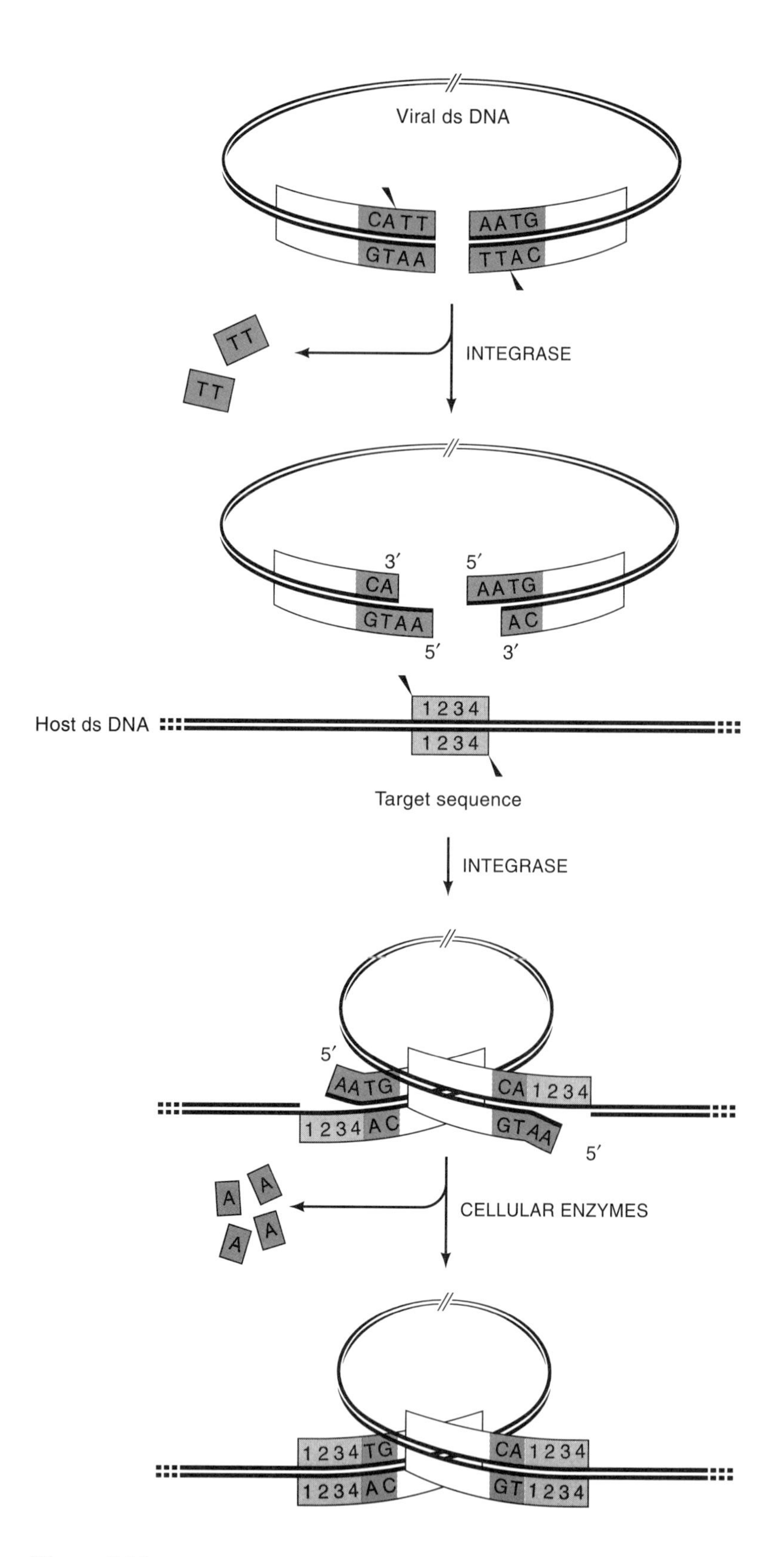

Figure 3.10
Structures of intermediates formed during the integration of the viral DNA into the host DNA. In a prefatory step, the termini of the full-length blunt-ended linear DNA are brought together and then cleaved by the endonuclease activity of the integrase protein to form recessed 3′ termini. In a second joining step, the recessed 3′ hydroxyl ends of the viral DNA are used by the integrase to attack phosphodiester bonds of the target DNA (arrows). The 3′ OH ends of the viral DNA are joined to the 5′ phosphates with displacement of the host 3′ OH. In the resulting intermediate, only one strand of the viral DNA is joined to the host DNA of each junction; two gaps and two unpaired protruding 5′ ends remain. Host enzymes are presumed to repair these discontinuities.

sequences at the termini, corresponding closely to the inverted repeat sequences; furthermore, these sequences must be present at both ends of the viral DNA molecule for either end to be cleaved, suggesting that the cleavage is a concerted reaction. Mutations introduced into the coding region for the IN protein showed that the cleavage requires the presence of the IN protein in the virion particle. Reconstructions of the reaction with purified IN protein confirm that this protein is the nuclease directly responsible for the reaction: purified IN protein is able to remove the terminal nucleotides from blunt-ended, linear DNA substrates in a highly sequence-specific reaction.

After cleavage, the trimmed DNA is joined to the target sequence. Experiments in which the reaction is performed *in vitro* have helped to determine the requirements of the joining reaction. The incubation of blunt-ended DNAs corresponding to the viral termini, a target DNA, and purified IN protein results in DNA joining. The integration reaction requires no added source of energy and apparently requires no small molecule cofactors other than suitable salt concentrations and divalent cation. Moreover, the energy of the phosphodiester bond in the viral DNA that is broken in the initial cleavage reaction is not apparently utilized. Precleaved DNA can be generated, purified, and used in the joining reaction; this DNA is just as active, or even more active, than the blunt-ended DNA. The energy for bond formation thus must come from the existing bonds in the target DNA. In essence, the reaction must be an exchange or strand transfer; as the 3′ hydroxyl group of the viral DNA is joined to the target, the 3′ hydroxyl group in the target is released.

Analysis of the exact structure of the products of the reaction carried out *in vitro* has helped define the mechanism. By using viral DNA substrates with radioactive label in alternate strands, researchers have determined which strand is linked to the target DNA. Only one strand of the viral DNA is covalently joined to the target; the 3′ hydroxyl of the viral DNA is linked to a 5′ phosphate of the target DNA, leaving a nick open in the target. Careful examination of the product showed that the 5′terminus of the viral DNA remains free and retains all the nucleotides present in the original blunt-ended DNA. Thus, the two 5′ terminal bases, left protruding by the removal of the TT dinucleotide, are unpaired and protruding from the joint.

The same joining reaction normally occurs simultaneously at each end of the viral DNA, with the joints being made at a precisely fixed distance apart in the target DNA; the spacing of the joints is characteristic of each particular virus and will ultimately determine the number of base pairs duplicated in the course of the reaction. When the structure is opened up, there are two single-stranded gaps in the target DNA flanking the viral DNA, and there are mispaired extended bases next to those gaps (Figure 3.9). Presumably, host enzymes are responsible for repairing the mispaired bases and filling in the gaps. Some recent work has led to the suggestion that the viral IN enzyme might be involved in aspects of the repair process, but this notion remains speculative.

c. Late Phase of the Life Cycle: Expression of the Integrated Provirus

Once the integrated provirus is formed, the early phase of the life cycle is ended. Integration marks a dramatic shift in style in the cycle. The events up to this time are uniquely viral reactions, quite removed from cellular host functions; the events after this time are more conventional and generally depend on the participation of the host machinery. The late stage is also different in that it is open-ended in time. Whereas the early stage of the life cycle is sharply terminated with integration, the late stage continues indefinitely, for the production of virus by the chronically infected cell will generally continue for the life of the cell.

Formation of viral mRNAs and genomic RNA The expression of the viral DNA to form viral mRNAs and genomic RNA is mediated by the host RNA polymerase II. The U3 region of the viral LTR contains a potent transcriptional promoter specifying the initiation of transcription at the U3/R boundary of the 5′ LTR. Retroviral promoters are among the most complex promoters studied. The promoter sequence can be divided broadly into two categories—core enhancer–like elements and a TATA box sequence (*Genes and Genomes,* chapter 8). The enhancer regions contain recognition sites for a great many DNA-binding proteins, only some of which have as yet been isolated and characterized. These promoters are exceptionally strong; a single integrated provirus may specify the formation of as much as 1 percent of the poly A+ RNA in the infected cell. In general, the promoters are constitutively active, though some notable exceptions exist: the MMTV promoter is highly regulated by glucocorticoids, and expression of the more complex retroviruses like HIV is regulated by both host factors and virus-encoded regulatory gene products. One of the most significant aspects of the strong transcriptional promoters carried by these viruses is the role played by the regulatory elements in their pathogenicity.

Transcription results in the formation of conventional 5′-capped precursor hnRNA in the nucleus, with synthesis extending from within the 5′ LTR through the whole provirus and through the 3′ LTR (Figure 3.11). The RNA is processed at the 3′ end, with a cleavage after the R/U5 boundary followed by polyadenylation. This unspliced transcript is then utilized in two ways: some of these molecules are exported from the nucleus for packaging into virions, and another

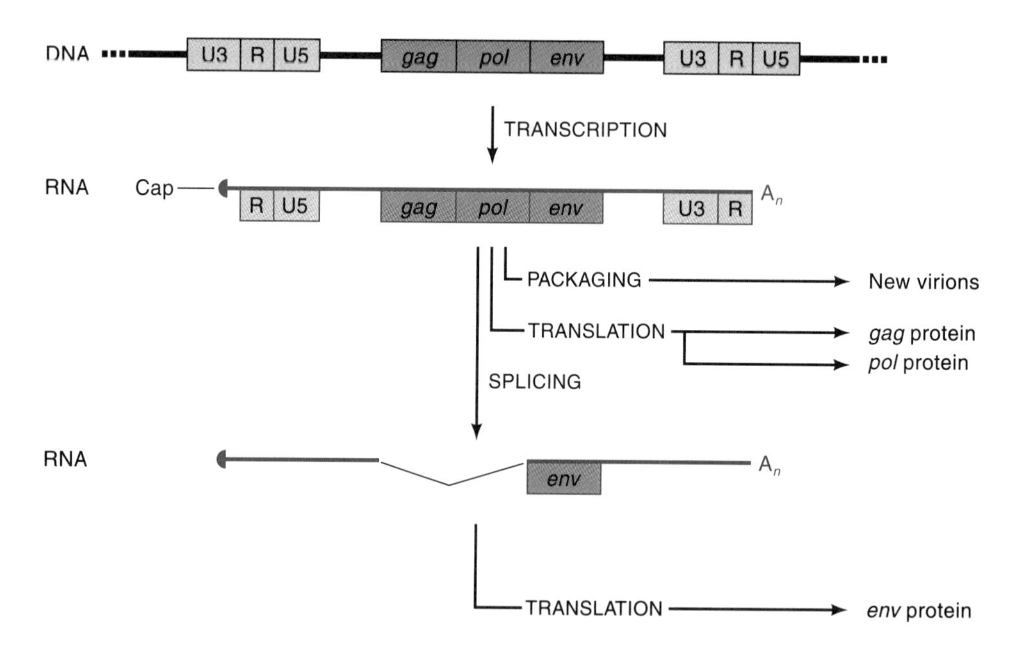

Figure 3.11
Expression of viral RNA from the integrated provirus. The integrated proviral DNA is shown with the two LTRs divided into the U3, R, and U5 regions. Transcription begins at the U3/R boundary of the 5′ LTR, and proceeds through the provirus. Cleavage and polyadenylation occur at the R/U5 boundary of the 3′ LTR. The resulting RNA contains direct repeats of the R region at the 5′ and 3′ ends. This RNA serves both as the genomic RNA for packaging into progeny virions, and as the mRNA for the formation of the *gag* and *pol* gene products. RNA splicing of a portion of these RNA molecules results in the removal of a sequence block as an intron and thus the formation of the mRNA for the *env* gene products.

portion is engaged by ribosomes in the cytoplasm, where it is translated to form the gag and gag-pol polyproteins. In addition, spliced RNAs are formed in the nucleus from the primary transcript. All retroviruses make at least one subgenomic mRNA, encoding the env protein, by excision of a single intron. In some cases, additional mRNAs are formed by alternative splicing.

A major mystery concerning retroviral transcription remains unresolved: how is transcription from the 5′ LTR favored strongly over transcription from the 3′ LTR, when both contain the exact same sequence? There is some evidence that the transcriptional activity of the 5′ LTR suppresses recognition of the 3′ LTR, but it is not clear if that is the whole explanation. There is a related mystery: why are the polyadenylation signals recognized at the 3′ LTR and not at the 5′ LTR? In many viruses, transcription through the 5′ LTR generates sequences that could be recognized for cleavage and polyadenylation, resulting in the formation of a very short transcript containing only the R region. Such transcripts are not formed or are not stable. Presumably, some aspect of the proximity of the sequence to the 5′ cap precludes its recognition, but the details of that explanation are not clear. These related issues become significant when considering how the leukemia viruses cause disease through the transcriptional activation of flanking protooncogenes.

Viral genes and genomic organization The genomes of all retroviruses include, at a minimum, three open reading frames between the two LTRs: from 5′ to 3′, these genes are *gag, pol,* and *env* (Figure 3.12). The *gag* gene encodes a polyprotein precursor that assembles to form the virion capsid or core; eventually, it is cleaved to form an array of mature proteins that are major constituents of the virion. The *pol* gene encodes several enzymes essential for the replication of the virus. The polyprotein product of the *pol* gene is formed in an unusual way and is also cleaved proteolytically to give rise to mature products. The *pol* gene products are present at relatively low abundance in the particles. Finally, the *env* gene encodes a glycoprotein present on the surface of infected cells and concentrated at high levels on the surface of virions. In most chronically infected cells, all these gene products are expressed constitutively and assembled continuously to generate progeny virions.

Formation and processing of polyprotein precursors As noted previously, most of the mechanisms involved in the expression of retroviral gene products are relatively conventional, mediated by normal host enzymes. But that is not to say that this phase of the retroviral life cycle is not without its odd aspects. The retroviral proteins are typically expressed in the form of large precursors, termed polypro-

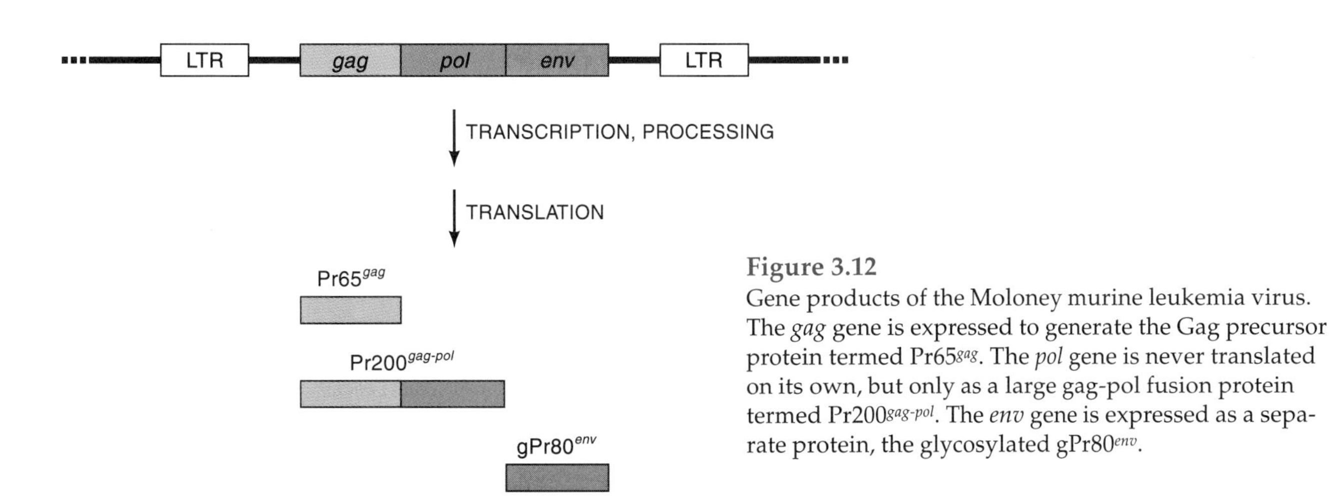

Figure 3.12
Gene products of the Moloney murine leukemia virus. The *gag* gene is expressed to generate the Gag precursor protein termed Pr65^{gag}. The *pol* gene is never translated on its own, but only as a large gag-pol fusion protein termed Pr200$^{gag\text{-}pol}$. The *env* gene is expressed as a separate protein, the glycosylated gPr80^{env}.

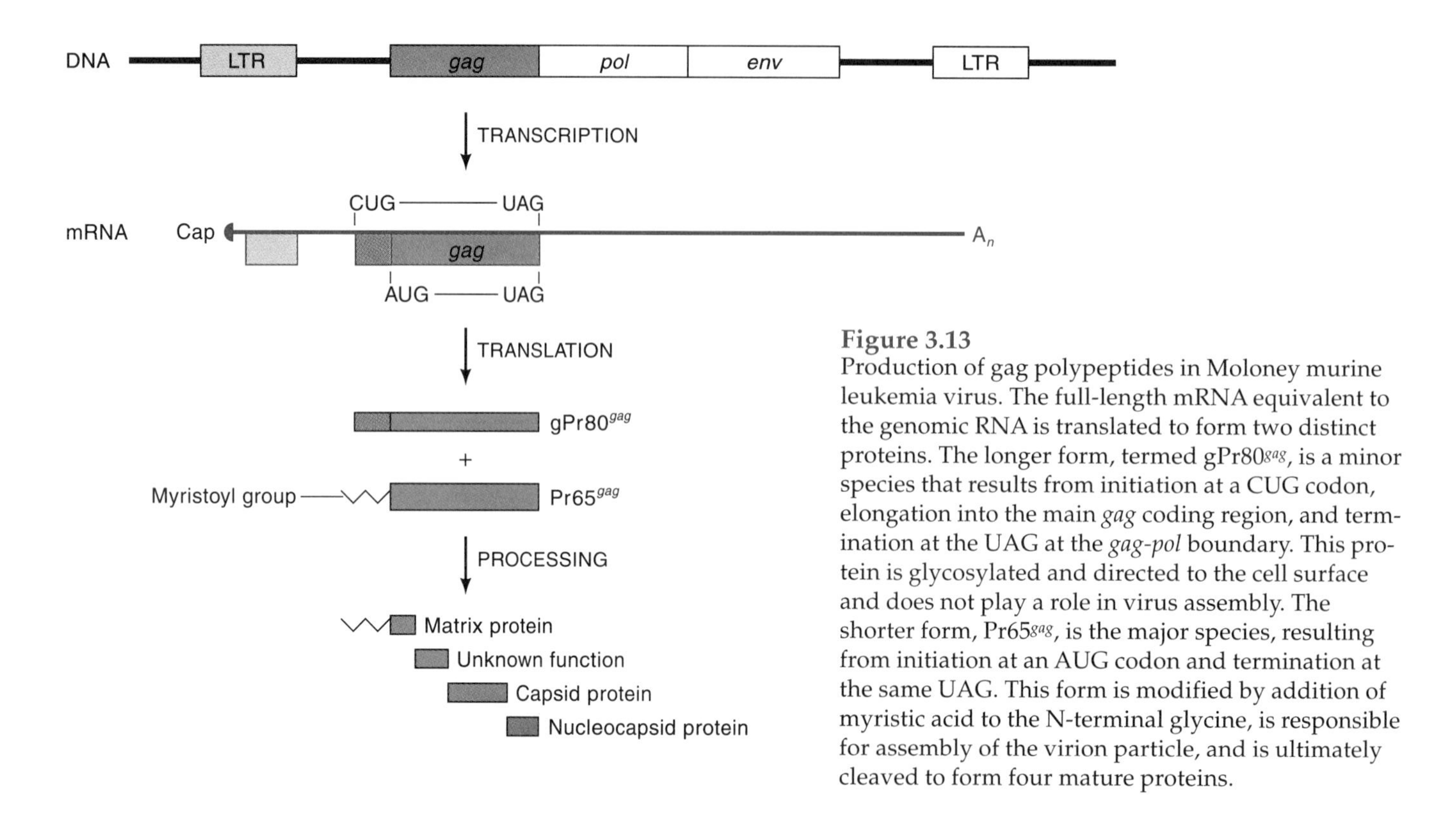

Figure 3.13
Production of gag polypeptides in Moloney murine leukemia virus. The full-length mRNA equivalent to the genomic RNA is translated to form two distinct proteins. The longer form, termed gPr80^{gag}, is a minor species that results from initiation at a CUG codon, elongation into the main *gag* coding region, and termination at the UAG at the *gag-pol* boundary. This protein is glycosylated and directed to the cell surface and does not play a role in virus assembly. The shorter form, Pr65^{gag}, is the major species, resulting from initiation at an AUG codon and termination at the same UAG. This form is modified by addition of myristic acid to the N-terminal glycine, is responsible for assembly of the virion particle, and is ultimately cleaved to form four mature proteins.

teins, containing several domains that are ultimately processed by proteolytic cleavages to form the mature protein products. In addition, the viruses break the "one mRNA, one protein" rule; two mechanisms permit the formation of two proteins from one mRNA.

Gag precursor: the MA, X, CA, and NC domains The *gag* gene products are formed by translation of the first open reading frame in the full-length, unspliced RNA that is indistinguishable in structure from the genomic RNA of the virus. Translation of the major gag protein begins with an AUG codon and proceeds to a terminator codon (Figure 3.13). The protein is cytoplasmic, soluble, and does not contain carbohydrate. In many viruses, the amino terminal methionine is removed, and the new amino terminal residue, a glycine, is decorated cotranslationally by the addition of a fatty acid, myristic acid, to its amino group. The modification may help direct the protein to the plasma membrane and is needed by some viruses for proper virion assembly.

Many of the murine and feline leukemia viruses encode an additional gag protein, a glycosylated version of gag. This protein is formed by initiating translation at a CUG codon upstream of the AUG at the major site of initiation. Translation from this upstream codon first forms a leader peptide and then subsequently proceeds in frame into the normal gag coding region, forming a larger protein that contains all the major gag sequences. The leader peptide, however, specifies that this protein is directed to the secretory pathway. The larger gag is transported through the endoplasmic reticuluum, glycosylated in the Golgi system, and deposited at the cell surface. Its lifetime on the surface is short. The larger gag is gradually cleaved into two fragments and then released into the extracellular medium. The protein's function is obscure; it is not present in the virion and is completely dispensable for replication of virus in tissue culture.

The major unglycosylated gag protein is responsible for assembling the virion particle under the plasma membrane. The expression of this protein alone in eukaryotic cells is sufficient to induce the formation of particles. As assembly occurs, the precursor protein is cleaved by a virus-encoded protease to form the mature proteins. From 5′ to 3′, or from N to C terminus, these proteins in the murine viruses are termed MA (for membrane associated or matrix), X (for protein of no known function), CA (for capsid), and NC (for nucleocapsid) (Figure 3.13). The list of the functions of these various proteins is certainly incomplete, and the position of each protein in the virion and its role in building or maintaining its structure are only vaguely known. In other retrovirus families, the cleavage patterns of the gag precursor are slightly different. In HIV-1, for example, there is no "X" protein between MA and CA, and there is an additional protein, so far termed p6, downstream of the NC protein. The functions of these "extra" cleavage products are even less well known than those of the more highly conserved products.

Gag-pol precursor: suppression of translational termination and frameshifting Although translation of the unspliced mRNA usually results in the formation of one or the other of the gag proteins, occasionally there can be a different outcome: formation of a larger gag-pol fusion protein. This is the only route by which the pol region is expressed, for it is never expressed on its own, independent of the gag region. Generally, the gag-pol protein is formed at a rather low level, at least in comparison with the level of the gag protein: typically, the gag-pol protein represents one-tenth or one-twentieth of the total gag production.

The gag-pol protein is formed through either one of two equally extraordinary mechanisms, each one utilized by different viruses. One class of viruses, including the murine and feline leukemia viruses, expresses the protein through translational suppression of termination at the end of the gag open reading frame (Figure 3.14). In these viruses, the gag and pol coding regions are in the same reading frame and are separated by only a single terminator codon, the rarely used UAG codon. Ribosomes initiating at the start of the *gag* gene read through this terminator at a relatively high frequency—roughly one-tenth or one-twentieth of the time—and continue on into the pol open reading frame to form the larger protein. Analysis of the amino acid sequence at the position corresponding to the terminator codon revealed that the ribosomes inserted a glutamine residue. The readthrough process can be recapitulated by *in vitro* translation systems, and research has shown that no viral proteins are required. Rather, sequences in the RNA—mostly those downstream of the terminator codon—are required to specify readthrough to the passing ribosomes. Mutagenesis of the region has shown that about 50 nt of sequence are most important, and analysis of such mutants and double mutants with changes predicted to conserve base pairing between certain sequence blocks has led to the proposal that a complex, folded, secondary structure is the likely feature mediating the readthrough; this structure is termed a **pseudoknot** (Figure 3.14). The position of

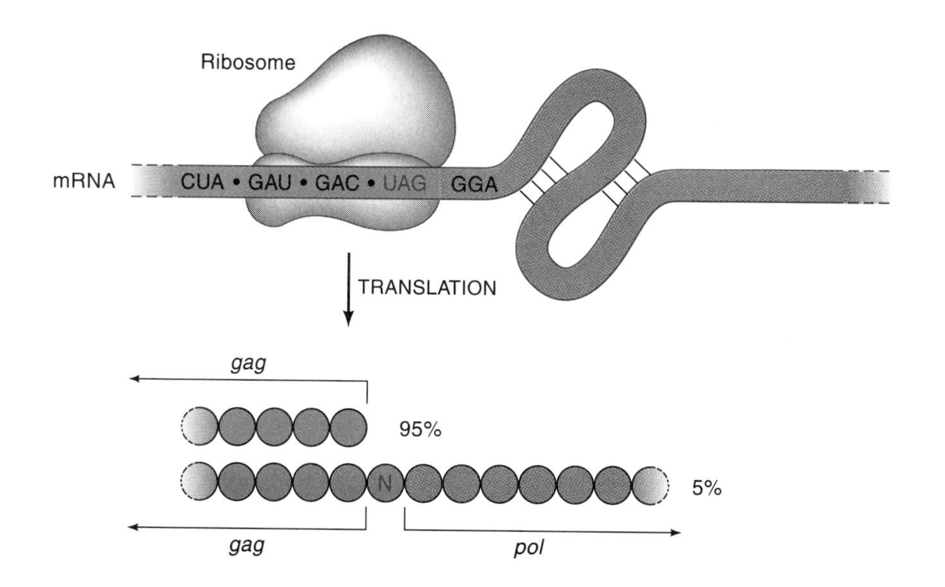

Figure 3.14
Suppression of translational termination at the *gag-pol* boundary of Moloney murine leukemia virus. The *pol* gene products are only synthesized in the form of a Gag-pol fusion protein by translational readthrough of the UAG terminator codon at the boundary of the *gag* and *pol* genes. The readthrough requires the presence of a folded secondary structure (a pseudoknot) in the RNA immediately downstream of the terminator codon.

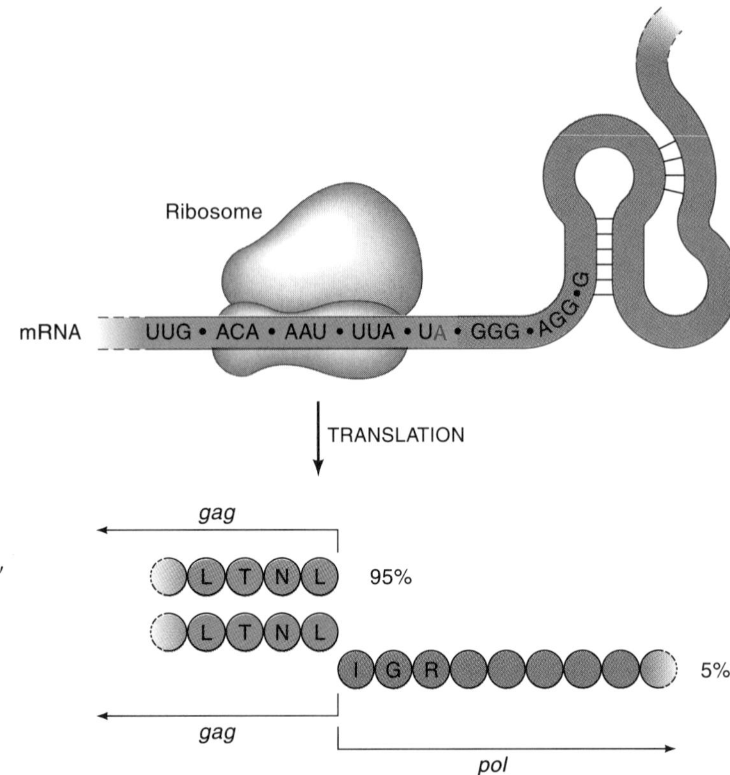

Figure 3.15
Translational frameshifting at the *gag-pol* boundary of Rous sarcoma virus. The *pol* gene products of this virus are only synthesized as a Gag-pol fusion protein by translational frameshifting immediately before the UAG terminator codon at the 3′ end of the *gag* gene. The frameshift, like readthrough in the murine viruses, requires a pseudoknot structure downstream of the frameshift site.

the pseudoknot relative to the terminator is important, but the event is not specific to a UAG codon; any of the three terminator codons, embedded in the proper flanking sequences, can be suppressed.

Another class of viruses uses a completely distinct mechanism to form their gag-pol protein. These viruses, including the avian leukosis viruses and the human immunodeficiency viruses, express their fusion protein by translational frameshifting (Figure 3.15). In these viruses, the pol coding region is in the -1 reading frame with respect to the gag coding sequence. As the ribosomes transit the gag sequence, translation shifts, at a low frequency, into the -1 reading frame before reaching the terminator codon at the end of gag; translation then proceeds into the pol region, passing through the terminator codon by reading the overlapping codons in the -1 frame. Analysis of the amino acid sequence of these gag-pol proteins showed that the frameshift occurs at a specific point in translation, at a so-called "shifty site" that includes a run of homopolymeric bases. The process, like translational readthrough, seems to be specified solely by the structural features in the RNA because no viral protein is required for the event, and normal ribosomal preparations from uninfected cells can recapitulate the frameshifting *in vitro*. The complete structural requirements for frameshifting remain undetermined, but there is evidence that here too a "pseudoknot," lying downstream of the frameshift site, may be important.

The various classes of retroviruses use quite diverse patterns of termination suppression and frameshifting to generate their gag-pol proteins (Figure 3.16). In some viruses, frameshifting is actually used twice to form a complete gag-pol protein. The MMTV genome contains separate open reading frames for the gag protein, for the protease, and for the pol products; each successive gene lies in the -1 reading frame with respect to the previous gene. The largest protein encoded by the virus is formed only when ribosomes undergo two frameshift events, one to switch from gag to protease, the second to switch from protease to

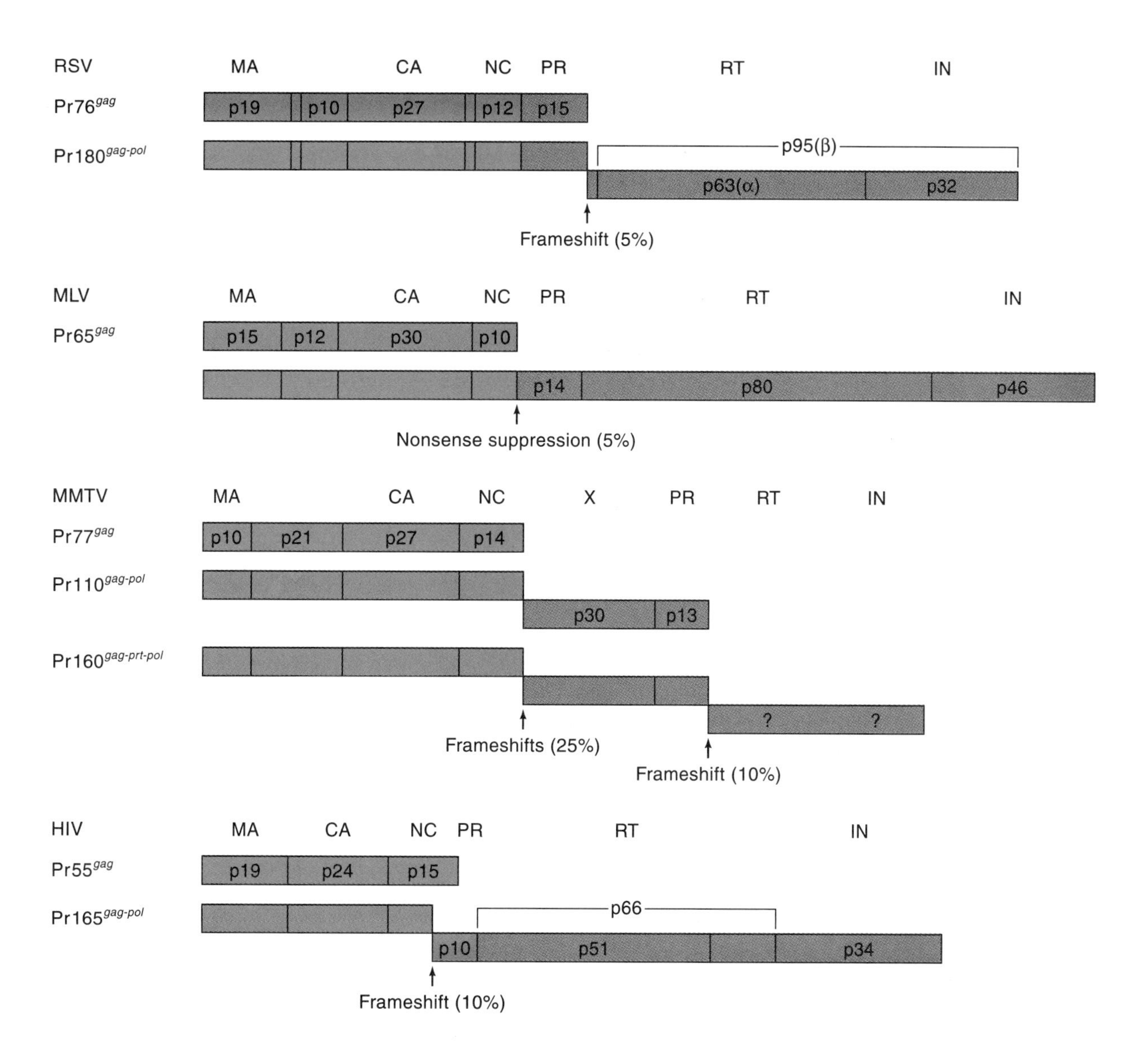

Figure 3.16
The organization of the *gag* and *pol* genes of various retroviruses. Though all the genomes contain related *gag* and *pol* sequences, the details of the mechanisms of expression, the sizes of the translation products, and the positions of the proteolytic cleavages are all different. In some viruses, two successive frameshifts are required to generate a complete gag-pol precursor protein.

pol. Each event utilizes independently functional sequences in the RNA. The frequency with which the full protein is formed is thus the product of the frequencies of the individual frameshifting events. In this setting, the individual frameshifts are extraordinarily efficient; one of the MMTV frameshifts occurs approximately 27 percent of the time that ribosomes pass down the RNA. This is a surprisingly efficient twist for the process of translation; normally, unscheduled frameshifting would be disastrous, and the process has evolved to limit frameshifting to extremely low frequencies. Although frameshifting has been found in other viruses, its use in cellular genes must be rare.

The pol *gene functions: the PR, RT, and IN domains* The *pol* gene of most retroviruses includes regions encoding several separate enzymes. These enzymes are initially formed as part of the large gag-pol precursor and are released as separate polypeptides only after assembly into the virion particle. However, not all pol proteins are handled in an identical way; rather, the pol portion of the precursor is processed into the various enzymes in different ways in different viruses. The processing patterns used by three viruses are shown in Figures 3.17, 3.18, and 3.19.

The N terminal proximal protein encoded in the pol region of the mammalian viruses is an unusual protease (PR). This protease is the enzyme responsible for the processing of both the gag and gag-pol polyproteins. Thus, the enzyme is apparently capable of cleaving itself from the precursor. The structure of this protein is known in atomic detail: the complete three-dimensional structure was determined by X-ray diffraction analysis of crystals of the protein from both HIV-1 and ALV. The enzyme is a homodimer of identical subunits, and there is evidence that dimerization is required for activity. This result suggests that the precursor itself must dimerize to activate the protease, perhaps during assembly into the virion, and that, once activated, the protease makes the many cleavages necessary to form the mature protein products of gag and pol. The enzyme shows an unusual sequence specificity, tending to cleave in stretches of hydrophobic amino acids, but with only a poorly defined consensus sequence. There is a limited structural similarity to other carboxypeptidases, and some inhibitors of the mammalian renins are also inhibitors of retroviral proteases. These and related compounds are under intense scrutiny as potential antiviral drugs for the treatment of AIDS.

The central domain of the mammalian retroviral *pol* gene encodes reverse transcriptase, the enzyme responsible for the synthesis of the double-stranded DNA version of the RNA genome. The RTs of different virus families are distinctively processed, yielding enzymes with distinctive subunit structures. The mammalian leukemia virus RTs are simple, monomeric enzymes, released from the precursor by proteolytic cleavages at either end. All the RTs have fairly similar overall enzymatic activities, and so the significance of the various subunit structures manifested by the different viral RTs is currently obscure.

All RTs known to date contain two major enzymatic activities: a DNA polymerase activity capable of using either RNA or DNA as template and a ribonuclease activity termed RNase H that is specific for RNA in RNA:DNA hybrid form. Both activities are essential for the complete synthesis of the double-stranded DNA genome as outlined in Section 3.2b, and the two together can account for all the reactions known to be required by the process. Broadly, the RNA-dependent DNA polymerase activity is used to synthesize the minus strand of the DNA product; the RNase H activity is responsible for removing the RNA genome whenever it appears in hybrid form; and the DNA-dependent DNA polymerase activity is responsible for synthesizing the second, plus, strand of the DNA product. The two activities are known to reside in two separate and separable parts of the RT molecule: the DNA polymerase in the N terminal portion and the RNase H activity in the C terminal portion. The activities are not normally separated by proteolysis in any of the known RTs. The two activities are likely to act in concert. Thus, during normal elongation of DNA on an RNA template, the DNA polymerase activity will extend the 3′ OH of the growing DNA chain, placing the RNA in RNA:DNA hybrid form, and the RNase H activity will then degrade the RNA behind the polymerase. At steady state, only a small stretch of sequence will be in hybrid form at any given time (Figure 3.7).

The third domain of the viral *pol* gene encodes the integrase function. In most viruses, this protein is released from the gag-pol precursor by a single cleavage (though in the avian viruses, only about half of the IN regions are

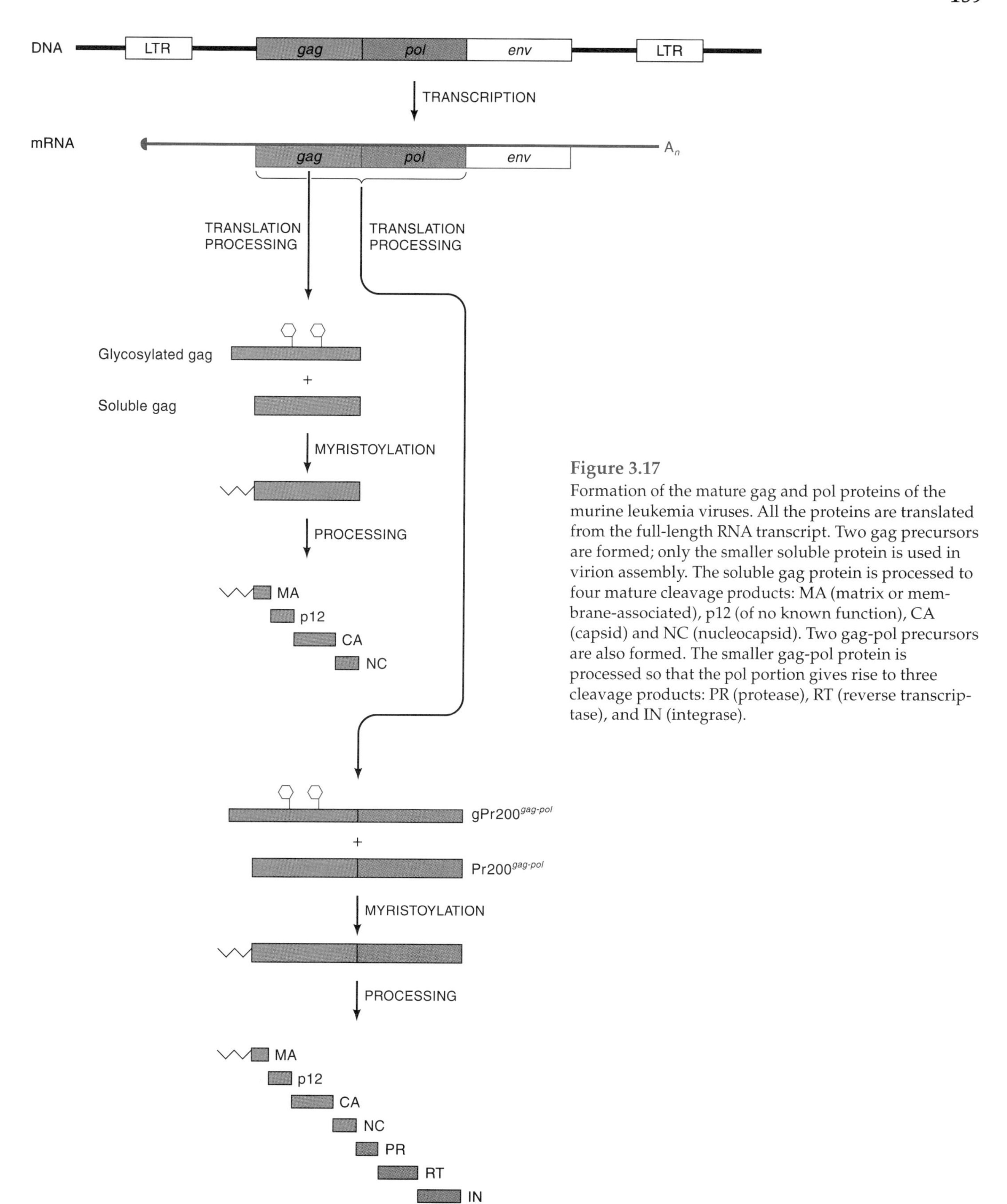

Figure 3.17
Formation of the mature gag and pol proteins of the murine leukemia viruses. All the proteins are translated from the full-length RNA transcript. Two gag precursors are formed; only the smaller soluble protein is used in virion assembly. The soluble gag protein is processed to four mature cleavage products: MA (matrix or membrane-associated), p12 (of no known function), CA (capsid) and NC (nucleocapsid). Two gag-pol precursors are also formed. The smaller gag-pol protein is processed so that the pol portion gives rise to three cleavage products: PR (protease), RT (reverse transcriptase), and IN (integrase).

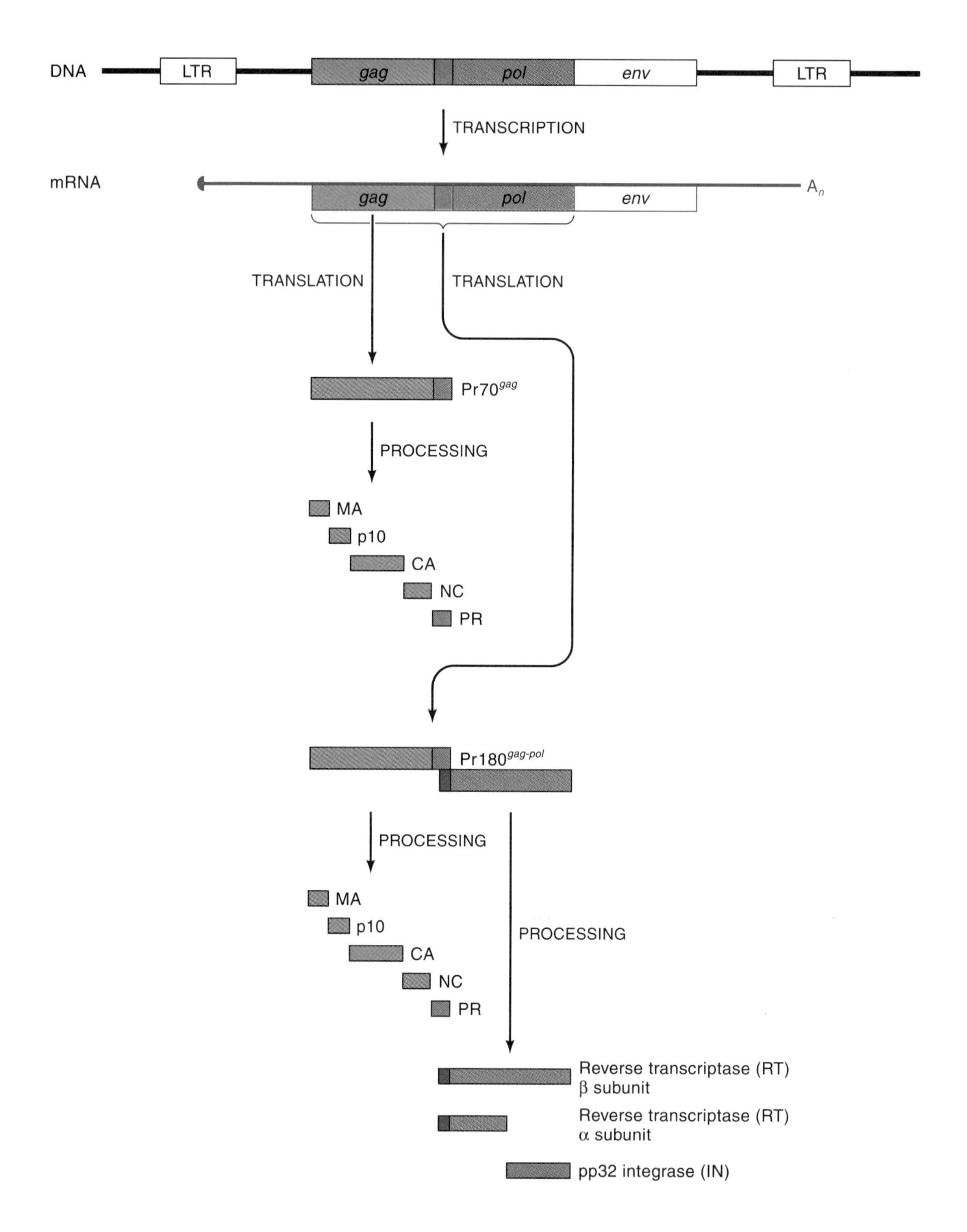

Figure 3.18
Formation of the mature gag and pol proteins of the avian leukosis viruses. All the proteins are translated from the full-length RNA transcript. The gag protein is processed to form five major products, including the PR (protease) from the C-terminus. The gag-pol precursor is processed so that the pol portion gives rise to a heterodimeric RT, consisting of one alpha and one beta subunit, and the pp32 IN (integrase). The cleavage of the beta subunit to form the alpha and IN products only occurs approximately half the time.

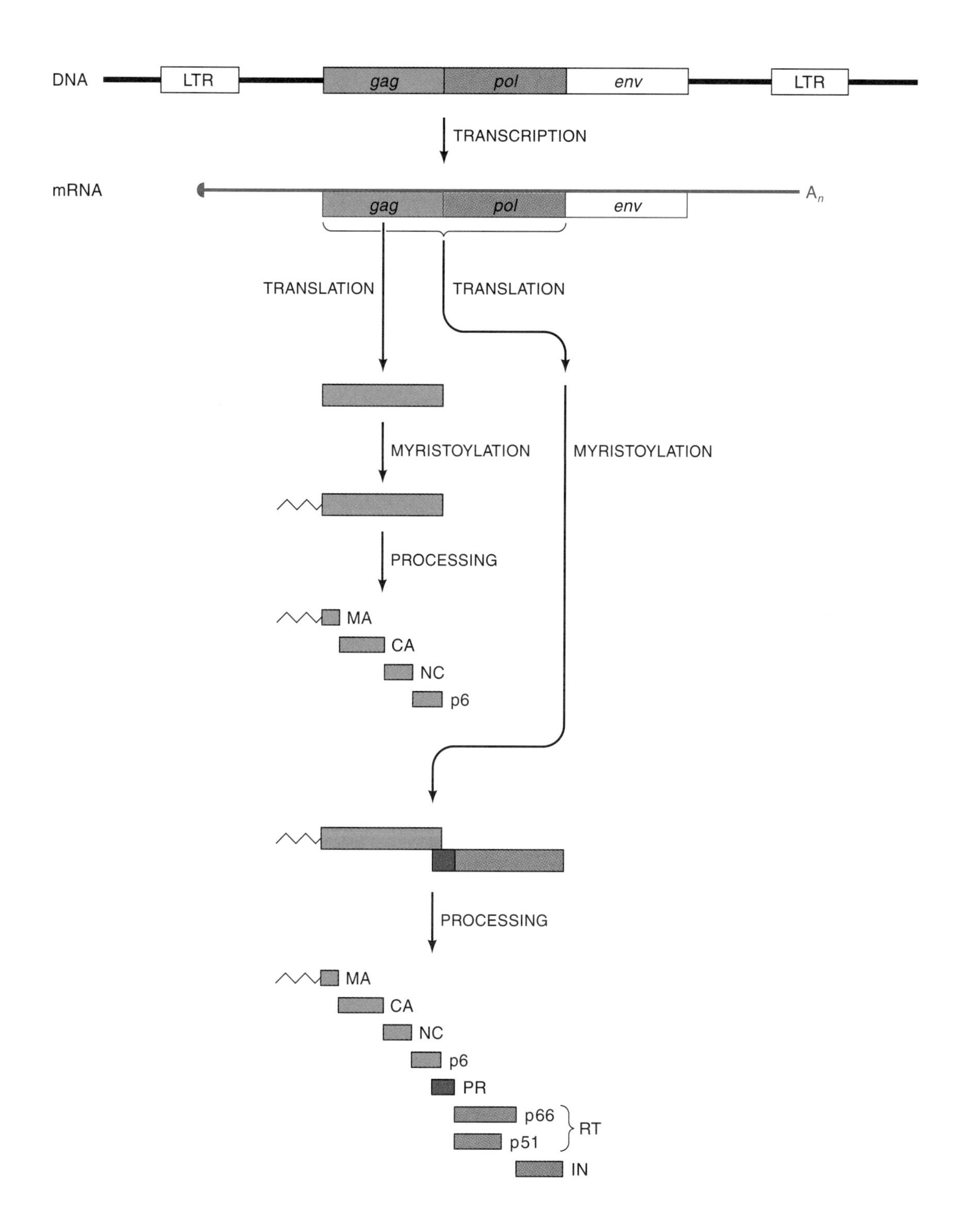

Figure 3.19
Formation of the mature gag and pol proteins of the human immunodeficiency virus type 1. The gag protein is processed to form four major products: MA, CA, NC, and p6. The gag-pol precursor is processed in a complex pathway: the pol portion is cleaved to give PR (protease) from the 5′ region, RT (reverse transcriptase) from the central region, and IN (integrase) from the 3′ region. Exactly half the p66 subunits of RT are cleaved to yield the p51 RT subunit, and the mature RT is a p66/p51 heterodimer. The fate of the remaining 15-kilodalton cleavage product is uncertain.

cleaved from RT; the rest are retained in the β-subunit of RT). The IN protein is essential for insertion of the viral DNA into the host chromosome. The isolated protein exhibits two major activities. First, it has an endonuclease activity with weak specificity for the DNA sequences at the tips of the full-length, linear, viral DNA; the nuclease releases the terminal dinucleotides from the 3′ ends (Section 3.2b). The activity is highest when the relevant sequence is positioned near the end of a linear DNA and relatively low when the same sequence is internal (i.e., far from any end). This "trimming" activity generates a new, recessed 3′ OH terminus at each end of the genome and determines the position on the viral genome that will ultimately be joined to the host DNA. Second, the IN protein has strand transferase activity, capable of joining the 3′ OH ends of the viral DNA to phosphates on a target sequence, with release of the 3′ OH previously linked to the phosphate. The reaction normally joins both ends of the linear viral DNA genome to the target DNA, leading to the formation of the integrated provirus. Purified IN protein alone, without added viral or host proteins, is sufficient to mediate integration of the viral DNA, albeit at low efficiency. It is possible, however, that the IN protein normally acts in the presence of the gag proteins, the reverse transcriptase, and possibly other proteins, and these proteins may facilitate or stimulate its activity *in vivo*.

The DNA sequence of the avian viral genome revealed that its *pol* gene is organized and processed rather differently from that of the mammalian viruses (Figure 3.17). First, the protease gene is not contained in the *pol* gene at all; it actually lies at the 3′ end of the *gag* gene, just upstream of the site of translational termination. Second, the avian viruses thus contain 10 to 20 times higher levels of the protease than the mammalian viruses. The avian viral RTs are dimeric. All the precurser proteins are cleaved at the N terminus, but only about half are cleaved at the C terminus. The resulting enzyme, the so-called α β-heterodimer, has an α subunit, the smaller product of complete cleavage at both ends, and a β-subunit, the larger product resulting from failure to cleave at the C terminus. The β-subunit thus retains the sequence content of the entire α-subunit fused to the C terminal domain, corresponding to the IN domain.

The gag-pol proteins of the human immunodeficiency viruses are processed in a third distinctive way (Figure 3.18). The HIV protease is encoded in the *pol* gene and is released in the same way as in the other mammalian viruses. The viral RTs are dimers, but of a still different sort than the avian RTs. Here cleavages at both the N and C termini of the RT region occur on all precursor molecules, but an additional cleavage occurs on half the subunits, releasing a small polypeptide from the C terminus. The result is a heterodimer of two subunits (p66 and p51) that have identical N termini but in which the smaller of the subunits (p51) is truncated at the C terminus. The HIV-1 reverse transcriptase has been purified from bacteria expressing the appropriate portions of the *pol* gene, crystallized, and analyzed by X-ray diffraction. The studies suggest that the HIV RT heterodimer is a highly asymmetric molecule in which the larger subunit is the active enzyme and the smaller subunit plays only a supporting role to help order the structure of the larger one. A detailed picture of the structure of this enzyme has been extraordinarily helpful in understanding its function.

Env precursor: the transmembrane glycoprotein on the virion surface The envelope proteins of all retroviruses, unlike the gag and gag-pol proteins, are translated from a spliced mRNA (Figure 3.20). The splicing joins a short leader sequence onto the coding region, with removal of a large intron containing the bulk of the gag and pol open reading frames. The splicing also removes the major determinant for the packaging of the viral genomic RNA into the capsid, ensuring that the envelope mRNA is only inefficiently packaged into virions. Incorporation of any such spliced mRNAs into the virion would be deleterious; their reverse transcription and integration would, in essence, result in the generation of a deleted

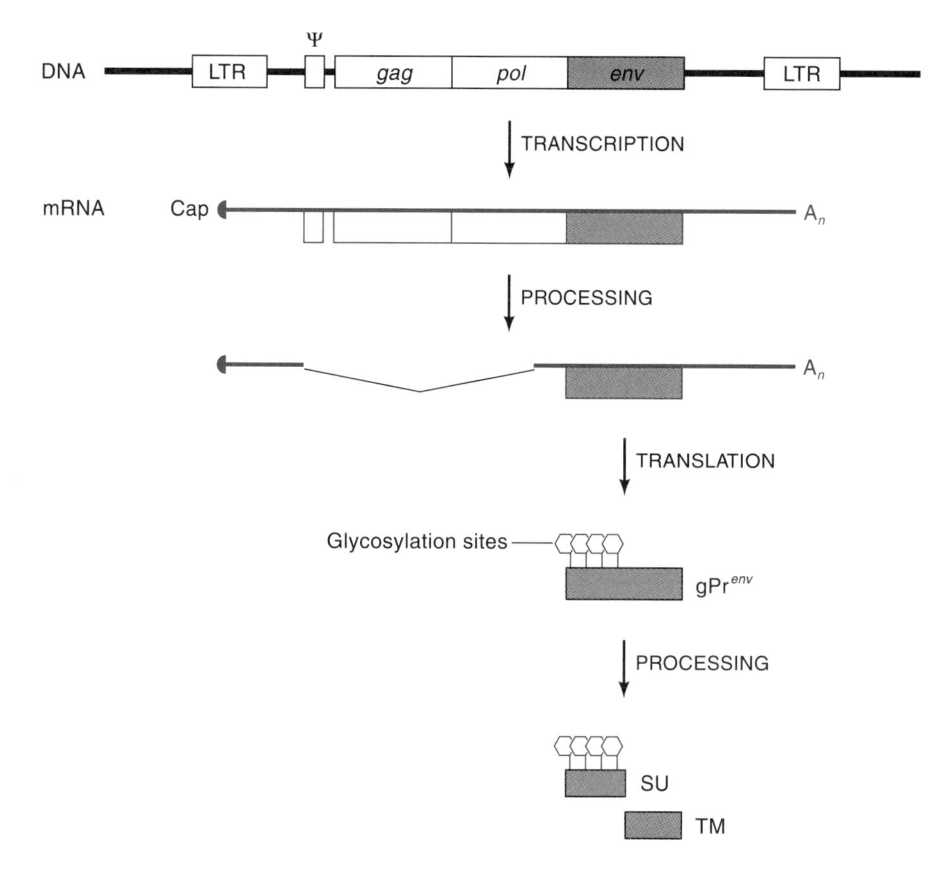

Figure 3.20
Formation of the *env* gene products. The env precursor protein is translated from a singly-spliced mRNA and is modified by addition of sugar moieties. The precursor is cleaved by host proteases in the Golgi to yield the mature SU (surface) and TM (transmembrane) products.

provirus that could not subsequently give rise to the needed viral proteins. In a normal situation, only a portion of the viral RNAs is spliced, producing about equal yields of unspliced and spliced mRNAs. This inefficient processing is determined not by a viral or host protein, but merely by inefficiently recognized splicing signals in the viral genome. The efficiency of the splicing reaction is apparently crucial to the healthy replication of the virus because mutants with abnormally high or low levels of splicing are severely impaired in replication and quickly give rise to variants with more appropriate levels of splicing.

The viral env mRNA is translated on the rough endoplasmic reticulum. The envelope protein contains an N terminal signal peptide, targeting the nascent protein to the membrane-bound signal recognition protein and initiating its translocation through the endoplasmic reticulum membrane. The protein is glycosylated cotranslationally by the addition of N-linked sugars at many Asn-X-Thr/Ser positions and in some cases by the addition of O-linked sugars. As the protein is transported from the ER through the Golgi compartments, the sugar chains are modified by trimming and decoration with high-mannose chains. In addition, the backbone of the protein is cleaved, probably by the newly characterized furin proteases, to generate a two-chain protein. These proteases, located in the Golgi, have specificity for dibasic amino acids, and are thought to be responsible for the cleavage of many host precursor proteins. The env protein is retained in the cell's plasma membrane by a single hydrophobic stretch of residues near the C terminus, acting as a stop transfer signal to fix the protein in the membrane. The protein then oligomerizes probably to form, in most cases, a trimer. The larger N terminal domain, termed SU for surface, lies fully on the outside of the membrane; the smaller C terminal domain, termed TM, anchors the protein in the membrane (Figure 3.20). The two chains are held together by noncovalent bonds and, in some cases, by disulfide linkages as well. The completed, fully modified protein is abundant in the membrane of infected cells. When virions are assembled, the protein is concentrated at the assembly site and incorporated into the particles for export. The envelope protein thus becomes the major, though not the only, membrane protein found on the virion surface.

The major role of the env protein is to target the virion to the appropriate cell receptor for virus infection. Because it binds specifically to the receptor, the env protein determines the species tropism of the virus—that is, the range of species in which the virus can grow—and, in some cases, the tissue tropism—that is, the range of tissues in which the virus replicates. Upon binding, the envelope protein is thought to undergo a conformational change, exposing the stretch of amino acids at the N terminus of the TM chain. These residues, sometimes termed the fusion peptide, seem to promote the fusion between the host membrane and the virion membrane. In some cases, that fusion event can occur at neutral pH and thus immediately after virus binding at the cell surface; in other cases, the reaction is triggered by low pH and occurs only after the virus has been internalized by uptake into the endosomes. In either case, the result of the fusion process mediated by the env protein is the delivery of the virion core, free of its lipid envelope, into the cytoplasm of the infected cell.

Assembly of progeny virions Examination of virus-producing cells by electron microscopy has provided the only available picture of assembly, albeit a rather crude one (Figure 3.21). Staining with antibodies linked to gold particles and examining thin sections in the electron microscope show under the membrane a patch of material that expands into a hemispherical structure, inducing curvature in the membrane. Antibodies specific for different viral proteins reveal that these patches are made up of gag proteins. As a particle grows, the membrane buds out from the cell, eventually forming a spherical shell connected to the cell only through a narrow stalk. Finally, the nascent virion is pinched off from the cell surface and released into the extracellular medium.

The analysis of naturally occurring variants and mutants generated by site-specific mutagenesis has revealed some of the requirements for assembly. The key "machine" for the budding process is the gag protein, which, in the form of the precursor, is transported to the host membrane—most often the plasma membrane, though for some viruses intracellular membranes—where aggregation occurs. The myristoyl group added cotranslationally to the N terminus promotes the targeting of the protein to membranes, though the mechanism of that targeting is obscure. The N terminal MA region (Figures 3.17, 3.18, and 3.19) is probably also important for membrane localization. As the protein accumulates, side-to-side contacts between the gag monomers generate patches, and the geometry of the contacts is thought to induce curvature in the membrane as the bud forms. Portions of the gag CA region are thought to be important for these contacts.

The expression of the gag protein alone in a cell is sufficient, in the complete absence of the viral env and pol products, to direct the formation of particles and their release from the cell. (Such particles, lacking the "hair" of the env protein on their surface, are sometimes termed "bald particles.") In addition, gag is sufficient to specify the incorporation of the viral genomic RNA into the particles; it alone can recognize, bind, and package viral RNAs containing the appropriate packaging signals. The NC region is the most critical for genome binding, though amino acid residues elsewhere in the precursor may also be important in the sequence specificity of the RNA binding activity.

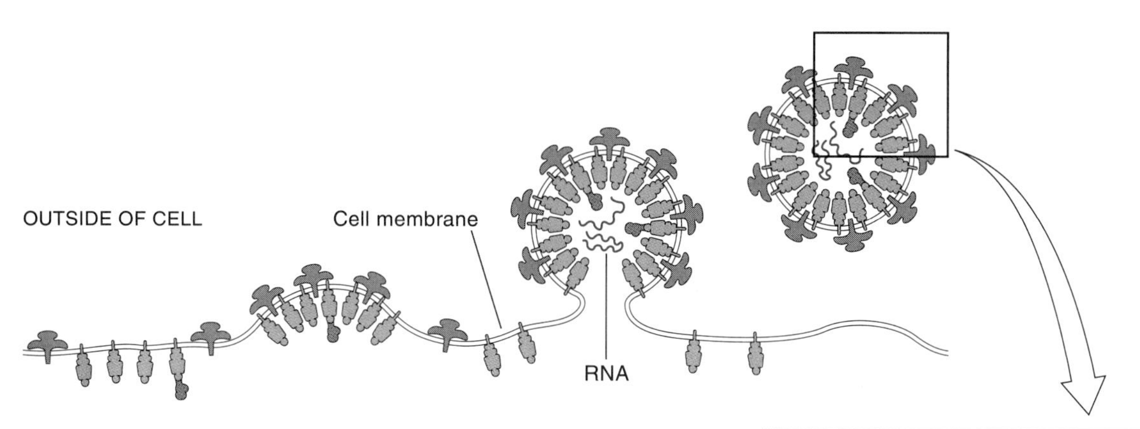

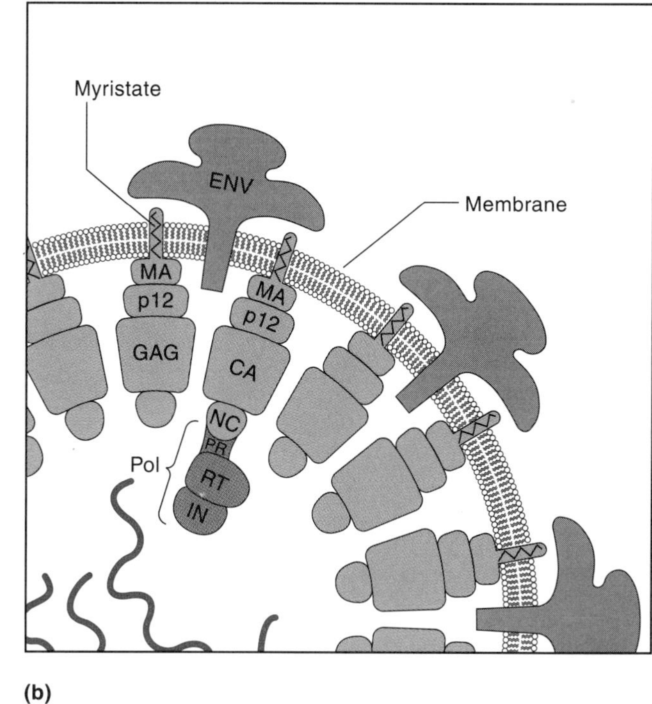

Figure 3.21
Assembly of the retroviral virion particle. The gag and gag-pol precursors are largely cytoplasmic and are retained in the membrane by their N-terminal fatty acid modification. The envelope protein is an integral membrane glycoprotein. The virion particle is formed by the side-to-side aggregation of the gag and gag-pol precursor proteins; as the particle grows, curvature is induced in the membrane. The env protein is attracted to the site of the bud, probably by contacts with gag. The RNA genome is incorporated, also through contacts with gag. At the end of the budding process the membrane is pinched off, and the gag and gag-pol precursors are processed to their mature products.

The proteolytic processing of the gag and pol proteins normally associated with assembly is not necessary for the assembly process; expression of gag in the absence of the protease still results in the production of particles. Thus, cells transformed with a construct expressing only the *gag* gene or with a complete provirus carrying a small mutation in the protease both release uncleaved particles that are stable though composed of uncleaved precursor proteins. However, processing is essential for viral replication because such particles are not at all infectious. The processing activates the infectivity of the particles for their successful entry into the next host cell.

The incorporation of pol proteins into virion particles is thought to be specified by the presence of gag sequences at the N terminus of the large fusion protein. Thus, the gag sequences of the gag-pol protein are not merely present to facilitate the expression of the *pol* gene, but rather play an active role in the proper handling of the pol products. The simplest notion is that the gag-pol precursor is incorporated into the growing virion as though it were a gag monomer with an extended C terminal tail; the assembly process might draw randomly on the pool of gag proteins, whether or not each molecule is in the form of a gag or a gag-pol protein. Consistent with this notion is the fact that the abundance of the gag-pol protein relative to gag protein in the final virion is roughly the same as seen intracellularly.

Stronger evidence in support of this simple model comes from studies of mutants in the *gag* region. Specific mutations can be introduced into the *gag* gene of a cloned proviral DNA; for example, deletion mutations can be created with restriction endonucleases within the MA, CA, or NC domains. Such mutants can then be tested in two assays: first, measuring the mutant gag protein's ability to mediate assembly, and, second, determining the corresponding mutant gag-pol protein's ability to be incorporated into particles being assembled from wild-type gag protein. In the first case, the mutants are analyzed for their ability to direct the release of particles when expressed in wild-type cells. In the second case, the mutants are tested for their ability to complement a cell line that already expresses a wild-type gag protein by providing the pol proteins to those empty virions. The result of such analyses is that there is a good, though not perfect, concordance between these two phenotypes. Thus, the same parts of the gag protein seem to be required for self-assembly and for the incorporation of gag-pol into virions. These results support the notion that the gag portion of the gagpol precursor protein is responsible for that incorporation.

The envelope protein seems to be a passive player in the assembly process. It is distributed throughout the plasma membrane, and only a fraction is concentrated at the sites of virion budding. The C terminal portion of the TM protein is probably responsible for interacting with the nascent particle. The standard model is that the cytoplasmic portion, or the portion actually embedded in the membrane, makes specific contacts with the gag protein as it assembles. At this time, the nature of those contacts remains unclear. As already noted, the env protein is not essential for forming particles per se, and the expression of the murine leukemia virus env gene in a cell already releasing "bald particles"—those made up only of gag protein—does not stimulate the rate of virion production.

One other component of mature virions is exceedingly important to their function: the viral genomic RNA. The incorporation of this RNA, and the exclusion of other cellular RNAs present in the cytoplasm, is thought to be mediated by specific protein-RNA interactions between the gag precursor protein and portions of the viral RNA. The sequence on the viral RNA recognized by the assembling virion, termed the ψ region, lies near the 5′ end, though the precise sequence and structure required are not known for any virus as of yet. A curious feature of the virion RNA is that two copies of the genome are incorporated into each virion particle. Soon after assembly, the two viral RNAs undergo a significant structural change, becoming complexed with the basic NC protein and condensing into a dimer structure held together by base pairing near the 5′ ends. The full significance of the dimer structure to the infectivity of the virus remains unknown, although both RNA chains can contribute to the formation of the proviral DNA (Section 3.1b).

d. Leukemogenesis: Insertional Mutagenesis

Most retroviruses are neither cytopathic nor cytotoxic but rather quite benign. This is because all the processes and steps of the life cycle normally occur without serious consequences to the host cell. Retroviruses usually enter a cell, synthesize the DNA genome, and integrate it into the chromosome without noticeable effect on cell physiology; the mere acquisition of an extra stretch of DNA is not a burden on the cell's resources. Even when the newly integrated provirus is strongly expressed to form viral mRNAs and proteins, serious consequences are rare. The expression itself may subvert up to 1 percent of the cell's synthetic effort, but most cells seem to be able to tolerate this expense without detectable effect. The budding process is not lytic or toxic, and the cell can continue to grow and shed virion particles throughout its normal lifetime. The only exceptions to this rule are the important immunodeficiency viruses, which can be rapidly cytolytic when strongly expressed. The mechanism of the cell killing is unknown, though there is some evidence that the high expression of the HIV-1 envelope protein may be at least in part responsible for cell death.

In spite of the relative harmlessness of most retroviruses during growth in cell culture, these viruses are very often oncogenic in animals. A variety of proposals was raised early in the study of these viruses to explain this discrepancy. One theory was that the viruses stimulated the immune system and enhanced the chances that rapidly dividing cells will form a tumor; another was that infection of T cells by the virus could result in autocrine activation of the T cell receptor. But the major route by which they cause disease is now firmly established: they act as insertional mutagens to activate the aberrant expression of selected endogenous cellular genes. When the affected gene can perturb the normal control of cell growth and division, the infected cell can be stimulated to divide inappropriately and can ultimately lead to the formation of a lethal tumor. This mechanism is broadly applicable across the known retrovirus families and is apparently important in tumorigenesis by retroviruses in mice, cats, and birds.

The insertion of retroviral DNA into the host genome is, to a first approximation, quite random, although there are certainly biases in the choice of target site. There is thought to be a broad bias for transcriptionally active regions of the chromosome and for DNAse-hypersensitive sites, and there is recent evidence for a small number of very strongly preferred sites, or "hot spots" for insertion. But these preferred sites, considered in aggregate, are likely to represent only a minority of the total insertion events in infected cells, and most are "pseudorandom." Thus, when it was discovered that the proviruses in leukemias were restricted to a small number of special loci not normally seen as preferred targets in cells infected in culture, it was apparent that these insertion events were special in that they correlated with the tumorigenic process. The process of tumorigenesis must have selected out those rare insertion events that lead to clonal outgrowth. The simple notion was that the insertion sites typical of leukemias had, in some way, activated neighboring host genes, thus leading to the aberrant growth.

In some systems, the provirus found in the tumor cell is not identical to the input virus, but has suffered substantial alterations. These changes may include simple deletions of portions of the genome that are involved in the activation process. In other cases, the alterations are complex and can be identified as substitution mutations arising in the virus before its insertion. Many tumors induced by the murine leukemia viruses, for example, are found to have acquired proviruses with substitutions in the envelope region. These viruses are recombinants between the input virus and endogenous proviral sequences resident in the host genome and arise during the spread of the virus in the animal. The new viruses, termed (for historical reasons) MCF viruses for mink cell focus-forming viruses, exhibit a significantly expanded host range by virtue of the new envelope sequences. Unlike the parental viruses, they can now infect cells of more diverse mammalian species and utilize distinct receptors. These viruses are also often more pathogenic than the input virus. The basis for the enhanced pathogenicity is unclear, but presumably the recombinants, being able to infect an expanded set of cell types, including those resistant to superinfection by the parental virus, can impact on a greater number of cells to initiate a tumor.

Endogenous oncogene activation Surveys of retrovirus-induced leukemias revealed that a small number of target loci are associated with leukemogenesis; to date, about 20 such loci have been identified in the different tumor types induced by the various replication-competent viruses. Many of these target genes have been found to be the very same ones previously identified through other means as protooncogenes, genes whose mutant forms are capable of initiating tumorigenic growth (Chapter 4). Most prominent among these genes are c-*myc*, c-*erb*B, and c-*myb*. Even more commonly, however, insertion sites have served to identify novel genes, such as those termed *int*-1, *int*-2, *Mlvi*-1 and -2, and *pim*-1. The analysis of these genes and their roles in the control of cell division constitutes a major field of investigation going on in many laboratories. There are several mechanisms by which proviral insertion can affect endogenous gene expression.

The simplest route by which a provirus can alter a target gene is by directly stimulating transcription into the gene (Figure 3.22a). The insertion of viral LTRs near the 5′ end of the gene can, in some cases, lead to high-level expression of hybrid mRNAs that initiate in the LTR and extend into the gene. In such "promoter insertions," the provirus is always oriented so that transcription from the LTR is in the same direction as that of the target gene. The hallmark of these events is the presence of RNAs that have viral 5′ ends, including the R and U5 regions of the LTR, joined to all or part of the adjacent gene's transcription unit. Generally, the insertion of a complete provirus does not lead to such mRNAs because the 5′ LTR leads only to the formation of viral mRNAs, and the 3′ LTR is relatively silent. But incomplete proviruses, especially those defective for either one or the other of the LTRs, will often drive formation of these aberrant transcripts. Depending on the point of insertion relative to the natural promoter for the gene, the hybrid mRNA may include all the normal mRNA sequence or only a 3′ terminal portion.

A more common mechanism for activation of gene expression by the retroviruses is the stimulation of transcription that initiates not in the virus but in nearby cellular sequences (Figure 3.22b). The initiation site may be the normal one for the cellular mRNA, or it may be at a completely novel position that does not act as a start site for the wild-type allele. With respect to the gene, the viral insertion can be very distant, either upstream or downstream, and in either orientation. Examples of all these types of insertions are found among insertional activations of the c-*myc* gene in avian leukosis–induced disease. In these events, the provirus is probably acting by bringing the powerful transcriptional enhancers of the LTR in proximity to the gene, and the mechanism is termed "enhancer insertion." Because the distances over which enhancers can act are very large (over 100 kb), locating the affected gene, even when the provirus has been identified and molecularly cloned, can be difficult. In some cases, a single insertion can activate more than one gene.

Retroviral insertions can mutate and inactivate host genes as well as activate them (Figure 3.23). In this way, retroviruses can act similarly to mobile genetic elements (*Genes and Genomes*, chapter 10). A retroviral insertion may directly disrupt the coding region of a gene by insertion into an exon. Alternatively, and perhaps more commonly, the insertion may block expression indirectly; insertions into introns may disrupt the formation of the primary transcript or the splicing of the transcript and result in the formation of a nonfunctional mRNA. Generally, disruption events are difficult to detect because the elimination of function of one allele of a host gene constitutes a recessive mutation that would not elicit any visible phenotype in a diploid cell. However, germ line insertion events—that is, insertions that become permanent parts of an animal's inherited genes—can be detected when homozygous individuals are generated by interbreeding. Several spontaneously generated mutations in the mouse germ line have been shown to have arisen by such retroviral insertions. In special cases, insertional inactivation can be detected: for example, insertions into the X-linked locus encoding hypoxanthine-guanine phosphoribosyltranferase can be scored because it is a single-copy gene even in diploid cells derived from males. Similarly, insertions into hemizygous genes, or loci that are heterozygous with distinguishable alleles, can often be observed. Insertions into the β2-microglobulin gene, for example, were detected after infection of heterozygous cells by selection for the mutant cells with allele-specific (allotypic) antisera and complement.

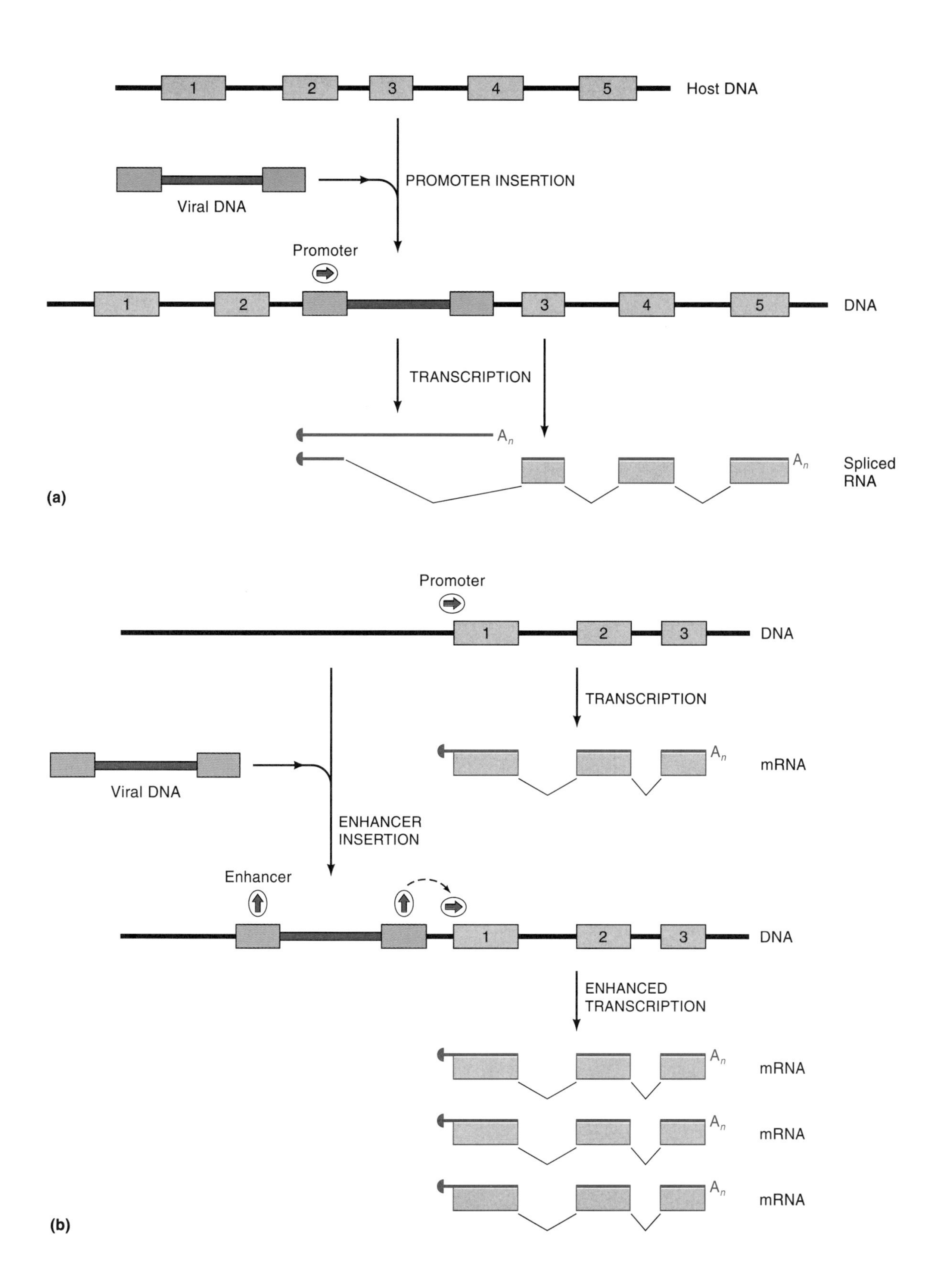

Figure 3.22
Insertional activation. (a) Activation by promoter insertion. In some cases, retroviral integration results in the expression of transcripts initiating in the promoter in either of the viral LTRs. (b) Activation by enhancer insertion. In other cases, retroviral integration results in increased rates of transcription initiating in the normal promoter of the host gene.

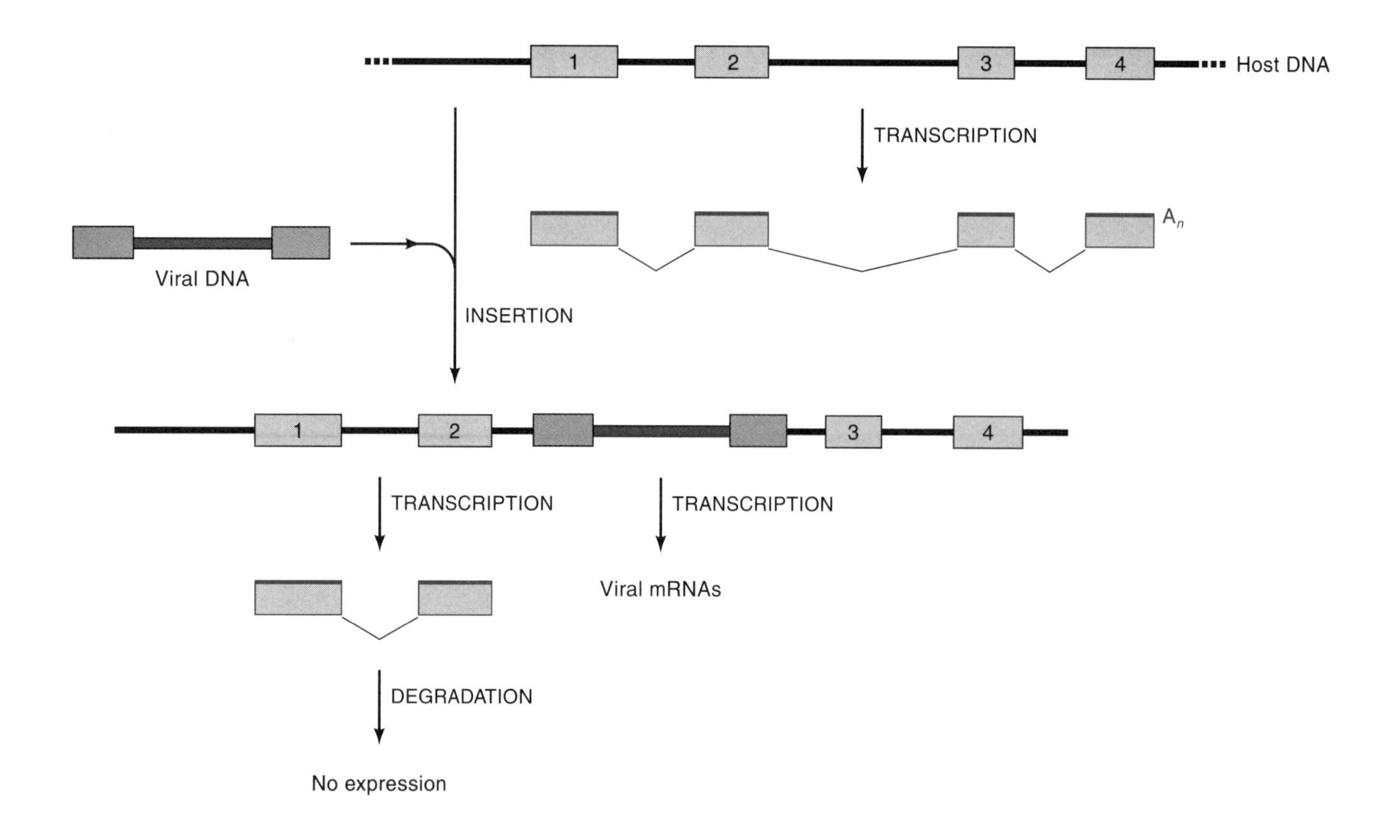

Figure 3.23
Gene disruption by retroviral insertion. Retroviral integration can block expression of the target gene. Even when insertion occurs in an intron sequence, the presence of the viral promoter and polyadenylation sites can affect the structure of the RNA formed at the locus such that the pre-mRNA is not successfully completed, processed, or exported from the nucleus.

3.3 Genetic Analysis of Retroviral Replication Functions

Much of our understanding of the role of the various viral gene products in the retroviral life cycle has come from analysis of viral mutants with defined alterations in the genome. Such studies can provide information at several levels. At the crudest level, analyzing the replication of mutants completely deleted for each coding region can quickly define the basic role of the affected gene product and its time of action. At an intermediate level, analysis of mutants with smaller mutations in restricted portions of the gene can define the location of various domains of the protein product. And, at the finest level, analysis of variant proteins containing substitutions of individual amino acids can define residues needed for catalysis and for contact with the protein's substrates.

In general, the retroviruses are uniquely straightforward among the RNA viruses as subjects for *in vitro* mutagenesis. Cloned copies of the integrated provirus can be obtained, and these cloned DNAs can be altered using all the power of molecular genetics. The mutant DNAs can be characterized precisely and then reintroduced into appropriate cells to establish a functional provirus. The subsequent analysis of phenotype proceeds as in the characterization of any mutant gene. But there is a slight twist compared with the analysis of classically derived mutant viruses: a mutant virus DNA introduced into cells by transfection initiates the viral life cycle at a different point from that of a normal infection. After DNA transfection, the first steps of the life cycle to occur are the expression of the viral genes and the formation of viral proteins. Thus, the effects of the mutation, if any, on the late stages of expression and virion assembly can be immediately determined in the transfected cell lines. If virion particles are successfully released from the transfected cells, the particles can be used to infect fresh cells. The effects of the mutation on the early stages of entry, reverse transcription, and integration can then be determined by analyzing infected cells. In these ways, all aspects of the life cycle can be studied after specific mutations are generated in cloned DNA.

a. Analysis of the Pol Region

A good example of a genetic analysis of a retroviral genome carried out by mutagenesis of cloned DNA is the definition of the domains of the *pol* gene of the Moloney murine leukemia virus. Starting with a proviral DNA cloned on a plasmid vector, researchers generated small deletions at various positions along the *pol* gene. They treated the plasmid DNA with selected restriction endonucleases to form linear molecules, exposed it to the exonuclease Bal31 to remove a small number of base pairs at the termini, and religated it to form circles. They recovered individual clones with specific deletions from bacteria and determined the extent of each deletion by sequence analysis. They established cell lines expressing each mutant DNA by cotransformation of NIH/3T3 mouse cells with a mixture of a given mutant provirus and a selectable DNA such as pSV2neo and recovered virus-producing lines after growth in the selective drug G418. Analysis of such lines showed that they all expressed the retroviral gag and env proteins normally and released virion particles. Further analysis of these particles revealed that the mutants fell into three classes based on their phenotypes and correlating perfectly with the position of the mutation (Figure 3.24).

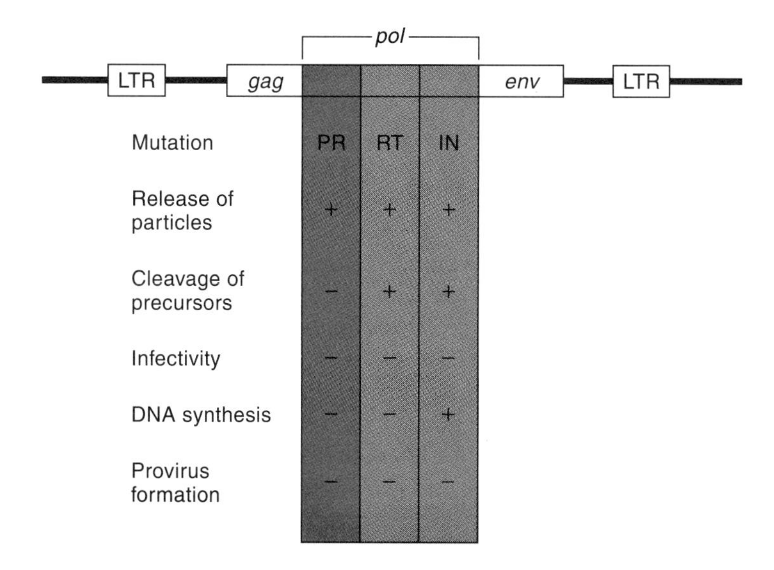

Figure 3.24
Phenotypes of mutant viruses with lesions in different parts of the *pol* gene. Entries under each domain of *pol* indicate the ability of an integrated mutant provirus defective in that domain to carry out the indicated portion of the life cycle.

Mutants with lesions in the 5′ proximal portion of the gene showed an informative phenotype: virus particles were formed and released at near normal levels, but the gag and gag-pol precursor proteins were not cleaved. The result led to the assignment of this region of the *pol* gene as encoding the viral protease (PR). Mutants with deletions in the central portion of the *pol* gene encoded particles devoid of reverse transcriptase activity, confirming that this central region encoded this crucial enzyme. Finally, mutants with lesions in the 3′ proximal portion of the *pol* gene showed a third distinct phenotype. Virus particles were released and were found to contain reverse transcriptase activity, but upon infection, the viruses were unable to establish the integrated proviral DNA in the host cell. Further studies helped define this region of the *pol* gene as encoding the integrase function.

b. Reverse Transcriptase

The retroviral reverse transcriptase is the defining enzyme for this class of viruses. In this sense, the enzyme's role in the life cycle is clearly known. But it is in fact a complex enzyme, and mutational studies have helped elucidate its structure and various functions.

Large deletions in the portion of the *pol* gene encoding reverse transcriptase are readily generated in a cloned provirus, and these mutant DNAs can be established in cells by transformation. The resulting cells express virions containing normal levels of genomic RNA and release them from the cell. Thus, reverse transcriptase is not necessary to assemble particles or encapsidate RNA. Such particles are devoid of RT activity and therefore uninfectious. When the mutant particles enter a cell, they do not direct the synthesis of any viral DNA. The input RNA is not detectably expressed, suggesting that the genomic RNA as introduced into the cell is not readily available for translation. Presumably, the RNA never leaves the virion core and is eventually degraded without generating any progeny virus.

Analysis of the effects of smaller mutations, such as short insertions or deletions, has allowed definition of RT subdomains. The effects of such mutations can be analyzed as for other mutations by introducing the mutant proviral DNAs into eukaryotic cells, recovering virus particles, and analyzing the enzyme contained within them. But in cases such as these, when the effects of the mutations need to be characterized at the enzymatic level, the analysis can more rapidly be performed on constructs that express only the enzyme of interest. Viral reverse transcriptase is, fortunately, a well-behaved enzyme and has been expressed as a separate gene product, free of all other *pol* gene products, in a large number of heterologous systems: *E. coli* and *B. subtilis* bacteria, yeast, insect cells, and even mammalian cells. The gene must be carefully engineered to include an appropriate promoter, an added ribosome binding site and initiator codon, and a terminator codon, all so as to produce a protein that mimics the natural enzyme formed by proteolytic cleavage. These constructs have allowed for the preparation of large quantities of the protein that exhibit the activities of the natural enzyme; both DNA polymerase and RNase H activities are readily demonstrated. Thus, the systems are ideally suited for the rapid screening of mutations introduced into the DNA.

The results of mutational studies have helped us understand the layout of the active sites in reverse transcriptase (Figure 3.25). Assays of a panel of mutants with small insertions for their DNA polymerase activity showed that mutations that destroy polymerase activity tend to group in the N terminal two-thirds of the protein; the mutations that have little or no effect are clustered in the C-terminal one-third. Assays of the same set of mutants for RNase H activity yielded a complementary picture. Here the mutations in the N terminal portion

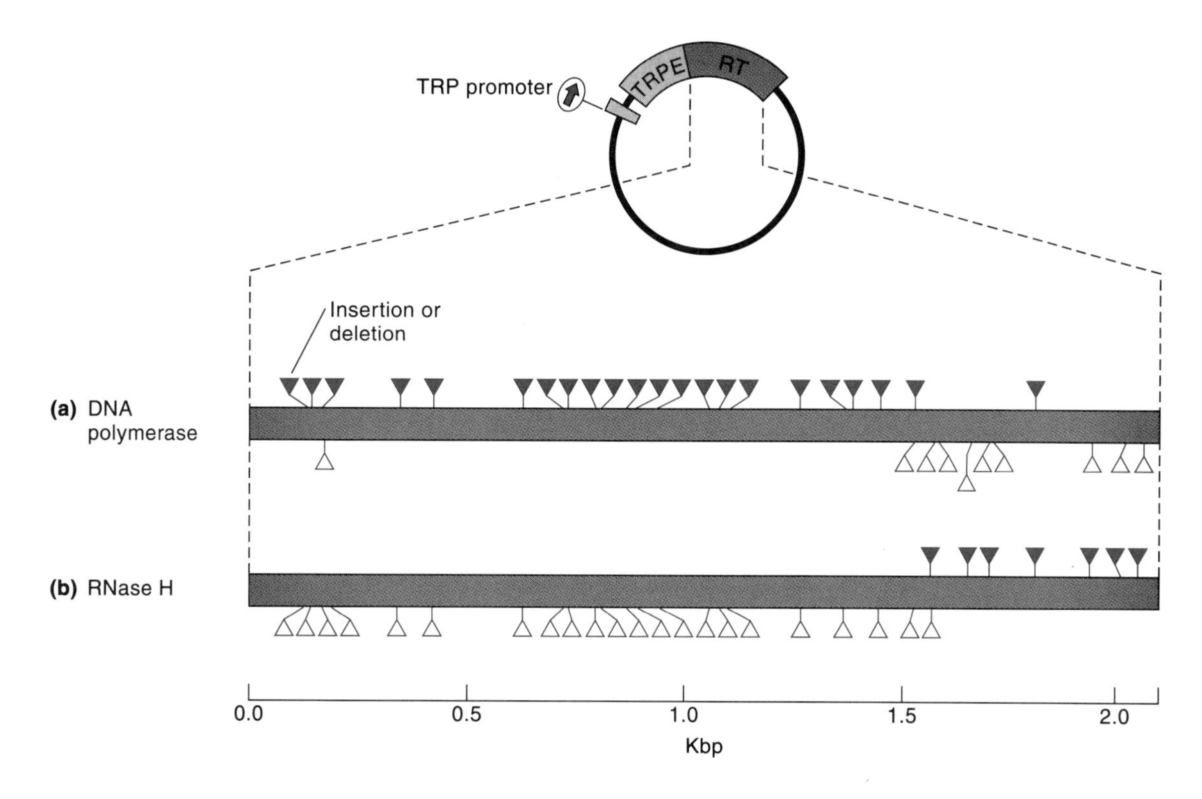

Figure 3.25
Definition of DNA polymerase and RNase H domains of reverse transcriptase. A series of small insertion and deletion mutations were generated in a construct expressing the Moloney murine leukemia virus reverse transcriptase, and each mutant protein was tested for DNA polymerase (a) and RNase H (b) activity. The scale gives the position in the viral genome as measured from the 5′ edge of the RT coding region. The position of each insertion is indicated by a triangle. Solid triangles indicate little or no activity, and open triangles denote wild-type or substantial levels of activity. (Adapted from N. Tanese and S. P. Goff. 1988. Domain structure of the Moloney murine leukemia virus reverse transcriptase. *Proc. Natl. Acad. Sci. USA* 85 1778.)

have little effect on RNase H activity, but those in the C terminal portion abolish the RNase H activity. In addition, the two activities were found to be fully separable and, at least to a first approximation, independent of one another: the separate expression of the N terminal region yields a protein with polymerase activity but no RNase H activity, and the expression of the C terminal region yields a protein with RNase H and no polymerase activity.

As the essential regions for enzyme activities became identified, sequence alignments between reverse transcriptases and other related enzymes helped define the conserved and more variable residues. These alignments helped identify the individual residues likely to be present at the active sites. Finally, the specific alteration of these individual residues and the analysis of their effects on enzyme activity has led to their firm assignment as providing groups critical for catalysis. The three-dimensional structure of the HIV-1 reverse transcriptase protein has been determined by X-ray diffraction analysis of crystals. The structure is remarkable in many ways. (See Figure 3.26.) The heterodimer is quite asymmetric, having no twofold axis of symmetry. The overall shape is similar to a human right hand, with the primer template thought to lie in the "palm," between "fingers" and the "thumb." In the near future, a coupling of mutational and structural studies will likely provide a detailed map of the active site.

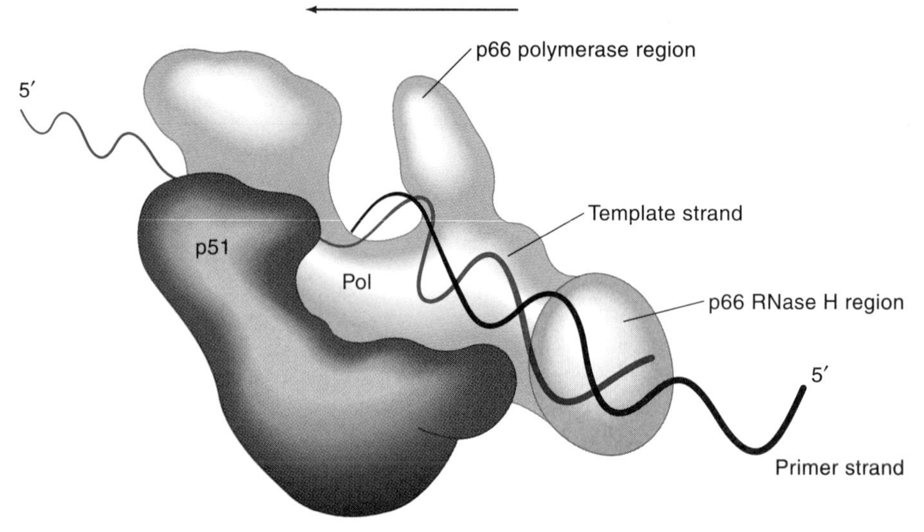

Figure 3.26
Model of the HIV-1 RT heterodimer bound to a template-primer. The p66 and p51 subunits are both included, with the polymerase and RNase H domains of the p66 subunit indicated. Elongation occurs toward the left.

c. Protease

Very detailed mutational studies have also been performed on the retroviral protease enzyme, powerfully fueled by a hope that the enzyme might provide a novel target for antiviral drugs. Both genetic and physical methods have successfully been applied to the study of this protein.

Gross deletions in the region of the *pol* gene that encodes the protease cause arrest of viral replication at a surprising time in the life cycle. Proviral DNAs lacking the protease function can still produce the gag, gag-pol, and env precursor proteins, and these proteins target appropriately to the cell membrane. Furthermore, the gag precursor can still direct the assembly of virion particles and their release from the cell surface. The particles contain the normal complement of RNA and seem in their physical properties to be wild-type, showing normal lipid content and near-normal buoyant density. But examination of the stable particles reveals that they contain the precursors of the gag and gag-pol proteins; the proteins are never cleaved. These immature particles are not infectious, and the block seems to be very early in the infection process, perhaps at the time of particle uncoating.

The sequence of the protease proteins, predicted from the gene's DNA sequence, shows similarity to other mammalian proteases. These other, generally larger proteases contain an internal direct repeat of a block of sequences that contain the active site. Structural analysis of these aspartyl proteases has shown that catalysis requires the concerted action of two aspartate residues in the repeat; evolutionary arguments suggest that the two blocks might have arisen by duplication of a primordial half-enzyme. The retroviral enzyme contains no such direct repeat but rather represents only one copy of the block; on that basis, researchers proposed early on that the enzyme might be a dimer, thereby recreating the active site geometry normally formed by the two parts of the nonviral enzymes. The alignments between the various enzymes also allow assignment of residues likely to be important for catalysis. Many of these assignments have been confirmed by extensive mutational analysis of the HIV protease. Using synthetic oligonucleotides, each of the 99 residues in the protein has been altered to one of several different amino acids, and the mutants' ability to catalyze their own processing and that of other substrates has been determined. In this way, a most complete knowledge of the essential parts of this enzyme has been developed (Figure 3.27). Inspection of the results shows that two large sequence

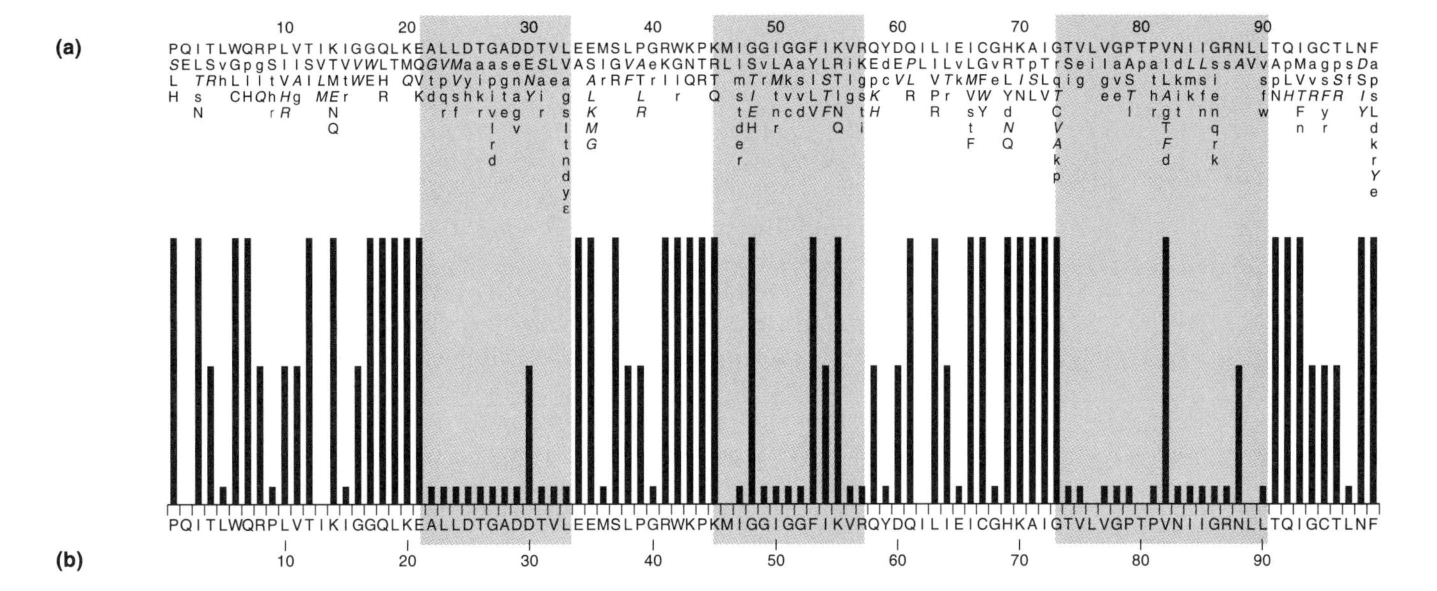

Figure 3.27
Mutational analysis of the HIV-1 PR region of the *pol* gene. (a) The top line presents the amino acid sequence of the wild-type protease, in single-letter code. The vertical column of amino acids underneath each position indicates the mutations generated at that position. The phenotypes are encoded: capital letter, wild-type or enzymatically active phenotype; italicized capital letter, intermediate phenotype; lowercase letter, negative phenotype. (b) The protease activity associated with nonconservative changes. Vertical black bars indicate the effect of nonconservative substitutions; very tall bars, wild-type or positive activity; short bars, no activity. When multiple phenotypes were recorded at one position, the most active phenotype is shown. Three regions of greatest sensitivity to mutation are indicated. (Reprinted with permission from D. D. Loeb et al., 1989. Complete mutagenesis of the HVI-1 protease. *Nature* 340 397–400. Copyright 1989 Macmillan Magazines Ltd.)

blocks are most critical for activity: residues 22–33 and 75–90 cannot be changed without severe effects.

The three-dimensional structure of the protease has been determined by analysis of X-ray diffraction patterns of protein crystals of the HIV-1 protease. Remarkably, the structure was first determined on protein that was completely synthesized *in vitro,* through purely chemical means. The structure that was found is indeed a homodimer, with a broad hydrophobic pocket able to bind substrate. These structures are being intensively studied to help direct the synthesis of inhibitors as antiviral drugs.

d. Integrase

Mutations in the 3′ portion of the *pol* gene helped define the function of this domain. Proviruses lacking this region are able to direct the normal assembly of virion particles, including cleavage of all the virion proteins. The virions contain viral RNA and normal levels of reverse transcriptase. These particles, like those deficient in other pol products, are also noninfectious, but the block to replication is at a distinct step in the life cycle. When these mutant virions are applied to sensitive cells, the particles enter the cell normally, and the entire reverse transcription process proceeds, yielding the full-length, double-stranded DNA product. The particles even enter the nucleus normally, as judged by the appearance of the two circular DNA forms, with either one or two tandem copies of the LTR.

But the DNA does not integrate into the host genome to form the provirus. Instead, the DNA simply persists for a few days as unintegrated forms and gradually disappears from the cultures as the cells continue to divide normally. The viral DNA does not replicate. Importantly, the mutant DNA genome is not efficiently transcribed to give rise to new viral mRNA and proteins. Thus, the unintegrated DNA is likely to be sequestered in some form that is inaccessible to the RNA polymerase II transcription machinery. The infection is thus abortive, with very little progeny virus produced.

Recently, the viral IN function has been expressed separately from the rest of the viral proteins in a number of heterologous systems, including *E. coli* and yeast. These IN proteins have displayed all the activities needed for integration *in vitro*, proving that the IN protein is sufficient, at least at low efficiency, for the entire process, including the trimming and the joining or strand transfer steps (Section 3.2.b). Mutagenesis of these expression constructs is just now being done, and several essential amino acid residues have been found. In general, most IN mutations seem to affect the trimming and joining reactions concordantly, suggesting that the structural requirements for these steps are very similar. The evidence suggests that there might be only one active site used for both reactions.

3.4 Replication-Defective Transforming Viruses

The viruses described previously—the leukemia viruses, in the main—are all replication competent, that is, they encode all the functions needed for their own replication. In contrast, replication-defective viruses have genomes derived from replication-competent viral genomes, but lacking one or more of the genes essential for replication. These defective viruses may have suffered a simple deletion of genetic material, but more often they actually contain sequences derived from the host cell in place of viral genes. These host sequences can confer on the virus important new biological properties; if the sequences code for an activated oncogene, the virus can be a potent **transforming virus**. The defective genomes retain all the *cis*-acting elements needed for replication (including the LTRs) and lack only some of the *trans*-acting genes. Thus, the defective genome can still be recognized by viral proteins and replicated normally so long as the essential replication proteins are provided. The usual means by which these defective genomes are replicated is along with a second virus (i.e., as one component in a mixture of two coreplicating genomes). The second genome is normally a replication-competent leukemia virus, and in this setting, the leukemia virus is said to "help" the defective virus or to act as a **helper virus**.

a. Transfer of Defective Genomes: Dependence on Helper Virus

To describe the replication of a mixture of two viral genomes in cell populations, I begin by considering a hypothetical cell carrying two proviruses integrated into the genome: one a replication-competent helper virus, the other a replication-defective (and perhaps transforming) virus. Such a cell will be a "producer cell," continuously releasing virion particles whose proteins the helper virus DNA encodes. Resident in the cell will be two distinct viral RNAs, namely, the unspliced RNAs of both the helper and the defective viruses, both suitable for packaging into particles. These RNAs will be randomly packaged into the particles in proportion to their abundance in the cell. Because two copies of the retroviral RNAs are packaged into the virions, the cell can, in principle, produce three classes of virions: those containing two copies of the helper virus genome, those containing two copies of the defective virus genome, and those containing one copy of the helper and one copy of the defective virus genome (Figure 3.28).

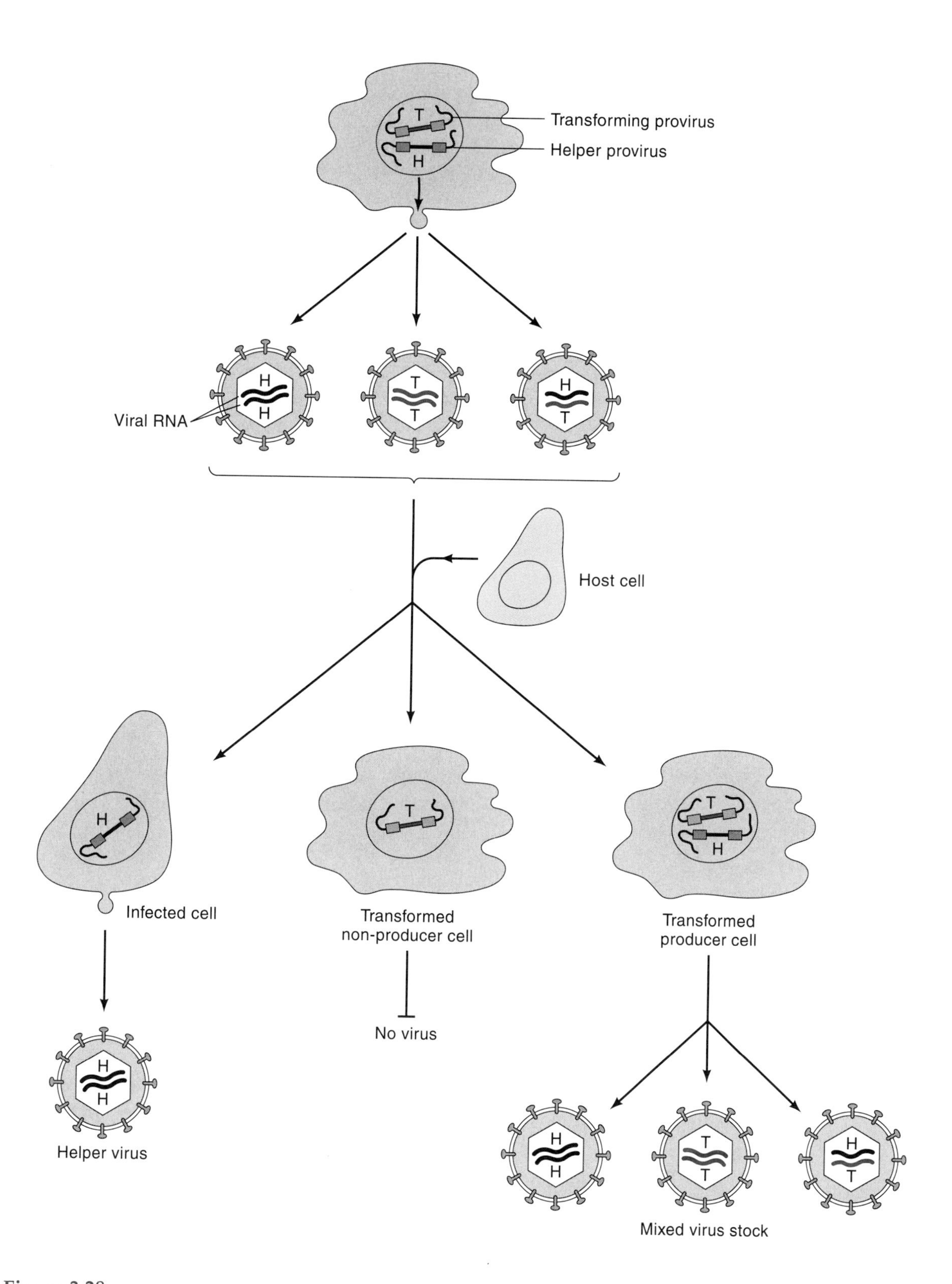

Figure 3.28
Transmission of replication-defective transforming retroviruses (T) in concert with replication-competent helper viruses (H). A cell line containing both proviral DNAs will express viral RNAs corresponding to both viral genomes. The helper virus proteins direct the formation and release of virion particles, and the particles encapsidate both viral RNAs, as homodimers and heterodimers. Upon infection of naive cells, three types of cells can be generated: cells carrying only the helper (H), cells carrying only the transforming virus (T), and cells carrying both (HT). HT cells are able to pass the transforming virus again.

When virions of these three types are applied to a naive, virus-sensitive cell, there are several different possible outcomes (Figure 3.28). Assume that the multiplicity of infection is low (i.e., that each cell in the infected cell population is exposed to only one virion). One possibility is that the cell will be infected by a particle containing only the helper genomes. In this case, the helper genome will be reverse transcribed into DNA and established in the cell as a provirus; the cell will then become a chronic producer of the helper virus. It will not be morphologically transformed because it did not receive a copy of the replication-defective transforming virus. A second possibility is that the exposed cell will be infected by a particle containing only replication-defective transforming virus genomes. In this case, the defective genome will be established in the cell, and the cell will thus become transformed. But because there are no helper genomes present to be expressed, the cell will not become a producer; no virions of any sort will be shed. The cell is said to be a transformed nonproducer cell. The virus genome in these cells is in a sense frozen in the cell, unable to spread further to neighboring cells in the population. The third possibility is that the exposed cell will be infected by a virion containing one copy each of the helper and defective virus RNAs. It is theoretically possible that both genomes could be copied into DNA and integrated into the host genome, but in practice, such an event is probably rare. The overall efficiency of the infection process is low by absolute standards, and hundreds or thousands of physical particles are probably required for each successful infection event. Thus, the odds that one virion would successfully contribute both of its RNA genomes to the cell are vanishingly small. The usual outcome of infection by such a particle is thus equivalent to infection by either one of the homodimeric particles. Either the defective genome will be copied and integrated, generating a transformed nonproducer cell, or the helper genome will be copied and integrated, generating a nontransformed producer cell. At relatively high frequency, one other event can occur: recombination between the two genomes during reverse transcription. Recombinant proviral DNAs are also formed from such infections, sometimes generating viruses with novel properties.

The provirus in transformed nonproducer cells can be mobilized at a later date by a second event: the infection of the cell by a new replication-competent helper virus. At that time, the new virus will establish itself in the cell, induce the release of virions, and offer the possibility of copackaging its own genome with that of the replication-defective transforming virus (Figure 3.29). Transformed nonproducer cells are important intermediates in the generation of stocks of transforming viruses that coreplicate with different helper viruses. The infection of these cells by various helper viruses is the simplest way to generate transforming viruses packaged in the coats encoded by different helper viruses.

A somewhat different set of events follows from a high multiplicity infection of a cell population by a mixed stock of virus. In such a case, a large proportion of the cells will be successfully infected by more than one virus particle; often the infected cell will receive at least one of each type of provirus, both the transforming virus and the helper virus. These cells will immediately become transformed and will also begin to release virus particles. The particles will package both viral genomes, and the cell thus becomes a transformed producer cell. It is equivalent to the original cell that generated the virus stock. In this way, the mixture of the two viruses can be propagated indefinitely through infected cell populations.

b. Genome Structure: Substitution of Cellular Genes for Viral Ones

The replication-defective transforming viruses are all substitution variants in which one or more sequence blocks derived from the host genome are substituted for viral sequences. The genomes of transforming retroviruses are quite di-

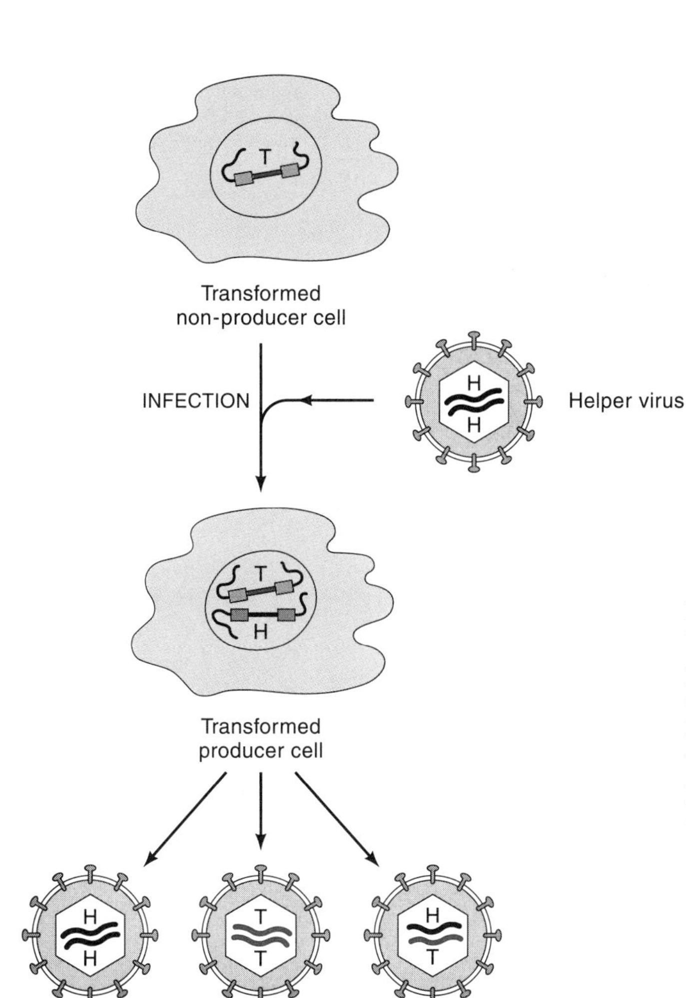

Figure 3.29
Rescue of a replication-defective transforming virus by infection with a helper virus. A transformed non-producer cell carrying only the transforming virus does not release virions. Infection of these cells by helper virus results in the insertion of a helper provirus, which then induces the formation of virion particles. The particles contain the RNA genomes of both the transforming virus and the helper.

verse in structure and mechanisms of expression, but they are constrained by the requirements for transmission mediated by helper viruses. Those requirements are that the RNA genome of the transforming virus be packaged into the helper virions, that the RNA be capable of serving as a template for reverse transcription, that the DNA be able to be integrated into the infected cell, and that the resulting provirus be able to be transcribed. All those requirements can be met by the retention of the two LTRs, the primer binding site for the tRNA primer, the packaging signals, and the polypurine tract for formation of the plus-strand primer. These sequences are all clustered at the two ends of the genome. Thus, the entire center of the genome—the region encoding the *gag, pol,* and *env* genes—is dispensable in a transforming virus and can be substituted with foreign sequences (Figure 3.30).

In the majority of the transforming viruses, a single block of novel sequences is substituted into the center of the genome, replacing, for example, part of *gag,* all of *pol,* and part of *env.* A very common means of expressing the foreign sequences is in the form of a "gag-X" fusion protein; that is, the new sequences are appended to a 5′ portion of retained gag sequences in the same translational reading frame so that the genome encodes a single large fusion protein. Often the splice acceptor for the envelope gene is deleted, and in this case, there may be only a single large RNA, corresponding to the full genome, that is formed from the provirus (Figure 3.30). In other cases, the new gene is expressed from a spliced subgenomic RNA, and translation initiates at its own internal AUG

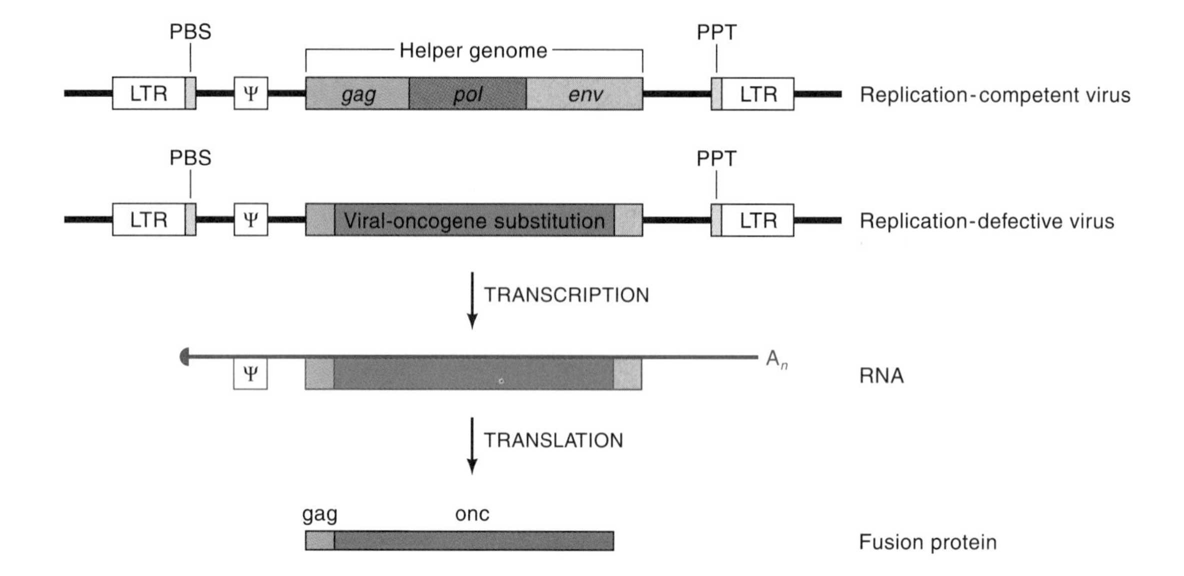

Figure 3.30
Genome of a prototypical replication-defective transforming virus compared to the genome of a parental replication-competent virus (top line). The replication-defective viruses (second line) retain the two LTRs, the primer binding site (PBS), the RNA packaging signal (Ψ), and the polypurine tract (PPT) of the parent. The central portion of the genome is replaced by sequences encoding a transforming protein (third line). In many cases, these novel sequences are expressed as a gag-onc fusion protein (bottom line).

codon. In other cases, much more complex structures have been found. Sometimes two new blocks of host sequences have been found in one virus. The process of forming new transforming viruses can clearly generate many different structures and allows evolutionary pressures to select for viruses that exploit many different methods of expression.

c. Origins: Model for Transduction

The creation of a novel transforming virus from a replication-competent helper virus is a process that occurs in nature, in an infected animal. We presume that an animal is infected with a virus and that during its many cycles of replication a rare transduction event occurs, detected only by virtue of the new properties conferred on the virus by the acquisition of the new sequences. That presumption has been confirmed only in one or two cases, in which the new virus arose in the laboratory, and the events are certainly so rare that they cannot be systematically studied. The nature of the event must be deduced from the structures of the known viruses and from limited experimental attacks on individual steps in the process.

The acquisition and transfer of a host gene by a retrovirus, although formally similar in its consequences to transduction by the bacteriophage, is thought to be mechanistically quite distinct. In nearly all cases, the newly acquired sequences in the transforming viruses correspond to cellular mRNAs rather than cellular genomic sequences: they consist of the spliced exons of a gene rather than the exons *cum* introns that are present in the DNA. This fact suggests that the acquisition process is complex. Analysis of the structures of the existing array of transforming viruses and some experimental work have led to a detailed model of a series of recombination events that can result in the formation of these viruses. The model involves recombination at two stages: one at the DNA level to form the 5′ junction between virus and host and one at the RNA level to form the 3′ junction (Figure 3.31).

Promoter insertion The first step in the transduction process is presumed to be the integration of the helper viral DNA in the vicinity of the gene to be transduced. The simplest initiating event would be the insertion of a provirus near the 5′end of the gene, oriented in the same transcriptional direction as the gene itself. This event might result in the activation of the gene; as such, this might be a common activating mutation that often leads to leukemia in a viremic animal. If the infected cell gives rise to an expanding tumor, there will be more cells potentially able to undergo the subsequent events.

Fusion by read-through or by deletion The next step in the process is presumed to be a spontaneous transcriptional read-through, fusing the provirus and the gene. Transcription initiating in the 5′ LTR of the virus must transit through the viral genome and then continue into the gene to form a stable mRNA that encodes a translatable version of the gene product. RNA splicing can remove portions of the viral and host sequences. Alternatively, deletions of DNA can occur, fusing viral sequences to the gene. If the gene is to be expressed as a gag-X fusion, the deletion determines the breakpoint of the fusion. A deletion of this sort in the DNA is supposed to be the most infrequent step in the overall transduction process and has the least experimental support. Cell lines in which such a deletion has occurred have not been isolated.

Transcription and splicing: packaging of a hybrid mRNA When the 5′ portion of the virus and the body of the gene have been fused, the locus will direct the formation of a hybrid transcript. The transcript will then be processed. The model presumes that the 3′ end of the transcript will be cleaved and polyadenylated at the normal position at the 3′ end of the host gene and that any complete introns contained within the gene's body will be excised. If the events that fused the viral to the gene's body actually joins the viral sequence to an exon, then all subsequent introns can be removed, forming a "normal" mRNA from the point of the fusion onward. If the fusion event joins the viral sequence to an intron, then only the 3′ half of that intron will be retained. The half intron will persist in the final mRNA and in the final genome of the transforming virus. Such a situation was found to have occurred in the transduction of several oncogenes, notably one termed "v-*fps*" present on an avian transforming virus.

The final mature mRNA expressed in the infected cell will generally result in the formation of a functional protein. In a simple example, the mRNA might be translated to form a gag-X fusion protein. The protein product, in the case of the acquisition of an oncogene, would presumably lead to the transformation of the cell. In addition, and more critically, the mRNA would be efficiently packaged into virions whenever the cell was infected by a helper virus. This packaging of the hybrid RNA is possible because all the signals for efficient packaging lie in the 5′ part of the RNA; there is in most viruses no need to retain the 3′ part. This packaging sets up the RNA for the second recombination step in the process.

Recombination during reverse transcription: cDNA formation The hybrid RNA packaged into virions is not yet a complete transducing virus because the RNA cannot itself be efficiently reverse transcribed and integrated into a newly infected cell. The RNA lacks the signals at the 3′ end—namely, the 3′ R sequence and the polypurine tract—that are required for its complete reverse transcription. But even without those sequences, the RNA can enter into the reverse transcription process and can generate a complete transducing virus. The addition of

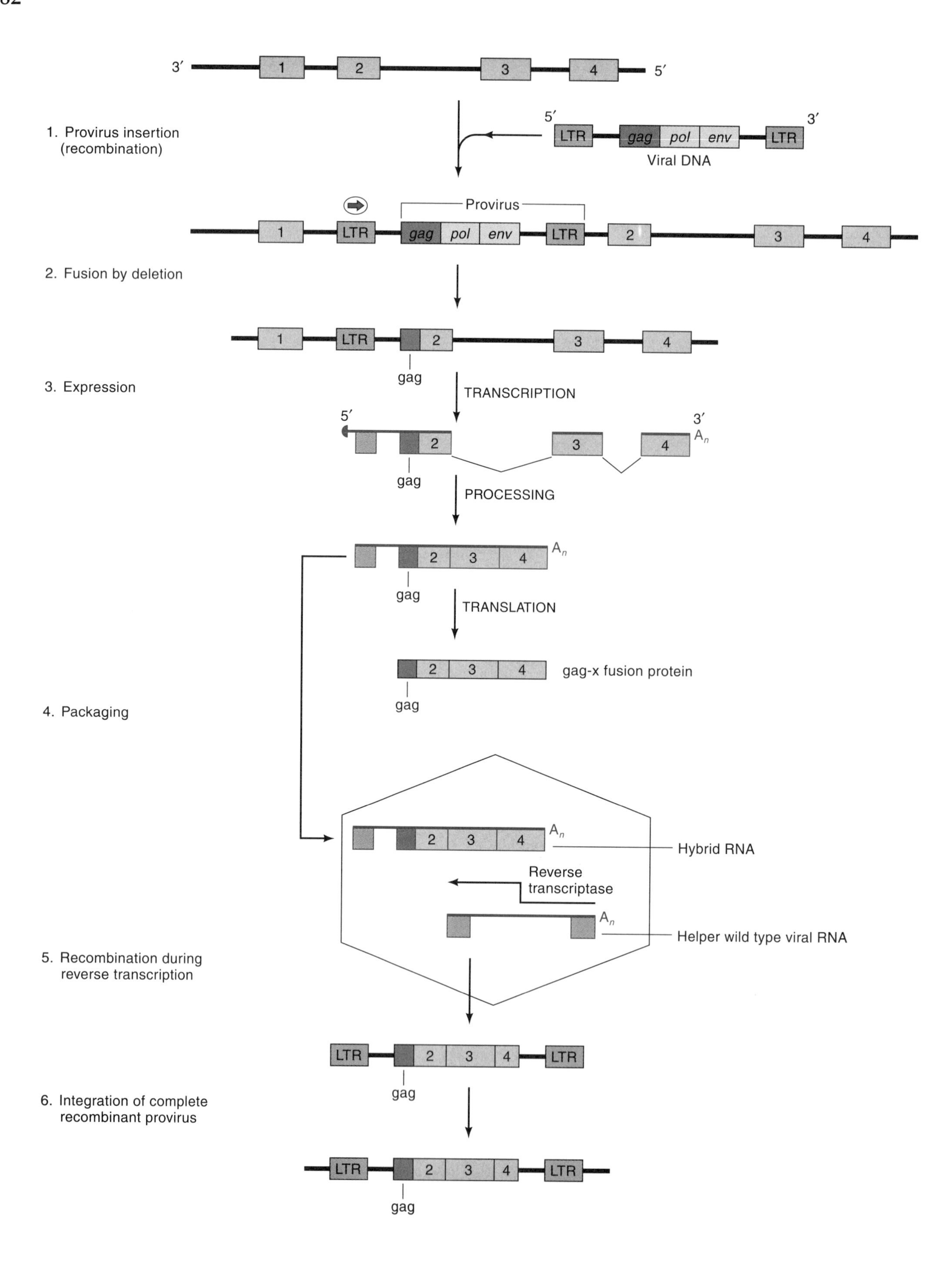
3′
1
2
3
4
5′
1. Provirus insertion (recombination)
5′
LTR
gag
pol
env
LTR
3′
Viral DNA
Provirus
1
LTR
gag
pol
env
LTR
2
3
4
2. Fusion by deletion
1
LTR
2
3
4
gag
3. Expression
TRANSCRIPTION
5′
3′
2
3
4
A_n
gag
PROCESSING
2
3
4
A_n
gag
TRANSLATION
2
3
4
gag-x fusion protein
gag
4. Packaging
2
3
4
A_n
Hybrid RNA
Reverse transcriptase
A_n
Helper wild type viral RNA
5. Recombination during reverse transcription
LTR
2
3
4
LTR
gag
6. Integration of complete recombinant provirus
LTR
2
3
4
LTR
gag

Figure 3.31 (*facing page*)
Model for the generation of a replication-defective transforming virus. Step 1: A replication-competent virus inserts its DNA near or within a cellular protooncogene. Step 2: Deletion of the 3′ portion of the provirus and a portion of the gene results in the formation of a fused transcription unit. Step 3: Transcripts initiating within the 5′ LTR are elongated into the gene, and any intact intron sequences are removed by splicing; the spliced hybrid RNA may encode a gag-onc fusion protein. It is also possible to generate the equivalent hybrid RNA by splicing read-through transcripts from an undeleted provirus; that is, without the deletion shown in step 2. Step 4: By virtue of the Ψ region present near its 5′ end, the spliced hybrid RNA is packaged into virions along with a helper viral RNA. Step 5: During infection of a new cell, reverse transcriptase switches templates from the helper to the hybrid RNA, placing a portion of the helper virus genome at the 3′ end of the oncogenic viral genome. Step 6: The completed viral DNA is integrated normally into the host genome. At this point, the new genome can be transmitted at high efficiency without further rearrangements.

---

a 3′ LTR to the proviral DNA containing the hybrid RNA sequence is one of the steps of transduction that can be re-created in cells in culture, and the step occurs with surprising efficiency.

The addition of a 3′ LTR is thought to occur during reverse transcription in a heterodimeric virion containing one copy of the helper virus genome and one copy of the hybrid RNA. In this model, reverse transcription normally starts at the primer binding site of either RNA and proceeds through the synthesis of the minus-strand strong stop DNA. That DNA then translocates normally, but can translocate only to the helper genome because only this RNA has a region of homology, in the form of the 3′ R region. The minus-strand DNA is then elongated. As its synthesis proceeds, the critical event happens: a "template switch" from the helper to the hybrid RNA, mediated by reverse transcriptase. In the course of synthesis, perhaps at a position where synthesis pauses, the DNA is extended not on the helper, but on the other RNA molecule; the transfer can be stimulated by a few bases of homology, though it can also occur at lower frequency without the benefit of any homology. Once the switch to the other template has been achieved, elongation continues on the hybrid RNA as usual. The rest of the reverse transcription process can proceed without complication to give a full-length, hybrid DNA genome with LTRs at both ends.

This last step in the generation of a transducing virus, unlike the earlier steps, can be reconstructed in cell culture. The frequency with which recombination occurs at the 3′ end of the virus can be tested by artificially generating a producer cell that packages a hybrid RNA containing a selectable marker gene. The 5′ end of a viral DNA is appended to a selectable 3′ gene sequence, and the DNA is introduced into a cell infected with wild-type virus. The virions the cell releases can be shown to carry the hybrid RNA. After infection of fresh cells with the virions, the marker gene is tranferred to a few cells; examination of individual recipients shows that each infection depends on a distinct recombination event that appends the 3′ end of the helper in the formation of each proviral DNA containing the hybrid RNA sequence. The frequency of such transfers suggests that these nonhomologous recombinations occur perhaps once in 10^4 reverse transcription events.

The net effect of these various steps is the transduction of a host gene and the generation of a new retrovirus with very different biological properties from the parental virus. The overall frequency is certainly low; only about 100 or so distinct viral transductions have ever been detected, in the form of the 100 known acute transforming viruses. The most important significance of these rare events has been the finger that they pointed at the cellular sequences responsible for the altered phenotype. The presence of host sequences on transforming viruses was the first, and remains numerically the most successful, route to the identification of oncogenes: genes capable of eliciting, or at least initiating, tumor formation. The study of these genes now constitutes several entire fields of research and touches on aspects of gene regulation, signal transduction, and development (see Chapter 4).

3.5 Human Retroviruses

The study of retroviruses was thrust from an academic study of tumorigenesis in animals into the clinical arena with the discovery of human pathogenic retroviruses. Some of the first such viruses to be discovered, the so-called human lymphotropic viruses HTLV-I and HTLV-II, cause intriguing and important diseases, but probably have a fairly limited range and distribution in people. The second family of viruses to be discovered, the human immunodeficiency viruses HIV-1 and HIV-2, are of profound clinical importance, causing one of the most frightening, publicized, and devastating pandemics of modern times: acquired immune deficiency syndrome, AIDS. These viruses have propelled retroviral biology into the public eye in an extraordinary way. Research into the replication of these viruses has expanded significantly in recent years, but the speed with which these viruses have been characterized depended on the vast body of information about animal retroviruses accumulated over the years and perhaps even more on the technologies of recombinant DNA. The rapid revelation of the structure and biology of these pathogens stands as perhaps the greatest single example of the worth of basic research in the ultimate understanding and resolution of medical problems.

a. HTLV-I and -II: Regulated Retroviruses

HTLV-I was the first human retroviral pathogen discovered; it remains a serious cause of disease today. The virus can induce a rare leukemia, termed adult T cell leukemia-lymphoma (ATLL), which is ultimately fatal after months to years of illness. The disease has several geographic foci, with highest incidence of disease in Japan, parts of the Carribean, and the southeastern United States. Although the epidemiology of the disease is clearly linked to infection by the virus, only a small proportion of those infected actually progress to disease. In addition to the leukemia, HTLV-I has been linked to an array of neurological symptoms, termed tropical spastic paraparesis.

The human leukemia virus HTLV-I is in many regards a typical replication-competent leukemia virus. The genome contains the conventional arrangement of *gag*, *pol*, and *env* genes and replicates via the mechanisms outlined previously. But it differs in two significant ways from the murine leukemia viruses: in its pathogenesis and in the expression of the integrated provirus. Both of these differences are probably due to the presence of two additional viral genes: the *tax* and *rex* regulatory genes. The products of these genes are synthesized from subgenomic, spliced mRNAs, and their coding regions are each contained in two separate exons that flank the *env* gene (Figure 3.32). The coding regions of *tax* and *rex* overlap extensively, and formation of the two proteins requires the translation of these sequences in two different reading frames.

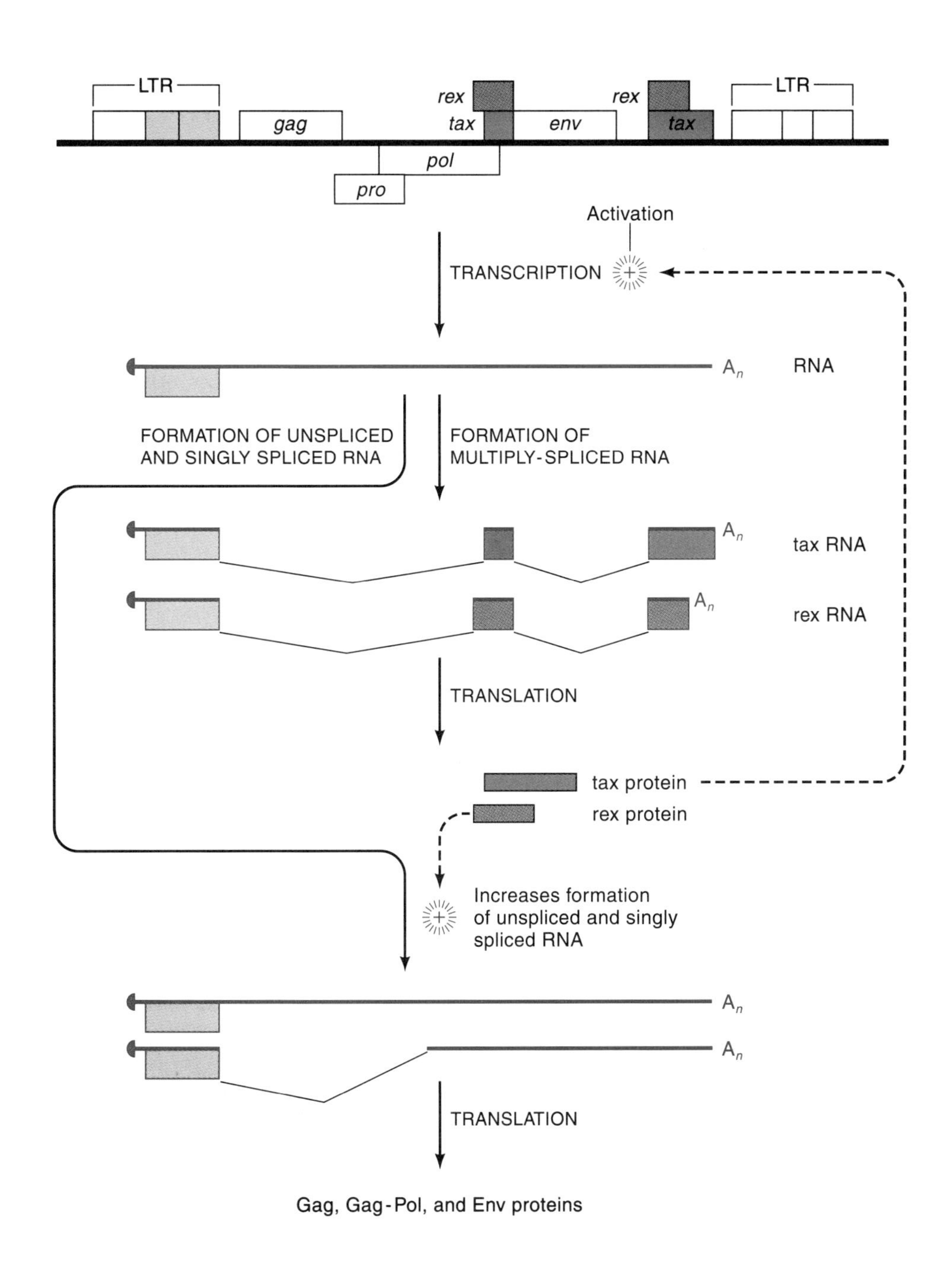

Figure 3.32
Genome structure and expression of the human T-cell lymphotropic virus type 1 (HTLV-I). In addition to the conventional structural genes, the HTLV-I genome includes two additional genes, *tax* and *rex*. The tax and rex proteins, each encoded in two exons, are translated from multiply spliced mRNAs. *Tax* increases transcription from the LTR; *rex* increases the formation of the larger unspliced and singly spliced mRNAs.

The tax *gene* When expression of the HTLV-I provirus occurs, the earliest cytoplasmic mRNAs to be formed are probably the subgenomic mRNAs that encode the two regulatory proteins tax and rex. Formation of these proteins is essential for the subsequent high-level expression of the provirus and the formation of its structural proteins. The *tax* gene product, named for its *trans*-activation function, stimulates the initiation of transcription from the viral LTR. Its expression is thus stimulatory for overall viral expression and can lead to the formation of very high levels of viral proteins. The tax protein is a small nuclear protein. Its mode of action is thought to be direct, through binding to DNA at a specific sequence in the LTR. With the help of still unidentified cellular transcription factors, the tax protein is thought to enhance recognition of the promoter by the host RNA polymerase II complex.

HTLV-I is unusual among the leukemia viruses in that it is capable of initiating transformation of cells when primary T cell populations are infected with virus *in vitro.* Thus, unlike Moloney MuLV, a long period of viremia *in vivo* is not required for transformation. Furthermore, there seems to be no common integration sites for the proviruses in various T cell tumors. These findings have led to the view that HTLV-I must carry its own oncogene, though it has no host cellular sequences.The present evidence is that the *tax* gene itself might be the transforming gene of the virus. Because the *tax* gene encodes a *trans*-acting factor, it has the capability of activating a number of host genes as well as the viral genes. Indeed, it has been demonstrated that the expression of the tax protein alone in cells is able to stimulate the expression of many host genes. Prominent among those tested to date are the genes for interleukin 2(IL2) and the IL2 receptor. The stimulation of expression of these genes may be the major mechanism of transformation. But many other host genes may act as important targets for the tax protein.

The rex *gene* The rex protein is also regulatory, but its action is quite different from that of the tax protein. The formation of the rex protein results in a perturbation of the ratio of spliced to unspliced viral RNAs; in particular, it increases the formation of unspliced genomic RNA and reduces the formation of spliced subgenomic RNAs. As rex protein is formed, it reduces the level of its own spliced mRNA and increases the level of mRNAs that encode the gag and gag-pol proteins. Conceptually, the protein is thus part of a switch that shifts virus expression from regulatory proteins to viral, structural proteins and the formation of virion particles.

The rex mode of action is still uncertain and controversial. Rex is a nuclear protein concentrated in the nucleolus. It seems to act posttranscriptionally, stabilizing and promoting the export from the nucleus of unspliced viral RNAs and singly spliced mRNAs like the env mRNA. The protein acts through specific target sequences in the RNA termed RREs, for rex-responsive elements; only RNAs that include an RRE are affected by rex. The RRE sequence forms a highly ordered secondary structure that includes a large and complex stem-loop. The rex protein is likely to bind directly to the RRE structure (based on similarity to the HIV-1 rev protein), and that binding is somehow able both to interfere with the splicing of the RNA and to facilitate its export from the nucleus. The relationship of the rex protein to the splicing machinery is under active investigation.

b. The AIDS Viruses: Regulated and Cytopathic Retroviruses

The most important known human retroviral pathogen is unquestionably HIV-1, the causative agent of AIDS. HIV-1 is a replication-competent retrovirus, and like all such viruses, it includes the conventional *gag, pol,* and *env* genes. Its mode of replication is also like that of any other retrovirus, as outlined previously. However, the virus has several unique properties that confer on it a potent pathogenicity.

The course of AIDS is long and complex. Upon exposure to the virus, a person first experiences a significant viremia, with a substantial number of blood cells likely to be infected and acquire proviruses. Thereafter, an immune response seems to reduce active viremia, but never to clear the virus completely from the body; infected cells persist. Most cells that harbor virus DNA are T cells, cells in the macrophage and monocyte lineages, and dendritic cells; all share the property of bearing the **CD4 protein**, an essential component of the receptor for the virus, on the cell surface. Over the course of as long as five to ten years, the number of CD4-positive T cells in the peripheral circulation decreases gradually, until the total CD4-positive T cell count drops from the normal level of about 800–1000 cells per cubic milliliter to less than 100–200 cells per milliliter. During this long asymptomatic phase, proviral DNA and viral RNA can be detected in the T cell population, but the proportion of the cells that carry virus at any given time is quite low. As the cell count drops, unrelated bacterial and viral infections occur. In severe AIDS, essentially no CD4-positive T cells remain, and the individual is severely immunocompromised. Eventually, most patients succumb to one of these infections. How these CD4-positive cells are killed and how the infection of only a few of the cells at any one time leads to the ultimate loss of the entire population remains obscure.

Receptor and tissue tropism HIV-1 is able to infect a very narrow class of cell types, including lymphocytes, monocytes, and dendritic cells, with the most significant restriction being at the level of virus entry. The basis for that narrow tropism became clear when the receptor for the virus was identified as CD4, a surface glycoprotein present on selected subclasses of hematopoietic cells. CD4-positive T cells are classically defined as the "helper" class of T lymphocytes, although exceptions to this rule are now known. Thus, HIV-1 efficiently enters only these cells, and the ultimate loss of CD4-positive cells would account for the effects on the immune system. The structure of the CD4 molecule is known, and mutational studies have shown that only a small portion is bound by the viral envelope protein. In cell culture, the virus induces the formation of multinucleate syncytia by cell-cell fusion. One hypothesis for the *in vivo* cytoxicity is that the expression of the viral envelope can induce similar cell-cell fusion, generating syncytial masses of cells that then die; in this scheme, one infected cell might be able to recruit neighboring uninfected cells into a dying syncytium. Other ideas center on the possibility that one viral gene product or another is toxic when expressed at some high threshold level.

The expression of the HIV-1 genes, like that of the HTLV-I, is highly regulated. The major players in that regulation are two regulatory genes termed *tat* and *rev*. These two gene products are translated from multiply spliced subgenomic RNAs and are each encoded by two exons located 5′ to *env* and within the 3′ part of *env* (Figure 3.33). Their effects are similar to those of the HTLV-I tax and rex proteins, though the mechanism of action of the tat protein product is probably different from that of the tax protein.

The tat *gene* The HIV-1 *tat* gene is essential for significant viral expression, increasing the transcription from the LTR by large factors, in some instances, hundreds of fold. The mechanism of action is remarkable, probably unique among transactivators studied to date, and still controversial. The region in the transcription unit required for *tat* action is termed the **tar element**, for tat-responsive region; the sequence lies in the first 59 nucleotides immediately downstream of the site of transcriptional initiation, within the LTR. The RNA encoded by *tar* is able to fold up into a specific stem-loop structure, and the tat protein is an RNA-binding protein that specifically recognizes the *tar* sequence. Mutational studies, in which nucleotides predicted to be important in the formation of the *tar* RNA secondary structure were systematically changed, have defined the sequences required for binding and further have shown that the protein must bind to RNA

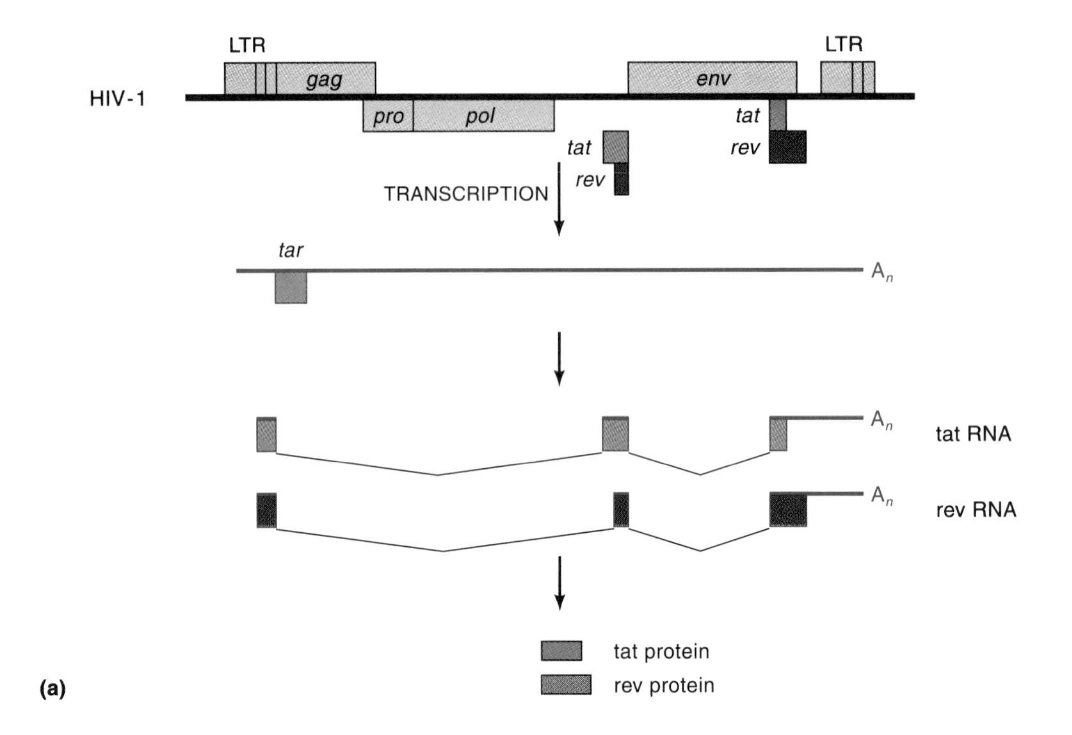
LTR
gag
LTR
env
HIV-1
pro
pol
tat
rev
tat
rev
TRANSCRIPTION
tar
A_n
A_n
tat RNA
A_n
rev RNA
tat protein
rev protein
(a)

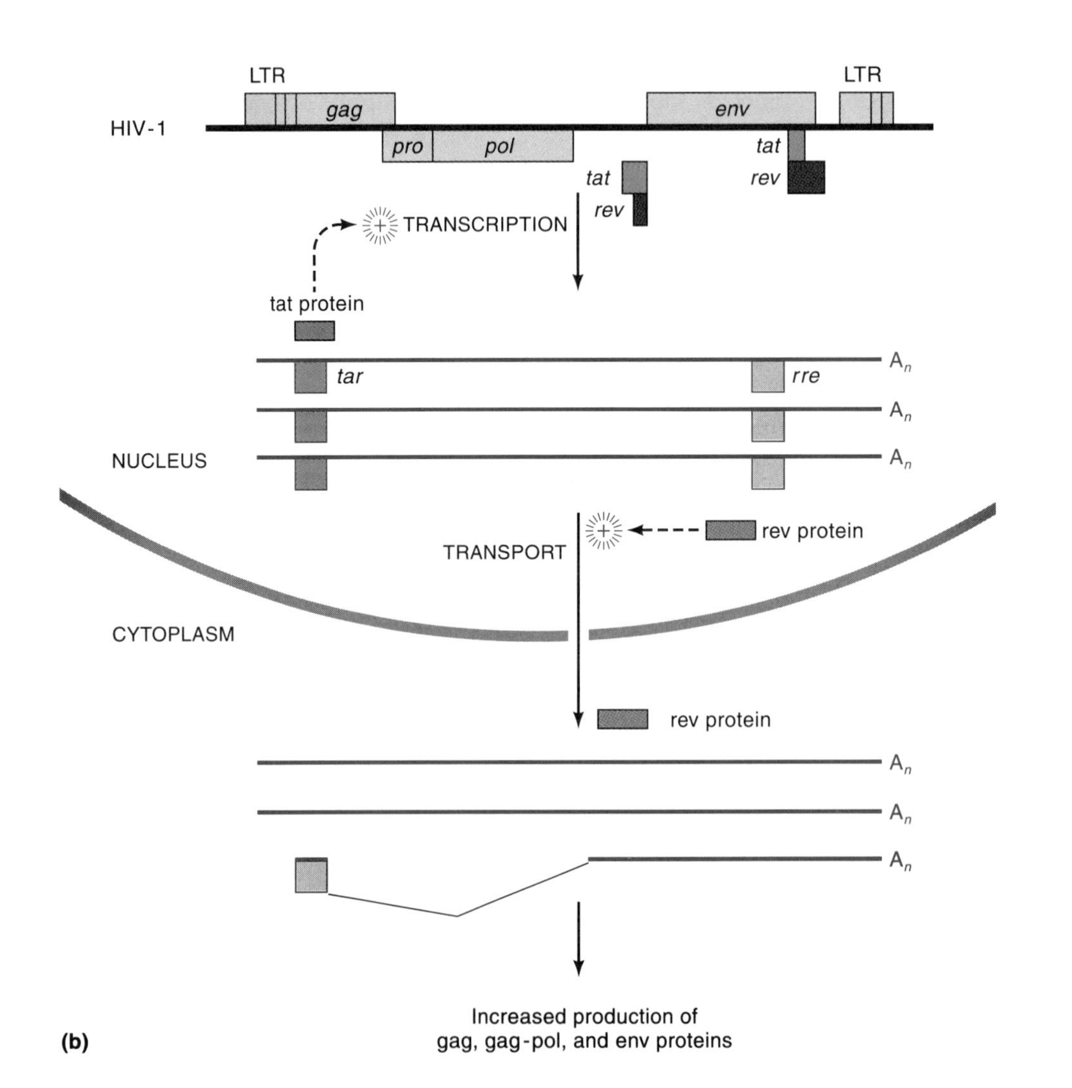
LTR
gag
LTR
env
HIV-1
pro
pol
tat
rev
tat
rev
TRANSCRIPTION
tat protein
tar
rre
A_n
A_n
NUCLEUS
A_n
TRANSPORT
rev protein
CYTOPLASM
rev protein
A_n
A_n
A_n
Increased production of
gag, gag-pol, and env proteins
(b)

Figure 3.33 *(facing page)*
Genome structure and expression of the human immunodeficiency virus type 1 (HIV-1). The HIV-1 genome includes many small open reading frames that encode several small proteins, all translated from multiply spliced small mRNAs. The functions of most are unknown. Important members of this group include *tat*, which acts through the *tat*-responsive element (*tar*) to increase transcription from the LTR; and *rev*, which acts through the *rev*-responsive element (RRE) to increase nuclear export of unspliced and singly spliced mRNAs. The net effect is a dramatic switch to increased expression of the virion structural proteins.

---

to transactivate. Such behavior would suggest that the protein acts post transcriptionally, perhaps to stabilize the RNA. But the binding of the tat protein seems to promote the initiation of transcription, the progression of transcription past the *tar* region, or both. An effect on initiation implies that the tat protein "reaches back" to the promoter and helps host factors initiate transcription; an effect on elongation implies that the bound protein must enhance RNA polymerase's ability to progress, perhaps by acting as an antiterminator.

The tat protein also interacts with one or more host proteins, but their role in the activation process is unclear. Tat might also activate host genes, but the identity of these genes and the consequences of their activation are not known. Researchers have noted that the addition of tat protein to the growth medium of cells in culture can *trans*-activate the LTR within those cells. Presumably, very low levels of tat protein, taken up by the cell, are sufficient for its activity. Thus, low levels of tat protein secreted by infected cells or released from dead cells might be able to *trans*-activate genes in many other cells.

The rev *gene* The rev protein, like the HTLV-I rex protein, alters the ratio of spliced to unspliced viral RNAs in the cytoplasm of infected cells. The 19 kDa protein is basic and localized in the nucleus and nucleolus. Before rev protein is formed, the major mRNAs formed and exported to the cytoplasm are multiply spliced mRNAs encoding only regulatory proteins; after rev protein accumulates, a substantial portion of the exported RNAs is unspliced genomic RNAs and singly spliced env mRNA formed at the expense of the doubly spliced mRNAs. Thus, rev protein lowers its own expression. The region required for its action is termed RRE, for *rev*-responsive element; *rev* has been shown to bind to the RRE region of RNA, recognizing a specific secondary structure. The mode of action is unclear, but a consensus is emerging that the primary effect may be on the export of the RNAs from the nucleus into the cytoplasm. There seem to be sequences located at several positions in the gag, pol, and env coding regions that act as negative *cis*-acting elements, blocking export in the absence of rev protein. The multiply spliced RNAs have lost these sequences and so can be exported without rev. When rev is expressed, and if RRE sequences are present on the RNA, then all RNAs are successfully exported. The mode of action of rev is probably similar to that of the rex protein of HTLV-I, even though there is little sequence similarity between the two proteins. The HTLV-I rex protein can activate expression of RNAs containing the HIV-1 RRE. Ultimately, we hope an understanding of rev action will provide many clues to the nature of RNA processing and export in uninfected cells.

Other viral genes The HIV-1 genome also encodes at least four other distinct viral gene products, formed by the translation of as many as 20 differentially spliced subgenomic mRNAs. The known genes include *vif,* for virion infectivity factor; *vpr*; *vpu*; *tev,* a hybrid of *tat* and *rev*; and *nef,* for negative regulatory factor. Most of these genes' functions are not known. In many cases, mutations introduced into these genes have little or no effect on the replication of the virus as detected during its growth in tissue culture. However, many of these genes may enhance the pathogenicity of the virus. Studies of mutations in the *nef* gene of the related simian immunodeficiency virus (SIV), for example, have shown that the *nef*-deficient viruses are unable to cause disease in infected monkeys.

c. Efforts at Antiviral Therapy for Retroviral Disease

The seriousness of the AIDS epidemic has encouraged a massive effort in many arenas to devise antiviral drugs that might inhibit the replication of HIV-1 in an infected individual. Each step in the life cycle provides a potential target for these antiviral agents. The following are a few of the current targets of therapy and the nature of the candidate inhibitors under evaluation at this time.

Viral entry A major class of inhibitors is aimed at blocking the virus-receptor interaction or subsequent steps in entry. The identification of the receptor as the CD4 molecule immediately provided one potential blocking molecule: a soluble version of the CD4 protein itself. The administration of high levels of soluble CD4, lacking the membrane anchor region, to virus or to cell cultures can indeed block infection by many cloned isolates of HIV-1. The clinical use of the molecule has been disappointing, however, because of the presence of resistant strains in the virus population. Other less specific molecules, especially polyanions such as dextran sulphate, have also been used to try to interfere with env-receptor interactions and so to block virus entry. These compounds have not been highly effective.

DNA synthesis A second major class of inhibitors, including all those currently in clinical use, are targeted at the reverse transcriptase. One group includes nucleoside analogues that, when phosphorylated to the triphosphate form, can act as chain terminators to block viral DNA synthesis. Examples include 3′-azido-3′-deoxythymidine (AZT, also called ziduvidine), dideoxyinosine (ddI), and dideoxycytidine (ddC). These drugs are moderately successful and show fair selectivity for the viral DNA polymerase activity of RT over that of the host DNA polymerases. Sometimes limited by toxicity and induced anemia, their use is ultimately limited only by the appearance of resistant strains of HIV. Using combinations of drugs may help avoid the generation of these resistant isolates. Another group includes nonnucleoside inhibitors such as nevirapine and a number of tetrahydroimidazo[4,5,1-jk][1,4]benzodiazepin-2(1H)-one and thione compounds (the TIBO series). These inhibitors act noncompetitively with the deoxynucleoside triphosphates and are very specific for HIV-1 over other viruses. Their detailed mode of action is unclear, although the binding site for these drugs on RT has been mapped in great detail. Another group includes RNase H inhibitors, mostly mimetics of polynucleotide substrates for the enzyme. To date, no safe, effective inhibitors of RNase H have been identified.

Expression Because HIV expression is so dependent on its regulatory proteins tat and rev, these gene products represent obvious targets for intervention. Several libraries of chemicals and extracts of natural products have been screened for inhibitors of tat or rev function or binding to RNA, and candidate compounds that block these steps have been found. The ultimate efficacy of these compounds *in vivo* remains unknown. The compounds may be quite useful in helping to dissect the functions of these proteins.

Assembly and virion activation Although the viral protease is not essential for virion assembly per se, its activity is required for forming infectious virions. The long history of protease inhibitors, including those specific for aspartyl proteases like renin, has facilitated analysis of compounds targeted to the viral protein. Several peptidelike inhibitors that block the viral protease with considerable specificity and potency have been prepared, and these compounds block virus production in culture but not entry of preexisting virions; thus, only immature, unprocessed proteins are formed. These compounds, like many peptide-based drugs, show a very short half-life in serum, however, and their utility *in vivo* may be limited.

The progress in the development of antivirals targeted to specific retroviral proteins is very rapid, and there is a broad optimism that several efficacious drugs will soon be available. There is little expectation that any of these drugs will effect a cure of the disease or eliminate the virus from the body. The hope is that the application of combinations of drugs in a carefully controlled regimen will arrest viral spread and control replication so that resistant variants do not arise. Such lifelong therapy may well be enough to extend the asymptomatic period, during which virus replication is occurring, to the length of a normal human life span.

3.6 Gene Transfer and Gene Therapy

One of the major attractions of the retrovirus life cycle to molecular biologists, geneticists, and even clinicians is its potential applications in gene transfer and gene therapy. The life cycle has evolved in answer to a specific problem imposed on the virus by the pressure to survive: how can a virus best endure in the face of a hostile environment and a hostile host's immune system? One answer to this problem, widely used by various viruses, is to establish latency and survive for some period of time without active replication. Retroviruses have developed perhaps the ultimate form of latency in establishing themselves in the host DNA as proviruses. That capability is the single most important feature of these viruses for those interested in gene transfer. In the last decade, many investigators have recruited retroviruses to help establish new genes in cells by infection. In this setting, portions of the retroviral genome are joined to a foreign gene to facilitate gene transfer, and viral proteins provided in trans are used to package the vector and actually carry out the transfer process. The altered retroviral genome that carries the gene is termed a retroviral vector, and the viral proteins are provided by a producer cell termed a helper cell (*Genes and Genomes,* section 5.7d).

a. Vector Genomes

A retroviral vector consists of a modified and truncated DNA copy of the viral genome, capable of accepting a foreign gene for the high-efficiency transfer of the gene from cell to cell. In general, most or all of the coding regions for the retroviral *gag, pol,* and *env* gene products are deleted, to be replaced by the foreign gene; and all the *cis*-acting regions required for transmission through the life cycle are retained. These requirements are most simply met by deleting the entire center of the genome, replacing it with the desired gene, and retaining the two LTRs, the primer binding site, the polypurine tract, and the packaging signals required for encapsidation of the viral RNA. Examples of various simple vectors carrying foreign genes are shown in Figure 3.34.

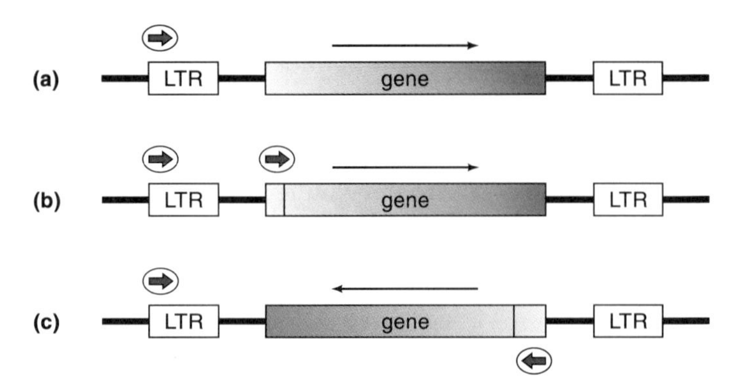

Figure 3.34
Structures of various retroviral vectors carrying a single gene, and various schemes for the expression of the inserted gene. (a) The simplest vectors specify transcription of a single mRNA from the viral LTR, which is translated to yield the protein. (b) Other vectors include an internal promoter that directs transcription of a subgenomic mRNA. (c) In other vectors, the promoter and gene are inserted in the opposite orientation from that of the transcription of the virus. In these cases, the subgenomic mRNA is translated to give the desired gene product, and the full-length viral RNA is only used as the genome for transfer to new cells.

Promoters The inserted gene can be expressed from either of two transcriptional promoters. The simplest scheme here is to use the viral LTR itself as the promoter to form mRNAs; those RNAs can serve both as the genomic RNA to be transmitted by infection and the mRNA for translation of the foreign protein. In these settings, the foreign gene is expressed like the *gag-pol* region of replication-competent viruses. An alternative scheme is to include an internal promoter inside the retroviral genome, in between the two LTRs, specifically for the formation of an mRNA encoding the foreign protein. In these cases, a provirus integrated into the host cell will express two RNAs: one longer RNA initiated in the LTR, suitable for packaging into virions for transmission, and one shorter RNA initiated at the internal promoter, not suitable for packaging but solely for the formation of protein. The gene can be inserted in either orientation with respect to the viral transcription.

Selectable markers Several features have been added to vectors to increase their utility and applicability. Perhaps the most important addition is the inclusion of a selectable marker, usually a drug resistance marker, within the vector, to be coexpressed along with the foreign gene. Various schemes have been devised to allow the simultaneous expression of both the selectable marker and the gene (Figure 3.35). One such is effectively to replace the *gag-pol* region with one gene and the *env* gene with the other. In these vectors, the signals for splicing of the *env*-equivalent gene are retained, with the intent that the two inserted genes will be coexpressed in the same way that gag-pol and env products are in a replication-competent virus. Another scheme is to place the two coding regions near each other on the genome so they can be translated from one unspliced mRNA. Translation can result in the synthesis of the 5′ proximal gene product, terminate, and then restart at the AUG of the 3′ proximal product. A third possibility is to express one of the genes—say, the marker—from its own internal promoter and express the other from the LTR.

Self-inactivating vectors Another twist in vector design is embodied in the so-called "self-inactivating" vector (Figure 3.36). These vectors, constructed in the

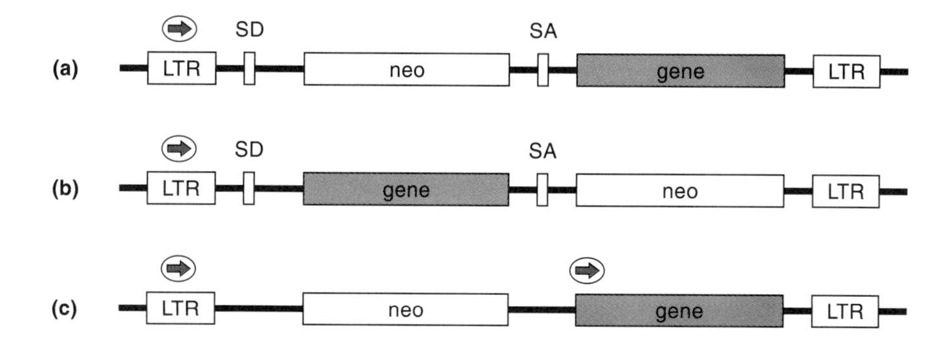

Figure 3.35
Structures of complex retroviral vectors. Vectors carrying two genes—a selectable marker like neomycin resistance (*neo*) and an unselected gene—can be engineered to express them in several ways. In one scheme, both genes can be expressed from transcripts initiating in the 5′ LTR. (a) One protein can be translated from an unspliced genomic RNA; (b) the other can be translated from a spliced subgenomic mRNA. (c) Alternatively, the downstream gene can be expressed from a separate internal promoter. Many other schemes are also possible.

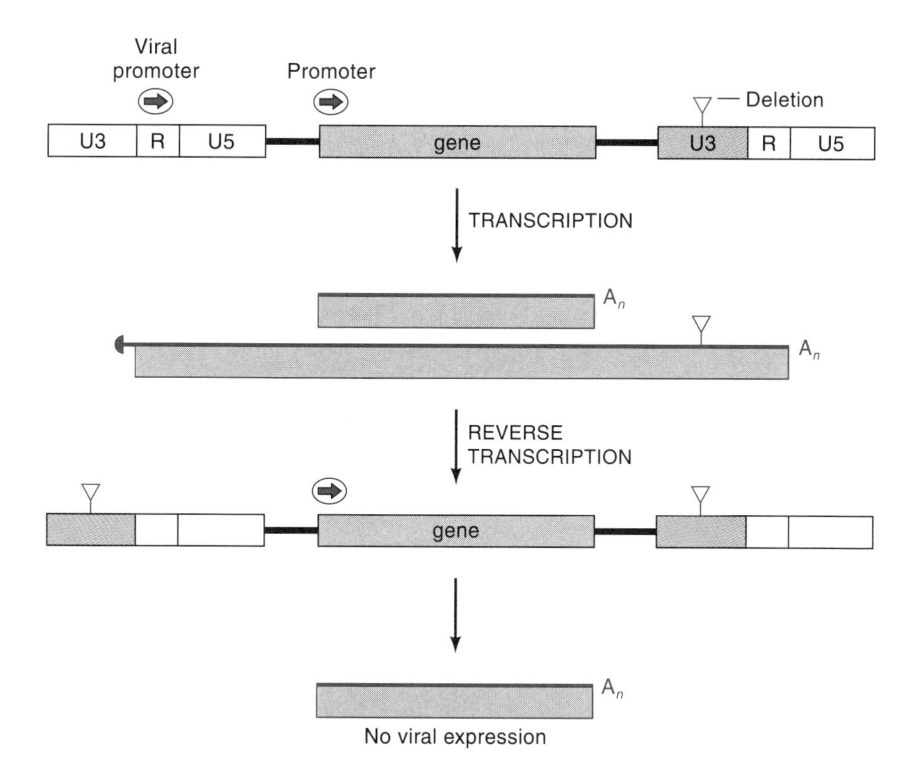

Figure 3.36
A "self-inactivating" vector. Proviral DNAs containing a deletion mutation in the 3′ LTR can be transmitted once from a producer cell to a recipient cell. During the process of reverse transcription of the mutant RNA, the deletion is copied so that it appears in both LTRs of the newly formed proviral DNA. The LTRs of the resulting integrated provirus do not contain functional transcriptional promoters; thus, no viral transcripts are formed. Any internal promoter, however, is retained and can direct transcription of transduced genes.

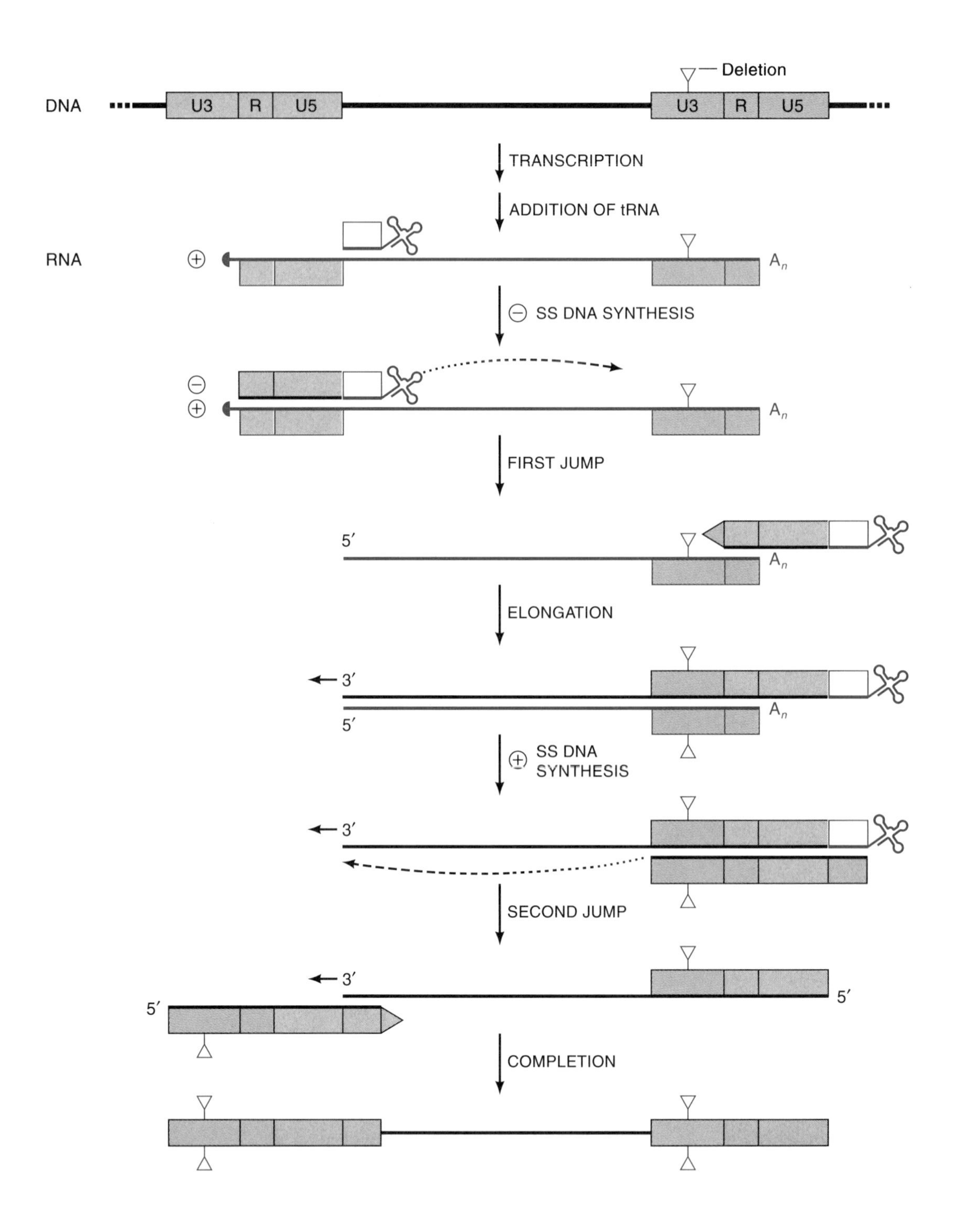

Figure 3.37
Transfer of a mutation in a 3′ LTR to both LTRs during a transcription-reverse transcription cycle. The crucial step is the translocation of the plus-strand strong stop DNA (the second jump), moving the mutation to the 5′ LTR. On completion of the provirus, both LTRs carry the mutation.

form of cloned, proviral DNAs, are built with a deletion in the U3 region of the 3′ LTR. Such a provirus, placed in a producer cell, can be expressed normally to form a vector genomic RNA; the transcription of the provirus is initiated in the 5′ LTR and depends as usual on the intact U3 region in that 5′ LTR. The RNA that is formed will contain the defective U3 region of the 3′ LTR at its 3′ end. When this RNA is packaged by the helper virus proteins, transmitted to the recipient cell, reverse transcribed, and integrated, a critical change occurs. During reverse transcription, the U3 sequence at the 3′ end of the RNA (the only copy of U3 in the RNA) is used to generate the U3 regions in both copies of the LTRs. When this DNA is integrated, the mutant U3 that was present in the donor cell only at the 3′ LTR will be present at both the 3′ and the 5′ LTR of the provirus produced in the recipient cell (Figure 3.37). Because the U3 region is essential for transcription, both LTRs will be silent, and the newly generated provirus will not be expressed. The provirus has thus been inactivated during the transmission process. However, if the vector includes an internal promoter driving expression of the foreign gene, that gene can be expressed. The net effect is to transfer an intact, functional transcription unit without the simultaneous transfer of the transcriptional signals of the viral vector. Because the newly transferred copy of the gene is not linked to the powerful transcriptional promoter and enhancers of the wild-type LTR, there is a much better chance that the gene will be expressed appropriately—that is, under the regulatory network of the associated promoter—than when the gene is imbedded in a transcriptionally active provirus.

b. Helper Viruses and Helper Cell Lines

Helper viruses, as described previously in the context of transforming virus replication, are viral genomes that provide the protein products necessary for forming virion particles; in turn, they mediate the transfer of vectors to new cells. The cell line expressing these viral proteins and shedding virion particles is termed a **helper cell** line. The conceptually simplest type of helper virus is a wild-type, replication-competent leukemia virus, and the correspondingly simplest type of helper cell line is a cell infected with wild-type virus that constitutively sheds virion particles (Figure 3.38). If such cells are induced to express a vector genome, the vector RNA will be packaged into the virions, and the application of these virions to fresh cells will result in the transfer of the vector's provirus into the host genome. The transfer is formally identical to the transfer of a replication-defective transforming virus to new cells by a helper virus. By using helper viruses of different host ranges, it is possible to transfer vectors to cells of many species. The murine amphotropic viruses and the avian spleen necrosis viruses are in use or under investigation for transfer to people.

A serious disadvantage of using wild-type viruses as helper viruses is the unwanted transfer to the recipient cell of the helper virus genome along with the vector. In many cases, it is distinctly undesirable to permit the cotransfer of the helper virus. For example, if the recipient cell population is to be used for successive infections, it must not become infected with helper, for the cell would become resistant to subsequent viral infections. Even more important, if the recipient cell population is derived from explanted tissue destined for return to an individual undergoing somatic gene therapy, the cells must not receive replication-competent virus; one must not introduce live virus into a patient.

The solution to this problem is the use of special helper cell lines that permit the generation of so-called helper-free virus stocks (Figure 3.39). A prototypical cell line is the ψ2 cell line. This cell contains a helper virus genome capable of expressing all the essential viral gene products, but it lacks the RNA packaging signals near the 5′ end of the genome. The ψ2 cells release "empty" virion particles;

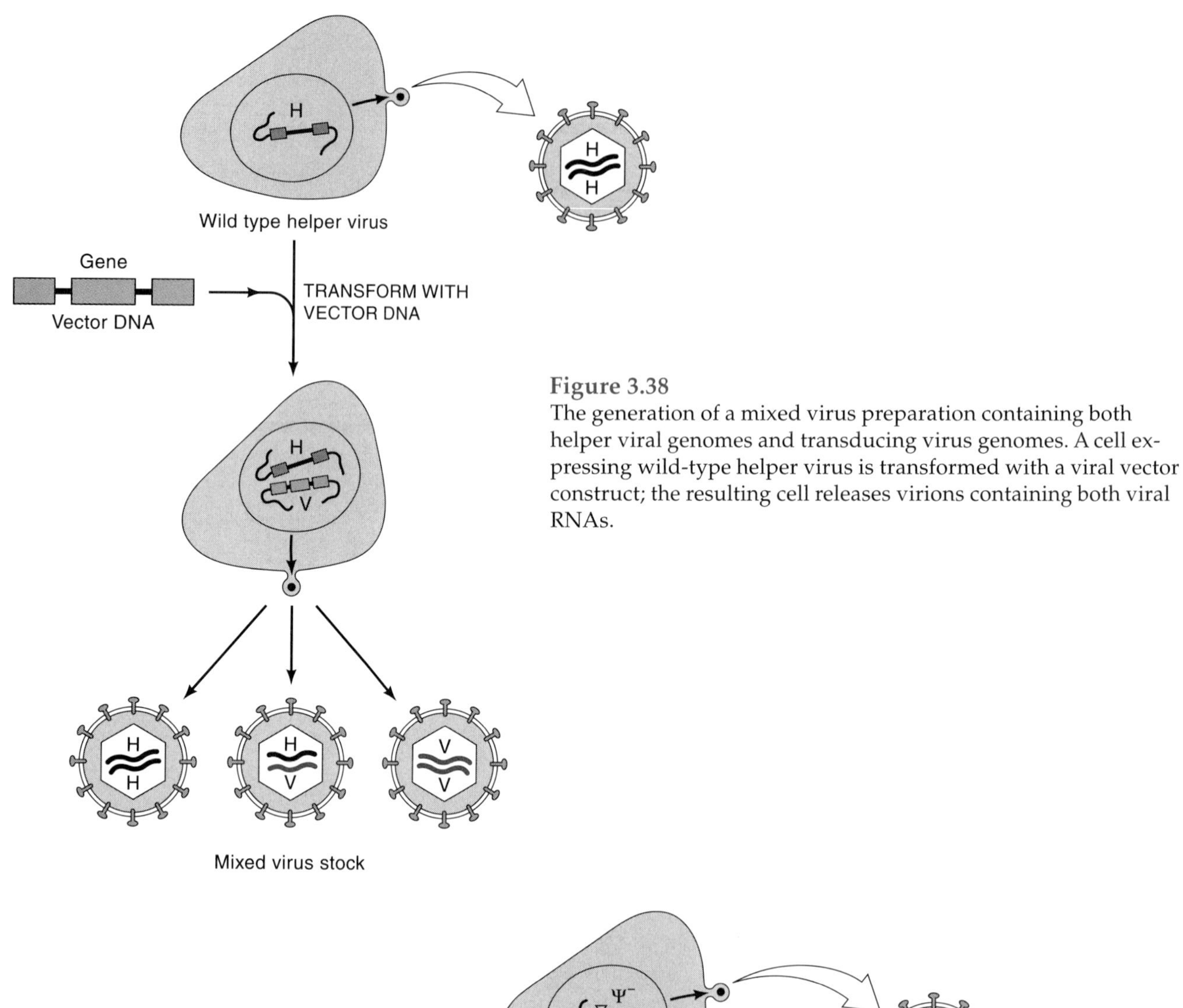

Figure 3.38
The generation of a mixed virus preparation containing both helper viral genomes and transducing virus genomes. A cell expressing wild-type helper virus is transformed with a viral vector construct; the resulting cell releases virions containing both viral RNAs.

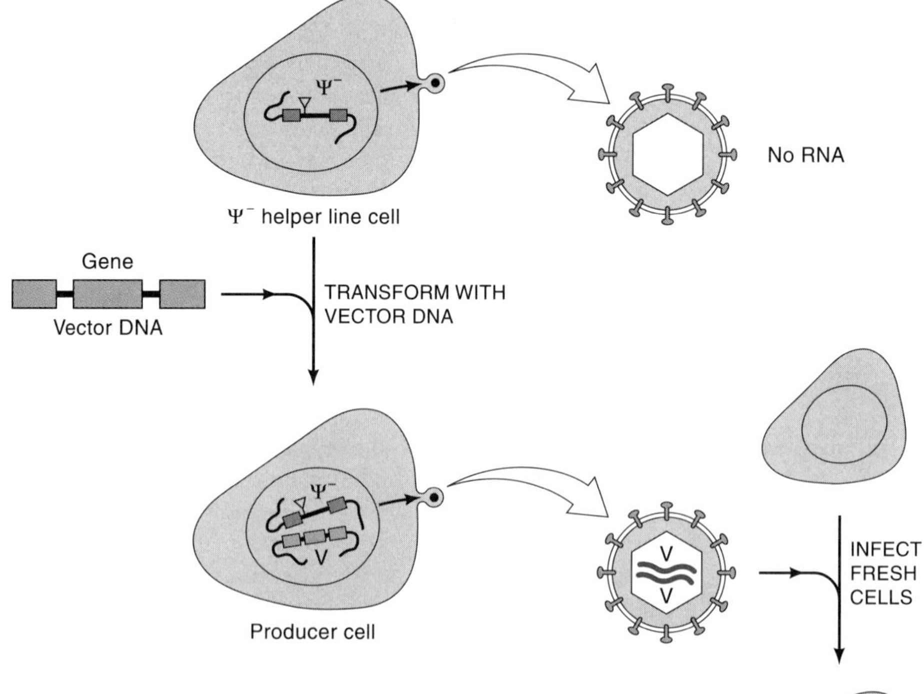

Figure 3.39
The generation of a "helper-free" virus preparation containing only transducing viral genomes. A cell line expressing a packaging-defective mutant provirus releases "empty" virion particles without viral RNA. When such a cell line is transformed with a viral vector construct, only the RNA of the viral vector is packaged into the particles. The infection of new cells by these particles results in the transduction of the viral vector without the transfer of helper virus.

that is, they assemble all the necessary viral proteins into virions, but they fail to incorporate substantial levels of the genomic RNA in those virions. (These virions do, however, include low levels of the viral RNA and significant levels of virus-related RNAs expressed from the virus-related DNAs in the genome.) When a vector construct that retains the packaging sequences is introduced into the ψ2 cell, its RNA is readily incorporated into the previously empty virions. Infection of fresh cells with these "helper-free" virions results in the transfer of only the desired vector construct and not the unwanted helper.

The ψ2 cells enormously expand the utility of retroviral vectors in gene transfer experiments. But there is still a significant problem with their use in certain applications. Because a low level of the helper RNA genome is packaged along with the vector RNA, there is the potential for recombination during reverse transcription to regenerate the packaging signal in the helper (Figure 3.40).

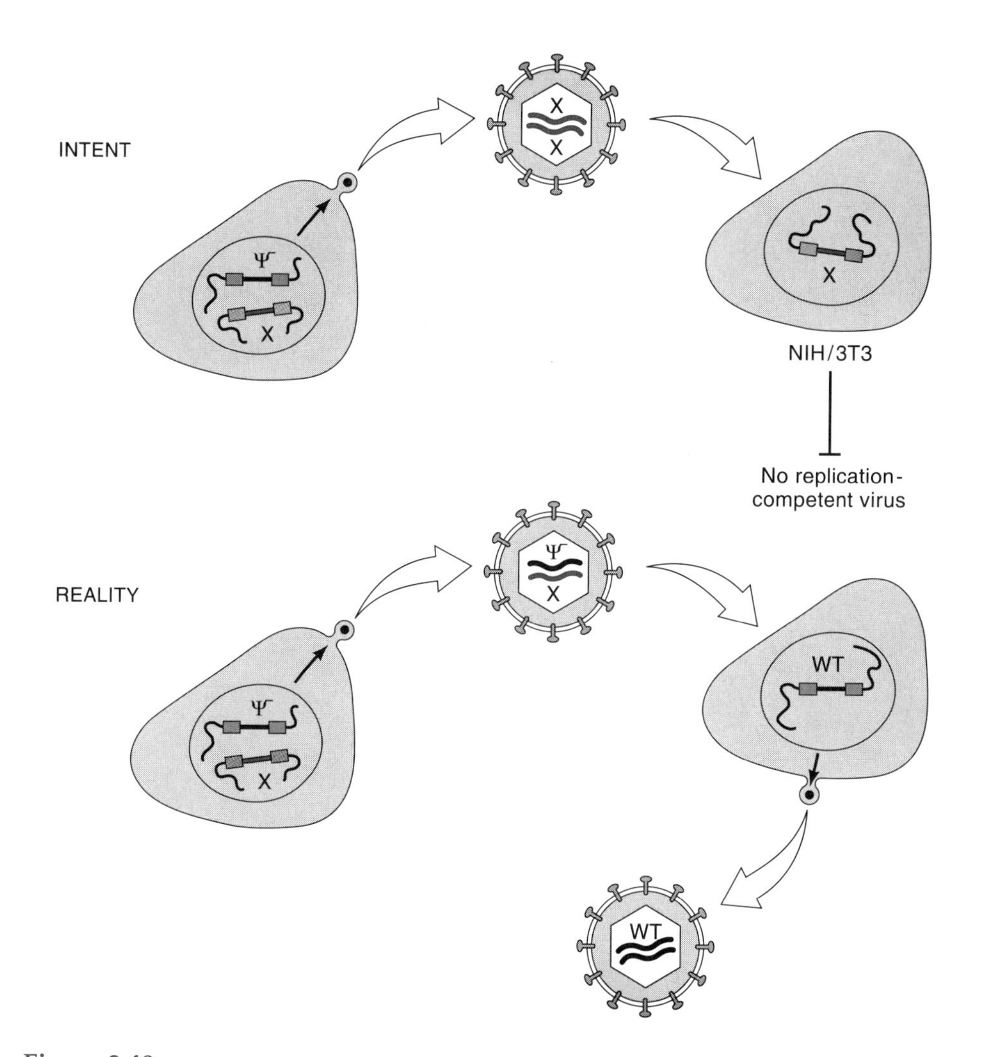

Figure 3.40
Formation of replication-competent virus by recombination between a packaging-defective mutant (Ψ^-) and a vector genome (χ). Packaging-defective cell lines containing a viral vector are intended to release particles containing only the vector RNA (top); infection with these virions should transfer only the vector genome. In reality (bottom), low levels of the helper viral RNA are packaged. On infection of new cells, recombination between the helper and vector results in the generation of a wild-type virus genome that can spread throughout the culture of infected cells.

The simplest means by which this might occur is that the 5′ proximal region of the vector is donated to the packaging-minus genome. Thus, wild-type virus may be formed at a low frequency during the infection of recipient cells. If many cells are being infected, and if there is a crucial need for the recipients to remain completely virus-free (as in human gene therapy), additional safety measures must be included to prevent the generation and spread of live virus in the cells.

The simplest way to reduce the frequency of live virus formation during gene transfer is to use helper cell lines in which the genes that provide the necessary virion proteins are physically separated from one another. In this way, all the necessary proteins can still be provided, but no single RNA molecule that might be packaged (even at a low level) and so act as a recombination parent in the formation of live virus is present in the helper cell. In most such helper cells, the *gag* and *pol* genes are inserted as one unit at one position in the genome, and the *env* gene is inserted as a separate unit at another position (Figure 3.41). Each cassette must induce the high-level expression of the viral gene products—so as to ensure high-level release of virion particles—but not encode RNAs capable of efficient pack-

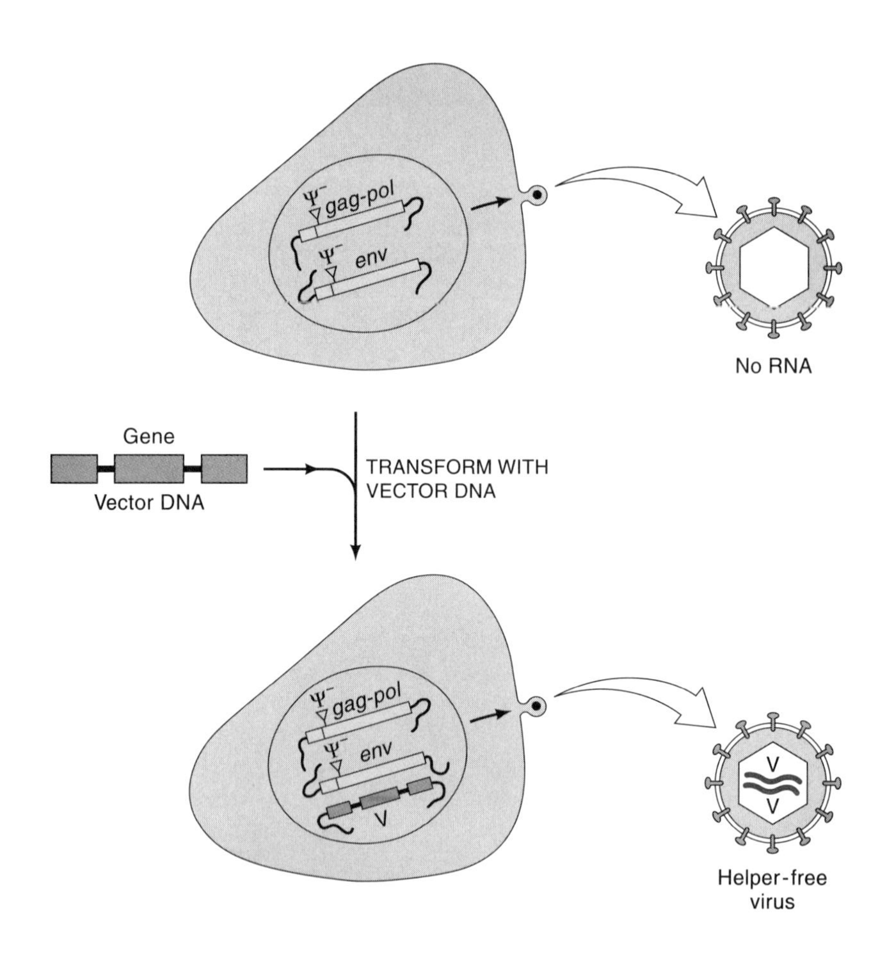

Figure 3.41
Prevention of formation of replication-competent virus. Packaging lines can be formed by introducing into a cell two "split" proviral DNAs: one expressing the *gag* and *gag-pol* genes, and another expressing the *env* gene. Both genomes are packaging-defective, so only empty virion particles are released. The subsequent introduction of a vector into these cells results in the packaging of the vector RNA into the particles for transfer into recipient cells. Even if low levels of either helper RNAs is packaged, there is no way for a single recombination event to create a wild-type viral genome.

aging into the particles. The usual design is to prepare two cassettes by modifying a proviral DNA. The packaging region and the 3′ LTR are removed from both cassettes. Alternative "unwanted" genes are removed from the two constructions. The cassettes can be modified by adding a substitute polyadenylation sequence at the 3′ end to help form discrete 3′ termini. Cells expressing both these separate constructs are ideal sources of virus for gene delivery, essentially free of the danger of forming live virus. Today such helper cells expressing the envelope genes of the murine-derived amphotropic viruses are used to transfer genes therapeutically into human cells.

c. Gene Transfer and Somatic Cell Therapy

Retroviral vectors have achieved their most advanced application to date in attempts to correct inherited genetic defects in human somatic tissues. Several inherited gene deficiencies could in principle be corrected by infecting a substantial number of somatic cells with a vector expressing the defective gene. Obviously, there are a number of considerations when candidate genes and the corresponding deficiency are evaluated. First, the defect must be correctable by the introduction of the missing gene at relatively late stages in the disease, certainly after diagnosis; the problem is that, for many genetic diseases, irreversible damage that might have occurred during development will not be corrected by the later introduction of the gene. Second, the problem must be corrected by the new expression of the gene in an accessible tissue and in a reasonable number of cells. Third, if the gene must be regulated, that proper regulation might have to be re-created by the transduced gene.

A few target diseases and the corresponding genes under consideration for correction today are listed in Table 3.3. Gene products whose expression in blood or in lymphoid tissues would be sufficient for correction are prime candidates for expression on retroviral vectors because lymphoid cells are readily accessible in the bone marrow, manipulated in culture, infected, and transplanted back into the circulation. The first application of the method has been the introduction of the adenosine deaminase (ADA) gene into the peripheral lymphoid cells of patients suffering from severe combined immune deficiency syndrome (SCIDS), a disease caused by an inherited defect in the gene. In this disease, the tissues most severely affected by the absence of the enzyme are lymphoid cells; their depletion seriously compromises the individual's ability to respond to infectious agents. Infection of these cells by retroviral vectors expressing the gene constitutively, followed by transfer of the infected cells back into the circulation, has re-

Table 3.3 ***Candidate Diseases for Correction by Gene Therapy***

Tissue	Disease	Gene
Bone marrow (ADA)	Severe combined immune deficiency syndrome (SCIDS)	Adenosine deaminase
	Hemoglobinopathies	β-globin
Liver	Phenylketonuria (PKU)	Phenylalanine hydroxylase
	Hypercholesterolemia	Low-density lipoprotein
	Hyperammonemia	Ornithine transcarbamylase
Muscle	Duchenne muscular dystrophy	Dystrophin
Lung	Cystic fibrosis	Cystic fibrosis transporter (CFTR)

sulted in detectable expression. Although it is controversial, significant clinical improvements have been claimed. Whether the treatment is cost-effective as compared with other therapies remains controversial.

Although the transduction of peripheral lymphoid cells results in only transient expression of the gene, due to the limited life span of these cells, transduction of hematopoietic stem cells could result in lifelong expression of the gene. These cells, a self-renewing population residing in the bone marrow, give rise to all classes of hematopoietic cells, including erythrocytes, macrophages, B cells, and T cells. The stem cells are present in bone marrow biopsies, are readily obtained, and can be grown, at least briefly, in culture before being reintroduced into a recipient. If these cells could be efficiently infected with a retroviral vector, a transduced gene would be present and potentially expressed in all hematopoietic cells. Though such experiments have been successfully carried out in mice, attempts in humans are just beginning. A crucial issue needing exploration in animal models is the level of expression of the transduced gene that can be achieved and the duration of that expression.

Other defects affecting other tissues will also be attacked soon through retroviral transduction. One plausible target is the liver, which can be readily infected after induction of cell division. These cells are attractive because expression there could result in the secretion of proteins into the serum. Skin fibroblasts are another accessible target, as are the endothelial cells lining the arteries. Here again the release of proteins both into the circulation and locally at the site of infection is envisioned.

d. Gene Transfer and Germ Line Therapy

The introduction of new genes into somatic cells in humans is now an accomplished historical fact. Although some ethical concerns about the widespread application of gene transfer technology to the treatment of genetic and other diseases remain, there is a growing acceptance and approval of the concept of introducing genes into the somatic cells of an affected individual. Obviously, such procedures may correct defects in the affected individual for a time, and if appropriate somatic cells are targeted, the correction can last the entire lifetime of that individual. But so long as somatic cells are the only targeted cells, the new gene will not be transmitted to the progeny of the treated individual, and the new genetic material will be lost at the end of life. In contrast, the introduction of new genes into germ cells—leading to transfer through sperm or egg—could initiate the transmission of the new genotype into an entire family line. The potential benefits and abuses of such technology are great, and the controversy surrounding the ethics of such procedures is profound. As of this writing, there is no practical means to introduce new DNA into human germ cells efficiently, and the controversy remains muted. But these technologies will likely be available in the future, and consideration of the methods and the ethics is important.

The simplest route for introducing new genes into mammalian germ lines is microinjection of naked DNA into a fertilized egg to create a so-called "transgenic" animal. This method is now routine in many mammalian species. After microinjection, the embryo is introduced into the uterus of a pseudopregnant female and allowed to implant and develop. A significant proportion of the animals born after this procedure carry the inserted DNA. Retroviruses have been used as an alternative route: very early embryos can be exposed to virus from a viremic mother so that a substantial proportion of the embryo's cells are infected. A fraction of the animals arising from these infected embryos will be capable of transmitting the provirus as a newly acquired gene to their offspring, and thereafter the provirus is inherited as a stable Mendelian allele.

A serious difficulty with all these methods of introducing DNA into the germ line is that the inserted DNA is positioned at random and therefore generally at inappropriate locations in the genome; the new DNA does not insert at the location of the resident copy of the gene. Retroviral transfer does not seem likely to provide a solution to this problem. The newly reverse-transcribed DNA does not normally insert via homologous recombination, and tests for increased rates of homologous recombination after infection with integrase-deficient viruses have not led to encouraging frequencies. For these reasons, among others, gene therapy using retroviral transduction is likely to be restricted to somatic cells. Each infected cell will take up the DNA at a different location in the genome, but the finite life span of the cell is likely to limit the effects of the insertion.

3.7 Conclusion

The retroviruses occupy a remarkable position in the history of molecular biology. They began at the very periphery of the field, drawing attention because they were oncogenic agents in animals. They moved onto the stage when their mode of replication was understood and into center stage when they provided help in identifying cellular oncogenes. Today they have become critically important objects of study to the clinician, both for their role in important human diseases and in curing genetic disorders. They will certainly continue to be powerful tools in the kit of the molecular biologist, permitting directed transduction of genes into cultured cells and tissues and ultimately into human beings.

Further Reading

General Review Articles

R. Weiss, N. Teich, H. Varmus, and J. Coffin (eds.). 1982. RNA Tumor Viruses, vol. 1. Cold Spring Harbor Press, Cold Spring Harbor, New York.

H. E. Varmus. 1983. Retroviruses, In J. Shapiro (ed.), Mobile Genetic Elements, pp. 411–503. Academic Press, New York.

R. Weiss, N. Teich, H. Varmus, and J. Coffin (eds.). 1985. RNA Tumor Viruses, vol. 2. Cold Spring Harbor Press, Cold Spring Harbor, New York.

H. Varmus and P. Brown. 1989. Retroviruses. In D. E. Berg and M. M. Howe (eds.), Mobile DNA, pp. 53–108. American Society for Microbiology, Washington, D.C.

J. M. Coffin. 1994. Retroviridae: The Viruses and Their Replication. In B. Fields et al. (eds.), Virology, 2nd ed., pp. 1437–1500. Raven Press, New York.

3.2b

E. Gilboa, S. W. Mitra, S. Goff, and D. Baltimore. 1979. A Detailed Model of Reverse Transcription and Tests of Crucial Aspects. *Cell* 18 93–100.

P. O. Brown, B. Bowerman, H. E. Varmus, and J. M. Bishop. 1987. Correct Integration of Retroviral DNA in Vitro. *Cell* 49 347–356.

S. P. Goff. 1990. Integration of Retroviral DNA into the Genome of the Infected Cell. *Cancer Cells* 2 172–178.

A. M. Skalka and S. P. Goff (eds.). 1993. Reverse Transcriptase. Cold Spring Harbor Press, Cold Spring Harbor, New York.

3.2c

E. Hunter and R. Swanstrom. 1990. Retrovirus Envelope Glycoproteins. *Curr. Top. Microbiol. Immunol.* 157 187–253.

J. W. Wills and R. C. Craven. 1991. Form, Function, and Use of Retroviral Gag Proteins. *AIDS* 5 639–654.

3.2d

B. G. Neel, W. S. Hayward, H. L. Robinson, J. Fang, and S. M. Astrin. 1981. Avian Leukosis Virus-Induced Tumors Have Common Proviral Integration Sites and Synthesize Discrete New RNAs: Oncogenesis by Promoter Insertion. *Cell* 23 323–334.

3.3a

N. Tanese and S. P. Goff. 1988. Domain Structure of the Moloney Murine Leukemia Virus Reverse Transcriptase: Separate Expression of the DNA Polymerase and RNAse H Activities. *Proc. Natl. Acad. Sci. USA* 85 1777–1781.

3.3b

D. D. Loeb, C. A. Hutchinson, M. H. Edgell, W. G. Farmerie, and R. Swanstrom. 1989. Mutational Analysis of the HIV-1 Protease Suggests Functional Homology with Aspartic Proteases. *J. Virol.* 63 111–121.

3.4

M. P. Goldfarb and R. A. Weinberg. 1981. Generation of Novel, Biologically Active Harvey Sarcoma Viruses via Apparent Illegitimate Recombination. *J. Virol.* 38 136–150.

H. E. Varmus. 1982. Form and Function of Retroviral Proviruses. *Science* 216 812–820.

A. Swain and J. M. Coffin. 1992. Mechanism of Transduction by Retroviruses. *Science* 255 841–845.

3.5

J. G. Sodroski, C. A. Rosen, and W. A. Haseltine. 1984. Trans-Acting Transcriptional Activation of the Long Terminal Repeat of Human T Lymphotropic Viruses in Infected Cells. *Science* 225 381–385.

G. Feuer and I. S. Y. Chen. 1992. Mechanism of Human T-Cell Leukemia Virus-Induced Leukemogenesis. *Biochim. Biophys. Acta* 1114 223–233.

3.6b

R. Mann, R. C. Mulligan, and D. Baltimore. 1983. Construction of a Retrovirus Packaging Mutant and Its Use To Produce Helper-Free Defective Retrovirus. *Cell* 33 153–159.

D. Markowitz, S. P. Goff, and A. Bank. 1988. A Safe Packaging Line for Gene Transfer: Separating Viral Genes on Two Different Plasmids. *J. Virol.* 62 1120–1124.

A. D. Miller. 1992. Retroviral Vectors. *Curr. Top. Microbiol. Immunol.* 158 1–24

CHAPTER 4

Oncogenes, Growth Suppressor Genes, and Cancer

TONY HUNTER

Cancer is a multistep disease, and the emergence of a tumor cell requires the accumulation of many—estimated to be between four and eight—genetic changes over the course of years. Two main types of mutation are responsible for the malignant phenotype. First, activating mutations convert normal cellular genes into **oncogenes**, which act in a dominant fashion and cause malignant transformation when introduced into normal vertebrate cells. The products of oncogenes are known as **oncoproteins**; these are altered forms of proteins that are normally involved in promoting cell growth. The first oncogenes were encountered as the transforming genes of tumor viruses. Subsequently, many oncogenes have been detected in the DNA of tumor cells and tissues. The second type of mutation implicated in tumorigenesis results in the inactivation of both alleles of a **growth suppressor gene**; the normal function of such genes, as the name implies, is to regulate cell growth in a negative fashion. The isolation and characterization of both oncogenes and growth suppressor genes, and the elucidation of their mechanisms of action, have relied heavily on the ability to isolate, characterize, and modify recombinant DNAs.

4.1 Oncogenes

The study of viruses that are able to cause tumors, **tumor viruses**, gave rise to the concept of oncogenes. Two types of tumor virus are known: one has DNA genomes (**DNA tumor viruses**) and the other contains RNA genomes (**RNA tumor viruses**); the latter are often referred to as retroviruses because their genomes replicate through a DNA intermediate (Table 3.1). Several different types of DNA virus cause cancer in animals, including papovaviruses (e.g., SV40 and polyomaviruses), papillomaviruses (e.g., bovine and human papillomaviruses), human adenoviruses, and herpesviruses (e.g., Herpes saimiri and Epstein-Barr virus [EBV]). The oncogenes of these viruses have been intensively studied, and a good deal has been learned about the molecular mechanisms of transformation, establishing a number of important principles. As I discuss in Section 4.3, the study of SV40, papilloma, and adenovirus oncogenesis has identified cellular proteins encoded by growth suppressor genes as their principal targets (Chapter 1). In contrast to DNA tumor virus oncogenes, whose origins are shrouded in the mists of evolutionary time, retroviral oncogenes are all derived from normal cellular genes, called **protooncogenes** (i.e., cellular genes with the potential to give rise to oncogenes). The studies of retroviral oncogenes have identified specific oncoproteins that act positively to convert cells into a tumorigenic state. In several instances, oncogenes derived from naturally arising tumors are related to retroviral oncogenes and were activated by similar mutations.

a. Retroviral Oncogenes

The archetypal tumor-causing retrovirus is Rous sarcoma virus (RSV), a chicken virus. Although the oncogenic properties of RSV have been known since 1911, not until 1970 was the viral genome shown to contain, in addition to the three genes *gag, pol,* and *env* required for propagation, sequences responsible for its transforming potential. In 1975, these sequences, termed the v-*src* gene, were found to have a cellular counterpart, the c-*src* gene. Subsequently, all retroviral oncogenes have also been assigned three-letter acronyms (v-*onc* genes). The initial characterization of the v-*src* gene required the painstaking isolation of sequences corresponding to the foreign, nonviral sequences in the RSV genome (Figure 4.1). This was accomplished by first synthesizing DNA complementary to the entire RSV genome with the endogenous viral reverse transcriptase (RT) and annealing it with an excess of RNA from the parental nontransforming virus; the unannealed DNA was presumed to contain the v-*src* sequence. This v-*src*-specific DNA "probe" was used to show that there are complementary DNA sequences in the chicken genome and that these chicken sequences are expressed in the form of a polyadenylated mRNA. Similar laborious experiments were carried out for two mammalian sarcoma viruses. Cross-hybridization of the oncogene probes for the three viruses suggested that each might have different cellular sequences, but definitive proof was lacking.

Three technical advances in the mid-1970s were essential to further progress in this field. First, the development of DNA blotting of restriction endonuclease digests of cellular DNA allowed demonstration of the nature of the sequences acquired by these oncogenic retroviruses. Second, the application of molecular

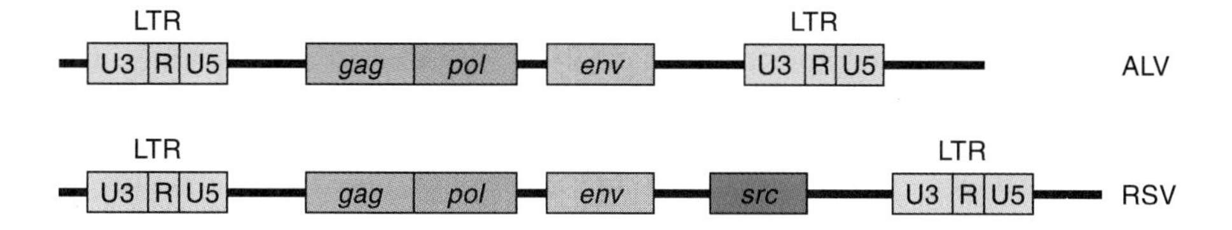

Figure 4.1
Comparison of the genomes of an avian leukosis virus (ALV) and Rous sarcoma virus (RSV) showing how cellular c-Src sequences were incorporated into ALV, which gives rise to RSV. LTR = long terminal repeat. The LTR contains an enhancer region, a promoter that initiates transcription at the start of R, and a polyadenylation signal that leads to addition of poly A at the end of R. U3, R, and U5 refer to sequences present in the 3′ untranslated (U3 + R) and 5′untranslated (R + U5) of the genomic RNA.

cloning techniques to the characterization of retroviral genomes led to the isolation of probes specific for the individual oncogenes. Third, rapid DNA sequencing techniques provided the complete nucleotide sequences of cloned retroviral genomes. Each of these approaches contributed to the realization that many different tumor-causing retroviruses had acquired different cellular sequences, although in some cases particular oncogenes were found in more than one virus.

b. Oncogenes from Tumor Cells

By the time it became apparent that there are multiple distinct retroviral oncogenes, another source of oncogenes was discovered (Table 4.1). Thus, DNA taken from certain human tumors and tumor cell lines was able to cause malignant transformation when transfected into cultured mouse fibroblast cells (NIH3T3 cells). This transforming activity can be transmitted serially from the primary transformants' DNA by repeated transfections. Although the initial recipient cell takes up a large amount of the foreign DNA (perhaps as much as a megabase), the amount of DNA derived from the original tumor cell that is present in the transformed recipient after three such transfection cycles is considerably less. Because human DNA has a characteristic set of short interspersed repeats that are distinct from the equivalent mouse repeats (*Genes and Genomes,* section 9.5), it is possible to detect and distinguish the human DNA in the transformed mouse cell using a probe for this human repeat. A common human DNA sequence found in a group of independently transformed cell clones isolated after trans-

Table 4.1 ***Genetic Changes Associated with Tumorigenesis***

Activation of protooncogenes into oncogenes
- DNA tumor virus oncogenes: origin unknown
- Retrovirus (RNA tumor virus) oncogenes: derived from cellular genes
- Tumor cell oncogenes
  - Mutations in coding sequences
  - Mutations leading to inappropriate regulation of expression
    - Retroviral promoter/enhancer insertion
    - Chromosomal translocation
    - Amplification and overexpression

Inactivation of growth suppressor genes

fection with a single tumor cell DNA is a presumptive tumor-derived transforming gene. In many cases, the responsible gene was cloned by screening a genomic library made from a secondary or tertiary transfectant with the human repeat probe. Comparison of these tumor-derived sequences with the corresponding sequences from normal human DNA revealed that the sequences derived from the tumors contained mutations of various sorts. Such transforming sequences have all been found to correspond to parts of recognizable cellular genes. The altered genes have been called **tumor oncogenes**, by analogy with the viral oncogenes. Although obtaining unequivocal proof that these mutated genes caused the original tumors is difficult, the fact that more than 20 percent of tumors contain mutant genes that can transform normal cells in culture strongly implies their involvement.

c. Diverse Origins of Oncogenes

There is no fundamental distinction between viral and tumor oncogenes; indeed, some have proven to be mutant forms of the same normal cellular gene. In addition to mutations in coding regions, oncogenic transformation can also occur as a result of changes in the regulation of gene activity (Table 4.1). For instance, the genomes of certain weakly oncogenic retroviruses lack their own oncogenes but can cause tumors by integrating adjacent to a cellular gene and activating its transcription and expression; this activation results from the juxtaposition of a viral promoter or enhancer and the cellular gene and is often referred to as **promoter/enhancer insertion** (Figure 4.2a and b) (see also Section 3.20). In such cases, the activated cellular gene can serve as an oncogene. Similar abnormal gene regulation is observed in many types of tumor cells at the borders of chromosomal translocation breakpoints. There are several examples where a gene that is normally transcriptionally silent is activated in a particular cell type when it is juxtaposed to a new locus by a chromosomal translocation (Figure 4.2c). Tumors are also known to arise when one or more protooncogenes is amplified, with a concomitant increase in expression of the protooncogene product. Members of gene families with oncogenic potential have been detected by a weak cross-hybridization signal after their amplification in tumors.

The cellular homologues of oncogenes are all highly conserved between species. Indeed, some protooncogenes, such as the c-*ras* gene (the "c-" prefix denotes the cellular gene corresponding to a v-*onc* gene), are conserved in organisms as distant as humans and yeast. This conservation implies that these genes play central roles in normal cellular metabolism. The fact that both retroviral and tumor oncogenes derive from such cellular genes raises the issue of how they are converted into transforming genes. A priori there are a number of possible activation mechanisms (Table 4.2), but two main themes predominate; both involve mutations. In the first case, the mutation does not affect the protein directly but

Table 4.2 ***Mechanisms of Protooncogene Activation***

Aberrant expression of an unaltered gene product

- Expression either at an improper time or in an inappropriate cell type (ectopic expression) of an unaltered gene product (e.g., due to regulatory mutations or to transposition or insertion of an active transcriptional control element in a region near the gene)
- Overexpression of an unaltered gene product in a cell where it is normally present (dosage hypothesis) (e.g., due to gene amplification)

Expression of an altered gene product

- Expression of a mutant protein that is deregulated as a result of point mutations or deletions in the protooncogene coding region

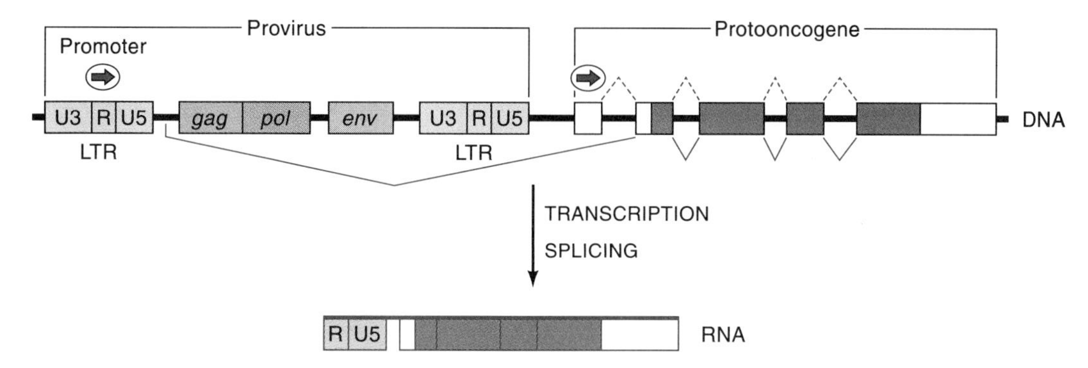

(a) Protooncogene activation by retroviral promoter insertion

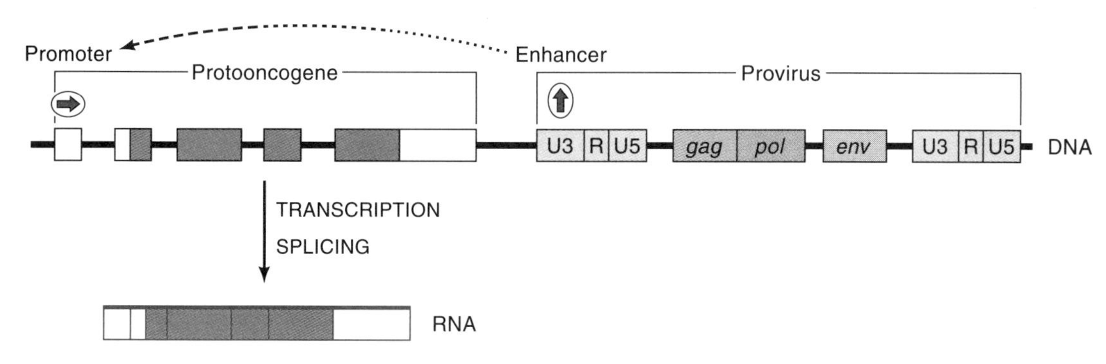

(b) Protooncogene activation by retroviral enhancer insertion

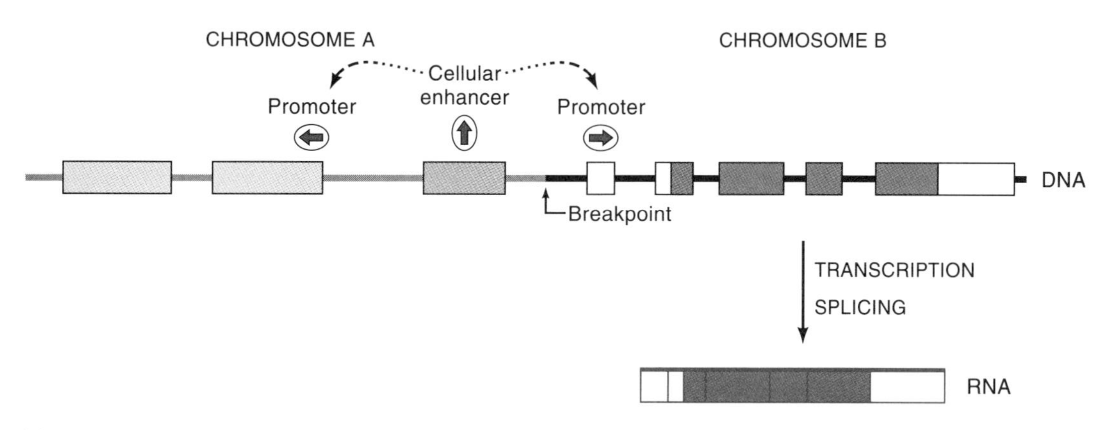

(c) Protooncogene activation by chromosomal translocation

Figure 4.2
(a) Protooncogene activation by retroviral promoter insertion. A transcript starting at the promoter in the LTR is extended through the viral genome into the adjacent cellular protooncogene. This RNA is spliced as indicated to give rise to a chimeric RNA containing viral (R + U5) sequences and the protooncogene coding exons (dark-colored regions) that encode the normal protooncogene product. (b) Protooncogene activation by retroviral enhancer insertion. The activity of the promoter for a cellular protooncogene is increased by the enhancer in the LTR of the provirus inserted downstream of the protooncogene. (c) Protooncogene activation by chromosomal translocation. The activity of the promoter for a cellular protooncogene on chromosome B is increased by the enhancer in another gene on chromosome A, which has been brought into juxtaposition by a chromosomal translocation.

leads to inappropriate levels of gene expression (e.g., at the wrong time, in the wrong place, or at excessive levels). For example, the inability to switch off expression of the rearranged c-*myc* gene in Burkitt's lymphoma cells may prevent their differentiation into normal B cells. In the second activation mechanism, mutations alter the protein itself. By and large, these changes seem to abrogate the cell's ability to regulate protein function.

d. Oncogenes, Protooncogenes, and Cancer

There are now more than 60 proven or presumptive oncogenes (Tables 4.3 and 4.4). There is direct genetic proof that the oncogene sequences found in tumor viruses are absolutely required for their oncogenic activity. The situation with the tumor oncogenes is less straightforward, and there can be only indirect evidence of their participation in the tumor phenotype. However, because carcinogenesis is a multistep process, it seems reasonable to propose that a mutation giving rise to a tumor oncogene detectable by DNA transfection could correspond to one of the steps. But in tumors where protooncogenes have been found to be amplified or rearranged, their involvement is less certain, although there are often suggestive correlations between the expression of the amplified or rearranged protooncogene and some aspect of the tumor phenotype. For example, in human neuroblastomas, where amplification of the N-*myc* gene is commonly observed, the degree of amplification correlates reasonably well with how far the disease has progressed.

To understand how the cumulative activation of oncogenes and inactivation of growth suppressor genes cause cancer, we need to appreciate how tumor cells differ from normal cells. The major characteristic of tumor cells is that they grow continuously in the absence of the exogenous growth factors required for the

Table 4.3 ***Oncoprotein and Growth Suppressor Protein Functions***

Oncoprotein	Number Known in Class
Growth factors	6
Ectopic expression leading to autocrine growth	
Receptor and nonreceptor protein-tyrosine kinases	16
Activated by deletion, point mutation, and fusion increasing protein kinase activity	
Receptors lacking protein kinase activity	4
Ectopic expression	
Membrane-associated G proteins and regulators	7
Activated by point mutation decreasing GTPase activity or increasing GTP exchange	
Cytoplasmic protein-serine kinases	5
Activated by deletion or ectopic expression	
Cytoplasmic regulators	3
Activated by deletion or overexpression	
Nuclear transcription factors	22
Activated by deletion, point mutation, fusion, and overexpression	
Cell cycle regulators	2
Activated by overexpression or truncation	
Antiapoptosis factors	1
Activated by overexpression	

Table 4.4 *Functions of Oncogene Products*

Oncogene	Oncoprotein	Cellular Location	Function
Class I: growth factors			
sis (V)	p28$^{env\text{-}sis}$	Secreted	PDGF-like growth factor
int-2 (T)	p34$^{int\text{-}2}$	Secreted	FGF-related growth factor (FGF-3)
hst (KS3)(T)	p22^{hst}	Secreted	FGF-related growth factor (FGF-4/K-FGF)
FGF-5 (T)	p27$^{FGF\text{-}5}$	Secreted	FGF-related growth factor
wnt-1 (T)	gp40$^{wnt\text{-}1}$	Membranes/secreted	Growth factor?
IL-3	p15$^{IL\text{-}3}$	Secreted	Cytokine
Class II: receptor and nonreceptor protein-tyrosine kinases			
ros (V)	P68$^{gag\text{-}ros}$	Membrane associated	Protein-tyrosine kinase
erbB (V)	gp68/74^{erbB}	Plasma/cytoplasmic membranes	EGF receptor protein-tyrosine kinase domain
*neu/c-erbB*2 (T)	p185^{neu}	Plasma membrane	Protein-tyrosine kinase
tel-PDGFR (T)	?	Cytoplasm?	Fusion between Tel, an Ets-like protein, and PDGFR protein-tyrosine kinase domain
fms (V)	gP180$^{gag\text{-}fms}$	Plasma/cytoplasmic membranes	Mutant CSF-1 receptor protein-tyrosine kinase
met	p65$^{tpr\text{-}met}$	Cytosol	Tpr-HGF receptor protein-tyrosine kinase domain fusion
trk (T)	p70$^{TM\text{-}trk}$	Cytosol	Tropopmyosin-NGF receptor protein-tyrosine kinase domain
kit (V)	P80$^{gag\text{-}kit}$	?	Truncated stem cell factor receptor protein-tyrosine kinase
sea (V)	gP155$^{env\text{-}sea}$	Plasma membrane	Truncated receptor protein-tyrosine kinase
ret (T)	p96^{ret}	?	Truncated receptor protein-tyrosine kinase
src (T)	p60$^{v\text{-}src}$	Plasma membrane	Protein-tyrosine kinase
yes (V)	P90$^{gag\text{-}yes}$	Plasma membrane?	Protein-tyrosine kinase
fgr (V)	P70$^{gag\text{-}fgr}$	?	Protein-tyrosine kinase
lck (T)	p56^{lck}	Plasma membrane	Protein-tyrosine kinase
fps/fes (V)	P140$^{gag\text{-}fps}$	Cytoplasm	Protein-tyrosine kinase
abl (V)	P160$^{gag\text{-}abl}$	Plasma membrane	Protein-tyrosine kinase
Class III: receptors lacking protein kinase activity			
mas (T)	p65^{mas}	Plasma membrane?	Angiotensin receptor
mpl (V)	P45$^{env\text{-}mpl}$	Plasma membrane?	Activated cytokine family receptor
int-3 (T)	p60$^{int\text{-}3}$	Plasma membrane?	Activated Notch receptor?
tan-1 (T)	p60$^{tan\text{-}1}$	Plasma membrane?	Activated Notch receptor?
Class IV: membrane-associated G proteins and regulators			
H-*ras* (V/T)	p21$^{H\text{-}ras}$	Plasma membrane	GTP binding/GTPase
K-*ras* (V/T)	p21$^{K\text{-}ras}$	Plasma membrane	GTP binding/GTPase
N-*ras* (T)	p21$^{N\text{-}ras}$	Plasma membrane	GTP binding/GTPase
gsp (T)	p41^{gsp}	Plasma membrane	Mutant form of Gα_s
gip (T)	p41^{gip}	Plasma membrane	Mutant form of Gα_i
dbl (T)	p66^{dbl}	Cytoplasm	GTP exchange factor for CDC42Hs
vav (T)	p91^{vav}	Cytoplasm?	SH2-GTP exchange factor
Class V: cytoplasmic protein-serine kinases			
raf/mil (V)	P90$^{gag\text{-}raf}$	Cytoplasm	Protein-serine kinase
mos (V)	p37$^{v\text{-}mos}$	Cytoplasm	Protein-serine kinase, cytostatic factor homologue
pim-1 (T)	p36$^{pim\text{-}1}$	Cytoplasm	Protein-serine kinase
cot/tpl-2 (V/T)	p52^{cot}	Cytoplasm?	Protein-serine kinase
akt (V)	P105$^{gag\text{-}akt}$	Cytoplasm/membranes?	Protein-serine kinase

Table 4.4 *(Continued)*

Oncogene	Oncoprotein	Cellular Location	Function
Class VI: cytoplasmic regulators			
crk (V)	$P47^{gag\text{-}crk}$	Cytoplasm	SH2/SH3 adaptor protein
shc	$p52/46^{shc}$	Cytoplasm?	SH2 adaptor protein
nck	$p47^{nck}$	Cytoplasm	SH2/SH3 adaptor protein
Class VII: nuclear transcription factors			
myc (V/T)	$P110^{gag\text{-}myc}$	Nucleus	bHLH-LZ transcription factor
N-*myc* (T)	$p66^{N\text{-}myc}$	Nucleus	bHLH-LZ transcription factor
L-*myc* (T)	$p64^{L\text{-}myc}$	Nucleus	bHLH-LZ transcription factor
lyl-1 (T)	$p29^{lyl\text{-}1}$	Nucleus?	bHLH transcription factor
tal1/scl (T)	$pp42^{tal\text{-}1}$	Nucleus	bHLH transcription factor
myb (V)	$P48^{gag\text{-}myb\text{-}env}$	Nucleus	Sequence-specific transcription factor
fos (V)	$p55^{v\text{-}fos}$	Nucleus	Mutant bZIP transcription factor; AP-1 subunit +Jun
jun (V)	$P65^{gag\text{-}jun}$	Nucleus	Mutant bZIP transcription factor; AP-1 subunit +Fos
maf (V)	$P100^{gag\text{-}maf}$	Nucleus?	Mutant bZIP transcription factor
erb-A (V)	$P75^{gag\text{-}erb\text{-}A}$	Nucleus	Dominant-negative mutant thyroxine receptor
pml-RAR (T)	$p110^{pml\text{-}RAR}$	Nucleus	Chimeric Pml-retinoic acid receptor α protein
evi-1 (T)	$p120^{evi\text{-}1}$	Nucleus?	Zinc finger transcription factor?
rel (V/T)	$p59^{v\text{-}rel}$	Nucleus/cytoplasm	Dominant-negative mutant NF-κB-related protein
bcl-3 (T)	$p46^{bcl\text{-}3}$	?	IκB family NF-κB inhibitor and transactivator
ets (V)	$P135^{gag\text{-}myb\text{-}ets}$	Nucleus	Mutant chimeric transcription factor
spi-1/PU-1 (T)	$p42^{spi\text{-}1}$	Nucleus	Ets family transcription factor
ski (V)	$P125^{gag\text{-}ski}$	Nucleus	Sequence-specific DNA-binding protein
E2A-pbx-1 (T)	$p77^{E2A\text{-}pbx\text{-}1}$	Nucleus	Chimeric E2A-homeobox transcription factor
trx (T)	?	Nucleus	Mutant *Drosophila* Trithorax-like homeobox protein
hox11 (T)	$p40^{hox11}$	Nucleus?	Overexpressed homeobox transcription factor
hox2.4 (T)	$p37^{hox2.4}$	Nucleus?	Overexpressed homeobox transcription factor
qin (V)	$P85^{gag\text{-}qin}$	Nucleus	Mutant forkhead family transcription factor
Class VIII: cell cycle regulators			
bcl-1 (T)	cyclin D1	Nucleus	Activating subunit of Cdk4, Cdk6
ΔN-cyclin A2 (T)	ΔN-cyclin A2	Nucleus	Mutant-activating subunit of Cdc2 and Cdk2
Class IX: antiapoptosis factors			
bcl-2 (T)	$p25^{bcl\text{-}2}$	Cytoplasmic membranes	Antagonizes apoptotic cell death

This table is not an exhaustive list, but gives the best characterized oncogenes.
(V) and (T) indicate viral and tumor cell oncogenes, respectively.
The approximate sizes of the proteins are given in kDa in the protein name (e.g., $p60^{v\text{-}src}$ is a 60 kDa protein). By convention, a capital P is used for viral oncoproteins that are fusion proteins with the viral Gag and Env structural proteins.
bHLH-LZ = basic region/helix-loop-helix/leucine zipper DNA-binding domain.
bHLH = basic region/helix-loop-helix DNA-binding domain.
bZIP = basic region/leucine zipper DNA-binding domain.

growth of normal cells. Tumor cells are also not subject to the normal growth inhibitory influence of surrounding cells. Furthermore, unlike normal cells, tumor cells usually do not require a substratum to grow on and have invasive properties that enable them to migrate between other cells within tissues.

Nondividing normal cells can be stimulated to grow by interaction of exogenous growth factors with cell surface receptors. Ligand binding to the receptors transduces a signal across the plasma membrane; the signal is transmitted to the nucleus, resulting in the induction of a set of more than 100 genes, the **early response genes**. Early response genes encode many types of protein, including transcription factors and secreted cytokines. Transcription factors encoded by early response genes are responsible for the induction of a second set of genes, which in turn initiate a gene expression cascade. This cascade propels the cell out of the resting state, sometimes known as G0, into the G1 phase of the cell cycle. Progression through the cell cycle is governed by a series of regulators that initiate and monitor the macromolecular synthesis events such as DNA replication that are required for cell division (Section 4.2i). Following cell division, the daughter cells can either commit to a further cell cycle or, if the appropriate growth conditions are not present, withdraw into the quiescent state. In tumor cells, the requirement for exogenous growth factors is largely superseded by intracellular signals generated by activated oncoproteins, and the normal growth inhibitory mechanisms are inoperative due to the loss of growth suppressor proteins, thereby allowing the tumor cell to grow in an unregulated fashion. Thus, to explain the phenotype of tumor cells in molecular terms, we need to understand the functions of oncoproteins and growth suppressor proteins.

4.2 The Functions of Oncogene Products

The seminal finding that pointed the way to understanding oncoprotein function was the discovery that the v-Src protein (also known as $pp60^{v\text{-}src}$) is a protein kinase. (Protein kinases are enzymes that phosphorylate particular amino acids in proteins using ATP as a phosphate donor.) This observation rapidly led to demonstrations that many retroviral oncogene products are protein kinases. In other instances, the subcellular location of the protein revealed likely functions for viral transforming proteins. For example, the v-*myc* and v-*fos* gene products are concentrated in the nucleus, suggesting that they may regulate the expression of other genes. Considerable progress in identifying oncogene functions and their relevance to cancer was made by comparing the predicted amino acid sequences of the oncoproteins with each other and with other known protein sequences. For instance, the fact that the v-*sis* gene encodes a protein related to a known growth factor emerged from such a comparison. This type of analysis has led to classification of the products of the dominant oncogenes into functional families (Table 4.3).

To provide some logic to the discussion of oncogene function, I have arranged the classes of oncogenes as follows: those whose products act outside the cell, those acting at the cell surface, those functioning in the cytoplasm, and ending up with the nuclear oncoproteins (Figure 4.3) (Table 4.4). Class I oncoproteins are growth factors and act on the same cells that produce them (an **autocrine** mechanism). Class II oncogenic proteins are protein-tyrosine kinases (i.e., they add phosphate to tyrosine residues in their substrate proteins), which may be either transmembrane proteins derived from growth factor receptors or cytoplasmic proteins that act as subunits of cell surface receptors. Class III proteins are cell surface receptors that lack protein kinase activity. Class IV oncoproteins are plasma membrane–associated signal-transducing proteins whose activity is linked to their GTP-binding and GTPase activities. Class V contains cytoplasmic protein-serine kinases, which may normally be involved in trans-

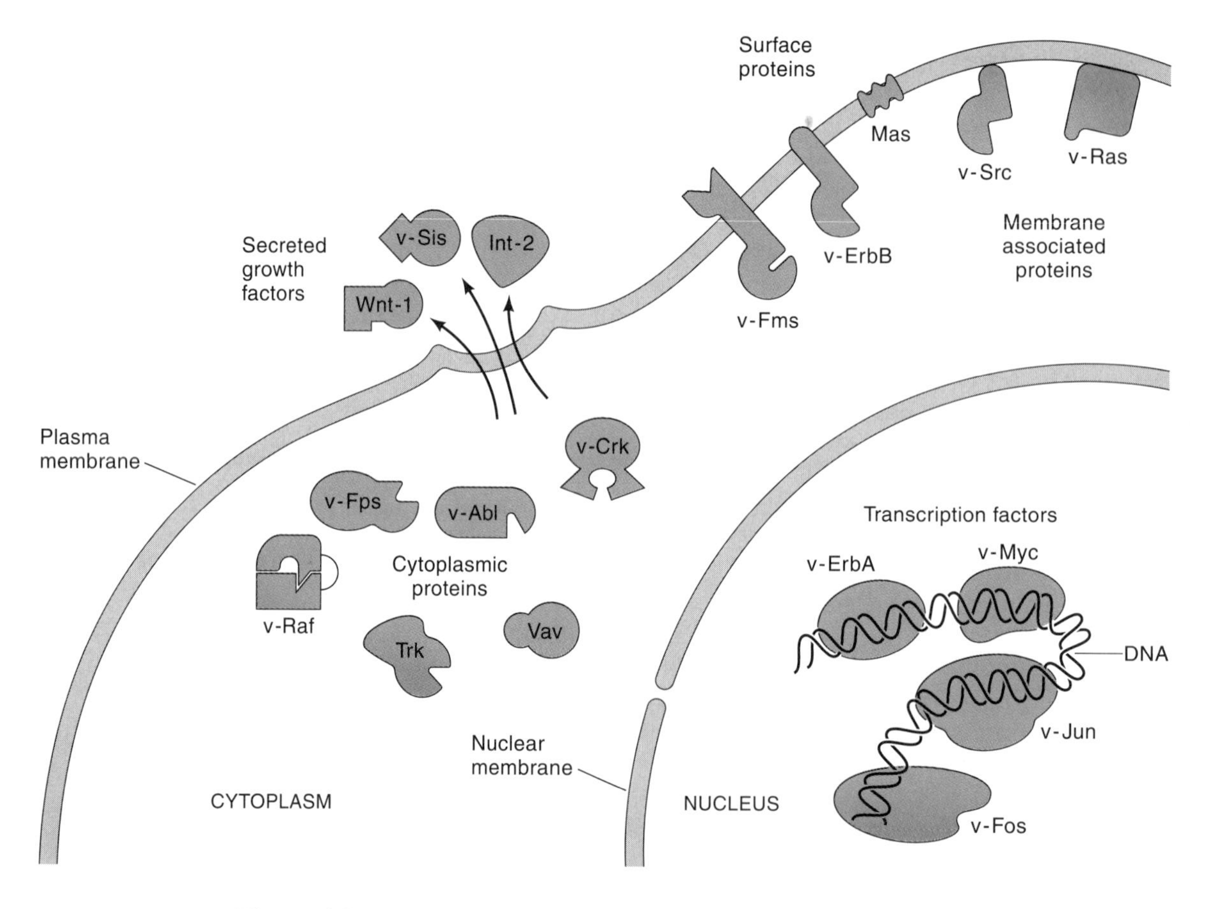

Figure 4.3
Subcellular localization of oncogene products. The subcellular locations of members of different classes of oncoprotein are shown in a schematic cell.

ducing signals to the nucleus. Class VI is comprised of proteins that function as cytoplasmic regulators. Class VII oncoproteins are nuclear proteins that regulate the expression of other genes directly or indirectly. Class VIII proteins regulate the cell cycle. Class IX proteins antagonize DNA damage–induced cell death known as **apoptosis**.

To illustrate the wide variety of mechanisms of transformation and oncogenic activation, I review each class. The products of the DNA tumor virus oncogenes and the growth suppressor genes are discussed separately. This review of oncoprotein function makes it evident how vital a part recombinant DNA technology has played. In general terms, this technology has made it possible to obtain complete predicted sequences of the oncogenic proteins. These sequences have proved valuable in determining the likely functions of these oncoproteins through comparison with the sequences of other proteins with known functions. Recombinant clones have also been indispensable in obtaining immunological reagents against these proteins. Finally, hypotheses regarding function and oncogenic activation mechanisms have been tested by making site-directed alterations in cloned oncogenes and expressing these mutants in cultured cells.

a. Class I: Growth Factors

Cell growth is governed by the availability of exogenous growth factors that act to stimulate growth through binding to specific cell surface receptors. When growth factors are withdrawn from cultured cells, they stop growing and enter a

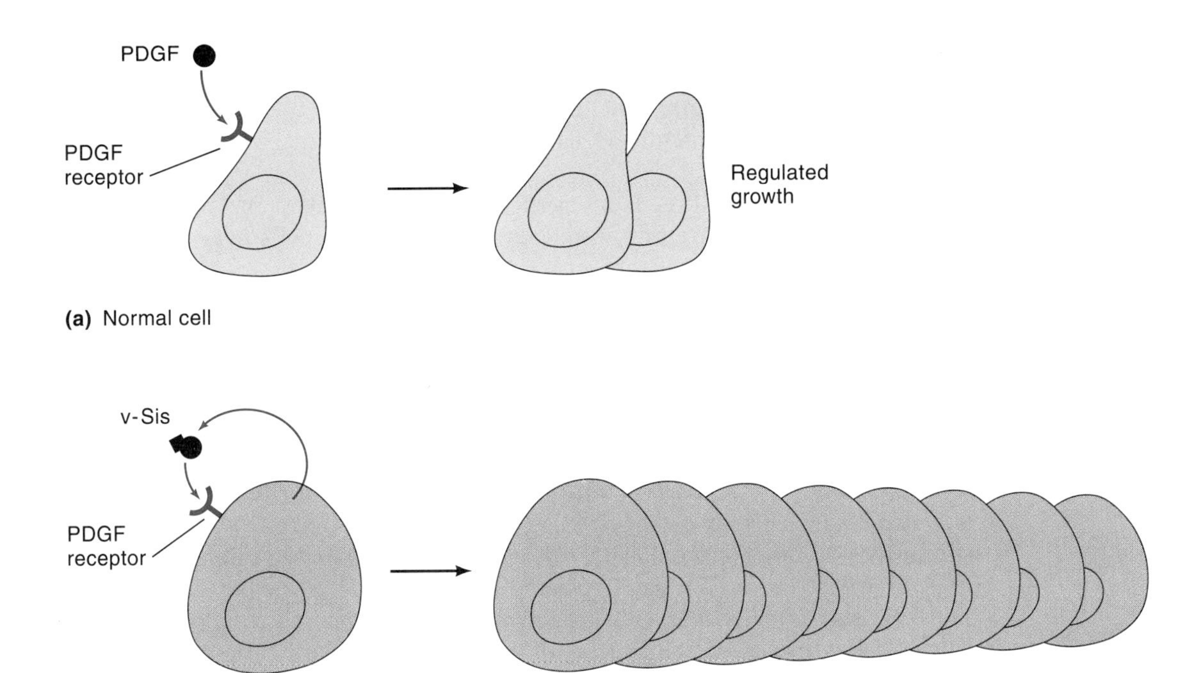

Figure 4.4
Contrast between normal cell and tumor cell growth regulation. (a) Normal cells show regulated growth in response to binding of platelet-derived growth factor (PDGF) to the PDGF receptor. (b) Cells transformed by v-Sis, a secreted PDGF-related growth factor, are continuously stimulated by the binding of v-Sis to the PDGF receptor (autocrine stimulation) leading to continuous unregulated growth.

quiescent state. For this reason, there was considerable excitement in the oncogene field when researchers found that the v-Sis protein, the product of the simian sarcoma virus (SSV) v-*sis* oncogene, was related to the B chain of platelet-derived growth factor (PDGF). This provided the first molecular connection between growth regulation and tumorigenesis. It is now established that the c-*sis* gene encodes the B chain of PDGF and that the PDGF receptor is expressed on the surface of fibroblasts and other mesenchymal cells. Indeed, the normal c-*sis* gene itself can transform fibroblasts that possess PDGF receptors on their surface. This means that the c-Sis protein need not be altered to create a transforming protein. Rather, transformation results from inappropriate expression of the PDGF B chain in cells that have the PDGF receptor. The PDGF receptor is one of a class of cellular growth factor receptors whose intrinsic protein-tyrosine kinase is activated upon ligand binding and that can be converted by mutation into oncoproteins (Section 4.2b).

Given the identity of the v-Sis protein with a potent growth factor, the obvious hypothesis is that the v-*sis* gene transforms through the production of a growth factor that drives the cells into continuous replication, through autocrine stimulation (Figure 4.4). There is a considerable body of evidence in favor of this idea. Thus, cells transformed by the v-*sis* gene secrete variable amounts of a PDGF-like protein with mitogenic activity, and there is a high level of cell-associated activity as well. Mutational analysis has shown that for the v-Sis protein to transform cells in which it is produced, it must enter the cellular secretion pathway, even if it is not actively secreted. Moreover, only cells that have PDGF receptors can be transformed by the v-*sis* gene. The puzzle is that treatment of cells bearing PDGF receptors with exogenous PDGF does not convert them into a malignant state; thus, there must be a difference between exogenous application of PDGF and chronic autogenous exposure. No direct activation of the c-*sis* gene in a human tumor has been reported, but several human tumor cell lines produce PDGF-like factors and show evidence for increased c-*sis* gene transcription.

A second type of Class I oncogene includes members of the fibroblast growth factor (FGF) family of genes; they also transform cells bearing FGF receptors via an autocrine mechanism. (There are several FGF receptors, each of which binds different members of the FGF family; all, however, are protein-tyrosine kinases.) One of these oncogenes, *hst* or K-FGF, was isolated from transfection studies with human tumor DNA. Another, *int*-2, was identified as one of the genes responsible for mouse mammary tumor virus–induced mammary tumors; in this instance, the *int*-2 gene is activated by the nearby insertion of the mouse mammary tumor virus (MMTV) genome. MMTV-induced mammary tumors also result from the activation of a third type of growth factor oncogene (*wnt*-1), which turns out to be a member of a large family of putative growth factors that are important in development. The *wnt*-1 gene transforms only cells of epithelial origin. The Wnt-1 receptor has not been identified, nor is it clear whether the Wnt-1 protein normally acts as a growth factor or a differentiation factor. Inasmuch as all the FGFs that have been tested can transform appropriate cells and that the epidermal growth factor (EGF) gene can transform EGF receptor-bearing cells, other growth factor genes will probably also prove to be oncogenes.

Most tumors secrete growth factors, which have been termed transforming growth factor (TGF). TGFs are not the products of the oncogenes in these tumors, but rather are induced indirectly. In appropriate combinations, TGFs can cause normal cells to adopt the properties of transformed cells, albeit reversibly. One of the TGFs, TGFα, is closely related to EGF and can bind to and activate the EGF receptor. Another TGF, TGFβ, is of interest because although it can stimulate the anchorage-independent growth of fibroblasts in the presence of TGFα, TGFβ inhibits the growth of many other cell types. TGFβ, itself a family of factors, can stimulate cells to express components of the extracellular matrix that are necessary to form a substratum for the growth of adherent cell types. There are also indications that TGFs may allow cells to escape their normal requirements for growth factors and an extracellular matrix and thereby contribute to the phenotypes of tumor cells.

b. Class II: Protein-Tyrosine Kinases

The addition of phosphate to a protein by a protein kinase can alter its activity. Therefore, that protein phosphorylation has turned out to be a major cellular regulatory mechanism is not surprising. For instance, glycogen phosphorylase activity is increased following phosphorylation by phosphorylase kinase. For protein activity to be regulated by phosphorylation, both a protein kinase and a protein phosphatase are required, the latter being needed to remove the phosphate and restore the protein to its initial state.

In a normal mammalian cell, most of the protein phosphorylation occurs on serine and threonine (90 percent and 10 percent, respectively), with only a small amount occurring on tyrosine. Although there are nearly as many protein-tyrosine kinases as protein-serine kinases, the paucity of phosphotyrosine in proteins can be explained by the presence of a high level of cellular protein-tyrosine phosphatase activity; moreover, protein-tyrosine kinases are generally much less abundant than the protein-serine kinases and are also tightly negatively regulated. Mutations that lead to the abolition of this negative regulation commonly result in the creation of oncogenes.

When the v-Src protein was discovered to be a protein kinase, only protein-serine/threonine kinases were known. Nevertheless, this property immediately suggested that the v-Src protein might cause transformation by phosphorylating cellular proteins and thereby altering their activity. The subsequent finding that the v-Src protein kinase activity was specific for tyrosine suggested that v-Src's transforming activity was linked to this uncommon type of phosphorylation. This view was strengthened by the discovery that several other oncogenic protein kinases are also specific for protein-tyrosine residues and that proteins in normal cells contain low but significant levels of phosphotyrosine. When two normal cellular proteins, c-Src and the EGF receptor, were found to have protein-tyrosine kinase activity, it seemed likely that there would be a large number of cellular genes that encode protein-tyrosine kinases.

Properties of protein-tyrosine kinases The normal functions of protein-tyrosine kinases are diverse. Many growth factor receptors have intrinsic protein-tyrosine kinase activity, which is increased severalfold upon binding their ligands; furthermore, the ensuing phosphorylation of target proteins on tyrosine is essential for the mitogenic response to growth factor binding. Tyrosine phosphorylation of proteins also influences the cell cycle, the architecture of the plasma membrane and the cytoskeleton, and secretion. Thus, the inappropriate activity of a protein-tyrosine kinase can, in principle, elicit both uncontrolled growth and altered cell morphology, each of which is characteristic of tumor cells.

The initial finding that v-Src had protein kinase activity relied on the detection of enzyme activity in immune complexes of v-Src isolated from RSV-transformed cells rather than with the purified protein. Thus, it was possible that this activity was not intrinsic to v-Src, but rather was due to an associated protein. The proof that v-Src had intrinsic protein-tyrosine kinase activity came from two directions. First, the protein resulting from the expression of the cloned v-*src* gene in *E. coli* had protein-tyrosine kinase activity; because *E. coli* lacks protein-tyrosine kinases, the activity could be attributed to v-Src itself. Second, the amino acid sequence predicted from the cloned v-*src* gene's nucleotide sequence was related to that of the catalytic subunit of the cAMP-dependent protein kinase. Subsequently, all the oncogenes encoding protein-tyrosine kinases have been found to be related to v-Src. Indeed, in several cases, the protein-tyrosine kinase activity of an oncogene product was predicted on the basis of its sequence relatedness to v-Src before it was shown biochemically.

We now know that all protein-serine and protein-tyrosine kinases share a domain of approximately 275 amino acids, which constitutes a catalytic domain (Figure 4.5). This domain contains several short amino acid sequences that are absolutely conserved among all the protein kinases and that serve as signatures for a protein kinase catalytic domain. The ATP-binding site is located in the N terminal half, and the substrate-binding site is in the C terminal half of this domain. Additional conserved sequences within the catalytic domain distinguish the protein-serine kinases from the protein-tyrosine kinases. The protein-serine and protein-tyrosine kinases can be divided into subfamilies based on the structures of both the catalytic domain and the sequences lying outside the catalytic domain. Most protein-tyrosine kinases are considerably larger than the minimal 30 kDa catalytic domain. The additional sequences serve a number of purposes, including regulation, subcellular targeting, and the determination of substrate specificity.

The oncogenes in Class II can be divided into two main groups based on the structure and subcellular localization of their products. The first class consists of protein-tyrosine kinase growth factor receptors that are membrane-spanning proteins expressed on the cell surface. The second group is made up of nonreceptor protein-tyrosine kinases. Several members of the *src* gene family protein-tyrosine kinases are in the second class; they bind to membranes through a

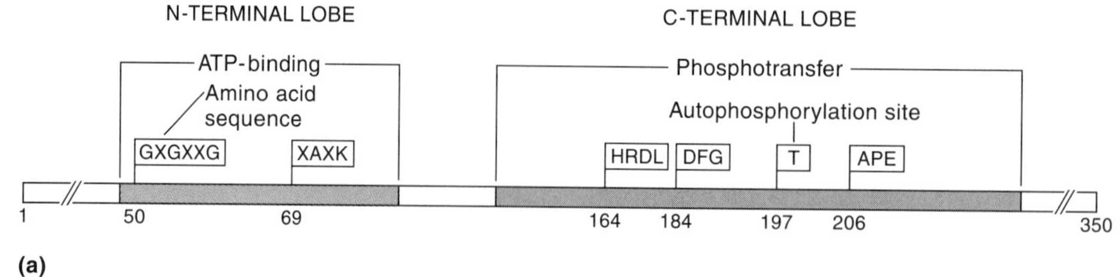

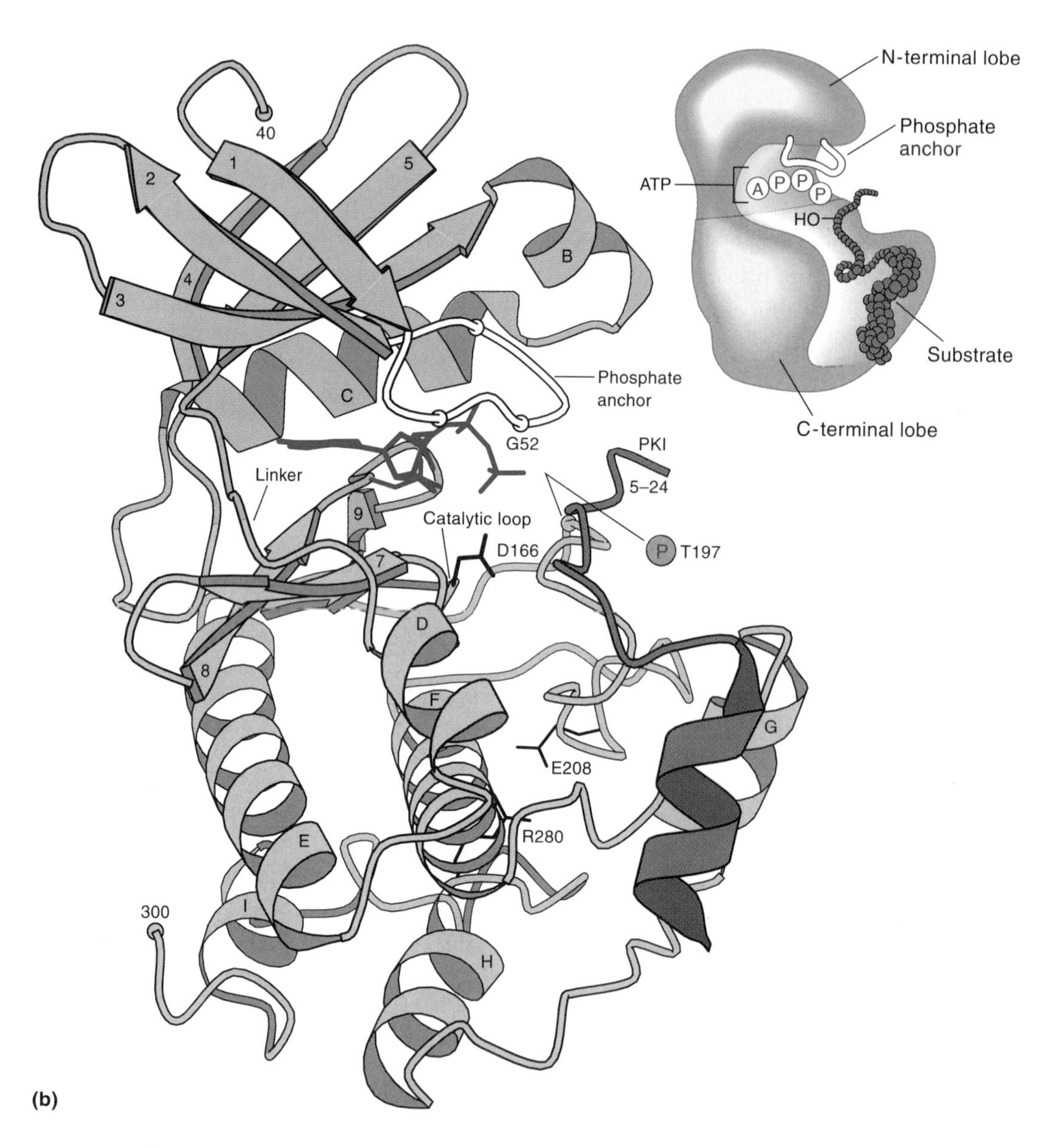

Figure 4.5
cAMP-dependent protein kinase catalytic domain structure. (a) Ribbon structure of the cAMP-dependent protein kinase catalytic subunit showing the two major functional domains and the key residues for catalysis. These include the glycine-rich phosphate anchor loop that binds ATP, and Asp 166 (the D in HRDL), the proposed catalytic base that initiates transfer of the γ-phosphate of ATP to the acceptor OH-group. (b) Three-dimensional structure of the cAMP-dependent protein kinase catalytic domain, showing the key conserved residues in the catalytic domain together with the catalytic loop and the ATP phosphate binding loop (adapted from Figure 3 in S. S. Taylor and E. Radzio-Andzelm. 1994. *Structure* 2 345.) Also shown is a bound peptide inhibitor, PKI, that mimics a bound substrate protein, as indicated in the inset, where the OH-group of the target serine is located close to the γ-phosphate of ATP and is in the correct position to accept the phosphate.

covalently attached lipid moiety and act as subunits of surface receptors that lack their own catalytic domain.

Oncogenes encoding growth factor receptors that are protein-tyrosine kinases The cloning and sequencing of the EGF receptor led to the unexpected discovery that the v-*erbB* oncogene, which is responsible for the leukemia induced by an avian erythroblastosis retrovirus and had previously been cloned and sequenced, was a truncated form of the EGF receptor. The EGF receptor has protein-tyrosine kinase activity. Several other oncogenes are now also known to be derived from cellular genes that encode growth factor receptors with protein-tyrosine kinase activity. These receptors are membrane-spanning proteins with a large, extracellular, glycosylated, N terminal ligand–binding domain, a single transmembrane domain, and a cytoplasmic region containing the protein kinase catalytic domain. Currently, 14 distinct families of receptors with protein-tyrosine kinase activity are known, but only 8 these have been implicated in oncogenesis either through the generation of oncogenes or by their overexpression in tumors.

When a growth factor binds to its receptor's extracellular domain, the cytoplasmic protein-tyrosine kinase domain is activated, leading to the tyrosine phosphorylation of cellular target proteins and the transmission of a mitogenic signal (Figure 4.6a). The exact mechanism by which activation of the receptor's protein-tyrosine kinase domain occurs is not fully understood, but ligand binding apparently triggers dimerization of the receptor in the plane of the membrane, presumably because a conformational change in the external domain exposes a dimerization site. As a result of dimerization, the cytoplasmic catalytic domains of two receptor molecules are brought into juxtaposition, thus allowing mutual phosphorylation. The phosphorylation that ensues occurs at regulatory "autophosphorylation" sites, some of which activate the protein-tyrosine kinase. Support for this model derives from the finding that a receptor whose protein-tyrosine kinase activity is abolished by mutation acts as a dominant suppressor of the wild-type receptor's activity when it is expressed inside cells. Presumably, this is because nonfunctional heterodimers containing one wild-type and one mutant receptor molecule are formed. Although the wild-type receptor can phosphorylate the mutant receptor in such dimers, the mutant receptor can neither phosphorylate nor activate the wild-type receptor.

Ligand-induced autophosphorylation also creates binding sites for a particular class of receptor protein-tyrosine kinase substrates that contain a so-called **SH2 domain** (Src Homology 2); proteins with SH2 domains bind to proteins at phosphorylated tyrosine residues. The binding of proteins with SH2 domains to activated receptors is an efficient mechanism for recruiting these protein substrates from the cytosol because the affinity of SH2 domains for phosphorylated tyrosines is in the nanomolar range, whereas non-SH2-containing substrates have affinities in the micromolar range. Many of these SH2 domain proteins are components of pathways that provide signals to the nucleus; I describe them more fully in "Targets for oncogenic protein-tyrosine kinases" later in this section.

Although the detailed regulation of each receptor's protein-tyrosine kinase is different, ligand-induced dimerization and substrate binding to autophosphorylation sites are general principles. Similarly, although the mutations that convert individual receptor protein-tyrosine kinases into oncoproteins are distinct, there are general mechanisms underlying oncogenic activation. In particular, regardless of whether the activated receptor-derived oncoprotein is membrane associated or cytosolic, the mutant proteins form dimers and are thereby constitutively activated for substrate phosphorylation.

At this point, a brief synopsis of five oncogenes derived from normal genes for growth factor receptors is useful to illustrate some of the different principles in the oncogenic activation of receptor protein-tyrosine kinases.

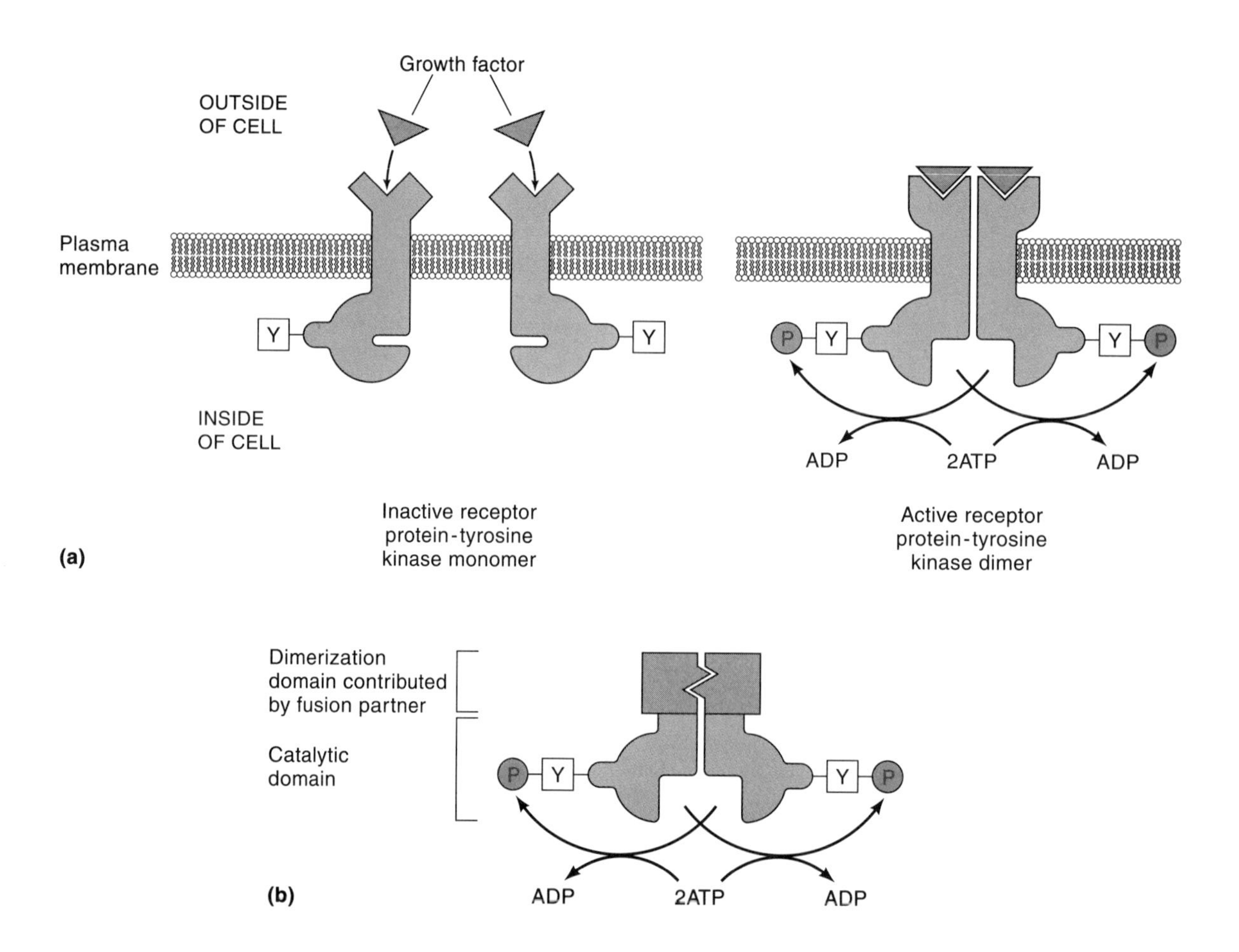

Figure 4.6
(a) Receptor protein-tyrosine kinases are activated by growth factor-induced dimerization. Growth factor binding to the extracellular ligand-binding domain of an inactive monomeric transmembrane receptor protein-tyrosine kinase causes a conformational change and thereby induces dimerization of the receptor. This results in a conformational change in both receptor cytoplasmic domains, which leads to transphosphorylation of one receptor subunit by the other. Transphosphorylation activates the receptor catalytic domains and also allows the cytoplasmic domains to bind and phosphorylate its substrates. (b) Oncogenic receptor protein-tyrosine kinases are constitutively activated dimers. One mechanism of oncogenic activation of receptor protein-tyrosine kinases is through fusion of the receptor cytoplasmic domain with a foreign N-terminal domain that has intrinsic dimerizing activity. Such receptor chimeras are therefore constitutively dimerized and activated in a ligand-independent fashion.

The **v-*erbB*** oncogene, which is present in two different avian erythroblastosis viruses (AEV), is derived from the chicken EGF receptor gene; besides being doubly truncated, it also harbors point mutations in the catalytic domain (Figure 4.7a). In addition to causing an erythroleukemia in chickens, these viruses can transform fibroblasts in culture. Both of the known forms of the v-ErbB protein lack almost the entire external ligand-binding domain and contain deletions of short sequences in the C terminal tail beyond the catalytic domain. Although the normal signal sequence is missing, the v-ErbB protein is glycosylated, and some of it becomes associated with the plasma membrane in a fashion topologically identical to the EGF receptor. However, much of the v-ErbB protein accumulates in the vicinity of the Golgi apparatus inside the cell. The v-ErbB protein has pro-

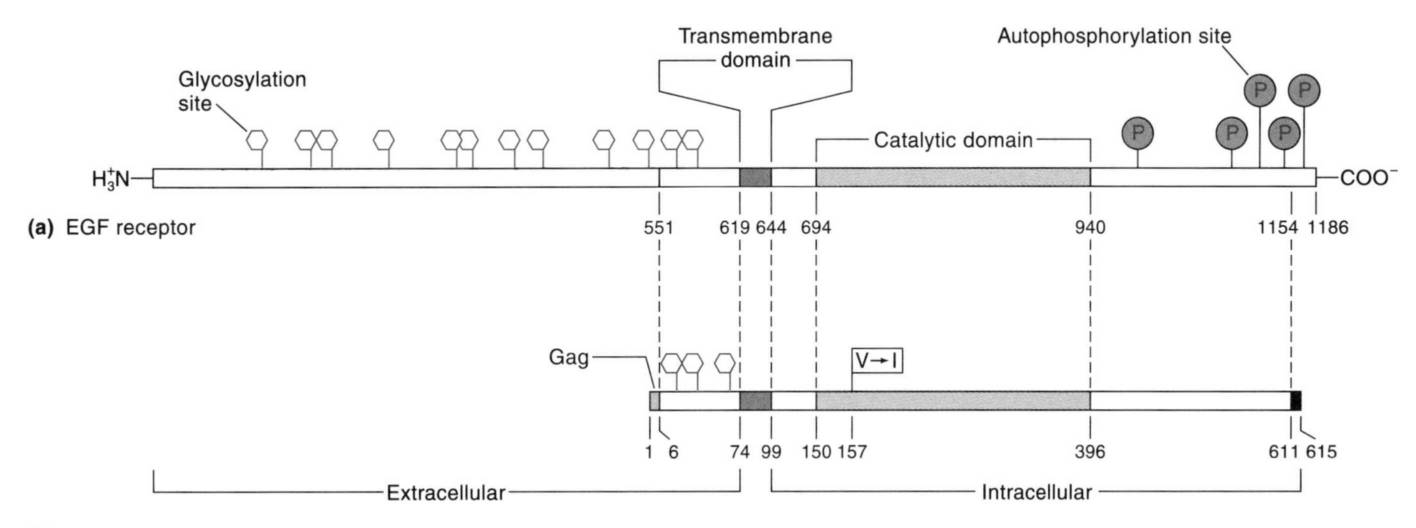

Figure 4.7
Oncogenic activation of the EGF receptor protein-tyrosine kinase by truncation. (a) The structure of the human EGF receptor. The positions of potential N-linked glycosylation sites in the extracellular ligand-binding domain (open hexagons), the transmembrane and catalytic domains, and the five mapped autophosphorylation sites (P) are shown. (b) The avian erythroblastosis virus (AEV-H) v-ErbB oncoprotein is derived from the chicken EGF receptor, which has a structure very similar to that of the human EGF receptor, by deletion of most of the extracellular domain and a short region at the C-terminus of the cytoplasmic domain, and by a point mutation in the catalytic domain in the ATP-binding site (indicated by a flag labelled V→I). Residues 1–6 are derived from the retroviral Gag protein, and residues 611–615 from out-of-frame *env* gene sequence.

tein-tyrosine kinase activity, although its activity is considerably weaker than that of the stimulated EGF receptor.

The key question is what changes are necessary to convert the EGF receptor gene into a transforming gene. The requirements for transformation by the v-*erbB* gene have been studied by site-directed mutagenesis of the cloned gene and also by isolating temperature-sensitive AEV mutants. These studies show that protein-tyrosine kinase activity is essential for transformation and that the v-ErbB protein has to reach the cell surface in order to be active for transformation. As to what mutations of the EGF receptor are required for oncogenic activation, the results are complex and depend on the cell type, the receptor origin, and the disease target. Thus, all that is required for erythroleukemogenesis is the N terminal truncation, but other mutations confer sarcomagenic potential. For instance, a 21 amino acid internal deletion in the C terminal tail near the catalytic domain as well as different point mutations in the ATP-binding lobe or in the phosphotransferase domain enable the protein to induce sarcomas. Although deletion of the very C terminus does not influence the protein's sarcomagenic potential, different deletions within the C terminal tail alter the disease potential *in vivo* and can have positive or negative effects on transformation of fibroblasts *in vitro*. The five major autophosphorylation sites lie in the **EGF receptor**'s C terminal tail beyond the catalytic domain, and the truncations in the two v-ErbB proteins eliminate at least one of these sites (Figure 4.7). Because these sites are important for the association and phosphorylation of v-ErbB substrates, the variable phenotypic effects of deletions in this region may be explained by which phosphorylation sites are removed and which substrates are crucial for a mitogenic response in different cell types.

The loss of most of the extracellular domain appears to be the most important event in converting the EGF receptor into an oncogene. The absence of this domain precludes EGF binding but apparently allows ligand-independent, constitutive dimerization, probably through the residual extracellular domain sequence. The activating mutations in the cytoplasmic domain all increase intrinsic catalytic activity, and the different mutations synergize to generate a more powerful transforming protein. The loss of the extracellular domain is the primary mutation, and the mutations in the cytoplasmic domain, which are selected for upon passaging the virus *in vivo,* are secondary. The accumulation of multiple independent activating mutations is typical of retroviral oncogene formation in general.

In summary, transformation by the v-ErbB oncoprotein involves mimicking an occupied EGF receptor by continuous expression of an unregulated protein-tyrosine kinase at the cell surface. When it transforms erythroid cells the v-ErbB protein probably usurps a mitogenic signal pathway normally initiated by tyrosine phosphorylation activated by an erythroid cell growth factor receptor, such as the erythropoietin receptor.

The ***neu/c-erbB2* oncogene** was first identified by transfection of DNA from N-ethylnitrosourea-induced glioblastomas into NIH 3T3 cells. Screening of the resulting clones of transformed cells for the presence of known oncogenes revealed the consistent presence of v-*erbB*-related sequences. This allowed the isolation of cDNA clones for both the oncogenic *neu* gene and its normal counterpart. The normal rat Neu protein is a 185 kDa surface protein with protein-tyrosine kinase activity. There is considerable sequence similarity with the EGF receptor throughout Neu, and although its C terminal regulatory region is less closely related, three autophosphorylation sites are conserved. The human homologue of the *neu* gene is known as c-*erbB2* or *HER2* due to the obvious relationship with the EGF receptor. In fact, human c-*erbB2* was originally cloned as a sequence related to the EGF receptor gene that was amplified in certain human tumor cell lines. Based on its structure, Neu, which is expressed on fibroblasts as well as many other cell types, is presumed to act as a surface receptor. The nature of the Neu ligand and its physiological role are still uncertain, but it is likely to be related to EGF.

In contrast to the major structural differences between the EGF receptor and v-ErbB protein, the normal Neu protein and the oncogenic version differ only by a single valine to glutamic acid change at position 664 (Val664Glu), eight residues in from the extracellular end of the transmembrane domain (Figure 4.8) The same mutation has occurred in the *neu* gene on multiple occasions, implying that it is the easiest way to activate the Neu protein. As a result of this single amino acid change, there is substantial ligand-independent dimerization accompanied by a five- to tenfold increase in the protein kinase activity of mutant Neu compared to that of the unliganded normal Neu protein. The mechanism by which this mutation elicits constitutive dimerization is unclear, but Glu 664 may be able to hydrogen bond with Glu 664 in a second Neu molecule, causing dimer formation via the two transmembrane domains. The only other amino acid substitutions at position 664 that result in activation are Val664Asp and Val664Gln, both hydrophilic replacements. Thus, both the positioning and the nature of the change in the transmembrane region are critical for activation of the Neu protein kinase. Interestingly, N terminal truncation of the *neu/c-erbB2* gene, like that of the EGF receptor, also results in oncogenic activation.

In summary, the mutant Neu oncoprotein transforms by acting as an unregulated protein-tyrosine kinase at the cell surface. This is consistent with the observation that transformation by the *neu* gene is blocked by a monoclonal antibody that clears the Neu protein from the surface. The fact that overexpression

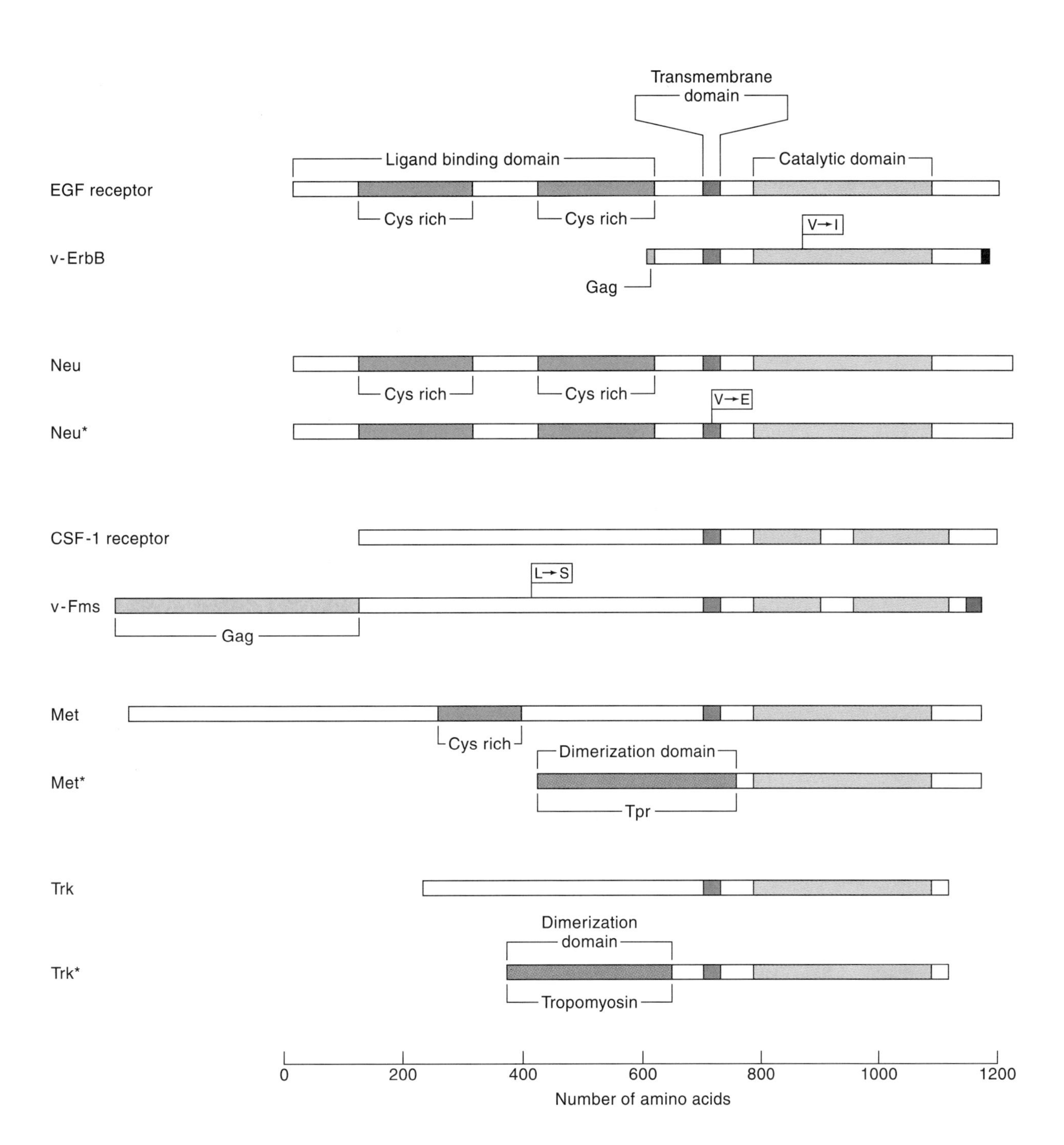

Figure 4.8
Structural rearrangements leading to oncogenic activation of receptor protein-tyrosine kinases. Examples of different mechanisms of mutational activation of receptor protein tyrosine-kinases are shown. The top structure in each pair is the normal cellular receptor protein-tyrosine kinase; the lower member is the oncogenic derivative. Activating mutations include truncations, fusions, and point mutations (indicated by flags V→I, V→E, L→S).

of the normal *neu/c-erbB2* gene transforms NIH 3T3 cells, coupled with the high level of expression of an apparently normal Neu/c-ErbB2 protein in many types of human tumor with amplified *neu/c-erbB2* genes, suggests that the activating mutation in the *neu* gene simply increases the protein kinase activity of the Neu protein rather than changing its substrate specificity.

The **v-*fms* oncogene** was first recognized as the oncogene of a feline sarcoma virus (FeSV). The v-Fms protein was shown to have protein-tyrosine kinase activity before the gene was cloned. The isolation and sequencing of the v-*fms* gene made it immediately clear that it belonged to the family of protein-tyrosine kinases, and its predicted structure indicated that it was derived from a receptor-like protein of unknown nature, with the interesting feature of having a large insert in the protein kinase domain. The clue to the identity of the c-Fms protein came from its high level of expression in macrophages, which suggested that c-Fms could be a receptor for a macrophage growth factor. Subsequently, researchers showed that the c-Fms protein is the receptor for colony-stimulating factor 1 (CSF-1), which is essential for macrophage differentiation and proliferation. Thus, the v-*fms* gene was derived from the cat CSF-1 receptor gene. A comparison of the sequences of the cat c-*fms* gene with that of the v-*fms* protein shows 10 scattered single amino acid changes and a striking divergence of the two protein sequences downstream of the catalytic domain starting 50 amino acids from the C terminus of c-Fms (Figure 4.8).

The v-*fms* gene has been studied extensively by site-directed mutagenesis, and the requirements for transformation have been determined by analyzing chimeric genes constructed using clones of the c-*fms* and v-*fms* genes. The v-Fms protein has to reach the cell surface and retain protein kinase activity to transform. The loss of a sequence that is rich in serine and contains a tyrosine four residues from the C terminus is responsible for the transforming activity of v Fms. The location of this tyrosine is reminiscent of the autophosphorylation sites in the EGF receptor and the regulatory tyrosine at the C terminus of the Src family protein-tyrosine kinases (see "Oncogenes derived from protein-tyrosine kinases that are not receptors"). Replacement of only the C terminal tyrosine in c-Fms with phenylalanine does not induce transforming activity; but in cells expressing CSF-1, the mutant protein with phenylalanine in place of tyrosine has considerably greater transforming activity than the wild-type CSF-1 receptor. By contrast, chimeras in which the C terminus of the normal c-Fms replaces the corresponding region of the v-Fms gene have reduced transforming activity. Moreover, substitution of the v-Fms C terminus by the corresponding segment containing phenylalanine suppresses the transforming phenotype less effectively than the normal C terminus. In aggregate, these data imply that the C terminus of the CSF-1 receptor has a negative regulatory function that requires the C terminal tyrosine, but there is no evidence that regulation by the C terminus involves phosphorylation of this tyrosine.

The fact that the v-Fms in another FSV also lacks the C terminus of c-Fms reinforces the conclusion that the loss of the C terminal region is important for the oncogenic activation of the c-*fms* gene. However, because the C terminal change is not sufficient for full transforming activity, one or more of the point mutations elsewhere in the v-*fms* gene must be required. The study of additional chimeric *fms* genes narrowed down the possibilities to two point mutations in the N terminal half of the ligand-binding domain. Researchers then used site-directed mutation to show that a leucine to serine change within this domain is responsible. This mutation does not affect CSF-1 binding, but apparently elicits constitutive ligand-independent dimerization. Subsequent random mutagenesis studies have shown that other point mutations in the c-*fms* sequence encoding the extracellular domain can also result in oncogenic activation in conjunction with the C terminal deletion.

In summary, the v-Fms oncoprotein appears to act as a ligand-independent protein-tyrosine kinase. Even though the v-Fms protein has an intact and functional ligand-binding site, CSF-1 binding does not cause a significant increase in v-Fms protein-tyrosine kinase activity *in vitro* or any change in growth properties of transformed cells. Indeed, the v-Fms protein accumulates in coated pits in the absence of CSF-1 and is internalized as if it were a liganded receptor. Because the normal CSF-1 receptor can transform fibroblasts provided they are expressing CSF-1, the v-Fms protein clearly mimics the normal CSF-1-stimulated receptor. In addition, because the CSF-1 receptor is expressed only in myeloid cells and macrophages, it must be able to trigger a normal fibroblast mitogenic pathway used by receptors like the EGF and PDGF receptors. Indeed, we now know that the CSF-1 receptor utilizes many of the same substrates as the EGF and PDGF receptors both in fibroblasts and in myeloid cells.

The ***trk* oncogene** was identified by transfection of human colon carcinoma DNA into NIH3T3 cells and cloned by the methods already outlined. The deduced amino acid sequence of the colon carcinoma *trk* oncogene indicated that it encodes a protein-tyrosine kinase and suggested that it is derived from a receptor protein-tyrosine kinase. A more extensive survey revealed that *trk* oncogenes occur in 25 percent of human papillary thyroid carcinomas. The normal *trk* protooncogene is expressed in an extremely restricted fashion *in vivo,* being present only in certain sensory ganglia in the central nervous system. The sequence of a full-length cDNA showed that the normal Trk protein is receptorlike, and when expressed in NIH3T3 cells, it encodes a 140 kDa glycoprotein with protein-tyrosine kinase activity. Subsequently, researchers identified nerve growth factor (NGF) as the ligand for the Trk protein, consistent with its distribution in sensory neurons.

Although the Trk oncoprotein is derived from a gene that encodes a 140 kDa cell surface receptorlike protein, it is itself a 70 kDa soluble cytoplasmic protein. The original isolate of the *trk* oncogene from a colon carcinoma was shown to be a chimeric gene resulting from a short interstitial deletion on human chromosome 1. It consists of the first seven coding exons from a nonmuscle tropomyosin fused to the *trk* gene at an exon just upstream of the transmembrane domain (Figure 4.8). Initially, the presence of the tropomyosin sequences in the Trk oncoprotein suggested that it might associate with the cytoskeleton. However, this is not the case because Trk is soluble and readily separable from cytoskeletal elements by cell fractionation. Surprisingly, despite the presence of a characteristic transmembrane sequence, Trk is not membrane associated; presumably, this results from the absence of a signal peptide. Although the new N terminal sequences are required for transformation by Trk, much of the tropomyosin sequence is dispensable. The tropomyosin domain cannot be replaced with sequences from actin, but other N terminal sequences also create transforming Trk proteins. Indeed, *trk* oncogenes from other sources have different N terminal sequences fused to the C terminal Trk sequence at different positions upstream of the catalytic domain. In each instance, the Trk oncoprotein contains the normal C terminus. The *trk* protooncogene itself is nontransforming even when expressed at high levels, but rearrangements of this gene that occur upon transfection readily give rise to transforming variants with altered N termini. Interestingly, some of these variants are membrane-associated proteins, implying that the soluble nature of Trk is not essential for its transforming activity. All these observations can be reconciled by the fact that the region of tropomyosin associated with the Trk C terminal segment is known to undergo a homotypic coiled-coil type of interaction resulting in tropomyosin dimers. Its presence in Trk presumably causes homodimerization, leading to activation of the catalytic domain. The same must be true for the other appended N terminal sequences that permit Trk to function as an oncoprotein.

A similar situation exists for the **Met oncoprotein**. The *met* oncogene, which was generated by mutagenesis of the human osteosarcoma cell line (HOS), was activated as a consequence of a translocation between the *met* gene on chromosome 7 and the *tpr* locus on chromosome 1. The normal Met protein, which is widely expressed, is a receptor for hepatocyte growth factor (HGF). The Met oncoprotein is a soluble 65 kDa protein with protein-tyrosine kinase activity, which, unlike the normal Met protein, is phosphorylated on tyrosine when isolated from the cell. The protein sequence encoded by the *tpr* locus joins the normal Met protein sequence 50 residues downstream of the transmembrane domain; thus, the chimeric Tpr-Met protein contains the Met catalytic domain and normal C terminus (Figure 4.8). The Tpr sequence does not provide a signal peptide; consequently, Tpr-Met is localized in the cytosol. The Tpr sequence is required for oncogenic activity, and the essential element is a "leucine zipper" motif, which allows Tpr-Met molecules to form homodimers, resulting in constitutive protein-tyrosine kinase activity. Mutation of leucines in the zipper region abolishes transforming activity. When the Tpr leucine zipper is replaced by leucine zippers from other proteins that homodimerize, oncogenic activity is retained. Thus, the ability of the foreign N terminal sequence to dimerize rather than a specific Tpr sequence is essential for oncogenic activation.

Thus, both the Trk and Met oncoproteins result from the replacement of an N terminal ligand-binding domain by a segment that promotes constitutive formation of homodimers, causing activation of the catalytic domain. In both cases, the product is soluble rather than membrane associated. Why the soluble Trk and Met oncoproteins transform is still unexplained because soluble forms of other activated protein-tyrosine kinases lack transforming activity.

In summary, although different receptor protein-tyrosine kinases are oncogenically activated by mutations that alter the extracellular, transmembrane, or cytoplasmic domains, the net outcome is the same, namely, that the oncoprotein undergoes spontaneous dimerization, resulting in ligand-independent constitutive protein-tyrosine kinase activity (Figure 4.6b) and thereby the activation of cellular mitogenesis pathways normally used by this type of receptor.

Oncogenes derived from protein-tyrosine kinases that are not receptors Oncogenic protein-tyrosine kinases of the second type are derived from cellular protein-tyrosine kinases that do not span the plasma membrane, although many of them are associated with cytoplasmic membranes and, in some cases, act as catalytic subunits for receptors that lack a catalytic domain. Currently, nine distinct families of nonreceptor protein-tyrosine kinase are known, of which three have given rise to oncogenes. One of these is the ***src* family of oncogenes**, the prototype being the **v-*src* oncogene** found in RSV.

As mentioned earlier, the v-*src* gene arose from the c-*src* protooncogene. The c-Src protein is a member of a family of closely related protein-tyrosine kinases with very similar structures. This family consists of the products of the c-*src*, c-*yes*, c-*fgr*, *fyn*, *lyn*, *lck*, *hck*, *blk*, and *yrk* genes. All these proteins are about 500 amino acids in length and have the same general organization in which the 250 residue catalytic domain near the C terminus is followed by a sequence of about 15 residues (Figure 4.9). There are two regulatory domains upstream of the catalytic domain: an SH2 domain and an **Src Homology 3 (SH3)** domain. The members of the Src family have about 70 percent identity over the approximately 420 residues spanning the SH2, SH3 domains and the catalytic domain. With the exception of the N terminal glycine residue, their N terminal 80 residues are almost completely divergent. This N terminal glycine is modified by covalent attachment of a myristoyl group through an amide linkage. Although these proteins have neither a signal peptide nor a transmembrane domain, the fatty acid substituent permits them to associate with the inner face of the plasma and other cytoplasmic membranes.

The physiological functions of the Src family protein-tyrosine kinases are not fully understood. The pattern of this gene family's expression, however, gives

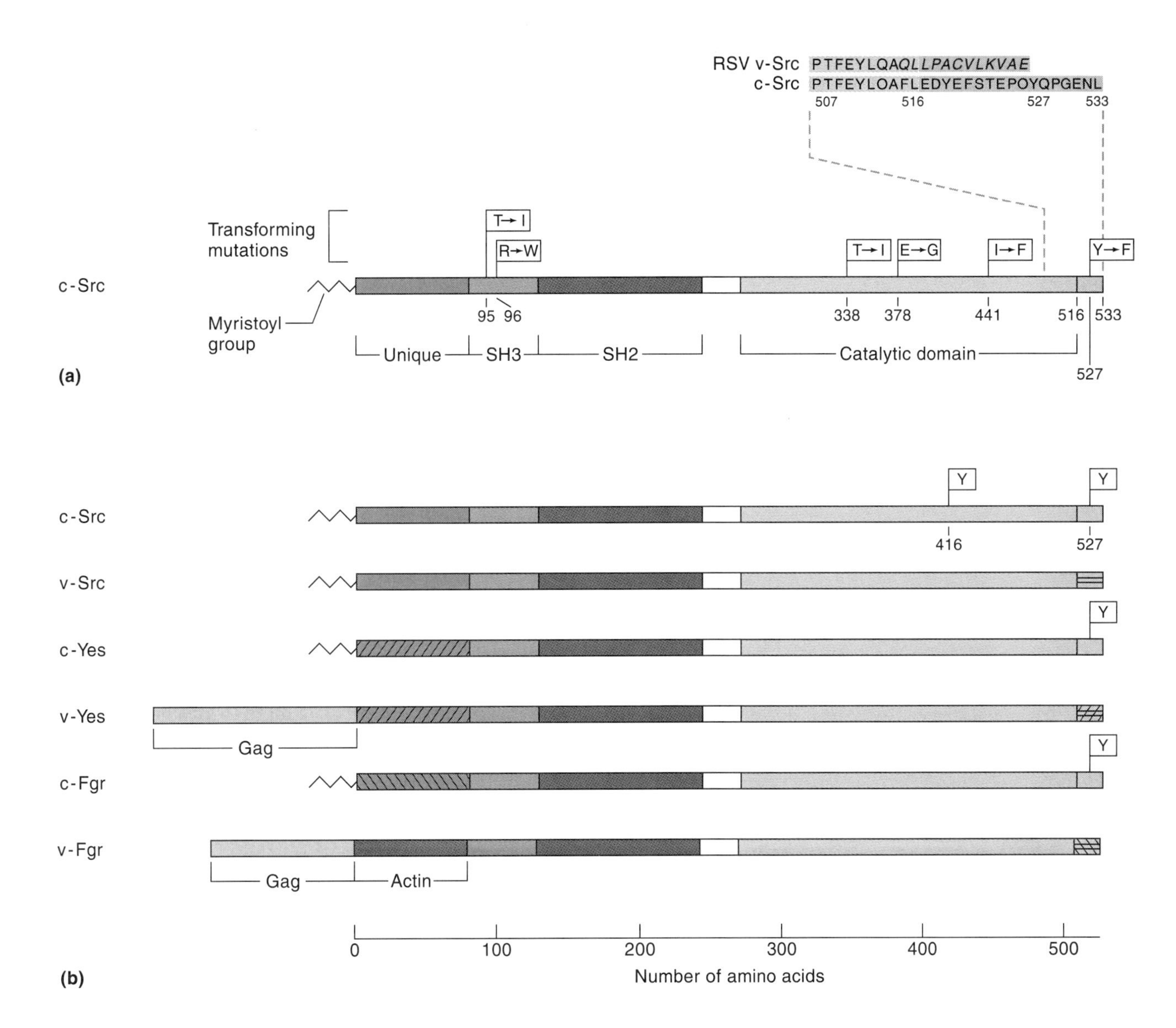

Figure 4.9
(a) Mechanism of oncogenic activation of c-Src by mutation. A schematic structure of the chicken c-Src protein is shown, with the functional domains indicated below. The RSV v-Src oncoprotein was derived from the chicken c-Src protein by replacement of the C-terminal 19 amino acids with 12 novel amino acids (shown in italics in gray shaded box) leading to loss of the C-terminal negative regulatory Tyr 527. The C-terminal boundary of the catalytic domain is Leu 516. Mutation of Tyr 527 to Phe by itself is sufficient to activate chicken c-Src into an oncoprotein. Other single-point mutations found in v-Src from different RSV strains that can also activate c-Src independently are indicated by flags (T→I, R→W, E→G, etc.) (b) Oncogenic activation of Src family protein-tyrosine kinases. As shown by the schematic structures, v-Yes and v-Fgr were derived from c-Yes and c-Fgr, respectively, by replacement of their C-terminal domains, including the regulatory tyrosine, by novel sequences, and by fusion of Gag or Gag + actin sequences upstream of the SH3 domain. The structures of c-Src and v-Src are shown for comparison.

some clues. Many of the genes are expressed in unique hematopoietic lineages. For instance, the Lck protein is present nearly exclusively in cells of the T cell lineage, whereas the Blk protein is expressed predominantly in B cells and Hck in granulocytes. However, c-Src is expressed in a somewhat different manner. It is present at high levels in megakaryocytes and platelets, but it is also present in nearly every cell type in the body. In particular, two alternatively spliced forms of c-Src mRNA are abundantly expressed in neurons of the central nervous system. These forms encode proteins with either 6 or 17 additional amino acids within the N terminal SH3 domain. Like c-Src, c-Yes and Fyn are also widely distributed. The recombinational "knockout" of *src* family genes in the mouse has shown that no one of these genes is essential for the organism's development or survival. However, mice lacking the c-*src* gene develop osteopetrosis, implying a defect in osteoclast function, and mice without the *lck* gene fail to develop T cells. Mice lacking the *fyn* gene have some defects in T cell signaling and a learning disability. Mice lacking both the *fyn* and c-*src* genes show a more severe phenotype than either single knockout, suggesting partial redundancy in the functions of c-Src and Fyn.

Even though mutant forms can incite continuous growth in transformed cells, the high levels of c-Src and other members of this family in terminally differentiated cells makes it unlikely that these protein kinases function simply to regulate cell growth. Their submembranous location suggests that they influence the cell's response to its external environment by transducing signals from transmembrane proteins, such as receptors. Thus, these proteins may act as subunits for surface receptors that lack an intrinsic protein kinase domain (Figure 4.10). The specific association of the Lck protein with CD4, the T lymphocyte surface protein, is consistent with this idea. Likewise, Fyn appears to associate with the T cell antigen receptor and Lyn with the B cell antigen receptor; in each case, the catalytic activity of these associated enzymes is increased when antigen binds to the respective receptors. As subunits of receptors, the Src family protein-tyrosine kinases are presumed to transduce signals for a growth factor receptor in dividing cells and a different type of receptor in differentiated cells. This dual role could explain the apparent paradox that these proteins can affect the growth of cells as well as be important in nondividing cells.

The function of c-Src itself is still something of an enigma. Based on the ability of v-Src to decrease intercellular communication through gap junctions, c-Src might play a role in intercellular communication. Because c-Src is present at high levels in many cells that are active in secretion (e.g., neurons, platelets, and chromaffin granule cells, where at least some of the protein is associated with secretory granules), researchers have proposed that c-Src may be involved in secretion. There are also indications that c-Src, c-Yes, and Fyn mediate signaling from the PDGF receptor protein-tyrosine kinase, suggestive of an important function for c-Src in mitogenesis.

One essential feature of the *src* family of protein-tyrosine kinases is that their catalytic activity is negatively regulated by phosphorylation of a tyrosine that lies just beyond the catalytic domain (Figure 4.10). Phosphorylation of c-Src Tyr 527 is catalyzed by the Csk nonreceptor protein-tyrosine kinase. Phosphorylation of Tyr 527 suppresses the catalytic activity of c-Src because the N terminal SH2 domain folds around and binds to phospho-Tyr 527 and either blocks access of substrates or induces a conformational change in the catalytic domain. When isolated from the cell, c-Src is nearly fully phosphorylated at Tyr 527 and is therefore largely in an inactive state. Presumably, the stimulus that turns on c-Src either inhibits the Tyr 527 protein kinase or else activates the phospho-Tyr 527 phosphatase. A conserved tyrosine is present at this position in each of the Src family protein kinases, and they all must be negatively regulated by phosphorylation at this residue.

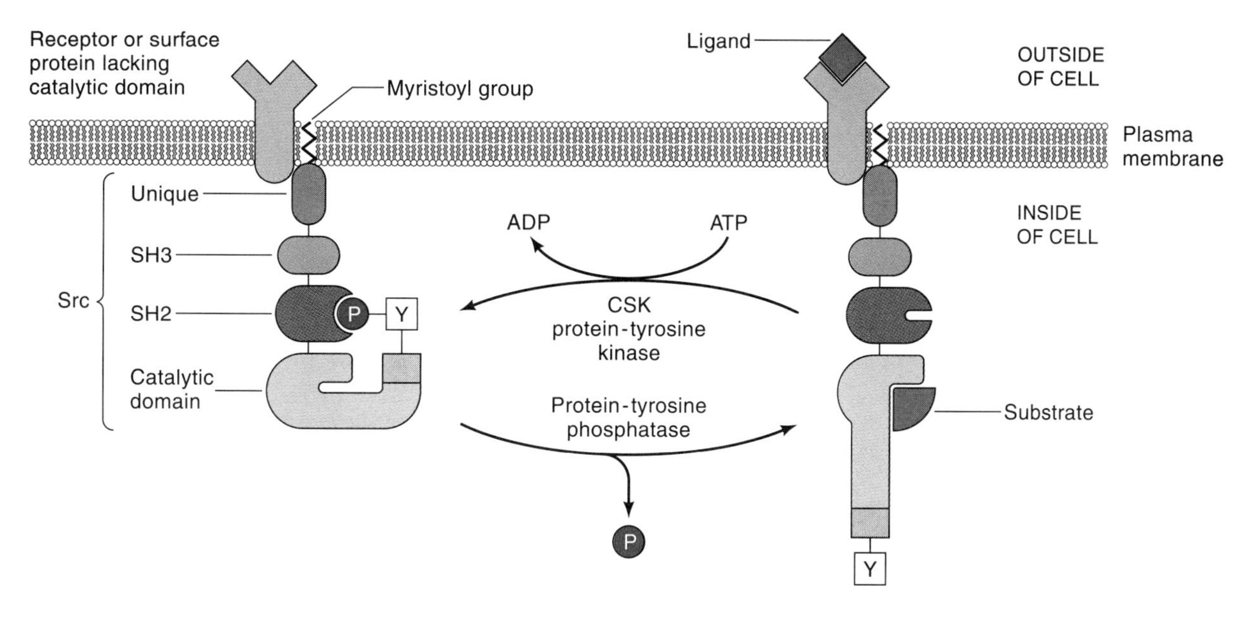

Figure 4.10
Regulation of Src family protein-tyrosine kinase activity. Src family protein-tyrosine kinases, phosphorylated at the C-terminal regulatory tyrosine by the CSK protein-tyrosine kinase and held in an inactive state by intramolecular interaction of the phosphorylated tyrosine with the SH2 domain, can associate with the cytoplasmic domains of surface receptors through their unique domain. Ligand binding to the receptor causes activation of the associated Src family protein-tyrosine kinase through dephosphorylation of the C-terminal tyrosine by a protein-tyrosine phosphatase, allowing phosphorylation of substrates.

The regulation of these normal cellular protein-tyrosine kinases has an important bearing on their oncogenic activation. As an example, compare c-Src and v-Src (Figure 4.9a). Apart from several scattered point mutations, chicken c-Src and v-Src proteins are very similar in structure over their 514 N terminal amino acids. However, the two proteins diverge completely at their C termini; the C terminal 19 amino acids of c-Src, starting at residue 515, are replaced in v-Src by 12 unrelated amino acids. The obvious consequence of this change is the elimination of the regulatory tyrosine. Indeed, as predicted by the negative regulation model, v-Src has much greater protein kinase activity than c-Src. Moreover, c-Src can be converted into a transforming protein simply by mutating Tyr 527 to phenylalanine, confirming that the major effect of the C terminal deletion is removal of the phosphorylation site.

Although the principal change in the conversion of the c-*src* gene into the v-*src* oncogene is elimination of the critical regulatory tyrosine, some point mutations in the body of the v-*src* gene are themselves sufficient to induce transforming activity. As is the case for the v-ErbB protein, these mutations were probably selected during RSV passage *in vivo* because their effects synergize with the loss of the C terminal tyrosine, resulting in more potent transforming viruses. Some of these mutations alter residues in the catalytic domain and presumably enhance catalytic efficiency directly through a conformational change. However, two of the mutations lie in the SH3 domain. Like SH2 domains, SH3 domains are protein interaction domains; but, in contrast to SH2 domains, which bind phosphotyrosine residues, SH3 domains interact with unphosphorylated, proline-rich sequences in other proteins. The activating mutations in the v-Src SH3 domain may therefore affect its binding to other proteins. Although mutations in the SH2 domain that prevent binding to phospho-Tyr 527 might be expected to be activating, such mutations generally inactivate v-Src, implying that SH2 domain function is important for the binding of v-Src to other proteins.

Two other members of the Src family have been found as retroviral oncogenes: v-*yes* and v-*fgr*. In both cases, the viral oncoproteins retain the complete catalytic domain of the normal cellular counterparts but are missing C terminal sequences, including the regulatory tyrosines (Figure 4.9b). Other Src family members can also be activated for fibroblast transformation by mutation of the C terminal regulatory tyrosine. Elevated expression of the normal Lck protein in certain Moloney murine leukemia virus–derived murine thymomas is also known to promote transformation. Perhaps overexpression of normal Lck protein in this particular cell type is sufficient to overwhelm the normal negative regulation, resulting in a transformed phenotype.

Site-directed mutation of Lys 295 in v-Src, a conserved residue in the catalytic domain ATP-binding site of all protein kinases, abolishes phosphorylating and transforming activity, thus proving that transformation requires tyrosine phosphorylation. Changing the N terminal glycine of v-Src to alanine, which prevents myristoylation and association with the membrane, also abrogates transforming activity, proving that membrane localization is essential for transformation. Perhaps certain critical substrates are present exclusively in the vicinity of the membrane.

Another oncoprotein derived from a nonreceptor protein-tyrosine kinase is **v-Abl**. The N terminal half of the c-Abl protein is organized very much like c-Src, with SH3 and SH2 domains lying just upstream of the catalytic domain (Figure 4.11). However, c-Abl has a long C terminal extension that contains a DNA-binding domain and an F-actin-binding domain. Even though one form of c-Abl is myristoylated at the N terminus, much of c-Abl is localized in the nucleus. The exact function of c-Abl is not known. Mice lacking the c-*abl* gene have a variety of wasting symptoms but are viable. The nuclear location of c-Abl implies that it normally phosphorylates nuclear proteins, and recent evidence suggests that c-Abl plays a role in regulating cell cycle progression through an association with the retinoblastoma (RB) protein in the G1 phase of the cycle. As a result of this association, the catalytic activity of c-Abl is suppressed.

Conversion of the c-*abl* gene into an oncogene can occur when it is incorporated into a retrovirus genome or by a chromosomal translocation. In both cases, the *abl* oncogene is expressed as a chimeric protein with foreign sequences at its N terminus (Figure 4.11). For the viral oncoprotein, v-Abl, the N terminal sequence corresponds to the viral Gag sequence that provides a myristoylation signal. As a result, the Gag-Abl protein is membrane associated. In addition, the SH3 domain in the Abl protein's N terminus is missing, but the SH2 domain is retained. The c-*abl* gene is also activated in chronic myelogenous leukemias (CML) that possess the **Philadelphia chromosome**. The Philadelphia chromosome arises from a chromosomal translocation that links the 3′ half of the c-*abl* gene on chromosome 9 with the 5′ half of the *bcr* locus on chromosome 22. As a result of the translocation, a segment from the N terminus of the Bcr protein is joined to c-Abl just upstream of the SH3 domain, resulting in a chimeric Bcr-Abl protein.

The v-Abl and Bcr-Abl proteins are phosphorylated on tyrosine when isolated from the cell, whereas c-Abl is not. This indicates that c-Abl protein is less active as a protein-tyrosine kinase *in vivo* than the oncogenic Abl proteins. Site-directed mutation of v-Abl indicates that the N terminal myristoyl group derived from Gag, the SH2 domain, and catalytic activity are all essential for transforming activity. Furthermore, unlike c-Abl, v-Abl is localized in the cytoplasm, a requirement for transformation. This suggests that transformation by

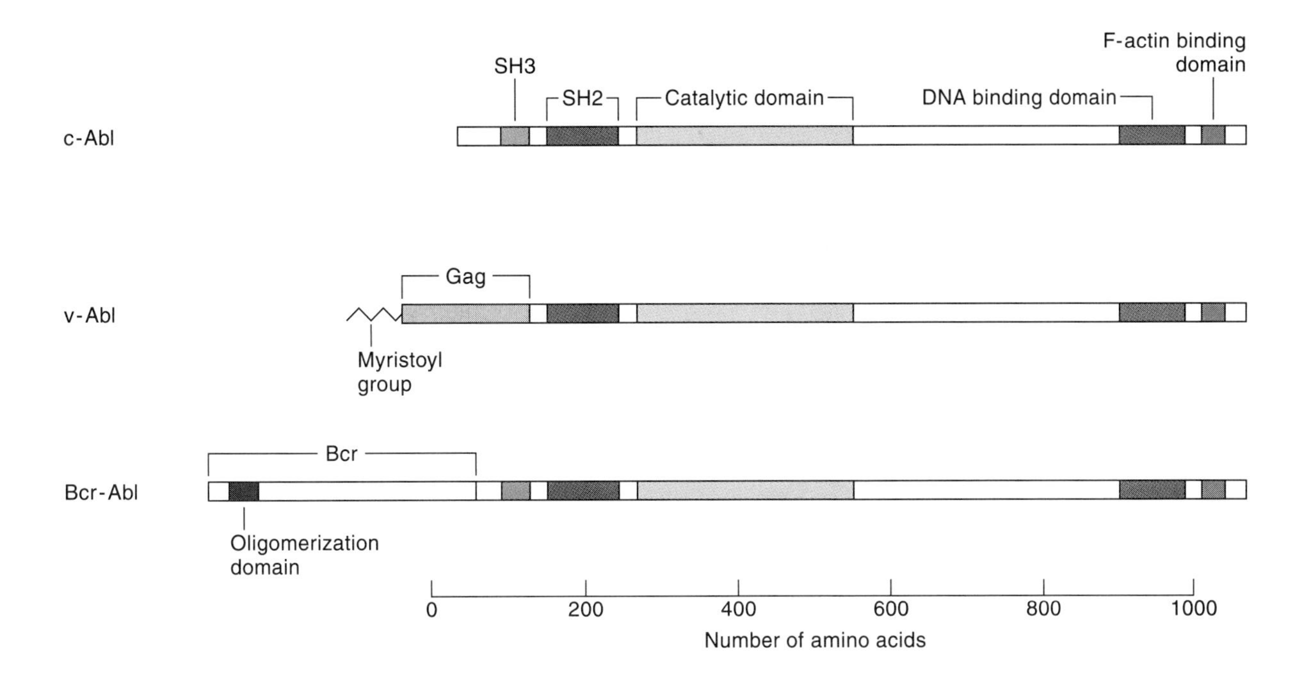

Figure 4.11
Structures of Abl oncoproteins. The schematic structure of the c-Abl protein-tyrosine kinase is shown compared to those of two c-Abl-derived oncoproteins. v-Abl is activated by deletion of the SH3 domain and the attachment of retroviral Gag sequences, which have an N-terminal myristoyl group. Bcr-Abl is activated by fusion of Bcr sequences near to the N terminus of c-Abl; the Bcr sequences include an oligomerization domain.

v-Abl involves more than simple deregulation of the functions performed by c-Abl.

One kind of mutation that results in oncogenic activation of c-Abl is deletion of the SH3 domain in the N terminus. This implies that the SH3 domain of c-Abl, like that of c-Src, is normally involved in negative regulation of activity. In contrast to c-Src, however, c-Abl is not normally phosphorylated on tyrosine; therefore, negative regulation of c-Abl catalytic activity does not require intramolecular interaction of the SH2 domain with a phosphotyrosine. Interestingly, the Bcr-Abl protein retains the SH3 domain, which implies that the appended N terminal Bcr sequence can override the negative regulatory function of the SH3 domain. Recent evidence indicates that this is because Bcr's N terminus can oligomerize through a coiled-coil type of interaction, suggesting that Bcr-Abl protein's oncogenic activity depends on oligomerization, as does that of receptor protein-tyrosine kinases.

Targets for oncogenic protein-tyrosine kinases The conversion of normal protein-tyrosine kinases into oncoproteins results from mutations that lead to constitutively elevated protein-tyrosine kinase activity. The key to understanding how these oncoproteins transform cells is the identification of their target substrates. As indicated previously, tyrosine phosphorylation is a rare event in normal cells, accounting for about 0.05 percent of phosphate esterified to protein. In cells transformed by oncogenic protein-tyrosine kinases, this value increases up to tenfold. Many proteins with increased amounts of phosphotyrosine can be identified in such cells, and these are presumably primary substrates for the protein-tyrosine kinases in question.

Table 4.5 ***Substrates for Signaling Protein-Tyrosine Kinases***

Enzymes	
PLC-γ1 and PLC-γ2	(SH2) (SH3)
GAP120	(SH2) (SH3)
Src family protein-tyrosine kinases	(SH2) (SH3)
SH-PTP1/HC-PTP/PTP1C	(SH2)
SH-PTP2/SYP/PTP1D	(SH2*)
PI3 kinase p85/p110	(SH2*)(SH3)
Vav (Dbl homology)	(SH2)(SH3)
FAK	Focal adhesion protein-tyrosine kinase
Adaptor and docking proteins	
IRS1	Docking protein
Shc	(SH2)
Crk	(SH2)(SH3)(SH3)
Nck	(SH2)(SH3)
Grb2	(SH2)(SH3) (not phosphorylated)
p62	Docking protein
Structural proteins	
Annexins I and II	
Ezrin	
Clathrin H chain	
Vinculin (focal adhesion)	
Talin (focal adhesion)	
Tensin (focal adhesion)	(SH2)
Paxillin	
AFAP110 (actin-binding protein)	
Cortactin	(SH3)
p120 (armadillo and β-catenin related)	
Connexin43	
Fibronectin receptor β subunit	

(SH2*): enzyme activity increases upon binding target phosphotyrosine sequence.

Within the past few years, many substrates for both normal and oncogenic protein-tyrosine kinases have been identified. Three different types of substrate can be distinguished: enzymes, adaptor proteins, and structural proteins (Table 4.5). Among the enzyme substrates are phospholipases, lipid kinases, nonreceptor protein-tyrosine kinases, protein-tyrosine phosphatases, and GTP exchange factors, all of which are either known or likely to be involved in growth factor signaling. Most of these proteins contain an SH2 domain (Figure 4.12), and they associate with activated protein-tyrosine kinase autophosphorylation sites through their SH2 domains prior to being phosphorylated (Figure 4.13a).

The activation or inhibition of enzymes by tyrosine phosphorylation is the most direct way that protein-tyrosine kinases can transduce signals. The best understood example of this principle is the γ-isoform of phospholipase C (PLC-γ). PLC-γ hydrolyzes membrane phosphoinositides, preferring phosphatidylinositol 4,5 diphosphate (PIP_2). The action of PLC-γ on PIP_2 generates two second messengers, diacylglycerol (DAG) and inositol trisphosphate (IP_3). DAG activates protein kinase C, a membrane-associated protein-serine kinase, which is itself involved in signaling (Figure 4.13b). IP_3 causes release of calcium from intracellular stores. Phosphorylation of PLC-γ on a specific tyrosine residue increases its catalytic activity, which in turn leads to elevated levels of DAG and IP_3 and stimulation of events downstream of these second messengers. Although

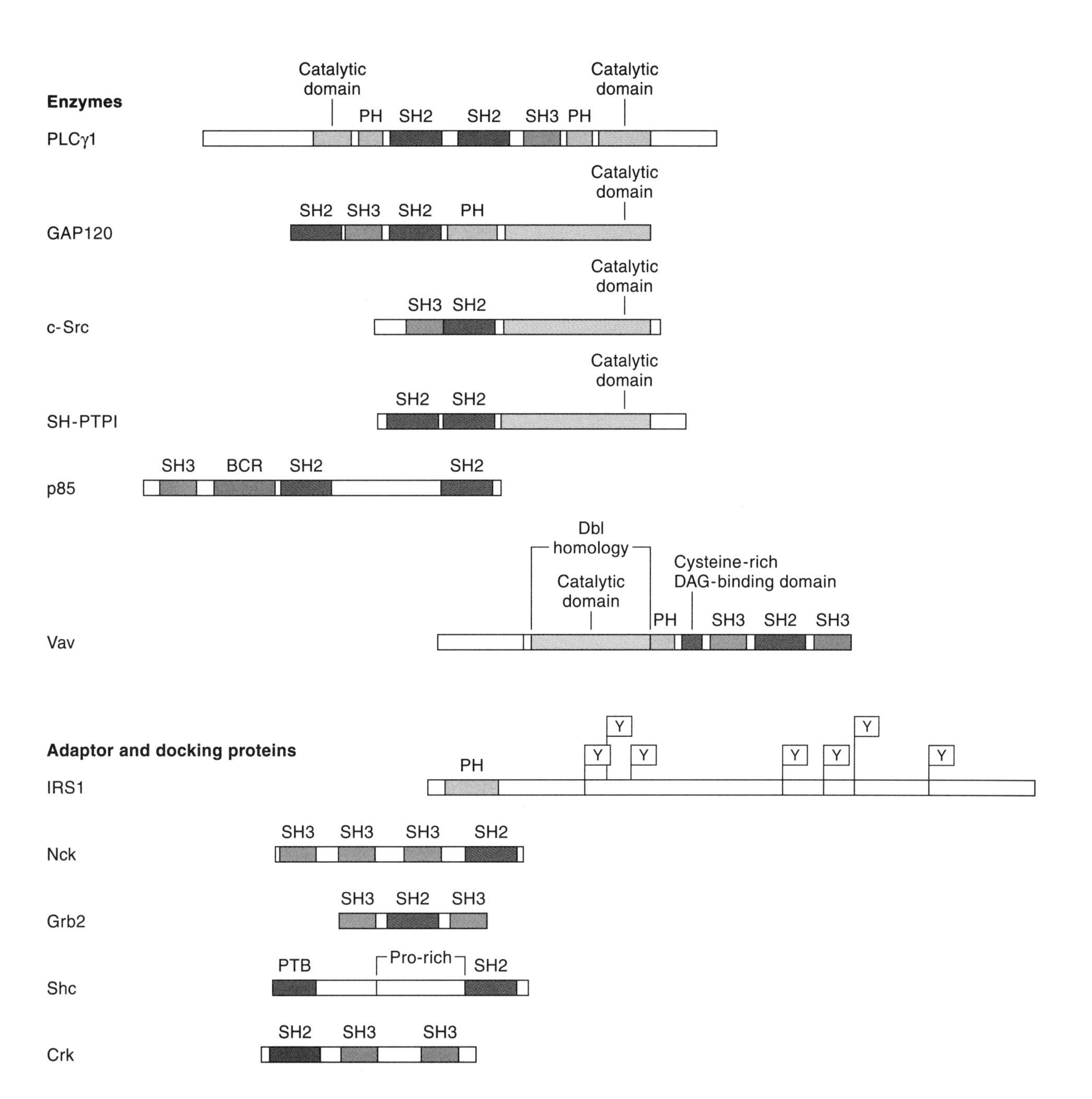

Figure 4.12
Targets for signaling protein-tyrosine kinases. Examples of structures for two of the three classes of target for signaling protein-tyrosine kinase are shown. The enzyme targets have a defined catalytic domain, except for p85, which is the regulatory subunit of PI3 kinase. The adaptor proteins consist primarily of protein interaction domains, including SH2 and SH3 domains. Additional domains are indicated: PH = pleckstrin-homology domain; PTB = phosphotyrosine-binding domain; Dbl homology = a region related in sequence to the Dbl protooncoprotein. The tyrosine phosphorylation sites (Y) in the IRS1 docking protein that can associate with SH2 domain proteins are shown as flags.

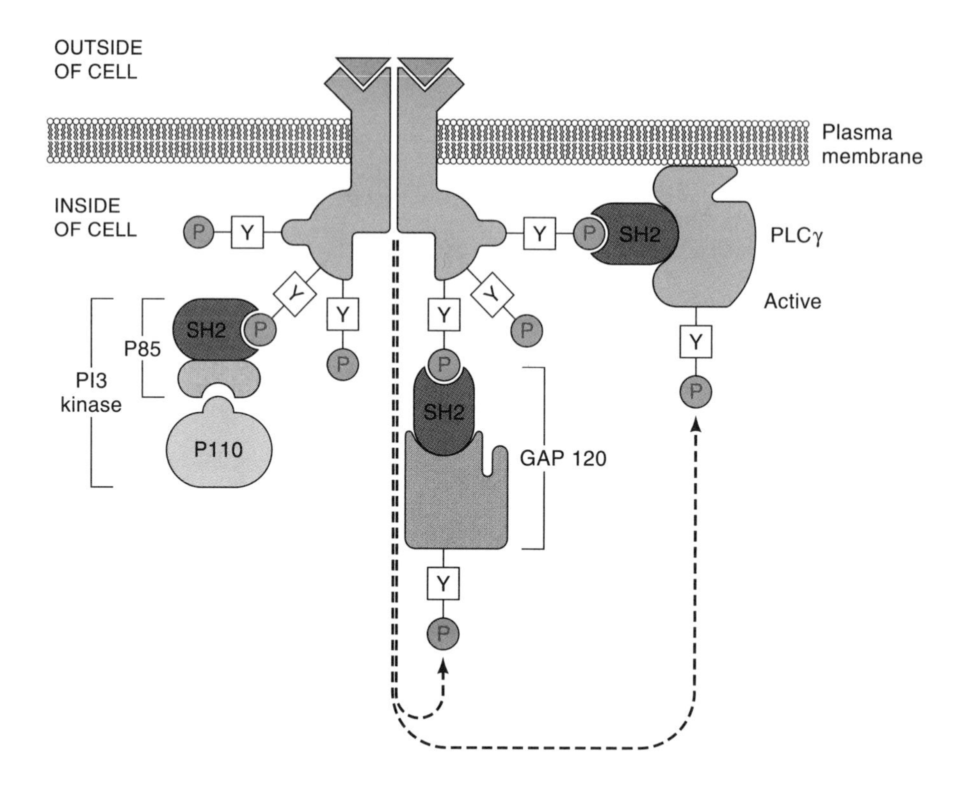
OUTSIDE
OF CELL
Plasma
membrane
INSIDE
OF CELL
P
Y
SH2
PLCγ
Active
P85
PI3
kinase
P110
GAP 120

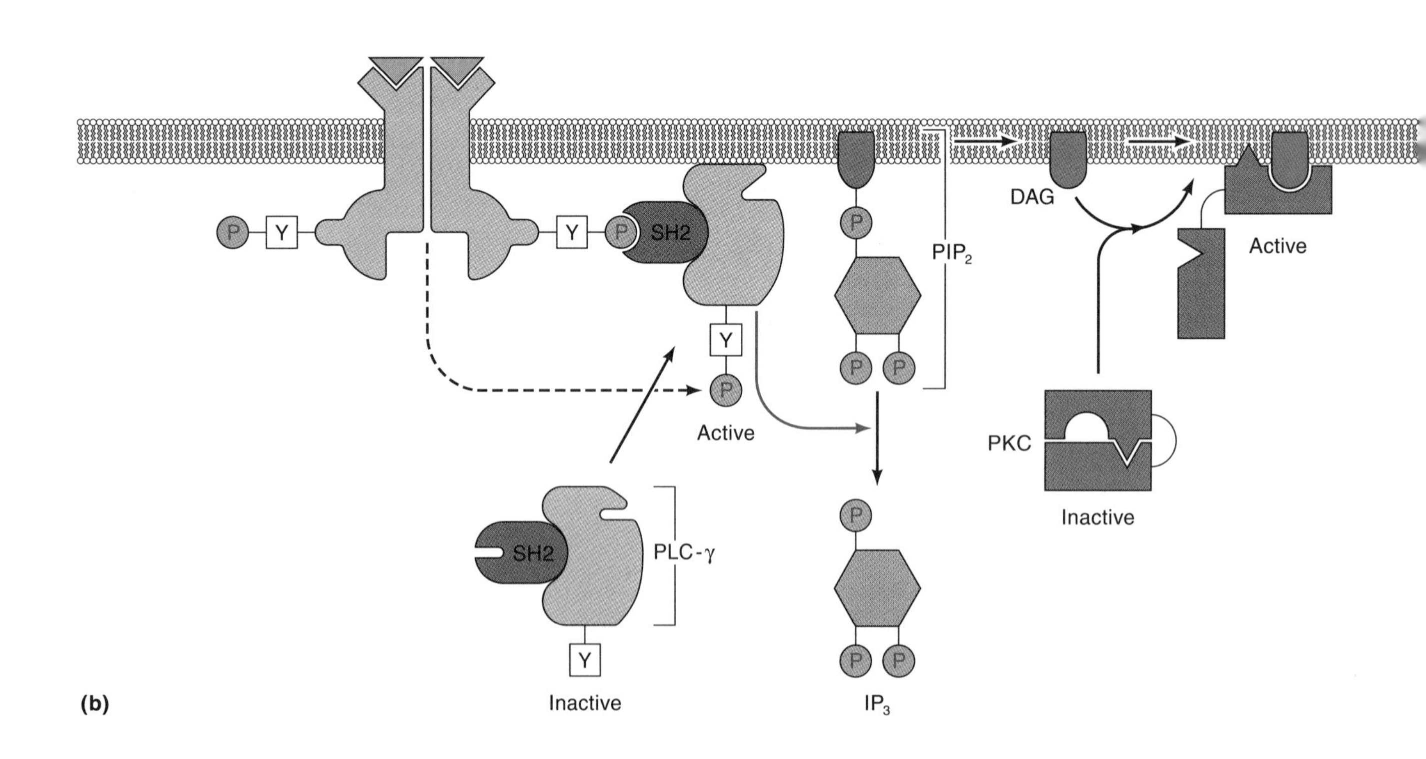
SH2
PIP$_2$
DAG
Active
PKC
Inactive
PLC-γ
(b)
Inactive
IP$_3$

Figure 4.13 (*facing page*)
(a) Recruitment of SH2-containing targets to an activated receptor protein-tyrosine kinase. Soluble cytoplasmic target proteins, such as phospholipase Cγ (PLC-γ), Ras GTPase activating protein (GAP120), and phosphatidylinositol 3′ kinase (PI3 kinase) bind via their SH2 domains to specific phosphotyrosines (P-Y) on an activated receptor protein-tyrosine kinase dimer. This is a high-affinity interaction and is a means of recruiting targets to activated receptors at the plasma membrane. After binding, target proteins can be phosphorylated by the receptor protein-tyrosine kinase at specific tyrosines, which leads to their activation. (b) Activation of PLC-γ by tyrosine phosphorylation. Once PLC-γ is bound to an activated receptor protein-tyrosine kinase via its SH2 domains, it is phosphorylated at a single tyrosine leading to activation of its catalytic domain. Activated PLC-γ, which may be released from the receptor, hydrolyzes phosphatidylinositol 4,5 diphosphate (PIP_2), a plasma membrane phospholipid, to generate diacylglycerol (DAG) and inositol 1,4,5 triphosphate (IP_3). DAG activates protein kinase C (PKC), a protein-serine kinase, and IP_3 stimulates release of calcium from intracellular stores.

---

it has not been shown that the activity of every enzyme substrate of this type is altered by tyrosine phosphorylation, this is likely to be true in most cases.

Substrates of the second type relay signals to enzymes involved in signal pathways but are not themselves enzymes. These proteins, called **adaptor proteins**, function by binding, through their SH2 domains, to phosphotyrosine sites on protein-tyrosine kinase and then generating a signal via another domain, commonly an SH3 domain (Figure 4.14a). The best example is the Grb2 adaptor protein, which contains a central SH2 domain flanked by two SH3 domains. The Grb2 protein, which acts downstream of receptor protein-tyrosine kinases, is highly conserved through evolution, and structural and functional homologues have been found in nematodes and flies. Grb2 is constitutively associated with a GTP exchange factor (called Sos) for the Ras protein through its SH3 domains. Following activation of a receptor protein-tyrosine kinase that has a Grb2-binding site, the Grb2-Sos complex binds to the appropriate phosphorylated tyrosine in the receptor's cytoplasmic domain and is thereby translocated to the inner face of the plasma membrane where Ras resides. Sos can then catalyze GTP exchange on Ras, resulting in its activation (Figure 4.14b). Activated Ras stimulates the Raf protein-serine kinase, the first member of a cascade of protein-serine kinases that signals to the nucleus (Figure 4.15). The Grb2 protein is not itself phosphorylated on tyrosine, but several other adaptor proteins are tyrosine phosphorylated by receptor protein-tyrosine kinases. The importance in oncogenesis of adaptor proteins as the mediators of protein-tyrosine activation is underscored by the fact that the c-*crk* gene, which encodes an adaptor protein, gave rise to the v-*crk* oncogene.

Substrates in the third category are structural proteins. Several such substrates are components of a protein skeleton under the plasma membrane, the same location as most of the signaling protein-tyrosine kinases. Tyrosine phosphorylation of the submembranous proteins may regulate cell shape and membrane architecture, although there is no direct proof that this is the case.

Substrates of all three types are constitutively phosphorylated on tyrosine in cells transformed by an oncogenic protein-tyrosine kinase. Given the normal functions of these proteins, one can readily appreciate how constitutive phosphorylation of the different signaling molecules by activated, oncogenic protein-tyrosine kinases could result in continuous cell growth and in the dramatic shape changes characteristic of transformed cells.

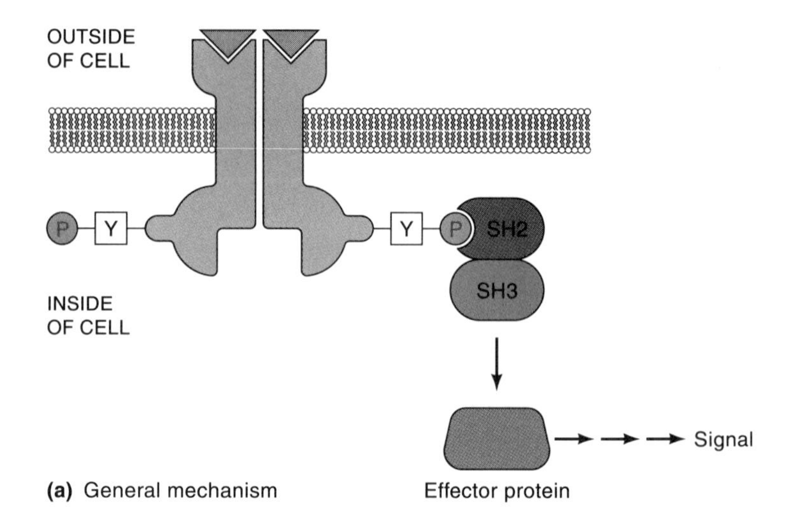

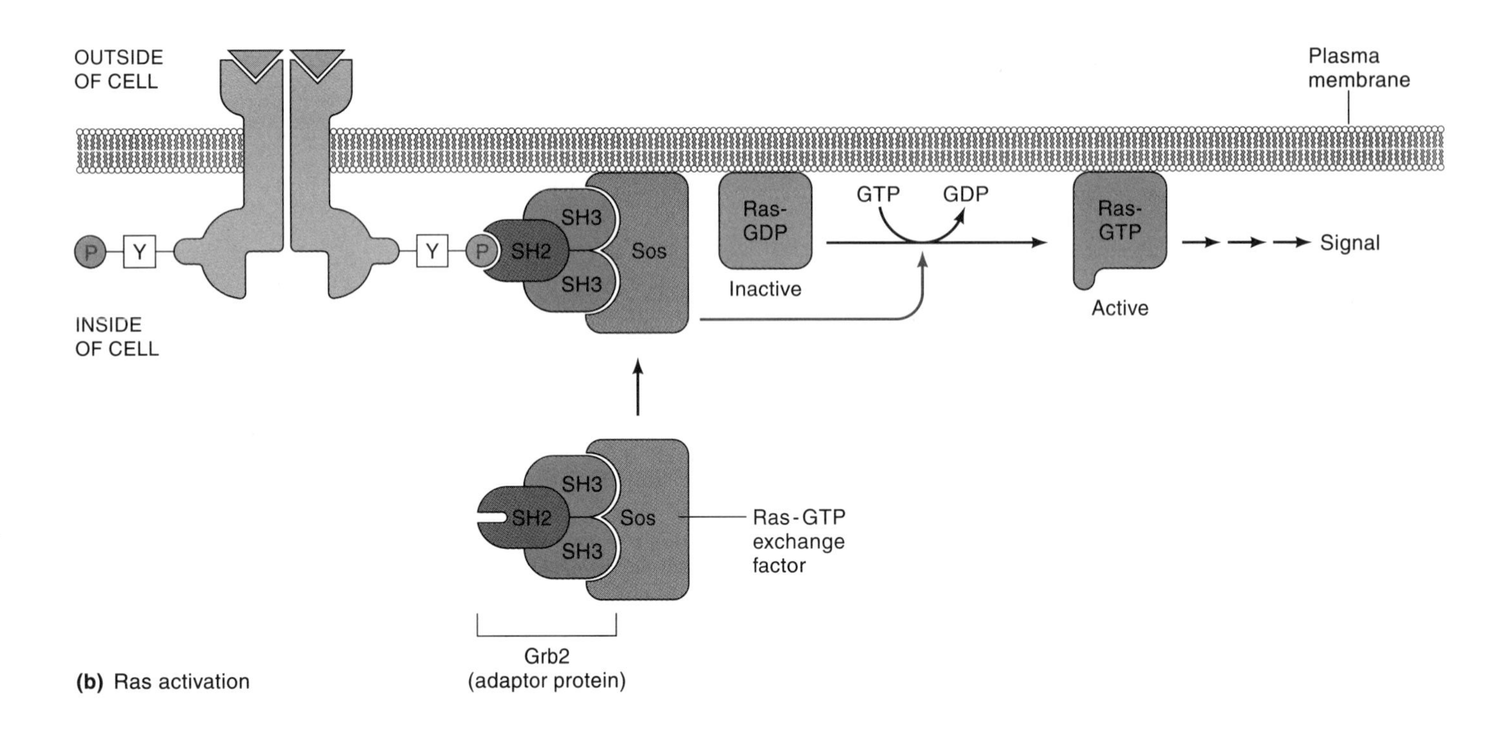

Figure 4.14
Receptor protein-tyrosine kinase signaling via adaptor proteins. (a) General mechanism. An adaptor protein, containing SH2 and SH3 domains, binds to an activated receptor protein-tyrosine kinase dimer via its SH2 domain; its SH3 domain(s) interacts with an effector protein either constitutively or inducibly. In this manner, the effector is translocated to the plasma membrane where it can initiate signaling. (b) Ras activation. Ras is activated by translocation of the adaptor protein Grb2 and the associated effector protein Sos to the plasma membrane. Ligand-dependent activation of a receptor protein-tyrosine kinase leads to autophosphorylation and creation of a binding site for the Grb2 SH2 domain (consensus binding site is P.Tyr.Xxx.Asn.Xxx, where Xxx is any amino acid). Grb2 is constitutively bound to Sos, a GTP exchange factor for Ras, through its SH3 domains, which interact with proline-rich sequences in the C-terminal part of Sos. The binding of the Grb2-Sos complex to the activated receptor results in juxtaposition of the Sos catalytic domain to the membrane where it can interact with inactive Ras-GDP, which facilitates exchange of GDP for GTP and the formation of active Ras-GTP.

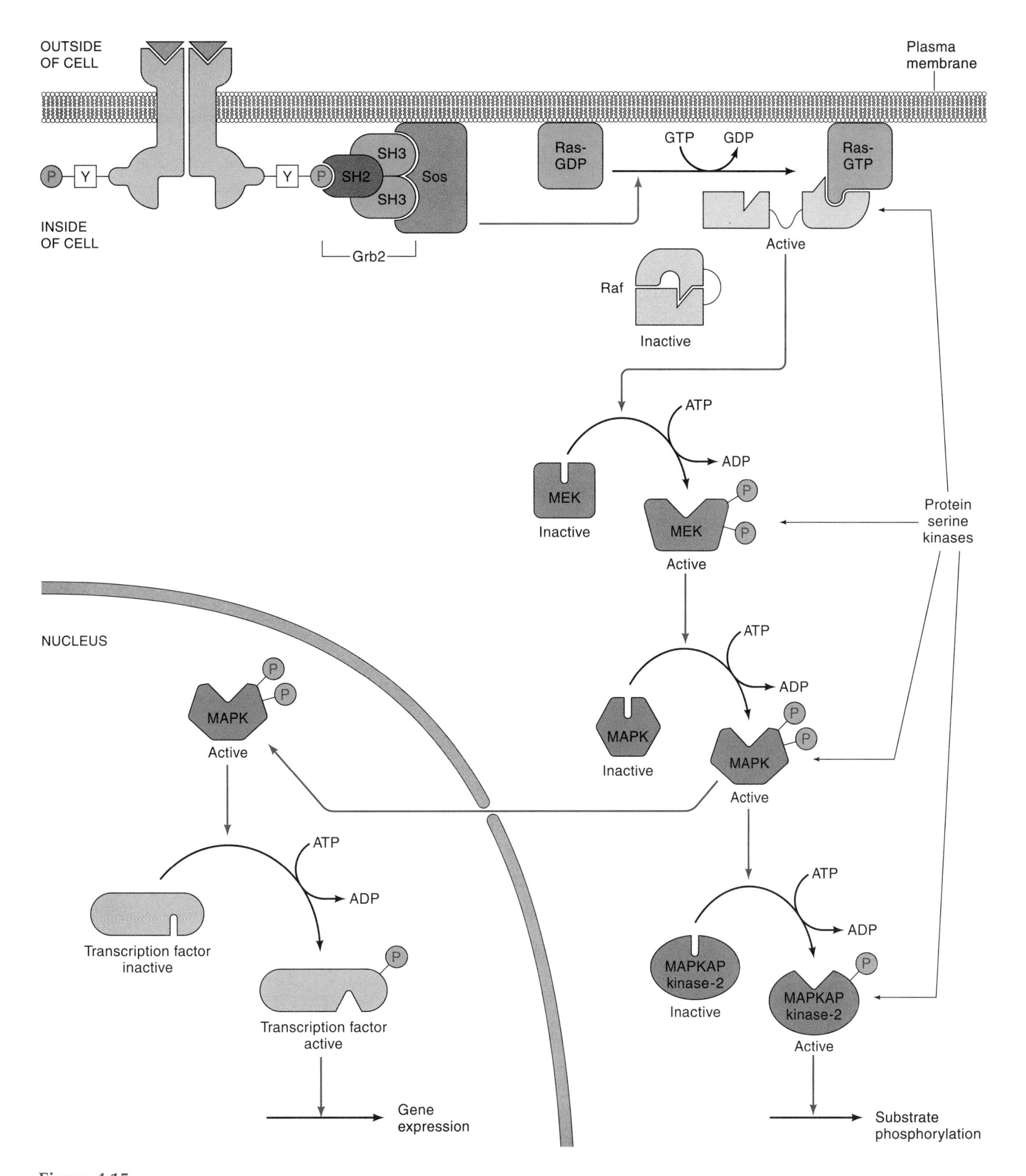

Figure 4.15
The MAP kinase pathway. Ras is converted into the active Ras-GTP state following translocation of Grb2-Sos to an activated receptor protein-tyrosine kinase. The Raf protein-serine kinase, which is normally in an inactive state in the cytoplasm, binds via its N-terminal regulatory domain to Ras-GTP, and this leads to activation of Raf in conjunction with a second unknown event (binding to Ras-GTP is not sufficient for full Raf activation). Raf then phosphorylates and activates MAP kinase kinase (MEK), a dual-specificity protein kinase, which in turn phosphorylates and activates MAP kinase. Activated MAP kinase phosphorylates cytoplasmic substrates, but also translocates into the nucleus where it phosphorylates and activates specific transcription factors and leads to gene expression. Activated MAP kinase can also phosphorylate and activate other protein-serine kinases, such as MAPKAP kinase 2, thus making a protein kinase cascade with four components. In principle, such a cascade can lead to great amplification of the original growth factor signal.

Oncogenic protein-tyrosine kinases in human cancer Several protein-tyrosine kinase oncogenes have been identified in human tumors. Mutant-activated forms of the cellular *trk* gene have been detected in a single colon carcinoma and frequently in papillary thyroid carcinomas. Mutant-activated forms of the cellular *ret* gene are also found in papillary thyroid carcinomas, and a mutant form of Ret, with a potentially activating mutation in the extracellular domain, may be responsible for the heritable multiple endocrine neoplasia type 2 (MEN2) syndrome. The c-*abl* gene is consistently found in a rearranged state as a result of a chromosomal translocation (Philadelphia chromosome) in CML; the resulting chimeric Bcr-Abl protein has constitutively high protein-tyrosine kinase activity (Figure 4.11). Although there is no formal proof for the involvement of this altered Abl protein in CML, the level of expression of the Bcr-Abl protein is well correlated with the stage of the disease, implying that it is important for the disease progression. Both the c-*erbB* (EGF receptor) and *neu*/c-*erbB2* genes are amplified in some human tumors, and the corresponding proteins are overexpressed. Amplified c-*erbB* is found in about 25 percent of glioblastomas multiform, and amplified *neu*/c-*erbB2* has been detected in a variety of tumors, including salivary adenocarcinoma and mammary and bladder carcinomas.

Given how easy it is to activate the protooncogenes in the *src* family by mutating the C terminal regulatory region, it is somewhat surprising that *src*-derived human tumor oncogenes are rare; the only known example is the activation of the *lck* gene in T cell leukemias as a result of its translocation to one of the T cell receptor genes. In one case, the translocated *lck* gene also contains activating point mutations, but these are not in the C terminal regulatory region. In colon carcinoma and some other types of tumor, c-Src and c-Yes protein-tyrosine kinase activity is commonly elevated. However, no mutations in the c-*src* and c-*yes* genes are found in such tumors, and the activation mechanism is unknown. Consequently, whether the activations of c-Src and c-Yes are significant events in human tumorigenesis is unclear.

c. Class III: Receptors Lacking Protein Kinase Activity

Serpentine receptors Although most receptor-derived oncoproteins are protein-tyrosine kinases, other types of cell surface receptor can also be converted into oncoproteins. For instance, **serpentine receptors** with seven transmembrane domains, such as the β-adrenergic receptor that binds and responds to adrenalin, can be activated into oncogenes. One example is the *mas* oncogene, which was isolated from human tumor DNA by the NIH3T3 cell transfection assay. Subsequently, the normal Mas protein was shown to be a receptor for angiotensin, which is present at high levels in serum and is made by many fibroblasts. In fact, the Mas oncoprotein is identical in sequence to the normal cellular Mas protein, which implies that ectopic synthesis of Mas in an inappropriate cell type is sufficient, in the presence of serum angiotensin, to induce transformation. A similar case is the serotonin 1c receptor, which is oncogenic when expressed in fibroblasts because of stimulation by serum serotonin. Other examples of serpentine growth factor receptors are the bombesin/GRP and the thrombin receptors. These receptors will probably prove to be oncogenic if expressed in the same cell as their ligands. All serpentine receptors are coupled to so-called G proteins (e.g., G_s and G_i), (Section 4.2d), which in turn are linked to effector enzymes such as adenylyl cyclase. A special set of G proteins is believed to couple to growth factor receptors of this type. One particular G protein that is activated by these receptors is G_q. G_q itself is coupled to the β-isozyme of PLC, which, like PLC-γ, hydrolyzes phosphoinositides to generate the second messengers DAG and IP_3. As might be expected, there are mutant forms of the α-subunit of G proteins, which are constitutively active and function as oncogenes.

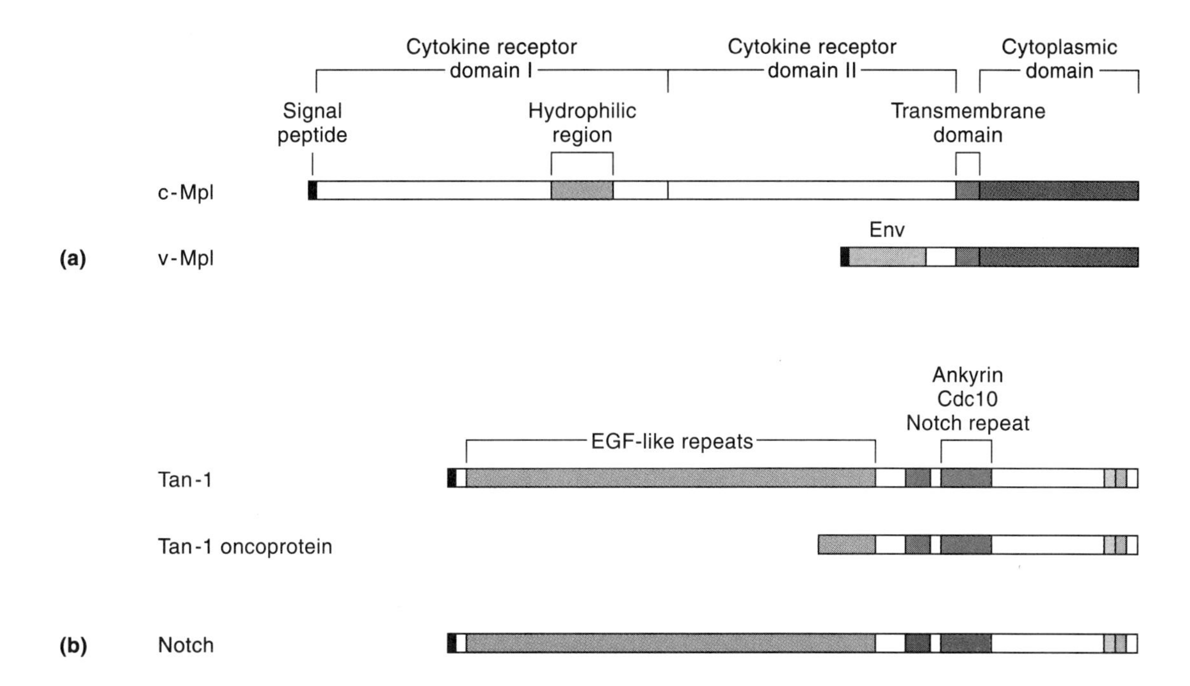

Figure 4.16
Oncogenic activation of cytokine and differentiation receptors. (a) As depicted by the schematic structures, c-Mpl, the cytokine receptor for the megakaryocyte growth factor, thrombopoietin, is converted into an oncoprotein by deletion of most of the extracellular ligand-binding domain and fusion with sequences from the N-terminal end of a retroviral Env protein. The chimeric protein is constitutively oligomerized and generates the same signals as c-Mpl following binding of ligand, which includes activation of members of the Jak family of nonreceptor protein-tyrosine kinases. (b) As indicated by the schematic structures, the Tan-1 oncoprotein is derived from Tan-1, a receptor in the Notch family of differentiation receptors, by N-terminal truncation of the normal Tan-1 protein. Through unknown mechanisms, this process leads to constitutive Tan-1 signaling, which is believed to block differentiation, as is the case for ligand-activated Notch. This signal involves the release of a cytoplasmic regulator of a transcription factor from the ankyrin/Cdc10/Notch repeat region of the cytoplasmic domain, which allows it to translocate into the nucleus. The structure of Drosophila Notch is shown for comparison.

Cytokine receptors The **v-*mpl* oncogene** of the myeloproliferative leukemia virus is derived from a gene that encodes a protein related to the cytokine receptor superfamily. This family includes receptors for interleukin 2, erythropoietin, and granulocyte colony-stimulating factor (G-CSF). The receptors have multiple subunits, which contain similar sequences in their extracellular domains, and appear to signal via activation of nonreceptor protein-tyrosine kinases, including members of the Src and Jak families. The c-Mpl protein is a subunit of a receptor for a recently identified cytokine that stimulates the growth and development of megakaryocytes. Although v-Mpl lacks most of the c-Mpl extracellular domain, the significance of its absence for oncogenic activity remains to be demonstrated (Figure 4.16a). Another cytokine receptor, the erythropoietin receptor, is activated by the spleen focus-forming virus (SFFV), which induces erythroleukemia in mice. The virus encodes a variant retroviral Env protein that associates with and activates the receptor, providing a continuous growth stimulus for erythroid precursor cells. These examples suggest that additional oncogenes may be derived from other cytokine receptor genes.

Receptors that control differentiation Two oncogenes, *int-3* and *tan-1*, are derived from genes encoding receptors in the Notch family (Figure 4.16b). Such receptors are normally involved in suppressing cellular differentiation during embryogenesis. Both Int3 and Tan1 oncoproteins lack the receptors' extracellular domains but retain the cytoplasmic domains, including a sequence motif known as the ankyrin repeats. These proteins may transform by providing a constitutive antidifferentiation signal through the ankyrin repeats.

d. Class IV: Membrane-Associated G Proteins and Their Regulators

The major oncogenes in this class are derived from two cellular gene families—the c-*ras* genes and G protein α-subunit genes, whose products act as signal transducers on the inner face of the plasma membrane.

The function of ras *genes* The ***ras* oncogenes** were first isolated from two murine sarcoma viruses, but most oncogenes detected by transfection studies with tumor cell DNA are also activated *ras* genes. The two v-*ras* genes are derived from different members of the c-*ras* family: the c-H-*ras* and c-K-*ras* protooncogenes. Tumor cell oncogenes derive from these two and the N-*ras* gene as well. All *ras* genes encode 21 kDa proteins. The proteins are modified by a farnesyl group attached to a cysteine four residues from the C terminus that targets them to the inner face of cytoplasmic membranes, including the plasma membrane.

Normal and oncogenic Ras proteins bind GTP or GDP equally tightly. Ras proteins also have GTPase activity; however, the normal c-Ras and activated Ras differ significantly in this activity, with activated Ras being fivefold to tenfold less active as a GTPase. These properties of Ras are a consequence of the presence in the N terminal two-thirds of the Ras proteins of a domain related to that found in proteins of the G protein α-subunit family. G proteins are heterotrimeric signaling proteins; their α-subunits bind GTP in response to activation of a serpentine cell surface receptor, and the α-subunit-GTP complex stimulates specific effector molecules, such as adenylyl cyclase. This signal is terminated upon the hydrolysis of the bound GTP to GDP by an intrinsic GTPase activity. The related Ras-GTP complex can also signal, and again the signal is terminated by hydrolysis of the bound GTP (Figure 4.17). Because the oncogenic forms of Ras have lower GTPase activity than normal c-Ras, the signal they generate persists.

The *ras* gene is highly conserved, and the human c-H-*ras* gene can even substitute for *S. cerevisiae* (budding yeast) *ras* genes. In yeast, the Ras proteins are positive activators of adenylyl cyclase and thereby regulate cyclic AMP levels. In vertebrates, however, there is good evidence that Ras does not activate adenylyl cyclase; rather, adenylyl cyclase is activated by the G_s α-subunit. Indeed, the level of cAMP is lower in v-Ras transformed cells than in the corresponding parental cells. This means that Ras must have another signaling function or functions.

One signaling pathway Ras regulates has been identified. It consists of a series of protein-serine kinases beginning with Raf (a protooncogene product), MAP kinase kinase (MEK), and, at the end of the pathway, mitogen-activated protein kinase (MAP kinase) (Figure 4.15). Each protein kinase phosphorylates the next protein kinase in line, thereby activating it. The pathway is activated by treatment with most mitogens, and the phosphorylated MAP kinase translocates into the nucleus, where it phosphorylates and activates specific transcription factors and consequently induces the expression of a set of early response genes. The GTP-bound form of Ras associates with the N terminal regulatory region of Raf and activates the Raf protein kinase. (The association with Ras is not, however, sufficient for Raf activation *in vitro,* suggesting that there must be another unknown step involved in Raf activation.) Consistent with these findings, cells transformed by v-Ras have constitutively elevated MAP kinase activity. Recent

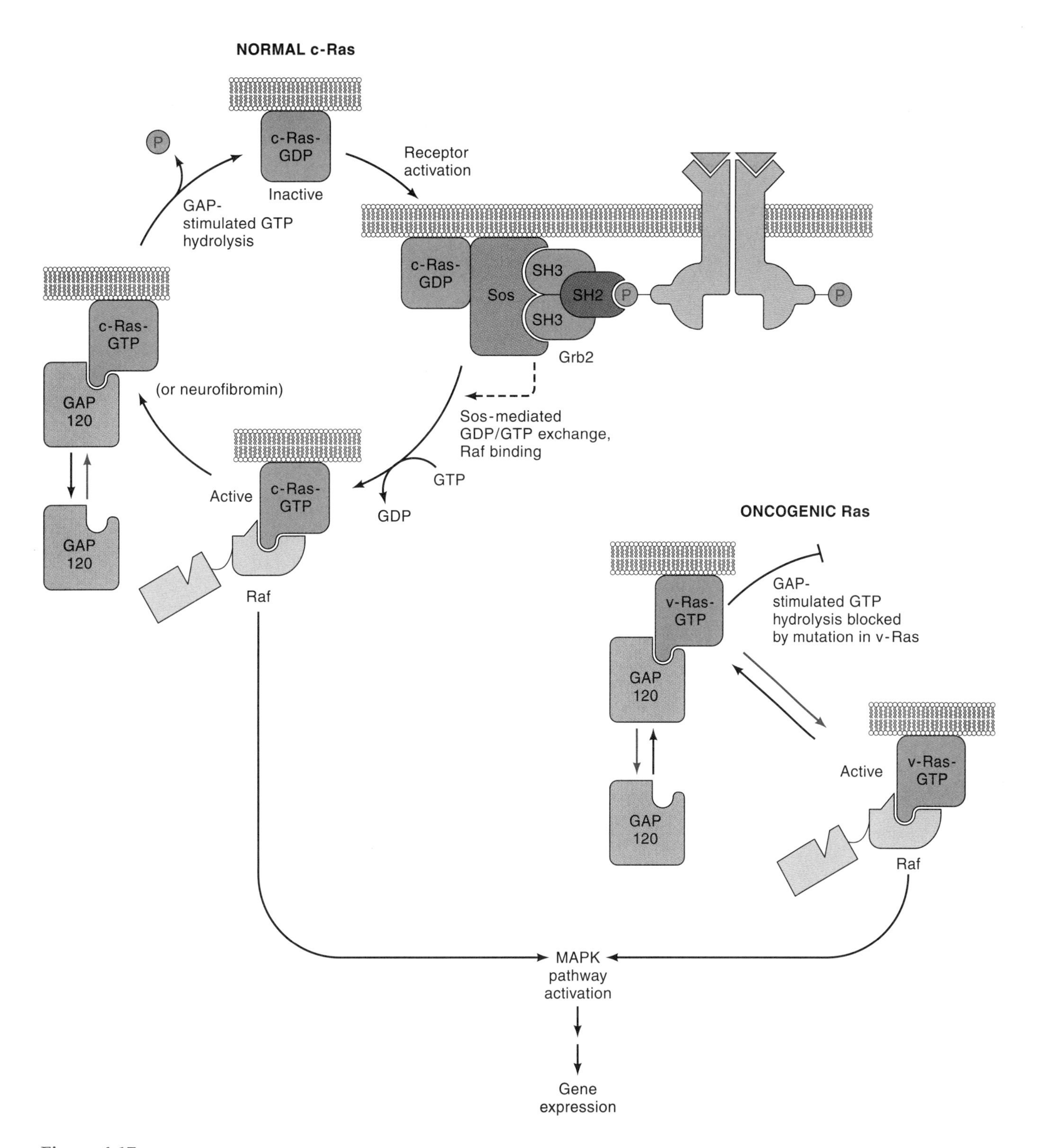

Figure 4.17
The Ras activation cycle. Normal c-Ras is converted from the inactive Ras-GDP state to the active Ras-GTP state by interaction with the Sos GTP exchange factor, which thus leads to the activation of the Raf protein-serine kinase. Active Ras-GTP is inactivated by interaction with GAP120 (or neurofibromin), which stimulates the intrinsic Ras GTPase activity and leads to the conversion of bound GTP to GDP. In the case of oncogenic mutant v-Ras, GAP120 can bind to v-Ras-GTP, but this does not stimulate GTP hydrolysis. In consequence, v-Ras, once bound to GTP, is continuously active, and therefore binds to and activates Raf constitutively. This leads, in turn, to the constitutive activation of the MAP kinase pathway.

evidence suggests that activation of the MAP kinase pathway is both sufficient and necessary for mitogenesis in fibroblasts, but Ras may still have additional targets.

Activated protein-tyrosine kinases encoded by oncogenes such as v-*src* also activate MAP kinase, and such oncogenes' ability to stimulate growth and transform cells depends on Ras function. This has been established in two ways. First, microinjection of an inhibitory monoclonal anti-Ras antibody into cells prevents their transformation by several oncogenic protein-tyrosine kinases. Second, expression of a Ras mutant in which Ser 17 is changed to asparagine blocks transformation by oncogenic protein-tyrosine kinases. This mutant has a dominant-negative phenotype, probably because it binds more tightly to GDP than GTP, and sequesters the Sos GTP exchange factor required for normal exchange of GDP for GTP. Because Raf is downstream of Ras, these findings place three different types of oncoprotein in a single signaling pathway. The c-Jun nuclear oncoprotein may also lie on this pathway (Section 4.2g). These examples emphasize the fact that mitogenic signaling involves a network of connected regulators, many of which drive cellular transformation when mutated into constitutively activated states.

The structure of ras *oncogenes* An intensive investigation of the precise alterations in c-Ras that lead to oncogenic activation was sparked by the finding that a point mutation at codon 12 of the c-H-*ras* gene in the T24 human bladder carcinoma cell line is sufficient to activate this gene. Subsequently, it became clear that mutations that change Gly 12, Gly 13, or Leu 61 to almost any other amino acid lead to activation. Moreover, the v-*ras* genes have the same types of point mutation as the tumor *ras* genes. The three-dimensional structures of c-H-Ras in both its GDP and GTP bound states are known and show that both Gly 12 and Leu 61 lie close to the GTP-binding pocket. Although they are not directly involved in binding, alteration of either residue might account for the decreased GTPase activity. The GTPase activity of c-Ras is stimulated by a number of cellular proteins, including a 120 kDa GTPase-activating protein (GAP120) and the product of the neurofibromatosis Type 1 gene, neurofibromin, a growth suppressor protein (Figure 4.17). Although the exact roles of GAP120 and neurofibromin in c-Ras function are not fully understood, they can in some circumstances down-regulate Ras signaling. GAP120 and neurofibromin also bind to oncogenic Ras proteins but do not stimulate their GTPase activity.

Between 20 percent and 30 percent of human tumors in which dominant oncogenes have been detected by transfection into normal fibroblast cell lines contain activated *ras* genes. Proof that these alterations are critical for tumorigenesis is still lacking, but there are interesting correlations between tumor type and the particular *ras* gene that is activated. For instance, activated c-K-*ras* genes are common in carcinomas, and oncogenic N-*ras* genes predominate in hematopoietic tumors. The logic behind this pattern is not yet apparent. Researchers have attempted to demonstrate a causal role for c-*ras* gene mutation in animal model systems. One study strongly indicated that the mutation of the c-H-*ras* gene is the initial step. In this case, ethylnitrosourea administration to 50-day-old female rats results in an almost 100 percent incidence of mammary carcinomas. This carcinogen has a half-life of only two hours in the animal, and yet the cells in more than 90 percent of the resultant mammary carcinomas are mutated at codon 12 of c-H-*ras*; a G to A base change in the second position of the glycine codon converts it into a glutamate codon. This transition is the type of mutation that ethylnitrosourea is known to make. Polymerase chain reaction (PCR) analysis demonstrated that mutation of codon 12 in c-H-*ras* is observable in mammary glands well before tumors are detected, indicating that the change is an early event in these mammary carcinomas. In skin papillomas, however, activating mutations in c-H-*ras* are apparently not sufficient to cause full malignant transformation because regressing papillomas still contain mutant c-H-*ras* genes.

G protein oncogenes Mutations that activate the G_s and G_i α-subunits occur in specific types of human tumor. Growth hormone–secreting pituitary tumors and a subset of thyroid carcinomas have mutations that lie in domains of α_s that are important for down-regulation of its activity by GTP hydrolysis. Both these cell types are unusual in that cAMP is mitogenic, and mitogens for these cells, such as thyroid-stimulating hormone, bind to serpentine receptors that stimulate adenylyl cyclase via G_s. Thus, mutations that activate α_s would be expected to drive the growth of such cells. Similar activating mutations are found in α_s in adrenal cortex tumors. G_i is coupled to phospholipases C and A_2, and the constitutive activation of these two enzymes, both of which generate second messengers involved in mitogenic responses, may account for the continuous growth of these cells.

Two other oncoproteins are included in this class because they are derived from GTP exchange factors that act on small G proteins in the Ras superfamily. Thus, the normal cellular counterpart of the Dbl oncoprotein is a GTP exchange factor for a Ras-related small G protein, Cdc42Hs, whose function is not known. The yeast homologue of Cdc42Hs plays a role in cell cycle control through activation of a protein-serine kinase, and the same may be true for Cdc42Hs, which can complement mutations in the yeast *cdc42* gene. Activation of Dbl into an oncoprotein requires the loss of N terminal sequences and leaves intact the central GTP exchange domain. This truncation may enhance the GTP exchange activity. Vav is an SH2/SH3 adaptor protein, which was also first identified by a human tumor oncogene. Normal Vav protein is expressed only in hematopoietic cells, and its function is essential for hematopoiesis. The central region of the Vav protein is related to the GTP exchange domain in Dbl and probably facilitates GTP exchange on a Ras-related small G protein. Oncogenic activation of Vav, like that of Dbl, requires the deletion of N terminal sequences and leads to constitutive exchange activity, consistent with a negative regulatory role for the N terminus. Vav also binds DAG and the tumor-promoting chemical, tetradecanoyl phorbol acetate (TPA).

e. Class V: Cytoplasmic Protein-Serine Kinases

The oncogenes in this class are derived from cellular genes that encode protein-serine kinases known or suspected to participate in transcytoplasmic signaling pathways. Two are retroviral oncogenes: v-*mil/raf* (*mil* is the chicken homologue of the mammalian *raf* gene) and v-*mos*. The proteins these genes specify contain sequences related to the canonical protein kinase catalytic domain and were suspected to be protein-tyrosine kinases when they were discovered. However, they display only protein-serine kinase activity *in vitro*, and mutation of key conserved residues in the catalytic domains leads to loss of kinase activity and transforming potential. There are several other oncogenic protein-serine kinases, including derivatives of known signaling protein-serine kinases. One example is protein kinase C, which constitutes a family of well-characterized protein-serine kinases that are activated by DAG and require calcium and phosphatidylinositol for activity. They are activated by many mitogens through the stimulation of phospholipase C, which generates DAG. Indeed, tumor promoters, such as TPA, which mimic DAG and activate protein kinase C, are mitogenic for some cell types. A mutant, oncogenic form of protein kinase C has also been found in cultured cells transformed by exposure to ultraviolet light.

Raf protein As discussed previously, c-Raf plays a role in mitogenic signaling downstream of c-Ras in the MAP kinase pathway. Upon activation, c-Raf phosphorylates and activates MAP kinase kinase (MEK), leading to activation of MAP kinase, which in turn activates specific transcription factors. The structure of the Raf protein is very similar to that of protein kinase C, which has an N ter-

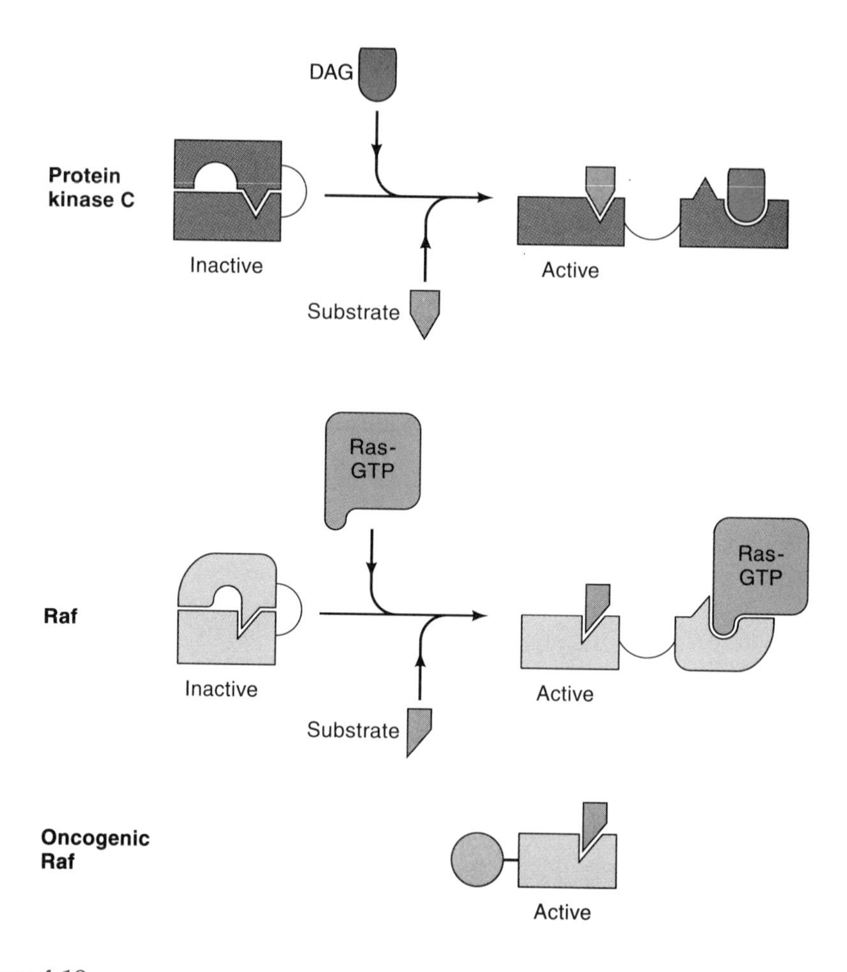

Figure 4.18
Parallels between activation of protein kinase C and Raf. Protein kinase C (PKC) is activated by the binding of its N-terminal regulatory domain, which contains two zinc-binding fingers, to DAG (and phosphatidylserine) in the membrane, thus allowing substrate binding and phosphorylation. Raf is activated by the binding of its N-terminal regulatory domain, which contains one zinc-binding finger, to Ras-GTP associated with the plasma membrane. In both cases, additional activating events are required, which may include phosphorylation. Oncogenic v-Raf is constitutively activated as a result of the loss of the N-terminal regulatory domain, which is replaced by retroviral Gag sequences.

minal regulatory domain containing two zinc-coordinated DAG-binding fingers and an N terminal pseudosubstrate sequence that interacts with the catalytic domain and suppresses activity. Binding of DAG dissociates the pseudosubstrate sequence and thereby reveals the enzyme's active site (Figure 4.18). The N terminal domain of c-Raf also seems likely to be a negative regulatory domain, and its interaction with the Ras-GTP complex may be part of the mechanism that relieves negative regulation. Also, the N terminal domain of c-Raf (expressed after transfection as an appropriately modified gene) acts in a dominant-negative fashion to suppress Ras signaling, presumably by sequestering Ras-GTP. Moreover, in every instance where oncogenically activated *raf* genes have been analyzed, N terminal sequences, including the Ras interaction domain, are missing. The C terminal part of Raf is constitutively active as a protein kinase and, when expressed *in vivo*, activates the MAP kinase pathway in the absence of Ras. The family of mammalian *raf* genes, c-*raf*, A-*raf*, and B-*raf*, all seem likely to have similar functions and be capable of oncogenic activation as a consequence of N terminal truncation.

Mos protein Although some single amino acid differences exist between v-Mos and c-Mos, the normal c-Mos protein is just as effective as v-Mos in cellular transformation assays. Mos proteins have serine-specific protein kinase activity *in vitro*, but their true function remained a mystery for a long time. The first clue to c-Mos function was the discovery that c-Mos is expressed exclusively in developing vertebrate germ cells. Later researchers found that oocyte maturation in *Xenopus* requires c-Mos. The protein is also a component of cytostatic factor, which maintains the activity of the maturation-promoting factor (MPF) in mature *Xenopus* oocytes that are blocked in the metaphase of meiosis II. The substrates that are phosphorylated in order for c-Mos to activate MPF are not known, although c-Mos is associated with polymerized tubulin in the oocyte and can phosphorylate tubulin *in vitro*. There is also strong evidence that c-Mos activates MAP kinase through the phosphorylation and activation of MAP kinase kinase. Indeed, microinjection of activated MAP kinase arrests cleavage of eggs in the same manner as c-Mos.

The mechanism of cellular transformation by Mos is not fully understood. Because c-Mos is normally produced only in developing germ cells, transformation probably results from ectopic expression. Somatic cells do not normally express c-Mos, and high levels of Mos are toxic to them. This, taken together with the association of c-Mos with microtubules, including spindle microtubules, and its role in the meiotic cell cycle, suggests that a critical concentration of Mos somehow triggers the mitotic cell cycle machinery into continuous operation. In addition, the ability of Mos to activate MAP kinase might provide the impetus for cells to progress through G1. No alterations in the c-*mos* gene have been detected in human tumors, but integration of an endogenous retroviral genome into the c-*mos* gene can cause oncogenic activation in some murine plasmacytomas.

f. Class VI: Cytoplasmic Regulators

The first example of an oncoprotein derived from a cytoplasmic regulatory protein was v-Crk. The oncogene of the CT10 avian sarcoma virus, v-*crk* encodes a Gag-Crk fusion protein. The v-Crk protein has an associated protein-tyrosine kinase activity, but the predicted amino acid sequence bears none of the hallmarks of other protein-tyrosine kinases. Instead, v-Crk has two regions related to noncatalytic sequences in the Src family protein-tyrosine kinases, specifically, the SH2 and SH3 domains (Figure 4.19). Mutagenesis studies demonstrated that both the SH2 and SH3 domains are required for v-Crk to transform. In transformed cells, v-Crk increases tyrosine phosphorylation and associates with a set of proteins that contain phosphotyrosine. This association is a property of the SH2 domain. Apparently, v-Crk signals through proteins bound to its SH3 domain, and some potential target proteins that bind to the v-Crk SH3 domain are known. v-Crk also associates with the c-Abl and Arg nonreceptor protein-tyrosine kinases, which probably accounts for the associated kinase activity and for the increase in phosphotyrosine-containing proteins in v-Crk-transformed cells. Indeed, v-Crk could be coupled to the same pathway as activated Abl proteins. The c-Crk protein differs from v-Crk by an additional C-terminal SH3 domain; the elimination of this domain is required for v-Crk transformation, implying that it normally plays a negative regulatory role.

Overexpression of two other adaptor proteins, Nck and Shc, elicits fibroblast transformation, although in these examples, tyrosine phosphorylation is not increased in the transformed cells. Overexpression may trigger normal signal

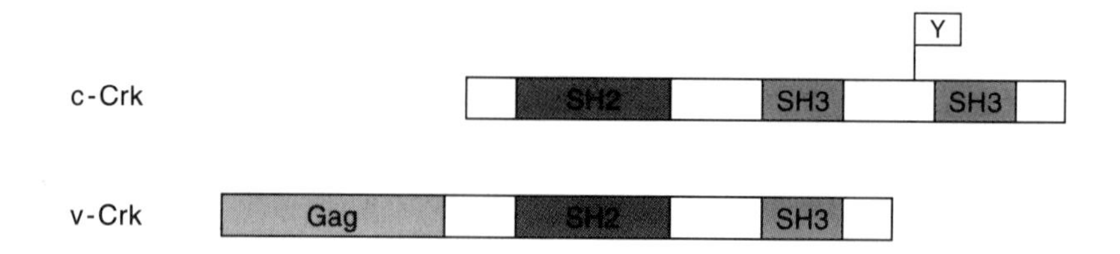

Figure 4.19
Activation of c-Crk into an oncoprotein. As depicted by the schematic structures, the adaptor protein c-Crk is converted into the v-Crk oncoprotein by fusion of retroviral Gag sequences to the N-terminus and by deletion of C-terminal sequences including an SH3 domain and a regulatory tyrosine Tyr 221 (shown by a flag, Y). In c-Crk Tyr 221 is normally phosphorylated and binds intramolecularly to the upstream SH2 domain, thus preventing both the SH2 and SH3 domains from binding to other proteins. The deletion of Tyr 221 in v-Crk means that both SH2 and SH3 domains are free to bind to other proteins, thus activating Crk-dependent signaling. One target for the v-Crk is the c-Abl protein-tyrosine kinase, which binds to the Crk SH3 domain. In the case of c-Crk, this leads to phosphorylation of Tyr 221 and inactivation of c-Crk. In the case of v-Crk, bound c-Abl may generate a constitutive signal.

pathways that lie downstream of Nck and Shc. The nature of the Nck-induced pathway is currently unknown, but Shc can associate with Grb2 if Shc is phosphorylated on tyrosine, and this can activate the Ras/MAP kinase pathway. As discussed earlier, the oncoprotein Vav could also be placed in this class because it contains two SH3 and one SH2 domains, as well as a domain related to the GTP-exchange region of Dbl. The purported Vav GTP exchange activity could be activated via tyrosine phosphorylation, and this phosphorylation presumably would require binding of the Vav SH2 domain to an activated protein-tyrosine kinase. Although Vav binds to activated receptor protein-tyrosine kinases when expressed in fibroblasts, the targets for the Vav SH2 domain in hematopoietic cells, where it is normally expressed, are not known.

g. Class VII: Nuclear Transcriptional Regulators

Changes in the pattern of gene expression are characteristic of tumor cells, and the early finding that the papovavirus-transforming proteins are located in the nucleus suggested that most oncoproteins would prove to be nuclear and affect gene expression directly. However, we now know that oncoproteins can function anywhere in the cell. Nevertheless, Class VII is the largest class of oncoproteins and includes proteins of both viral and tumor origin. Many of the tumor oncogenes in this class are associated with chromosomal breakpoints in human leukemias.

The progenitors of most Class VII oncoproteins are sequence-specific DNA-binding proteins known (or suspected) to act as transcriptional regulators. Many different types of transcription factor with distinct DNA-binding domains can be activated into oncoproteins, including DNA-binding domains in the helix-loop-helix, leucine zipper, homeodomain, and zinc finger families.

A number of general principles operate in the oncogenic activity of these proteins. In several instances, overexpression of the normal protein results in transformation, presumably through increased expression of specific target genes. This is true for c-Myc and c-Fos, which are early response genes, whose transcription is induced by mitogen treatment. Because the products of these genes are proposed to be transcriptional activators involved in a gene expression cascade required for exit from G0 into G1, imagining how their overexpression could trigger sustained proliferation is easy.

For other proteins, mutations are required for the oncogenic phenotype; in most cases, the activating mutations leave the DNA-binding function unaffected but alter transcriptional activation or regulatory domains. Such activating mutations can result either in increased or decreased transcriptional activity. In general, increased activity leads to the expression of genes that are normally involved in cell growth. In those instances where activity is decreased, the normal transcription factors appear to be required for the induction of differentiation genes, and the modified oncoproteins may bind to the response elements in the target genes and prevent differentiation by blocking the binding of the normal counterpart.

Transformation by ectopic expression or overexpression of nuclear oncoproteins One of the best examples where overexpression of a normal transcription factor causes transformation is the Myc protein. The v-*myc* oncogene was initially isolated from the MC29 avian retrovirus, which induces myelocytomatosis in chickens. The MC29 v-Myc protein contains several mutations when compared to the normal chicken c-Myc and is also fused to Gag sequences at its N terminus. However, these mutations may not be essential for v-Myc to transform because increased expression of the normal c-Myc protein is causally implicated in many types of leukemia. For instance, oncogenic activation of c-*myc* expression can occur by insertion of a viral transcriptional control element so that its expression is no longer regulated normally; this happens in the majority of avian leukosis virus-induced bursal lymphomas in chickens. In Burkitt's lymphoma, there are characteristic chromosomal translocations that involve the c-*myc* gene locus on chromosome 8 (Table 4.6). In each case, the translocation partner, which can be

Table 4.6 ***Human Oncogenes Activated by Chromosomal Translocation***

Disease	*Translocation*	*Gene Product*
Overexpression of normal protein by fusion with immunoglobulin or T cell receptor gene enhancer wequences		
Burkitt lymphoma (B cell leukemia)	t(8:14), t(8;22), t(2;8)	c-Myc (IgH and IgL)
B cell chronic lymphocytic leukemia	t(11;14)	Bcl1/PRAD1/cyclin D1; G1 cyclin (IgH)
B cell follicular lymphoma	t(14;18)	Bcl2; antiapoptosis factor (IgH)
B cell chronic lymphocytic leukemia	t(14; 19)	Bcl3; IκB family member (IgH)
Pre–B cell acute leukemia (pre–B ALL)	t(5; 14)	Interleukin 3 cytokine (IgH)
T cell acute leukemia (T-ALL)	t(10; 14)	Hox11; homeodomain transcription factor (TCR)
T cell acute leukemia (T-ALL)	t(7; 19)	Lyl1; bHLH transcription factor (TCR)
T cell acute leukemia (T-ALL)	t(1; 14)	Tal1; bHLH transcription factor (TCR)
T cell acute leukemia (T-ALL)	t(11; 14)	Ttg1-Rhom1; zinc finger transcription factor (TCR)
Expression of chimeric proteins containing fused sequences from the genes on either side of breakpoint		
Chronic myelogenous leukemia (CML)	t(9; 22)	Bcr-Abl; activated Abl protein-tyrosine kinase
Acute lymphocytic leukemia (ALL-1)	t(4; 11), t(11; 19), t(1; 11)	Fusion of human Trithorax-like protein (Trx) with various sequences
Acute promyelocytic leukemia (APL)	t(15; 17)	Pml-RARα; chimeric nuclear regulator
Acute lymphoblastic (pre-B) leukemia	t(1; 19)	E2A-Pbx1; chimeric transcription factor
Acute myeloid leukemia (AML-1)	t(8; 21), t(3;21)	Fusion of Aml1 human Runt-like protein with various sequences
Acute myeloid leukemia (AML)	t(6; 9)	Dek-Can; unknown nuclear function
Acute myeloid leukemia (AML)	t(16; 21)	Fusion of Erg, an Ets-related protein, with an unknown partner
T cell acute leukemia (T-ALL)	t(7; 9)	Tan1; truncated Notch receptor homologue
Ewing sarcoma (EWS)	t(11; 22)	Fusion of Fli1, an Ets-related protein, with an Ews locus protein
Myxoid liposarcoma	t(12; 16)	Fusion of Chop, a bZIP transcription factor, with Tls, and Exs-related protein

on chromosome 2, 14, or 22, is an immunoglobulin locus. As a result, the translocated c-*myc* gene is transcriptionally regulated like an immunoglobulin gene. Similar translocations are observed in murine plasmacytomas. Cell fusion experiments demonstrate that the normal c-*myc* gene is switched off in mature B cells while, as would be expected from the immunoglobulin nature of its control element, the translocated gene continues to be expressed. This implies that the inability to down-regulate the expression of the c-*myc* gene may be a crucial component in the maintenance of the transformed state of these lymphomas and their failure to differentiate to mature B cells.

The c-*myc* gene is amplified in several human leukemia cell lines. The increased gene copy number may allow partial escape from the normal regulatory control. During the screening of human tumors for alterations in the c-*myc* gene, researchers discovered a number of genes similar to c-*myc*. In these cases, the gene in question was detected despite its weak homology to authentic c-*myc* because it was amplified. One such gene is called N-*myc* because it is amplified in neuroblastomas; another is termed L-*myc* because it is amplified in some small cell lung carcinomas. The N-*myc* gene has properties akin to those of the c-*myc* gene in oncogene cooperation assays. N-*myc* and L-*myc* genes have only small patches of homology with c-*myc*, but interestingly, in each case, these regions are related to the same c-*myc* segments, which include the DNA-binding domain. As is commonly the case with c-*myc*, N-*myc* and L-*myc* are amplified rather than rearranged in tumors.

The discovery that c-*myc* is rapidly induced when quiescent cells are treated with many growth factors led to the idea that c-Myc is a cell cycle regulator. However, once induced, c-Myc protein drops to a low level and then does not vary during subsequent cell cycles in continuously growing cells, indicating that if c-Myc is a cell cycle regulator, changes in its level are not required. Nevertheless, experiments with an estrogen receptor–c-Myc fusion protein whose activity is regulated by estrogen indicate that c-Myc alone is sufficient to drive quiescent cells into the cell cycle when overexpressed. Treatment of cells with antisense oligonucleotides to ablate c-*myc* mRNA also suggests that c-Myc is required for progression through G1. Synthesis of these diverse findings into a coherent model for c-Myc action depends on understanding c-Myc structure and the discovery of an additional protein that interacts with c-Myc.

The nuclear location of c-Myc suggested that it would be a transcriptional regulator. There is now strong evidence that this is the case (Figure 4.20a). The c-Myc protein can dimerize with itself and other proteins through its basic region/helix-loop-helix/leucine zipper domain. The c-Myc homo- and heterodimers can both bind to specific DNA sequences; c-Myc also possesses a discrete transcriptional activation domain. However, c-Myc acts only as a transcriptional activator in the form of a heterodimer with Max, a smaller protein that forms a dimer with c-Myc through their helix-loop-helix/leucine zipper domains. The c-Myc-Max heterodimer activates transcription of reporter genes bearing multiple copies of DNA-binding sites specific for c-Myc. Overexpression of the c-Myc protein is sufficient to increase expression of such reporter genes, presumably because Max is present in a large excess. The c-Myc:Max ratio is probably the critical factor in determining whether c-Myc induces gene expression because Max homodimers act as repressors. However, the picture is complicated because Max has other partners that can displace c-Myc, such as Mad. Indeed, the level of Mad rises as cells differentiate while that of c-Myc falls; together this effectively reduces the level of active c-Myc-Max complex. Unexpectedly, overexpression of normal c-Myc in the absence of growth factors leads to apoptosis of fibroblasts. Thus, c-Myc function may be critical for the decision whether to grow or die.

The genes that c-Myc-Max normally induces to permit entry into the cell cycle are not yet known. The transcriptional activating activity of c-Myc is undoubtedly central to its transforming function because mutations that disrupt DNA binding or remove the transactivation domain abolish transforming activity. Presumably, the target genes for transformation and cell cycle progression are largely the same. The transactivation activity of c-Myc is increased by phosphorylation of the transactivation domain. However, one of the activating phosphorylation sites is commonly mutated in c-Myc, encoded by translocated c-*myc* genes in Burkitt's lymphoma, suggesting that transformation may require attenuated transactivation activity. Perhaps this reduces the Myc-mediated apoptotic response, allowing tumor cell survival.

The v-Fos oncoprotein was first discovered as the product of the v-*fos* oncogene in the FBJ murine sarcoma virus. Like c-*myc*, c-*fos* is an early response gene. The protein encoded by c-*fos* has two functional domains, a basic region/leucine zipper domain that enables it to form heterodimers with Jun family proteins and bind DNA and a C terminal transcriptional activation domain (Figure 4.20b). Thus, Fos-Jun heterodimers can bind to specific DNA sequences and activate transcription of associated promoters. Although v-Fos harbors several mutations when compared to c-Fos, the normal c-Fos protein can transform nearly as efficiently as v-Fos *in vivo* and *in vitro*. The main difference between c-Fos and v-Fos is that v-Fos is missing the normal C terminus; this domain has the important role of down-regulating c-*fos* transcription to terminate mitogenic induction. Mutational analysis shows that transcriptional activation by Fos is important for transformation. However, the target genes for v-Fos-induced transformation are not yet known.

Mutational activation of transcription factors One of the best studied examples of a nuclear oncoprotein where mutation is required is v-Jun. The v-*jun* gene, the oncogene of the ASV17 avian sarcoma virus, arose from the chicken c-*jun* gene. The c-Jun protein has an N terminal transcriptional activation domain and a C terminal basic region/leucine zipper DNA-binding domain. It can homodimerize or heterodimerize with other Jun or Fos family members and form active transcription factors. There are two structural differences between v-Jun and c-Jun: a deletion of 27 amino acids near the N terminus and three point mutations near the basic region (Figure 4.20b). The N terminal deletion is most important for transformation, although the functional consequence of this deletion is not fully understood. Although v-Jun retains transactivating activity, phosphorylation of two serines in the transactivation domain, whose phosphorylation is needed for maximal transactivating activity, is reduced in v-Jun, apparently because the missing amino acids contain part of the binding site for the Jun protein kinase. Thus, although v-Jun transformation requires its transactivation function, transactivation activity apparently has to be attenuated, as is the case for c-Myc.

Another nuclear oncoprotein activated by mutation is v-Myb, the oncogene of the avian myeloblastosis virus (AMV). In this case, the c-Myb protein is expressed at high levels in erythroid, myeloid, and lymphoid progenitor cells. Targeted inactivation of the c-*myb* gene in the mouse shows that it is essential for hematopoiesis, presumably because it is required for the proliferation of hematopoietic progenitor cells. The c-Myb protein has sequence-specific DNA-binding activity. It has an N terminal DNA binding domain and a transactivation domain near the C terminus. Activation of c-Myb into an oncoprotein requires the loss of both N and C terminal sequences, leaving the DNA-binding and transactivation domains intact (Figure 4.20c). The c-Myb protein has low intrinsic transactivating activity, apparently because the N and C termini interact, thus masking both the DNA-binding and transactivation domains; elimination of this interaction generates a constitutively active transcription factor. The target genes for v-Myb transformation are unknown, although it can cooperate with other transcription factors such as C/EBP in activating model promoters.

Figure 4.20
Structures of oncogenic transcription factors and their normal cellular counterparts. (a) Myc and Max. c-Myc forms a heterodimer with Max through the interaction of their helix-loop-helix-leucine zipper domains, which binds to the indicated DNA target sequence via the basic regions, thereby allowing the transactivation domain to stimulate gene expression. (b) Jun and Fos. The c-Jun protein is converted to the v-Jun oncoprotein by fusion of retroviral Gag sequences to the N terminus, three point mutations (indicated by open triangles), which each contribute to oncogenic activity, and deletion of 21 amino acids (δ) near the transactivation domain, which is the binding site for the Jun protein kinase that phosphorylates two serines leading to increased transactivating activity. The c-Fos protein is converted into the v-Fos protein by deletion of C-terminal sequences, which include part of the C-terminal transactivation domain and also sequences needed for repression of c-Fos expression. Fos and Jun family members form heterodimers via their leucine zippers, which bind to the indicated DNA target sequence via the basic regions, which allows the transactivation domains to stimulate gene expression. v-Jun and v-Fos appear to have the same DNA-binding specificity as their normal counterparts. (c) Myb. The c-Myb protein is converted to the v-Myb oncoprotein by the fusion of retroviral Gag sequences into the first of the three repeats in the DNA-binding domain, loss of a C-terminal negative regulatory domain, and four point mutations (indicated by open triangles). Myb binds to the indicated DNA target sequence and stimulates gene expression. v-Myb appears to have the same DNA-binding specificity as its normal counterpart. (d) ErbA. The c-ErbA protein (thyroxine receptor) is converted into the v-ErbA oncoprotein by fusion of retroviral Gag sequences, 11 point mutations (indicated by open triangles), and loss of a short sequence from the C-terminus. As a result v-ErbA retains DNA-binding activity, but lacks thyroxine-binding activity. c-ErbA forms a heterodimer with RXR, and this binds to the indicated target DNA sequence. In the presence of bound thyroxine this heterodimer stimulates gene expression. v-ErbA heterodimerizes with RXR and binds to the same target sequences, but cannot stimulate transcription, and thus blocks expression of differentiation genes normally stimulated by thyroxine through binding to c-ErbA. (e) Rel. The c-Rel protein is converted into v-Rel oncoprotein by loss of a short sequence at the N terminus, and deletion of part of the transactivation domain at the C terminus. c-Rel forms heterodimers with other NF-κB family members, such as p50, and binds to the indicated target DNA sequence and stimulates gene expression. v-Rel can also heterodimerize with NF-κB family members, but this heterodimer, either sequestered in the cytoplasm or bound to the same target DNA sequences, cannot stimulate transcription, and thus blocks expression of differentiation genes.

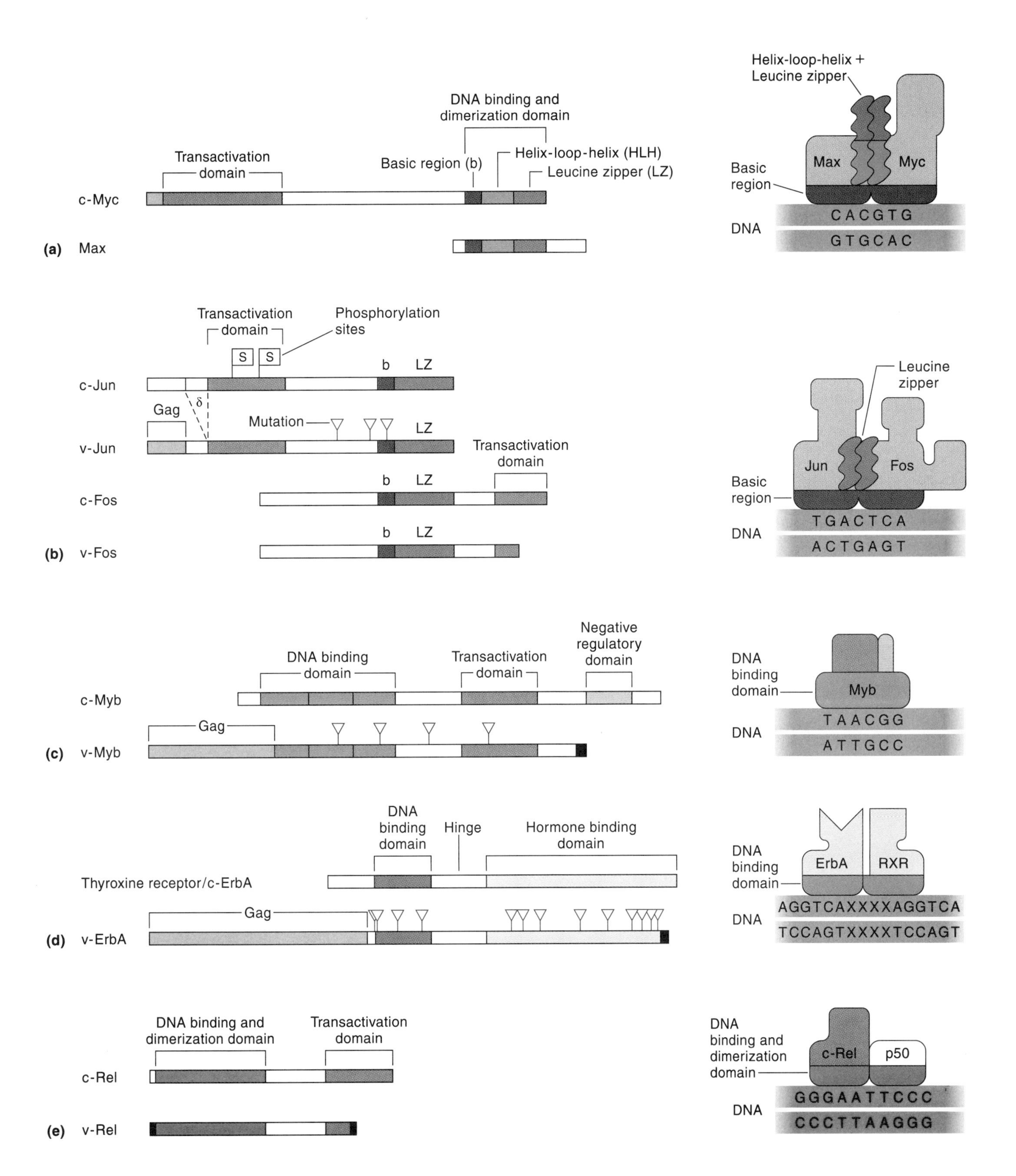

Helix-loop-helix +
Leucine zipper
DNA binding and
dimerization domain
Transactivation
domain
Basic region (b)
Helix-loop-helix (HLH)
Leucine zipper (LZ)
c-Myc
(a) Max
Basic
region
Max
Myc
DNA
CACGTG
GTGCAC
Transactivation
domain
Phosphorylation
sites
S
S
b
LZ
c-Jun
Gag
δ
Mutation
LZ
v-Jun
Transactivation
domain
b
LZ
c-Fos
b
LZ
(b) v-Fos
Leucine
zipper
Jun
Fos
Basic
region
DNA
TGACTCA
ACTGAGT
Negative
regulatory
domain
DNA binding
domain
Transactivation
domain
c-Myb
Gag
(c) v-Myb
DNA
binding
domain
Myb
DNA
TAACGG
ATTGCC
DNA
binding
domain
Hinge
Hormone binding
domain
Thyroxine receptor/c-ErbA
Gag
(d) v-ErbA
DNA
binding
domain
ErbA
RXR
DNA
AGGTCAXXXXAGGTCA
TCCAGTXXXXTCCAGT
DNA binding and
dimerization domain
Transactivation
domain
c-Rel
(e) v-Rel
DNA
binding and
dimerization
domain
c-Rel
p50
DNA
GGGAATTCCC
CCCTTAAGGG

Nuclear oncoproteins where mutation decreases transcriptional activity The v-Rel and v-ErbA oncoproteins are examples where there are mutations that decrease the transcriptional activity of these factors. In both cases, researchers believe these proteins act to suppress transcriptional induction by the corresponding wild-type proteins, thus blocking the expression of genes needed for differentiation.

The v-*erbA* gene is the second oncogene of AEV. It is not an oncogene in its own right, but cooperates with v-*erbB* in inducing erythroleukemia in chickens. The clue to c-ErbA function came from an unexpected direction: the cloning of the glucocorticoid receptor, which is a glucocorticoid-regulated transcription factor. The sequence of the DNA-binding domain of the glucocorticoid receptor proved to be related to part of c-ErbA, which suggested that c-ErbA might also be the receptor for a small organic molecule. Subsequent work showed that c-ErbA is the receptor for the thyroid hormone, thyroxine, and that c-ErbA is a thyroxine-regulated transcription factor. Thyroxine is involved in the differentiation of several cell types, but its importance in this context is that it is required for the differentiation of red cell precursors.

There are several differences between v-ErbA and c-ErbA, including point mutations in the zinc finger DNA-binding and C terminal hormone-binding domains and a deletion of several amino acids at the C terminus (Figure 4.20d). The v-ErbA protein retains DNA-binding activity and binds with the same sequence specificity as c-ErbA, but as a result of the C terminal deletion, v-ErbA cannot bind thyroxine or stimulate transcription when bound to DNA. The loss of hormone binding and transactivation functions is critical for v-ErbA to assist v-ErbB in blocking erythroid cell differentiation and thus promoting growth. Like other members of the steroid receptor superfamily, c-ErbA binds to its target DNA sequence as a heterodimer with another member of the superfamily called RXR. In the presence of thyroxine, this complex stimulates expression of target genes, in this case, genes involved in red cell terminal differentiation such as carbonic anhydrase and the red cell anion transporter. Apparently, v-ErbA blocks the expression of such genes simply by binding to the c-ErbA regulatory elements in their promoters, thus preventing the binding of the thyroxine-c-ErbA-RXR complex, which would normally lead to the transcription of these genes. In this sense, v-ErbA is acting in a so-called dominant-negative fashion to repress that activity of its normal cellular counterpart.

v-*rel* is the oncogene of avian reticuloendotheliosis virus, which causes a lymphoid leukemia in chickens. The progenitor of v-Rel, c-Rel, is a member of a family of transcription factors called the NF-κB family, which bind to DNA as homo- or heterodimers. When compared to c-Rel, v-Rel retains the N terminal DNA-binding/dimerization domain, but has suffered a C terminal deletion in the transactivation domain (Figure 4.20e). Evidence suggests that the loss of the C terminus diminishes v-Rel-transactivating activity. Thus, a plausible model for v-Rel transformation of B cells is that it blocks the transcription of genes needed for B cell differentiation, which are normally transcribed in response to binding of members of the NF-κB family, including c-Rel. Among such genes is the immunoglobulin κ light chain gene, which was the first gene shown to require NF-κB activity. The v-Rel protein could act in a dominant-negative fashion either by binding to NF-κB binding sites or by sequestering other members of the NF-κB family in inactive complexes.

The ability of v-ErbA and v-Rel to block differentiation of hematopoietic cells illustrates an important principle in oncogenesis: that tumor cells often represent cells blocked at a stage in differentiation when they would normally be dividing to generate more precursor cells. Thus, simply by blocking differentiation, one will get an expanded population of precursor cells that continue to divide. The converse is that if one can induce these cells to differentiate, they will stop dividing.

Chimeric nuclear oncoproteins There are several examples where chromosomal translocations in human leukemias result in the fusion of two genes, and this fusion encodes a chimeric protein (Table 4.6). In several cases, these chimeras are comprised of two distinct nuclear proteins, each of which is thought to regulate transcription in its own right. There is also one example of a viral oncoprotein that is a fusion of two nuclear proteins, Myb and Ets. (Table 4.4). The Myb-Ets oncoprotein causes erythroleukemia in chickens, and mutation of either partner in the fusion abolishes transforming activity. A virus designed to express the two halves of this chimeric protein separately does not cause disease; in the few infected chickens that develop leukemia, a rearranged virus is found in which the *myb* and *ets* genes are once again fused. Thus, in this case, one can be certain that the fusion is essential for transforming activity.

Two examples of oncogenic chimeras in human leukemias are E2A-Pbx1 and Pml-RAR. E2A-Pbx1 is a fusion between Pbx1, a putative transcription factor in the homeodomain family, and E2A, a member of the basic region/helix-loop-helix family. It is implicated in a type of pediatric pre–B cell leukemia with a t(1;19) translocation (Figure 4.21a). In the fusion protein, the E2A partner has lost its DNA-binding domain but retains its transactivation domain; the Pbx1 partner retains its homeobox DNA-binding domain. A virus expressing E2A-Pbx1 transforms fibroblasts in culture and causes myeloid leukemia in mice. E2A-Pbx1 activates transcription of reporter genes with a Pbx1-binding site through the E2A transactivation domain. E2A is normally expressed at high levels in pre–B cells

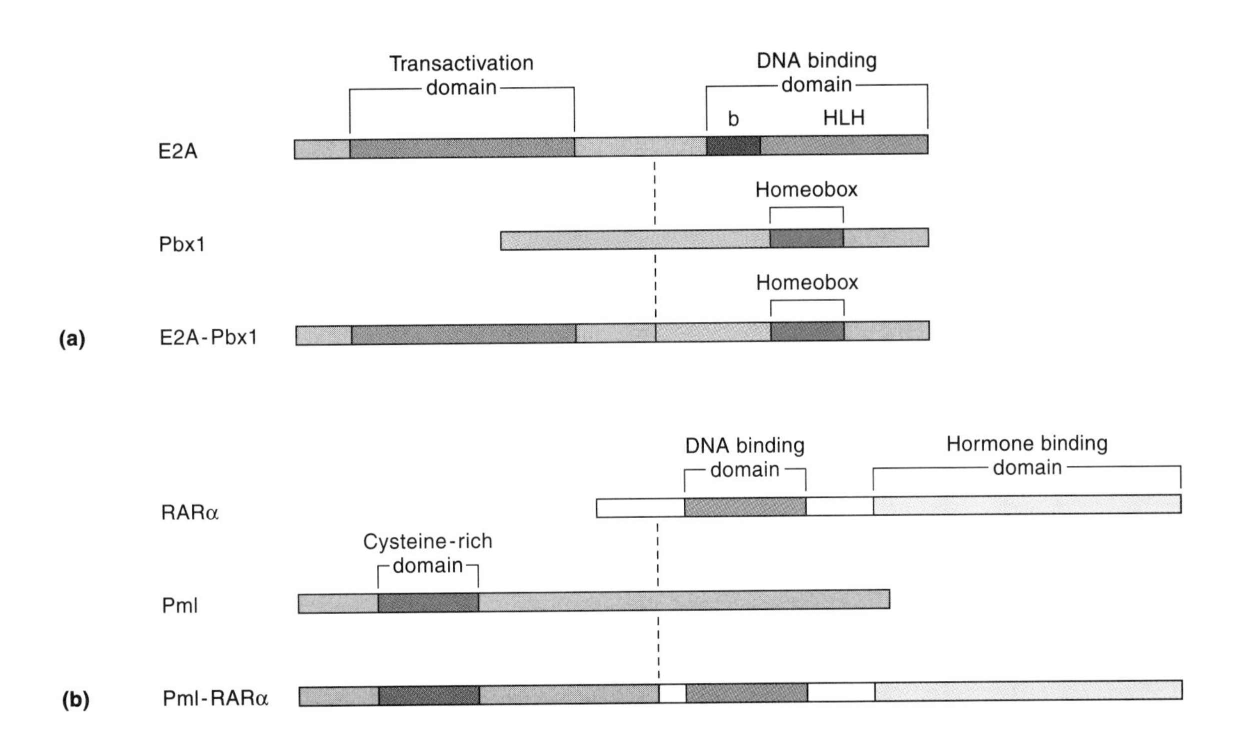

Figure 4.21
Oncogenic chimeric transcription factors. (a) E2A-Pbx1. As indicated by the schematic structures, the chimeric E2A-Pbx1 oncoprotein results from a fusion of the E2A and Pbx1 genes, which creates a chimeric gene encoding a protein with the transactivation domain of the E2A transcription factor and the homeobox DNA-binding domain of Pbx1. (b) Pml-RARα. The chimeric Pml-RARα oncoprotein results from a fusion of the Pml and retinoic acid receptor α(RARα) genes, which creates a chimeric gene encoding a protein with the N-terminal domain of Pml and the DNA-binding and hormone-binding domains RARα.

and is needed for expression of the immunoglobulin κ light chain gene. In contrast, Pbx1 is not normally expressed in pre–B cells. Because the *E2A-PBX-1* fusion gene is under the control of the E2A promoter, it is expressed at high levels in pre–B cells, and this leads to transcription of genes to which the Pbx1-binding domain can bind that are not normally expressed in pre–B cells. Exactly which target genes are needed for E2A-Pbx1 to transform, however, is not yet known.

The *PML-RAR* gene is a fusion between Pml, a nuclear protein of uncertain function, and the retinoic acid receptor (RAR) α-protein, which is a retinoic acid–regulated transcription factor in the steroid receptor superfamily (Figure 4.21b). The *PML-RAR* gene is created by a t(15;17) translocation found in acute promyelocytic leukemia. The Pml-RAR fusion protein retains most of the Pml protein and the DNA-binding and retinoic acid–binding domains of RARα. One model for how Pml-RAR transforms promyelocytes is that it binds to the same promoter regulatory elements recognized by RARα, and blocks RARα-dependent transcription needed for differentiation. However, Pml-RAR and RARα exhibit very similar retinoic acid–stimulated transcriptional activity on reporter genes. Thus, it seems unlikely that inhibition of RARα function accounts for the oncogenic potential of Pml-RAR. An alternative model is that the fusion of Pml with RARα abrogates a normal Pml function required for differentiation of promyelocytes. This idea is supported by the demonstration that Pml is normally localized to specific nuclear substructures, but this structure is disrupted in acute promyelocytic leukemia cells. Retinoic acid treatment of such cells causes reformation of these structures and differentiation of the cells. Excitingly, retinoic acid treatment of patients with this disease also causes differentiation of the leukemia cells *in vivo* and remission, and this is now being used as a therapy. Thus, Pml-RAR appears to act in a dominant negative fashion by inhibiting the formation of a specialized nuclear structure needed for promyelocyte differentiation.

h. Class VIII: Cell Cycle Regulators

One outcome of the recent explosion of information about the cell cycle is the realization that some of the fundamental cell cycle regulators are directly involved in oncogenesis. Because the continuous cycling of cells is a hallmark of cancer, the principal emphasis in studying the abnormal growth of cancer cells has been on how genetic changes stimulate quiescent cells in G0 to enter the cell cycle. However, deregulation of other decision points in the cell cycle can apparently also cause unbridled cell division and play an important role in tumorigenesis.

The eukaryotic cell cycle The major control points in the eukaryotic cell cycle have been defined largely by the genetic analysis of the cell cycle in fission and budding yeast (Figure 4.22). However, the conserved nature of cell cycle regulatory mechanisms in eukaryotes allows us to extrapolate to vertebrate cells. The major control points are the G1 to S phase transition, at which cells become committed to DNA synthesis, and the G2/M boundary, where cells enter mitosis. In yeast, both decisions are controlled by the activity of a single protein-serine kinase, Cdc2, in conjunction with one of two different but related classes of Cdc2-activating subunits known as cyclins. The so-called **G1 cyclins** regulate the G1 to S transition. The **mitotic cyclins**, first discovered as proteins whose abundance oscillates in invertebrate embryonic cell cycles, regulate the G2 to M transition.

In higher eukaryotes, the simple picture of two Cdc2-regulated cell cycle decision points derived from studies with yeast proves inadequate. There are more than ten distinct cyclin genes in the mammalian genome and at least six cyclin-

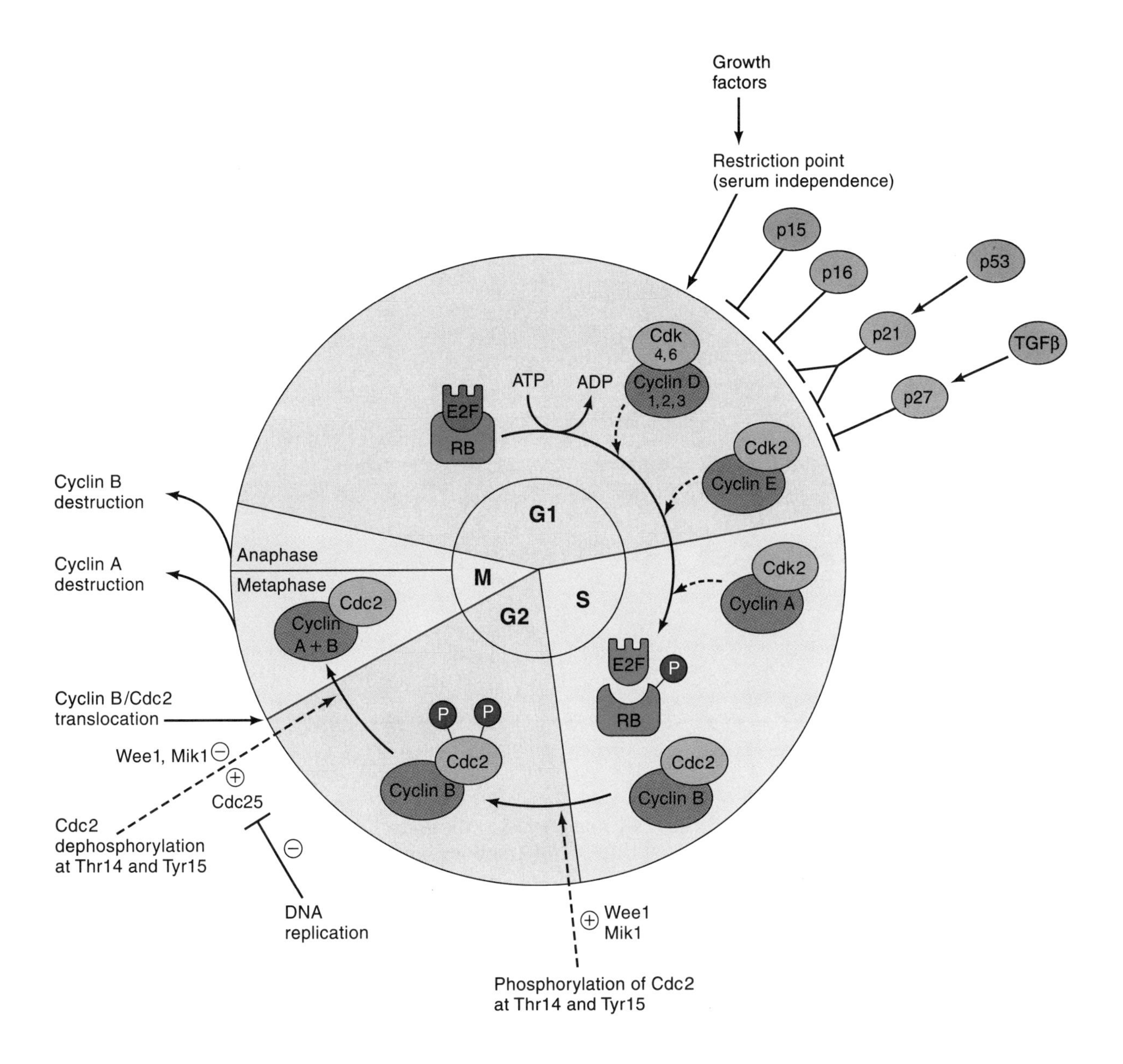

Figure 4.22
Cell cycle regulation by cyclin-Cdk protein kinases. Progression through the four phases of the cell cycle (G1, S, G2, and M) is governed by the successive formation, activation, and inactivation of a series of heterodimeric complexes, which contain a cyclin-dependent kinase (Cdk) bound to an activating cyclin subunit. In G1, cyclin D-Cdk4 (or Cdk6) complexes appear first and phosphorylate retinoblastoma (RB)-E2F complexes and lead to the release of the E2F transcription factor, which stimulates expression of genes encoding proteins needed for S phase, such as DNA polymerase α. Cyclin E-Cdk2 complexes appear next and are required for entry into S phase. The activity of the G1 cyclin-Cdk complexes is negatively regulated by a series of small inhibitor proteins, whose expression is controlled by DNA damage and exogenous stimuli. p15 and p16 are related Cdk inhibitors that act on cyclin D-Cdk complexes, and p21 and p27 are related Cdk inhibitors that act on both cyclin D-Cdk and cyclin-Cdk2 complexes. Cyclin A2-Cdk2 complexes form at the end of G1 and are needed for progression through S phase. Cyclin B-Cdc2 complexes form during G2, but are held in an inactive state through phosphorylation of Thr 14 and Tyr 15 in Cdc2 by the Wee1 and Mik1 protein kinases. Dephosphorylation of these sites at the end of G2 by the Cdc25 phosphatase triggers entry into mitosis (M phase). Inactivation of the cyclin A2-Cdc2 and cyclin B-Cdc2 complexes through destruction of the cyclins is required for exit from M phase into the next G1.

Table 4.7 *Cyclins and Cyclin-Dependent Kinases*

Cyclin	Cell Cycle Function	Cdk Partners
Cyclin A1	Meiosis	?
Cyclin A2	S and G2/M	Cdk2; Cdc2
Cyclin B1	G2/M Cdc2	
Cyclin B2	G2/M Cdc2	
Cyclin C	?	?
Cyclin D1 (Bcl1, PRAD1)	Late G1	Cdk4; Cdk6
Cyclin D2	Late G1	Cdk4; Cdk6
Cyclin D3	Late G1	Cdk4; Cdk6
Cyclin E	G1/S	Cdk2
Cyclin F	?	
Cyclin G	?	
Cyclin-Dependent Kinases	**Cell Cycle Function**	**Cyclin Partners**
Cdc2 (Cdk1)	G2/M	Cyclin A2; cyclin B1/2
Cdk2	G1/S	Cyclin A2; cyclin E
Cdk3	?	?
Cdk4	Late G1	Cyclin D1; cyclin D2; cyclin D3
Cdk5	?	?
Cdk6	Late G1	Cyclin D1; cyclin D2; cyclin D3

dependent kinase (Cdk) partners for these cyclins (Table 4.7). In principle, this could provide a vast number of combinations, but only certain cyclin-Cdk complexes are found, and there are clear restrictions on which Cdk can partner which cyclin. The multiplicity of known cyclin-Cdk complexes implies that, in addition to the G1/S and G2/M cell cycle transitions defined from work on the yeast cell cycle, there are probably many regulated cell cycle transitions in the vertebrate somatic cell cycle. The current belief is that cyclin D complexed to Cdk4, or Cdk6 regulates a transition in late G1. Cyclin E/Cdk2 appears to regulate the G1/S transition, and cyclin A2/Cdk2 is needed for progression through S phase. Cyclin A2/Cdc2 is required for the G2/M transition, but cyclin B1/Cdc2 (and possibly cyclin B2/Cdc2) is the main effector at this step.

As yet, the critical substrates for the cyclin-Cdk complexes are not well defined. However, many substrates are known for the activated cyclin B1-Cdk2 complex, including lamin, a protein whose phosphorylation is essential for nuclear lamina breakdown. Also, the retinoblastoma (RB) growth suppressor gene product is a key cyclin/Cdk substrate in late G1, and there is good evidence that RB, if present, has to be phosphorylated for cells to progress into S phase (Figure 4.23). Cyclin/Cdk-mediated phosphorylation is thought to release transcription factors and other nuclear proteins that are specifically bound to the unphosphorylated form of RB (and RB-related proteins), thus allowing induction of genes needed for progression into S phase. Among these transcription factors, E2F has received considerable attention because it may be a regulator of a number of genes, such as cyclin A2 and DNA polymerase α, whose expression is induced in late G1. An RB-related protein (p107), the p53 protein, replication origin-binding proteins, and the replication factor RPA are other likely targets for cyclin-Cdk complexes in G1 and S phases.

Cell cycle regulation and oncogenesis Several experimental observations imply a very direct connection between the cell cycle regulatory components and onco-

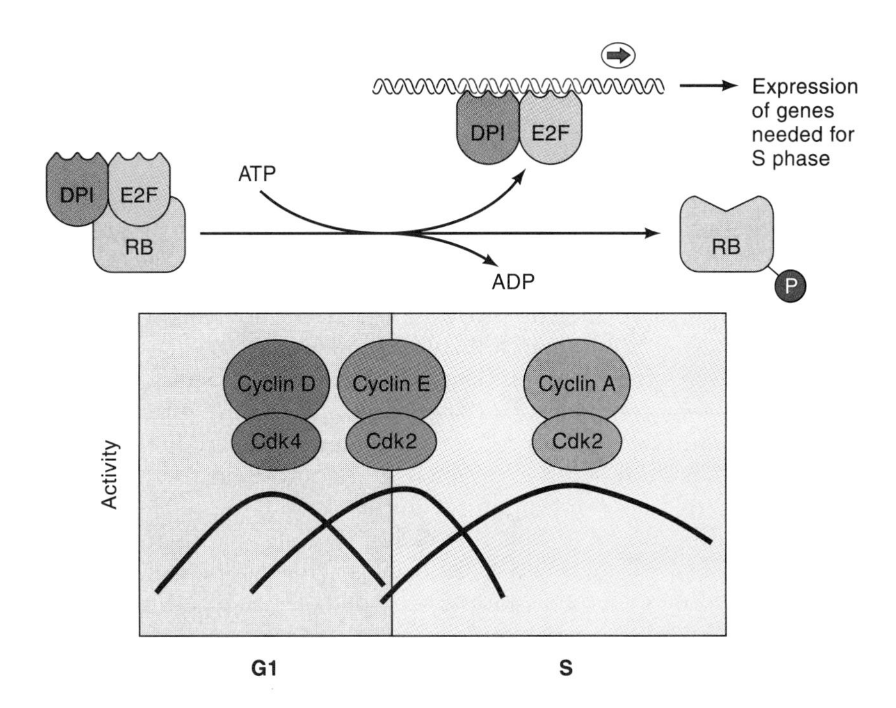

Figure 4.23
Phosphorylation of RB by cyclin-Cdk kinases is needed for progression from G1 into S phase. In early G1, RB is bound to E2F-DP1 heterodimers and other proteins, thus inactivating them. Successive activation of cyclin D-Cdk4 (or Cdk6), cyclin E-Cdk2, and cyclin A2-Cdk2 complexes results in sustained RB phosphorylation during late G1 and S phases. Phosphorylation of RB leads to the release of E2F-DP1 (and other bound proteins), allowing it to bind to its target gene regulatory sequences and induce transcription of genes whose expression is needed for S phase.

genesis, as the following few examples illustrate. Cyclin A2 is implicated in tumorigenesis in several ways. Thus, the adenovirus-transforming protein E1A acts through its association with a complex that contains the RB-related protein p107, cyclin A2, and Cdk2. This complex is formed when E1A displaces the active form of the E2F transcription factor from a less active E2F-p107-cyclin A2-Cdk2 complex. Second, hepatitis B virus sequences are integrated into the cyclin A2 gene in a human hepatoma that expresses an altered, stabilized cyclin A protein.

Cyclin D1 is overexpressed in at least three types of human tumor: in some parathyroid tumors, cyclin D1 expression is deregulated by translocation of the *cyclin D1* gene to the parathyroid hormone gene promoter; cyclin D1 expression is altered by translocation of the *cyclin D1* gene to the immunoglobulin heavy chain locus in one type of B cell lymphoma; cyclin D1 is part of an amplified locus and is overexpressed in some breast and esophageal carcinomas. Three additional lines of evidence suggest that cyclin D1 can act as an oncoprotein. First, overexpression of cyclin D1 in fibroblasts shortens the G1 phase, and such cells, although not morphologically transformed, form tumors in nude mice. Second, cyclin D1 cooperates with oncogenic Ras in transforming primary rat fibroblasts. Third, transgenic mice expressing cyclin D1 under control of the MMTV transcriptional regulatory region develop mammary carcinomas after lactation when the MMTV promoter is activated.

The activity of cyclin-Cdk complexes during the cell cycle is regulated in part by protein inhibitors. The fact that both copies of the *p16* gene, which encodes a 16 kDa inhibitor of cyclin D/Cdk4 activity, are disrupted in a small fraction of human carcinomas suggests that the inactivation of cell cycle inhibitory protein genes can play a role in tumorigenesis. Such genes can now be included in the category of growth suppressor genes (Section 4.4).

i. Class IX: Antiapoptosis Factors

Excessive DNA damage or exposure to inappropriate growth conditions can trigger the induction of a cell death program known as apoptosis. The Bcl2 oncoprotein, first identified in a human follicular B cell lymphoma at a translocation breakpoint with the immunoglobulin heavy chain locus, acts to prevent the apoptotic response. Overexpression of Bcl2 in transgenic mice prevents lymphocytes that have not undergone a productive immunoglobulin or T cell receptor gene rearrangement from being eliminated by apoptosis in the usual manner during development, thus resulting in accumulation of these cells. This increase in immature lymphocytes in turn leads to a high frequency of lymphomas. The exact function of Bcl2 is unknown. It associates with cytoplasmic membranes and may act as a component of an antiprotease pathway to block apoptosis. Bcl2 has recently been found to interact with a Ras-related protein, suggesting it may be part of a signaling system. Because apoptosis is a means of eliminating cells with potentially deleterious mutations, such as those that contribute to tumor progression, other oncoproteins that block apoptosis will probably be found.

4.3 DNA Tumor Virus Oncoproteins and the Cell Cycle

The transforming proteins of the small DNA tumor viruses of the polyoma, papilloma, and adenovirus families were extremely valuable in elucidating cell cycle regulators. For example, p53 was first identified through its association with the SV40- and adenovirus-transforming proteins, and the association of adenovirus E1A with RB provided a vital clue to RB function. We now know that a major activity of these DNA tumor virus–transforming proteins is to trigger entry of quiescent cells into the cell cycle upon infection. This is critical for viral proliferation because the viruses rely on cellular DNA replication functions for the replication of their genomes. In large part, the viral-transforming proteins act by blocking the activity of growth suppressor proteins. Thus, p53 is inactivated either by its sequestration as a complex with the transforming protein or by induction of its degradation. As discussed in Chapter 1, the binding of SV40 large-T, E1A, and E7 to RB results in the release of a series of proteins that are normally freed by phosphorylation at the G1/S boundary.

There are some exceptions to the rule that DNA virus–transforming proteins inactivate growth suppressor proteins. For instance, the polyomavirus middle-T protein, which is associated with cytoplasmic membranes, binds to and activates the c-Src protein-tyrosine kinase and also associates with protein phosphatase 2A to modulate its substrate specificity. The association with c-Src is important because this leads to the phosphorylation of middle-T, thereby creating binding sites for PI-3 kinase and the adaptor protein Shc. As already described, binding of the latter activates the Ras signaling pathway through its association with Grb2/Sos. More details of the association of DNA virus–transforming protein with cellular proteins and the consequences of these interactions are in Chapter 1.

4.4 Growth Suppressor Proteins

The realization that some of the mutations involved in tumorigenesis result in the inactivation of both alleles of a cellular gene has had an enormous impact on our understanding of this process. There are over 40 heritable predispositions to-

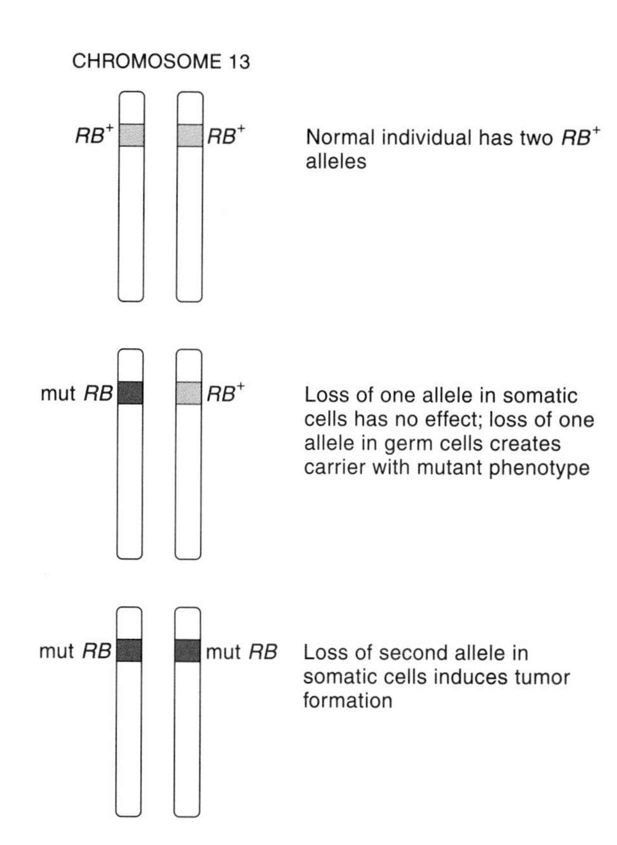

Figure 4.24
Loss of both copies of a growth suppressor gene plays a role in tumorigenesis. Hereditary retinoblastoma is caused by the inheritance of a mutated copy of the *RB* gene on chromosome 13q1.4 (shown schematically). Inactivation of the second normal copy of the *RB* gene in a somatic cell in the retina of such an individual leads to tumor formation. In sporadic retinoblastoma, both copies of the *RB* gene have to be inactivated in the same cell to cause tumor formation. The loss or inactivation of both copies of the *RB* gene plays a role in several types of human cancer.

ward specific types of cancer, implying the occurrence of mutations carried in germ line cells. The discovery that a macroscopic deletion of chromosome 13q14 is commonly present in one such case, families with heritable retinoblastoma, suggested that the disease arises as the result of inactivation of the second allele. Sporadic retinoblastoma, which is a rare disease, might then be a consequence of a new mutation in one allele of the same gene (the retinoblastoma gene), followed by mutational inactivation of the second allele (Figure 4.24). Investigation of polymorphic markers on chromosome 13 in inherited retinoblastoma tumor cells revealed that in many cases heterozygosity had been lost for either the whole or part of chromosome 13. This loss of heterozygosity can arise in several ways, including nondisjunction of sister chromatids at mitosis, mitotic recombination, and gene conversion. The loss of heterozygosity has now become a key to identifying novel growth suppressor genes of this type and to confirming the role of a particular locus in carcinogenesis.

The functions of several growth suppressor genes have now been established, and it is worth reviewing the best understood cases (Table 4.8); additional discussion of these genes is found in Chapter 1.

a. Retinoblastoma Protein

RB is a growth suppressor protein commonly inactivated by mutation in human tumors. The identification of the *RB* gene from genetic studies and its mapping to chromosome 13q14 allowed positional cloning of the gene. The cloned gene was used to demonstrate that, in patients genetically predisposed to retinoblastoma, retinal tumor cells carry inactivating mutations in both copies of the *RB* gene. The existence of cells lacking RB indicates that it is not essential for cell division, but several lines of evidence suggest a role for RB in cell cycle control. Thus, exogenously expressed RB suppresses the tumorigenicity of certain cell lines lacking RB, and overexpression of RB inhibits the growth of some normal cell lines.

Table 4.8 *Functions of Growth Suppressor Proteins*

Protein	Disease	Location	Function
DCC	Colon carcinoma	Plasma membrane	Adhesion molecule with Ig-like and fibronectin Type III repeats
VHL	von Hippel-Lindau (germ line mutation)	Nucleus	Inhibitory subunit of RNA polymerase II elongation factor
Neurofibromin	NF-1 (germ line mutation)	Cytoplasm	Activates Ras GTPase, thus inactivating Ras
Merlin	NF-2 (germ line mutation)	Cytoskeleton?	Ezrin-related putative cytoskeletal protein
APC	Adenomatous polyposis coli (germ line mutation)	Cytoplasmic?	Catenin-binding protein that may be involved in a cadherin adhesion receptor-signaling pathway
PKR	Leudemia?	Cytoplasm	Double-stranded RNA-dependent protein-serine kinase
RB	Retinoblastoma (germ line mutation)	Nucleous	Binds/inactivates transcription factors in unphosphorylated state
WT-1	Wilms tumor (germ line mutation)	Nucleus	Zinc finger DNA-binding protein with Egr1-like binding specificity that acts as a transcriptional repressor
p53	Multiple cancers (germ line mutation)	Nucleus	Transcription factor that can activate or repress; involved in DNA damage response pathway (Mutant forms cannot transactivate and act dominantly byt sequestering wild-type p53 in complexes, thus inactivating it.)
MSH	HNPCC (germ line mutation)	Nucleus?	Protein involved in mismatch DNA repair
MLH1	HNPCC (germ line mutation)	Nucleus?	Protein involved in mismatch DNA repair
p16	Carcinomas	Nucleus?	Inhibitor of cyclin D/Cdk4 activity

An important step in understanding RB function was the demonstration that RB is phosphorylated in a manner that is regulated by the cell cycle; a significant increase occurs at the G1/S boundary (Figure 4.23). In addition, differentiated and nondividing cells contain predominantly hypophosphorylated RB. Microinjection of hypophosphorylated RB in G1 blocks cells from entering S phase, confirming a major site of action at the G1/S boundary. In growing cells, one or more of the cyclin-Cdk complexes phosphorylates and thereby inactivates RB during G1, allowing progression into S phase.

RB binds to a number of cellular and viral proteins, including the SV40 large-T, papilloma E7, and adenovirus E1A transforming proteins. These viral proteins stimulate cells to enter the cell cycle by displacing bound cellular proteins from RB, and these proteins are also released by cyclin/Cdk phosphorylation of RB at the G1/S boundary. Among the cellular proteins that bind to RB is the transcription factor E2F. RB binds to the transactivation domain of E2F, suppressing E2F-mediated transcriptional activation. Thus, the phosphorylation-dependent release of E2F can cause the induction of genes regulated by E2F.

b. p53

Because researchers inadvertently used a mutant form of the *p53* gene rather than a normal gene in early experiments, they initially thought p53 was a dominant oncoprotein. However, wild-type p53 is actually a growth suppressor protein that inhibits the growth of both normal and transformed cells when overexpressed. Most human tumors have no wild-type copy of the *p53* gene and express either no p53 or a mutant form. Li-Fraumeni families have a predisposi-

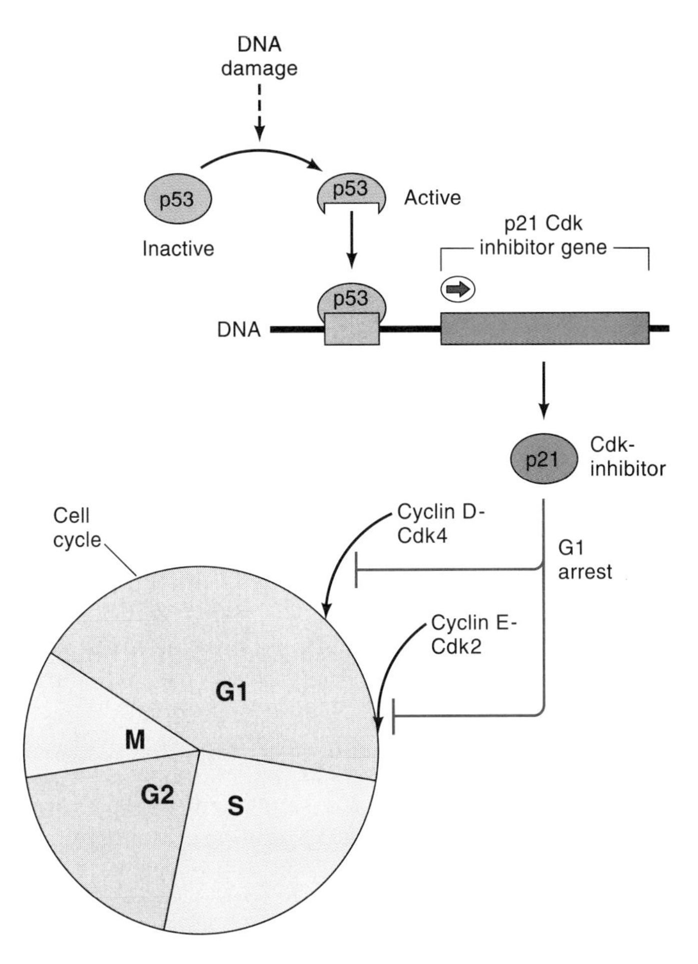

Figure 4.25
DNA damage in G1 induces cell cycle arrest in a p53-dependent fashion. DNA damage in G1 leads to the activation of the p53 transcription factor, which binds to a regulatory element in the p21 Cdk inhibitor gene and induces its expression. As p21 accumulates, it binds to and inactivates cyclin D-Cdk4 (or Cdk6) and cyclin E-Cdk2 complexes, thus blocking progression through G1 and into S phase. This allows the cell time to repair DNA damage before starting DNA replication.

tion to breast cancer; diseased family members have a germ line mutation in the *p53* gene, and both copies of the *p53* gene are mutated in their tumors. Wild-type p53 is a sequence-specific DNA-binding protein that can activate the transcription of some genes and repress others. This property is believed to account for its role in controlling normal cell proliferation. The mutant p53s have increased stability and therefore accumulate to higher than normal levels. Moreover, although mutant p53 lacks transcriptional regulatory activity, it can suppress wild-type p53 activity by sequestering the wild-type p53 in a multimeric complex, thus acting in a dominant-negative fashion.

Mice lacking p53 develop normally, indicating that p53 is not essential for orderly development or for cell division. However, such animals develop tumors at an early age. A striking feature of cells lacking p53 is that they fail to arrest in the G1 phase of the cycle in response to DNA damage, suggesting that p53 functions in a DNA damage control system (Figure 4.25). The absence of this checkpoint may result in genetic instability and the accumulation of mutations, leading to oncogenesis. A few target genes whose transcription is believed to be part of this damage control function have been identified. One of these encodes a 21 kDa protein inhibitor of the cyclin-Cdk complexes that are required for the

G1/S transition. The induction of this potent inhibitor could explain how p53 induces growth arrest in G1. Thus, radiation-induced G1 arrest requires the presence of wild-type p53 and results in the accumulation of the 21 kDa inhibitor, but irradiation of cells lacking functional p53 fails to induce the inhibitor or G1 arrest. If the DNA damage is too great to repair, cells undergo apoptosis, leading to cell death. This is a specialized means of eliminating cells, in this case, cells with potentially deleterious mutations. Damage-induced apoptosis requires wild-type p53 function, and thus tumor cells lacking p53 cannot activate this apoptotic pathway. This is of practical significance because anticancer drugs that induce DNA damage fail to induce apoptosis in tumor cells lacking p53.

c. WT-1

The loss of *WT-1* gene function is implicated in the etiology of Wilms tumor, a childhood kidney tumor. A putative growth suppressor gene was mapped to chromosome 11p13, and the *WT-1* gene was cloned from this locus. As expected, the *WT-1* gene was mutated in individuals with hereditary Wilms tumor. The nucleotide sequence of *WT-1* predicted that it encoded a zinc finger–containing protein. WT1 actually contains four zinc fingers in a C terminal domain, and this region bears striking similarity to the DNA-binding domain of Egr1, a protein induced early after exposure of resting cells to serum factors. WT1 indeed binds DNA with a sequence specificity very similar to that of Egr1. However, whereas Egr1 induces transcription when its response element is placed upstream of a basal promoter linked to a reporter gene WT1 represses transcription of the same reporter gene and antagonizes Egr1 induction. Moreover, WT1 also represses transcription of reporter genes driven by natural promoters, such as those for *EGR*-1, *IGF*-2, and *PDGF*-A chain. This implies that the normal role of WT1 is to suppress the expression of genes whose products are needed for the continued growth of the metanephric blastema cells in the developing kidney, thereby allowing them to differentiate into components of the nephron. In the absence of functional WT1, sustained expression of such genes precludes differentiation, and cells continue to grow, ultimately resulting in tumor formation.

d. Neurofibromin

In patients with susceptibility to Type 1 neurofibromatosis, the *NF-1* gene carries a germ line mutation. When this gene was cloned, the predicted product, called neurofibromin, proved to be related in its central region to the catalytic domain of the GTPase-activating protein GAP120 that acts on Ras. This immediately suggested that neurofibromin has Ras GAP activity, as indeed it does (Figure 4.17). Malignant Schwannomas (tumors derived from Schwann cells) arising in Type 1 neurofibromatosis patients lack neurofibromin function and have increased levels of Ras-GTP complexes, even though these cells also contain wild-type GAP120. This implies that neurofibromin plays a critical role in the negative regulation of Ras function in Schwann cells and that these cells are transformed through activation of the Ras signaling pathway. The inactivation of neurofibromin may also be involved in sporadic tumors because certain human melanoma and neuroblastoma cell lines show loss of neurofibromin protein.

e. Growth Suppressor Genes and Colon Cancer

There are likely to be many cellular genes encoding growth suppressors. Their inactivation in tumors can lead to their identification through loss-of-heterozygosity analysis. Once these genes are known, studying them often reveals fundamental information about normal growth control. A few genes associated with colon cancer illustrate this point (Figure 4.26a). The *DCC* gene, which is inactivated in a fraction of human colon carcinomas, has a structure that predicts a cell surface receptor protein, with motifs in its extracellular domain resembling those characteristic of cell adhesion molecules, such as N-CAM. The DCC protein is present on the surface of differentiated cells of the intestine, consistent with a role for DCC in intestinal epithelial cell differentiation. Although we do not know whether DCC functions in cell-cell interactions, the *APC* (adenomatous polyposis coli) gene product interacts with α- and β-catenins, cytoplasmic subunits of the cadherin family of calcium-regulated receptors, that are involved in cell-cell interactions through formation of adherens junctions. β-catenin is related to a *Drosophila* gene called *armadillo* that plays a role in defining morphological segments during embryogenesis and is required for *wingless,* the *Drosophila* homologue of Wnt1, to generate a positive signal. In mammals, Wnt1 stimulates the expression of catenins and cell adhesion, suggesting that it interacts with this system. One possible interpretation of these data is that the APC protein, which is cytoplasmic, is required for a signal pathway that tells the cell it is in contact with surrounding cells and leads to growth inhibition. Thus, the loss of APC function may prevent the normal growth inhibitory signal from affecting surrounding cells in the colon.

Mutations in the *HNPCC* gene on chromosome 2p16 have been detected in families with heritable nonpolyposis colon carcinomas and may account for more than 15 percent of all human colon carcinoma. This gene encodes a homologue of the bacterial MutS and the budding yeast MSH proteins, which are required for the repair of mismatches in DNA. A second *HNPCC* gene on chromosome 3p22 encodes a homologue of the bacterial MutL and the budding yeast MLH1 proteins. The loss of critical repair functions in colonic epithelial cells, which are exposed to continual mutagenic insults, might be expected to lead to a high frequency of mutations (Figure 4.26b). The finding of frequent alterations in microsatellite sequences in tumors from HNPCC patients is consistent with this view.

The principle underlying the inherited predispositions to HNPCC differs from that for genes like retinoblastoma because the mutation is not in a growth suppressor gene. However, the loss of a DNA mismatch repair system will result in an increased mutation frequency, and this in turn could lead to the mutational inactivation growth suppressor genes and mutational activation of protooncogenes. Perhaps mutations in other DNA repair genes will prove to be the cause of other heritable predispositions to cancer. For instance, in the case of Bloom syndrome, where multiple different cancers are found, there is an increased rate of recombination and sister chromatid exchange, which may lead to the unmasking of recessive mutations. The *p53* gene can also be placed in this category because its product is required for arrest in G1 in response to DNA damage preliminary to repair (Figure 4.26b).

4.5 Cooperation Between Genetic Events in Tumorigenesis

As mentioned, tumorigenesis in humans involves multiple steps. Yet many tumors contain oncogenes that can transform laboratory cell lines into a malignant state in a single step. The nature of the tester cell lines can apparently explain this paradox. When tumor oncogenes are assayed by transfection into primary cells taken straight from an animal rather than into well-established, continuously propagated laboratory cell lines, they generally have no detectable effect. Only when one uses particular combinations of oncogenes are primary cells con-

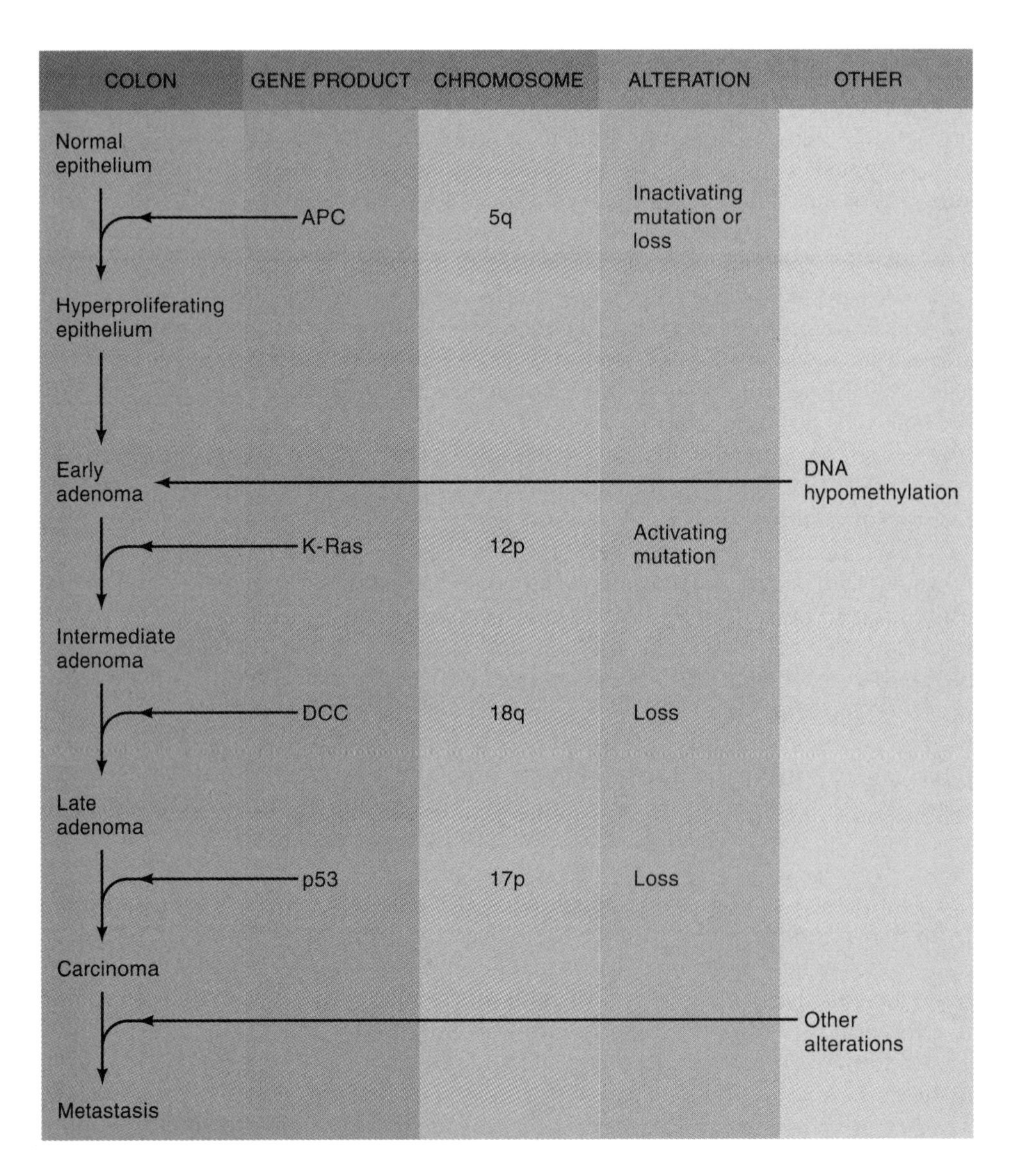

Figure 4.26
Genetic changes leading to colon carcinoma. (a) Genetic model of colorectal tumorigenesis. The genetic changes that accumulate during the progression of a normal colonic epithelial cell into a full-blown metastatic colon carcinoma cell are depicted. APC, DCC, and p53 are the products of growth suppressor genes, located at the indicated chromosomal positions. The observed changes commonly occur in the indicated order, but this is not absolutely fixed. The changes occur during the course of many years. (b) Responses of normal and mutated colon cells to DNA damage. A normal colonic cell undergoes cell cycle arrest in response to DNA damage, and this leads either to repair of the damaged

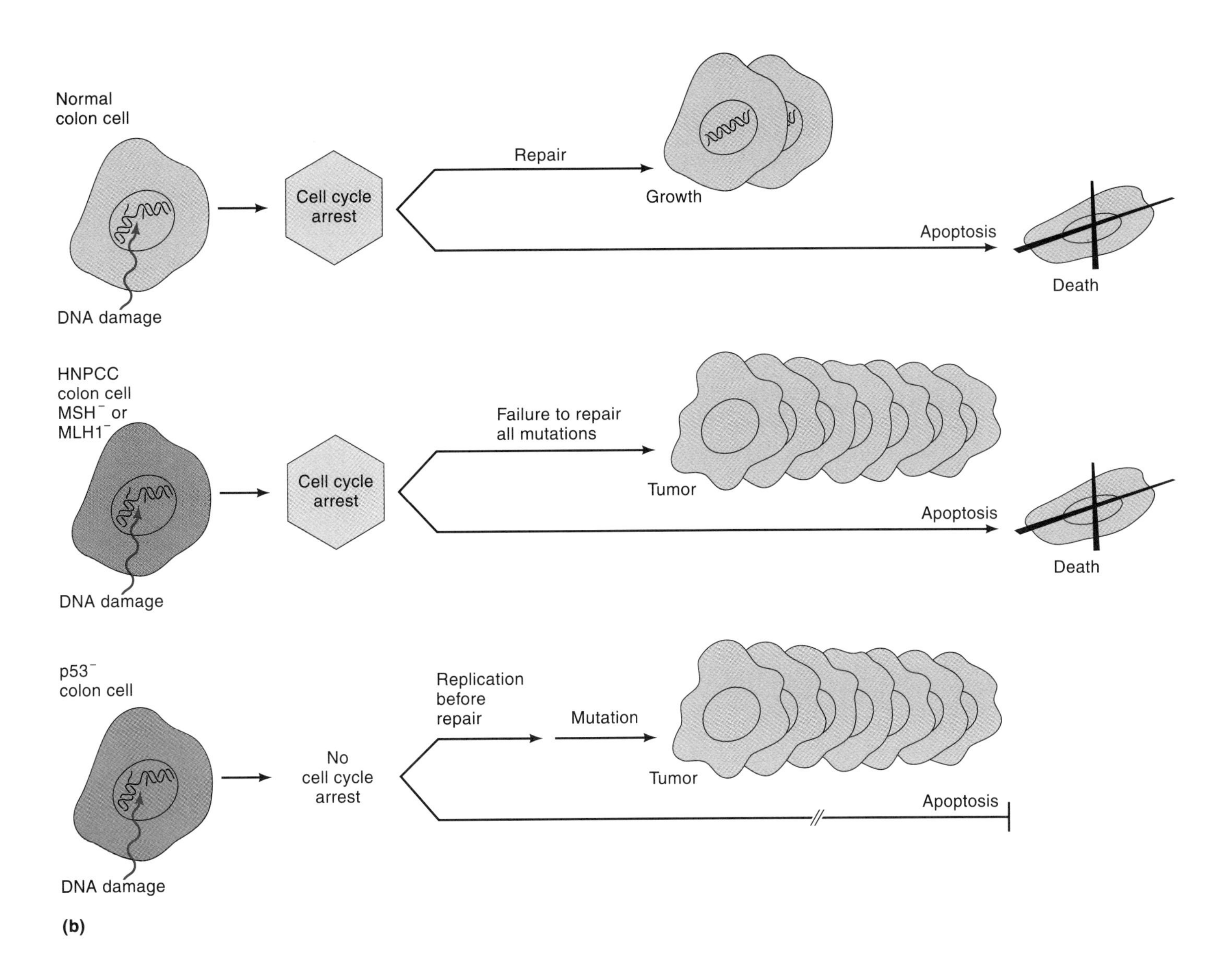

DNA and renewed growth or to apoptosis (programmed cell death). In an HNPCC colonic cell, in which both copies of the *MSH* or *MLH1* DNA repair genes are inactivated, not all DNA damage can be repaired even though there is damage-induced cell cycle arrest and apoptosis. Eventually, some cells replicate their damaged DNA, thus propagating mutations, which ultimately leads to tumor formation. In a p53-deficient colonic cell, DNA damage does not induce cell cycle arrest and the damaged DNA is replicated, which leads to propagation of mutations. In addition, there is a defect in the apoptotic response to damaged DNA, which also requires p53.

verted into tumor cells. Thus, certain combinations of oncogenes can transform primary rodent fibroblasts or kidney cells in culture, whereas individually the same oncogenes are impotent.

The specific combinations of oncogenes that are able to cooperate led to the concept that nuclear oncoproteins cooperate best with cytoplasmic oncoproteins (Table 4.9), although this is by no means a hard-and-fast rule. For instance, the nuclear Myc protein cooperates with the cytoplasmic membrane–associated Ras protein to transform rat embryo fibroblasts. Based largely on studies in rodent adherent cells, the "nuclear" category includes Myc, N-Myc, L-Myc, Fos, Jun, adenovirus E1A, polyomavirus and SV40 large-T proteins, papillomavirus E7, and HTLV-I Tax. The "cytoplasmic" category includes H-Ras, K-Ras, N-Ras, Src (and in avian bone marrow cells, other protein-tyrosine kinases as well), and polyomavirus middle-T antigen. The nuclear/cytoplasmic classification reflects the phenotypes induced by the two types of oncoprotein as well as their location. Although there are exceptions, expression of the nuclear oncoproteins by themselves does not generally alter cell morphology or growth factor or anchorage requirements, but it may "immortalize" cells and prevent normal senescence. Immortalization may be important, considering that even without any cell death, a tumor needs 30 cell doublings to reach a mass of 1 kg, which is approaching the normal limit in the number of doublings before a cell senesces. Conversely, the cytoplasmic oncoproteins can reduce growth factor requirements, induce cell shape changes, and lead to anchorage-independent growth; but they do not immortalize cells.

a. Oncogenes in Transgenic Mice

Compelling evidence for oncogene cooperation comes from studies with transgenic mice. Transgenic mouse strains harboring individual oncogenes expressed from tissue-specific or ubiquitous promoters generally show a strongly enhanced level of oncogenesis, but depending on the system, this is often delayed into adulthood. Moreover, the tumors that arise are monoclonal, indicating that additional events are required for tumorigenesis. In contrast, when two appropriate strains of oncogene-bearing mice are crossed, the incidence of tumors is often greatly accelerated, demonstrating that the oncogenes expressed in these mice can cooperate when expressed in the F1 generation.

Table 4.9 ***Cooperation Between Oncogenes***

Group I Oncogene	Group II Oncogene
Confers anchorage independence for growth	Immortalizes or establishes cells as permanent lines
Alters cell morphology	
Increases growth factor and protease secretion	Increases sensitivity to growth factors
Encodes a cytoplasmic protein	Encodes a nuclear protein
Activated by structural mutations	Activated by regularory mutations
Examples:	Examples:
N-*ras*	*myc*
K-*ras*	*myb*
H-*ras*	*ski*
src	*fos*
neu	Polyoma virus large-T
Polyoma virus middle-T	Adenovirus E1A

The analyses of transgenic mouse strains carrying a single oncogene are instructive because they reveal what sort of cooperating oncogenes can be selected for *in vivo,* in contrast to the combinations of oncogenes that work in cell culture experiments. The *in vivo* system has the added advantage that it is sometimes possible to target expression of an oncogene to a particular cell type not available in a laboratory cell line.

The relatively long latency observed for tumorigenesis in mice expressing a single oncogene can be explained by the necessity for additional events before a frank tumor is evident. Examination of the subsequent events in tumors from oncogene-bearing transgenic mice confirms conclusions deduced from *in vitro* studies and identifies new oncogenes. For instance, B cell lymphomas occurring in mice carrying a *myc* transgene under the control of the Eμ immunoglobulin heavy chain enhancer (Eμ-*myc* mice), which is expressed largely in B cells, have activated *ras* genes. However, when Eμ-*myc* mice are infected with Moloney murine leukemia virus, there is an accelerated incidence of B cell lymphoma, and the *pim-1* oncogene is commonly activated by a retroviral insertion in the leukemic cell. Interestingly, in mice expressing the *pim-1* gene from the Eμ enhancer, the c-*myc* and N-*myc* genes are the ones commonly activated in lymphomas that arise following retroviral infection. This system is reciprocal; in both cases, there is activation of a cytoplasmic protein-serine kinase, Pim1, and a nuclear transcription factor, c-Myc or N-Myc. However, tumors from infected Eμ-*myc* and Eμ-*pim-1* mice also reveal retroviral activation of previously unidentified oncogenes. One such gene, *bmi-1,* is activated at a reasonably high frequency in Eμ-*myc* mice. Bmi1 appears to be a chromatin-binding protein and is related to a family of proteins implicated in the regulation of gene expression; although its function is not known, Bmi1 might be expected to lie on the same pathway as Pim1.

b. The Mechanisms of Cooperation

In some cases, cooperation may reflect the cooperating oncogene's ability to help overcome the influence of normal surrounding cells. There are several striking *in vivo* and *in vitro* examples where the suppressive effects of an excess of normal cells on the growth and progression of cells expressing oncogenes have been noted. The mechanism of suppression is not known, but it could involve the secretion of growth inhibitory cytokines by the normal cells, inhibitory signals delivered via cell-cell contact, or the direct intercellular transmission of a negative regulator. Another role for one of the two oncogenes in a cooperating pair is to block the apoptotic response induced by the other oncogene. For instance, the adenovirus E1A transforming protein elicits many aspects of the transformed phenotype but also induces an apoptotic response through p53, and one role of the cooperating E1B-transforming protein is to block the E1A-induced apoptotic response.

At the single cell level, cooperating oncogenes probably constitutively activate distinct mitogenic signal pathways; moreover, the cooperating oncogenes may very well affect different signal transduction cascades that converge at the level of transcription to effect either induction and/or repression of gene expression. This would predict that oncoproteins lying on the same signal pathway would fail to cooperate, as is true for Ras and Src. Even in cases where the ultimate target of cooperation is nuclear, cytoplasmic oncoproteins may be more effective than nuclear oncoprotein, because signals initiated in the cytoplasm can branch and lead to two or more independent nuclear signals. Thus, at least three nuclear oncogenes might be required to achieve the same end result as one nuclear and one cytoplasmic oncoprotein. In addition, other aspects of the transformed phenotype, such as cell shape, invasiveness, and anchorage-independent growth, may require signals generated in the cytoplasm. Indeed, because most if not all cytoplasmic oncoproteins lead to changes in protein phosphorylation, cytoplasmic phenotypes may be directly induced by cytoplasmic phosphorylation events. Cytoplasmic oncoproteins may also be very effective in triggering key cell cycle transitions, particularly the G2 to M transition, which clearly involves changes in protein phosphorylation. In general, two cytoplasmic oncoproteins do not cooperate. Perhaps this is because of feedback control between pathways, where stimulation of one pathway would inhibit the other oncoprotein-activated pathway or because most of the cytoplasmic oncoproteins activate the same pathway or set of pathways. The growth suppressor proteins may normally act as negative regulators of the pathways that are activated by oncoproteins. For instance, neurofibromin is a negative regulator of Ras. Normally, these growth suppressor proteins are powerful enough to interdict the stimulatory signal provided by the oncoproteins, and thus tumor cells are strongly selected for inactivation of genes encoding growth suppressor proteins.

It is not clear how many of the multiple events needed to generate a cancer can be explained by the activation of oncogenes and the inactivation of growth suppressor genes that are currently known. Additional genetic events are needed for tumor angiogenesis or metastasis, and genes that may account for some of these steps are being studied. Lesions in apoptotic pathways leading to cell death appear to be important in tumorigenesis, and the loss of p53 function may be critical in this regard (Figure 4.26b). Ultimately, we are unlikely to understand fully how the multiple genetic changes cooperate until we know more about the different cellular signal pathways, their mutual interactions, and their nuclear targets.

4.6 Tumorigenesis in Humans

The multistep nature of tumorigenesis is well documented in humans. The incidence of human cancer follows a curve related to the fifth power of age, suggesting there could on average be as many as five stages before the disease is manifest. The exact nature of the multiple steps involved may differ in every cancer, but many of them are clearly genetic events. In colon and esophageal carcinomas, where a clear disease progression can be found and analyzed by biopsy, multiple genetic changes of increasing number can be detected during tumor progression (Figure 4.26a). Each type of cancer has genetic changes that are characteristic, although some events, like the inactivation of p53, are found in many types of cancer. About 20 percent of human tumors contain mutant genes detectable as oncogenes in transfection assays. This relatively low proportion could reflect limitations in the assay system, which is particularly sensitive to oncogenes that can transform fibroblasts and could miss other types of oncogenes. Nevertheless, the loss of growth suppressor genes may turn out to be more important than the generation of dominantly acting oncogenes. We do not know if genetic changes have to occur in a required order to generate a given tumor type. There is some evidence for a preferred order in esophageal cancer, but the accumulation of a set of genetic changes that in combination confer the tumor phenotype may be more important. Each genetic change must give that cell a selective growth advantage and allow the acquisition of an additional mutation, which in turn provides an additional growth advantage. This stepwise selection process continues until a full-blown tumor cell arises.

The types of mutation observed in human cancer cells vary widely. Most tumors have an aneuploid chromosome complement, and some of the mutations, such as those inactivating p53, MSH, and MLH1, may affect DNA damage repair systems, thus increasing the frequency of chromosomal aberrations and mutations responsible for tumor progression. Chromosomal telomeres, which are made up of a repeated DNA sequence, shorten with age, and shortening below the limit required for chromosomal stability may be one cause of cellular senescence. Telomerase, the enzyme that lengthens telomeres, is normally present at very low levels in somatic cells but is readily detectable in tumor cells. The expression of telomerase may reverse telomere shortening and contribute to the immortal phenotype of tumor cells. How telomerase expression is increased in tumor cells is not clear, but this could be a result of mutation.

Increasingly, characteristic chromosomal translocations are being found in human tumors. In leukemias derived from B and T lymphocytes, translocations generally arise as a result of an abortive rearrangement of one of the immunoglobulin or T cell receptor loci bringing the transcriptional enhancer element of the immunoglobulin or T cell receptor gene close to a protooncogene, thereby causing overexpression or ectopic expression. In the loss of growth suppressor genes, nondisjunction of sister chromatids at mitosis, mitotic recombination, and gene conversion have all been observed. Carcinogens are mutagens, and many carcinogens cause point mutations; one example is ethylnitrosourea, which causes mutation of codon 12 in the c-H-*ras* gene. However, small and large deletions are also frequently found. The amplification of protooncogenes in tumors is very common and occurs through aberrant replication and recombination. Many types of mutation occur during DNA replication, and even mutations caused by carcinogens, which can occur in quiescent cells, can be fixed only by replication. Thus, the genetic events that result in continuous growth are very important in allowing the accumulation of additional tumorigenic mutations.

4.7 Viruses and Human Cancer

Human cancer does not, in general, have a viral etiology. There are, however, four instances where viruses may be involved. Although there are no acutely transforming human retroviruses, adult human T cell leukemia is caused by a retrovirus: human T cell leukemia virus I (HTLV-I). This virus lacks an oncogene but has an unusual mode of transcriptional regulation compared to most other retroviruses. Besides the three normal retroviral structural genes, HTLV-I encodes several additional regulatory proteins. One of these, Tax, acts positively on the virus's own transcriptional control element. Tax may also act, fortuitously, on certain cellular protooncogenes and activate their transcription. This and the fact that the viral transcriptional signals are efficiently used only in T cells can explain the specificity of the virus in causing T cell lymphomas. HTLV-II, a virus closely related to HTLV-I, is implicated as the causal agent of hairy cell leukemia. HIV, the causal agent of AIDS, is a retrovirus related to HTLV-I and II. HIV also encodes a *trans*-acting transcription factor, Tat; but Tat expression leads ultimately to the death of infected T cells and hence immune deficiency, rather than transformation.

Epstein-Barr virus (EBV), a herpesvirus, is implicated in Burkitt lymphoma and nasopharyngeal carcinoma in some areas of the world. For several reasons, however, the role the virus plays seems likely to be ancillary rather than causal. First, some Burkitt lymphoma cell lines, in addition to translocations of the c-*myc* gene, contain dominant oncogenes that are detectable by transfection but do not correspond to EBV sequences. Second, although some herpesviruses are tumorigenic (e.g., Herpes saimiri causes tumors in monkeys), EBV infection of its normal host cell, the B cell, does not cause transformation. Instead, EBV infection provides a permanent growth stimulus, which may increase the chances of a second event such as translocation of the c-*myc* gene to an immunoglobulin locus, activation of a protooncogene, or inactivation of a growth suppressor gene.

Infection with hepatitis B virus (HBV) leads to an increased risk of hepatocellular carcinoma. HBV itself, however, lacks an oncogene and does not transform cells in culture. HBV DNA sequences can be found integrated in some hepatocellular carcinomas; but unlike avian leukosis virus, which activates the c-*myc* gene via insertional mutagenesis, no common sites of integration have been identified. In one case, HBV integration yielded a mutated and consequently nondegradable form of cyclin A2, which may have played a role in tumorigenesis. In general, however, the mechanism of HBV involvement remains obscure.

The most significant virus involvement in human cancer is in cervical carcinoma. More than 80 percent of cervical carcinomas have sequences from one or another strain of human papillomavirus (HPV), and infection by certain strains of HPV engenders a high risk of cervical carcinoma. Because papillomaviruses are recognized DNA tumor viruses, one is tempted to speculate that their transforming genes are a causal element. The HPV sequences present in cervical carcinoma almost always include viral sequences encoding the E6 and E7 proteins, which inactivate p53 and RB, respectively, and can transform cells in culture (see Chapter 1).

4.8 Conclusions and Prospects

The accumulation of mutations that activate protooncogenes into oncogenes and inactivate growth suppressor genes definitely plays a fundamental role in human cancer. For many oncoproteins and growth suppressor proteins, we are beginning to glean an understanding of the molecular events associated with malignant transformation. How this will help us prevent and treat the disease is an important question. In terms of prevention, an obvious lesson is the avoidance of exposure to mutagens in the environment. Also, where inherited predisposition to cancer is suspected, genetic screening can determine whether an individual is a carrier of the mutant growth suppressor gene. With respect to treatment, the first applications of our knowledge of oncogenes will be for diagnostic purposes. For instance, reagents that can detect activated *ras* genes and proteins are available; therefore, we can determine whether a particular tumor contains an activated *ras* gene. We may then correlate this information with the effectiveness of particular chemotherapeutic regimes. In the long run, we may be able to devise specific drugs to combat specific oncogenes. Again the *ras* gene is leading the way; drugs that block the essential C terminal farnesylation of Ras proteins have been developed, and these can reverse the phenotype of cells transformed by an oncogenic *ras*. Drugs that inhibit the activity of oncogenic protein-tyrosine kinases are also being developed.

Enormous progress toward an understanding of cancer has been made since the discovery of the *src* gene in 1975. This knowledge is beginning to be used to provide better cancer diagnosis and therapy.

Further Reading

H. Varmus and A. J. Levine (eds.). 1983. Readings in Tumor Virology. Cold Spring Harbor Laboratory Press, Cold Spring Harbor, New York.

R. A. Weinberg (ed.). 1989. Oncogenes and the Molecular Origins of Cancer. Cold Spring Habor Laboraory Press, Cold Spring Harbor, New York.

G. M. Cooper. 1990. Oncogenes. Jones and Bartlett, Boston.

J. M. Bishop. 1991. Molecular Themes in Oncogenesis. *Cell 64 225–248.*

T. Hunter. 1991. Cooperation Between Oncogenes. *Cell* 64 249-270.

E. Solomon, J. Borrow, and A. D. Goddard. 1991. Chromosome Aberrations and Cancers. *Science* 254 1153–1161.

G. M. Cooper. 1992. Elements of Human Cancer. Jones and Bartlett, Boston.

A. Murray and T. Hunt. 1993. The Cell Cycle. W. H. Freeman, New York.

H. Varmus. 1993. Genes and the Biology of Cancer. W. H. Freeman, New York.

L. Lanfrancome, G. Pelicci, and P. G. Pelicci. 1994. Cancer Genetics. *Curr. Opin. Genet. Develop.* 4 109–119.

4.2a

S. A. Aaronson. 1991. Growth Factors and Cancer. *Science* 254 1146–1153.

4.2b

L. C. Cantley, K. R. Auger, C. Carpenter, B. Duckworth, A. Graziani, R. Kapeller, and S. J. M. Soltoff. 1991. Oncogenes and Signal Transduction. *Cell* 64 225–248.

J. Schlessinger and A. Ullrich. 1992. Growth Factor Signaling by Receptor Tyrosine Kinases. *Neuron* 9 383–391.

T. Pawson and J. Schlessinger. 1993. SH2 and SH3 Domains. *Curr. Biol.* 3 434–442.

G. A. Rodrigues and M. Park. 1994. Oncogenic Activation of Tyrosine Kinases. *Curr. Opin. Genet. Develop.* 4 15–24.

P. van der Geer, T. Hunter, and R. A. Lindberg. 1994. Receptor Protein-Tyrosine Kinases and Their Signal Transduction Pathways. *Annu. Rev. Cell Biol.* 10 251–337.

T. Pawson. 1995. Protein Modules and Signalling Networks. *Nature* 373 573–580.

4.2e

Y. Nishizuka. 1992. Intracellular Signalling by Hydrolysis of Phospholipids and Activation of Protein Kinase C. *Science* 258 607–614.

4.2g

M. L. Cleary. 1991. Oncogenic Conversion of Transcription Factors by Chromosomal Translocations. *Cell* 66 619–622.

B. Lewin. 1991. Oncogenic Transcription Factors. *Cell* 64 303–312.

4.2h

T. Hunter and J. Pines. 1991. Cyclins and Cancer. *Cell* 66 171–174.

T. Hunter and J. Pines. 1994. Cyclins and Cancer II: Cyclin D and CDK Inhibitors Come of Age. *Cell* 79 573–582.

4.2i

E. A. Harrington, A. Fanidi, and G. I. Evan. 1994. Oncogenes and Cell Death. *Curr. Opin. Genet. Develop.* 4 120–129.

4.4

C. J. Marshall. 1991. Tumor Suppressor Genes. *Cell* 64 313–326.

R. A. Weinberg. 1991. Tumor Suppressors. *Science* 254 1138–1146.

P. W. Hinds and R. A. Weinberg. 1994. Tumor Suppressor Genes. *Curr. Opin. Genet. Develop.* 4 135–141.

4.5

J. M. Adams and S. Cory. 1991. Transgenic Models of Tumor Development. *Science* 254 1161–1167.

4.7

H. zur Hausen. 1991. Viruses and Human Cancers. *Science* 254 1167–1173.

4.8

J. F. Hancock. 1993. Anti-Ras Drugs Come of Age. *Curr. Biol.* 3 770–772.

CHAPTER 5

Mapping Markers and Genes in the Human Genome

RAY WHITE

5.1 Introduction

Recombinant DNA technology revolutionized human genetics and made possible fruitful analyses of the more than 3000 disorders attributable to familial inheritance of altered genes. The resulting molecular analysis of human mutations is revealing the underlying mechanisms of complex biological systems. However, despite remarkable successes, most of the mutant genes causing this wide spectrum of human diseases still await cloning and analysis. If mutant genes that predispose members of some families to psychiatric disorders such as schizophrenia and manic-depressive illness could be identified and cloned, for example, they would surely offer insights into the metabolic events that influence emotions and intellectual processes.

But new information about basic mechanisms in human physiology is already coming from studies of an increasing variety of cloned genes, as illustrated in various chapters in this book. For example, molecular analysis of the hemoglobinopathies, including mutations responsible for producing defective hemoglobin in thalassemia, has provided key contributions to our understanding of RNA transcription regulation and mRNA processing. Similarly, mutations in the gene encoding the receptor for low density lipoprotein (LDL) have yielded insights into the several steps in receptor-mediated endocytosis. Like mutations at the hemoglobin locus, mutations in the LDL receptor gene are identified by virtue of an inherited disease. In this case, a molecular defect leads to high cholesterol concentrations in plasma and an early onset of atherosclerosis, a syndrome known as familial hypercholesterolemia. Analysis of the mutations elucidated the processing and glycosylation of the receptor protein as part of its transport to the cell surface, the attachment of the receptor molecule to the coated pits of the cell membrane, and the specific interaction of the receptor with its ligand, the LDL particle. Fundamental studies of this genetic system laid the foundation for our nascent understanding of the genetic basis of risk for cardiovascular disease and of the importance of cholesterol and other blood lipids in atherosclerosis.

DNA markers have played, and continue to play, vital roles in the cloning of medically important genes. Successful efforts over the past few years to identify genes associated with cystic fibrosis, neurofibromatosis Types 1 and 2 (inherited disorders involving nerve tissue), X-linked muscular dystrophies, and Huntington disease (a fatal neurological syndrome) are a few cases in point.

a. The Genetics of Cancer: DNA Markers as Tools for Discovery

Two lines of evidence—the mutagenic activity of most carcinogens and numerous mutations identified in known oncogenes—indicate that carcinogenesis is primarily a genetic phenomenon (see Chapter 4). Additional support for the genetic hypothesis comes from DNA studies of rare families where some members inherit a mutation that almost invariably leads to a specific type of cancer. The genetic defects in these families are proving to be of major importance to understanding the molecular basis of cancer in all humans because the same genes may be altered by somatic (i.e., noninherited) mutation in the general population.

Figure 5.1 shows the pedigree of a family with an inherited syndrome called adenomatous polyposis coli (APC). Individuals who have inherited the family's mutant version (allele) of the *APC* gene can show a variety of phenotypic abnormalities, but the most important in terms of cancer predisposition is the appearance in the colon of multiple polyps—benign epithelial tumors called adenomas—at an early age (Figure 5.2). The polyps in family members who carry the APC mutation are precursors to colon cancer; one or more of these polyps will progress to malignancy by midlife unless the colon is removed.

Because this and other evidence indicates that the *APC* gene controls an early step in the development of colon cancer, it became a major target of cloning efforts. Success was achieved in 1991 (Section 5.6b). However, cloning of this "cancer gene" was possible only because linkage of the *APC* locus to known DNA markers initially confirmed its approximate location on chromosome 5q and opened the door to an intensive search of the critical region.

b. The Problem: Identifying and Cloning an Unknown Gene That Harbors a Deleterious Mutation

Most mutant human genes are identified solely by their phenotypic effect; that is, an inherited disease may be defined clinically without any biochemical evi-

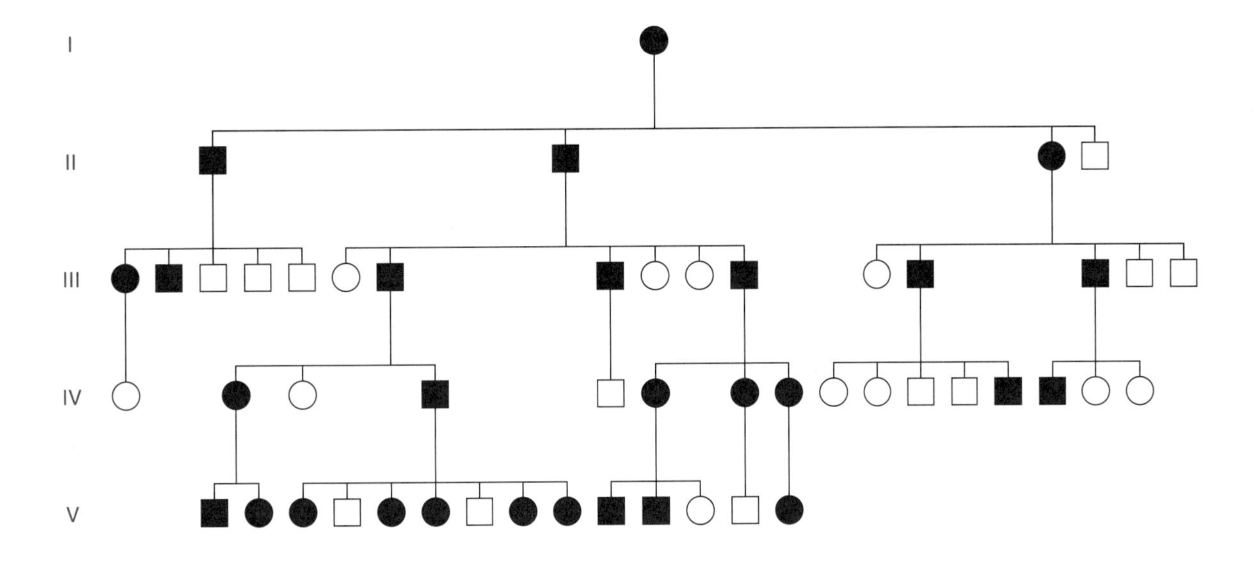

Figure 5.1
Abbreviated pedigree of a large kindred that segregates a mutant gene causing Gardner syndrome (GS), a form of familial polyposis coli. Individuals who exhibit symptoms of GS or who have died of colon cancer are indicated by filled symbols. Appearance of the disease in all generations indicates inheritance of a dominant mutation. (Adapted from D. Barker, M. McCoy, R. Weinberg, M. Goldfarb, M. Wigler, R. Burt, E. Gardner, and R. White. 1983. A Test of the Role of Two Oncogenes in Inherited Predisposition to Colon Cancer. *Mol. Biol. Med.* 1 201.)

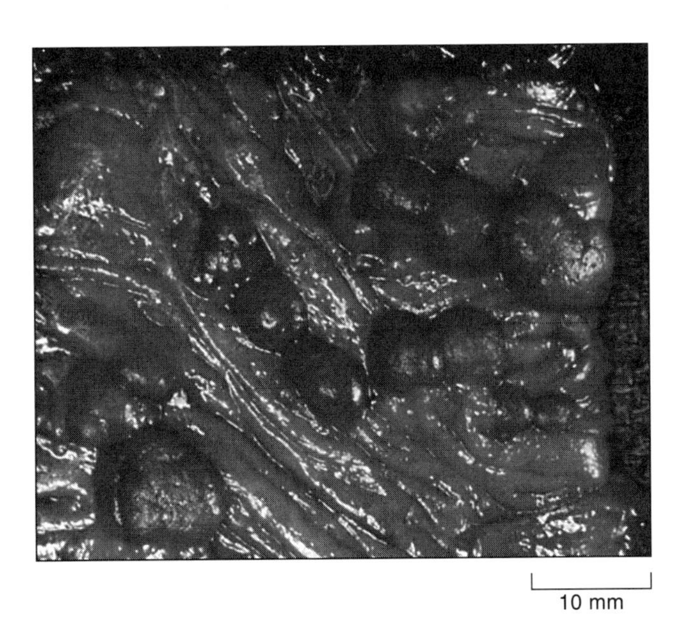

Figure 5.2
Colonic tissue from a patient with Gardner syndrome. Multiple polyps are evident. (Photograph courtesy of Dr. Randall Burt, University of Utah School of Medicine.)

dence as to the nature of the altered protein product. Genes in this category cannot be cloned through the standard approaches of molecular biology because no biochemical "handles" exist. Such handles might be an abundant mRNA or a known amino acid sequence that can be used to design hybridization probes for screening genomic libraries or for preparing an antibody that could detect an antigen that might be expressed from a cloned gene. Lacking such handles, researchers need an alternative approach that does not depend on biochemical knowledge of the gene product.

Positional cloning One answer to this problem is to combine classical family studies with modern DNA technology to map the unknown gene to a chromosomal location. The premise of this approach is that, if the gene can be mapped with sufficient precision, it can eventually be identified and cloned. Specifically, if the mutation can be mapped to a particular 1 megabase pair (mbp) region of DNA, only a few genes within that segment are likely to be expressed (i.e., produce mRNA) in affected tissue and thus represent "candidate genes" for the disease. Additional functional or genetic tests performed on these candidate genes may reveal which one is associated with the disease. Successful cloning of a number of genes that harbor mutations causing inherited diseases has verified this **positional cloning** approach for identifying unknown genes (Table 5.1).

Linkage Linkage studies in families can localize a mutant gene with a high level of precision. Genetic "markers" embedded throughout the genome allow investigators to track the inheritance of specific segments of individual parental chromosomes. The mutant allele of a gene will usually be inherited along with particular alleles of one or more markers that are present on the same chromosomal segment (Figure 5.3).

The rationale for linkage determination is based on recombination during meiosis, when homologous chromosomes interchange segments of DNA. The frequency of exchange between two points on a chromosome is approximately proportional to the physical distance separating them. Therefore, if a DNA marker is close to a particular gene on a chromosome, the two are unlikely to be separated by recombination. Consequently, if within a family, the same allele of a polymorphic marker is usually or always inherited along with a disease, the mutant gene causing that disease probably lies in the same region. Because marker alleles may sometimes also be coinherited with a disease simply by chance, statistical analyses have been developed to estimate, from the data, the likelihood of "linkage" due to physical proximity. The closer the marker is to the

Table 5.1 ***Some Genetic Disorders for Which the Responsible Genes Were Mapped with the Aid of Linkage Studies and Subsequently Identified by Positional Cloning and/or Investigation of Candidate Genes in the Region***

Disease	Chromosomal Location	Gene Name (alternate)
Acoustic neuroma (NF Type 2)	22q11.21-q13.1	*NF2* (*ACN*)
Adenomatous polyposis coli; Gardner syndrome	5q21-22	*APC* (*GS, FPC*)
Alport syndrome	Xq22	*COL4A5* (*ATS; ASLN*)
Chronic granulomatous disease	Xp21.1	*CYBB* (*CGD*)
Cystic fibrosis	7q31.2	*CFTR* (*CF*)
Duchenne and Becker muscular dystrophies	Xp21.2	*DMD* (*BMD*)
Huntington disease	4p16.3	*HD*
Hyperkalemic periodic paralysis; paramyotonia congenita	17q23.1-q25.3	*SCN4A* (*HYPP; NAC1A*)
Martin-Bell syndrome (fragile X site)	Xq27.3	*FRAXA* (*FMR1*)
Multiple endocrine neoplasia, Type 2; medullary thyroid carcinoma	10q21.1	*MEN2A* (*MEN2*)
Myotonic dystrophy	19q13.2-q13.3	*DM*
Neurofibromatosis, peripheral (Type 1)	17q11.2	*NF1* (*VRNF*)
Retinoblastoma; ?osteosarcoma	13q14.1-q14.2	*RB1*
Supravalvular aortic stenosis	7q11.2	*ELN*
von Hippel-Lindau syndrome	3p26-p25	*VHL*

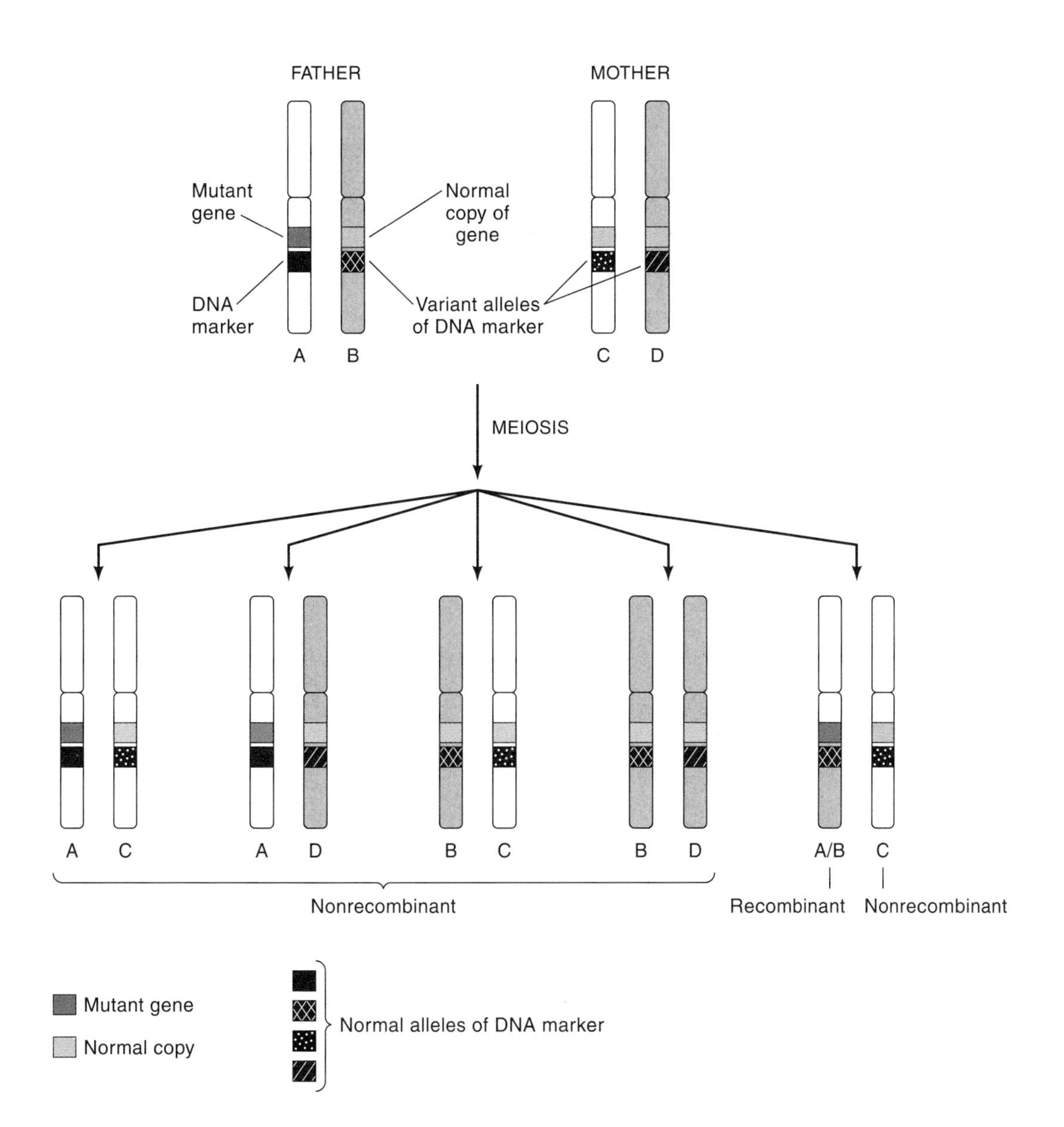

Figure 5.3
Linkage between a disease gene and a polymorphic marker. The relevant segments from two chromosomal homologues (A or B from the father, C or D from the mother) are inherited intact by four of the five children. The defective allele (dark color) has not been separated during meiosis from the marker allele (black) carried on the father's A chromosome. Normal alleles at the disease locus are indicated in light color, and patterns designate alleles at the marker locus that differ slightly from the marker carried on the father's A chromosome. The fifth child has inherited a recombinant chromosome from his father; this child carries both the disease allele and a different version of the marker than was originally on the A chromosome. The farther the marker allele lies from the disease gene on a chromosomal segment, the more frequently recombinants are likely to be observed in a family. (Adapted from R. White and J.-M. Lalouel. 1988. Chromosome Mapping with DNA Markers. *Sci. Am.* 258 44.)

gene and the larger the family sample, the more precise will be the localization by these algorithms.

For the linkage approach to be effective, however, the human genomic "map" must be covered with a large number of genetic markers. A major effort toward this end has culminated in the definition of thousands of new genetic markers, arrayed on chromosomal maps at intervals of approximately one megabase. Such maps became feasible when it was discovered that an essentially unlimited number of sequence variations in DNA—polymorphisms that have no deleterious consequences—occurs in human genomes. These polymorphisms are found not only in the genes coding for variant proteins, such as blood-type antigens, but at many "anonymous" locations in the genome.

5.2. Variations in DNA Sequence: The New Genetic Markers

a. Restriction Fragment Length Polymorphisms

Many genetic markers are defined by variations in DNA sequence that affect cleavage sites for restriction endonucleases. Thus, a specific recognition sequence may be present on one chromosome and absent at the allelic position on the homologue. The difference can be detected indirectly by determining the lengths of the DNA fragments produced by digestion of genomic DNA with a restriction endonuclease (Figure 5.4). If the site is absent, a long fragment will be produced from that region (i.e., because sites that *are* cleaved by the enzyme are farther apart on the chromosome); if the site is present, the fragment from that location will be shorter. However, digestion of complex mammalian DNA with a restriction endonuclease usually produces more than a million DNA fragments. Therefore, the restriction fragment lengths associated with a specific locus must be determined by hybridizing digested, denatured DNA to a specific labelled

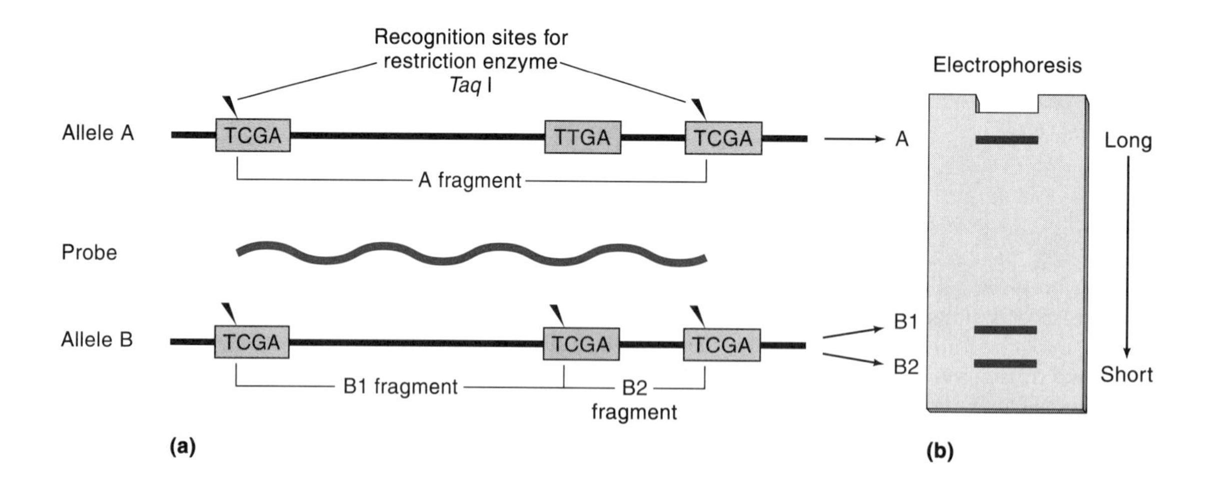

Figure 5.4
Single-site DNA polymorphism. (a) On allele A, the recognition sequence for the restriction endonuclease *TaqI* (TCGA) has been altered to TTGA at the locus homologous to the probe DNA. (b) The difference is reflected on a Southern blot. Electrophoresis has separated two shorter fragments (from the B allele) from the A fragment, which is longer because the enzyme did not cut it at the altered site and therefore it migrates more slowly through the gel. (Adapted from R. White and J.-M. Lalouel. 1988. Chromosome Mapping with DNA Markers. *Sci. Am.* 258 44.)

DNA probe after the fragments are separated by size on an electrophoretic gel and transferred to a membrane by DNA blotting (*Genes and Genomes,* section 6.1b). A useful restriction fragment length polymorphism (RFLP) marker depends on having both a probe complementary to the DNA segment and a variable restriction endonuclease site (polymorphism) within that segment.

A DNA sequence variant can be used to follow the inheritance of a specific region of a chromosome (see Figure 5.3). If only one of the two allelic segments on a homologous pair of parental chromosomes contains the middle cleavage site for the restriction endonuclease (therefore yielding shorter fragments), the length of the fragment produced by the child's DNA identifies which of the two segments was inherited from that parent. (The child's DNA will also contribute a DNA fragment containing the locus in question from the other parent's chromosome, but the respective parental contributions can usually be distinguished.) DNA markers are especially useful in genetic analyses because they are **codominant** (one allele can be detected in the presence of another). This circumstance is not a feature of all phenotypically polymorphic systems: in the ABO blood group system, for example, the O allele cannot be detected in the presence of either the A or the B allele.

Methylated cytosines (^{me}C) in human DNA are good places to look for variation because they are "hotspots" for mutation. In bacterial systems, deamination of cytosine bases in DNA is fairly frequent, and the cell has evolved defenses to reduce the mutational burden. Specific glycosylases recognize and remove the deoxyuracil produced by deamination (*Genes and Genomes,* section 2.3b). Deamination of methylcytosine produces thymidine, which cannot be distinguished from the normal nucleoside and thus is immune to this mechanism.

In mammalian DNA, most of the methylcytosine is found in CG dimer sequences. Therefore, assuming polymorphism frequencies to be proportional to mutation frequencies, we might expect restriction endonucleases that have CG dinucleotide sequences in their recognition sites to be especially effective in revealing DNA sequence polymorphisms. In fact, two such enzymes, *Msp*I and *Taq*I, are approximately twice as efficient in this respect as enzymes whose recognition sequences lack CG dinucleotides.

b. Oligonucleotide Probes

Not all base pair differences can be detected with restriction endonucleases. An alternative method for detecting DNA polymorphism takes advantage of the sensitivity to the presence of mispaired bases of the melting temperature of double-stranded DNA segments. An adenine paired with a guanine, for example, will destabilize the DNA helix in that region.

Probes as short as 12–20 nucleotides hybridize stably to single-stranded genomic DNA under appropriate conditions. However, if there is an internal base pair mismatch (if the homology is not complete), the melting temperature of the annealed product will be reduced by a few degrees. Therefore, in principle, a variant DNA sequence can be detected directly with an oligonucleotide probe because a hybridized oligonucleotide can detach from a DNA strand that differs from it by 1 bp at a lower temperature than that required to detach the oligonucleotide from a perfectly complementary DNA sequence.

The shorter the oligonucleotide, the greater the destabilizing effect of a mismatch. However, an additional complication arising from the complexity of human DNA limits the length of probes. A simple calculation indicates that a particular 12 bp sequence is expected to occur some $3 \times 10^9/4^{12}$ or 179 times in the genome. Thus, a probe that is too short will occur often and will not be specific for a single locus. A similar calculation indicates that an oligonucleotide must be at least 16 bp long to have a reasonable chance of being unique in the genome. In practice, oligonucleotides 17–19 bp long are used.

A homozygous DNA sample of normal sequence will remain hybridized at a critical melting temperature only with the normal oligonucleotide. Similarly, a homozygous variant DNA sample will remain hybridized only with the variant oligonucleotide. If a heterozygous DNA sample hybridizes with either the normal or the variant oligonucleotide, two kinds of duplex can result: one stable at the critical melting temperature and one that is unstable. Therefore, each genotype can be detected in the presence of the other. This marker system, like that based on RFLP, is codominant. Because the differences in melting temperature are small, discrimination may be difficult—depending on the particular base pair involved. Recent work suggests that results are clearer when an unlabelled probe with a one-base mismatch is added to the hybridization mixture with the labelled probe to compete for binding to the one-mismatch site.

c. Amplification of the Region

Because detection of DNA sequence variation by DNA blotting is a contest between signal and noise, a technique that permits amplification of a specific genomic region *in vitro* is of particular interest. Such is the polymerase chain reaction (PCR) (see *Genes and Genones,* Section 7). Oligonucleotide primers that flank the region to be amplified are synthesized; at each PCR cycle, the primed region will be doubled, yielding a geometric increase in its representation. In the first cycle of amplification, two strands of indeterminate length are synthesized; but on the next cycle, when priming occurs on the just synthesized strand, the new strand will have a discrete termination site defined by the first primer. In principle, 20 cycles should yield a millionfold amplification; in practice, the efficiency is somewhat reduced, but amplifications of many thousands are possible. The amplified DNA becomes a robust substrate either for detecting DNA sequence variation or for cloning the specific region.

d. Human DNA Sequence Variants as Genetic Markers

DNA sequence variation in the human genome is predominantly of two kinds: (1) change, or insertion or deletion of a single base, as illustrated in Figure 5.4, and (2) variation in the number of tandem sequence repeats at a locus (Figure 5.5). The markers derived from these two types of structural alterations differ in their variability, or heterozygosity, within the population. Only two alternative fragment lengths occur with a single base pair change because the cleavage site is either present or absent; the respective sizes of the two fragments will be the same in all members of the population. Similarly, detection of particular single base changes with oligonucleotide probes reveals only the presence or absence of an alteration. By contrast, many different restriction fragment lengths are possible at loci having the tandem repeat structure.

Single-base polymorphism How frequently must a site variation occur in the population before we classify it as a polymorphic marker? Almost any base might sport a variant in one or two people in the world. The accepted convention defining a "polymorphism" is its occurrence in at least 1 percent of the population. With this figure limiting the calculations, base pair change polymorphisms are estimated to occur every 200–500 bp in the human genome, on average. However, the range is wide, with some regions showing very little variation. The entire X chromosome seems to have an average variant frequency that is severalfold lower than that in other chromosomes; for example, the region encoding the Factor VIII protein, a coagulation factor that when defective causes hemophilia A, contains stretches of thousands of bases with no detectable variation.

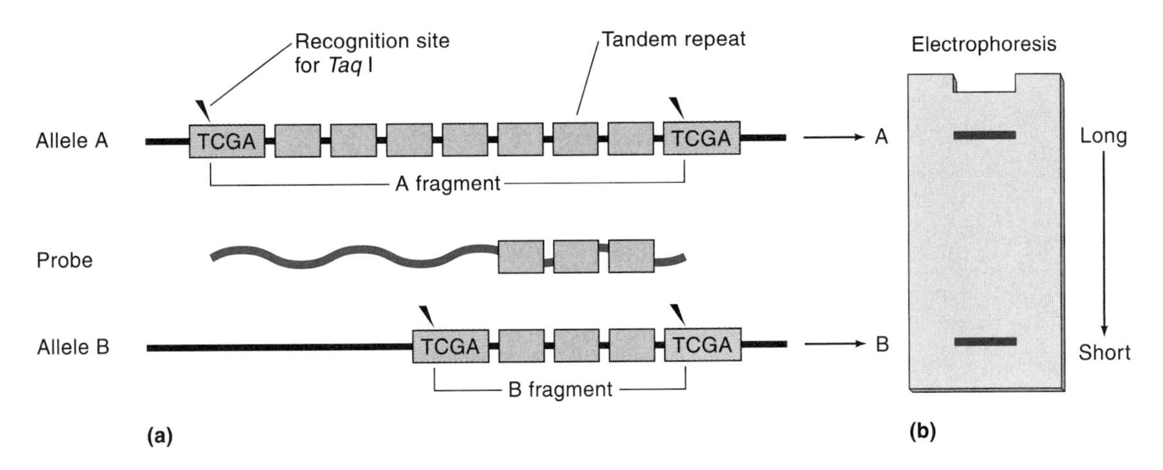

Figure 5.5
VNTR polymorphism. (a) Restriction endonucleases produce DNA fragments of different lengths, as with site polymorphisms, but in this case, the *Taq*I cutting sites are separated by a varying number of repeats of a short sequence of bases (represented here by gray boxes). A probe containing the repeat unit(s) hybridizes to the locus. (b) Autoradiography of a Southern blot reveals the difference in molecular weight between the two alleles. (Adapted from R. White and J.-M. Lalouel. 1988. Chromosome Mapping with DNA Markers. *Sci. Am.* 258 44.)

VNTR markers Variations in the number of tandem repeats of a short sequence of nucleotides at a given locus are a rich source of polymorphic markers. The first anonymous DNA marker examined in human DNA revealed many different fragment lengths in a set of unrelated individuals; almost every person tested was heterozygous, and more than 50 different fragment lengths were subsequently detected at that locus. Several similar polymorphisms were later discovered at the insulin, Harvey-*ras*, ζ-globin, and myoglobin loci. Sequencing at the insulin locus revealed a region containing more than a hundred tandem repeats of a 14 bp sequence; different alleles contain different numbers of repeats of the sequence. We now call such systems **variable number of tandem repeat** (**VNTR**) loci. As one would expect for this type of polymorphism, which involves insertion or deletion of copies of the repeat unit and thus increased or decreased distances between cleavage sites, these variants typically can be revealed with any of several restriction endonucleases.

"Families" of VNTR loci exist within the human genome; hybridization of genomic DNA with a specific VNTR sequence at reduced stringency (a temperature that allows annealing of sequences that are less than perfectly matched) often reveals a number of related but dispersed loci. It is now apparent that this method can be used for screening genomic libraries to yield new VNTR loci. The large number of highly variable loci already detected by particular probe sequences in this manner generates **genetic "fingerprints"** unique to each individual. The fingerprints have forensic as well as genetic applications.

The fact that VNTR loci are members of defined sequence families and identifiable by partial homologies to particular oligonucleotide probes has been important for developing new markers for the genomic map. Probes derived from the consensus sequences can be used to screen human genomic libraries and to identify clones that in turn can reveal polymorphisms at unique loci. With this protocol as the paradigm, many loci with multiple alleles have been identified and subsequently mapped to chromosomal regions.

Sequence-tagged site markers Another class of markers for genomic maps consists of dinucleotide, tetranucleotide, or other very short sequences that occur in tandem arrays throughout the genome, together with the unique sequences that flank each array. Such loci have come to be known as **microsatellites** or **simple sequence repeats** (SSRs). Like VNTRs, they tend to be highly polymorphic. However, unlike the polymorphisms that are detected with restriction endonucleases, microsatellite loci are identified by the unique sequences flanking their core repeats. Thus, hybridization probes for these loci are frequently relatively short, synthesized oligonucleotides and therefore do not need to be maintained as clones in living bacteria. SSR probes can be synthesized at will for mapping or other purposes from information stored in data bases. Because their sequences are known, microsatellite loci are physical markers for the genome; thus, they help bring physical and genetic maps of chromosomes into closer relationship. Nearly all DNA markers now being developed are of the SSR type.

Genotypic analysis with polymorphic SSR markers requires technology different from the DNA blotting used for RFLPs. A locus containing an SSR region is identified, for example, by probing a genomic library with a $(CA)_n$ repeat sequence. Then the sequence of the DNA on either side of the repeats in the isolated clone is determined, and unique sequences are chosen to define the primers to be synthesized for PCR amplification of genomic DNA at that locus. One primer of the pair is tagged with a radioisotope. The PCR product from the DNA of the individual being screened is electrophoresed on a polyacrylamide gel and autoradiographed. Size markers indicate the length of the amplified fragments, and the number of repeats present in each allele of the test sample can be calculated.

Heterozygosity A parent must be heterozygous at a marker locus, and the two alleles must be distinguishable if the inheritance of a particular allele is to be followed. Otherwise, it is not possible to tell which of the two homologous chromosomal regions containing the marker has been passed to a child. Whether an individual is heterozygous at a particular locus depends on chance, but chance is strongly influenced by the number of alleles present in the population as a whole. If there are 20 allelic variants in the population, many more people will be heterozygous than if only two alleles are present in the genetic pool. Thus, the difference in the frequency between single site and tandem repeat polymorphism is important to their usefulness as genetic markers.

Because markers based on single base pair changes are detectable with restriction endonucleases in only two allelic forms (the cleavage site is either present or absent), heterozygosity for any given single base pair polymorphism is likely to be found in no more than 50 percent of the population; the remaining individuals can provide no information on inheritance. The situation is somewhat more promising if several base pair changes can be identified in the same region using various enzymes; but because each substitution must usually be determined separately, the additional work is considerable.

When marker systems are derived from tandem repeat loci, the number of repeating elements within the restriction fragment of each allele will determine fragment length. Because the number of repeats can vary a great deal, there are often many different alleles in the population, and most individuals are likely to be heterozygous. This consideration makes VNTRs and SSR markers especially valuable for mapping loci associated with genetic diseases because almost all individuals in the rare and often small families transmitting a defective allele will provide linkage information.

DNA sequence variants are abundant. About 0.5 percent of the bases in DNA exhibit frequent polymorphism in human populations, so a variation is likely to occur every 200 to 500 bp. This means that the total number of frequently variant sites in the human genome should be on the order of 3×10^9 bp $\times$ 1/200 $-$ 500 bp = 1.5×10^7 to 6×10^6. From so large a pool, as many markers as desired should be detectable with currently available techniques.

5.3 Analysis of Genetic Linkage in Humans

a. Genetic Mapping

There are various reasons for constructing a human genetic map. One is the importance of positional information in confirming or denying the identity of a particular gene as the source of a phenotype. Are two genes, defined according to different operational criteria, actually different genes, or are they the same? The logic of mapping can often resolve that issue. If the genes are the same, they must map to the same chromosomal location; if they map to different places in the genome, they are separate genes.

Map location can also provide important clues to function. For example, dramatic insights into the genetics of cancer have been obtained from the observation that some oncogenes are located at or near the breakpoints of tumor-specific chromosomal translocations. This juxtaposition has implicated the cytogenetic events in the activation of the oncogenes.

The new generation of DNA markers for the human genome removes many of the difficulties inherent in earlier family studies. It is now feasible to construct genetic linkage maps of human chromosomes—maps that locate anonymous DNA markers, cloned genes of known sequence or function, and genes known only by their mutant phenotype. The maps will ultimately provide a set of refined investigative tools for localizing any gene.

b. The Linkage Approach

Linkage analysis is based on the principle that inheritance of a disease implies the inheritance of the specific chromosomal region in which the defective gene is located. One can follow the inheritance pattern of that chromosomal segment in a family either by observing the individual members with respect to disease phenotype or by identifying a particular allele of a DNA "marker" sequence embedded in the region. For mapping purposes, the region should contain at least one known genetic marker in addition to the disease gene.

As mentioned earlier, normal meiotic recombination exchanges segments of the parental chromosomes. If the breakpoint of an exchange occurs between a marker and a disease gene, it will separate the marker allele from the mutant allele responsible for the disease (Figure 5.6). It follows that if the gene locus and the marker locus are not in physical proximity, alleles originally on the same parental chromosome will often fail to be inherited together. However, if a marker and the gene both lie within the same small region of a chromosome, a recombination event between them is unlikely; the two loci are said to be **linked** because the two original alleles on a parental chromosome will almost always be inherited together. The probability of an exchange (crossing-over) occurring between the gene and the marker locus at meiosis is a function of the **recombination frequency** (or θ), a parameter expressed as a fraction or a percentage that reflects the observable incidence of crossing-over events within the interval bounded by the two loci.

The chromosomal region associated with a disease gene can carry alleles of one or several genetic markers that show an identical or nearly identical pattern of inheritance as that of the disease. Such a pattern establishes the colocation of marker(s) and mutant gene within the same segment of the same chromosome. If we know the chromosomal location of the marker, we also know the chromosomal location of the disease gene. Classically, the markers consisted of electrophoretic or antigenic variations in the products of known genes, but today markers consisting simply of variations in the sequence of DNA bases or in the number of tandem repeats of an oligonucleotide are the most widely used.

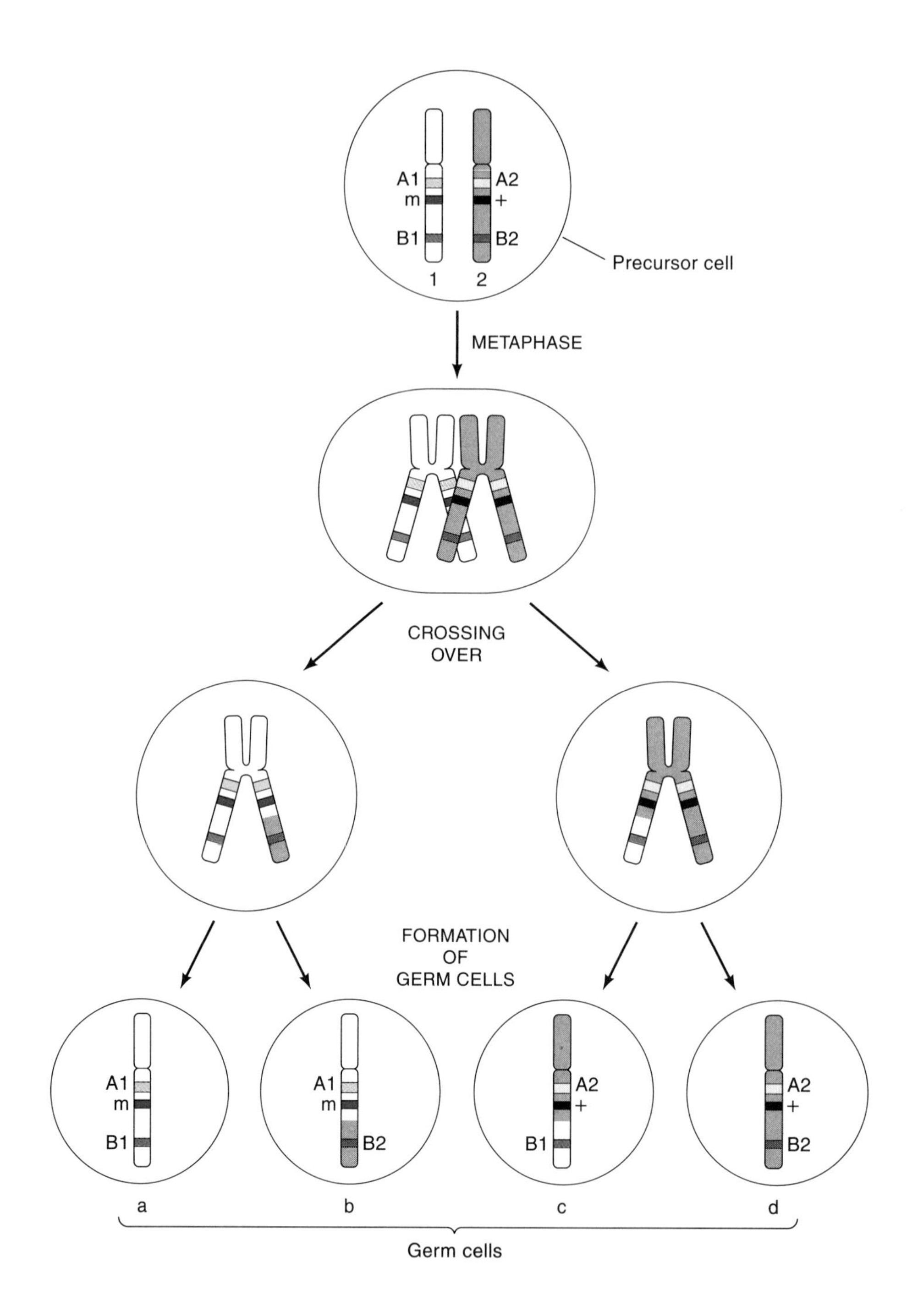

Figure 5.6
Meiosis and recombination. In this diagram, haploid spermatozoa or ova resulting from meiosis are at risk for bearing a disease-causing allele (m) of a gene whose normal allele is designated +. During the crossing-over stage, segments of DNA are exchanged between homologues of all autosomes and between parts of the X and Y chromosomes. In the chromosome depicted here, crossing-over takes place between marker loci A and B, with the result that two germ cells (a and d) carry the parental combinations of alleles, and two (b and c) contain recombinant chromosomes. In cell b, the mutant gene is still found with allele 1 at locus A but is now joined by allele 2 at locus B. A low frequency of observed recombinants between the disease gene and marker A among many meioses would indicate that the gene and this marker are closely linked. (Adapted from R. White and J.-M. Lalouel. 1988. Chromosome Mapping with DNA Markers. *Sci. Am.* 258 43.)

c. Analytical Complexities

Two issues are usually primary for analyzing human genetic linkage. What is the recombination frequency between pairs of markers (are the markers in fact linked)? What is the linear order of the gene(s) and markers? For each of these problems, **likelihood analysis** is the statistical approach most often used for estimating parameters and determining statistical significance. The need for likelihood analysis stems from two essential characteristics of the human system: (1) the phase of the parental alleles (i.e., which alleles are associated with which of the two homologous chromosomes) is very often unknown, and (2) many individuals in a sample set will be only partially informative—heterozygous at two out of three loci, for example.

Likelihood analysis provides a quantitative expression of the level of support offered by a specific data set for a hypothesis relative to some other hypothesis. Its ability to determine relative support for each hypothesis becomes important when the data set does not eliminate the alternative hypothesis. The question then becomes: which hypothesis is the more favored and by how much?

The likelihood approach requires calculation, under each hypothesis, of the probability of the data set. The analysis is based on the proposition that the relative likelihood of a hypothesis is proportional to the probability given to the observed data set. The hypothesis that gives the higher probability to the observed data is the hypothesis with the higher relative likelihood. Furthermore, that hypothesis is favored over another in direct proportion to the ratio of the calculated probabilities of the data set under each hypothesis: the **relative likelihood** of each hypothesis is proportional to the **ratios of the probabilities** that each predicts for the observed data. For example, if the probability of a certain data set would be 10^{-4} under hypothesis A and 10^{-5} under hypothesis B, we would say that the relative likelihood of hypothesis A over hypothesis B is 10.

d. Estimating Recombination Frequency

Recombination frequency can be estimated in most genetic systems from an "observed" recombination frequency that is determined by counting the recombinant chromosomes in a data set and dividing by the total. However, one aspect of human genetic data gives even this simple operation an intrinsically probabilistic flavor. As mentioned previously, the "phase" of the alleles over the parental chromosomes is often unknown (i.e., one cannot deduce with certainty which alleles are located on the same chromosome of a pair of homologues in the parents).

For example, in Figure 5.7a, we are unable to tell whether the A1 and B5 alleles are carried on the same maternal homologue or whether A1 is linked to B3. The distinction is important because the maternally derived chromosome of individual II-1, for example, whose genotype is (A1 B5), will be a recombinant if the maternal phase is (A1 B3/A2 B5) and a nonrecombinant if the maternal phase is (A1 B5/A2 B3). There is no direct means of determining whether the maternal chromosome in individual II-1 is recombinant or nonrecombinant for purposes of linkage analysis.

For the data set illustrated here, we can nevertheless calculate a probability of observing the (A1 B5) inheritance pattern as a function of recombination frequency. Note that the paternal phase is irrelevant because the father is a homozygote and does not contribute recombination information in this example. Because we do not know the maternal phase and either phase is equally likely, we assign a probability 1/2 to each phase and calculate the probability of the phase in the progeny chromosome under both possibilities. The probability of

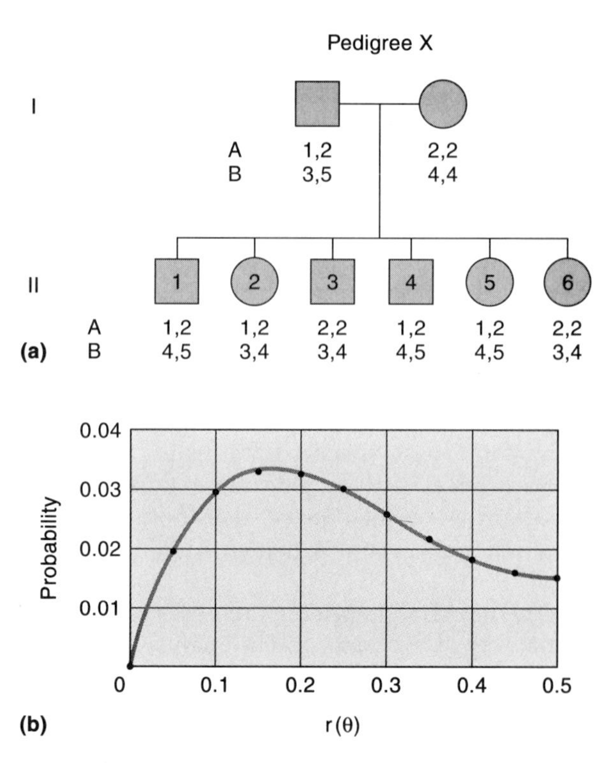

Figure 5.7
(a) Two-generation pedigree showing genotypes for two markers (A and B) under each symbol. The alleles are, for A, 1, and 2 and for B, 3, 4, and 5. The phase of the paternal chromosomes is unknown. If DNA had been available from the father's parents, the phase of his chromosomes would be known, and linkage analysis at the marker loci would be more straightforward. (b) Likelihood values for paternal recombination frequency. P is the probability of obtaining the observed data in generation II, assuming values for r, the recombination frequency, between 0 and 0.5.

the data as a function of r (recombination frequency) is proportional to $1/2(1 - r)r^5 + 1/2r(1 - r)^5$.

Therefore, we can obtain an estimate of recombination frequency for the family in Figure 5.7a under the principle of likelihood analysis by repeatedly determining the recombination frequency giving the highest probability to the observed data. Figure 5.7b shows the results of such a calculation for values of r between 0 and 0.5. The value of r that gives the highest probability at this level of resolution is 0.15; this value of r is called the **maximum likelihood** estimate because it is the most likely value of r indicated by the data.

With the small sample sizes characteristic of human genetic studies, a possibility exists that cosegregation of alleles from two loci may be due to chance alone. The relative likelihood that an observed inheritance pattern is due to linkage, as opposed to chance, is obtained by calculating the probability of observing the data given the value of $r = r_{max}$ (the maximum likelihood value of r) versus the probability of observing the data at a value of r = 0.5, the value of r for unlinked markers. The ratio of these probabilities provides the relative likelihood of the two hypotheses.

For the family data shown in Figure 5.7, the maximum likelihood value of r, 0.15, has an associated probability 0.0333; the probability of the data set at r = 0.5 is 0.0156. The relative likelihood of linkage, given by the ratio of the two probabilities, is 2.13/1 in favor of linkage in this example.

How large must the likelihood ratio be for us to conclude that the markers are indeed linked? A likelihood ratio of 1,000/1 is considered adequate to support the linkage hypothesis if only a few marker loci are being tested. The primary source of error will be the chance cosegregation of a pair of markers that are in fact not physically close to one another. The probability of this happening is independent of the size of the genome and is in fact 1/2 raised to the power of the number of meioses. This means that in ten meioses, for example, the chance that a pair of markers on different chromosomes would completely cosegregate is about 1/1000. The more meioses one has in a data set, the smaller the probability that chance cosegregation will cause misleading results in a linkage study.

For convenience in combining data from different laboratories, a convention has been adopted for expressing the likelihood ratio in terms of its logarithm. This allows results to be added together. The **logarithm of the odds** is expressed as a **LOD** score. A LOD score of three, therefore, is accepted as sufficient evidence of linkage for most purposes.

e. Multilocus Analysis

Human genetic data are often only partially informative. Therefore, it is useful to be able to combine results from overlapping genomic intervals into an overall estimate of recombination frequency for a specific interval. This requires an expression for recombination frequency in one interval in terms of recombination frequencies within the smaller intervals it encompasses.

Consider, for example, a region with three marker loci, A, B, and C, arrayed in the order ABC. Assume that there are a number of meioses informative for A and B and for B and C that are not informative for recombination between A and C. The recombination fractions rAB and rBC cannot simply be added together to give the rAC fraction because some recombinants in rAB may be associated with a second recombination event within the BC interval; such chromosomes will not score as recombinants in interval AC.

Assume that recombination events in AB are independent of those in BC (i.e., a recombination event in AB has no influence over the likelihood of recombination in BC); it then appears that $rAC = rAB + rBC - 2(rAB \times rBC)$. In fact, however, this assumption is incorrect for a number of genetic systems; recombination in adjacent intervals is often depressed somewhat by mechanisms that are not understood. This phenomenon—recombination in one interval reducing the chance of recombination in an adjacent interval—is called **interference**. However, the error introduced by the assumption of interference is small compared to the error expected as a consequence of the limited sample sizes available in human genetics. Therefore, such analyses can explicitly assume that recombination events are independent of one another.

What does complicate matters is the fact that double exchanges within an interval do not give rise to observed recombinants. Linkage analysis is simplified when there is an expression for the average number of actual recombinational events in an interval, rather than just the odd number of exchanges that result in observable recombination between the two markers that define the interval. Such an expression is called a **mapping function**. The "map distances" derived with such a function have the property of additivity. Several mathematical formulations have been developed for mapping purposes, the first by J. B. S. Haldane in 1919. Genetic distances derived in this way are expressed in **centiMorgans (cM)**. For values of θ less than 0.06, centiMorgans are roughly equivalent to θ times 100 (e.g., θ equal to 0.05 is equivalent to 5 cM).

f. The Problem of Gene Order

The most important characteristic defining a genetic map is the sequential order of the genes and other markers along a chromosome. Although recombination frequencies can be expected to change as additional data are obtained and to vary between the sexes, it would be convenient if the linear order could be estimated with sufficient accuracy that it would be unlikely to change as new meioses and new markers are added to the data base. The most important parameter to estimate, therefore, is the most likely gene order; the statistical support for that order as opposed to others is a crucial reflection of the strength of the map.

Simple comparisons of data in human sample sets often lead to contradictions. For example, if the recombination frequencies among three marker loci in different, nonoverlapping sample sets are 5 percent between A and B, 10 percent between B and C, and 15 percent between A and C, the estimated order would be ABC. However, the variance in such estimates can be high, given the limited sample sizes available for human studies; a different population sample may well give a different result.

The reliability of map order determination is markedly enhanced by developing inheritance data for all defined loci on the same sample set and analyzing the loci jointly, three at a time. Qualitatively, if the order is ABC, a recombinant between A and B will also be a recombinant between A and C, but not between B and C. If the order is BAC, a recombinant between A and B will almost always be a recombinant between B and C. This approach reveals order almost directly if the sample set is fully characterized at all loci.

All chromosomes from a three-locus analysis will fall into one of four classes: nonrecombinant, recombinant in AB and AC, recombinant in AC and BC, or recombinant in AB and BC. For any particular recombinant class, two orders will be consistent with a single exchange, and one will require a double exchange. In the previous example, the recombinant in AB and AC is a single-exchange chromosome under either order ABC or ACB. Under the order BAC, however, a chromosome recombinant in both AB and AC would be a double exchange.

The order of markers is most often defined by determining which order requires the smallest number of exchanges. The number of exchanges required to generate the observed recombinant chromosomes in a given cross can be determined for each possible order of markers. For any given data set, an explanation for the generation of each possible order—ABC, ACB, and BAC in this example—will ordinarily require different numbers of recombination events. Because multiple exchanges are less likely than single exchanges, the game in determining order is to distribute the progeny chromosomes into the four recombinant classes and determine which group is represented by the fewest chromosomes. This will be the double-exchange class; it will reveal the gene order directly.

Estimates of the relative likelihoods of particular gene orders are much stronger if they are derived from multilocus analysis because a hypothesis with an incorrect gene order will demand several recombinational exchanges in adjacent intervals, a highly improbable scenario. Likelihood analysis will reveal the relative support for the three possible orders of loci A, B, and C; orders that require double exchanges will give low probabilities to the data set. However, the unlikely possibility of occurrence of several exchanges within a small region can be taken into account quantitatively only if several loci are scored jointly.

Some information, albeit much less, is contributed by the recombination frequencies themselves. In practice, owing to the phase uncertainties in human data sets, it is useful to obtain a quantitative statement of the support for a particular order over the next most likely. This is accomplished by calculating the likelihoods of the data sets under the two hypotheses for order; the more likely hypothesis is the one that gives the higher probability to the data set.

g. Combining Data from Different Pedigrees

As the examples have indicated, small nuclear families by themselves do not provide adequate support for hypotheses about gene order or recombination frequencies. For linkage studies that involve DNA markers only—placing new markers on a map, for example—it is appropriate to consider the genotypes determined for each family as an independent trial and to combine the data by multiplying the probabilities obtained for each family for each specification of the parameters.

However, for linkage studies in families with a genetic disease, this approach has pitfalls. Specifically, what we think of as a single genetic disease can sometimes be the result of mutations in any of several different genes. Xeroderma pigmentosa, for example, involves several complementation groups that are likely to map to different chromosomes. Recently, an X-linked form of spastic paraplegia was shown to result from lesions at either of two different loci. If several families under study for a given disorder are in fact transmitting defective alleles of different genes, the signals are likely to be canceled out and a false negative result obtained when we combine the linkage data. One of the best ways to be certain that the same gene is being examined is to obtain extended pedigrees that are by themselves sufficiently large to give a strong linkage signal; the disease is virtually certain to be the consequence of the same genetic defect within a single pedigree.

5.4 Linkage Maps of Human Chromosomes

For the linkage approach to be efficient, it is useful to construct a map of linked markers for each chromosome. Such maps allow the selection of markers that will provide complete coverage of each chromosome. Thus, the distances between mapped markers should be such that a new disease gene or marker lying in any region will show evidence of linkage with at least one of the already mapped markers flanking that region. Linkage maps are constructed on the basis of genotypic information from a panel of reference families whose pedigree structures contain three living generations and a large sibship. The parameters established by the maps are useful for multilocus analysis of disease genes, even though the reference families are not afflicted. In contrast, the sample sizes available from families carrying the diseases are often quite small. They cannot by themselves reveal the order and recombination frequencies among markers that are critical to establishing the precise location of a disease gene. Mapped sets of markers for human chromosomes also provide an opportunity to study the behavior of normal human chromosomes at meiosis.

a. Family Panel for Linkage Mapping

Almost a thousand progeny chromosomes in defined family structures are required to make reliable estimates of linkage distances of the order 1 percent. The work of linkage mapping is most efficient if there are many children in each of the families that contributes DNA to the reference panel; fewer parents then need to be typed. Ambiguities with respect to phase can be resolved if grandparents can also be sampled. More than 50 such "ideal" families have been identified and sampled in Utah, and immortalized cell lines from the majority of them have been contributed to a DNA archive established in Paris by the Human Polymorphism Study Center (CEPH). The DNA reference panel CEPH maintains is widely available to investigators. Its usefulness will increase as the inheritance patterns of more and more chromosomal regions are characterized in laboratories throughout the world.

b. Linkage Maps of Autosomes

Primary maps of all human autosomes—maps that cover nearly the entire chromosome with linked marker loci at intervals of approximately 20 cM, on average—have already been constructed, and high-resolution maps, with markers spaced at 1–2 cM, are essentially complete. The primary linkage map of human chromosome 7, a medium-sized autosome, is illustrated in Figure 5.8a. Because

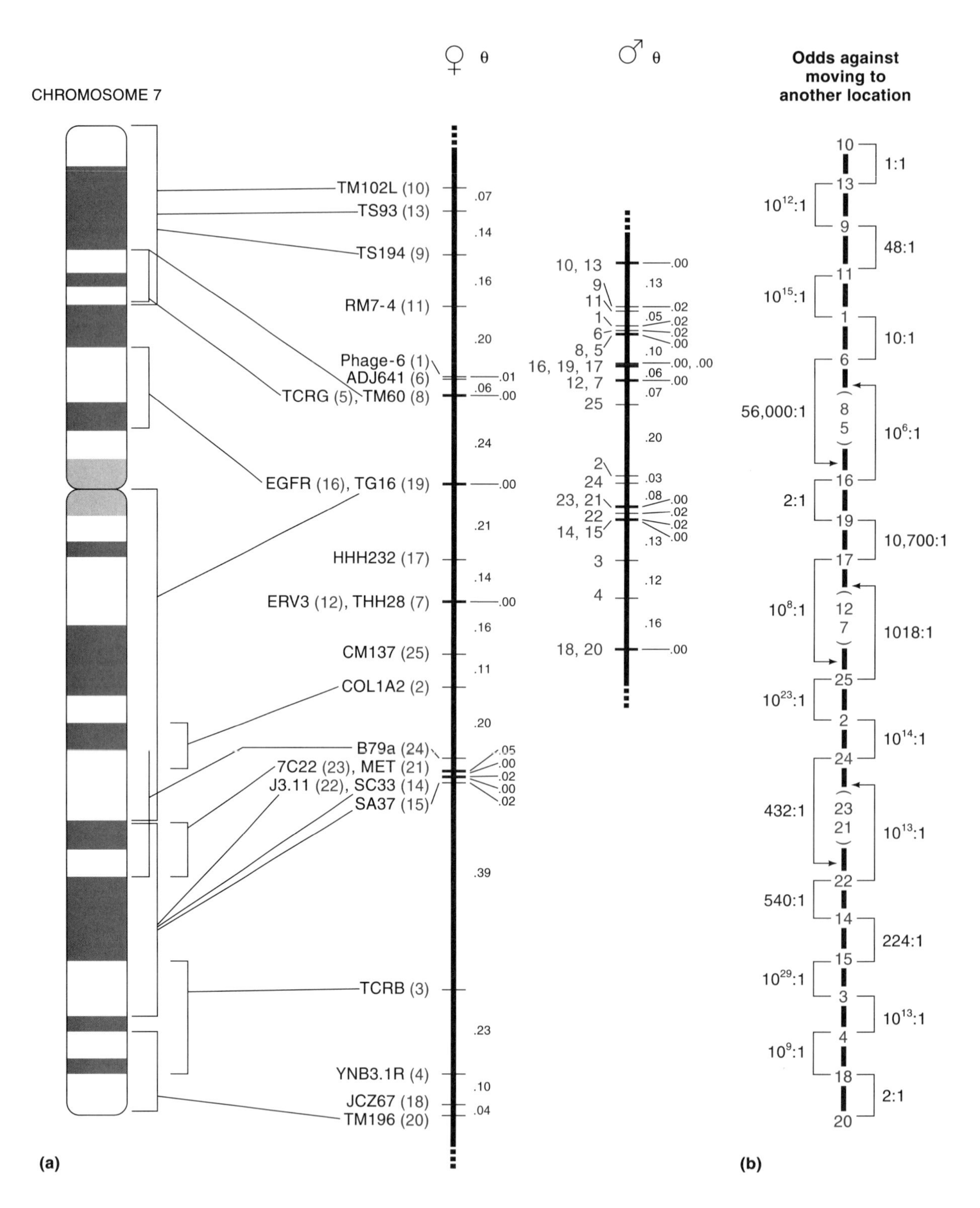

Figure 5.8
(a) Sex-specific genetic maps of chromosome 7. The maps, scaled in centiMorgans, are derived from recombination frequencies (θ) through the Haldane mapping function. Locus numbers on the male map correspond to the numbers in parentheses after each marker on the female map. Approximate physical locations of markers, where known, are indicated on the karyogram. (b) Final order of markers, showing odds against inversion of adjacent loci (brackets) or against placement in an alternative interval (arrows). Loci in parentheses could not be ordered with available data. (From G. M. Lathrop, P. O'Connell, M. Leppert, Y. Nakamura, M. Farrell, L.-C. Tsui, J.-M. Lalouel, and R. White. 1989. Twenty-five Loci Form a Continuous Linkage Map of Markers for Human Chromosome 7. *Genomics* 5, 866–873.)

several of the markers have been localized on the chromosome by physical as well as genetic methods (*Genes and Genomes,* section 7.4b), it is clear that virtually the entire chromosome is spanned by the linkage map. Because recombination events in many regions of the genome are more frequent in female than in male meioses, at least one large "gap" is apparent in the female map shown in Figure 5.8a. When data from males and females are combined, however, the coverage of the chromosome is nearly complete for purposes of disease studies. Markers shown on this map (first TCRB, later MET) first suggested and then confirmed localization of the cystic fibrosis gene to chromosome 7, a gene subsequently identified and cloned.

The first problem in constructing a linkage map is establishing the order of the markers. As already described, establishing the relative likelihood of one order compared to another requires calculating the probability for each of two orders in a particular data set by analyzing the segregation patterns of all the markers jointly.

In principle, gene order could be determined by calculating the likelihood of all possible orders and choosing the most favored—the maximum likelihood order. However, for 15 marker loci, there would be $15!/2 = 6 \times 10^{11}$ possible orders. In practice, therefore, a useful strategy is to eliminate from consideration orders that are possible but have a very low probability. Many possible gene orders can be eliminated by simple three-locus analyses.

These considerations make it appropriate when representing the genetic linkage map of a chromosome to indicate the support offered by the data for the most favored order relative to the several next most likely. Although different maps are required to represent male and female recombination frequencies, a single gene order should be consistent for both sexes. Figure 5.8b indicates the support obtained from combined linkage data for the illustrated order of the markers on chromosome 7.

Recombination frequencies are not strictly proportional to physical distance. Although in a given region it is appropriate to expect that recombination frequencies will increase with physical distance, the ratio of one of these two parameters to the other can vary significantly from region to region. Some of the most intriguing variation is in the differences between recombination frequencies observed in male versus female meioses. As the genetic map in Figure 5.8 shows for chromosome 7, the recombination frequencies observed in female meioses seem uniformly higher than those seen in males. Because the available analytical computer programs allow independent estimation of the female and male map distances for each interval, we can determine whether the data offer significant support for sex-specific differences from one interval to the next. On chromosome 7, no regions showing a significant excess of male recombination have been observed.

However, for chromosome 11 and some other chromosomes, including 1, 9, 10, 12, 15, 16, 17, and 19, the result is different (Figure 5.9). The recombination frequency observed in males within the region of chromosome 11p between HBBC and HRAS1 considerably exceeds that of females. Furthermore, in an adjacent interval, HBBC to PTH/CALC1, the female map distance is slightly greater than the male map distance. This difference in the ratio of sex-specific map distances for adjacent intervals was found to be significant in two independent studies. Moreover, when the region between HBBC and HRAS1 is subdivided by internal markers, the male-specific increase in map distance appears to be distributed over the entire interval, a result that argues against the presence of a single male-specific "hotspot" for recombination somewhere in the HBBC-HRAS segment.

The biochemical basis for the distinction between female and male recombination frequencies is unclear, but an intriguing conclusion can be drawn. Because the *physical* distance between the marker loci must be the same in both

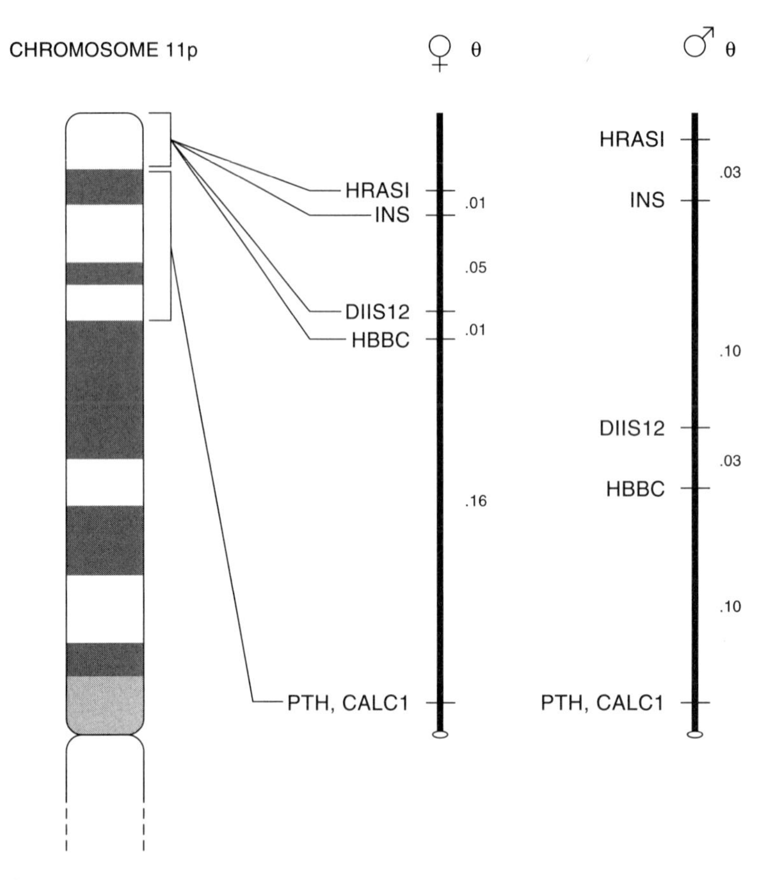

Figure 5.9
Physical localizations and genetic maps for five loci on chromosome 11p. Male and female maps are scaled to genetic distance in centiMorgans by the Haldane mapping function. Recombination fractions (θ) are given in each interval. The loci defined by the genetic systems indicated on the maps are PTH = parathyroid hormone; CALC1 = calcitonin 1; HBBC = β-globin complex; D11S12 = anonymous marker; INS = insulin; and HRAS1 = Harvey-*ras* protooncogene. Recombination fractions (θ) are shown in each interval. (Adapted from R. White and J.-M. Lalouel. 1987. Genetic Maps, 4th ed. S. O'Brien, ed. Cold Spring Harbor Laboratory, Cold Spring Harbor, New York.)

males and females (the same chromosome will often be inherited by females from males and vice versa), the frequency of crossing-over within a given physical distance in DNA must vary in males, females, or both.

c. The X Chromosome and Linkage

The X chromosome has excited wide interest because of the large number and variety of known X-linked diseases. This situation reflects the fact that X is the only human chromosome carrying a substantial number of essential genes that is regularly in a haploid, or hemizygous, condition (in males). As a result of the interest, several laboratories have contributed to the development of markers for this chromosome. These markers have been tested in a large set of human families, and the genotypic data have been compiled to construct a map that spans the entire chromosome between the telomeres.

An intriguing finding emerging from such studies is that the X and Y chromosomes contain homologous regions—the pseudoautosomal regions—that regularly undergo recombinational exchange. (Over the majority of their lengths, the X and Y chromosomes neither recombine nor show extensive homology.) Genes in the pseudoautosomal regions of X and Y, where recombination does take place, are not sex-linked; that is, the traits encoded by those genes are not transmitted in a hemizygous fashion to male offspring. A much higher frequency of recombination has been observed in the pseudoautosomal region in males than in females. This higher frequency of recombination is consistent with the idea that each chromosome tetrad is obliged to undergo a recombination event at meiosis. If this recombination event is constrained within a small region, the recombination frequency per unit of DNA becomes very high.

5.5 Mapping and Isolating Genes Responsible for Human Diseases

a. Linkage Mapping in Disease Families

The most efficient way to map an unknown gene by linkage analysis is by reference to a map containing markers at intervals of less than 20 cM on each chromosome. Such maps are most useful if they are constructed with markers that are informative in most people. Then it is possible to derive subsets of markers spaced so that linkage of a disease locus ought to be detectable with at least one subset. Furthermore, linkage analysis is much more sensitive if multiple markers mapped to small chromosomal regions can be examined jointly; as much as 50 percent more information on linkage can be obtained this way. However, before such maps became widely available, the urgency to locate genes associated with important genetic diseases led to a "shotgun" approach: searching for linkage with essentially unmapped markers. This approach proved successful in a number of instances. It was facilitated by the chromosomal translocations that occasionally accompany an inherited disease; the breakpoints indicate regions where important genes may be disrupted, and linkage studies can focus on those regions.

There is a large element of chance, both in the initial detection of linkage and in the progress toward cloning that follows. The search for the cystic fibrosis (CF) gene is a good example: within weeks of the initial finding of linkage to markers on chromosome 7, researchers discovered other markers even closer to the putative gene (i.e., linked with smaller recombination percentages). Limiting the scope in this way brought the region to be searched in reach of techniques for studying large fragments of DNA, and within two years, several genes in the region were identified as candidates. However, cloning and sequencing the correct gene from among the candidates and identifying the mutations responsible for CF in patients took two years more.

Once a disease gene has been located to within a recombination frequency of a few percent from a mapped marker, the challenge is to identify which of the genes present within that region is actually the mutated gene causing the disease. How close physically is the disease locus to the marker? How many possible coding regions does such a distance represent? Current estimates suggest that a 1 percent recombination frequency (1 cM) corresponds, on the average, to approximately one million base pairs. However, the lack of strict proportionality between physical and genetic distance mentioned earlier indicates that a 1 percent recombination frequency could represent as few as several hundred thousand or as many as several million base pairs. How many genes might be contained within such a region? Estimates of the total number of genes imbedded within and expressed from the human genome are in the vicinity of 100,000 or 1 per 30 kbp of DNA on average—a figure reasonably consistent with the average length of genes thus far characterized.

In a specific tissue such as liver or kidney, however, as few as 10,000 genes may be expressed. Therefore, if one can identify a specific tissue within which the mutant gene is expressed, the number of potential candidate genes may be limited, on the average, to only one per 300 kbp of DNA or ten per three million base pairs. A variety of methods, other than linkage analysis, must then be applied to identify the sought after gene.

b. Identifying and Isolating the Gene

Having identified the relevant region of a chromosome by means of linked markers and narrowed the area of search by developing close flanking markers, researchers usually try to identify a number of cloned segments of DNA from that region. These clones serve as probes to search for submicroscopic deletions, to identify new polymorphic markers that may be closer to the gene, and to scan cDNA libraries in the hope of finding expressed sequences from within the region.

Most cloning approaches involve developing a DNA source of limited complexity that contains the chromosomal region of interest. With overlapping clones, researchers can search for mutations that might be causing the disease under investigation. Base pair substitutions detected by sequencing are difficult to correlate with a disease at this stage because, as we have seen, such variants occur normally throughout human DNA every 200–500 base pairs. In a one-megabase region, for example, we expect 2000–5000 base pair variants, far too many to test, although more discriminating tests in the neighborhoods of specific genes might be feasible. But deletions are not frequent in human DNA and have proven highly useful in pinpointing the locations of some previously unknown genes. The genes responsible for retinoblastoma (a rare cancer of the eye) and Duchenne muscular dystrophy were located in this way.

Detecting deletions Disease-causing mutations in a gene or in critical neighboring sequences are sometimes physical deletions. Deletions are likely to range in size from a few base pairs to millions of base pairs; at the high end of the scale, a deletion can be detected visually, as a missing, often translocated, segment in a metaphase chromosome. Smaller deletions will affect the lengths of restriction fragments so that standard DNA blotting can be used to scan a large region of DNA in many affected individuals. Larger (but still submicroscopic) deletions should affect the length of the very large restriction fragments produced by endonucleases whose recognition sequences do not occur very often in the human genome. Scanning such restriction fragments, which are as large as several megabases and can be resolved by pulsed-field gel electrophoresis (*Genes and Genomes,* section 6.5), facilitates discovery of deletions associated with genetic disease.

Confirming identification Disease-causing mutations will often have occurred at the level of single base pairs, creating in-frame stop codons or frameshifts, for example. Sequencing and comparing exonic material in affected and unaffected individuals is thus usually necessary to confirm the gene's involvement. Conclusive evidence of the role of a particular gene in an inherited disease involves demonstrating disease-specific aberrations in a number of unrelated patients. The cloned genes listed in Table 5.1, all originally mapped with the aid of linkage analysis, were identified through a variety of molecular strategies.

5.6 DNA Sequence Variations as Markers in Somatic Cells

Genetic markers can trace somatic genetic events as well as the meiotic events observed in family studies. This is particularly important for analysis of the somatic cell mutations that are critical in carcinogenesis. Before recombinant DNA technology was available, geneticists studying chromosomes showed that specific translocations are characteristic of tumor cells in certain hematopoietic malignancies. DNA markers permit, in addition, characterization of the somatic mutations occurring during development of solid tumors. The story of the retinoblastoma and APC genes illustrates what can be achieved.

a. The Retinoblastoma Gene

Researchers hypothesized early on that a predisposition for a childhood tumor of the eye, retinoblastoma, was the result of an inherited mutant gene and that appearance of the tumor required a second, somatic genetic event. Note that if inheritance of the mutant gene was sufficient to cause the tumor, then tissue derived from retinoblast cells—all of which would be carrying the germ line mutation—would become generally tumorous. As it is, however, such tissue sports only the few independent tumor clones characteristic of inherited retinoblastoma.

The nature of the second event was not obvious, and here DNA markers became crucial to the story. It was reasonable to hypothesize that the inherited lesion is—from a cell's point of view—a recessive mutation; that is, the normal allele carried on the chromosomal homologue is able to complement the defective function of the mutant allele, and no tumor forms. By this model, tumor clones would emerge if the normal allele were lost from one or more retinal cells by some genetic accident (Figure 5.10). One possibility would be point mutations in the normal allele. Alternatively, larger scale events such as major deletions or even loss of the entire chromosome due to a nondisjunctional accident at mitosis could inactivate the normal allele. Although DNA markers offered no immediate hope of identifying point mutations in the normal gene, they could detect large-scale chromosomal loss.

Chromosome 13, band q14, was implicated as the likely site for the retinoblastoma (*RB*1) gene because a microscopically observable deletion in one chromosome 13, always involving the q14 band, was present in the cells of about 5 percent of individuals with retinoblastoma. If the two chromosomes 13 were differentially marked by several DNA markers distributed over the chromosome, any event that caused the loss of a significant chromosomal segment would also eliminate the marker alleles carried on that segment. Thus, loss of alleles was detected by comparing DNA prepared from tumor cells with DNA from some other tissue of individuals hosting the tumor (Figure 5.11). Apparent loss of the entire chromosome 13 through nondisjunction or some similar mechanism occurs in the genesis of more than 50 percent of retinoblastoma tumors; such events are readily detected. Careful examination of several exceptional tumors revealed that occasionally yet another mechanism—**mitotic recombination**—can also rid a cell of one of its alleles. Although well characterized in *Drosophila* and in fungi, mitotic recombination had previously been hard to demonstrate in mammalian systems.

Retinoblastoma can also occur in the absence of an inherited mutant gene, as is true of approximately half the cases. It was of some interest, then, to find that loss of one of the chromosome 13 homologues was prevalent even in these "sporadic" tumors. Thus, mechanisms similar to those operating in familial

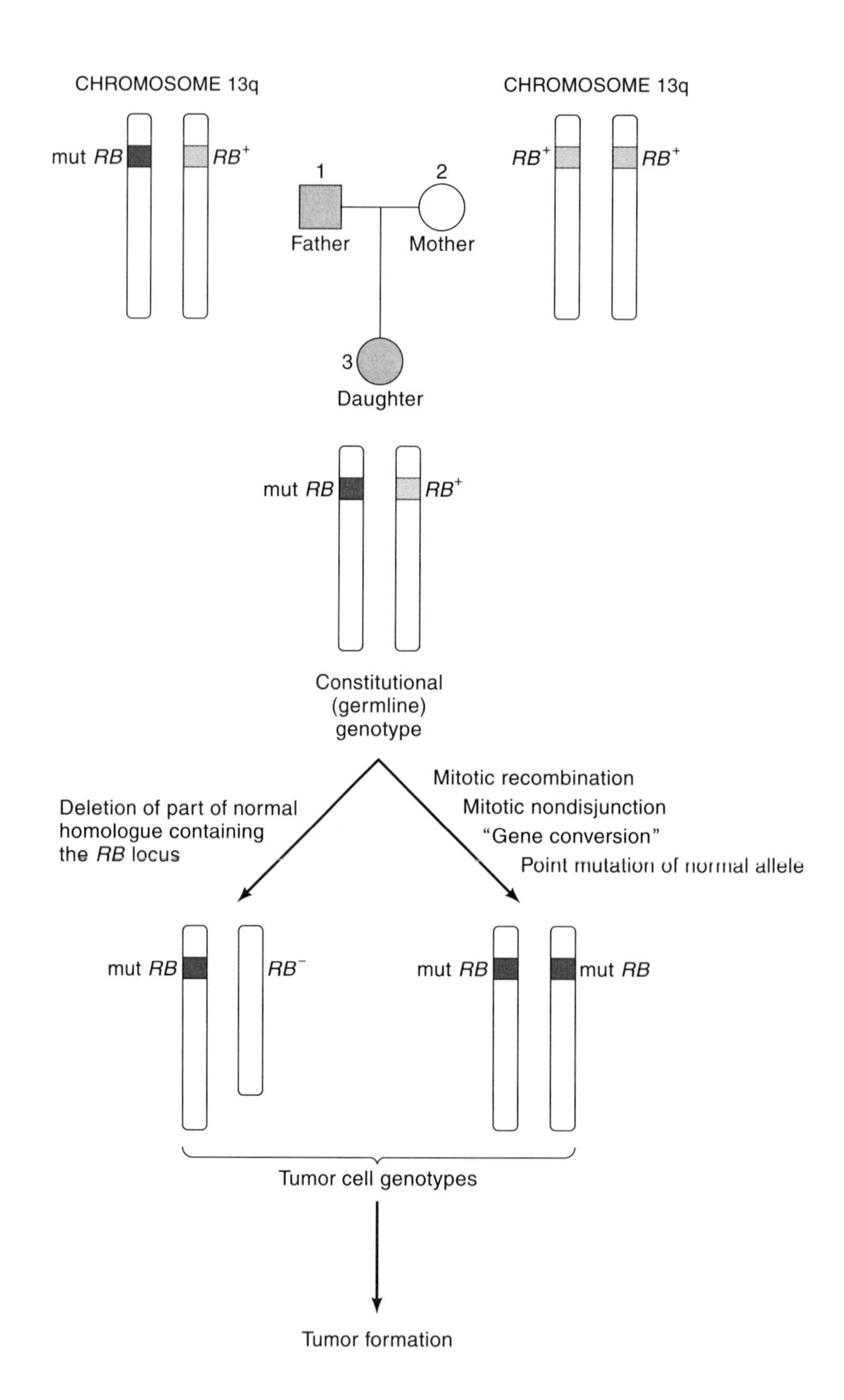

Figure 5.10
Models for expressing a mutant tumor suppressor gene in retinoblastoma. Colored boxes denote the location on chromosome 13q of the *RB*1 locus (mut *RB* = mutant allele; RB^+ = normal gene). Individuals 1 and 3, who have inherited the mutant allele, each have also inherited one normal chromosome 13 to compensate. However, if a retinal precursor cell (retinoblast) undergoes a somatic event during the child's embryonic life, such as mitotic recombination or nondisjunction—resulting in homozygosity for the mutant allele—or deletion of the normal homologue of the *RB*1 gene, a tumor clone will emerge in early childhood. (Adapted from R. White, D. Barker, T. Holm, J. Berkowitz, M. Leppert, W. Cavenee, R. Leach, and D. Drayna. 1983. Approaches to Linkage Analysis in the Human. In Banbury Report No. 14, Recombinant DNA Applications to Human Disease, p. 248. Cold Spring Harbor Laboratory, Cold Spring Harbor, New York.)

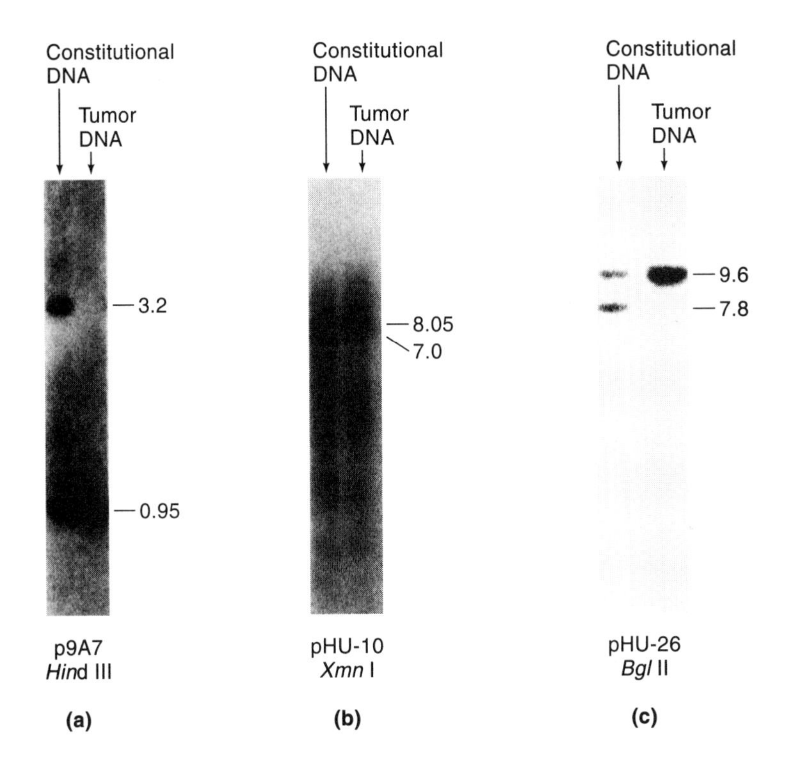

Figure 5.11
Reduction to homozygosity as a consequence of loss of one chromosome 13 homologue with reduplication of the remaining (mutant) one. Autoradiographs of DNA from normal (constitutional) cells and tumor cells from a patient with retinoblastoma show alleles at loci on chromosome 13 defined by three different probe/enzyme systems. One allele is absent at all three loci in the tumor cells. Numbers to the right of each panel indicate the size of each allele in kbp. (From W. Cavenee et al. 1983. Expression of Recessive Alleles by Chromosomal Mechanisms in Retinoblastoma. *Nature* 305 780. Reproduced by permission.)

retinoblastoma were apparently at work in the sporadic tumors, but the primary mutational event in the latter was associated with somatic rather than germ line cells. However, because somatic inactivation of both alleles of the retinoblastoma gene (*RB*1) in a formerly normal retinal precursor cell requires a very rare combination of events (two mutational "hits" on a pair of allelic loci within the same cell), patients with the noninherited form of retinoblastoma rarely develop more than one tumor, but predisposed children often have tumors in both eyes.

An interesting corollary to the retinoblastoma story arose from epidemiological studies that followed the fate of individuals who had inherited a mutant retinoblastoma gene but had survived their initial retinoblastoma tumors. Additional tumors frequently arose later in life in these individuals, and prominent among them were (otherwise rare) osteosarcomas. Loss of a chromosome 13 homologue was likewise characteristic of these tumors. Further study provided evidence that chromosome 13 loss may be characteristic of osteosarcoma as well as retinoblastoma.

Such mechanisms subsequently have been shown to operate in other tumors as well: loss of a chromosome 11 homologue is prevalent in the genesis of Wilms tumor and some rhabdomyosarcomas, and loss of a chromosome 22 homologue is prevalent in the genesis of acoustic neuromas.

DNA markers, therefore, have been instrumental in showing that the inherited genetic lesion in retinoblastoma, and in several other tumor types as well, is a recessive mutation that is not revealed until additional mutational events inactivate or expel the normal homologue. Tumors then form when cellular control mechanisms are disrupted by complete loss of the gene product. Genes like *RB*1 have therefore come to be known as "tumor suppressors" (Section 4.4a).

b. The APC *Story*

One of several recent successes of the mapping paradigm was the identification in 1991 of *APC*, the gene whose mutant forms cause adenomatous polyposis coli (Figures 5.1 and 5.2). If untreated, this inherited condition, which manifests itself in early adulthood, inevitably leads to colon cancer because each of the numerous polyps typical of APC represents a precursor to a malignant tumor. Prevention consists of surgical removal of the colon from affected individuals. Before the gene was localized and marker systems became available to identify carriers in APC families, all possible carriers were monitored by frequent colonoscopies, with the anxiety, expense, and discomfort that entailed.

Although such families are rare, their contributions to localization of *APC* by linkage studies were instrumental in making possible the eventual isolation of a gene whose identity and behavior have far-reaching implications for the population at large. This is because colon cancer occurs in at least 5 percent of the U.S. population, and a mutation in *APC* within a colonic epithelial cell appears to contribute to carcinogenesis even in sporadic (presumably noninherited) cases.

Over a period of years beginning several decades ago, a study was made of a multigenerational Utah kindred in which numerous individuals developed Gardner syndrome (GS), defined by polyposis associated with a variety of other manifestations such as osteomas, desmoid tumors, and epidermoid cysts. Pedigree analysis showed that this syndrome was transmitted in an autosomal dominant fashion. Whether GS and classical APC were manifestations of lesions in the same gene was unknown.

In 1987, in two independent laboratories, researchers undertook linkage studies in GS and APC families using a DNA marker that had been mapped by physical methods (such as *in situ* hybridization) to chromosome 5q21–22. They were following a clue suggested by a published report describing a patient with GS; the patient's cells all contained one abnormal chromosome 5 that was seen upon microscopic examination to be missing some material. Logic suggested that the patient's disease might be associated with loss of one or more specific genes from the deleted region. Linkage analyses in a number of families showed that APC and GS were both linked to a marker on the long arm of chromosome 5, and the effort to find the gene(s) began in earnest.

Within a year, a series of new markers for the 5q21-22 region had been developed using DNA from human × rodent hybrid cells in which the only human content was derived from chromosome 5. These markers were ordered into a genetic map by genotypic analysis of a reference panel of DNAs from 59 three-generation pedigrees. Then linkage analyses among APC and GS families showed that one of the new markers was much closer to the putative gene than the original 5q marker, an estimated 3000 kbp away, and indicated that APC and GS were probably caused by the same gene.

Soon it was discovered that alleles of markers in the same region were often missing in colonic adenomas and carcinomas of patients without APC. These allelic losses suggested that alteration of a tumor suppressor gene in that region might contribute to development of sporadic as well as inherited forms of colon cancer. The region commonly lost in tumor cells appeared to be centered around a polymorphic locus detected by a DNA fragment called c5.71. Subclones of this fragment were tested for cross-hybridization with DNA from other animal species, on the grounds that important genetic information is usually conserved during evolution and can identify regions likely to contain expressed sequences. This strategy revealed, at first, a previously unknown gene labelled *MCC* (for **m**utated in **c**olon **c**ancer) that was found to have been altered in several sporadic colon carcinomas. *MCC*, because it lay in the region known to contain the *APC* locus, became a candidate gene for the inherited polyp syndrome. However, careful examination of *MCC* in APC patients failed to reveal any inherited mutations.

Several large DNA fragments from the region around c5.71 were ordered into a map by means of overlapping sequences, and the search for *APC* continued. At this molecular level, small deletions affecting one of the fragments—one deletion 100 kbp long and the other 260 kbp, the first lying within the second—were identified in two unrelated APC patients. *MCC* was shown to be outside the 100 kbp region deleted from one chromosome 5 allele in both patients. However, three genes *were* present in that 100 kbp portion of normal chromosomes.

Each of these three genes was studied for inactivating mutations in other APC patients by a method known as **single-strand conformation polymorphism (SSCP)** analysis. This technique can often detect single-base substitutions, some of which might be capable of inactivating a gene or rendering the gene product nonfunctional. DNA sequencing of aberrations revealed by SSCP identified potentially inactivating mutations (i.e., sequence anomalies that would result in a shift in the reading frame or in premature stop codons) in one of the three new candidate genes in several patients, and thus the APC gene was identified. It turned out to be very large, with a coding sequence of more than 8500 bp, arranged in 15 exons. Alternative splicing, both within the gene and in the 5′ leader region, is an intriguing feature; some evidence indicates that although *APC* is expressed in all human tissues tested, different forms of the spliced transcript may be unique to certain tissues.

The mutations that have been identified to date in many APC families make definitive, early (even prenatal) diagnosis possible for members of those families. Linkage analysis and mutational studies have also detected inherited mutations in *APC* in some families with a high incidence of colon cancer but without polyposis. Early evidence suggests that in these families, the inherited mutation has occurred close to the 5′ end of the gene (where transcription begins); the implications of the effect of location on the phenotypic consequences of a given mutation have yet to be fully worked out, but as the number and variety of characterized *APC* mutations increases, the picture may become clearer.

What does *APC* do when it is functioning normally? We are now seeking to find out. The DNA sequence of the cloned gene shows little homology to other known genes and thus does not indicate the presence of specific functional domains (e.g., transmembrane regions characteristic of receptor proteins or cyclic AMP-binding sites). However, the amino acid sequence encoded by *APC* does contain a region capable of forming coiled, dimeric structures with itself or with similar peptides. Most of the mutations found so far in APC families have occurred in a "downstream" region of the gene, a location that apparently permits expression of a truncated form of the protein product with the dimerizing region intact. The shortened peptide could presumably form complexes with normal proteins and might "poison" their activity.

On the basis of the early evidence, *APC* normally functions as a tumor suppressor, in that it appears to have a role in controlling the growth and differentiation of epithelial cells. Even before polyps appear in individuals who have inherited a mutant allele, there is some evidence of abnormality in the colonic mucosa. Might the *APC* product be involved in recognition of a differentiation factor, perhaps a cell surface receptor? The cloned gene now provides the molecular biologist with the tools necessary to find out (see also section 4.4e).

5.7 Conclusion: Genetics and Human Variation

From disease predisposition to musical or athletic ability, individual humans show a wide range of phenotypic variation. Geneticists would like to understand the contribution of germ line inheritance to both "normal" variation and medical problems. There is ample basis for suspecting that much of individual variation is encoded in our genes; two copies of each of 100,000 or more genes, a significant proportion of which exist in multiple allelic forms, provide combinatorial possibilities sufficient to support enormous variation. However, it is not always easy to determine whether a variation in human phenotype reflects genetic difference or the influence of different experiences and environments.

Much of the work of human genetics therefore consists of determining whether or to what extent specific features are genetically encoded. Characteristics most urgently in need of classification are the diseases and disease predispositions. At present, diseases are accepted as a mixture of genotype and environment; the respective environmental and genetic components are hard to separate because of variations affecting them both.

In experimental plant and animal systems, the distinction between environment and genetics can be assessed by raising a crop or breed of genetically identical individuals, placing them in a controlled environment, and examining their phenotypes. Genotypes of interest are produced by forming, through controlled matings or construction of transgenic lines, an appropriate population or strain. Such approaches are not applicable to the human. However, human matings are abundant in nature. Because DNA markers provide a way to track specific genes through families, identifying individuals of any appropriate genotype, including those with the same genotype at a relevant locus, should be possible.

Although the human system has few of the advantages that make some organisms particularly amenable to genetic investigation—short generation time, ease of manipulation, and large numbers of progeny for scoring, for example—it is studied intensively because of its special relationship to the investigator. Thus, the human has emerged as the best characterized of any mammalian system and as a useful system for basic scientific investigations in its own right. Although the difficulties of experimentation in the intact organism have in the past impeded work in human genetics, modern *in vitro* technologies (cell culture and recombinant DNA) now permit direct experimentation and allow a focus of resources on many aspects of human biology and genetics. In consequence, the molecular biologist has fertile and unprecedented research opportunities.

As recently as 1980, the concept of mapping the human genome with markers derived from anonymous polymorphisms rather than from genes encoding variant protein products was visionary. Only 12 years later, a comprehensive genetic linkage map became a reality. Collaboration between the CEPH consortium and the U.S. National Institutes of Health produced useful maps of all chromosomes. Together, these maps contain 1416 marker loci, including 279 genes and expressed sequences. In addition, a French collaborative group ordered 813 newly characterized $(CA)_n$ repeat loci into linkage groups that together span a distance corresponding to about 90 percent of the estimated length of the human genome. The markers and genotypic data used to construct these maps are the fruit of work in numerous laboratories over the intervening years.

Telomeric loci, historically elusive, now define physical and genetic endpoints on at least seven chromosomal arms: 2q, 4p, 7q, 8p, 14q, 16p, and 16q. An estimated 92 percent of the autosomal genomic length and 95 percent of the X chromosome appear to be covered by the published genetic maps, and marker density is such that genes responsible for many rare inherited diseases may now be amenable to linkage mapping even when pedigree resources are limited.

The pace of identification of previously unknown genes is accelerating rapidly, in large part because mapping resources and technologies have reached such a high level of sophistication. Characterization of newly cloned genes can yield novel information about human biology in general and disease processes in particular. Clarification of the functions of these genes may in turn open doors to development of therapies for many diseases that are not yet well understood.

Further Reading

General References

R. White and J.-M. Lalouel. 1987. Investigation of Genetic Linkage in Human Families. *Adv. Human Genet.* 16 121–228.

R. White and J.-M. Lalouel. 1988. Sets of Linked Genetic Markers for Human Chromosomes. *Annu. Rev. Genet.* 22 259–279.

R. White and C. T. Caskey. 1988. The Human as an Experimental System in Human Genetics. *Science* 240 1483–1488.

R. White and J.-M. Lalouel. 1988. Chromosome Mapping with DNA Markers. *Scientific American* 258 40–48.

5.2

D. Botstein, R. White, M. Skolnick, and R. Davis. 1980. Construction of a Genetic Linkage Map in Man Using Restriction Fragment Length Polymorphisms. *Amer. J. Hum. Genet.* 32 314–331.

R. White, M. Leppert, T. Bishop, D. Barker, J. Berkowitz, P. Callahan, T. Holm, and L. Jerominski. 1985. Construction of Linkage Maps with DNA Markers for Human Chromosomes. *Nature* 313 101–105.

Y. Nakamura, M. Leppert, P. O'Connell, R. Wolff, T. Holm, M. Culver, C. Martin, et al. 1987. Variable Number of Tandem Repeat Markers for Human Gene Mapping. *Science* 235 1616–1622.

Y. Nakamura, M. Carlson, K. Krapcho, M. Kanamori, and R. White. 1988. New Approach for Isolation of VNTR Markers. *Amer. J. Hum. Genet.* 43 854–859.

J. Weber. 1990. Informativeness of Human $(dC\text{-}dA)_n$ ° $(dG\text{-}dT)_n$ Polymorphisms. *Genomics* 7 524–530.

5.3

G. Lathrop, J.-M. Lalouel, C. Julier, and J. Ott. 1984. Strategies for Multilocus Linkage Analysis in Humans. *Proc. Natl. Acad. Sci. USA* 81 3443–3446.

G. Lathrop, J.-M. Lalouel, C. Julier, and J. Ott. 1985. Multilocus Linkage Analysis in Humans: Detection of Linkage and Estimation of Recombination. *Amer. J. Hum. Genet.* 37 482–498.

G. Lathrop, J.-M. Lalouel, and R. White. 1986. Construction of Human Linkage Maps: Likelihood Calculations for Multilocus Linkage Analysis. *Genet. Epidemiol.* 3 39–52.

5.4

G. Lathrop, P. O'Connell, M. Leppert, Y. Nakamura, M. Farrall, L.-C. Tsui, J.-M. Lalouel, and R. White. 1989. Twenty-five Loci Form a Continuous Linkage Map of Markers for Human Chromosome 7. *Genomics* 5 866–873.

J. Dausset, H. Cann, D. Cohen, M. Lathrop, J.-M. Lalouel, and R. White. 1990. Centre d'Étude du Polymorphisme Humain (CEPH): Collaborative Genetic Mapping of the Human Genome. *Genomics* 6 575–577.

J. C. Murray et al. 1994. A comprehensive human linkage map with centimorgan density. *Science* 265 2049–2054.

D. Adamson et al. (the Utah Marker Development Group). 1995. A collection of ordered tetranucleotide-repeat markers from the human genome. *Am. J. Hum. Genet.* 57 619–628.

5.5

M. Koenig, E. Hoffman, C. Bertelson, A. Monaco, C. Feener, and L. Kunkel. 1987. Complete Cloning of the Duchenne Muscular Dystrophy (DMD) cDNA and Preliminary Genomic Organization of the DMD Gene in Normal and Affected Individuals. *Cell* 50 509–517.

J. Riordan, J. Rommens, B. Kerem, N. Alon, R. Rozmahel, Z. Grzelczak, J. Zielenski, et al. 1989. Identification of the Cystic Fibrosis Gene: Cloning and Characterization of Complementary DNA. *Science* 245 1073–1080.

R. Cawthon, R. Weiss, G. Xu, D. Viskochil, M. Culver, J. Stevens, M. Robertson, et al. 1990. A Major Segment of the Neurofibromatosis Type 1 Gene: cDNA Sequence, Genomic Structure, and Point Mutations. *Cell* 62 193–201.

M. Wallace, D. Marchuk, L. Andersen, R. Lethcer, H. Odeh, A. Saulino, J. Fountain, et al. 1990. Type 1 Neurofibromatosis Gene: Identification of a Large Transcript Disrupted in Three NF1 Patients. *Science* 249 181–186.

D. Viskochil, A. Buchberg, G. Xu, R. Cawthon, S. Stevens, R. Wolff, M. Culver, et al. 1991. Deletions and a Translocation Interrupt a Cloned Gene at the Neurofibromatosis Type 1 Locus. *Cell* 62 187–192.

5.6

A. Knudson, H. Hethcote, and B. Brown. 1975. Mutation and Childhood Cancer: A Probabilistic Model for the Incidence of Retinoblastoma. *Proc. Natl. Acad. Sci. USA* 72 5116–5120.

W. Cavenee, T. Dryja, R. Phillips, W. Benedict, R. Godbout, B. Gallie, A. Murphree, L. Strong, and R. White. 1983. Expression of Recessive Alleles by Chromosomal Mechanisms in Retinoblastoma. *Nature* 305 779–784.

W. Cavenee, M. Hansen, M. Nordenskjold, E. Kock, I. Maumenee, J. Squire, R. Phillips, and B. Gallie. 1985. Genetic Origin of Mutations Predisposing to Retinoblastoma. *Science* 228 501–503.

S. Friend, R. Bernards, S. Rogelj, R. Weinberg, J. Rapaport, D. Albert, and T. Dryja. 1986. A Human DNA Segment with Properties of the Gene That Predisposes to Retinoblastoma and Osteosarcoma. *Nature* 323 643–646.

W. Bodmer, C. Bailey, J. Bodmer, H. Bussey, A. Ellis, P. Gorman, F. Lucibello, et al. 1987. Localization of the Gene for Familial Adenomatous Polyposis on Chromosome 5. *Nature* 328 614–616.

M. Leppert, M. Dobbs, P. Scambler, P. O'Connell, Y. Nakamura, D. Stauffer, S. Woodward, R. Burt, J. Hughes, E. Gardner, M. Lathrop, J. Wasmuth, J.-M. Lalouel, and R. White. 1987. The Gene for Familial Polyposis Coli Maps to the Long Arm of Chromosome 5. *Science* 238 1411–1413.

Y. Nakamura, M. Lathrop, M. Leppert, M. Dobbs, J. Wasmuth, R. Wolff, M. Carlson, et al. 1988. Localization of the Genetic Defect in Familial Adenomatous Polyposis Within a Small Region of Chromosome 5. *Am. J. Hum. Genet.* 43 638–644.

M. Orita, Y. Suzuki, T. Sekiya, and K. Hayashi. 1989. Rapid and Sensitive Detection of Point Mutations and DNA Polymorphisms Using the Polymerase Chain Reaction. *Genomics* 5 874–879.

J. Groden, A. Thliveris, W. Samowitz, M. Carlson, L. Gelbert, H. Albertsen, G. Joslyn, et al. 1991. Identification and Characterization of the Familial Adenomatous Polyposis Coli Gene. *Cell* 66 589–600.

G. Joslyn, M. Carlson, A. Thliveris, H. Albertsen, L. Gelbert, W. Samowitz, J. Groden, et al. 1991. Identification of Deletion Mutations and Three New Genes at the Familial Polyposis Locus. *Cell* 66 601–613.

K. Kinzler, M. Nilbert, L.-K. Su, B. Vogelstein, T. Bryan, D. Levy, K. Smith, et al. 1991. Identification of FAP Locus Genes from Chromosome 5q21. *Science* 253 661–665.

K. Kinzler, M. Nilbert, B. Vogelstein, T. Bryan, D. Levy, K. Smith, A. Preisinger, et al. 1991. Identification of a Gene Located at Chromosome 5q21 That Is Mutated in Colorectal Cancers. *Science* 251 1366–1370.

I. Nishisho, Y. Nakamura, Y. Miyoshi, Y. Miki, H. Ando, A. Horii, K. Koyama, et al. 1991. Mutations of Chromosome 5q21 Genes in FAP and Colorectal Cancer Patients. *Science* 253 665–669.

5.7

NIH/CEPH Collaborative Mapping Group. 1992. A Comprehensive Genetic Linkage Map of the Human Genome. *Science Reprint Series,* AAAS (66 pages) (expanded from material published in *Science* 258 67–86).

J. Weissenbach, G. Gyapay, C. Dib, A. Vignal, J. Morissette, P. Millasseau, G. Vaysseix, and M. Lathrop. 1992. A Second-Generation Linkage Map of the Human Genome. *Nature* 359 794–801.

V. A. McKusick and J. S. Amberger. 1993. The Morbid Anatomy of the Human Genome: Chromosomal Location of Mutations Causing Disease. *J. Med. Genet.* 30 1–26.

CHAPTER 6

Molecular Genetics of the Hemoglobin Genes

HAIG H. KAZAZIAN, JR. and STYLIANOS ANTONARAKIS

Hemoglobin has had a major role in mammalian biology for over 100 years. As the protein that carries and releases oxygen in red blood cells, it is central to the life process. Because the analysis of this protein has served as a paradigm in many biological studies, its molecular genetics has been singled out for special discussion in this book. But because it is a highly specialized protein and very abundant in a single cell type, some aspects of its biology are not generalizable to most other proteins.

Several factors have been important to the acquisition of knowledge about hemoglobin. First, it is the major protein of red blood cells. Adult hemoglobin, hemoglobin A, with its two protein components, the α- and β-globin chains, makes up about 90 percent of the protein of the red blood cell; so it can be easily purified. Second, its tissue source is readily accessible; a sample of blood contains 5×10^9 red cells per cubic millimeter. Third, the mRNAs encoding the globin chains of adult hemoglobin make up a very large part (90–95 percent) of the total mRNA in young red blood cells, called **reticulocytes**. Fourth, the globin genes are small and simple compared to other mammalian genes, thereby furthering the analysis of gene expression. Fifth, several human disease phenotypes resulting from mutations in globin genes were readily identified; the analysis of these mutations that altered sequences either within the adult globin genes or at some distance from them on the same chromosome provided the first comprehensive information about sequences important for gene expression *in vivo*.

Hemoglobin was the first protein to be crystallized, and it was the first to have its three-dimensional structure solved. Lysates of reticulocytes served as important tools for general studies of protein synthesis. Globin mRNAs were among the first mRNAs to be isolated. The β-globin complementary DNA was the second cDNA cloned after the one encoding chorionic somatomammotropin. The human globin gene clusters were the first to be mapped and subjected to extensive nucleotide sequence analysis. A key regulatory region controlling an entire gene cluster or gene family, the so-called **locus control region** (LCR) was first observed with the β-globin gene cluster. Mutations producing abnormal hemoglobins were the first to be comprehensively analyzed at the protein level. Later, the first detailed analysis of a genetic disease at the gene sequence level was carried out for β-thalassemia, a disorder of β-globin production. In the past few years, numerous transcription factors specific for certain cell types have been discovered. One of the first of these was an erythrocyte factor, first discovered because of its action on globin genes. Deletions of subtelomeric sequences have recently been shown to affect the α-globin genes, producing an unusual form of α-thalassemia associated with mental retardation.

Other surprises of general biological interest probably await further investigations of globin gene regulation. The mechanisms involved in coordinate regulation of α- and β-globin gene expression and in the switching of expression from one α- or β-like gene to another during development still need to be resolved. Because of the large body of information currently available on hemoglobin biology, we focus on topics that illustrate general principles of human biology.

6.1 The Globin Genes and Proteins

a. Hemoglobin Protein Structure

Normal human hemoglobins are tetramers containing two each of two different polypeptide chains, α or α-like and β or β-like. Adult hemoglobin contains two identical α-globin chains and two identical β-globin chains. Each of the four globin chains contains a heme group in which an iron atom is bound to a histidine; the iron atom remains in the ferrous state (Fe^{2+}) in both deoxyhemoglobin and oxyhemoglobin.

Different kinds of hemoglobins are made during human development. During the first six weeks of human embryonic development, the hemoglobin consists of two α-like chains, ζ, and two β-like chains, ε, the tetramer being $\zeta_2\varepsilon_2$. The major hemoglobin of the fetus (weeks 6–35 of gestation) consists of two α-globin chains and two new β-like chains, γ; this protein, called hemoglobin F, is designated $\alpha_2\gamma_2$. In the adult, the major component (98 percent) is hemoglobin A ($\alpha_2\beta_2$), and the minor component (2 percent) is hemoglobin A_2 ($\alpha_2\delta_2$). Globin chains present in embryonic, fetal, and neonatal life are shown in Figures 6.1 and 6.2.

The **primary structure** of each globin chain is its amino acid sequence: 141 amino acids in the α- and ζ-chains and 146 in the β-, γ-, δ-, and ε-chains. Interactions between adjacent amino acids along the chain produce both helical and β-pleated sheet **secondary structures** (*Genes and Genomes,* chapter 1). Approximately 75 percent of native hemoglobin is in the α-helical form (Figure 6.3).

X-ray crystallography demonstrated that the three-dimensional or **tertiary** conformations of the α- and β-subunits are similar. The rodlike α-helix is interrupted at specific locations in the hemoglobin chains by nonhelical segments that allow folding. The β-globin chain has eight helical segments, A through H, and the secondary structure of α-globin corresponds to that of β-globin except

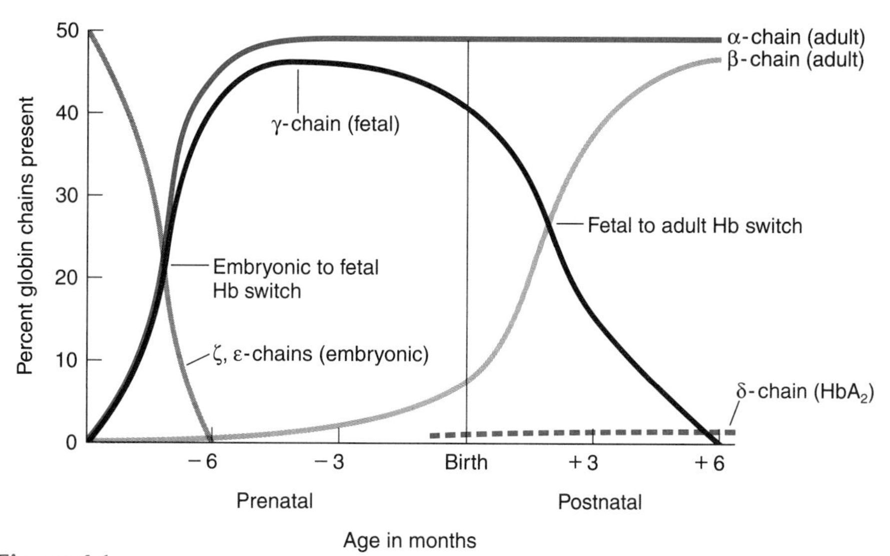

Figure 6.1
Qualitative and quantitative changes in globin chains during human development. Note the ζ-to-α-chain switch and the ε-to-γ-chain switch at 6–8 weeks of fetal life and the γ-to-β-chain switch at about 32 weeks of fetal life. This latter switch is relative in that small amounts of β-chains are produced as early as 6 weeks of fetal life. (Adapted from H. F. Bunn and B. G. Forget, eds., 1986. Hemoglobin: Molecular, Genetic and Clinical Aspects, p. 68. W. B. Saunders Co., Philadelphia.)

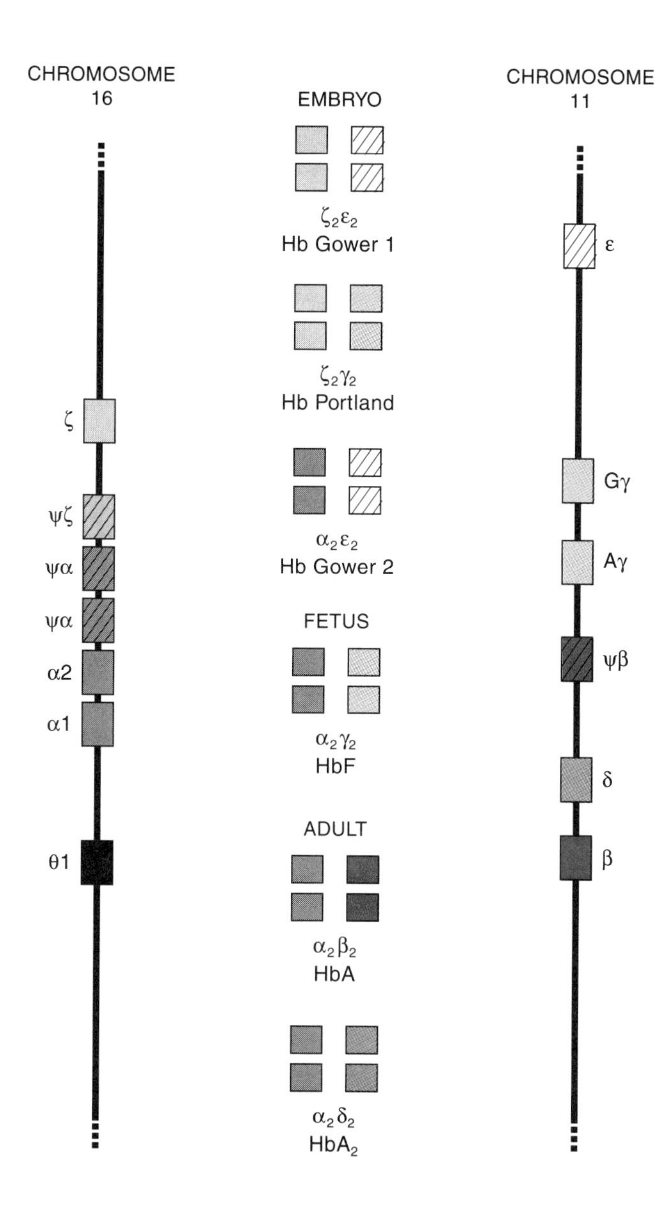

Figure 6.2
Organization of the α- and β-globin gene clusters. In the left column is the α-globin cluster located at the tip of the short arm of chromosome 16 (16p) with a ζ gene, two α genes, and up to four pseudogenes. The functional status of the θ_1 gene is unknown. In the right column is the β-globin cluster located near the tip of the short arm of chromosome 11 (11p) with functional ε-, $^{G}\gamma$-, $^{A}\gamma$-, δ-, and β-globin genes and one pseudogene. Hemoglobin types formed by the gene products at different stages of development are shown in the center column. The embryonic hemoglobins were originally named after the geographic places in which they were discovered. (Adapted from D. G. Weatherall et al. 1989. The Hemoglobinopathies. In C. R. Scriver, A. L. Beaudet, W. S. Sly, and D. A. Valle, eds., The Metabolic Basis of Inherited Disease, 6th ed., pp. 2281–2339. McGraw-Hill, New York.)

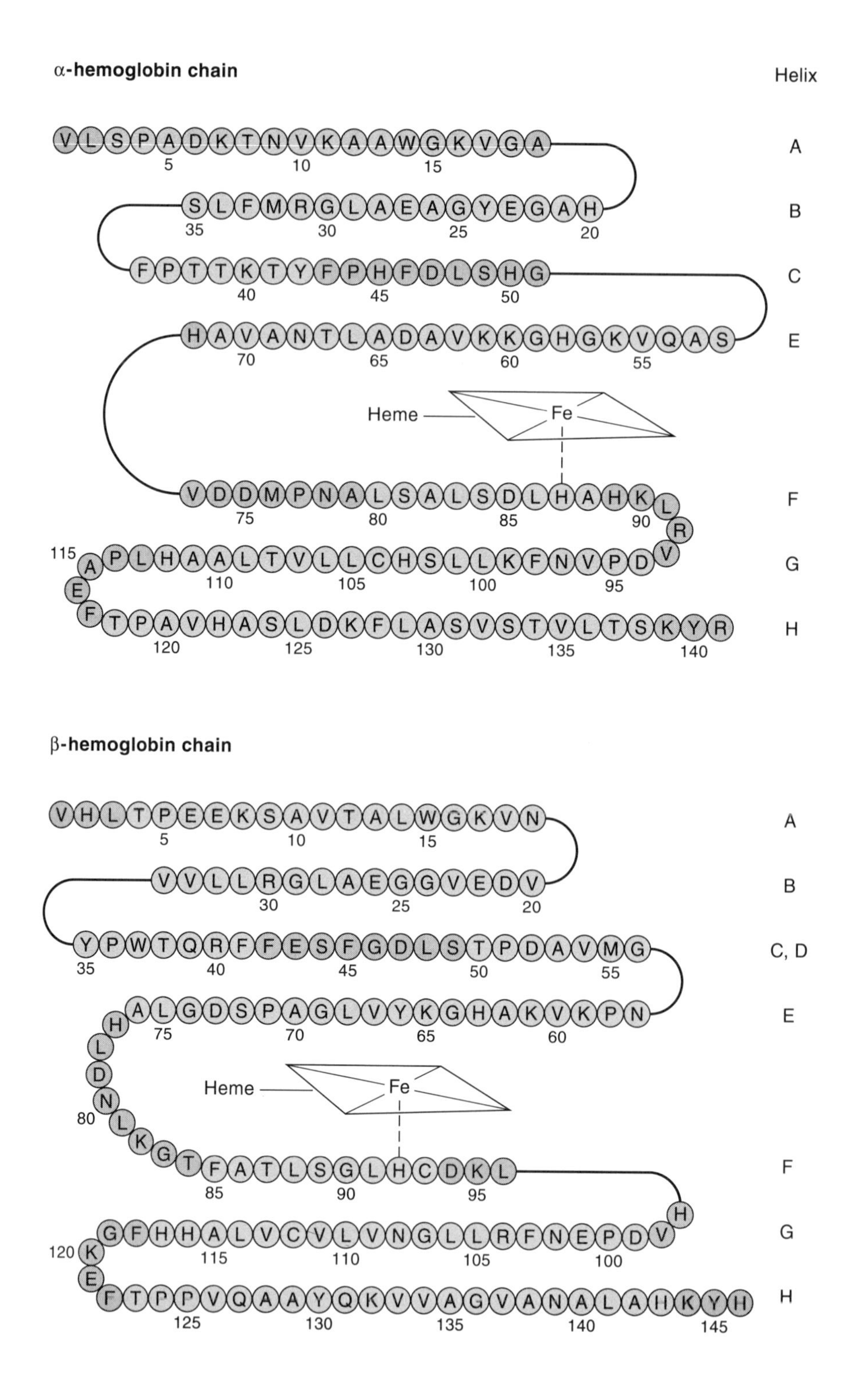

Figure 6.3
Schematic representation of the primary and secondary structures of the α-globin (top) and β-globin (bottom) chains. Residues in an α-helix configuration are in pink; non-helical residues are in gray. Heme groups and charged residues are also shown. (From M. Murayama. 1971. Molecular Mechanism of Human Red Cell (with HbS) Sickling. In R. M. Nalbandian, ed., Molecular Aspects of Sickle Cell Hemoglobin, pp. 3–19. Courtesy of Charles C Thomas, Publisher, Springfield, Illinois.)

for the absence of residues forming the D helix (Figure 6.3). Amino acids with charged side groups (e.g., lysine and glutamic acid) lie on the external surface; uncharged residues tend to be within the interior of the molecule. The heme iron is covalently linked to the histidine residue at position 8 of the F helical segment (F8), which is at position 87 in the α- and 92 in the β-chain. Mutations altering these histidines may reduce the stability of the heme within its globin pocket, producing an unstable protein, or they may increase the stability of the oxidized iron state, resulting in **methemoglobin**.

The crystallographic analysis of hemoglobin revealed its **quaternary structure**, which refers to the relationship of the four globin subunits to each other, showing that the heme groups associated with each of the four globin chains fit into surface clefts equidistant from one another. The tetramer forms an oblate spheroid with a diameter of 5.5 nm and a single axis of symmetry (*Genes and Genomes*, figure 1.36). The four subunits forming a tetramer are labelled α_1, α_2, β_1, and β_2. (Note that these numerical designations have no relationship to the numbering of the genes. For example, the α_1 subunit may be derived from either of the two α-globin genes.) There is no contact between the two β-chains in hemoglobin, but each α-chain touches both β-chains. Bonds across the $\alpha_1\beta_1$ interface are stronger than those at the $\alpha_1\beta_2$ interface. Mutations that cause substitutions of amino acids that form contact points between the α_1 and β_1 globin chains can markedly alter specific functional properties of the hemoglobin. Hemoglobin's quaternary structure changes markedly, depending on whether the heme groups are associated with oxygen (oxyhemoglobin) or not (deoxyhemoglobin). This difference in the quaternary structures accounts for many of the observed changes in the physical properties of the hemoglobin.

b. Functional Properties

To fulfill its physiologic role, hemoglobin must bind oxygen with a certain affinity. One measure of oxygen affinity is P_{50} (the partial pressure of oxygen in mm of Hg required for 50 percent saturation of hemoglobin). Thus, hemoglobin with an increased P_{50} has a decreased affinity for oxygen (Figure 6.4). Oxygen affinity

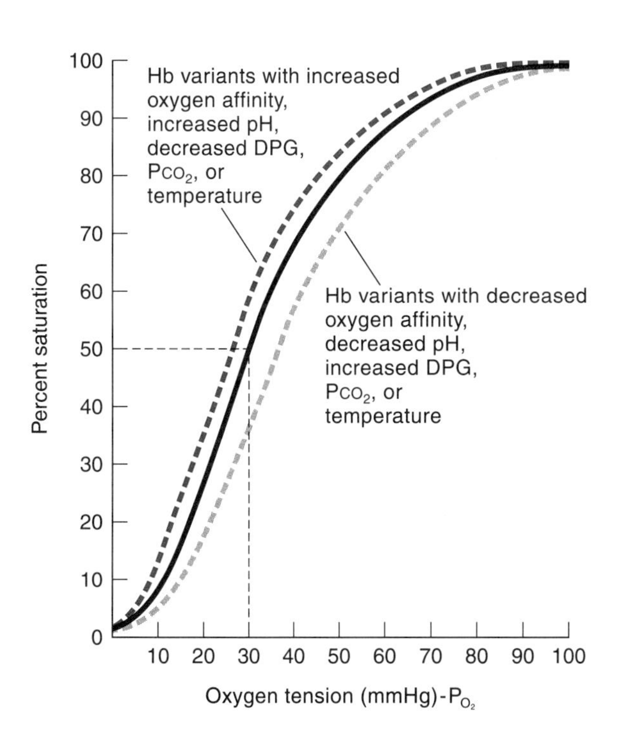

Figure 6.4
The oxyhemoglobin dissociation curve and the effect of different factors on oxygen affinity. Note the effects of pH, 2,3-DPG concentration, P_{CO_2}, and temperature on the percent saturation of hemoglobin at various levels of P_{O_2} (oxygen tension). (Adapted from J. A. Phillips and H. H. Kazazian, Jr. 1990. Haemoglobinopathies and Thalassaemias. In A. E. H. Emery and D. L. Remoin, eds., Principles and Practice of Medical Genetics. Churchill Livingstone, Edinburgh.)

is affected by a number of environmental factors, including temperature, pH, concentration of organic phosphate (DPG), and the partial pressure of carbon dioxide (P_{CO_2}).

The sigmoid shape of the hemoglobin's oxygen dissociation curve reflects heme-heme interaction (i.e., successive oxygenation of each heme group in the tetramer increases the oxygen affinity of the remaining unoxygenated heme groups). The heme-heme interaction that occurs on oxygenation results from the decreased atomic radius of the heme iron that allows the iron atom to fit into the plane of the heme group's porphyrin ring. A series of conformational changes in the protein, each affecting the other heme groups, amplifies this alteration. The sigmoid curve representing the association and dissociation of oxygen from hemoglobin at various partial pressures of oxygen has great physiologic importance because it enables large amounts of oxygen to be bound or released with a small increase or decrease in oxygen tension. In contrast to hemoglobin A, the abnormal hemoglobins H (β_4) and Barts (γ_4) lack subunit interaction and have a hyperbolic rather than a sigmoid oxygen dissociation curve; as a consequence, oxygen is not released at physiologic oxygen tensions.

The Bohr effect is a change in the oxygen affinity of hemoglobin as a function of the pH. Thus, oxygen dissociates from oxyhemoglobin more readily as the pH decreases. This effect is beneficial at the tissue level, where the lower pH promotes oxygen release. The uptake of oxygen in the lungs is enhanced by the opposite changes in pH and elevated carbon dioxide.

Red cells have unusually high concentrations of 2,3-diphosphoglycerate (2,3-DPG). One molecule of 2,3-DPG binds to deoxyhemoglobin in a pocket formed by specific β-chain residues (1, 2, 82, and 143 of both β-chains). Binding of 2,3-DPG to deoxyhemoglobin is important because it stabilizes this form of hemoglobin in preference to the oxygenated form, thereby lowering the oxygen affinity of the molecule. The γ-chain of hemoglobin F lacks the β^{143}-histidine residue, and the resulting decrease in the binding of 2,3-DPG to hemoglobin F accounts for the increased oxygen affinity of fetal red cells compared to that of adult red cells. By contrast, elevated levels of 2,3-DPG, which occur when individuals move to high altitudes where the P_{O_2} is low, promotes more efficient dissociation of oxygen from oxyhemoglobin.

c. The Globin Gene Families

The general structure of globin genes Each globin gene has three coding regions, or **exons**, and two intervening sequences, or **introns** (Figure 6.5). The exons are roughly 200 bp in size (about average for a mammalian exon). The first intron is 150–200 bp, and the second is usually about 850 bp. The 5′-most 60 bp of the first exon and the 3′-most 150 bp of the last exon are not translated. This exon/intron structure has been conserved during evolution, and all vertebrates and many invertebrates have oxygen-carrying proteins whose gene structures are very similar to those of the human globin genes. The vertebrate α- and β-globin genes occur in two clusters, the α-globin gene cluster and the β-globin gene cluster (Figure 6.2).

The a-globin gene cluster The α-globin gene cluster occupies about 25 kbp very close to the telomere of the short arm of chromosome 16. The cluster contains a single ζ gene that is functional in all individuals (ζ_2); another embryonic ζ gene is functional in some individuals (ζ_1), but in others, it contains a mutation that abolishes its function and makes it a pseudogene ($\varphi\zeta_1$). The ζ_2 gene (which is closest to 16pter) is located about 150 kbp from the telomere of chromosome 16 in some individuals, but that region appears to be polymorphic because, in other

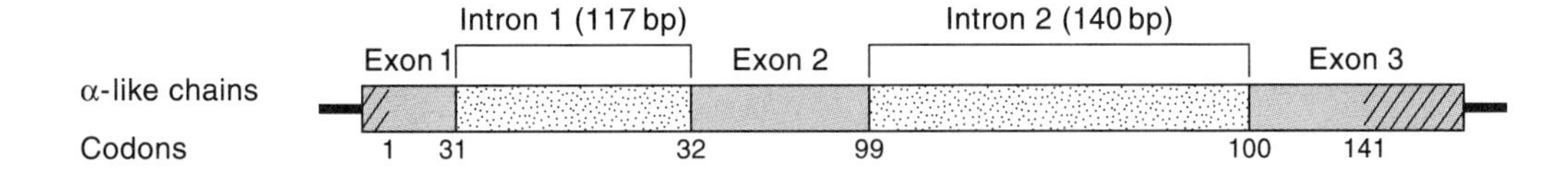

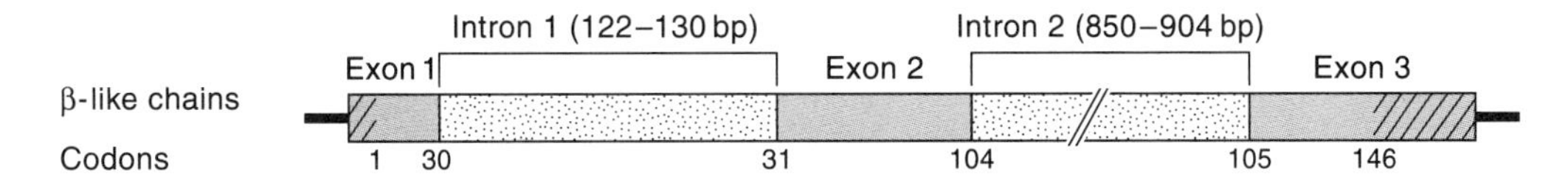

Figure 6.5
General structure of α-like and β-like globin genes. The α-like genes are roughly 0.8 kbp in size and contain 3 exons and 2 introns; the β-like genes are roughly 1.5 kbp in size and have the same exon/intron structure.

individuals, the distance to the telomere can be up to 430 kbp. Besides two nonfunctional α-pseudogenes ($\varphi\alpha_2$, $\varphi\alpha_1$), there are two functional α-genes, α_2 and α_1, and a recently found α-like gene, θ1, whose sequence suggests it is functional but whose polypeptide product has not yet been found.

There are several other interesting features about the α-globin gene cluster. A number of 4 kbp blocks of sequence homology occur around the two α-genes; one pair of blocks includes the two α-genes and their 5′ flanking regions, and another pair of blocks occurs in the 3′ flanking regions of the α_2 gene and the $\varphi\alpha_1$ pseudogene (Figure 6.6). These blocks of homology suggest that an original single α-gene was first duplicated and later gave rise to additional gene copies. These duplications probably took place by nonallelic pairing of homologous regions and unequal crossing-over (see Figure 6.6). Unequal crossing-over is an important mechanism that reduces or increases the number of copies of a gene segment in the genome. Examples of this phenomenon have been found in both globin gene clusters.

Under normal conditions, α- and β-globin chains are produced at the same rate in erythroid cells (see Section 6.2a). α-Thalassemia is a condition in which α-globin synthesis in erythroid cells is reduced relative to that of β-globin synthesis. Nearly all instances of α-thalassemia are due to gene deletions, a large fraction of which is secondary to unequal crossing-over events within the α-globin homology blocks (Figure 6.6). The two commonly found chromosome 16s carrying a single α-globin gene are the result of different kinds of crossing-over events, each of which has occurred a number of times in human history.

An additional feature of the α-gene cluster is the unequal expression of the two α-globin genes. The α_2 gene (leftward gene) is expressed about two and a half to three times more efficiently than the α_1 gene (rightward gene) (see Section 6.2a). Another feature of the α-globin gene cluster is its locus control region (α-LCR), which controls the expression of all the α-like genes in the cluster. The α-LCR lies in a region of open chromatin within the intron of another gene about 40 kbp 5′ to the cluster.

The β-globin gene cluster The β-globin gene cluster is a 50 kbp region at the distal end of the short arm of chromosome 11. It contains one embryonic gene (ε), two fetal genes ($^{G}\gamma$ and $^{A}\gamma$), one major adult gene (β), one minor adult gene (δ), and one pseudogene ($\varphi\beta_1$). The homology blocks within the β-globin cluster are

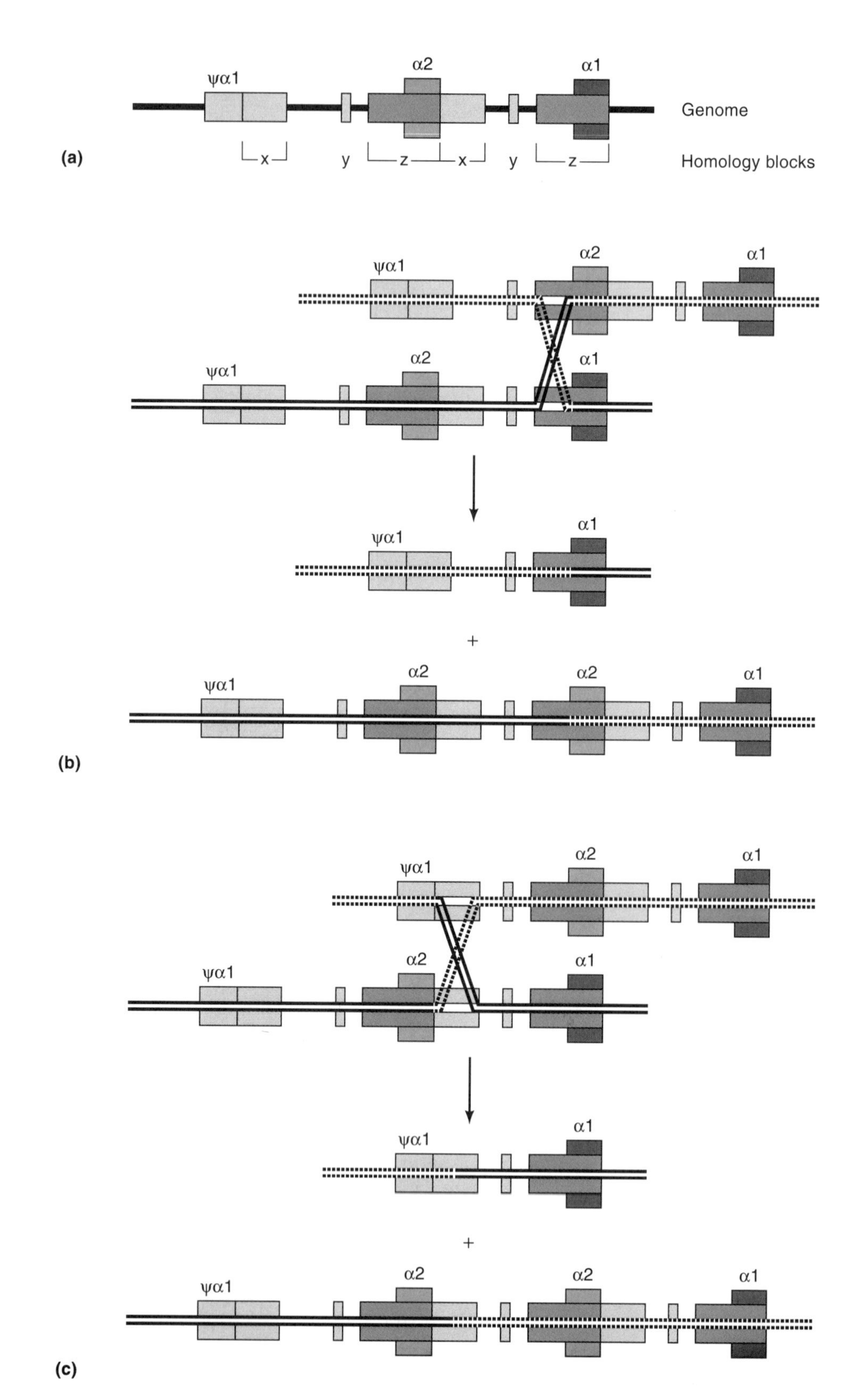

ψα1
α2
α1
Genome
(a)
x
y
z
x
y
z
Homology blocks
ψα1
α2
α1
ψα1
α2
α1
ψα1
α1
+
ψα1
α2
α2
α1
(b)
ψα1
α2
α1
ψα1
α2
α1
ψα1
α1
+
ψα1
α2
α2
α1
(c)

Figure 6.6 *(facing page)*
Homologous recombination in the α-globin gene cluster. (a) The X, Y, and Z homology blocks in the DNA near the α-globin genes are shown. (b) Nonallelic pairing and crossing-over in the Z homology block produces a 3.7 kbp deletion and results in a chromosome carrying a single functional α gene (α1). The reciprocal product of this crossing-over event is a chromosome with 3 α genes, which has also been observed in humans. (c) Nonallelic pairing and crossing-over in the X homology block produces a chromosome with a 4.2 kbp deletion and a single α gene (α1). (Adapted from D. J. Weatherall et al. 1989. The Hemoglobinopathies. In C. R. Scriver, A. L. Beaudet, W. S. Sly, and D. A. Valle, eds., The Metabolic Basis of Inherited Disease, 6th ed., pp. 2281–2339. McGraw-Hill, New York.)

---

generally limited to the protein coding regions themselves; for the most part, the intron sequences differ greatly even among genes whose protein sequences are highly homologous. An exception to this rule is the structure of the $^{G}\gamma$ and $^{A}\gamma$ genes. These genes, which encode proteins that differ by a single amino acid, have significant homology at the nucleotide level beginning 5′ to the transcription start site and extending through nearly all the second intron sequences. This homology is most probably a consequence of **gene conversion**, a phenomenon by which one sequence (in this case, the upstream $^{G}\gamma$ gene) is used to correct a second sequence (here the downstream $^{A}\gamma$ sequence) (see Figure 6.7).

A locus control region for the β-globin gene cluster (β-LCR) is located 10–30 kbp 5′ to the ε-globin gene. As implied by the nomenclature, the β-LCR controls the expression of the entire β-globin gene cluster.

d. Hemoglobin Ontogeny

The globin genes are expressed at different times and in different relative amounts during human development (Figure 6.1). In addition, as a rule, they are expressed in a tissue-specific fashion. The ε- and ζ-chains are expressed in the embryonic yolk sac, α- and γ-chains are synthesized during fetal life in fetal liver cells, but late in fetal life and after birth, the adult α- and β-chains are produced in bone marrow cells. However, small amounts of γ-globin have been found in yolk sac cells, and β-chains are produced in small quantities in liver as early as six weeks of fetal life, indicating that the timing and tissue specificity of globin gene expression are not absolute.

The sequence in which the various globin chains appear is helpful in understanding the timing of the onset of clinical manifestations of the hemoglobinopathies and thalassemias. For example, a deficiency of α- and γ-chain synthesis or the occurrence of α- or γ-variants with abnormal functions should be observed at birth, but a deficiency of β-chains may not cause symptoms until several months of age. The levels of β-chain variants, such as occurs in hemoglobin S of individuals with sickle cell disease, also increase progressively during the first months of life, so the onset of the clinical manifestations of sickle cell disease may be delayed until the latter half of the first year. In Section 6.2, we discuss what is known about the molecular basis of **hemoglobin switching** during development, the process by which synthesis of one globin chain stops and is replaced by expression of another gene, (e.g., how the production of the γ-chain is replaced by the β-chain).

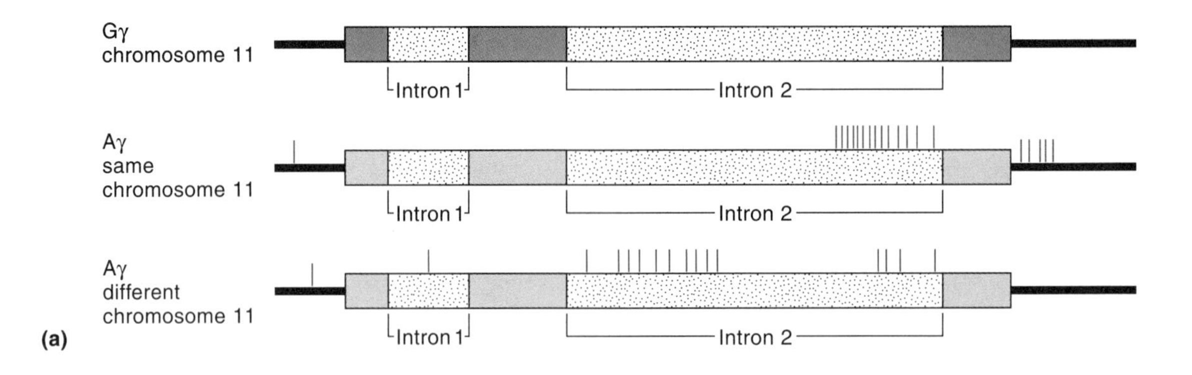

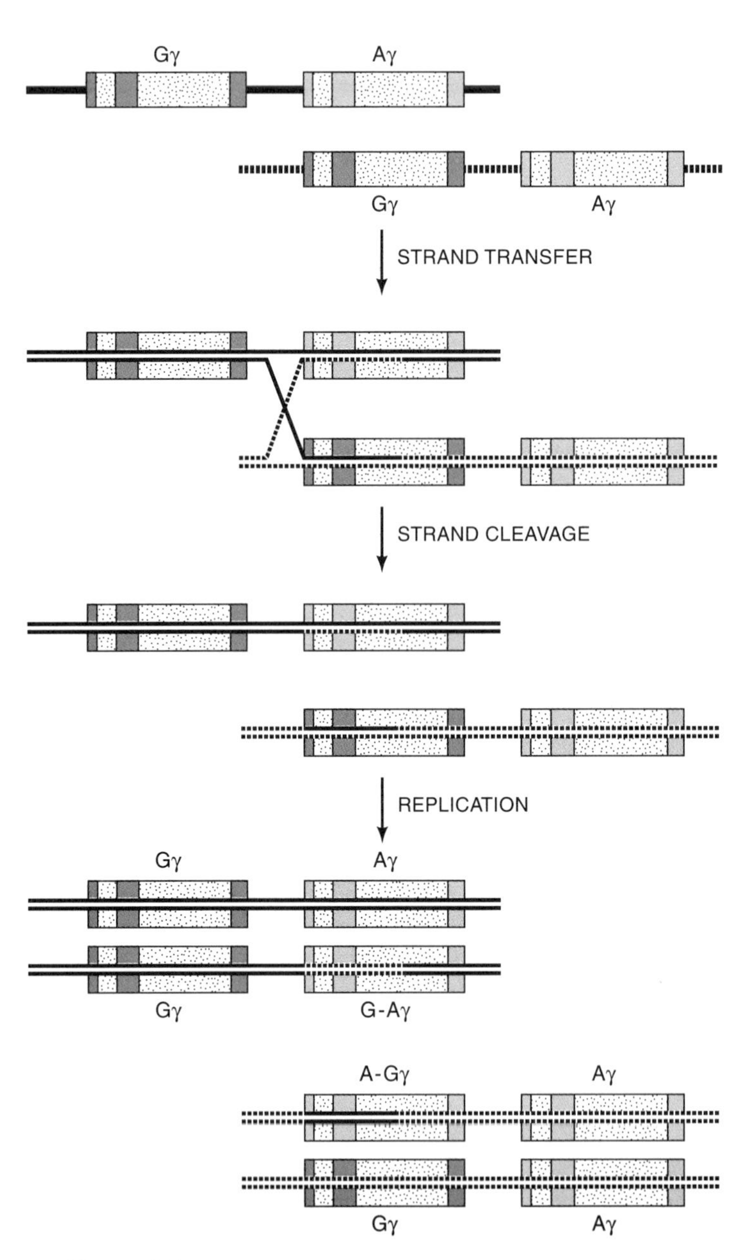

Figure 6.7
Gene conversion in the γ-globin genes. (a) The nucleotide sequences of the roughly 1.1 kbp 5′ and within the $^{G}\gamma$ and $^{A}\gamma$ genes on the same chromosome 11 are identical, but sequences 5′ and 3′ to this region have diverged. Other $^{A}\gamma$ genes differ significantly from this $^{A}\gamma$ gene throughout intron 2. (b) The best interpretation of these data is that the sequence of the upstream $^{G}\gamma$-globin gene was used to correct the sequence of the downstream $^{A}\gamma$-globin gene through a recombinational event between a $^{G}\gamma$-globin gene aligned with an $^{A}\gamma$-globin gene. This event is essentially a double recombination over a DNA distance of 1 kbp or less.

6.2 Expression of Globin Genes

a. Coordinate Regulation of α- and β-Globin Chain Synthesis

By the 1960s, considerable indirect evidence indicated that there are two α-globin loci and only a single β-globin locus. First, in an individual with a variant α-globin chain (e.g., α^G), the variant hemoglobin ($\alpha_2^G\beta_2^A$) made up 15–25 percent of the total adult hemoglobin. In contrast, in an individual carrying a variant β-globin chain (e.g., β^S), the abnormal hemoglobin S ($\alpha_2\beta_2^S$) made up 40–50 percent of the total hemoglobin. Second, although there were only two β-thalassemia states, there were four distinct α-thalassemia states of increasing severity (deficiency states corresponding to the lack of one, two, three, or all four genes at the two α-globin loci).

Considering that early erythroid cells (reticulocytes) can express each of the four α-globin genes and the two β-globin genes and that the hemoglobin tetramer has equal numbers of α- and β-chains, it is important to understand how the expression of the two β-globin genes and the four α-globin genes is regulated to produce equal numbers of chains. The answer turns out to be more complicated than expected and is discussed here because it is still the only complete description of coordinate regulation of unlike polypeptides of a heteromeric protein. In addition, it provides an informative illustration of how progress depends on available experimental methods.

Early experimental approaches By 1968, a great deal was known about the mechanisms of mRNA translation, including the following: (1) protein synthesis can be studied in cell-free extracts, including ones made from reticulocytes, (2) polyribosomes and nascent polypeptide chains can be isolated, and (3) a number of drugs inhibit various steps in translation. Incubating rabbit reticulocytes with a ^{3}H-amino acid found predominantly in one globin chain and a ^{14}C-amino acid present more frequently in the other allowed the determination of the number of ribosomes on each globin mRNA. The answer was a big surprise. It was expected that two mRNAs encoding proteins of the same size, such as the globin mRNAs, would be associated with the same number of ribosomes. Instead, there are three to four ribosomes on the α-mRNA and five to six per β-mRNA. Inasmuch as α-mRNA is translated at the same rate as β-mRNA, the ribosome-binding data indicate that translation initiation on α-mRNAs occurs more slowly than on β-mRNAs.

In 1971, the relative numbers of α- and β-globin mRNA molecules in rabbit reticulocytes were counted. This was accomplished by adding inhibitors of peptide bond formation to reticulocytes incubated with the ^{3}H- and ^{14}C-amino acids mentioned previously and measuring the size of the β- and α-polyribosomes. Then the ratio of the newly synthesized α- and β-globin was measured after separating the α- and β-globin polypeptides from the isolated total globin by column chromatography (Figure 6.8). The inhibitors, which greatly reduced peptide bond formation and total globin synthesis, nevertheless resulted in the accumulation of globin mRNAs that were fully packed with ribosomes. The relative amounts of α-mRNAs to β-mRNAs obtained under these conditions turned out to be not one or two, but about one and a half. The amount of each of the globin mRNAs was also estimated by solution hybridization techniques (which had been developed by the mid-1970s). The result was the same: an α-mRNA to β-mRNA ratio of 1.5 in reticulocytes. From all these experiments, researchers concluded that coordinate regulation of globin chain synthesis from two α-globin loci and a single β-globin locus resulted from a combination of (1) down-regulation of α-mRNA production and (2) lowered translation of each α-mRNA molecule resulting from reduced initiation.

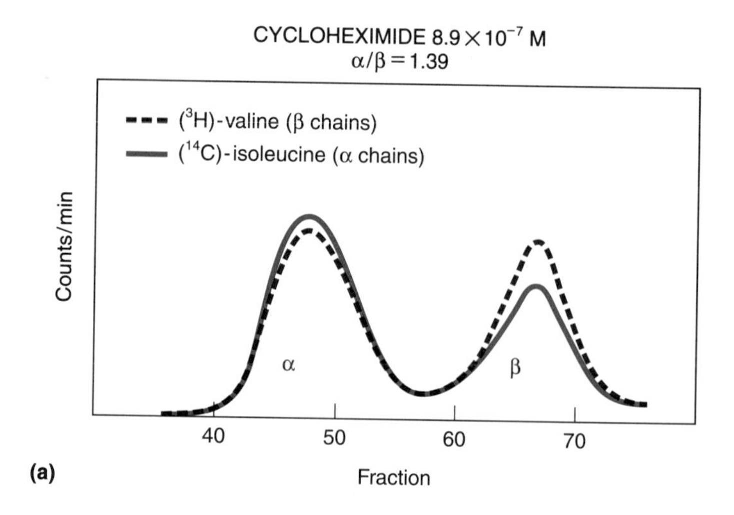

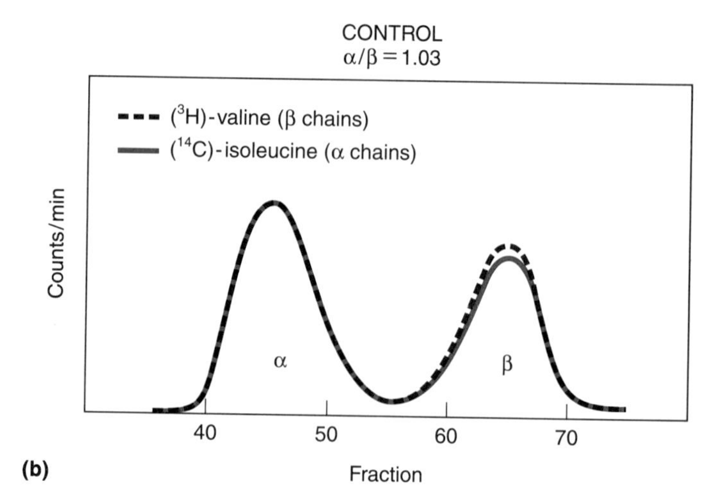

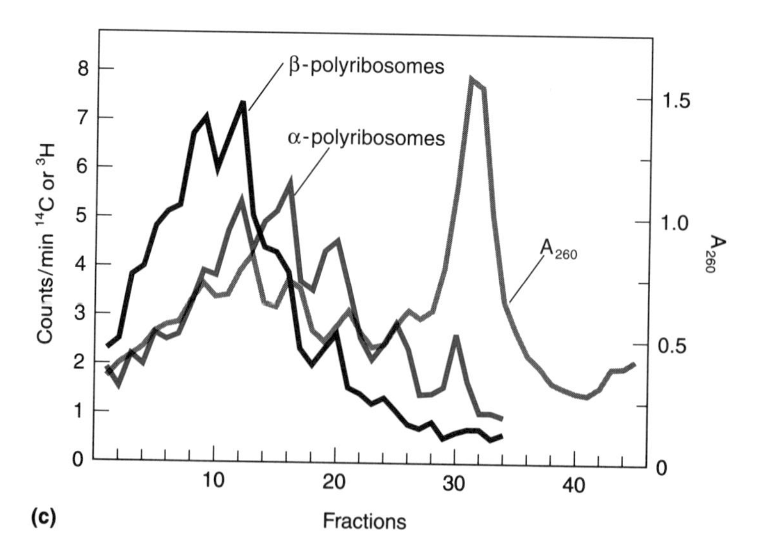

Figure 6.8
The number of mRNA molecules is about 1.5 times that of β-mRNA molecules in rabbit reticulocytes. (a) Reticulocytes incubated with (^{14}C)-isoleucine and (^{3}H)-valine either with or without cycloheximide, an inhibitor of peptide bond formation. Each nascent α-chain contains, on average, 2.4 isoleucine residues and 4.2 valine residues; each nascent β-chain has 0.23 isoleucines and 9 valines. Thus, the (^{3}H) radioactivity is principally derived from nascent β-chains, and the (^{14}C) radioactivity is mainly due to α-chains. (a) The number of nascent α-chains exceeds that of nascent β-chains when the cells are incubated with cycloheximide and protein synthesis is inhibited. (b) The result was confirmed by analysis of the completed α- and β-globin chains synthesized during the incubation. (c) Polyribosomes were separated by centrifugation of the ribosomal fraction in a sucrose density gradient and globin chains were separated by carboxymethylcellulose column chromatography. (Adapted from H. F. Lodish. 1971. Alpha and Beta Globin Messenger Ribonucleic Acid: Different Amounts and Rates of Initiation of Translation. *J. Biol. Chem.* 246 7131–7138.)

Table 6.1 *Coordinate Regulation of α- and β-Globin Chain Synthesis*

Gene	α_2	α_1	β	α/β Ratio 2.0
Relative mRNA concentration	1.1	0.4	1.0	1.5
Relative translation rate	0.7	0.7	1.0	0.7
Globin synthesis/gene	0.75	0.25	1.0	1.0

Direct quantification of each globin mRNA With the advent of the recombinant DNA era, the globin gene clusters were characterized, and measuring the quantity of the two α-globin mRNAs using specific, distinguishing DNA probes derived from their 3′-untranslated regions became possible. With these α_2- and α_1-globin gene probes and reticulocyte mRNA from an individual carrying a single normal α_1-gene and a single variant α_2-gene, Steven Liebhaber and Y. W. Kan showed that the synthesis of mRNA from the variant α_2-gene was two and a half times that of the α_1-gene. Later through quantification of the synthesis of specific α-globin mutants encoded by each α-globin gene, researchers showed that the α_2- and α_1-globin mRNAs are translated at similar rates and that α_2-globin chains are synthesized at two and a half to three times the rate of α_1-globin chains. These data demonstrate a surprising complexity in the mechanisms used to achieve coordinate regulation of α- and β-globin synthesis, a summary of which is shown in Table 6.1.

A similar question relates to the unequal synthesis of δ- and β-globin chains. Fifty times more β-globin than δ-globin chains are produced in adults. In this case, however, the DNA sequence in the 5′ promoter region of the δ-globin gene differs from that in the β-globin gene promoter, and these differences greatly reduce the ability of the δ-gene to be transcribed.

b. Transcriptional Control of Globin Gene Expression

The regulation of expression of the β-globin cluster of genes provides a model for understanding human gene expression during development. As mentioned earlier, the various globin genes are expressed in erythroid cells in a developmental sequence. The mechanisms of switching expression from one specific gene to another are not clearly understood. The genes of the α-like globin gene cluster are arranged on chromosome 16 in the order of their expression: embryonic (ζ) and adult (α). The β-like globin genes are arranged on chromosome 11 also in the order of their expression: embryonic (ε), fetal ($^G\gamma$ and $^A\gamma$), and adult (δ and β). The importance of regulatory DNA sequences within the β-globin gene cluster (***cis*-acting elements**) and proteins that bind to those elements but are encoded by genes elsewhere in the genome (***trans*-acting factors**) has been recently revealed.

The search for regulatory sequences in the immediate 5′ and 3′ flanking sequences of the β-like globin genes revealed enhancer elements located upstream, within, and downstream of the genes. These *cis*-acting elements are responsible for correctly regulated expression in erythroid cells. Their functions were assessed in transgenic mice carrying the human β- or γ-globin genes and their neighboring enhancer sequences (see Chapter 11). In these mice, transgene expression was tissue-specific and developmentally appropriate. However, the levels of expression were very low (i.e., less than 1 percent of the endogenous mouse globin genes). Therefore, neighboring regulatory sequences, although conferring tissue and developmental specificity, are not sufficient to regulate normal levels of β-globin gene expression.

The locus control region Patients carrying partial deletions within the β-globin gene cluster provided clues to the location of additional, previously unsuspected, regulatory elements. For example, deletions that eliminate sequences upstream of the β-globin gene completely silence its expression even when the gene itself is intact. DNAse I–hypersensitive sites (in which the chromatin can be digested by low concentrations of DNAase I) are present in a region about 20 kbp upstream of the ε-globin gene. Ligation of this region of DNAase I hypersensitivity to the β- or α-globin genes or even heterologous genes and introduction of these constructs into transgenic mice or cultured erythroid cells result in high-level, erythroid-specific expression equivalent to the level of expression of the endogenous mouse globin genes. The level of expression also depends on the number of introduced gene copies. For example, in one study, the level of expression of the β-globin transgene containing the DNAase I–hypersensitive sites was, on average, 110 percent of the endogenous mouse β-globin gene per gene copy in virtually all transgenic mice. In contrast, the omission of the DNAse I–hypersensitive sites results in expression in only one-third of the animals and at levels only 0.5 percent on a per gene basis of the endogenous mouse β-globin gene. This region of DNAse I hypersensitivity is termed the locus control region; it consists of four elements: HS1 to HS4, located between 6 and 20 kbp upstream of the ε-globin gene (Figure 6.9). As mentioned earlier, an LCR is also present about 40 kbp upstream of the α-globin gene cluster.

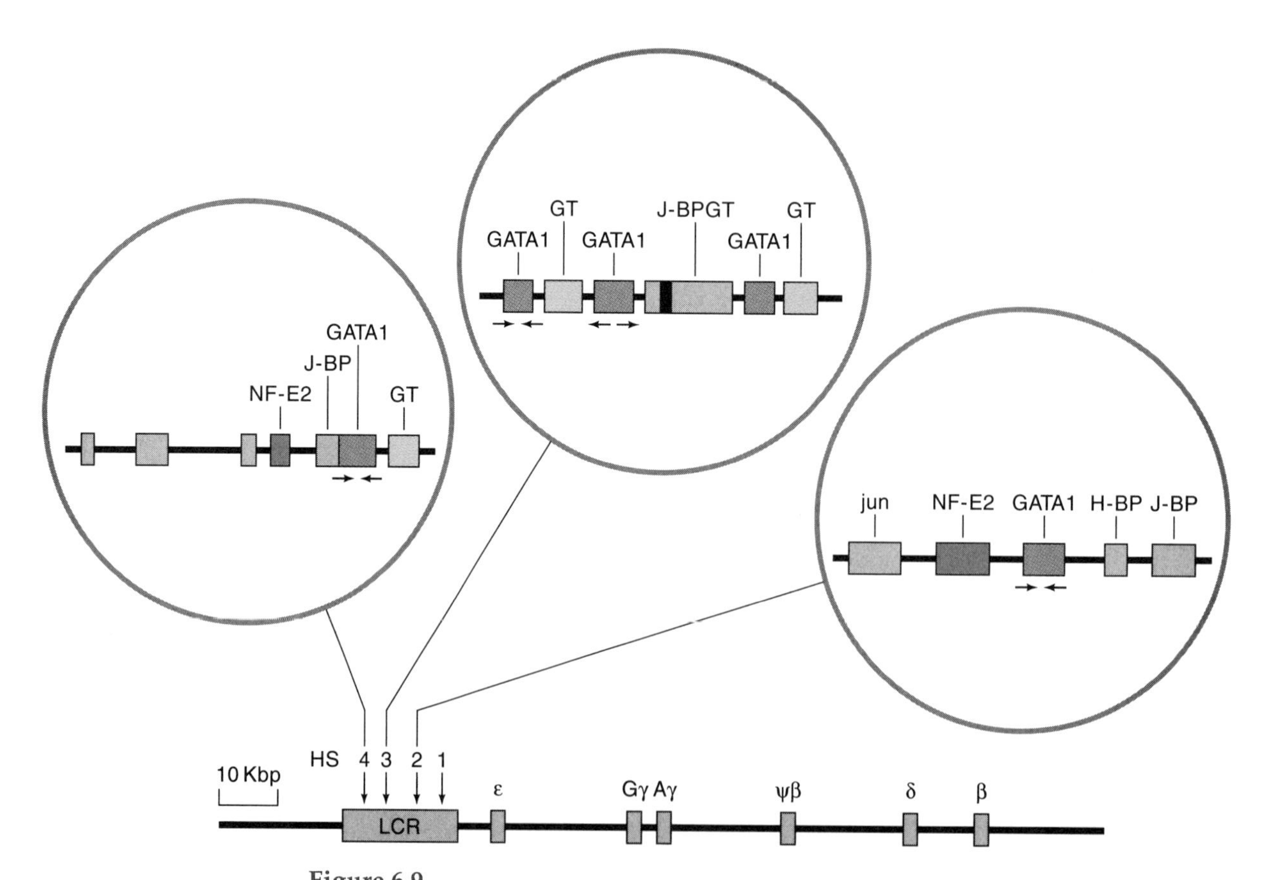

Figure 6.9
An LCR (locus control region) is located 6–20 kbp 5′ to the ε-globin gene. Four DNAse I hypersensitive sites (HS1 to HS4) are located within the LCR, and within HS2, HS3, and HS4 are consensus sequences for binding of several ubiquitous transcription factors and the tissue-specific factors, GATA-1, and NF-E2.

All four HS elements are needed to obtain normal levels of expression. HS2 is perhaps the most important because it alone can increase expression of a linked β-globin gene about 40 percent in transgenic experiments. Although HS2 is important for increasing β-globin gene expression *in vivo,* this sequence by itself is not an efficient enhancer of globin gene expression. Molecular dissection of the various HS elements revealed that their full activity is manifested by small segments of 0.5 to 1 kbp per HS. Figure 6.9 illustrates several of the identified DNA sequences found within the HS elements of the LCR and known to interact with specific DNA-binding regulatory proteins. These sequences include binding sites for the Jun/Fos family of DNA-binding proteins and the ubiquitous H-BP, J-BP, Sp1, and TEF-2 factors. More interesting are the erythroid-specific GATA1 and NF-E2 proteins. There are additional, not yet identified, regulatory factors also involved in the function of the LCR. Extensive sequence homology between the human, mouse, and goat LCR also underscores its importance in β-globin gene expression. A naturally occurring deletion that specifically affects the LCR is known to silence the intact β-globin gene. This 35 kbp deletion, which causes a "Hispanic" type of γδβ thalassemia, eliminates DNA sequences upstream of the ε-globin gene, including HS2 to HS4, and renders the γ-, δ-, and β-globin genes nonfunctional.

Function of the LCR The LCR is a novel regulatory region with two probable functions. The first is the organization of the globin gene cluster into an "active chromatin" domain that permits regulatory factors to gain access to individual genes. Thus, the Hispanic γδβ-thalassemia deletion results in an altered chromatin structure throughout more than 100 kbp in the β-globin gene cluster as revealed by a change in the sensitivity to DNAse I digestion (see Figure 6.10). Second, the LCR acts as an enhancer of transcription of the genes in the β-globin gene cluster, as illustrated in experiments with transgenic mice.

The developmental regulation of expression of the individual genes of the human β-globin gene cluster has been studied extensively in transgenic mice (Figure 6.11). The human ε-globin transgene is not expressed in transgenic mice.

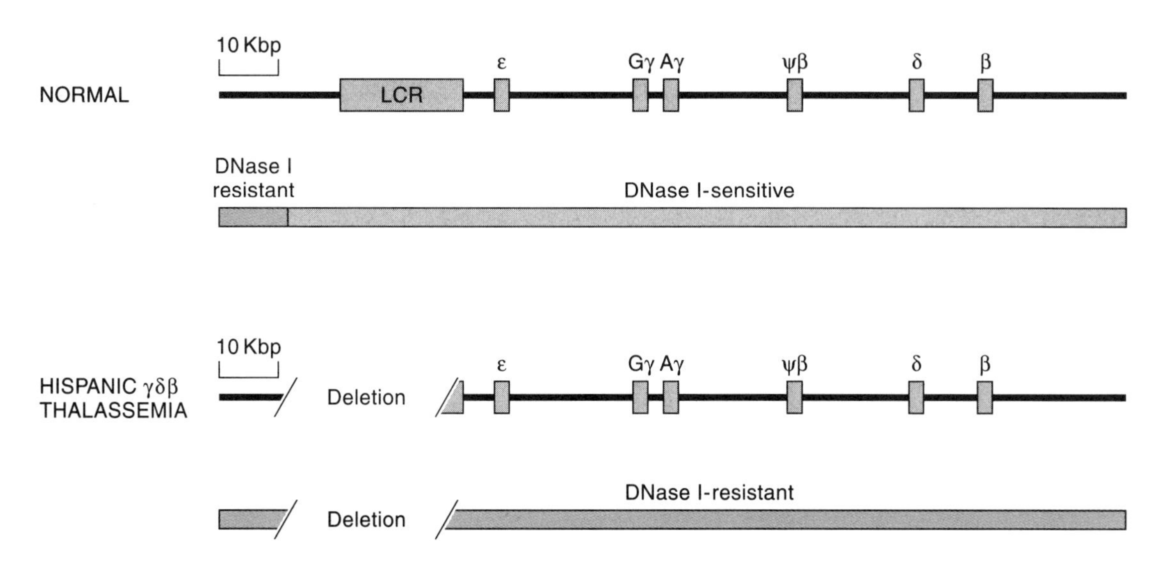

Figure 6.10
Deletion of the LCR of the β-globin gene cluster inactivates the β-globin gene. The LCR maintains an "open chromatin" structure. The 35-kbp deletion within the β-globin gene cluster that is associated with the Hispanic γδβ-thalassemia makes the whole β-globin gene cluster DNAse I resistant. In the normal chromosome the β-globin gene cluster is DNAse I sensitive. The β-globin gene 3′ to the Hispanic deletion is completely inactive.

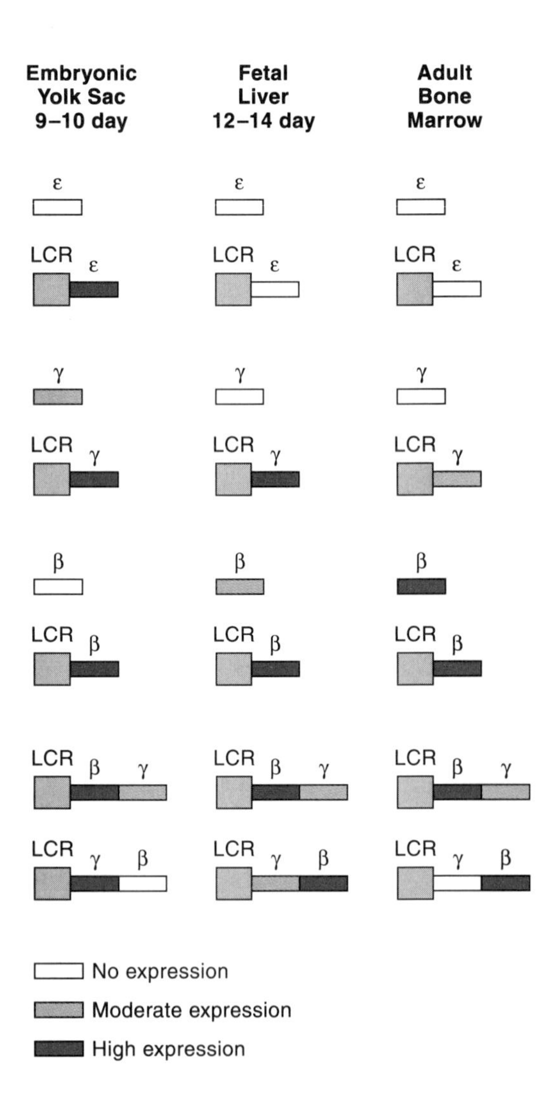

Figure 6.11
The effect of the LCR on expression of globin transgenes in mice. The LCR effect on ε-, γ-, and β-globin expression is complex, but developmental and tissue-specific expression is appropriate with an LCR-γβ transgene.

However, when the ε-globin gene is linked to the LCR, it is highly expressed in the embryonic stage (yolk sac, days 9–10) and silenced in the fetal liver (days 12–14) and adult bone marrow. Sequences 200 to 300 nucleotides 5′ to the ε-globin gene play a role in silencing its expression. This silencer may have an important function in switching from ε- to γ-globin gene expression.

When the human γ-globin gene is put into the mouse genome as a transgene, it is expressed during the embryonic stage and is silent in fetal liver and adult bone marrow. When human LCR sequences are added to the γ-globin gene, it is expressed in all developmental stages. However, when the construct contains the LCR and both the γ- and β-globin genes, the γ-gene is silent in adult mice, and the β-globin gene is expressed in fetal liver and adult bone marrow. This result suggests that the globin genes may compete for the LCR effect on expression, whatever it may be. In some transgenic mice, in which only one or two copies of the LCR-γ-gene construct are integrated, the γ-globin gene is silenced in adult bone marrow; this suggests the existence of a γ-globin gene silencer that shuts off expression of this gene independent of the presence of the β-globin gene. This element has not yet been identified.

The human β-globin transgene alone is expressed only in definitive erythroid cells (fetal liver and adult bone marrow) and is silent in the yolk sac. When the β-globin gene is linked to the LCR in a transgene, the β-globin gene loses its developmental control and is expressed at all stages. In an LCR-γ-β transgene, the β-globin gene is not expressed in the yolk sac, again suggesting competition between the globin genes for LCR action. Moreover, in an LCR-β-γ transgene, the β-globin gene is expressed in the same manner as in the LCR-β construct, suggesting that the proximity of the gene to the LCR is also important for the developmental regulation of expression.

Switching globin gene expression The current model for switching expression from one β-like gene to another is schematically shown in Figure 6.12. The LCR maintains the chromatin of the whole gene cluster in an "open" configuration accessible to transcription factors. The ε-globin gene, the closest to the LCR, responds to its influence and is transcribed in the primitive cells of the yolk sac. The other globin genes are not expressed, possibly because the ε-globin gene competes for the important available transcription factors. After the embryonic period, the ε-globin gene is shut off through action of its autonomously regulated silencer. This silencing cannot be overcome by the LCR enhancer function. Subsequently, the γ-globin genes are efficiently expressed in the fetal liver, presumably because of their proximity to the LCR. The β-globin gene remains relatively inactive during this period because its genomic position is disadvantageous in the competition for enhancer factors. Later, the β-globin gene attains high levels of expression in the adult bone marrow, but it is less clear how this activation occurs. It is not clear whether the silencing of the γ-globin gene in the

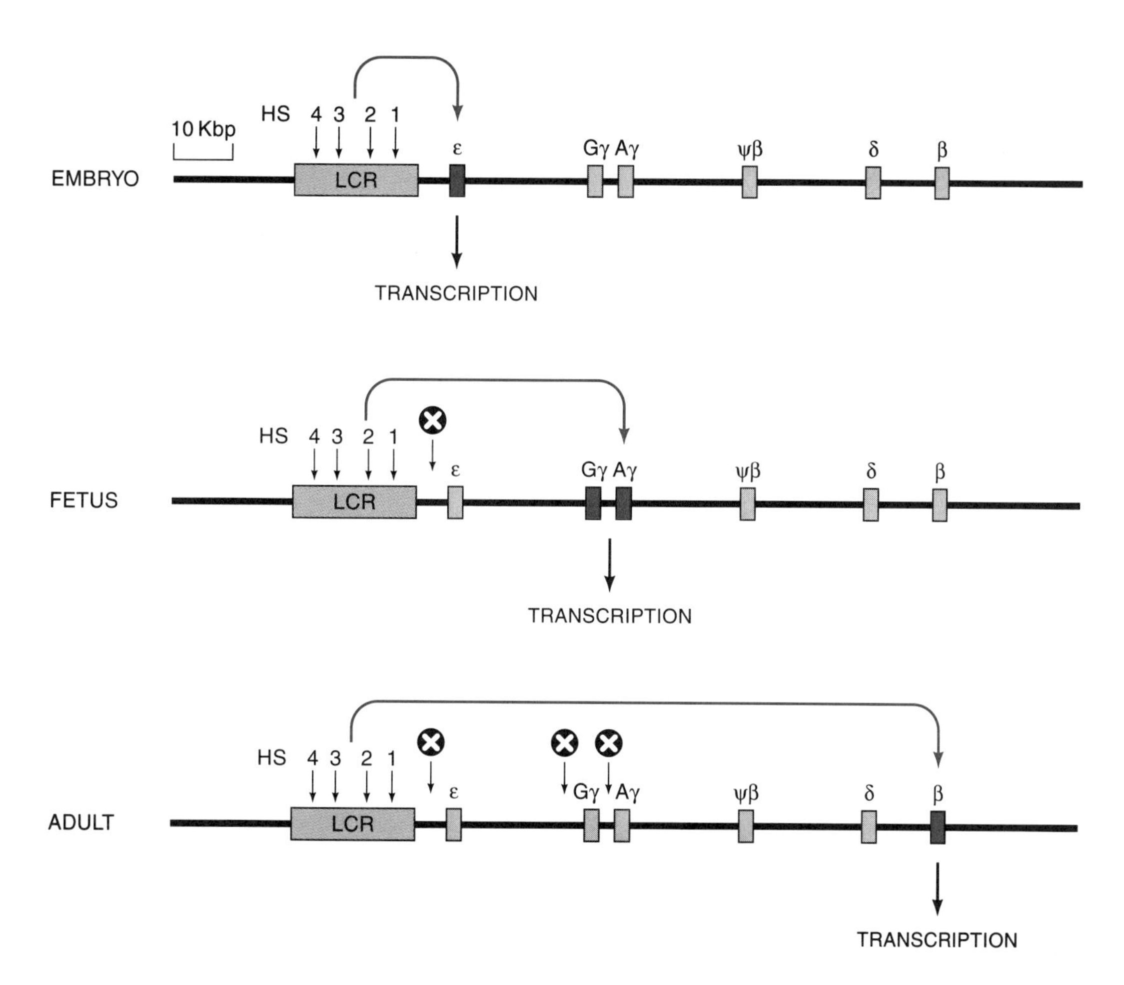

Figure 6.12
Model for human globin gene switching in the β gene cluster. The LCR is thought to act on the most proximal globin genes, which are not independently silenced. Thus, in the embryo it acts on the ε gene; in the fetus it acts on the γ genes because the ε gene is silenced; and in the adult where the ε and γ genes are silenced, the LCR acts on the adult β gene.

postnatal liver is regulated autonomously through silencer sequences; naturally occurring mutations in the 5′ flanking sequences of the γ-globin gene suggest that such silencing elements do exist. A more detailed understanding of globin gene developmental regulation and switching will undoubtedly emerge in time, and the model presented here is likely to be modified accordingly.

Hereditary persistence of fetal hemoglobin Most deletions in the β-globin cluster affect more than the β-globin gene itself and produce complicated phenotypes. Among these are the deletions associated with **hereditary persistence of fetal hemoglobin (HPFH)** and δβ-thalassemia. Elucidation of the mutations that cause HPFH was particularly important because these mutations provided insights into the control of gene expression and hemoglobin switching. Deletions of the δ- and β-globin genes that leave the γ-globin genes intact are usually responsible for these disorders. In certain situations, mutations in the β-globin gene cluster result in the production of substantial amounts of γ-chains but no δ- or β-chains. In the conditions known as HPFH, γ-globin synthesis is much greater than in δβ-thalassemia, and this accounts for the milder phenotype in the former condition.

HPFH can also be caused by point mutations in the distal promoter sequences of either the $^{G}\gamma$ or $^{A}\gamma$ genes (Figure 6.13). These mutations result in overexpression of the affected gene and are referred to as "up promoter" mutations. A β-globin gene cluster containing an HPFH point mutation will produce γ-globin of either the $^{G}\gamma$ or $^{A}\gamma$ type at about 30 percent of the rate expected for adult β-globin but will produce a concomitant decrease in the number of adult

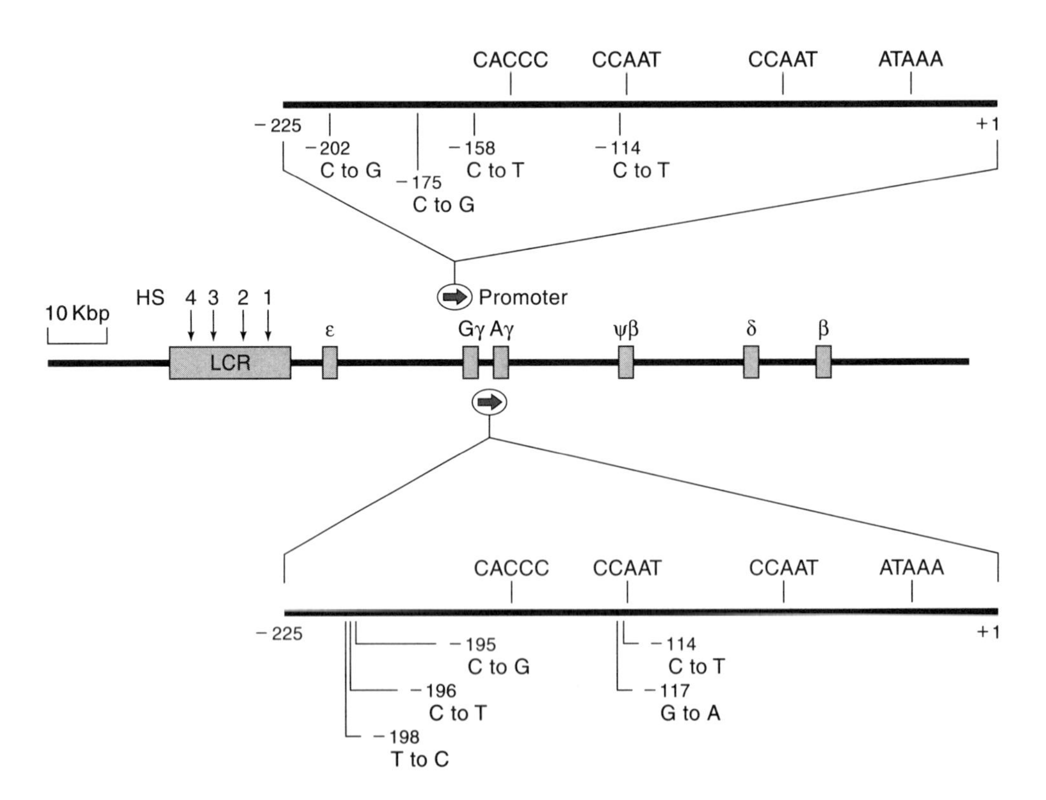

Figure 6.13
Point mutations that produce a form of hereditary persistence of fetal hemoglobin (HPFH). These mutations (called "up promoter" mutations) occur 114 to 202 nucleotides upstream of one or the other γ-globin genes. Each mutation alters expression of its neighboring gene; for example, overexpression of the $^{G}\gamma$-globin gene results from a C to G mutation at position −202, 5′ of the start site (position +1) of $^{G}\gamma$-gene transcription. Mutations shown are a subset of known mutations producing HPFH.

β-globin chains (i.e., about 70 percent of normal). Thus, these mutations seem to affect the normal switch from γ- to β-globin production within the affected β-globin cluster.

Nondeletion HPFH mutations are thought to disrupt a binding site for one or more transcription factors whose normal function is to turn off γ-globin expression during the switch from fetal to adult globin (see Figure 6.13 for details of these mutations). For example, the C to T mutation at base pair −114 in either the $^{G}\gamma$ or the $^{A}\gamma$ gene, which has been found in different patients, is in the distal CCAAT-box, an important promoter element for many eukaryotic genes. However, little is known about the exact pathogenetic mechanism by which these particular point mutations act.

Recently, the C to T mutation at base pair −117 of the $^{A}\gamma$ gene, two nucleotides upstream of the distal CCAAT-box, was introduced into transgenic mice. It resulted in persistence of γ-globin gene expression and a concomitant decrease in β-globin gene expression. In addition, there was diminished binding of the transcription factor GATA1 to the γ-globin gene promoter in these mice, suggesting that the GATA1 protein may act as a negative regulator of γ-globin gene expression in adults. Other HPFH mutations have been shown to interfere with the binding of other proteins such as Sp1 and CP1, ubiquitous transcription factors that are not erythroid cell specific.

Erythroid-specific transcription factors The regulation of gene expression in the β-globin gene cluster is mediated by *trans*-acting factors (i.e., proteins that bind to DNA and to each other and control the expression of specific genes). These proteins can be categorized as either ubiquitous factors required for the activation of most genes or erythroid-specific factors that are involved only in the regulation of expression of erythroid genes, including the β-globin gene cluster. A description of several ubiquitous *trans*-acting factors is included in *Genes and Genomes,* chapter 8.

The search for erythroid-specific proteins that recognize specific DNA (*cis*-acting elements) in the β-globin gene cluster has uncovered relatively few such factors. The protein that has been best studied is GATA1, which recognizes a consensus WGATAR motif (where W is A or T, and R is a purine) in proximity to the majority of globin and other genes expressed in erythroid cells (erythropoietin, erythropoietin-receptor, pyruvate kinase, glycophorin B). The gene encoding GATA1 protein maps to the short arm of the human X chromosome and is a member of a larger gene family, two other members of which, GATA2 and GATA3, have been identified. GATA1 protein contains a DNA-binding domain that consists of two zinc finger structures and a DNA activation domain. It strongly activates the transcription of minimal promoters containing the appropriate binding site. Mutational knockout of the GATA1 gene in embryonic stem cells by homologous recombination and the subsequent introduction of these cells into mouse blastocysts results in chimeric mice with blocked erythroid differentiation. These mice have severe anemia with no contribution of erythroid stem cells to mature erythrocytes, and they often die *in utero.* There are several GATA1 binding sites in the promoter sequences of the GATA1 gene itself, suggesting that the factor may autoregulate its own expression.

Another erythroid-specific factor is NF-E2. Its cellular distribution is similar to that of GATA1, and its binding site closely resembles the Jun/Fos site. The gene encoding this protein has been recently cloned. The protein binds to sequences in the HS2 region of the LCR of the β-globin gene cluster and is a basic region-leucine zipper transcription factor.

A third erythroid-specific factor is EKLF, the erythroid Krüppel-like factor, which is named for its strong homology to the zinc finger region of the *Drosophila* gap protein Krüppel. The factor binds specifically to the distal promoter sequence CACACCC at about −90 to the transcription start site of the β-globin gene. β-thalassemia mutations in the sequence (Section 6.3b) greatly reduce both the binding of the DNA to the zinc fingers of the protein and the transactivation of a reporter gene by EKLF.

6.3 Molecular Basis of Hemoglobinopathies

A mutation in one or another of the globin genes can produce either a distinctive alteration in the hemoglobin structure, a hemoglobin variant, or decreased production of the protein, resulting in thalassemia. Prior to the recombinant DNA era, numerous different hemoglobin variants were found, mostly through changes in the electrophoretic property of the protein. Many more rare variants that do not have an altered charge undoubtedly remain to be discovered. Since 1980, most mutations in globin genes have been found through examination of the genes themselves, and the molecular basis of the thalassemias has been unraveled.

a. Hemoglobin Variants

Hemoglobin variants result from mutations within a coding region that change the sequence or number of nucleotides or from a recombination that leads to fusion of two different globin genes. Mutation can cause the substitution, addition, or deletion of one or more amino acids in the polypeptide sequence of the affected globin. Most commonly, single base changes result in single amino acid substitutions, for example, hemoglobin S ($\beta^{6\ \text{Glu-Val}}$). Rarely, single base changes produce shortened chains due to premature termination of translation ($\beta^{145\ \text{Tyr-Termination}}$) or elongated chains. Elongated globins result when the termination codon undergoes mutation to a codon for an amino acid, such as the UAA to CAA change in the termination codon of the α_2-globin gene in Hemoglobin Constant Spring. Two other mutations in the termination codon of the α_2-globin gene have been found, and, in all three, the α-chains have 31 additional amino acid residues (the next in-frame stop codon occurs 31 codons downstream) and differ from each other only at residue 142.

Over 400 hemoglobin variants are now known (Table 6.2). Nearly all of these are very rare and have been found only in the heterozygous state. A few hemoglobins, S, C, D, and E, are common and often occur in a homozygous state. The number of β-variants (232) is about twice that of the α-variants (126), even though there are two α-loci and a single β-chain locus. The single β-locus makes detection of β-variants easier because the variant hemoglobin in the heterozy-

Table 6.2 *Known Variants of Hemoglobin*

		Abnormal Properties			
Globin Chain	Total Variants *n*	Clinically Silent *n* (%)	Unstable *n* (%)	Abnormal Oxygen Affinity, *n* (%)	Ferric Hemoglobin *n* (%)
α	126	102 (81)	16 (12)	6 (5)	2(2)
β	232	122 (52)	68 (30)	39 (16)	3 (1)
γ	38	36 (95)	1 (3)	—	1 (3)
δ	15	15 (100)	—	—	—
Total	411	275 (66)	85 (22)	45 (11)	5 (1)

Source: Modified from H. F. Bunn and B. G. Forget, Hemoglobin: Molecular, Genetic, and Clinical Aspects. Philadelphia: W.B. Saunders, 1986.
Note: The percentages may be greater than 100 because some variants have more than one abnormal property. Fusion variants are not included.

gote is 40–50 percent of the total adult hemoglobin, whereas, in an α-variant affecting only one-quarter of the α-genes, the variant hemoglobin generally accounts for only 15–25 percent of the total. Thus, many of the α-variants may have gone undetected. Also, the percentage of β-chain variants associated with abnormal physical properties (47 percent) is twice that of the α-chain variants (19 percent), presumably because the percentage of variant hemoglobin is much greater with a β-chain variant than it is with an α-chain variant.

Variants due to gain or loss of nucleotides Single base deletions and additions can cause a frameshift in the normal ribosome-reading process. If the frameshift occurs close to the 3′ end of the coding region of the gene, a polypeptide product may still be observed. For example, in α^{Wayne}, a single base deletion in codon 139 causes insertion into the α-chain of abnormal residues at positions 139, 140, and 141 and three additional residues, 142, 143, and 144, prior to the occurrence of a termination codon in the new reading frame. Deletions of three or multiples of three nucleotides in the coding region cause deletion of one or more amino acids from the affected globin. There are 13 examples of this type of mutation, and all are β-globin variants.

Other deletions of gene segments may be due to crossing-over after mispairing of homologous DNA regions in meiosis. For example, when the β-globin gene of one chromosome 11 mispairs with the δ-globin gene of the other chromosome 11, and crossing-over occurs between the mispaired genes, a δβ-fusion globin results. This fusion globin is called a Lepore globin (Figure 6.14). The other recombination product of this crossing-over event is an anti-Lepore globin, a βδ-

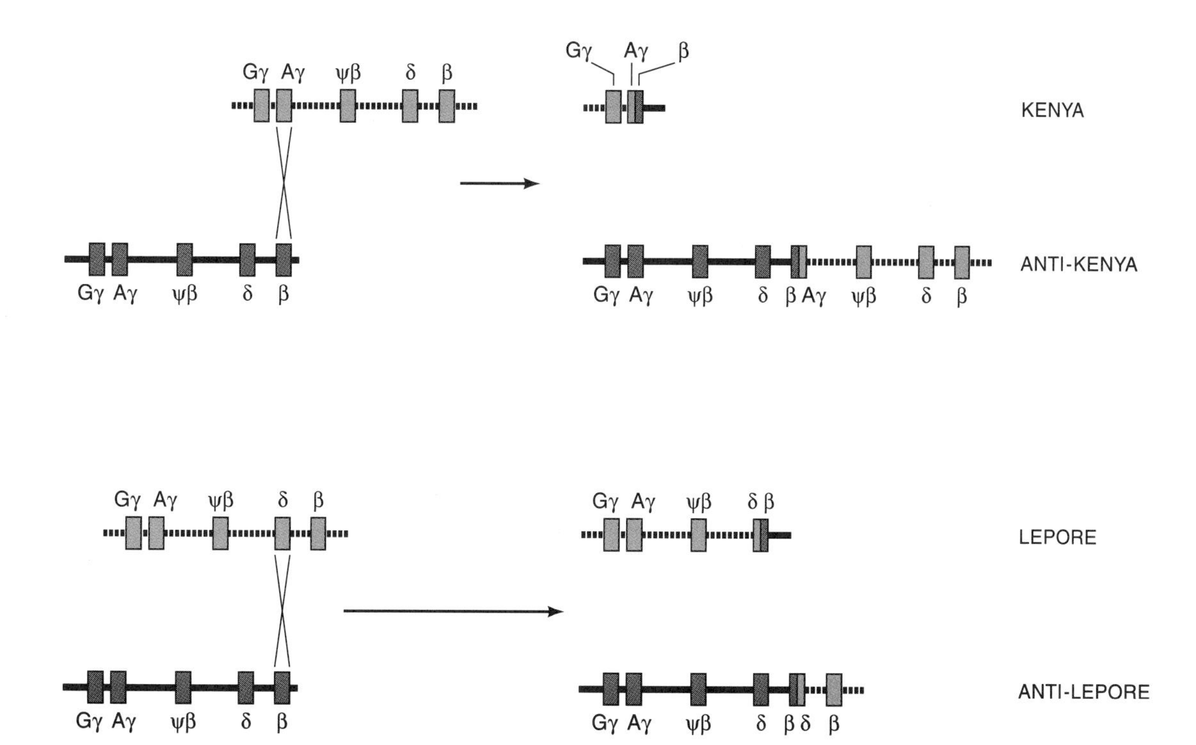

Figure 6.14
Unequal crossing-over in the β-globin gene cluster produces Lepore, anti-Lepore, and Kenya globins. The $^{A}\gamma$ gene mispairs with the β gene and crossing-over produces a Kenya globin; anti-Kenya globin has not yet been observed. The δ gene mispairs with a β gene and crossing-over produces Lepore and anti-Lepore globins.

fusion globin. Similar crossing-over between mispaired $^{A}\gamma$ and β-globin genes produces a Kenya globin (γβ-fusion chain). Because the Lepore genes (δβ-fusions) retain the weak δ-globin promoter, they are expressed at low levels and are a cause of β-thalassemia.

Variants with amino acid substitutions The majority of these mutants arises from a single base substitution that results in a single amino acid substitution. However, many of these amino acid substitutions, even some that produce abnormal physical properties in the variant hemoglobins, are clinically silent and were detected by population screening. Other substitutions cause clinical symptoms even in the heterozygous state through one of a variety of mechanisms: (1) instability of the tetramer; (2) deformation of the hemoglobin three-dimensional structure; (3) inability to maintain iron in the ferrous state; (4) alteration of the residues that make important interactions with heme, 2,3-DPG, or at αβ contact sites; (5) other abnormalities of hemoglobin's properties. Indeed, the hemoglobin variants have long been a primer on how amino acid substitutions in a protein can result in a large variety of phenotypes that vary from no effect on the individual to severe diseases, each with different features.

The location of the amino acid changed by the mutation can often be correlated with the resultant clinical phenotype. Unstable hemoglobin variants are caused by several types of changes in the primary sequence that affect the secondary, tertiary, or quaternary structure. These substitutions tend to be at residues in the molecule's interior, at contact points between the chains, at residues that interact with the heme groups, or when a proline residue replaces another amino acid within an α-helical region, thereby disrupting the helix—for example, hemoglobins Genova (β^{28}[B10]$^{Leu\text{-}Pro}$) and Abraham Lincoln (β^{32}[B14]$^{Leu\text{-}Pro}$). Hemoglobin Philly (β^{35}[C1]$^{Tyr\text{-}Phe}$) is also unstable, secondary to a missing hydrogen bond normally found between the α_1 and β_1 subunits. Many other unstable hemoglobins result from mutations affecting residues that bind heme or that alter the hydrophobic heme cleft—for example, hemoglobins Gun Hill ($\beta^{91\text{–}95[Leu\text{-}His\text{-}Cys\text{-}Asp\text{-}Lys]\text{-}0}$) and Hammersmith ($\beta^{42 Phe\text{-}Ser}$).

Substitutions on the surface of the molecule usually do not affect the tertiary structure or heme-heme interaction, but they may permit molecular interactions that decrease solubility under certain conditions, for example, hemoglobin S ($\beta^{6\ Glu\text{-}Val}$). As a consequence of the valine substitution at the β^{6}-position, the deoxyhemoglobin S aggregates into fibers, thereby inducing the sickling property of the erythrocyte. The substitution of tyrosine for the histidine at F8 that covalently binds the iron molecule or the histidine at E7 that forms a hydrogen bond with oxygen results in increased stability of the ferric, or oxidized, iron state seen in Met hemoglobins—for example, M_{Boston} (α^{58}[E7]$^{His\text{-}Tyr}$) and M_{Iwate} (α^{87}[F8]$^{His\text{-}Tyr}$). A substitution at an $\alpha_1\beta_1$ contact point, such as β99—hemoglobin Kempsey (β^{99}[G1]$^{Asp\text{-}Asn}$)—disturbs the heme-heme interaction, causing increased oxygen affinity and polycythemia (increased numbers of red cells in the blood).

b. β-Thalassemias

Decreased synthesis of one or another of the globin chains causes **thalassemia**. Although most globin gene mutations produce either a hemoglobin variant or a thalassemia, a few do both (i.e., lead to reduced production of an altered globin chain).

Characteristics of the disease The cause of β-thalassemia is defective synthesis of β-globin chains in red cell precursors. It was the first genetic disorder to be characterized at the molecular level. Severe β-thalassemia major is characterized by a pronounced anemia requiring blood transfusions every three to four weeks, leading to the development of massive iron overload and death by the fourth decade. Although α-thalassemia is caused largely by deletions of the α-globin gene (Section 6.3c), β-thalassemias are generally caused by point mutations in the β-gene or its promoter. Clinically, we recognize two β-thalassemia states: trait and disease. Most individuals with the β-thalassemia trait carry one normal β-globin gene and one gene that is either poorly expressed or not expressed at all (i.e., a β-thalassemia gene). Such people are essentially in good health and have a normal life span.

Individuals with two defective β-globin genes usually have what is termed β-thalassemia major, but gradations of severity exist. These result not only from differences in the mutations producing the disorder, but also from differences in modifying genes and sequences. In some cases, people with two defective β-globin genes are discovered only as an incidental observation in middle age. Three separate effects produce the anemia of β-thalassemia major: markedly reduced red cell production, red cell lysis, and low concentrations of hemoglobin in red cells. The first two are due to the accumulation and consequent precipitation of free α-chains resulting in damage to the red cell membranes; the last is caused by the reduced β-globin production.

The alleles β-thalassemia genes occur at a significant frequency (two or more per hundred individuals) in many parts of the world, including the Mediterranean basin, Southeast Asia, India, Africa, and Indonesia. We believe β-thalassemia has its particular geographic distribution because individuals carrying the trait have less morbidity when infected with malarial parasites than do normal individuals lacking the trait. This has been well documented for α-thalassemia; even the silent carrier state of α-thalassemia confers some protection against malaria. Because of the distribution of β-thalassemia alleles and their incidence, any mutation that produces β-globin deficiency in an individual residing in a malarial-infested region is probably subject to positive selection, and its gene frequency will increase. Alleles with high gene frequencies in a population group are thought to have originated earlier than rarer alleles. The high-frequency β-thalassemia alleles clearly postdate the divergence of the human races; each ethnic or racial group has its own specific battery of alleles. This suggests that malaria as a significant endemic disease and the etiology of genetic selection also postdate racial divergence.

β-thalassemia mutations were studied extensively from 1980 to 1986 by a three-step procedure. First, β-globin gene clusters containing β-thalassemia genes were subjected to haplotype analysis (Section 6.4) to determine which β-thalassemia genes had not been previous characterized. Second, this screening process was followed by cloning and sequencing of the mutant β-globin genes. Third, the expression of the mutant genes was analyzed. Transfection of a construct containing the SV40 72 base pair enhancer sequence and a β-globin gene resulted in efficient transient expression of β-globin mRNA (Figure 6.15). Using this assay, researchers could determine the amount of β-globin mRNA produced relative to the α-globin mRNA transcribed from a cotransfected α-globin gene. Thus, the ratio of β- to α-globin RNA served to detect and measure defects in the transcribability of either the β-globin or α-globin genes. Researchers could also determine whether the structure of the β-globin mRNA was normal or altered as a consequence of abnormal splicing or defective cleavage of the β-globin transcript. By 1987, the most common mutations in the affected ethnic groups, about 40 in number, had been discovered, and most of these were documented in the transient expression assay.

At that time, amplifying regions of the β-globin gene by PCR, starting with genomic DNA from white blood cells, and directly sequencing the amplified products also became possible. By 1993, another 80 uncommon alleles had been characterized using this approach in selected individuals known to lack the alleles common in the relevant ethnic group. Thus, of the roughly 240 β-globin alleles known to produce clinical phenotypes, about 50 percent are β-thalassemia

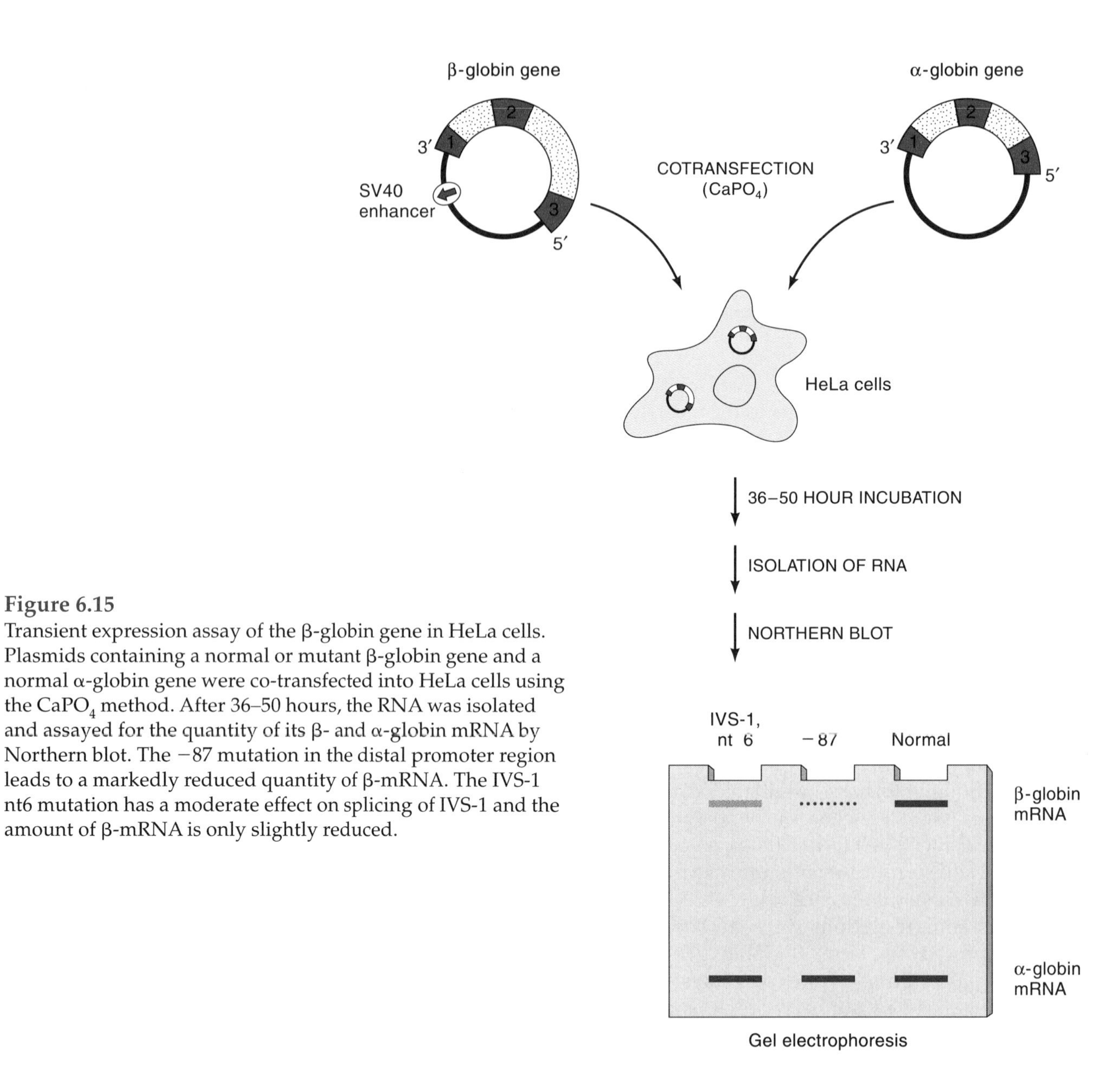

Figure 6.15
Transient expression assay of the β-globin gene in HeLa cells. Plasmids containing a normal or mutant β-globin gene and a normal α-globin gene were co-transfected into HeLa cells using the $CaPO_4$ method. After 36–50 hours, the RNA was isolated and assayed for the quantity of its β- and α-globin mRNA by Northern blot. The −87 mutation in the distal promoter region leads to a markedly reduced quantity of β-mRNA. The IVS-1 nt6 mutation has a moderate effect on splicing of IVS-1 and the amount of β-mRNA is only slightly reduced.

alleles; the "abnormal hemoglobin" alleles account for the remaining 50 percent.

The spectrum of β-thalassemia alleles has been determined in a wide variety of population groups, including African Americans, Asian Indians, Italians and Greeks, Spanish, Turks, Egyptians, Israelis, Kurdish Jews, Chinese, Japanese, and Thais. In most affected populations, such as Mediterranean peoples, Chinese and Southeast Asians, Asian Indians, and African-Americans, only a few ethnic group–specific alleles account for slightly over 90 percent of β-thalassemia genes. For example, among Chinese from Kuontung province, 4 alleles account for about 90 percent of β-thalassemia genes. About 15 rare alleles account for the remaining 10 percent of β-thalassemia genes among Chinese. Although about 35 alleles are observed in diverse regions of the Mediterranean basin, allele frequencies vary greatly from one country to another. For example, the codon 39 (C-T) allele, the most common in the region among thalassemia

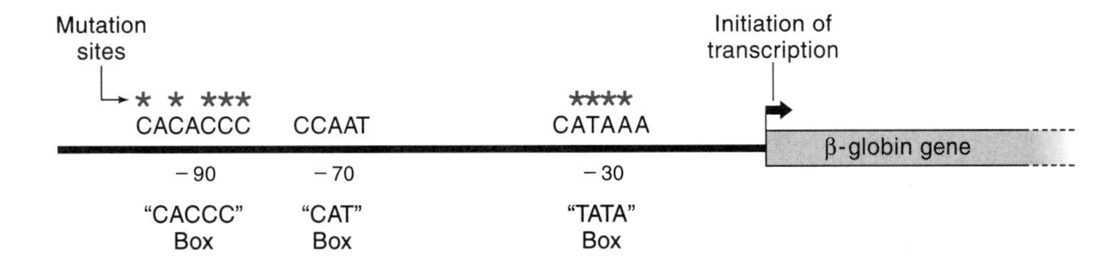

Figure 6.16
Nucleotide substitutions in the promoter region of the β-globin gene that produce β-thalassemia. Note that a large number of mutations have been observed in the CACACCC distal promoter sequences at −85 to −90 from the transcription start site, and in the "TATA" box at −30. No mutations have been found in the "CCAAT" box at −70 from the cap site.

genes, has a frequency that varies from 70 percent in Spain to only 1 percent in Egypt.

Because so many alleles are found in each region, most individuals with β-thalassemia major carry two different alleles and are called **genetic compounds**. True homozygotes who carry two copies of the same allele are in the minority. A wealth of general information about nucleotide sequences important for gene expression *in vivo* has been derived from studying the β-thalassemia mutations.

Transcription mutants Many alleles that affect transcription of the β-globin gene have been observed in the TATA-box (the sequence CATAAA located about 30 bp upstream of the transcription start site) and in the proximal and distal CACACCC sequences at base pairs −90 and −105 upstream of the gene, respectively (Figure 6.16).

In vitro transcription studies had implicated the TATA region as important for efficient transcription and defining the start site of transcription. Mutations are known at the ATAA residues from base pairs −31 to −28; these are generally associated with mild clinical phenotypes and reduced transcription initiating at position +1. However, ethnic variation in phenotype is observed. Thus, African American homozygotes with the −29 A to G mutation at base pair −29 are very mildly ill or even completely normal, and the only known Chinese homozygote for this mutation has severe disease. This difference in clinical phenotype is thought to result from increased transcription of the $^{G}\gamma$ gene resulting from a C to T substitution at base pair −158 of the $^{G}\gamma$ globin gene (Figure 6.13). This "up promoter" substitution, present in the chromosome bearing the African American mutation at base pair −29 but absent from the Chinese chromosome carrying the same base pair −29 β-globin allele, results in a greater production of $^{G}\gamma$ globin in the black versus the Chinese homozygote.

Several mutations causing relatively mild disease occur at each of the five C residues in the proximal CACACCC sequence (base pairs −92 to −86). A so-called "silent carrier" mutation has been identified at the 3′ most C (base pair −101) of the distal CACACCC sequence. A silent carrier has normal hematologic values, but an individual who has both a silent carrier allele and a severe allele can be severely affected. Therefore, a silent carrier is at risk for severely affected offspring if he or she mates with a carrier of a severe allele. Even though numerous mutations affecting transcription are known, none has been found in the CCAAT-box at base pair −70. Although the CCAAT-box was one of the first regions to be implicated in *in vitro* studies of β-globin gene promoter activity, this region's importance in transcriptional regulation *in vivo* is unclear.

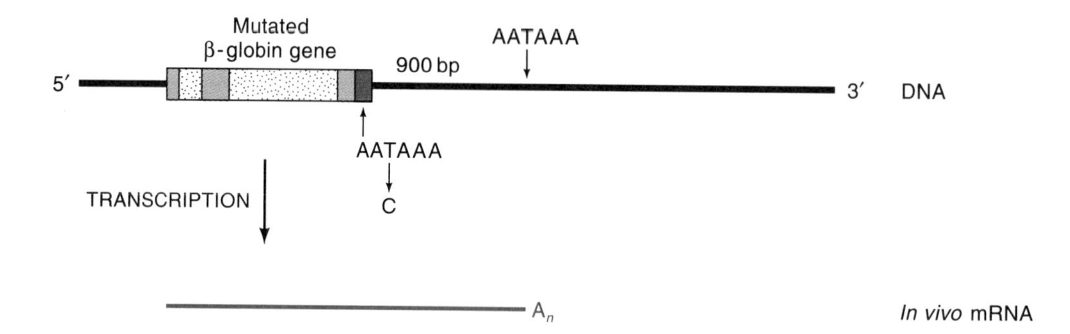

Figure 6.17
Mutations in the mRNA cleavage and polyadenylation signal sequence disrupt proper cleavage of the β-mRNA transcript. Elongated transcripts that terminate at RNA cleavage signals downstream of the β-globin gene have been observed *in vivo.*

Cap site mutant The start site for transcription at position +1 is also a site at which the primary transcript is modified by a **cap**. A 7-methylguanosine is linked through an unusual triphosphate to the 5′ end of the RNA, and such a cap appears critical for efficient translation of mRNA. Analysis of a number of genes indicated that any 5′ terminal nucleotide, not just the commonly seen A residue, can serve as the site of RNA capping. The A to C mutation observed at position +1 of the β-globin gene has a very mild effect; a homozygote has the hematologic values of a mild β-thalassemia carrier, and heterozygotes are "silent carriers." This mild mutation may reduce the transcription rate, or it may reduce capping with a secondary effect on translation. However, the combination of this mutation with a severe β-thalassemia allele can produce severe β-thalassemia.

RNA cleavage and polyadenylation mutants Eukaryotic genes have at their 3′ ends the sequence AATAAA, or AAUAAA in the mRNA. Nascent RNA transcripts are cleaved and poly A tails are added about 10–20 nucleotides beyond the AAUAAA sequence. Several mutations have been found in the AATAAA sequence of the β-globin gene, some of which have been studied in the transient expression assay (Figure 6.17). In these instances, only a small percentage of the RNA transcripts is cleaved properly; nearly all transcripts are cleaved only after transcription has proceeded between 1 and 3 kb beyond the normal AAUAAA signal to AAUAAA signals downstream. The elongated transcripts can be translated properly into β-globin, but they are unstable, and their concentrations are only about 10 percent of normal.

Mutants affecting RNA splicing Most mutations affecting RNA splicing occur at key sequences at or near the splice junctions. For splice donor sites (5′ ends of introns), the consensus sequences include the last three nucleotides of the exon and the first six nucleotides of the intron; for splice acceptor sites (3′ end of introns), they are the last ten nucleotides of the intron and the first nucleotide of the exon. In the β-globin gene, mutations at the donor site of an intron have been observed at base pair −3, −1, +1, +2, +5, and +6 from the splice site. Mutations at −1, +1, and +2 are severe, a mutation at +6 is especially mild, and mutations at +5 produce diverse clinical phenotypes. Mutations at −2 and −1 of the acceptor site are severe, and mutations at base pair −3 are moderately severe. One mutation, a C-A substitution at base pair −8, which lies within the polypyrimidine tract, has been observed.

Nucleotide substitution can also lead to the introduction of a consensus splice site sequence within either an intron or an exon (Figure 6.18). These muta-

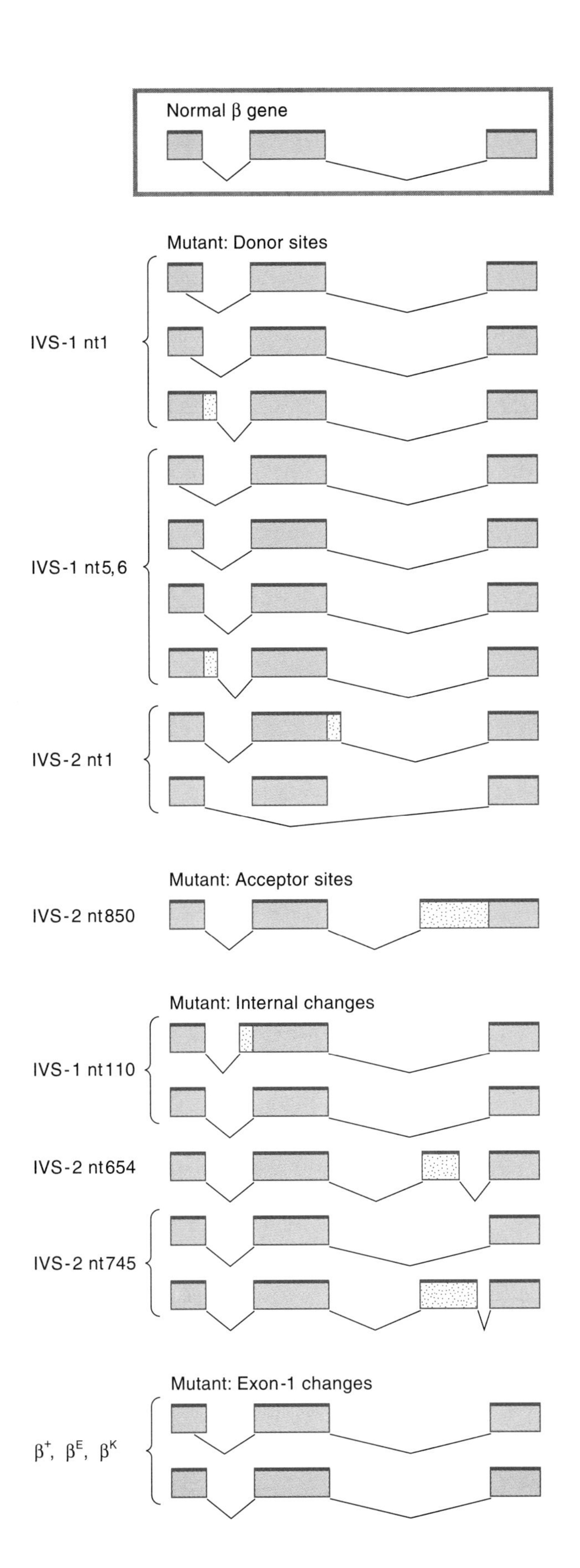

Figure 6.18
Abnormal RNA processing in β-thalassemia. The splicing pattern of the normal β-globin gene is shown in the top box. Solid colored boxes represent exons, and thin dark-colored lines show splicing at donor (5′ end) and acceptor (3′ end) sites. Open boxes are intron or exon sequences that are included in the abnormal final processed mRNA molecule. Note that mutations in intron 2 that create abnormal donor splice sites lead to splicing from that abnormal site to the normal acceptor splice site. In addition, a cryptic acceptor site at nucleotide 580 of IVS 2 is activated and used with the normal donor site. Mutations in codons 26 to 28 of exon 1 activate a cryptic donor site and lead to splicing from that site to the normal acceptor site. In addition, they reduce splicing at the normal donor and acceptor sites.

tions are very interesting because their location and effect need to be explained in any proposed mechanism of splicing. Mutations of this type produce new donor sites roughly 100, 150, and 200 nucleotides from the acceptor site of intron 2. The first two lead to reduced production of normal mRNA, and the latter completely eliminates normal splicing. Thus, donor splice sites created within intron 2 must compete with the normal 5′ donor splice site during splicing.

Mutations of this type also activate the cryptic donor splice site in codons 24–27 of exon 1 and are generally mild alleles (Figure 6.18). A cryptic splice site is a sequence that is quite similar to a normal splice site sequence but is never used under normal circumstances. When a mutation in a cryptic site (e.g., codons 24–27) makes the sequence resemble the normal splice site more closely, it can be used for splicing, although at a low efficiency. A cryptic site can also be activated when a mutation completely abolishes function of the normal splice site (e.g., at base pair +1 or +2 of the intron 1 donor splice site).

Translation mutants About one-third of the 120 known β-thalassemia mutations affect translation of the RNA. A few of these alter the initiation codon for translation, ATG in DNA. Others produce chain termination (nonsense mutations) at various sites in the coding region of the gene. Most of the translation mutants are frameshift alleles caused by insertion or deletion of one, two, four, or seven base pairs in an exon coding for β-globin. Among a wide variety of human diseases inherited as autosomal recessive disorders, very few *de novo* mutations have been found; the affected individual's two parents are nearly always carriers of the disorder. Three of the translation mutations—conversion of codon 39 to a nonsense codon (by change of CAG to TAG), conversion of codon 121 to a nonsense codon, and a frameshift at codon 64—have been observed as *de novo* mutations, meaning that each has been seen in a child neither of whose parents carried the allele. Recognition of these *de novo* mutations in β-thalassemia occurs because the β-globin genes of a very large number of affected patients have been studied, and those rare, hematologically normal individuals who have borne affected children are targeted for intensive study.

Earlier we mentioned gene conversion between the two γ-globin genes within a chromosome. Gene conversion can also occur between alleles, an event termed **interallelic gene conversion**. The conversion or correction usually takes place over one to a few kilobase pairs and, although an infrequent event, may have an important role in evolution. Interallelic gene conversion has been postulated to explain a number of phenomena observed within the β-globin gene cluster, including the presence in a single ethnic group of a specific mutation, such as the codon 39(C-T) allele, in two or more β-globin genes containing sequence differences both upstream and downstream of the mutation.

Deletions producing β-thalassemia Although most deletions in the β-globin gene cluster affect more of the cluster than just the β-globin gene and produce more complex phenotypes, a few deletions affect only the β-globin gene and produce β-thalassemia. One in particular accounts for about one-third of β-thalassemia genes in Asian Indians. As mentioned earlier, δβ-fusion genes, referred to as Lepore hemoglobin genes, arise after crossing-over due to mispairing of the δ-globin gene of one chromosome 11 with the β-globin gene of the other (Figure 6.14). These deletions also produce β-thalassemia because the δβ-fusion genes are transcribed at a reduced rate.

Other deletions that leave the β-globin gene intact yet silence its expression are of particular interest and led to the discovery of the locus control region described in Section 6.2b.

Mutations in unknown genes In 1984, a silent carrier β-thalassemia mutation that was not linked to the β-globin cluster was reported. The mutation had been passed from a parent to two offspring, each of whom had received a different β-globin gene cluster from that parent. The β-globin genes of the silent carrier parent were shown to be normal. Moreover, the mutation is presumably in another gene located elsewhere in the genome, possibly a transcription factor that is important in β-globin gene expression. Molecular analyses of a large number of β-thalassemia cases in many laboratories indicate that about 1 percent of cases is caused by mutations in unknown genes distant from the β-globin gene cluster. Genes encoding erythroid-specific transcription factors are excellent candidate genes for these unknown β-thalassemia genes.

c. α-Thalassemias

Because mutations can affect either one or both α-genes on each of the two homologous chromosome 16s, individuals can have any of four α-thalassemia states: silent carrier (three functional genes), α-thalassemia trait (two functional genes), hemoglobin H disease (one functional gene), or α^0-thalassemia (no functional gene). The first two conditions are of negligible clinical importance, but the latter two are disorders with significant morbidity and mortality.

Structure of α-thalassemia genes In contrast to the molecular basis of β-thalassemia, more than 95 percent of α-thalassemia genes are deletions of either one or both α-globin genes. Recall that there are X, Y, and Z homology blocks in 4 kbp arrays in the α-globin clusters (Figure 6.5). Single-gene deletions are of either the rightward type, involving the Z-blocks (3.7 kbp deletions), or the leftward type, involving the X-blocks (4.2 kbp deletions). Both types of single-gene deletions are due to mispairing of homologous sequences within Z- or X-blocks followed by crossing-over. The reciprocal chromosomes bearing three α-globin genes have also been observed. The 3.7 kbp and 4.2 kbp deletions have occurred on different occasions because haplotype analysis shows that each is present on a number of different chromosome backgrounds. The rightward 3.7 kbp deletion has a high frequency in American blacks (about 30 percent of all chromosome 16s in this population carry only one α-globin gene), but because chromosomes with deletions of both α-genes are rare in blacks, only individuals lacking one or two of the four possible α-genes are seen in this population. In contrast, deletions involving both α-globin genes are prevalent in Southeast Asia, South China, and the Philippines. Homozygosity for these chromosomes leading to α^0-thalassemia and neonatal death is quite common in these regions of the world.

Nondeletion α-thalassemia alleles are very uncommon, but several are now known. Perhaps the most interesting, $\alpha^{\text{Constant Spring}}$, was characterized about 1970 and is a mutation in the termination codon of the α_2-globin gene. The product of the mutated gene is a 172 amino acid polypeptide instead of the normal 141 residue protein and is markedly deficient because the corresponding mRNA is highly unstable.

α-Thalassemias associated with mental retardation Rare cases of α-thalassemia are associated with mental retardation. Two different genomic alterations have been found in such individuals: a microdeletion and a defective gene encoding a *trans*-acting factor. Individuals with the microdeletion syndrome inherit, from one parent, a large deletion encompassing at least 120 kbp including the α-globin gene cluster; presumably, the deletion also affects a gene located close to the globin cluster that is important in mental development. Other, similarly affected individuals, all males, carry a defective X chromosome gene thought to encode a protein important for the expression of several genes in the 16p13.3 chromosomal region.

6.4 DNA Polymorphisms in the Globin Gene Clusters and Their Use in Analysis of Disease-Producing Mutations

a. RFLPs

Nucleotide substitutions are normally found at between 1 in 10^2 to 10^3 nucleotides in flanking and intronic DNA in the human genome. Thus, normal variation in DNA sequence is extensive and occurs in a number of forms. These include nucleotide substitutions leading to alterations in susceptibility to cleavage by restriction endonucleases (**restriction fragment length polymorphisms**), insertion or deletion of one or more repeats within a tandem array of short, repeated sequences of 15–50 nucleotides (**variable number of tandem repeats**), variation in the number of CA dinucleotides repeated at a locus (**short sequence repeats**), and length variation in the 3′ ends of Alu sequences (see Chapter 5).

The first frequent DNA polymorphism in the human globin genes, a variation detected by the presence or absence of a *Hpa*I restriction endonuclease site some 6 kbp 3′ to the β-globin gene, was detected in 1978. The polymorphism has a strong racial predilection, being present almost exclusively in blacks. In addition, in African Americans, the site is present in about 95 percent of β^A-bearing chromosomes but is absent in about 70 percent of β-globin gene clusters carrying a β^S-globin gene (Figure 6.19). Further study of the distribution of β^S-globin

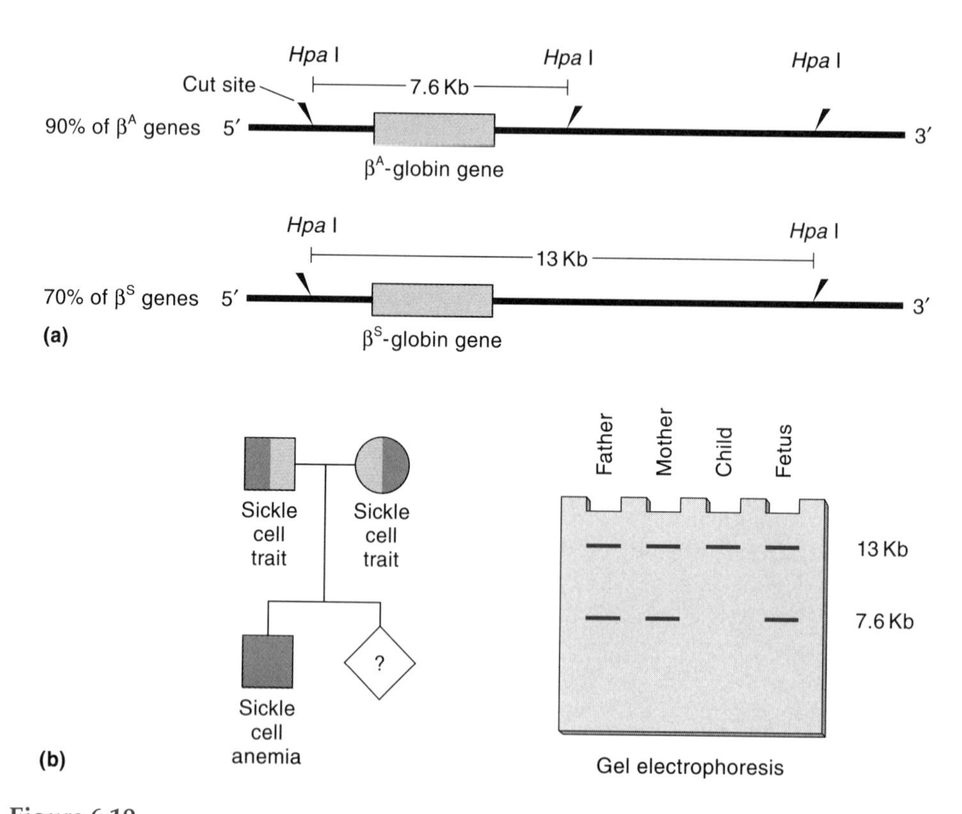

Figure 6.19
Use of *Hpa* I polymorphism 3′ to the β-globin gene in prenatal diagnosis of sickle cell anemia. (a) The *Hpa* I site 3′ to the β-globin gene is present on 95 percent of β^A-bearing chromosomes and is associated with a 7.6 kpb *Hpa* I fragment. The site is absent from 70 percent of β^S-bearing chromosomes, thus yielding a 13 kbp *Hpa* I fragment. (b) Both members of the couple shown carry B^S and, thus, the sickle cell trait and a 7.6 kbp/13 kbp genotype. Their child affected with sickle cell anemia has a 13 kbp/13 kbp genotype, thereby establishing that the β^S genes of both parents are linked to the 13 kbp genotype. DNA analysis of fetal cells of a subsequent pregnancy gave a 7.6 kbp/13 kbp genotype.

genes and the chromosome backgrounds on which they occur strongly suggested that there were at least three independent origins for the β^S-mutation in Africa: in the countries of Benin, Senegal, and the Central African Republic. The earliest β^S-mutation presumably occurred in Benin on an uncommon β-globin cluster lacking the polymorphic *Hpa*I site, and that chromosome accounts for 70 percent of all β^S-chromosomes. This polymorphism was used for the first prenatal DNA diagnostic test in couples whose fetuses were at risk for sickle cell anemia. When both members of a couple are heterozygous for the *Hpa*I polymorphism, there is a very high probability that the chromosome containing the *Hpa*I cleavage site bears the β^A-allele; the chromosome lacking the *Hpa*I site carries the sickle cell mutation. Thus, the couple can produce a fetus homozygous for both the lack of the restriction site and the sickle β^S-allele.

Soon other RFLPs were found in the β-globin gene cluster, and these were used to determine the chromosome background on which disease-producing mutations occurred. Analysis of the pattern of polymorphic restriction sites around the 50 kbp of the β-globin cluster demonstrated that the number of distinctive patterns was small, on the order of 10, even though the possible number of patterns, considering seven commonly studied polymorphic sites each having two alleles, was 2^7, or 128. The pattern of polymorphic restriction sites in a restricted region of a chromosome is called a **haplotype** (Figure 6.20), after the term used for serum markers in the human leukocyte antigen (HLA) system.

b. Haplotypes

In order to determine the haplotypes of the two β-globin gene clusters carried by an individual, researchers need to study the polymorphisms of interest in the individual and his or her parents. A key requirement for a haplotype analysis is that the polymorphisms making up the haplotype should all be in a small region of a chromosome so that a meiotic crossing-over event within the region would be extremely rare. In other words, the DNA encompassing the polymorphisms of interest should be inherited as a unit.

The introduction of the DNA haplotype concept in the early 1980s greatly aided the rapid characterization of β-thalassemia alleles. In 1978–1980, four β-thalassemia alleles were characterized by cloning the β-globin genes from patients and sequencing the key regions of the gene. Haplotype analysis of β-thalassemia chromosomes then became a useful screening procedure for new alleles because it indicates which β-thalassemia genes are likely to be novel and therefore worth characterizing. With the advent of PCR in 1987, haplotype analysis became much less valuable as a screening procedure because a gene could be sequenced without the need for cloning and subcloning the DNA fragments. We discuss it here because it remains useful for analyzing alleles in populations and for prenatal diagnosis of genetic disease.

How is haplotype analysis applied to screening for mutations? In each ethnic group studied (e.g., Chinese), haplotype analysis based on the seven common polymorphic restriction endonuclease sites identified about ten haplotypes in normal β-globin clusters. When β-thalassemia genes from an ethnic group were studied, the thalassemic β-globin gene clusters generally had the same haplotypes and in the same frequency as did the normal β-globin clusters. Because the expected recombination frequency within the 50 kbp region of the β-globin gene cluster making up a haplotype is 1 in 2000 meioses, these findings led to the following hypothesis: once formed, mutations to β-thalassemia are only rarely associated with another haplotype through recombination. Therefore, researchers predicted that the number of β-thalassemia alleles present in each ethnic group would be about ten, corresponding to the number of β-thalassemia haplotypes. If there were only one or two β-thalassemia alleles in an ethnic group, one would expect to find only a small number of β-thalassemia

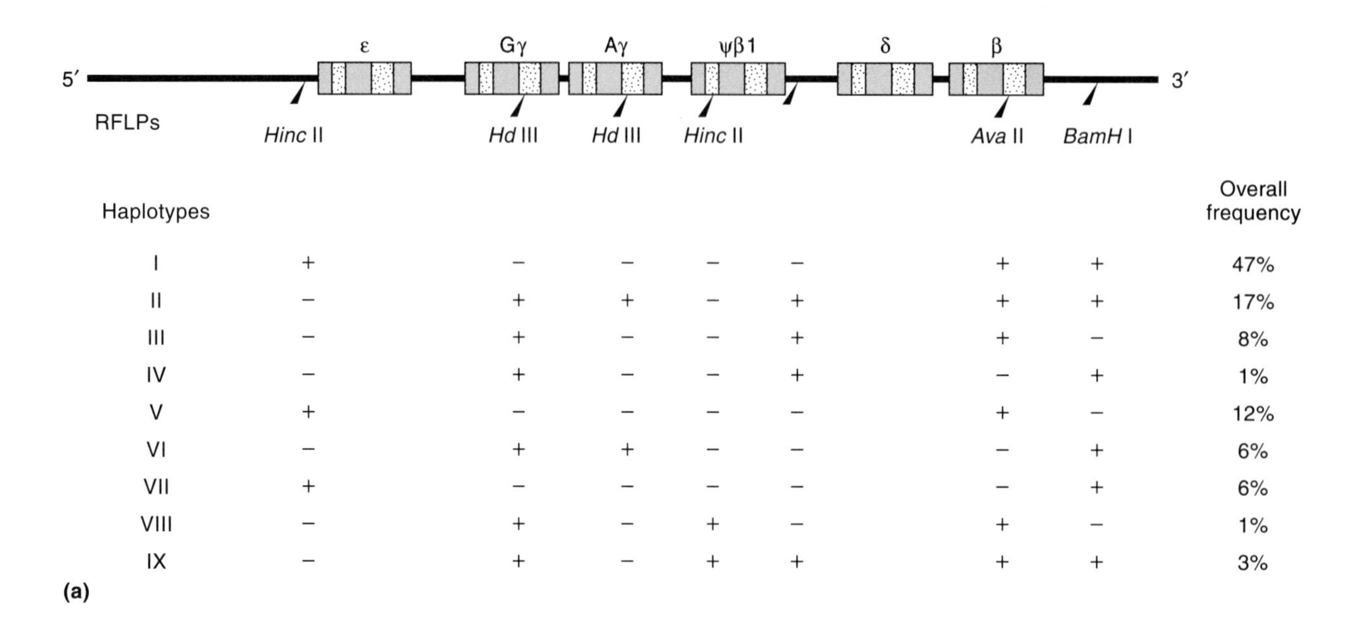

Haplotypes	Hinc II	Hd III	Hd III	Hinc II		Ava II	BamH I	Overall frequency
I	+	−	−	−	−	+	+	47%
II	−	+	+	−	+	+	+	17%
III	−	+	−	−	+	+	−	8%
IV	−	+	−	−	+	−	+	1%
V	+	−	−	−	−	+	−	12%
VI	−	+	+	−	−	−	+	6%
VII	+	−	−	−	−	−	+	6%
VIII	−	+	−	+	−	+	−	1%
IX	−	+	−	+	+	+	+	3%

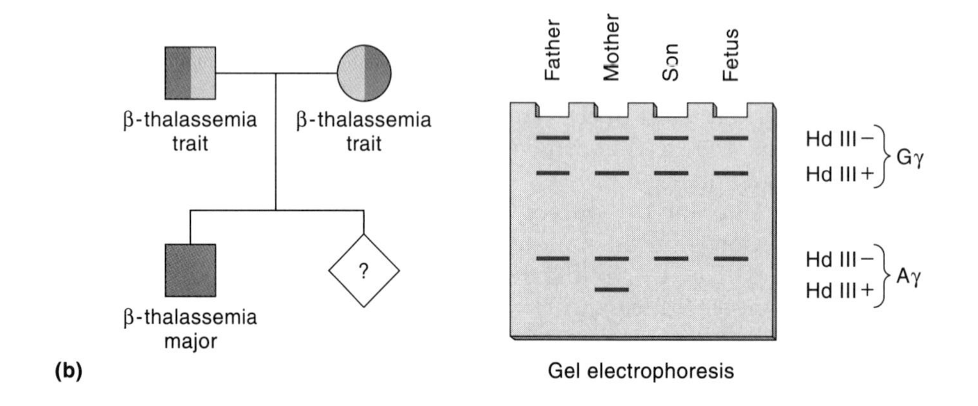

Figure 6.20
Common β-globin gene cluster haplotypes defined by analysis of seven RFLPs and their use in prenatal diagnosis. (a) Polymorphic restriction endonuclease sites are shown as solid black wedges and the presence (+) or absence (−) of a site is noted for the seven commonly used sites. Nine haplotypes common in most ethnic and racial groups for both β^{A}-bearing and $\beta^{\text{thalassemia}}$-bearing chromosomes are shown. (b) Tracking β-thalassemia genes in a couple at risk. The common RFLPs (two shown) are analyzed in the parents and in their child with β-thalassemia. Haplotypes for the two β-thalassemia and two normal β-globin genes are determined in the family and prenatal diagnosis is carried out in the subsequent pregnancy.

haplotypes in that group. Moreover, most of the β-thalassemia genes found in association with a particular haplotype would be a different allele from the allele making up most of the β-thalassemia genes on another haplotype. In general, these predictions turned out to be true.

First, researchers determined the haplotypes of the β-thalassemia chromosomes from about 20 patients within an ethnic group. Then they analyzed the mutation associated with one β-globin gene from each haplotype. With this procedure, there was a high probability (about 80 percent) that the same mutation would not be analyzed more than once. Although the association of a particular mutation with a particular haplotype was by no means completely exclusive or inclusive, the procedure was so useful that haplotype analysis became a model for analysis of other recessive genetic diseases, such as α-thalassemia (α-globin defects), phenylketonuria (phenylalanine hydoxylase defects), familial hypercholesterolemia (LDL receptor defects), and cystic fibrosis (cystic fibrosis transmembrane regulator defects).

A corollary of these observations is that when the same mutation appears in two different ethnic groups on strikingly different chromosome backgrounds, two independent origins of the mutation is the likely explanation. About 15 percent of the β-thalassemia mutations recur in this way, and at least two of them have had three independent origins. Similarly, finding the β^S-globin mutation present mainly on three different haplotypes, each of which predominates in a different geographic region of Africa, suggests that the β^S-mutation occurred on three separate occasions in different parts of that continent.

c. Prenatal Diagnosis of the Thalassemias

Using RFLPs Using the *Hpa*I polymorphism 3′ to the β-globin gene, researchers were able to carry out prenatal diagnosis of sickle cell anemia in many couples at 25 percent risk of having a child with the disease. As other RFLPs in the gene cluster were found, prenatal diagnosis of both sickle cell anemia and β-thalassemia using linkage analysis became feasible in the vast majority of couples at risk. After a family study in which both parents and either a previous child or the two sets of grandparents were studied to determine their RFLP types, the normal β^A-bearing gene cluster and the mutant β-thalassemia-bearing cluster could be tracked through the family using one or more informative RFLPs (see Figure 6.20).

Using PCR Although most of the important mutations producing β-thalassemia were characterized by 1984, methods for their detection were both cumbersome and time-consuming, making direct mutation analysis in the context of rapid diagnosis very difficult. In 1987, the introduction of the PCR using the thermostable enzyme *Taq* polymerase produced a dramatic change in the detection of β-thalassemia. The β-globin gene can be amplified by PCR from one million to ten million–fold *in vitro* in about three hours as two nonoverlapping fragments (*Genes and Genomes,* section 7.9). PCR, and the detection methods used in conjunction with it, mean that the β-globin mutations present in any couple at risk and their fetus can be easily and directly detected; RFLP analysis of families is no longer required. The fact that each ethnic group has only four to six mutations that account for 90–95 percent of the β-thalassemia genes in that group has been a substantial help to prenatal diagnosis of the disease by direct mutation analysis.

One of the first methods used to detect a mutation in a PCR product was **restriction endonuclease analysis**. Many mutations alter a 4 or 5 bp restriction endonuclease site that is quite frequent in the genome. Because the DNA fragments produced by the presence or absence of such frequent restriction endonuclease sites were too small to be detected by Southern blotting, use of this type of analy-

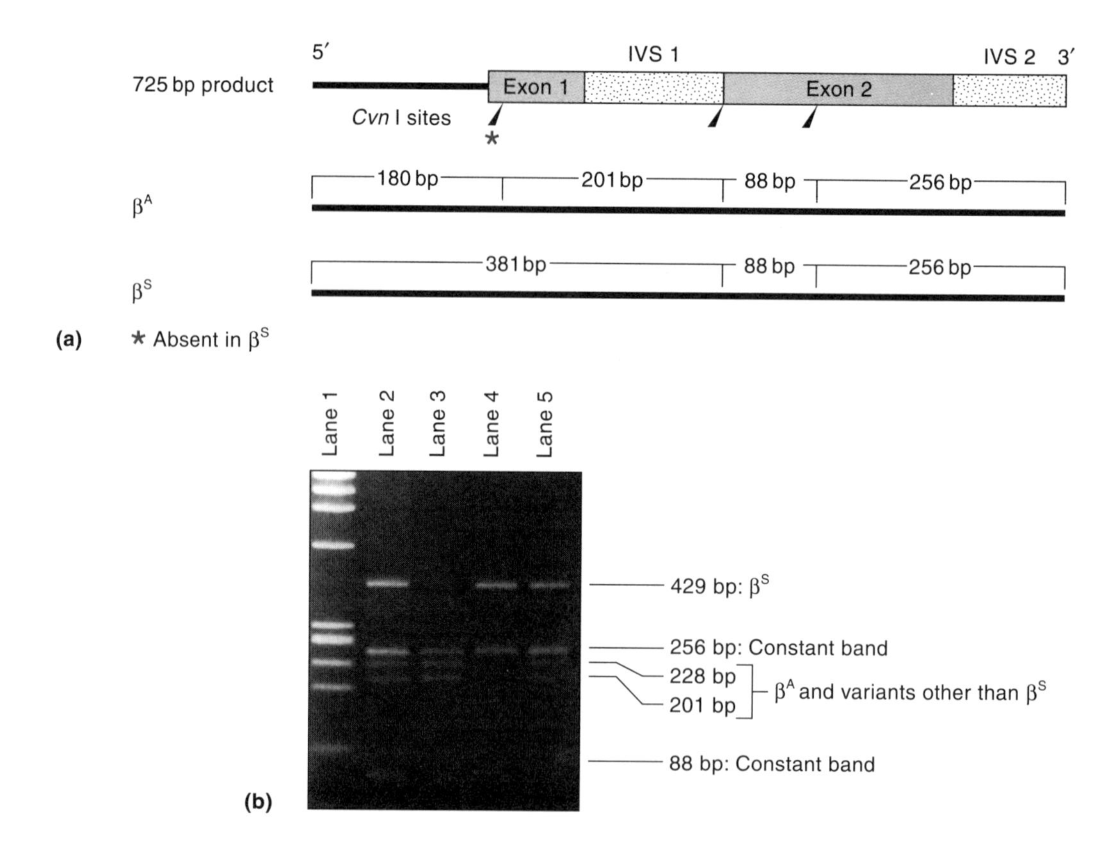

Figure 6.21
Prenatal diagnosis of sickle cell anemia using restriction endonuclease digestion of a PCR fragment. (a) *Cvn* I restriction map of the 5′-end of the β^A- and β^S-globin genes. Because the β^S mutation in codon 6 eliminates a *Cvn* I site, the β^A and β^S genes are easily distinguished by agarose gel electrophoresis of the *Cvn* I-digested PCR fragment containing the relevant portion of the β-globin gene. (b) *Cvn* I endonuclease digestion patterns of normal (AA) (lane 3), sickle trait (AS) (lanes 2 and 5), and sickle cell anemia (SS) (lane 4) individuals. Lane 1 contains a molecular weight marker.

sis for mutation detection prior to PCR was quite limited. But virtually all alterations of restriction endonuclease sites can be detected in a conveniently sized PCR fragment because amplification provides sufficient amounts so that the fragment as well as its cleavage products can be easily detected on electrophoresis gels by staining techniques. Many β-thalassemia mutations and the β^S-globin mutation are detected in this manner (Figure 6.21).

Another common method for detecting mutations in PCR fragments utilizes **allele-specific oligonucleotides** (ASOs). Eighteen to 20 nucleotide long oligonucleotides specific for either the mutant sequence in question or for the normal sequence are synthesized. These oligonucleotides commonly differ at a single nucleotide in the sequence. The oligonucleotides are then tagged with a radioactive isotope or biotin and hybridized to a filter-bound PCR product that contains the potential mutation site (a **dot blot**). In a **reverse dot blot** procedure, the oligonucleotides representing a number of mutant and normal sequences are bound to the filter, and the PCR fragment of the β-globin gene is labelled with biotin and hybridized to the many mutant sequences at once (Figure 6.22). In both procedures, detection of hybridized biotin depends on a simple enzymatic assay, and in the dot blot procedure, radioactivity is detected by autoradiography.

A variety of other methods that also depend on PCR fragments is commonly used for direct mutation analysis and prenatal diagnosis of the thalassemias.

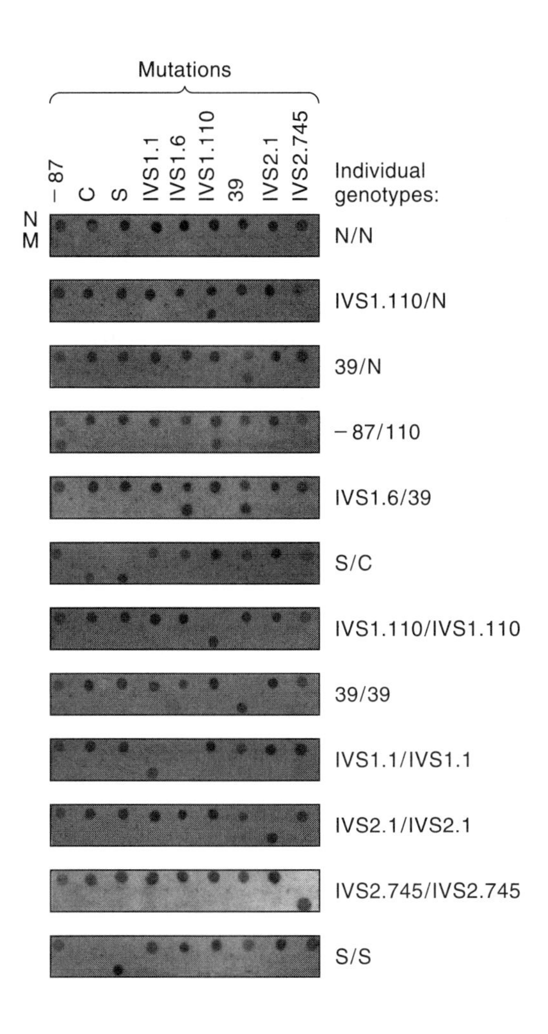

Figure 6.22
Reverse dot blot analysis of 9 β-globin gene mutations (β^S, β^C, and $\beta^{\text{-thalassemia}}$). In this technique, oligonucleotides specific for the normal and various mutant genes are covalently attached to a filter and hybridized to PCR products which encompass the genomic region of mutations encoded by the oligonucleotides. The normal (N) and mutant (M) probes for each mutation are dotted, respectively, on the upper and lower rows of each strip and in the same order on all strips. In this example, the PCR fragments were labelled with biotin during their synthesis by attaching biotin to one or both of the PCR primers. After hybridization and washing under specific conditions worked out for the oligonucleotides, the presence of biotin on the dots is assayed by an enzymatic colorimetric reaction. A normal individual shows a signal with all normal probes but not with any mutant probe. From the right, the strips are normal, heterozygous for mutation IVS-1.110; heterozygous for the C-to-T mutation in codon 39; genetic compounds for −87 and IVS-1.110, IVS-1.6, codon 39 (C–T), and β^S and β^C; and homozygous for IVS-1.110, codon 39 (C–T), IVS-1.1, IVS-2.1, IVS-2.745, and β^S. (From A. Maggio et al. 1993. Rapid and simultaneous typing of hemoglobins, hemoglobin C, and seven Mediterranean β-thalassemia mutations by covalent reverse dot-blot analysis: Application to prenatal diagnosis in Sicily. *Blood* 81 239–242.)

They include direct DNA sequencing to find the mutation, chemical cleavage of the DNA at the mutation site, and separation of normal and mutant double-strand DNA, making use of their different melting characteristics on gradient gel electrophoresis under denaturing conditions. All of these methods are relatively simple and quite useful for analysis of the small β-globin gene. In fact, the ease with which any known point mutation can be detected in a human DNA sample today is a major surprise to those investigators who have worked in human genetics for the past 20 years.

By 1992, prenatal diagnostic programs for β-thalassemia had played a substantial public health role in Italy, Greece, Cyprus, Israel, China, Taiwan, and the United States. The incidence of the disease at birth has dropped significantly in these countries; the decrease ranges from about 50 percent in the United States to 90 percent in Italy and Greece. Prevention of the disease has been accomplished through several steps. First, couples are screened for the β-thalassemia trait by simple hematological analyses before the birth of their first child. If both members have the trait, the couples are then counseled about their one in four risk of having an affected child and are offered prenatal diagnosis with each pregnancy. Most couples carrying an affected fetus decide to terminate the pregnancy, usually within the first trimester. Although preventing the disease through selective abortion of affected fetuses is abhorrent to many and less than optimal for most individuals, it has been effective in alleviating substantial suffering. Moreover, prenatal diagnosis has also had a positive effect. Before prenatal diagnosis for the disease was available, these couples had significantly reduced their childbearing to avoid bringing an affected child into the world. After prenatal diagnosis became available, the number of healthy children born to couples at risk increased dramatically.

Further Reading

6.1

J. L. Slightom, A. E. Blechl, and O. Smithies. 1980. Human Fetal $^{G}\gamma$ and $^{A}\gamma$ Globin Genes: Complete Nucleotide Sequences Suggest That DNA Can Be Exchanged Between Duplicated Genes. *Cell* 21 627–634.

D. J. Weatherall and J. B. Clegg. 1981. The Thalassemia Syndromes, 3rd ed. Blackwell Scientific Publications, Oxford.

H. F. Bunn and B. G. Forget. 1986. Hemoglobin: Molecular, Genetic, and Clinical Aspects. W.B. Saunders, Philadelphia.

M. F. Perutz. 1987. Molecular Anatomy, Physiology, and Pathology of Hemoglobin. In G. Stamatoyannopoulos, A. W. Nienhuis, P. Leder, and P. W. Majerus (eds), The Molecular Basis of Blood Diseases, pp. 127–170. W.B. Saunders, Philadelphia.

A. O. M. Wilkie, D. R. Higgs, V. J. Buckle, N. K. Spurr, N. Fishel-Ghodsian, I. Ceccherini, W. R. A. Brown, and P. C. Harris. 1991. Stable Length Polymorphism of up to 260 kb at the Tip of the Short Arm of Chromosome 16. *Cell* 64 595–606.

6.2a

T. Hunt, T. Hunter, and A. Munro. 1968. Control of Haemoglobin Synthesis: A Difference in the Size of the Polysomes Making α and β Chains. *Nature* 220 481.

H. F. Lodish. 1971. Alpha and Beta Globin Messenger Ribonucleic Acid: Different Amounts and Rates of Initiation of Translation. *J. Biol. Chem.* 246 7131–7138.

S. H. Orkin and S. C. Goff. 1981. Duplicated Human α-Globin Genes: Their Relative Expression as Measured by RNA Analysis. *Cell* 24 345–351.

S. A. Liebhaber, F. E. Cash, and S. K. Ballas. 1986. Human α-Globin Gene Expression: The Dominant Role of the α_2 Locus in mRNA and Protein Synthesis. *J. Biol. Chem.* 251 15327–15333.

6.2b

D. Tuan, W. Solomon, L. S. Quilang, and I. M. London. 1985. The "β-Like Globin" Gene Domain in Human Erythroid Cells. *PNAS* 32 6384–6388.

F. Grosveld, G. Blom van Assendelt, D. R. Greaves, and G. Kollias. 1987. Position-Independent High-Level Expression of the Human β-Globin Gene in Transgenic Mice. *Cell* 51 975–985.

D. R. Higgs, W. G. Wood, A. P. Jarman, J. Sharpe, I. Lida, I.-M. Pretorius, and H. Ayyub. 1990. A Major Regulatory Region Located Far Upstream of the Human α-Globin Gene Locus. *Genes Dev.* 4 1588, 1601.

S. H. Orkin. 1990. Globin Gene Regulation and Switching: Circa 1990. *Cell* 63 665–672.

T. M. Townes and R. R. Behringer. 1990. Human Globin Locus Activation Region (LAR) Role in Temporal Control. *TIG* 6 219–223

N. Dillon, D. Talbot, S. Philipsen, O. Hanscombe, P. Fraser, S. Pruzina, M. Lindenbaum, and F. Grosveld. 1991. The Regulation of the Human β-Globin Locus. *Genome Analysis* 2 99–118.

G. Stamatoyannopoulos. 1991. Human Hemoglobin Switching. *Science* 252 383.

E. Epner, C. G. Kim, and M. Groudine. 1992. What Does the Locus Control Region Control? *Current Biology* 2 262–264.

N. C. Andrews, H. Erdjument-Bromage, M. B. Davidson, P. Tempst, and S. H. Orkin. 1993. Erythroid Transcription Factor NF-E2 Is a Haemopoietic-Specific Basic-Leucine Zipper Protein. *Nature* 362 722–728.

6.3

S. H. Orkin, H. H. Kazazian Jr., S. E. Antonarakis, S. C. Goff, C. D. Boehm, J. P. Sexton, P. G. Weber, and P. V. J. Giardina. 1982. Linkage of β-Thalassemia Mutations and β-Globin Gene Polymorphisms with DNA Polymorphisms in the Human β-Globin Gene Cluster. *Nature* 296 627–631.

R. Treisman, S. H. Orkin, and T. Maniatis. 1983 Specific Transcription and RNA Splicing Defects in Five Cloned Beta-Thalassaemia Genes. *Nature* 302 591.

F. S. Collins, C. J. Stoeckert, G. R. Serjent, B. G. Forget, and S. M. Weissman. 1984. ^{G}Gamma-Beta (+) Hereditary Persistence of Fetal Hemoglobin: Cosmid Cloning and Identification of a Specific Mutation 5′ to the $^{G}\gamma$ Gene. *PNAS* 81 4894–4898.

S. H. Orkin and H. H. Kazazian Jr. 1984. The Mutation and Polymorphism of the Human Beta Globin Gene and Its Surrounding DNA. *Ann. Rev. Genet.* 18 131.

D. R. Higgs, M. A. Vickers, A. O. M. Wilkie et al. 1989. A Review of the Molecular Genetics of the Human α-Globin Gene Cluster. *Blood* 73 1081–1104.

H. H. Kazazian Jr. 1990. The Thalassemia Syndromes: Molecular Basis and Prenatal Diagnosis in 1990. *Sem. in Hemat.* 27 209–228.

6.4

Y. M. Kan and A. M. Dozy. 1978. Polymorphism of DNA Sequence Adjacent to Human β-Globin Gene: Relationship to Sickle Cell Mutation. *PNAS* 75 5631–5635.

P. R. F. Little, G. Annison, S. Darling, R. Williamson, L. Cambra, and B. Modell. 1980. Model for Antenatal Diagnosis of Beta-Thalassemia and Other Monogenic Disorders by Molecular Analysis of Linked DNA Polymorphisms. *Nature* 285 144–147.

S. E. Antonarakis, C. D. Boehm, P. J. V. Giardina, and H. H. Kazazian Jr. 1982. Nonrandom Association of Polymorphic Restriction Sites in the β-Globin Gene Cluster. *PNAS* 79 137–141.

A. Maggio, A. Giambona, S. P. Cai, J. Wall, Y. W. Kan, and F. F. Chehab. 1993. Rapid and Simultaneous Typing of Hemoglobin S, Hemoglobin C, and Seven Mediterranean β-Thalassemia Mutations by Covalent Reverse Dot-Blot Analysis: Application to Prenatal Diagnosis in Sicily. *Blood* 81 239–242.

CHAPTER 7

Generation of Antigen Receptor Diversity

JIANZHU CHEN and FREDERICK W. ALT

7.1 Overview of the Immune System

The **immune system** protects vertebrates from infections by foreign agents such as bacteria, viruses, parasites, tissues, and macromolecules and self-abnormalities such as cancer cells. The immune system's nearly unlimited capacity to recognize any potentially harmful substance, or **antigen**, relies on the enormous

diversity of **antigen receptors** expressed by the specific cells of the immune system (**lymphocytes**). (See **Genes and Genomes**, Section 10.6c.) During an **immune response** against a particular antigen, the immune system can specifically recognize and eliminate the invading antigen from the body. The **specificity** of this recognition results in large part from the facts that each lymphocyte expresses a unique antigen receptor and that only those lymphocytes whose receptors recognize the foreign antigen are activated and recruited into the immune response. The system's enormous diversity is generated through mechanisms that produce vast numbers of lymphocytes, each of which recognizes a unique spectrum of antigens. The generation of lymphocyte antigen receptor diversity involves a series of **gene rearrangement** events as well as a **somatic hypermutation** process.

In this chapter, we focus on the mechanism and control of these genomic rearrangements and mutational events and how they lead to diversification of antigen receptor repertoires. The methodologies for gene cloning, characterization, and modification have been critical in the development of our ideas and understanding of the immune system. Particularly important is the ability to construct model substrates that permit ready assays to study the mechanisms of genomic rearrangements that are so central to the creation of the complex repertoires of antigen receptors. Equally significant is the extent to which the creation of transgenic animals and directed alterations of genes *in vivo* have made it possible to examine the physiological importance of various regulatory and mechanistic features of the immune system. (See Chapter 1.1.)

The specific effector cells of the immune response are **T** and **B lymphocytes** (Figure 7.1). T lymphocytes effect **cell-mediated immunity**, which is particularly effective against fungi, parasites, viral-infected cells, cancer cells, and foreign tissues. B lymphocytes mediate **humoral immunity** through secretion of soluble **immunoglobulin** (**Ig**) molecules referred to as **antibodies**. Antibodies are important in the clearing of foreign macromolecules, reacting against extracellular phases of bacterial and viral infection and, upon binding to foreign antigens on infected cells, activating other elimination pathways such as the complement pathway.

T cells mediate cellular immunity through their surface antigen receptors or **T cell receptors (TCR)**, which recognize foreign antigens on the surface of the target cells. The interaction of antigen and antigen receptor (cell-cell contact) leads to the activation of lytic enzymes in **cytotoxic T cells** and the lysis of target cells. Other types of T cells, termed **helper T cells**, help other T cells or B cells respond to antigen through antigen-mediated cell-cell interactions and the secretion of soluble factors termed **cytokines**. B cells recognize soluble or surface antigens through their surface antigen receptor or immunoglobulin. This interaction leads to the activation of B cells, which subsequently secrete immunoglobulin molecules with the same specificity as that of their surface Ig receptor. The secreted antibodies bind to the antigen and activate downstream effector pathways that lead to elimination of the complexes or lysis of the reactive cells. Thus, B cells can express immunoglobulins in both membrane-bound and secreted form. When expressed on the B cell surface, they serve as antigen receptors; when secreted, they serve as **effector** molecules. T cell antigen receptors are expressed only in a membrane-bound form.

a. Immunoglobulin Structure

A monomeric immunoglobulin molecule is comprised of two identical **heavy** and two identical **light** polypeptide chains (**HC** and **LC**, respectively), linked by disulfide bonds (Figure 7.2). Both polypeptide chains fold into distinct structural **domains** that are approximately 110 amino acids long and share high degrees of homology. Each domain consists of two layers of antiparallel β-pleated sheets

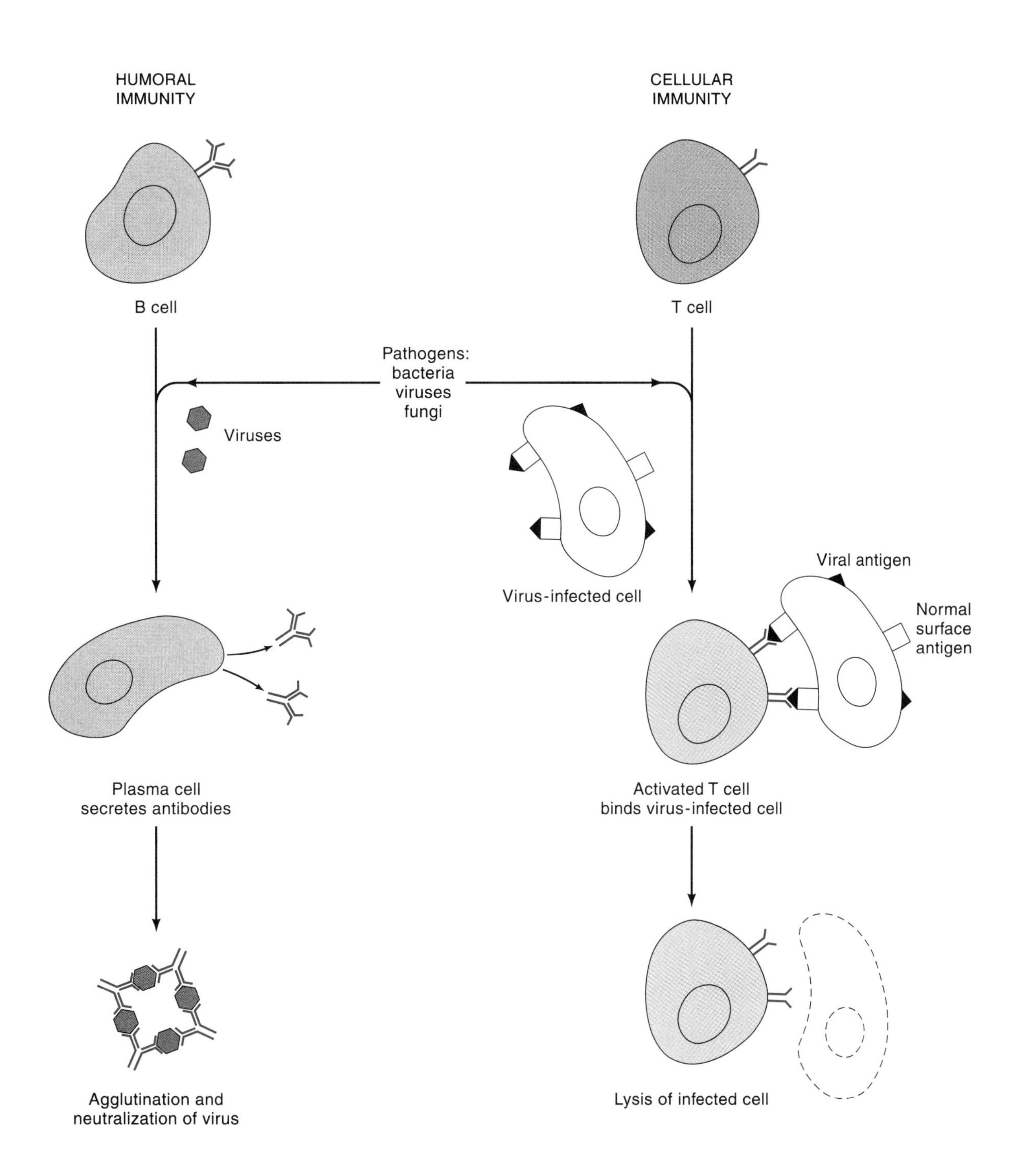

Figure 7.1
The immune system. (Adapted from L. E. Hood, I. L. Weissman, W. B. Wood, and J. H. Wilson. 1984. Immunology, 2nd ed. p. 4. Benjamin/Cummings, Menlo Park, California.)

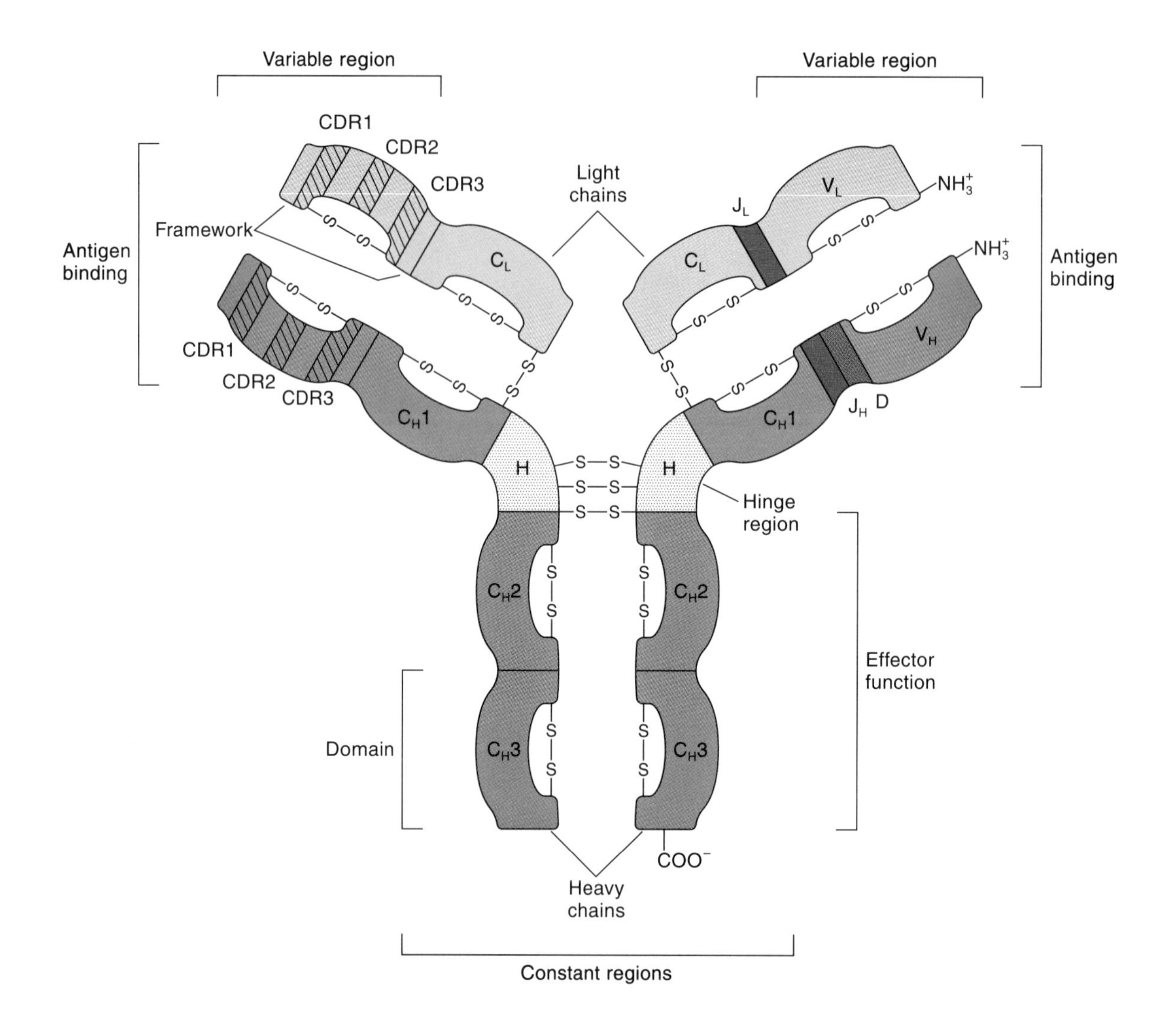

Figure 7.2
An IgG molecule. (From G. Rathbun, J. Berman, G. Yancopoulos, and F. W. Alt. 1989. Organization and Expression of the Mammalian Heavy-Chain Variable Region Locus. In T. Honjo, F. W. Alt, and T. H. Rabbitts, eds. Immunoglobulin Genes. p. 64. Academic Press, San Diego.)

and an internal disulfide linkage. This conserved three-dimensional structure is known as the **immunoglobulin fold** (Figure 7.3). The HC contains three to five domains, depending on the HC **class** (or **isotype**). The LC contains two domains. The membrane-bound form of the HC protein is inserted into the cell membrane through a transmembrane sequence at its carboxyl terminal end. Ig LC proteins do not have a transmembrane region.

The carboxyl terminal domains of Ig HC and LC of a particular class have constant amino acid sequences and are referred to as **constant (C) regions** (Figure 7.2). In contrast, the amino terminal domain of both Ig HC and LC are variable in amino acid sequence from species to species of antibody molecule and are referred to as **variable (V) regions** (Figure 7.2). The V regions of HC and LC interact to form the **antigen-binding site**, which determines the antigen specificity of a particular Ig molecule. Variable region domains consist of four subregions of relatively constant amino acid sequence (**framework regions**) interspersed with three subregions of highly variable amino acid sequence (**hyper-**

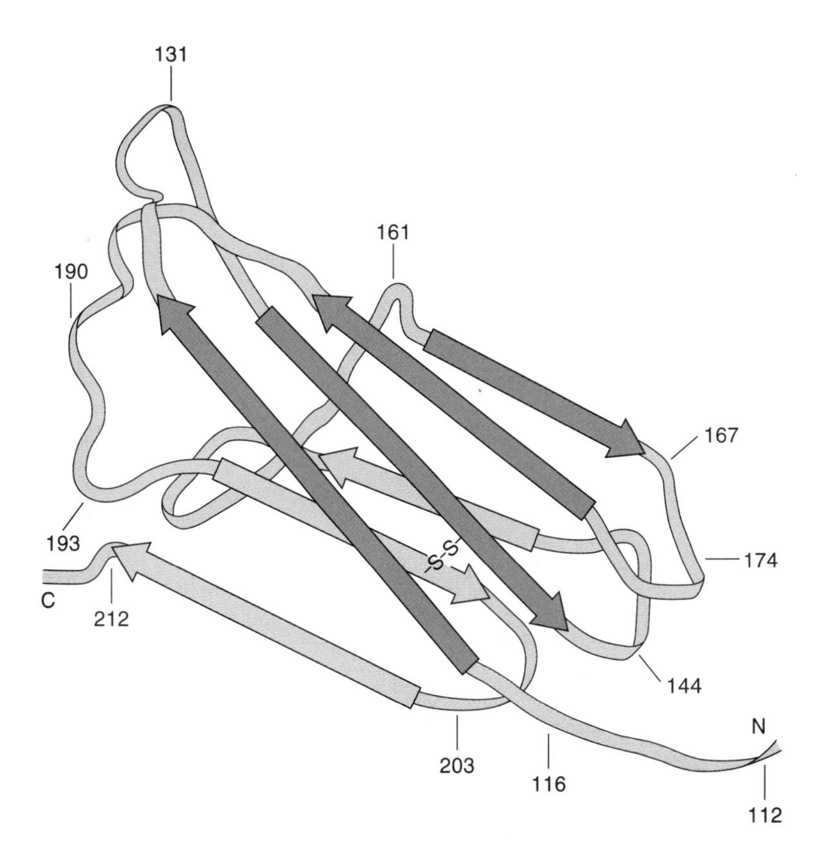

Figure 7.3
The immunoglobulin fold. Represented is a schematic diagram of the three-dimensional structure of a constant region domain of an IgG-κ light chain. The numbers represent particular amino acid positions, the arrows represent β-pleated sheets and point from the N to the C terminus; the solid line (SS) indicates disulfide linkage. (From A. B. Edmundson, K. R. Ely, E. E. Abola, M. Schiffer, and N. Panagiotopoulos. 1975. Rotational Allomerism and Divergent Evolution of Domains in Immoglobulin Light Chains. *Biochemistry* 14 3954.)

variable regions) (Figure 7.4). The hypervariable regions of HC and LC proteins directly contact the antigen. The contacting surface is complementary to the three-dimensional surface of the bound antigen; thus, the hypervariable regions are referred to as **complementarity-determining regions (CDRs)** (Figures 7.2 and 7.4).

Ig molecules are further divided into classes and subclasses (subtypes), depending on the reactivity to specific antisera or homologies of C region amino acid sequences. In mice, Ig classes include **IgM**, **IgD**, **IgG**, **IgE**, and **IgA** (Figure 7.5), and their HCs are denoted by their respective Greek letters: μ, δ, γ, ε, and α. IgG subclasses include IgG3, IgG2a, IgG2b, and IgG1. These different classes and subclasses of Ig are determined by the C regions of the HC. The C regions mediate different immunologic **effector functions** of the Ig, including placental transfer, binding to the cell surface Fc receptors, and complement fixation. Thus, HC isotypes and subtypes determine the effector functions of humoral immunity. Two LC isotypes, κ and λ, are present in mammals; however, no functional differences have been identified for the LC isotypes.

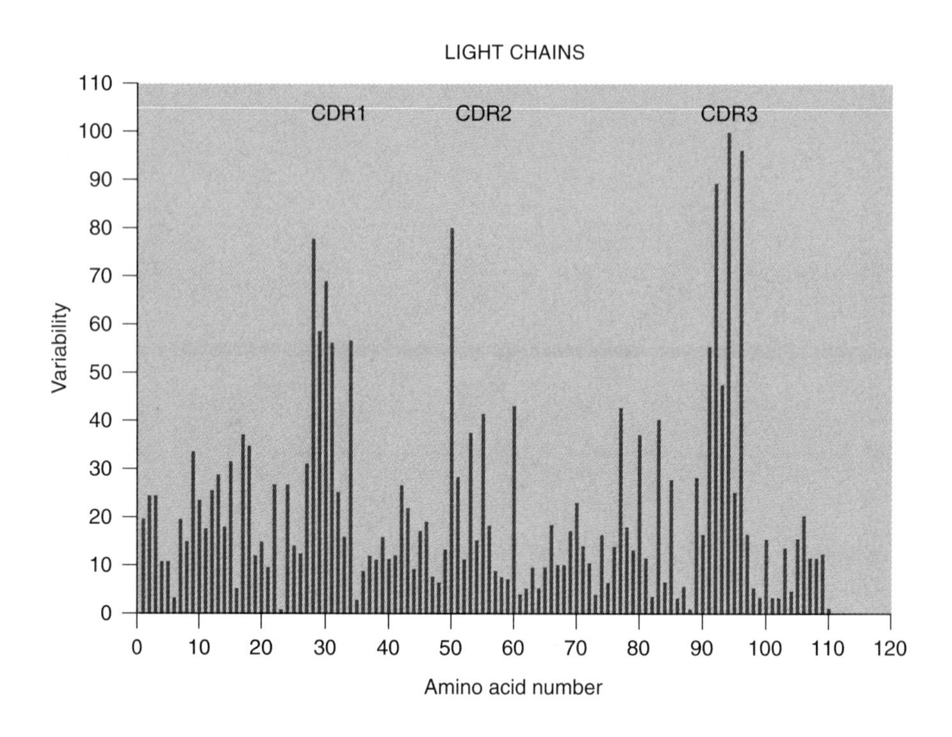

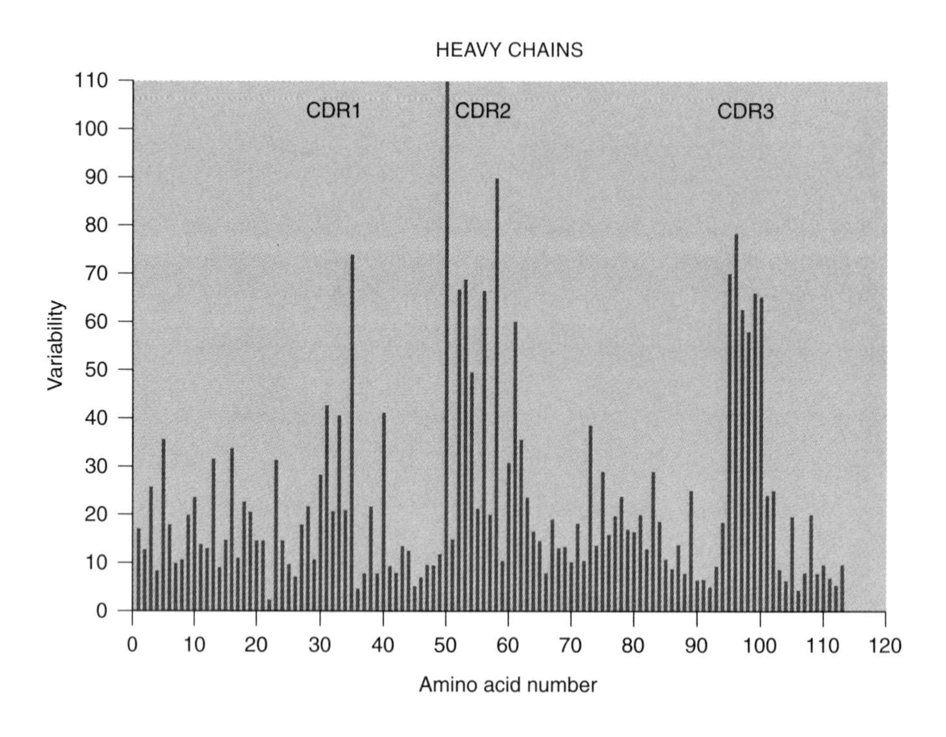

Figure 7.4
Regions of amino acid variability in the variable region of Ig HC and LC molecules. (From E. A. Kabat, T. T. Wu, M. Reid-Miller, H. M. Perry, and K. S. Gottesman. 1987. Sequences of Proteins of Immunological Interest, 4th ed. p. 730. U.S. Department of Health and Human Services, Public Health Service, National Institutes of Health, Bethesda, Maryland.)

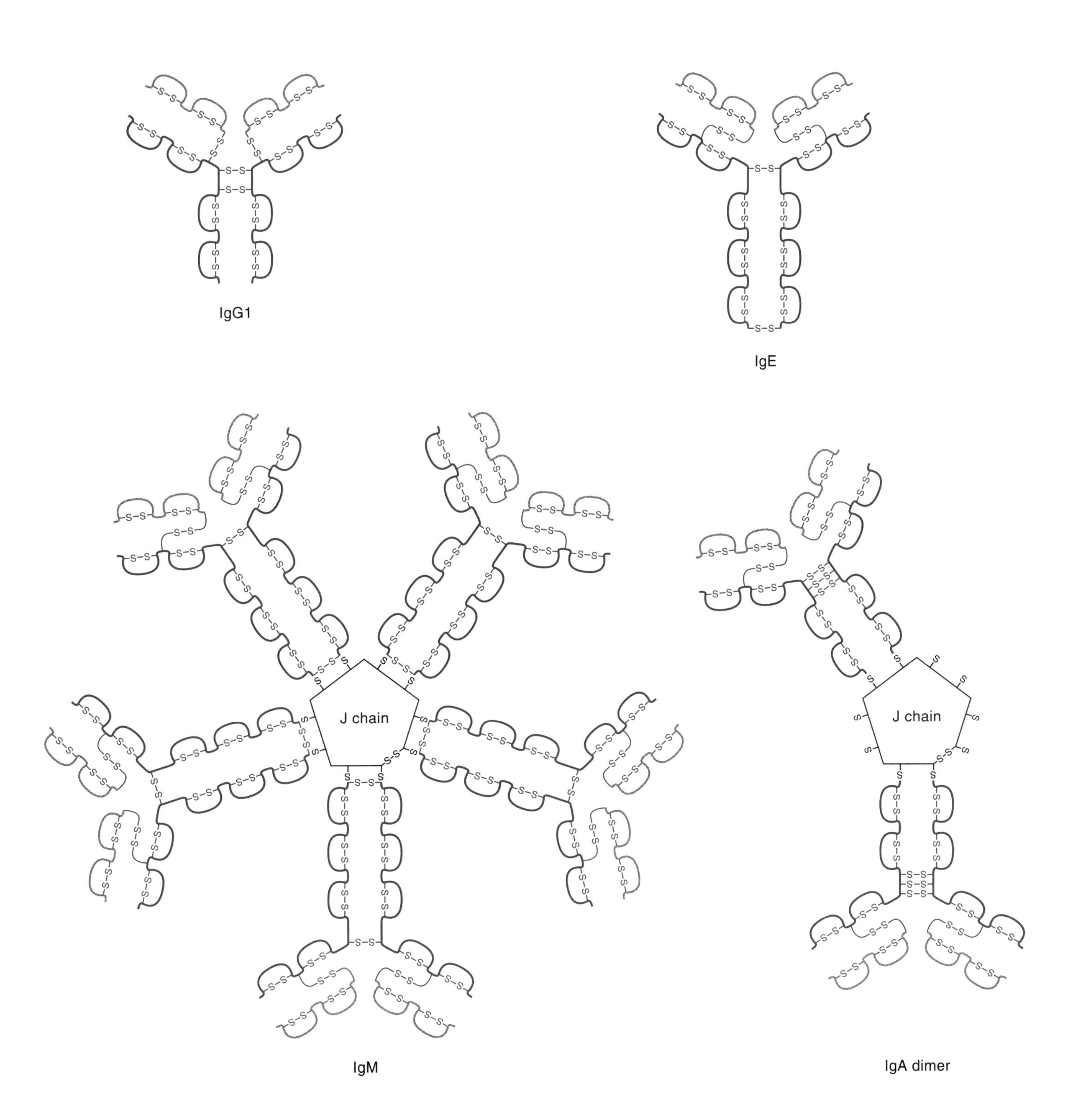

Figure 7.5
Structure of secreted Ig molecules. The structures of human IgG1, IgE, IgM, and IgA are diagrammed. Note that IgM and IgA are secreted as polymeric forms, linked by a J chain, of the basic Ig monomer. Mouse Igs are similar in structure. (Adapted from J. Gally. 1973. Structure of Immunoglobulins. In M. Sela, ed. The Antigen. p. 209. Academic Press, New York.)

b. T Cell Receptor Structure

The antigen receptors expressed by T cells are heterodimers of αβ or γδ TCR chains (Figure 7.6). Similar to Ig HC and LC, each of the four T cell receptor chains is organized into two immunoglobulin fold domains. The carboxyl terminal domains of the four chains are relatively constant in amino acid sequence (constant regions), but the amino terminal domain of each type of chain has a variable amino acid sequence (variable region) among individual TCR. The combination of the variable regions of α- and β-chains or γ- and δ-chains forms the antigen-binding sites. Unlike Ig molecules, however, both chains of the TCR heterodimer contain transmembrane domains, and TCR exist only in membrane-bound form. Furthermore, TCR do not bind soluble antigens; rather, they bind antigenic peptides complexed with the self **major histocompatibility complex (MHC)** molecules on the target cell surface (Figure 7.7). Thus, T cell antigen recognition is **MHC restricted**; the antigen-binding portions of TCR variable regions must contact both antigen and MHC molecules.

c. Somatic Assembly of Ig and TCR Genes

The variable regions of Igs and TCRs form the antigen-binding sites; therefore, the diversity and specificity of variable regions determine the diversity and specificity of the immune system. To generate the enormous receptor diversity required by the immune system, vertebrates have evolved somatic diversification mechanisms to maximize the coding potential that can be generated from a limited number of germ line gene segments. Both B and T cells generate enormous diversity through mechanisms associated with the assembly of the Ig and TCR genes from component germ line gene segments. The variable region of the antigen receptors is encoded by multiple gene segments denoted variable (**V**), diversity (**D**), and joining (**J**) segments. During early lymphocyte development, these gene segments are assembled into complete V(D)J antigen receptor variable region genes through a process referred to as **V(D)J recombination**. The V(D)J recombination process, which is uniquely found in developing lymphocytes, has built-in features that generate antigen receptor diversity.

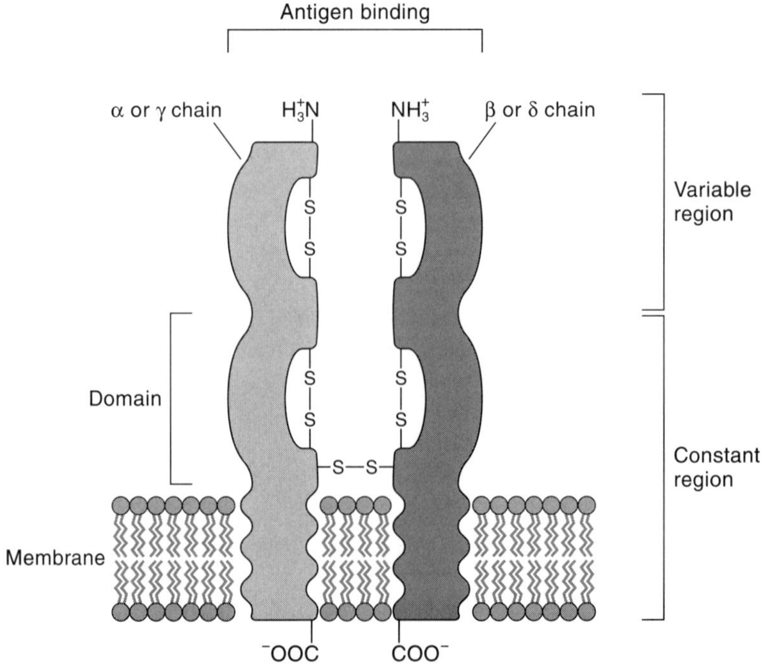

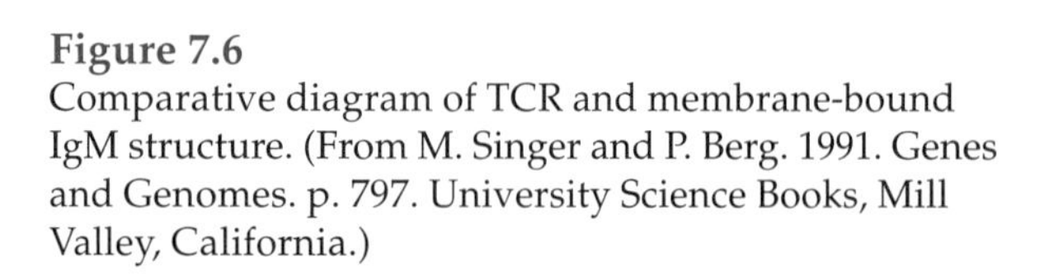
Figure 7.6
Comparative diagram of TCR and membrane-bound IgM structure. (From M. Singer and P. Berg. 1991. Genes and Genomes. p. 797. University Science Books, Mill Valley, California.)

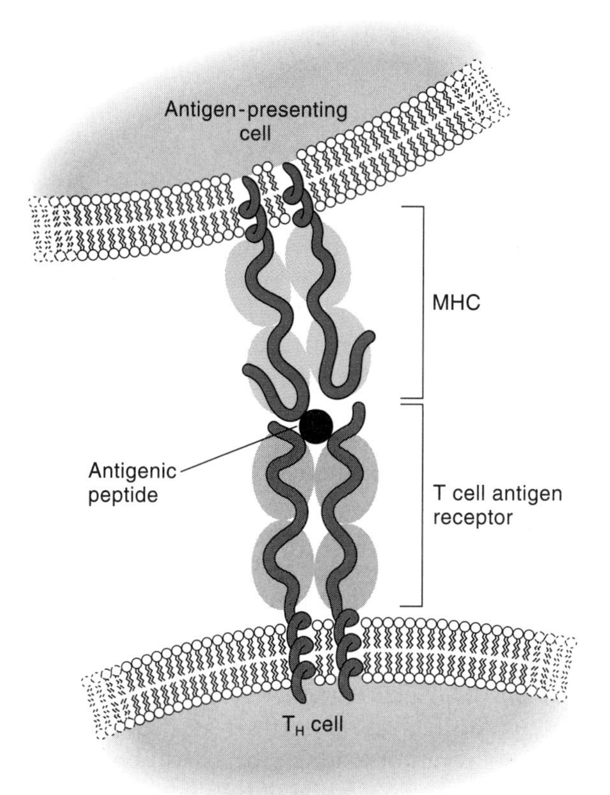

Figure 7.7
TCR interaction with antigen-MHC complexes. (From M. Singer and P. Berg. 1991. Genes and Genomes. p. 799. University Science Books, Mill Valley, California.)

The Ig HC variable region gene is assembled from V_H D_H and J_H segments. Ig LC genes of either the κ- or λ-type are encoded only by V_L and J_L gene segments. The V_H and V_L segments encode the bulk of the variable regions, approximately the 96 N terminal amino acids. In particular, the V_H and V_L encode framework regions 1–3 and CDR 1 and 2 (Figure 7.8). Thus, the occurrence of numerous germ line HC and LC V gene segments contributes to germ line–encoded diversity of antigen-binding specificities. The remainder of the HC variable region is encoded by the D_H (usually 1–3 amino acids) and the J_H segment (12 amino acids). The $V_HD_HJ_H$ junctional region encodes CDR3, and framework 4 is encoded by the J_H (Figure 7.8). Thus, the combinatorial assortment of particular V_H D, and J_H segments is another important source of germ line–encoded variable region diversity. Likewise, the LC CDR3 is encoded by the V_LJ_L junctional sequences, and framework 4 is encoded by the J_L (Figure 7.8). Because the HC and LC CDR3s (as well as the analogous region of the TCR) are encoded by junctional sequences, they are particularly rich sources of diversity. In this context, diversification of these regions is further enhanced by somatic mechanisms that modify their coding capacity through the addition or loss of nucleotides.

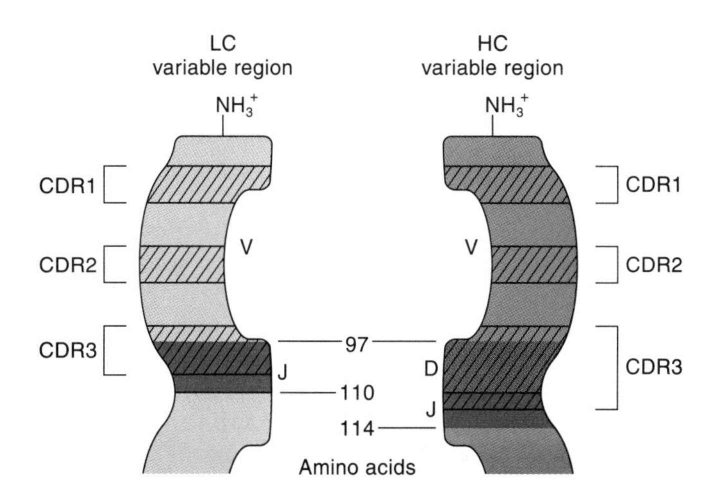

Figure 7.8
Role of V, D, and J gene segments in encoding CDR1, 2, and 3 regions of Ig variable region genes. (Adapted from A. K. Abbas, A. H. Lichtman, and J. S. Pober. 1991. Cellular and Molecular Immunology. p. 78. W. B. Saunders, Philadelphia.)

d. Overview of B Cell Differentiation

Primary B cell differentiation B Lymphocytes are generated from hematopoietic stem cells in the fetal liver and adult bone marrow. The **differentiation** process proceeds through many stages and involves well-ordered molecular events of Ig gene rearrangement and expression (Figure 7.9). At the **pro-B** cell stage, Ig HC genes are assembled and expressed. The expression of Igμ HC, which is the first HC isotype expressed during B cell development, marks the transition of pro-B to **pre-B** cells. Pre-B cells undergo Ig LC gene rearrangement and expression, resulting in the expression of IgM on **immature B** cells. If the antigen receptor expressed on immature B cells reacts with antigens normally found in the host organism (i.e., is **self-reactive**), interaction with cognate antigen will lead to the **deletion** or **anergy** of the immature B cell (self-selection). If the antigen receptor is not self-reactive, immature B cells will mature by expressing both IgM and IgD on the cell surface and migrate to peripheral lymphoid organs such as spleen, lymph node, and Peyer's patches, where they can undergo further, antigen-dependent differentiation events.

Antigen-dependent B cell differentiation During their lifetime in the periphery, **mature B** cells may encounter foreign antigen through their surface receptors. The interaction of antigen with antigen receptor leads to the activation and further differentiation of mature B cells (Figure 7.9). Some of the cells differentiate into short-lived (a few days) **plasma cells**, which secrete large quantities of antibodies. Some of the cells differentiate into long-lived **memory cells**, which can initiate a rapid immune response subsequent to a reencounter with the same antigen. During memory B cell generation, Ig variable regions undergo **somatic hypermutation**, resulting in an increase of receptor affinities (**affinity maturation**). This form of diversification operates over the entire assembled variable region gene; thus, it is a mechanism by which all of the CDR sequences can acquire somatic diversification. During this differentiation process, activated B cells may undergo **class switch** recombination (Figure 7.9), which is different from V(D)J recombination, to express Ig classes other than the initial IgM and IgD. Because different classes of antibodies have different effector functions, class switching maximizes the efficiency of antibodies in removal of specific antigens.

e. Clonal Selection and Allelic Exclusion

Millions of B cell clones, each with a unique antigen receptor specificity, are fed into the peripheral lymphoid tissues each day. A specific immune response can be generated by a specific foreign antigen because only those B cells that happen to express receptors with high enough affinities for that antigen are activated to secrete antibodies and to undergo additional downstream differentiation events such as somatic mutation, heavy chain class switching, and memory B cell generation (Figure 7.10). This process is known as **clonal selection**. Thus, although the antibody repertoire is enormously diverse, each immune response is antigen specific.

For clonal selection to be operative, individual B cells must express only a single Ig HC and a single Ig LC variable region gene and, thereby, a homogenous set of antigen receptors. Although both chromosomal sets of Ig HC genes and both sets of Ig κ and λLC genes can potentially be rearranged to form unique variable regions, individual B cells functionally rearrange and express only a single HC and LC gene (allele). This phenomenon is called **allelic exclusion**. With respect to Ig LC genes, B cells also express either κ or λ, but not both—a phenomenon referred to as **LC isotypic exclusion.**

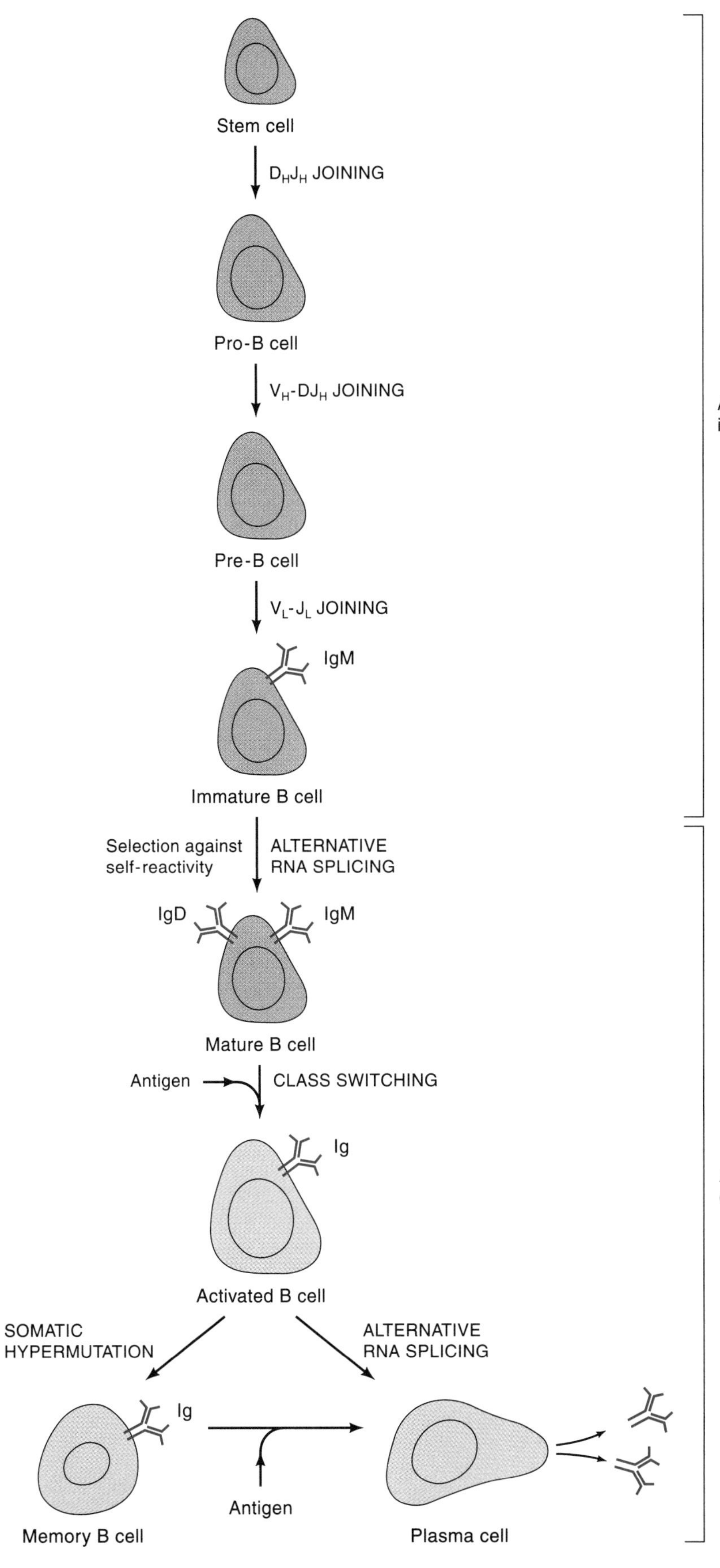

Figure 7.9
A model of B cell differentiation.

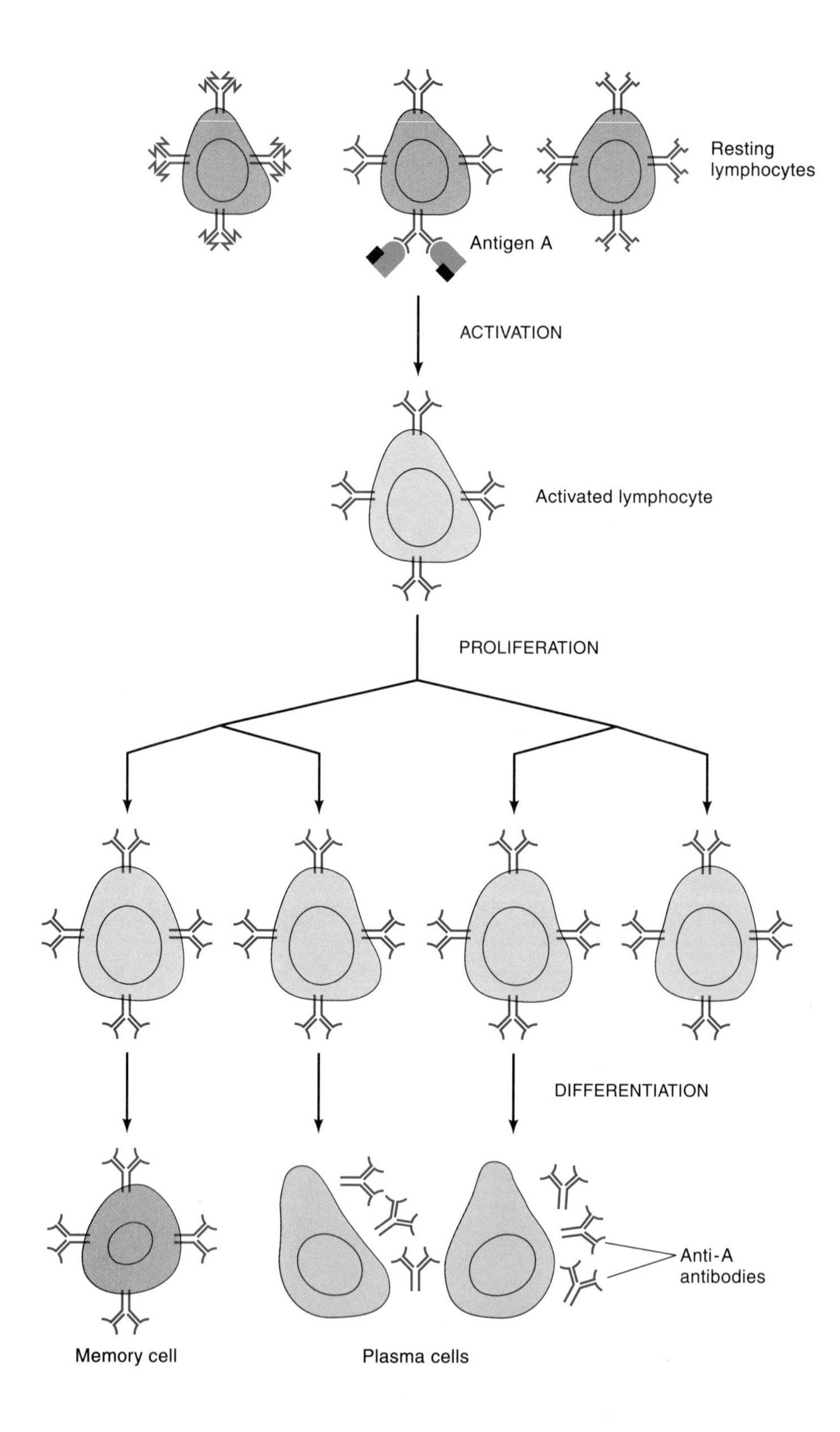

Figure 7.10
The clonal selection process. (From A. K. Abbas, A. H. Lichtman, and J. S. Pober. 1991. Cellular and Molecular Immunology. p. 11. W. B. Saunders, Philadelphia.)

7.2 Cell and Animal Models To Study B Cell Differentiation

a. Cell Lines

Various tumors and transformed cell lines have provided homogenous populations of cells that represent models for studies of the various stages of B cell differentiation.

Precursor stages The Abelson murine leukemia virus (A-MuLV) is a replication-defective retrovirus that specifically transforms normal fetal liver or adult marrow B cell precursors in culture. The transformed lines generated by this approach usually represent the pro- and pre-B stages (Table 7.1). Such lines often continue to express V(D)J recombination activity during propagation in culture, as evidenced by their ability to assemble endogenous HC and sometimes LC variable region genes. Analyses of the pattern of endogenous Ig HC and LC gene rearrangements in these lines helped to provide early insights into the normal sequence of endogenous Ig gene rearrangements. Analyses of the ongoing V(D)J recombination process in A-MuLV transformed pre-B lines has also been useful for elucidating aspects of the mechansim and control of this process, for example, by using the lines as hosts for substrates designed to assay the V(D)J recombination mechanism. However, most transformed precursor B cell lines do not represent normal pro- or pre-B cells in that they do not tend to differentiate fully in culture and often express levels of V(D)J recombination activity that are orders of magnitude lower than those of normal precursor lymphocytes. Recent work suggests that expression of the V-*abl* gene product introduced in their transformation may block their overall differentiation potential. The use of temperature-sensitive versions of this virus has provided a promising method to generate permanent cell lines that will further differentiate at the nonpermissive temperature.

More recently, B cell precursors at various stages of differentiation have been isolated directly from bone marrow by flow cytometry sorting on the basis of the expression of particular sets of cell surface markers (Table 7.1). The analysis of Ig gene rearrangement and expression in these purified subsets has confirmed the general picture of early B cell development provided by transformed cell lines and has allowed the process to be even more finely characterized (Figure 7.9).

Table 7.1 ***Different Stages of B Cells and Their Corresponding Cell Lines***

B Cell Stage	Surface Marker	Cell Line	Usage
Pro-B	$B220^+$, $CD43^+$, IL-$2R\alpha^-$	A-MuLV transformants	B cell differentiation
Pre-B	$B220^+$, $CD43^-$, IL-$2R\alpha^+$	Long-term bone marrow culture	V(D)J recombination
Immature B	$B220^+$, IgM^+, IgD^-	Lymphoma	B cell activation
Mature B	$B220^+$, IgM^+, IgD^+	Leukemia	Structure and expression of rearranged Ig genes Class switching
Plasma cell	$B220^+$, IgM^-, IgD^-	Myeloma Plasmacytoma Hybridoma	Ig structure Structure and expression of rearranged Ig genes Somatic hypermutation Monoclonal antibodies

Nontransformed B cell precursors also can be propagated in cell culture systems that depend on the use of appropriate stromal cell lines and/or the presence of added pre-B cell growth factor, interleukin-7. Recent work indicates that pre-B cells can even be differentiated from totipotent embryonic stem (ES) cell lines by employing variations of such cultures. These latter methods may offer powerful approaches for future efforts to elucidate the factors that control very early B cell differentiation.

Mature stages Among the earliest available lines were myelomas and plasmacytomas, which represent terminally differentiated Ig-secreting cells. Such lines secrete homogenous Ig molecules from uniquely rearranged Ig HC and LC genes and thus were particularly important for studies of antibody structure and for early studies of the structure and expression of rearranged Ig genes (Table 7.1). Fusion of plasmacytomas that have lost endogenous Ig expression with normal splenic B lymphocytes allows the generation of permanent hybrid cell lines, referred to as hybridomas, which produce the antibody encoded by the HC and LC genes of the normal B cell partner (Table 7.1). This process, which has been of far-reaching significance to both experimental and applied immunology, allows the generation of hybridoma cell lines that secrete a single species of specific antibody, for example, following an immunization. The antibody secreted by these clonal cell lines is referred to as a "monoclonal" antibody. The hybridoma technology has been extremely important for studying mechanisms of Ig diversification, in particular for following secondary mechanisms such as somatic mutation and HC class switching that occur during a specific immune response.

Normal mature murine B cell populations, for example, as prepared from the adult spleen, can be activated in culture to proliferate and secrete Ig by treatment with nonspecific ("polyclonal") activators such as bacterial lipopolysaccharide (LPS) or by cross-linking their surface Ig receptors. Such polyclonal activation systems have been particularly useful for studying the molecular events involved in B cell activation and the effects of lymphokines on Ig secretion and class switching.

b. Transgenic and Mutant Animals

Some of the most powerful approaches to study aspects of B cell differentiation involve the use of transgenic and mutant mice. The use of such animals has been of fundamental importance for many of the studies described here.

Ig transgenic mice Rearranged Ig genes introduced into mice germ lines often are expressed in a proper tissue-specific manner as a result of the strong tissue-specific regulatory sequences associated with them. Thus, the introduction of preassembled Ig HC or LC transgenes into mice leads to the generation of B cell populations of which most members express the HC or LC encoded by the transgenes. Such methods have been useful for studying how Ig HC or LC gene expression effects particular developmental regulatory events, such as transitions from pro-B to pre-B cells and allelic exclusion. The generation of Ig transgenic mice that harbor transgenes that encode autoreactive antibodies, for example, has also been important in studies of the cellular mechanisms underlying immunological tolerance. Furthermore, transgenic Ig gene constructs containing germ line V region gene segments are assembled into variable region genes during B cell differentiation if they contain the necessary *cis*-acting regulatory sequences. The use of such transgenic V(D)J recombination substrates has allowed physiological assays of the ability of various *cis*-elements, such as transcriptional enhancers, to regulate V(D)J recombination during normal lymphocyte differentiation.

Mutant mice Several naturally occurring mouse mutations have been useful for providing models to elucidate certain aspects of B cell differentiation. However, the introduction of germ line mutations by the method of gene-targeted mutation of embryonic stem cells has been particularly powerful for analyzing the development of the immune system because mice do not need their immune systems to survive if they are maintained in pathogen-free environments. Therefore, even mutations that completely block immune system generation can be maintained in mouse lines. Such lines can then be further analyzed by introduction of additional germ line mutations or complementing transgenes into the mutant background through breeding of appropriate mouse lines. We discuss this powerful approach in the context of a variety of mutations that affect V(D)J recombination and/or B cell development.

7.3 Organization and Expression of Ig Genes

a. IgH Locus

The murine IgH locus is located on chromosome 12. There are 4 J_H segments, 13 characterized D_H segments, and a few hundred to several thousand V_H segments, depending on the particular mouse strain (Figure 7.11). In germ line DNA, the 4

H chain locus (chromosome 12)

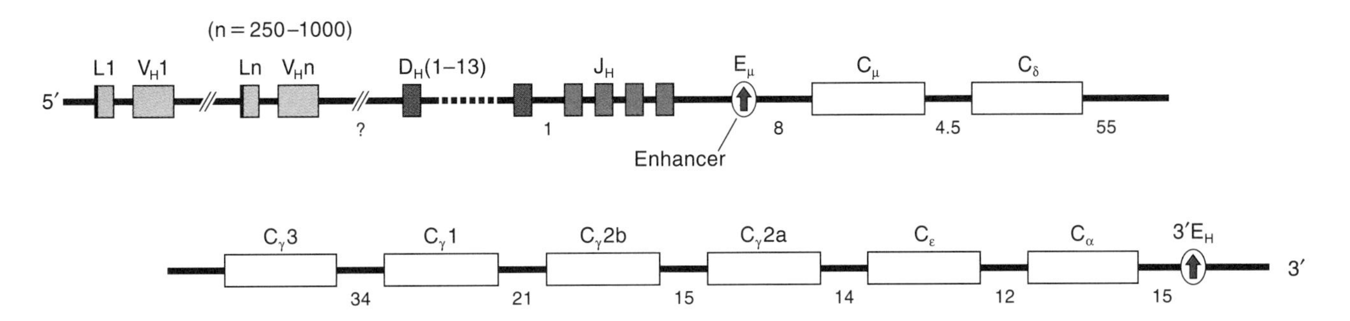

κ chain locus (chromosome 6)

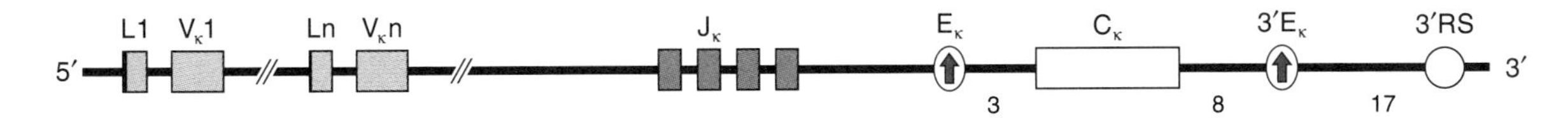

λ chain locus (chromosome 16)

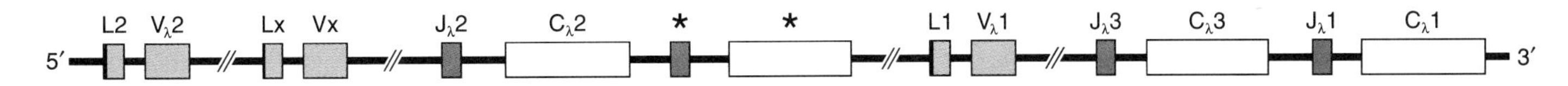

Figure 7.11
Organization of mouse Ig gene families. (Adapted from A. K. Abbas, A. H. Lichtman, and J. S. Pober. 1991. Cellular and Molecular Immunology. p. 77. W. B. Saunders, Philadelphia.)

J_H segments lie within 2 kb of one another and are about 8 kb 5′ of exons that encode the Cμ gene. The 13 D segments are spread over approximately 80 kb, and the most 3′ D is about 1 kb upstream of the most 5′ J_H segment. The V_H gene segments are located in a region that encompasses several megabases and begins approximately 100–200 kb upstream from the J_H segments.

The V_H gene segments have been grouped into 14 families based on nucleotide sequence homology. Murine V_H gene families are generally organized into overlapping clusters such that certain V_H genes or families can be described as J_H proximal and others as J_H distal. In the mouse, the position of the V_H genes on the chromosome seems to correlate, to some extent, with their frequency of rearrangement, with J_H proximal families being rearranged more frequently than J_H distal families. V_H gene segments have been found to occur in only one transcriptional orientation, which is the same orientation as the J_H and C_H exons.

Murine V_H families are highly polymorphic in different inbred strains. However, although the absolute number of V_H gene segments may vary greatly among different mouse strains, the complexity of the V_H locus (number of functional V_H segments that encode unique sequences) is probably similar for all and probably is less than 100. Thus, strains with larger numbers of V_H gene segments appear to have acquired them from recent gene duplication events and therefore have many copies of the same V_H gene. In addition, there are many pseudogenes among the V_H genes due to the presence of in-frame stop codons or the absence of target sequences for the V(D)J recombination activity.

The C_H gene locus encompasses approximately 200 kb (Figure 7.11). Following assembly of a complete V(D)J variable region gene, transcription is initiated from a promoter that lies upstream of the V_H gene segment and proceeds downstream through the four major exons of the Cμ gene. This transcription is influenced by a strong enhancer element that lies in the intron between J_H4 and Cμ (referred to as the **intronic IgH enhancer** or **Eμ** element). Several additional transcriptional enhancer elements have been identified in the region at the far 3′ end of the HC locus, although it is not yet known if they have a role in normal Ig HC gene expression. Recent evidence suggests the existence of additional elements that influence Ig HC gene transcription, but these remain to be identified.

Mature B cells express membrane IgM and IgD as antigen receptors. Activation of B cells leads to their terminal differentiation into plasma cells. These short-lived plasma cells secrete large quantities of antibodies to neutralize foreign antigens. In plasma cells, the level of Ig gene transcription is dramatically increased. Termination of Ig HC gene transcripts just 3′ of the Cμ4 exon generates a primary transcript that is processed to yield an mRNA that encodes the secreted form of the μ protein (μ_s) (Figure 7.12). The coding sequences that specify this form lie in the extreme 3′ portion of the Cμ4 exon; the C terminal amino acids encoded by this region form a domain that interacts with another protein, called the J chain, to allow formation of pentameric IgM molecules (Figure 7.5). There are two additional small exons 3′ of the Cμ4 exon that encode a C terminus transmembrane domain. Thus, if transcription continues through these exons, RNA processing mechanisms utilize an in-phase splice donor site, thereby deleting the most 3′ Cμ4 region encoding the secretory sequences and appending the downstream exons to generate an mRNA that encodes the membrane-bound form of the protein (μ_m) (Figure 7.12). Regulation of the generation of these two different types of primary transcripts allows B cells first to express mostly the membrane-bound form of the μ protein and, subsequent to activation, to express high levels of the same molecule as a secreted form.

The Cδ gene lies approximately 4.5 kb downstream of the Cμ gene and is generally expressed simultaneously with Cμ in mature B cells through a differential RNA-processing mechanism similar to that employed to generate the μ_m and μ_s mRNA sequences (Figure 7.13). The remainder of the C_H genes lie far downstream of the Cμ gene (Figure 7.11). The Cγ3 gene is approximately 55 kb

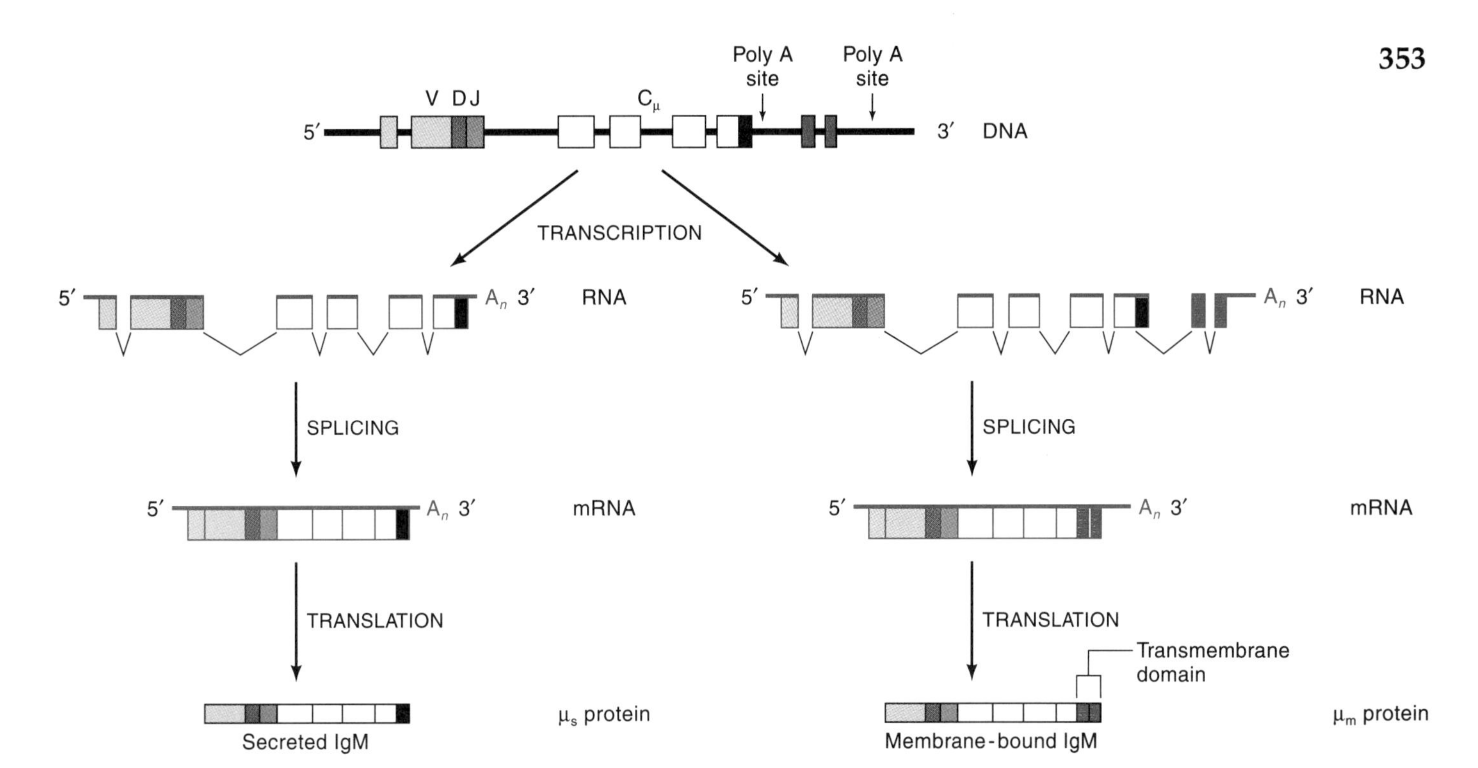

Figure 7.12
Control of membrane-bound and secreted IgM production. (Adapted from L. E. Hood, I. L. Weissman, W. B. Wood, and J. H. Wilson. 1984. Immunology, 2nd ed. p. 95. Benjamin/Cummings, Menlo Park, California.)

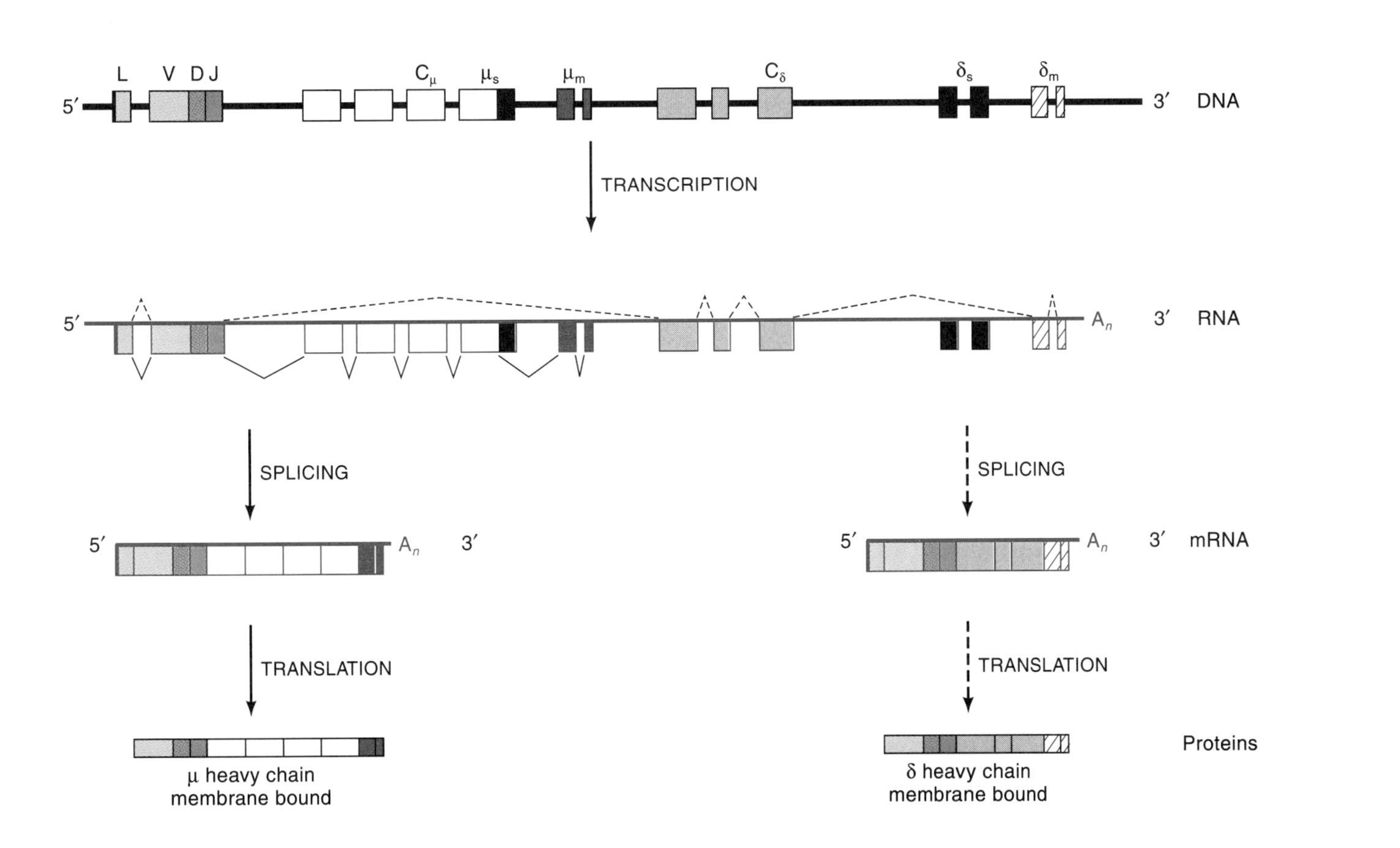

Figure 7.13
Control of IgM and IgD expression. (Adapted from L. E. Hood, I. L. Weissman, W. B. Wood, and J. H. Wilson. 1984. Immunology, 2nd ed. p. 94. Benjamin/Cummings, Menlo Park, California.)

downstream of Cδ on the chromosome, followed by the Cγ1, Cγ2b, Cγ2a, Cε, and Cα genes, which occur at increasingly closer intervals. Expression of these downstream C_H genes generally occurs only in activated B cells. It depends on a different form of B cell–specific recombination referred to as **HC class switch recombination** and juxtaposes the downstream C_H gene to the J_H locus with the deletion of all the intervening C_H genes. All downstream HC genes are organized similarly to the Cμ gene in the sense that they can encode either membrane-bound or secreted protein forms by generating transcripts with alternative 3′ coding sequences.

b. κ-Locus

In the mouse, κ-LCs make up nearly 95 percent of the serum LCs. The murine κ-LC locus lies on chromosome 6. At least 200 Vκ gene segments (of which many may be pseudogenes) lie 5′of four functional Jκ segments that, in turn, lie about 3 kb 5′ of the Cκ coding exon (Figure 7.11). The murine Vκ gene segments have been grouped into 14 families. Vκ gene segments exist in both transcriptional orientations relative to the Jκ and Cκ exons, although the functional significance of this organization, if any, is unknown.

Following assembly of a VκJκ variable region gene, transcription initiated from the Vκ promoter generates a primary transcript that is processed to form the mature VκJκCκ mRNA sequence. Two transcriptional enhancer elements have been identified in the vicinity of the Cκ gene, one in the Jκ-Cκ intron (**intronic κ-enhancer** or **Eκ**) and one about 8 kb downstream of Cκ, referred to as the **3′ κ-enhancer** (**3′ Eκ**) (Figure 7.11). Currently, both seem to be involved in κ-LC rearrangement and/or expression, perhaps in a stage-specific manner.

c. λ-Locus

The murine λ-LC locus is located on chromosome 16. There are 3 Vλ gene segments and 4 Jλ-Cλ complexes in which a Jλ lies just 5′ to each Cλ exon (Figure 7.11). The murine λ-locus encompasses 200 kb and is organized into two duplicated units: Vλ2-Vx-Cλ2-Cλ4-Vλ1-Cλ3-Cλ1 (Figure 7.11). In the mouse, only 5 percent of the serum LCs are of λ-isotype, of which 90 percent are Cλ1. The relatively low contribution of the λ-genes to the LC repertoire probably results from the lower complexity of the V region gene segments. Thus, in other mammalian species, such as humans, a greater complexity of Vλ sequences is associated with a much larger contribution to the serum Ig levels. Other than the promoter elements associated with the Vλ sequences, transcriptional control elements involved with λ-LC gene expression have been defined in much less detail than for the HC and κ-LC loci.

7.4 Mechanism of Antigen Receptor Variable Region Gene Assembly

a. Sequences of Germ Line and Rearranged V, D, and J Segments

Our general understanding of the V(D)J recombination mechanism has derived in large part from early studies that compared the structure of germ line Ig variable region gene segments to that of the Ig V region gene segments fused to form completely assembled variable region genes in B-lineage cells. Originally, such analyses involved laborious constructing and screening of genomic DNA li-

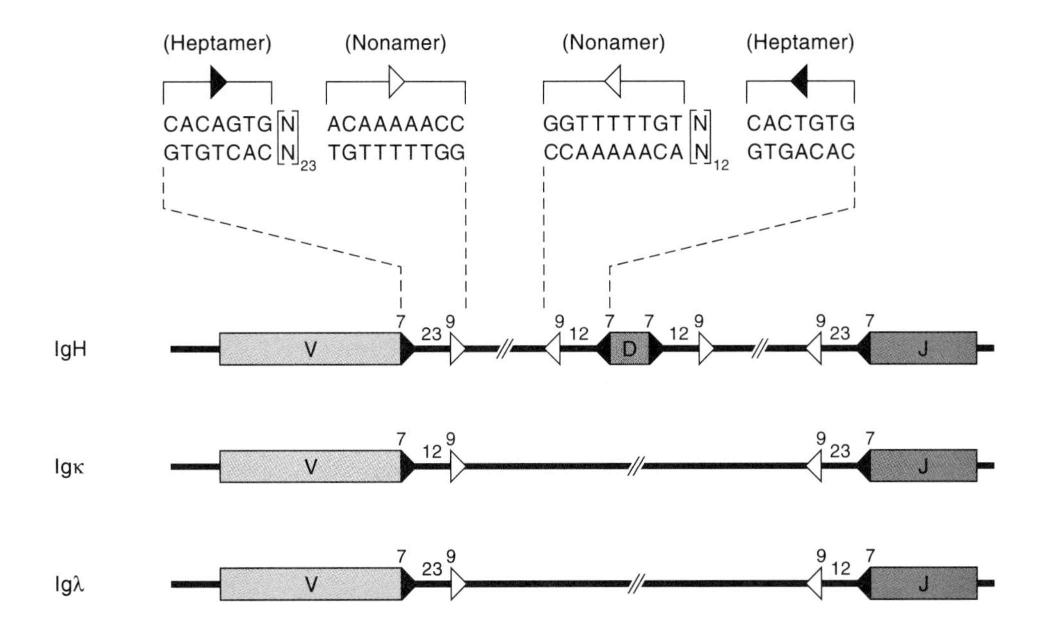

Figure 7.14
Recombination signal sequences at V, D, and J gene segments of IgH, κ, and λ gene loci. (From R. Lansford, A. Okada, J. Chen, E. M. Oltz, K. Blackwell, F. W. Alt, and G. Rathbun. 1996. Mechanism and Control of Immunoglobulin Gene Rearrangement. *Molecular Immunology,* 2nd ed. B. D. Hames and D. M. Glover, eds. IRL Press.)

braries for each analyzed B-lineage cell line (e.g., myelomas and plasmacytomas) to obtain cloned genes. Nowadays, large numbers of endogenous (or substrate) rearrangements can be rapidly isolated and analyzed by polymerase chain reaction (PCR) methodologies. However, the original cloning and sequencing analyses revealed a number of fundamental aspects of this unique site-specific recombination process.

Recombination signal sequences Immediately adjacent to their recombination point, each germ line variable region gene segment is flanked by a set of conserved sequences termed **recombination signal sequences** (**RSS**). The RSS consist of a conserved heptameric sequence (CACAGTG), a spacer of 12 or 23 bp, and an AT-rich nonamer sequence (related to ACAAAAACC) (Figure 7.14). The conservation of these sequences 3′ of all V gene segments on both sides of all D segments and 5′ of all J segments suggest that they are the target sequence for the recombination process. Examination of the RSS sequences between V_L and J_L sequences indicated that they had, respectively, spacers of 12 and 23 bp. This finding led to the prediction of the **12/23 rule**, which states that two variable region gene segments can be joined only if one contains a 12 and the other a 23 bp spacer.

The 12/23 rule Early studies demonstrated that both J_H and V_H segments are flanked by RSS with 23 bp spacers; this is one of the findings that led to the prediction of an additional gene segment, the D segment, for HC variable region genes. Once D segments were identified, researchers found, as predicted, that they are flanked on both their 5′ and 3′ sides by RSS with 12 bp spacers—which permits D to J_H and V_H to DJ_H joining in accordance with the 12/23 rule (Figure 7.14). Subsequent analyses of TCRβ variable region genes showed that the J segments were flanked by RSS with 12 bp spacers, and the Vβ segments were flanked by RSS with 23 bp spacers. Yet, DNA sequence analyses suggested the presence of D segments and led to the prediction of D segments flanked on one side with RSS containing 23 bp spacers and on the other with RSS containing 12 bp spacers to permit joining to the Jβ and Vβ, respectively, Such D segments were subsequently identified, adding further support for the 12/23 rule, which was later confirmed in detail by recombination substrate studies.

The 12 or 23 nucleotides, respectively, permit one or two complete turns of the DNA helix, aligning the heptamers and nomamers of both types of RSS in the same rotational position on the DNA molecule. This configuration may allow spatial relationships between putative heptamer- and nonamer-binding factors that optimize binding or interaction and, therefore, efficient recombination. In any case, the 12/23 joining reaction provides direction to the V(D)J recombination process. Thus, assembly of Ig HC variable regions necessarily involves D to J_H and V_H to DJ_H joining because V_H segments cannot join to the J_H segments directly. In contrast, the 12/23 rule specifies that V_L segments will join directly to J_L.

Junctional diversification Examination of multiple independent joints involving the same germ line element indicated that the joining process is not necessarily precise with respect to the coding ends. The sequence variations observed in these regions is referred to as **junctional diversity**. In particular, nucleotides could be lost from the potential coding sequences, and nucleotides could be added to junctional regions by *de novo* mechanisms (Figure 7.15).

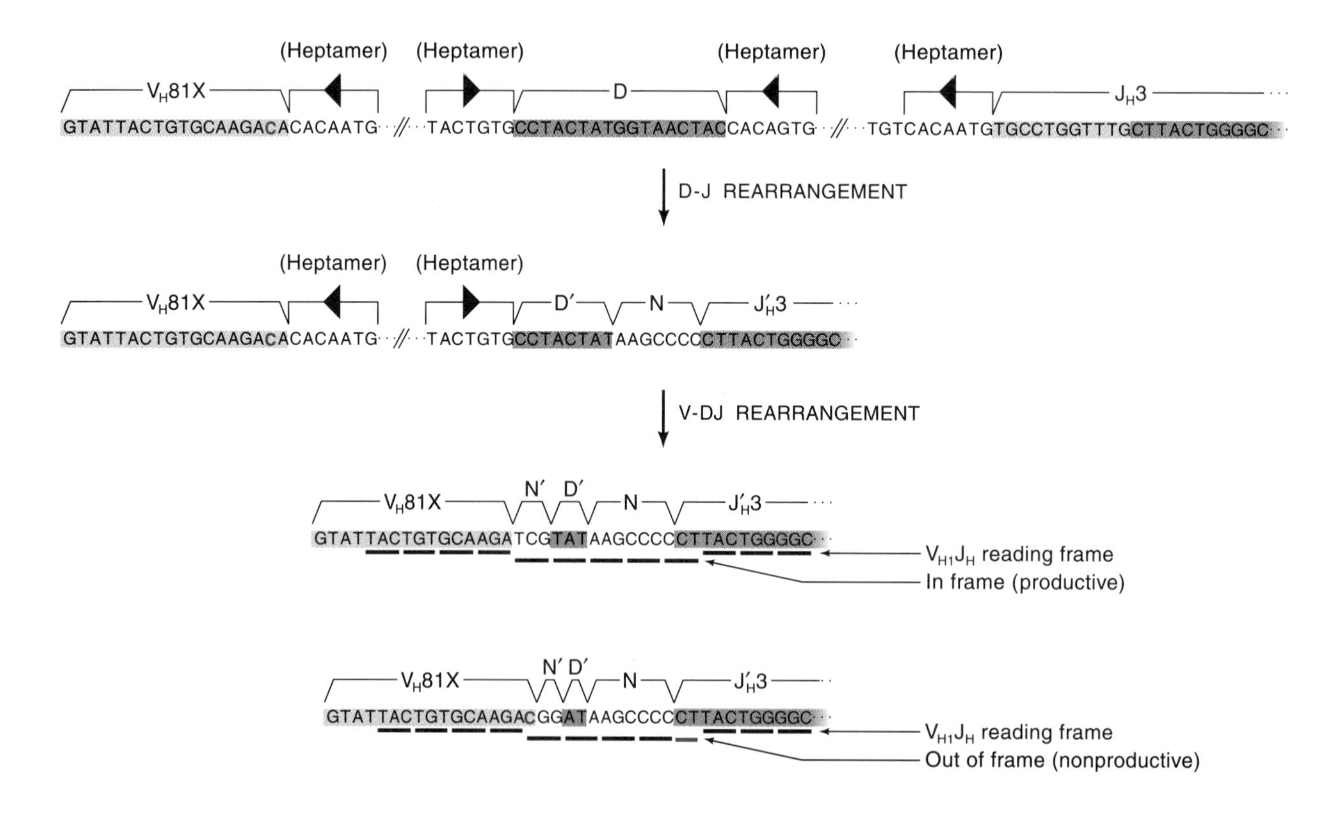

Figure 7.15
Junctional diversities and in-frame and out-of-frame rearrangements. (Adapted from S. V. Desiderio, G. D. Yancopoulos, M. Paskink, E. Thomas, M. A. Boss, N. Landau, F. Alt, and D. Baltimore. 1984. Insertion of N Regions into Heavy-Chain Genes Is Correlated with Expression of Terminal Deoxytransferase in B Cells. *Nature* 311 754.)

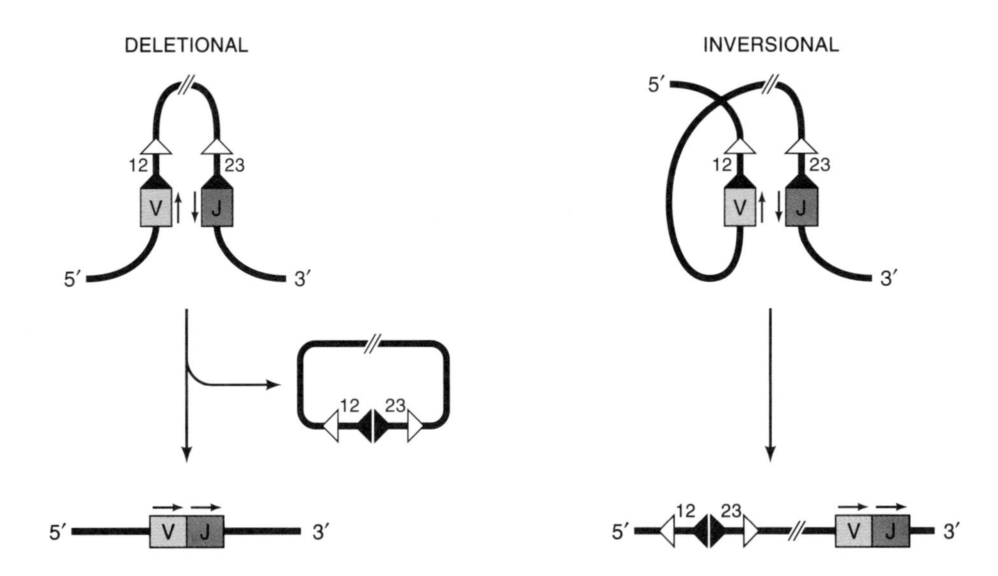

Figure 7.16
Inversional and deletional joinings. (Adapted from F. Alt and D. Baltimore. 1982. Joining of Immunoglobulin Heavy Chain Gene Segments: Implications from a Chromose with Evidence of Three D-JH Fusions. *Proc. Natl. Acad. Sci. USA* 79 4121.)

Deletional and inversional joints Analyses of Ig-producing cell lines demonstrated that joining of the coding sequences that lie in the same transcriptional orientation results in deletion of the intervening DNA sequences (**deletional joining**) (Figure 7.16). However, researchers also observed that V(D)J recombination can occur between gene segments with the opposite transcriptional orientation relative to each other (e.g., some Vκ gene segments). In this case, the recombination event leads to the inversion of the intervening DNA segment (**inversional joining**) (Figure 7.16). Analyses of such inversion events demonstrated that the two participating RSS were perfectly joined (no loss of nucleotides), but the coding ends often exhibited variable loss of nucleotides. These findings led to several predictions that were later verified by more sophisticated methods, including: (1) the V(D)J recombination process is initiated by the generation of precise double-strand breaks at the border of coding and RSS sequences, (2) the RSS and coding ends are differentially handled, RSS sequences being precisely joined and coding ends being subject to extensive junctional diversification (modification), and (3) with respect to deletional joining, the intervening DNA is deleted as a circle (Figure 7.17).

b. V(D)J Recombination Substrates

The most efficient way to study the mechanism of an enzymatic reaction is to reconstitute the reaction *in vitro*. Prior to the recent accomplishment of *in vitro* V(D)J recombination, much of our understanding of the V(D)J recombination process has relied on the introduction of **V(D)J recombination substrates** into cell lines. In such assays, vectors containing portions of variable region gene sequences plus their flanking RSS (or just the RSS linked to non-V region DNA sequences) are introduced into cell lines, such as A-MuLV-transformed pre-B cells, that express V(D)J recombination activity. The introduced V(D)J recombination substrates undergo specific V(D)J recombination events that can subsequently be recovered for analysis. Such vectors have been developed in two general forms, chromosomally integrated and extrachromosomal.

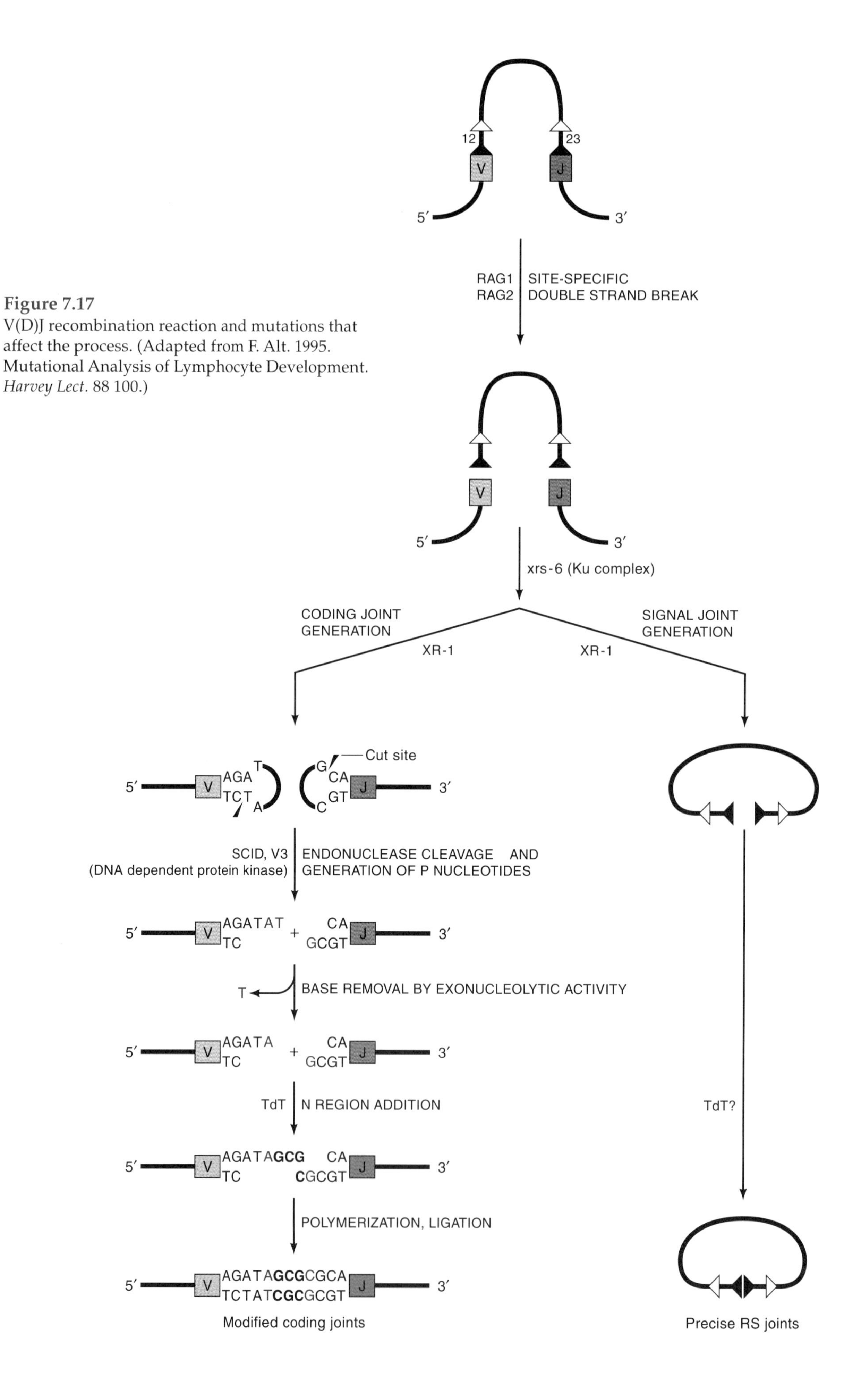

Figure 7.17
V(D)J recombination reaction and mutations that affect the process. (Adapted from F. Alt. 1995. Mutational Analysis of Lymphocyte Development. *Harvey Lect.* 88 100.)

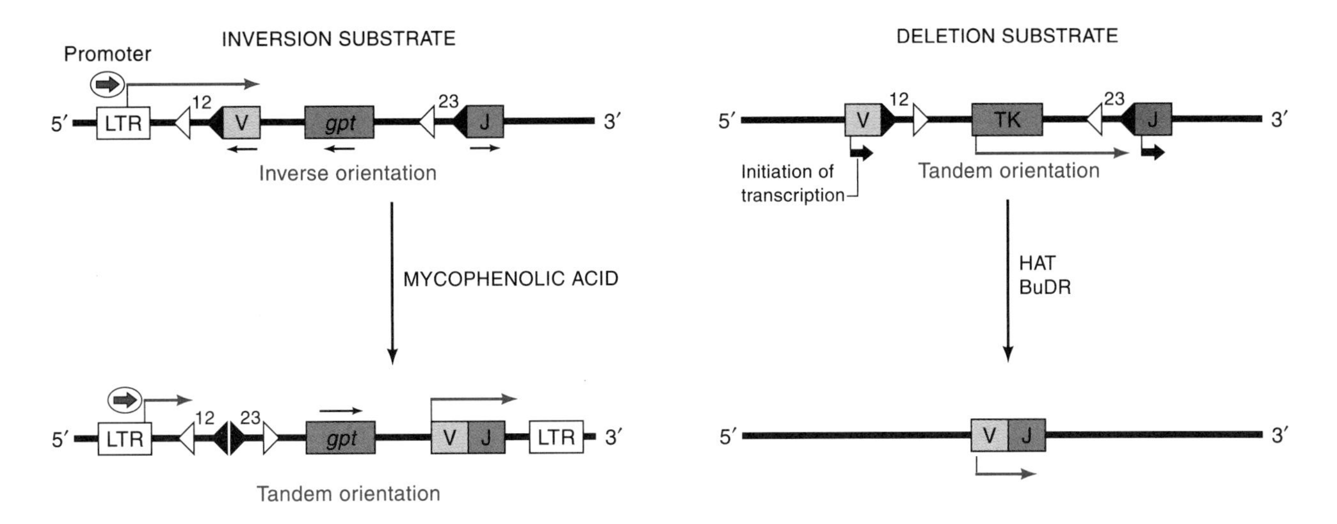

Figure 7.18
Two types of chromosomally integrated recombination substrates. (Adapted from F. Alt, T. Blackwell, and G. Yancopoulos. 1987. Development of the Primary Antibody Repertoire. *Science* 238 1084.)

Chromosomally integrated recombination substrates The first V(D)J recombination substrates were stably introduced into the genome of A-MuLV-transformed pre-B cell lines. The most versatile form of these vectors exploited the inversional mode of V(D)J joining to activate a selectable marker gene—such as the guanidine phosphoribosyltransferase (*gpt*) gene, which, when expressed, confers a drug resistance to the mammalian cell (Figure 7.18). In these substrates, variable region gene segments containing complementary RSS (with 12 and 23 bp spacers) were positioned in inverted orientation on either side of the selectable marker gene. The selectable marker gene was inserted in inverse orientation relative to a transcriptional promoter so that it was not functionally expressed unless it was inverted. Such vectors were then introduced into pre-B cell lines either in the context of a retroviral vector or by stable transfection methods. Following the generation of clones in which the vector was integrated, cells that had undergone rearrangements could be selected on the basis of the drug resistance conferred by the expression of the inverted selectable marker gene. In some cell lines with high levels of V(D)J recombination activity, rearrangement of the introduced substrates occurred at high enough levels so that DNA-blotting analysis of mass populations or nonselected subclones was sufficient to identify substrate rearrangements. These types of vectors and analyses have been important for assaying mechanistic and regulatory aspects of the reaction. Most important, such vectors have provided a method for cloning genes that encode specific components of the V(D)J recombination reaction.

Extrachromosomal recombination substrates A major innovation with respect to V(D)J recombination substrate technology was the transient introduction of recombination substrates (Figure 7.19). These substrates contain target variable region gene segments (or just RSS) upstream and downstream of a bacterial transcriptional termination signal. A bacterial chloroamphenicol transacetyltransferase gene, which when expressed confers chloroamphenicol resistance (*cam*^{r}) to bacteria, lies downstream. A bacterial promoter lies upstream of the 5′ variable region gene segment (RSS). These substrates are introduced transiently into

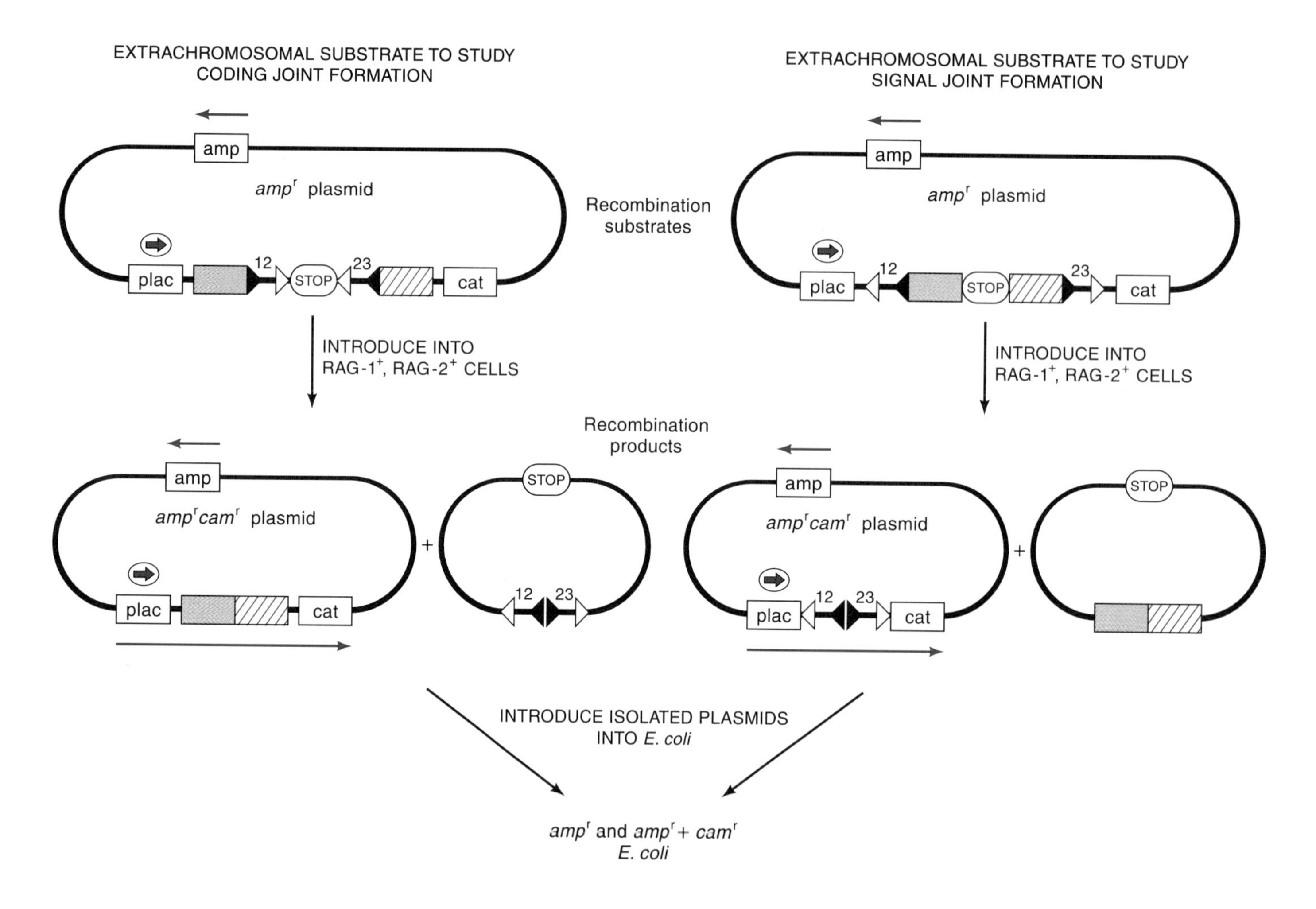

Figure 7.19
Two types of extrachromosomal recombination substrates. (Adapted from J. E. Hesse, M. R. Lieber, M. Gellert, and K. Mizuuchi. 1987. Extrachromosomal DNA Substrates in Pre-B Cells Undergo Inversion or Deletion at Immunoglobulin V-(D)-J Joining Signals. (*Cell* 49 776, as modified by R. Lansford, A. Okada, J. Chen, E. M. Oltz, K. Blackwell, F. W. Alt, and G.Rathbun. 1996. Mechanism and Control of Immunoglobulin Gene Rearrangement. *Molecular Immunology,* 2nd ed. B. D. Hames and D. M. Glover, eds. IRL Press.)

mammalian cells where, if present, V(D)J recombination activity removes the transcriptional termination signal. Thus, when specifically recombined segments are recovered from the mammalian cells and introduced into bacterial cells, V(D)J recombination can be scored by generation of chloroamphenicol-resistant colonies. Because the vectors also incorporate a constitutively active ampicillin resistance gene (*amp*^{r}), an index of the V(D)J recombination activity can be estimated by comparing the number of *amp*^{r}/*cam*^{r} bacterial colonies and *amp*^{r} colonies detected (Figure 7.20).

The transient vectors were designed to assay for coding sequence joints, RSS joints, and, in the case of inversion vectors, both types of joints (Figure 7.19). Analysis of the RSS joints by nucleotide sequencing across the junctions in isolated rearranged products generates an index of the fidelity of the joining process. These substrates have been particularly useful for dissecting mechanistic details of the V(D)J recombination reaction, for example, the influence of RSS

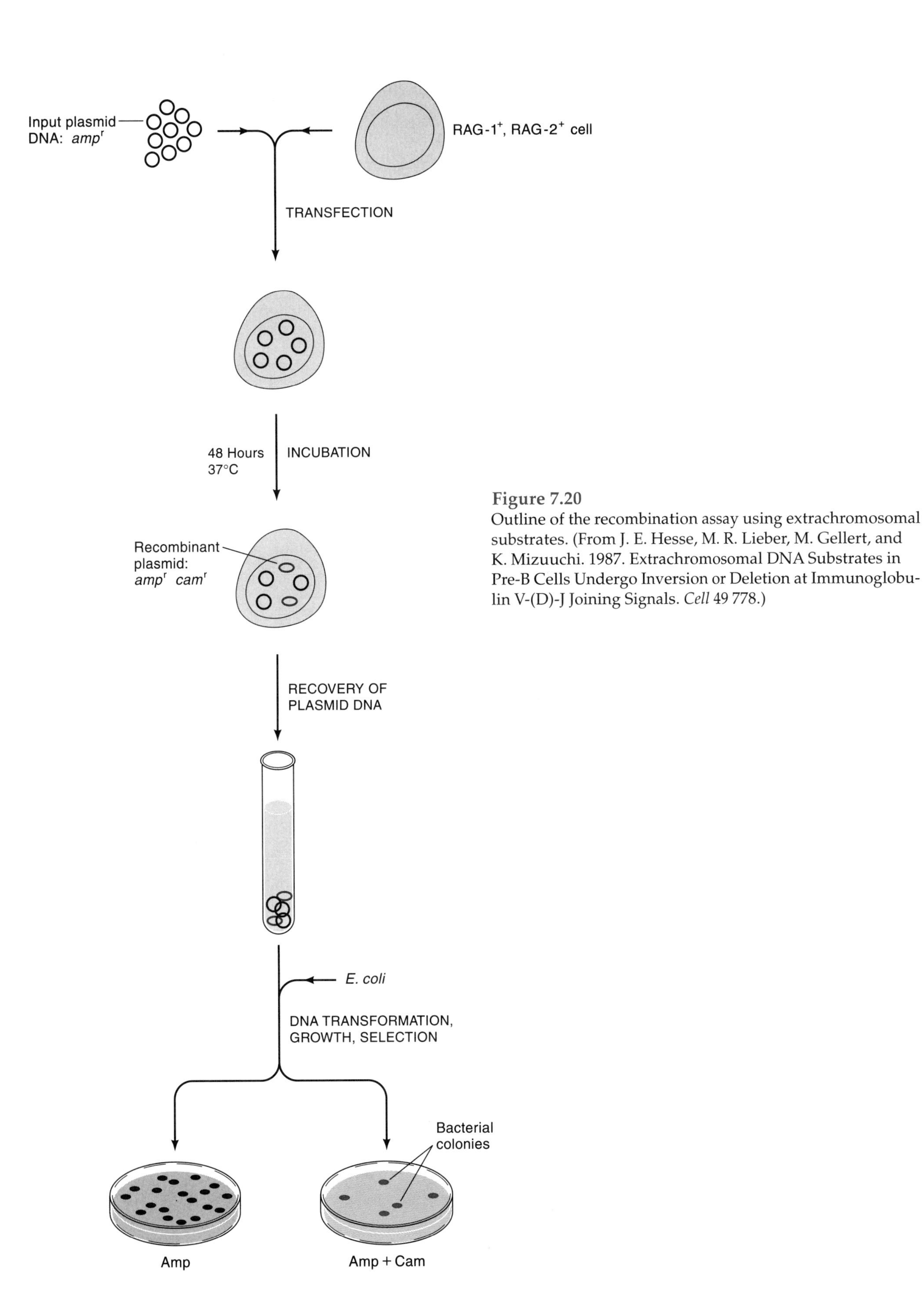

Figure 7.20
Outline of the recombination assay using extrachromosomal substrates. (From J. E. Hesse, M. R. Lieber, M. Gellert, and K. Mizuuchi. 1987. Extrachromosomal DNA Substrates in Pre-B Cells Undergo Inversion or Deletion at Immunoglobulin V-(D)-J Joining Signals. *Cell* 49 778.)

and coding sequences (by systematic modification of these sequences in a given substrate) and the identification of novel joining pathways such as **hybrid joints** (where RSS join coding segments) and **open and shut joints** (where the RSS are separated from their associated coding segment and then reattached, often with loss or addition of nucleotides) (Figure 7.21). A particular advantage of the transient recombination substrates is the ability to recover and analyze large numbers of independent recombination events.

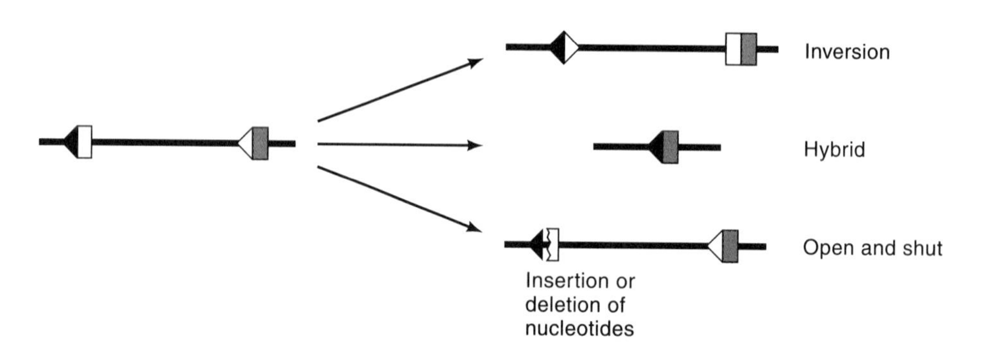

Figure 7.21
Normal and aberrant products of V(D)J recombination. (From M. Gellert. 1992. V(D)J Recombination Gets a Break. *Trends Genet.* 8 410.)

c. The V(D)J Recombination Mechanism

A detailed scheme that presents current understanding of the V(D)J recombination mechanism is presented in Figure 7.17. The V(D)J recombinase system and its cognate RSS appear highly conserved in even the most primitive vertebrate species analyzed, such as sharks; however, V(D)J recombination has not been identified in invertebrate species, suggesting that it evolved relatively recently.

The initial step in the V(D)J recombination reaction involves the recognition of the participating RSS sequences. Several lines of evidence indicate that two RSS or at least one complete RSS plus a heptamer are required to initiate the V(D)J recombination reaction. Thus, for example, open and shut joints are not observed with a single RSS, but they are observed with a single RSS plus certain mutant heptamer sequences (Figure 7.21). Mutational analyses of heptamer and nonamer sequences in stable or transient recombination substrates demonstrated that significant variations from the consensus heptamer can be tolerated but that conservation of the three coding segment–proximal bases of the heptamer is required for efficient rearrangement (Figure 7.22). Apparently, the nonamer can tolerate even greater variations from the consensus sequence and still

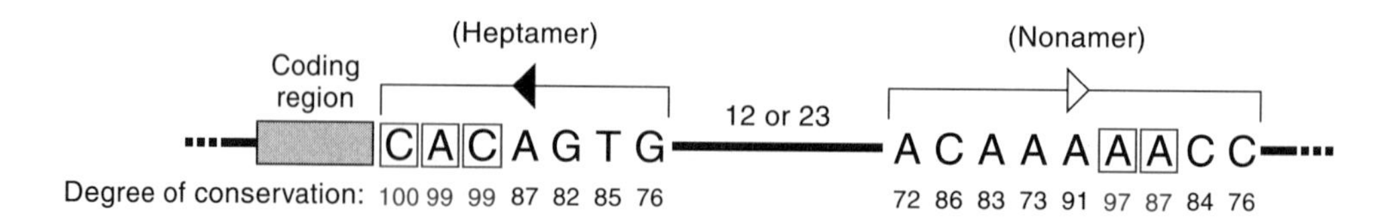

Figure 7.22
Conservation of recombination signal sequences. (From M. Gellert. 1992. V(D)J Recombination Gets a Break. *Trends Genet.* 8 409.)

function to promote relatively efficient rearrangement if in the presence of a consensus heptamer. Genes that encode proteins that bind to isolated RSS, heptamer, or nonamer sequences have been isolated. However, some such binding activities appear to be artifactual, and no specific functions for any of the encoded proteins in the context of V(D)J recombination have yet been identified.

Because of the less stringent requirement for precise nonamer sequences in V(D)J recombination, sequences flanked by a complete RSS can be joined to sequences flanked by a heptamer alone. These rearrangements may have important physiological functions. For example, **V_H gene replacement**, which could be a way of salvaging nonfunctional V(D)J rearrangements during normal lymphocyte development, can occur between the RSS of a germ line V_H gene segment and the heptamer conserved in the third framework region of most murine and human V_H gene segments (Figure 7.23). Similarly, heptamer-mediated rearrangements can result in the deletion of Cκ genes so often observed in λ-producing B cell lines. Conversely, because typical heptamer sequences occur frequently in other places in the genome, inappropriate V(D)J recombination events may lead to translocations observed in certain lymphomas. Thus, the recognition of heptamer (or even complete RSS) must not be the only factor that guides the recombinase to the appropriate substrate variable region gene segments. We explore this concept more fully in the context of the regulation of the reaction (see Section 7.5).

The initiation of the V(D)J recombination reaction appears to depend on the specific recognition of the RSS sequences followed by the introduction of a precise **double-strand break** (**DSB**) at the border of both participating RSS and coding sequences (Figure 7.17). As outlined previously, this process was initially inferred from the analysis of inversional joints in which the RSS heptamers were nearly always precisely ligated to each other. More recently, the unfused DSB intermediates of rearranging TCR and Ig RSS sequences have been isolated. These

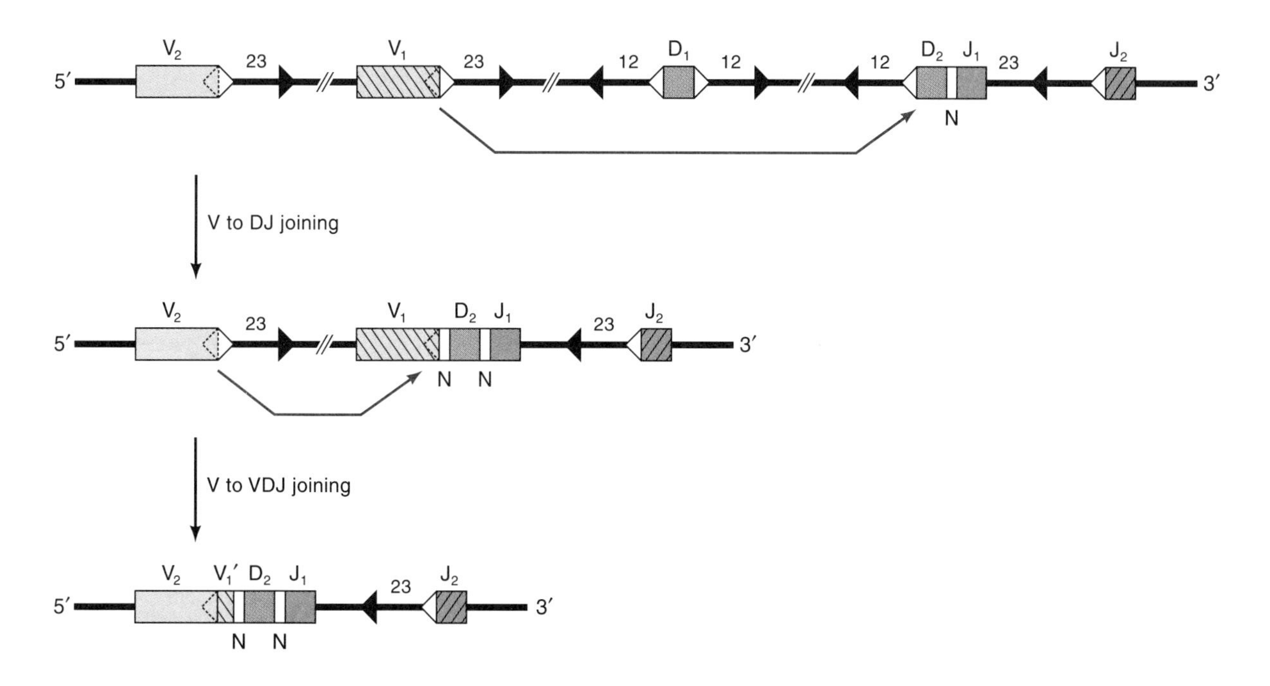

Figure 7.23
V gene replacement. The dotted-lined triangles in the V regions represent the heptamers conserved in the third framework region. (Adapted from F. W. Alt. 1986. Antibody Diversity: New Mechanism Revealed. *Nature* 322 772.)

intermediates are blunt, sensitive to exonuclease, and phosphorylated at their 5′ ends. The enzymatic activities responsible for the double-stranded cleavage are encoded in the *RAG* 1 and *RAG* 2 genes (see below). Despite the identification of unfused DSB intermediates, researchers generally assume that the RSS sequences are usually precisely ligated to each other, heptamer to heptamer, in an early step in the process. In deletional joining pathways, this leads to the liberation of the intervening DNA from the chromosome as a circle. Such excised circles from both Ig and TCR loci have, in fact, been identified in lymphoid cell lines and tissues.

One important difference in the structures of the liberated RSS and coding sequences is that the ends of the latter appear to be **hairpin** structures (Figure 7.17). This difference may provide, at least in part, a mechanistic basis for the differential processing reactions involved in coding and RSS joints. Initially coding end hairpins were identified only in the context of a murine mutation (*scid*) that affects ability to generate normal coding sequence joints, but they are known to be the product of the RAG 1 RAG 2 mediated cleavage. However, a potential to generate a hairpin structure during normal coding joint formation was proposed even before their identification. The proposal accounted for several unusual features of the V(D)J reaction (such as P nucleotide formation). If we assume that most coding ends are generated in the form of hairpins, an endonuclease activity would be required to open them for further processing. An endonucleolytic scission at the apex of the hairpin would generate the equivalent of a blunt DSB at the heptamer border. Endonucleolytic scissions interior to the apex of the hairpin would result in an overhanging end, in the form of an inverted repeat, that could be subjected to further processing events (Figure 7.24).

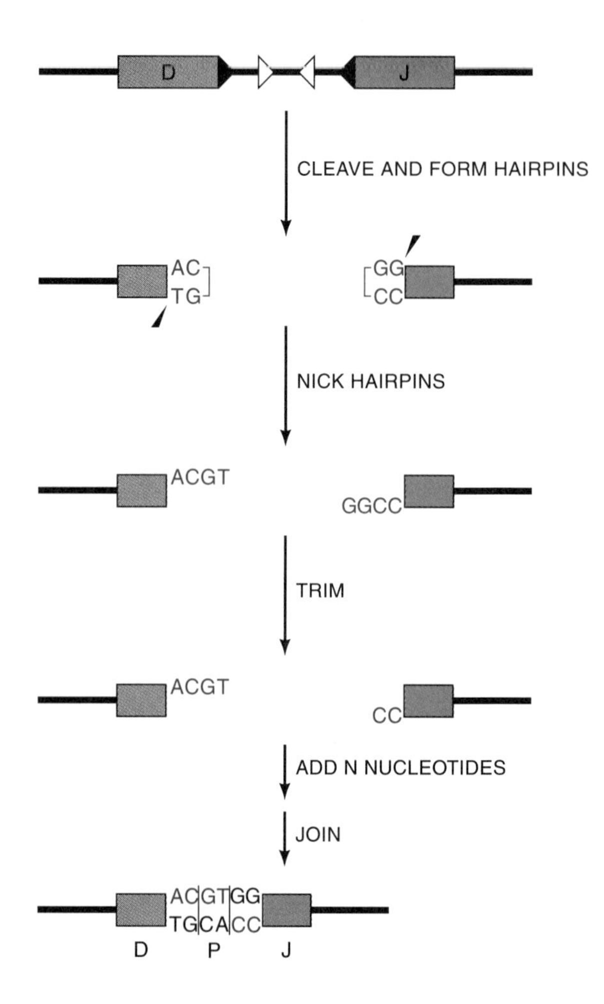

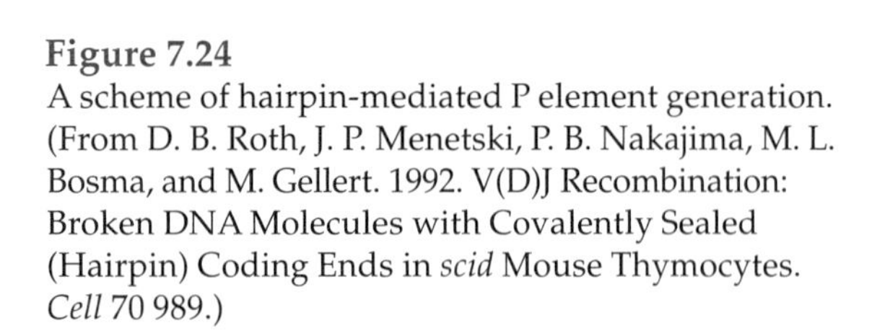
Figure 7.24
A scheme of hairpin-mediated P element generation. (From D. B. Roth, J. P. Menetski, P. B. Nakajima, M. L. Bosma, and M. Gellert. 1992. V(D)J Recombination: Broken DNA Molecules with Covalently Sealed (Hairpin) Coding Ends in *scid* Mouse Thymocytes. *Cell* 70 989.)

Nucleotide deletion and addition at junctional regions Comparison of the junctional sequences of germ line versus assembled variable region gene-coding segments revealed both loss and addition of nucleotides at the junctions (Figure 7.15). Such activities can lead to extensive modifications of the potential CDR3 coding sequences; therefore, they are a very important source of variable region diversity. The mechanism of nucleotide loss is unknown, although an exonucleolytic activity has been speculated. Such exonucleolytic activity might work on overhanging ends or on flush ends (Figure 7.24). However, other potential mechanisms relating to the recombinational resolution of hairpin sequences are also conceivable.

The extra nucleotides found at junctions fall into two general categories. Non–germ line encoded nucleotides, referred to as **N regions**, can have an essentially random sequence and are normally less than 10 bp long. **P (palindromic) nucleotides**, which are found adjacent to intact coding sequences, form palindromes with the terminal sequences of the coding regions and are usually only 1 or 2 bp long (Figure 7.25).

N regions are added to V(D)J junctional regions by the enzyme **terminal deoxynucleotidyl transferase (TdT)** (Figure 7.17), a lymphoid-specific enzyme that adds nucleotides onto the 3′ ends of DNA sequences in a template-independent manner. N regions may also be added to the RSS joints. Because TdT is generally expressed only at the pro-B cell stage in adult mice, N regions usually are present only at the DJ_H and V_HD junctions. Ig LC variable genes, which are assembled at the pre-B cell stage when TdT expression declines, generally do not have N region additions. N regions are prominent in the junctional regions of all the various TCR variable region gene segments generated in adult mice. Fetal repertoires generally have few N regions because of limited TdT expression in developing fetal lymphocytes. Thus, the developmental and stage-specific expression of TdT results in the generation of qualitatively different V(D)J joints. Although only speculative at this point, antigen receptors encoded by such differential rearrangements may have important developmental stage-specific functions.

P nucleotides appear to result from the opening, interior to their apex, of hairpin regions at the liberated coding ends. This results in a short, palindromic, overhanging flap that, if incorporated into the final coding region sequences, generates the short, inverted repeats of several base pairs (Figures 7.17, 7.24, 7.25). The **severe combined immunodeficiency (*scid*)** mutation of mice results, when homozygous, in an almost complete absence of mature T and B cells. Analyses of V(D)J recombination in precursor lymphocytes from homozygous *scid* mutant mice (**SCID mice**) showed that the defect is due to an inability to join liberated coding ends as constrasted with RSS ends (Figure 7.17). The cleaved coding sequence intermediates of TCRδ variable region gene rearrangements in the thymus of SCID mice appear to accumulate as covalently closed hairpins. In this context, exceptionally long P nucleotides of up to 15 bp have been found in junctions of V(D)J rearrangements from SCID lymphoid cells. Thus, the SCID defect may lie in the inability to resolve the hairpin structures normally; nicking in regions more distal to the apex of the hairpin would result in the abnormally long P sequence. The *scid* gene may encode a factor that is either directly or indirectly required for normal nicking or, when mutated, interferes with normal nicking.

Homology-mediated joining Analyses of coding joints that occur in the absence of N regions (e.g., in endogenous joints in fetal lymphocytes or joints within recombination substrates assayed in TdT-deficient cell lines) have revealed the presence of **homology-mediated joining** of coding segments. Homology-mediated joining occurs when several complementary bases exist at the ends of the rearranging segments. These apparently pair to generate joints with characteristic ambiguous nucleotides that could be derived from either of the participating

$TdT^{+/-}$

	V_H81X	P	N	P	D	P	N	P	J_H3
Germ line	TGT GCA AGA CA								AC TAC TTT GAC TAC TGG
22a2-4	TGT GCA AGA		GCCCCCA		TAACTGG		T		AC TAC TTT GAC TAC TGG
22a2-10	TGT GCA AGA				TGATTACGAC				TAC TTT GAC TAC TGG
22a2-12	TGT GCA AGA C		GG		TACTATGATTACGAC	G	G		TTT GAC TAC TGG
22a2-13	TGT GCA AGA C				GATTACGAC	G			AC TAC TGG
22a2-3	TGT GCA AGA CA	T	AG		AGTATGGTAAC		CC	T	AC TAC TTT GAC TAC TGG
22a2-11	TGT GCA AGA CA	T	AG		AGTATGGTAAC		CC	T	AC TAC TTT GAC TAC TGG
22b2-4	TGT GCA AG		GC	GG	CTATGGTAACT		GAGG	GT	AC TAC TTT GAC TAC TGG
22b2-1	TGT GCA AGA C				GGTTACT				GG
22b2-7	TGT GCA AGA CA	T	A		CTATGATGGT			T	AC TAC TTT GAC TAC TGG
22a2-6	TGT GCA AGA CA		GGG		ACTATAGGTACGAC	G	AAGGGC		TGAC TAC TGG
22a2-9	TGT GCA AGAC				TATAGGTACGAC				TTT GAC TAC TGG
22b2-2	TGT GCA AGA CA		GG		CTACTATAGGTACG		CG		GAC TAC TGG
22b2-3	TGT GCA AGA CA		GG		CTATAGGTACGAC	GT			C TAC TGG
22b2-11	TGT GCA A			GG	CCTACTATAGGTACGAC	GT			AC TTT GAC TAC TGG
22b2-13	TGT GCA AGA		GGGG		AGGTACG		TG		C TAC TTT GAC TAC TGG
22b2-5	TGT GCA AGA CA	T			TACTACGGTAGTAGCT				C TAC TGG
22b2-12	TGT GCA A		TAT		TACGGC			T	AC TAC TTT GAC TAC TGG
22a2-1	TGT GCA AGA CA		G		GGGCTAC				GG
22a2-5	TGT GCA AGA C								C TTT GAC TAC TGG

$TdT^{-/-}$

	V_H81X	P	N	P	D	P	N	P	J_H3
Germ line	TGT GCA AGA CA								AC TAC TTT GAC TAC TGG
84a2-3	TGT GCA AGA CA	TG			CTGGGAC				TAC TTT GAC TAC TGG
84b2-4	TGT GCA AG				GAC				TAC TGG
84b2-11	TGT GCA AGA C				TGGGAC				TTT GAC TAC TGG
84b2-16	TGT GCA AGA C				TAACTGGGAC				TAC TTT GAC TAC TGG
84b2-19	TGT GCA A				ACTGGGAC	GT			AC TTT GAC TAC TGG
84a2-12	TGT GCA AGA C				TATGATTACGAC				TAC TGG
84a2-17	TGT GCA AGA CA	TG			TGATTACGAC				TTT GAC TAC TGG
84b2-7	TGT GCA AGA CA	TG			ATTACGAC	GT			AC TAC TTT GAC TAC TGG
84a2-6	TGT GCA AG				GTAAC			T	AC TAC TTT GAC TAC TGG
84a2-13	TGT GCA AGA CA	TG			TGGTAACTAC				TTT GAC TAC TGG
84b2-1	TGT GCA AGA C				TGGTAAC			T	AC TAC TTT GAC TAC TGG
84b2-2	TGT GCA AGA C				GGTAACT				TT GAC TAC TGG
84b2-3	TGT GCA AGA C				GGTAACTA				AC TAC TGG
84b2-13	TGT GCA AGA CA	TG			TGGTAACTAC				TTT GAC TAC TGG
84b2-17	TGT GCA AGA				TGGTAAC			T	AC TAC TTT GAC TAC TGG
84b2-18	TGT GCA AGA CA	T			ATGGTAACTAC				TTT GAC TAC TGG
84b2-8	TGT GCA AGA C				CTACTATGGT				GAC TAC TGG
84a2-4	TGT GCA AGA CA			G	CCTAGTATGGTAACTAC				TGG
84a2-7	TGT GCA AGA CA	TG			TATGGTAACTAC				TTT GAC TAC TGG
84a2-8	TGT GCA AGA				TAGTATGGTAACTAC	G			AC TAC TGG
84b2-14	TGT GCA AG				TATGG				GAC TAC TGG
84b2-20	TGT GCA AGA C				TAGTATGGTAACTAC				TTT GAC TAC TGG
84b2-21	TGT GCA				GTATGGTAAC			T	AC TAC TTT GAC TAC TGG
84a2-21	TGT GCA AGA CA	TG			GTATGGTAACT				TT GAC TAC TGG
84a2-1	TGT GCA AGA CA	T			ACTATAGGTACGAC				TAC TGG
84a2-5	TGT GCA AGA				TAGGTACGAC				TAC TTT GAC TAC TGG
84a2-9	TGT GCA AG				TACTATAGGTACGAC				TAC TGG
84a2-11	TGT GCA AGA C				TATAGGTA				AC TAC TGG
84a2-14	TGT GCA AGA CA				CTATAGGTACGAC	GT			AC TTT GAC TAC TGG
84b2-6	TGT GCA AGA CA	T			AGGTA				GAC TAC TGG
84a2-10	TGT GC			AA	TTTATTACTACGGTAGTAGC			T	AC TAC TTT GAC TAC TGG
84a2-18	TGT GCA AGA C				TACTACGGTAGTAG				AC TAC TTT GAC TAC TGG
84a2-2	TGT GCA AG			A	TAGACAGCTCGGGCTAC				AC TGG
84a2-15	TGT GCA AGA				C			T	AC TAC TTT GAC TAC TGG
84b2-9	TGT GCA AGA C				T				TT GAC TAC TGG
84b2-10	TGT GCA AGA C				GG			GT	AC TAC TTT GAC TAC TGG

Figure 7.25
The role of TdT in N region addition. (top section) A set of Ig heavy chain V(D)J junctions from mice heterozygous for an inactivating mutation of the TdT gene. (lower section) A set of Ig heavy chain V(D)J junctions from mice homozygous for the TdT inactivation mutation. Note the lack of N nucleotides in junctions. (From T. Komori, A. Okada, V. Stewart, and F. W. Alt. 1993. Lack of N Regions in Antigen Receptor Variable Region Genes of TdT-Deficient Lymphocytes. *Science* 261 1173.)

coding segments (Figure 7.25). Analyses of V(D)J rearrangement junctions formed in the absence of TdT, which obscures and/or eliminates the use of homologies, indicates that homology-mediated joining occurs quite frequently, but apparently is not absolutely required for the joining process. In addition, not all homologies appear to promote the reaction with equal efficiency, and the extent to which they are employed may depend on the surrounding sequences. Although homology-mediated joining can restrict variable region diversity, it may have evolved to form a specific variable region gene and thereby circumvent the wasteful selection process. Such "programmed" variable region genes have been suggested to be important for establishing a repertoire responsive to specific antigens during early development.

Completion of the reaction The V(D)J recombination reaction is presumably completed by steps dependent on, for example, DNA polymerase and ligase activities that repair the DSB intermediates generated by the processes already outlined (Figure 7.17). However, the reaction as presented is probably oversimplified because many additional activities have been implicated in similar site-specific recombination processes in lower organisms. Insight into such additional activities has been obtained from studies that focus on mammalian DSB repair pathways as discussed in the following section.

d. Genes and Activities Involved in V(D)J Recombination

As outlined previously, the V(D)J recombination process involves a series of distinct steps that must be mediated by numerous components. Collectively, the reaction components are referred to as **the V(D)J recombinase**. By analogy to other site-specific recombination systems, the V(D)J recombinase is likely a complex (or collection) of proteins. These proteins include some that are specifically found in lymphoid cells; thus, the *RAG*1 and *RAG*2 proteins provide for RSS recognition and specific double-strand endonucleolytic activity. Other participating proteins are likely to be more generally expressed in a variety of cell types; these generally expressed activities could carry out the more routine aspects of the reaction such as repair and ligation. The generally expressed activities are probably recruited by lymphoid-specific components to carry out their particular functions in the context of V(D)J recombination. A variety of different approaches has now led to the identification of both specific and general components.

Recombination-activating genes All necessary lymphocyte-specific components of the V(D)J recombination reaction can be provided by the synergistic activities of the proteins encoded by the **recombination-activating genes (*RAG*1 and *RAG*2)**. Isolation of the *RAG*1 and *RAG*2 genes was one of the major breakthroughs in our understanding of the V(D)J recombination process and lymphocyte development. The isolation procedure depended on the fact that V(D)J recombination activity has been found only in cell lines that represent precursor lymphocytes. The activity is not present in nonlymphoid cells or cell lines. Thus, the *RAG* genes were cloned on the basis of their ability to activate selectively an inversional V(D)J recombination substrate that had been introduced into a fibroblast cell line. When activated by specific inversion, this vector drives expression of a marker gene (bacterial guanosine phosphoribosyl transferase gene, *gpt*); the transferase confers resistance to mycophenolic acid on the fibroblast cells. Transfection of genomic DNA from a human B cell lymphoma line into the fibroblast cell line and selection for *gpt* expression led to the isolation of a resistant cell clone in which the *gpt* gene was activated by inversional rearrangement of the V(D)J recombination substrate (Figure 7.26). The transfected DNA se-

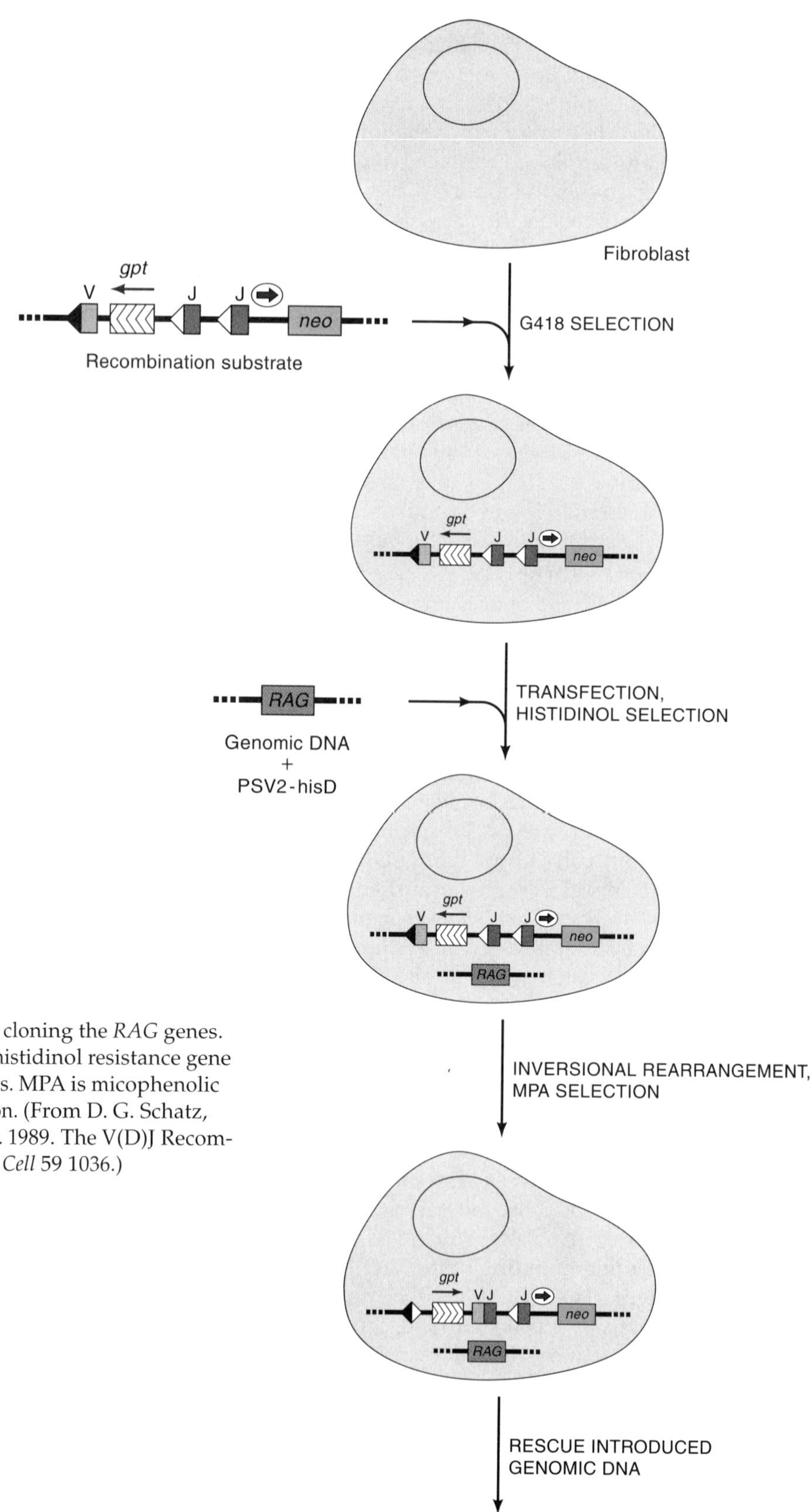

Figure 7.26
Genomic transfection strategy for cloning the *RAG* genes. The pSV2-his plasmid encodes a histidinol resistance gene driven by SV2 regulatory elements. MPA is micophenolic acid for selection for *gpt* expression. (From D. G. Schatz, M. A. Oettinger, and D. Baltimore. 1989. The V(D)J Recombination Activating Gene, *RAG-1. Cell* 59 1036.)

quences were rescued by molecular tagging and secondary transfections of fibroblast cells. Ultimately, a genomic clone that conferred V(D)J recombination activity to the fibroblast cells was isolated. The gene encoding this activity was initially named the recombination-activating gene-1 (*RAG*1); it is transcribed into an approximately 7 kb mRNA that has an open reading frame encoding a protein of 120 kDa.

Surprisingly, a *RAG*1 cDNA expression vector conferred V(D)J recombination activity at only a very low level. Yet when the *RAG*1 cDNA was transfected into fibroblasts along with the *RAG*1 genomic clone, high levels of activity were generated. This initially puzzling result led to the notion that the original *RAG*1-containing genomic clone contains another gene in addition to *RAG*1 whose activity is required to provide V(D)J recombination activity. Further examination of this clone revealed that it did, in fact, contain an additional gene located 4 kb from the *RAG*1 gene and faced in the opposite transcriptional orientation (Figure 7.27). This gene specifies a 2.2 kb mRNA with an open reading frame encoding a

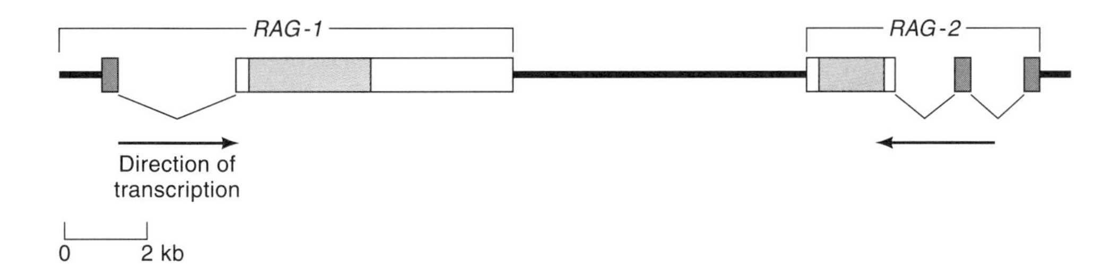

Figure 7.27
The *RAG* genomic locus. (From D. G. Schatz, M. A. Oettinger, and M. S. Schlissel. 1992. V(D)J Recombination: Molecular Biology and Regulation. *Ann. Rev. Immunol.* 10 366.)

60 kDa protein. Simultaneous transformation of nonlymphoid cells with *RAG*1 and *RAG*2 expression vectors results in a high level of V(D)J recombination activity.

The *RAG*1 and *RAG*2 genes encode nuclear phosphoproteins that do not share substantial homology with any other known proteins. However, these proteins do contain a number of motifs found in other proteins and implicated in transcription and/or recombination (e.g., topoisomerases). The *RAG*1 and *RAG*2 genes were so named because it was not proven whether they in fact encoded specific components of the V(D)J recombinase or encoded other factors that were somehow involved in activating the actual V(D)J recombinase proteins. This issue has been resolved by recent *in vitro* (V(D)J recombination experiments which demonstrate that both of the *RAG* gene products play primary roles in the V(D)J recombination process.

The *RAG* genes are coexpressed only in precursor B and T cells or their cell line counterparts. Furthermore, expression of transfected *RAG* genes under the control of inducible promoters indicates that V(D)J recombination activity and RAG protein levels correlate quite precisely. Furthermore, mice that are homozygous for mutations that inactivate either the *RAG*1 or *RAG*2 gene are completely blocked in their ability to generate mature B and T cells because they are unable to initiate the V(D)J recombination process. This leads to a blockade in T and B cell development precisely at the precursor stages (pro-B and pro-T cells) at which antigen receptor gene rearrangement normally begins (Figure 7.28). Thus, these animals have a complete **severe combined immune deficiency** (no B or T cells and no functioning specific immune system).

*RAG*1- and *RAG*2-deficient mice that have been produced by gene "knockouts" in embryonic stem cells do not have any other defects and can survive and reproduce if kept in a specific pathogen-free environment to prevent infections. These findings suggest that the *RAG* genes evolved to function specifically in the context of the immune system. Additional experiments have demonstrated the

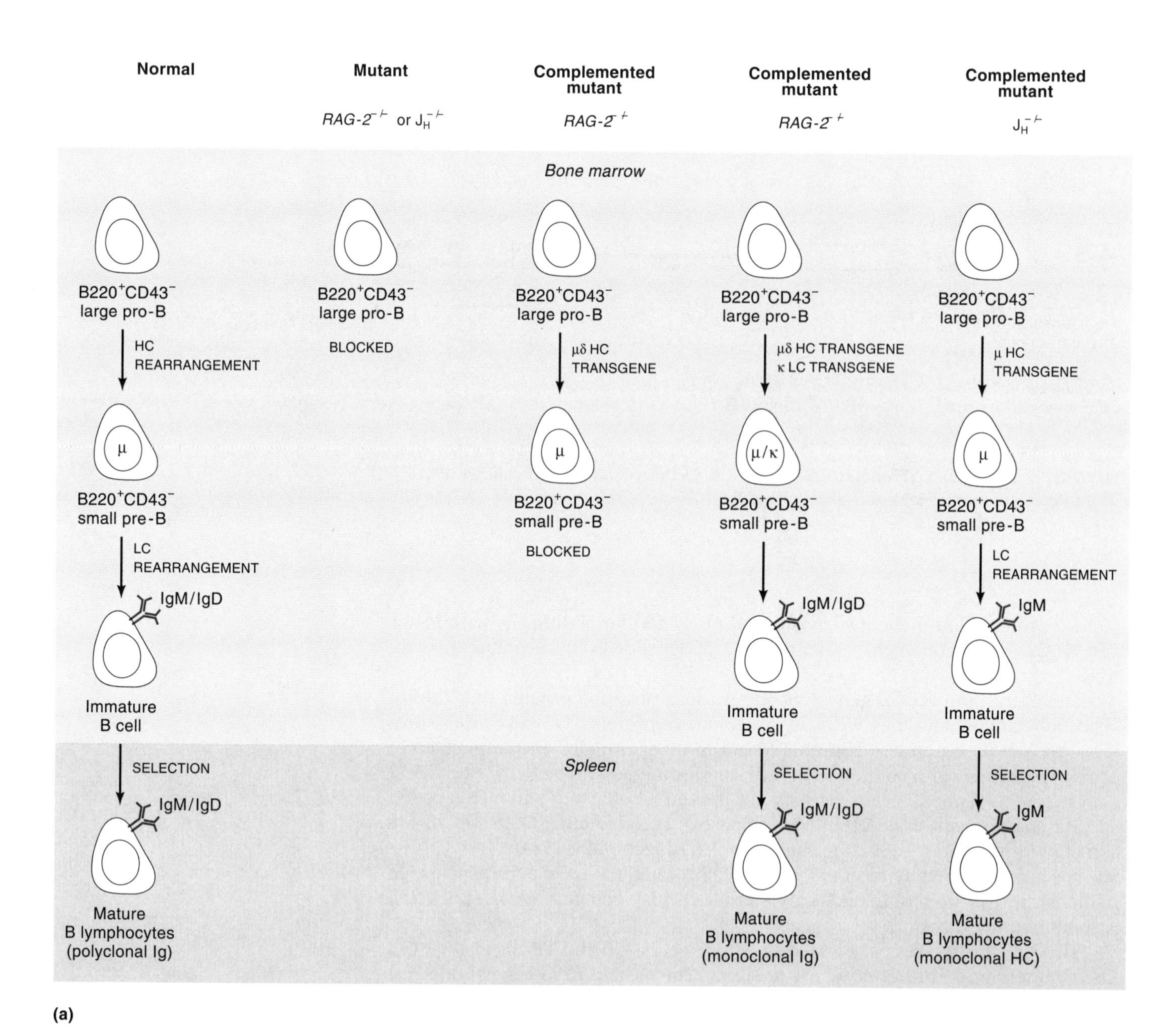

Figure 7.28
Restoration of lymphocyte development by complementation of *RAG* mutant mice. (a) Rearrangement of Ig genes. (b) Rearrangement of TCR genes. (Adapted from F. Alt. 1995. Mutational Analysis of Lymphocyte Development. *Harvey Lect.* 88 100.)

specificity of the *RAG*1 and *RAG*2 defects. Thus, if a rearranged Ig HC transgene is introduced into *RAG*-deficient mice, pro-B cell development proceeds to the pre-B stage, but no further (Figure 7.28a). Introduction of both HC and LC transgenes into the *RAG*-deficient mice leads to the generation of mature B cell populations in the periphery, all of which express the antigen receptor encoded by the HC and LC transgenes. Introduction of transgenes encoding the two chains of the TCR also rescues T cell development in the *RAG*-deficient background (Figure 7.28b). Together these findings further indicate that *RAG* gene products

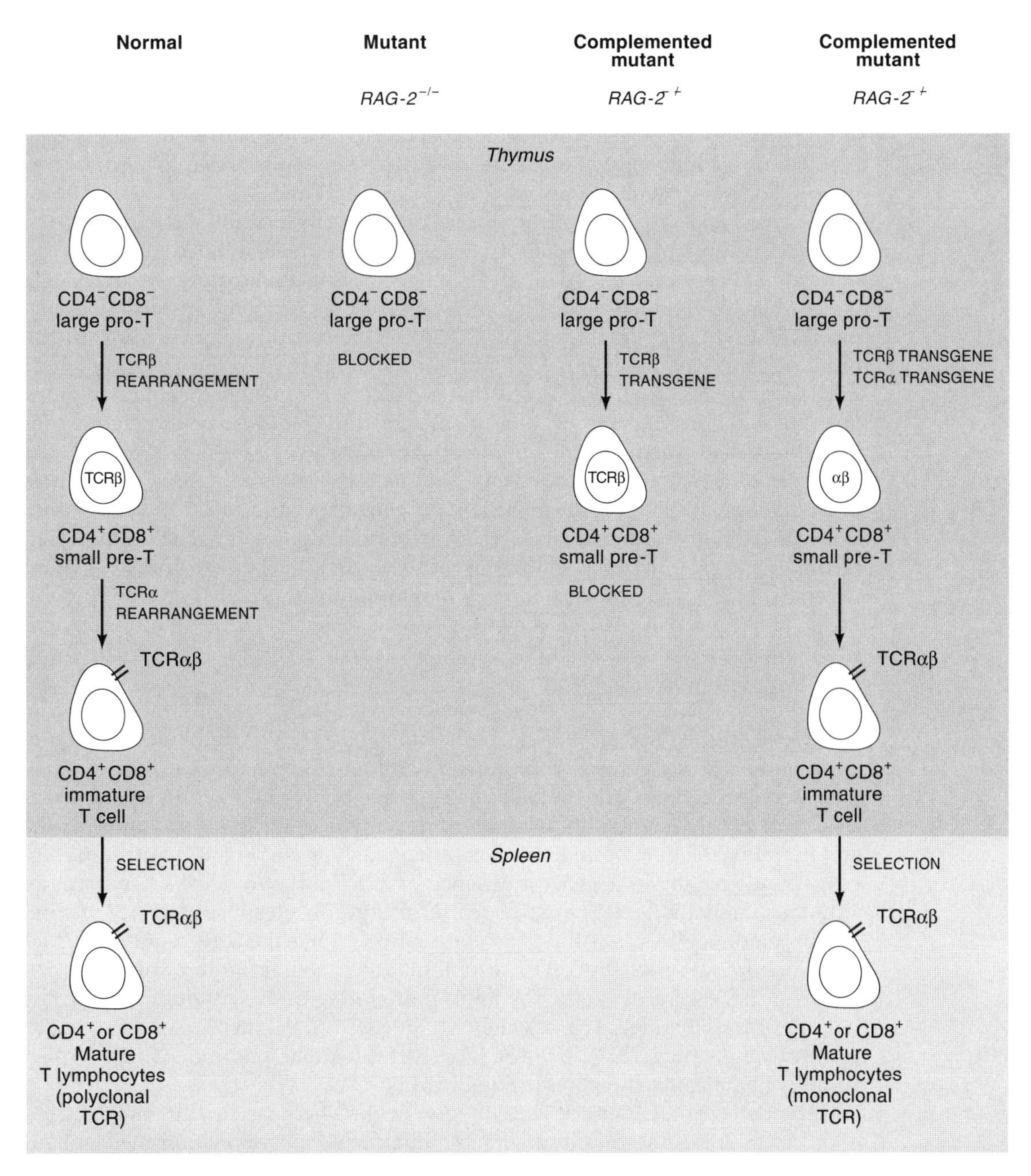

(b)

probably are required for lymphocyte development only with respect to their role in the initiation of the V(D)J recombination process. The issue of whether *RAG*1 and *RAG*2 proteins are important for later B cell–specific processes such as somatic mutation and HC class switching has been considered but not yet tested.

In conclusion, *RAG*1 and *RAG*2 gene products are required to initiate the V(D)J recombination reaction. *In vitro* V(D)J recombination assays have now confirmed that *RAG*1 and *RAG*2 are involved in recognizing the RSS sequences and in introducing the specific DSB. They also may recruit other activities needed for V(D)J recombination. Ongoing mutational analyses of the various *RAG* gene functional domains should provide more definitive evidence with respect to additional functions.

Terminal deoxynucleotydl transferase The nontemplated addition of nucleotides at the junction of variable region gene segments was proposed to be effected by **terminal deoxynucleotidyl transferase (TdT)**. This proposal was based on the known activity of the enzyme and its specific expression pattern in lymphocyte precursors. A large body of evidence now has been accumulated to support this notion. For example, although transfection of *RAG*1 and *RAG*2 genes into fibroblasts, which lack TdT, still allows these cells to perform V(D)J recombination of introduced substrates, the junctions recovered in the recombined substrates do not contain N regions. If a TdT expression vector is included along with the *RAG* genes, then V(D)J junctions containing N regions are formed. The most convincing evidence to support the role of TdT in N region addition came from studies of mice that are homozygous for a mutation that inactivates the TdT gene. Such mice generate mature B and T cells that are nearly completely lacking N regions at V(D)J junctions, but the V(D)J junctions produced in their heterozygous mutant littermates contained the characteristic pattern of N regions (Figure 7.25). Analyses of these mice indicated that TdT activity is not essential for V(D)J recombination, lymphocyte development, or generation and function of any other cell type. However, when expressed, its activity qualitatively modifies the V(D)J recombinational junctional sequences. Notably, in addition to adding N region diversity, TdT expression also appears to inhibit homology-mediated V(D)J joinings. Thus, expression of TdT contributes to the diversity of the repertoire in two different ways. In summary, TdT is a nonessential, tissue-specific participant in the V(D)J recombination reaction.

Activity affected by the murine scid *mutation* The V(D)J recombination reaction apparently recruits generally expressed activities, such as those used in some DNA repair reactions. The first evidence for the involvement of more general activities in V(D)J recombination came from analyses of mice homozygous for the *scid* mutation. The *scid* mutation is spontaneously occurring and autosomal recessive and results in a general absence of mature B and T cells. Analyses of V(D)J recombination of either endogenous or introduced substrates in SCID precursor lymphocytes show that the *scid* mutation, unlike the *RAG* mutations, does not block initiation of the V(D)J recombination reaction. Rather, this mutation causes a marked blockage of the joining of coding ends without affecting the joining of the liberated RSS segments (Figure 7.17). As mentioned previously, some evidence suggests that the SCID defect impairs the normal ability to open coding end hairpins, thereby blocking further reaction.

The defect in SCID mice is relatively "leaky" because the liberated coding ends can be joined at low frequency by illegitimate recombination mechanisms to generate what appear to be relatively normal coding joints. Thus, as such homozygous *scid* mutant mice age, they do accumulate both B and T lymphocytes. By contrast, in homozygous *RAG*-deficient mice, the defect is not leaky. Consequently, because *RAG*-deficient mice cannot initiate the V(D)J recombination process and specific coding ends are not generated, there are no ends to be joined by any mechanism.

Mutations in yeast that affect recombination also frequently affect the DNA repair process, particularly the repair of double-stranded DNA breaks. A striking finding was that homozygous *scid* mice not only have a V(D)J recombination defect, but they are also generally sensitive to ionizing radiation and to radiomimetic drugs. Thus, these animals have a defect in the ability to repair DSBs. This suggested that the *scid* mutation affects the activity of a generally expressed gene product needed to repair both randomly occurring DSBs and the specific DSBs introduced during V(D)J recombination. The inability of SCID precursor lymphocytes to form coding joints probably leads to frequent generation of lethal double-stranded breaks in the DNA of developing lymphocytes that attempt V(D)J recombination. This may explain the difficulty experienced in trying to complement the homozygous *scid* mutation with functionally rearranged Ig or TCR transgenes as compared to the successful complementation of the *RAG*-deficient phenotype with similar transgenes (Figure 7.28).

Activities affected in Chinese hamster ovary cell DNA repair mutants An independent approach to identify genes active in both V(D)J recombination and DNA repair involved screening a large panel of mutant Chinese hamster ovary (CHO) cell lines with known defects in either excision repair (UV sensitivity) or DSB repair (ionizing radiation sensitivity). Such lines had been further subdivided into a number of different complementation groups with respect to their DSB repair defect based on somatic cell fusion analyses. These lines were tested for their ability to rearrange introduced V(D)J recombination substrates following the introduction of *RAG*1 and *RAG*2 expression vectors to promote V(D)J recombination activity in the nonlymphoid cell lines. Not surprisingly, none of the excision repair mutants was defective in their ability to promote V(D)J recombination. However, three independent DSB repair mutants were also defective in V(D)J recombination.

One of these mutant CHO lines, V-3, is defective in V(D)J recombination in much the same way as was found with the homozygous *scid* mutation. Thus, joining of coding sequence is blocked, but RSS joining is normal (Figure 7.17). Cell fusion studies established that the hamster V-3 and murine *scid* mutations were in fact in the same complementation group, suggesting that they affect the same gene. The gene affected by the *scid* mutation has recently been identified; it encodes the catalytic subunit of the DNA-dependent protein kinase (DNA-PK).

Two additional CHO mutant cell lines, *xrs*6 and *XR*1, are defective in V(D)J recombination but in ways different from what had previously been observed. When provided the *RAG* gene activities, these two cell lines can initiate V(D)J recombination properly (with DSBs at the RSS), but they are drastically impaired in forming both coding and RSS joints (Figure 7.17). Researchers had previously shown that the ionizing radiation sensitivity of the *xrs*6 mutants could be rescued by introducing human chromosome 2 into the mutant CHO line. Likewise, the X-ray sensitivity of *XR*1 can be corrected by introducing human chromosome 5. Besides complementing the radiation sensitivities of *xrs*6 and *XR*1, introducing the appropriate human chromosomes also restores the ability of these mutant cell lines to perform V(D)J recombination. This indicates that there are genes that encode activities essential for both DSB repair and V(D)J recombination and that the V(D)J recombination process recruits proteins that are used to repair other kinds of DSBs.

The gene affected by the *XR*1 mutation has been identified and is unrelated to any previously characterized genes. However, the human gene that complements the *xrs*6 mutation appears to encode a subunit of a protein first identified as an autoantigen and referred to as the **Ku protein**. The Ku protein, which is a complex of 70 and 86 kDa subunits, binds nonspecifically to DNA ends, after which it serves as the DNA-binding subunit of DNA-PK. Several lines of evidence implicate Ku in the *xrs*6 defect. First, the *xrs*6 mutant lacks the end-binding activity associated with Ku. Second, the *xrs6* mutation maps to a small subregion of human chromosome 2, the same region to which the gene encoding the 86 kDa subunit of Ku has been localized. Finally, introduction of cDNA constructs that can express the 86 kDa subunit of Ku into the *xrs6* cells restores DNA end-binding activity and simultaneously results in substantial restoration of the cell's X-ray resistance and ability to undergo V(D)J recombination.

The function of the DNA-PK components in DSB repair and V(D)J recombination is presently unknown. However, researchers speculate that Ku's ability to bind to the ends of DNA at double-strand breaks protects the free ends generated during the V(D)J recombination process (Figure 7.17). Alternatively, and not mutually exclusively, Ku protein binding to the free ends may activate downstream steps, by activating the DNA-dependent protein kinase catalytic subunit—which has among its substrates transcription factors, RNA polymerase, and topoisomerase II. Identification of DNA-PK subunits as generally expressed components of the V(D)J recombination reaction should allow additional insights into this process and perhaps facilitate identification of additional components.

e. Overview of the Generation of Primary Repertoire Diversity

The primary antigen receptor repertoire is generated during the differentiation of B cells from hematopoietic stem cells. The process is thought to be independent of antigen contact or other selective forces. The nearly unlimited diversity of Ig specificities in mice is generated in large part as a result of the V(D)J recombination process. As we have discussed, four factors are involved. (1) Many germ line HC and LC variable region gene segments encode the different V region amino acid sequences of CDR 1 and 2. (2) The HC variable region is made up of V_H, D_H, and J_H segments; the LC gene is made up of V_L and J_L segments. The junctional region of these segments encodes CDR3. Therefore, the joining process increases diversity by combinatorial assortment of the different germ line gene segments. (3) The joining process creates junctional diversity in the form of deleted and added coding sequences; this junctional diversification greatly amplifies the diversity of CDR3. (4) Any HC can theoretically associate with any LC to form a complete Ig molecule; therefore, the combinatorial association of different HC and LC variable regions in a given B cell further increases the diversity of the primary repertoire. A theoretical estimate of the contribution of these various factors to the murine primary Ig and TCR variable region gene repertoire is indicated in Table 7.2. The antibody repertoire, but not the TCR repertoire, is further modified through somatic hypermutation mechanisms during antigen-stimulated clonal proliferation.

Despite the evolution of numerous mechanisms to diversify the Ig variable region repertoire, other factors exist that may actually lead to restrictions in this process, at least in newly generated lymphocytes. Most notably, hundreds to thousands of V_H gene segments are spread over many million base pairs in the mouse genome. Yet, the most J_H-proximal V_H segments in the mouse are preferentially utilized in V_H to DJ_H rearrangements in developing pro-B cells. The mechanism that accounts for this preference is unknown, although various models have been proposed, including one-dimensional tracking of the V(D)J recombinase, differences in RSS (or adjacent) sequences, and/or the presence of elements that differentially regulate V_H gene segment accessibility for recombination. In any case, although proximal V_H segments are used preferentially, essentially all V_H segments appear to be rearranged at some frequency, indicating mechanisms that promote the utilization of even the most distal segments.

Table 7.2 ***Sequence Diversity in T Cell Receptor and Immunoglobulin Genes***

	Immunoglobulin		TCR $\alpha:\beta$		TCR $\gamma:\delta$	
	H	κ	α	β	γ	δ
Variable (V) segments	250–1000	250	100	25	7	10
Diversity (D) segments	13	0	0	2	0	2
Joining (J) segments	4	4	50	12	2	2
Ds read in all frames	Rarely	—	—	Often	—	Often
N region additions and P nucleotides	*V–D, D–J*	None	*V–J*	*V–D, D–J*	*V–J*	*V–D₁, D₁–D₂, D₁–J*
Variable region combinations	62,500–250,000		2500		70	
Junctional combinations	$\sim 10^{11}$		$\sim 10^{15}$		$\sim 10^{18}$	

Source: Adapted from M. M. Davis and P. J. Bjorkman, T-cell antigen receptor genes and T-cell recognition. *Nature* 334 (1988), p. 397.

Although this biased rearrangement process results in a biased primary repertoire, the V_H gene repertoire is normalized to be more representative of all V gene segments when the B cells migrate to the periphery—most likely by mechanisms that operate at the level of cellular selection. The preferential rearrangement process can nevertheless significantly affect the fetal antibody repertoire, which, as previously discussed, is further biased by the absence of TdT activity. One possible significance of the biased usage of V_H gene segments and limited junctional diversity in fetal life is that the encoded variable regions may be directed against important foreign antigens or may be involved in reacting with other antibodies to influence the establishment of the antibody repertoire.

f. Productive and Nonproductive Rearrangements

An important prerequisite to understanding the issues associated with the control of rearrangement is the notion that the rearrangements can be either functional or nonfunctional with respect to their ability to encode Ig (or TCR) proteins (Figure 7.15). The reading frame of an HC or LC mRNA sequence is determined by the ATG translation start site at the 5′ end of the V gene sequence; the reading frame of the variable region sequence relative to the C region sequence is determined by the reading frame of the J region sequence, which is spliced in frame with the C region sequence. Because of the imprecision of the V(D)J joining process (junctional diversification), approximately two of three V gene sequences (assuming random insertion and deletion of nucleotides) will be out of frame with respect to the J region to which they are joined. For out-of-frame rearrangements, the encoded mRNA sequences will not be translated into a functional polypeptide. Such rearrangements are referred to as nonproductive.

With HC variable region genes, the reading frame of D is theoretically neutral and could be used in all three reading frames. This is in fact the case for TCR D segments. However, most murine D_H segments have translational termination codons in at least one of their reading frames. Therefore, such D_H segments cannot be used in that frame unless the termination codon is deleted by the junctional diversification mechanisms. Finally, as already noted, developing B cells may have ways to increase the frequency of functional (in-frame) rearrangements, at least one of which could be homology-mediated joining pathways.

7.5 Control of V(D)J Recombination

For clonal selection to work, B lymphocytes must functionally assemble and express only one set of immunoglobulin heavy and light chain genes. The V(D)J recombination process can also lead to recombination events that result in chromosomal rearrangements of genes and sequences other than those associated with the Ig and TCR loci. Such unwanted rearrangements occasionally lead to detrimental effects such as activation of cellular oncogenes. Thus, the V(D)J recombination is tightly controlled so as to ensure that the proper rearrangements occur only in lymphocytes, that the rearrangements are distinctively different in B and T cells, that rearrangements occur at the correct developmental stage, and that only a single productive rearrangement takes place for a given antigen receptor locus in each developing lymphocyte (allelic exclusion).

a. Cell Type–Specific Control of V(D)J Recombination

V(D)J recombination occurs exclusively within the B and T lymphoid cell lineages. V(D)J recombination is limited to lymphocytes through the control of V(D)J recombinase expression. This results from the highly restricted expression of the *RAG*1 and *RAG*2 genes only in developing B and T cell precursors. Thus, strict regulation of the expression of the *RAG*1 and *RAG*2 genes ensures that V(D)J recombination occurs exclusively in specific lymphoid precursor cell types. However, there must be an additional level of control because, although expression of *RAG*1 and *RAG*2 in nonlymphoid cells initiates V(D)J recombination of introduced recombination substrates, the endogenous Ig and TCR loci are not rearranged. The differential accessibility of the introduced versus the endogenous variable region gene segments in nonlymphoid cells suggests that substrate accessibility is important for directing the activity of the V(D)J recombination system once it is expressed.

b. A Common V(D)J Recombinase and Accessibility Control

V(D)J recombination in B and T lymphoid cells is mediated by a common recombination mechanism. This is evident from the fact that Ig and TCR gene rearrangements are mediated by the conserved RSS. Moreover, although endogenous TCR genes are not rearranged in pre-B cell lines, TCR recombination substrates introduced into the chromosomes of pre-B cell lines are rearranged as efficiently as introduced Ig gene recombination substrates. The finding that the *RAG* genes are expressed in both precursor B and T cells and that targeted mutation of *RAG* genes in mice blocks initiation of the V(D)J rearrangement process of both Ig and TCR loci further supports the notion of a common V(D)J recombination system (Figures 7.9, 7.17, and 7.28).

In spite of a common V(D)J recombinase, the complete rearrangement of Ig loci occurs only in B-lineage cells and the complete rearrangement of TCR loci only in T-lineage cells. Therefore, the lineage specificity of this process must be controlled by modulating the availability of the substrate Ig and TCR variable region gene segments in a lineage-specific manner. This concept is generally referred to as regulated substrate gene segment accessibility. The accessibility of various gene segments involved in the respective rearrangements may be responsible for additional modes of regulation of V(D)J recombination in developing lymphocytes. Thus, the accessibility of the V, D, and J DNA segments has been correlated with a number of factors, including transcription, hypomethylation, and the state of the target germ line loci's chromatin structure.

c. Developmental Stage-Specific Control of V(D)J Recombination

Lymphocyte differentiation from hematopoietic stem cells to effector cells goes through many well-defined stages (Figure 7.9). Even before the rearrangement of their endogenous antigen receptor loci, lymphoid progenitors already express lineage-specific surface proteins, such as B220 for B-lineage cells and Thy-1 for T-lineage cells. Although the expression of these markers suggests early commitment of progenitor cells to the B or T cell lineages, the rearrangement of either the Ig or TCR loci completes this commitment.

RAG gene expression activates the expression of V(D)J recombinase and, as a result, permits progression of lymphocytes through their differentiation pathways. Signals that initiate *RAG* expression at the pro-lymphocyte stage are critically important for this process. However, neither *cis-* nor *trans*-regulatory elements for controlling *RAG* gene transcription have been identified. Expression of complete Ig or TCR molecules on the cell surface of immature lymphocytes can lead to the generation of signals that down-regulate *RAG* gene expression and thereby V(D)J recombination activity (**feedback regulation**). The molecular events that mediate signals generated on the cell surface to block *RAG* gene transcription in the nucleus remain to be elucidated.

d. Control of Ordered Ig Gene Rearrangement

Ordered HC and LC gene rearrangement The major steps of B cell differentiation have been divided into a series of distinct stages based on the expression of a set of characteristic stage-specific surface markers: the pro-B cells (B220$^+$, CD43$^+$, IL-2Rα^-), the pre-B cells (B220$^+$, CD43$^-$, IL-2Rα^+), and the immature B cells (B220$^+$, IgM$^+$) (Figure 7.9 and Table 7.1). The ordered V(D)J rearrangement process and the expression of Ig HC and LC gene proteins appears to be intimately associated with the progression of B lineage cells through this defined pathway.

During normal precursor B cell differentiation, the rearrangement and expression of Ig HC genes generally occur before those of Ig LC genes (Figure 7.9). Early evidence favoring an ordered rearrangement pathway came from studies of A-MuLV-transformed pro-B cell lines. Such lines were found to have undergone, at a minimum, D to J_H rearrangements on both HC alleles but usually had not undergone any LC gene rearrangement. In addition, pro-B cells, purified from the mouse bone marrow on the basis of their characteristic surface marker expression (Figure 7.9 and Table 7.1) were also found to have HC rearrangements but few LC gene rearrangements. Thus, HC gene rearrangement is initiated at the pro-B cell stage. The expression of the Igμ HCs from V_HDJ_H rearrangements assembled at the pro-B cell stage appears to promote the differentiation of pro-B cells to the pre-B stage, where most LC rearrangement occurs.

Additional evidence supporting this proposed pathway comes from the analyses of mice that harbor targeted mutations that block developing B cell ability to form functional HC gene rearrangements (such as the *RAG* gene mutations or the targeted deletion of J_H segments; Figure 7.28). Such mutations block B cell differentiation at the pro-B cell stage. Introduction of a functionally expressed Igμ HC transgene into the *RAG*-deficient background promotes B cell differentiation from the pro-B to pre-B cell stage. In contrast, similar introduction of transgenic Ig LC expression vectors into the *RAG*-deficient background does not promote any further B cell differentiation. Introduction of Igμ HC expression vectors into the J_H mutant background leads to the rearrangement of endogenous light chain genes and generation of pre-B cells in the bone marrow and B cells in the periphery. Thus, rearrangement and expression of Igμ HC genes appear to promote the differentiation of pro-B to pre-B cells.

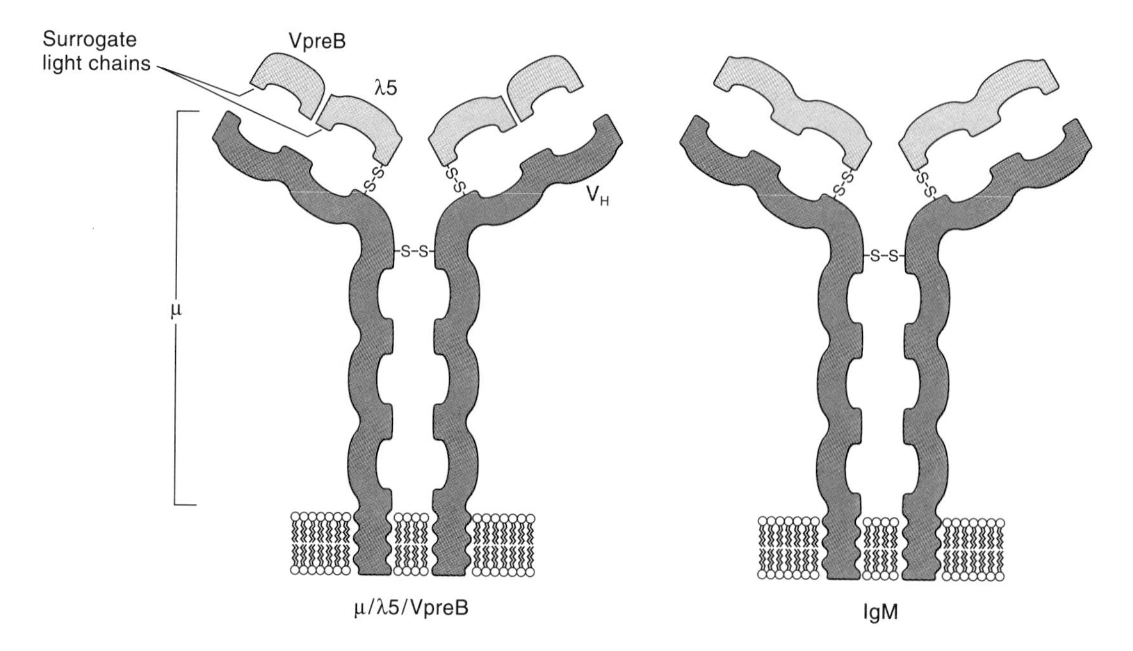

Figure 7.29
A schematic comparison of IgM and μ-surrogate LC complex. (Adapted from F. Melchers, H. Karasuyama, D. Haasner, S. Bauer, A. Kudo, N. Sakaguchi, B. Jameson, and A. Rolink. 1993. The Surrogate Light Chain in B Cell Development. *Immunol. Today* 14 64.)

Early experiments using vectors encoding either the membrane-bound (μ_m) or secreted (μ_s) form of the μ protein suggested that expression of the μ_m but not μ_s protein promotes pro-B to pre-B cell differentiation and the onset of LC gene rearrangement. This notion was tested through the creation of mice with a mutation that enables B cells to express μ_s but not μ_m chains. In such mice, B cell development was indeed blocked at the pro-B cell stage. How the μ_m chain functions to promote this differentiation event is still unknown. But various lines of evidence suggest that μ_m chains perform this function by forming a complex with surrogate LC molecules present in the pro-B cells (Figure 7.29) and that this complex, perhaps after reaching the cell surface, generates signals that mediate further differentiation events. We discuss this issue in more detail in the context of the regulation of allelic exclusion.

Ordered HC and LC gene rearrangement is apparently achieved through a temporal difference in accessibility of the two loci. This may be achieved by differential activities of transcriptional enhancers. Thus, the Eμ enhancer is active at least as early as the pro-B cell stage, but Eκ enhancer is apparently not active until pre-B cell stage. The temporal activation of such enhancer elements or other such control elements during B cell development may be involved in making the associated loci accessible for V(D)J recombination.

Expression of a μ heavy chain transgene in *RAG*-deficient pre-B cells induces the appearance of cells that have increased expression of germ line κ LC transcripts. Consequently, the rearrangement and expression of μ heavy chains are thought to initiate LC gene rearrangement by promoting differentiation of pro-B cells to the pre-B stage, at which time the κ LC locus becomes accessible for rearrangement. However, a low level of κ LC gene rearrangement has been observed in mutant mice that cannot produce a functional μ HC (or μ_m HC). Such rearrangements might be due to the developmental block imposed by the failure to produce μ HC. However, these findings have also been interpreted as suggesting that a minor B cell differentiation pathway exists in which LC gene assembly precedes that of HC genes.

Ordered DJ_H and V_H DJ_H rearrangement At the IgH locus, joining of D to J_H generally occurs prior to joining of V_H to D during B cell differentiation. This was deduced from finding that A-MuLV-transformed pre-B cell lines almost always have DJ_H rearrangements on both alleles and less frequently have appended V_H segments to these DJ_H complexes. Analyses of HC gene rearrangements in purified pro-B cell subsets from murine bone marrow confirm that the ordered joining sequence of D to J_H and then V_H to DJ_H is the predominant route of Ig HC gene assembly in normal differentiating B-lineage cells (Figure 7.9). This ordered assembly of the Ig HC variable region genes from the component gene segments may be important for controlling this process, especially by allowing for allelic exclusion through regulation of the V_H to DJ_H joining step. In this context, it is notable that murine T-lineage cells often show DJ_H rearrangements of their endogenous Ig HC loci but do not generate V_HDJ_H rearrangements, indicating that regulation of this step is also important in the lineage-specific control of V(D)J recombination.

κ- versus λ-rearrangement Individual B cells generally express either κ or λ LCs in their Ig, but not both. This phenomenon is referred to as **LC isotype exclusion**. The mouse has many more Vκ than Vλ gene segments; correspondingly, approximately 95 percent of the serum Ig contains κ light chains. In humans, the κ- and λ-variable region gene loci are both complex, and κ and λ LCs are found at comparable levels in the serum. However, both murine and human pre-B cells appear to rearrange their κ-loci in preference to their λ-loci. Thus, κ LC–expressing B cells and cell lines normally have their λ-locus in germ line configuration, but λ-expressing B cells and cell lines have either aberrant κ-gene rearrangements on both alleles or deletions in their Cκ loci. In this deletional rearrangement, the Cκ gene is specifically lost when Vκ sequences or Jκ-Cκ intronic heptamer sequences are joined to an RSS that lies far downstream of the Cκ gene (Figure 7.30). Such Cκ deletions, which are very frequent in λ-producing human B cells, may be important in the generation of B cells with λ-rearrangements.

Two models exist to explain the general findings just outlined. One suggests an ordered pathway in which rearrangement of κ-genes precedes that of λ-genes so that λ-rearrangement takes place only after two aberrant κ-rearrangements

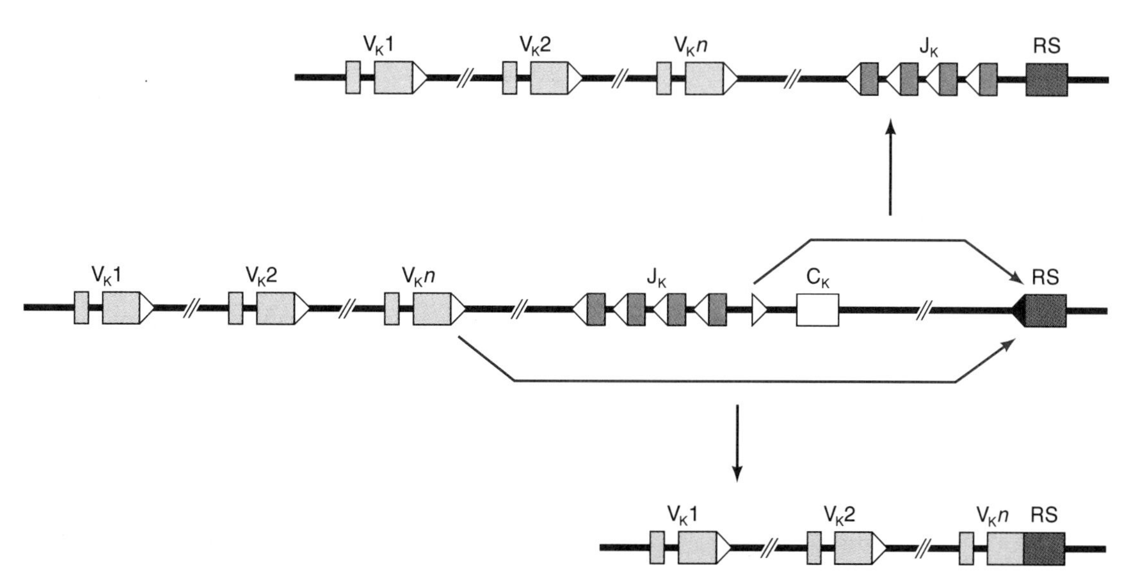

Figure 7.30
3′RS rearrangement at the murine κ locus.

have occurred. The other suggests a stochastic mechanism in which κ- and λ-genes are rearranged during the same developmental period but by which κ-genes have a much higher probability of rearrangement than λ-genes. Thus, a differential probability of rearrangement might result from differential accessibility or differences in RSS. These two possibilities remain unresolved. However, that mice in which the κ-loci are functionally inactivated by targeted deletions of either the Jκ or Cκ region generate nearly normal numbers of B cells, all of which express λ LC in their Ig. In these mice, λ-expressing B cells are generated approximately ten times more efficiently than in normal mice without a substantial increase of pre-B cell pool size. This is consistent with the possibility that inactivation of the κ-locus may increase the probability of λ-gene rearrangements in a given pre-B cell.

e. Allelic Exclusion

HC allelic exclusion Murine B cells potentially can form two different Ig HCs by assembling two different functional V_HDJ_H sequences at their two HC alleles. They can also form at least six different Ig LC genes by assembling functional LC variable regions at their different allelic κ- and λ-gene loci (Figure 7.11). Thus, an individual B cell could theoretically express multiple species and specificities of Ig. However, over 99 percent of B cells rearrange and express only a single HC and LC gene, resulting in the expression of a single specificity of Ig per cell. These phenomena are called HC **allelic exclusion** and LC allelic (and isotype) exclusion. Allelic exclusion was first observed through the use of antibodies that recognized amino acid differences between the constant regions encoded by different Ig LC or HC alleles. By using such antibodies, individual B cells were found to express only one or the other, but not both, of their Ig HC or LC genes. The allelic exclusion process is thought to be regulated, at least in large part, at the level of Ig variable region gene assembly, that is, developing B cells must have a feedback mechanism to monitor functional HC and LC gene rearrangements.

In pro-B cells, both HC alleles are rearranged to the DJ_H stage. Pre-B and mature B cells must have one productive V_HDJ_H rearrangement; the other HC allele is either frozen at the DJ_H rearrangement stage (60 percent) or has undergone a nonproductive V_HDJ_H rearrangement (40 percent). Various considerations based on the ordered pathway of HC variable region gene assembly, the presence of "frozen" DJ_H rearrangements, and the ratio of DJ_H to V_HDJ_H rearrangements on the nonproductive allele have led to a model proposing that HC allelic exclusion is regulated at the V_H to DJ_H rearrangement stage. Thus, an initial productive V_HDJ_H rearrangement feeds back to prevent V_H to DJ_H rearrangement on the second allele (Figure 7.31). If the initial V_HDJ_H rearrangement is nonproductive, developing B cells still have a chance to assemble a functional rearrangement on the second allele. Because both productive and nonproductive rearrangements are usually transcribed, the feedback signal for allelic exclusion must be an Igμ polypeptide expressed from the productive rearrangement (Figure 7.31).

The feedback effect of μ heavy chain protein on HC gene rearrangement was demonstrated in transgenic mice. Expression of μ_m but not μ_s chains from human and mouse transgenes resulted in an inhibition of endogenous HC gene rearrangement and generation of B cell populations that expressed HCs from predominantly the transgene, rather than the endogenous HC loci. In mice that are heterozygous for a mutation that deletes the exon specifying the μ_m domain, productive rearrangement and expression of the mutant allele that expresses only the μ_s protein fails to exclude the rearrangement and expression of the wild-type allele. Thus, expression of the μ_m protein is required to mediate allelic exclusion as well as for further B cell differentiation.

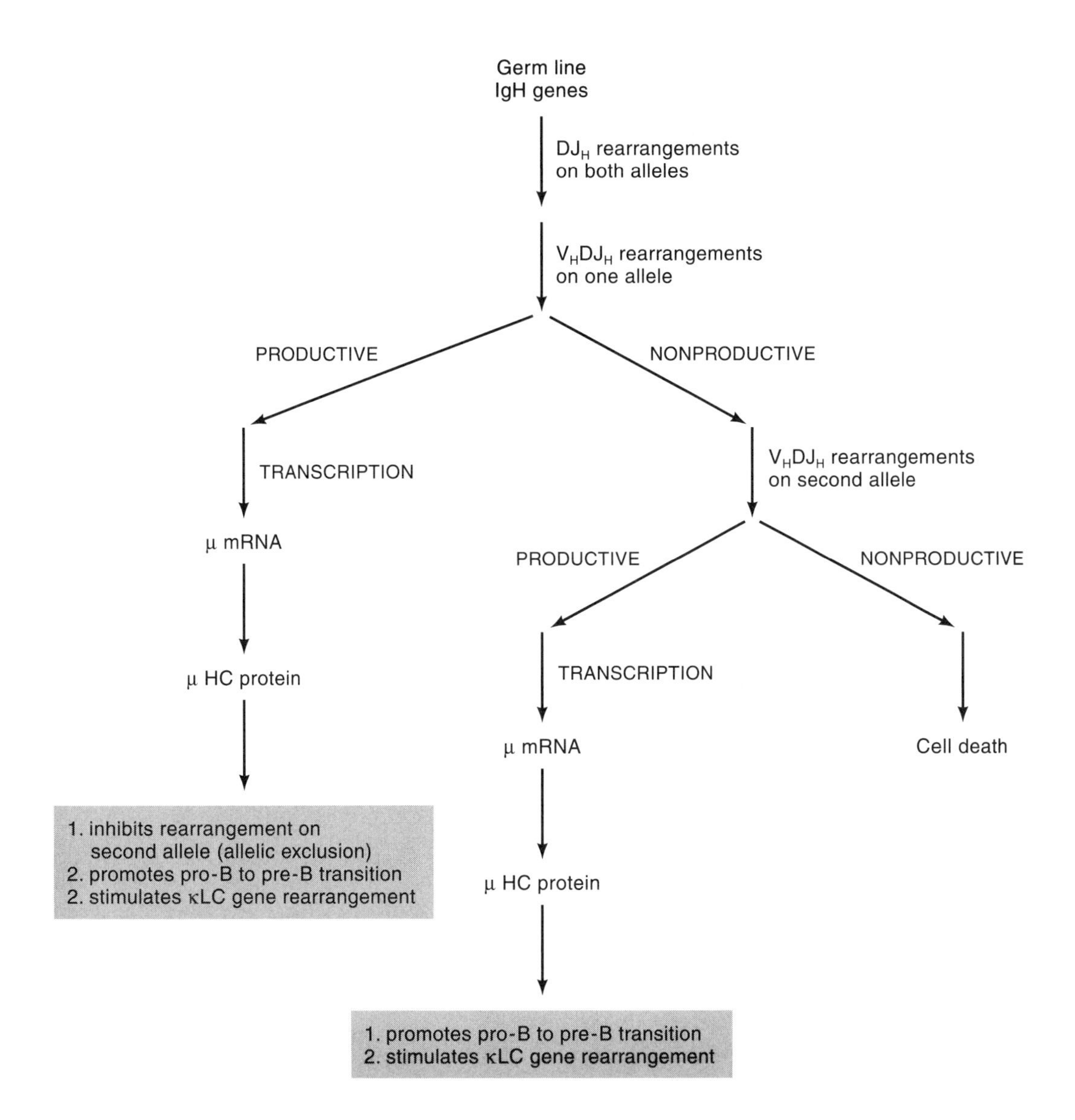

Figure 7.31
Ordered rearrangement and expression of IgH loci. (Adapted from A. K. Abbas, A. H. Lichtman, and J. S. Pober. 1991. Cellular and Molecular Immunology. p. 81. W. B. Saunders, Philadelphia.)

The μ_m HC protein expressed at the pre-B cell stage is linked by disulfide bonds to a λ LC constant region-related molecule termed λ5 and noncovalently to a λ LC variable region-related molecule termed Vpre-B (Figure 7.29). The λ5 and Vpre-B molecules are encoded by nonrearranging genes linked to the λ LC locus. The proteins they encode are termed surrogate light (SL) chains. Mice that are homozygous for a targeted mutation that inactivates the λ5 gene are blocked in B cell differentiation at the pro-B cell stage, although the block is leaky, and mature B cells accumulate in the periphery with age. This finding further supports the notion that a complex of μ_m with the surrogate light chains interacts with putative ligands and signals a block in V_H to DJ_H rearrangement on the second allele while promoting the differentiation of pro-B cells to the pre-B stage.

The μ_m-SL chain complex was initially found on the surface of pre-B cell lines and the later stage of normal pre-B cells. However, pre-B cells generated in μ_m-transgene-complemented *RAG*-deficient mice did not have detectable surface μ_m or $\lambda 5$ expression, although both μ_m and SL chains were found intracellularly. Further analysis showed that pro-B cells from *RAG*-deficient mice already express surrogate light chains on the cell surface, probably in association with a complex of glycoproteins. In normal mice, most pre-B cells express μ_m but not surrogate light chains, and only a minor population coexpresses μ_m and SL chains intracellularly. Thus, whether the initial signal from the complex is generated on the cell surface or intracellularly is still not clear. Furthermore, the nature of the signal generated to effect HC allelic exclusion is not known. Because pre-B cells continue to express V(D)J recombinase for LC gene rearrangement, HC allelic exclusion is likely controlled through the accessibility of V_H and/or DJ_H segments on the second allele. This suggests that the signal emanating from the complex of μ_m and the SL light chains ultimately affects the accessibility of the unrearranged V_H and/or DJ_H segment.

Light chain allelic and isotype exclusion Most B cells rearrange and express only one functional LC gene. Occasionally, B cells have two productive LC gene rearrangements (i.e., both encode proteins); however, only one of the encoded LCs associates with the cellular HC to form Ig (i.e., only one is actually expressed as a functional antibody). LC allelic exclusion also appears to occur through a feedback inhibition mechanism. In this case, the signal is proposed to result from the formation and surface deposition of a complete IgM. The IgM complex may cause shutdown of endogenous LC gene rearrangement through an accessibility mechanism such as that proposed for control of HC allelic exclusion. However, as already described, expression of surface IgM may effect LC allelic exclusion by causing the down-regulation of *RAG* gene expression and, correspondingly, the cessation of V(D)J recombination.

Unsolved problems related to allelic exclusion Given that it takes a significant amount of time from the formation of a functional HC or LC gene to the generation of the signals that cause the cessation of further rearrangement, a correlate of the feedback model is that a differentiating B cell precursor must rearrange only one of its HC or LC loci during that time period. How such a process could be achieved remains an enigma. The presence of limiting levels of recombination or accessibility factors is possible but seems unlikely given the apparent efficiency of the V(D)J rearrangement process. Another possibility is that only one V(D)J recombination event can occur during a cell cycle, for example, by occurring at a unique location in the nucleus. Regarding the potential cell cycle preference of V(D)J recombination, it is notable that the stability of the RAG2 protein is dramatically affected by phosphorylation catalyzed by cyclin-dependent kinases (CDKs) such as $p34^{cdc2}$. Thus, the V(D)J recombinase activity could be restricted to a particular interval of the cell cycle.

f. Factors That Correlate with V(D)J Recombinational Accessibility

The accessibility of the regions involved in recombination has been correlated with transcription. Thus, transcription of most Ig and TCR variable region gene segments precedes their rearrangement (Figure 7.32). For example, the endogenous J_H locus is actively transcribed in *RAG*-deficient pro-B cells. In addition, at least some germ line V_H gene segments are transcribed in pro-B cell lines that undergo V_H to DJ_H rearrangement. These observations led to the suggestion that transcription somehow makes one or both of the recombining segments more accessible. However, there is no compelling evidence that all V_H genes, particularly

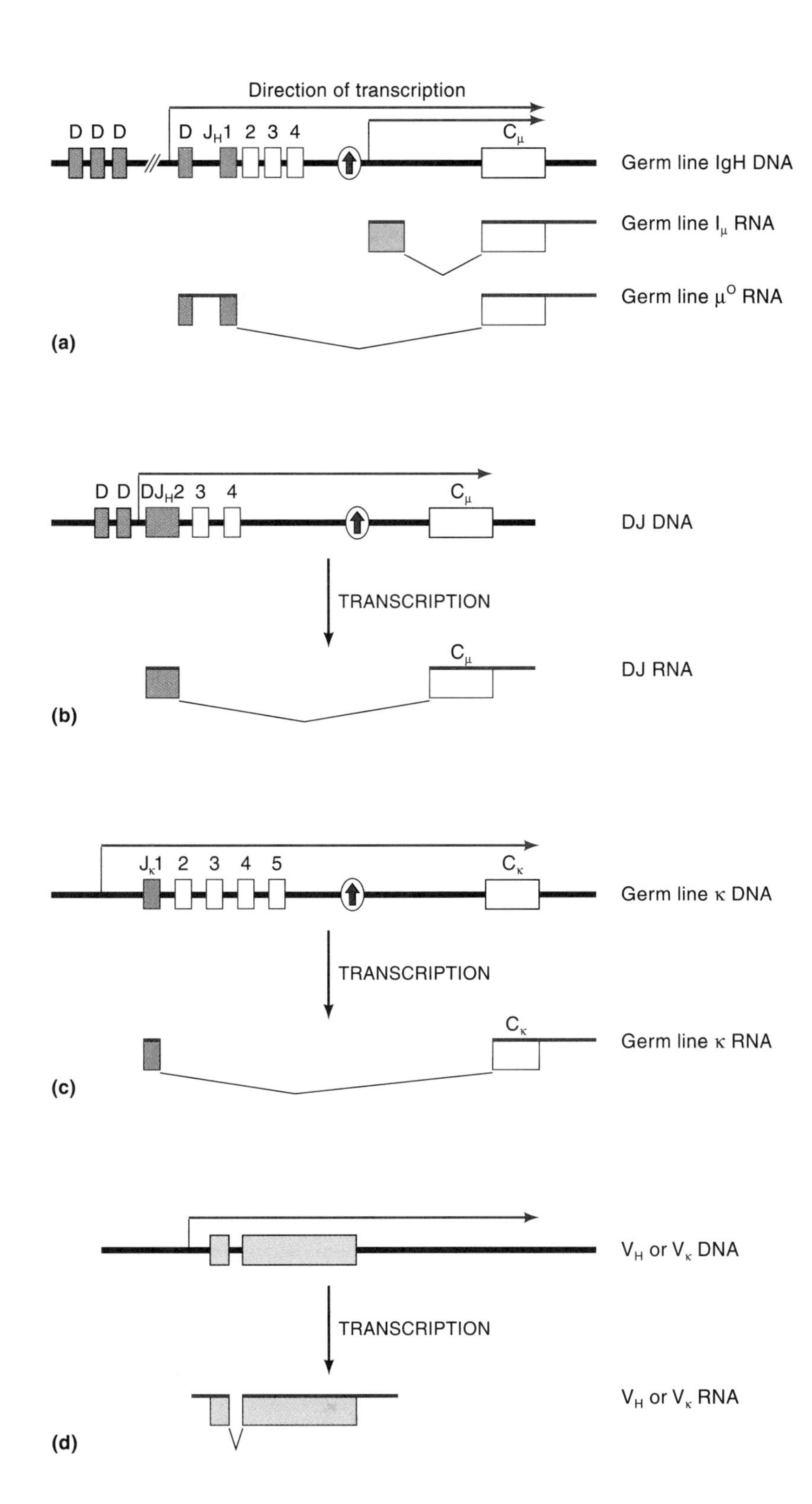

Figure 7.32
Germ line transcription and transcripts. (From D. G. Schatz, M. A. Oettinger, and M. Schlissel. 1992. V(D)J Recombination: Molecular Biology and Regulation. *Ann. Rev. Immunol.* 10 373.)

the 3′ V_H genes that are most frequently rearranged, are transcribed prior to or during V(D)J rearrangement in pro-B cells. Thus, the overall importance of transcription with respect to V(D)J recombination in pro-B cells is not clear. Nevertheless, transcription induction at a particular Ig locus often correlates with rearrangement induction at that locus. For example, lipopolysaccharide treatment of pre-B cell lines induces transcription of the germ line Jκ-Cκ locus followed by Vκ to Jκ rearrangement. Also, the rearrangement of a particular TCR Vγ gene segment is clearly preceded by transcription of that segment during fetal thymic development. However, some recent studies with recombination substrates suggest that transcription of V region gene segments is not always sufficient for target rearrangement, implying the existence of additional controlling elements, some of which may act negatively.

Transcriptionally active loci tend to be hypomethylated, and transcriptionally inactive loci tend to be hypermethylated. **Hypomethylation** of variable region gene loci also has been correlated with accessibility for V(D)J recombination. In cell lines, methylated recombination substrates are less efficiently rearranged than hypomethylated ones. In mice, rearrangement occurs predominantly on hypomethylated recombination substrate transgenes compared to hypermethylated copies integrated at the same chromosomal locus. The J_H locus in *RAG*-deficient pro-B cells is transcribed and hypomethylated and undergoes rearrangement upon the introduction and expression of *RAG* genes. Replacement of the Eμ enhancer with a *neo*^{r} gene leads to the generation of pro-B cells in which the J_H locus is hypermethylated and not transcribed; such cells are substantially unable to undergo J_H rearrangements. However, some data indicate that all hypomethylated loci may not be highly accessible for rearrangement. Thus, hypomethylation, like transcription, may be important but not necessarily sufficient fully to promote V(D)J recombinational accessibility.

Cis-*acting elements that mediate V(D)J recombinational accessibility* The accessibility model for controlling V(D)J recombination implies the existence of *cis*-acting regulatory elements that control this process. Direct evidence that transcriptional **enhancer** elements function as V(D)J recombinational enhancers comes from studies with recombination substrates and analysis of targeted mutations. Thus, transcriptional enhancer elements introduced into transgenic recombination substrates promote V(D)J rearrangement in a lineage- and stage-specific fashion. Furthermore, rearrangement of recombination substrates containing TCR Vβ, Dβ, and Jβ segments linked to the Ig Cμ gene depends on the inclusion of transcriptional enhancer elements, both in cells capable of carrying out V(D)J recombination and in transgenic mice (Figure 7.33). In cell lines, any active enhancer, such as the SV40 enhancer, appears able to promote V(D)J recombination. In addition, *in vivo*, the expression and rearrangement of Eα- or Eβ-containing substrates is regulated in the same developmental manner as the endogenous TCRα and TCRβ loci during thymocyte differentiation. These results suggest that enhancer elements associated with endogenous Ig or TCR loci can stimulate both transcription and V(D)J recombination.

To test directly the role of transcriptional enhancer elements in accessibility, researchers either deleted or replaced the Eμ or Eκ intronic enhancer elements with a *neo*^{r} gene in ES cells via gene-targeted mutation. Following reconstitution of mice from such mutated ES cells, J_H or Jκ rearrangement in differentiating precursor lymphocytes was evidently impaired. However, loci with deletions in Eμ or *neo*^{r} gene replacements were not completely blocked in their ability to rearrange the J_H segments, suggesting the existence of additional *cis*-acting elements that can promote V(D)J recombination. More recently, gene-targeted deletions of the TCRβ enhancer element have shown that rearrangement of its associated locus also depends on the integrity of this element.

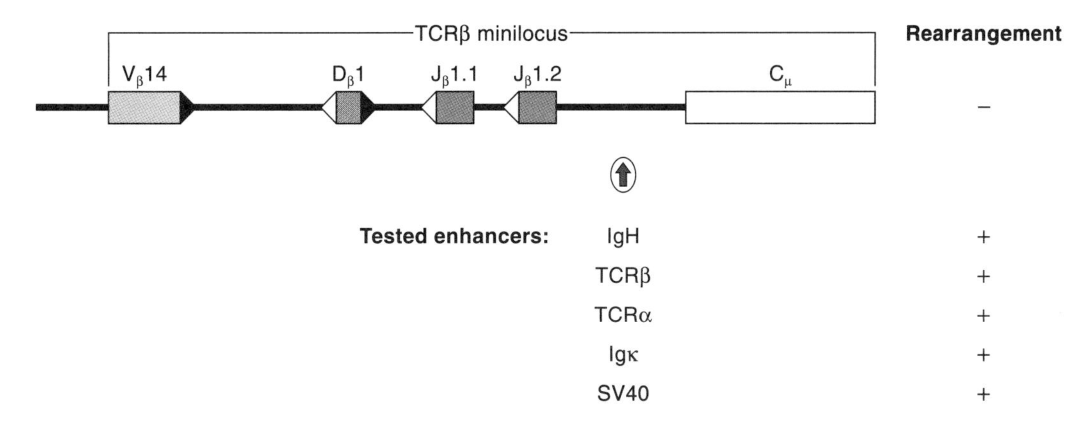

Figure 7.33
Rearrangement status of TCRβ minilocus with various enhancer elements. (From R. Lansford, A. Okada, J. Chen, E. M. Oltz, K. Blackwell, F. W. Alt, and G. Rathbun. 1966. Mechanism and Control of Immunoglobulin Gene Rearrangement. *Molecular Immunology,* 2nd ed. B. D. Hames and D. M. Glover, eds. IRL Press.)

Molecular basis of accessibility Transcriptional enhancer elements clearly play a role in V(D)J recombinational accessibility of endogenous Ig and TCR loci. One possible mechanism by which they may act is through their transcriptional enhancing activity. However, various lines of evidence indicate that germ line transcripts or their encoded polypeptides are not directly involved in V(D)J recombination. Transcription per se could potentially confer recombination accessibility by promoting necessary interactions between recombination and transcription complexes or by altering DNA structure. Notably, transcription has been shown to enhance transposase-mediated site-specific recombination in bacteria. Transcriptional enhancers may mediate V(D)J recombinational accessibility in a manner independent of their ability to enhance transcription, for example, by serving as entry sites for the recombination factors or by leading to a general opening of the surrounding chromatin. Hypomethylated DNA apparently assumes a different configuration than methylated DNA with respect to the organization of nucleosomes in chromatin. In the latter context, the Eκ element also can target tissue- and stage-specific demethylation of Ig LC miniloci following transfection into B cell lines.

Finally, additional DNA elements, yet to be defined, likely play a significant role in directing V(D)J recombinational accessibility. Some recent experiments suggest the existence of elements that influence accessibility in a negative fashion, analogous to the way silencers affect transcription. Likewise, some long-range elements, such as those we discuss in the context of Ig HC class switching (Section 7.6c), probably operate to promote and regulate accessibility of selective V gene segments within the very large region over which V(D)J recombination occurs.

g. Overlapping Regulatory Strategies in Precursor B and T Cell Differentiation

Developing B and T lymphocytes share many strategies for regulating V(D)J recombination and progression of the cells through their respective pathways to the stage where surface receptors are expressed. Like B cells, T cells develop through a series of stages that are well defined by the expression of various surface markers (Figure 7.28b). The pro-thymocyte stage is also referred to as the

double-negative (**DN**) stage because pro-thymocytes do not express either the CD4 or CD8 coreceptor molecules. The prethymocyte stage is referred to as the **double-positive** (**DP**) stage because prethymocytes express both CD4 and CD8. Finally, prethymocytes give rise to thymocytes, which, through the MHC restriction process, express either CD4 (Class II MHC restricted helper T cells) or CD8 (Class I MHC restricted cytotoxic T cells). These cells are referred to as **single-positive** (**SP**) thymocytes.

Assays of TCR gene rearrangements in purified thymocyte populations, cell lines, and mutant mice indicate that, analogous to ordered Ig HC and LC gene rearrangement, TCRβ variable region genes are assembled and expressed before those of TCRα genes during the differentiation of α/βT cells. V(D)J recombination is initiated at the pro-thymocyte stage at the TCRβ locus; *RAG*-1 or *RAG*-2 deficient mice, as well as SCID mice and mice harboring deletions of the TCRβ locus, are blocked in T cell development at this stage (Figure 7.28).

The TCRβ variable region genes are assembled from Vβ, Dβ, and Jβ segments via an ordered pathway in which Dβ is joined to Jβ followed by appendage of Vβ segments to the Dβ-Jβ complex. Studies that analyzed the effects of already assembled TCRβ functional transgenes on thymocyte development provided the first evidence that TCRβ variable region gene assembly is also allelically excluded. As is true for HC variable region gene assembly, the production of a functional TCRβ chain provides the signal for feedback regulation of the Vβ to DJβ rearrangement step at the second allele. In addition to regulating allelic exclusion of TCRβ gene rearrangement, the TCRβ protein, which also resides on the surface of the pro-T cell in association with a surrogate TCR chain (the pre-T cell receptor α chain or pTα), promotes the development of pro-thymocytes to the prethymocyte stage.

TCRα variable region genes are assembled from component Vα and Jα gene segments at the prethymocyte stage. Thus, the expression of TCRβ chains appears to mediate the onset of TCRα gene rearrangement. However, just as some LC gene rearrangement does occur in Ig HC–deficient mice, so can some TCRα gene rearrangement take place in TCRβ-deficient mice. The interpretation of this finding is subject to the same general types of considerations raised for LC rearrangement in HC-deficient mice. However, one significant difference between the regulation of Ig V_LJ_L and TCR VαJα assembly is that the latter process does not appear to be subject to strict allelic exclusion. The functional significance of the lack of allelic exclusion at the TCRα locus is still a matter of speculation.

Expression of functionally rearranged TCRα genes leads to the generation of the mature αβ TCR on the surface of T cells and cellular selection processes that ultimately result in the generation of MHC-restricted, SP T cells. Notably, cross-linking of the complete TCR on the surface of immature, TCR-expressing DP thymocytes leads to down-regulation of *RAG* gene expression, suggesting that this feedback process ultimately fixes the rearrangement patterns in mature, peripheral T cells.

7.6 Modification of the Primary B Cell Antibody Repertoire

The primary antibody repertoire generated in the fetal liver and adult bone marrow through V(D)J recombination is further modified as immature B cells progress along the differentiation pathway in the bone marrow and peripheral lymphoid organs. The driving force for these modifications is the interaction between self-reactive or foreign antigens and antigen receptors expressed on immature and mature B cells. If the antigen is a macromolecule indiginous to the host, its binding to IgM on immature B cells or to IgM and IgD on mature B cells leads to elimination of the B cell and reduction in the complexity of the primary antibody repertoire. If the antigen is a foreign macromolecule, its binding to IgM and IgD on mature B cells leads to B cell activation. During an immune response, the primary antibody repertoire can be further diversified and the antigen specificity refined by **somatic hypermutation**. Furthermore, immunoglobulins can be functionally modified through **class switching**.

a. Self-Reactive B Cell Selection

Assembly of variable region gene segments is a random process in terms of V, (D), and J gene segment combinations. Among the seemingly unlimited numbers of variable region genes generated, some may encode immunoglobulins that bind macromolecules normally produced by the organism itself while the rest encode immunoglobulins specific for non-self-molecules. To prevent **autoimmunity**, that is, immune reactions against host structures, self-reactive B cells have to be eliminated shortly after their generation. This process operates at the cellular level and occurs first at the IgM-expressing, immature B cell stage in the bone marrow and then at the IgM- and IgD-expressing mature B cell stage in the periphery. The fate of reactive immature and mature B cells depends on whether the self-macromolecules encountered are soluble or membrane bound.

Clonal deletion Interaction of membrane-bound self-antigen with IgM on immature B cells in the bone marrow tends to result in the death and therefore **clonal deletion** of the immature B cells (Figure 7.34). Thus, B cells fail to develop in mice treated with anti-IgM antibody from birth. In addition, transgenic mice whose B cells express IgM specific for H-2K^b or H-2K^k Class I MHC molecules in animals whose MHC type is either H-2k^b or H-2K^k, respectively, have markedly reduced numbers of B cells and a complete absence of peripheral B cells that express the transgenes.

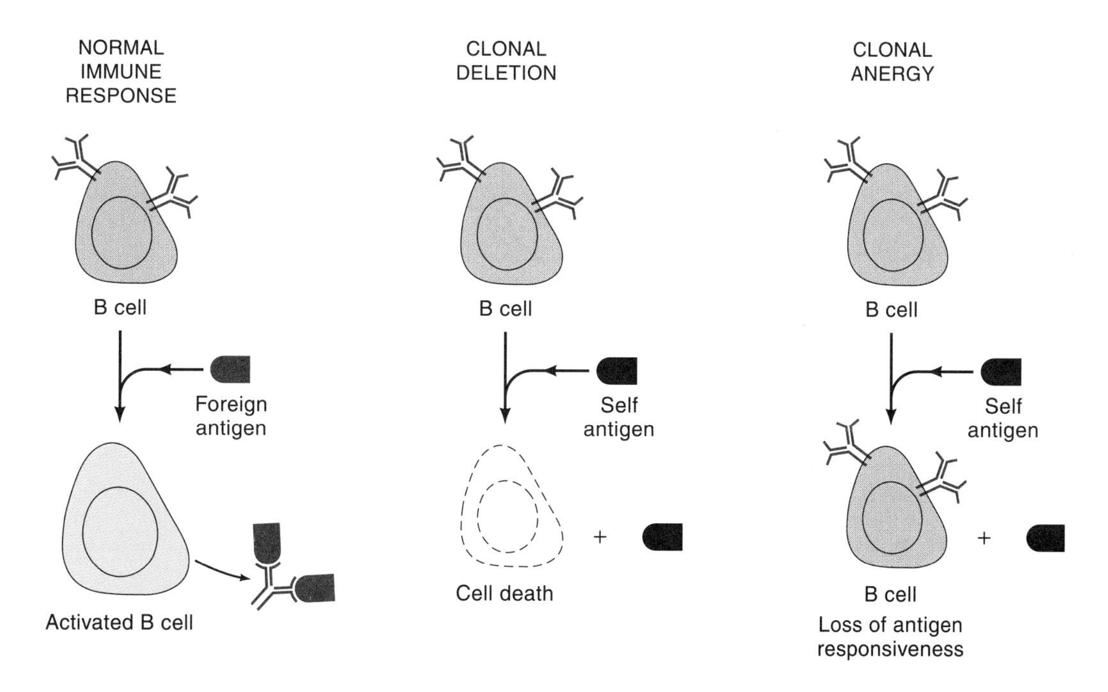

Figure 7.34
Comparison of B cell activation and selection. (Adapted from A. K. Abbas, A. H. Lichtman, and J. S. Pober. 1991. Cellular and Molecular Immunology. p. 210. W. B. Saunders, Philadelphia.)

In another transgenic system, the clonal B cell deletion process has been shown to proceed through two sequential events: **arrested development** and **cell death**. In double transgenic mice expressing membrane-bound hen egg lysozyme (mHEL) and IgM and IgD specific for the lysozyme (Figure 7.35), immature B cells are initially blocked of further differentiation because they fail to express adhesion molecules and receptors important for migration and activation. These arrested B cells remain in the bone marrow and die after a few days. Cell death probably occurs through **apoptosis** because the expression of *bcl*-2, a gene whose expression confers resistance to apoptosis in many cell types, prolongs the survival of the arrested B cells in the bone marrow.

Clonal anergy Binding of soluble self-antigen to IgM and IgD on mature B cells in the periphery results in the functional inactivation of mature B cells (**clonal anergy**) (Figure 7.34). Thus, in transgenic mice expressing IgM and IgD directed against lysozyme, the presence of soluble lysozyme down-modulates surface IgM, but not IgD, expression on B cells. Other antigens, such as single-stranded

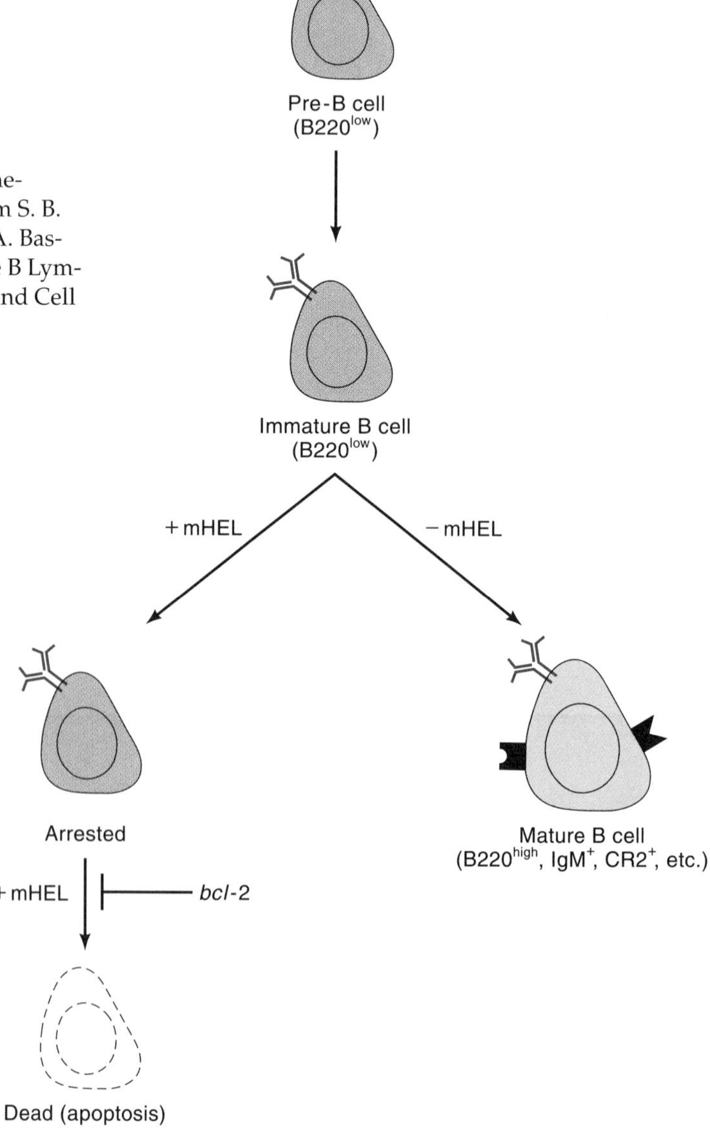

Figure 7.35
Proposed pathway for clonal deletion in the anti-membrane-bound hen egg lysozyme (mHEL) transgenic system. (From S. B. Hartley, M. P. Cooke, D. A. Fulcher, A. W. Harris, S. Cory, A. Basten, and C. C. Goodnow. 1993. Elimination of Self-Reactive B Lymphocytes Proceeds in Two Stages: Arrested Development and Cell Death. *Cell* 72 332.)

DNA (ssDNA) in transgenic mice expressing surface Ig specific for ssDNA, do not affect reactive B cells phenotypically. However, both types of self-antigens render reactive B cells functionally inactive. These B cells are unable to transduce signals for protein tyrosine phosphorylation, proliferation, and differentiation into antibody-secreting plasma cells through their remaining surface antigen receptors. Such anergic cells occupy normal anatomical sites in the lymphoid organs and survive for some time in the body. The signals that lead to clonal anergy and the physiological changes the anergic B cells undergo are not known. The initial phase of receptor down-modulation is probably mediated through endocytosis, but the sustained down-modulation probably involves changes in cellular processes such as posttranslational processing of immunoglobulin molecules.

For both clonal deletion and anergy, the self-antigen has to be present at a sufficiently high concentration. If the antigen receptor occupancy on immature and mature B cells is below 5 percent, the presence of self-antigen has no detectable effect on reactive B cell development and function (**clonal ignorance**). This has been observed in both antilysozyme and anti-H-$2K^k$ or H-$2K^b$ transgenic mice when lysozyme or H-$2K^k$ or H-$2K^b$ expression is too low. The absence of an anergic response when the self-antigens are present at low concentration or the B cells have too low an affinity for abundant self-antigens may be physiologically significant.

Receptor editing Most self-reactive immature B cells are eliminated through deletion and anergy. Some self-reactive transgenic B cells can survive by changing the antigen receptor expressed. In transgenic mice producing B cells specific for ssDNA or H-$2K^k$ or H-$2K^b$, antigen binding to IgM leads to expression of the *RAG* genes and thus activation of V(D)J recombination and the rearrangement and expression of endogenous LC genes. Surface IgM consisting of endogenous LC and transgenic HC may no longer bind to the self-antigen, and the resultant B cells can thus survive the selection process. Such **receptor editing** probably occurs during normal B cell differentiation but has been detected only in self-reactive transgenic mice.

Autoantibodies and autoimmunity Normally, self-reactive B cells are effectively censored. However, if self-reactive B cells are incompletely eliminated or anergic B cells recover, due to either genetic or environmental factors, they can produce **autoantibodies**. Such autoantibodies may cause autoimmune disease by eliminating or functionally mimicking normal physiologically important molecules or by leading to the deposition of immune complexes. For instance, Graves' disease, an autoimmune disease of hyperthyroidism, is usually caused by autoantibodies mimicking thyroid-stimulating hormone (TSH) by binding to TSH receptors on thyroid epithelial cells. Systemic lupus erythematosus is an autoimmune disease in which autoantibodies and their complexes contribute to the clinical manifestation of glomerulonepheritis and arthritis.

b. Somatic Hypermutation

Surviving non-self-reactive B cells reside in the secondary lymphoid organs. During an immune response, binding of foreign antigen to IgM and IgD receptors on mature B cells initiates a series of signal transduction events mediated by calcium mobilization, phospholipid turnover, and protein tyrosine phosphorylation. With T cell help, **activated B cells** undergo rapid proliferation and further differentiation into short-lived **plasma cells** and long-lived **memory B cells** (Figure 7.9). During this differentiation process, the primary antibody repertoire may be further diversified by **somatic hypermutation**.

The idea that somatic mutation generates antibody diversity was proposed in the 1950s as an alternative to the hypothesis that such diversity is germ line encoded. The first convincing evidence for somatic mutation was obtained from examination of amino acid sequences of the λ LC produced by mouse myeloma tumors. Since then, the cloning and sequencing of specific variable region genes from B cells and hybridomas derived from mice before and after immunization has unequivocally established that the primary antibody repertoire can undergo extensive somatic hypermutation during an immune response.

Target of hypermutation Comparison of variable region gene sequences from both HC and LC genes before rearrangement, after rearrangement but before antigen stimulation, and after immunization has shown that mutations are highly localized in the variable region genes (Figure 7.36). They are introduced at similar frequencies in V_HDJ_H and V_LJ_L regions regardless of whether they represent productive or nonproductive rearrangements. However, they usually do not occur in unrearranged V gene segments and occur less frequently in rearranged DJ_H complexes. In the case of nonproductive rearrangements, mutations tend to be spread somewhat randomly over the entire variable region DNA, whereas in productively rearranged genes, they are more concentrated in the CDRs. This finding is interpreted to mean that (1) the somatic mutation process is relatively random with respect to the variable region gene sequence and (2) cellular selection amplifies the recovery of cells in which the mutations result in increased affinity for the immunizing antigen.

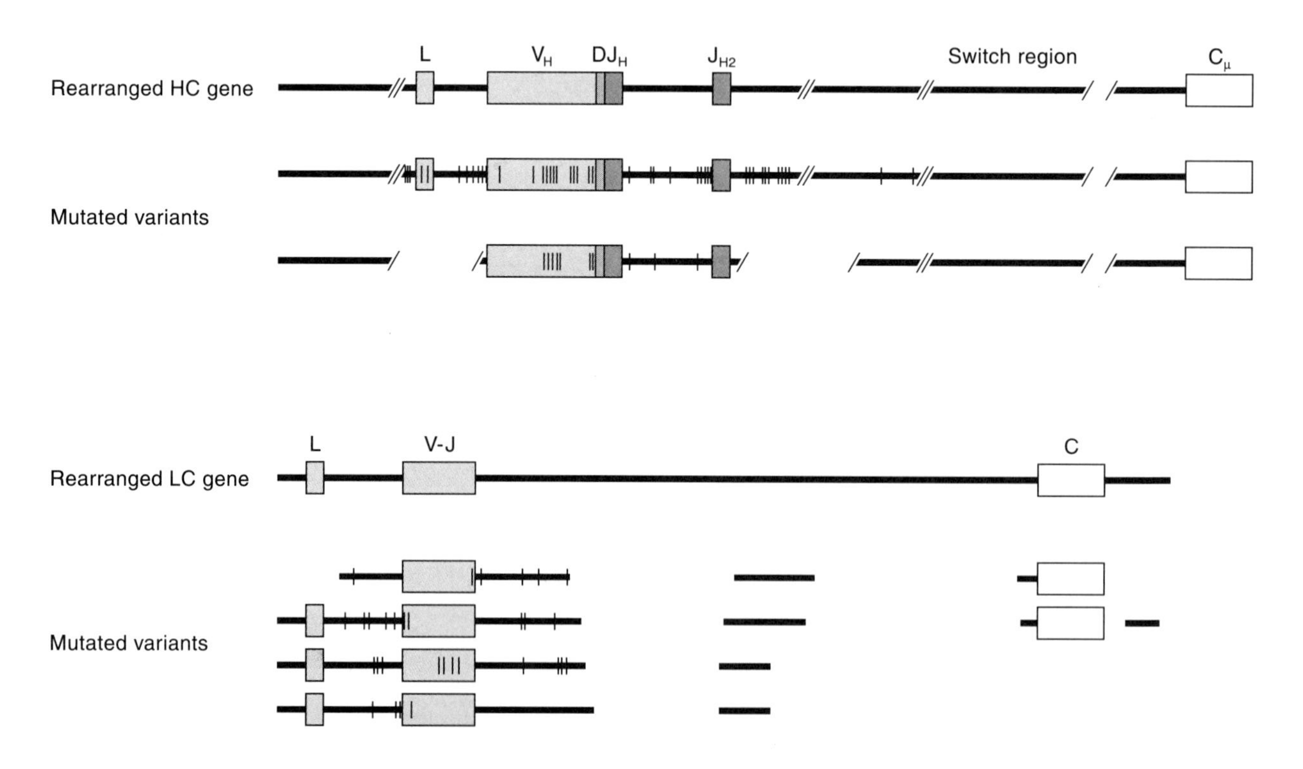

Figure 7.36
Rearranged HC and LC genes and their mutated variants. (Adapted from S. Kim, M. M. Davis, E. Sinn, P. Patten, and L. Hood. 1981. Antibody Diversity: Somatic Hypermutation of Rearranged VH Genes. Clusters of Point Mutations Are Found Exclusively Around Rearranged Antibody Variable Genes. *Cell* 27 573; and P. J. Gearhart and D. F. Bogenhagen. *Proc. Natl. Acad. Sci. USA* 80 3441.)

Although somatic hypermutation can occur throughout the variable region gene, its frequency greatly diminishes approximately 1 kb from each end. Moreover, these mutations are rarely found in C region genes. The highly localized occurrence of these mutations on completely assembled variable region genes suggests that rearrangement may generate a *cis*-acting signal for the targeting of mutation. Alternatively, targeting may be through *cis*-acting elements brought into proximity but outside of rearranged variable regions themselves. Such *cis*-acting elements apparently are not brought in through class switching because mutations are found in both IgM and IgG variable regions.

Transgenic mice have been used in attempts to identify the putative *cis*-acting elements involved in somatic mutation. The transgenic constructs contained a rearranged HC or LC gene in the presence or absence of various possible *cis*-acting elements, for example, transcriptional control elements. Hybridomas were made before and after immunization of the transgenic animals and the nucleotide sequences of endogenous HC and/or LC genes as well as that of the transgenes were determined. Alternatively, the transgenes in selected populations of activated B cells were sequenced directly by PCR amplification without prior generation of hybridomas. These types of analyses showed that the V_H promoter and Eμ enhancer elements are not sufficient to target somatic hypermutation in the HC locus. Thus, the putative *cis* signals do not appear to be within the rearranged gene or its immediately flanking sequences. By contrast, both the intronic and 3′ κ enhancers, but not the Vκ promoter, appear to be required for efficient somatic mutation of the κ locus.

Nature of mutation Somatic hypermutations are predominantly point mutations, some small deletions, occasionally additions, and a few possible gene conversion events. Both replacement mutations, which result in amino acid changes, and silent mutations, which do not lead to amino acid changes, are present. In some cases, mutations are repeatedly observed at the same spot; these are apparently not due to cellular selection because of their occurrence in nonproductive rearrangements, for example, and probably represent mutational hot spots.

The clonal origin of hybridomas derived from individual mice after a single antigen stimulation can be established by their shared unique V(D)J junctional sequences and identically rearranged nonproductive alleles. Comparison of mutations in the variable region genes of these related hybridomas demonstrated that mutations are accumulated stepwise during clonal expansion (Figure 7.37). Furthermore, related hybridomas can be reconstructed into a genealogical tree according to the accumulation of mutations. If we assume that cells are doubling every 17 to 18 hours and the mutations occur continuously, the mutation rates calculated from the genealogical trees are on the order of 10^{-3} to 10^{-4}/base pair/generation. Theoretical modeling of the population dynamics suggests a mutation rate of 10^{-3}/base pair/generation if the presence of silent substitutions (nonselectable mutations) in the progeny cells is to be explained.

Possible mechanisms The extremely localized high rate of somatic hypermutation suggests the existence of a special mutator mechanism, although its molecular basis remains a matter of speculation. One possibility is error-prone DNA repair. In this case, mutations would be introduced as errors in either DNA replication or repair and fixed upon replication in the progeny. DNA polymerase β has an apparently high error rate of 1.5×10^{-3} and may play a role in DNA repair. Another considered mechanism is gene conversion, that is, the acquisition of new sequence arrangements by recombination with a related but different sequence. Although it is unlikely to be a major mechanism, some support for this hypothesis comes from the existence in the genome of pseudovariable gene segments that could serve as donor sequences and the few sequence changes that

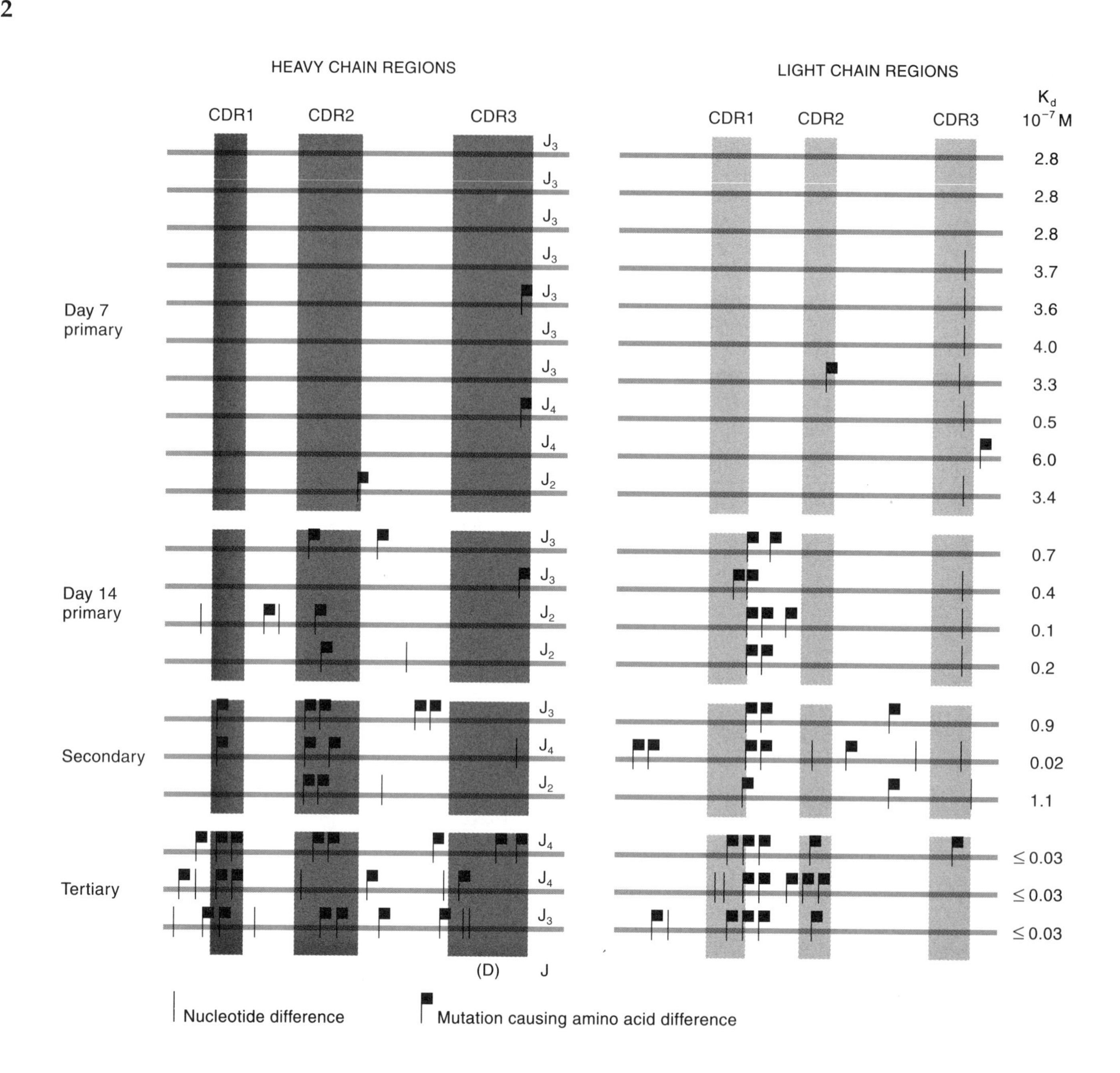

Figure 7.37
Stepwise accumulation of mutations in Ig variable region genes. (From C. Berek and C. Milstein. 1987. Mutation Drift and Repertoire Shift. *Immunol. Rev.* 96 26.)

can be accounted for by conversion during somatic hypermutation. Gene conversion is a plausible mechanism because it is the main mechanism for generating primary antibody diversity during B cell differentiation in the bursa of the chicken. Directed mutation mechanisms have also been proposed because repeats and/or palindromes occur at high frequency in the variable region genes. This may account for some mutational hot spots. In any event, in the absence of an easily manipulable system, the elucidation of the mechanism of somatic hypermutation remains a significant challenge. Because λ LC diversification in Ilial Peyer's patches in sheep appears to occur by somatic mutation, analysis of this system may shed light on the mechanisms of somatic hypermutation in other systems.

Germinal centers and somatic hypermutation During a humoral response to a T cell–dependent antigen, antigen-activated B cells proliferate in the periphery of the periarteriolar lymphoid sheath (PALS) and then either differentiate into foci of antibody-secreting cells or initiate the formation of so-called germinal centers in the lymphoid follicles (Figure 7.38). Somatic mutation is detectable at day 7 after immunization and peaks at about day 14, corresponding exactly to the kinetics of germinal center formation. Germinal centers form due to the rapid proliferation of activated B cells. In the germinal center's "dark zone," primary B cell blasts, termed centroblasts, lose their surface Ig and proliferate to give rise to progeny, centrocytes, that are nondividing and express surface Ig (Figure 7.39). Centrocytes interact with follicular dendritic cells, and some of them differentiate into memory B cells and possibly plasma cells. The rest die through apoptosis. Sequence analysis of isolated germinal center B cells' variable region genes demonstrated that somatic mutation occurs during memory B cell generation in the germinal center but not in the B cells that form antibody-secreting foci.

Affinity maturation After somatic hypermutation, the binding affinity of the mutated variable regions for the original antigen may be the same, increased, decreased, or completely abolished. Furthermore, mutated variable regions may gain significant binding affinities for different antigens or even self-antigens. Thus, after somatic mutation, B cells must undergo another round of selection to enrich for those bearing antigen receptors with higher affinity for the original

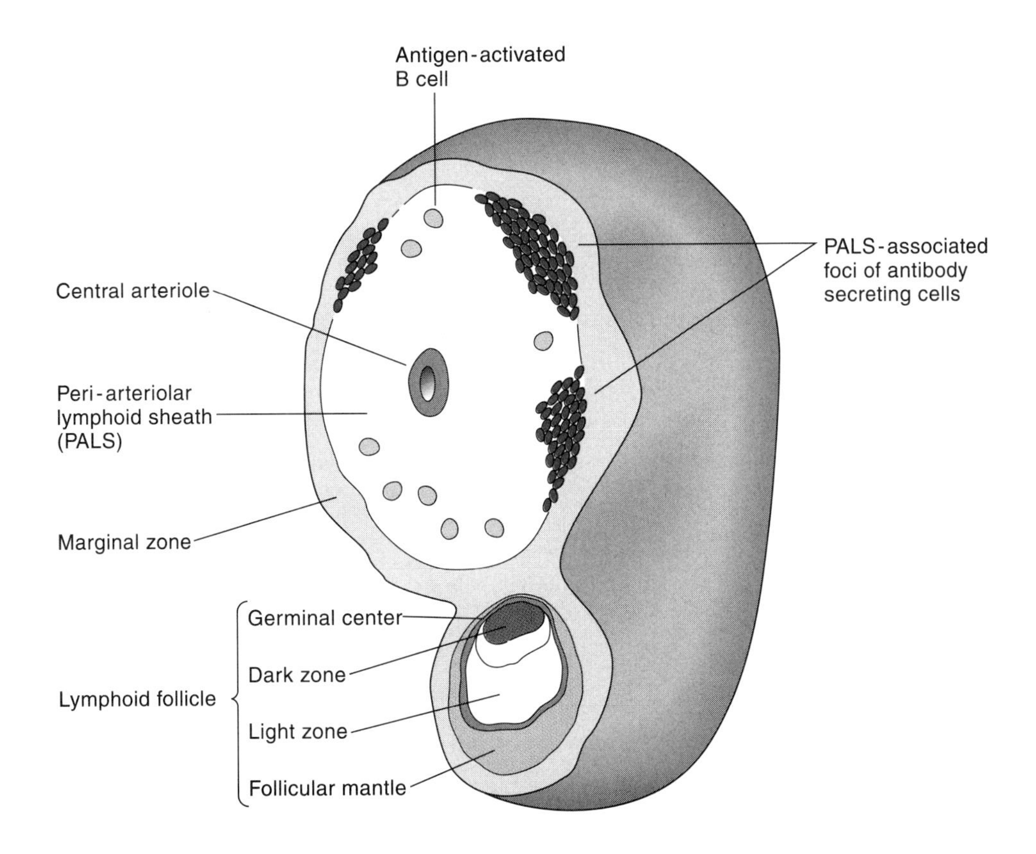

Figure 7.38
Cross-section of a white pulp area of a lymph node. (Adapted from G. Kelsoe and B. Zheng. 1993. Sites of B Cell Activation *in Vivo*. *Curr. Opin. Immunol.* 5 419.)

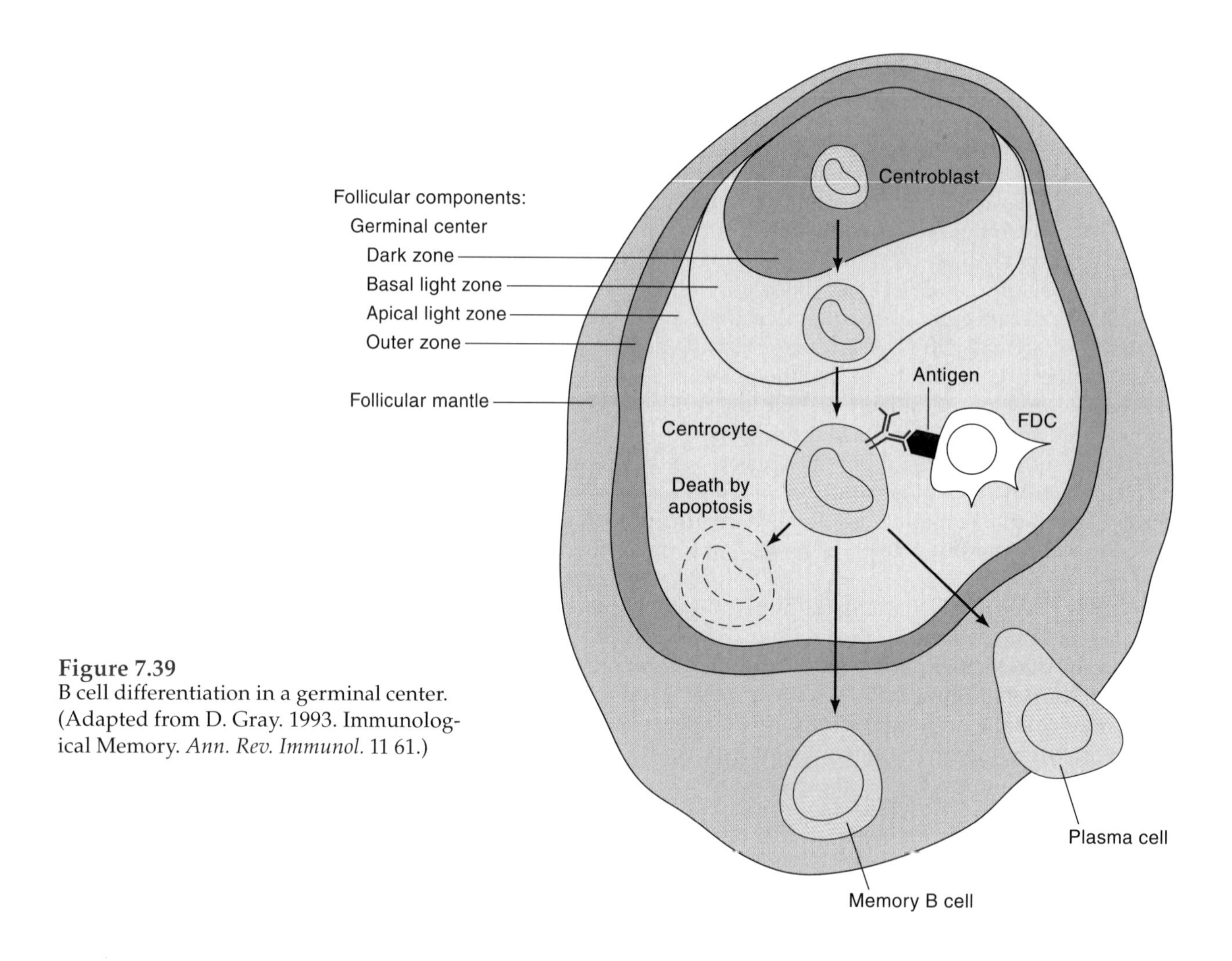

Figure 7.39
B cell differentiation in a germinal center. (Adapted from D. Gray. 1993. Immunological Memory. *Ann. Rev. Immunol.* 11 61.)

antigen. The rest of the B cells, including those bearing self-reactive antigen receptors, must be eliminated. This selection process occurs through the interaction of surface Ig with antigen trapped on the surface of follicular dendritic cells in the germinal center. Presumably, B cells bearing receptors with a decreased or qualitatively altered affinity do not receive any signal and die through apoptosis. B cells bearing receptors with an increased affinity to the antigen receive a signal and differentiate into long-lived memory B cells. When these cells encounter the same antigen again, they rapidly differentiate into antibody-secreting plasma cells, yielding an increase in overall antibody affinity. Thus, the somatic mutation process indirectly leads to **affinity maturation** of antibodies.

Hypermutation control Somatic hypermutation is clearly initiated by antigen stimulation of B cells through their cell surface Ig receptors. However, many other signals, especially those involved in B and T cell collaboration, are probably required to complete the mutation process after initial B cell activation. Thus, interaction between **CD40** expressed on the B cell surface and **CD40 ligand (CD40L)** expressed on the activated T cells is required for the germinal center formation and the generation of memory B cells. Mice harboring a targeted mutation in the genes encoding either CD40 or CD40L generate virtually phenotypically normal populations of T and B cells. However, they do not generate germinal centers after immunization with T cell–dependent antigens. It would be of interest to know if somatic mutation occurs in these mutant mice. Similarly, full activation of the somatic mutation process may also require interaction of the B7 surface antigen on activated B cells with the surface proteins CD28 and CTLA-4 on activated T cells. The cessation of somatic mutation seems to correlate with the reexpression of Ig on centrocytes. Analogous to IgM signaling for cessation of V(D)J recombination in immature B cells, Ig expressed on centrocytes may signal cessation of somatic mutation in the germinal center.

c. Heavy Chain Class Switching

HC class switch recombination (CSR) is a B cell–specific process that allows a clonal B cell population to produce antibodies that retain their initial antigen specificity while acquiring a different effector function. Class switching involves an apparently B cell–specific **recombination/deletion** mechanism that is distinct from V(D)J recombination. As a result of HC CSR, a different and downstream C_H segment is juxtaposed to the V(D)J sequences with the accompanying deletion of Cμ and intervening segments (Figure 7.40). This process occurs by recombinational events between **switch region** (**S region**) sequences that reside in the introns upstream of the various C_H coding elements. Class switching therefore involves recombination between the Sμ sequence of Cμ and the S sequence associated with a downstream C_H gene. Thus, the downstream C_H gene is appended to the variable region transcription unit. Transcription from the V_H promoter then generates HC mRNA sequences in which the same HC variable region initially expressed with Cμ is expressed with the new C_H gene. This recombination/deletion event usually occurs by an intrachromosomal process with the intervening DNA generally deleted as a circle (Figure 7.40).

Just as for early V(D)J recombination, most of our knowledge of the mechanism of the HC CSR process is derived from comparative analyses of the structures of the C_H loci in germ line cells and cell lines (plasmacytomas or hybridomas) representing B cells that have switched to expression of downstream C_H genes (Table 7.1). The isolation of circular excision products from splenic B cells that had undergone CSR events has allowed further examination of the process in normal cells. Some B-lineage cell lines undergo HC CSR in culture; analyses of these lines has also provided some insights into the mechanism and control of this process. Such lines have also been used as recipients for CSR substrates analogous to those employed to study V(D)J recombination. These types of studies provide the best evidence for the B cell specificity of this reaction. Thus, CSR was observed when switch recombination substrates were introduced into certain B-lineage cell lines but not when they were introduced into a limited number of nonlymphoid lines. However, because the frequency of CSR events in most cell lines is relatively low and the target S region sequences are complex, only limited mechanistic insights have been obtained with this approach. No genes or gene products have been specifically implicated in CSR.

S region sequences CSR is distinct from the V(D)J recombination process in many ways. First, it occurs at a late stage of B cell development (i.e., during antigen-driven B cell differentiation) (Figure 7.9). Second, the reaction involves recombination between the S region sequences that lie in intronic regions; thus, the recombination event does not affect coding potential. Each C_H gene, except Cδ, which does not undergo normal class switching, is preceded on the chromosome by an S region. The S regions range from 2.5 to 10 kb in length and lie approximately 2 kb upstream of their respective C_H coding exons (Figure 7.40). S regions are mainly imprecise, tandemly repeated sequences with varying degrees of homology to each other. Thus, the Sμ region is most homologous to the Sα and Sε regions (associated with the Cα and Cε coding segments, respectively) and less so to the Sγ region upstream of Cγ.

CSR occurs randomly within the S regions, in contrast to the site specificity of the V(D)J recombination process. No consensus sequence for recombination within the S regions has been conclusively identified. In addition, although several factors that bind to S regions have been described, their roles in the CSR

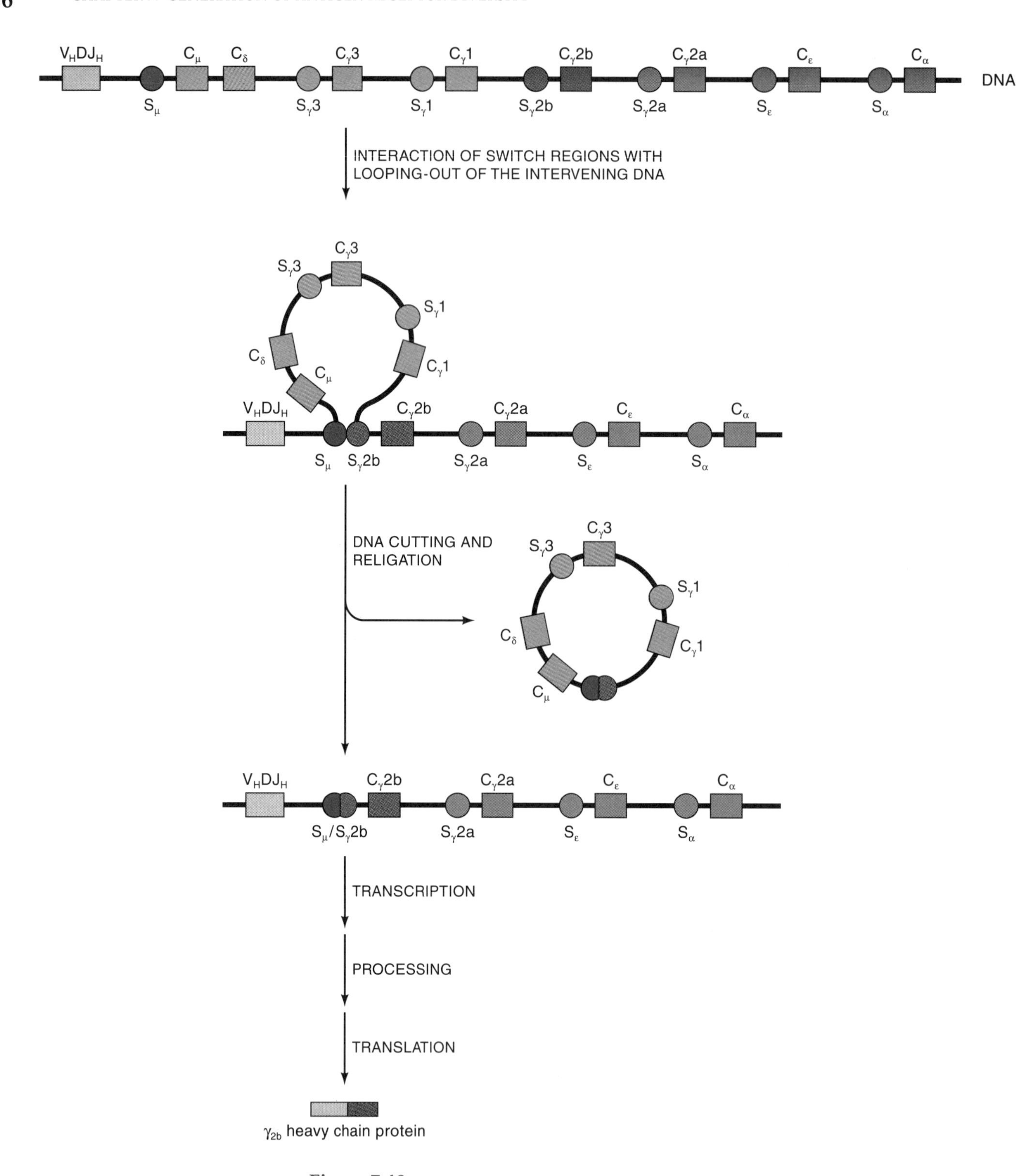

Figure 7.40
Deletion model of class switch recombination. (Adapted from S. Luztker and F. Alt. 1989. Immunoglobulin Heavy Chain Class Switching. In D. Berg and M. Howe, eds. Mobile DNA. p. 669. American Society for Microbiology, Washington, D.C.)

process, if any, remain to be determined. The repetitive nature of S regions suggests that a general recombination process is employed in this reaction. However, although the Sμ and Sε regions are the most homologous at a sequence level, switching to Cε is the most infrequent form of class switching. Therefore, S region–specific recombination systems remain a possibility. In any case, there

Table 7.3 *Properties of Thymus-Dependent and Thymus-Independent Antigens*

Property	Thymus Dependent	Thymus Independent	
	Protein	Lipopolysaccharide, *Brucella abortus*	Polymeric Antigens, Especially Polysaccharides
Antibody response in			
Athymic mice	No	Yes	Yes
T cell–depleted cultures	No	Yes	No or reduced
Features of antibody response			
Isotype switching	Yes	No	No (usually)
Affinity maturation	Yes	No	No
Secondary response (memory B cells)	Yes	No	No
Polyclonal B cell activation	No	Yes	No

Source: Adapted from A. K. Abbas, A. H. Lichtman, and J. S. Pober. 1991. B cell Activation and Antibody Production. In Cellular and Molecular Immunology. W.B. Saunders, Philadelphia, p. 199.

obviously must be mechanisms to ensure B cell specificity; perhaps the choice of S region sequences that recombine with Sμ is related to the control of locus accessibility.

Class switching regulation Evidence that CSR is a directed process came initially from studies of B-lineage cells or cell lines that had undergone CSR. In a large percentage of such cells that had undergone CSR, a switch to the same C_H region occurred on both chromosomes. Because only one of the chromosomes of a given B cell has a functional V(D)J rearrangement, cellular selection seemed an unlikely basis for the enrichment of cells switched to a particular C_H gene; indeed, some form of directed CSR appeared more likely. Cellular studies have further shown that directed HC CSR is regulated by external factors, the **cytokines** or **lymphokines** secreted by activated T cells and other accessory cells of the immune system. Class switching generally appears to occur in B cells that have been activated normally in the context of a specific immune response.

B cell activation can be effected in two general ways, one independent of and one dependent on T cells (Table 7.3). Antigens with high molecular weights and repeating structures, such as bacterial lipopolysaccharide, stimulate B cells in a T cell–independent fashion. A classical way to mimic T cell–independent B cell activation pathways *in vitro* is to treat B cells with a polyclonal activator such as lipopolysaccharide. This treatment results in B cell proliferation, IgM secretion, and, ultimately, class switching and secretion of predominantly IgG_{2b} and IgG_3 (Figure 7.41). Treatment of splenic B cells with LPS plus the T cell factor interleukin-4 also stimulates B cell proliferation and IgM secretion, but the B cells switch preferentially to secretion of IgG_1 and (in an interleukin-4 concentration–dependent fashion) to IgE (Figure 7.41). Other lymphokines and/or activation treatments lead to directed switching to other classes. For example, treatment of splenic B cells with LPS plus transforming growth factor β or interferon γ leads to switching to IgA and IgG_{2a}, respectively.

Most antigens activate B cells only with help from T cells. T cell–dependent B cell activation results from antigen-mediated T/B cell interactions that involve additional surface molecules, in particular, interaction of the CD40 ligand on activated T cells with CD40 on B cells. B cells in culture can also be activated by treatment with soluble CD40 ligand, activated T cell membranes, or anti-CD40 antibodies. In humans, an immune deficiency disease, X-linked immune defi-

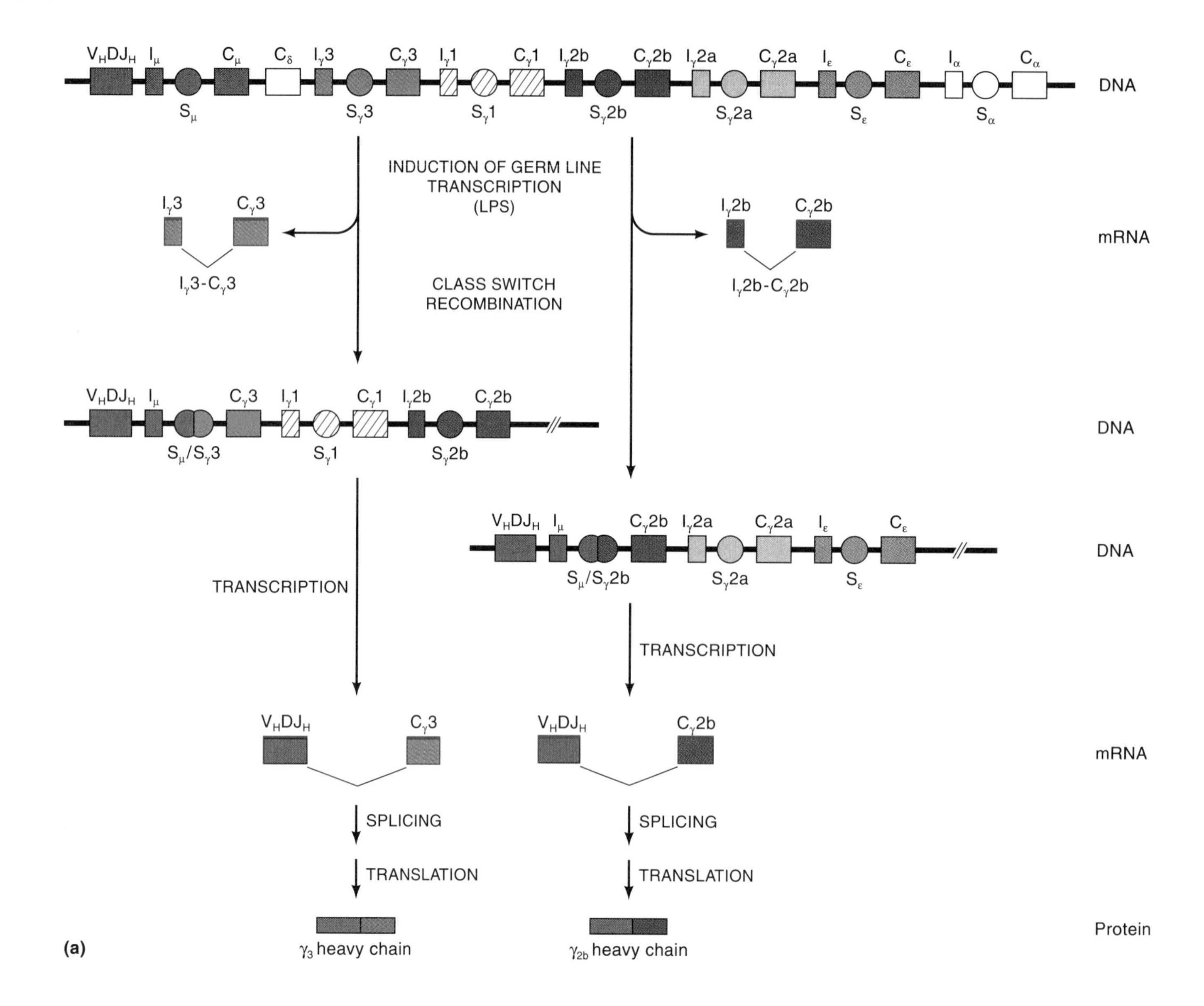

Figure 7.41
Accessibility model of class switch recombination. (From Regulation of Class Switch Recombination of the Immunoglobulin Heavy Chain Genes. 1995. In T. Honjo, F. W. Alt, and T. H. Rabbitts, eds. Immunoglobulin Genes, 2nd ed. Academic Press, New York.)

ciency with hyper IgM, results from a mutation in the gene encoding the CD40 ligand and impairs B cells' ability to undergo T cell–dependent class switching. Mice engineered to lack CD40 or CD40 ligand expression by targeted mutation of either of these genes lose their ability to mount responses to T cell–dependent antigens and, as mentioned previously, do not form germinal centers. *In vitro,* lymphokines can influence CSR promoted by the CD40/CD40 ligand pathway just as they do in a T cell–dependent pathway (Table 7.3).

Class switching *in vivo* is strongly affected by the type of antigen, the route of immunization, and, as a result, the types of lymphokines the B cells encounter during the response. For example, interleukin-4 is required for IgE production in response to helminthic parasite infections *in vivo.* Mice deficient in interleukin-4, as a result of ablation of the gene, lack serum IgE and do not respond to parasitic infections by secreting IgE. IgE is normally involved in allergic responses via binding to Fc receptors on mast cells and basophils. The γ-interferon produced following viral infections enhances IgG_{2a} production but inhibits class switching

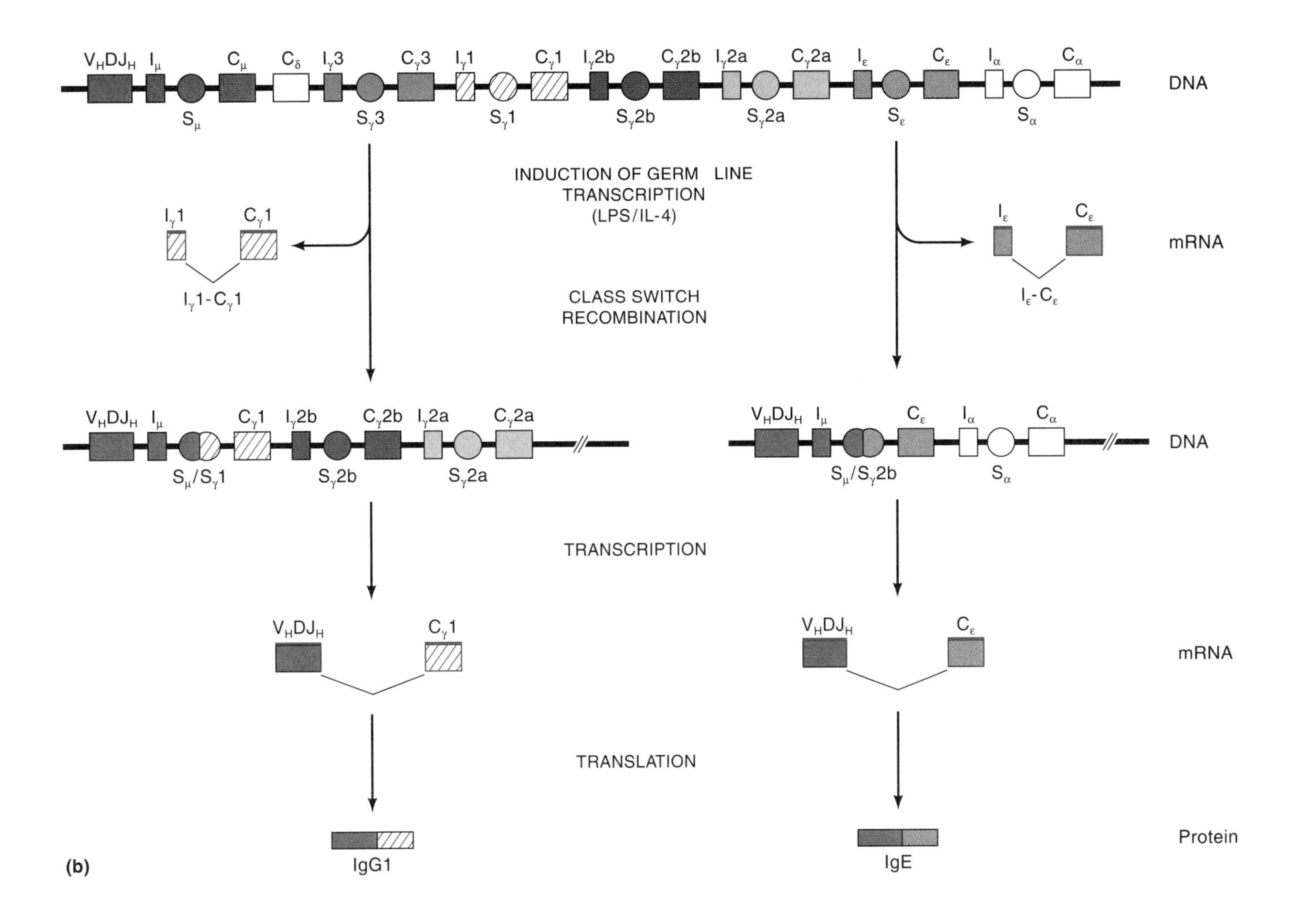

to IgE and IgG$_1$; conversely, interleukin-4 inhibits switching to IgG$_{2a}$, indicating an intricate regulation of class switching by lymphokines. Interleukin-5 and transforming growth factor-β are important for IgA production and therefore may be particularly important for regulating mucosal immunity. Thus, directed class switching is a means of regulating the humoral immune response to produce antibodies with effector functions most appropriate for eliminating the particular foreign antigen encountered.

Control of class switch recombination Directed class switching is thought to be controlled by the accessibility of the recombining sequences in much the same way as is proposed for control of V(D)J recombination. However, formal proof of this mechanism and evidence for the existence of a common class switch recombinase is lacking. The CSR accessibility model is based on the observation that a given germ line C_H gene is specifically transcribed prior to the CSR event. Moreover, mitogen and lymphokine treatments specifically induce transcription of unrearranged C_H regions prior to the induced CSR event, suggestive of a cause-and-effect relationship. For example, splenic B cells cultured in the presence of lipopolysaccharide express germ line γ_{2b} and γ_3 transcripts, but splenic B cells cultured with lipopolysaccharide plus interleukin-4 express γ1 and ε germ line transcripts while suppressing expression of the γ2b and γ3 genes (Figure 7.41).

Transcription of unrearranged or the germ line type of C_H regions initiates at a sequence that is 5′ of the I exon, proceeds through the S region, and terminates downstream of the involved C_H segments (Figure 7.42). For several of the C_H genes, these "germ line" transcripts initiate at promoters that respond to lymphokines. For example, the promoter upstream of the Cε region contains both lipopolysaccharide and interleukin-4 responsive elements.

The primary "germ line" C_H gene transcripts are processed to yield mature transcripts in which the I exon is spliced to the C_H exon (Figure 7.42). In terms of overall structure, these transcripts resemble a mature HC mRNA except that the I exon is found in place of the variable region exon. The "germ line" C_H transcripts do not appear capable of encoding proteins because I exons contain multiple translation stop codons, some of which are in frame with the C_H exon coding sequences. Therefore, they are considered "sterile" transcripts. Yet the "germ line" transcription units are conserved for all C_H genes (except Cδ) and in multiple mammalian species. The conservation of these noncoding transcription units and the structure of their processed products suggest they may play a regulatory role in class switching.

The apparent inability of "germ line" transcripts of C_H genes to encode protein products has made researchers consider a number of other possible functions for them. One is that the act of transcription through the target S regions renders them accessible for CSR just as transcription promotes homologous recombination in yeast. Another, not mutually exclusive, possibility is that the primary "germ line" C_H transcripts or their processed products play a direct role in the CSR process. For example, the RNA transcripts might interact directly with the S region sequences to form more recombinogenic structures. Another scenario speculates that interaction of the processed transcripts with the genomic I region and downstream sequences (e.g., C_H sequences), presumably mediated by protein factors, renders the intervening S region more accessible.

A final speculative role for "germ line" C_H transcripts is that they may serve as substrates for transsplicing reactions that would allow the variable region gene sequences expressed with the Cμ transcription unit to be appended to the downstream C_H gene at the RNA level. This would allow an activated B cell to express simultaneously receptors for IgM and the downstream isotype. Indirect

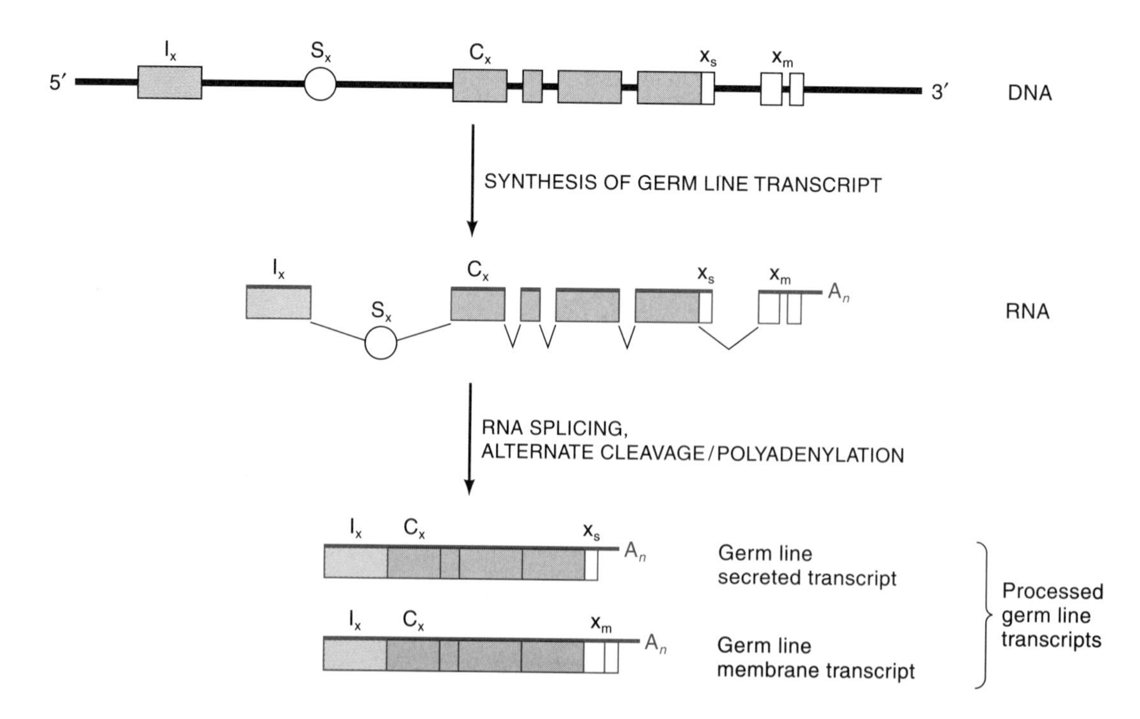

Figure 7.42
Synthesis and processing of germ line transcripts. (From P. Rothman, S. Li, and F. Alt. 1989. The Molecular Events in Heavy Chain Class-Switching. *Sem. Immunol.* 1 71.)

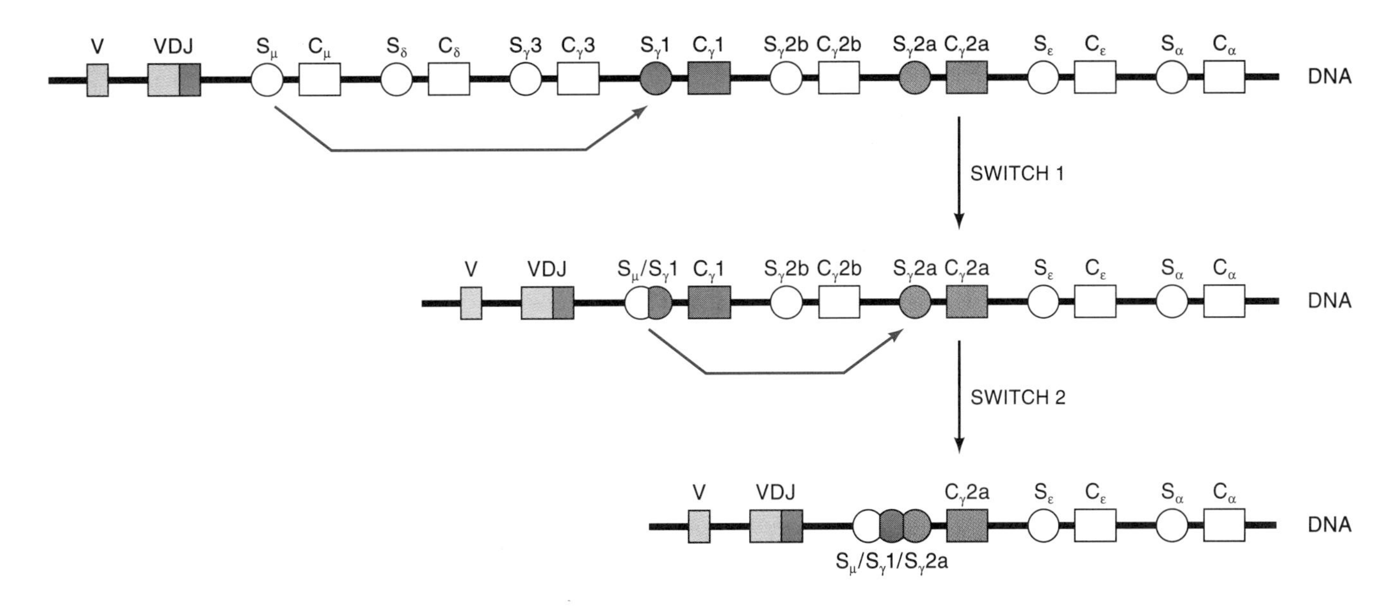

Figure 7.43
Successive class switch recombinations. (From L. E. Hood, I. L. Weissman, W. B. Wood, and J. H. Wilson. 1984. Immunology, 2nd ed. p. 103. Benjamin/Cummings, Menlo Park, California.)

evidence for this proposed transsplicing mechanism has been obtained, but the utility of such a mode of class switching in B cell physiology is not clear.

The role of C_H transcription in CSR Some of the proposed functions for "germ line" C_H gene transcription are being tested by analysis of targeted mutations. The activity of targeted mutations in several different C_H regions confirms that the integrity of the I exon and its promoter is essential for efficient class switching to a given C_H gene. This strongly supports the proposed regulatory role of these sequences in CSR. For example, deletion of the I exon and promoter associated with $C_{\gamma 1}$ leads to a block in the ability of B cells to switch to the $C_{\gamma 1}$ region both *in vivo* and *in vitro* following treatment with lipopolysaccharide and interleukin-4. However, this mutation does not affect switching to C_H regions that lie 5′ or 3′ of the $C_{\gamma 1}$ on the chromosome, including the Cε gene (Figure 7.41), indicating the specificity of the affected control element for $C\gamma_1$.

B cells can undergo sequential CSR. Thus, once a B cell has switched to a given downstream C_H region such as $C_{\gamma 1}$, additional stimulation can lead to switching from $C_{\gamma 1}$ to a downstream region such as Cε (Figure 7.43). In this case, switch recombination occurs between the hybrid $S\mu/S_{\gamma 1}$ S region and the Sε region, further suggesting the relatively nonsequence specificity of the recombination reaction. In any case, the $I_{\gamma 1}$ exon gene–targeted mutational experiments described previously indicate that sequential CSR is not a required mechanism for switching to further downstream C_H genes.

Another experiment showed that targeted replacement of the Iε exon and promoter with a lipopolysaccharide inducible promoter (a V_H promoter plus the Eμ element) led to the generation of B lymphocytes that switch to Cε *in vitro* following treatment with lipopolysaccharide alone, a phenomenon not observed in normal B cells. Because the switching levels in the mutant cells were very low compared to normal, transcription per se through the S region can likely promote CSR, but the integrity of the I region and/or promoter is probably necessary for efficient switching.

Together, these types of experiments demonstrate the importance of "germ line" transcription for normal targeting of the CSR process. However, the exact function of this process awaits the analysis of additional mutations that will allow evaluation of the relative roles of the "germ line" promoters, the I exons, the transcription process, and the encoded primary and processed transcripts.

The IgH 3′ enhancer regulatory locus How the complex regulation of the various C_H transcription units is differentially effected is largely unknown. As already mentioned, some of these promoters do contain specific lymphokine responsive elements. An additional control region lying downstream of the Cα region at the most 3′ end of the C_H locus has been implicated in this process. Transient transfection assays indicate that a DNA segment lying 15 kb downstream of the Cα gene generates strong transcription-enhancing activity in plasmacytoma cells. The element responsible for this activity is termed the 3′HC enhancer (3′E_H) (Figure 7.12). Its location and apparent specificity for more mature B-lineage cell stages led to the speculation that 3′E_H functions in some aspect of HC gene expression in activated B cells. Conceivably, it could enhance Ig gene expression in plasma cells or be involved in regulating the HC CSR process.

The potential role of the 3′E_H sequences was tested by replacing a 5 kb region surrounding this sequence with a *neo*^{r} gene. This mutation, when homozygous, had no obvious effects on V(D)J recombination or expression of the Cμ gene. However, it led to marked defects in the ability of activated B cells to induce "germ line" transcripts from and to switch to certain C_H genes, including Cγ3, Cγ2b, Cγ2a, and Cε. Notably, the Cγ3 gene, which is severely affected, lies over 100 kb 5′ to the 3′E_H mutation. It is also significant that switching to other C_H genes, such as $C_{\gamma 1}$, remained relatively unaffected. Although the precise mechanism by which the elimination of the 3′E_H segment affects class switching is unknown, the mutation may disrupt the regulation of "germ line" C_H gene transcription and, as a result, accessibility for CSR. Such a putative regulatory element may function similarly to so-called **locus control regions (LCRs)**, which regulate the expression of embryonic, fetal, and adult globin genes during development. Both the genomic structure of the region at or downstream of the 3′EH mutation (e.g., the occurrence of four nuclease hypersensitivity sites within this region) and the ability of DNA sequences from this region to confer position- and copy number–independent expression in transfection assays support this notion.

7.7 Conclusion

The immune system's capacity to respond specifically to any foreign antigen requires the generation of an enormously diverse repertoire of effector molecules. The primary repertoire is generated from multiple gene segments by the V(D)J recombination process, through which the coding capacity of limited numbers of gene segments is maximized by combinatorial joining of independent gene segments and the random addition and deletion of nucleotides at the joints. Because of the randomness of the V(D)J recombination process, some of the antigen receptors generated are destined to react with the self-molecules. Lymphocytes expressing self-reactive antigen receptors are eliminated at various stages of lymphocyte differentiation, thus restricting the primary antigen receptor repertoire. In contrast, lymphocytes with antigen receptors specific for a foreign antigen are selectively expanded to give rise to the specificity of the immune response. For B lymphocytes, the repertoire for a given antigen can be further refined by somatic mutation and selection processes after encountering the antigen. Thus, the immune system's antigen receptor diversity is shaped by the V(D)J recombination process's intrinsic tendency to generate as diverse a repertoire as possible at the molecular level and the selection for non-self-reactive and antigen-specific lymphocytes at the cellular level.

Further Reading

7.1

A. Nisonoff, J. R. Hopper, and S. B. Spring. 1975. The Antibody Molecule. Academic Press, New York.

S. Tonegawa. 1983. Somatic Generation of Antibody Diversity. *Nature* 302 575–581.

M. M. Davis and P. J. Bjorkman. 1988. T-Cell Antigen Receptor Genes and T-Cell Recognition. *Nature* 334 395–402.

A. Rolink and F. Melchers F. 1991. Molecular and Cellular Origins of B Lymphocyte Diversity. *Cell* 68 1081–1094.

R. Lansford, A. Okada, J. Chen, E. M. Oltz, K. Blackwell, F. W. Alt, and G. Rathbun. 1996. Mechanism and Control of Immunoglobulin Gene Rearrangement. In *Molecular Immunology,* 2nd ed., B. D. Hames and D. M. Glover (eds.). IRL Press.

7.2

G. Kohler and C. Milstein. 1975. Continuous Cultures of Fused Cells Secreting Antibody of Predefined Specificity. *Nature* 256 495–497.

N. Rosenberg, D. Baltimore, and C. D. Scher. 1975. In Vitro Transformation of Lymphoid Cells by Abelson Murine Leukemia Virus. *Proc. Natl. Acad. Sci. USA* 72 1932–1936.

H. Bluthmann and P. Ohashi (eds.). 1994. Transgenesis and Targeted Mutagenesis. Academic Press, San Diego.

7.3

T. Honjo and F. W. Alt. 1995. Immunoglobulin Genes. Academic Press, New York.

7.4

N. Hozumi and S. Tonegawa. 1976. Evidence for Somatic Rearrangement of Immunoglobulin Genes Coding for Variable and Constant Regions. *Proc. Natl. Acad. Sci. USA* 73 3628–3632.

P. Early, H. Huang, M. Davis, K. Calame, and L. Hood, L. 1980. An Immunoglobulin Heavy-Chain Variable Region Gene Is Generated from Three Segments of DNA: V_H, D, and J_H. *Cell* 19 981.

J. E. Hesse, M. R. Lieber, M. Gellert, and K. Mizuuchi. 1987. Extrachromosomal DNA Substrates in Pre-B Cells Undergo Inversion of Deletion at Immunoglobulin V-(D)-J Joining Signals. *Cell* 49 775–783.

D. G. Schatz, M. A. Oettinger, and D. Baltimore. 1989. The V(D)J Recombination Activating Gene, RAG-1. *Cell* 59 1035–1048.

M. A. Oettinger, D. G. Schatz, C. Gorka, and D. Baltimore. 1990. RAG-1 and RAG-2, Adjacent Genes That Synergistically Activate V(D)J Recombination. *Science* 248 1517–1523.

F. W. Alt, E. M. Oltz, F. Young, J. Gorman, G. Taccioli, and J. Chen. 1992. VDJ Recombination. *Immunol. Today* 13 306–314.

M. Gellert. 1992. V(D)J Recombination Gets a Break. *Trends Genet.* 8 408–416.

P. Mombaerts, J. Iacomini, R. S. Johnson, K. Herrup, S. Tonegawa, and V. E. Papaloannou. 1992. RAG-1-Deficient Mice Have No Mature B and T Lymphocytes. *Cell* 68 869–877.

Y. Shinkai, G. Rathbun K.-P. Lam, E. M. Oltz, V. Stewart, M. Mendelsohn, J. Charron, M. Datta, F. Young, A. M. Stall, and F. W. Alt. 1992. RAG-2 Deficient Mice Lack Mature Lymphocytes Owing to Inability to Initiate V(D)J Rearrangement. *Cell* 68 855–867.

J. Chen and F. W. Alt. 1993. Gene Rearrangement and B-Cell Development. *Cur. Opi. Immunol.* 5 194–200.

D. C. Van Gent, J. F. McBlane, D. A. Ramsden, M. J. Sadofsky, J. E. Hesse, and M. Gellert. 1995. Initiation of V(D)J recombination in a cell free system. *Cell* 81 925–934.

G. Taccioli and F. W. Alt. 1995. Genes involved in V(D)J Recombination and DNA Repair: Potential Targets for Autosomal SCID Mutations. *Cur. Opi. Immunol.* 7 436–440.

7.5

G. Yancopoulos, T. Blackwell, H. Suh, L. Hood, and F. Alt. 1986. Introduced T Cell Receptor Variable Region Gene Segments Recombine in Pre-B Cells: Evidence That B and T Cells Use a Common Recombinase. *Cell* 44 251–259.

R. R. Hardy, C. E. Carmack, S. A. Shinton, J. D. Kemp, and K. Hayakawa. 1991. Resolution and Characterization of Pro-B and Pre-Pro-B Cell Stages in Normal Mouse Bone Marrow. *J. Exp. Med.* 173 1213–1225.

D. Kitamura and K. Rajewsky. 1992. Targeted Disruption of μ Chain Membrane Exon Causes Loss of Heavy-Chain Allelic Exclusion. *Nature* 356 154–156.

D. Kitamura, A. Kudo, S. Schaal, W. Muller, F. Melchers, and K. Rajewsky. 1992. A Critical Role of λ5 Protein in B Cell Development. *Cell* 69 823–831.

M. Malissen, J. Trucy, E. Jouvin-Marche, P.-A. Cazenave, R. Scollay, and B. Malissen. 1992. Regulation of TCR α and β Gene Allelic Exclusion During T-Cell Development. *Immunol. Today* 13 315–322.

P. Mombaerts, A. R. Clarke, M. A. Rudnicki, J. Iacomini, S. Itohara, J. J. Lafaille, L. Wang, Y. Ichikawa, R. Jaenisch, M. L. Hooper, and S. Tonegawa. 1992. Mutations in T-Cell Antigen Receptor Genes α and β Block Thymocyte Develpment at Different Stages. *Nature* 360 225–231.

M. Groettrup and H. von Boehmer. 1993. A Role for a Pre-T-Cell Receptor in T-Cell Development. *Immunol. Today* 14 610–614.

7.6

C. Kocks and K. Rajewsky. 1989. Stable Expression and Somatic Hypermutation of Antibody V Regions in B-Cell Developmental Pathway. *Ann. Rev. Immunol.* 7 537–559.

C. C. Goodnow. 1992. Transgenic Mice and Analysis of B-Cell Tolerance. *Ann. Rev. Immunol.* 10 489–518.

Y.-J. Liu, G. D. Johnson, J. Gordon, and C. M. MacLennan. 1992. Germinal Centers in T-Cell-Dependent Antibody Response. *Immunol. Today* 13 17–21.

R. Coffman, D. Lebman, and P. Rothman. 1993. The Mechanism and Regulation of Ig Isotype Switching. *Adv. Immunol.* 54 229–270.

D. Gray. 1993. Immunological Memory. *Ann. Rev. Immunol.* 11 49–77.

S. B. Hartley, M. P. Cooke, D. A. Fulcher, A. W. Harris, S. Cory, A. Basten, and C. C. Goodnow. 1993. Elimination of Self-Reactive B Lymphocytes Proceeds in Two Stages: Arrested Development and Cell Death. *Cell* 72 325–335.

CHAPTER 8

Biosynthesis of Intercellular Messenger Peptides

MAXINE SINGER and PAUL BERG

One of the most amazing feats of evolution is the intricate coordination of cellular growth, organization, and function in multicellular organisms, a coordination that depends on the sending and receiving of myriad intercellular signals. Although direct contacts between cells are important, much intercellular intercourse is accomplished by chemical messengers. These communication networks function across great distances such as those between the brain and peripheral muscles and also at very close quarters as at the synapses between nerve cells. Many different kinds of molecules serve as messengers, and many of them are polypeptides.

Table 8.1 *Some Messenger Peptides*

Peptide	Type	Number* of Amino Acids	Site of Synthesis	Primary Target
α-Pheromone	Mating factor	13	*S. cerevisiae*	a-Type Yeast Cells
Mellitin	Toxin	26	Bees	In Venom
Insulin	Hormone	21 plus 30	Pancreatic β-cells	Many cell types
Adrenocorticotropin	Hormone	39	Anterior pituitary	Adrenal cortex
Oxytocin	Hormone	9	Hypothalamus	Uterus
Vasopressin	Hormone, neurotransmitter	9	Hypothalamus	Pituitary, blood vessel, kidney
Enkephalin	Opioid	5	Nerve cells	Nerve cells
β-Endorphin	Opioid	31	Pituitary, hypothalamus	Not yet clear
Epidermal growth factor	Growth factor	53	Salivary glands and other cells	Many cell types
Somatomedins	Growth factor	70	Liver	Many

*The size and sequence of these molecules is very similar but not necessarily identical from one species to another.

Historically, peptides that circulate in the blood and coordinate cellular functions across long distances were designated "endocrine hormones," and those active over short distances, like those separating synapses between nerve cells, were characterized as "neurotransmitters." The two were studied by different kinds of biologists. More recently, important similarities between the two kinds of **messenger peptides** became apparent, and many peptides that probably serve both kinds of function were recognized. Thus, the distinction between peptide hormones and peptide neurotransmitters is fading.

Besides the classical peptide hormones and neurotransmitters, various factors that stimulate cell growth are also intercellular messengers, as are the opioid peptides. Table 8.1 lists some of the more familiar peptide messengers of vertebrates as well as a few that occur in invertebrates and fungi. The latter include, for example, the mellitin of honeybees, the bag cell peptides of snails, and the pheromones of yeast. Some of the familiar peptide messengers probably have behavioral effects in addition to their well-known physiological consequences: vasopressin, for example. Still others, such as the bag cell peptides, seem to be specifically involved in eliciting inherent behavioral patterns. Altogether, there is a panoply of messenger peptides whose combined effects allow complex collections of cells to function as a single organism or single-cell eukaryotes to mate.

The primary translation products of genes encoding messenger peptides are much longer than the peptides themselves and often contain sequences for several active peptides. These translation products, the **preproproteins**, are modified, processed, and transported intracellularly in complex ways that eventually produce biologically active peptides and release them at appropriate times into the cell's environment.

8.1 The Special Properties of Intercellular Messenger Peptides and Their Genes

a. The Communications Network

Regardless of whether they function over long or short distances, most messenger peptides act in a fundamentally similar way (Figure 8.1). First, the peptide is

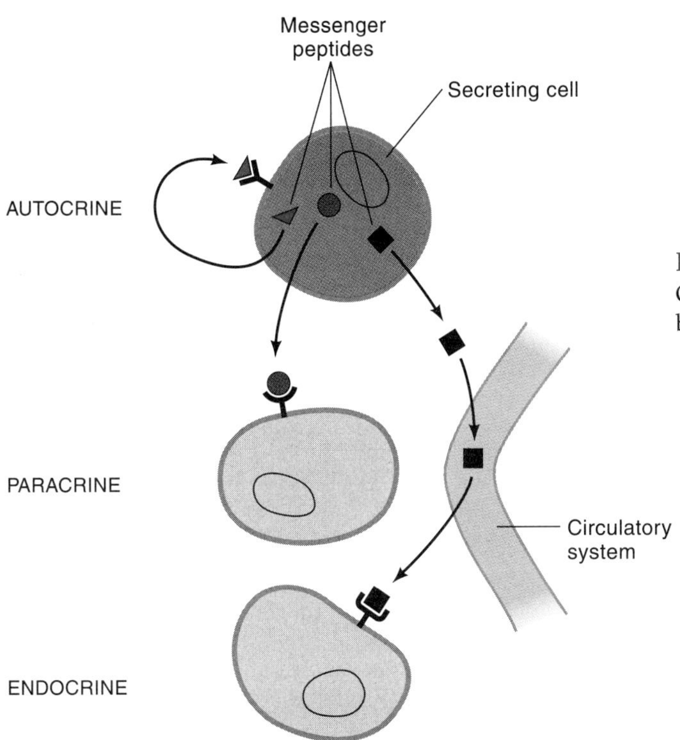

Figure 8.1
Overall scheme for intercellular communication by messenger peptides.

synthesized in a specific cell type, usually in response to a specific extra- or intracellular signal; sometimes the signal is provided by another chemical messenger. Next, the peptide is secreted by its cell of origin. It may then travel a short distance to affect a neighboring cell (a **paracrine** system, also referred to as an endocrine synapse) or even the cell that secreted it (an **autocrine** system, also referred to as an endocrine presynaptic receptor), or it may enter the circulatory system, ultimately to affect cells in distant organs (an **endocrine** system).

Cells prepared to receive a particular message do so by an interaction between the messenger peptide and a matching receptor on the cell surface. A cell that lacks a receptor specific for a given messenger peptide is not in the communications network defined by that peptide. For example, adrenocorticotropin is synthesized in the anterior pituitary and then circulates in the blood. Although all cells are bathed in the hormone, many are unaffected because they lack the correct receptors. Specific receptors do, however, occur on the cells of the adrenal cortex, the target of adrenocorticotropin; the hormone is bound, binding releases an intracellular signal, and the adrenal cortex cell responds by synthesizing cortisol. In this way, the messenger elicits a precise response from a particular cell type. The same peptide may also find an appropriate receptor on quite a different cell type, and the response of the second cell type may be distinctive. Adrenocorticotropin, for example, is also "sensed" by fat cells, which respond by releasing stored fatty acids. Thus, there is a coordinated response to the release of a particular hormone, and many distant and distinctive cell types are brought into a single network. Additional levels of coordination can occur if more than one hormone is synchronously released.

When messenger peptides work over short distances, another factor contributes to the specificity of response, namely, the proximity of the recipient cell. Coordination in nervous tissue is thus governed by the topology of the network of nerve cells as well as by receptor specificity. Neurotransmitters tend to be relatively rapidly degraded, thereby limiting their ability to affect distant cells. One consequence of this is that the same neurotransmitter may be used for communication between different, geographically isolated sets of neurons.

b. Synthesis of Messenger Peptides

Genes encoding messenger peptides Some messenger peptides are quite short. Thyrotropin releasing hormone (TRH), for example, is a tripeptide, and the enkephalins are only five amino acids long. In principle, then, these peptides might be synthesized directly from single amino acids by a series of enzymatically catalyzed condensations; the tripeptide glutathione and the *Bacillus brevis* pentapeptide antibiotic gramicidin S are constructed in just that way. However, all neuroendocrine messenger peptides known are instead encoded by genes and synthesized through the standard route involving mRNA and ribosomes. As a general rule, the genes for messenger peptides encode long preproproteins. Although inactive themselves, preproproteins yield active neuroendocrine peptides upon proteolytic processing.

Each of the genes encoding preproproteins is associated with sequences that are critical for gene transcription in a specific cell type, at a particular time, or in response to another signal (*Genes and Genomes,* chapter 8). Some of the genes include alternative polyadenylation sites or allow for alternative splicing. Thus, a particular gene may yield quite different messenger peptides in different cell types.

The primary translation products and their processing The structure of a preproprotein is made up of several elements (Figure 8.2). The "pre" portion is an

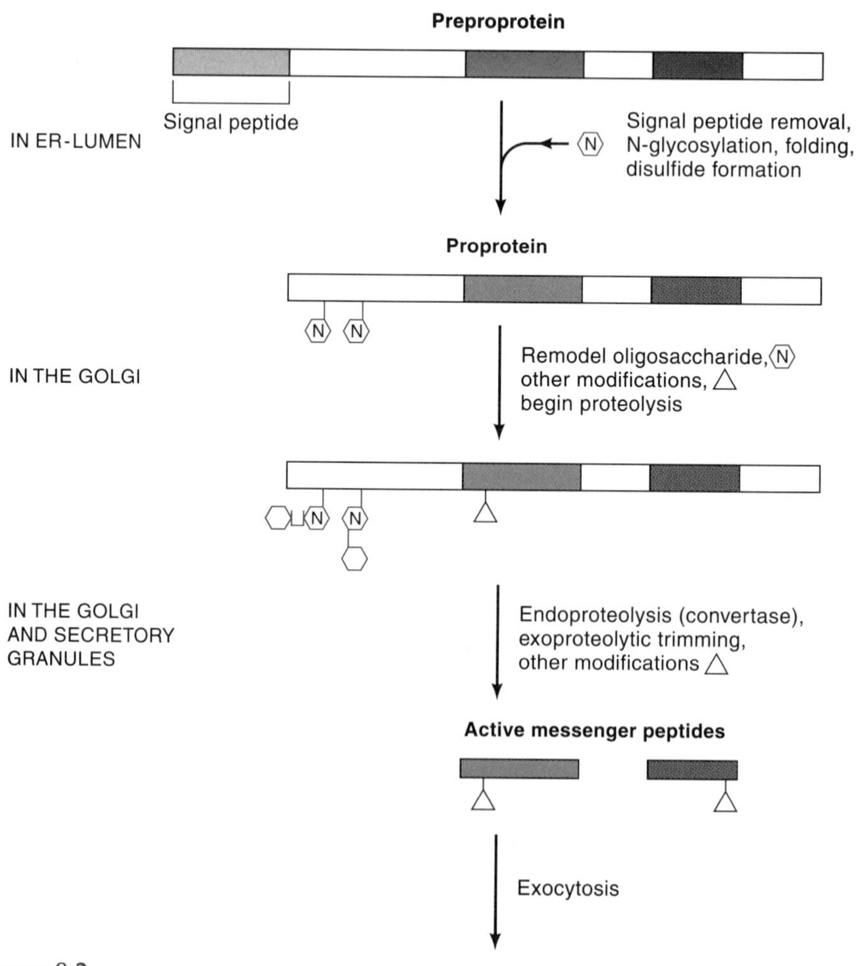

Figure 8.2
Preproteins are modified in various ways and processed by proteolysis to yield active messenger peptides. Each step in the pathway takes place in a particular intracellular site.

amino terminal signal peptide, a feature common to nearly all proteins destined to be secreted or targeted to special intracellular membrane compartments. As described in *Genes and Genomes,* section 3.10b, the signal peptide, in conjunction with the signal recognition particle (SRP), directs the cotranslational insertion of the nascent polypeptide into the membrane of the endoplasmic reticulum (ER). The subsequent removal of the approximately 20 amino acid long signal peptide in the lumen of the ER by a specific endopeptidase yields a **proprotein** and is the first step in converting the preproprotein to an active messenger peptide. In a few instances, an internal amino acid sequence rather than an amino terminal segment acts as a signal to direct a nascent polypeptide to the ER without an accompanying cleavage.

Messenger peptides are embedded in the longer proprotein precursors, which also contain polypeptide regions that will not be included in the active peptides, the "pro" segment(s). Thus, proproteins are subject to proteolytic processing and usually contain special amino acid sequences that are required for specific proteolysis. For example, "pro" segments that are amino terminal, as many of them are, tend to be hydrophobic and end in a sequence that provides a signal for cleavage of the "pro" segment from the rest of the polypeptide. In addition, proproteins often include short amino acid sequences that are targets for covalent modifications such as N-glycosylation, amidation, acetylation, phosphorylation, and sulfation.

Proteolytic processing is required to produce active peptides from a proprotein. In general, processing includes endoproteolytic cleavages to yield relatively small peptides followed by exoproteolysis to trim amino acids from the ends of the immediate precursors of the messenger peptides. The messenger peptide may represent only a very small part of the proprotein, and one proprotein frequently produces more than one active messenger peptide; such a proprotein is termed a **polyprotein**.

The multiple posttranslational events required to form a functional messenger peptide provide regulatory opportunities. Active messenger peptides are not produced, for example, unless the correct processing enzymes are available. Different processing enzymes sometimes occur in different cell types, allowing a single proprotein to yield different active peptides, depending on its environment. Moreover, some of the intermediates formed in the process may also have biological activity. Thus, the intricate intercellular communications network is subject to very fine regulation and modulation.

The relation between processing events and intracellular location Each of the processing events associated with the conversion of a proprotein to an active messenger peptide takes place in a particular cellular location. This requires the precursor to be transported to a particular site at each step of the overall process. Such transport may depend on covalent modifications, signals contained in the primary amino acid sequence, or protein folding, any or all of which may direct the maturing polypeptide to the next specific site.

The ER is the first site most nascent preproproteins encounter (Figure 8.3). Besides the removal of the signal peptide, N-glycosylation of specific arginine residues occurs in the ER. The next stop is the Golgi apparatus; only properly folded and assembled proteins are exported from the ER to the Golgi. Vesicles that bud off from the ER carry the modified proprotein to the Golgi and there fuse with the entry face (or *cis* Golgi cisternae). During passage from the *cis* to the *trans* Golgi (or exit face) through the intermediate Golgi cisternae, additional covalent modifications occur, and further proteolytic processing begins. Precursors or already mature peptides are packaged into secretory granules (also called secretory vesicles) as they exit from the *trans* Golgi. Additional proteolytic processing and covalent modifications can occur in the vesicles. Once they are in their mature form, messenger peptides are transported in the vesicles to appropriate positions within the cell. There they may be stored and released only upon

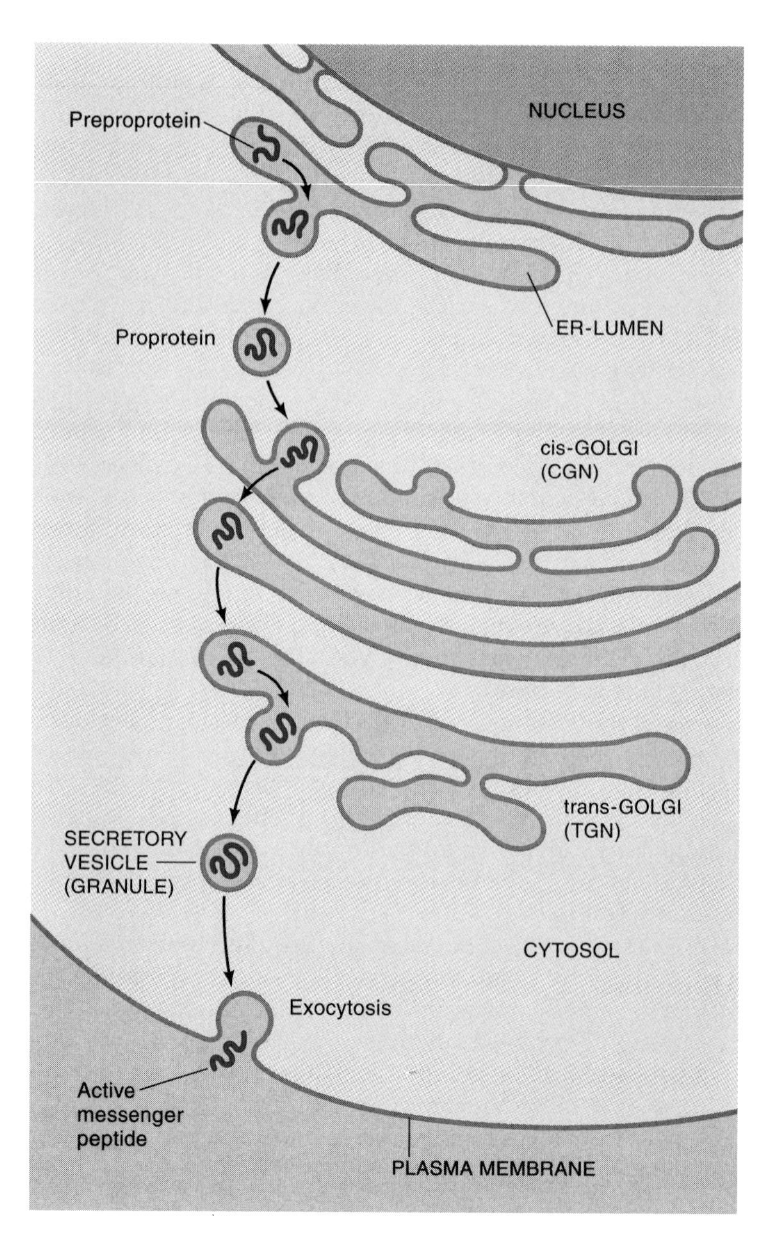

Figure 8.3
Schematic drawing showing the intracellular sites through which a preproprotein must travel to produce a secreted messenger peptide (not to scale).

receipt of an appropriate signal. For example, in peptidergic neurons, the secretory granules move down the axon in preparation for release at the synapse.

Regulated and constitutive secretory pathways Storage of messenger peptides in secretory granules permits control of the time of release. The maturation process for messenger peptides destined for controlled secretion is termed the **regulated secretory pathway.** This pathway is distinguished from the **constitutive secretory pathway** by which other polypeptides, destined for direct transport to the plasma membrane either for immediate secretion or for insertion into the membrane (e.g., cell surface receptors), are processed. These latter molecules are also often cleaved out of protein precursors during their maturation. However, even though precursor transport from one membrane compartment to another as well as the processing events are similar in both secretory pathways, the constitutive secretory pathway is active in all cell types, but regulated secretion occurs only in specialized cells. This implies that the particular proteins and enzymes required for regulated secretion will have a highly specific distribution among cell types. Indeed, the messenger peptides discussed in this chapter are synthesized specifically in those cells of multicellular organisms referred to as endocrine or neuroendocrine cells. By the time messenger peptide precursors emerge in vesicles at the *trans* Golgi stack, they have been sorted away from those proteins destined for the constitutive secretory pathway or other cellular locations. Thus, different types of vesicles are associated with the constitutive and regulated secretory pathways.

8.2 The Biochemistry of Cotranslational and Posttranslational Modifications and Proteolytic Processing

a. Proteolytic Processing

Proteolysis of a proprotein is an essential step in generating active messenger peptides. This occurs in two stages. First, endoproteolytic cleavages yield polypeptides of intermediate size between the proprotein and the final peptide. Second, exopeptidases trim the intermediate size polypeptides down to the final length. The available data indicate that endoproteolysis may begin in the Golgi apparatus and continue in the secretory vesicles. However, the details vary from one proprotein to another and from one cell type to another.

Endoproteases: the convertases Identification of the enzymes that catalyze endoproteolysis of proproteins was not easy. The activities are present in low concentration, and other enzymes, with similar specificities, occur. Purification of the specific endoproteases is, in principle, more clear-cut if subcellular organelles such as the Golgi complex or secretory vesicles are first purified. However, obtaining the organelles free from contamination is a challenge; for example, minor contamination with material from lysosomes, which contain abundant proteases, can readily obscure the **converting enzymes**—or **convertases**—as the specific enzymes are called.

The specificity of convertases was first deduced from the observed cleavage sites in the proproteins. Proprotein substrates frequently contain runs of two or more consecutive basic amino acids (e.g., Arg-Arg or Lys-Arg) or a single arginine, at which cleavage occurs. A variety of trypsinlike endoproteases with the required specificity and apparent location were identified in cells over the course of many years of work in many laboratories. Some of these convert particular proproteins to the expected messenger peptides *in vitro*. However, convincing proof that these enzymes were specific convertases, active as such in cells, was lacking. Moreover, the properties of these enzymes failed to account for the fact that, in cells, many potential proprotein cleavage sites may be left intact and that different sets of potential cleavage sites on the same proprotein may be cleaved in different cell types.

Clarification of the nature of the specific endoproteases began to emerge from work on the biosynthesis of the yeast (*S. cerevisiae*) intercellular mating factor, α-pheromone. This work illustrates the advantages of an experimental approach that combines biochemistry with the methods of classical and molecular genetics.

The Kex2 *convertase of* S. *cerevisiae* The α-pheromone is a peptide produced from the precursor, prepro-α-pheromone (Section 8.4). After removal of the signal peptide and N-glycosylation in the ER, additional processing occurs in the Golgi and secretory vesicles. *S. cerevisiae* cells with a mutation in the gene called *KEX2* do not process the precursor. Biochemical and genetic studies of the Kex2 protein demonstrated that it is a Ca^{2+}-requiring serine protease. Kex2 has, near its carboxy end, a transmembrane domain followed by a tail (Figure 8.4). The enzyme is integral to yeast intracellular membranes, and the tail protrudes into the cytoplasm; in particular, the enzyme is found in the "late" Golgi, that is, closer to the *trans* than the *cis* face. The large N terminal region of the protein includes the catalytic region and is homologous, over about 300 amino acids, to the bacterial proteases known as subtilisins. Kex2 cleaves most efficiently on the car-

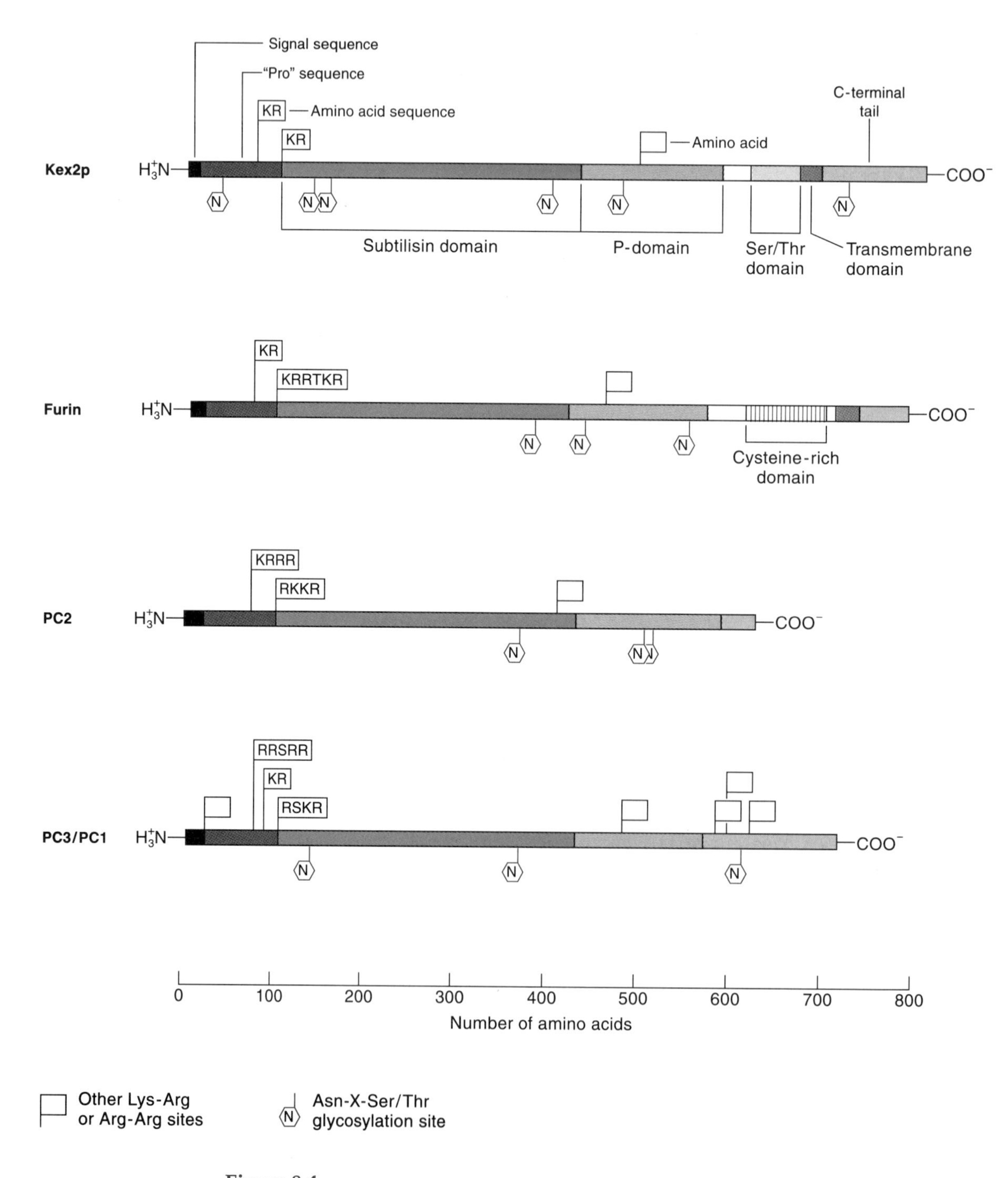

Figure 8.4
Structures of the preproproteins that are precursors to the convertases. Kex2p and furin each have a transmembrane domain and are associated with the Golgi membranes. PC3/PC1 and PC2 have no transmembrane domain.

boxy side of the second basic amino acid in a peptide of structure -X-Lys-Arg-Y or -X-Arg-Arg-Y. Moreover, when the cloned *KEX2* gene is expressed in mammalian cells, it produces a protease that can cleave mammalian proproteins to the proper messenger peptides. Thus, Kex2 appears to be a true convertase.

The discovery and characterization of Kex2 led to a search for homologous mammalian proteins. Two approaches were successful. First, sequence data banks were searched for homologous proteins (or genes). The computer searches proved fruitful. They indicated homology between the active site region of Kex2 and the predicted protein product, **furin**, of a human gene of then unknown function. The name furin had been chosen because the gene itself is located upstream of the gene for the *fes/fps* oncogene and was termed *FUR* (for *fes/fps* upstream region). Second, it was found that cells expected to carry out proprotein processing express mRNAs whose sequences encode two additional proteins with homology to the active site regions of Kex2 and, as it turned out, furin. Probes were generated from these mRNAs by the polymerase chain reaction and used to clone full-length cDNAs for the proteins (Figure 8.4). By now, the genes encoding furin, PC3/PC1, and PC2 (PC for proprotein convertase) have been studied in many organisms including mice, humans, *D. melanogaster,* and *C. elegans.*

Furin The mRNA encoding furin (also called PACE for paired basic amino acid cleaving enzyme) is widely distributed in all mammalian cell types examined, as indicated by RNA blot analysis and *in situ* hybridization. Cells that had been transfected with an expression vector carrying furin cDNA were used for detailed analysis of the cellular localization and properties of mammalian furin. The fluorescence pattern obtained after treating the cells with a fluorescent-tagged furin antibody showed that furin is localized to the Golgi complex. Cell fractionation experiments combined with immunoblotting after gel electrophoresis under denaturing conditions confirmed the association of furin with membranes and revealed two immunoreactive polypeptides corresponding to profurin and furin. Enzymatic activity depends on removing the "pro"-sequence, a cleavage that appears to be autocatalytic (Figure 8.4); cleavage at the sequence Arg-Thr-Lys-Arg, at the carboxy end of the pro segment, is consistent with data demonstrating that furin cleaves other substrates preferentially after Arg-X-Lys(or Arg)-Arg, where X can be a variety of amino acids. Furin is a Ca^{2+}-dependent enzyme, as is Kex2.

Two properties of furin suggest that it is not necessarily essential to the synthesis of all messenger peptides. First, although furin is found in many cell types, convertases specific for the regulated secretory pathway might be expected to be found exclusively in endocrine or neuroendocrine cells. Second, proproteins that are handled by the regulated secretory pathway are frequently cleaved after simple pairs of basic amino acids such as Lys-Arg. In contrast, proproteins processed by the constitutive secretory pathway are frequently cleaved after the Arg-X-Lys(or Arg)-Arg that furin prefers. Thus, the weight of current evidence suggests that furin may participate in the constitutive secretory pathway, but not necessarily in the regulated secretory pathway. Proteins such as plasma membrane receptor precursors, for example, the insulin receptor, are among the likely substrates for furin in the constitutive pathway.

PC3/PC1 and PC2 All available data indicate that PC3/PC1 and PC2 play major roles in the formation of messenger peptides in the regulated secretory pathway. Thus, in contrast to the distribution of furin, PC3/PC1 and PC2 appear to be localized to the neuroendocrine and endocrine cells that secrete messenger peptides through the regulated secretory pathway. Messenger RNAs for these endoproteases are found in such cells by *in situ* hybridization and in cell extracts by RNA blotting, but not in cells that have only the constitutive secretory pathway.

These proteases have no transmembrane domain (Figure 8.4) and are localized primarily in secretory vesicles. Depending on the particular cell type, PC3/PC1 or PC2 may predominate. Both enzymes hydrolyze polypeptides on the carboxy side of a pair of basic amino acids, either Lys-Arg or Arg-Arg. The surrounding amino acids are of some importance, and, depending on what they are, the rate of hydrolysis can be diminished or the sequence can be resistant to cleavage. PC3/PC1 and PC2 appear to hydrolyze at somewhat different efficiencies, depending on the particular cleavage site. Cotransfection of an expression vector containing either PC3/PC1 or PC2 cDNA with one expressing a preproprotein permits cleavage of the preproprotein to the expected messenger peptides, even in cells that do not normally process messenger peptides and do not contain the regulated secretory pathway.

Genes encoding prohormone convertases and the proteins they predict *KEX*2, *FUR*, and the genes encoding mammalian PC3/PC1 and PC2 belong to a gene family and encode related proteins (Figure 8.4); the human genes are all on different chromosomes. Each of the genes predicts a preproprotein with an amino terminal signal sequence; thus, each polypeptide should be cotranslationally transported into the ER and from there into the Golgi and, in some instances, the secretory granules. Although furin contains a transmembrane domain, neither PC3/PC1 nor PC2 appears to have one; they do have a carboxy terminal amphipathic helical domain that may help direct them to the secretory granules. Processing of the preproconvertases, including removal of the "pro"-segments, provides another junction at which regulation of the secretory process may occur. Moreover, the occurrence of the enzymes in specific cell types is probably at least in part a consequence of regulation at the level of transcription, as has been demonstrated for *PC*2 (human).

Other endoproteases In some instances, cleavage of a proprotein may occur after a single basic residue instead of a stretch of two or more basic amino acids. In such cases, cleavage of proproteins may be catalyzed by trypsinlike enzymes such as the kallikreins or by aspartyl proteases. Kallikreins are abundant in secretory granules in, for example, submaxillary glands and have been identified in mouse and rat intermediate pituitary glands. Renin, a pepsinlike aspartyl protease, is also abundant in submaxillary glands; it is a secreted protein and is formed from a preprorenin by a mechanism similar to that used in the biosynthesis of insulin (Section 8.5b). Renin plays an important role in a cascade of proteolytic reactions important for regulating blood pressure.

Exoproteases The final trimming of precursors to yield active messenger peptides occurs in secretory granules and is catalyzed by specific enzymes such as dipeptidyl amino peptidase and carboxypeptidase. The exoproteolytic dipeptidyl amino peptidases are integral membrane proteins within the secretory pathway and can trim peptides at their amino terminal ends.

Although the carboxypeptidases isolated from secretory granules of different mammalian tissues appeared to be distinctive and were given different names, they are likely all encoded by a single gene. In recognition of this, the name **carboxypeptidase H** (or **CP-H**) is now widely used. CP-H is characterized, regardless of cell type, by its location in secretory granules and its biochemical properties. Both soluble and membrane-bound forms of CP-H occur in secretory granules, but after solubilization by detergent in the presence of high salt concentrations, the properties of the membrane-bound and soluble forms are similar. Generally, there is two to three times as much membrane-bound as soluble enzyme. Trimming of peptide precursors by CP-H generally removes the carboxy terminal basic amino acids that remain after convertase cleavage. The enzyme encoded by the yeast *KEX*1 gene appears to carry out a similar function in the maturation of α-pheromone.

b. Glycosylation

Proproteins are often glycosylated at asparagine residues in the context Asn-X-Ser(or Thr), where X can be any amino acid: **N-glycosylation**. Some proproteins, for example, the precursor to human erythropoietin, may also undergo **O-glycosylation** at serine or threonine residues. The two kinds of glycosylation, besides being biochemically distinctive, also occur at different locations in the secretory pathway.

N-glycosylation The first step in N-glycosylation of nascent polypeptides is the transfer of a complex, branched oligosaccharide from the lipid dolichol pyrophosphate to an asparagine residue (*Genes and Genomes,* figure 3.63). The reaction is catalyzed by N-oligosaccharyl transferase, an enzyme located in the ER lumen. This N-glycosylation is a cotranslational modification, occurring as the polypeptides enter the ER lumen. Later, the N-linked oligosaccharide will be remodeled by a stepwise series of specific glycosyl hydrolases and glycosylation reactions that occur either in the ER or as the polypeptide passes through the Golgi. Different proteins acquire different final N-linked oligosaccharides, depending on their primary and secondary structures and the presence or absence of specific oligosaccharide processing enzymes in the cell carrying out the synthesis. A specific inhibitor of N-glycosylation, tunicamycin, has been used to study the effect of N-glycosylation on the activity of normally N-glycosylated polypeptides.

Purified N-oligosaccharyl transferase (from dog pancreas) has three subunits. Two of these had been previously characterized as ER-localized proteins of unknown function and called ribophorin I and II. The ribophorins as well as the third subunit have typical "membrane spanning" domains that apparently anchor them in the ER membrane so that the bulk of the protein is within the lumen and a smaller portion is in the cytosol. Considerable evidence suggests that the cytosolic portion is intimately associated with the synthesizing ribosomes on the surface of the ER. Thus, there is a one-to-one molar ratio of ribophorins and ribosomes, the ribophorins and ribosomes can be chemically cross-linked, and antibodies to the ribophorin cytosolic tails appear to block the targeting of ribosomes to the ER.

O-glycosylation Unlike N-glycosylation, O-glycosylation of serines or threonines occurs after polypeptide synthesis is complete and mainly after the polypeptide reaches the Golgi. In the first step, UDP-N-acetylgalactosamine donates its N-acetylgalactosamine to the hydroxyl of the amino acid. Subsequent steps add different monosaccharides. These steps were studied in connection with the biosynthesis of proteoglycans and mucins, and their applicability to peptide messenger synthesis is now being established.

c. α-Amidation

Many messenger peptides require a carboxy terminal α-amide for biological activity (Figure 8.5). Amidation involves neither free NH_3 nor transamidation. Rather, the amide group is derived from a glycine that abuts the carboxy end of the messenger peptide segment in the proprotein. Immediately following the glycine at the carboxy side is at least one basic amino acid. Thus, the sequence X-Gly-Arg or X-Gly-Lys in a proprotein is a possible site of amidation. The proprotein precursor of the messenger peptide generally requires prior endoproteolytic cleavage and exoproteolytic trimming to provide the carboxy terminal glycine. Like many posttranslational modifications of messenger peptides, the amidation reaction appears to be an ancient process; the bee hormone mellitin as well as many vertebrate messenger peptides all acquire their carboxy terminal amides in the same way.

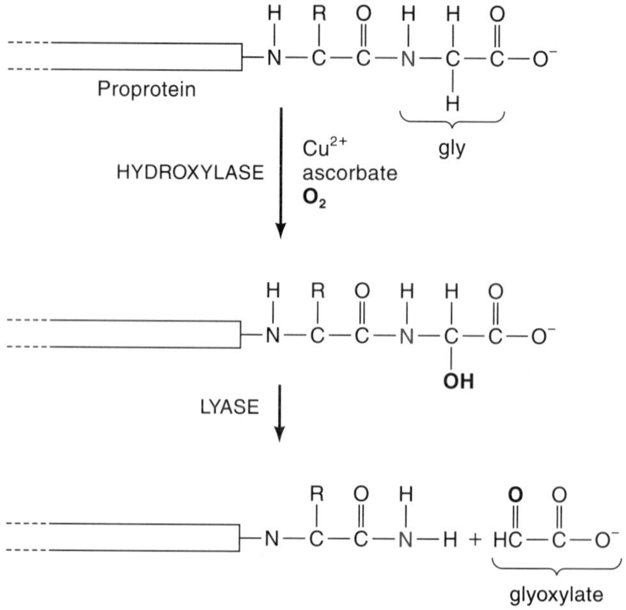

Proprotein

HYDROXYLASE Cu^{2+} ascorbate O_2

gly

LYASE

glyoxylate

Figure 8.5
The α-amidation of the carboxy terminus of messenger peptides. The amide nitrogen is derived from the glycine residue at the carboxy terminus in the proprotein or intermediate polypeptide.

Enzymatic mechanism α-Amidation results from the action of an enzyme called **peptidylglycine α-amidating monooxygenase** (abbreviated **PAM**) or **α-amidating enzyme**, encoded by a single gene in mammalian species. The overall reaction is illustrated in Figure 8.5 and has two discrete steps. In the first step, the terminal glycine is oxidized to peptidyl-α-hydroxyglycine with the involvement of ascorbate and O_2 and a requirement for Cu^{2+}; additional products are dehydroascorbate and water. In the second step, the peptidyl-α-hydroxyglycine, which is relatively stable at neutral pH, is cleaved to give the peptidyl-α-amide and glyoxylate. The reaction mechanism has been confirmed by stoichiometric analysis of reactants and products and by the demonstration, using isotopes, that the hydroxyl O (in the peptidyl-α-hydroxyglycine) and the amide N are derived from the O_2 and the peptide glycine, respectively.

The enzyme PAM occurs within secretory granules in a wide variety of cell types and is a bifunctional enzyme. A region in the amino terminal half of the molecule catalyzes the oxidation of the glycine, and a region in the carboxy terminal half catalyzes the lyase (or cleavage) reaction. In some types of cells, the PAM molecule itself is subject to proteolysis, thereby separating the molecule into two independent and active enzymes. Following the catalytic domains, toward the carboxy end of the PAM molecule, is a transmembrane domain that permits PAM to insert into the membrane of a secretory granule. The short carboxy terminal region of the enzyme projects into the cytoplasm. Proteolytic processing at pairs of basic amino acids can cleave off the transmembrane and cytoplasmic domains, leaving a soluble form of PAM in the secretory granules. The distribution of membrane-bound and soluble forms of the enzyme is tissue specific.

Besides the multiple forms of PAM produced by posttranslational proteolysis, additional forms are also generated by tissue-specific alternative splicing of the primary transcript of the single mammalian PAM gene. The gene has more than 26 exons and covers at least 150 kbp of DNA. One exon encodes the amino acids between the two functional regions of the enzyme and includes the basic dipeptide that is the cleavage site at which the two regions can be separated. Messenger RNAs lacking this exon yield a protein that cannot be cleaved. Moreover, because the various mRNAs yield proteins that can be differentially processed, even more forms of the active enzyme can be produced. Particular forms of the mRNA and particular posttranslational processing reactions appear to be preferred in different tissues, although the physiological significance of the many resulting forms of PAM remains to be discovered.

d. Phosphorylation

Some peptide messengers have phosphate groups on one or more serine residues. Phosphorylation is presumed to be carried out by a kinase with either ATP or GTP serving as phosphate donor, although the responsible enzyme(s) have not been identified or located within cells. A possible candidate, one of a large number of so-called **casein kinases** (because they phosphorylate casein efficiently *in vitro*), was found associated with the Golgi lumen in rat liver and mammary gland. Comparison of the phosphorylated serines in a number of peptides suggests that the phosphorylating enzyme favors serines in the context Ser-X-Glu, although this is not the preferred sequence for known casein kinases. In any case, extents of phosphorylation vary from one tissue to another, and some isolated messenger peptides are known to be mixtures of phosphorylated and unphosphorylated molecules. Clearly, a lot remains to be learned about the phosphorylation of messenger peptides.

e. Sulfation

Tyrosine residues in some active messenger peptides are sulfated, as are specific tyrosines in some other secreted and plasma membrane proteins. The reaction is catalyzed by **tyrosylprotein sulfotransferase**, and 3′-phosphoadenosine-5′-phosphosulfate donates the sulfate. The enzyme resides in the membranes of the *trans* Golgi; its catalytic site faces the Golgi lumen. Studies with synthetic peptides indicate that acidic amino acids on the amino side of a tyrosine are important for enzyme activity. Thus, the sequence Glu-Tyr-Gly is a typical sulfation site.

Appropriate tyrosine sulfation can be essential for activity of a messenger peptide. For example, all active forms of cholecystokinin are sulfated at a tyrosine residue six amino acids from the carboxy terminus. Removal of the sulfate residue reduces hormone activity by several hundred–fold. Unlike phosphorylation, sulfation does not seem to be reversible; thus, it is unlikely to regulate the activity of a messenger peptide.

f. N-Acetylation

Some, but not all, messenger peptides are acetylated at the amino terminus. Moreover, not all of a particular messenger peptide may be acetylated in any particular tissue. The effect of acetylation on biological activity is also variable, with some peptides being more and others less active when the N-acetyl group is present.

N-acetyltransferases occur in many different tissues and cell compartments, where they may modify the ε-amino groups of internal lysine residues as well as the terminal α-amino groups of a variety of polypeptides. Acetyl coenzyme A is the donor of acetyl groups. The N-acetyltransferase found in significant amounts in the secretory granules of certain cell types appears to be responsible for N-acetylation of messenger peptides; whether it is a distinctive form of the enzyme compared to the activities in other cellular compartments is unclear. It is likely that the presence or absence of the enzyme and acetyl coenzyme A controls whether or not acetylation occurs. Moreover, the sequence of the first few amino terminal amino acids of a peptide influences the efficiency of N terminal acetylation.

8.3 The Secretory Pathway

Nascent polypeptides directed to the ER by interaction of their amino terminal signal peptides with the SRP have one of several destinations: the membranes of the secretory network itself (e.g., the localization of furin in the Golgi), direct transport to the plasma membrane either for immediate secretion (the constitutive secretory pathway) or for insertion into the membrane, transport to lysosomes, or storage in secretory granules for further modifications and controlled release from the cell (the regulated secretory pathway). Almost all messenger peptides synthesized in mammalian cells follow the regulated secretory pathway. The secretion of α-pheromone by yeast cells, however, appears to proceed through a constitutive pathway with little, if any, storage of peptide for later, regulated release.

The regulated and constitutive pathways have many common steps. Consequently, pathway analysis has been facilitated by complementary biochemical investigations using cell-free preparations from yeast and mammals and the investigation of yeast mutants. Prior to the development of the biochemical and genetic approaches to the study of secretory pathways, and continuing to this day, electron microscope studies using mammalian cells reveal the general outlines of the pathway, including the structure of the rough endoplasmic reticulum, the Golgi, and the role of secretory granules. The summary given in this section only hints at the breadth and depth of the experiments. It barely indicates the novel experimental approaches used to acquire an understanding of what long seemed to be intractable problems. Studies of membranes have, in general, been experimentally challenging, and the interplay between genetics and biochemistry has been the key to unambiguous analysis.

a. Targeting to the Endoplasmic Reticulum

The mechanism whereby the SRP recognizes and directs appropriate nascent polypeptides to the endoplasmic reticulum is described in *Genes and Genomes,* section 3.10b. Therefore, we pick up the story as the ribosome attaches to the membrane through a special receptor, after which the nascent polypeptide's signal sequence enters the ER membrane, and the SRP dissociates.

Entering the ER lumen The SRP brings the nascent polypeptide and associated ribosomes in contact with the cytosolic surface of the ER by interaction with the **signal recognition particle receptor** (**SRP-R**) in the membrane. Other proteins integral to the ER membrane are also critical for binding the nascent polypeptide/SRP/ribosome complex and translocating the polypeptide across the ER membrane.

The currently favored model suggests that a cluster of several different proteins within the membrane constitutes a **translocation site**, and some of these form a functional **channel** (or pore) within the membrane. Several such proteins have been identified in *S. cerevisiae,* some of which are encoded by genes in the so-called *SEC* group; the *SEC* genes were identified by the isolation of temperature-sensitive mutant yeast cells whose phenotypes at the nonpermissive temperature suggested they were defective in general secretory processes. Proteins important for translocation were also identified in mammalian cells; these could be chemically cross-linked to nascent secretory polypeptides in the ER membranes. In several instances, functional homologies have been demonstrated between yeast and mammalian proteins; and, where the corresponding protein structures are known (usually from the sequence of cloned genes), structural conservation is also often apparent (Table 8.2). The precise order and way that these proteins together contribute to translocation is now beginning to be under-

Table 8.2 *Some Proteins of the Secretory Pathway*

Site	Yeast Protein	Mammalian Protein	Function*
ER membrane	Sec62p	NEM sensitive	Between targeting and translocation
	Sec63p	NEM sensitive	
		TRAM	As above
	Sec61p	P37 or Sec61p	Translocation channel
Inside the ER lumen	Sec11p	Signal peptidase	
	Kar2p	BiP	
		PDI	Protein disulfide isomerase
ER to Golgi	Erd2p	KDEL receptor	ER retention
	Sec21p	β-COP	Coatomer
	Sec23p	GAP	Vesicle formation
	Sar1p	ARF (Rab)	GTPase, vesicle formation
	Sec12p	GRF	Vesicle formation
	Ypt1p	Rab1	GTPase, vesicle docking
	Bet2p	Geranylgeranyl transferase	Directs association with membranes
	Sec17p	α-SNAP	Docking at Golgi
	Sec18p	NSF	Docking at Golgi
	Sec13p	G-protein	β-Subunit
Golgi to plasma membrane	Sec4p		GTPase, vesicle docking

Note: The table lists only proteins mentioned in the text. Many additional proteins important for the secretory pathway have been identified.
*See the text for more detail.

stood, but much remains to be clarified. For example, they may act in concert or in an ordered sequence of events. Moreover, several of them may have similar functions but act on specific classes of peptide precursors.

*SEC*62 and *SEC*63 encode proteins (Sec62p and Sec63p) that span the ER membrane, interact with one another (and several other proteins), and appear to function soon after targeting of the nascent polypeptide/SRP/ribosome complex to the ER membrane but before actual translocation begins (Figure 8.6). Translocation requires ATP, and the membrane channel may be formed from

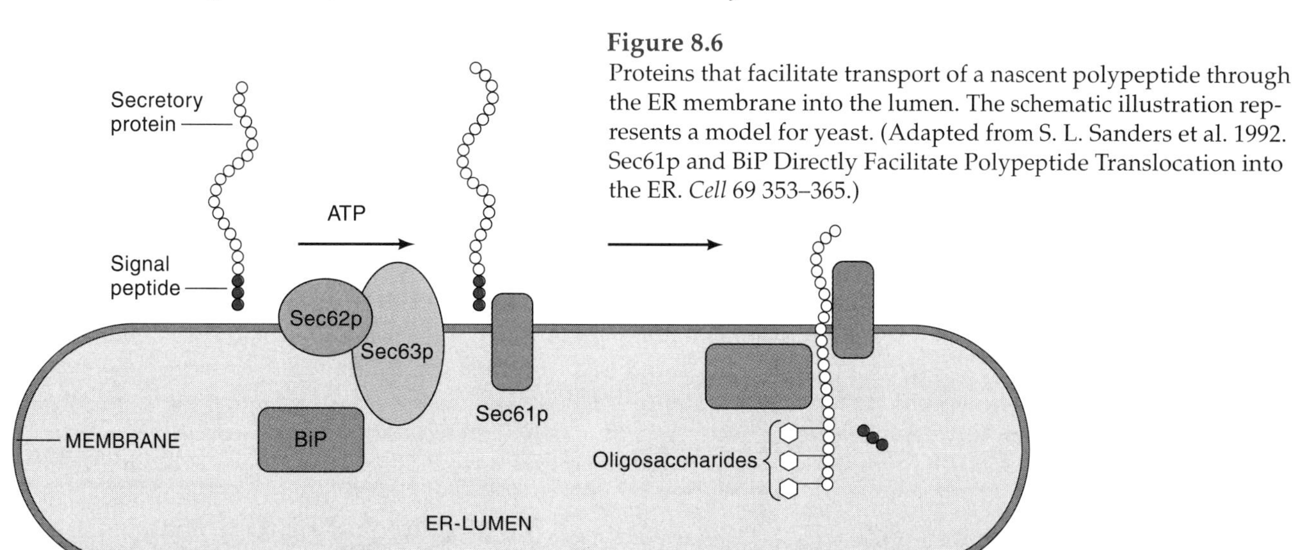

Figure 8.6
Proteins that facilitate transport of a nascent polypeptide through the ER membrane into the lumen. The schematic illustration represents a model for yeast. (Adapted from S. L. Sanders et al. 1992. Sec61p and BiP Directly Facilitate Polypeptide Translocation into the ER. *Cell* 69 353–365.)

Sec61p, which is known to be an integral membrane protein. Thus, at the nonpermissive temperature, *SEC*61 mutants show diminished processing of a variety of secretory and membrane proteins, and appropriate chemical treatment cross-links Sec61p to translocating polypeptides.

In mammalian cells, a group of integral ER membrane proteins play roles similar to those of the yeast Sec proteins. Thus, mammalian cells contain a protein with significant homology to Sec61p that binds tightly to 80S ribosomes attached to nascent polypeptides; it may form the ribosome binding site that captures the nascent polypeptide/ribosome complex when the SRP is released. Release of the nascent polypeptide from the SRP requires GTP hydrolysis and probably involves particular SRP subunits that are GTP binding proteins. This theme, the involvement of small GTP binding proteins and GTP hydrolysis, recurs each time the polypeptide moves from one membrane to another in the secretory pathway. Another mammalian protein, identified by its sensitivity to N-ethylmaleimide, may be functionally similar to the Sec62p/Sec63p complex. Another, the glycoprotein TRAM (for translocating chain-associating membrane), can be chemically cross-linked to nascent polypeptide. TRAM and the N-ethylmaleimide-sensitive protein appear to function in concert to shepherd the nascent polypeptide across the ER membrane.

Unique and definitive data have been contributed by studying synthetic lipid bilayers to which ER membranes along with bound ribosomes are fused. Such constructs allowed the direct demonstration of a translocating channel (Figure 8.7). Thus, when a translocating polypeptide plugs a channel, ions cannot pass through; if the polypeptide is released from ribosomes by the addition of puromycin and allowed to pass through the membrane, ions can then pass through the membrane, and their passage is detectable by conductance measurements. Ribosomes appear to keep the channels open because they must be retained on the membrane to observe conductance. If they are released with high salt concentrations, the channels appear to close, and ions no longer flow through.

Inside the ER lumen Nascent and completed polypeptides are modified in at least three different ways inside the ER lumen: the signal peptide is cleaved off; the polypeptides are subject to conformational modeling, including the formation of appropriate disulfide bonds; and N-glycosylation occurs. The mammalian protein BiP (for binding protein) and its yeast homologue, Kar2p, are found within the ER lumen, where they appear to have multiple possible functions, including facilitating translocation and folding the nascent protein. Thus, mutations in the yeast *KAR*2 gene, which encodes Kar2p, interfere markedly with translocation. Kar2p also interacts with Sec63p, but the significance of this is not yet clear. BiP (and Kar2p) is a member of the so-called **heat shock 70** class of proteins, also termed molecular **chaperones**. It can form complexes with a variety of polypeptides, including nascent chains in the ER lumen, and ATP is generally required for dissociation of the complex. In the ER lumen, BiP as well as other chaperone proteins are probably important for maintaining nascent chains in configurations required for correct further transport and processing.

The enzyme **protein disulfide isomerase** (**PDI**) is also within the ER lumen, and in substantial amounts. As its name implies, this enzyme catalyzes the intramolecular exchange of one disulfide bond for another. Thus, it facilitates formation of those disulfide bonds needed for folding proteins into stable and functional conformations. PDI belongs to a family of proteins, defined by the presence of one or more amino acid segments homologous to the active site of thioredoxin, the protein involved in a variety of redox reactions in all organisms. Several such proteins have been identified in the ER lumen in mammals.

N-oligosaccharyl transferase (Section 8.2b) and signal peptidase are also within the ER lumen. Both appear to be anchored in the membrane, with their catalytic sites in the lumen. One of the five subunits of the signal peptidase,

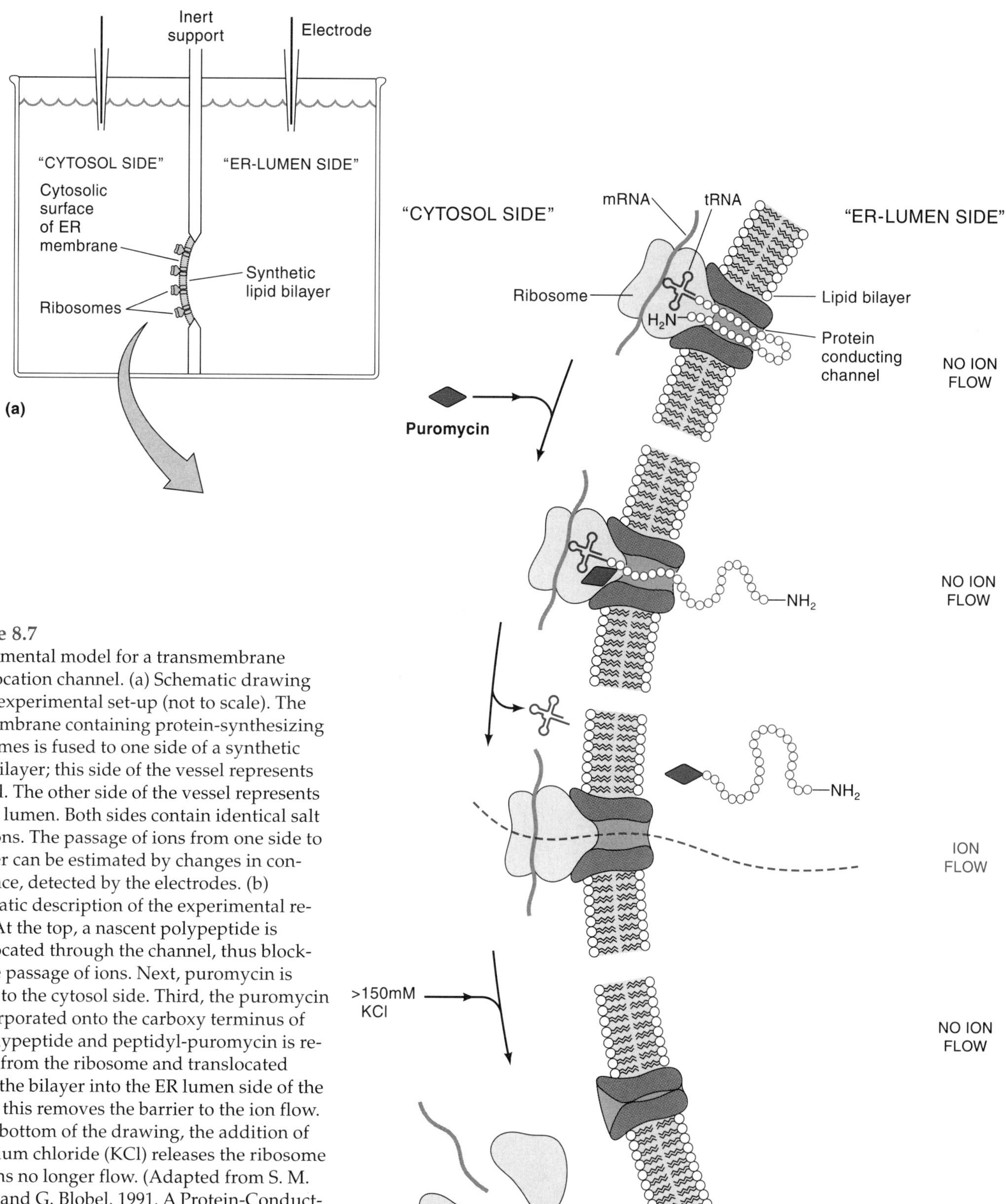

Figure 8.7
Experimental model for a transmembrane translocation channel. (a) Schematic drawing of the experimental set-up (not to scale). The ER membrane containing protein-synthesizing ribosomes is fused to one side of a synthetic lipid bilayer; this side of the vessel represents cytosol. The other side of the vessel represents the ER lumen. Both sides contain identical salt solutions. The passage of ions from one side to another can be estimated by changes in conductance, detected by the electrodes. (b) Schematic description of the experimental results. At the top, a nascent polypeptide is translocated through the channel, thus blocking the passage of ions. Next, puromycin is added to the cytosol side. Third, the puromycin is incorporated onto the carboxy terminus of the polypeptide and peptidyl-puromycin is released from the ribosome and translocated across the bilayer into the ER lumen side of the vessel; this removes the barrier to the ion flow. At the bottom of the drawing, the addition of potassium chloride (KCl) releases the ribosome and ions no longer flow. (Adapted from S. M. Simon and G. Blobel. 1991. A Protein-Conducting Channel in the Endoplasmic Reticulum. *Cell* 65 371–380.)

Sec11p, is similar in amino acid sequence to some subunits of the mammalian enzyme.

b. Transfer of Proteins from One Membrane to Another

General considerations Messenger peptide precursors, like precursors to other secreted polypeptides and the integral constituents of the Golgi and secretory granules, are transported in special vesicles from one membrane complex to another. The first such trip is from the ER to the *cis* Golgi. Subsequent transits, from one Golgi cisternae to another and from the *trans* Golgi network to the cell surface, are biochemically similar. The processes involve discriminating between those molecules to be transported and those to be retained, budding of a transport vesicle containing the precursor from the donor membrane, identifying the target system, transporting the vesicle to the specific target, and fusing with the target membranes (Figure 8.8). Although we here present the description of the

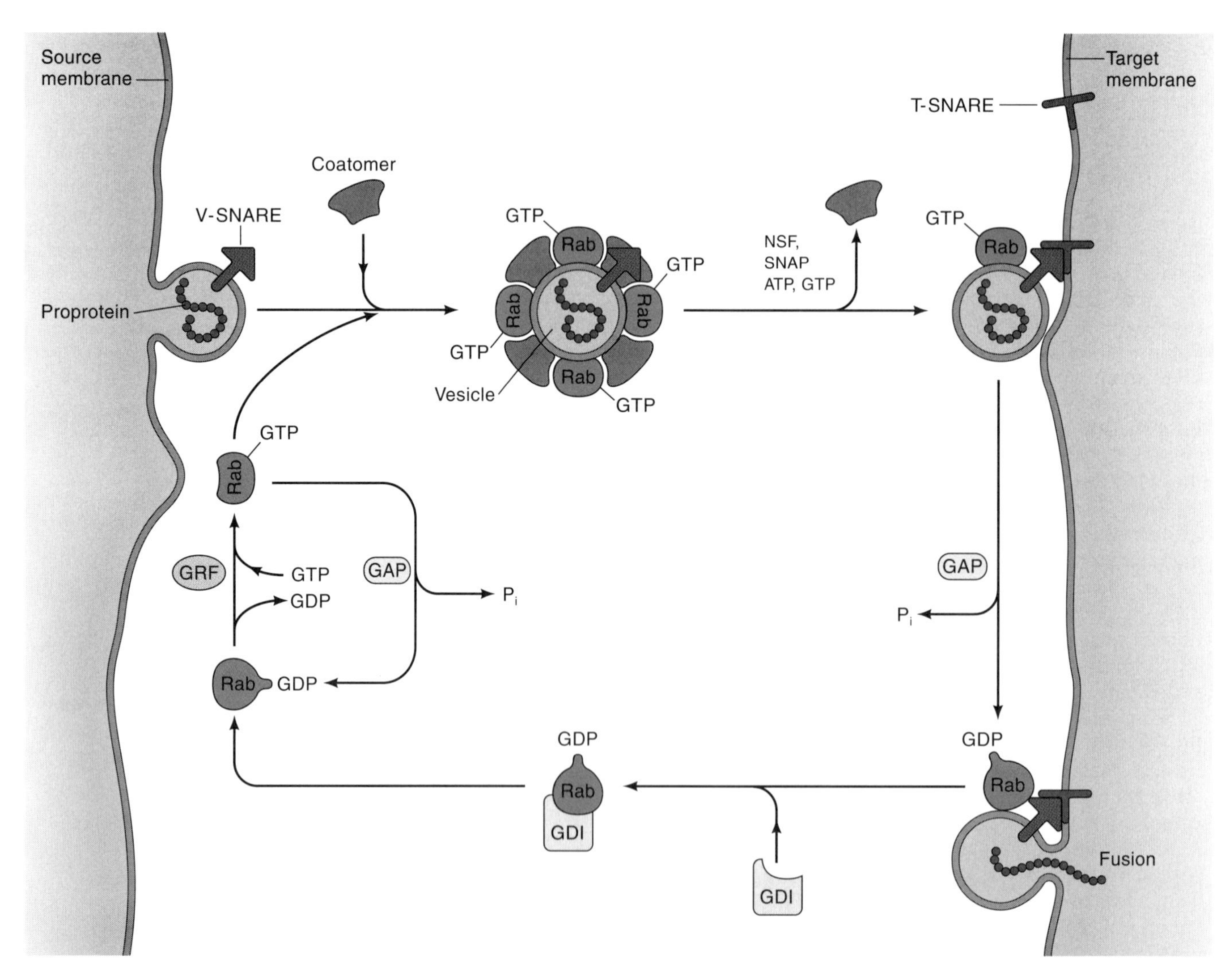

Figure 8.8
Vectorial vesicular transport. The schematic drawing outlines the general features of budding from the source membrane (e.g., the ER) and targeting and fusion to the target membrane (e.g., the *cis* Golgi). The various proteins are described in the text. Note that the term Rab is used as a general name for small GTP-binding proteins (e.g., ARF, Sar1p, etc.)

ER to *cis* Golgi transit first and in the most detail, the initial discoveries and many of the important observations were made in studies on transit through the Golgi stacks.

Selecting molecules for transport What discriminates between proteins destined to be shipped out of the ER and those to remain as residents? At least two properties of the proteins appear to be important. First, to be exported from the ER, proteins must be folded into appropriate tertiary and quarternary structures. Thus, mutant proteins containing altered amino acids that interfere with proper folding are generally retained in the ER. Second, proteins carrying a carboxy terminal tetrapeptide Lys-Asp-Glu-Leu (KDEL) in mammals or His-Asp-Glu-Leu (HDEL) in *S. cerevisiae* are marked for residence in the ER. When KDEL (or HDEL) is removed from an ER resident protein, the protein goes through the whole pathway and is secreted. Contrarily, if KDEL (or HDEL) is added to a normally secreted protein, the protein localizes in the ER. It is not clear whether KDEL (or HDEL) functions directly to retain proteins in the ER. What is clear is that when such proteins are carried to the Golgi, they are shipped back by a retrograde pathway to the ER. How is it possible to tell that a resident ER protein spent some time in the Golgi? Such proteins carry telltale modifications such as remodeled oligosaccharides and phosphates, modifications known to occur exclusively in the Golgi, even though they reside in the ER. Moreover, in yeast mutants blocked in transport from the ER to the Golgi (e.g., *Sec*18), such modifications do not occur.

The mechanism of the retrograde pathway is poorly understood. However, a yeast gene, ***ERD*2 (ER-retention defective)**, appears to encode a receptor that resides in the Golgi, binds HDEL, and can then move to the ER; a human gene encodes a Golgi protein with extensive homology to Erd2p. Besides the retrograde pathway, some proteins may be retained in specific membranes or compartments by virtue of yet uncharacterized structural features.

Budding of vesicles As pits in the ER membrane that contain the precursors to be transported begin to bud on the outside of the ER, they are coated by protein complexes (Figure 8.8). Such multisubunit 700–800 kDa **coatomers** were first discovered and characterized in mammalian Golgi, where they contain seven distinct subunits. Their counterparts in *S. cerevisiae* are also 700–800 kDa coatamer complexes containing at least six different polypeptide components, some of which are encoded by previously identified *SEC* genes. One polypeptide component of the yeast coatomer cross-reacts with antibody to β-COP, a major component of mammalian coatomers. Antibodies prepared against yeast Sec21p made in *E. coli* were used to purify Sec21p from yeast subcellular fractions and to localize the 105 kDa Sec21p in the coatomer. Moreover, yeast cells with a mutant *Sec*21 gene are unable to accumulate stable buds, and introduction of a cloned, wild-type *SEC*21 gene into such cells restores their ability to accumulate stable buds.

Budding from the ER (and, consequently, secretion) is also inhibited by mutations in the yeast *SAR*1 gene. *In vitro* experiments demonstrated that interaction between Sar1p and Sec12p, which occurs within the ER membrane, is required for vesicle formation at the ER membrane; these proteins may help define and organize the budding site. Sar1p (21 kDa) is one of a large number of ubiquitous, small GTPases (often referred to as small GTP-binding proteins) similar to the prototypical mammalian Ras protein (Chapter 4). Related GTPases are essential in mammalian secretory pathways; they are known by a bewildering variety of names such as **Rab** and **ARF** (**ADP-ribosylation factor**). Biochemical fractionation of cellular components and electron microscopy in conjunction with the use of specific antibodies have localized different small, homologous, monomeric GTP-binding proteins to specific cellular locations. Sar1p and ARF (ARF homologues occur in yeast) seem to be involved in the budding of vesicles, and Rab proteins participate in the interaction between a vesicle and its target membrane. The yeast proteins Ypt1p and Sec4p, members of the Rab protein family, are involved in targeting vesicles to a recipient membrane. Sec4p is associated with vesicles that have already left the Golgi and will fuse with the plasma membrane. The mammalian homologue of Ypt1p, Rab1, is associated with the ER, and the related Rab3a is located on synaptic vesicles.

How the small GTPases act in the secretory pathway is not precisely clear, but we know a good deal about their properties. Generally, they are 21–25 kDa in size. They bind GTP and, in the presence of a **GTPase-activating protein** (**GAP**), hydrolyze the GTP to GDP. Thereafter, they can exchange the GDP for another molecule of GTP, a reaction facilitated by another group of proteins, the **guanine nucleotide releasing factors** (**GRFs**). There are significant differences in the conformation of the small GTP-binding proteins, depending on whether GTP or GDP is bound. The active forms of the well-characterized members of this group of proteins are those bound to GTP; in that form, they appear to catalyze the fusion of vesicle and target membranes. They are capable of specific interactions with additional proteins, usually effectors of additional biochemical reactions.

The importance of the GTPases to secretion is consistent with the fact that GTP hydrolysis is required at many steps in the secretory pathway, including the formation and docking of vesicles at various membranes. At each step, a different GTP-binding protein appears to function. Thus, GTP hydrolysis is associated with the role of Sar1p during vesicle formation at the ER membrane and with the role of Ypt1p (the Rab 1 homologue) and other similar proteins during targeting or fusion to the *cis* Golgi. Moreover, presence of the nonhydrolyzable GTP analogue GTP-γ-S leads to the accumulation of filled transport vesicles that are unable to dock at the next membrane in the pathway. Similarly, vesicles accumulate rather than fuse with their target membranes in the presence of antibodies to a Rab protein and at elevated temperatures in yeast mutants with temperature-sensitive genes for Rab homologues.

For those GTPases that function upon interaction of a vesicle with its target membrane, GTP hydrolysis may accompany target recognition and/or binding, fusion of the two membranes, or release of the GTPase for recycling; the details are currently being studied. The resulting protein, now inactive and bound to GDP, is retrieved from the membrane by a cytosolic protein called GDI (guanine nucleotide displacement inhibitor). It is returned to its organelle of origin in a process that couples exchange of its GDP for GTP (Figure 8.8). Recent experiments indicate that besides the small, monomeric, GTP-binding proteins, other guanine nucleotide–binding proteins, the **heterotrimeric G proteins**, also play important roles at various steps in the secretory pathway. One possibility is that the heterotrimeric G proteins are GRFs, facilitating the dissociation of GDP from the monomeric GTPases and thus the exchange of GDP for GTP to regenerate active monomeric protein.

The GTP-binding proteins (GTPases) must be directed to distinct membrane compartments. Their association with membranes depends on the presence of lipid moieties added posttranslationally. Thus, some of the GTP-binding proteins contain geranylgeranyl groups (20 carbon isoprenoid derivatives) in thioether linkage to cysteine near the carboxy terminus. Others, such as the ARF proteins, are myristylated at their amino termini, another modification known to direct proteins to membranes. Experiments in which the carboxy terminal regions of several Rab proteins were exchanged, by exchanging the 3′ sequences of their cloned genes, demonstrated that association of the proteins with specific membranes is directed by this carboxy end of the polypeptide.

c. Transport from the Endoplasmic Reticulum to the Golgi

Fusion of vesicles with the cis *Golgi* The picture derived from mammalian cell biology suggests that the "docking" of the coated vesicle on the target Golgi membrane is facilitated by a multicomponent complex including the protein **NSF (N-ethylmaleimide-sensitive fraction)** and several **SNAPs (soluble NSF attachment proteins)**. This role of NSF/SNAP in membrane targeting was first discovered in studies of intra-Golgi vesicle transport. The proteins are now known to be necessary for the fusion of ER-derived vesicles to the Golgi and also for fusion to other target membranes at various points in the general secretory pathway. The SNAPs bind to a receptor in the target membrane and perhaps to another in the vesicle membrane, and NSF binds to the SNAPs. In a reconstituted system, efficient docking is inhibited by N-ethylmaleimide, and unbound vesicles accumulate. The protein receptors in the vesicle and target membranes are called v-SNAREs and t-SNAREs (for SNAP receptors), respectively, and their role in intracellular membrane targeting and fusion is beginning to be studied in detail. Interaction between "matching" v-SNAREs and t-SNAREs plays a role in assuring that vesicles dock at the appropriate membrane. Earlier work on proteins involved in the fusion of synaptic vesicles with synaptic plasma membranes identified proteins that are now recognized as prototypical SNAREs: VAMPs or synaptobrevins for v-SNAREs and syntaxins for t-SNAREs. The complex formed at the Golgi membrane facilitates membrane/membrane fusion but in as yet poorly understood ways.

Certain yeast mutations are different from those already described in that the cell cytoplasms accumulate 50 nm vesicles, unattached to any membrane, just as vesicles accumulate in the reconstituted system in the presence of N-ethylmaleimide. All available evidence indicates that these are vesicles that bud off the ER but are unable to fuse to the *cis* Golgi. Two such mutations, in the genes *SEC*18 and *SEC*17, are known, for example, to encode homologues of the mammalian proteins NSF and α-SNAP, respectively; and Sec18p can substitute for NSF in reconstituted, *in vitro* mammalian systems. Similarly, certain yeast mutants identify proteins that are likely to function as v-SNAREs and t-SNAREs.

The coatomer is generally shed when the transport vesicle initiates its attachment to the recipient membrane, leaving an uncoated particle to undergo fusion. Hydrolysis of the GTP by the ARF (or other Rab) on the transport vesicle probably also occurs in association with attachment. The coatomer and GTP-binding protein are thereafter presumably recycled. Fusion itself seems to be accompanied (or preceded) by release of the NSF/SNAP/receptor complex and hydrolysis of the ATP bound to NSF.

Within the Golgi Once inside the Golgi, peptide precursors undergo further modifications, including the beginning of proprotein proteolysis and remodeling of the core oligosaccharides. Those modifications may take place in one or more of the Golgi compartments as the precursor makes its way from the *cis* Golgi, between cisternae, into the *trans* Golgi compartment, thence into secretory granules. Transport of the precursor from each subcompartment to the next is mediated by transport vesicles much like those constructed for the trip from the ER to the Golgi. Indeed, most of the coat and particle components already described, or homologous and analogous proteins, function in similar ways in these later steps in the pathway. For instance, ARF and the COP proteins are the only proteins needed to form the vesicle coats required for transport in the Golgi. Uncoating of the vesicle in preparation for fusion with the target membrane is associated with hydrolysis of the GTP bound to ARF.

The *trans* Golgi appears to be the site at which precursors are sorted among different vesicles and thereby for different fates: regulated or constitutive secretory pathways, lysosomes, or plasma membrane. Until this point in the pathway, the "default" path has been to proceed to the next step; only proteins destined for positions in the membranes or lumens of the pathway itself are previously diverted. Most of the messenger peptides discussed in this chapter are destined for regulated secretion. As such, they leave the *trans* Golgi in secretory vesicles often also called secretory granules. One exception is the yeast α-pheromone, which appears to follow a constitutive pathway (Section 8.4)

d. Incorporation of Proteins into Secretory Granules and Exocytosis

Sorting The sorting of secreted messenger peptides from other proteins in the secretory pathway, such as those destined for constitutive secretion or retention in the Golgi, appears to start in the *trans* Golgi as the peptides are packaged into secretory granules for storage and regulated release. The signals specifying sorting and packaging are not understood. It is widely believed that the properties of the messenger peptide precursors are such that they can selectively aggregate under the low pH and high Ca^{2+} conditions prevailing in the *trans* Golgi and secretory granules and that such aggregation is required for incorporation into granules. The proteins become densely packed within the secretory granules themselves, making these granules relatively distinctive (and facilitating their identification and purification). Thus, the granules containing messenger peptides differ from the synaptic vesicles containing nonpeptide neurotransmitters, although both are storage vesicles for regulated secretion.

Secretory granules The secretory granules that form at the *trans* Golgi membrane differ from the coatomer-coated transport vesicles that function in earlier steps in the secretory process. Moreover, the granules that package peptides for regulated secretion are different from those that package materials for other destinations or nonpeptide neurotransmitters. One clear difference is that those granules that package messenger peptide precursors also contain the enzymes required for the final steps in messenger peptide maturation. However, there is no definitive description of the packaging events at the *trans* Golgi; this topic is being actively investigated and is the subject of some controversy.

The earliest granules packaged at the *trans* Golgi, what might be called immature granules, may have one structure, and the mature granules, those ready to release their peptide contents, another. The immature granules may have coats constructed from **clathrin**, a protein first discovered in association with the plasma membrane, where it forms the cagelike lattice found in the coated pits involved in endocytosis (clathrate means a latticelike structure and is derived from a Latin word meaning lattice). Immunocytochemical techniques suggest the presence of clathrin at the *trans* Golgi. Each clathrin unit contains three **clathrin heavy chains** of 180 kDa, each associated with one of several different **clathrin light chains** of 30–40 kDa.

The proteins called **adaptins**, which link the clathrin cage to the membrane and facilitate the assembly of the clathrin units, may play an indirect role in selecting polypeptides for inclusion in the secretory granules. Adaptins occur in oligomeric complexes made up of two adaptin chains (of which several occur) and several other proteins. The adaptins found at the *trans* Golgi are different from those found at the plasma membrane. This could distinguish the clathrin-coated vesicles that form during endocytosis at the plasma membrane from those that form at the *trans* Golgi and coat secretory granules. At least one adaptin is homologous to the β-COP protein of the coatamer, suggesting similarities in the way various kinds of vesicles form and then fuse with their target membranes.

Exocytosis Granule formation initiates at the *trans* Golgi membrane where the protein cargo of the vesicles begins to accumulate, forming a core. The vesicle containing the core then buds from the membrane and is pinched off. As with every budding and pinching of vesicles in the secretory pathway, this process involves a number of proteins. Some granules for regulated secretion are stored in the cytoplasm close to the cell membrane, linked to the cytoskeletal actin chains. Others appear to bind directly to the cytoplasmic surface of the plasma membrane.

Secretion in the regulated pathway requires an appropriate signal from outside the cell. Generally, the signal is transmitted as a change (increase) in cytosolic Ca^{2+} concentration. Upon exocytosis, the granule membrane fuses with the plasma membrane. This fusion, like the membrane fusions that occur earlier in the secretory pathway, appears to require ATP and a battery of proteins analogous or in some cases identical to those active in the earlier membrane fusions, for example, the small GTP-binding proteins. After release of the granule's contents, the secretory granule membrane can recycle to the *trans* Golgi. Exocytosis is accompanied by disruption of the actin cytoskeleton, presumably as part of the mechanism that frees the granules to move to the plasma membrane. Thus, for example, antibodies to fodrin, the protein that cross-links actin, inhibit secretion.

8.4 The α-Pheromone of *Saccharomyces cerevisiae*

Communication between individual haploid yeast cells is a prelude to mating, and only cells of the two different mating types, MATa or MATα, mate effectively (*Genes and Genomes,* section 10.6b). The product of mating is a MATa/MATα diploid cell. Mating type recognition depends on distinctive, small, peptide hormones called **pheromones** that are secreted by cells of the MATa or MATα type, respectively, as well as on cell surface receptors specific for the pheromone of the opposite mating type. Binding of a pheromone to its cognate receptor on a cell of the opposite type initiates cellular changes that permit mating. Thus, although yeast is not a multicellular organism, intercellular communication plays an important role in its biology.

The 13 amino acid long α-pheromone peptide is produced in MATα cells from a preproprotein encoded in the yeast genome, as described in Section 8.4a. The a-pheromone peptide is also encoded in a precursor larger than its 12 amino acids, but it is produced by a process distinct from that typically associated with preproprotein precursors and is not described here.

a. The Genes for α-Pheromone

Two genes encoding prepro-a-pheromones occur in *S. cerevisiae, MF*α1 and *MF*α2. The *MF*α1 gene was cloned by transfecting α-pheromone-deficient mutant yeast cells with a library of yeast genomic fragments in a yeast plasmid vector and selecting for cells that acquired the ability to mate. In addition, both the *MF*α1 and *MF*α2 genes were detected and cloned by screening a yeast genomic library with a DNA probe whose sequence was deduced from the amino acid sequence of α-pheromone. The genes are not related to any sequence within the yeast mating locus *MAT*α, as expected from genetic experiments that had shown them to be controlled by but not contained within that locus. Expression of the cloned *MF*α1 gene in transformed cells does, however, require the presence of the α1 gene product of MATα locus (*Genes and Genomes,* section 10.6b). Neither gene contains any introns.

The protein sequence encoded by *MF*α1 is 165 amino acids long with a 19 amino acid hydrophobic signal sequence immediately following the initiating

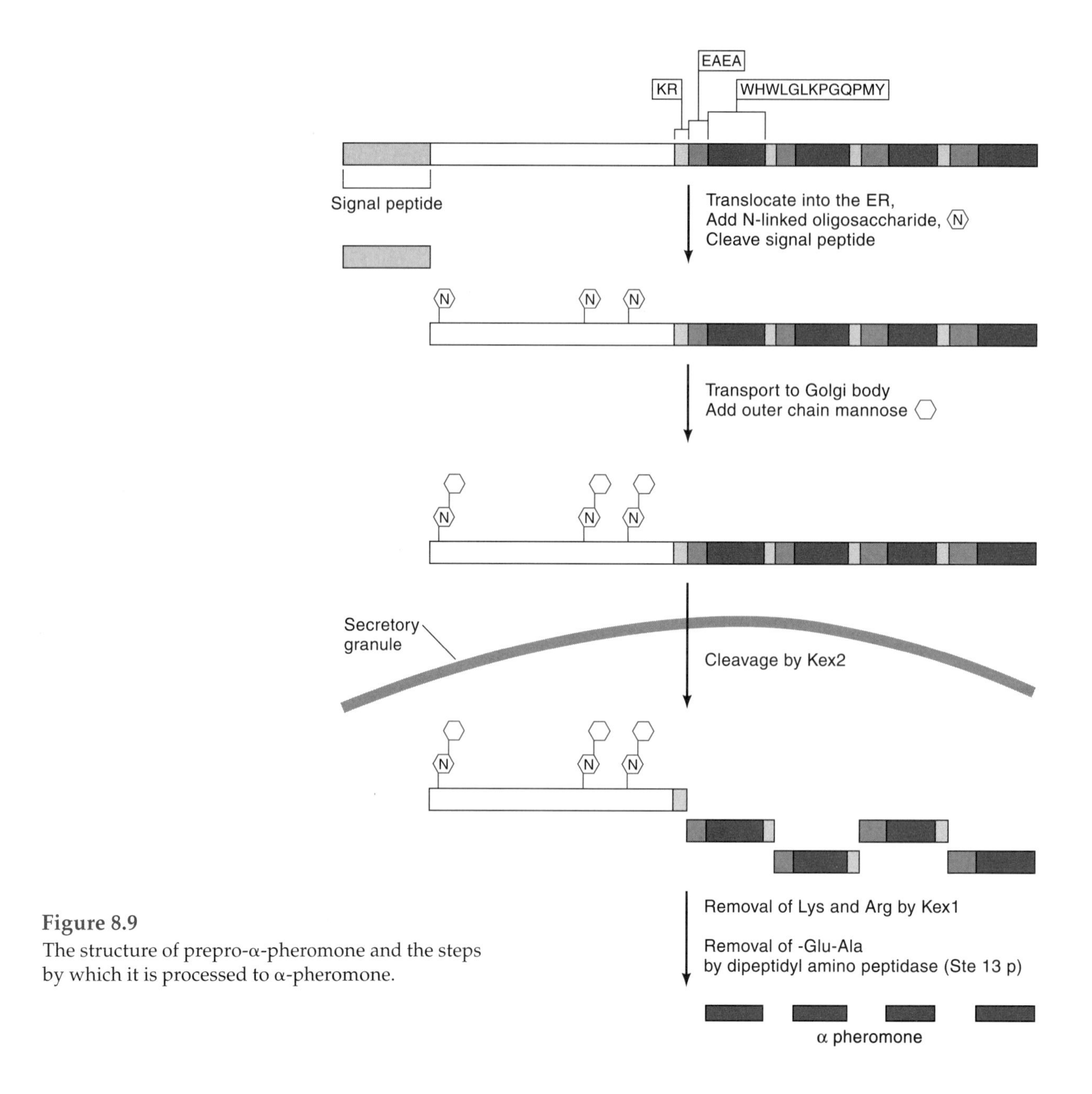

Figure 8.9
The structure of prepro-α-pheromone and the steps by which it is processed to α-pheromone.

methionine (Figure 8.9). This is followed by an approximately 60 amino acid long hydrophilic region, the "pro" segment. Thereafter, there are four tandem copies of a 21 amino acid long peptide, each beginning with an amino terminal Lys-Arg dipeptide and ending with a carboxy terminal α-pheromone sequence. Thus, processing of α1-preproprotein can yield four molecules of α-pheromone. *MF*α2 is very similar, but it encodes only two copies of α-pheromone in a 120 amino acid long precursor. Most of the α-pheromone in yeast cells is produced by *MF*α1. Cells carrying mutations in both genes cannot mate.

b. The Secretory and Processing Pathway

The processing of prepro-α-pheromone is intimately associated with its passage through the endoplasmic reticulum, the Golgi, and the secretory granules (Figure 8.9). Secretion of the mature 13 amino acid long α-pheromone follows the constitutive secretory pathway. Indeed, all secretion from yeast cells appears to proceed through a constitutive pathway with little if any storage of polypeptides for later release. Although this chapter emphasizes regulated secretion, we include a discussion of the α-pheromone because it stresses the conservation, throughout the eukaryotes, of the strategies used in secretory processes.

At least 25 genes in yeast are known to produce products essential to the secretory process; some of these, including the various *KEX* and *SEC* genes, are described in Sections 8.2a and 8.3, respectively. This section describes the biochemical modifications that convert prepro-α-pheromone to the active α-pheromone and those aspects of the secretory pathway that are typical of secretion in yeast.

The signal peptide region of prepro-α-pheromone facilitates binding of the translating ribosome to the endoplasmic reticulum and translocation of the nascent polypeptide across the membrane into the ER lumen. Within the ER, the signal peptide is removed by cleavage after residue 19 (alanine) and N-glycosylation occurs at Asn-X-Ser (or Thr) sites in the "pro" region. Yeast cells also appear to have an alternative way of targeting proteins to the secretory pathway. Thus, when the SRP is nonfunctional because of the deletion of genes encoding SRP constituents, cells still grow, albeit inefficiently. In addition, certain fully translated proteins can enter the ER *in vitro* even in the absence of the SRP docking component. Such entry into the ER requires ATP, and the prepro-α-pheromone must be unfolded. One of the **chaperonins** (an Hsp70 or heat shock protein 70) plays a role in keeping the molecule in an open conformation.

Once in the Golgi, the Kex2 convertase begins to hydrolyze the pro-α-pheromone, although a substantial portion of the proteolysis occurs later in the secretory pathway. Thus, for example, a large percentage of pro-α-pheromone remains uncleaved in cells carrying a mutation in the *SEC7* or *SEC4* genes, which block, respectively, transport between the Golgi cisternae or from the Golgi to the plasma membrane. Cleavage by Kex2 occurs on the carboxy side of the Lys-Arg dipeptides at the amino end of each of the repeated 21 amino acid long segments (Figure 8.9). The products include three molecules of the 21 residue peptide and a fourth, 19 residue long peptide derived from the carboxy terminus of pro-α-pheromone. This fourth peptide lacks the dibasic, carboxy terminal dipeptide found on the other three. Thereafter, the carboxypeptidaselike enzyme, Kex1, removes the carboxy terminal lysine and arginine from the three 21 residue long peptides. In cells lacking a functional *KEX*1 gene, secretion of α-pheromone is much reduced; some α-pheromone is produced from the carboxy terminal copy of the peptide because there is no amino terminal Lys-Arg to remove.

The final step in the production of active α-pheromone is the removal, by dipeptidyl amino peptidase, of the six amino terminal residues from the 19 amino acid long intermediates. Dipeptidyl amino peptidase A is encoded by the *STE*13 gene; mutations in *STE*13 cause sterility in MATα yeast cells, and the cells secrete nonfunctional α-pheromone precursors carrying extra amino terminal amino acids.

Processing of α-pheromone is completed in the secretory granules, as evidenced by the presence of fully mature α-pheromone in cells carrying a *Sec*1 mutation, which blocks fusion of the secretory granules with the plasma membrane. Prior to or concomitant with fusion, the granule binds to the plasma membrane, the Sec4p-bound GTP is hydrolyzed to GDP, and the Sec4p-GDP complex is released.

8.5 Insulin: One Gene, Two Polypeptides, One Protein

Many heteropolymeric proteins are the product of the interaction of several polypeptides, each encoded by a separate gene. But others, like insulin and some other messenger peptides, are the product of interaction between separate polypeptides encoded by a single gene. Thus, although most polyproteins yield several distinct and independently acting messenger peptides, preproinsulin yields two peptides that evolved to act in concert.

a. The Synthesis of Insulin from Preproinsulin

Preproinsulin is processed to yield the two polypeptides that comprise the active hormone (Figure 8.10). Of the two chains in vertebrate insulins, A is usually 21 and B usually 30 amino acids long. The two are held together by two disulfide bonds. The fact that both peptides A and B are produced from a single proprotein precursor was known well before it was possible to analyze gene structures. Thus, proinsulin was isolated from an insulin-rich tumor of pancreatic islet cells and found to contain the sequences of both the A and B chains flanking another, connecting peptide segment called C. Proinsulin itself is not very active as a hormone, but the presence of C facilitates interaction between the A and B segments and thus disulfide bond formation. Thereafter, the C portion is removed by proteolysis at the basic amino acids that mark C off from both A and B. Active insulin is the product of this proteolysis. Proteolysis depends on more than just the presence of the dibasic amino acids; for example, if the histidine at residue 10 in the B chain (B10) is changed to an aspartic acid, proinsulin is not cleaved at the B/C junction.

b. The Preproinsulin Multigene Family

The preproinsulin gene The preproinsulin gene is specifically transcribed in the β-cells of the islets of Langerhans in the mammalian pancreas. The coding region begins with a typical signal peptide: a group of large hydrophobic amino acids followed by several somewhat smaller neutral amino acids. The B chain segment starts with the first phenylalanine immediately after the signal peptide. The end of the reading frame coincides with the carboxy end of the A peptide.

Most vertebrate insulin genes have two introns, one within the 5′ untranslated region of the primary transcript and one interrupting the C peptide coding region. Most vertebrates have only a single (haploid) insulin gene, but rats and mice (but not all rodents) have two nonallelic insulin genes. One of these has only a single intron in the usual position in the 5′ untranslated region; the second gene has the two typical introns. Both genes are functional. Because the single gene in other vertebrates has two introns, the one-intron gene of some rodents probably evolved by duplicating an ancestral two-intron gene and precisely excising the second intron. One way this might have occurred is through an RNA intermediate that had already lost one intron through splicing (see *Genes and Genomes*, section 10.7).

Other genes in the preproinsulin family Insulinlike genes appear to be quite ancient, occurring in invertebrates including insects and even sponges, which separated from other metazoans about 1.6 billion years ago. A variety of insulinlike polypeptides occurs in vertebrates, and the similarity of their genes indicates that they probably had a common ancestral gene.

Relaxin is a hormone produced in the corpus luteum during pregnancy that is active during parturition. Like insulin, it is a heterodimeric protein, and the A (22 amino acids) and B (25 amino acids) peptides are derived from a prorelaxin by proteolytic removal of a connecting C peptide. Although quite different from insulin in amino acid sequence (the homology is only about 20 percent), the two have identically placed cysteine residues and very similar tertiary structures.

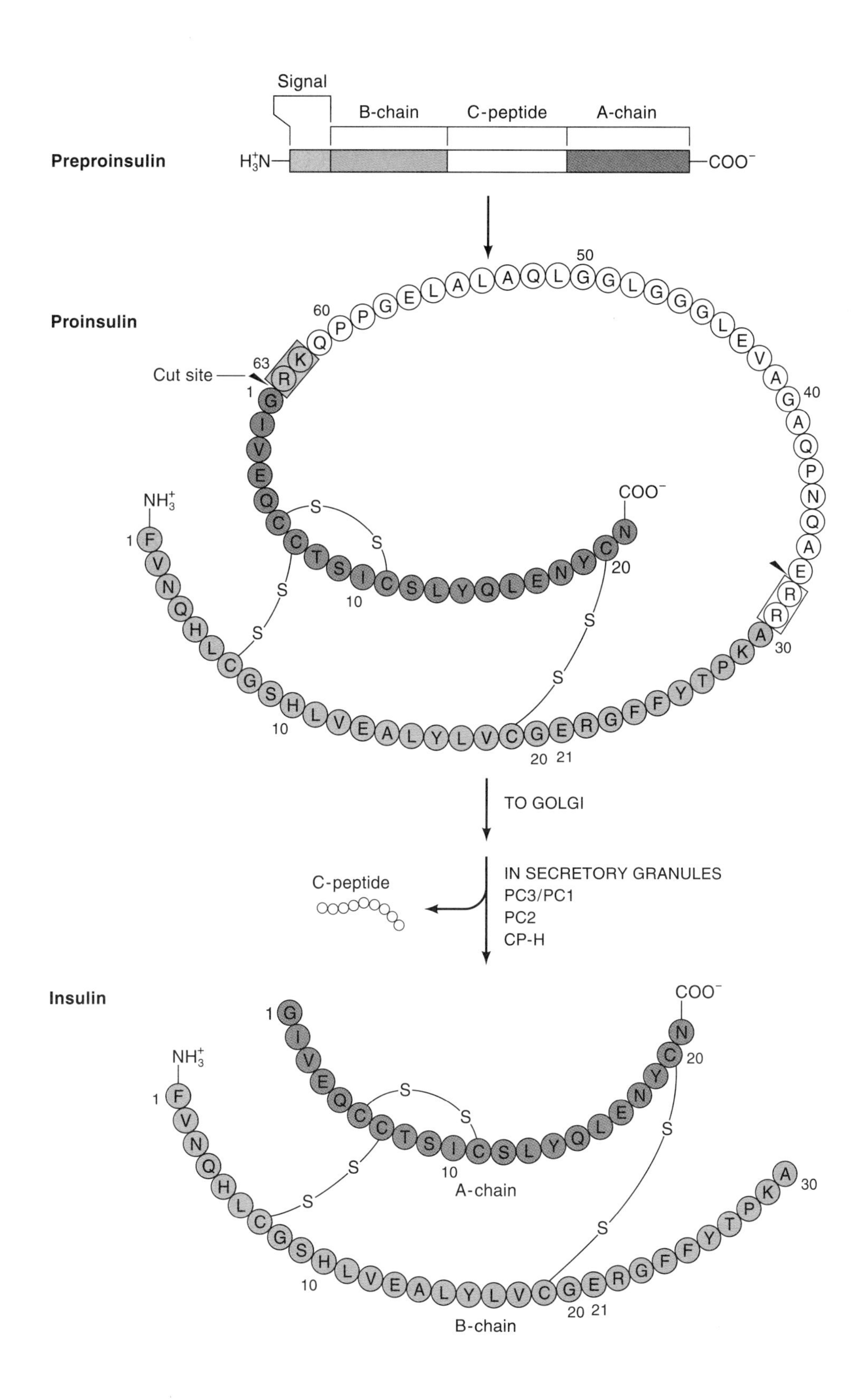

Figure 8.10
The processing of preproinsulin. The structure of a typical mammalian preproinsulin (porcine) is shown at the top, followed by the products of processing: proinsulin and mature insulin.

Moreover, the gene encoding relaxin is organizationally similar to the preproinsulin gene; it encodes a preprorelaxin and has an intron in the C peptide region. The extensive differences between the primary structures of the two genes and the encoded preproproteins make it difficult to conclude that they derive from an ancient common precursor (relaxin occurs even in such primitive forms as sharks), but it is not unlikely.

Two messenger peptides that are growth factors, **insulinlike growth factors I and II** (**IGF I and IGF II**), also called **somatomedins** (see Table 8.1), are much more homologous to insulin than is relaxin. Somatomedins are synthesized in liver and other tissues in response to growth hormone and are direct stimulators of growth and cell division in many somatic cells. Each IGF is a single polypeptide chain about 70 amino acids long with three disulfide bonds. It is as though the portions analogous to the A, B, and C peptides of insulin remain together rather than being separated by cleavage; the amino and carboxy portions of the IGFs are more homologous to the B and A peptides of insulin, respectively, than is the central portion to the C region of proinsulin, and the pairs of dibasic amino acids are missing. But the homologies are great enough so that somatomedins bind weakly to insulin receptors and thereby have a weak insulinlike activity. The reverse is also true; the binding of insulin to specific somatomedin receptors is likely to account for its growth-promoting activity. These interactions are also fostered by the structural similarities between the insulin, the IGF I, and IGF II receptors.

The insulinlike gene family illustrates the diverse ways related coding sequences may be used. Changes in coding sequences yield products with different and highly specific physiological effects. Changes in coding sequences can also alter the potential for processing and again yield related but distinctive polypeptides. Moreover, each of the genes must evolve in concert with the genes encoding their specific cell surface receptors as well as with DNA segments that direct the tissue-specific expression of those receptors.

Processing of proproteins in the insulin gene family Processing the preprohormones of the insulin gene family requires removing the signal peptides upon entry into the ER lumen, forming disulfide bonds within the lumen, and making internal cleavages at the carboxy side of dibasic dipeptide segments at subsequent sites in the regulated secretory pathway (Figure 8.10). The process is not rapid; it takes about 40 minutes for radiolabelled amino acids to appear in mature insulin after they are added to pancreatic islets *in vitro.* Endoproteolysis of proinsulin appears to begin in secretory granules in pancreatic β-cells. Demonstration of this localization depended on two monoclonal antibodies, one specific for the carboxy terminal region of the insulin B chain and one specific for proinsulin; very little cross-reactivity was observed with either antibody. Cells were reacted with one or the other antibody, and the antibodies were visualized with a complex of protein A bound to gold particles (protein A is a bacterial protein that specifically binds immunoglobulin G, IgG). Electron microscope observations indicated that proinsulin is in the Golgi and in immature secretory vesicles (possibly clathrin coated), but most of the insulin is found in mature, uncoated, dense secretory granules (Figure 8.11). Further, PC3/PC1 and PC2 convertases are localized, by immunocytochemistry, in the secretory granules in the β-cells of the pancreatic islets of Langerhans, but not in the exocrine pancreas cells that secrete trypsin, chymotrypsin, and pancreatic amylase.

The endoproteolytic cleavages can occur only in cells that contain the regulated secretory pathway and secretory granules; as described earlier, other cell types do not make the required cleavages, although they contain furin, which could, but does not, make the required cleavages. Thus, neither proinsulin nor prorelaxin is properly processed when the cloned genes are expressed from transfected recombinant vectors in typical laboratory lines of primate kidney

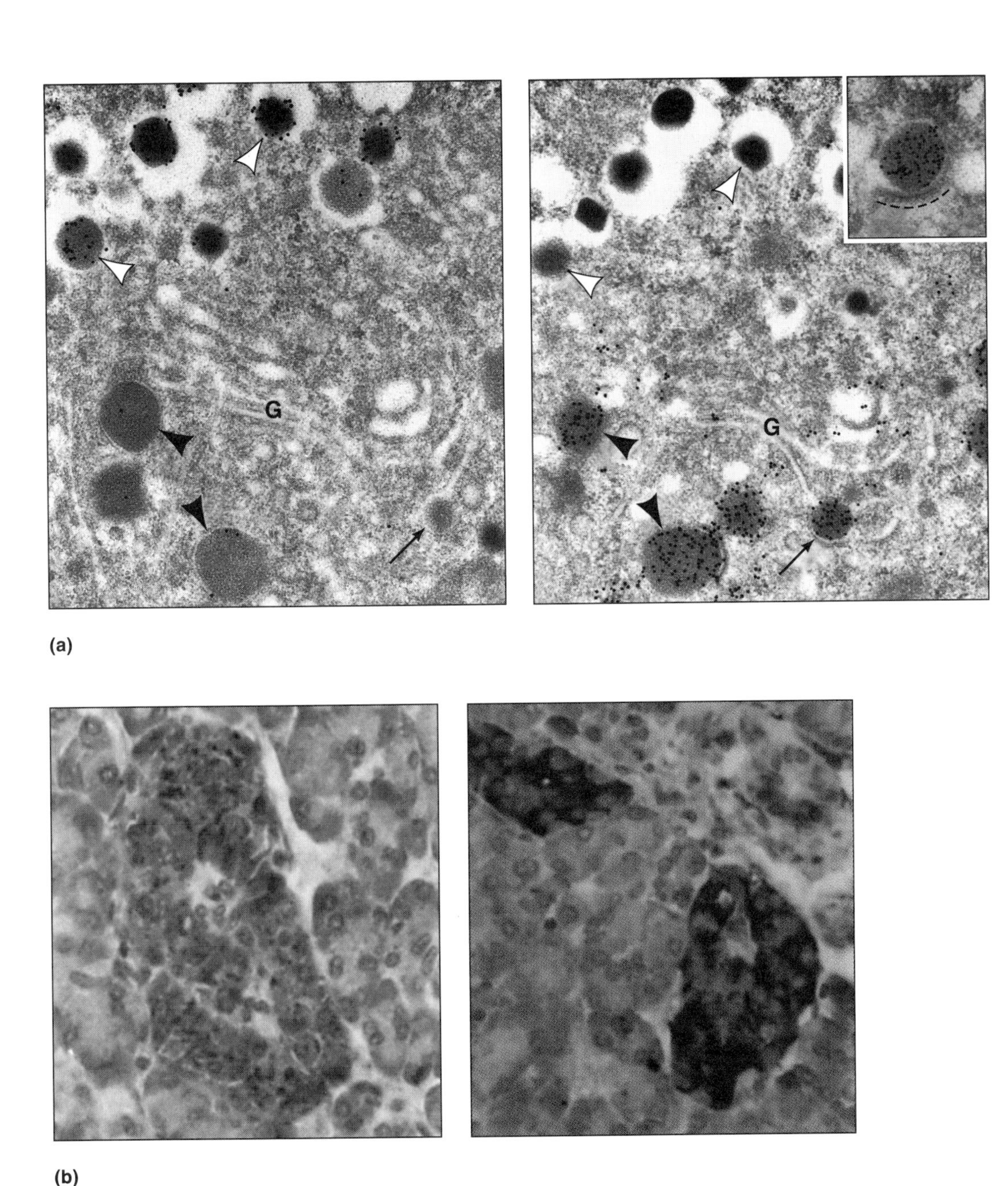

Figure 8.11
Processing of proinsulin by PC3/PC1 and PC2 in secretory granules in β-cells in the islets of Langerhans. (a) Electron microscope images of consecutive serial sections marked by monoclonal antibodies to insulin (left) or proinsulin (right). The visible dark spots are formed by gold particles. Open arrowheads point to mature secretory granules that react with antibody to mature insulin; filled arrowheads point to immature, clathrin-coated vesicles that stain mainly with antibody specific for proinsulin. (b) Sections of human pancreas marked by reaction with antibody to mature PC2 (left) or PC3 (right). Visualization is by means of avidin biotin complexes carrying alkaline phosphatase, the color being the result of chromogen generation after phosphatase action on a specific substrate (indole-fast red). (Part a from L. Orci et al. 1987. Proteolytic Maturation of Insulin Is a Post-Golgi Event Which Occurs in Acidifying Clathrin-Coated Secretory Vesicles. *Cell* 49 865–868; part b from S. P. Smeekens et al. 1992. Proinsulin Processing by the Subtilisin-Related Proprotein Convertases Furin, PC2, and PC3. *Proc. Natl. Acad. Sci. USA* 89 8822–8826. Reproduced with permission.)

cells that contain only the constitutive secretory pathway. For example, prorelaxin rather than relaxin is secreted from such cells, presumably through the constitutive pathway. However, if the gene encoding PC3/PC1 convertase is cotransfected into such cells along with the preprorelaxin gene, then prorelaxin is properly processed to relaxin, which is secreted into the medium surrounding the cells. Not only do the correct cleavages occur at the junctions between the B and C and C and A chains, but the B chain is correctly trimmed of its carboxy terminal Lys-Arg by carboxypeptidase. Similar experiments with preproinsulin confirm that PC3/PC1 and PC2 process the proprotein in pancreatic β-cell secretory granules. Thus, on the basis of colocalization as well as biochemical activity, PC3/PC1 and PC/2 appear to be responsible, in the pancreas, for proinsulin processing.

Additional experiments in which either the PC2 or the PC3/PC1 gene or both were cotransfected with the preproinsulin gene and the processing products analyzed suggest that PC3/PC1 may tend to cleave first at the B/C junction, with PC2 tending to act later, at the junction between the C and A peptides. Such specificities are likely to depend on the primary amino acid sequence surrounding the dibasic dipeptide as well as the conformation of the proprotein chain and the properties of the convertase. In most mammals, the B/C junction has a Lys-X-Arg-Arg sequence and the C/A junction, an X-X-Lys-Arg (X can be various amino acids). If these two sites are changed to a typical furin target sequence, Arg-X-Lys(or Arg)-Arg, by site-specific mutagenesis of a preproinsulin cDNA, then cleavage occurs even in nonendocrine cells, presumably by furin.

The two carboxy terminal arginines remaining on the B polypeptide after convertase action must be removed to form the mature B peptide. This is catalyzed by CP-H within the secretory granules, and secretion of CP-H along with insulin and the C peptide is observed.

8.6 Multiple Active Peptides from Single Genes by Differential Posttranslational Processing

a. Vasopressin and Oxytocin

The two polypeptide hormones vasopressin and oxytocin (Table 8.1 and Figure 8.12) are each derived from a preproprotein that is synthesized in specialized neurosecretory cells in the hypothalamus. Vasopressin has several physiological effects. It functions in the constriction of arteries, thus increasing blood pressure; it is an antidiuretic, decreasing water loss from the kidney; it brings about release of adrenocorticotropin from the anterior pituitary; and it causes behavioral changes. Oxytocin promotes milk release and uterine contractions during parturition.

The preprovasopressin and preprooxytocin genes are independently regulated, and each is expressed in a different set of hypothalamic neurons. After the proproteins are formed, they are packaged into secretory granules and transported, intracellularly, along axons toward the posterior pituitary. During transport, proteolytic cleavage of each proprotein yields two products: the active hormone and a protein called neurophysin (Figure 8.11). The neurophysins associated with vasopressin and oxytocin, respectively, are different; neither has any known function except to accompany the hormone to axonal termini in the pituitary, where the neurophysin is released. Vasopressin also may act as a local hormone, helping to stimulate the release of adrenocorticotropin from the anterior pituitary.

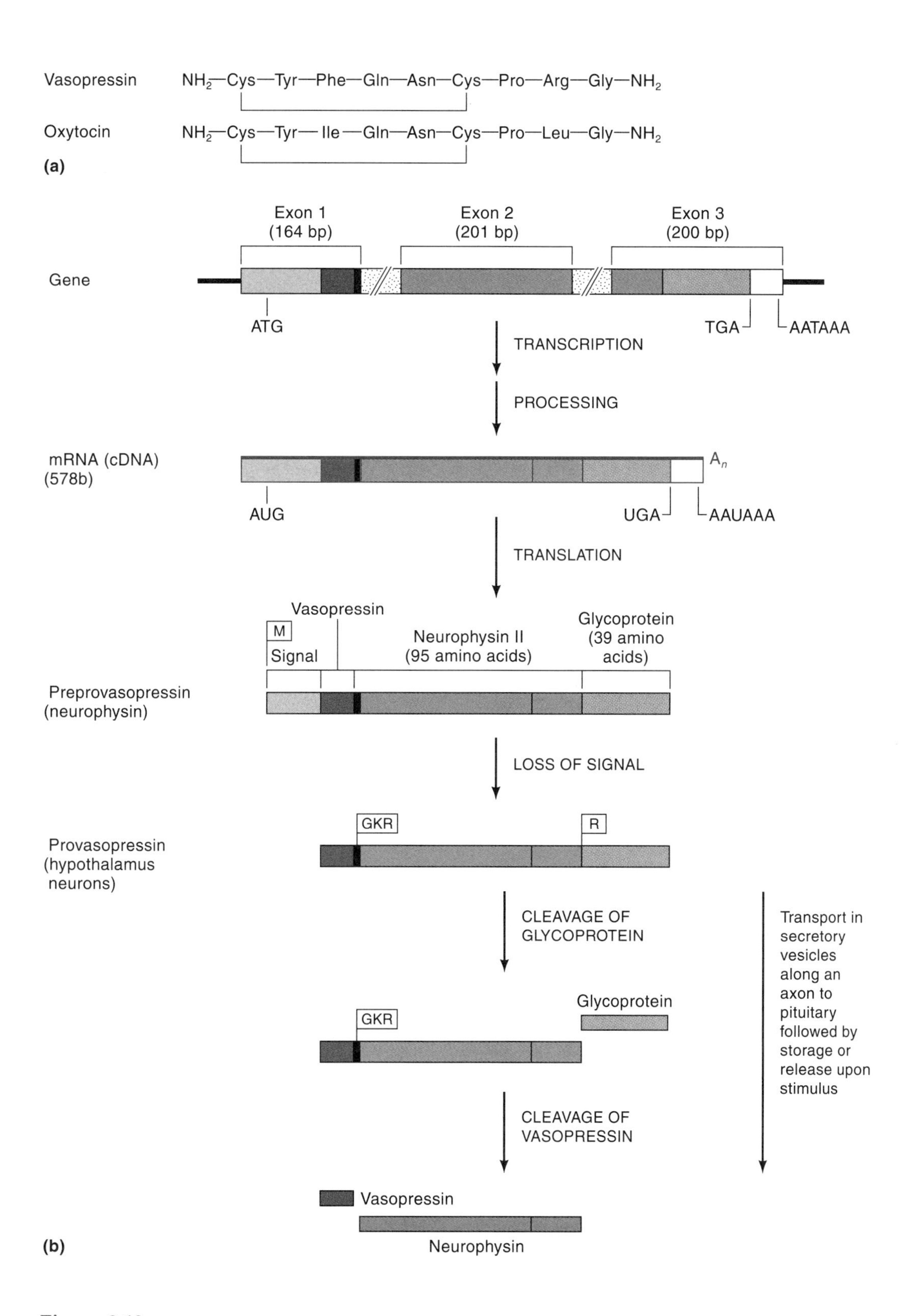

Figure 8.12
(a) Mammalian oxytocin and vasopressin. (b) The preprovasopressin gene and the synthesis of vasopressin. The diagram is based on a gene isolated from rats and a cDNA from cows.

Cloning the genes for preprovasopressin and preprooxytocin The amino acid sequence of each hormone has been known for years. Moreover, genetic and biochemical experiments indicated the association of each of the hormones with a neurophysin and also that each hormone-neurophysin combination is derived from a single polyprotein. Therefore, scientists were not surprised to find that when hypothalamic mRNA is translated *in vitro,* the products precipitated by antibodies to either vasopressin or oxytocin were long, single polypeptides. Analysis of the tryptic peptides derived from these immunoprecipitable products demonstrated that both the hormone and the corresponding neurophysin were present in each case. However, the translation products were even longer than needed to contain the two known components. Enriched mRNA was then used to help select appropriate cDNA clones from a hypothalamic library, and the clones were used in turn to isolate genomic clones. The primary nucleotide sequences of the cDNAs and genes are completely consistent with the polypeptide sequences determined earlier. Moreover, the organizations of coding sequences in the two preproprotein genes are very similar. The main difference between the two, besides specific codon changes, is the presence in preprovasopressin but not preprooxytocin of a segment encoding a third polypeptide, a glycopeptide (Figure 8.12). Two introns interrupt the coding sequences, and the nucleotide sequence of exon 2 is identical in both (bovine) preproproteins. The two genes very likely represent the duplication of a single common ancestor. Notably, the three exons more or less coincide with the coding regions specifying the three vasopressin gene products.

Processing and transport The preprovasopressin predicted from the gene structure has the customary signal peptide, and the three polypeptide domains are marked off by basic amino acids, suggesting a typical polyprotein processing mechanism. Moreover, the tripeptide Gly-Lys-Arg separates the vasopressin and neurophysin II segments; this glycine is the source of the amide group on the carboxy terminal glycine of vasopressin (see Section 8.2c). It is not known at which point the processing occurs during transport of secretory granules from cell body to axonal termini. Also unknown is the function, if any, of the glycopeptide, although it is not complexed with vasopressin during transport to axonal termini as is neurophysin II.

b. Opioid Peptides and Related Hormones

A partial list of the opioid peptides that have been isolated from mammalian brains, pituitary, and adrenal glands is given in Table 8.3. Regardless of their

Table 8.3 **Some Opioid Peptides**

Gene	Opioid	Sequence*
POMC	β-Endorphin	YGGFMTSEKSQTPLVTLFKNAIIKNAHKKQ
PEA	Met-enkephalin	YGGFM
	Leu-enkephalin	YGGFL
PDY	Dynorphin A	YGGFLRRIRPKLKWDNQ
	Leu-enkephalin	YGGFL
	Dynorphin B	YGGFLRRQFKVVT (also called leumorphin)
	α-Neoendorphin	YGGFLRKYPK
	β-Neoendorphin	YGGFLRKYP

*The one-letter abbreviations for amino acids are used here.

length, all opioid peptides have the same four amino acids at their amino termini followed by either a methionine or a leucine. Thus, they all represent extensions, at the carboxy end, of the two simplest opioid peptides, the pentapeptides **met-enkephalin** and **leu-enkephalin**. Moreover, all the known opioid peptides are derived from three preproproteins encoded, respectively, by three genes (Figure 8.13). Different opioids are produced in different cells and tissues, and control over where and when each opioid accumulates is exercised on at least two levels: transcription of the appropriate gene and proteolytic processing of the proprotein gene product. The situation is quite complex because opioid peptides are found not only in neural tissue but in such sites as germ line cells, testis, mesodermal cell lineages in many parts of mouse embryos, and embryonic cultured fibroblasts.

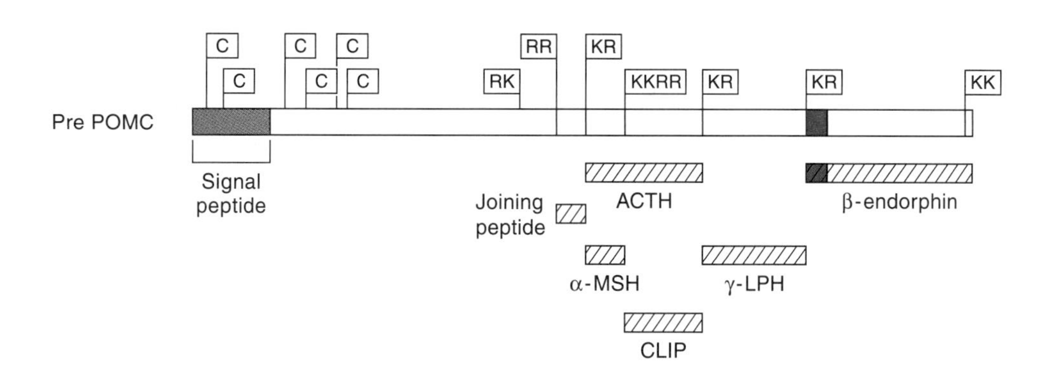

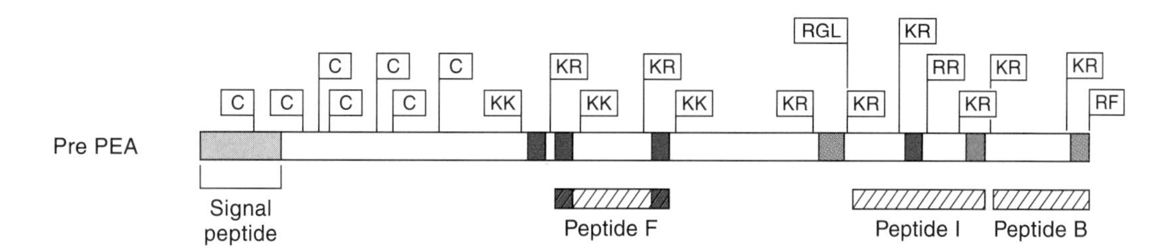

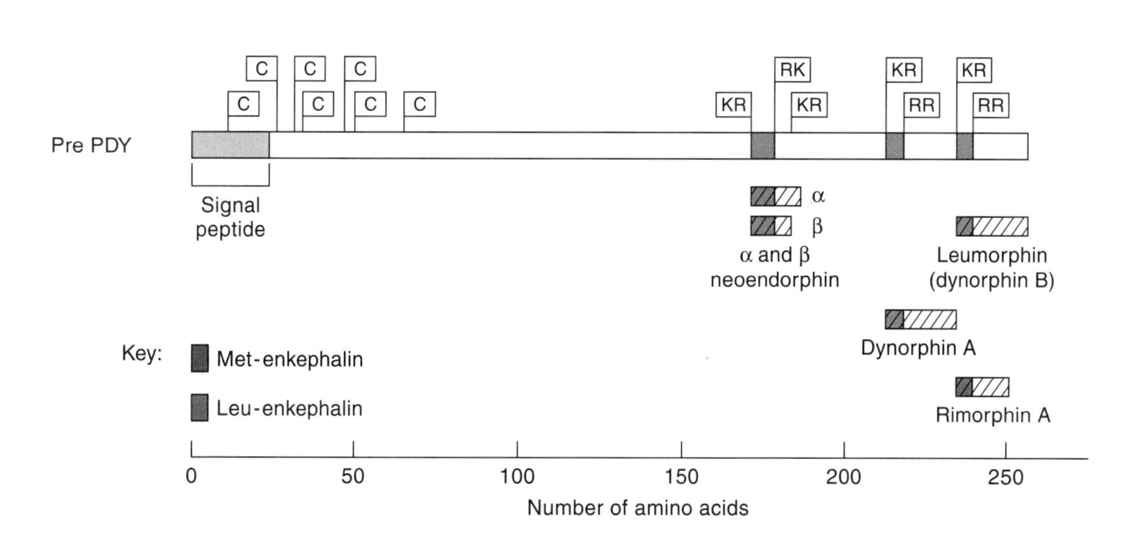

Figure 8.13
Many opioid peptides are encoded in three genes.

The POMC gene The polyprotein **proopiomelanocortin (POMC)**, also known as **proACTH/endorphin (PAE)**, is precursor to the hormones adrenocorticotropin (ACTH) and γ-lipotropin as well as to several opioid peptides, including β-endorphin (Figures 8.13 and 8.14). It is found in both anterior and intermediate lobes of the pituitary gland and in the hypothalamus as well as in other parts of the central nervous system and, among nonneuronal tissues, in the gut and testicular germ line cells as well as in many tissues of mouse embryos. POMC itself is inactive, and its processing to active hormones involves cleavage at pairs of basic amino acids.

Homologous genes and cDNAs that encode POMC in cows, mice, and humans have been cloned and sequenced; a single active gene occurs in most mammals, and there is one pseudogene in mice. Cloning of all these genes from genomic libraries was facilitated by the marked homology of the coding sequences in different species. In one initial approach, the critical probe was obtained by purification of poly A^+ RNA from ER-associated polysomes isolated from a mouse pituitary tumor cell line that secretes copious adrenocorticotropin and β-endorphin. Purification was monitored by immunoprecipitation of the products of *in vitro* translation of the RNA after addition of antibody to adrenocorticotropin. Fractionation by size gave a preparation of active mRNA about 1.4 kb long; 75 percent of the protein synthesized *in vitro* from this mRNA was immunoprecipitated. A double-stranded cDNA prepared from this still impure RNA yielded, upon cleavage with endonuclease *Hae*III, an abundant fragment of 140 bp. Its abundance suggested that it might represent part of the POMC mRNA, and the fragment was cloned in *E. coli* after adding *Hind*III linkers. The nucleotide sequence of the cloned fragment predicted a portion of the γ-lipoprotein amino acid sequence. Once this portion was in hand, it served as a probe to clone full length cDNAs and ultimately the POMC gene.

Besides the POMC coding sequences, the gene encodes a 26 amino acid long signal sequence; thus, the primary translation product is a preproopiomelanocortin. One intron occurs in the 5′ untranslated region, and another interrupts the coding sequence shortly after the end of the signal peptide segment. Alu sequences occur in both intervening sequences in the human gene.

Processing of proopiomelanocortin The processing of POMC occurs in a complex and incompletely understood tissue- and cell-specific manner. We give only a few examples here to provide insight into the range of distinctive pathways that occur (Figure 8.14). The picture that has emerged is the consequence of several experimental approaches: biochemical, immunocytochemical staining followed by electron microscopy, and tracking of the translation products of cloned cDNAs linked to appropriate promoters.

N-linked oligosaccharides are added to POMC in the ER. Additional modifications, including phosphorylation of a serine near the carboxy terminal (by a casein kinase–type enzyme), take place in the Golgi. When genes encoding PC3/PC1 or PC2 or both and the POMC gene are simultaneously expressed from recombinant vectors in nonendocrine cells that do not normally express these genes, the POMC is selectively cleaved at appropriate sites. This suggested that one or both of these enzymes is responsible for endoproteolytic cleavage of POMC.

In corticotropes (POMC-expressing cells in the anterior pituitary lobe), some POMC molecules at the *trans* Golgi are already cleaved at the Lys-Arg dipeptides bordering the ACTH segment, although the bulk of such cleavages occurs in secretory granules. Cleavage at an additional dibasic site (Figure 8.14) yields the so-called **joining peptide** (of unknown function). All these cleavages are likely to be made by PC3/PC1 because that is the major convertase expressed in corticotropes. Carboxypeptidase trimming and α-amidation (of joining peptide) also occur in the secretory granules. Aminopeptidase appears to function in the

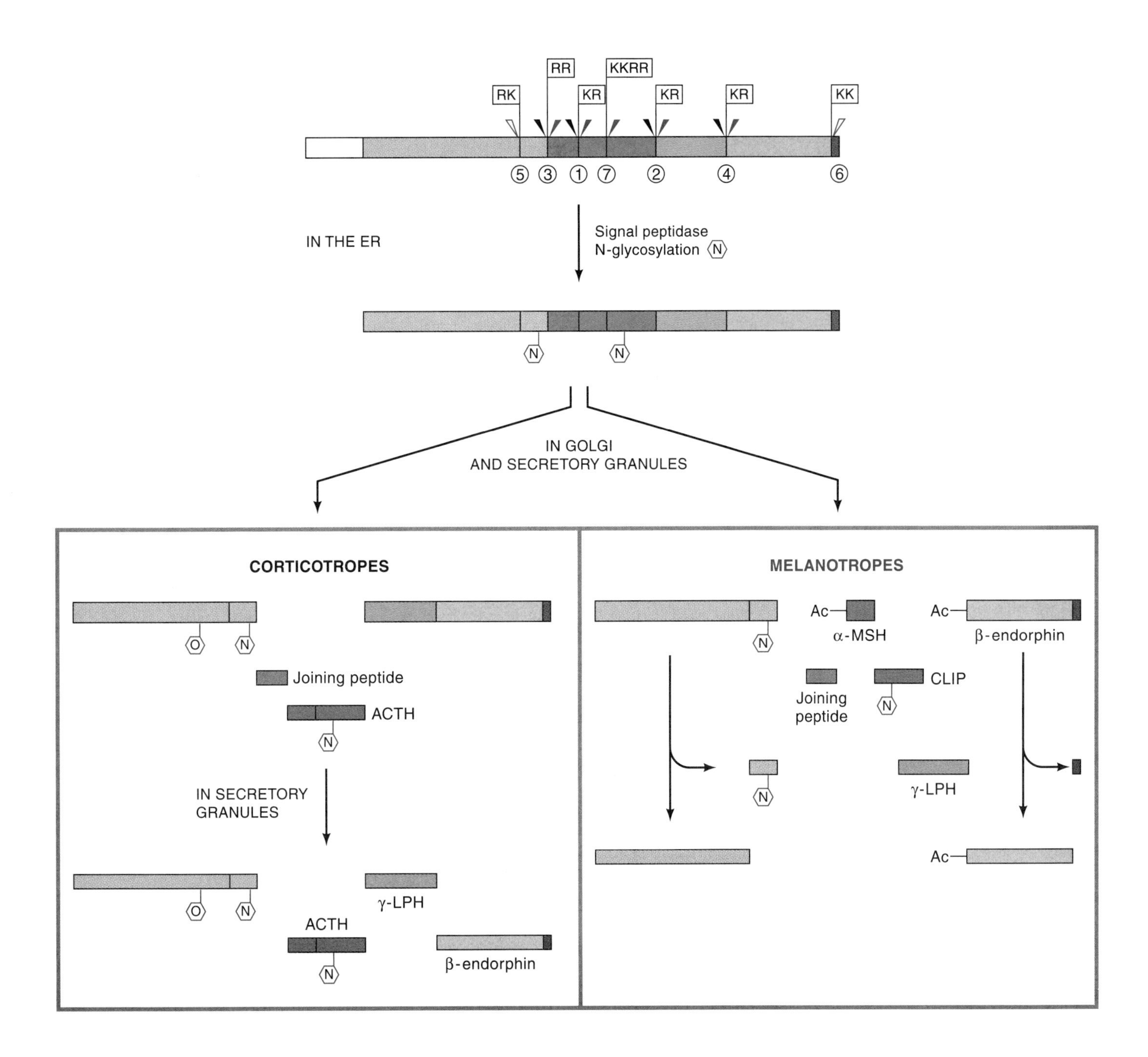

Figure 8.14
Preproopiomelanocortin (POMC or PAE) is processed in tissue-specific ways to yield a collection of messenger polypeptides. On the schematic drawing of prePOMC at the top, black arrowheads show reactions occurring in both the anterior lobe of the pituitary gland (corticotropes) and in the intermediate lobe (melanotropes); colored arrowheads show reactions occurring primarily in the intermediate lobe; open arrowheads indicate inefficiently cleaved sites. The size of POMC and some polypeptides varies slightly from one organism to another. (See R. E. Mains and B. A. Eipper. 1990. The Tissue Specific Processing of Pro-ACTH/Endorphin: Recent Advances and Unsolved Problems. *Trends Endocrinol. Metab.* 1 388–394.)

maturation of POMC, although its significance, if any, is not clear. Thus, amino terminal residues are missing from some percent of some of the peptide products (e.g., the amino terminal serine of ACTH). Small amounts of β-endorphin and γ-lipotropin are also produced in corticotropes.

The main endoproteolytic cleavages observed in corticotropes occur at a distinct subset of the available dibasic dipeptide sequences: Lys-Arg or Arg-Arg (marked sites 1, 2, 3, and 4 on Figure 8.14). Other sites containing Arg-Lys (5), Lys-Lys (6), and Lys-Lys-Arg-Arg (7) are uncleaved. What is the explanation for this specificity? In part, it could be the sensitivity of the convertase toward amino acids surrounding the cleavage site. Also, an O-linked oligosaccharide (see Figure 8.14) that is added in corticotropes and appears to inhibit some cleavages is likely to provide at least a partial explanation. The selection of cleavage sites appears to be regulated because the processing of proPOMC in the melanotropes of the intermediary lobe of the pituitary is different from the events in the corticotropes. In part, this could be because PC2 is more abundant in melanotropes than is PC3/PC1. Also, some POMC molecules lacking the O-linked oligosaccharide occur in melanotropes, and these are cleaved, albeit slowly, at site 5 (Arg-Lys), which is closest to the O-glycosylation site. The sole Lys-Lys (site 6) is also cleaved slowly in melanotropes. But the cell type specificity of processing is most markedly illustrated by the fact that the Lys-Lys-Arg-Arg (site 7) in the ACTH region is efficiently cleaved in melanotropes. As a consequence, these cells secrete **α-melanocyte stimulating hormone** (**α-MSH** or **α-melanotropin**) and the peptide called **CLIP** (**corticotropinlike intermediate lobe peptide**). Another example of cell-type-specific modification is the α-N-acetylation of ACTH and β-endorphin, which occurs in melanotropes but not in corticotropes.

Besides the cell-type specificity of POMC processing in adult mammals, there is a marked developmental regulation of the endoproteolytic processing steps in the anterior pituitary corticotropes, at least in rats. Thus, endoproteolysis of POMC is relatively inefficient in newborn animals compared to adults, where almost all the precursor is cleaved. Moreover, the KKRR (site 7) in ACTH, which, as mentioned previously, is inefficiently cleaved in adult corticotropes, is frequently processed in newborns. This effect appears to be influenced by glucocorticoids.

Enkephalins and the processing of preproenkephalin A The five amino acids at the amino end of β-endorphin represent the structure of the opioid peptide met-enkephalin: Tyr-Gly-Gly-Phe-Met. In principle, an additional or alternative processing step for POMC could yield this opioid in those tissues where it is found, including brain, sympathetic ganglia, and neurons in the intestines. However, POMC and its gene are not the source of enkephalins. Enkephalins and several somewhat longer opioid peptides that include enkephalin segments are especially abundant in the adrenal cortex. The discovery of the longer peptides suggested that they, rather than β-endorphin, might be precursors to met- and leu-enkephalin, and this suggestion was confirmed when the structures of the genes and cDNAs for **preproenkephalin A** (**prePEA**) and **preprodynorphin** (**prePDY**), the distinctive polyprotein precursors of the enkephalins, were determined.

The abundant mRNA for enkephalin in bovine adrenal medulla and a human adrenal tumor, a pheochromocytoma, was a key to cloning the prePEA gene. However, because the polyprotein precursor of enkephalin had not previously been characterized and antibodies were not available, the mRNA and its cDNA could not be identified by *in vitro* translation. Therefore, a pentadecadeoxynucleotide probe whose structure was deduced from the structure of opioid peptides themselves was synthesized. The choice of likely codons was narrowed by considering the known frequencies of codon usage in mammals and by inspecting the sequences encoding met-enkephalin in the POMC gene, assuming that similar codons might be used. The probe identified an mRNA about 1.5 kb long in polyadenylated RNA from bovine adrenal medulla and from the pheochromocytoma (in pituitary cell cytoplasm, the same probe anneals to the shorter POMC mRNA). Investigators partially enriched the putative mRNA by gradient centrifugation and prepared cDNA and cloned it in a plasmid vector. Upon screening the resulting cDNA library with the pentadecanucleotide, they selected colonies containing homologous cDNA. They then used the clones to select the corresponding gene segments from a genomic library. Finally, they obtained the human prePEA cDNA and gene and deduced the primary translation product (Figure 8.13).

PEA includes six copies of met-enkephalin and one of leu-enkephalin. In addition, there are coding sequences corresponding to several of the larger opioid peptides (e.g., F, I, and B) that had been isolated from bovine adrenal medulla and contain enkephalin pentapeptides within their sequences. Pairs of basic amino acids flank most of the opioid peptides, as expected if processing of PEA is accomplished by the typical endoproteases. PrePEA encodes an amino terminal signal peptide (24 amino acids), and there is a lengthy peptide of unknown significance between the signal peptide and the first met-enkephalin segment. Also, like prePOMC, the prePEA gene has an intron in the 5′ untranslated region just before the ATG initiation codon and another shortly after the end of the segment encoding the signal peptide; in addition, a third intron interrupts the noncoding region.

Opioid peptides from preprodynorphin The characterization of prePOMC and prePEA and their genes accounted for the biosynthesis of many but not all known opioid peptides. The existence of at least one additional polyprotein was therefore postulated, one that might produce **dynorphin** or **β-neoendorphin** or **α-neoendorphin** or all three, and a search was undertaken. Oligodeoxynucleotides corresponding to the carboxy terminal pentapeptide of dynorphin (Lys-Trp-Asp-Asn-Gln) were synthesized and used as probes against a cDNA library prepared from hypothalamic poly A^+ mRNA. The clones obtained were, however, significantly shorter than the approximately 2.6 kb mRNA that annealed to the probe on RNA blots, indicating that they did not represent full length cDNAs. Another oligodeoxynucleotide whose sequence precisely matched that encoding the dynorphin carboxy terminal pentapeptide in the isolated short cDNA was synthesized. This was annealed to hypothalamic poly A^+ mRNA, and double-stranded cDNA was synthesized after the synthetic primer was extended with reverse transcriptase. The resulting mixed population of cDNA clones yielded clones representing the complete mRNA, and these were used as probes to clone the human gene encoding the three opioid peptides—the preprodynorphin gene (Figure 8.13).

However, besides encoding dynorphin and the two neoendorphins, the **preprodynorphin** (*prePDY*) gene predicted the existence of two additional opioid peptides, assuming that pairs of basic amino acids mark off potential cleavage sites. In fact, by the time the cloning and sequencing of the *prePDY* gene was complete, one of the predicted opioid peptides, rimorphin, had been isolated from posterior pituitary lobes, a tissue in which dynorphin and α-neoendorphin also occur. The opioid activity of the second predicted product, leumorphin (also called dynorphin B), was confirmed using a chemically synthesized peptide even before it was identified as a naturally occurring component of the anterior pituitary. Each of the peptides, β-neoendorphin, dynorphin, and leumorphin, has a leu-enkephalin sequence at its amino terminus; thus, all the opioid peptides contained in PDY are of the leu-enkephalin class.

Comparison of the POMC, PEA, and PDY genes and gene products There are striking similarities between the three preproprotein genes that together encode the known opioid peptides (Figure 8.13). After each signal peptide, there is an amino terminal region that is processed to a peptide of unknown significance, and all three polyproteins have a similarly placed cluster of cysteine residues within this peptide. The carboxy terminal portion of each polyprotein is processed to multiple hormones or opioid peptides. In the three genes, the 5′ untranslated portion of the transcription unit is interrupted by an intron whose acceptor site is very close to the ATG start codon. Another intron interrupts each gene shortly beyond the signal peptide region. The *prePEA* and *prePDY* genes, but not *prePOMC,* have an additional, similarly placed intron in the 5′ noncoding region. Moreover, there are striking similarities between the nucleotide and amino acid sequences of prePEA and prePDY in all species. Thus, these two genes are likely to share a common ancestor and also a common ancestor with the *POMC* gene. Internally, each of the three genes shows signs of evolution by amplification. For example, of the six met-enkephalin sequences in PEA, four are encoded in the sequence TATGGGGGCTTCATG.

Cell-specific regulation of opioid and hormone production Given the complex, specific actions of the neuropeptides and hormones, that their synthesis is regulated in a complex manner should not be surprising. Although a detailed description of the neurobiology is not within the scope of this book, a few general points are pertinent to gene expression. Most important, the production of specific neuroendocrine peptides in particular tissues or cell types is regulated on at least two levels: transcription and processing. Thus, for example, in the anterior pituitary, transcription of *prePOMC* is down-regulated by glucocorticoid hormones, and appropriate *cis*-acting sequences occur in the 5′ flank of the gene (*Genes and Genomes,* section 8.3f). Interestingly, expression of the gene in the intermediate lobe of the pituitary is not affected by glucocorticoids. The explanation for the difference appears to be the absence of glucocorticoid receptors in intermediate lobe pituitary cells.

Differential transcriptional regulation of the three genes is frequent. The *prePEA* gene, for example, is transcribed relatively abundantly in the adrenal medulla, and high levels of *prePOMC* gene transcripts are associated with the pituitary. In some cells, all three genes, *prePEA, prePDY,* and *prePOMC,* are silent. *PrePEA* transcripts (and even peptides) are found in nonneuronal, mesodermal cells. This is especially striking in rat embryos, where prePEA mRNA (but not prePDY or prePOMC mRNAs) is found in substantial amounts in primordial mesodermal tissues such as kidney. Upon terminal differentiation of the embryonic cells, the prePEA mRNA disappears. The amount of enkephalin detected in such tissues does not necessarily reflect the amount of prePEA mRNA, and the significance of the findings is not clear.

The three genes, *prePOMC, prePEA,* and *prePDY,* can together produce around 20 different peptides with enkephalin activity as well as several products with no opioid action. But the relative amounts of the particular peptides produced by any one of the three is distinctive in different cell types. This directly reflects differential proteolytic processing and, in particular, differential cleavage at the pairs of basic amino acids. For example, the processing of POMC proceeds in steps, some of which occur in the intermediate but not the anterior pituitary lobes (as discussed earlier in this section). Thus, a different set of peptides is generated at each site (Figure 8.14). Processing in the hypothalamus is different still.

In addition, β-endorphin is relatively abundant in the hypothalamus and the anterior pituitary, which means that little cleavage occurs at the Lys-Lys pair near its carboxy end (Figure 8.14). A short version of β-endorphin (amino acids 1–27) is more abundant in the midbrain and the intermediate pituitary, indicating that cleavage does occur in some tissues. Similarly, differences occur in the processing of PEA and PDY in different tissues. These differences are likely important to the specific and quantitative aspects of neuropeptide action. Generally, smaller peptides like met- and leu-enkephalin appear to be less potent than longer peptides that include enkephalin segments at their amino termini. Moreover, different carboxy extensions on the same type of enkephalin appear to alter the interaction with receptors. By specific interaction with different receptors, the closely related peptides can exert multiple different effects.

8.7 Multiple Active Peptides from a Single Gene by Alternative Splicing: Calcitonin and CGRP

The previous section illustrated one way different biologically active peptides may be generated from a single gene and a single transcription unit, namely, by posttranslational processing of a polyprotein. In that case, tissue-specific production of one or another peptide from a single proprotein depends on distinctive proteolysis. An alternative mechanism for generating different active peptides from a single gene and single transcription unit is by alternative splicing of the primary transcript to yield different mRNAs (*Genes and Genomes,* section 8.5f). Here, tissue-specific production of one or another product is regulated at the level of RNA processing. The synthesis of **calcitonin** and **calcitonin gene related peptide** (**CGRP**) in the thyroid and nervous system, respectively, is a relevant example. This system again illustrates how molecular analysis of genes and their RNA transcripts actually can lead to the discovery of previously unknown polypeptides; the existence of CGRP was proven only after it had been postulated from the DNA and RNA structures.

a. Two mRNAs from One Gene

Calcitonin is a 32 amino acid long, α-amidated messenger peptide whose primary physiological effect is to inhibit osteoclasts, thereby protecting bone, particularly from calcium deficiencies that can occur during pregnancy and lactation. The clue that calcitonin gene expression involves something unusual came from analysis of a thyroid tumor in rats. The tumor itself produced calcitonin, but upon serial transplantation, calcitonin biosynthesis sometimes decreased by as much as tenfold. Annealing of a calcitonin cDNA probe to RNA from both types of tumor indicated that the decrease in calcitonin production is not associated with a lower level of complementary mRNA. However, it is associated with a change in the size of the homologous RNA. In particular, besides the 1050 nucleotide long calcitonin mRNA typical of the cytoplasm of normal thyroid and of the original, high-producing tumor, the low-producing tumors contain another homologous polyadenylated cytoplasmic RNA that is about 200 nucleotides longer. Clones of cDNAs for both types of RNA and the single rat calcitonin gene were isolated and analyzed (Figure 8.15).

The two cDNAs (mRNAs) are identical in the 5′ noncoding regions and for 227 nucleotides after the start codon; thereafter, they are not homologous, although both have open reading frames. When investigators compared the sequence of the calcitonin gene with that of the two cDNAs, they saw that the distinctive 3′ portions of the two cDNAs are encoded by two different exons. Moreover, the 3′ end of the common region coincides with the 5′ splice donor site of an intron. Splicing of that site to the next exon yields an mRNA encoding calcitonin; a polyadenylation signal at the 3′ end of the exon corresponds to the polyadenylation site of calcitonin cDNA. Alternatively, splicing the same 5′

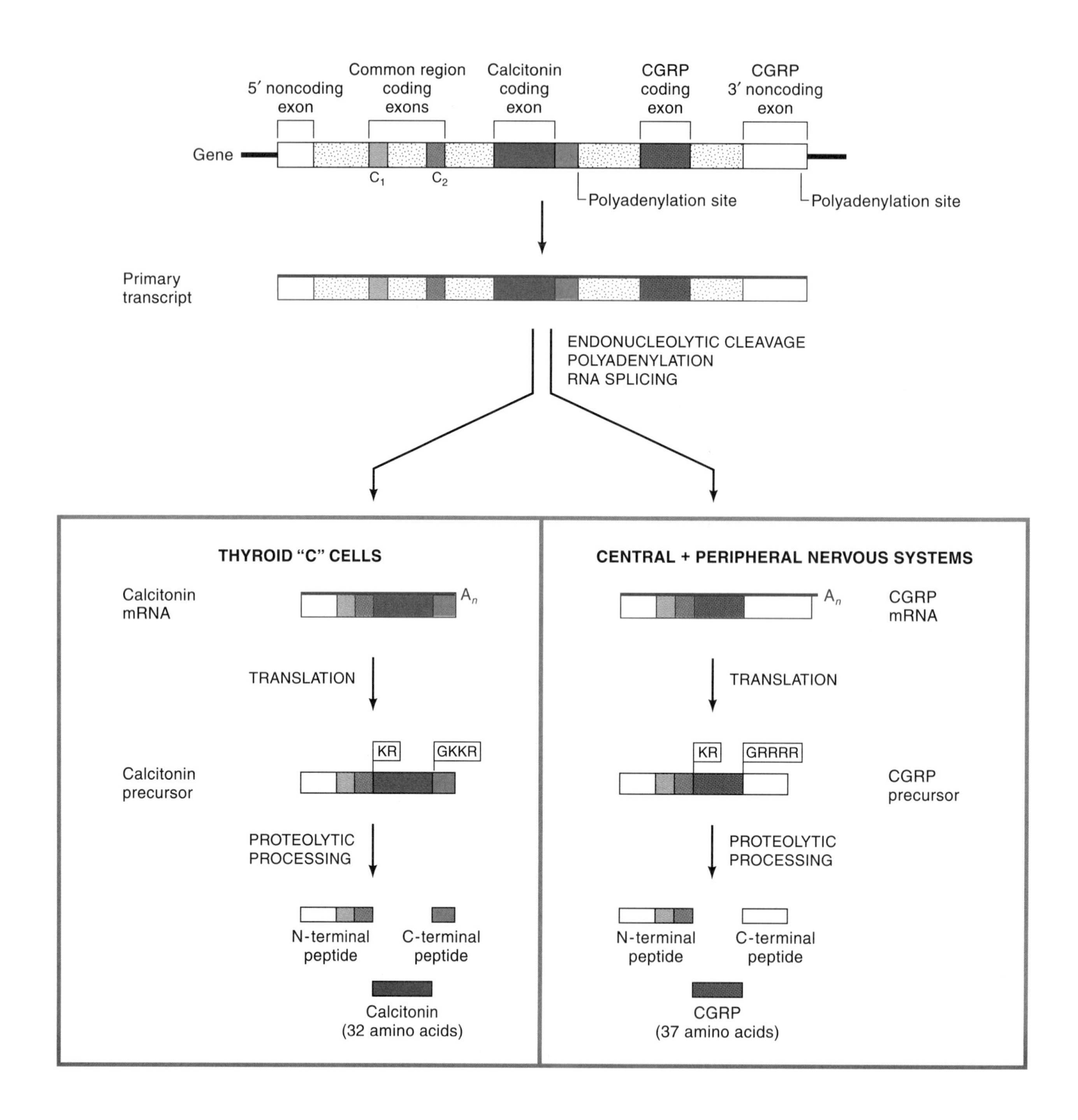

Figure 8.15
Differential processing of the primary transcript of the calcitonin/CGRP gene yields different messenger peptides in different tissues.

donor site to an exon farther downstream deletes the calcitonin exon and yields an mRNA encoding a previously unknown polypeptide, CGRP. To generate CGRP mRNA, an additional intron must be removed from the primary transcript, and the final 3′ noncoding exon provides the polyadenylation signal for CGRP mRNA. Although this discussion focuses on two mRNAs, they are not the only mRNAs known to be produced by alternative splicing of the primary transcript of the calcitonin/CGRP gene.

Translation of each of the mRNAs produces a precursor polypeptide that must then be processed to calcitonin or CGRP. The nucleotide sequences predict that processing is by proteolysis at pairs of basic amino acids at the amino termini of the calcitonin and CGRP segments. Near their carboxy termini, each of the peptide precursors contains a glycine followed by several basic amino acids; removal of these amino acids is followed by peptide α-amidation (Section 8.2c). As predicted, a single long nuclear transcript containing both the calcitonin and CGRP coding sequences is demonstrable. Thus, alternative splicing must account for the formation of the two different mRNAs (*Genes and Genomes*, section 8.5f).

b. Synthesis of Calcitonin and CGRP in Different Tissues

The observations made on the rat thyroid tumor led to the identification of a new messenger peptide, CGRP. Thus, probes containing unique regions of the two cDNAs were prepared; one annealed exclusively to the calcitonin mRNA and the other to CGRP mRNA. When extracts of normal rat tissues were screened with these probes, calcitonin and CGRP mRNAs were found in thyroid C cells in a ratio of about 98 to 1, but the mRNA in brain was exclusively CGRP mRNA. This marked tissue specificity was confirmed by an antibody raised to a chemically synthesized peptide corresponding to the 14 carboxy terminal residues of CGRP. The antibody reacted with central nervous system tissues and peripheral neurons in a pattern distinct from that of any previously known neuropeptide. It failed to react with thyroid tissue at all. Thus, the distributions of calcitonin and CGRP are different. CGRP (now called α-CGRP because another, related messenger peptide called β-CGRP and encoded in another gene has been discovered) plays important roles in mammalian physiology. It is found at developing, presynaptic motor neuron termini, where it modulates the postsynaptic response of acetylcholine receptor channels. More generally, it effects, for example, cardiovascular neuromuscular junctions, pain perception, and gastrointestinal function.

Regardless of the tissue, the primary calcitonin/CGRP gene transcripts are full length. This suggested that the alternative splicing modes might be specified in different cell types. To approach this question, investigators generated lines of transgenic mice carrying the calcitonin/CGRP gene under the control of the mouse metallothionein I promoter. The gene was, as expected, transcribed in many different cell types, including tissues that do not normally express the gene. Nonneuronal tissues, including nonneuronal brain cells, produced mainly calcitonin mRNA, and neuronal cells produced mainly CGRP mRNA. Thus, neurons apparently must have specific competence for the splices leading to CGRP mRNA, a competence lacking in other cell types. Similarly, when epithelial cells (human HeLa cell line) or teratocarcinoma cells (the F9 mouse cell line that has many properties of neuronal cells) were stably transformed with the rat calcitonin/CGRP gene, calcitonin mRNA predominated in the nonneuronal cells and CGRP mRNA in the F9 cells (greater than 94 percent in each case).

What specifies the alternative splicing modes? Two models have been considered: splicing specificity may be determined by the initial selection of one or the other polyadenylation site or by the choice of splice site itself. When Hela and F9 cells were transformed with a calcitonin/CGRP gene in which the normal calcitonin polyadenylation site in exon 4 was replaced by the distinctive CGRP polyadenylation site in exon 6 (and the exon 6 site was left unchanged), the relative amounts of calcitonin mRNA and CGRP mRNA in the two cell types remained unchanged. Thus, splicing specificity is apparently not determined by the polyadenylation site, but by choice of splice site. Splice site selection could, for example, result from an inhibition of exon 3/exon 4 splicing or an activation of the exon 3/exon 5 splicing in neuronal cells. Some experiments suggest that no special, neuronal cell–specific factors are likely to be required for exon 3/exon 5 splicing. Thus, when either HeLa or F9 cells are transfected with a calcitonin/CGRP gene in which the 3′ splice acceptor site at the beginning of exon 4 is deleted, both cell types use the 3′ acceptor site at the beginning of exon 5. This observation indicates that, at least in HeLa cells, everything required for the splicing to form CGRP mRNA is available. In contrast, F9 cells are unable to produce calcitonin mRNA when transfected with a calcitonin/CGRP gene in which the 3′ splice acceptor site at the beginning of exon 5 is deleted; this would be consistent with the presence, in neuronal cells, of a factor that inhibits splicing at the acceptor site used to produce calcitonin mRNA.

8.8 Messenger Peptides and Fixed Behavioral Patterns in *Aplysia*

Most eukaryotes display certain fixed behavioral patterns that are inherited and not generally influenced by learning. At least two genetic components appear necessary for such instinctive and rigidly programmed performances. First, the correct network of neurons and effector cells such as muscle cells must be laid down during development. Second, a system that restricts the response of the network to particular external or innate developmental stimuli must exist. There are hints that some of the neuroendocrine peptides described earlier in this chapter participate in such innate behavioral patterns. However, the ability to study the molecular genetics of mammalian stereotypical behavior is now very limited. For one thing, such patterns are often perturbed in very complex organisms by intersecting behavioral systems or by learned responses. For another, the central nervous systems are extraordinarily complex. Fortunately, some quite simple invertebrates also display fixed behaviors. Analysis of the messenger peptides that participate in such systems reveals a striking similarity to the features typical of mammalian genes for neuropeptides.

a. Fixed Behavioral Patterns in Aplysia

In the marine snail *Aplysia californica,* the correlation between gene structure, nerve cell function, and behavior is relatively simple. The snail has only about 2×10^4 nerve cells, and most of them are enormous—up to a millimeter in diameter. Many of these neurons are polyploid and contain vast quantities of DNA. From one snail to another, each neuron is connected to others in the same way, and this pattern is reproducibly identifiable. Moreover, specific neurons are associated with defined, stereotypical behaviors such as courtship, mating, and egg laying. The genetic principles likely to underlie all these patterns are exemplified by egg laying.

After copulation and internal fertilization of eggs within the reproductive duct, *Aplysia* lay their eggs onto surfaces in the sea. A million or more eggs are extruded in a long string from the genital aperture by contraction of the muscles surrounding the duct. The snail assists extrusion by waving its head, thereby

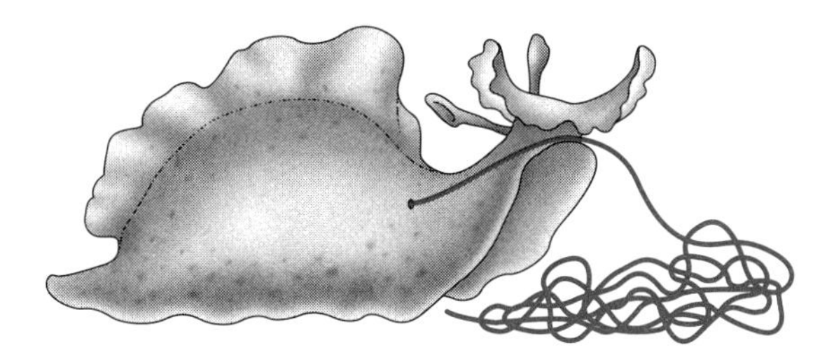

Figure 8.16
Schematic drawing of an *Aplysia* laying eggs. (Adapted from L. J. Jung and R. H. Scheller. 1991. Peptide Processing and Targeting in the Neuronal Secretory Pathway. *Science* 251 1331, Figure 1B.)

pulling the string along the body (Figure 8.16). The string becomes a tangled mass, held together by a sticky mucus secreted by the mouth, and one final wave of the head deposits the mass on a solid surface. While all this is going on, the snail stops eating and walking, and its heart and respiratory rates increase.

The 2×10^4 *Aplysia* neurons are collected in four head ganglia and one abdominal ganglion. At the top of the abdominal ganglion are two clusters, each containing about 400 neurons called **bag cells** (Figure 8.17). The bag cells are always fired prior to spontaneous egg laying, and extracts of bag cells, when injected into virgin snails, elicit the whole egg-laying performance even though the snail's eggs are unfertilized. Four physiologically active peptides that influence egg-laying behavior have been purified from the bag cell extracts. They are **egg-laying hormone** (**ELH**), **α-bag cell factor**, **β-bag cell factor**, and **γ-bag cell factor**.

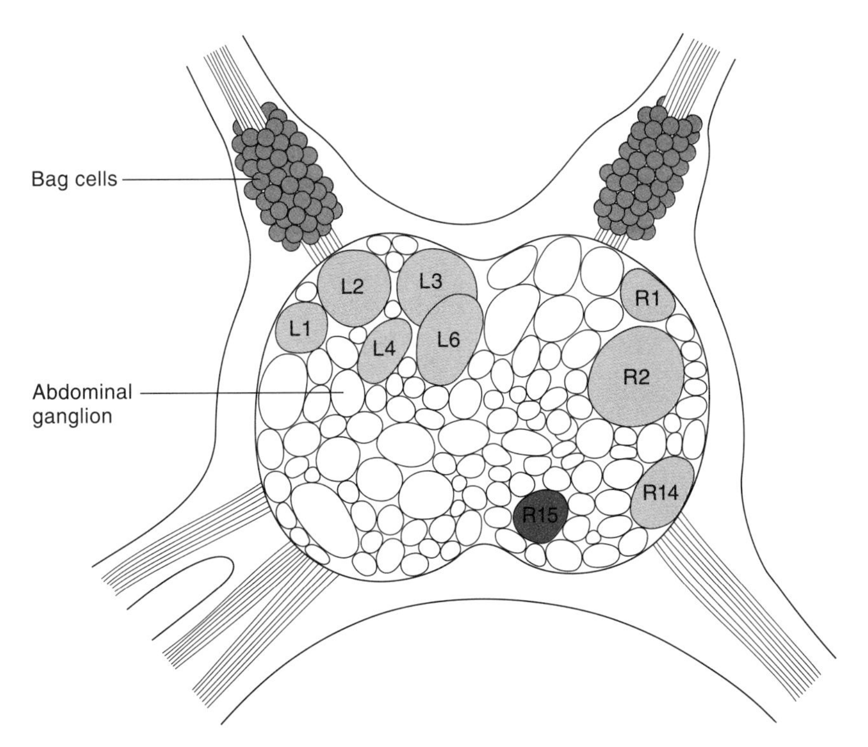

Figure 8.17
The abdominal ganglion of *Aplysia*. (Adapted from L. J. Jung and R. H. Scheller. 1991. Peptide Processing and Targeting in the Neuronal Secretory Pathway. *Science* 251 1334, Figure 5.)

The signal that initiates egg-laying behavior must relate somehow to fertilization and accompanying events in the reproductive duct. This or a subsequent signal must lead to the firing of bag cells and ultimately to the secretion of the peptides. Release of ELH from bag cells results in the excitation of neuron R15 as well as several other cells in the lower left quadrant of the abdominal ganglion; ELH also travels to the reproductive duct via the circulatory system and induces contraction of the duct muscles and release of oocytes. Thus, ELH is both a local neurotransmitter and a hormone. The bag cell factors also excite neurons in the abdominal ganglion and have, in addition, autocrine effects.

Another set of messenger peptides is secreted by the atrial gland, an exocrine tissue of *Aplysia's* reproductive organs; these peptides are secreted into the oviduct. The atrial gland peptides are related in structure to the bag cell peptides. Although at least some of them stimulate egg laying when injected into nonlaying *Aplysia,* their physiological function(s) is unclear. There is speculation that they may act as pheromones, thereby influencing mating behavior among individual *Aplysia.*

b. The Genes Encoding Egg-Laying Proteins

Preproproteins About half the protein made in bag cells during the egg-laying season is proELH, and the concentration of preproELH mRNA is correspondingly high. Differential screening of an *Aplysia* genomic library using cDNAs to the mRNA yielded a set of clones that contain sequences expressed in bag cells but not in other abdominal ganglion or nonneural cells. Under relatively relaxed annealing conditions, these clones also hybridize with cDNA prepared from atrial gland mRNA. However, under stringent conditions, the cloned genomic segments divided into two groups; one hybridizes with bag cell mRNA and the other with atrial gland mRNA. Several genes were then cloned using the cDNAs as probes.

The nucleotide sequences of the segments revealed that one of the genes encodes a preproELH, and others encode polyprotein precursors of atrial gland peptides. Figure 8.18 shows the structures of preproELH and one of the atrial gland peptide precursors, preproA, as deduced from the structures of the genes (these genes do not contain introns). The start of the polypeptides was taken to coincide with the methionine codons at the 5′ ends of the long open reading frames. Confidence in these conclusions came from *in vitro* translation experiments in which each of the corresponding mRNAs yielded precursor proteins of the predicted size. Both preproproteins are analogous in structure to vertebrate neuroendocrine peptide precursors, including an approximately 30 amino acid long typical signal peptide segment; the processing sites are also similar. Thus, the amino acid sequences of the known peptide hormones are flanked at the amino side by one or more basic amino acids. At the carboxy side, they are followed by the residues Gly-Lys-Arg, which provide for the carboxy terminal α-amidation of the peptides.

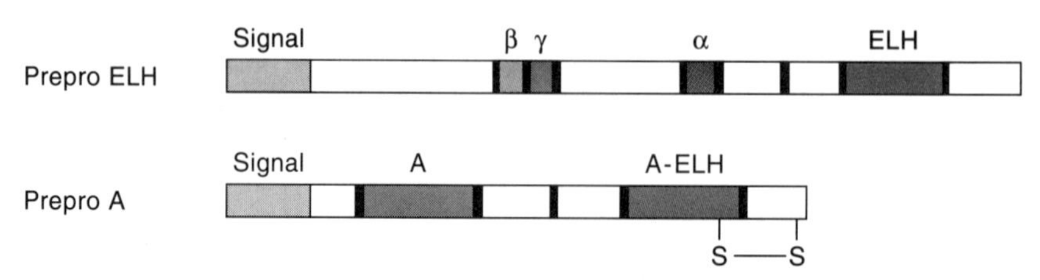

Figure 8.18
Aplysia genes for preproELH and preproA, shown here in the form of the encoded preproproteins.

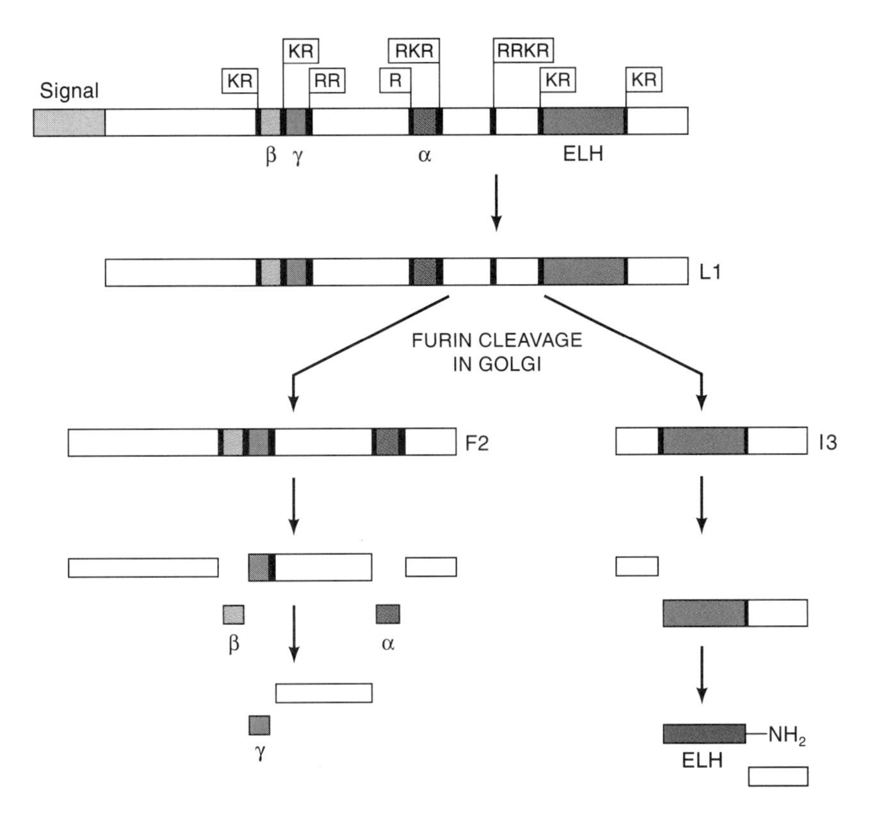

Figure 8.19
The processing of preproELH. After removal of the signal sequence in the ER, the proELH is cleaved by furin (at the Arg-Arg-Lys-Arg sequence) in the Golgi. The two products, F2 and I3, are sorted into different vesicles at the *trans* Golgi membrane and are processed to produce the α, β, and γ BCPs and the mature, amidated ELH, respectively. Cleavage by furin is relatively rapid (half-life of the proELH in the Golgi is 1–2 hours), but the formation of ELH takes about 4 hours and formation of the BCPs more than 20 hours. Moreover, much more of ELH is produced than of BCPs.

Additional peptides predicted by the genes The arrangement of familiar proteolytic cleavage sites suggested that the 271 amino acid long preproELH could be processed to yield several peptides besides ELH itself, and the suggestion has been confirmed. Several of these were previously identified as possible nonsynaptic neurotransmitters in the abdominal ganglion; others were identified after the structure of the gene predicted their existence. Thus, of the predicted **bag cell peptides** (**BCP**), the α-, β-, and γ-BCPs are known. The *A* gene also predicted previously unknown peptides such as the so-called A-ELH complex, which is now known to exist and be biologically active.

PreproELH processing The modification and processing of preproELH and the other *Aplysia* preprohormones, as well as secretion, depend on the same kind of vectorial passage through the network of intracellular membranes and vesicles as has already been described for yeast and mammals (Figure 8.19).

The signal sequence of preproELH is removed upon insertion of the nascent chain into the ER. Before leaving the Golgi, the 242 residue long proELH is cleaved into two parts: an amino terminal (F2, 18 kDa) and a carboxy terminal (I3, 8 kDa) portion. This cleavage occurs at a sequence of four basic amino acids, Arg-Arg-Lys-Arg (residues 181–184), and is relatively rapid; the half-life of proELH in the Golgi is about 30 minutes. That furin catalyzes the cleavage is very likely because (1) the site is the consensus furin target and (2) furin resides in the *trans* Golgi. Subsequent cleavages appear to occur in the post-Golgi secretory vesicles. Experiments with the carboxylic acid ionophore monensin confirm this model. In the presence of monensin, the Golgi cisternae swell, and the normally acidic pH within the lumen is neutralized; proteins do not leave the Golgi, and post-Golgi proteolytic cleavage of many prohormones cannot occur. In the presence of monensin, however, proELH is still cleaved into the two large fragments, although further processing does not take place.

Immunocytochemical analysis using specific antibodies and both light and electron microscopy has demonstrated that the two products of the first cleavage of proELH are sorted into separate and distinctive vesicles upon exiting the *trans* Golgi (Figure 8.20). This separation is associated with distinctive subsequent pathways for the ELH and bag cell peptides produced from the carboxy and amino terminal products, respectively. The shorter carboxy terminal fragment is packaged, at the *trans* Golgi, into relatively small vesicles with dense cores. Within these vesicles, the fragment is further processed, apparently in a stepwise manner, presumably by PC3/PC1- or PC2-like enzymes. Two successive cleavages take place at the Lys-Arg residues surrounding the ELH coding region. After the second cleavage, at the carboxy side, the two basic amino acids can be removed by CP-H, leaving a terminal glycine for the α-amidation of the carboxy terminus of the 36 amino acid long active ELH. Vesicles containing the maturing and matured ELH appear both in the cell body and in the nerve cell processes. Release of ELH after an appropriate stimulus delivers some of the hormone into the vascular system and some into abdominal ganglion, where it modulates the activity of specific ganglion neurons. The endocrine functions of ELH may explain why a higher concentration of ELH compared to the BCPs is advantageous.

Post-Golgi vesicles containing the 18 kDa amino terminal fragment F2 appear to be relatively large (compared to those containing the 8 kDa carboxy terminal fragment I3), and processing of the amino terminal fragment in these vesicles is relatively slow, perhaps because they contain only limiting amounts of the processing enzymes or have unfavorable conditions (e.g., pH). Significant numbers of these larger vesicles may fuse with lysosomes, leading to the degradation of the precursor molecules rather than productive processing to active molecules. Thus, one consequence of the separate vesicular sorting of the two portions of proELH is the production of significantly more ELH than α-, β-, and γ-BCP; the ratio is more than five to one. Thus, even though all these peptides are derived from a single proprotein, their relative amounts can be regulated. After stimulation of bag cells, the secreted BCPs act in an autocrine fashion to further influence bag cell activity and in a paracrine fashion on cells in the abdominal ganglion.

Those amino terminal fragments that are effectively processed are cleaved at five positions, again in an ordered series of reactions. All these cleavages are blocked in the presence of monesin, confirming that the very first cleavage of proELH is distinctive. One of these cleavages is after the single arginine flanking the amino terminal residue of α-BCP; cleavage is likely to be by a trypsinlike enzyme. Trimming by CP-H occurs in some instances, including at the carboxy end of α-BCP; forms of α-BCP containing nine, eight, or seven amino acids are known, and the shortest peptide is about 30 times more active physiologically than the longest. Overall, processing of the amino terminal portion of proELH is a much slower process than processing of the carboxy terminal portion, on the order of a day compared to about four hours.

The processing of proA illustrates a different mechanism for regulating secretion of active peptides. One of the products of proA is the A-ELH complex: two peptides derived from the carboxy end of proA are held together by inter-

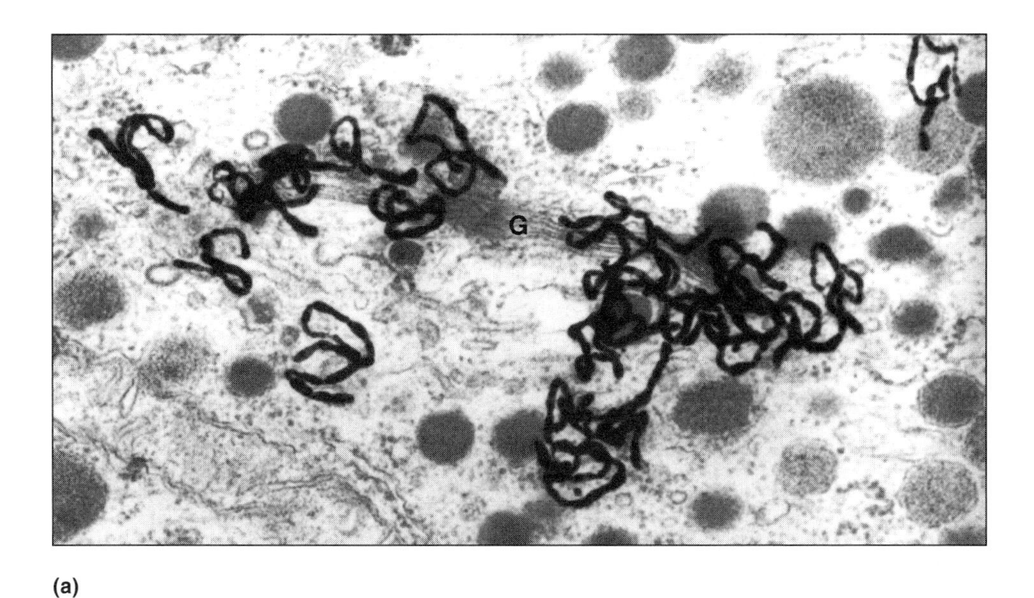

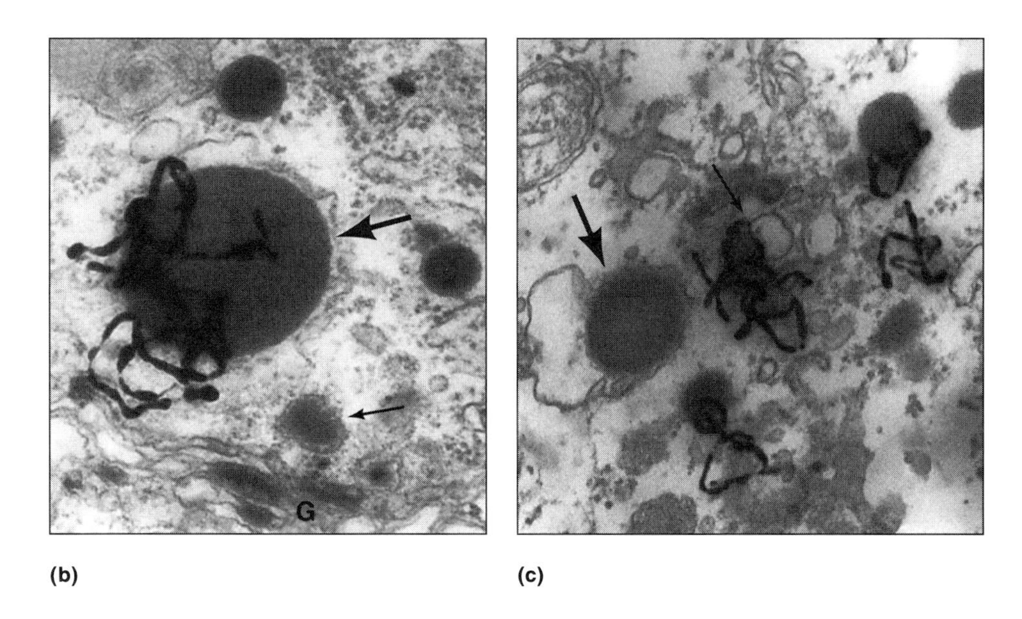

Figure 8.20
Electron micrographs of bag cell vesicles showing sorting of ELH and BCPs into different vesicles. The pictures show thin sections of ^{3}H-labelled cells. Note that in proELH almost all the isoleucine is in the carboxy terminal third and thus labels the ELH, while all the phenylalanine is in the amino terminal two-thirds and labels the BCPs. (a) After pulse labeling with ^{3}H-Ile, ^{3}H grains are over the Golgi (G). (b) After a pulse-chase experiment using ^{3}H-Phe, the grains are over large (large arrow) but not small (small arrow) immature vesicles. (c) After a pulse with ^{3}H-Ile and a chase, only the immature small vesicles are labelled. (From W. S. Sossin, J. M. Fisher, and R. H. Scheller. 1990. Sorting Within the Regulated Secretory Pathway Occurs in the *Trans* Golgi Network. *J. Cell. Biol.* 110 1–12. Reproduced by permission.)

molecular disulfides (as in insulin). Although the A-ELH complex is biologically active, its significance is not now understood. About 13 percent of the A-ELH complex can be cleaved at a Leu-Leu dipeptide (presumably by a reninlike enzyme) in one of the two chains, rendering the complex inactive. Such reactions illustrate the kind of fine-tuning that can occur under different conditions or perhaps in different cells.

Similarities between the preproA *and* preproELH *genes* Close inspection of the two genes depicted in Figure 8.18 reveals that they are remarkably similar. This should not be too surprising because, as noted earlier, they hybridize to one another under all but the most stringent conditions. Thus, the *preproELH* gene is very much like *preproA* except that it has a large insert (240 bp encoding 80 amino acids) in the region equivalent to that encoding the A peptide (or the *preproA* gene has a deletion). After the interruption, the sequence of *preproELH* is closely homologous to that of *preproA.* Within the 80 codon insertion, two of the small bag cell peptides, β and γ, are marked off by processing signals. The α-BCP is included in the A-like region of the *preproELH* gene. Thus, the 5′ portions of the two genes are remarkably similar. Similarly, both genes resemble one another in the 3′ half. The region encoding ELH on the *preproELH* gene is matched by homologous regions on *preproA,* but in the latter case, a missense mutation bars formation of ELH. Moreover, the homology extends to the long stretch between the A segment and the ELH segment of the two genes as well as into the carboxy terminal regions.

These two genes (and several other *Aplysia* genes as well) constitute a gene family; their similarities are so close that it seems certain they evolved by amplification and mutation from a single ancestral DNA segment. Together they provide a beautiful example of how specialized functions evolve from gene multiplication (*Genes and Genomes,* section 9.4b). The evolution of the coding sequences must have proceeded hand in hand with the evolution of regulatory regions that ensure expression of the particular genes in the specific cells, a question that remains to be studied.

c. The Development and Differentiation of Nerve Cells

Cloned genes are a powerful adjunct to classical cytological techniques such as immunocytochemistry. Consider, for example, the question of whether ELH is synthesized in nerve cells other than bag cells. Synthesis may be at an extremely low level but still of great significance. The use of cloned genes in a variety of ways markedly improves the sensitivity of methods, enabling the study of such questions. For example, specific probes tagged with fluorescent markers, including monoclonal antibodies, can be prepared and used to determine the cellular location of peptides and proteins in thin sections of tissues. Also, a labelled probe containing the sequence of the gene can be annealed to thin sections to mark cells that are synthesizing a specific RNA; even a single positive cell surrounded by many cells that are not expressing the gene can be detected. These sorts of experiments demonstrated that ELH is made in many nerve cells, including some in three of the head ganglia of *Aplysia.* When the techniques are applied to tissue samples from various embryonic stages, the developmental history of ELH-producing cells is displayed. Expression of the *preproELH* gene is detectable in ectodermal cells in very early *Aplysia* larvae. These cells appear to line the body wall. Later in development they migrate from the body wall to take up residence in their adult positions within the nervous system.

8.9 Comments

Increasing numbers of genes that encode more than one active polypeptide are being recognized. In some instances, different transcriptional start sites are used to produce different polypeptides. In others, alternative splicing or the use of alternative polyadenylation signals accounts for the versatility. These possibilities are also reflected in the expression of genes encoding messenger peptides. But most characteristically, versatility among the messenger peptide genes is provided by posttranslational processing of the primary translation product, the preproprotein, into multiple active peptides. The processing includes proteolysis and the modification of amino acid side chains. And, just as transcription or modification of primary transcripts is often regulated in distinctive ways in different cell types or at different times during development, so too is the posttranslational processing of preproproteins. Thus, evolutionary processes led to a variety of ways to make the most of the material at hand. Duplication of genes followed by independent evolution of regulatory processes and coding sequences gave rise to multiple, related genes whose products function in different ways, in different places, and at different times. Similarly , even within a single gene, reiteration of coding sequences allowed the synthesis of multiple copies of a messenger peptide or the independent evolution of the multiple copies to produce related but distinctive, active peptides.

Besides the important stories of the individual genes for messenger peptides, the secretory pathways themselves command great interest. Although highly complex and involving many proteins in a very highly organized train of membranes, the structures and processes appear to be quite ancient, as old as the earliest eukaryotes and perhaps even older because related structures and processes occur not only in plants but in prokaryotes as well. Considering the diversity of gene structure and mechanisms of gene expression adopted during evolution, there is a remarkable similarity among the genes, messenger peptides, and processes involved in the production of the peptides in diverse organisms. Moreover, these structures and processes themselves proved versatile during evolution and gave rise to such related but distinctive processes as the constitutive and regulated secretory pathways.

The evolution of messenger peptides must proceed in concert with the evolution of their corresponding cell surface receptor molecules because a messenger without a hospitable receptor is of no consequence. Similarly, the evolution of processing enzymes and preproproteins must be interrelated. Although this interdependent selective pressure is in one sense restrictive, it is also notable for the flexibility it provides. The extraordinary versatility of the neuroendocrine system reflects a rich evolutionary history and a diversity of functional opportunities.

Caution is required before concluding that all contemporary genes for neuroendocrine peptide precursors had some common origin. It is equally plausible in principle that selective pressure forced disparate precursors to adopt the same general strategy either independently or by joining related DNA segments to different peptide coding regions. Nevertheless, the evidence for some element of common and ancient origin is strong; yeast and mammals utilize similar genetic systems and processing pathways to produce intercellular messenger peptides. It is thus tempting to believe that such systems appeared at a very early stage of eukaryote evolution.

Late note: The complex process by which proproteins are transported from the endoplasmic reticulum to the Golgi complex remains under active investigation.

Recent results suggest that the description given in this chapter may require modification. Readers are referred to the following two papers: (1) H. R. B. Pelham (1994), "About Turn for the Cops?" in *Cell 79,* 1125–1127 and (2) L. Orcie, A. Perrelet, M. Ravazzola, M. Amherdt, J. E. Rothman, and R. Schekman (1994), "Coatomer-rich Endoplasmic Reticulum," in *Proceedings of the National Academy of Science USA,* 91:11924–11928.

Acknowledgments

The authors thank C. Zioudrou, R. E. Mains, and B. A. Eipper for their comments on a very early draft of this chapter and S. Pfeffer for a critical reading of the manuscript.

Further Reading

8.1

E. Herbert and M. Uhler. 1982. Biosynthesis of Polyprotein Precursors to Regulatory Peptides. *Cell* 30 1–2.

D. T. Krieger. 1983. Brain Peptides: What, Where, and Why? *Science* 222 975–985.

8.1a

J. M. Fisher and R. H. Scheller. 1988. Prohormone Processing and the Secretory Pathway. *J. Biol. Chem.* 263 16515–16518.

T. G. Sherman, H. Akil, and S. J. Watson. 1989. The Molecular Biology of Neuropeptides. Elsevier, Amsterdam. (This is vol. 6, no. 1 of the Discussions in Neuroscience series.)

K. Siddle and J. C. Hutton (eds.). 1990. Peptide Hormone Action, a Practical Approach. IRL Press, Oxford.

8.1b

J. C. Hutton and K. Siddle (eds.). 1990. Peptide Hormone Secretion, a Practical Approach. IRL Press, Oxford.

D. F. Steiner. 1991. The Biosynthesis of Biologically Active Peptides: A Perspective. In L. D. Fricker (ed.), Peptide Synthesis and Processing, pp. 1–15. CRC Press, Boca Raton, Florida.

M. K. Bennett and R. H. Scheller. 1993. The Molecular Machinery for Secretion Is Conserved from Yeast to Neurons. *Proc. Natl. Acad. Sci. USA* 90 2559–2563.

Y. P. Loh (ed.). 1993. Mechanisms of Intracellular Trafficking and Processing of Proproteins. CRC Press, Boca Raton, Florida.

8.2

8.2a

D. Julius, A. Brake, L. Blair, R. Kunisawa, and J. Thorner. 1984. Isolation of the Putative Structural Gene for the Lysine-Arginine-Cleaving Endopeptidase Required for Processing of Yeast Prepro-α-Factor. *Cell* 37 1075–1089.

L. D. Fricker. 1988. Carboxypeptidase E. *Ann. Rev. Physiol.* 50 309–321.

R. S. Fuller, R. E. Sterne, and J. Thorner. 1988. Enzymes Required for Yeast Prohormone Processing. *Ann. Rev. Physiol.* 50 345–362.

G. Thomas, B. A. Thorne, L. Thomas, R. G. Allen, D. E. Hruby, R. Fuller, and J. Thorner. 1988. Yeast *KEX*2 Endopeptidase Correctly Cleaves a Neuroendocrine Prohormone in Mammalian Cells. *Science* 241 226–230.

P. A. Bresnahan, R. Leduc, L. Thomas, J. Thorner, H. L. Gibson, A. J. Brake, P. J. Barr, and G. Thomas. 1990. Human *fur* Gene Encodes a Yeast KEX2-Like Endoprotease That Cleaves Pro-β-NGF *in Vivo*. *J. Cell Biol.* 111 2851–2859.

S. P. Smeekens and D. F. Steiner. 1990. Identification of a Human Insulinoma cDNA Encoding a Novel Mammalian Protein Structurally Related to the Yeast Dibasic Processing Protease Kex2. *J. Biol. Chem.* 265 2997–3000.

D. F. Steiner. 1991. Prohormone Processing Revealed at Last. *Curr. Biol.* 1 375–377.

C. Brenner and R. S. Fuller. 1992. Structural and Enzymatic Characterization of a Purified Prohormone-Processing Enzyme: Secreted, Soluble, Kex2 Protein. *Proc. Natl. Acad. Sci. USA* 89 922–926.

S. Ohagu, J. LaMendola, M. M. LeBeau, R. Espinosa III, J. Takeda, S. P. Smeekens, S. J. Chan, and D. F. Steiner. 1992. Identification and Analysis of the Gene Encoding Human PC2, a Prohormone Convertase Expressed in Neuroendocrine Tissues. *Proc. Natl. Acad. Sci. USA* 89 4977–4981.

N. G. Seidah and M. Chrétien. 1992. Proprotein and Prohormone Convertases of the Subtilisin Family. *Trends Endocrinol. and Metab.* 3 133–140.

D. F. Steiner, S. P. Smeekens, S. Ohagi, and S. J. Chan. 1992. The New Enzymology of Precursor Processing Endoproteases. *J. Biol. Chem.* 267 23435–23438.

C. A. Wilcox, K. Redding, R. Wright, and R. S. Fuller. 1992. Mutation of a Tyrosine Localization Signal in the Cytosolic Tail of Yeast Kex2 Protease Disrupts Golgi Retention and Results in Default Transport to the Vacuole. *Mol. Biol. Cell* 3 1353–1371.

8.2b

M. C. Pascale, M. C. Erra, N. Malagolini, F. Serafini-Cessi, A. Leone, and S. Bonatti. 1992. Post-Translational Processing of an O-Glycosylated Protein, the Human CD8 Glycoprotein, During the Intracellular Transport to the Plasma Membrane. *J. Biol. Chem.* 267 25196–25201.

S. Silberstein, J. Kelleher, and R. Gilmore. 1992. The 48-kDa Subunit of the Mammalian Oligosaccharyltransferase Complex Is Homologous to the Essential Yeast Protein WBP1. *J. Biol. Chem.* 267 23658–23663.

M. D. Snider. 1992. A Function for Ribophorines. *Curr. Biol.* 2 443–445.

8.2c

B. A. Eipper and R. E. Mains. 1988. Peptide α-Amidation. *Ann. Rev. Physiol.* 50 333–344.

N. S. Rangaraju, J. F. Xu, and R. B. Harris. 1991. Pro-Gonadotropin-Releasing Hormone Protein Is Processed Within Hypothalamic Neurosecretory Granules. *Neuroendocrinol.* 53 20–28.

D. A. Stoffers, L. Ouafik, and B. A. Eipper. 1991. Characterization of Novel mRNAS Encoding Enzymes Involved in Peptide α-Amidation. *J. Biol. Chem.* 266 1701–1707.

B. A. Eipper, D. A. Stoffers, and R. E. Mains. 1992. The Biosynthesis of Neuropeptides: Peptide α-Amidation. *Annu. Rev. Neurosci.* 15 57–85.

8.2d

W. B. Huttner. 1988. Tyrosine Sulfation and the Secretory Pathway. *Ann. Rev. Physiol.* 50 363–376.

C. Niehrs and W. B. Huttner. 1990. Purification and Characterization of Tyrosylprotein Sulfotransferase. *EMBO J.* 9 35–42.

C. Niehrs, M. Kraft, R. W. H. Lee, and W. B. Huttner. 1990. Analysis of the Substrate Specificity of Tyrosylprotein Sulfotransferase Using Synthetic Peptides. *J. Biol. Chem.* 265 8525–8532.

W. Huttner, C. Niehrs, and C. Vannier. 1991. Bind or Bleed. *Curr. Biol.* 1 309–310.

8.2e

T. R. Gibson and C. C. Glembotski. 1985. Acetylation of α-MSH and β-Endorphin by Secretory Granule-Associated Acetyl Transferase. *Peptides* 6 615–620.

8.3

8.3a

R. B. Freedman. 1984. Native Disulphide Bond Formation in Protein Biosynthesis: Evidence for the Role of Protein Disulphide Isomerase. *Trends Biochem. Sci.* 9 438–441.

S. M. Simon and G. Blobel. 1991. A Protein-Conducting Channel in the Endoplasmic Reticulum. *Cell* 65 371–380.

R. E. Dalbey and G. von Heijne. 1992. Signal Peptidases in Prokaryotes and Eukaryotes—A New Protease Family. *Trends Biochem. Sci.* 17 474–478.

M. J. Getting and J. Sambrook. 1992. Protein Folding in the Cell. *Nature* 355 33–45.

J. Nunnari and P. Walter. 1992. Protein Targeting and Translocation Across the Membrane of the Endoplasmic Reticulum. *Curr. Opin. Cell Biol.* 4 573–580.

T. A. Rapaport. 1992. Transport of Proteins Across the ER Membrane. *Science* 258 931–936.

S. L. Sanders and R. Shekman. 1992. Polypeptide Translocation Across the Endoplasmic Reticulum Membrane. *J. Biol. Chem.* 267 13791–13794.

R. Gilmore. 1993. Protein Translocation Across the Endoplasmic Reticulum: A Tunnel with Toll Booths at Entry and Exit. *Cell* 75 589–592.

A. E. Johnson. 1993. Protein Translocation Across the ER Membrane: A Fluorescent Light at the End of the Tunnel. *Trends Biochem. Sci.* 18 456–458.

R. B. Kelly. 1993. Much Ado About Docking. *Curr. Biol.* 3 474–476.

K. E. S. Matlack and P. Walter. 1993. Shedding Light on the Translocation Pore. *Curr. Biol.* 3 677–679.

R. I Morimoto. 1993. Chaperoning the Nascent Polypeptide Chain. *Curr. Biol.* 3 101–102.

R. B. Freedman, T. R. Hirst, and M. F. Tuite. 1994. Protein Disulphide Isomerase: Building Bridges in Protein Folding. *Trends Biochem. Sci.* 19 331–336.

8.3b

H. R. B. Pelham. 1989. Control of Protein Exit from the Endoplasmic Reticulum. *Annu. Rev. Cell Biol.* 5 1–23.

M. Bomsel and K. Mostov. 1992. Role of Heterotrimeric G Proteins in Membrane Traffic. *Mol. Biol. of the Cell* 3 1317–1328.

M. Hosobuchi, T. Kreis, and R. Schekman. 1992. *SEC21* is a Gene Required for ER to Golgi Protein Transport That Encodes a Subunit of a Yeast Coatomer. *Nature* 360 603–605.

S. R. Pfeffer. 1992. GTP-Binding Proteins in Intracellular Transport. *Trends Cell Biol.* 2 41–46.

N. K. Pryer, L. J. Wuestehube, and R. Shekman. 1992. Vesicle-Mediated Protein Sorting. *Annu. Rev. Biochem.* 61 471–516.

J. E. Rothman and L. Orci. 1992. Molecular Dissection of the Secretory Pathway. *Nature* 355 409–415.

R. Schekman. 1992. Genetic and Biochemical Analysis of Vesicular Traffic in Yeast. *Curr. Opin. Cell Biol.* 4 587–592.

J. Armstrong. 1993. Two Fingers for Membrane Traffic. *Curr. Biol.* 3 33–35.

C. Barlowe, C. d'Enfert, and R. Schekman. 1993. Purification and Characterization of SAR1p, a Small GTP-Binding Protein Required for Transport Vesicle Formation from the Endoplasmic Reticulum. *J. Biol. Chem.* 268 873–879.

C. Barlowe and R. Schekman. 1993. *SEC*12 Encodes a Guanine-Nucleotide-Exchange Factor Essential for Transport Vesicle Budding from the ER. *Nature* 365 347–349.

L. Orci, D. J. Palmer, A. Amherdt, and J. E. Rothman. 1993. Coated Vesicles Assembly in the Golgi Requires Only Coatomer and ARF Proteins from the Cytosol. *Nature* 364 732–734.

P. A. Takizawa and V. Malhotra. 1993. Coatomers and SNAREs in Promoting Membrane Traffic. *Cell* 75 593–596.

G. Warren. 1993. Bridging the Gap. *Nature* 362 297–298.

T. Yoshihisa, C. Barlowe, and R. Shekman. 1993. A Requirement for a GTPase Activating Protein in Vesicle Budding from the ER. *Science* 259 1466–1468.

M. Zerial and H. Stenmark. 1993. Rab GTPases in Vesicular Transport. *Curr. Opin. Cell Biol.* 5 613–620.

I. Mellman (ed.). 1994. Membranes and Sorting. *Curr. Opin. Cell Biol.* 6 497–560.

8.3c

H. F. Lodish. 1988. Transport of Secretory and Membrane Glycoproteins from the Rough Endoplasmic Reticulum to the Golgi. *J.Biol. Chem.* 263 2107–2110.

I. Mellman and K. Simons. 1992. The Golgi Complex: In Vitro Veritas? *Cell* 68 829–840.

J. P. Lian and S. Ferro-Novick. 1993. Bos1p, an Integral Membrane Protein of the Endoplasmic Reticulum to Golgi Transport Vesicles Is Required for Their Fusion Competence. *Cell* 73 735–745.

C. E. Machamer. 1993. Targeting and Retention of Golgi Membrane Proteins. *Curr. Opin. Cell Biol.* 5 606–612.

M. Moya, D. Roberts, and P. Novick. 1993. *DSS4-1* Is a Dominant Suppressor of *sec4*-8 That Encodes a Nucleotide Exchange Protein that Aids Sec4p Function. *Nature* 361 460–463.

P. Novick. 1993. Friends and Family: The Role of the Rab GTPases in Vesicular Traffic. *Cell* 75 597–601.

T. Söllner, S. W. Whiteheart, M. Brunner, H. Erdjument-Bromage, S. Geromanos, P. Tempst, and J. E. Rothman. 1993. SNAP Receptors Implicated in Vesicle Targeting and Fusion. *Nature* 362 318–324.

N. Calakos, M. K. Bennett, K. E. Peterson, and R. H. Scheller. 1994. Protein-Protein Interactions Contributing to the Specificity of Intracellular Vesicular Trafficking. *Science* 263 1146–1149.

S. Ferro-Novick and R. Jahn. 1994. Vesicle Fusion from Yeast to Man. *Nature* 370 191–193.

J. E. Rothman and G. Warren. 1994. Implications of the SNARE Hypothesis for Intracellular Membrane Topology and Dynamics. *Curr. Biol.* 4 220–233.

8.3d

J. H. Keen. 1990. Clathrin. *Ann. Rev. Biochem.* 59 415–438.

S. G. Miller and H.-P. Moore. 1990. Regulated Secretion. *Curr. Opin. Cell Biol.* 2 642–647.

M. J. Rindler. 1992. Biogenesis of Storage Granules and Vesicles. *Curr. Opin. Cell Biol.* 4 616–622.

S. M. Hurtley. 1993. Membrane Proteins Involved in Targeted Membrane Fusion. *Trends Biochem. Sci.* 18 453–455.

R. B. Kelly. 1993. Storage and Release of Neurotransmitters. *Cell* 72/*Neuron* 10 *Review Supplement* 43–53.

P. Melancon. 1993. G Whizz. *Curr. Biol.* 3 230–233.

H. R. B. Pelham and S. Munro. 1993. Sorting of Membrane Proteins in the Secretory Pathway. *Cell* 75 603–605.

T. C. Südhof, P. De Camilli, H. Niemann, and R. Jahn. 1993. Membrane Fusion Machinery. *Cell* 75 1–4.

8.4

8.4a

J. Kurjan and I. Herskowitz. 1982. Structure of a Yeast Pheromone Gene (*MFα*): A Putative α-Factor Precursor Contains Four Tandem Repeats of Mature α-Factor. *Cell* 30 933–943.

A. Singh, E. Y. Chen, J. M. Lugovoy, C. N. Change, R. A. Hitzeman, and P. H. Seeburg. 1983. *Saccharomyces cerevisiae* Contains Two Discrete Genes Coding for the α-Factor Pheromone. *Nucleic Acids Res.* 11 4049–4063.

J. Kurjan. 1985. α-Factor Structural Gene Mutations in *S. cerevisiae*: Effects on α-Factor Production and Mating. *Mol. Cell. Biol.* 5 787–796.

8.4b

D. Julius, L. Blair, A. Brake, G. Sprague, and J. Thorner. 1983. Yeast α Factor Is Processed from a Larger Precursor Polypeptide: The Essential Role of a Membrane-Bound Dipeptidyl Aminopeptidase. *Cell* 32 839–852.

R. S. Fuller, R. E. Sterne, and J. Thorner. 1988. Enzymes Required for Yeast Prohormone Processing. *Ann. Rev. Physiol.* 50 345–362.

G. L. Bush, A.-M. Tassin, H. Fridén, and D. M. Meyer. 1991. Secretion in Yeast: Purification and *in Vitro* Translocation of Chemical Amounts of Prepro-α-Factor. *J. Biol. Chem.* 266 13811–13814.

8.5

8.5a

P. Lomedico, N. Rosenthal, A. Efstratiadis, W. Gilbert, R. Kolodner, and R. Tizard. 1979. The Structure and Evolution of the Two Nonallelic Rat Preproinsulin Genes. *Cell* 18 545–558.

I. Sures, D. V. Goeddel, A. Gray, and A. Ullrich. 1980. Nucleotide Sequence of Human Preproinsulin Complementary DNA. *Science* 208 57–59.

S. P. Smeekens, A. G. Montag, G. Thomas, C. Albiges-Rizo, R. Carroll, M. Benig, L. A. Phillips, S. Martin, S. Ohagi, P. Gardner, H. H. Swift, and D. F. Steiner. 1992. Proinsulin Processing by the Subtilisin-Related Proprotein Convertases Furin, PC2, and PC3. *Proc. Natl. Acad. Sci. USA* 89 8822–8826.

8.5b

A. Ullrich, C. H. Berman, T. J. Dull, A. Gray, and J. M. Lee. 1984. Isolation of the Human Insulin-Like Growth Factor Gene Using a Single Synthetic DNA Probe. *EMBO J.* 3 361–364.

M. B. Soares, E. Schon, A. Henderson, S. K. Karathanasis, R. Cate, S. Zeitlin, J. Cirgwin, and A. Efstratiadis. 1985. RNA-Mediated Gene Duplication: The Rat Preproinsulin I Gene Is a Functional Retroposon. *Mol. Cell. Biol.* 5 2090–2103.

Y. W.-H. Yang, M. M. Rechler, S. P. Nissley, and J. E. Coligan. 1985. Biosynthesis of Rat Insulin-Like Growth Factor II. *J. Biol. Chem.* 260 2578–2582.

L. Orci, M. Ravazzola, M.-J. Storch, R. G. W. Anderson, J.-D. Vassalli, and A. Perrelet. 1987. Proteolytic Maturation of Insulin Is a Post-Golgi Event Which Occurs in Acidifying Clathrin-Coated Secretory Vesicles. *Cell* 49 865–868.

A. Robitzki, H. C. Schröder, D. Ugarkovic, K. Pfeifer, G. Uhlenbruck, and W. E. G. Müller. 1989. Demonstration of an Endocrine Signaling Circuit for Insulin in the Sponge *Geodia cydonium*. *EMBO J.* 8 2905–2909.

D. Marriott, B. Gillece-Castro, and C. M. Gorman. 1992. Prohormone Convertase-1 Will Process Prorelaxin, a Member of the Insulin Family of Hormones. *Mol. Endocrinol.* 6 1441–1450.

C. J. Rhodes, B. Lincoln, and S. R. Shoelson. 1992. Preferential Cleavage of des-31,32-Proinsulin Over Intact Proinsulin by the Insulin Secretory Granule Type II Endopeptidase: Implications of a Favored Route for Prohormone Processing. *J. Biol. Chem.* 267 22719–22727.

8.6

8.6a

D. Richter. 1983. Vasopressin and Oxytocin Are Expressed as Polyproteins. *Trends Biochem. Sci.* 8 278–281.

H. Schmale, S. Heinsohn, and D. Richter. 1983. Structural Organization of the Rat Gene for the Arginine Vasopressin-Neurophysin Precursor. *EMBO J.* 2 763–767.

R. Ivell and D. Richter. 1984. The Gene for the Hypothalamic Peptide Hormone Oxytocin Is Highly Expressed in the Bovine Corpus Luteum: Biosynthesis, Structure and Sequence Analysis. *Embo J.* 3 2351–2354.

8.6b

S. Nakanishi, Y. Teranishi, Y. Watanabe, M. Notake, M. Noda, H. Kakidani, H. Jingami, and S. Numa. 1981. Isolation and Characterization of the Bovine Corticotropin/β-Lipoprotein Precursor Gene. *Eur. J. Biochem.* 115 429–438.

M. Comb, P. H. Seeburg, J. Adelman, L. Eiden, and E. Herbert. 1982. Primary Structure of the Human Met- and Leu-Enkephalin Precursor and Its mRNA. *Nature* 295 663–666.

S. Legon, D. M. Glover, J. Hughes, P. J. Lowry, P. W. J. Rigby, and C. J. Watson. 1982. The Structure and Expression of the Preproenkephalin Gene. *Nucleic Acids Res.* 10 7905–7918.

P. L. Whitfeld, P. H. Seeburg, and J. Shine. 1982. The Human Proopiomelanocortin Gene: Organization, Sequence, and Interspersion with Repetitive DNA. *DNA* 1 133–143.

M. Comb, H. Rosen, P. Seeburg, J. Adelman, and E. Herbert. 1983. Primary Structure of the Human Proenkephalin Gene. *DNA* 2 213–229.

S. Horikawa, T. Takai, M. Toyosato, H. Takahashi, M. Noda, H. Kakidani, T. Kubo, T. Hirose, S. Inayama, H. Hayashida, T. Miyata, and S. Numa. 1983. Isolation and Structural Organization of the Human Preproenkephalin B Gene. *Nature* 306 611–614.

D. Liston, G. Patey, J. Rossier, P. Verbanck, and J.-J. Vanderhaegen. 1984. Processing of Pro-Enkephalin Is Tissue-Specific. *Science* 225 735–737.

R. E. Mains and B. A. Eipper. 1990. The Tissue Specific Processing of Pro-ACTH/Endorphin: Recent Advances and Unsolved Problems. *Trends Endocrinol. and Metab.* 1 388–394.

S. Benjannet, N. Rondeau, R. Day, M. Chrétien, and N. G. Seidah. 1991. PC1 and PC2 Are Proprotein Convertases Capable of Cleaving Proopiomelanocortin at Distinct Pairs of Basic residues. *Proc. Natl. Acad. Sci. USA* 88 3564–3568.

J. Korner, J. Chun, D. Harter, and R. Axel. 1991. Isolation and Functional Expression of a Mammalian Prohormone Processing Enzyme, Murine Prohormone Convertase 1. *Proc. Natl. Acad. Sci. USA* 88 6834–6838.

B. A. Thorne, O. H. Viveros, and G. Thomas. 1991. Expression and Processing of Mouse Proopiomelanocortin in Bovine Adrenal Chromaffin Cells: A Model System to Study Tissue-Specific Prohormone Processing. *J. Biol. Chem.* 266 13607–13615.

J. Silberring, M. E. Castello, and F. Nyberg. 1992. Characterization of Dynorphin A–Converting Enzyme in Human Spinal Cord. *J. Biol. Chem.* 267 21324–21328.

A. Zhou, B. T. Bloomquist, and R. E. Mains. 1993. The Prohormone Convertases PC1 and PC2 Mediate Distinct Endoproteolytic Cleavages in a Strict Temporal Order During Proopiomelanocortin Biosynthetic Processing. *J. Biol. Chem.* 268 1763–1769.

8.7

8.7a

J. Allison, L. Hall, I. MacIntyre, and R. K. Craig. 1981. The Construction and Partial Characterization of Plasmids Containing Complementary DNA Sequences to Human Calcitonin Precursor Polyprotein. *Biochem. J.* 199 725–731.

M. G. Rosenfeld, C. R. Lin, S. G. Amara, L. Stolarsky, B. A. Roos, E. S. Ong, and R. M. Evans. 1982. Calcitonin mRNA Polymorphism: Peptide Switching Associated with Alternative RNA Splicing Events. *Proc. Natl. Acad. Sci. USA* 79 1717–1721.

M. G. Rosenfeld, J.-J. Mermod, S. G. Amara, L. W. Swanson, P. E. Sawchenko, J. Rivier, W. W. Vale, and R. M. Evans. 1983. Production of a Novel Neuropeptide Encoded by the Calcitonin Gene via Tissue Specific RNA Processing. *Nature* 304 129–135.

M. I. Sabate, L. S. Stolarsky, J. M. Polak, S. R. Bloom, I. M. Varndell, M. A. Ghatei, R. M. Evans, and M. G. Rosenfeld. 1985. Regulation of Neuroendocrine Gene Expression by Alternative RNA Processing. *J. Biol. Chem.* 260 2589–2592.

8.7b

E. B. Crenshaw III, A. F. Russo, L. W. Swanson, and M. G. Rosenfeld. 1987. Neuron-Specific Alternative RNA Processing in Transgenic Mice Expressing a Metallothionein-Calcitonin Fusion Gene. *Cell* 49 389–398.

S. E. Leff, R. M. Evans, and M. G. Rosenfeld. 1987. Splice Commitment Dictates Neuron-Specific Alternative RNA Processing in Cacitonin/CGRP Gene Expression. *Cell* 48 517–524.

R. B. Emeson, F. Hedjran, J. M. Yeakley, J. W. Guise, and M. G. Rosenfeld. 1989. Alternative Production of Calcitonin and CGRP mRNA Is Regulated at the Calcitonin-Specific Splice Acceptor. *Nature* 341 76–80.

M. Zaidi, B. S. Moonga, P. J. R. Bevis, Z. A. Bascal, and L. H. Breimer. 1990. The Calcitonin Gene Peptides: Biology and Clinical Relevance. *Critical Rev. Clinical Lab. Sciences* 28 109–174.

S. Minvielle, S. Giscard-Darteville, R. Cohen, J. Taboulet, F. Labye, A. Jullienne, P. Pivaille, G. Milhaud, M. Moukhtar, and F. Lasmoles. 1991. A Novel Calcitonin Carboxyl-Terminal Peptide Produced in Medullary Thyroid Carcinoma by Alternative RNA Processing of the Calcitonin/Calcitonin Gene-Related Peptide Gene. *J. Biol. Chem.* 266 24627–24631.

G. J. Ademoa and P. D. Baas. 1992. A Novel Calcitonin-Encoding mRNA Is Produced by Alternative Processing of Calcitonin/Calcitonin Gene-Related Peptide-I Pre-mRNA. *J. Biol. Chem.* 267 7943–7948.

B. Lu, W.-M. Fu, P. Greengard, and M.-M. Poo. 1993. Calcitonin Gene-Related Peptide Potentiates Synaptic Responses at Developing Neuromuscular Junction. *Nature* 363 76–79.

8.8

8.8a

R. H. Scheller, B. S. Rothman, and E. Mayeri. 1983. A Single Gene Encodes Multiple Peptide-Transmitter Candidates Involved in a Stereotyped Behavior. *Trends Neurosci.* 6 340–345.

J. M. Fisher, W. Sossin, R. Newcomb, and R. H. Scheller. 1988. Multiple Neuropeptides Derived from a Common Precursor Are Differentially Packaged and Transported. *Cell* 54 813–822.

L. J. Jung and R. H. Scheller. 1991. Peptide Processing and Targeting in the Neuronal Secretory Pathway. *Science* 251 1330–1335.

8.8b

R. Newcomb, J. M. Fisher, and R. H. Scheller. 1988. Processing of Egg-Laying Hormone (ELH) Precursor in the Bag Cell Neurons of *Aplysia*. *J. Biol. Chem.* 263 12514–12521.

W. S. Sossin, J. M. Fisher, and R. H. Scheller. 1990. Sorting Within the Regulated Secretory Pathway Occurs in the *Trans*-Golgi Network. *J. Cell Biol.* 110 1–12.

L. J. Jung and R. H. Scheller. 1991. Peptide Processing and Targeting in the Neuronal Secretory Pathway. *Science* 251 1330–1335.

M. K. Bennett and R. H. Scheller. 1993. The Molecular Machinery for Secretion Is Conserved from Yeast to Neurons. *Proc. Natl. Acad. Sci. USA* 90 2559–2563.

L. J. Jung, T. Kreiner, and R. H. Scheller. 1993. Prohormone Structure Governs Proteolytic Processing and Sorting in the Golgi Complex. *Rec. Progr. in Hormone Res.* 48 415–436.

8.9

A. Driouich, L. Faye, and L. A. Staehelin. 1993. The Plant Golgi Apparatus: A Factory for Complex Polysaccharides and Glycoproteins. *Trends Biochem. Sci.* 18 210–214.

M. K. Bennett and R. H. Scheller. 1993. The Molecular Machinery for Secretion Is Conserved from Yeast to Neurons. *Proc. Natl. Acad. Sci. USA* 90 2559–2563.

S. M. Bajjalieh, K. Peterson, R. Shinghal, and R. H. Scheller. 1992. SV2, a Brain Synaptic Vesicle Protein Homologous to Bacterial Transporters. *Science* 257 1271–1273.

CHAPTER 9

Regulation of *Drosophila* Development by Transcription Factors and Cell-Cell Signaling

THOMAS B. KORNBERG and MATTHEW P. SCOTT

9.1 The Developmental Program of *Drosophila*

Developmental biology encompasses the many disciplines that seek to understand how organisms grow and how their cells respond to the changing environments that development creates. This field was born during the last decades of the nineteenth century when meticulous descriptions of embryogenesis began to plot how various plants, vertebrates, and invertebrates mature. During the past decade, our understanding of animal and plant development has progressed rapidly as the application of modern molecular genetics, revolutionary recombinant DNA technologies, and classical genetics has transformed this previously descriptive science into one that can now trace the outlines of the underlying molecular mechanisms that define the developmental program.

Critical to the explosive progress in this field have been the increasingly efficient ways in which the genes that direct developmental programs can be identified and isolated. Such studies of the genes that regulate the development of the fruit fly *Drosophila melanogaster* have been the most successful and most informative. In this chapter, we provide a general overview of the genetic approaches used to identify these genes in *Drosophila,* the molecular techniques that have made it possible to isolate and characterize them, and some of the properties of a few of the regulatory proteins that these genes encode.

a. Egg, Embryo, Larva, Adult

Female flies store sperm after mating and dispense one sperm to each maturing egg. The newly laid fly egg already has evident polarity, with respiratory appendages and other asymmetries that invariably predict, in wild-type flies, the orientation of the embryo that will develop. The egg develops very rapidly: after about 22 hours, a larva has formed with a complete nervous system, musculature, and epidermal structures. *Drosophila* is a bit unusual in having a **syncytial** form of development in the early embryo. The first 14 divisions after the pronuclei fuse are nuclear divisions in a common cytoplasm (Figure 9.1). The embryo takes full advantage of the opportunity to move macromolecules without membrane interference. After the divisions are complete, cell membranes form and the embryo briefly is in a cellular blastoderm stage. It is a monolayer of about 6000 cells with yolk inside, and by this time, many important regulatory genes are differentially active within the visually indistinguishable cells. **Gastrulation** begins immediately to create multiple cell types and layers. The first cells to invaginate are the mesoderm cells, which move to the interior as a block. Next, individual cells release from a large region of the ectoderm and singly move to the interior as neuroblasts, which will give rise to the entire nervous system. Cells located at each end of the embryo move to the interior to form the endoderm, and organogenesis follows. The epidermal cells secrete the cuticle, the exoskeleton, of the first larval stage.

When the larva hatches at about 22 hours, it has the ambition of a glutton. Its breathing tubes protrude through its tail, allowing it to burrow head first in food—in the lab, made of yeast, cornmeal, and molasses and, in the wild, mainly yeast. Under favorable conditions, the larva grows so fast that after only a day, it is bursting at its seams and must molt to allow growth to continue. A pulse of the steroid hormone ecdysone triggers molting. The second instar larva, as voracious as the first, molts again after another day. At each molt, the same epidermal cells secrete the new cuticle while destroying the old one. The third instar larva remains as such until the ripe old age of two days (for a total of five since fertilization). At this time, the larva has an inclination to wander, leaving the safe haven of the food and emerging on a pilgrimage about 8 hours long—in the lab, up and around the side of a glass vial. After identifying a favorable anchoring spot, it attaches itself to the glass, and the cuticle darkens and shortens over a few hours to form the puparium, encased in the modified third instar larval cuticle. Four hours after puparium formation, the epidermis has retracted from the outer cuticle and secreted a prepupal cuticle. The prepupa is distinct from the pupa in having no external legs or other appendages.

Some cells are sequestered during embryogenesis, reserved as precursors of adult tissues. Among these are the small clusters of cells that will form **imaginal disks**, the primordia of the eyes, wings, legs, halteres, antennae, other head structures, and genitalia of the adult fly (Figure 9.2). During larval development, the imaginal disk cells grow at a constant rate and eventually form flattened sacs consisting of tens of thousands of cells. In addition to the epithelial cells of the disks, cells that will later form the muscles of the appendages migrate to the

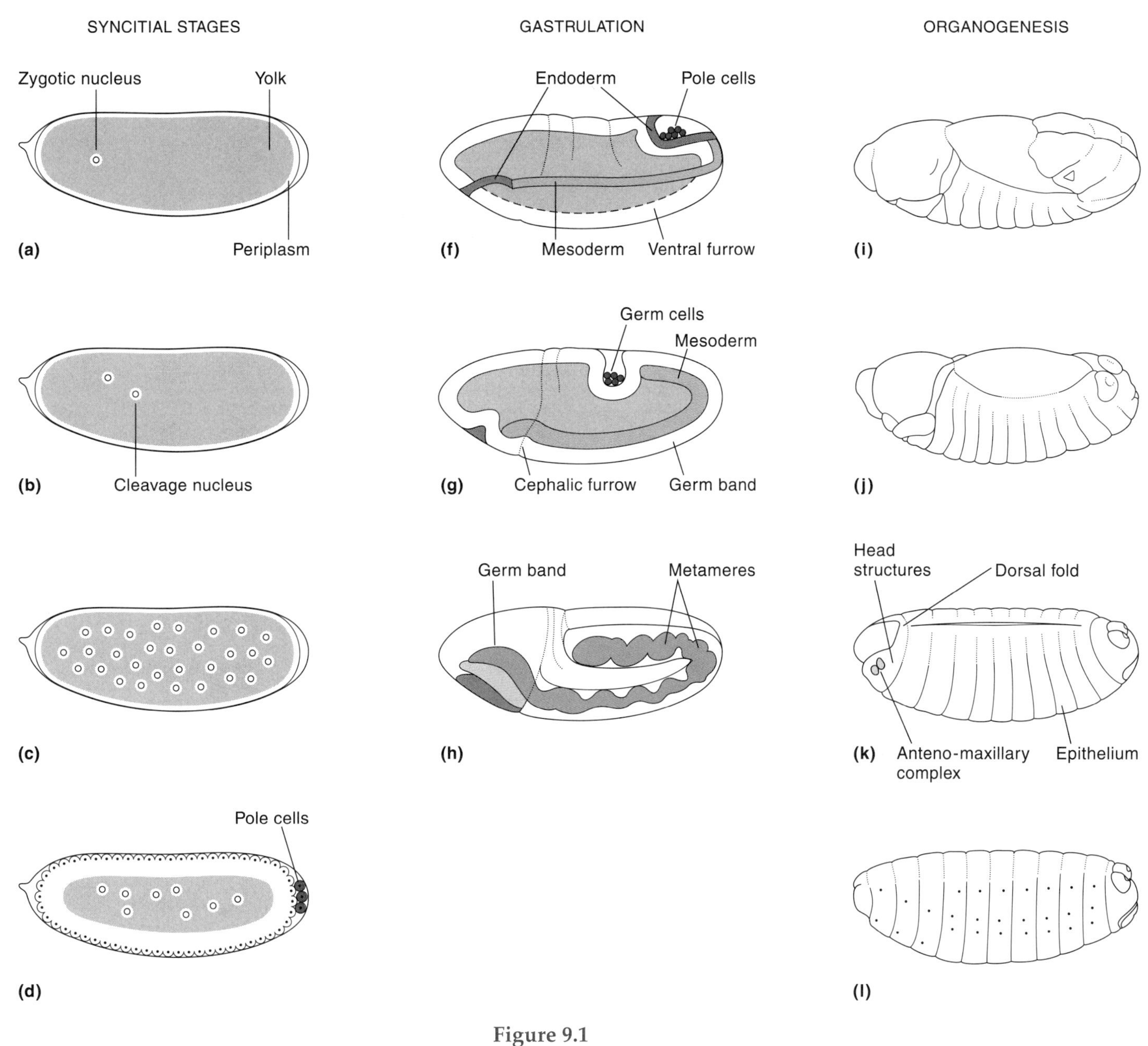

Figure 9.1
Development of *Drosophila* from fertilization to the segmented embryo. In these side views, the anterior is to the left and the ventral side is down. (a)–(e) Syncytial stages; the nuclei divide without cytokinesis and migrate to the periphery. (f)–(h) Gastrulation; the cellularized blastoderm initiates a set of regulated cell divisions that are coordinated with cell movements along the surface and invaginations into the interior that reorganize the embryo and generate the primitive gut, germ layers, and metameres. (i)–(l) Organogenesis; continuing divisions and shuffling of the embryo cells internalize the head structures, cover the surface with an epithelium, and mature the internal organs.

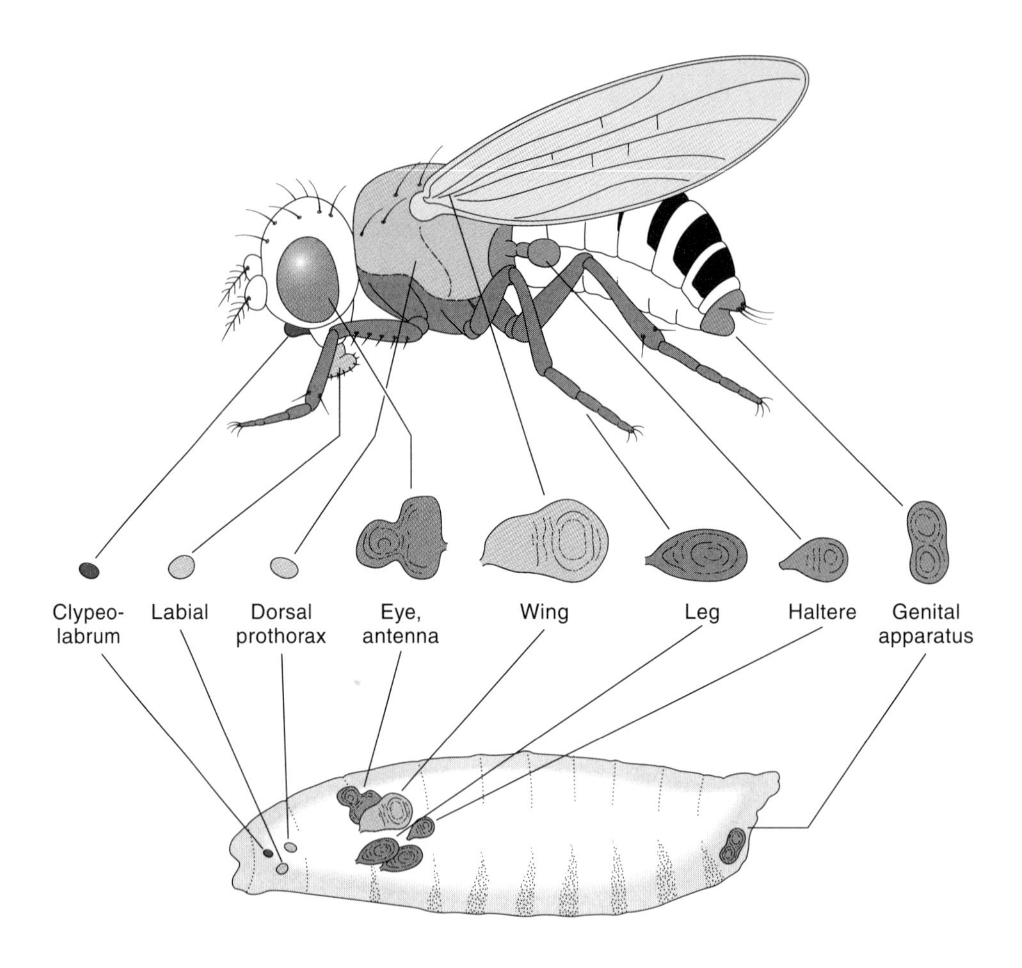

Figure 9.2
The imaginal disks of a *Drosophila* larva. These organs are nourished during the larval period while tethered to the larval epidermis, and generate the adult head, thoracic and genital epithelia after metamorphosis. (From V. Hartenstein. 1993. Atlas of Drosophila Development. Cold Spring Harbor Press, Cold Spring Harbor, New York.)

disks. Together these cells remain attached to the larval epidermis until metamorphosis, when they spread out to replace the dying larval cells. To form appendages, the cell sheets extend and elongate. The cells that had been at the center of the disks extend the most, to form the most distal parts of the appendage. Those cells located closer to the disk's connection to the larval body wall epidermis form more proximal structures. Abdominal segments form from clusters of cells called **histoblast nests**. Thus, the surface of the adult is a patchwork quilt, stitched together with pieces contributed by 17 imaginal disks (the odd one is the genital disk; all others come in pairs) and 14 histoblast nests. Internal organ precursor cells are set aside during embryonic stages as well, often interspersed with larval cells. Most larval cells become terminally differentiated and polyploid; adult precursors remain diploid.

Pupa formation occurs about 12 hours after puparium formation, when a new pupal cuticle is secreted. This new encasement leaves the pupa cosily within the pupal cuticle, the prepupal cuticle, and, on the outside, the puparium. Metamorphosis of internal tissues follows with the near complete breakdown of many tissues, such as muscles, and the partial reconstruction of others, such as the brain.

Four days after puparium formation, the adult emerges from the pupa. About another half day is required before the flies are fertile. Thus, the complete life cycle is about ten days at 25°C.

b. The Embryo of Drosophila melanogaster

The *Drosophila* egg is relatively large (500 mm × 180 mm) and has an asymmetric shape that clearly distinguishes its anterior and posterior poles and dorsal and ventral surfaces. However, remarkably few of the materials that make up the egg are sequestered in restricted locations. Most of the RNA, protein, and other components are distributed uniformly within either the egg membranes or cytoplasm. Although there is little to suggest the presence of a blueprint for subsequent development, the processes that initiate after fertilization reproducibly position the embryo within this structured egg.

Fertilization initiates rapid changes in the composition and organization of the *Drosophila* egg (Figure 9.1). The meiotic process of the mother, which had been arrested in meiosis I, completes its production of a maternally derived pronucleus and three nonfunctional polar bodies. Within 20–30 minutes of fertilization, replication of both the chromosomes of the maternal pronucleus and sperm and fusion of the two pronuclei in the interior of the embryo signal the beginning of a series of mitotic divisions that proceed without cytokinesis. These mitoses occur synchronously, with a cycle time of approximately 12 minutes. As the nuclei increase in number, they migrate outward. They eventually populate the surface of the zygote, leaving the yolk-filled egg interior almost completely devoid of nuclei. One hundred thirty minutes after fertilization, the approximately 4000 nuclei stop dividing, and in a period of about 40 minutes, membranes grow between the nuclei to subdivide the egg periphery into individual cells. This first cellular state is called the **cellular blastoderm**.

Two distinct populations of cells exist in the cellular blastoderm of the *Drosophila* embryo. At the posterior tip, a cluster of cells that formed during the period of the last few syncytial divisions becomes physically isolated from the other cells of the embryo. These will become the **germ cells** of the animal. If they are removed surgically or do not form in mutant embryos, the adults that develop will have no germ cells and will be sterile. The remaining nuclei distributed around the periphery have no morphological features that distinguish them from each other. Yet genetic, molecular, and other experimental approaches reveal that the processes that partition the embryo into different germ layers and segments have already begun when the concerted cell movements of gastrulation begin, approximately three hours after fertilization. In outline, gastrulation begins with the invagination of a sheet of mesoderm precursors along the ventral midline. Ectodermal cells still on the outside of the embryo move toward this ventral midline, and the ventral band of cells, called the germ band, extends around the posterior end of the embryo until the segmental primordia are wrapped in a nearly complete U shape, with tail structures behind the head. During this process, the ectoderm takes on a salt-and-pepper character, with presumptive neuroblast cells leaving the ectodermal cell sheet and moving as single cells into the interior.

Determination of cell fates depends on differential activation of the zygotic genome along the anterior-posterior (A-P) and dorsal-ventral (D-V) axes and on functions provided by RNA and proteins in the egg cytoplasm inherited from the mother. The maternal substances are critical to forming the proper pattern in the embryo.

9.2 Maternal Influences on Developmental Events

a. Anterior-Posterior Segmentation and Dorsal-Ventral Polarization

The embryo activates its own genome to control segmentation, but the initial events of embryogenesis, as well as some of the subsequent functions, employ RNA and protein molecules provided to the egg by the mother during oogenesis. The necessity for maternal products is revealed by mutations that when homozygous in the mother alter her progeny's pattern of development. This genetic condition contrasts with the essential segmentation genes, which provide their contributions from the genome of the zygote and therefore depend upon both the maternal and paternal chromosome contributions. When the **zygotic segmentation genes** are mutant, there are dramatic effects such as changes in the number and character of body segments, but several key features of the embryonic pattern are unaffected. None of the zygotic segmentation mutants alters the relationship of the dorsal and ventral structures of the embryo. Nor do any of the zygotic lethal mutants alter the underlying A-P polarity of the embryo. The maternally provided products set up the initial polarity of the embryo along both axes.

The many genes required to establish the A-P and D-V polarities of the embryo are expressed by the mother during oogenesis, and the products of these genes are provided to the egg as part of the maternal dowry. Female flies homozygous for mutations in some of these genes develop normally and are perfectly viable, but they are sterile (Figure 9.3). They cannot provide all the necessary gene products (RNA or protein) to their progeny. Such mutations are classified as having **maternal-effects**, and they include flies unable to produce eggs at all, flies that make visibly defective eggs, and flies that make eggs that appear normal but are unable to develop properly.

If a gene is required for both oogenesis and development of the zygote, it will not be possible to obtain homozygous mutant females: they will die for lack of the gene product and will never have the opportunity to produce eggs. To test whether a gene required for development of the adult is also needed for oogenesis, scientists produce mosaic female flies that have a single functional gene in most of their tissues, allowing survival, but no functional gene in their germ line cells (Figure 9.4). The embryos can then be assessed for developmental defects due to their mother's (germ line) genetic constitution.

Thus, there are three classes of genes for embryonic development: those required exclusively maternally, those required exclusively zygotically, and those required during both oogenesis and zygotic development. The latter include all the "housekeeping" genes needed for basic cellular functions, which will be essential but not necessarily of prime relevance to embryonic patterning. The *Drosophila* egg contains a rich supply of components that allow the embryo to develop through the syncytial stages with little or no transcription of the zygotic genome. Mutations in many genes that provide products to the egg during oogenesis have a female sterile phenotype, but most of the genes have no role in pattern formation. Genetic screens seeking to identify those functions involved specifically in embryonic axis formation are difficult and laborious because they must sift through many irrelevant female sterile lines.

Screens to identify all genes required to orient the embryonic axes require two steps. The first is the generation of many individual lines, each with a mutagenized chromosome. In order to produce viable females so their progeny can be examined, the newly mutant chromosomes cannot carry any zygotic lethal mutations. To eliminate such chromosomes from the screen, investigators cross F2 progeny of mutagenized males to each other to generate stocks of flies that can produce offsping that are homozygous for a particular chromosome. The second step involves screening the progeny of these stocks to identify lines whose ho-

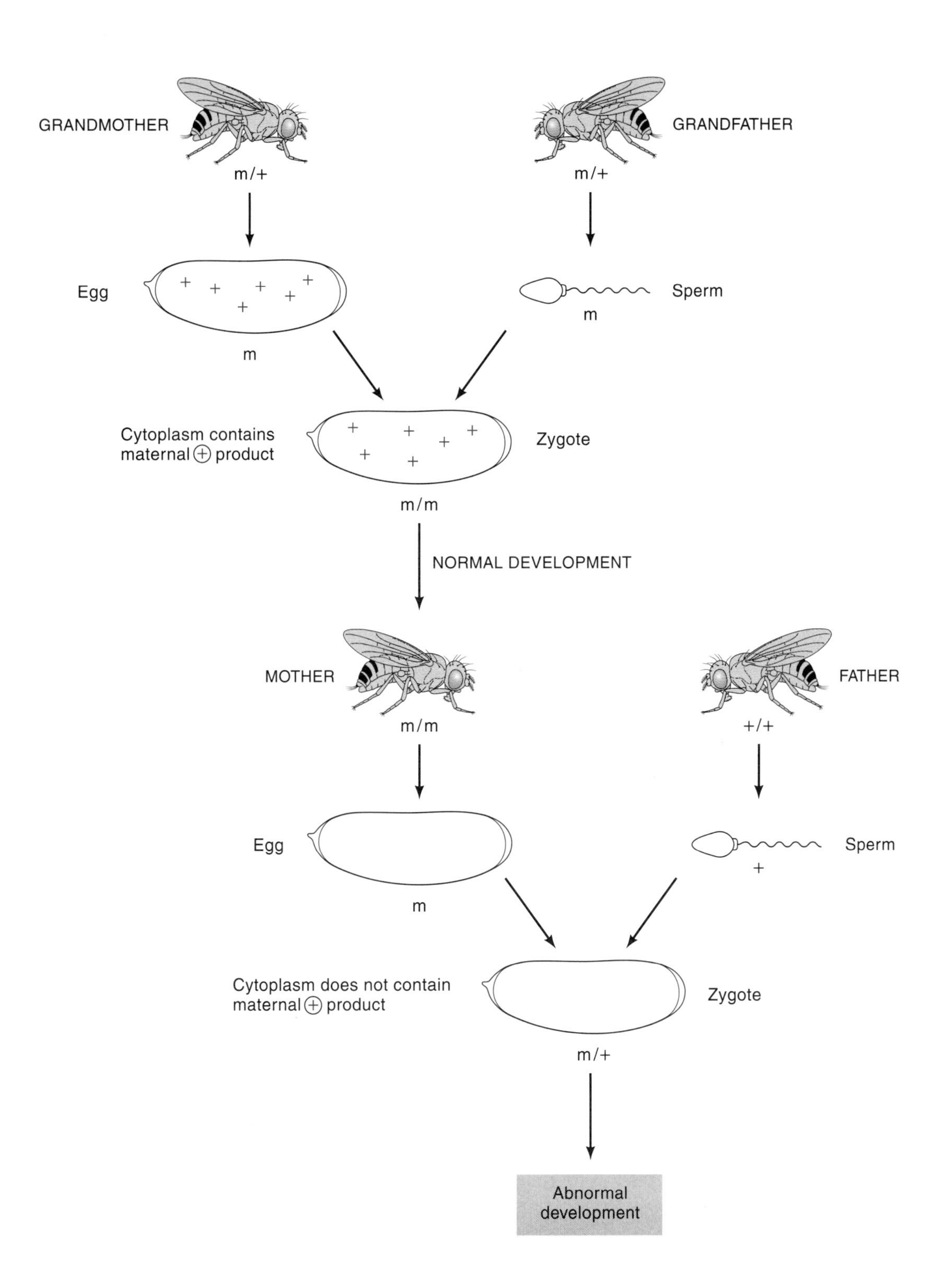

Figure 9.3
Recessive maternal-effect mutations manifest their effects in the progeny of affected females. These mutations do not affect the development of heterozygous (m/+) or homozygous (m/m) animals, but the normal gene product (+) must be supplied by the mother for the embryo to survive. A paternally contributed wild-type gene cannot compensate for this type of maternal deficit. (From B. Alberts et al. 1994. Molecular Biology of the Cell, 3rd ed. p. 1082. Garland, New York.)

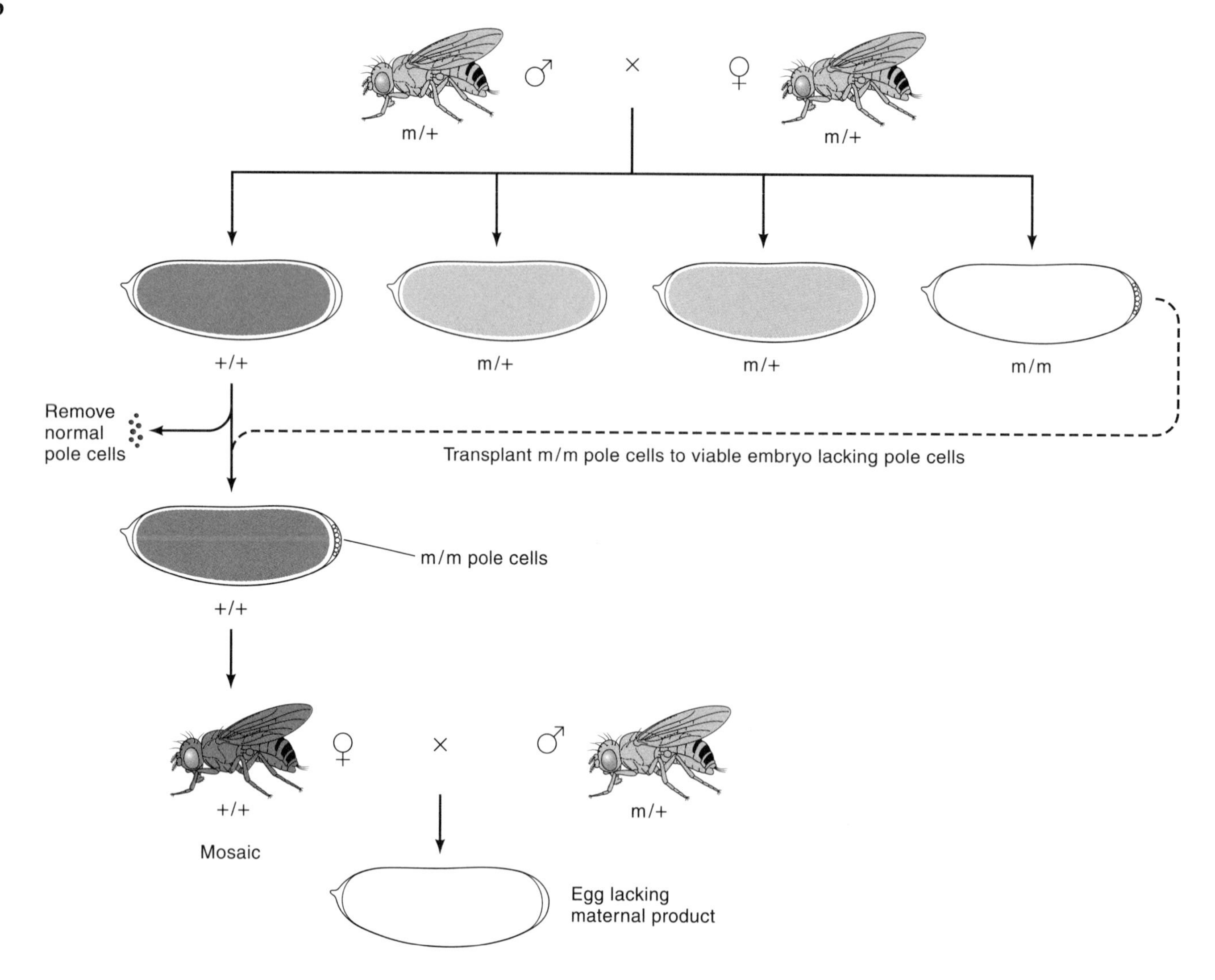

Figure 9.4
Embryos without functional germ cells can be rescued by transplantation. Micropipets can be used to transfer germ cells from embryos that cannot survive but have functional germ cells (m/m; white) to embryos that can develop but lack functional germ cells. The surviving animals are genetically mosaic because their somatic cells and germ cells have different genotypes. (From P. Lawrence. 1992. The Making of a Fly. p. 36. Alden Press, Oxford, England.)

mozygous females fail to generate viable progeny and whose embryos die with abnormal patterns. Two-step screens for such maternal effect lethals have been conducted for each of the major autosomes and for the X chromosome. From extensive screens of this nature, researchers estimate that approximately 30–40 genes are needed to endow the egg with the capacity to generate normal embryonic patterns. Interestingly, most of the isolated mutants affect either the A-P or the D-V pattern of the embryo. The patterning mechanisms that organize the two major axes of the *Drosophila* egg initially function independently of each other.

b. Maternal Dowry Regulation of Embryonic A-P Pattern Polarization

The mechanism that generates the A-P axis has three components; these have been called the anterior system, the posterior system and the terminal system.

Eighteen genes that staff these systems are known, the lack of any one of which leads to a **maternal lethal** phenotype. ("Maternal lethal" means that if the female fly is homozygous, her progeny will die. "Female sterile" has some of the same implication, except this designation may also mean females that fail to produce eggs at all.) Study of these mutant phenotypes and cloning and characterization of many of these genes reveals that the three systems operate through different molecular mechanisms, but fundamentally, all encode position in a similar manner. Each produces a **concentration gradient** of a protein that has the capacity to regulate a transcriptional process in a concentration-dependent manner.

One of the key regulatory proteins of the anterior system is encoded by the maternal effect gene *bicoid* (*bcd*). It produces one of the exceptional molecules in the maternal dowry that is localized during oogenesis. ***In situ* hybridization** has been used to show the location of *bcd* mRNA, which is tightly sequestered at the anterior tip of the oocyte during oogenesis. This is a method in which labelled copies of a gene sequence (a probe) are applied to a tissue sample under conditions that encourage annealing of the probe to complementary sequences in the tissue. If the tissue contains both chromosomal DNA and mRNA and the preparation of the tissue did not denature the DNA, then probe hybridization under conditions that favor RNA:DNA hybrid formation will identify regions in the tissue that contain RNA sequences complementary to the DNA probe. *In situ* hybridization probes may be prepared in a number of different ways, depending upon the nature of the tissue sample and other parameters. If the probe is prepared with radioactive nucleotides, hybridized probe can be detected by autoradiographic methods in which photographic emulsion is layered over the sample after hybridization. If the probe is prepared with certain derivatized bases that can be recognized with antibodies directed against the unusual nucleotides, then the hybridized probe can be detected by immunohistochemical detection methods. Such methods allow for highly sensitive detection with sufficient spatial resolution to distinguish areas of RNA accumulation within actively transcribing nuclei. Sequences in the 3′ untranslated region of the *bcd* mRNA have been implicated in its anterior localization in a process requiring the participation of the products of three other genes: *exuperentia, swallow,* and *staufen.*

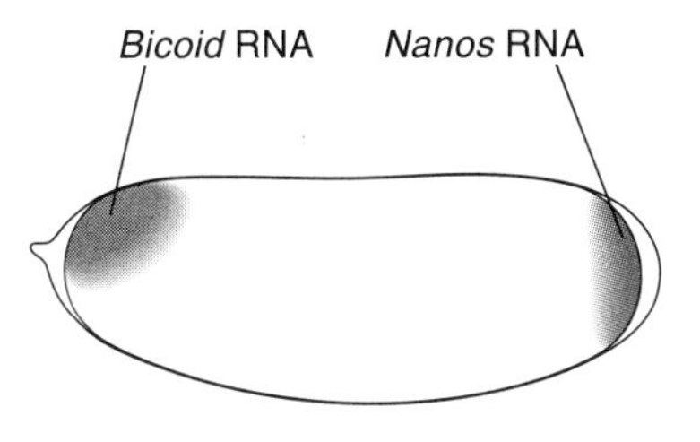

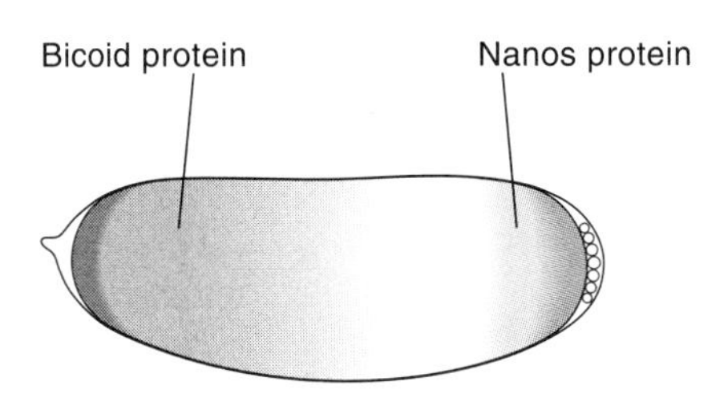

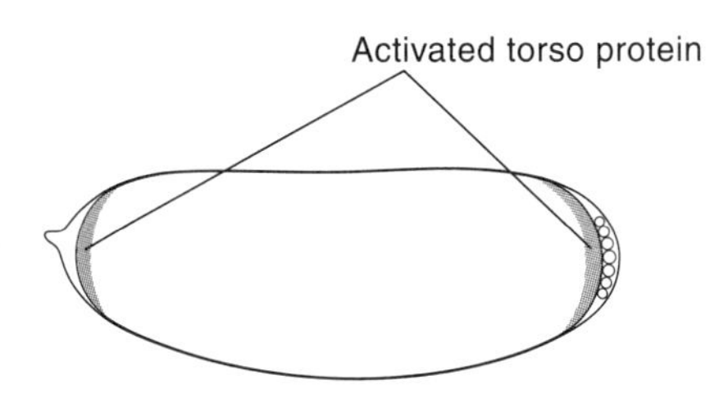

Figure 9.5
Localized determinants at the ends of the *Drosophila* embryo control its A/P polarity. RNA sequestered at the anterior pole (*bicoid* mRNA) and posterior pole (*nanos* mRNA) of mature eggs generates gradients of protein after fertilization. The uniformly distributed *torso* receptor is activated only at the poles.

RNA locations determined by *in situ* hybridization are usefully compared to protein distubutions determined with antibodies, for example, to see whether a protein may be encoded in one location and move to another (Figure 9.5). Bcd protein is produced in the egg, but only after fertilization, and its diffusion away from the anterior pole generates a concentration gradient during the precellular stages of embryogenesis. The formation of the gradient takes advantage of the lack of cell membranes in the early embryo.

The instructional capacity of the *bcd* gradient has been demonstrated in two ways. The amount of *bcd* mRNA sequested at the anterior tip of an egg can be varied by changing the number of copies of *bcd* genes in the mother (Figure 9.6). With increasing numbers of copies of the *bcd* gene, the mother provides the egg with more copies of *bcd* mRNA, and the A-P proportionality of the embryo changes in response to the different amounts of *bcd* mRNA, presumably due to altered levels of Bcd protein. One copy of the *bcd* gene in the mother results in an embryo in which the segments form in a more anterior location than if the mother has two copies of the *bcd* gene; embryos produced by mothers with four copies of *bcd* form segments in a more posterior location; embryos produced by mothers with six copies of *bcd* form segments in an even more posterior location.

Bcd protein is a transcriptional regulator that is thought to act as a **morphogen** by eliciting distinct transcriptional responses from its several targets in response to different concentrations of protein. Among its targets are the segmentation genes *buttonhead, orthodenticle, and empty spiracles,* each of which acts in head development, the gap genes *hunchback* and *giant,* and the pair-rule gene *even-skipped*. Expression of these target genes is sensitive to the level of Bcd protein, and the promoter of each gene has sites to which the Bcd protein can bind.

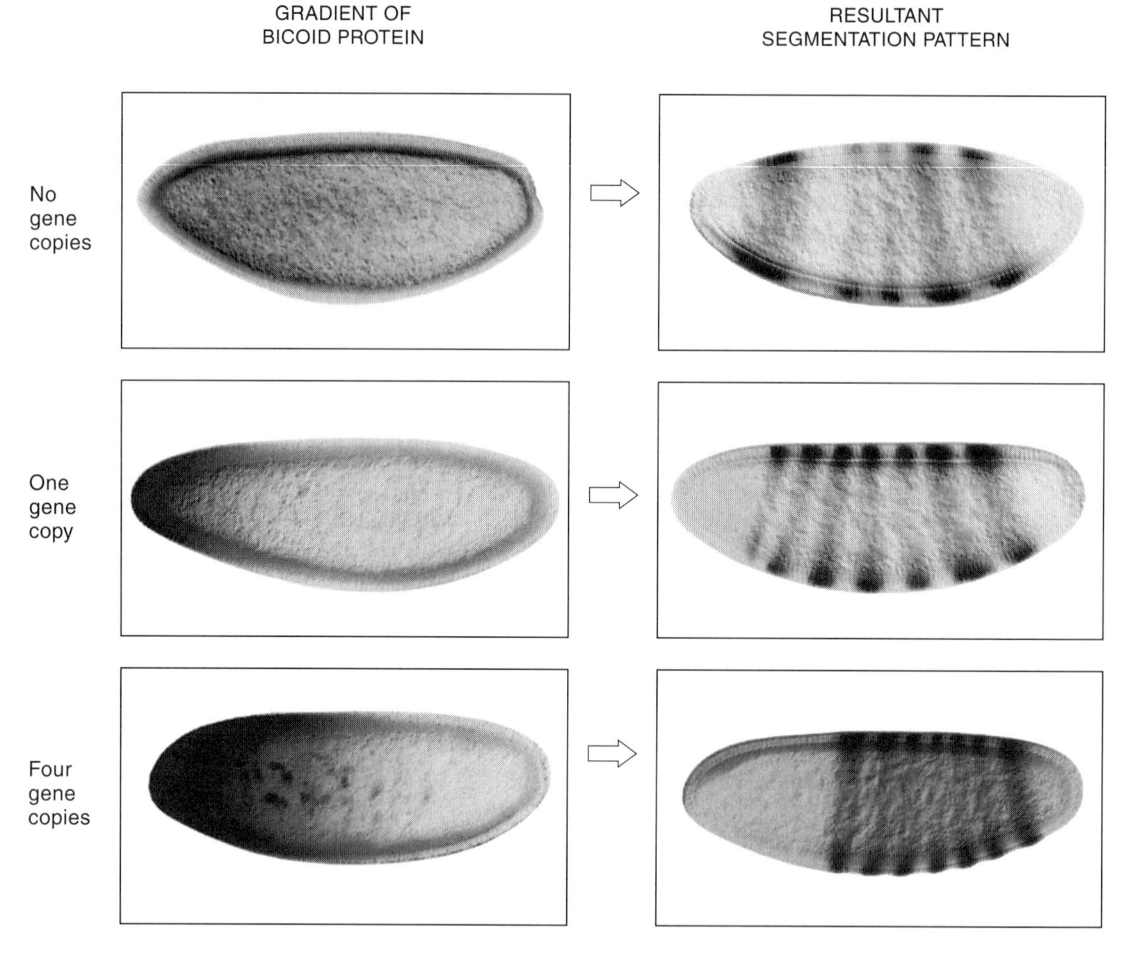

Figure 9.6
The organizing influence of the bicoid protein is revealed by affecting its distribution. Normally, females have two copies of the bicoid gene that together produce bicoid RNA that is deposited in the developing egg. If the number of functional bicoid genes is reduced or increased, the amount of bicoid mRNA in the embryo changes proportionally, as does the distribution of bicoid protein. In response, the pattern of stripes of pair-rule genes, such as *fushi tarazu,* changes. (From Alberts et al. 1994. Molecular Biology of the Cell, 3rd ed. p. 1087. Garland, New York.)

We discuss the targets further in Section 9.3.

The definition of a morphogen requires that the response to it be concentration dependent. In the case of *bicoid,* each target gene is responsive in a different spatial pattern. Some targets become active only in response to high levels of bicoid protein (i.e., in the most anterior region of the embryo); other target genes are activated at lower concentrations of bicoid and therefore are active in more posterior regions. Why the genes respond to different bicoid concentrations is not fully worked out, but possible mechanisms include: (1) different affinity binding sites for bicoid at different targets, (2) different numbers of binding sites at different targets, and (3) different interactions with other proteins at different targets. All three of these mechanisms may apply to the interaction between bicoid and its targets. Sites to which Bcd protein can bind with high affinity respond to lower concentrations of Bcd at more posterior locations along the Bcd concentration gradient than do sites with a lower affinity. Target genes activated in the more anterior regions have low affinity binding sites for *bcd,* whereas target genes activated in the more posterior regions have sites that bind *bcd* with

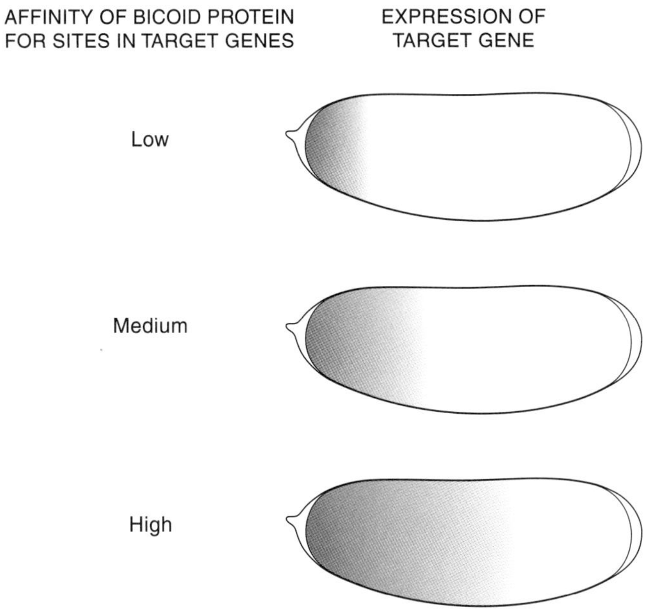

AFFINITY OF BICOID PROTEIN FOR SITES IN TARGET GENES

EXPRESSION OF TARGET GENE

Low

Medium

High

Figure 9.7
Concentration-dependent activation of bicoid target genes. The mechanism that translates the gradient of bicoid protein into differential responses involves the variable affinity of bicoid protein for regulatory sequences in the bicoid target genes. Regulatory sequences with low affinity can be activated only where the concentration of bicoid protein is high—the most anterior locations. Conversely, regulatory sequences with high affinity can be activated where the concentration of bicoid protein is low—the more posterior locations.

higher affinity. By this mechanism, the broad monotonic gradient of Bcd protein is transduced into the smaller domains of expression of its targets (Figure 9.7).

In the gene system that controls posterior development, *nanos* RNA is tightly sequestered at the posterior pole. Sequences in the 3′ untranslated region of the *nanos* mRNA have been implicated in its posterior localization, a process that requires the function of at least seven genes (*cappuccino, oskar, spire, staufen, tudor, valois,* and *vasa*). Nanos protein is produced only after fertilization and is distributed in a gradient whose peak is at the posterior pole. Nanos protein is thought to target specific mRNAs for **translational repression** and/or degradation. Sequence elements responsive to *nanos* have been found in the 3′ untranslated regions of *bcd* and *hunchback* (a gap segmentation gene) mRNAs, and *nanos*-dependent repression of these transcripts allows abdominal development to proceed. The concentration of *nanos* protein probably provides an instructional measure of distance from the posterior pole by proportionally reducing the concentration of *hunchback* RNA.

Despite the tight association of *bcd* and *nanos* RNAs with the anterior and posterior poles of the embryo, respectively, *bcd* and *nanos* functions are not primarily responsible for the development of the extreme anterior and posterior regions of the embryo. Rather, a separate **"terminal" system** of maternally provided products is required for the anterior head and for the tail (Figure 9.8). In contrast to the anterior and posterior systems, the terminal system has no determinants known to be localized in the embryo. The terminal pathways are thought to be initiated by a signaling pathway that is controlled by a **receptor tyrosine kinase, *torso***. Torso protein is synthesized after fertilization and distributed evenly within the plasma membrane of the embryo. Activation of the *torso* receptor must be localized to the embryo poles. The product of the gene *torsolike,* a novel protein, has recently been identified as the best candidate for the ligand. The *torsolike* signal emanates from one or several of the immediately adjacent somatic cells. Gene products implicated in the terminal system signal transduction pathway include a serine-threonine kinase homologue (*l[1]pole hole*), a tyrosine phosphatase homologue (*corkscrew*), *ras1,* a positive regulator of *ras1* (*son of sevenless*), and a MAP kinase homologue (*Dsor1*) (Figure 9.8). Ongoing studies promise to identify additional components of the *torso* signal transduction pathway and ultimately to reveal how activation of *torso* leads to position-dependent responses of target genes in the terminal regions.

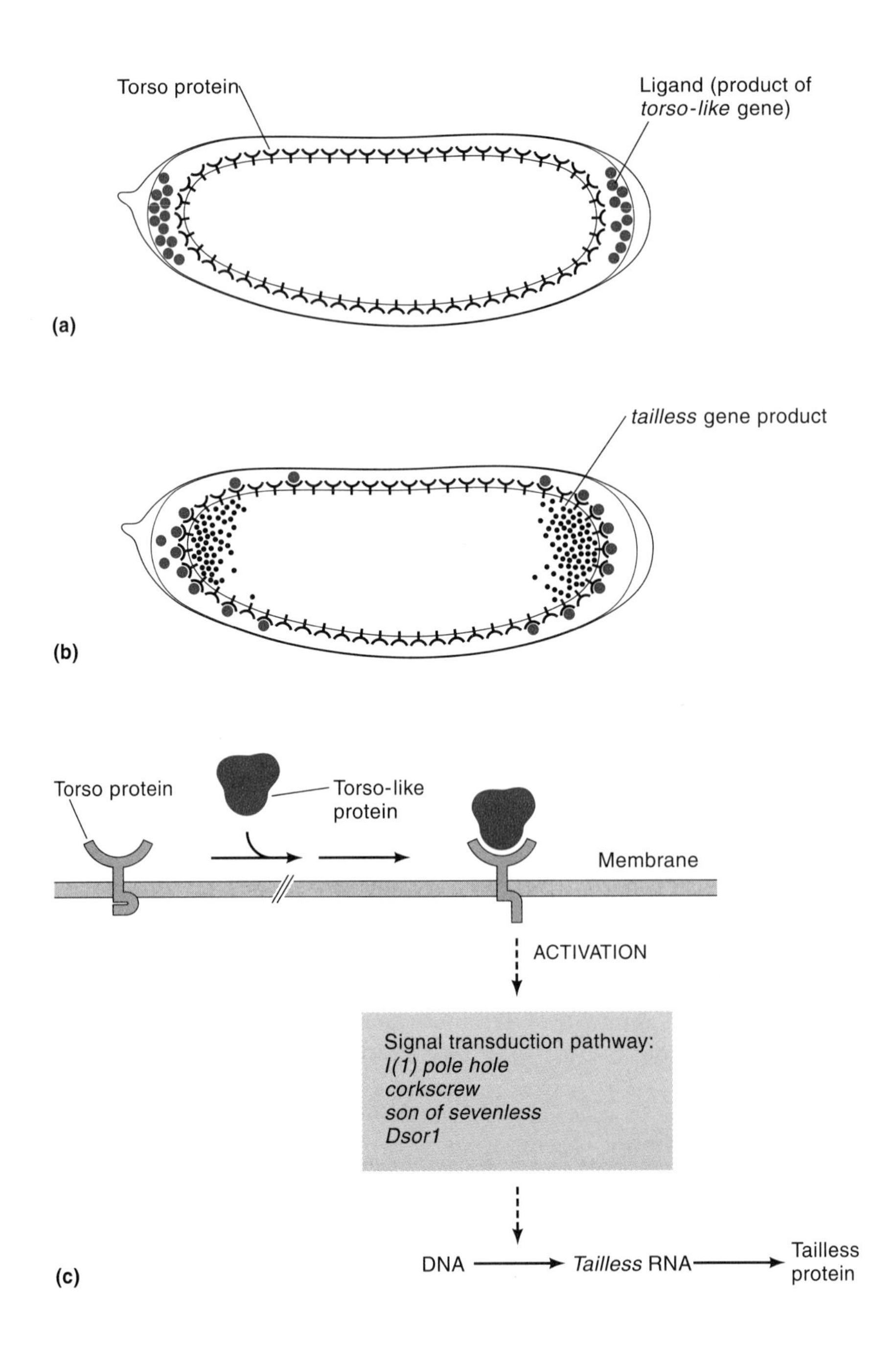

Figure 9.8
The organization of the embryo termini is controlled by the six terminal group genes. (a) Activation of the *torso* receptor at the anterior and posterior poles by the torso-like ligand. (b) The activation initiates a signal transduction pathway. (c) The pathway leads to the transcription of genes such as *tailless.*

In summary, the A-P axis is established during the first several hours after fertilization of the *Drosophila* embryo by three independent systems, two of which rely on prior localization of determinants during oogenesis (*bicoid* for the anterior system and *nanos* for the posterior system), and one of which relies upon localized activation of a signal transduction pathway (the terminal system). These systems operate during the period of rapid nuclear divisions, prior to cellularization of the embryo. In the anterior and posterior systems, positional information is thought to be encoded in the form of protein concentration gradients and to be elaborated by response elements in target nucleic acids that differ subtly in their relative affinities. The mechanism through which *torso* receptor activation leads to position-dependent gene activation at the embryo poles is less well understood.

c. D-V Pattern Formation

Patterning along the D-V axis also involves a protein gradient, although it is generated in a manner quite different from the protein gradients that pattern the A-P axis. Unlike the *bicoid* and *nanos* determinants, which are sequestered at the poles of the embryo, the two principal factors that determine the D-V axis are initially distributed uniformly in the embryo (Figure 9.9). These are the protein products of the *Toll* and *dorsal* genes, which are maternal effect lethal genes without whose function embryos develop only dorsal structures. Therefore, both genes are needed to promote ventral development; dorsal development is the default state. The two genes are among a large set of genes isolated because mutations alter dorsal-ventral patterning. Mutations in many of the genes, including dorsal, have maternal effects; mutations in other genes act exclusively in the zygote.

In situ hybridization with probes for *Toll* and *dorsal* mRNA reveal that both genes are expressed primarily during oogenesis. Antibodies directed against the Toll and Dorsal proteins reveal Toll to be distributed around the embryo periphery and Dorsal to be restricted to the cytoplasm in young embryos. In syncytial blastoderm stage embryos, however, Dorsal protein relocates to nuclei in restricted locations. The concentration of nuclear-localized Dorsal protein is greatest along the ventral midline, and it declines in more lateral and dorsal regions. This nuclear localization gradient of dorsal protein establishes the embryo's D-V axis.

Dorsal protein is a member of the **Rel** family of transcription factors, and its presence in nuclei signals its recruitment to regulate a battery of target genes. Known targets include *twist, snail, rhomboid, tolloid, decapentaplegic,* and *zerknüllt.* These genes specify different regions of the D-V pattern, and they are collectively responsible for the differentiation of the mesoderm, the neuroectoderm, the lateral and dorsal ectoderm, and the most dorsal structure, the amnioserosa. Expression of these genes correlates exactly with the regions they specify. The genes *twist* and *snail* are expressed in the most ventral regions of the embryo that will produce mesodermal tissues; *rhomboid, tolloid,* and *decapentaplegic* are expressed in the lateral regions that will produce the neuroectoderm and ectoderm; *zerknüllt* is expressed in the most dorsal region. These target genes are thought to respond to different concentrations of the Dorsal protein. The embryo is exquisitely sensitive to levels of Dorsal protein. If the level of Dorsal protein is low, then the embryo produces more dorsal structures and fewer ventral structures, and probes reveal a ventral shift in the pattern of expression of the *dorsal* target genes. Conversely, if the level of Dorsal protein is high, then embryos develop with an excess of ventral structures, and the expression pattern of the *dorsal* target genes shifts dorsally. Thus, *dorsal* has the properties of a morphogen.

The gradient of Dorsal protein is produced in response to the localized activation of Toll. Toll is a transmembrane receptor that appears to be activated in a graded manner by an **extracellular ligand**. The distribution of this extracellular ligand presumably determines the subsequent distribution of *dorsal,* with its highest concentration along the ventral midline and diminishing amounts in more lateral and dorsal locations. Toll mediates a signal from surrounding non–germ line cells, called follicle cells, to the embryo. Follicle cells are also the secretory cells that produce the eggshell. Although several maternal effect genes

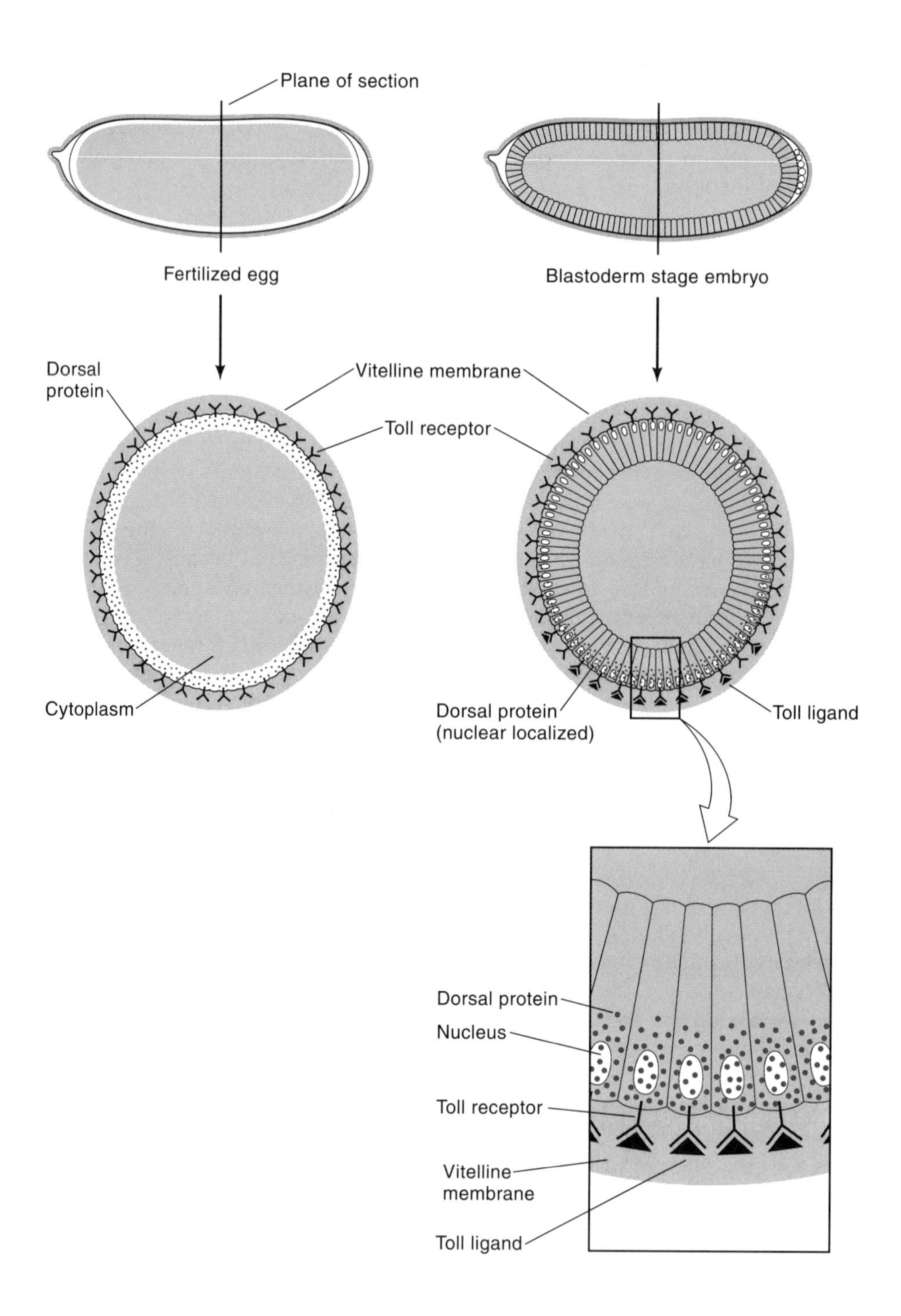

Figure 9.9
The D-V axis is organized when the *Toll* receptor is activated along the ventral midline. Represented in cross-section, the drawing on the left shows the *Toll* receptor (Y) and the dorsal protein (dots) uniformly distributed in the egg membrane and peripheral cytoplasm, respectively. Binding of the Toll ligand (filled triangles) causes the dorsal protein to translocate to the nucleus in the immediate vicinity of the activated *Toll* receptor and in a concentration gradient whose peak coincides with the ventral midline.

whose function is required to produce the appropriately distributed Toll ligand have been identified (e.g., *fs(1)K10, squid, cappucino, spire, gurken, cornichon, torpedo, easter, pelle, gastrulation defective,* and *snake*), *spätzle* appears to encode a ligand for the Toll receptor. The ligand was identified both biochemically in extracts capable of inducing dorsal localization and genetically. The other genes may help produce, process, release, or receive the ligand.

9.3 Anterior-Posterior Patterning

a. Larval Body Segmentation

Segmentation partitions cells along the embryo A-P axis into groups whose descendents will generate the head, thoracic, and abdominal segments of the embryo, larva, and adult. Although these segments first become visible as morphologically distinct units at approximately ten hours of embryonic development, cells have clearly been committed to particular segmental lineages by the cellular blastoderm stage (at three hours of development). These commitments can be demonstrated in two ways.

If a tiny beam (20 mm in diameter) of ultraviolet laser light is used to kill small numbers of cells in different regions of the cellular blastoderm, many of the surviving animals have defects (Figure 9.10). Defect location corresponds

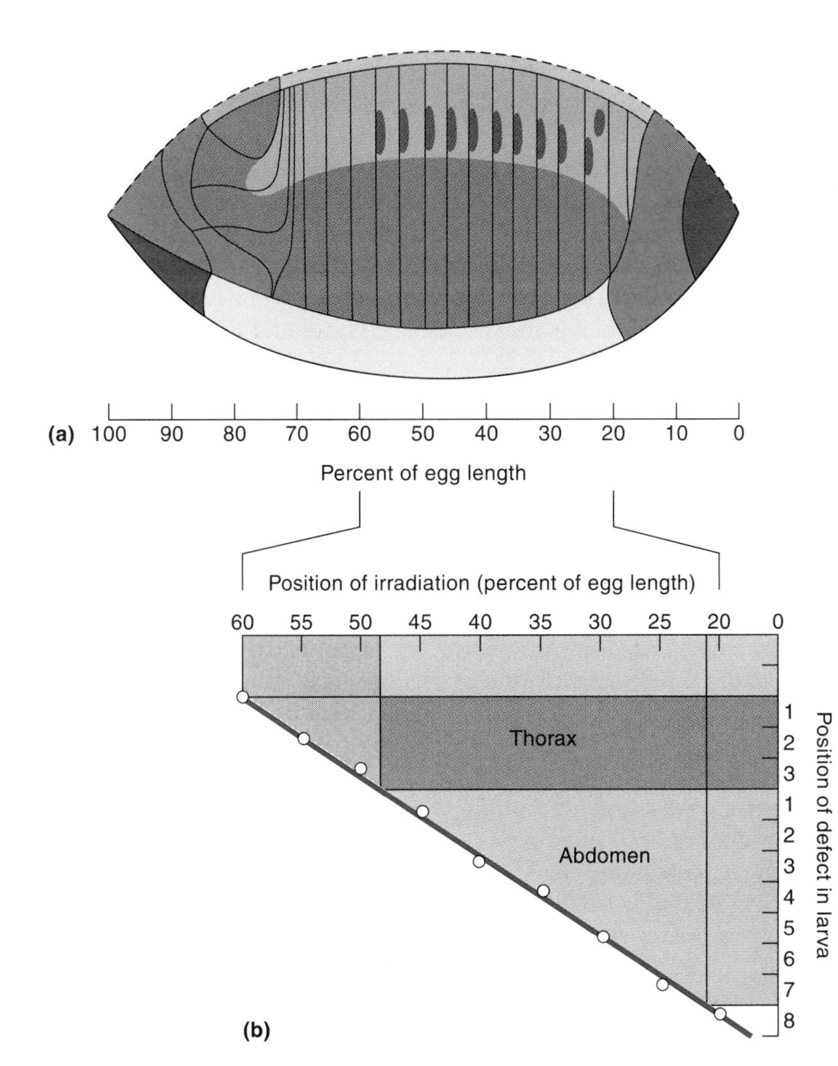

Figure 9.10
Fate map of the *Drosophila* blastoderm. (a) Lateral representation of a blastoderm stage *Drosophila* embryo indicates the approximate position of the various internal and external organs. The scale indicates percentage of egg length, from posterior to anterior. (b) Killing blastoderm cells by focusing UV light on particular regions along the length of the embryo reveals a good correlation between the location of the irradiated cells and the defects that subsequently arise in the embryo. (Part a from V. Hartenstein. 1993. Atlas of Drosophila Development. p. 5. Cold Spring Harbor Press, Cold Spring Harbor, New York; part b from M. Lohs-Schardin et al. 1979. A Fate Map for the Larval Epidermis of *Drosophila melanogaster. Dev. Biol.* 73 250.)

closely to the irradiation site. For instance, killing cells in the lateral area of the blastoderm from 20 to 60 percent egg length (0 percent = posterior pole and 100 percent = anterior pole) results in localized defects in the larval epidermis. By correlating the position where progenitors of the different thoracic and abdominal segments are affected by the light-induced killing, investigators can show that the cells that will generate the segments are equally spaced along the A-P axis and that the progenitor cells for each segment occupy a transverse stripe about three to four cells wide. These experiments reveal where the cells that will produce different segments originate on the blastoderm surface and demonstrate that, in the cellular blastoderm, the processes have already begun to restrict the fate of the cells.

Molecular markers for determining segmental cell fates A dramatic and more direct demonstration that cells have begun to receive their developmental assignments at the cellular blastoderm stage comes from efforts to identify the genes that establish the identity of the different regions. To understand how the fly builds itself by using a group of genes to define its various parts, one must first understand that the developmental unit of importance is a region called the **parasegment**, a region related to but not identical with the segment (Figure 9.11). Before segments form, parasegment borders partition the embryo into segment-sized units. This process occurs during the last stages of cellular blastoderm formation and just as gastrulation commences. Only several hours later do the segment borders form at a position posterior to each parasegment border.

Parasegment border formation can be seen directly by visualizing the expression of the *engrailed* gene. This gene is expressed in young gastrulating embryos in transverse stripes of cells, one cell wide, just posterior to the parasegment border. Its expression can be visualized by the technique of *in situ* hybridization. When an *engrailed* probe is applied to gastrulating embryos, a striking pattern of expressing cells is observed (Figure 9.11). Stripes one cell wide spaced approximately every four cells along the A-P axis create a zebralike pattern that defines the anterior portion of each parasegment. The *engrailed* gene is one of a set of genes whose expression in stripes reveals, at early stages, the metameric organization of the later embryo. However, the genes are more than just early markers: they are the genes that control cell fates and organize a multicellular pattern.

The homeotic genes In 1894, William Bateson described homeotic mutants as animals where one part of the body develops as a copy of another part. Often called transformations, these switched fates are really replacement phenomena. Groups of cells choose an inappropriate developmental path, apparently confused about their location along the A-P axis. The genes that do the switching of fates, called **homeotic genes**, were first studied in detail in *Drosophila.*

Fourteen *engrailed* stripes form in the cellular blastoderm and mark the presence of 14 parasegments. These parasegments are assigned to their distinct developmental fates by a small set of homeotic genes that reside in one of two locations, either in the Antennapedia Complex (ANT-C) or the Bithorax Complex (BX-C) of genes (Figure 9.12). For instance, in the ANT-C, the *Deformed* (*Dfd*) gene is responsible for the development of parasegment 1, *Sex combs reduced* (*Scr*) is responsible for parasegments 2 and 3, and *Antennapedia* (*Antp*) is responsible for parasegments 4 and 5. Parasegments 5–13 require the genes of the BX-C, *Ultrabithorax* (*Ubx*) in parasegments 5 and 6, *Abdominal-A* (*abd-A*) in parasegments 7, 8, and 9, and *Abdominal-B* (*abd-B*) in parasegments 10–13 (Table 9.1). The pattern in each part of the body along the A-P axis is therefore determined by a different homeotic protein or, in some cases, a combination of proteins.

Parasegments 5 and 6, where *Ubx* is expressed, will produce the portion of the embryo extending from the posterior second thoracic segment (pT2) to the

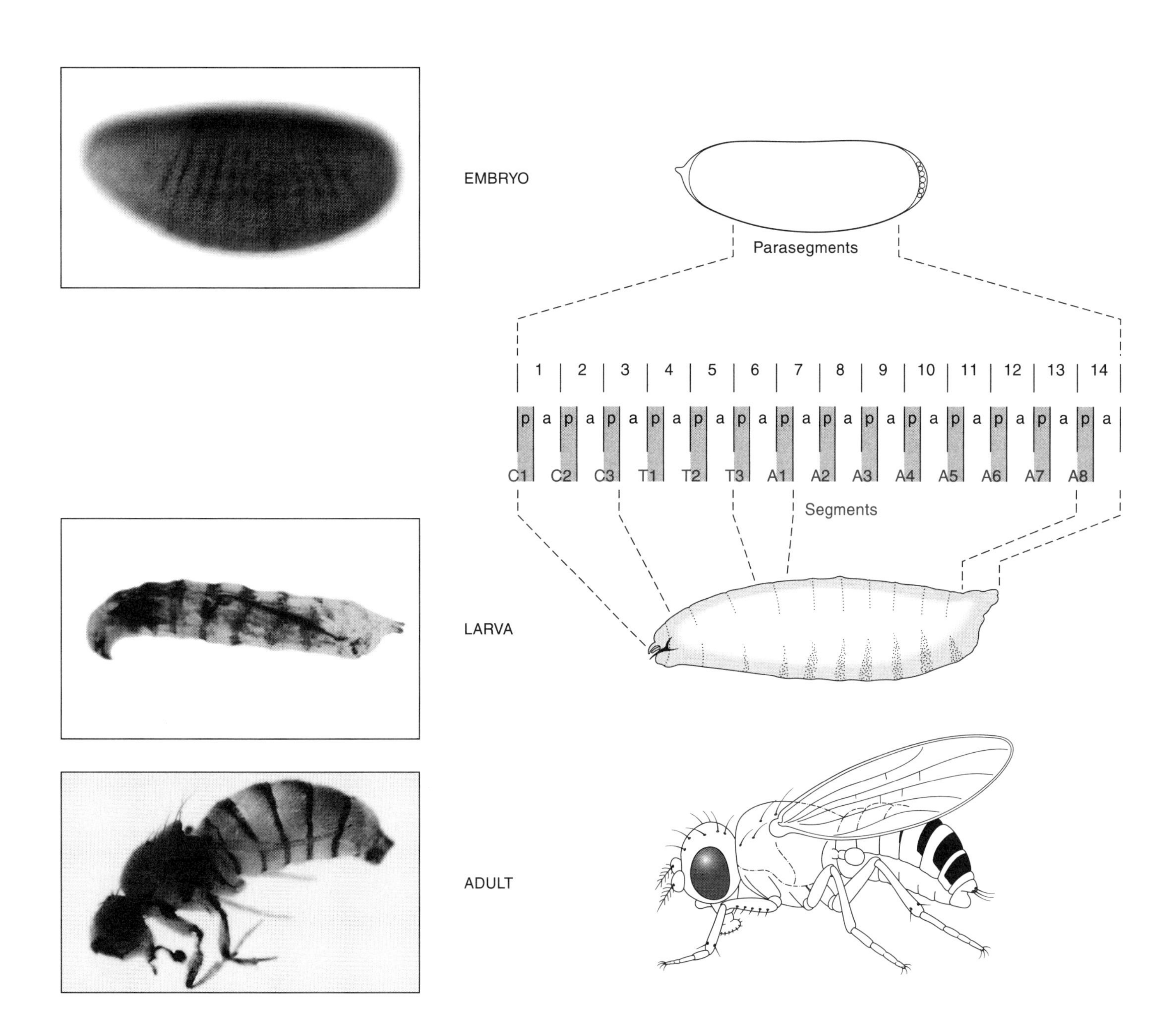

Figure 9.11

The metameric organization of *Drosophila.* Subdivision of the embryo along the A-P axis is first manifested when genes such as *engrailed* are expressed in stripes. There are 14 stripes of *engrailed*-expressing cells along the main trunk region of the embryo; these mark the anterior part of each of the 14 parasegments. After 10 hours of embryonic development, the stripes of *engrailed*-expressing cells mark the posterior part of each segment, and this pattern of expression is retained in both the larval and adult forms. The embryo was hybridized with a probe for the *engrailed* mRNA; the larva and adult are from a strain of *Drosophila* containing the regulatory sequences of *engrailed* coupled to the gene encoding the enzyme β-galactosidase, whose activity can be detected histochemically.

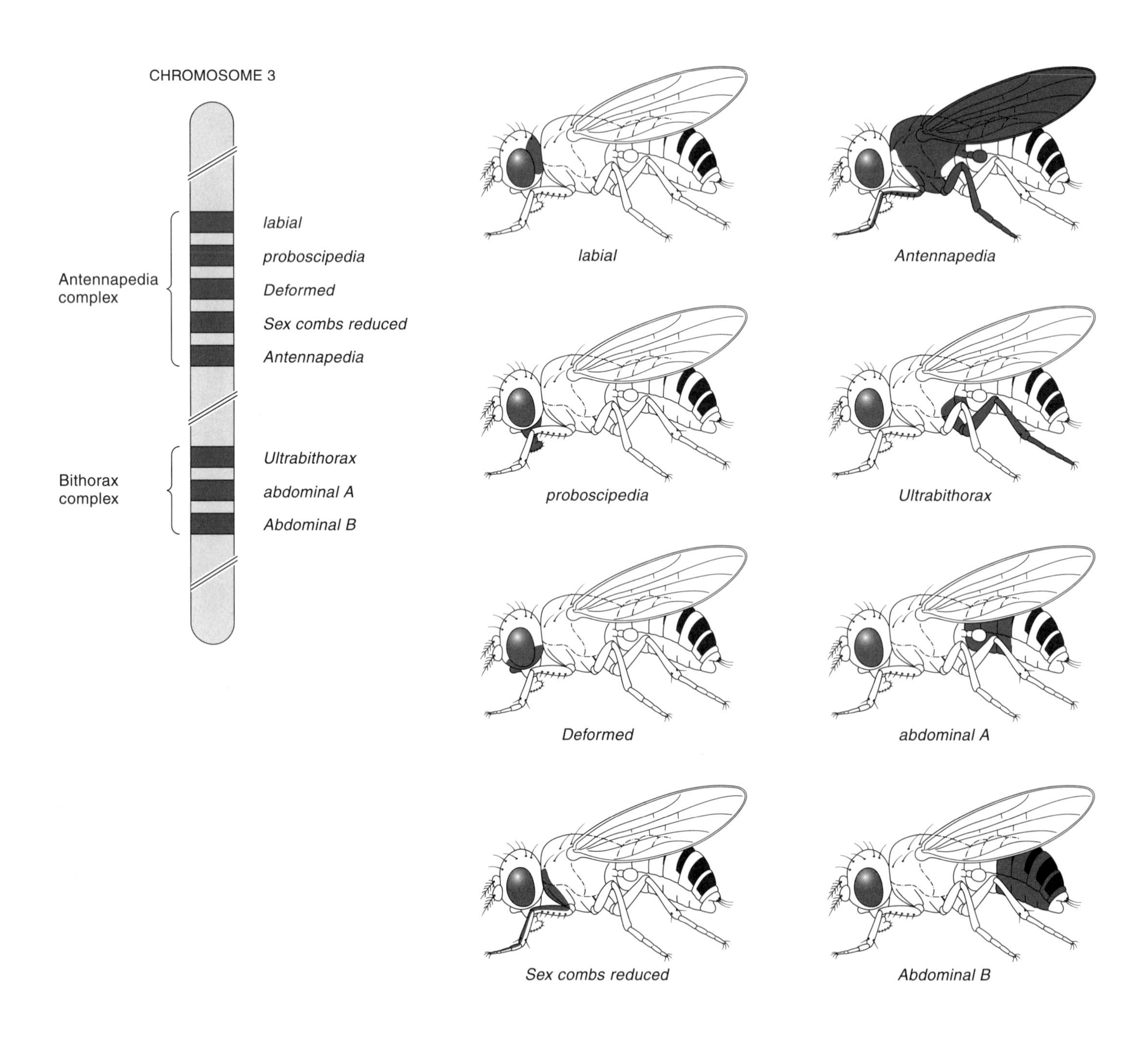

Figure 9.12
Chromosomal location and patterns of expression of the homeotic genes. The homeotic complex in *Drosophila* is split, but the two parts retain the correlation between the proximodistal sequence of genes and the A-P sequence of expression domains. The shaded areas in the drawings of the flies indicate the expression domains of the genes in the homeotic complex.

Table 9.1 ***Homeotic Genes and Their Sites of Action***

Gene Name	Site of Action
Labial	Parasegment 0
Proboscipedia	Parasegment
Deformed	Parasegment 1
Sex combs reduced	Parasegments 2, 3
Antennapedia	Parasegments 4, 5
Ultrabithorax	Parasegments 5, 6
Abdominal-A	Parasegments 7–12
Abdominal-B	Parasegments 10–13

anterior part of the first abdominal segment (aA1), including the intervening third thoracic segment (aT3 and pT3). Mutations that alter the *Ubx* gene display a variety of phenotypes, depending on how they affect *Ubx* function. Those that eliminate *Ubx* function altogether arrest development during the embryonic period and produce embryos that are normal except that parasegments 5 and 6 have been transformed to look like parasegment 4. Several fascinating mutations alter *Ubx* function only in adult tissues. For instance, two mutant alleles of *Ubx*, *bithorax* (*bx*) and *postbithorax* (*pbx*), transform the anterior and posterior halves of the haltere, respectively, into corresponding regions of the wing. Such "homeotic" transformations affect the musculature, the pattern of ennervation, and the pattern and types of bristles and hairs that decorate the cuticle surface. Flies with *bx pbx* have duplicated second thoracic segments (one at the expense of the third thoracic segment) and therefore two complete sets of wings (Figure 9.13).

Due in part to the remarkable phenotypes that mutant animals display, such homeotic mutants have long captured the fascination of biologists. They are capable of placing whole body parts in inappropriate locations—in effect, in the

(a)

(b)

Figure 9.13
(a) A normal fly, with two wings. (b) A four-winged fly produced by combining bithorax and postbithorax mutations. (Courtesy of E. B. Lewis.)

case of *Ubx*, replacing halteres with wings. The phenotype of *Ubx* mutants suggests that the role of *Ubx* is to instruct cells in parasegments 5 and 6 to choose and follow a developmental pathway appropriate to their location and that these cells may otherwise choose among alternative and essentially equivalent pathways. These alternative pathways require the action of the other homeotic genes in either the ANT-C or BX-C.

Isolation of the *Ubx* gene provided a means to discover when and where the *Ubx* gene is expressed. *In situ* hybridization of probes prepared with the *Ubx* gene sequences reveals that cellular blastoderm stage *Drosophila* embryos express *Ubx* predominantly in the region where the progenitor cells of parasegments 5 and 6 are located (Figure 9.12). Similar experiments using antibodies to detect Ubx protein reveal an identical distribution. Because *Ubx* function is required to instruct these cells to develop according to their position in the embryo, the processes that partition the embryo into developmentally distinct segmental units have apparently already defined the relevant locations by this stage.

The other genes of the BX-C, *abd-A* and *abd-B*, affect the more posterior abdominal segments. Their mutant phenotypes indicate that these genes are responsible for the segmental identity of the segments posterior to those affected by mutations in *Ubx*. The realms of action of these three homeotic genes are essentially complementary and nonoverlapping. The function of *Ubx* extends from the posterior second thoracic segment (pT2) to the first abdominal segment (A1), *abd-A* function extends from A2 to A4, and *abd-B* function extends from A5 to A9. As already described, inactivation of *ubx* transforms T3 to T2. Inactivation of *abd-A* transforms A2-A4 to A1. Inactivation of *abd-B* transforms A5-A9 to A4. Remarkably, the specificity of these phenotypes is such that double and triple mutant animals have additive phenotypes. Although animals that lack any of the BX-C functions do not survive beyond the embryo stage, the cuticle of embryos deficient for the entire BX-C does form and can be analyzed to understand what developmental program each of the segments followed. All segments of BX-C-deficient embryos develop patterns characteristic of T2.

Because the segments of these embryos follow identical developmental paths, the genes of the BX-C are apparently responsible for diversifying the different segments. The phenotypes of these homeotic mutants suggest that the groups of cells whose descendents will make these body parts choose between **alternative developmental pathways** during the course of normal development and imply that these decisions are taken and/or executed by the products of the homeotic genes. It follows that the cells whose developmental program can be switched between alternative fates share an inherent homology and that the different homeotic genes that direct their ultimate diversification function in similar capacities.

BX-C-deficient animals develop with many identical segments, but they retain the capacity to create the normal number of individual metameres despite their inability to diversify them (Figure 9.14). Mutations in the genes that are responsible for placing the *engrailed* stripes in their appropriate locations (so-called **segmentation genes**) or mutations in the *engrailed* gene itself kill affected animals during embryogenesis. In contrast to animals that lack BX-C function and develop the normal number of segments, each of which has a normal appearance (albeit one that is inappropriate to the location), animals lacking segmentation gene functions have abnormal and incomplete segments.

Thus, the genes that organize the *Drosophila* embryo can be classified as one of two types. The segmentation genes partition the embryo into homologous units that are inherently similar. Mutations in these genes alter the segmentation process by affecting either the number of segments or the basic organization within each one. The homeotic functions elaborate different developmental pathways in each metamere. In effect, homeotic mutants misassign developmental programs to the functionally separate developmental units.

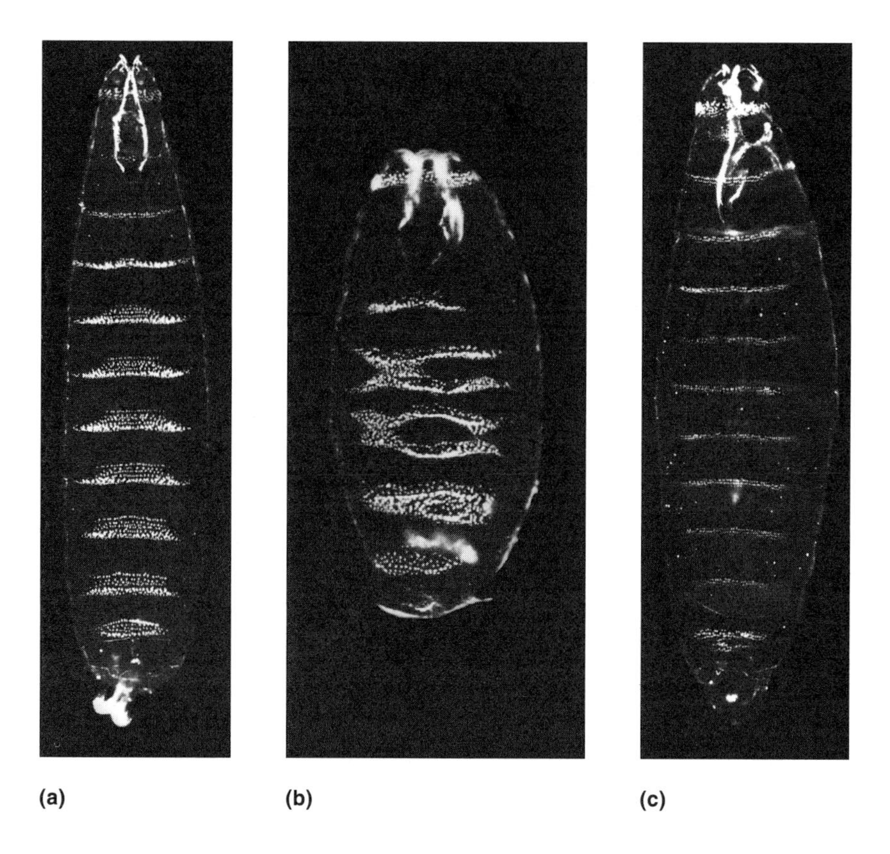

Figure 9.14
The effects of mutations that inactivate the segmentation gene *engrailed* or the BX-C. (a) The ventral cuticle of a normal *Drosophila* larva shown in darkfield illumination. (b) A similar view of a larva mutant for the *engrailed* gene. (c) A similar view of a mutant larva deleted for most of the BX-C. Note the misshapen segments in the *engrailed* mutant and the transformation of the parasegments posterior to PS5 to a PS5 appearance in the BX-C mutant. (Parts a and b courtesy of E. Wieschaus. From C. Nüsslein-Volhard and E. Wieschaus. 1980. Mutations affecting segment number and polarity in *Drosophilia. Nature* 287 797; part c courtesy of Alberts et al. 1994. p. 1087. Molecular Biology of the Cell, 3rd ed. Garland, New York.)

Systematic genetic screens for segmentation genes One of the key developments that has aided the identification and characterization of the segmentation and homeotic genes has been the facility with which mutant forms of the genes can be recognized and studied. Although the first mutant alleles in homeotic genes were discovered more than 70 years ago and although mutant alleles in most of the homeotic gene loci and in several of the segmentation genes were discovered and described over the ensuing 60 plus years, their isolation was for the most part serendipitous and sporadic. Flies appeared spontaneously in fly cultures that, because of their interesting and unusual phenotypes, were saved and characterized. We know now that almost every one of the spontaneous alleles was caused by the insertion of a transposable element in the regulatory regions of the affected gene. Interesting though these phenotypes are, they represent the consequences of incomplete gene inactivation and so fail to reveal a complete and accurate picture of the role of the affected genes. All the mutants were recognized by virtue of their adult phenotype, yet we now know that each affected segmentation or homeotic gene provides an essential function, without which the animals die as embryos.

Systematic efforts to isolate additional alleles of the homeotic genes initially met with little success for several reasons. Because extant mutants produced viable flies, few systematic screens for lethal alleles of the genes were attempted. This was due in part to researchers' lack of appreciation that the same homeotic transformations that affect the adult fly would also affect embryonic and larval structures in a similar way or that such transformations could be easily seen. Once simple methods to fix and mount embryo and larval cuticles for examination by light microscopy were developed, the universe of genes that are required to pattern the embryo became accessible to genetic analysis. No longer were these analyses limited to the study of spontaneous or other rare alleles that affected the expression of these genes in unusual ways.

Thorough genetic characterization of a locus can be accomplished with a standard but necessarily elaborate screen for lethal alleles in the region of interest. In a diploid organism like *Drosophila*, most mutations that would cause lethality if homozygous (present with another mutant gene copy) or if hemizygous (without any normal or mutant gene copy) do not markedly affect the development or behavior of the animal when heterozygous (present with a normal gene copy). For most of the genome, deletion of a small portion (on the order of 0.5–1.0 percent) affects a fly in only minor ways as long as one normal copy of each affected gene is present. Screens among the progeny of mutagenized flies for mutant chromosomes that fail to support life when hemizygous can be used to "saturate" a region for lethal alleles. Such screens involve several steps (Figure 9.15). First, the progeny of the pool of mutagenized germ cells must be isolated in order to assay each mutagenized chromosome of interest individually.

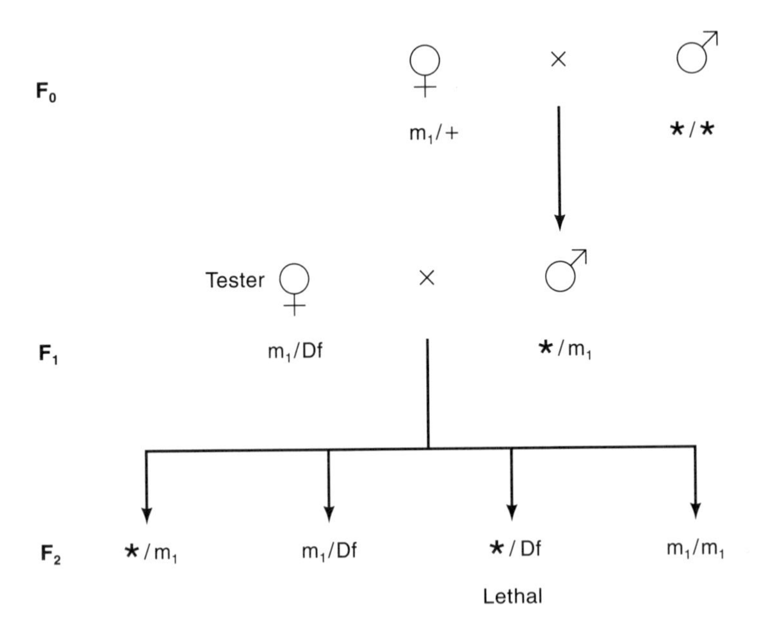

Figure 9.15
Experimental scheme to screen for lethal mutations. Such screens are conducted separately for chromosomes 1, 2, and 3. As an example of a method used to identify mutations on chromosome 2, males are mutagenized (e.g., with a dose of X-rays or with a chemical mutagen) and mated with virgin females that have genetically marked second chromosomes. Their male progeny, which carry one mutagenized chromosome 2 (*) and one of the marked chromosomes from the mother (e.g., m_1), are mated individually with "tester" females. If lethal mutations in a region of a known deletion are sought, the tester females carry the deletion of interest (Df) balanced with a marked chromosome (e.g., m_1). Examination of the F_2 progeny identifies the progeny of a mating that fails to produce viable */Df flies, and the chromosome that carries the mutation of interest can be isolated as */m_1 males and females.

In practice, F1 male progeny of a mutagenized parent are isolated and tested individually. Such males will have a haploid set of mutagenized chromosomes, and generally, only one of his chromosomes will be tested in a particular genetic screen. Each F1 male is separately mated with tester virgin females to place his mutagenized chromosome of interest in *trans* to the deletion-bearing chromosome. This step must be carried out so that if a lethal mutation in the region of the deletion has been induced, it is possible to recover the mutagenized chromosome in siblings that provide a heterozygous (and viable) environment for the mutant gene. Ultimately, F2 males and females of the appropriate genotype must be recovered and mated in order to preserve each mutant chromosome for further study. Their laborious nature generally limits such screens to fewer than 10,000 mutagenized chromosomes. Nevertheless, with frequencies approximating 0.01 percent per gene (one mutant in an average gene per 1000 chromosomes) for the recovery of mutations in a gene of average size, such screens can be used to identify essential functions in specific regions of the genome.

Such screens have been used successfully to characterize a number of important genetic loci, such as the BX-C and the ANT-C. However, such screens are necessarily directed to specific chromosomal regions for the purpose of isolating particular alleles. Because the number of loci in the genome responsible for regulating the processes of developmental organization is small relative to the total number of genes in the genome, such screens rarely uncover previously unknown loci of interest to the developmental biologist. Searching the entire genome for new alleles with a recessive lethal phenotype is a task of enormous magnitude. Nevertheless, ingenious screens that have succeeded in revealing many interesting mutants that bear lethal alleles of genes required during the organizational processes of embryonic development have been devised. These screens proceeded in several stages.

Drosophila has two large autosomes of approximately equal size (chromosomes 2 and 3), an autosome (chromosome 4) that is approximately 2 percent the size of the other autosomes, and an X chromosome that is approximately half the size of the large autosomes. The Y chromosome is heterochromatic and carries only male fertility functions. Each of the three larger chromosomes was analyzed separately in a process that involved several separate steps. For each of the large autosomes, 4000–5000 individual fly lines were generated, each of which carried a different lethal mutation on the autosome under study. This number was chosen to achieve a distribution of mutants such that there was a reasonable expectation that each of the estimated 2000 genes on the chromosome would be represented at least once. Next, each lethal line, represented by heterozygous males and females, was characterized to determine when the homozygous mutant progeny die. Approximately one-third of the mutants died as embryos. *Drosophila* oogenesis provides a rich dowry of maternally supplied material to the embryo, and almost all embryonic lethal mutants survive long enough to produce an epidermal cuticle. As a consequence, examination of dead embryos reveals cuticle patterns that reflect the basic organization of the embryo. One of the advantages of studying embryonic pattern formation in *Drosophila* is the ease with which mutations that lead to an alteration of embryonic pattern can be analyzed. Embryonic development results in a simple metameric pattern of 11 repeating units, the 3 thoracic and 8 abdominal segments of the hatching larva. This cuticular pattern of the larval integument, the hypoderm, is well marked with structures indicating position and polarity. The pattern may be dramatically altered by mutations, the changes indicating the overall changes in the spatial organization of the embryonic material (Figure 9.16).

Many specialized cuticular structures decorate the surface of the *Drosophila* embryo. These include specialized sense organs, segmentally repeating patterns of small hairs arranged in rows (called denticle belts), each of which is unique to a particular segment, and structures unique to the head and tail. Most embry-

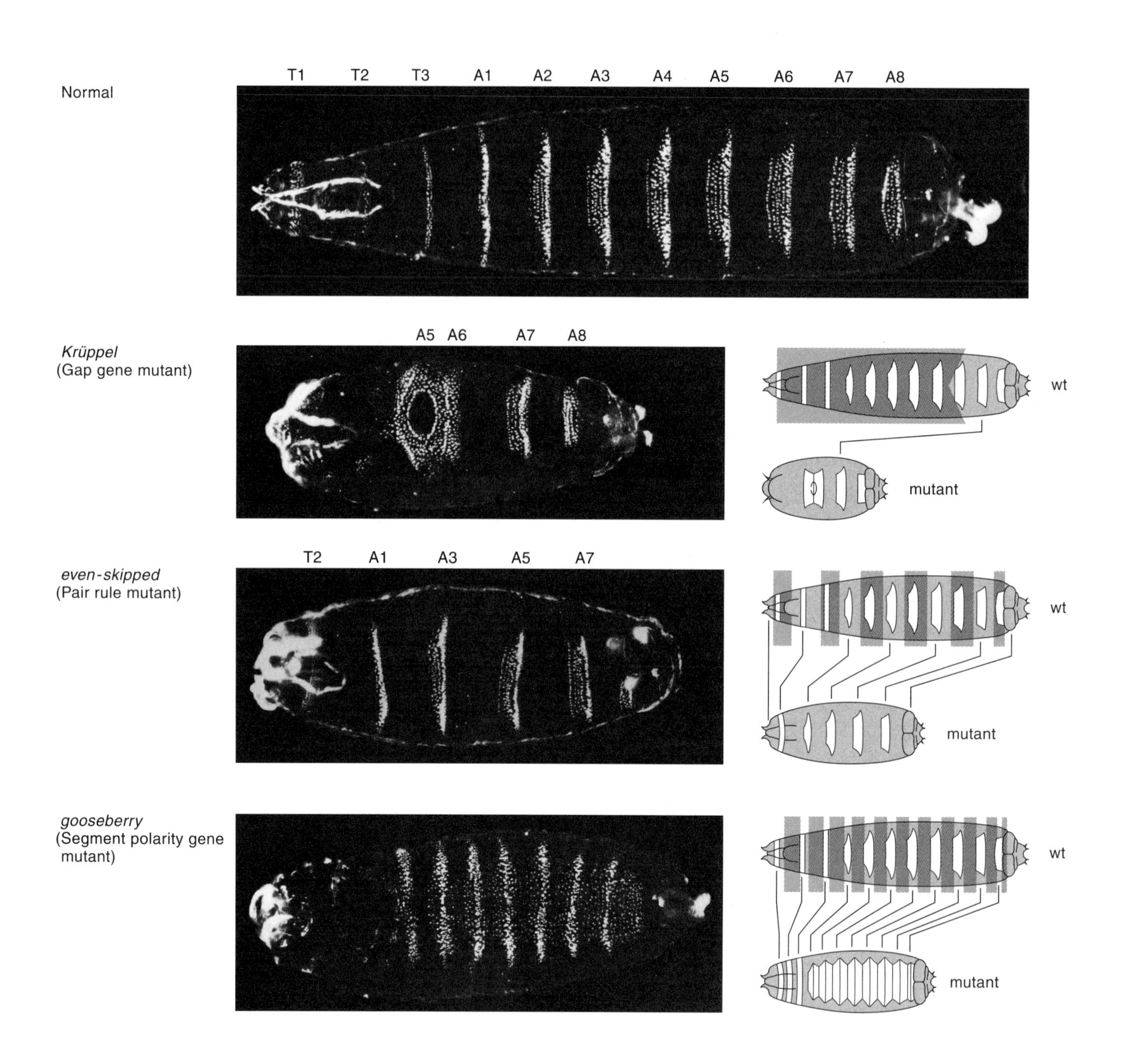

Figure 9.16
Examples of the phenotypes of the three types of segmentation gene mutants. Shown are larvae typical of the normal state, a gap gene mutant (*Krüppel*), a pair-rule mutant (*even-skipped*), and a segment polarity gene mutant (*gooseberry*). The drawings below depict with shading the areas that are either deleted or are replaced with mirror-image duplications in the mutants. (From C. Nüsslein-Volhard and E. Wieschaus. 1980. Mutations affecting segment number and polarity in *Drosophilia. Nature* 287 796.)

onic lethal mutants die without affecting these patterns, but in a small proportion of the mutants, these cuticular structures are arranged in abnormal patterns. In some, no rows of denticles form. In others, the patterns of small hairs in the denticle rows are abnormally arranged. In others, the number of denticle belts is reduced. Among almost 15,000 lethal lines isolated on the X, second, and third chromosomes, about 35 different genetic loci were identified whose function is needed to determine the proper number of segments or to arrange the features of the segments in a normal way. This collection of genes probably represents most of the genes of this type in *Drosophila.*

Segmentation gene classes Mutants in 18 genes reduce the number of segments (Figure 9.16). Among these "segmentation genes," 7 reduce the number of segments by one-half. These are collectively called the **pair-rule** group. Mutants in 11 reduce the number of segments to a greater extent, generating embryos with fewer than six broad and poorly formed segments. Because these patterns are missing broad sections of the segmental pattern, these mutants are called the **gap** genes. The phenotypes suggest that the segmentation process is sequential and that intermediate stages subdivide the embryo into progressively smaller units. The gap genes would act first, followed by the pair-rule genes. This is now known to be correct.

In addition to the segmentation mutants that reduce the number of segments, there are 17 that produce animals with the normal number of segments but fail to pattern the structures appropriately in each of the segments. These are called the **segment polarity** genes.

b. Elaborating the A-P Pattern in the Syncytial Blastoderm

The broad concentration gradients of Bicoid and Nanos protein that form along the length of the embryo activate a cascade of cross-regulating genes during the syncytial blastoderm stages (see also *Genes and Genomes,* section 8.3g). All these genes encode transcription factors whose targets include other members of the cascade. Those gap genes expressed in the middle of the embryo, the central gap genes, include *hunchback, Krüppel, knirps,* and *giant.* They are expressed exclusively in restricted domains that will ultimately give rise to groups of contiguous segments along the A-P axis. Gap gene deployment is accomplished in two steps. First, *hunchback* transcription is activated anteriorly by *bicoid,* and translation of maternally inherited *hunchback* transcripts is blocked posteriorly by *nanos.* The mechanism of *nanos* action is not known. Second, the resulting gradient of Hunchback protein activates the transcription of the *Krüppel, knirps,* and *giant* gap genes in a series of broad domains, presumably by providing a series of concentration thresholds that independently regulate each gene. Krüppel, Knirps, and Giant proteins are themselves thought to form short-range morphogenetic gradients that help refine their respective expression domains. Such mutual interactions are thought to involve competition by activators and repressors for overlapping sites in the regulatory regions of the respective genes (Figure 9.17).

In summary, a set of six transcription factors is synthesized in restricted domains along the A-P embryonic axis in response to maternal signals. In the central part of the embryo, these domains are defined in part by broad gradients of Bicoid and Nanos protein. Subsequent input from cross-regulatory interactions among the gap gene products helps refine the relative placement of these domains along the A-P axis.

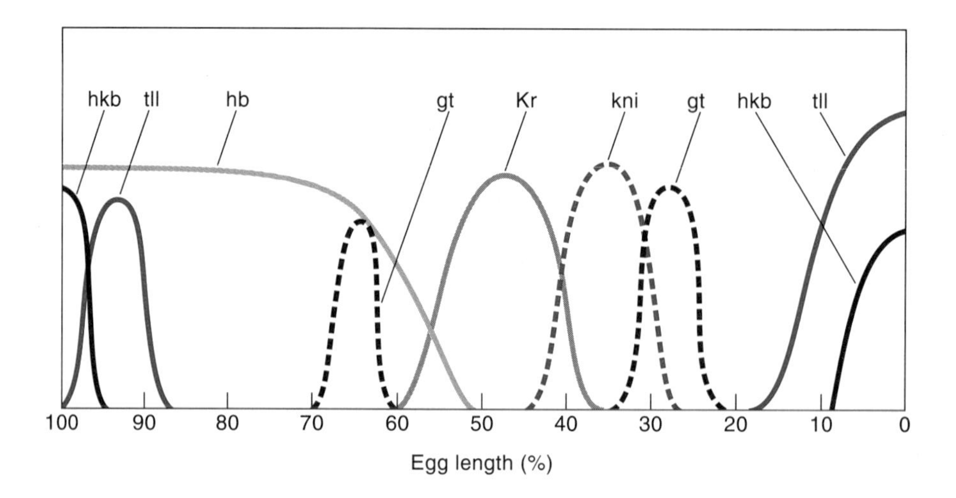

Figure 9.17
The approximate domains of expression of the gap segmentation genes. The regions of *huckebein* (hkb), *tailless* (tll), *hunchback* (hb), *Krüppel* (Kr), *knirps* (kni), and *giant* (gt) expression have graded boundaries and overlap. The additive effects of their gene products help to refine the expression patterns of the downstream pair-rule genes. (From P. Lawrence. 1992. p. 58. The Making of a Fly. Alden Press, Oxford, England.)

c. Refining the Pattern in the Cellular Blastoderm: Regulating the Pair-Rule Genes

The broad bands of gap gene expression along the A-P axis lead to a more refined and higher resolution pattern of pair-rule and segment polarity gene expression during the last minutes of the syncytial blastoderm stage. For the pair-rule genes, expression patterns describe seven transverse stripes that each correspond to two segment intervals along the body axis; their expression patterns bring the first signs of metamerization. Each segment primordium consists of about four cells along the A-P axis, so the pattern of pair-rule transcription will be four cells on, four cells off, and so on down the embryo (Figure 9.18). Because the expression pattern of each pair-rule gene has a similar periodicity but different registration, unique combinations of pair-rule gene products are expressed in each cell. For example, the pair-rule genes *fushi tarazu* (*ftz*) and *even-skipped* (*eve*) are transcribed in perfectly complementary striped patterns; together they cover all the segments. Cells making *ftz* RNA do not make *eve* RNA. Other pair-rule gene transcription patterns align with neither *ftz* nor *eve* and therefore overlap with one or both of them (Figure 9.18). The striped patterns are dynamic; as development proceeds, stripes of the different pair-rule gene products may become subdivided, shrink to narrower stripes covering fewer cells, or new stripes may appear between the original ones.

Primary pair-rule genes Two types of mechanisms are thought to be involved in activating the striped domains of pair-rule gene expression. For the so-called primary pair-rule genes, *hairy*, *runt*, and *eve*, different stripes are independently regulated by separate DNA sequence elements: **stripe response elements**. These sequence elements are utilized by subsets of gap gene products expressed in overlapping domains in the embryo (Figure 9.19). Combinations of gap gene protein products either activate, in the regions of expression, or repress, in the interstripe regions. The best studied example of gap gene regulation of pair-rule gene expression is the element responsible for expression of *eve* stripe 2. In this

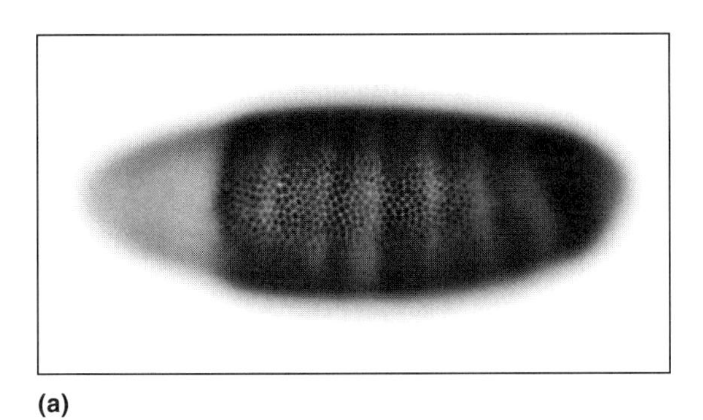
(a)

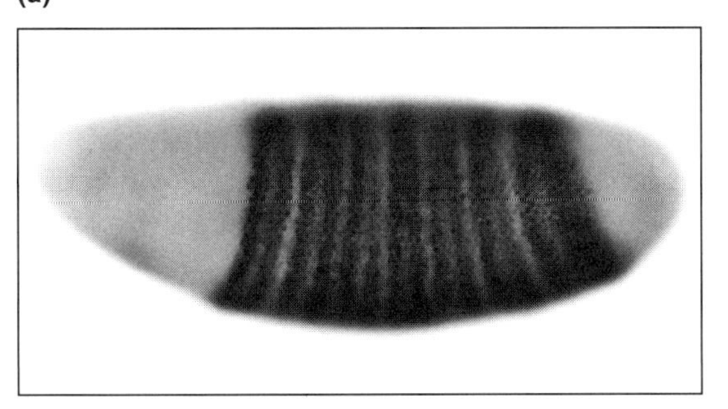
(b)

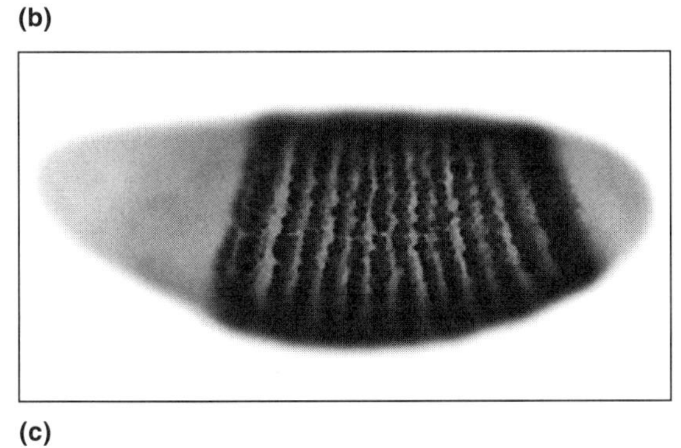
(c)

Figure 9.18
Expression patterns of the *eve* and *ftz* pair-rule genes. The photographs reveal the non-overlapping domains of *eve* (light) and *ftz* (dark) expression as they narrow and sharpen during the blastoderm period. (From P. Lawrence. 1992. p. 109. The Making of a Fly. Alden Press, Oxford, England.)

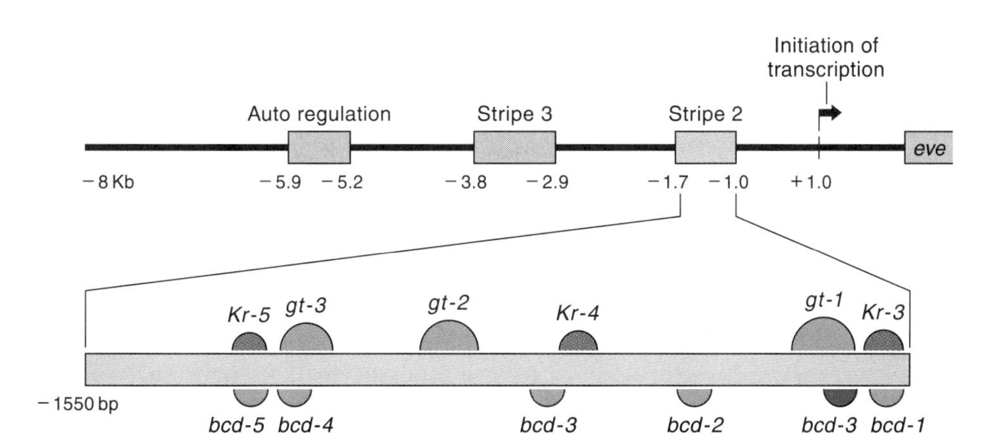

Figure 9.19
Organization of *eve* regulatory region. The horizontal line represents the 5′flanking sequences of the *eve* gene. Discrete regulatory elements within this sequence initiate *eve* transcription in stripes 2 and 3. An autoregulatory element that controls maintenance and refinement during gastrulation is also present in this region. Within an essential portion of the *eve* stripe 2 regulatory element, binding sites for bcd and hb activator proteins are indicated below the enlarged segment, and binding sites for the gt and Kr repressor proteins are indicated above it. Note the close proximity of the activator and repressor sites. (From A. Small et al. 1992. Regulation of Even-Skipped Stripe 2 in the *Drosophila* Embryo. *EMBO J.* 11 4048.)

480 bp element, closely linked binding sites for Hunchback, Krüppel, Giant, and Bicoid proteins have been mapped (Figure 9.19). Activation is thought to be mediated by the combined actions of *hunchback* and *bicoid.* The limits of expression are thought to be defined by the repressive effects of *giant* anteriorly and by *Krüppel* posteriorly. The element is in the OFF state in the anterior embryo because the high concentration of giant protein apparently overcomes the activating influence of Hunchback and Bicoid. Krüppel has the same effect in regions posterior to the stripe. Thus, the ON state of the enhancer occurs only where the activators outweigh the effects of any repressors. The exact mechanisms of transcriptional activation and repression are as poorly understood for these stripe elements as they are for most other eukaryotic genes, but the arrangement of binding sites suggests competition for sites between activators and repressors. Presumably, other *eve* (and *hairy* and *runt*) stripes respond to other combinations of gap and maternal regulators, each stripe forming due to its own stripe response element. Some stripe elements that respond to a regulator in part of the embryo where the regulator is not abundant are found to contain large clusters of binding sites for the regulator. A low concentration of regulator is made functional by the large number of sites.

Secondary pair-rule genes Studies of the relationships among the pair-rule genes indicate that expression patterns of the so-called secondary pair-rule genes (*ftz, paired, odd-paired,* and *odd-skipped*) depend upon the prior expression of *eve, runt,* and *hairy.* Among these secondary pair-rule genes, regulation of *ftz* is best understood. In contrast to the organization and regulation of *eve* and *hairy,* control of all seven of the *ftz* stripes is mediated through a single regulatory element. The expression of *ftz* is initially activated in a broad domain extending 10–70 percent of egg length, and repression in the interstripe regions generates the striped pattern. A 669 bp "zebra element" in the *ftz* regulatory region has been shown to contain binding sites for a multitude of activating and repressing transcription factors, only a few of which are known proteins. The positive regulatory sites can mediate expression throughout most of the germ band, and the negative regulatory sites transform a continuous pattern of gene expression into discrete stripes. The exact mechanisms involved are not understood.

Expression domains of *eve* and *ftz* change shape continuously during the course of cellularization and gastrulation. Initially, the stripes are broad, with diffuse and overlapping borders, and the distribution of protein within each stripe is bell-shaped. These stripes subsequently narrow as the expression of each gene is extinguished in single rows of cells in the regions of overlap. Sharply defined expression domains of each gene remain, leaving rows of cells expressing neither gene in between. The stripes of *ftz* and *eve* expression are no longer symmetric, but have anterior margins that express strongly and are sharply defined; the posterior margins are less well defined. Normally, the stripes of expression of each gene narrow to the same extent and so remain equal in width and evenly spaced. However, in mutant embryos that either partially inactivate *eve* or hyperactivate *ftz,* the spacing of the stripes becomes unequal. For instance, in *eve* mutant embryos, *ftz* stripes remain symmetric, they lack sharp borders, and the metameres that subsequently form are spaced unevenly. These observations suggest that the regular spacing of the segmental primordia that is characteristic of normal embryos depends upon the mutually antagonistic activities of Eve and Ftz proteins.

In summary, transcription factors produced in successive temporal waves and in discrete spatial locations lead to the spatially periodic expression of a set of seven pair-rule genes. The regularity of the final pattern is in part a consequence of the finely tuned mechanisms that produce Eve and Ftz proteins in functionally equivalent quantities. This process transduces a pattern established by at least five different genes (*bicoid, giant, hunchback, Krüppel,* and *knirps*) to a more refined one directed by only two. Presumably, any irregularities in the spacing of the initial domains of expression of the pair-rule genes can be corrected by these mutually antagonistic interactions.

d. Refining the Pattern After Cellularization: Regulating the Segment Polarity Genes

The segment polarity genes constitute the machinery for the last step of the sequential control of successively finer aspects of segmental patterning. These genes assume control of the segmentation process as cellularization of the embryo is completed and gastrulation commences. In contrast to the gap and pair-rule gene products, all of which are transcription factors that function in nuclei in the absence of cell membranes, segment polarity gene products are diverse in character and mediate communication between the newly formed cells of the embryo. Some of the segment polarity gene products are nuclear transcription factors (Engrailed [En], Gooseberry [Gsb], Cubitus interruptus [Ci], and Sloppy-paired [Slp]); others are protein kinases (Fused [Fu] and Shaggy [Sgg]), membrane associated (Patched [Ptc] and Armadillo [Arm]), or secreted proteins (Wingless [Wg] and Hedgehog [Hh]) (Table 9.2). In contrast to the gap and pair-rule segmentation genes, whose roles in patterning the cells of the *Drosophila* ectoderm terminate early, the contributions of the segment polarity genes to patterning the ectoderm continue after the embryonic stages. Their roles in regulating growth and pattern in the imaginal disks is particularly important.

The segment polarity genes are expressed in various different patterns in the newly cellularized embryo. Several (*fu, sgg,* and *arm*) are expressed ubiquitously or in broad domains. Others (*ci, en, wg, hh, ptc,* and *gsb*) are expressed in segmentally reiterated stripes; *en* and *hh* are expressed in the most anterior cell of each of the *ftz* and *eve* stripes; *ptc* and *ci* are expressed in cells that do not express *en; wg* is expressed in the cells that are anterior and adjacent to the cells that express *en; gsb* is expressed in the cells that express *en* and *wg.* Initially, these patterns of expression are a consequence of a combination of positive and negative influences

Table 9.2 ***Segment Polarity Genes and Their Products***

Gene Name	Product
wingless	Secreted signal
lines	?
engrailed	Homeodomain transcription factor
zeste white 3/shaggy	Ser/thr kinase
dishevelled	Novel protein
hedgehog	Secreted signal
cubitus interruptus	Zinc finger transcription factor
fused	Ser/thr kinase
armadillo	β-catenin-like cell junction protein
smooth	membrane protein
arrow	?
patched	Multiple transmembrane domain protein
naked	?
costal-2	?
sloppy paired	Forkhead/HNF3 class transcription factor

from the pair-rule gene products. However, the role of the pair-rule genes is transitory, at least in part because their products disappear; and once the periodicity of segment polarity gene expression has been established shortly after gastrulation commences, the segmental pattern becomes the sole province of the segment polarity genes. At this stage of development, interactions among the segment polarity gene products themselves determine the expression patterns of the segment polarity genes.

Mutant embryos that either lack or misexpress one of the segment polarity genes have abnormal segments. Depending upon the particular mutant, the pattern of naked cuticle and belts of denticles in each segment changes, either losing or rearranging portions of denticle belts or naked cuticle (Figure 9.16). The identity of the cells and the pathway of differentiation they follow depends upon which of the segment polarity genes they express. Interestingly, for many of the mutants, the changes in the segmental patterns extend beyond the cells that normally express the mutated gene, indicating that signaling among the cells plays a critical role in patterning.

e. The Cell-Cell Signaling System in the Cellularized Embryo

As we have described, the early *Drosophila* embryo takes advantage of its syncytial organization to move macromolecules through its cytoplasm unimpeded by cell membrane boundaries. This is best illustrated by the manner in which the Bcd and Nos protein morphogens become distributed in the embryo. These proteins establish the location of the various head, thoracic, and abdominal segments through a concentration-dependent process: their distributions from tightly localized sites of synthesis at the poles of the embryo arise by simple diffusion.

Once the embryo cellularizes, the opportunity to create such instructional gradients by simple diffusion is lost, and other mechanisms must substitute. These mechanisms establish relative position within and between metameres by providing a way for the embryo cells to communicate with each other. The segment polarity genes encode the components of these signaling pathways. Although this group of genes represents what is certainly an incomplete list of the functions involved in these processes, their genetic interactions and the properties of the proteins they encode provide valuable hints about the underlying mechanisms. The phenotypes of mutations in the segment polarity genes also suggest signaling pathways because both cell fates and cell polarities are altered (Figure 9.20).

Once segment polarity gene expression achieves independence of the pair-rule genes, the interdependence of the segment polarity genes assumes preeminent importance in the pattern-determining process. In the 6–12 hour embryo, the cells that express *en* do so continuously only if their anterior neighbors across the parasegmental border express *wg*. Similarly, the stripes of cells in these embryos that express *wg* do so continuously only if their posterior neighbors express *en*. The genes *wg* and *en* become dependent upon each other for their continued and stable expression. These are two examples of how cells at particular locations along the A-P axis selectively express segment polarity genes by processes that depend upon the identity of the adjacent anterior or posterior cells.

Communication between *en*- and *wg*-expressing cells that foster such interactions are thought to involve most of the other segment polarity genes. Although the details of the signal transduction pathways are not yet understood in detail, the response of the *wg*-expressing cells to their *en*-expressing neighbors seems to involve both the patched protein, which is an integral membrane protein that has multiple membrane-spanning domains and could be a receptor at

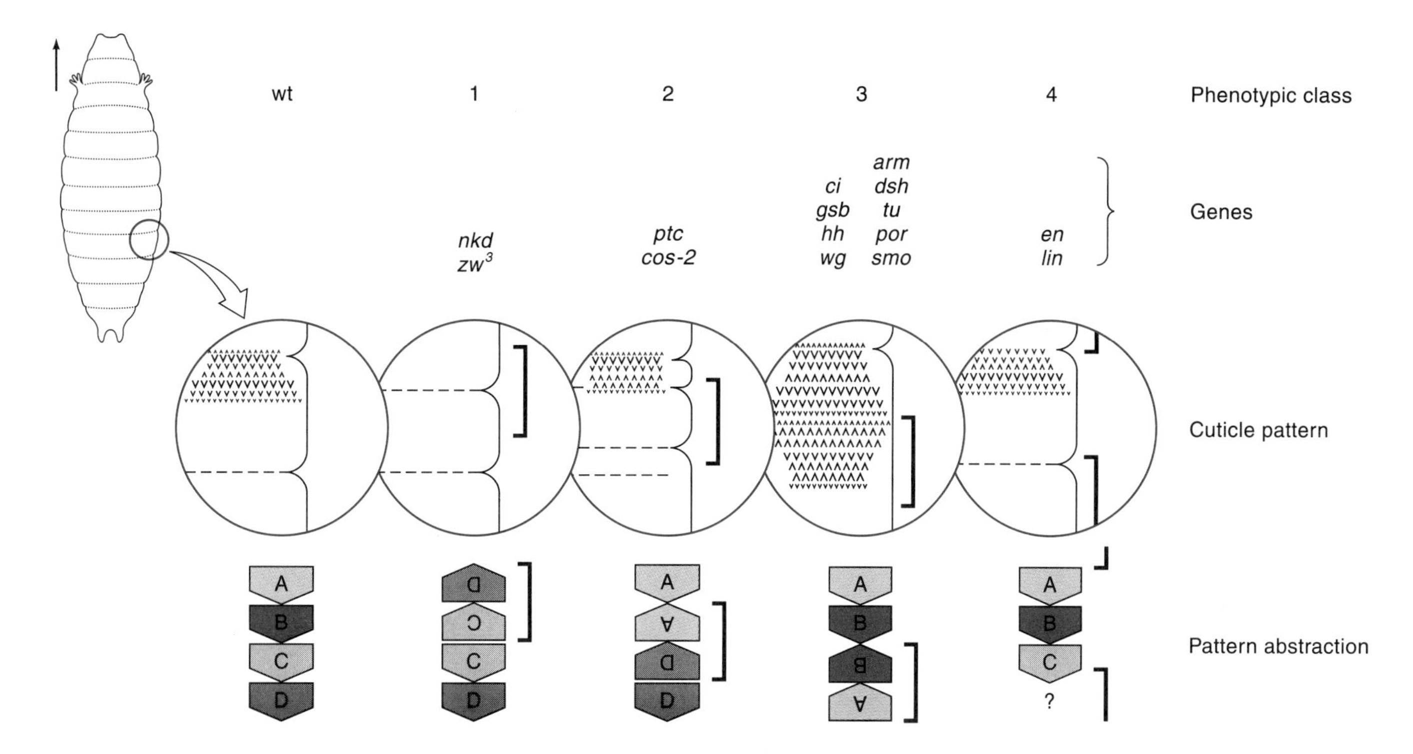

Figure 9.20
The pattern of denticles that distinguishes different parts of larval segments along the A-P axis has characteristic polarized rows in the anterior and most posterior parts of each segment and naked cuticle between (enlargements in circles). Larvae homozygous for lethal mutations in segment polarity genes survive long enough to produce the cuticular patterns, and the patterns reveal alterations in the fates of cells along the A-P axis. The patterns are shown in the circles and abstracted as A, B, C, and D. The changes in the mutants reveal changes in cell fate, with cells in a certain position making structures characteristic of a different part of the wild-type embryo. In addition, the polarities of the denticle rows are sometimes changed, indicating that both cell fates and cell polarities are controlled. For example, in ptc mutants, the cells in the central part of each segment make structures characteristic of anterior segment, and in a mirror-image duplication with respect to the normal structures. In the largest group of mutants, the posterior naked cuticle is replaced by denticle-covered cuticle characteristic of the anterior. Segment boundaries are pattern elements, too, and some mutants have duplicated or changed locations of segment boundaries.

the cell surface, and the Ci protein, which is likely to be a transcription factor (Figure 9.21). The patched protein is thought to be a constitutive negative regulator of its own and *wg* transcription, through a poorly understood pathway leading from the cell surface to the nucleus. This negative effect of Patched is blocked by the Hedgehog signal coming from adjacent cells. Neither *wg* nor *ci* is expressed in cells that express *en*. Across the parasegmental border, the *en*-expressing cells use the Armadillo protein, which is thought to associate with cell junctions, the Dishevelled protein, whose function is unknown, and the Zeste white 3 (also known as Shaggy) protein, a putative protein kinase, in the order shown in Figure 9.21 in their response to their *wg*-expressing neighbors. Other components of these signal transduction pathways likely include other segment polarity gene products as well as ubiquitous factors with more general functions.

The molecules that convey the signals between *wg*- and *en*-expressing cells are the Wg and Hh proteins (Figure 9.21). Wg is secreted from the cells in which

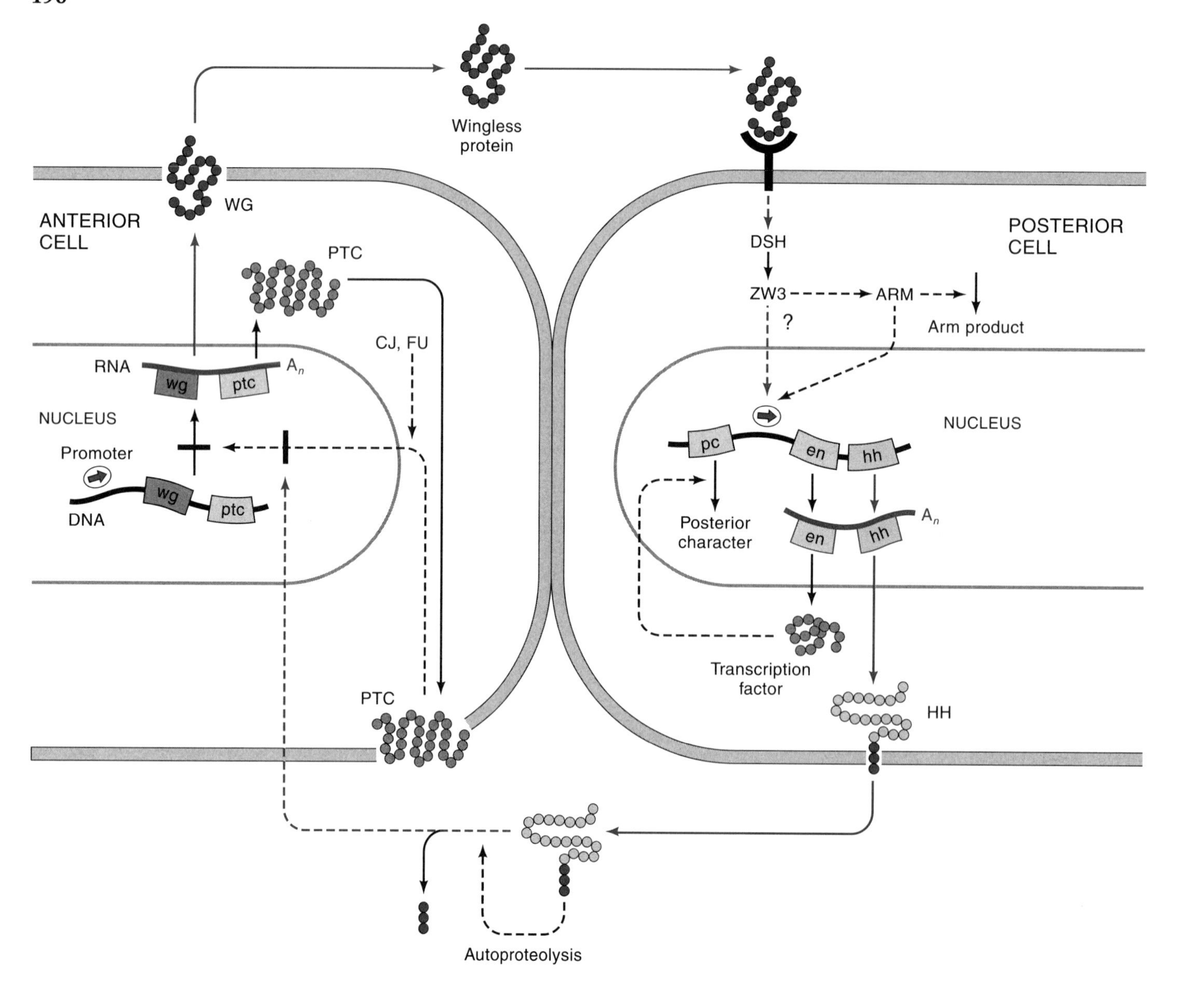

Figure 9.21
Diagram of the two signaling pathways in the posterior half of each *Drosophila* embryonic segment primordium. Each cell represents a belt of cells running around the embryo. The anterior belt of cells makes the secreted signaling protein Wingless (WG); the posterior belt cells make the signaling protein Hedgehog (HH). The two signaling proteins reinforce each others' synthesis, each protein acting through a signal transduction pathway to cause persistent transcription of the gene encoding the other signal. WG binds to a receptor and then acts through a series of proteins including the ZW3 kinase and the ARM b-cateninlike protein. The transmitted signal causes transcription of the transcription factor Engrailed (EN), which in turn causes persistent transcription of the hedgehog gene. EN also has other functions in conferring upon cells their posterior character. The secreted HH protein induces transcription of the wingless gene and another gene called *Patched.* The product of *patched* (PTC) is a transmembrane protein that, indirectly, represses transcription of its own gene and wingless. HH induces transcription of wingless and patched by opposing the function of PTC. The activating function of HH also requires the zinc finger protein CI and the kinase FU. Overall, the two signals maintain different fates in the two adjacent cell types.

it is made, and it then is taken up by neighboring cells, or it precipitates on the extracellular matrix. Wg moves two or three cells from its source, traversing the width of the *en* stripe. The correlation between Wg distribution and *en* expression is excellent, and it supports the proposition that Wg itself carries the signal that determines whether or not a cell will express *en*. Hh is produced by all the cells that express *en*, and in embryos, its expression depends upon *en*. Hh is secreted from these cells and is distributed among the cells in which it is expressed as well as in the neighboring cells. Although its activity is not understood, the Hh protein that diffuses across the parasegmental border is thought to effect the enhancement of *ptc* expression and the maintenance of *wg* expression.

9.4 The Roles of Transcription Factors in Early Development

The successful application of genetic and recombinant DNA methodologies to the study of developmental processes has revealed the preeminent role played by transcription factors in *Drosophila* development. In the early embryo, the A-P and D-V axes are generated by gradients of the *bicoid* and *dorsal* transcription factors, respectively. Segment identity is later controlled by other transcription factors, the proteins encoded by the homeotic genes. Embodied in these transcription factors and the sites upon which they act are the mechanisms that regulate the basic developmental processes in *Drosophila*. As we have seen, maternally provided transcription factors bind to specific DNA sequences to effect their regulatory functions, as in the regulation of *hunchback* by bicoid protein. The gap genes, *hunchback, knirps,* and *Krüppel,* encode proteins belonging to the **zinc finger–containing family** of DNA-binding proteins. The product of the *hairy* pair-rule gene belongs to the **helix-loop-helix** family. Most of the others, including all the homeotic genes, belong to the **homeodomain** family.

a. The Homeodomain Transcription Factor Superfamily

As described previously, the phenotypes of the homeotic mutants suggest that the groups of cells controlled by homeotic genes choose between alternative developmental pathways during the course of normal development. The cells whose developmental program can be switched between alternative fates therefore share an inherent homology, and the different homeotic genes, which direct the ultimate diversification, build upon the homology in directing cells into variant pathways of pattern formation. E. B. Lewis suggested that the homeotic genes might share a common ancestry several years before any were isolated.

The notion that the present-day homeotic genes might have arisen by duplication and divergence gained immediate acceptance when their isolation and characterization revealed, in each, a conserved **homeobox** region. The homeobox is defined as the 180 bp sequence of DNA. The homeodomain is the 60 amino acid sequence encoded by the homeobox. Homeodomains are now known to be **DNA-binding domains** that recognize specific DNA sequences. They may also serve in protein-protein contacts. The homeodomain has since been found to be present in a large superfamily of eukaryotic regulatory proteins. Homeodomains can be classified into a large number of families that extend across vast evolutionary distances. The homeobox family now numbers more than 500, continues to grow, and includes genes from organisms broadly representative of both the plant and animal kingdoms. In plants, the first homeobox genes were identified because they regulate leaf development. In yeast and other fungi, homeobox genes regulate mating type. To be in plants and animals, homeodomains must be at least a billion years old, barring horizontal exchange of genes. Although it is now clear that most members of this gene family have developmental roles unrelated to homeosis, the homeobox remains a hallmark of homeotic genes, and several of these genes appear to have retained both their precisely ordered tandem arrangement in the genome and their developmental roles in axial patterning across vast evolutionary time (Figure 9.22).

The impressive conservation of nucleotide sequences of the homeoboxes has fostered the efficient and rapid isolation of homeobox-containing genes from diverse sources. Even more impressive is the conservation of the protein segment encoded by the homeobox, the homeodomain. For instance, the *Drosophila Antp* and human *HoxB7* homeodomains differ at only 1 of 61 residues, and 16 other homeodomains from fly, human, sea urchin, mouse, rat, and frog, related by sequence to the *Antp* homeodomain, differ at fewer than 7 residues. Such extraordinary conservation must indicate conserved functions.

b. Homeodomain Structure

X-ray crystallography and nuclear magnetic resonance studies of homeodomains in solution have been used to analyze the structure of homeodomain proteins. The first three proteins were analyzed as complexes with DNA. They were found to have virtually superimposable homeodomain backbone structures, a striking finding because the three proteins, one from yeast (MATα2) and two from flies (Engrailed and Antennapedia), are about as different in primary sequence as any two of the known homeodomains. The yeast and fly proteins share only about 15 amino acids out of 61. Nonetheless all three proteins are folded into three α-helical regions joined by sharp turns (Figure 9.23). The second and third helices are comparable to two α-helices in the **helix-turn-helix** configuration of certain bacterial DNA-binding proteins including λ-repressor and cro (*Genes and Genomes,* section 8.6a). Helix 3 sits in the major groove of the DNA and makes base-specific contacts now known to be critical for binding site recognition. A notable feature of homeodomains as compared to other helix-turn-helix proteins is an N terminal region adjacent to the first α-helical region that touches the minor groove of the DNA. This part of the homeodomain is also important in determining the specific developmental effects of homeodomain proteins in flies.

Nearly all homeodomains have four amino acids in common, serving as a sort of diagnostic tag. However, homeodomain proteins span a vast range of sequences, and some proteins with a homeodomain structure are not likely to be recognized as such merely by examining the primary sequence. Furthermore, there are several variant forms of "homeodomains" with added "loops" or adjacent conserved sequences, making it difficult to arrive at a strict definition of the word. As more proteins are studied, a gradation of forms from homeodomains of the types now recognized to rather different DNA recognition structures will probably emerge.

c. Functions of Homeodomain Transcription Factors

Homeoboxes have been especially useful because the conserved stretch in the encoded protein is continuous, and the highly conserved block of DNA sequence can be used to isolate other homeobox genes by low stringency cross-hybridization (*Genes and Genomes,* section 6.6) or polymerase chain reaction techniques (*Genes and Genomes,* section 7.9). Different degrees of sequence conservation make it possible to group the homeodomain proteins in classes.

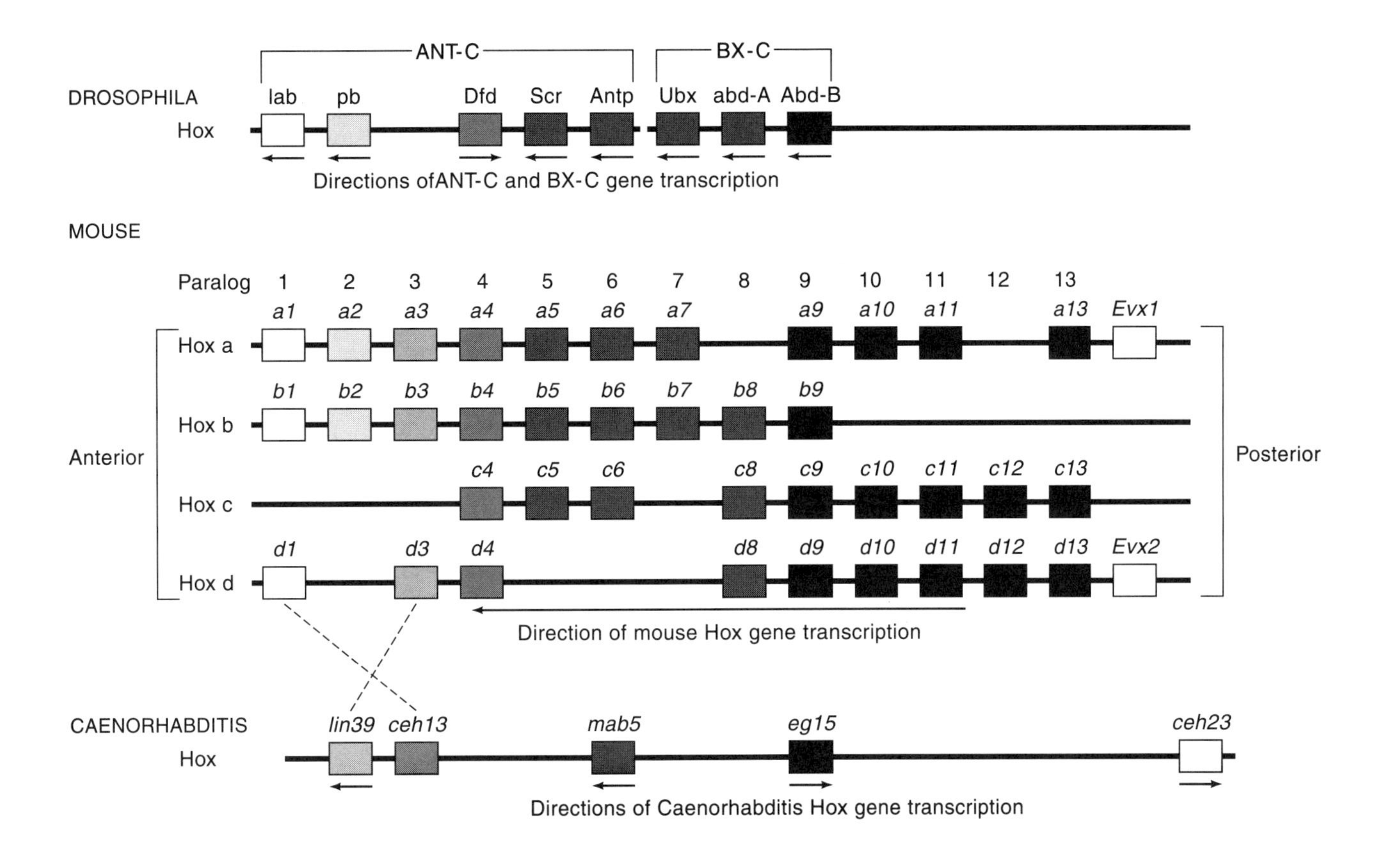

Figure 9.22
The organization of the HOX complexes in three animals, the fly *Drosophila,* the nematode worm *Caenorhabditis,* and mouse *Musculus,* are compared. Each box represents a single gene; arrows indicate the direction of transcription. Each gene encodes a transcription factor with a homeodomain DNA-binding domain. It is presumed that a common ancestor to all the organisms had a single complex with genes expressed in the anterior of the organism located at one end and genes in the posterior at the other end. The reason for this organization is unclear, but it has apparently persisted with few exceptions for about a half-billion years. The expression patterns of the genes reflect their sites of action: mutants show defects in the regions where the genes are normally expressed. The fly complex has divided during evolution into two parts, the Antennapedia complex (ANT-C) and the bithorax complex (BX-C). Other insects have the complex intact. In mammals, different types of change have occurred. Four copies of the complex have formed (a–d) by at least two duplication events, and several copies of the rightmost genes exist in comparison to a single representative of these genes in flies and worms. The different types of mammalian genes have been assigned "paralog numbers" 1–13. Some paralogs are missing from some copies of the mammalian HOX complex, and two HOX complexes have *Evx* genes, which are a distinct type of homeobox gene. The worm has a much simpler complex in one copy. The shading of the boxes indicates similarities between homeodomain protein sequences. The protein product of the Paralog 4 gene in each mammalian complex is more closely related to the fly DFD protein than to the other mammalian proteins, for example. The paralog 6 group is related to *mab5* and *Antp.* The degree to which the protein similarities correspond to evolutionarily conserved functions of the genes remains to be learned. However, it appears that distinct parts of the A-P body axis in most or all animals are patterned by conserved classes of transcription factors.

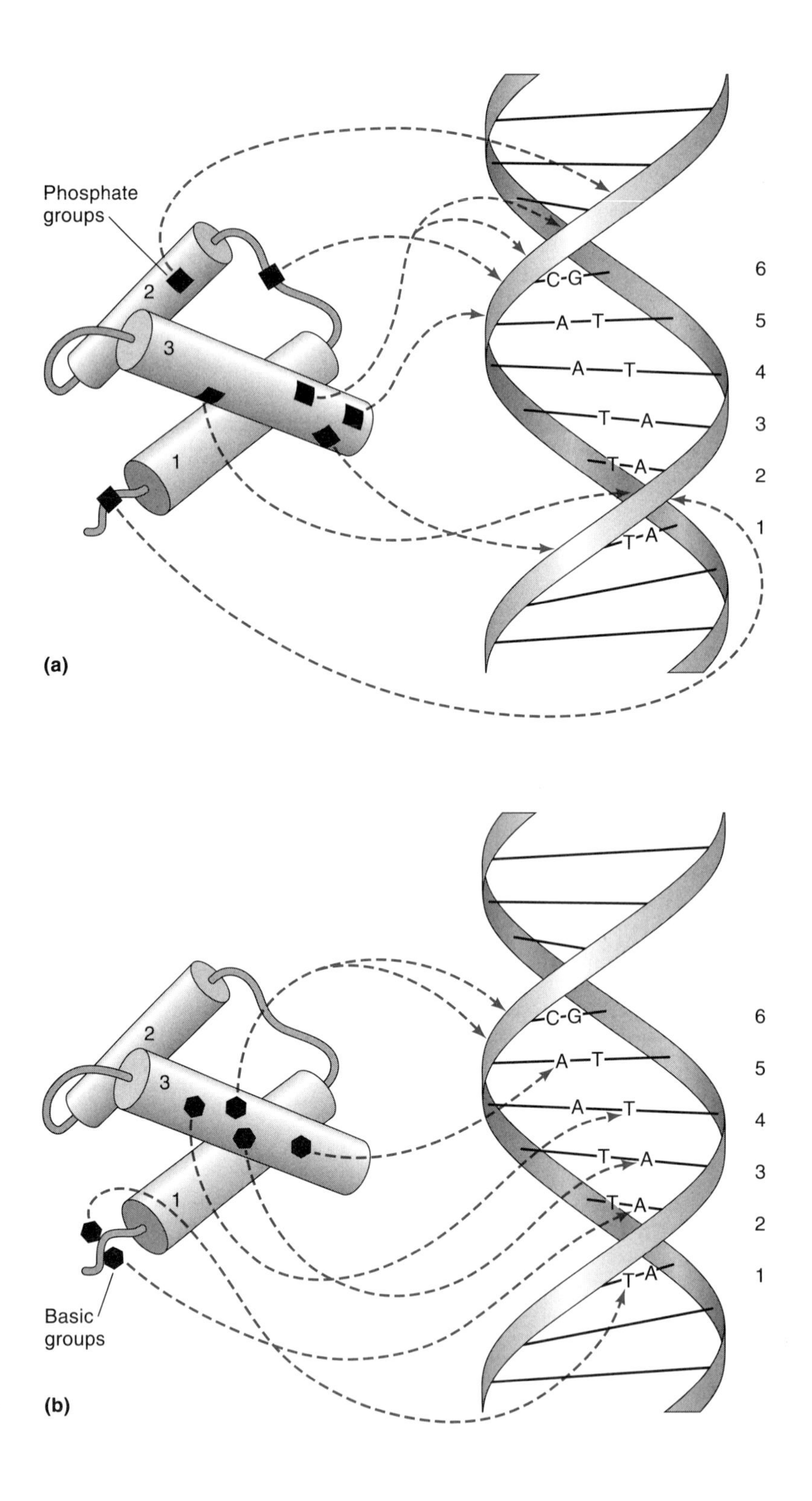

Figure 9.23
The homeodomain makes contact with both the phosphate backbone and the bases of DNA. (a) On the left are cylindrical representations of the three α-helices of the engrailed homeodomain as they contact the phosphate backbone (right). (b) The base-specific contact of DNA. Lines and arrows indicate the specific areas of contact.

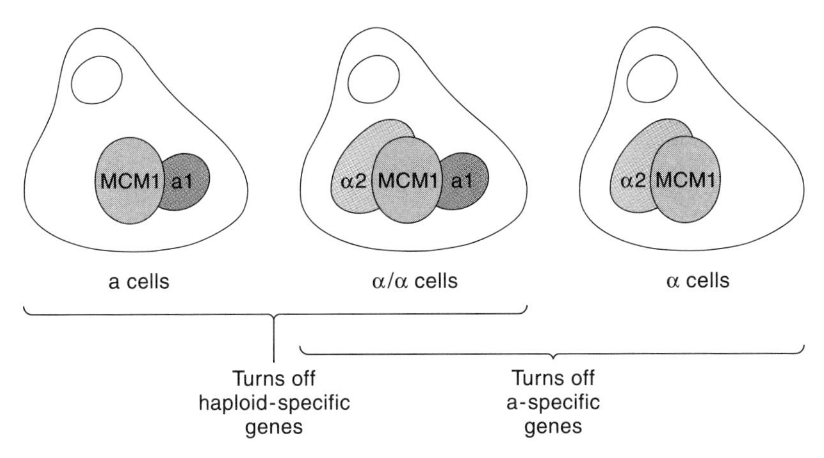

Figure 9.24
Key regulatory proteins in three different types of cells of the yeast *Saccharomyces cerevisiae.* All contain the accessory protein MCM1. A cells contain the a1 homeodomain protein but not the a homeodomain protein, α cells contain α2 but not a1. α/α diploids contain all three proteins.

Homeodomain protein control of yeast mating types The yeast *S. cerevisiae* switches between two alternative **mating types** called a and α. Programmed DNA rearrangements whose occurrence is regulated by the cell cycle and cell lineage control mating type. A cassette-switching mechanism (*Genes and Genomes,* section 10.6b) controls the type of protein produced at the MAT locus, gene conversion events swapping one sequence for another. Alternative proteins are encoded by MAT, depending on which cassette sequence is resident at the locus. The two best studied MAT products are homeodomain proteins called a1 and α2. The MAT proteins provide an excellent example of how a small number of key transcription factors can coordinate the activities of an array of genes whose expression controls differentiation (see *Genes and Genomes,* figure 10.59).

The yeast MAT system provides useful ideas about how other homeodomain proteins might work. The MATα2 protein forms alternative heteromeric complexes (Figure 9.24). It works as a tetramer in diploids composed of two copies of α2 joined with two copies of α1. In haploid α-type cells, a different tetramer with two copies of α2 associates on the DNA and perhaps in solution with two copies of a protein called MCM1. The α2/a1 tetramer is a repressor of haploid-specific genes; the α2/MCM1 tetramer has different target specificity, repressing a-specific genes but not haploid-specific genes. In the absence of α2, the a1 protein has no discernible function. Each protein complex gains its target specificity from the combinatorial action of two protein types, but in different ways. Both the α2 and a1 proteins are homeodomain proteins that bind DNA sequences, and the haploid-specific genes contain adjacent binding sites for the two proteins. MCM1 confers specificity of binding to α2 in a different way, by setting the spacing of the two molecules of α2 as well as by itself recognizing a DNA sequence (Figure 9.25). MCM1 touches α2 protein at a site N terminal to the homeodomain, and a1 touches α2 on the C terminal side of the homeodomain. Taken together, the results demonstrate how protein-DNA and protein-protein interactions confer specificity of target gene selection.

The POU group Originally named for the first three genes found to have these sequences, *pit1* and *oct1* of mammals and *unc86* of nematodes, these **POU proteins** have diverse functions in regulating transcription and development (*Genes and Genomes,* section 8.6a, figure 8.98). One of the POU proteins, Oct1, is thought

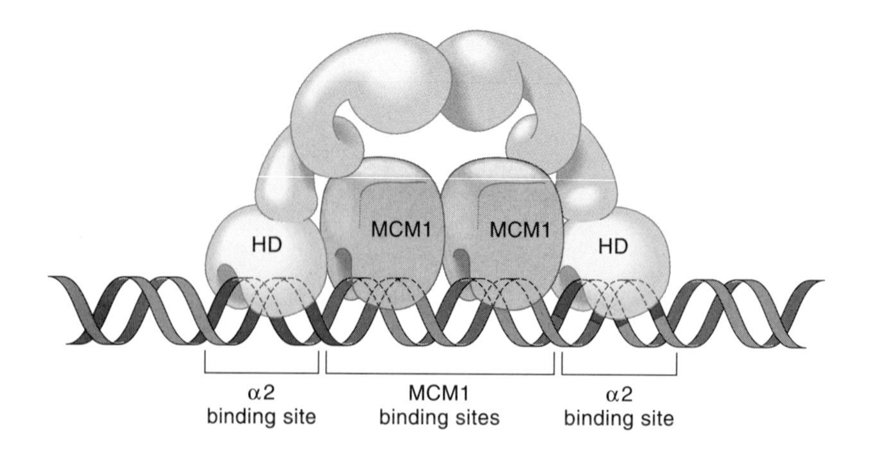

Figure 9.25
Model of α2 and MCM1 bound to a target promoter. The positioning of MCM1 binding sites between the α2 binding sites and the contacts between MCM1 and α2 accurately position this protein complex at the appropriate site on the chromosome.

to be required for transcription of histone genes and is, as expected, found in all cells. A very similar protein, Oct2, is expressed primarily in developing lymphocytes, where it activates immunoglobulin gene expression.

The POU domain is striking because only part of it is a classical homeodomain, although with a distinctive sequence that defines a POU homeodomain class. The remainder, the so-called POU-specific domain, is located upstream of the homeodomain and is composed of two parts, called A and B. Recent structural studies of the POU-specific domain have led to an astonishing finding: it is extremely similar to the structure of the DNA-binding portion of the repressor protein of bacteriophage λ, perhaps the most intensively studied transcription factor (Figure 9.26; see also *Genes and Genomes,* section 3.11e). If we assume, as is reasonable, that other POU domains have similar structures, we are left with a picture of two helix-turn-helix motifs joined in a collaboration to bind to specific DNA sequences. Neither the POU-specific domain nor the POU homeodomain binds DNA very well alone. The combinatorial binding presumably confers great specificity, much as the combination of the yeast homeodomain proteins a1 and α2 have much greater binding specificity and affinity than either protein alone.

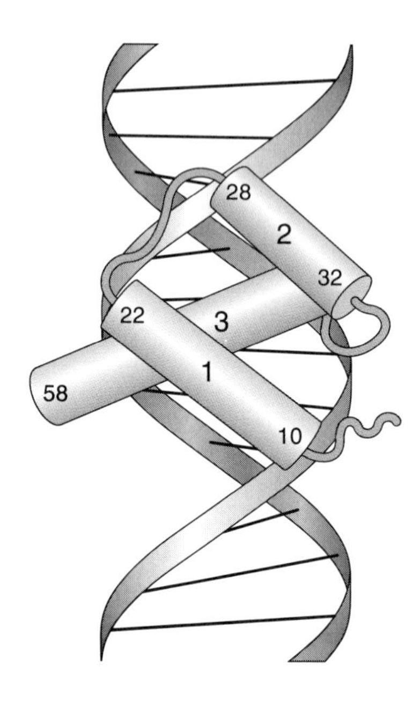

Figure 9.26
The POU domain of the mammalian transcription factor Oct1 has three α-helices (cylinders 1, 2, and 3) whose orientation and contact with DNA are similar to the DNA-binding domain of the Engrailed protein (a homeodomain) and of the prokaryotic 1 and 434 repressor proteins.

Pax genes Like the POU group, the **Pax group** shares a distinctive DNA-binding domain, called the paired domain, in addition to a characteristic type of homeodomain. This second domain is called the paired domain not because it is duplicated but because it was first found in the fly segmentation gene *paired.* The paired domain is about 130 amino acids long and includes a region that may form a helix-turn-helix structure. Although the POU-specific domain has yet to be found without its accompanying POU homeodomain, the paired domain typical of Pax proteins is also found in proteins that have no homeodomain at all.

The Pax group has been shown to be involved in developmental regulation in flies, fish, mice, and humans, presumably as a transcriptional regulator. The *Pax1* gene, which contains a paired domain but no homeodomain, is altered in three alleles of the mouse gene *undulated.* The mutations cause vertebral column defects; the expression of *Pax1* in the prevertebral column is consistent with the phenotype.

Mouse and human mutations have been mapped to the *Pax3* gene. The mouse mutations, called *Splotch,* have severe developmental defects. Heterozygotes are deaf and have pigment abnormalities, but homozygotes exhibit far worse problems, including spina bifida, exencephaly, and neural crest derivative defects. The human syndrome, observed only in heterozygotes, presumably because homozygotes do not survive, is called Waardenburg syndrome Type 1. The phenotype is similar to that of the *Splotch* mice.

There are also known mutations in the human *Pax6* gene. These heterozygous *Aniridia* patients partially or completely lack the iris of the eye. Similarly, homozygous *Small eye* mice, without a functional *Pax6* gene, completely lack eye and nose structures. These results provide striking confirmation of the usefulness of finding vertebrate homologues (and more distant relatives) of fly segmentation genes.

The HOM and HOX complexes: conservation in all animals We previously discussed the clustered homeotic genes that determine the fates of fly segments. When E. B. Lewis first described the BX-C, he emphasized how it influenced his views of gene structures and relationships. Subsequently, in work done over several decades and continuing today, he demonstrated the importance of the BX-C in organizing development. There were three main thrusts to his conclusions. (1) Each part of the animal along the A-P body axis requires a different homeotic gene activity, as we have discussed. (2) Loss of a gene function causes transformations of more posterior body parts into copies of more anterior parts. The activity of a gene where it is normally inactive causes anterior body parts to develop like copies of posterior body parts. (3) The order of the genes along the chromosome corresponds to the order of the body parts they affect along the A-P axis.

Today we are able to say, still fresh with astonishment, that these three rules apply remarkably well to mammalian development and probably to most animals. Although the exact structures formed are obviously drastically different in a fly, a nematode, and a mouse, homologous homeotic gene clusters in these disparate life forms control differentiation along the A-P axis (Figure 9.22). Distinct differentiation pathways are presumably regulated by controlling the transcription of downstream "target" genes.

The homeobox was discovered as a region of cross-hybridizing DNA among the genes in the ANT-C. Twelve genes in the ANT-C and BX-C contain homeoboxes. The pair-rule segmentation gene *ftz* and the maternally active gene *bicoid* also encode homeodomain proteins. There are 2 closely related *zen* genes, 1 of which is necessary for dorsal-ventral patterning. We mentioned some of the 5 ANT-C homeotic genes, *labial,* and *proboscipedia, Dfd, Scr, Antp,* and the 3 BX-C homeotic genes, *Ubx, abd-A,* and *Abd-B,* reaching the final tally of 12 (Table 9.1). The colorful names are historical and have limited relation to gene functions. Mutations in *labial* affect the labial (mouthpart) structures. Mutations in *proboscipedia* can transform the proboscis into legs. Mutations in *Deformed* distort the head, and those in *Sex combs reduced* transform first legs into second legs, which is seen as the reduction of a row of bristles called the sex comb, which normally marks anterior (first) male legs. Some *Antennapedia* mutations transform antennae into legs. Mutations in *Ultrabithorax* can cause a transformation of the third thoracic segment into the second, and because the second thoracic segment makes up the bulk of the thorax, the flies become bithoracic to an extreme degree. The two abdominal genes affect segment fates in the abdomen. The names can be misleading in that *Antp* mutations, for example, that transform antennae into legs, suggest a role of *Antp* in antennal development. In fact, the gene normally acts in the thorax to control leg development, and the effects of the mutations are due to gene activation in the head, where it normally is inactive.

The two homeotic gene clusters of flies were probably originally one, based on the mammalian and nematode data and on analyses of a beetle complex. The full cluster is sometimes referred to as the HOM complex, but it may be simplest to refer to the homologous complexes of all animals as the **HOX complexes**. The order of ANT-C genes along the chromosome, conceptually joined to the BX-C genes, largely agrees with the order of affected body segments. Each gene is transcribed in, and acts in, a different part of the body. In some body regions, more than one gene is expressed and functional.

The homologous mouse and human complexes were discovered by cross-hybridization with fly probes. Most of the sequence similarity is limited to the homeobox, but there are small regions of additional homology in other parts of the coding regions. The mouse and human HOX complexes appear to have been included in two large chromosome duplication events involving many genes in addition to the homeotic genes, resulting in four copies of the complex called A, B, C, and D (Figure 9.22). Each complex contains at least nine genes, more than the eight types in the pooled fly complexes. If the genes are aligned by homeodomain sequence, a *labial*-like set of mammalian genes, a *Dfd*-like set, an *Abd-B*-like set, and a set of four similar to the *Antp, Scr, Ubx,* and *abd-A* genes, all of which have rather similar sequences, can be seen. Comparable members of different mammalian clusters are called **paralogs** and are given corresponding numbers. The *labial*-like paralogs in all four clusters are given ones, for example, hence A1, B1, and D1. (The C cluster does not have a *labial* type gene.) The "extra" mammalian genes are of the *Abd-B* type and, appropriately, are located at the posterior end of the complex. Each cluster seems to have lost some of its 13 paralogs. There are 38 HOX-type genes altogether in the four clusters (Figure 9.22). Curiously, all mammalian genes are transcribed in the same direction, so one can speak of the posterior end (for where the genes work in the animal) synonymously with the 5′ end (for the direction of transcription of all the genes).

The nematode worm *Caenorhabditis elegans* cluster of homeotic genes has also been characterized. It contains at least five genes and is organized similarly to the fly and mammalian clusters (Figure 9.22). Mutants lack parts of the animal and exhibit homeotic transformations. The worm homeotic genes offer clear examples of combinatorial regulation: two proteins expressed in the same cell can mutually cancel out their effects or can work together to direct a fate not seen with either protein alone. Because nematode cell lineages are so well understood, researchers can observe regulation by homeotic genes of the fates, differentiation, and migration of single cells. The similarity in the clustered homeobox genes of flies, worms, and mammals strongly suggests that similar clusters will be found in all animals.

Mutations of HOX genes have been constructed using gene-targeted gene disruption in embryonic stem cell lines followed by creation of heterozygous mice from the stem cells. When these mice are bred to each other, homozygotes are generated, and the phenotypes can be examined. The resulting homeotic phenotypes demonstrate astonishing parallels between the insect and mammalian gene functions (Figure 9.27). In one striking instance, due to loss of function of *Hoxc-8* (a *Scr-Antp-Ubx-abd-A* type gene), the ribs are transformed into more anterior ribs, and vertebrae that normally lack ribs can gain them. As in the insect mutants, loss of function leads to development of anterior structures in more posterior regions. Knockouts of other *Hox* genes have given rise to different defects, including ear, cranial nerve, and hindbrain disruptions due to loss of *Hoxa-1* (a labial group gene) and homeotic transformations of the cervical vertebrae due to mutations in *Hoxb-4.*

If mouse homeotic genes are to be like those of flies in other respects, activation of genes where they ought to be off should also cause transformations. This is in fact the case and usually follows the fly rule: more anterior structures are formed with posterior characteristics. Thus, additional vertebrae are seen in the head of mice when *Hoxa-7* or *Hoxd-4* (a *Scr-Antp-Ubx-abd-A* type gene) is expressed in the head region, where it is normally off. There may, however, be exceptions to the A-P rules. Thus, an extra rib phenotype similar to the effect of loss of *Hoxc-8* function is seen if *Hoxc-6* is expressed at high levels in the posterior,

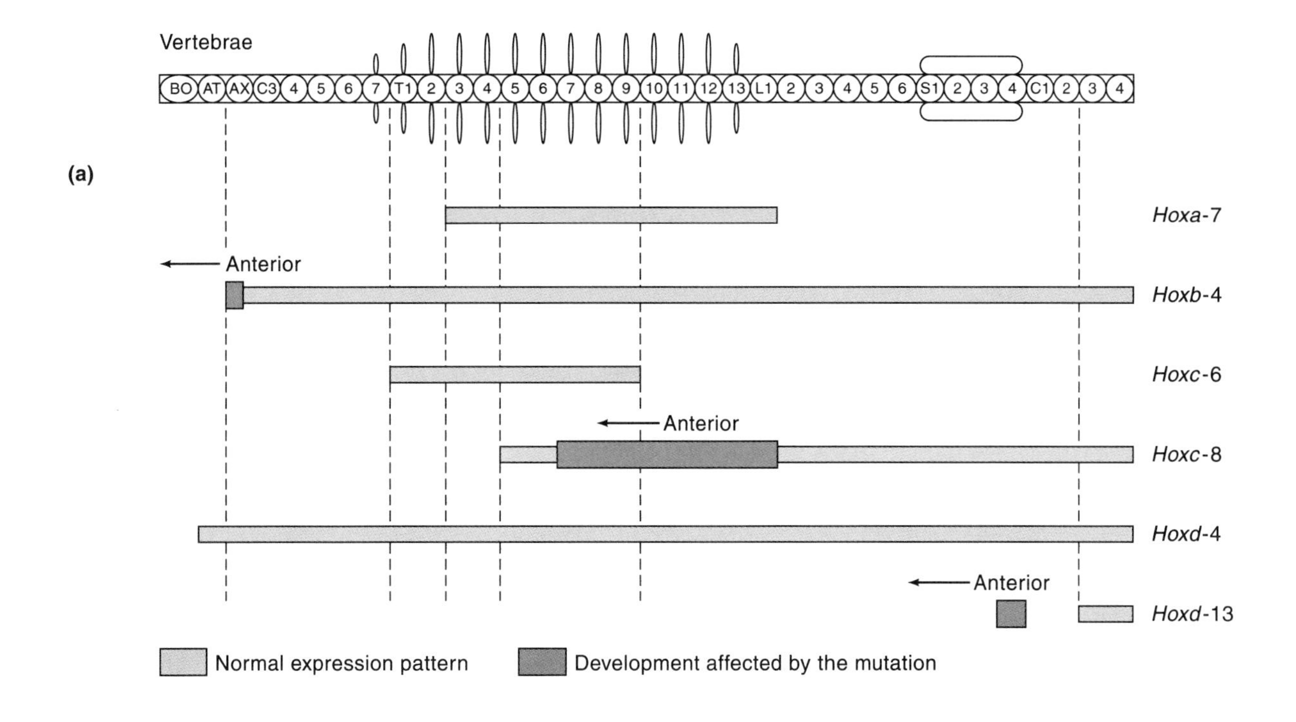

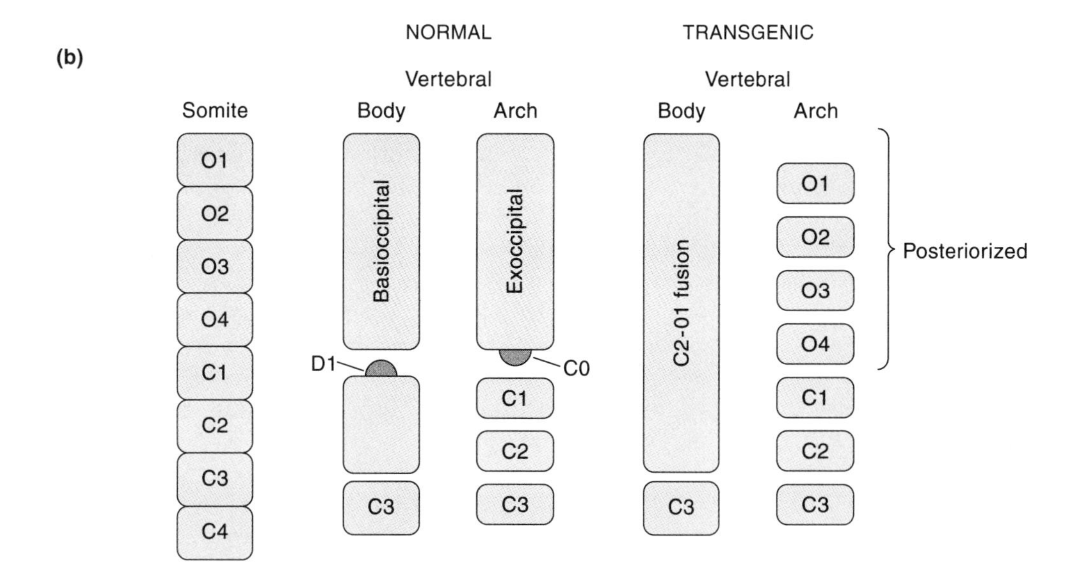

Figure 9.27
Examples of homeotic phenotypes that arise when HOX genes are not expressed normally. (a) Mice normally develop 14 pairs of ribs, but mutants lacking a functional HOXC-8 gene have an extra pair of ribs attached to the sternum. These extranumery ribs are produced by the L1 vertebra and are a striking example of the anterior transformation that alters affected tissues. This figure shows the anterior limits of expression of 6 HOX genes (vertical dashed lines) and the areas most affected by mutations in HOXB-4, HOXC-8, and HOXD-13. (b) Ectopic expression of HOXB-4 in a more rostral pattern than normal causes a transformation of the occipital bones (O1–4) toward a more posterior phenotype, into structures that resemble cervical vertebrae. Vertebrae are composed of two parts, vertebral body (Body) and neural arch (Arch), and normally only C2 and C1 are fused. However in mutants with ectopic HOXB-4 expression, the vertebral bodies from C2-O1 are also fused. No axis dens (D) or occipital condyle (CO) formed in the mutants, and the exoccipital arch between O1–4 did not fuse. (Part a from H. Moellic et al. 1992. Homeosis in the mouse induced by a null mutation in the HOX 3.1 gene. *Cell* 69 256; part b from T. Lufkin et al. 1992. Homeotic transformation of the occipital bones of the skull by ectopic expression of a homeobox gene. *Nature* 359 840.)

where it is normally off, as though it is interfering with the function of a more posterior gene.

Insects and mammals are thought to be separated from their last common ancestor by about 600 million years. Nematodes are probably even more distant, yet they too have a complex composed of at least four homeobox genes that can be linked, by sequence and order along the chromosome, to the fly and mammalian genes. The HOX and HOM genes have an Antennapedia class homeodomain and are dedicated to similar roles in vastly different animals. Homeodomains per se must be very ancient because yeast, plants, and animals probably have a billion years of evolution separating them.

Upstream and downstream of HOX/HOM genes Now the questions become how the Hox genes control differentiation and how differential Hox expression is set up in the first place. In fact, there are two major questions about the functions of HOX/HOM genes: how are they regulated and how do they regulate? As we have seen, we know much about how the segmentation genes initiate homeotic gene expression in flies. The gap genes play a particularly important role in controlling where the homeotic genes are expressed along the A-P axis. The patterns are further refined by direct action of the pair-rule transcription factors, with additional contributions from at least some of the segment polarity genes. Similar questions for vertebrate systems await investigation. For example, how do *Hox* genes control digit differentiation, and how is differential expression in the limb bud or hindbrain established?

In flies, at least three other groups of genes appear to be involved in **maintaining** proper patterns of homeotic gene transcription long after the initiating factors have disappeared. First, some of the products of homeotic genes cross-regulate, in the most common instance, with posterior-acting genes preventing transcription of more anterior-acting genes. Some of the genes also **autoregulate** positively or negatively. Second, a group of genes of the **Polycomb class** keeps homeotic genes off where they should be off. Mutations in any of the ten or so genes in this class do not change the initial patterns of homeotic gene transcription, but allow extra expression of *HOM* genes to occur later. The extra expression leads to homeotic transformations similar to those caused by gain-of-function mutations in the homeotic genes themselves. The third group of genes, the **trithorax class**, provides activator functions that appear to counterbalance the actions of Polycomb group genes. Trithorax group genes help maintain patterns of homeotic gene expression. The Trithorax protein itself is a zinc finger protein closely related to a human protein implicated as a leukemia protooncogene. Another member of the trithorax group is closely related to a yeast protein that, as part of a protein complex, offsets the repressive effects of chromatin to allow transcriptional activation. Human homologues of this protein also exist. Little is known as yet about the molecular mechanisms of either the activator or repressor group, although they both clearly affect transcription, and each group contains some DNA- or chromatin-binding proteins.

If there is some progress on regulation of homeotic genes (although little as yet in vertebrates), there is less in understanding how they control differentiation. Homeotic genes must coordinate a large number of cellular processes, such as production of specialized products, cell division, cell movements, cell surface attributes, cell asymmetries, and so on. Only a few genes that are directly regulated by the homeotic transcription factors have been found, but there are good reasons to believe that many such **"target" genes** exist. The targets that have been found are mostly powerful regulators of development in their own right, so the extraordinary influence of the homeotic genes is in part due to their ability to coordinate the activities of subservient regulators. Among the target genes are, for example, secreted signaling proteins of the **transforming growth factor β (TGFβ)** and **Wnt** classes. Identification of target genes and studies of how the homeotic proteins regulate their specific targets and not those under the control of other homeotic genes are key to understanding how differentiation is controlled.

Homeotic proteins seem unlikely to work alone. The yeast MAT and POU proteins are good examples of how combinatorial actions of proteins lead to increased potency and specificity. Recently, one excellent candidate for a **cofactor protein** that interacts with some of the homeotic proteins has been found. The fly gene is called *extradenticle,* and it encodes a homeodomain protein. The protein is remarkably similar to a mammalian protein called Pbx1, first identified as a key factor in a chromosome rearrangement that causes leukemia in humans. Mutations of *extradenticle* do not change where HOM genes are transcribed, but instead alter the effects of some genes in their normal domains of action. Thus, homeotic and Extradenticle proteins may act together upon target genes. Indeed, researchers recently showed directly that target gene expression is altered, sometimes by loss of activation and sometimes by loss of repression, in *extradenticle* mutants.

Homeotic genes are master regulators, but cells respond to homeotic genes in the context of cell position in the embryo and cell history. Homeotic genes are in many instances expressed in all cells of a given body segment, yet the cells do different things according to their positions. They may all contribute thoracic structures, but their individual contributions differ. Thus, cells must integrate information about their sex from the sex determination pathway, about their tissue type from regulators such as myogenin, about their position from segment polarity genes, and so forth. Homeotic transcription factors work in the context of all these other regulators, and a major job ahead is learning how all this information is integrated to create a whole, functioning animal.

9.5 Molecular Methods in *Drosophila* Developmental Genetics

a. Gene Isolation and Molecular Cloning

Key to the explosive progress that has transformed our understanding of many developmental mechanisms has been the increasing facility with which the genes involved can be characterized genetically, cloned, and analyzed. A complete understanding of a gene's role and function in a developmental process requires that we know when and where the gene is expressed, what its product is, what activity the product has, and the context in which the product interacts. Having the gene in recombinant form and having mutant forms of the gene that provide clues to its normal function are essential to such studies. As described previously, genetic screens have succeeded in identifying most of the genes involved in patterning the *Drosophila* embryo. The challenge of isolating these genes has been met with a variety of approaches.

Two basic approaches to cloning genes in *Drosophila* are to (1) identify and isolate a chromosomal region that is altered in a mutant line or (2) isolate DNA related by sequence homology to a known gene. The first approach can proceed in several ways. Mutants can be induced with a mutagen that physically marks the mutational event in such a way as to facilitate its molecular isolation. Such methods have been developed using the insertional transposon called the **P element**. Mutations that physically break the chromosome can be accurately mapped cytologically, and if a closely linked DNA sequence is available, neighboring and overlapping recombinant clones can be isolated in successive steps from the initial clone (see Section 9.5d). The second approach involves molecular hybridization under conditions in which related and nonidentical sequences can hybridize to probes.

b. The P Element Transposon

Vectors and mutagens P elements are described in *Genes and Genomes,* section 10.2, but we briefly review their biology and applications here. When flies caught in the wild are crossed with laboratory flies, a peculiar phenomenon called **hybrid dysgenesis** is often observed. Fly progeny appear to have been mutagenized. The most unusual aspect of the effect is the asymmetry of the mutagenesis with respect to the sexes of the crossed flies. If the female is the laboratory fly, progeny exhibit genetic defects of many sorts. If the female is the wild fly, no dysgenesis is seen.

The basis of these strange events is a small transposable element called the P element, which produces its own transposase and jumps to new chromosomal sites whenever it finds itself in the absence of an inhibitor. Inhibitor builds up gradually in any cells that have P elements. The arrangement seems to be logical because if flies bearing P elements survive, they will eventually build up enough inhibitor to prevent transposition. The immobilized P elements then remain quiescent in surviving flies, which presumably have P elements in harmless insertion sites. The wild flies are believed to be in just such a state. Laboratory flies, by contrast, contain no P elements, apparently because when the laboratory stocks were established decades ago, P elements were not prevalent in the wild flies. Most present-day laboratory flies contain no P elements and therefore no inhibitor.

Returning to the asymmetry of the cross results, we can see a sensible reason for dysgenesis occurring only when male flies carrying P elements (wild flies) are crossed to female flies (lab flies) that do not. The male donates P-bearing chromosomes to an egg cell environment lacking inhibitor. Transposase is synthesized from the intact P elements, the "music starts to play," and the P elements "play musical chairs," inserting into new sites and potentially wreaking genomic havoc. In contrast, the eggs from a female wild fly contain P elements but also inhibitor. Regardless of which males fertilize the egg, there will be no music and no jumping.

Some P elements are internally deleted and therefore unable to produce the transposase—to play the music. Others have defective ends and cannot move from one chromosomal site to another—they cannot "hear" the music. The former provide useful tools because when transposase-producing elements are removed from the cell, the remaining P elements will not jump, allowing stable inserts to be maintained even without inhibitor. The latter class of defective elements, without two good ends, are useful tools because they stay put. Thus, one of these "deaf" inserts, located at a known genomic site and linked to easily followed genetic markers, can be used as a source of transposase and then removed to stabilize remaining P inserts that cannot play. In other words, the experimenter can turn the music on and off, controlling the timing of transposition.

P elements have been turned into extraordinarily powerful tools for developmental genetics. They can be used as mutagens, and the inserts can be used as a molecular tag to clone the mutated gene. They can be used as vectors to carry DNA into the chromosomes. They can be used as sensors of spatial gene expression, allowing the identification of genes with developmentally interesting transcription patterns.

The transposase gene can be interrupted or replaced with large chunks of foreign DNA sequence, up to at least 40 kbp, and the new vector used to carry foreign or engineered genes into the fly genome. The element is then defective because it does not encode a functional transposase, but it can be complemented by providing other intact P elements or a copy of the transposase gene in *trans.* In the absence of transposase, P elements and DNA inserted into them remain stable, permanent, and functional parts of the genome. Most inserts are single copies of the transposon.

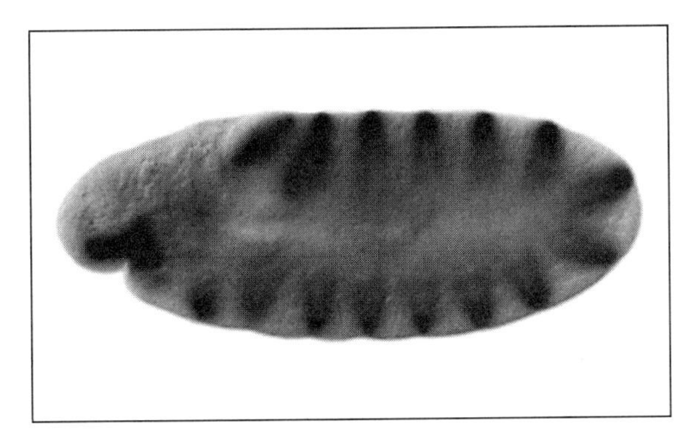

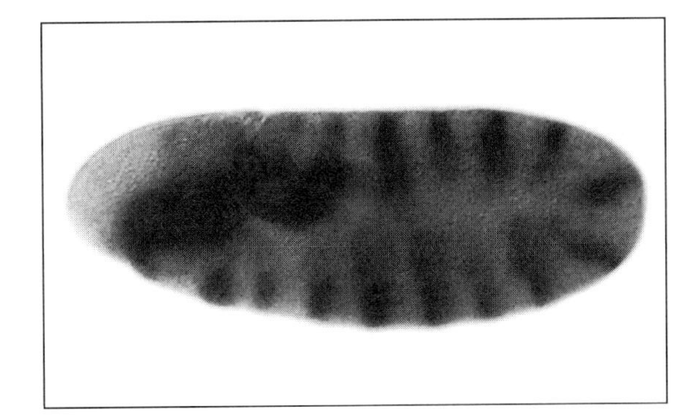

Figure 9.28
Pattern of expression of the segment polarity gene *hedgehog* revealed by in situ hybridization to *hedgehog* mRNA (left) or by histochemical detection of β-galactosidase (right). The embryo on the left is a normal embryo; the one on the right has an enhancer trap insertion upstream of the *hedgehog* transcription unit.

Another important application is the use of P elements for making animals mosaic for a certain gene function. The P element vector is used to import yeast recombinase systems into flies. The yeast recombinase system can be used to delete a gene flanked by sites upon which the recombinase acts. By doing this during development of the animal, researchers create a mosaic animal that contains the gene in some cells but not others. This sort of mosaic is important for analyses of which cells require a particular gene, for example, in a signaling situation.

Enhancer trapping Many important regulatory genes of the types we have described are transcribed in precise spatial and temporal patterns. The logic can be turned around to imply that genes expressed in such patterns are likely to be important regulators. This rationale has stimulated work on methods for detecting genes expressed in suggestive patterns during development. Using a variation on a standard *E. coli* approach, scientists can fuse the *E. coli lacZ* gene to random genes in the *Drosophila* genome. Separate lines of flies are generated, each with a *lacZ* insert in a different location. Then *lacZ* expression can be observed by staining embryos or other stage animals with antibody or the color-generating substrate X-gal (Figure 9.28). The key to the procedure is to use a P element carrying a transcriptionally silent, or nearly silent, *lacZ* gene. The *lacZ* gene is located within the P element in a position that brings it under the control of the quite weak P element promoter. When the P element inserts near a transcriptional enhancer in the genomic DNA, its promoter and therefore *lacZ* is brought under the influence of the enhancer, and β-galactosidase is produced in a pattern reflecting the normal pattern of gene expression at the insertion site. The method is referred to as **enhancer trapping** because it detects the function of transcriptional enhancers.

Thousands of lines of flies have been examined for patterns of *lacZ* expressions. Experience has shown that accurate β-galactosidase patterns are often seen with previously characterized genes and with genes newly discovered using the enhancer-trapping approach. The result has been an explosion of knowledge about genes under spatial and temporal control. For example, one can screen for genes expressed in specific cells in the developing eye, wing, nervous system, or blood cells. The pattern of β-galactosidase does not always precisely report the full and normal pattern of the insertion site gene, presumably because the insert position sometimes allows a response to only some of the regulatory elements or the insert itself damages the gene. Success is prevalent, however, and enhancer trapping has been added as a major experimental advantage of *Drosophila.*

If a P(*lacZ*) insert is found to have an interesting pattern, the gene can be mapped and cloned, and new mutations can be made as follows. First, the location of the P element is determined by *in situ* hybridization to polytene chromosomes (see Section 9.5c). Because the P element seves as a marker for the chromosomal DNA at the site of its insertion, that DNA region can be obtained by preparing a genomic DNA library from the fly stock and screening it with a P element probe or by using a procedure called "plasmid rescue." Many P elements have been engineered to contain a ColE1 origin of DNA replication and a drug resistance marker such as ampicillin resistance. Genomic DNA from the flies of interest is digested with a restriction enzyme that does not cut within the P element, so the P element will be included in a larger fragment of DNA that incorporates flanking genomic DNA. Ligation of the whole genomic DNA mixture under appropriate conditions leads to DNA circles, but only the P element–bearing circle will be able to replicate in *E. coli* on ampicillin media. The ligated genomic DNA is therefore transformed into bacteria, and ampicillin-resistant colonies are selected. The successful isolation of genomic DNA surrounding the P element is confirmed by hybridizing the isolated clones to polytene chromosomes from a fly stock that has no P elements. Only the flanking genomic sequences will hybridize, revealing their origin, which will, if all has gone well, be the same as the original P element insertion site. More genomic DNA can then be isolated from adjacent regions by chromosome walking, and the region can be searched for transcripts using RNA blots. Any transcript found is hybridized *in situ* to whole mount embryos to determine whether the expression pattern reflects that of the P element *lacZ* gene.

Finding a transcript in an interesting pattern is only the first step. It is crucial to prove that the newly discovered gene has a function where it is transcribed, a function relevant to the developmental process, such as eye development, that is under study. Sometimes the insertion will already have created a mutation in the target site gene. If not, the P(*lacZ*) insert can be used to produce a mutation in the insertion site gene. Thus, the P element can be mobilized by providing transposase from another P element in *trans.* Some of the resulting transposition events will delete DNA from the insertion site, creating mutations that delete some or all of nearby transcripts. These mutants can be examined for their ability to support, for example, proper eye development. Because the mutations are isolated in heterozygotes, even essential genes can usually be deleted and later tested in the homozygous condition. Genes can be transcribed in remarkably precise and intriguing patterns without having an easily detectable function, so these tests are important and worthwhile.

c. The Polytene Chromosomes

The *Drosophila* genome project began in about 1910 when T. H. Morgan started mapping genes to assemble a genetic representation of the chromosomes. Decades of genetic analysis by Morgan and his associates provided an amazing and detailed picture of chromosome behavior and structure, all prior to the discovery of the most powerful tool of chromosome cytology: the giant **polytene chromosomes** of the *Drosophila* larval salivary gland. The polytene chromosomes provided stunning confirmation of the genetic results, including, for example, great detail about chromosome rearrangements, while adding an unparalleled visual physical map of the euchromatic part of the genome.

Polytene chromosomes have a banded appearance similar to the labels used for machine pricing of supermarket boxes (Figure 9.29). The dark bands are regions of compact chromatin; the lighter "interbands" are regions of more extended chromatin. In flies, the largest extended salivary gland chromosome is more than 0.5 mm long, enormous compared to the same chromosome viewed in metaphase. Five chromosome arms carry most of the fly genome; the arms average about 1000 bands each, all visible with a light microscope. Chromosome length is due to partial decondensation of the chromatin. Chromosome breadth (3–5 μm), about the same as the length of a metaphase chromosome, is due to the alignment of many copies of each double helix. The polytene chromosomes are really a bundle of chromosomes produced by DNA replication without cell division. The near perfect alignment of the individual chromosomes gives rise to the distinctive banded pattern. Because the banding pattern is so reproducible and the pattern of one part of the genome is distinguishable, with practice, from others, the bands are a physical map of the genome. An inverted section of the chromosome can easily be recognized by a group of bands found in a flipped pattern. Deletions, translocations, and insertions can also be detected if they are sufficiently large, typically more than 50–100 kbp.

Not all of the genome is equally replicated, so the polytene chromosomes give an incomplete picture of chromosome structure. Heterochromatin, including all of the Y chromosome and about half of the X, is not overreplicated. Ribosomal RNA genes are replicated for a few rounds, but not for the ten or so doublings that create the 1000 strands of polytene euchromatin. An average band contains about 21.6 kbp of DNA in each of about 1000 parallel aligned copies. The 5000 bands therefore contain a complexity of about 100,000 kbp, accounting for most of the complexity of the *Drosophila* genome.

This high-resolution band pattern makes polytene chromosomes a superb link between molecular biology and cytology. The link required more than facile cytogenetics. There had to be a way to map cloned DNA or purified RNA. *In situ* hybridization was developed for this purpose. Chromosomes are denatured with an alkaline treatment in a way that preserves band morphology. Radioactively labelled or chemically labelled nucleic acid probes are hybridized to the chromosomes and detected by autoradiography or enzymatic methods (Figure 9.29). The resolution is about as good as the banding pattern.

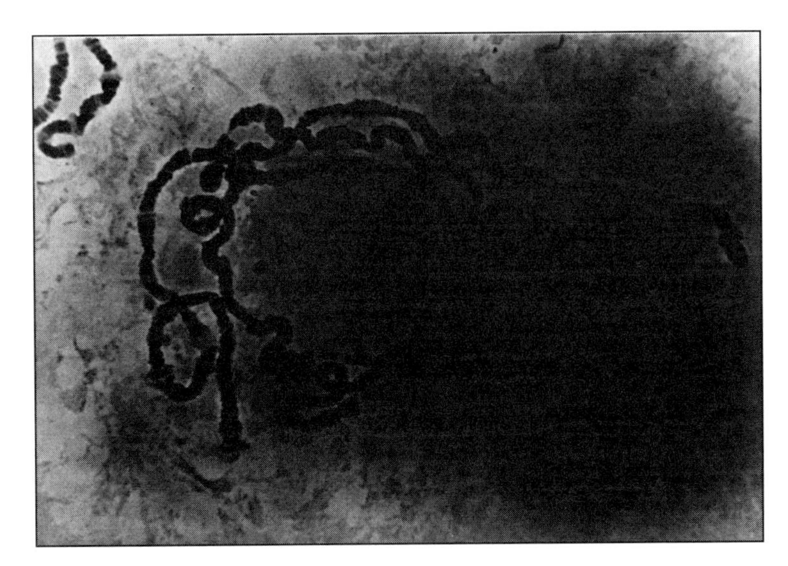

Figure 9.29
Giant polytene chromosomes from the nucleus of a larval salivary gland. This preparation was hybridized in situ with a probe for a P element transposon, and the single site of hybridization (arrow) indicates the position of transposon insertion.

d. Cloning Strategies

General cloning strategies are described in *Genes and Genomes,* sections 6.6 and 7.1–7.5, but several procedures take advantage of the special properties of *Drosophila* and deserve further description.

Positional cloning The first genes to be cloned were accessible because their mRNAs are abundant and easily isolated. The mRNA could be hybridized to libraries of genomic DNA and the corresponding gene isolated. This procedure is useless for most genes because the mRNAs are rare or impossible to identify. The vast wealth of genetic mapping information in *Drosophila* makes it desirable to clone genes according to their positions on the chromosome rather than from any knowledge of the genes' products. The first application of positional cloning was to certain *Drosophila* genes encoding enzymes and regulatory proteins, using a procedure called **chromosome walking**.

Chromosome walking isolates overlapping fragments of genomic DNA from a library containing the entire genome. Each newly isolated fragment is used to identify the next one by hybridization, thus allowing a representation of a stretch of chromosome to be assembled from a series of restriction maps. A starting point is needed and is usually chosen from available serendipitously identified DNA clones near the gene of interest. *In situ* hybridization to the giant polytene chromosomes allows determination to within accuracies of about 100 kbp where a cloned piece of DNA is located, and data bases of the mapped DNA clones have been established. The advantages of the large chromosomes extend beyond providing a starting point, making it possible to determine the orientation of the cloned DNA relative to known positions on the chromosome. Using chromosomes carrying inversions or translocations, *in situ* hybridization with different clones will reveal whether a particular cloned DNA fragment is to the left or right of a chromosome breakpoint. If the breakpoint causes a mutation in the gene of interest, all the better, because mapping the breakpoint within the cloned DNA will provide a good clue about where the gene is located.

Sometimes obtaining a good starting point for a walk is difficult. In these instances, the direct approach of microdissection of polytene chromosomes has been useful. A minute fragment of chromosome, usually on the order of 1 percent of the genome, is dissected, and the DNA is fragmented and isolated as a small library of clones (*Genes and Genomes,* figure 6.38). Individual clones can be hybridized to polytene chromosomes to confirm their origin and then used to initiate a walk or to isolate large genomic clones that can be linked by cross-hybridization into an overlapping series.

Mapping point mutations to identify the gene of interest Mapping breakpoints or point mutations to a particular piece of cloned DNA provides a different and very useful assessment of where the gene is located. In addition to mapping breakpoints using *in situ* hybridization to polytene chromosomes, there are now powerful methods for detecting point mutations. In one such method, PCR is used to amplify a short region of genomic DNA, such as a region encoding a candidate transcript. The PCR product is denatured to make single-stranded DNA and then run on a gel that allows some secondary structure to form. The two strands will migrate differently because their complementary sequences will fold differently. In parallel, the same experiment is done, but instead of using homogeneous wild-type DNA, DNA from a mutant animal is used. Many point mutations have a sufficient effect on single-strand conformation and folding to detect on these "single-strand-conformation-polymorphism" (SSCP) gels. Often both strands migrate differently from the corresponding strands from wild-type animals. Heterozygotes also work as a source of DNA, and in the best instances, four bands are seen: the two bands of wild-type DNA and the two bands of mutant DNA. Not all point mutations cause detectable changes in band migration on SSCP gels, but if a large enough number of alleles is tested, it is often possible to find a change.

Finding the transcript that corresponds to the gene of interest Having reached the vicinity of a gene, one must identify transcripts and determine which is the right one. Either pieces of cloned DNA can be used to probe RNA blots, or total complex cDNA made from a relevant source of mRNA can be used to probe the cloned DNA. Both experiments lead to the detection and mapping of candidate transcripts. If information is available about when or where the gene of interest is transcribed, further clues about which transcript is the right one can be obtained by determining temporal and spatial patterns of transcription for each candidate RNA.

RNA blots using RNA from staged animals or *in situ* hybridization to mRNA in whole mount embryos often provide very useful information. For example, when the first homeotic gene, *Ultrabithorax,* was cloned, its transcripts were initially discovered as bands on RNA blots. Because the exons corresponding to the transcripts were found to be distributed over a region of the chromosomes where numerous *Ubx* mutations map, there was strong reason to believe the transcripts were the correct ones. However, a more dramatic confirmation came from *in situ* hybridization studies demonstrating the distribution of the transcripts. They were found to be spatially controlled and at their highest levels in exactly the body segments where *Ubx* is known to be required. Once a transcript appears to be the right one, the full extent of its exons can be mapped in detail by S1 nuclease analysis, RNAse protection, and cDNA cloning.

Proving a candidate transcript is the right one The most overwhelmingly compelling proof for having correctly correlated a genetic function with a transcript is a "rescue" experiment in which a copy of the cloned gene is shown to substitute for the function of the normal gene. A genomic DNA fragment encompassing the candidate transcription unit and joined to its presumed regulatory DNA is introduced into the genome by P element transformation. Then conventional crosses are done to bring the transgene into a homozygous mutant that would normally die or have some other phenotype. If the mutant phenotype is not observed, the introduced gene must contain all the necessary information. The gene can be trimmed down further to identify the minimum piece that will work.

9.6 Perspectives

Drosophila has been extraordinarily useful for discovering how initially undifferentiated cells are organized into all the complex tissues of an animal. About 6000 genes have been identified by mutation, and the protein sequences of over 1300 are known. The wealth of information has revealed many new families of proteins, such as the homeodomain family, and nearly every family so revealed has proved to have relatives in other animals. Thus, the progress of *Drosophila* developmental genetics has been beautifully linked to exciting findings about the development of mouse, worm, frog, and human embryos and to elegant studies of gene regulation in yeast, bacteria, and viruses.

Despite all the excitement, an enormous amount remains to be learned. The relationships between cell biology and developmental biology are crucial. How does differential gene expression lead to changes in cell behavior, metabolism, and structure? What other signals remain to be discovered? How are cell fates remembered? How do developmental clocks work? What controls healing and regeneration? How does aging occur? Studies of development will continue to bring into focus the roles of new classes of proteins and will reveal novel relationships among the protein types already found.

Drosophila is likely to be important for learning about mammalian proteins, given the high success rate to date in finding similarities between fly and mammalian protein structures and, most remarkably, functions. As genome projects accelerate, taking a protein known only from its sequence and finding out what it does will be increasingly important. Also, human geneticists are identifying large numbers of genes demonstrably critical for healthy human life. Humans are the only organisms to screen themselves, and the screen involves vast numbers of individuals. Even subtle changes in behavior or thought no animal can tell us will be detected. Human genetics is less useful for subsequent analysis both because it is often difficult to get the requisite variety of alleles and because many types of experiments cannot ethically be performed. To learn the function of a novel protein, powerful genetic approaches are needed to assign functions to proteins. A protein found either to be important to a disease in humans or through biochemical methods or as a mystery sequence or as a component of a novel organelle is often usefully studied by analyzing homologous or similar proteins in other creatures. It is possible to identify and mutate corresponding bacterial, yeast, worm, fly, or mouse genes to learn something about their normal activity.

The use of genetically accessible organisms for identifying gene functions is at least as important as their usefulness in attaching functions to previously described proteins. Biochemical and genetic studies of bacteria provided most of what we know of metabolic pathways, with the genetics providing the gene discovery tool and biochemistry providing information about the enzymatic and regulatory functions of proteins. Similarly, the nematode *Caenorhabditis,* the fly *Drosophila,* and the mouse offer the opportunity to saturate genomes with mutations in every gene needed for a particular function, be it a metabolic pathway, a developmental event, or a behavior. In this way, pathways are assembled and novel proteins are discovered.

In the late 1970s, it seemed unlikely that studies of developmental processes in invertebrates would yield results directly transferable to vertebrates; now such extrapolations are not only frequent, but expected. This change in thinking has profoundly altered our views of developmental processes. The findings have raised a set of new questions most easily described by a computer analogy: is only the hardware conserved but not the software? Many of the classes of proteins involved in development, the transistors and buttons, are conserved, much as enzymes are. But are the regulatory relationships and higher organization of how the components are used—the software—also conserved? To what extent? Answering this question addresses the unity of life as much or more than determining whether all organisms use the Krebs cycle.

Further Reading

General Reviews

J. A. Campos-Ortega and V. Hartenstein. 1985. The Embryonic Development of *Drosophila melanogaster.* Springer-Verlag, Berlin.

M. Akam. 1987. The Molecular Basis for Metameric Pattern in the *Drosophila* Embryo. *Development* 101 1–22.

M. P. Scott and S. B. Carroll. 1987. The Segmentation and Homeotic Gene Network in Early *Drosophila* Development. *Cell* 51 689–698.

P. W. Ingham. 1988. The Molecular Genetics of Embryonic Pattern Formation in *Drosophila. Nature* 335 25–34.

M. Kessel and P. Gruss. 1990. Murine Developmental Control Genes. *Science* 249 374–379.

P. A. Lawrence. 1992. The Making of a Fly: The Genetics of Animal Design. Blackwell Scientific Publications, London.

M. Bate and A. Martinez-Arias. 1993. The Development of *Drosophila melanogaster,* vols. 1 and 2. Cold Spring Harbor Laboratory Press, Cold Spring Harbor, New York.

9.2a

W. Driever and C. Nüsslein-Volhard. 1988. The *Bicoid* Protein Determines Position in the *Drosophila* Embryo in a Concentration-Dependent Manner. *Cell* 54 95–104.

W. Driever and C. Nüsslein-Volhard. 1988. A Gradient of *Bicoid Protein* in *Drosophila* Embryos. *Cell* 54 83–93.

R. Chasan and K. V. Anderson. 1989. The Role of Easter, an Apparent Serine Protease, in Organizing the Dorsal-Ventral Pattern of the *Drosophila* Embryo. *Cell* 56 391–400.

R. Chasan, Y. Jin, and K. V. Anderson. 1992. Activation of the Easter Zymogen Is Regulated by Five Other Genes To Define Dorsal-Ventral Polarity in the *Drosophila* Embryo. *Development* 115 607–616.

R. Geisler, A. Bergmann, Y. Hiromi, and C. Nüsslein-Volhard. 1992. *Cactus,* a Gene Involved in Dorsoventral Pattern Formation of *Drosophila,* Is Related to the IkappaB Gene Family of Vertebrates. *Cell* 71 613–621.

D. St. Johnston and C. Nüsslein-Volhard. 1992. The Origin of Pattern and Polarity in the *Drosophila* Embryo. *Cell* 68 201–219.

9.2b

W. Driever and C. Nüsslein-Volhard. 1989. The *Bicoid* Protein Is a Positive Regulator of *hunchback* Transcription in the Early *Drosophila* Embryo. *Nature* 337 138–143.

G. Struhl, K. Struhl, and P. M. Macdonald. 1989. The Gradient Morphogen Bicoid Is a Concentration-Dependent Transcriptional Activator. *Cell* 57 1259–1273.

J. L. Brown, S. Sonoda, H. Ueda, M. P. Scott, and C. Wu. 1991. Repression of the *Drosophila fushi tarazu* (*ftz*) Segmentation Gene. *EMBO J* 10 665–674.

9.2c

C. Hashimoto, K. L. Hudson, and K. V. Anderson. 1988. The *Toll* Gene of *Drosophila,* Required for Dorsal-Ventral Embryonic Polarity, Appears to Encode a Transmembrane Protein. *Cell* 52 269–279.

H. J. Doyle, R. Kraut, and M. Levine. 1989. Spatial Regulation of *zerknüllt*: A Dorsal-Ventral Patterning Gene in *Drosophila. Genes Dev.* 3 1518–1533.

Y. T. Ip, R. Kraut, M. Levine, and C. A. Rushlow. 1991. The Dorsal Morphogen Is a Sequence-Specific DNA-Binding Protein That Interacts with a Long-Range Repression Element in *Drosophila. Cell* 64 439–446.

J. Jiang, D. Kosman, Y. T. Ip, and M. Levine. 1991. The Dorsal Morphogen Gradient Regulates the Mesoderm Determinant Twist in Early *Drosophila* Embryos. *Genes Dev.* 5 1881–1891.

D. Stein, S. Roth, E. Vogelsang, and C. Nüsslein-Volhard. 1991. The Polarity of the Dorsoventral Axis in the *Drosophila* Embryo Is Defined by an Extracellular Signal. *Cell* 65 725–735.

E. L. Ferguson and K. V. Anderson. 1992. Decapentaplegic Acts as a Morphogen To Organize Dorsal-Ventral Pattern in the *Drosophila* Embryo. *Cell* 71 451–461.

Y. T. Ip, R. E. Park, D. Kosman, E. Bier, and M. Levine. 1992. The Dorsal Gradient Morphogen Regulates Stripes of Rhomboid Expression in the Presumptive Neuroectoderm of the *Drosophila* Embryo. *Genes Dev.* 6 1728–1739.

Y. T. Ip, R. E. Park, D. Kosman, K. Yazdanbakhsh, and M. Levine. 1992. Dorsal-Twist Interactions Establish Snail Expression in the Presumptive Mesoderm of the *Drosophila* Embryo. *Genes Dev.* 6 1518–1530.

F. Sprenger and C. Nüsslein-Volhard. 1992. Torso Receptor Activity Is Regulated by a Diffusible Ligand Produced at the Extracellular Terminal Regions of the *Drosophila* Egg. *Cell* 71 987–1001.

9.3a

M. Lohs-Schardin, C. Cremer, and C. Nüsslein-Volhard. 1979. A Fate Map for the Larval Epidermis of *Drosophila melanogaster*: Localized Defects Following Irradiation of the Blastoderm with an Ultraviolet Laser Microbeam. *Dev. Biol.* 73 239–255.

9.3b

H. Jäckle, U. B. Rosenberg, A. Preiss, E. Seifert, D. C. Knipple, A. Kienlin, and R. Lehmann. 1985. Molecular Analysis of *Krüppel,* a Segmentation Gene of *Drosophila melanogaster. Cold Spring Harbor Symp. Quant. Biol.* 50 465–473.

D. C. Knipple, E. Seifert, U. B. Rosenberg, A. Preiss, and H. Jäckle. 1985. Spatial and Temporal Patterns of *Krüppel* Gene Expression in Early *Drosophila* Embryos. *Nature* 317 40–44.

P. M. Macdonald and G. Struhl. 1986. A Molecular Gradient in Early *Drosophila* Embryos and Its Role in Specifying the Body Pattern. *Nature* 324 537–545.

U. Gaul and H. Jäckle. 1987. Pole Region-Dependent Repression of the *Drosophila* Gap Gene *Krüppel* by Maternal Gene Products. *Cell* 51 549–555.

M. J. Pankratz, M. Hoch, E. Seifert, and H. Jäckle. 1989. *Krüppel* Requirement for *knirps* Enhancement Reflects Overlapping Gap Gene Activities in the *Drosophila* Embryo. *Nature* 341 337–340.

M. Hoch, C. Schröder, E. Seifert, and H. Jäckle. 1990. Cis-Acting Control Elements for *Krüppel* Expression in the *Drosophila* Embryo. *EMBO J.* 9 2587–2595.

D. Weigel, G. Jürgens, M. Klingler, and H. Jäckle. 1990. Two Gap Genes Mediate Maternal Terminal Pattern Information in *Drosophila. Science* 248 495–498.

9.3c

S. B. Carroll and M. P. Scott. 1986. Zygotically-Active Genes That Affect the Spatial Expression of the *fushi tarazu* Segmentation Gene During Early *Drosophila* Embryogenesis. *Cell* 45 113–126.

K. Howard and P. Ingham. 1986. Regulatory Interactions Between the Segmentation Genes *fushi tarazu, hairy,* and *engrailed* in the *Drosophila* Blastoderm. *Cell* 44 949–957.

P. M. Macdonald, P. Ingham, and G. Struhl. 1986. Isolation, Structure and Expression of *even-skipped*: A Second Pair-Rule Gene of *Drosophila* Containing a Homeobox. *Cell* 47 721–734.

S. B. Carroll, A. Laughon, and B. S. Thalley. 1988. Expression, Function, and Regulation of the *hairy* Segmentation Protein in the *Drosophila* Embryo. *Genes Dev.* 2 883–890.

S. B. Carroll. 1990. Zebra Patterns in Fly Embryos: Activation of Stripes or Repression of Interstripes? *Cell* 60 9–16.

M. J. Pankratz, E. Seifert, N. Gerwin, B. Billi, U. Nauber, and H. Jäckle. 1990. Gradients of *Krüppel* and *knirps* Gene Products Direct Pair-Rule Gene Stripe Patterning in the Posterior Region of the *Drosophila* Embryo. *Cell* 61 309–317.

S. Baumgartner and M. Noll. 1991. Network of Interactions Among Pair-Rule Genes Regulating *paired* Expression During Primordial Segmentation of *Drosophila. Mech. of Devel.* 33 1–18.

D. Stanojevic, S. Small, and M. Levine. 1991. Regulation of a Segmentation Stripe by Overlapping Activators and Repressors in the *Drosophila* Embryo. *Science* 254 1385–1387.

J. Treisman, E. Harris, and C. Desplan. 1991. The Paired Box Encodes a Second DNA-Binding Domain in the Paired Homeo Domain Protein. *Genes Dev.* 5 594–604.

P. Zuo, D. Stanojevic, J. Colgan, K. Han, M. Levine, and J. L. Manley. 1991. Activation and Repression of Transcription by the Gap Proteins Hunchback and Krüppel in Cultured *Drosophila* Cells. *Genes Dev.* 5 254–264.

S. Small, A. Blair, and M. Levine. 1992. Regulation of Even-Skipped Stripe 2 in the *Drosophila* Embryo. *EMBO J.* 11 4047–4057.

9.3d

S. DiNardo, J. M. Kuner, J. Theis, and P. H. O'Farrell. 1985. Development of Embryonic Pattern in *Drosophila melanogaster* as Revealed by Accumulation of the Nuclear *engrailed* Protein. *Cell* 43 59–69.

F. Rijsewijk, M. Schuermann, E. Wagenaar, P. Parren, D. Weigel, and R. Nusse. 1987. The *Drosophila* Homolog of the Mouse Mammary Oncogene *int-1* Is Identical to the Segment Polarity Gene *wingless. Cell* 50 649–657.

S. DiNardo, E. Sher, J. Heemskerk-Jorgens, J. Kassis, and P. O'Farrell. 1988. Two-Tiered Regulation of Spatially Patterned *engrailed* Gene Expression During *Drosophila* Embryogenesis. *Nature* 332 604–609.

P. W. Ingham, N. E. Baker, and A. A. Martinez. 1988. Regulation of Segment Polarity Genes in the *Drosophila* Blastoderm by *fushi tarazu* and *even skipped. Nature* 331 73–75.

M. P. Weir, B. A. Edgar, T. Kornberg, and G. Schubiger. 1988. Spatial Regulation of *engrailed* Expression in the *Drosophila* Embryo. *Genes and Devel.* 1194–1203.

M. van den Heuvel, R. Nusse, P. Johnston, and P. Lawrence. 1989. Distribution of the *wingless* Gene Product in *Drosophila* Embryos; a Protein Involved in Cell-Cell Communication. *Cell* 59 739–749.

S. Eaton and T. B. Kornberg. 1990. Repression of *ci-D* in Posterior Compartments of *Drosophila* by *engrailed. Genes and Devel.* 4 1068–1077.

J. Heemskerk, S. DiNardo, R. Kostriken, and P. H. O'Farrell. 1991. Multiple Modes of Engrailed Regulation in the Progression Towards Cell Fate Determination. *Nature* 352 404–410.

U. Grossniklaus, P. R. Kurth, and W. J. Gehring. 1992. The *Drosophila sloppy paired* Locus Encodes Two Proteins Involved in Segmentation That Show Homology to Mammalian Transcription Factors. *Genes Dev.* 6 1030–1051.

T. Tabata, S. Eaton, and T. B. Kornberg. 1992. The *Drosophila hedgehog* Gene Is Expressed Specifically in Posterior Compartment Cells and Is a Target of *engrailed* Regulation. *Genes Dev.* 6 2635–2645.

J. P. Vincent and P. H. O'Farrell. 1992. The State of Engrailed Expression Is Not Clonally Transmitted During Early *Drosophila* Development. *Cell* 68 923–931.

9.3e

T. Kornberg. 1981. *Engrailed*: A Gene Controlling Compartment and Segment Formation in *Drosophila. Proc. Natl. Acad. Sci. USA* 78 1095–1099.

A. Martinez-Arias and P. W. Ingham. 1985. The Origin of Pattern Duplications in Segment Polarity Mutants of *Drosophila melanogaster. J. Embryol. Exp. Morphol.* 87 129–135.

A. Martinez-Arias, N. E. Baker, and P. W. Ingham. 1988. Role of Segment Polarity Genes in the Definition and Maintenance of Cell States in the *Drosophila* Embryo. *Development* 103 157–170.

J. E. Hooper and M. P. Scott. 1989. The *Drosophila patched* Gene Encodes a Putative Membrane Protein Required for Segmental Patterning. *Cell* 59 751–765.

Y. Nakano, I. Guerrero, A. Hidalgo, A. Taylor, J. R. S. Whittle, and P. W. Ingham. 1989. The *Drosophila* Segment Polarity Gene *patched* Encodes a Protein with Multiple Potential Membrane Spanning Domains. *Nature* 341 508–513.

N. H. Patel, E. Martin-Blanco, K. G. Coleman, S. J. Poole, M. C. Ellis, T. B. Kornberg, and C. S. Goodman. 1989. Expression of *engrailed* Proteins in Arthropods, Annelids, and Chordates. *Cell* 58 955–968.

R. G. Phillips, I. J. H. Roberts, P. W. Ingham, and J. R. S. Whittle. 1990. The *Drosophila* Segment Polarity Gene *patched* Is Involved in a Position-Signalling Mechanism in Imaginal Discs. *Development* 110 105–114.

P. W. Ingham, A. M. Taylor, and Y. Nakano. 1991. Role of the *Drosophila patched* Gene in Positional Signalling. *Nature* 353 184–187.

R. Nusse and H. Varmus. 1992. *Wnt* Genes. *Cell* 69 1073–1087.

P. W. Ingham and A. Hidalgo. 1993. Regulation of *wingless* Transcription in the *Drosophila* Embryo. *Development* 117 283–291.

E. Siegfried, T.-B. Chou, and N. Perrimon. 1993. *Wingless* Signalling Acts Through *zeste-white 3,* the *Drosophila* Homologue of *glycogen synthase kinase-3,* To Regulate *engrailed* and Establish Cell Fate. *Cell* 71 1167–1179.

G. Struhl and K. Basler. 1993. Organizing Activity of Wingless Protein in *Drosophila. Cell* 72 527–540.

9.4a.

A. Laughon and M. P. Scott. 1984. Sequence of a *Drosophila* Segmentation Gene: Protein Structure Homology with DNA-Binding Proteins. *Nature* 310 25–31.

W. McGinnis, R. L. Garber, J. Wirz, A. Kuroiwa, and W. J. Gehring. 1984. A Homologous Protein-Coding Sequence in *Drosophila* Homeotic Genes and Its Conservation in Other Metazoans. *Cell* 37 403–408.

W. McGinnis, M. S. Levine, E. Hafen, A. Kuroiwa, and W. J. Gehring. 1984. A Conserved DNA Sequence in Homoeotic Genes of the *Drosophila* Antennapedia and Bithorax Complexes. *Nature* 308 428–433.

M. P. Scott and A. J. Weiner. 1984. Structural Relationships Among Genes That Control Development: Sequence Homology Between the *Antennapedia, Ultrabithorax,* and *fushi tarazu* Loci of *Drosophila. Proc. Natl. Acad. Sci. USA* 81 4115–4119.

M. P. Scott, J. W. Tamkun, and G. W. Hartzell III. 1989. The Structure and Function of the Homeodomain. *BBA Rev. Cancer* 989 25–48.

W. J. Gehring. 1992. The Homeobox in Perspective. *Trends Biochem. Sci.* 17 277–280.

9.4b

G. Otting, Y. Q. Qian, M. Müller, M. Affolter, W. Gehring, and K. Wüthrich. 1988. Secondary Structure Determination for the *Antennapedia* Homeodomain by Nuclear Magnetic Resonance and Evidence for a Helix-Turn-Helix Motif. *EMBO J.* 7 4305–4309.

M. Billeter, Y. Qian, G. Otting, M. Müller, W. J. Gehring, and K. Wüthrich. 1990. Determination of the Three-Dimensional Structure of the Antennapedia Homeodomain from *Drosophila* in Solution by 1H Nuclear Magnetic Resonance Spectroscopy. *J. Mol. Biol.* 214 183–197.

C. R. Kissinger, B. Liu, B. E. Martin, T. B. Kornberg, and C. O. Pabo. 1990. Crystal Structure of an Engrailed Homeodomain-DNA Complex at 2.8 Å Resolution: A Framework for Understanding Homeodomain-DNA Interactions. *Cell* 63 579–590.

C. Wolberger, A. K. Vershon, B. Liu, A. D. Johnson, and C. O. Pabo. 1991. Crystal Structure of a MATa2 Homeodomain-Operator Complex Suggests a General Model for Homeodomain-DNA Interactions. *Cell* 67 517–528.

A. F. Schier and W. J. Gehring. 1992. Direct Homeodomain-DNA Interaction in the Autoregulation of the *fushi tarazu* Gene. *Nature* 356 804–807.

9.4c

E. B. Lewis. 1951. Pseudoallelism and Gene Evolution. *Cold Spring Harbor Symp. Quant. Biol.* 16 159–174.

E. B. Lewis. 1963. Genes and Developmental Pathways. *Am. Zool.* 3 33–56.

E. B. Lewis. 1978. A Gene Complex Controlling Segmentation in *Drosophila. Nature* 276 565–570.

P. W. Ingham and R. Whittle. 1980. Trithorax: A New Homeotic Mutation of *Drosophila melanogaster* Causing Transformations of Abdominal and Thoracic Segments. *Molec. Gen. Genet.* 179 607–614.

T. C. Kaufman, R. Lewis, and B. Wakimoto. 1980. Cytogenetic Analysis of Chromosome 3 in *Drosophila melanogaster*: The Homeotic Gene Complex in Polytene Chromosomal Interval 84A, B. *Genetics* 94 115–133.

R. E. Denell, K. R. Hummels, B. T. Wakimoto, and T. C. Kaufman. 1981. Developmental Studies of Lethality Associated with the Antennapedia Gene Complex in *Drosophila melanogaster. Dev. Biol.* 81 43–50.

I. M. Duncan. 1982. Polycomblike: A Gene That Appears To Be Required for the Normal Expression of the Bithorax and Antennapedia Gene Complexes of *Drosophila melanogaster. Genetics* 102 49–70.

G. Struhl. 1982. Genes Controlling Segmental Specification in the *Drosophila* Thorax. *Proc. Natl. Acad. Sci. USA* 79 7380–7384.

G. Struhl and D. Brower. 1982. Early Role of the esc+ Gene Product in the Determination of Segments in *Drosophila. Cell* 31 285–292.

W. Bender, M. Akam, F. Karch, P. A. Beachy, M. Peifer, P. Spierer, E. B. Lewis, and D. S. Hogness. 1983. Molecular Genetics of the Bithorax Complex in *Drosophila melanogaster. Science* 221 23–29.

M. P. Scott, A. J. Weiner, T. I. Hazelrigg, B. A. Polisky, V. Pirrotta, F. Scalenghe, and T. C. Kaufman. 1983. The Molecular Organization of the *Antennapedia* Locus of *Drosophila. Cell* 35 763–776.

E. Hafen, M. Levine, and W. J. Gehring. 1984. Regulation of *Antennapedia* Transcript Distribution by the Bithorax Complex in *Drosophila. Nature* 307 287–289.

A. D. Johnson and I. Herskowitz. 1985. A Repressor (MAT a2 Product) and Its Operator Control Expression of a Set of Cell Type Specific Genes in Yeast. *Cell* 42 237–247.

F. Karch, B. Weiffenbach, M. Peifer, W. Bender, I. Duncan, S. Celniker, M. Crosby, and E. B. Lewis. 1985. The Abdominal Region of the Bithorax Complex. *Cell* 43 81–96.

G. Struhl and R. A. White. 1985. Regulation of the *Ultrabithorax* Gene of *Drosophila* by Other Bithorax Complex Genes. *Cell* 43 507–519.

J. Wirz, L. Fessler, and W. J. Gehring. 1986. Localization of the *Antennapedia* Protein in the *Drosophila* Embryo and Imaginal discs. *EMBO J.* 5 3327–3334.

I. M. Duncan. 1987. The Bithorax Complex. *Ann. Rev. Genet.* 21 285–319.

R. Balling, U. Deutsch, and P. Gruss. 1988. *Undulated,* a Mutation Affecting the Development of the Mouse Skeleton, Has a Point Mutation in the Paired Box of *Pax-1. Cell* 55 531–535.

R. G. Clerc, L. M. Corcoran, J. H. LeBowitz, D. Baltimore, and P. A. Sharp. 1988. The B-cell-Specific Oct-2 Protein Contains POU Box- and Homeo Box-Type Domains. *Genes Dev.* 2 1570–1581.

G. Gibson and W. J. Gehring. 1988. Head and Thoracic Transformations Caused by Ectopic Expression of *Antennapedia* During *Drosophila* Development. *Development* 102 657–675.

C. Goutte and A. D. Johnson. 1988. a1 Protein Alters the DNA Binding Specificity of Alpha-2 Repressor. *Cell* 52 875–882.

W. Herr, R. A. Sturm, R. G. Clerc, L. M. Corcoran, D. Baltimore, P. A. Sharp, H. A. Ingraham, M. G. Rosenfeld, M. Finney, G. Ruvkin, and H. R. Horvitz. 1988. The Pou Domain: A Large Conserved Region in the Mamallian *pit-1, oct-1, oct-2* and *C. elegans unc-86* Gene Products. *Genes Dev.* 2 1513–1516.

J. A. Kennison and J. W. Tamkun. 1988. Dosage-Dependent Modifiers of *Polycomb* and *Antennapedia* Mutations in *Drosophila. Proc. Natl. Acad. Sci. USA* 85 8136–8140.

J. W. Mahaffey and T. C. Kaufman. 1988. The Homeotic Genes of the Antennapedia Complex and the Bithorax Complex of *Drosophila.* In G. M. Malacinski (Ed.), Developmental Genetics of Higher Organisms: A Primer in Developmental Biology, pp. 329–360. Macmillan, New York.

R. A. Sturm and W. Herr. 1988. The POU Domain is a Bipartite DNA-Binding Structure. *Nature* 336 601–604.

R. Balling, G. Mutter, P. Gruss, and M. Kessel. 1989. Craniofacial Abnormalities Induced by Cctopic Expression of the Homeobox Gene Hox-1.1 in Transgenic Mice. *Cell* 58 337–347.

D. Duboule and P. Dollé. 1989. The Structural and Functional Organization of the Murine HOX Gene Family Resembles That of *Drosophila* Homeotic Genes. *EMBO J.* 8 1497–1505.

A. Graham, N. Papalopulu, and R. Krumlauf. 1989. The Murine and *Drosophila* Homeobox Gene Complexes Have Common Features of Organization and Expression. *Cell* 57 367–378.

S. Hayashi and M. P. Scott. 1990. What Determines the Specificity of Action of *Drosophila* Homeodomain Proteins? Cell 63 883–894.

X. He, M. N. Treacy, D. M. Simmons, H. A. Ingraham, L. W. Swanson, and M. G. Rosenfeld. 1989. Expression of a Large Family of POU-Domain Regulatory Genes in Mammilian Brain Development. *Nature* 340 35–41.

E. S. Monuki, G. Wienmaster, R. Kuhn, and G. Lemke. 1989. SCIP: A Glial POU Domain Gene Regulated by Cyclic AMP. *Neuron* 3 783–793.

H. A. Ingraham, V. R. Albert, R. P. Chen, 3. E. B. Crenshaw, H. P. Elsholtz, X. He, M. S. Kapiloff, H. J. Mangalam, L. W. Swanson, M. N. Treacy, and M. G. Rosenfeld. 1990. A Family of POU-Domain and Pit-1 Tissue-Specific Transcription Factors in Pituitary and Neuroendocrine Development. *Annu. Rev. Physiol.* 52 773–791.

H. A. Ingraham, S. E. Flynn, J. W. Voss, V. R. Albert, M. S. Kapiloff, L. Wilson, and M. G. Rosenfeld. 1990. The POU-Specific Domain of Pit-1 Is Essential for Sequence-Specific, High Affinity DNA Binding and DNA-Dependent Pit-1-Pit-1 Interactions. *Cell* 61 1021–1033.

T. C. Kaufman, M. A. Seeger, and G. Olsen. 1990. Molecular and Genetic Organization of the Antennapedia Gene Complex of *Drosophila melanogaster. Adv. Genet.* 27 309–362.

J. Simon, M. Peifer, W. Bender, and M. O'Connor. 1990. Regulatory Elements of the Bithorax Complex That Control Expression Along the Anterior-Posterior Axis. *EMBO J.* 9 3945–3956.

M. Tanaka and W. Herr. 1990. Differential Transcriptional Activation by Oct-1 and Oct-2: Interdependent Activation Domains Induce Oct-2 Phosphorylation. *Cell* 60 375–386.

D. G. Wilkinson and R. Krumlauf. 1990. Molecular Approaches to the Segmentation of the Hindbrain. *Trends Neurosci.* 13 335–339.

O. Chisaka and M. R. Capecchi. 1991. Regionally Restricted Developmental Defects Resulting from Targeted Disruption of the Mouse Homeobox Gene hox-1.5. *Nature* 350 473–479.

T. Dick, X. H. Yang, S. L. Yeo, and W. Chia. 1991. Two Closely Linked *Drosophila* POU Domain Genes Are Expressed in Neuroblasts and Sensory Elements. *Proc. Natl. Acad. Sci. USA* 88 7645–7649.

D. J. Epstein, M. Vekemans, and P. Gros. 1991. Splotch (Sp2H), a Mutation Affecting Development of the Mouse Neural Tube, Shows a Deletion Within the Paired Homeodomain of Pax-3. *Cell* 67 767–774.

R. E. Hill, J. Favor, B. Hogan, C. Ton, G. F. Saunders, I. M. Hanson, J. Prosser, T. Jordan, N. D. Hastie, and V. Van Heyningen. 1991. Mouse Small Eye Results from Mutations in a Paired-Like Homeobox-Containing Gene. *Nature* 354 522–525.

M. Kessel and P. Gruss. 1991. Homeotic Transformations of Murine Vertebrae and Concomitant Alteration of Hox Codes Induced by Retinoic Acid. *Cell* 67 89–104.

T. Lufkin, A. Dierich, M. LeMeur, M. Mark, and P. Chambon. 1991. Disruption of the Hox-1.6 Homeobox Gene Results in Defects in a Region Corresponding to Its Rostral Domain of Expression. *Cell* 66 1105–1119.

R. Paro and D. S. Hogness. 1991. The Polycomb Protein Shares a Homologous Domain with a Heterochromatin-Associated Protein of *Drosophila. Proc. Natl. Acad. Sci. USA* 88 263–267.

M. G. Rosenfeld. 1991. POU-Domain Transcription Factors: Pou-er-ful Developmental Regulators. *Genes Dev.* 5 897–907.

G. Ruvkun and M. Finney. 1991. Regulation of Transcription and Cell Identity by POU Domain Proteins. *Cell* 64 475–478.

S. Stern and W. Herr. 1991. The Herpes Simplex Virus Trans-Activator VP16 Recognizes the Oct-1 Homeo Domain: Evidence for a Homeo Domain Recognition Subdomain. *Genes Dev.* 5 2555–2566.

M. N. Treacy, X. He, and M. G. Rosenfeld. 1991. I-POU: A POU-Domain Protein That Inhibits Neuron-Specific Gene Activation. *Nature* 350 577–584.

C. P. Verrijzer, J. A. van Oosterhout, W. W. van Weperen, and P. C. van der Vliet. 1991. POU Proteins Bend DNA Via the POU-Specific Domain. *EMBO J.* 10 3007–3014.

C. Walther and P. Gruss. 1991. Pax-6, a Murine Paired Box Gene, Is Expressed in the Developing CNS. *Development* 113 1435–1449.

B. Zink, Y. Engström, W. J. Gehring, and R. Paro. 1991. Direct Interaction of the *Polycomb* Protein with *Antennapedia* Regulatory Sequences in Polytene Chromosomes of *Drosophila melanogaster. EMBO J.* 10 153–162.

O. Chisaka, T. S. Musci, and M. R. Capecchi. 1992. Developmental Defects of the Ear, Cranial Nerves and Hindbrain Resulting from Targeted Disruption of the Mouse Homeobox Gene Hox-1.6. *Nature* 355 516–520.

A. Franke, M. DeCamillis, D. Zink, N. Cheng, H. W. Brock, and R. Paro. 1992. Polycomb and Polyhomeotic Are Constituents of a Multimeric Protein Complex in Chromatin of *Drosophila melanogaster. EMBO J.* 11 2941–2950.

J. G. Heuer and T. C. Kaufman. 1992. Homeotic Genes Have Specific Functional Roles in the Establishment of the *Drosophila* Embryonic Peripheral Nervous System. *Development* 115 35–47.

H. Le Mouellic, Y. Lallemand, and P. Brûlet. 1992. Homeosis in the Mouse Induced by a Null Mutation in the *Hox3.1* Gene. *Cell* 69 251–264.

T. Lufkin, M. Mark, C. P. Hart, P. Dollé, M. LeMeur, and P. Chambon. 1992. Homeotic Transformation of the Occipital Bones of the Skull by Ectopic Expression of a Homeobox Gene. *Nature* 359 835–841.

W. McGinnis and R. Krumlauf. 1992. Homeobox Genes and Axial Patterning. *Cell* 68 283–302.

R. A. Pollock, G. Jay, and C. J. Bieberich. 1992. Altering the Boundaries of *Hox3.1* Expression: Evidence for Antipodal Gene Regulation. *Cell* 71 911–923.

M. P. Scott. 1992. Vertebrate Homeobox Gene Nomenclature. *Cell* 71 551–553.

J. Simon, A. Chiang, and W. Bender. 1992. Ten Different Polycomb Group Genes Are Required for Spatial Control of the abdA and AbdB Homeotic Products. *Development* 114 493–505.

D. L. Smith and A. D. Johnson. 1992. A Molecular Mechanism for Combinatorial Control in Yeast: MCM1 Protein Sets the Spacing and Orientation of the Homeodomains of an Alpha2 Dimer. *Cell* 68 133–142.

J. W. Tamkun, R. Deuring, M. P. Scott, M. Kissinger, A. M. Pattatucci, T. C. Kaufman, and J. A. Kennison. 1992. Brahma: A Regulator of *Drosophila* Homeotic Genes Structurally Related to the Yeast Transcriptional Activator SNF2/SWI2. *Cell* 68 561–572.

M. N. Treacy, L. I. Neilson, E. E. Turner, X. He, and M. G. Rosenfeld. 1992. Twin of I-POU: A Two Amino Acid Difference in the I-POU Homeodomain Distinguishes an Activator from an Inhibitor of Transcription. *Cell* 68 491–505.

A. M. Tymon, U. Kües, W. Richardson, and L. A. Casselton. 1992. A Fungal Mating Type Protein That Regulates Sexual and Asexual Development Contains a POU-Related Domain. *EMBO J.* 11 1805–1813.

N. Assa-Munt, R. J. Mortishire-Smith, R. Aurora, W. Herr, and P. E. Wright. 1993. The Solution Structure of the Oct-1 POU-Specific Domain Reveals a Striking Similarity to the Bacteriophage Lambda Repressor DNA-Binding Domain. *Cell* 73 193–205.

9.5b

G. M. Rubin and A. C. Spradling. 1982. Genetic Transformation of *Drosophila* with Transposable Element Vectors. *Science* 218 348–353.

A. C. Spradling and G. M. Rubin. 1982. Transposition of Cloned P Elements into *Drosophila* Germ Line Chromosomes. *Science* 218 341–347.

G. M. Rubin and A. C. Spradling. 1983. Vectors for P Element-Mediated Gene Transfer in *Drosophila. Nucl. Acids Res.* 11 6341–6351.

H. J. Bellen, C. O'Kane, C. Wilson, U. Grossniklaus, R. K. Pearson, and W. J. Gehring. 1989. P-Element-Mediated Enhancer Detection: A Versatile Method To Study Development in *Drosophila. Genes and Devel.* 3 1288–1300.

E. Bier, H. Vaessin, S. Shepherd, K. Lee, K. McCall, S. Barbel, L. Ackerman, R. Carretto, T. Uemura, E. Grell, L. Y. Jan, and Y. N. Jan. 1989. Searching for Pattern and Mutation in the *Drosophila* Genome with a P-*lacZ* Vector. *Genes and Devel.* 3 1273–1287.

U. Grossniklaus, H. J. Bellen, C. Wilson, and W. J. Gehring. 1989. P-Element-Mediated Enhancer Detection Applied to the Study of Oogenesis in *Drosophila. Development* 107 189–200.

C. Wilson, R. K. Pearson, H. J. Bellen, C. J. O'Kane, U. Grossniklaus, and W. J. Gehring. 1989. P-Element-Mediated Enhancer Detection: An Efficient Method for Isolating and Characterizing Developmentally Regulated Genes in *Drosophila. Genes and Devel.* 3 1301–1313.

9.5c

M. Ashburner. 1967. Patterns of Puffing Activity in the Salivary Gland Chromosomes of *Drosophila.* I. Autosomal Puffing Patterns in a Laboratory Stock of *Drosophila melanogaster. Chromosoma* 21 398–428.

K. I. Sidén, R. Saunders, L. Spanos, T. Majerus, J. Treanear, C. Savakis, C. Louis, D. M. Glover, M. Ashburner, and F. C. Kafatos. 1990. Towards a Physical Map of the *Drosophila melanogaster* Genome: Mapping of Cosmid Clones Within Defined Genomic Divisions. *Nucleic Acids Res.* 18 6261–6270.

F. C. Kafatos, C. Louis, C. Savakis, D. M. Glover, M. Ashburner, A. J. Link, K. I. Sidén, and R. Saunders. 1991. Integrated Maps of the *Drosophila* Genome: Progress and Prospects. *Trends Genet.* 7 155–161.

9.5d

W. Bender, P. Spierer, and D. S. Hogness. 1983. Chromosomal Walking and Jumping To Isolate DNA from the *Ace* and *rosy* Loci and the Bithorax Complex in *Drosophila melanogaster. J. Mol. Biol. 168 17–33.*

CHAPTER 10

Manipulating Protein Structure

CYNTHIA N. KISER and JOHN H. RICHARDS

10.1 General Considerations

The ability to design and synthesize molecules is crucial to understanding the fundamental relationship between the structure of a molecule and its properties and function. Since the preparation of urea by Friedrich Wöhler in 1828, synthetic chemistry has improved at an ever-increasing pace. Today the chemical synthesis of any small organic molecule, however complex its structure, is almost within the ability of the best synthetic chemistry. Indeed, most of the new pharmaceuticals that have transformed large areas of clinical medicine, including agents effective in treating hypertension, neurological and psychiatric disorders, infectious diseases, cancer, and a host of other human maladies, have arisen from these abilities. In the realm of biological chemistry, the ability to synthesize and determine the three-dimensional structure of simple dipeptides played an essential role in establishing some fundamental structural principles such as the planarity of the peptide bond. Exploiting recent advances in the synthesis of the complex macromolecules of living systems, biological chemists are now trying to determine the relationship between the amino acid sequence of a protein and its three-dimensional structure and functional properties. These questions can now be posed with unprecedented breadth and answered with hitherto unattainable insights.

For example, what structural features determine the stable three-dimensional folding of a protein? How important are the packing of the hydrophobic core, specific hydrogen bonds, and electrostatic interactions? What aspects of three-dimensional proteins influence the exquisite specificity with which they bind ligands, for example, in the binding of an antigen by an antibody, the binding of a peptide hormone by a receptor, or the recognition of a specific base sequence in DNA by a transcription factor? How do the many subunits interact in an allosterically modulated process such as during the acquisition and release of oxygen by hemoglobin? What features of an enzyme play central roles in catalysis? Is it just those few residues whose side chains directly contact the substrate, or are other, more remote aspects of the enzyme's structure such as a helix-dipole moment or flexibility in a particular region of the protein also crucial to function? What is the mechanism by which light causes the separation of charge in the photosynthetic reaction center and thereby initiates the process of photosynthesis?

These questions are now being addressed. The results are often surprising, revealing for the first time many hitherto unrecognized and subtle structural features that are essential for a protein's efficient functioning in living systems. One can also ask how the proteins of today evolved their function and, in those cases where they function less than optimally, try to improve their efficiency by introducing various mutations affecting structure and function. As our knowledge of the rules that relate structure to function expands, we can envision a time when proteins or protein-analogues can be designed to carry out useful processes not now possible with any known protein—as, for example, enzymatic catalysis that joins two peptide fragments (Section 10.3h).

a. Synthesis of Altered Proteins

Genetic engineering approaches Today the best and most widely used method of obtaining protein for study is by coopting a cell to carry out its efficient synthesis. DNA coding for the target protein joined to appropriate transcription and translation signals is introduced into a suitable host organism (*Genes and Genomes,* section 7.8), and the protein of interest is purified from the cell extract. Alterations in the amino acid sequence, produced by changing the coding sequence of the DNA, are limited to the 20 naturally occurring amino acids.

A much wider range of protein variants containing an unnatural amino acid at the position of an introduced stop codon can be prepared *in vitro.* Briefly, by using chemical synthesis, researchers can artificially charge a tRNA that recognizes a stop codon with an unnatural amino acid and then use it in an *in vitro* translation system (Figure 10.1). Though of considerable power, the use of an *in vitro* system limits this approach to the synthesis of relatively small amounts of the variant proteins. Moreover, the introduction of the unnatural amino acid uniquely at the position of a stop codon restricts the number and position of such amino acids investigators can incorporate in any one variant.

Chemical synthesis Chemical synthesis of proteins is also now feasible. Coupling of the amino acid building blocks, starting from a carboxy (C) terminal residue anchored to a solid matrix, is generally accomplished using one of several well-defined chemical schemes. Suitably protected derivatives of the 20 naturally occurring amino acids are available commercially, and the synthesis is generally carried out using an automated synthesizer to facilitate the repeated cycles of activation and coupling. Chemical protein synthesis is not restricted to the 20 amino acids found in living systems, and scientists have exploited this in order to ask detailed questions about the relation of amino acid side chain structure to function. Despite its versatility, chemical synthesis is not the most com-

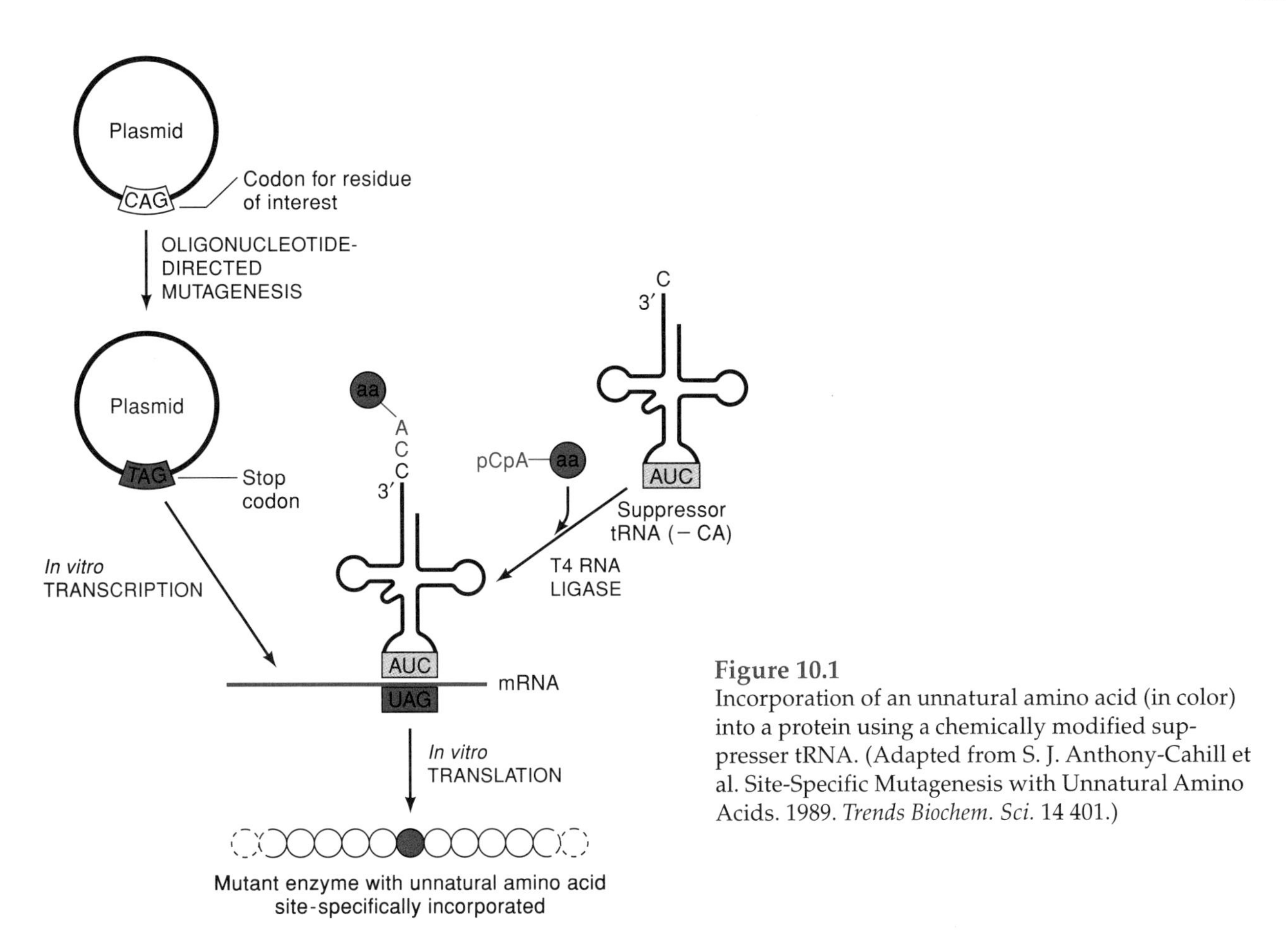

Figure 10.1
Incorporation of an unnatural amino acid (in color) into a protein using a chemically modified suppresser tRNA. (Adapted from S. J. Anthony-Cahill et al. Site-Specific Mutagenesis with Unnatural Amino Acids. 1989. *Trends Biochem. Sci.* 14 401.)

mon method of obtaining protein for further analysis because of the severe limit on the size of the protein that can be synthesized. Use of a solid support minimizes the loss of material during the washings between steps, but even assuming a 99 percent yield at each step, overall yield for a 50 amino acid protein is only 60 percent. With the best available techniques, only proteins of 100 residues or smaller can be obtained with a reasonable effort.

Similar limitations apply to chemical synthesis of large oligonucleotides, but the existence of DNA ligases allows these smaller synthetic units to be joined into a DNA chain of essentially any length. Protein synthesis awaits the development of enzymes or chemical procedures that will allow the joining, under mild conditions, of presynthesized oligomers. Such an advance would enormously expand the power and versatility of chemical synthesis and would usher in an era in which the synthesis of proteins (and variants) of several hundred residues and an essentially limitless variety of molecular structures would become feasible.

b. Mutational Strategies

Specific mutations The application of recombinant DNA techniques for protein production allows for two fundamentally different experimental strategies. In the first, specific mutant proteins are produced one at a time, and their properties are ascertained. (This approach is the direct analogue of structure/function studies of organic molecules and potential pharmaceutical agents.) Many techniques, such as site-directed or cassette mutagenesis, can be employed to create specific mutants; the mutants thereby created can vary by as little as a single or by several amino acids or can involve more complex alterations such as substitution of entire domains to create chimeric proteins. However minor or extensive the change introduced by this approach, one creates and assesses the properties of a single mutant protein.

Families of mutants In the alternative strategy, families of mutant proteins can be generated and screened for those that possess particular properties. Proteins that confer a selectable phenotype on their host organism, thereby facilitating the rapid screening of large numbers of mutant proteins to select the most interesting for further analysis, are especially amenable to this approach. For example, studies of proteins conferring antibiotic resistance are facilitated by the fact that mutations in key catalytic or binding residues alter the host's ability to grow in the presence of one or more related drugs. In another example, indirect selection methods have greatly facilitated studies of transcription factors. The mutated transcription factor is introduced into a cell containing a reporter construct. A gene conferring some easily assayed phenotype is fused to the DNA elements through which the transcription factor acts. In the most common construct, an active factor triggers the synthesis of β-galactosidase, which catalyzes the hydrolysis of a chromogenic substrate, X-gal (5-bromo-1-chloro-3-indoyl-β-D-galactopyranoside), to give an easily detectable blue color.

Scientists can employ a variety of synthetic approaches to create mutant families. For example, they can introduce all possible amino acids at a given site (termed site saturation) or concentrate mutagenesis at a small number of sites. Alternatively, they can create random mutations at all sites in a protein or within domains of interest. Use of an appropriate methodology will give rise to virtually any series of mutants that wit can devise; the only limitation is that the numbers of potential mutants escalates rapidly. For example, if one uses the site saturation approach, the possibilities grow at the rate of 20^n where n is the number of sites being saturated. For 20 sites, this will be about 10^{26} possibilities; for a protein of 20 residues, this would require almost 400 kg of sample that would contain only one molecule of each of the 10^{26} variant proteins. This highlights the importance of having a rapid technique for selecting bacteria possessing proteins with altered activity. It also illustrates the importance of choosing a mutagenesis strategy compatible with the analysis being done: incomplete, random mutagenesis to identify important residues in an unanalyzed protein or site saturation to dissect the role of a previously identified critical residue.

Nonspecific mutagenesis coupled with phenotypic screening does not restrict the design of the variants to be studied and allows very rapid acquisition of general information about the role of various amino acid residues in determining a protein's functional characteristics. It also can yield quite astonishing surprises; sometimes a sequence that sophisticated theoretical analysis might have predicted to behave in a particular way turns out to confound all the predictions and displays a completely unexpected personality. Rather than avoiding such surprises, the adventurous will seek them out. They can stimulate new insights far more original and powerful than those gained by the more conservative creation and study of single mutants where the labor involved inherently militates against making mutants that would be deemed fanciful or quixotic by existing principles of structure/function relationships.

10.2 Protein Folding and Stability

a. Overview

A canon of modern molecular biology holds that the folding of a chain of amino acids to generate a native protein is uniquely determined by the sequence of amino acids. Other agents, such as chaperonins, may become involved at intermediate stages between the unfolded chain and native protein to prevent aggregation of partially folded intermediates. However, no additional machinery seems to be required to direct folding actively along a specific pathway to the most thermodynamically stable three-dimensional structure. For proteins synthesized in a precursor form, such as preproinsulin or preproelastase, the conformation adopted is generally the one in which the precursor is most stable. This may or may not be the most stable conformation for the processed, functional protein; thus, the processed protein may be difficult or impossible to reconstitute in a native form because some of the essential features that directed folding have been removed during the processing.

The stability of most folded proteins is sufficient that they do not spontaneously unfold or denature under the conditions of their normal environments, although there is not generally a large margin of additional stability. Thus for organisms that normally exist between 30 and 40°C, a temperature rise to 50 or 60°C will result in the denaturation of many of the organism's proteins; but proteins in thermophiles that normally live in hot springs may, for example, be stable at temperatures even in excess of 100°C. Furthermore, the thermodynamic energy of importance in determining the stability of a native protein reflects the *difference* in free energy between the folded, native protein and the unfolded, denatured protein. (There are many forms of a denatured protein—from forms that may still possess significant secondary structure to a random coil in which almost all elements of higher-order structure are absent.) Accordingly, assessing the effect of a mutation on protein stability requires analyzing how the change will alter the free energy of *both* the folded and unfolded forms.

The relative stability of the folded compared to the unfolded molecule depends on complex interactions that include electrostatic salt bridges, hydrogen bonding, the effect of moving hydrophobic side chains from an aqueous environment in the unfolded protein to a hydrophobic environment in the closely packed interior of the folded protein (the **hydrophobic effect**), interactions between helix dipoles, and so forth. The sum of these effects commonly amounts to 0.5–1 kcal per residue. Mutagenic experiments that seek to determine the relative contribution of any single residue should always be analyzed with caution because a change that converts, for example, a serine to an alanine will not only remove a hydrogen bond but also alter the hydrophobic character of that region of the protein. Furthermore, apparently simple changes, although often being accommodated with only minimal change in the overall structure, may, under other circumstances, cause major alterations to the three-dimensional structure. Accordingly, careful structural analysis, for example by X-ray diffraction, plays an important role in interpreting changes in stability caused by mutations and in relating such changes to the fundamental forces involved in folding and stability.

Hydrophobic effects Examination of the crystal structures of a wide variety of proteins reveals that, in spite of large differences in their overall shapes, the forces that determine folding fall into just a few broad categories. Probably the most important of these is the hydrophobic effect. Upon folding, hydrophobic side chains pack together in the interior of the protein, from which water is generally excluded. Although van der Waals contacts provide some favorable energy effects upon such packing, the dominant energy gain from burying hydrophobic side chains arises from the increase in entropy when ordered water molecules surrounding the hydrophobic side chains of the random coil are released to bulk solvent upon folding of the protein. The energy thus gained is roughly proportional to the surface area of the side chain transferred from water to the protein's hydrophobic core: 25–30 cal $mol^{-1}Å^{-2}$.

Hydrogen bonding Very specific hydrogen bonds also play important roles in determining and stabilizing protein structures. Common secondary structural motifs such as α-helices and β-sheets involve hydrogen-bonding interactions between backbone amide hydrogens and carbonyl oxygens. Polar amino acid side chains are also often involved in hydrogen-bonding interactions with each other and with backbone amide groups. Each hydrogen bond contributes approximately 1.3 kcal mol^{-1} (this value can vary significantly) to the stability of the folded protein. However, because of the entropic factors involved in the specific orientation of the donor and acceptor in the folded protein and the hydrogen bonds that the unfolded protein can form with solvent water, the net stabilization gain from hydrogen bond formation upon folding can be a relatively small difference between two large energy terms (the sums of the energies of the hydrogen bond interactions in the unfolded state and native protein).

Disulfide bonds Covalent bonds between the sulfur atoms of two cysteine residues fasten polypeptide chains together and thereby stabilize the native form of a protein. Such disulfide bonds often hold together separate polypeptide chains whose backbones were initially part of a single covalent chain that was cleaved by proteolytic processing (e.g., the A and B chains of insulin). Disulfide bonds are particularly prevalent in thermostable proteins.

The stabilizing effect of a disulfide bond rests less in its energy of formation than in the constraints that, once formed, it imposes on the conformations accessible to the unfolded protein. Unfolded proteins can assume a vast number of conformations (with resulting high entropy); during folding, the number of possible conformations becomes increasingly restricted until only one remains. Such conformational restriction represents a large loss of entropy. A covalent disulfide bond linking two residues that may be far apart in the primary sequence constrains these residues to be neighbors in both the folded and the unfolded form. Thus, the presence of the disulfide bond significantly reduces the entropy of the unfolded state without appreciably altering the entropy of the folded protein and thereby lowers considerably the unfavorable loss of entropy upon folding. These considerations further suggest that the thermodynamic stabilization conferred by a disulfide bond should increase as the number of residues between the two cysteines increases.

Backbone flexibility Smaller but significant changes in the difference in entropy between the unfolded and folded states can occur when the backbone flexibility of a residue is altered. For example, the cyclic nature of proline restricts rotation about the C-N bond. Therefore, substitution of proline for another residue limits the conformations available around that site. If this distorts the conformation in this region of the normally folded protein, such a substitution can destabilize a protein. But if the substitution is introduced at a residue whose conformation in the folded protein happens to be just that of proline, no difference in the conformational energy of the folded protein would result. Instead, the entropy of the unfolded protein would be lowered relative to that of the normal protein because the mutant would have fewer conformations available. Accordingly, substitution of proline can lead to a net stabilization of the folded protein. Replacing an alanine with a glycine will also alter the entropy balance between folded and unfolded states, in this case, raising the entropy of the unfolded state because of the high degree of flexibility glycine introduces into the unfolded backbone. Having only hydrogens as C_α substituents, glycine frees rotations of the backbone from constraints that would be imposed by the van der Waals interactions of bulkier side chains with other parts of the unfolded polypeptide. Thus, introducing glycine can stabilize an unfolded protein relative to its folded state. Moreover, loss of the normal side chain when alanine is replaced by glycine could cause loss of the favorable van der Waals interactions, leaving a destabilizing void in the folded protein.

b. Hydrophobic Interactions

Hydrophobic core The salient features of protein interiors were studied by varying seven residues in the hydrophobic core of the amino (N) terminal domain of the λ-repressor (see *Genes and Genomes,* section 3.11e). This experimental approach demonstrates the power of the combination of cassette mutagenesis to generate many random mutants within a particular region and subsequent screening or selection to identify the mutants of most interest. Three or four of the core residues were simultaneously varied, and mutants that could protect their hosts from infection by lytic λ-phage were isolated. Expression of the repressor domain was controlled by an inducible *tac* promoter in the plasmid (*Genes and Genomes,* section 7.8b). The *tac* promoter is sufficiently leaky that even under noninducing conditions, native and other fully functional mutant repressors can protect the host from infection. Partially active mutant repressors must be present in higher concentrations to repress lytic infection of their host. Such partially active mutants can be recognized by selecting for protection from infection in the presence of IPTG, which induces the *tac* promoter, thereby increasing the amount of λ-repressor present.

After screening 75,000 colonies, investigators identified 31 fully active and 62 partially active mutants. Acceptable replacements for the core residues (Leu 18, Val 36, Met 40, Val 47, Phe 51, Leu 57, and Leu 65) were all drawn from a small subset of mostly apolar amino acids: Ala, Cys, Thr, Val, Ile, Leu, Met, and Phe. This finding emphasizes the importance of hydrophobic interactions in the core in determining protein stability (Figure 10.2). (Though threonine is not

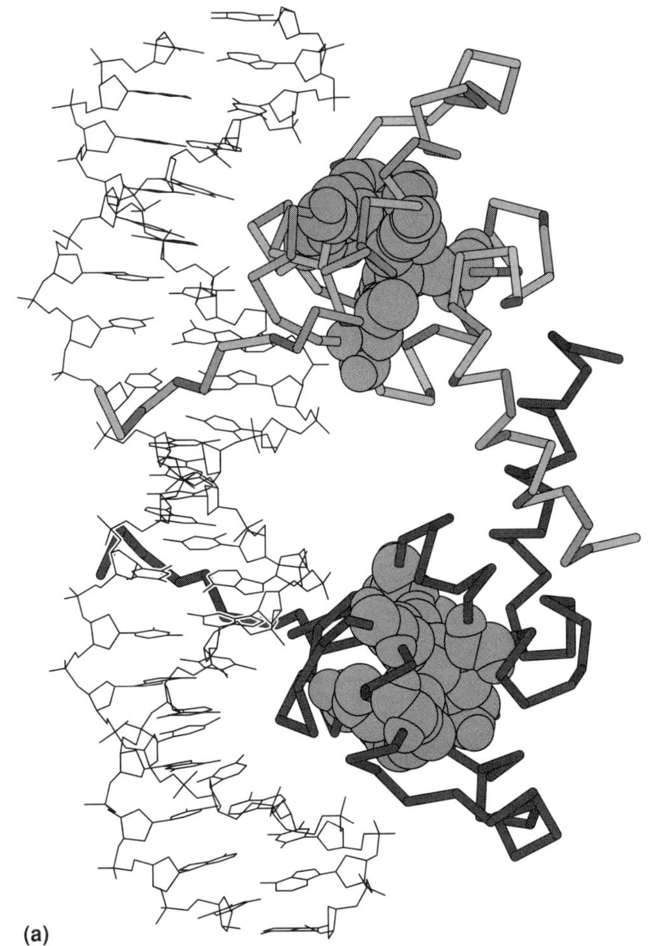

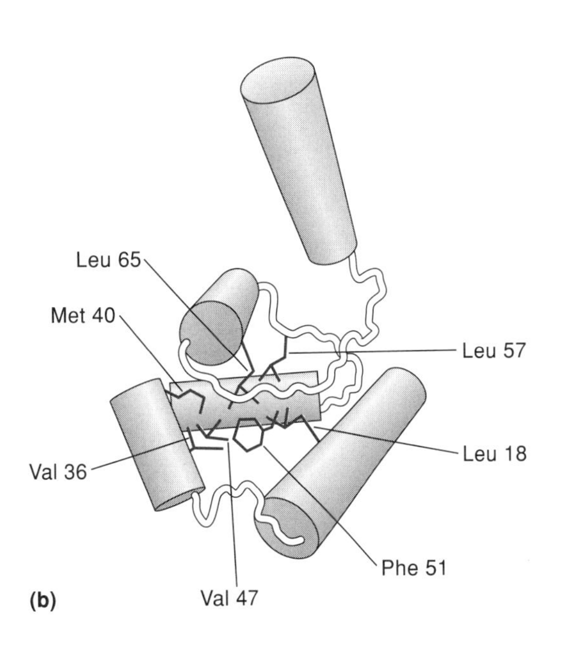

Figure 10.2
(a) A dimer of N terminal domains of λ-repressor bound to the operator DNA. The backbone of the dimer is shown in gray and the van der Waals surfaces of the core residues, 18, 36, 40, 47, 51, 57, and 65, are shown in color. (b) The core residues in one monomer. Compared to (a), the complex has been rotated 90° about the axis of the DNA such that the DNA is behind the plane of the page. (Adapted from W. A. Lim and R. T. Saur. 1989. Alternative Packing Arrangements in the Hydrophobic Core of λ-Repressor. *Nature* 339 31.)

strictly apolar, its hydroxyl group is frequently seen to hydrogen bond to the protein backbone, and its methyl group can occupy a hydrophobic pocket.) The calculated volume of the residues occupying the interior core in each fully or partially active mutant was generally found to be close to that of the wild-type repressor: 543 $Å^3$. Core volumes in active repressors ranged from approximately 40$Å^3$ larger than wild type to approximately 60$Å^3$ smaller. This is equivalent to between about two extra or three fewer methylene groups in the protein interior of the mutants relative to the native repressor.

Side chain polarity and volume are not the only criteria for activity. Many inactive mutants had core volumes well within the range of those observed for active mutants. Other, more specific packing interactions also play a vital role in establishing core stability. Such compatibility in packing interactions is manifest when one compares the sequence of one of the fully active mutants (Val36-Leu40Val47), with that of one of the partially active mutants (Leu36Val40Val47). Both have the same composition and total packing volume, the only change being alteration of the sequence order of Val and Leu. Thus, detailed steric considerations influence the stability and functionality of mutants, indicating the importance of high-resolution structural data for understanding the forces governing these energetic factors.

Additional interest focuses on a comparison of acceptable amino acids at one site as a function of changes in neighboring residues. Allowing changes in each of two adjacent residues rather than in only one of the pair often increases the number of permissible substitutions because of compensating changes that preserve the overall tight packing of the hydrophobic core. In the λ-repressor mutagenesis study, altering nearby residues increased the range of acceptable substitutions in many but not all situations.

Packing interactions T4 lysozyme is one of the last genes expressed from the phage genome after infection. The enzyme causes the lysis of the bacterial host by catalyzing the hydrolysis of the polysaccharides of the bacterial cell wall. The protein consists of two globular domains and a long α-helix that acts as a hinge between these domains. T4 lysozyme is also one of the first proteins to have been studied by site-directed mutagenesis.

The N terminal region of the protein packs tightly against the long helix, which puts Ile 3 into a hydrophobic environment. To assess the contribution to the protein's overall stability made by this hydrophobic interaction, researchers replaced Ile 3 with 13 different amino acids. The thermal stability of most of the mutant proteins correlated well with the free energy associated with transfer of the amino acids from water to organic solvents such as ethanol and n-octanol or N-methyl acetamide; this free energy serves as one of the quantitative measures of the hydrophobicity of an amino acid. Not surprisingly, mutants containing the large aromatic residues, especially Trp and Tyr (Ile3Trp and Ile3Tyr), were less stable than expected from the large energies of transfer (Figure 10.3). This is because, as demonstrated by X-ray structural analysis, the replacement of Ile with the much larger Tyr in a hydrophobic core causes relatively large, destabilizing changes in the local structure. Indeed, the tyrosine side chain cannot be accommodated in the interior of the protein and swings out into the solution, leaving a cavity in the hydrophobic core. In such a situation, not only does the protein not gain the free energy expected from transfer experiments, but the cavity left in the core is itself destabilizing, so the overall effect is appreciable (T_m wild-type 64.7°C; T_m Ile3Tyr mutant 58.8°C).

An alternative method of assessing the contribution of an individual residue to the hydrophobic core depends on calculating its free energy of transfer from a fully solvated form to that found within the folded protein. This approach involves first calculating the surface areas of the side chain that are exposed to solvent in both the unfolded and folded forms and then multiplying these by an

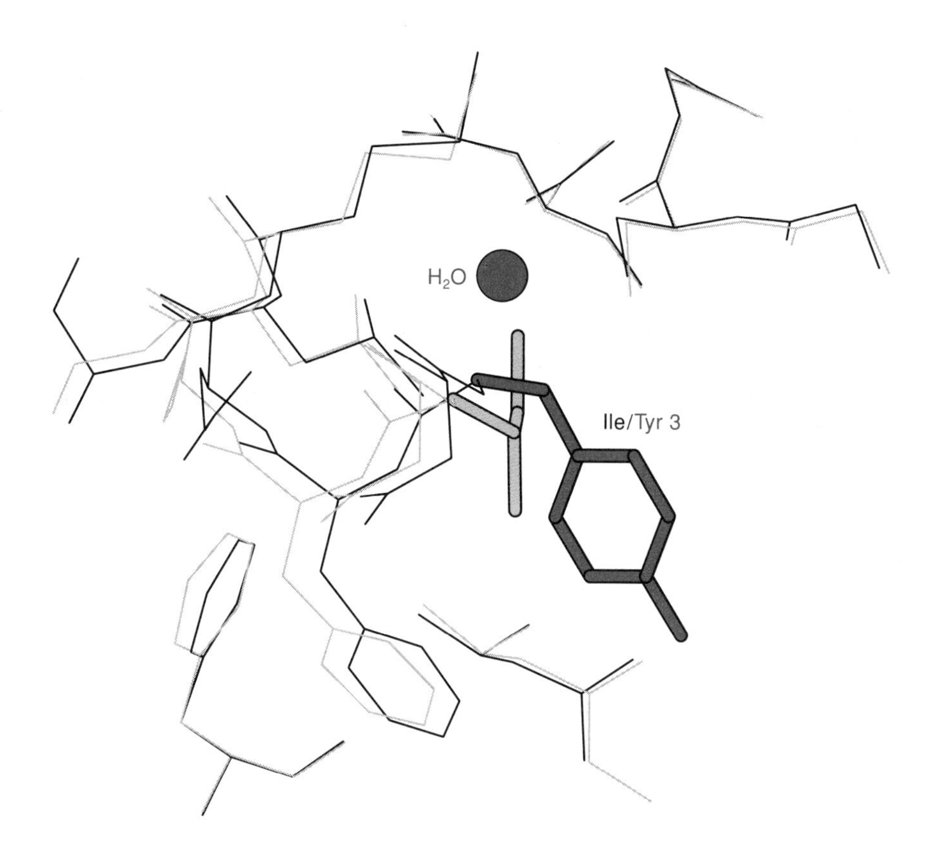

Figure 10.3
The structure of the Ile 3 to Tyr mutant (black) superimposed on that of the wild-type T4 lysozyme (gray). The solvent molecule is present only in the mutant structure. The tyrosine (color) cannot fit in the space between the three convergent helices occupied by isoleucine 3 (light color). Instead, it twists out onto the exterior of the protein and the space originally occupied by Ile 3 contains a single molecule of water. (Adapted from M. Matsumura et al. 1988. Hydrophobic Stabilization in T4 Lysozyme Determined Directly by Multiple Substitutions of Ile 3. *Nature* 334 407.)

energy term calculated in molar free energy per Å^2. (Solvent-exposed surface areas for the unfolded form are generally calculated from data on model peptides such as Ala-X-Ala, where X is the amino acid of interest, because these more closely approximate the area of residue X in an unfolded polypeptide chain than do the measurements of the surface area of the free amino acid itself.) Application of this approach to mutants of lysozyme at Ile 3 gave data in good quantitative agreement with the free energy of folding determined by thermal denaturation.

Cavity filling That the lower thermostability of the Ile3Tyr mutant is partly a consequence of the creation of a cavity in the interior of the protein suggests the possibility of improving the thermal stability of proteins containing such cavities by filling these cavities with hydrophobic side chains. One such cavity in wild-type T4 lysozyme is only partially filled by Leu 133. In the mutant Leu133Phe, the cavity is indeed filled by the Phe side chain, but no concomitant increase in thermal stability results. Detailed examination of the structure of the Leu133Phe mutant showed that, in order to fit into and fill the cavity, the Phe side chain has to adopt an unfavorably strained configuration; the strain energy from this distortion offsets the increased stabilization expected from the favorable hydrophobic interactions associated with filling the cavity. This result emphasizes the need for extremely careful analysis of even subtle structural aspects of a mutant before reaching firm conclusions about the origins of the observed changes in properties.

c. Hydrogen Bonding

As already mentioned, the hydrogen bond energy of importance is the *difference* between the sum of the energies for all the interactions of the unfolded protein with itself and with water and the sum of the energies of all such interactions in the folded protein. This usually represents a relatively small difference between two rather large energies. Nevertheless, hydrogen bonds contribute appreciably to the stability of a folded protein.

Importance of existing hydrogen bonds The contributions of hydrophobic interactions and hydrogen bonds to the stability of RNase T1 were estimated to be approximately equal. The energetic contributions of individual hydrogen bonds to the stability were estimated by replacing amino acids with hydrogen-bonding side chains with residues having similar side chains that lack the ability to participate in hydrogen bonding. Five substitutions of Tyr to Phe, three of Ser to Ala, and four of Asn to Ala were made. The Tyr to Phe and Ser to Ala changes retain the carbon frameworks but replace a hydroxyl group with a hydrogen; in addition, the volume of the side chain changes slightly, a factor that is more pronounced in the significantly less conservative substitution of Ala for an Asn. Changes in the free energy of stabilization were calculated from experimental determinations of the reversible melting temperature of each mutant; corrections were applied for changes in the hydrophobicity of the side chains. Stabilization-free energies ranged from +0.02 to -2.89 kcal mol^{-1}, from which the free energies for individual hydrogen bonds of 0–3.2 kcal mol^{-1} were calculated. If an average hydrogen bond energy of 1.3 kcal mol^{-1} is used, the 86 intramolecular hydrogen bonds inferred from the RNase T1 crystal structure contribute a total of 112 kcal mol^{-1} to protein stability. For comparison, summation of the free energies of transfer for the buried residues leads to estimates of between 98 and 132 kcal mol^{-1} stabilization energy from the hydrophobic effect. Thus, the contributions to stabilization by hydrogen bonding and the hydrophobic effect are roughly equal in RNase T1.

Additional assessments of hydrogen-bonding contributions to stability have also been made with T4 lysozyme. In this protein, Thr 157 resides on a surface loop, and, as seen in the X-ray structure, its hydroxyl group hydrogen bonds to the backbone nitrogen of Asp 159 (Figure 10.4). The importance of this hydrogen bond to the protein's thermal stability was first inferred from the temperature sensitivity of a Thr157Ile mutant found during screening of randomly generated mutants. This observation documents the ability of random mutagenesis to reveal global questions and to highlight important features that may not be apparent even from careful examination of available structural evidence. The finding was followed up using specific mutagenesis. The results demonstrate the powerful complementarity of the two approaches.

When other substitutions at Thr 157 were studied, the least destabilizing mutants were those with amino acids that allow hydrogen bonding to the Asp 159 backbone; for example, Asn, Asp, and Ser provide hydrogen-bonding side chains, and Gly creates room for binding an additional water molecule that can form a hydrogen bond. Substitutions that do not provide hydrogen-bonding partners are much more destabilizing, most notably Phe and Ile, which destabilize the folded protein by 2.4 and 2.9 kcal mol^{-1}, respectively. These results emphasize the importance of pairing all potential hydrogen-bonding groups in the

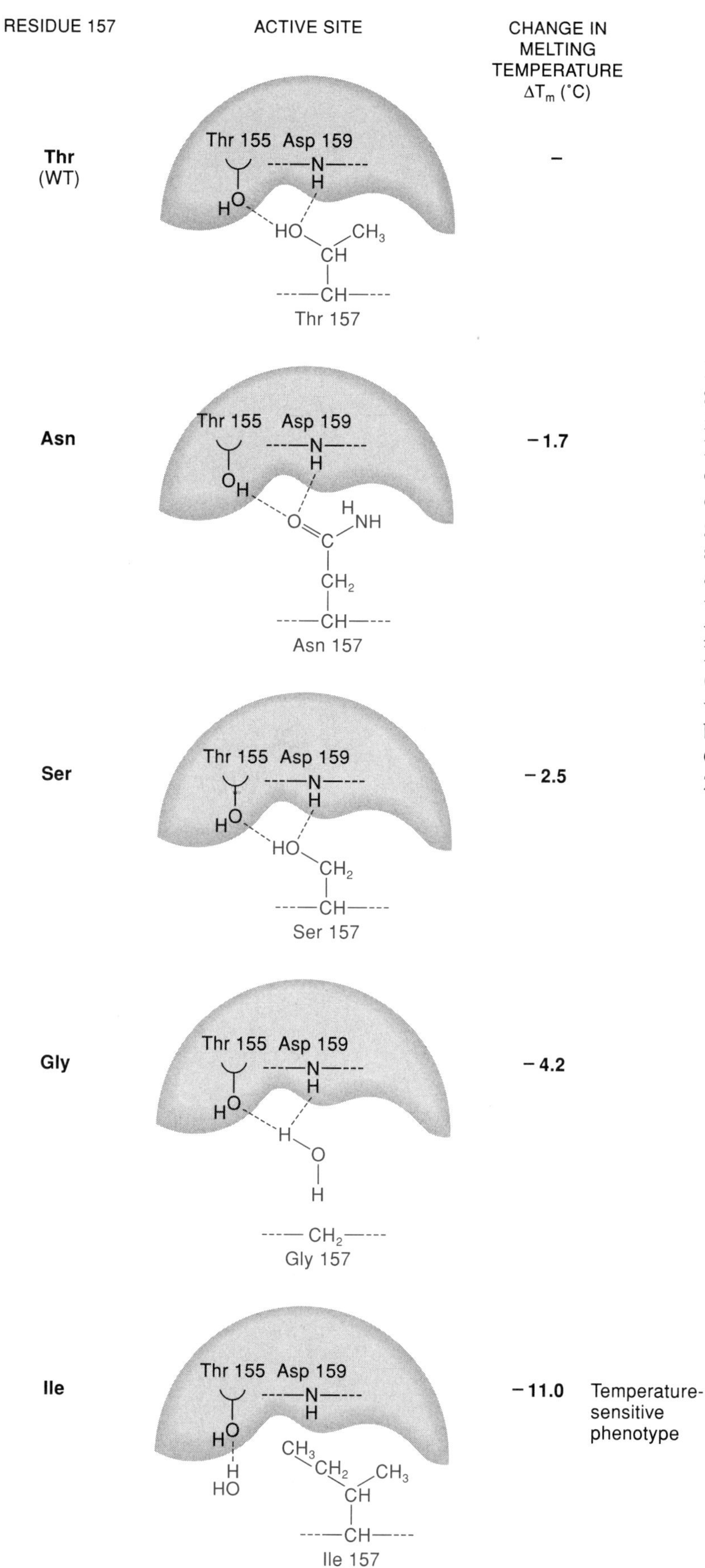

Figure 10.4
Schematic illustration of the interactions displayed by five representative amino acids at position 157 of T4 lysozyme and the melting temperature changes in each mutant protein. In wild-type lysozyme, the γ-oxygen of Thr 157 participates in a network of hydrogen bonds that is conserved in the Ser 157 and Asn 157 structures. In the case of the Gly 157 structure, X-ray analysis shows that a water molecule serves to retain the hydrogen bond network, giving a relatively stable protein. (The position of Thr 157 in the wild-type T4 lysozyme structure is shown in color in Figure 10.5.) (Adapted from J. A. Bell et al. 1990. Approaches Toward the Design of Proteins of Enhanced Thermostability. In W. G. Laver and G. M. Air, eds. Use of X-Ray Crystallography in the Design of Antiviral Agents. p. 236. Academic Press, San Diego.)

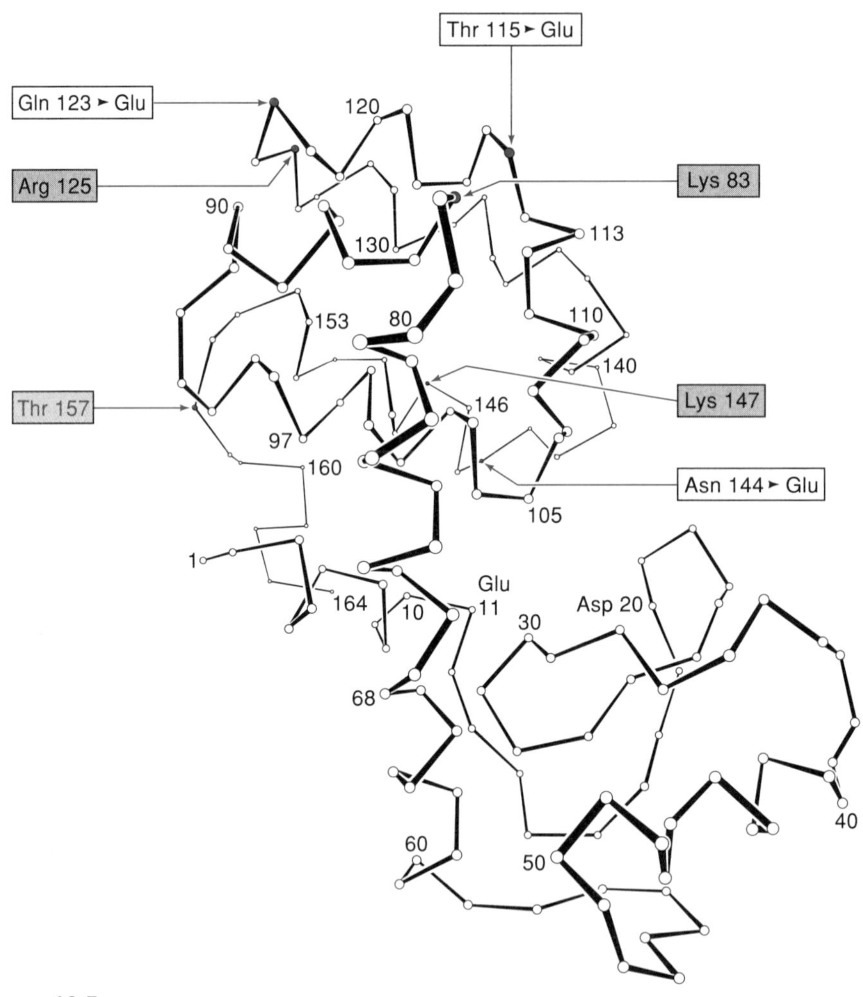

Figure 10.5
Backbone of T4 lysozyme showing residues used to investigate the role of hydrogen bonds in protein stability. The position of Thr 157 is shown in color. Mutations of Gln 123, Thr 115, and Asn 144 to Glu (white boxes) are intended to provide salt bridge partners for Arg 125, Lys 83, and Lys 147 (gray boxes), respectively. (From S. Dao-pin et al. 1991. Contributions of Engineered Surface Salt Bridges to the Stability of T4 Lysozyme Determined by Directed Mutagenesis. *Biochem.* 30 7143.)

folded protein. In the unfolded state, groups capable of hydrogen bonding do so with solvent molecules; upon folding, such intermolecular interactions are generally replaced by intramolecular hydrogen bonds of similar energy, and the ordered water is released to the bulk solvent. When protein folding demands breaking hydrogen bonds with solvent without replacement by intramolecular hydrogen bonds, much potentially stabilizing bond energy is lost.

Increasing stability by hydrogen bonds The experiments already described as well as many other analogous studies in which favorable interactions are deleted have confirmed the importance of hydrogen bonding in stabilizing a folded protein. Converse experiments, aimed at increasing stabilization by introducing new hydrogen bonds have, however, been much less successful. For example, in an attempt to add hydrogen bonds to the surface of T4 lysozyme, researchers introduced glutamate residues within hydrogen-bonding distance of preexisting lysine or arginine side chains. In two cases (Gln123Glu and Asn144Glu), the hydrogen bonds were intended to be between two residues separated by one turn of an α-helix; a third attempt (Thr115Glu) was predicted to bridge the ends of two different helices (Figure 10.5). Double mutations, in which the putative partner for the salt bridge was removed, were also constructed. The first mutation (Gln123Glu) stabilizes the protein by 0.4 kcal mol^{-1}, but the increased stability shows none of the pH dependence that would have been diagnostic for electrostatic interactions, and no hydrogen bond is apparent in the crystal structure. The other intrahelix mutation (Asn144Glu) stabilizes the protein by 0.2 kcal mol^{-1}, but the crystal structure reveals that the designed hydrogen bond to Lys 147 is present only as a minor conformer. The major conformer in the mixture contains a bound water molecule that forms a hydrogen bond with Arg 48. In the third case (Thr115Glu), thermal stability increases as the pH increases from 2.0 to 6.5; this is the predicted behavior for a glutamic acid residue, which should develop a negative charge at higher pH and thus be able to accept a hydrogen bond. However, the stability of the Thr115Glu/Lys83Met control increases by a similar amount over the same pH range, indicating that the increase in stability results not from salt bridge formation but from interaction of the negatively charged glutamate side chain with the positive charge at the N terminus of the helix 115–123. Comparisons of crystallographic thermal parameters in the wild-type and mutant proteins showed that the side chains expected to be held in salt bridges instead remain quite mobile, indicating that the oppositely charged residues interact only weakly. These observations lead to the conclusion that the entropic cost of localizing a solvent-exposed, charged side chain in the proper conformation for hydrogen bonding is greater than the energy that could be gained from the formation of the salt bridge. Thus, although salt bridges can make significant contributions to overall protein stability, their introduction into proteins does not afford a general route to increasing protein stability because of the need for the rest of the protein to hold the charged residues in the proper, rigid alignment for hydrogen bonding.

d. α-Helices

Helix capping Ordered cylinders of amino acids in which hydrogen bonds form the glue are called α-helices. Within such a helix, each amide group donates a hydrogen bond to the carbonyl oxygen that lies about 4 residues away in the previous helical turn. (Because of the pitch of the α-helix, there are in fact 3.6 residues per turn.) At the ends of a helix, there is no apparent acceptor for the amide hydrogens on the three amino acids closest to the N terminus and no donor for the carbonyl oxygens on the three amino acids closest to the C terminus. But structural studies reveal that these potential hydrogen-bonding groups frequently interact with side chain or backbone partners in other parts of the protein.

For example, the N terminal residue of each of the two α-helices in RNase from *B. amyloquefaciens* (barnase) is a threonine so situated that the oxygen of the side chain hydroxyl acts as an acceptor for the amide hydrogen in the first turn of the helix (Figure 10.6). Systematic changes have been made in the residues at the helix N terminal caps to study the contribution of each portion of the threonine side chain to the stability of the helix and ultimately to that of the protein itself. Two mutants (Thr6Ser and Thr26Ser) in which threonines were changed to serines were created, thereby removing γ-methyl groups that point away from the center of the helix in the folded protein. These methyl groups make stabilizing van der Waals contacts with other portions of the protein, and thus their removal destabilizes the protein (by 0.22 kcal mol^{-1} for Thr6Ser and 0.56 kcal mol^{-1} for Thr26Ser) (see also the discussion of the hydrophobic effect in Section 10.2b).

In other mutants, the hydrogen-bonding hydroxyl was replaced with a methyl group (Thr to Val), or substituents on the β-carbon were removed (Thr to Ala). Lastly, the β-carbon and its two substituents were deleted entirely (Thr to Gly). Of these substitutions, valine caused the greatest destabilization (2.31 kcal mol^{-1} for Thr26Val) despite the similarity in overall size of threonine and valine.

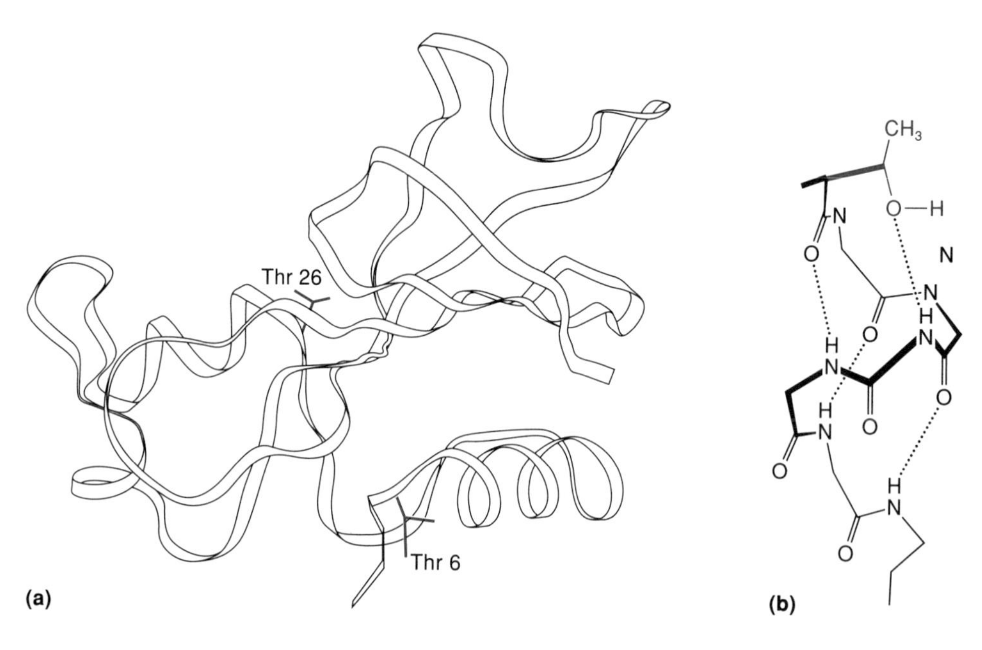

Figure 10.6
(a) A ribbon representation of the backbone of barnase showing Thr 6 and Thr 26 (color) at the N termini of their respective helices. (b) Detail of the barnase helix that starts with Thr 6, showing the hydrogen bonds between the γ-hydroxyl of Thr 6 and the backbone nitrogen of residue 9. (Adapted from L. Serrano and A. R. Fersht. 1989. Capping and α-Helix Stability. *Nature* 342 297.)

Removal of both the methyl and hydroxyl groups had a similar destabilizing effect (2.53 and 2.11 kcal mol^{-1} for Thr6Ala and Thr26Ala, respectively). The change to glycine caused the least destabilization (1.34 kcal mol^{-1} for Thr6Gly and 1.58 kcal mol^{-1} for Thr26Gly); glycine can create a pocket that could allow solvation of the N terminal amide. In contrast, replacing the threonine hydroxyl with a methyl group (Thr to Val) precludes the easy entry of any additional water. This causes the loss of hydrogen bonds to the N terminal amide hydrogen of the helix, thereby destabilizing the folded protein relative to an unfolded form in which the amide hydrogen can interact with water.

In addition to these relatively conservative substitutions, more drastic changes were made in barnase by substituting Asn, Gln, Asp, and Glu for these threonines. Compilations of amino acids in known helices suggest a distinct preference for Asn at the N terminal capping residue; the acidic residues Asp and Glu tend to occur near, not at, the N terminus of the helix. In these two barnase helices, substitutions of either Thr 6 or Thr 26 by both Asn and Gln were found to be destabilizing (1.3 kcal mol^{-1} and 1.7–1.9 kcal mol^{-1}, respectively). Substituting the acidic residue Glu for these threonines was significantly less destabilizing (0.05–0.27 kcal mol^{-1}), and the mutation Thr6Asp actually stabilized the protein (-0.11 kcal mol^{-1}). Comparison of the effect of Thr6Asn with that of Thr6Asp shows that the Asp carboxylate contributes significantly more to stability than the isomorphous carboxamide of Asn (a difference of 1.38 kcal mol^{-1}), probably because of the stabilizing interaction of the anionic Asp with the helix dipole.

Helix dipole Because of the uncompensated amides at the N terminus, the uncompensated carbonyls at the C terminus, and the inherent dipoles of each hydrogen-bonding interaction along the backbone of an α-helix, the helix itself has a dipole. This dipole's positive and negative ends lie at the N and C termini, respectively. Thus, introducing positively charged groups at or near the C termi-

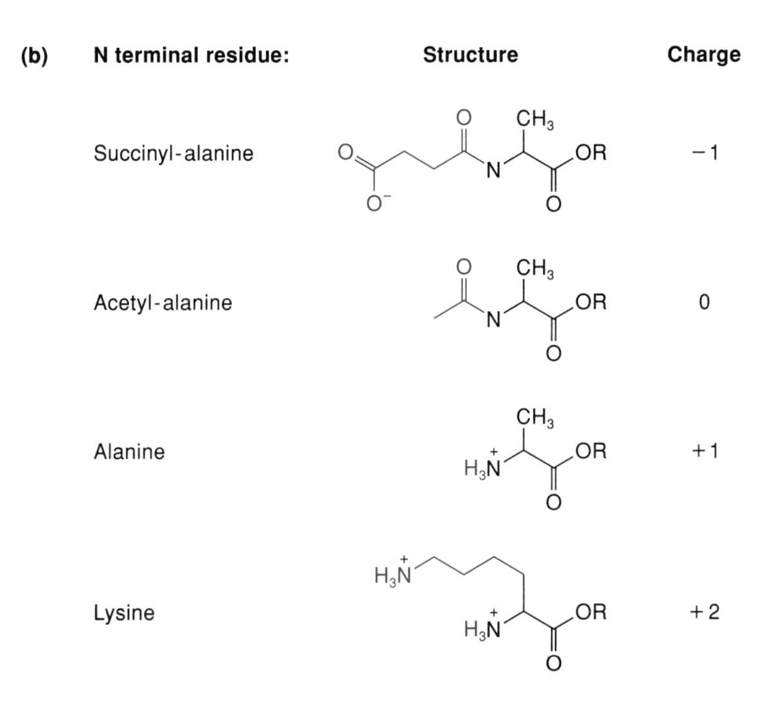

Figure 10.7
(a) Sequence of the C peptide of RNase A. Residues whose charged side chains interact favorably with the helix dipole (Glu 2 and His 12) are shown in color. (b) Structures of the amino acids substituted for the usual N terminal Lys of the C peptide.

nus and negatively charged groups at or near the N terminus is expected to exert a stabilizing effect on a helix because of the favorable interactions between these charges and the charges inherent in the helix dipole.

RNase A is an RNA endonuclease composed of 124 amino acids. The enzyme can be cleaved specifically by the protease subtilisin into two fragments, the S peptide (residues 1–15) and the S protein (residues 21–124). Although neither fragment is active alone, together they form a noncovalent complex that has full ribonuclease activity. Both the S peptide and a shorter derivative, the C peptide (residues 1–13) show significant α-helical structure at 0°C, and C peptide has been used as a model peptide to study α-helix structure (Figure 10.7). The small size of the C peptide makes chemical synthesis of analogues very convenient.

Circular dichroism observations at 0°C show that the native C peptide has significant α-helical structure. The helicity is pH dependent, as expected if charge interactions are important to α-helix formation Thus, decreasing the pH from 5 to 2 leads to a dramatic decrease in α-helix formation correlated with the protonation of side chain carboxylate. The effect is associated with neutralization of the negative charge on Glu 2, which eliminates the stabilizing anion near the N terminus. Confirmation of this assignment came from the study of a synthetic peptide with a Glu2Ala substitution; this variant has a significantly lower helical content than does native C peptide, and the helical content is independent of pH over the range pH 1 to 6.

The helicity of the C peptide drops dramatically as the pH is raised from 6 to 8, correlating with titration of the protonated His 12 cation near the C terminus; the cation can interact favorably with the negatively charged C terminal end of the helix dipole. In confirmation of this interpretation, the helical content of a C peptide in which His 12 is replaced by Ala (His12Ala) is unchanged as the pH is raised from 6 to 12.

As emphasized earlier, chemical synthesis allows great flexibility in the choice of amino acid substitutions. Thus, a series of C peptide analogues with a progressive change in charge at the N terminus was synthesized: succinyl-Ala, charge -1; acetyl Ala, charge 0; Ala, charge +1 (because of the N terminal ammonium cation); Lys, charge +2 (the normal residue). As predicted by the helix dipole hypothesis, the negatively charged succinyl-Ala provides the most favorable interaction with the positively charged N terminus of the α-helix; of the series, this peptide has the highest degree of helicity.

Further confirmation of the contribution of electrostatic interactions to helix stability comes from studies of the effect of increasing salt concentration on helix formation. The screening of charges by increased salt concentrations should decrease the magnitude of the electrostatic interaction between the helix dipole and a charged amino acid side chain. Indeed, raising the salt concentration decreases helix formation by the succinyl-Ala peptide and increases helix formation by the N terminal Lys peptide.

Reconstituted RNase A made by combining the 15 amino acid S peptide with the 104 amino acid S protein can be used to help determine whether the stability of a particular domain has any effect on the overall stability of the protein itself. The reconstituted molecule contains two noncovalently joined domains and lacks five residues of the native protein that form a loop joining the N terminal helix to the body of the molecule. Studies of RNase A analogues reconstituted from synthetic 15 amino acid S peptides containing the same variant sequences as those described for the 13 amino acid C peptides answer the question affirmatively; the stabilities of the reconstituted RNase A analogues parallel the stabilities of their C peptide helices.

Suitably placed negatively charged residues have also been used to stabilize α-helices in T4 lysozyme. In this context, investigators introduced aspartate residues near the N termini of five different helices, choosing sites for these mutations so as to avoid unstable interactions in the resulting proteins. To assess the relative importance of hydrogen bonding compared to electrostatic interactions, they introduced asparagine substitutions into the same sites. In general, Asn substitutions cause little or no difference in protein stability. In contrast, introducing negatively charged Asp increases protein stability by 0.6–1.3 kcal mol^{-1}. Comparison of the X-ray structures of mutants containing Asp and Asn at each site showed that hydrogen-bonding interactions are essentially identical for both substitutions. Accordingly, differences in stability are attributable to the favorable interaction between the negatively charged carboxylate and the positive end of the helix dipole.

e. Disulfide Bonds

Increasing thermostability A number of proteins containing disulfide bonds display extreme thermal stability. These bonds are thought to contribute to thermodynamic stability by decreasing the entropy gained when the protein unfolds. Introducing new disulfide bonds into a protein therefore could present an attractive potential approach to increasing thermal stability.

In some cases, introducing new disulfide bonds into certain positions does increase the mutant's stability compared to the normal protein. At other positions in the same protein and in other proteins, new disulfide bonds are, however, destabilizing. Important factors leading to increased stability include introduction of cysteines at sites that, in the native protein, lie within normal distances for disulfide bonds and designing the new bonds so that the cysteines have torsion angles similar to those seen in naturally occurring proteins. Thus, the new disulfide should itself have a favorable geometry, and its introduction should not provoke strain or distortion. Other, more subtle, factors may also be important. For example, if a region's flexibility is essential to function, then the rigidity imposed by a disulfide could destroy normal protein activity.

These factors have been exploited in the successful stabilization of T4 lysozyme by added disulfide bonds. Wild-type T4 lysozyme contains two cysteines that do not normally form a disulfide bond. Because these cysteines could create ambiguities in disulfide formation by newly introduced cysteines, a pseudo-wild-type background was created for these studies; the two wild-type cysteines were replaced by the double mutation Cys54Thr/Cys97Ala, which has no effect on the melting temperature or activity of the enzyme.

The effects of introducing various pairs of cysteines into positions in T4 lysozyme where they might be expected to form stable disulfide under oxidizing conditions are summarized in Figure 10.8a. One of these mutants ties the N terminal helix to the C terminal domain by disulfide formation between a new cysteine at residue 3 (Ile13Cys) and the wild-type cysteine at 97. (This is formally a double mutant Ile3Cys/Cys54Thr, so only one disulfide is possible.) Another disulfide between residues 9 and 164 pins the N terminal domain to the C terminal amino acid. A third substitution creates between residues 21 and 142 a disulfide that spans the active site. Compared to the wild type, these substitutions decrease the melting temperatures (T_m) of reduced protein by 1.9–6.5°C and increase the T_m of oxidized protein by 4.8–11°C. Mutants containing combinations of these disulfides were then constructed; the increases in stabilities associated with multiple disulfide bonds approximated the sums of the increases from the individual disulfides (Figure 10.8b). For example, in a lysozyme with three added disulfide bonds, the T_m increased by 23.4°C over that of the wild-type enzyme with no disulfides.

Not all new disulfides conferred greater stability to T4 lysozyme, highlighting the subtle combination of factors that determine the results of substitutions. For example, two attempts to add stabilizing disulfide linkages within the C terminal domain actually destabilized the protein; T_m for the oxidized mutant containing cysteines at residues 90 and 122 was lower by 0.5°C compared to wild type, and the T_m for the Cys127/Cys154 mutant was 2.4°C lower. In these two cases, the loops formed by the disulfides are much smaller than those formed by the highly stabilizing disulfide bonds between residues 9–164 and 3–97. The smaller loops are expected to lower the entropy of the unfolded state less than the larger loops and therefore be less stabilizing.

Comparison of the stabilizing and destabilizing disulfides suggests additional factors that may be important in designing stabilizing disulfides. For example, the substitution of cysteine for other amino acids may itself not be entirely benign (if it unfavorably alters van der Waals and hydrogen bond interactions in the region of the replacement). Examination of naturally occurring disulfide bonds emphasizes several stringent geometric constraints that influence their stability; these include a narrow range for optimal distances and torsional angles. Theoretical estimates suggest a high value for the strain energy associated with disulfide formation between residues 127–154, and measurements of the equilibrium constant between the oxidized and reduced form of this mutant confirm that formation of this disulfide bond is difficult.

Two of the most stabilizing disulfides, those between residues 9–164 and 21–142, are located on flexible parts of the molecule as judged by crystallographic thermal parameters. By contrast, the two destabilizing disulfides are in a region of close-packed α-helices, where high rigidity presumably restricts severely the protein's ability to adapt to disulfide formation. Theoretical estimates of the strain energy associated with bond formation between residues 21–142 (which spans the cleft) were, in fact, much higher than those estimated for bond formation between residues 127–154. But the model used in these calculations

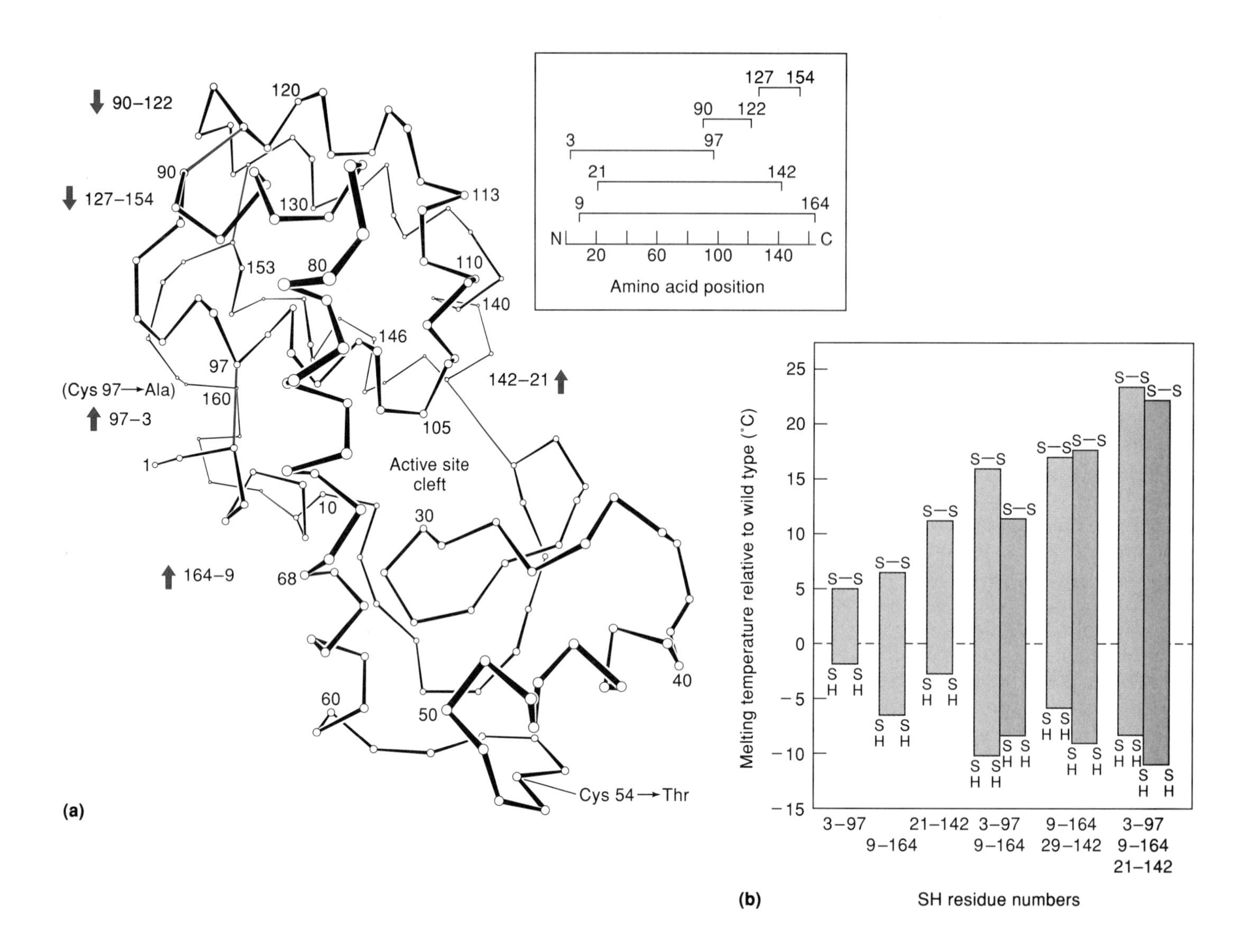

Figure 10.8
(a) View of the α-carbon backbone of T4 lysozyme showing the location of the five engineered disulfide bonds. Arrows indicate whether the disulfide bond stabilizes or destabilizes the protein. Also shown are the positions of the two native cysteines, Cys 54 and Cys 97, that are replaced by threonine and alanine, respectively, to give the pseudo-wild-type protein to which the mutants were compared. The insert shows the sizes of the loops formed by the various disulfide bridges. (b) Melting temperatures (T_m) of single, double, and triple disulfide-bonded lysozymes relative to the pseudo-wild-type at pH 2.0. The colored bars show the changes in the observed melting temperatures of the oxidized and reduced forms of the mutant lysozymes. The gray bars for the multiply bridged proteins correspond to the sums of the changes in T_m for the constituent singly bridged lysozymes. (From M. Matsumura et al. 1989. Stabilization of Phage T4 Lysozyme by Engineered Disulfide Bond. *Proc. Natl. Acad. Sci. USA* 86 6563.)

overlooks the possibility of the hinge-bending motions known to occur in T4 lysozyme which are likely to be of significant importance to function.

Preventing enzyme inactivation by disulfide bonds Disulfide bonds have been introduced into subtilisin in an attempt to protect the protein from irreversible inactivation; once unfolded, wild-type subtilisin does not regain its native state even when returned to conditions under which the folded protein is stable. Disulfide bonds were introduced at sites that link a wide variety of different types of structural elements, their bond strengths were measured, and their ability to protect subtilisin from inactivation was assessed. Unfortunately, there was no correlation between the strength of the disulfide bond, as measured by reduction equilibrium with glutathione, and stability against irreversible inactivation. One possible explanation for this observation is that irreversible inactivation does not involve an equilibrium between folded and unfolded protein. Thus, the presence of a disulfide bond, although thought to increase stability of the folded state by selectively destabilizing the unfolded state, has no influence on the process of irreversible inactivation that may be kinetically rather than thermodynamically controlled. In an extreme example of a disulfide bond's inability to protect subtilisin from inactivation, a disulfide bridge introduced between positions 148 and 243 actually has less resistance to thermal inactivation than its reduced counterpart.

Although the use of disulfides to increase thermal stability of subtilisin has proven unrewarding, at least one benefit has come from this study: determination of the crystal structures of four altered subtilisins. These reveal some surprising and instructive features. The torsion angles of the atoms of the disulfide bond differ appreciably in some cases from the angles predicted by modeling. Additionally, the molecules show a variety of accommodations made by different types of secondary structures due to the addition of a new covalent bond. For example, in two mutants where the disulfides linked α-helices to β-strands, concerted movements are seen in both secondary structural elements. In each case, the α-helix is more disrupted than the β-sheet, but structural perturbations propagate through the strand into the sheet. The structured water surrounding the protein undergoes numerous changes; in some cases, the side chain alterations leave hydrophobic cavities that are filled with disordered water molecules. Although such cavities might be thought to be energetically unfavorable, their existence in these mutants shows them to be less destabilizing than alterations in the protein's local geometry that would fill the cavities with amino acid side chains.

Controlling protein function using disulfide bonds Another novel use of added disulfides is for regulating protein activity. One of the T4 lysozyme mutants discussed previously places cysteines on opposite sides of the active site cleft (residues 21 and 142). Under reducing conditions, the mutant retains 68 percent of the activity of the wild-type enzyme; in the absence of reducing agents, the disulfide bond forms spontaneously and prevents the substrate's access to the active site, thereby abolishing all catalytic activity. Thus, two appropriately placed cysteines can provide a switch for regulating activity in response to the redox conditions of the environment.

Disulfides can also prevent protein motions that are, in some systems, an essential feature of function. Troponin C is one of the structural proteins in skeletal and cardiac muscle (Figure 10.9). Conformational changes are induced in troponin C by calcium binding and transmitted to other components of the thin filaments, thereby triggering muscle contraction. Although the exact structural changes are unknown, one hypothesis suggests that binding calcium causes helices B and C to move away from helices A and D. To test this postulate, scientists introduced two cysteines within disulfide-bonding distance of each other, one in

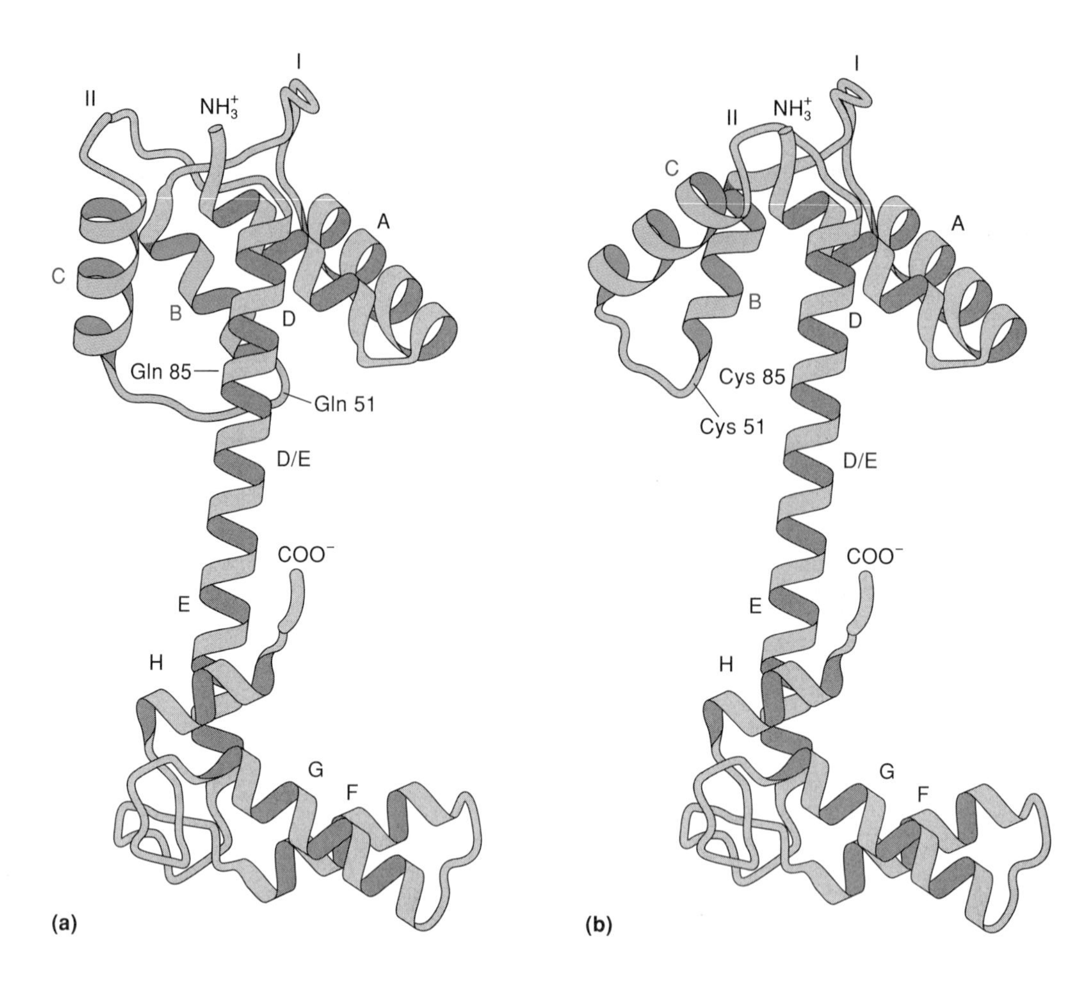

Figure 10.9
(a) The crystal structure of the inactive form of troponin C. (b) Binding of two Ca^{2+} ions to loops I and II is thought to cause helices B and C to move away from the central D/E helix. To test this hypothesis, Gln 51 and Gln 85 were changed to cysteines. Under conditions that allow disulfide bond formation, the mutant troponin C cannot trigger muscle contraction, presumably because movement of the B and C helices is prevented. (Adapted from C. O. Herzber, J. Moult, and M. N. G. James. 1986. A Model for the Ca^{2+}-Induced Conformational Transition of Troponin. *J. Biol. Chem.* 261 2641.)

the loop between helices B and C and one in a position on helix D. Although the alterations are 20Å from the calcium-binding site, calcium binding is significantly reduced in the oxidized mutant protein. Additionally, the ATPase activity of the thin filaments reconstituted with the oxidized mutant is no longer regulated by calcium ions; calcium regulation, is, however, normal if the two cysteines are carboxymethylated to prevent disulfide bond formation. This supports the notion that movement of the troponin C helices plays an essential functional role and suggests that introducing disulfide bonds may provide a general test for the importance of conformational changes to protein function.

f. Entropic Stabilization

Adding disulfide bonds increases protein stability by restricting the conformational freedom available to the unfolded molecule. Other methods of increasing protein stability also rely on an analogous restriction of an unfolded protein's entropy. This can be accomplished, for example, by replacing glycine with alanine or introducing proline in place of some other amino acid.

Because glycine lacks a side chain, it provides an unusually flexible site in a protein backbone. Such flexibility may be necessary to achieve a special twist in the folded protein; in such cases, retaining the glycine is essential to normal protein folding. Similarly, conserving glycine is essential in regions of very tight packing, where no additional side chain carbons can be sterically tolerated. In other sites, however, glycine can be replaced, for example, by alanine, without insult to the folded protein, but with a significant reduction in the flexibility and therefore the entropy of the unfolded protein. This should stabilize the folded relative to the unfolded protein. This hypothesis was tested by a Gly77Ala mutation at a site in T4 lysozyme where the crystal structure suggested that the alanine methyl side chain could easily be accommodated. The substitution causes small, local changes in the structure but has no effect on enzymatic activity; it does, however, stabilize T4 lysozyme toward both reversible and irreversible thermal inactivation (at pH 6.5, the increase in free energy of folding is 0.4 kcal/mol^{-1}).

As noted earlier (Section 10.1a), a suppressor tRNA charged with an unnatural amino acid can be used in an *in vitro* translation system to incorporate the unnatural amino acid at the site of a suppressor codon in mRNA. With this approach, sterically hindered amino acid analogues were incorporated at position 82 of T4 lysozyme to explore the hypothesis that rotational constraint of the backbone of the unfolded protein increases the relative stability of the folded form (Figure 10.10). Incorporation of α-aminoisobutyric acid, an amino acid with two α-methyl groups, does indeed increase the melting temperature by 1°C; an important aspect of the stabilizing effect of this substitution is that the angles (ϕ, φ) corresponding to the lowest energy conformation of this analogue are very similar to the angles observed for Ala 82 in wild-type T4 lysozyme. Substitution with a pipecolic acid, a cyclic six-membered analogue of proline, would also be expected to decrease the entropy of the unfolded protein; however, the T_m of such a lysozyme variant is 2°C lower than that of the wild type. This may reflect the much poorer match between the ϕ, φ angles in the normal folded protein and those of pipecolic acid. Unfortunately, *in vitro* protein synthesis systems do not yield sufficient protein for X-ray crystallographic study.

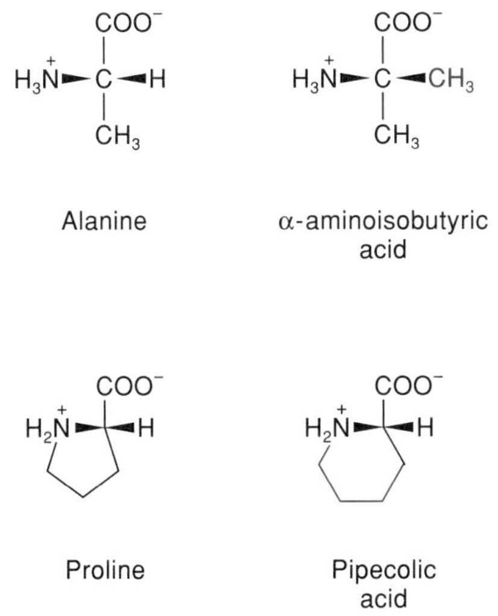

Alanine

α-aminoisobutyric acid

Proline

Pipecolic acid

Figure 10.10
Alanine, proline, and the sterically hindered analogs, α-aminoisobuteric acid and pipecolic acid, introduced into T4 lysozyme using chemically charged suppressor tRNAs.

g. Kinetics of Folding and Residues that Direct Folding Pathways

Folding pathway of RNase T1 In addition to studying the equilibrium thermodynamics between folded and unfolded proteins, we are making progress in our knowledge of the pathways of protein folding and how sequence influences this important process. In many cases, folding is such a highly cooperative phenomenon that discrete intermediates cannot be identified; in other cases, bi- and triphasic kinetics have been observed, leading to efforts to dissect the physical processes associated with each phase. RNase T1, a single-domain protein that can fold reversibly under a wide variety of conditions, is amenable to this study. Kinetics of folding show three discrete phases: the first is very fast and is completed within one second; the second and third are much slower and take from minutes to hours to complete. Folded RNase T1 contains two *cis* prolines, suggesting that, in the two slow steps, the two X-Pro amide bonds of Pro 39 and Pro 55 isomerize from the *trans* to the *cis* conformations. In the completely unfolded protein, these residues would be expected to adopt the *trans* conformation. The acceleration of the intermediate folding phase by prolyl-peptide isomerase (PPIase) indicates that the isomerization of at least one of these two prolines could be a rate-determining step in the folding pathway.

Replacing Ser 54-Pro 55 with Gly 54-Asn 55 does not affect the overall stability of the protein but does lead to the loss of the intermediate slow folding phase. Similar attempts to replace Pro 39 were unsuccessful, indicating that Pro 39 is essential to RNase T1 stability. The final slow folding phase in the Ser54Gly/Pro55Asn mutant is tentatively associated with isomerization of Pro 39 and is surprisingly slow compared to other *trans-cis* isomerizations. This phase can be significantly accelerated by adding chaotropic agents such as guanidinium hydrochloride. This observation led to the hypothesis that the nativelike regions in the folding intermediate hinder the isomerization of Pro 39, trapping this residue in the transconformation. The crystal structure of RNase T1 shows that Trp 59 makes van der Waals contact with Pro 39; moreover, the fluorescence of Trp 59 is influenced by this contact. A Trp59Tyr mutation does not affect the activity, stability, or triphasic kinetic character of the folding pathway, but does appreciably accelerate the slowest folding step (putatively, the Pro 39 isomerization) such that the rate of this step overlaps that of the intermediate phase (putatively, the Pro 55 isomerization). Although the rates of the two proline isomerizations are nearly identical in the Trp59Tyr mutant, they are distinguished by very different responses to PPIase. The isomerization rate of Pro 55 is accelerated in the presence of PPIase, but that of Pro 39 remains unchanged.

Temperature-sensitive folding mutants of phage P22 tailspike protein Studies of the folding of the tailspike protein of phage P22 have also yielded information about the process of protein folding, especially about the assembly of multimeric proteins. *In vivo,* the trailspike protein mediates the adsorption of the phage to its *Salmonella* host. When assembly of the phage is blocked by a mutation in the major coat protein, soluble trimers of the tailspike protein accumulate in the bacterial cytoplasm. The folding of the tailspike protein shows three distinct phases (Figure 10.11): first the newly synthesized monomer must fold into a conformation that is competent to associate with two more subunits. These three subunits form a protrimer which folds further to give the mature trimeric tailspike; the protrimer can be distinguished from the mature trimer by the sensitivity of the protrimer to proteases.

Temperature sensitive mutants of tailspike protein have been found where the protein, instead of folding into soluble trimers, accumulates as large, insoluble aggregates known as inclusion bodies. Often when foreign proteins are overexpressed in bacteria, they partition into inclusion bodies. In some cases, the formation of inclusion bodies can be used to the researcher's advantage since the protein of interest can be partially purified by separating the inclusion bodies from the rest of the cellular proteins by centrifugation. The inclusion bodies are then solubilized by adding a denaturing agent and the protein is recovered in nearly pure form by refolding. However, some proteins do not refold readily (rapidly and in high yield) from denaturants; so inclusion body formation can be a major obstacle to overproduction of some proteins.

Many of the temperature-sensitive tailspike mutants were inferred to involve defects in the folding process (tsf mutants) since the protein produced when these mutants were grown at the permissive temperature (30°C) have melting temperatures very similar to that of the wild-type tailspike protein (88°C) and all have melting temperatures more than 40°C higher than the temperature at which they show a temperature-sensitive phenotype *in vivo.* Since the mutant proteins, once folded, are stable at high temperature, the temperature sensitivity must be in the folding step.

The crystal structure of the P22 tailspike protein shows an unusual β-helix fold with the tsf mutations clustered in the body and 'dorsal fin' of the protein. This information, combined with studies of the kinetics of folding, suggests that the first stage of folding of the tailspike protein involves folding this β-helix into an intermediate conformation that is able to associate with two similar subunits to give the protrimer. Tsf mutations interfere with the formation of this competent intermediate and the improperly folded subunits associate in nonproductive ways and eventually accrete into insoluble inclusion bodies. The interdigitated C terminal domain (Figure 10.11c) provides a possible explanation for the observed protrimer maturation step: although the three subunits associate rapidly to form the protrimer, rearranging the C terminal strands to their final form might be expected to be a relatively slow process. Once the C terminal domain has folded it would be very difficult to dissociate—explaining the high melting temperature of both the mature wild type and tsf mutant proteins.

Genetic screens for suppressers of the tsf folding defects turned up two mutants, Val331Gly and Ala334Val, that can suppress a number of different folding defects. Neither of these suppresser mutations appreciable affect the stability or function of the folded protein; the T_m of the purified proteins is unchanged and the mutant tailspike proteins have the same ability as the wild type to bind phage heads and mediate infection of the *Salmonella* host. Both suppresser mutations, however, strongly influence the partitioning of maturing tailspike protein between native, soluble trimers and inclusion bodies. Pulse labeling experiments at 39°C show that while wild type tailspike protein partitions to give approximately 35% mature trimers and 65% aggregated inclusion bodies, the single-site temperature sensitive mutants produced 100% aggregated inclusion bodies. However, the folding of proteins containing both the temperature sensitive folding mutation and the suppresser mutation resembled that from the wild type protein, yielding 35% soluble trimers at 39°C. Impressively, proteins containing only the suppresser mutations fold much more readily at high temperature than does the wild type (60% soluble trimers for tailspike protein with both suppresser mutations compared to 35% of the native protein at 39°C). Because inclusion body formation can be a formidable obstacle to the overexpression of foreign proteins in bacteria, researchers hope the investigations into the factors governing partitioning of the P22 tailspike protein between aggregation and productive folding pathways will give general insight into inclusion body formation.

10.3 Protein Function

Once they have folded to their unique and exquisite three-dimensional structures, proteins carry out an extraordinary variety of functions. They are the molecular machines of life, catalyzing every metabolic interconversion from the apparently simplest such as the hydration of carbon dioxide (by carbonic anhydrase) to those as complicated as protein synthesis (catalyzed by a ribosome, the

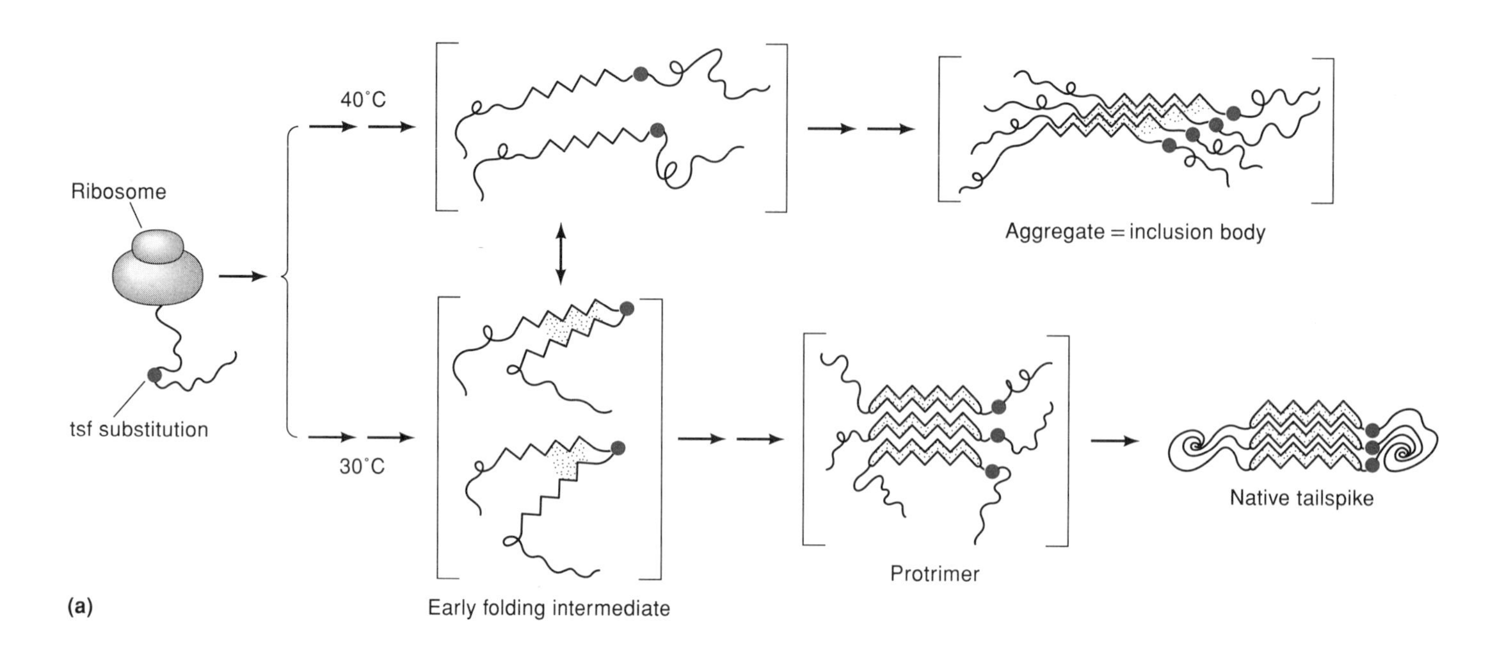
40°C
Ribosome
tsf substitution
30°C
Aggregate = inclusion body
Early folding intermediate
Protrimer
Native tailspike
(a)

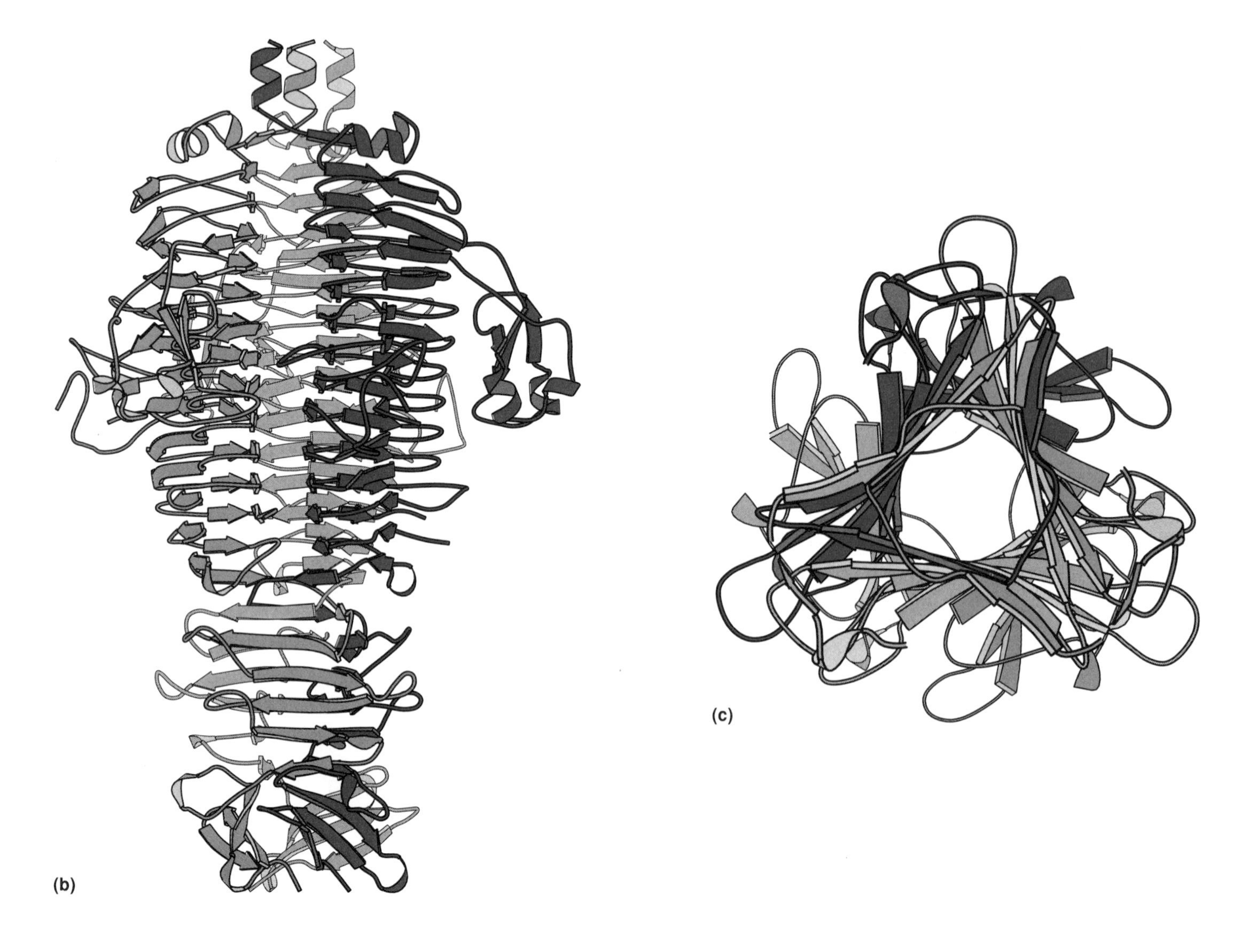
(b)
(c)

Figure 10.11 (*facing page*)
(a) Proposed pathway for the intracellular folding and assembly of the P22 tailspike protein. At permissive temperatures, newly synthesized chains fold to a conformation competent for specific chain-chain interactions leading to the formation of protrimers. These incompletely folded protrimers then finish folding to yield the extremely stable native tailspike. At higher temperatures, the folding mutation is thought to interfere with the early folding into an assembly-competent intermediate. The misfolded intermediates accumulate and aggregate nonspecifically to form large, insoluble inclusion bodies. (b) Crystal structure of the mature P22 tailspike protein. The main body of the trimer consists of three β-helices, each with a small extension called the dorsal fin. (c) A view up the central axis of the tailspike trimer showing how the C terminal domains of the monomers fold around each other in mature structure. (From A. Mitraki et al. 1991. Global Suppression of Protein Folding Defects and Inclusion Body Formation. *Science* 253 55; D. P. Goldenberg and T. E. Creighton. 1994. A Fishy Tail of Protein Folding. *Current Biol.* 4 1027.)

molecular machine that coordinates the interplay of charged tRNAs, mRNA, and proteins). *E. coli* ribosomes are composed of 55 different polypeptides along with three ribosomal RNAs (which may themselves contribute to catalysis). Proteins also transduce environmental signals (e.g., the conversion of light into the sensation of vision, the sensing of nutrient concentration resulting in chemotaxis, and the detection of neurotransmitter binding leading to the opening and closing of ion channels in a precisely timed manner, thereby creating a signal along a neuron's axon). Proteins convert energy from the hydrolysis of ATP into muscle action. Proteins on an individual's cell surfaces give all its tissues a unique identity so that the immune system can recognize self from nonself.

Alterations in protein structure by mutagenesis now play a central role in learning how all these functions depend on the structures that mediate them. Mutagenic studies of virtually every class of proteins have been undertaken and have given insights, in an ever-increasing number of cases, into those structural elements that are crucial for function and how they work together to create a molecular machine. Because of the enormous and rapidly expanding scope of this field, no single chapter can possibly consider the manifold systems that have been studied by mutagenesis. We concentrate on catalysis, limiting the discussion to a few examples that demonstrate the essence and potential of this approach.

a. Classical Enzymology and Kinetics

Before the development of site-specific mutagenesis, enzyme mechanism was studied by many different techniques, including the kinetics of enzymatic reactions as a function of pH, temperature, substrate structure and concentration, and in the presence or absence of molecules that might influence the enzyme's catalytic activity. Great ingenuity allowed the chemical modification of selected amino acids at or near the active site, and the results of such experiments revealed important aspects of protein function. Biophysical studies using nuclear magnetic resonance, sometimes with labelling of specific residues, also played a significant role. But site-specific mutagenesis fundamentally changed the rules of the game. The ability to obtain relatively large amounts of structurally altered enzymes opened entirely new avenues and, together with three-dimensional structure determination by single crystal X-ray diffraction and, more recently, by multidimensional nuclear magnetic resonance techniques, has brought new insights into how enzymes work. The latest step in this saga is the attempt to alter the activity of a native enzyme so that it can perform useful new tasks, an effort referred to as protein engineering.

In a very general way, an enzymatic reaction involves two steps: recognition and binding of a specific substrate or group of substrates, followed by catalysis of changes in covalent bonds. Several factors contribute to the enzyme binding substrate. There is a complementarity between the substrate and the active site of the enzyme that includes an essentially hand-in-glove steric fit between the shape of the substrate and that of the active site. It also involves charge interactions (e.g., a positive charge on the substrate interacting with a negative charge on the enzyme), hydrogen bonding, and van der Waals stabilization (hydrophobic regions of a substrate binding in hydrophobic regions of the enzyme). In catalysis, several of functional groups in the amino acid side chains of the protein (or even the amides of the peptide backbone) act together in a highly collaborative process to catalyze the breaking of the old and the making of the new bonds.

In thinking about binding and catalysis, one can overemphasize the so-called "active site" and regard the rest of the protein as an essentially rigid molecule. In fact, enzyme flexibility may be necessary to let substrate in and product out; movements of the groups involved in catalysis also are essential. In addition, binding of an appropriate molecule may trigger dramatic changes in shape that are critical to function. Such changes in shape occur, for example, when an effector binds on the outside surface of a cell to a receptor protein inserted across the cell membrane; the shape change is thereby communicated across the membrane into the cell's interior, where it triggers appropriate responses to the external signal.

Substrate binding and dissociation are generally relatively rapid compared to bond rearrangements although this is by no means always true. In some important cases, product dissociation is, in fact, rate determining. This substrate-binding step is customarily described by the following equation:

$$E + S \underset{k_2}{\overset{k_1}{\rightleftharpoons}} E \cdot S$$

where *E*, *S*, and *E·S* are enzyme, substrate, and enzyme-substrate complex, respectively. In the simplest cases, this process can be considered a rapid equilibrium that precedes any rearrangement of covalent bonds, and it can be described by an appropriate equilibrium constant K_S:

$$\frac{[E][S]}{[E \cdot S]} = K_s = \frac{k_2}{k_1}$$

Substrate binding is followed by the catalytic steps in which old bonds are broken and new ones formed to generate, sometimes after many steps, the products of catalysis, which then leave the enzyme relatively rapidly compared to the rate of the bond rearrangement. These events can often be described by a kinetic constant k_{cat} that defines the rate of the slowest step:

$$E \cdot S \xrightarrow{k_{cat}} E + P$$

where P represents the product(s). The overall rates of many enzymatic processes can be quantitatively described by classical Michaelis-Menten kinetics, in

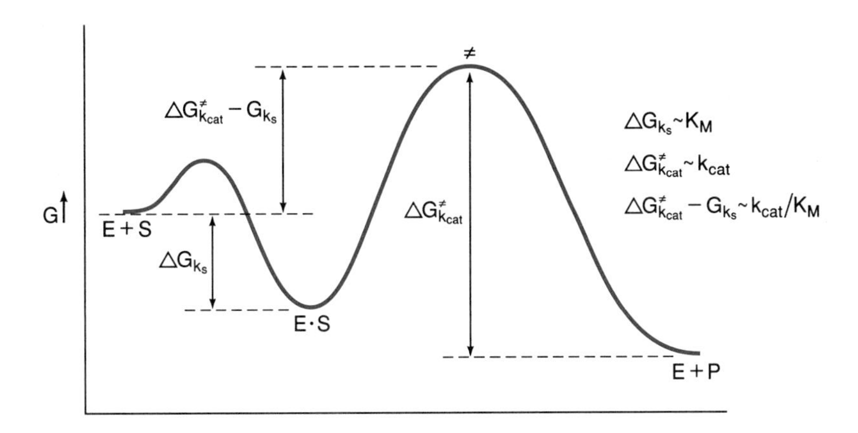

Figure 10.12
Free energy diagram showing the relative energies of the enzyme and free substrate, the enzyme-substrate complex, and the enzyme and released products.

which the equilibrium constant K_S is replaced by a somewhat more complicated constant, K_M (the Michaelis constant):

$$K_M = \frac{k_2 + k_{cat}}{k_1}$$

In situations where the catalytic process (k_{cat}) is slow relative to the rate of dissociation of the E · S complex (k_2):

$$K_M = K_S$$

K_M is often taken as a measure of an enzyme's affinity for substrate and k_{cat} as a measure of the speed with which the enzyme catalyzes the bond rearrangements in the rate-determining step. In this scheme, the overall rate of an enzymatic reaction can be expressed as the following function of the two constants just discussed:

$$\text{Rate} = \frac{k_{cat}[E]_o[S]}{K_M + [S]}$$

Figure 10.12 illustrates the energetics of a simple case. In considering the overall conversion of the free enzyme and substrate to the transition state, the slowest step is associated with the maximum increase in free energy.

$$\Delta G^{\ddagger}_{k_{cat}} - \Delta G_{K_S}$$

Rate and equilibrium constants are exponentially related to the corresponding free energies as follows:

$$K_S = \exp(-\Delta G_{K_S}/RT)$$

$$k_{cat} = \text{const} \cdot \exp(-\Delta G^{\ddagger} k_{cat}/RT)$$

hence

$$\exp(\Delta G_{K_S} - \Delta G^{\ddagger}_{k_{cat}})/RT \approx kcat/K_S$$

The ratio k_{cat}/K_S, or more generally, k_{cat}/K_M, reflects a combination of the enzyme's ability to bind substrate and its ability to catalyze the necessary bond rearrangements. Thus, the ratio provides a useful measure of the enzyme's overall effectiveness in catalyzing the conversion of substrates to products.

In many cases, the kinetic parameters K_M (or K_s), k_{cat}, and k_{cat}/K_M reflect very complex phenomena well beyond the scope of this chapter. However, this discussion, though vastly oversimplified, provides a useful and necessary introduction for understanding the case studies described in the following sections.

b. Serine Protease Families

Both the formation and cleavage of amide bonds play central roles in the metabolic interconversions carried out by every living organism. Cells synthesize proteins and degrade them when no longer needed; the released amino acids can be salvaged for further protein syntheses. During digestion, an organism breaks down proteins from its food supply so that it can assimilate the component amino acids. Many physiological processes require activation of an inactive precursor protein by cleavage of specific peptide bonds to produce an active product. Examples of such activation occur in the mechanisms that control blood pressure, account for homeostasis in blood clotting, provide the complement cascade crucial to the immune response, and allow a single sperm to penetrate the coating surrounding an ovum during fertilization. Thus, cleavage of the amide bonds in proteins is a central and ubiquitous process.

There are several families of proteases. Some cleave a single amino acid (the exopeptidases) from either the C terminus (the carboxypeptidases) or N terminus (the amino peptidases). Others, the endopeptidases, cleave peptide bonds within the interior of a polypeptide chain. Some of these, the acid proteases, function in an acidic environment such as the stomach (pH 1–2). Other endopeptidases (the group with which we are especially concerned) act under mildly basic conditions (pH 7–8) within the small intestine or biological fluids such as blood. This family of endopeptidases, called the serine proteases and typified by trypsin, is commonly found both in eukaryotes such as vertebrates (ourselves importantly included) and in some prokaryotes (e.g., *Streptomyces*). The trypsin-like serine proteases share a highly conserved protein fold as well as a common catalytic apparatus and mechanism; in the first step of the hydrolysis reaction, they all use the hydroxyl functionality of a serine residue to attack the carbonyl group of the amide bond to be hydrolyzed (the scissile bond). They also share other features, such as a histidine and an aspartate that, together with serine, participate importantly in catalysis (the catalytic triad). Another subgroup of serine proteases, typified by subtilisin, uses an essentially identical catalytic triad but is evolutionarily distinct from the trypsin family; members of this subgroup possess a common three-dimensional architecture that differs profoundly from that of the trypsin family. The trypsin and subtilisin families of serine proteases thus provide one of the clearest and most fascinating examples of convergent evolution at the molecular level.

c. Substrate Specificity in the Trypsin Family

The various members of the trypsin family catalyze the cleavage of peptide bonds (Figure 10.13). Trypsin itself cleaves the peptide bond immediately, following a positively charged side chain—usually arginine or lysine. This specificity-determining residue (which becomes the C terminus of the cleaved protein) is called the **P_1 residue**. Chymotrypsin cleaves after large hydrophobic or aromatic residues, and elastase shows a preference for small hydrophobic P_1

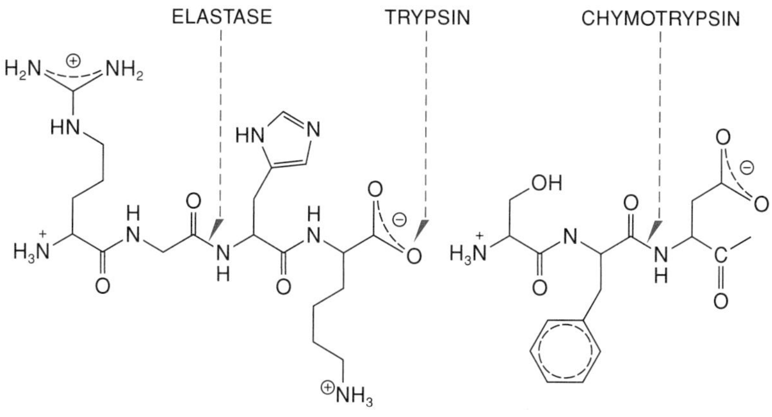

ELASTASE TRYPSIN CHYMOTRYPSIN

Arginine-Glycine-Histidine-Lysine Serine-Phenylalanine-Aspartate

Figure 10.13
This peptide illustrates the cleavage preferences of the various members of the trypsin family of serine proteases. Elastase cleaves the amide bond immediately following a small amino acid such as glycine or alanine. By contrast, chymotrypsin prefers large hydrophobic side chains in the P_1 position. Trypsin is shown cleaving the amide bond following a positively charged lysine residue. (From R. M. Stroud. 1974. A Family of Protein-Cutting Proteins. *Sci. Am.* July 75.)

residues. Other members of the trypsin family show specificity for more than a single residue; for example, factor Xa of the blood-clotting cascade prefers to cleave after the sequence -Ile-Glu-Gly-Arg-X-. These differences in substrate specificity reflect interactions between the side chains of the amino acids that line the so-called **S_1 pocket** of the enzymes and the side chain of the P_1 residue on the substrate. The S_1 pockets in the trypsin family are formed by a common protein fold that involves the extension of two β-strands from the core of the enzyme (Figure 10.14). But, the S_1 pockets of trypsin, chymotrypsin, and elastase differ in the side chains that project from the backbone into the interior.

The three digestive enzymes, trypsin, chymotrypsin, and elastase, present an ideal system for inquiry into the factors that influence substrate specificity. In a simplistic view, interchanging those side chains of the three enzymes that project into the S_1 pocket should convert the specificities of the resulting variants. However, a number of more subtle factors operate, including interactions between the S_1 side chains and other neighboring residues, and the experimental results are far less straightforward than the simple notion predicts. They dramatize the complexity of the interplay between many different residues in the enzyme and emphasize the extended cooperativity between protein active sites and distant surrounding structures in controlling the protein function.

Components of substrate binding in trypsin The peptide frameworks of the S_1 binding pockets of trypsin and chymotrypsin are superimposable. The major difference between them is the presence of either aspartate (Asp 189) or serine (Ser 189) at the base of the S_1 binding pocket of trypsin or chymotrypsin, respectively. (The numbering of these residues derives from a serine protease consensus sequence equivalent to the longest known such enzyme. Residues at analogous three-dimensional positions are given the same number wherever they fall in the linear sequence of amino acids. The relative order of residues remains the same, and the convention allows for deletions in the linear sequence of particular enzymes to achieve this useful simplification in nomenclature.) The negative

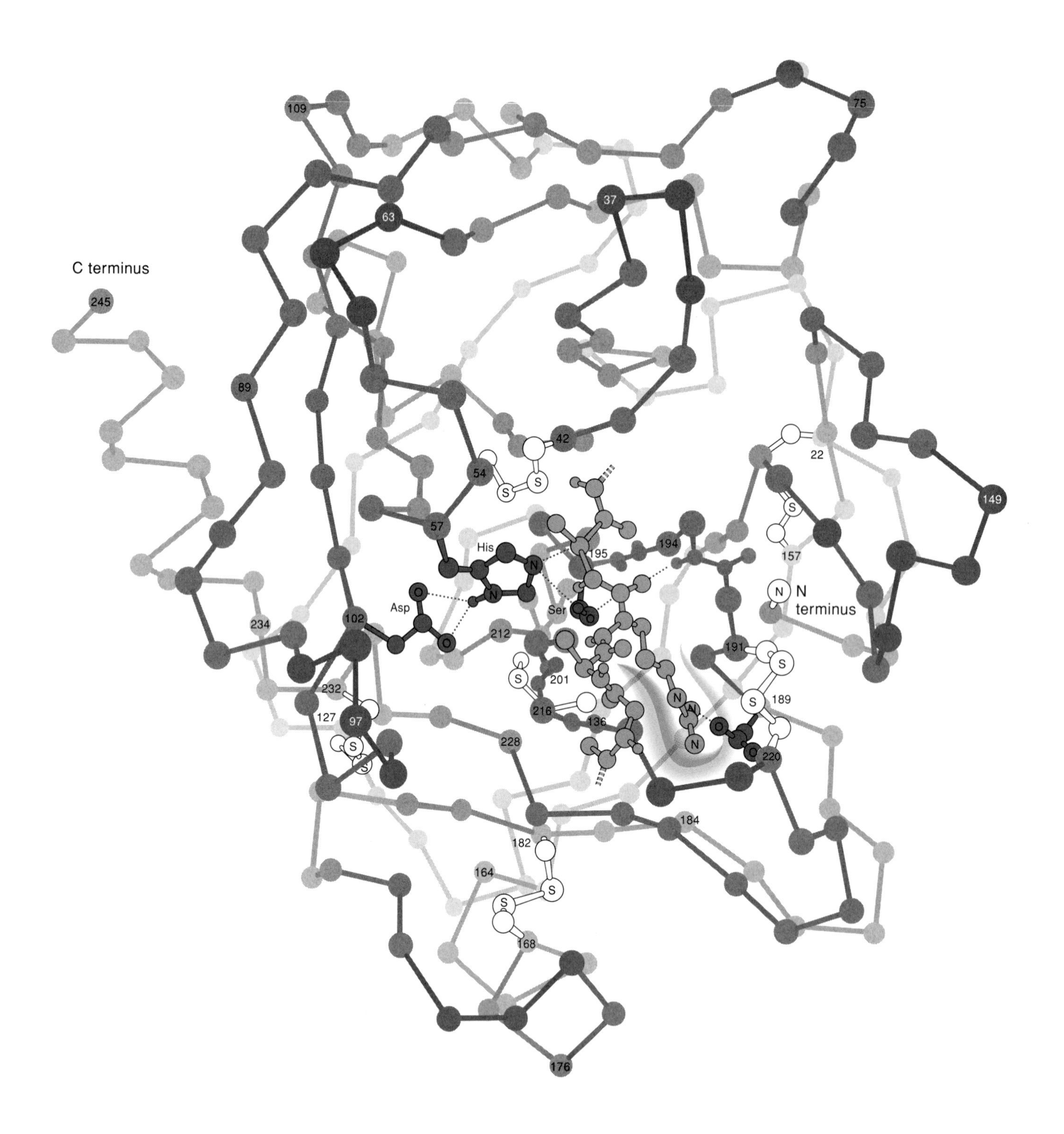

Figure 10.14
Trypsin with a substrate peptide (light color) modeled into the active site. Binding of a basic residue to Asp 189 (color) at the base of the S_1 binding pocket (surface shown in light color) aligns the next peptide bond with the active site serine, making it susceptible to nucleophilic attack. Side chains of the catalytic triad are shown in color. (Adapted from R. M. Stroud. A Family of Protein-Cutting Proteins. *Sci. Am.* July 75.)

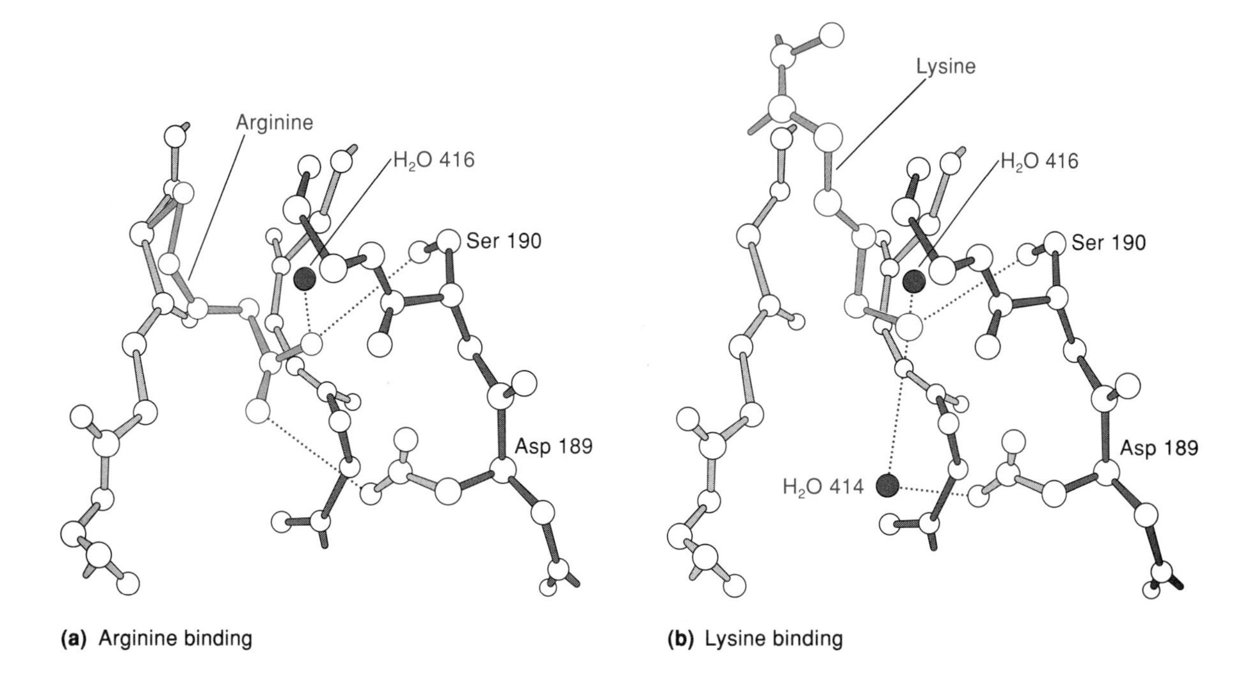

Figure 10.15
Details of the binding interactions in the S_1 specificity pocket of trypsin. (a) An arginine in the P_1 position (color) can form hydrogen bonds directly to the Asp 189 and Ser 190 side chains at the base of the S_1 pocket and to a bound water molecule numbered 416. (b) A lysine residue in the same position (light color) still forms hydrogen bonds directly to that water molecule and to Ser 190, but hydrogen-bonding interactions with Asp 189 are mediated by an additional bound water molecule, 414. (Adapted from C. Brandon and J. Tooze. 1991. Introduction to Protein Structure. p. 240. Garland Publishing, New York.)

charge associated with Asp 189 within an otherwise hydrophobic environment permits stabilizing electrostatic interactions between the enzyme and typical trypsin substrates (i.e., Lys or Arg at P_1). The importance of this negative charge to trypsin activity is reflected in the approximately 10^5-fold decrease in activity of the isosteric Asp189Asn mutant on lysyl and arginyl peptide substrates.

The geometry of the binding pocket also significantly influences substrate specificity. With native trypsin, the arginyl side chain of the substrate extends to the base of the pocket, and hydrogen bonds directly to Asp 189 (Figure 10.15a). For lysyl substrates, a water molecule bridges the gap between the ammonium ion of the lysine side chain and the carboxylate of Asp 189 (Figure 10.16b). As expected, therefore, arginyl residues provide better substrates (by a factor of 4.2) than do lysyl residues. If one tries to enhance trypsin activity for lysyl substrates by moving the negative charge up from the base of the pocket, as in an Asp-189Glu mutant, one lowers the enzyme's ability to bind *both* lysyl and arginyl substrates (an increase in K_M by 40–50-fold). More surprisingly, one also slows the catalytic rate, k_{cat}, by 4-fold for lysyl substrates and 70-fold for arginyl substrates. Though overall a significantly poorer enzyme for both substrates, the Asp189Glu mutant does show a reversal of the substrate specificity of the native enzyme; that is, there is a greater reduction in the activity for arginyl than for lysyl substrates, so the mutant has an 8.3-fold preference for lysyl over arginyl substrates. However, the overall lower effectiveness of the enzyme for either substrate emphasizes the delicate balance that nature has achieved in the evolution of enzymatic catalysts. Structural changes that, on simplistic grounds, should lead to predictable alterations in function very often require far more subtle analysis to understand quantitatively.

Because of the often unexpected properties of designed mutations, applying strategies that create families of mutants whose functions can be quickly assessed by either selection or screening for particular activities is especially useful. Using this approach, researchers made a broad survey of residues acceptable at the bottom of the trypsin specificity pocket; the assay relied on a genetic screen for active trypsin. Bacteria cannot normally metabolize arginine β-naphthylamide. Consequently, bacteria auxotrophic for arginine cannot grow on media containing arginine β-naphthylamide as the only source of arginine. However, trypsin (not a native bacterial enzyme) can cleave the amide bond of this derivative, freeing the vital arginine. Accordingly, arginine auxotrophs that secrete an active trypsin can grow on media containing arginine β-naphthylamide as the only potential source of arginine. Random mutagenesis created a group of mutants including all possible combinations of residues at positions 189 and 190 (site saturation mutagenesis), and arginine auxotrophs expressing these altered trypsins were selected for their ability to grow on media containing arginine β-naphthylamide. The results confirmed the necessity of a negative charge at the base of the pocket (either Asp or Glu at 189 or 190). The Asp189Ser/Ser190Asp and Asp189Ser/Ser190Glu mutations were about as effective as Asp189Glu but considerably less effective than the native enzyme. Also, these mutants show no preference for lysyl over arginyl substrates. The nature of the neighboring amino acids also matters; only small side chains are permissible adjacent to the Asp or Glu.

The Ser 190 hydroxyl group in native trypsin plays more of a role in facilitating binding of lysyl substrates than in overall catalysis. Binding interactions with a lysyl side chain involve hydrogen bonds to two water molecules (as just noted), one of which is held in place by the Ser 190 hydroxyl. Loss of this anchoring hydroxyl lowers the specific activity of trypsin for lysyl substrates, but binding of arginyl substrates, which depends largely on direct contacts with Asp189, is relatively unaffected. Thus, a Ser190Gly mutant has a 40-fold preference for arginyl over lysyl substrates.

Designing substrate specificities The electrostatic interaction between Asp 189 in the enzyme's S_1 pocket and the Lys or Arg side chain of the substrate should dominate trypsin specificity. The prediction that a simple alteration from a negative to a positive charge in the base of the binding pocket (e.g., in an Asp189Lys mutant) can create an enzyme that specifically cleaves after glutamate or aspartate residues rests on the assumption that the ammonium cation of lysine at 189 would occupy essentially the same position as that occupied by the carboxylate anion of aspartate. Although lysine would extend farther than aspartate from the pocket's base, the side chains of the proposed substrates for the Asp189Lys mutant would be correspondingly shorter, so the interaction could involve a simple reversal of charges. Nature is, however, less accommodating, and the properties of such a mutant are surprising. Whereas the Asp189Lys mutation does abolish activity toward basic substrates, the mutant shows no increase in proteolysis involving acidic residues at P_1. If anything, the mutant behaves more like chymotrypsin, preferring substrates with large hydrophobic side chains. Thus, the Asp189Lys trypsin, like chymotrypsin (but not as efficiently) cleaves after phenylalanyl, tyrosyl, and leucyl residues, although not after tryptophan. In fact, trypsin itself possesses a low level of catalytic activity for substrates with aromatic residues at P_1, though this weak activity is generally masked by the much more vigorous activity toward substrates with basic side chains. Because neither trypsin nor chymotrypsin cleaves after leucyl residues, this activity of the mutant, although weak, is nonetheless novel.

Computer modeling of the Asp189Lys mutant shows that the lysine side chain is long enough to extend completely across the pocket's base such that the

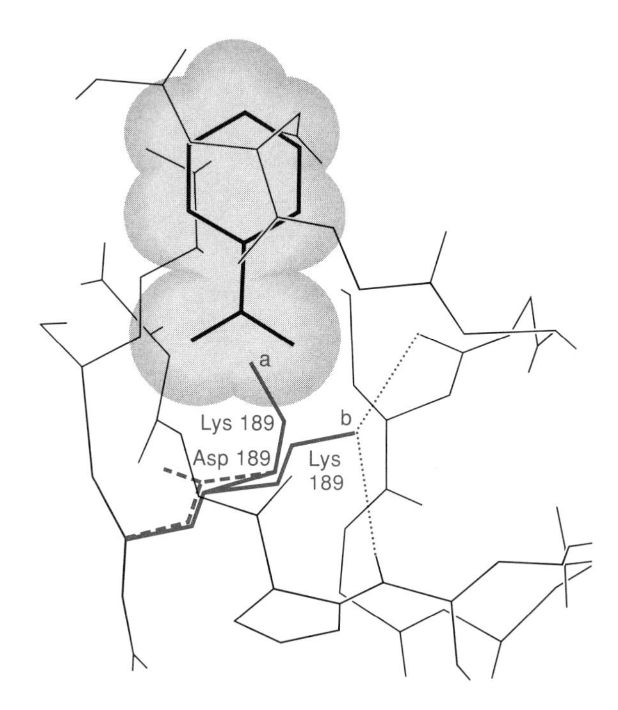

Figure 10.16
Computer model showing the S_1 binding pocket of the Asp189Lys trypsin mutant. The gray surface of the benzamidine inhibitor delineates the interior of the protein pocket. The wild-type recognition residue, Asp 189, is shown as a dashed line, and two alternate positions of lysine at position 189 (color) are labelled a and b. The original expectation was that Lys 189 would extend into the S_1 binding pocket, yielding a pocket with a positively charged amino group at its base (a). (b) If, however, the lysine side chain extends across the base of the binding pocket and forms hydrogen bonds with two backbone carbonyls, the base of the binding pocket consists of the methylene groups of the lysine side chain, giving a large hydrophobic binding surface. (From L. Graf et al. 1987. Selective Alteration of Substrate Specificity by Replacement of Aspartic Acid-189 with Lysine in the Binding Pocket of Trypsin. *Biochem.* 26 2619.)

ammonium ion is positioned on the far side of the groove in a hydrophilic region where it can hydrogen bond with two nearby carbonyl oxygens (Figure 10.16). Such a structure effectively hides the positive charge of the lysine, leaving only the four hydrophobic methylene groups of its side chain lining the base of the S_1 pocket. This hydrophobic character can account for the observed chymotrypsin-like preference for substrates with large hydrophobic side chains. This result again emphasizes the potential complexities and presently unforeseeable outcomes of what at first glance appear as entirely straightforward modifications of protein function by mutagenesis.

Switching specificities within a family of proteases Trypsin and chymotrypsin share a protein fold and a high degree of sequence conservation in their specificity pockets (Figure 10.17). Thus, a reasonable conclusion would be that the enzyme's substrate preferences arise largely from the few differences in the amino acids that line the pocket. Most obviously, chymotrypsin has Ser 189 in a position analogous to the Asp 189 of trypsin. However, the Asp189Ser mutant of trypsin does not mimic chymotrypsin, but rather is a sluggish protease, an equally poor catalyst for substrates with either basic or hydrophobic residues at P_1. Even if the other three binding pocket residues that differ between trypsin

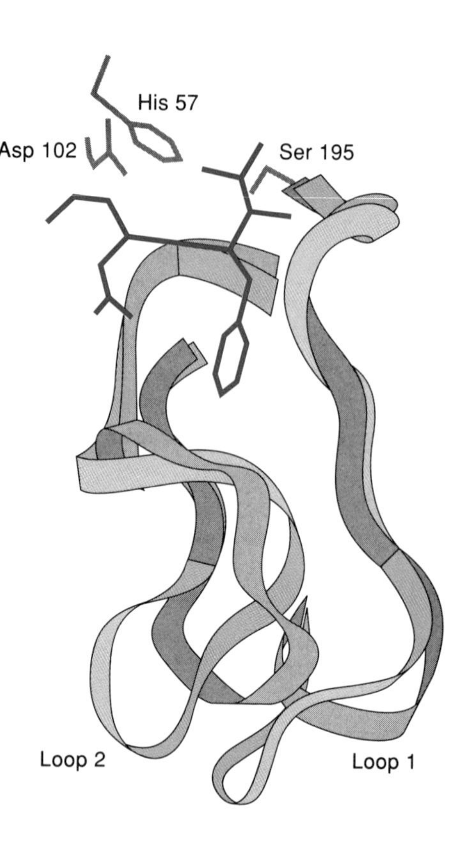

Figure 10.17
Loops 1 and 2 which connect the β-strands forming the S_1 binding pockets of trypsin (color) and chymotrypsin (gray). A phenylalanine substrate fragment (color) is bound, and the catalytic triad is shown in gray. (From J. J. Perona and C. S. Craik. 1995. Structural Basis of Substrate Specificity in the Serine Proteases. *Protein Sci.* 4 352.)

and chymotrypsin are changed to those of the chymotrypsin pocket (the three changes are Asp189Ser, Glu192Asp, and addition of threonine corresponding to a Thr 219 that is in chymotrypsin's S_1 pocket but absent in trypsin), the result is an enzyme with activity on hydrophobic substrates only slightly improved compared to native trypsin and still four orders of magnitude less than the activity of chymotrypsin.

More extensive comparisons of families of trypsin and chymotrypsin sequences around the S_1 pocket show that the two loops that join the β-strands making up the pocket are highly conserved in sequence and structure. Although neither loop makes direct contact with substrates, conservation suggested an important role in determining function. To test the hypothesis, researchers synthesized hybrid enzymes in which chymotrypsin loops were incorporated into a trypsin containing the chymotrypsin S_1 binding pocket. Substituting loop 1 alone gave no increase in chymotrypsin activity, but substituting loop 2 alone caused a modest increase in activity toward substrates with hydrophobic residues at P_1. Substituting both loops, led to a dramatic increase (approximately 1000-fold) in chymotrypsinlike activity and a preference for chymotrypsin over trypsin substrates by a factor of 10^2–10^3, comparable to the preference shown by the native enzyme. Nevertheless, chymotrypsin itself is more active than the hybrid enzyme on such substrates.

These intriguing and complex results again dramatize the subtlety with which structure mandates function. Even protein regions that seem removed from direct interaction with substrate, and therefore might appear to be excess baggage, only filling space and joining the working regions, can contribute significantly to a fully functional protein.

d. Binding Specificity in Subtilisin

Whereas the trypsin family of proteases encompasses a relatively narrow range of substrate specificities, the members of the subtilisin family act on a broad set of substrates having both small and large hydrophobic and positively and negatively charged residues at P_1. Such a difference is consistent with the biological roles of these families. The trypsinlike enzymes are involved in mammalian physiology, which often requires very specific cleavages of peptide bonds in carefully regulated cascades. In contrast, the subtilisin-related enzymes often operate as nutritional opportunists, hydrolyzing proteins and peptides in the bacterial environment, thereby providing amino acid nutrients for growth. But even in these promiscuous proteases, cleavage efficiency varies significantly, depending on the substrate. For example, native subtilisin from *Bacillus amyloliquefaciens* hydrolyzes peptide bonds with increasing efficiency, as measured by k_{cat}/K_M, as the size and hydrophobicity of the P_1 residue increase. Generally, size and hydrophobicity correlate, and mutagenesis strategies can reveal their interplay. To this end, the Gly 166 at the base of the subtilisin S_1 binding site was systematically altered and the catalytic efficiency of the resulting mutants assessed on a panel of substrates with different amino acids in the P_1 position.

Hydrophobic interactions in substrate-enzyme binding As expected, increases in the size of residue 166 in subtilisin lead to preferences for substrates with P_1 side chains of decreasing size. Optimally, the side chain of residue 166 and the P_1 side chains together occupy about 160 ±30 Å^3, and every 120 Å^3 of volume in excess of this value decreases the transition state binding energy (which is directly related to k_{cat}/K_M) by approximately 2.6 kcal/mol. (For comparison, a leucyl side chain occupies about 100 Å^3).

Up to this threshold, where steric crowding becomes dominant, the catalytic efficiency on a given substrate increases as the volume of the side chain at position 166 increases so that the space is entirely filled by the combination of enzyme and substrate side chains. Moreover, increasing the hydrophobicity of isosteric side chains in the enzyme increases activity. For example, cysteine is considerably more hydrophobic than serine, so the Gly166Cys mutant becomes catalytically more efficient relative to the Gly166Ser mutant as the size and hydrophobicity of the substrate P_1 side chain increase from alanine to phenylalanine. If only size mattered, the marginally larger cysteine should somewhat decrease the relative efficiency for a phenylalanine substrate. Interestingly, glycine is conserved at position 166 in subtilisins from many bacterial species, thus ensuring a broad range of substrate specificity. This suggests an evolutionary advantage in secreting an enzyme that can hydrolyze peptide bonds between many different amino acids even at the expense of decreased efficiency on certain substrates.

Electrostatics in subtilisin substrate binding In addition to the hydrophobic effects just discussed, subtilisins also employ complementary electrostatic interactions between substrate and enzyme. X-ray crystallographic studies show two somewhat distinct modes of P_1 side chain binding into the S_1 site. With amino acids such as phenylalanine, the large hydrophobic side chain projects its aromatic ring toward the Gly 166 at the base of the S_1-binding pocket. In contrast, the side chain of lysine lies across this cleft and positions its terminal ammonium cation adjacent to Glu 156, which lies at the edge of the cleft (Figure 10.18). In addition to the electrostatic attraction, the ammonium cation forms hydrogen bonds with the Glu 156 carboxylate. By adding a complementary negative charge at position 166, investigators tried to redirect lysine binding so that the positively charged side chain projected deeply into the binding cleft. Mutations at 166 (to Glu, Gln, and Ser) were coupled with a wide variety of complementary changes at 156 (Glu156Asp, Asn, Gln, Met, Ala, Gly, Lys, Arg). Researchers then measured how active these mutants are toward essentially isosteric substrates with differently charged side chains at P_1 (e.g., glutamate vs. glutamine, lysine vs. methionine). Increasing the positive charge at the S_1-binding site (as in Glu156Gln/Gly166Lys)

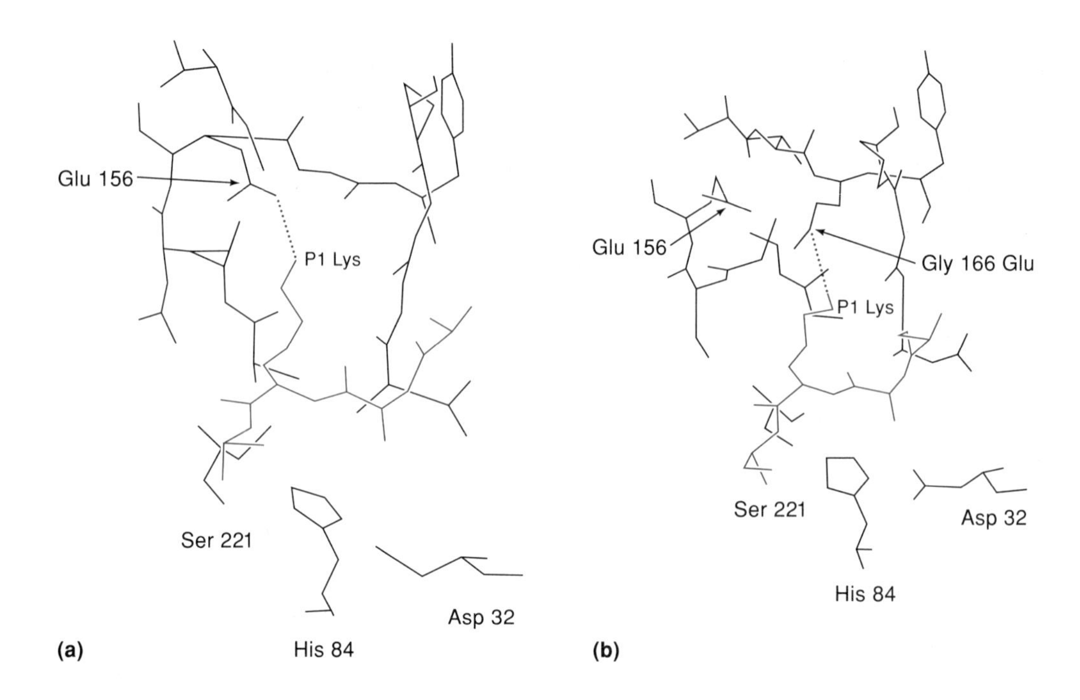

Figure 10.18
A model of the catalytic triad and S_1 binding pocket of subtilisin containing a substrate (color) with lysine in the P_1 position. (a) In the wild-type enzyme, lysine lies across the S_1 binding cleft and makes a hydrogen bond with Glu 156 (black arrow). (b) It has been proposed that in a Gly166Glu mutant the negative charge at the base of the binding pocket could provide an alternate binding mode in which the P_1 lysine side chain extends into the binding pocket. (Adapted from J. A. Wells et al. 1989. Designing Substrate Specificity by Protein Engineering of Electrostatic Interactions. *Proc. Natl. Acad. Sci. USA* 84 1220.)

increases the catalytic efficiency (k_{cat}/K_M) on substrates with Glu at P_1 by a factor of 1900 over the wild type. Efficiency on the analogous neutral substrate with glutamine at P_1 also increases, but much less dramatically. (These substrates were succinylated peptides; so whatever the residue at P_1, the negative charge on the succinate carboxyl may account for some of the increase by virtue of non-specific electrostatic interaction between substrate and enzyme.) Substrates with methionine at P_1 are also better substrates for the Glu156Gln/Gly166Lys mutant. In anticipated contrast, the mutant is less effective (by a factor of sevenfold) on substrates with lysine at P_1. Mutants containing charged residues at both 156 and 166 afford roughly additive increases in activity on substrates with opposite charges on P_1, suggesting that these two relatively close sites interact only weakly with each other. As would be expected for effects arising from electrostatic interactions, differences in substrate preference are most pronounced at low ionic strengths.

Specificity differences among members of the subtilisin family Subtilisins from different species of bacteria show differing activity profiles toward various substrates (Figure 10.19). For example, subtilisin from *Bacillus lichenformis* is much more active with glutamate at P_1 than is subtilisin from *Bacillus amyloliquefaciens.* Although these two subtilisins share a folding pattern, they differ at 86 out of a total of 275 residues. Only 2 of these differing residues, Glu156 and Tyr217, can come into direct contact with the substrate; a third binding pocket residue, Gly169, is within 7 Å. Yet these few amino acid changes can alter the activity pro-

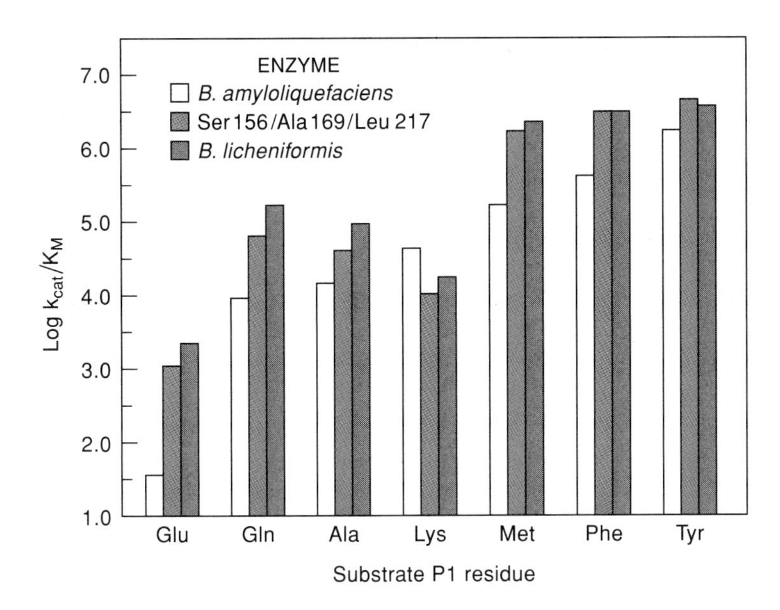

Figure 10.19
Specificity profiles of three subtilisins. Log(k_{cat}/K_M) on substrates with different P_1 residues are shown for each of three enzymes: white bars, *B. amyloliquefaciens;* gray bars, *B. licheniformis;* and colored bars, *B. amyloliquefaciens* triple mutant containing the binding pocket residues of *B. licheniformis,* Glu156Ser, Tyr217Leu, and Gly169Ala. (From J. A. Wells and D. A. Estell. 1988. Subtilisin—An Enzyme Designed to be Engineered. *Trends Biochem. Sci.* 13 292.)

files. To graft a *B. lichenformis* specificity onto a *B. amyloliquefaciens* background, scientists made three changes in the binding pocket, both singly and in combination: Glu156Ser, Tyr217Leu, and Gly169Ala.

As already described, the residue at position 156 contacts the P_1 side chain of the substrate. A Glu156Ser mutation shifts the specificity of *B. amyloliquefaciens* subtilisin toward that of *B. lichenformis* subtilisin, namely, a decrease in affinity for positively charged and an increase for negatively charged residues in P_1. Residue 217 lies near the $P_{1'}$-leaving group (the residue that constitutes the amino part of the scissile amide bond). A Tyr217Leu mutation increases the k_{cat} of the *B. amyloliquefaciens* enzyme almost to that of *B. lichenformis.* This increase may result from a reduced steric hindrance toward the leaving group when leucine replaces tyrosine. A Gly169Ala mutation has no significant effect on the enzyme's preference for various substrates but does lower K_M (increase binding affinity) for all substrates tested. This may arise from a general change in the hydrophobicity of S_1 or may reflect more subtle alterations in substrate binding mediated by changes in the position of the nearby carbonyl oxygen of residue 152.

The kinetic effects are essentially additive, indicating that the residues involved act almost independently. Thus, effects in the double mutant Glu156Ser/Tyr217Leu roughly equal the sum of the effects of the single mutations. Moreover, the introduction of the third mutation, Gly169Ala, essentially adds the functional effects of that single mutation to the double one. The triple mutant of *B. amyloliquefaciens* subtilisin substantially mimics the substrate specificity of *B. licheniformis,* indicating that the differences in substrate preferences of these two enzymes are primarily due to differences in their S_1 binding sites. The small differences in specificity that remain between the triple mutant and the *B. licheniformis* enzyme must be due to subtle differences outside the S_1 binding pockets of the two enzymes. As the work on substrate specificity within the trypsin family of enzymes so clearly revealed, many subtle features, by no means entirely localized to the obvious residues in the S_1-binding pocket, act in concert to create differing substrate specificities in evolutionarily and structurally related enzymes.

e. Catalysis by Serine Proteases

As noted earlier, the subtilisins and trypsinlike serine proteases provide a striking example of convergent evolution. These two families of enzymes occur in widely different phyla and have very different overall three-dimensional structures. Nevertheless, both families use the same functional side chains to form almost identical catalytic active sites. The key catalytic functionality is the hydroxyl group of a serine residue. The hydroxyl oxygen attacks the carbon in the carbonyl group of the scissile amide bond (Figure 10.20). A tetrahedral species is thus formed, followed by expulsion of the C terminal polypeptide segment. The resulting acyl-enzyme intermediate is a covalent ester between the serine hydroxyl group of the enzyme and the carboxyl group on the N terminal peptide fragment. Water binds to active sites and attacks the serine ester. The reaction proceeds through a second tetrahedral transition state and the N terminal peptide fragment is released from the enzyme.

The serine hydroxyl is activated when an adjacent histidine imidazole ring (which may be many residues removed from the catalytic serine in the primary amino acid sequence) accepts a proton from the hydroxyl group to form an imadazolium cation; the cation is itself stabilized in a charge relay network by an adjacent carboxylate anion provided by an aspartate residue. These three residues (Ser, His, Asp) together form the catalytic triad (Figure 10.21). They occur in a consensus numbering scheme: His 57, Asp 104, and Ser 195 in the trypsin family and Asp 32, His 64, and Ser 221 in the subtilisin family. In the tetrahedral intermediate, the carbonyl oxygen becomes a so-called oxyanion, which is stabilized by strong hydrogen-bonding interactions in an **oxyanion hole** formed by the backbone amides of Gly 193 and Ser 195 in the trypsin family and by the backbone amide of Ser 221 and the side chain amide of Asn 155 in the subtilisin family.

Catalytic triad The highly cooperative nature of the catalytic triad becomes manifest on analysis of the activity of mutants in which these residues are altered, either singly or in combination. For example, substituting the subtilisin active-site serine, histidine, or aspartate with alanine results in a decrease in k_{cat} by a factor of 2×10^6, 2×10^6, or 3×10^4, respectively; replacement of all three

---

Figure 10.20 (*facing page*)
Hydrolysis of a peptide bond catalyzed by a serine protease. The enzyme binds substrate (1), forming an enzyme-substrate complex (2). The catalytic serine attacks the adjacent peptide bond, proceeding via a tetrahedral transition state (3) to yield an acyl-enzyme intermediate in which the active site serine is covalently attached to the N terminal peptide fragment (4) and the C terminal peptide fragment is released into the solvent. The acyl-enzyme intermediate then binds water (5), which attacks the ester bond between the enzyme serine and the peptide. Proceeding via a second tetrahedral transition state (6), the covalent attachment between the enzyme and substrate is broken (7) and the second product is released (8).

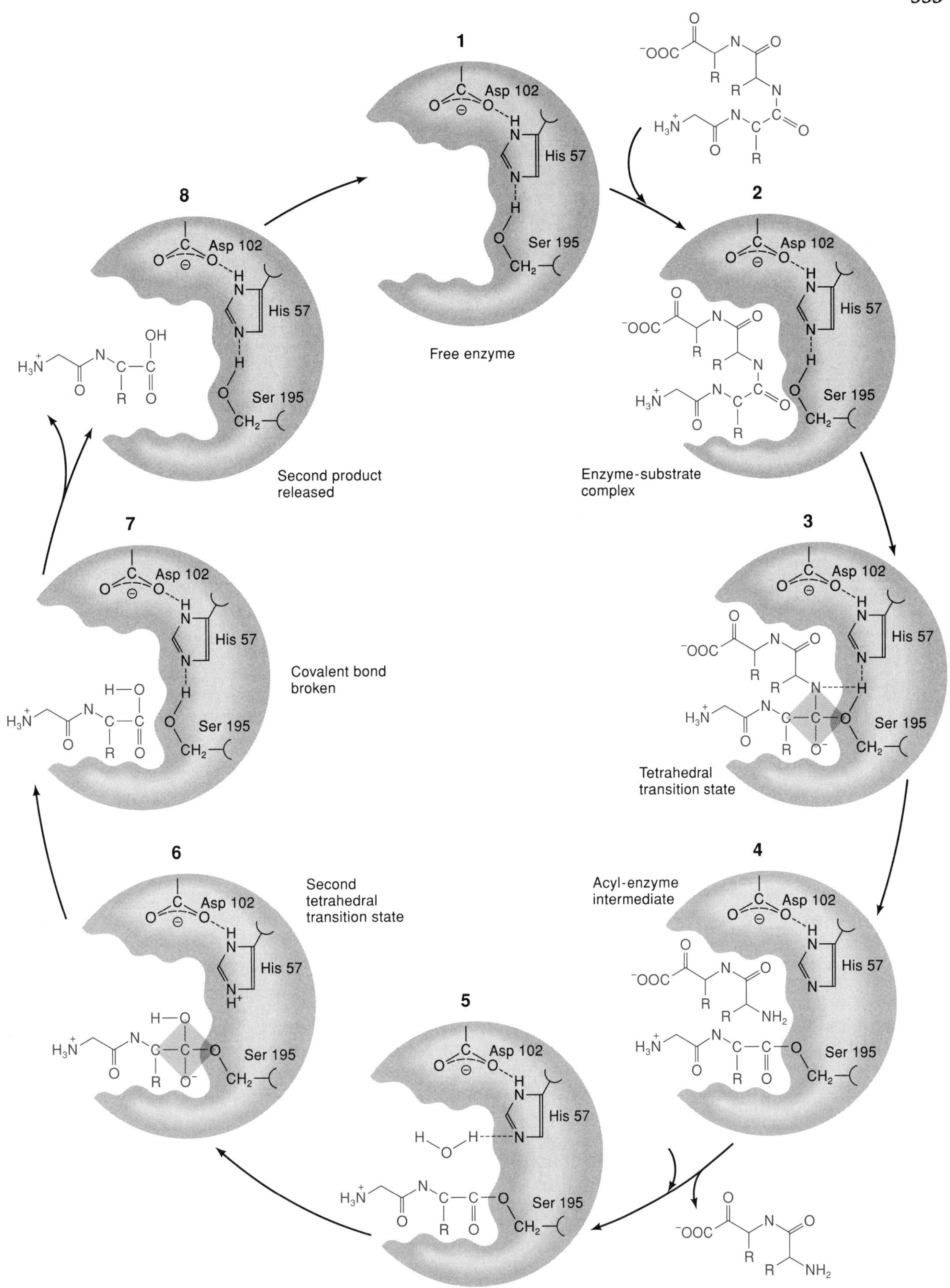

1
Asp 102
His 57
Ser 195
Free enzyme
2
Enzyme-substrate complex
3
Tetrahedral transition state
4
Acyl-enzyme intermediate
5
6
Second tetrahedral transition state
7
Covalent bond broken
8
Second product released

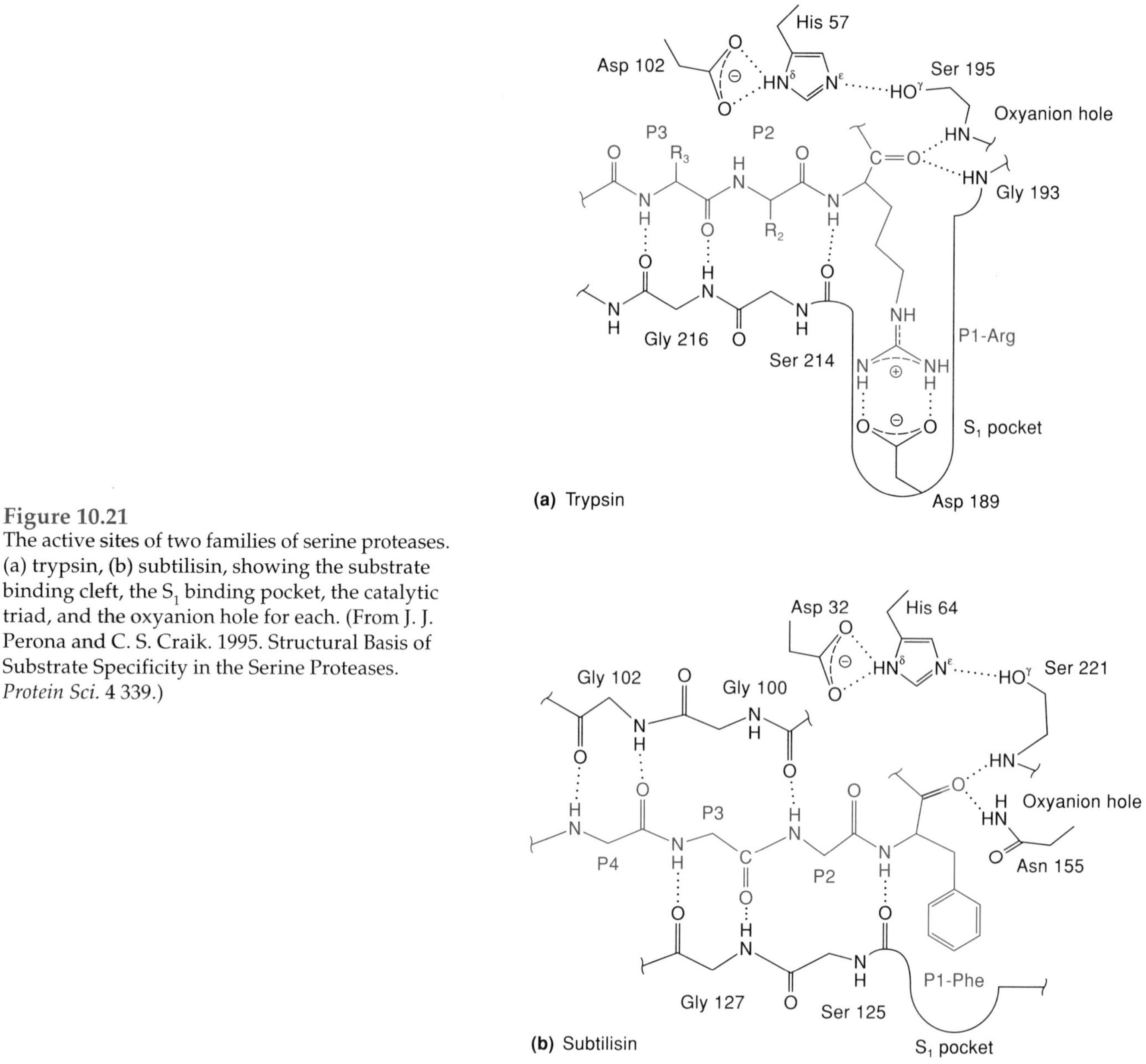

Figure 10.21
The active sites of two families of serine proteases. (a) trypsin, (b) subtilisin, showing the substrate binding cleft, the S_1 binding pocket, the catalytic triad, and the oxyanion hole for each. (From J. J. Perona and C. S. Craik. 1995. Structural Basis of Substrate Specificity in the Serine Proteases. *Protein Sci.* 4 339.)

residues by alanine decreases activity by a factor of 2×10^6. Intriguingly, substitution of either the histidine or serine has an essentially similar devastating effect on catalytic activity as substitution of all three residues. However, the Asp32Ala substitution, which leaves the serine-histidine dyad unchanged while removing the polarizing and orienting effect of the aspartate, yields an enzyme that is 100 times more active than the other mutants. Further evidence for the role of Asp 32 in orienting the imidazole ring of His 64 is that the presence of the carboxylate of Asp 32 in the His64Ala mutant actually hinders the formation of the negatively charged tetrahedral intermediate; k_{cat} for the double mutant His64Ala/Asp32Ala is 10 times higher than that for the single mutant His64Ala.

Histidine's role in the catalytic triad can be satisfied in an alternate and fascinating way: by a histidine residue at the P_2 position of the substrate. In a mutant such as His64Ala, the imidazole ring of a substrate with histidine at P_2 can adopt a position close to that occupied by the His 64 imidazole ring in the native enzyme. Consequently, the His46Ala mutant shows significantly greater activity when the substrate has histidine at P_2. Although substrates that are unable to provide a replacement histidine to the catalytic triad may bind equally well, they cannot assist in their own hydrolysis.

Oxyanion hole Enzymatic catalysis involves the cooperation of many factors, some of which may be very subtle. For example, although mutations such as Asp32Ala/His64Ala/Ser221Ala reduce activity by 2×10^6-fold, this protein, with all three residues of the catalytic triad converted to inert alanines, carries out amide hydrolysis almost 10^3-fold more efficiently than the nonenzymatic reaction at pH 7.5. Note that the oxyanion hole remains intact in the triple mutant and may account for the remaining activity. A test of this possibility would be difficult in the trypsin family because the polypeptide backbone provides both of the amides that form the oxyanion hole. This feature of catalysis can be tested more easily with a subtilisin, where Asn 155 provides one of the two oxyanion-binding amides. Replacing Asn 155 with Thr, Glu, Asp, or His results in somewhat less favorable binding of substrate (small increase in K_M) but significantly poorer catalysis (k_{cat} reduced by 200–4000-fold). Unexpectedly, however, residual activity is actually increased by replacing Asn 155 in the Ser221Ala mutant. Thus, for Ser221Ala, $k_{cat} = 3.4 \times 10^{-5}$ sec^{-1}, down by 6×10^6 from the wild type $k_{cat} = 5.9$ sec^{-1}. For Ser221Ala/Gln155Gly, $k_{cat} = 1.8 \times 10^{-4}$ down by about 3×10^{-5} from kcat for the wild type enzyme.

A mutant such as Ser221Ala, which lacks an essential aspect of catalysis like the serine hydroxyl and yet still shows appreciable activity, may operate by an entirely different mechanism than the native enzyme. This is likely to be the situation in the case just described. In the absence of a hydroxyl group from the enzyme, hydrolysis probably starts with a direct attack by water to yield a tetrahedral intermediate lacking any covalent bonds between enzyme and substrate. This intermediate could then collapse to give, directly, the free carboxylate and ammonium ion products of proteolysis. Moreover, a direct attack by water cannot come from the same direction as one by the serine hydroxyl; the methyl group in the Ser221Ala mutant will prevent solvent access from this side. Accordingly, the oxyanion of the tetrahedral intermediate will not project into the normal oxyanion hole of which Asn 155 is a part. Indeed, in the Ser221Ala mutant, Asn 155 might interfere sterically with an attack by water from a new direction, and removing the Asn 155 side chain, as in the double mutant Asn155Gly/Ser221Ala, might then ease the access of water to the scissile amide group.

Additional interactions involved in catalysis Structural analysis of subtilisin bound to substrate and to transition state analogues indicates that the P_1 residue penetrates approximately 1 Å farther into its binding site in the transition state than in the E·S complex, forming a new hydrogen bond between the P_1 amide nitrogen and the carbonyl oxygen of nearby Ser 125. Further evidence for involvement of P_1 binding in the reaction comes from comparing the effect of different P_1 residues on substrate binding (measured by K_M) and on catalysis (measured by k_{cat}). As the P_1 residue is altered, K_M changes up to 300-fold, whereas k_{cat} changes by 1700-fold. Additionally, extending substrates to include the $P_{2'}$ through P_4 residues leads to further increases in catalytic efficiency—not because of improved K_M but because of a 200-fold increase in k_{cat}. Over millennia, enzymes have evolved a great many tricks to enhance catalytic efficiency. The experiments described here illustrate how dominant contributions of a few groups to an enzyme's catalytic process can be dissected and how more subtle but substantial effects can be revealed.

f. Triose Phosphate Isomerase—A Perfect Enzyme

Triose phosphate isomerase (TIM) catalyzes the interconversion of R-glyceraldehyde-3-phosphate (GAP) and dihydroxyacetone phosphate (DHAP) in glycolysis. These molecules are the products of fructose-1,6-diphosphate hydrolysis, and only GAP can go forward in the metabolic pathway.

TIM presents an especially intriguing example of catalysis because the enzyme appears to have achieved evolutionary perfection. It accelerates the rate of isomerization by almost ten orders of magnitude (10^{10}) over the rate of simple base catalysis by, for example, an acetate ion. Furthermore, kinetic analysis of the individual steps reveals that the slowest, and therefore rate-limiting, step for the conversion of GAP to DHAP is the diffusion of GAP into the enzyme's active site; for the reverse conversion of DHAP to GAP, the rate-limiting step is the release of the GAP product. (The equilibrium between DHAP and GAP favors DHAP by a factor of about 20, so the DHAP to GAP conversion rate must be slower than the reverse process rate by this factor.)

Substrate binding by TIM As noted earlier, any enzymatic conversion begins with binding of substrate to the enzyme's active site. The phosphate groups on GAP and DHAP dominate binding, and changes in these groups (e.g., substitution by sulfate) can abolish binding. In contrast, the enzyme accepts a large variety of changes elsewhere in GAP or DHAP, and such substrate variants are vigorous competitive inhibitors. The enzyme binds the phosphate as the dianion, and hydrogen bonds are formed to four backbone amides (Gly 171, Ser 211, Gly 237, and Gly 233). In general, amide nitrogens are considered to have about a half unit of positive charge. In this case, two of the four amides probably possess somewhat more positive character because they are located at the positive end of short helices that are aimed straight at the phosphate (Figure 10.22). The

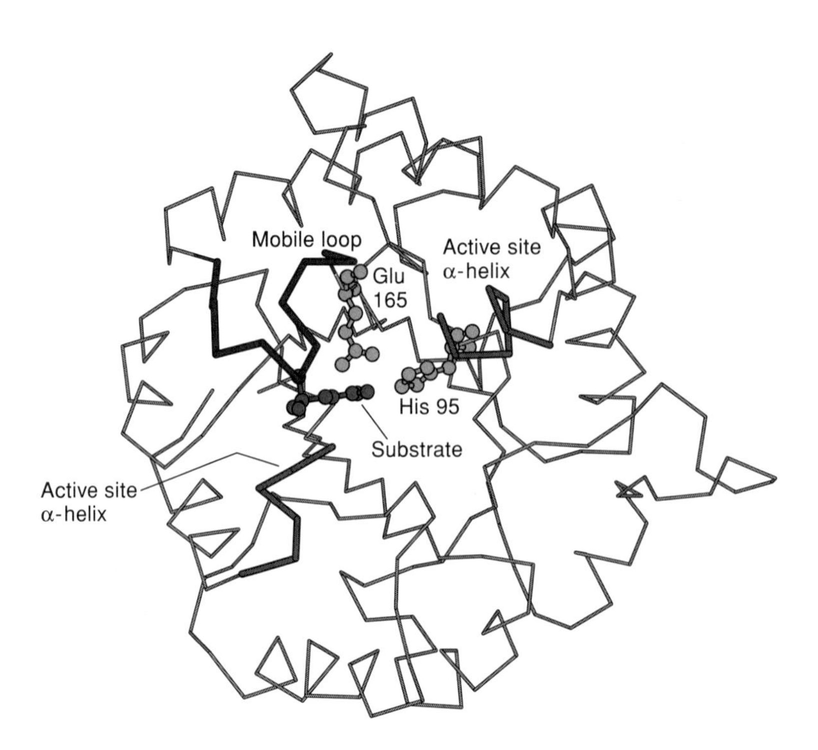

Figure 10.22
One subunit of triosephosphate isomerase showing the active site residues Glu 165 and His 95 (light color) The two α-helices aimed at the active site are shown in gray and the mobile loop that closes over the bound substrate (color) is black. (Adapted from J. R. Knowles. 1991. Enzyme Catalysis: Not Different, Just Better. *Nature* 350 123.)

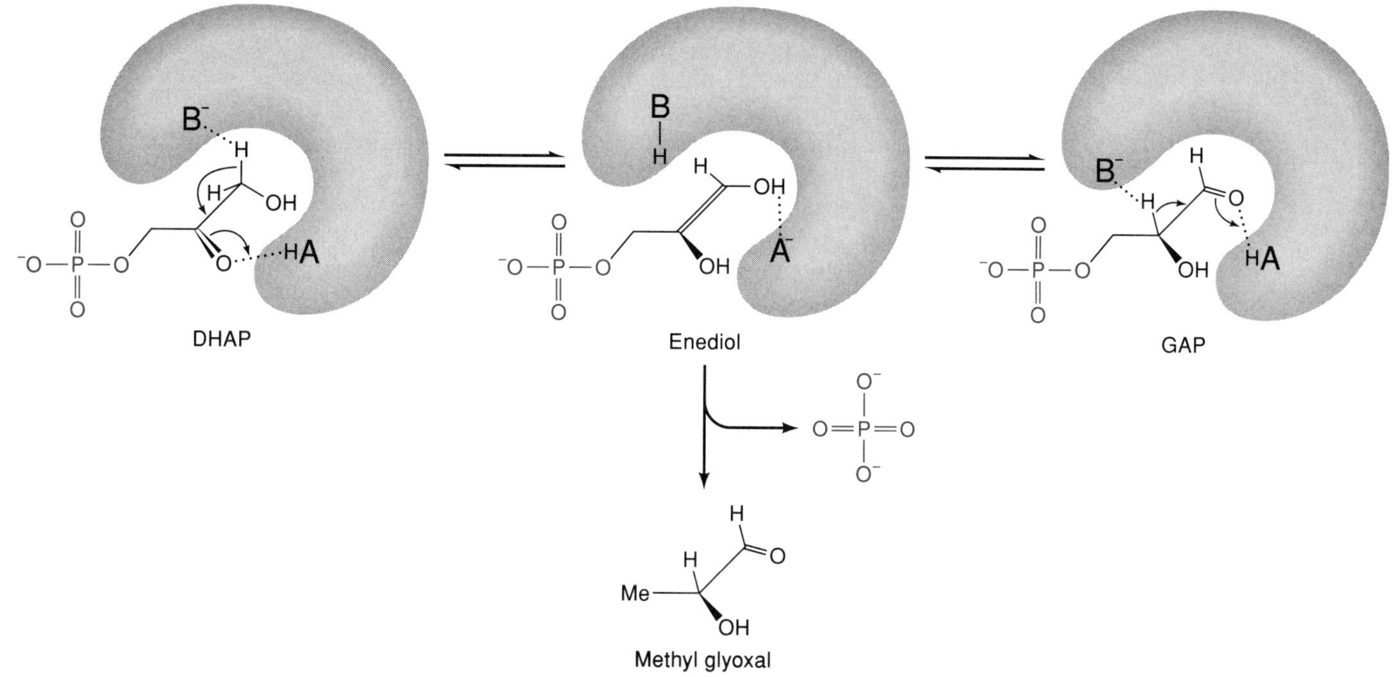

Figure 10.23
The reaction catalyzed by TIM, showing the proposed enediol intermediate and the methyl glyoxal elimination product produced by a TIM mutant missing four residues of the lid covering the substrate binding site. (Adapted from J. R. Knowles. 1991. Enzyme Catalysis: Not Different, Just Better. *Nature* 350 121.)

enzyme might have used positively charged residues such as arginine or lysine to bind and neutralize the phosphate dianion, but such side chains can have considerable flexibility. Consequently, the precise localization of the positive charge required for substrate binding might significantly decrease the entropy change on substrate binding (thereby disfavoring binding). In contrast, dependence on the backbone amides, which are inherently well ordered and relatively inflexible, appears to be more effective overall for this enzyme.

Comparison of crystal structures of TIM with and without bound substrate shows that a loop consisting of residues 166–176 moves in a concerted fashion to close like a lid over bound substrate, in accordance with the binding or dissociation step being rate limiting. A small section of this lid, residues 170–173, forms a looplet that snaps down to provide the final amide hydrogen bond (Gly 171) to the phosphate. Deletion of residues 170–173 causes very little distortion in the loop because the two ends of this looplet are only 5 Å apart; however, in the absence of the looplet, the loop cannot fully envelop the substrate. Although such a deletion causes only a slight increase in K_M, it has an almost catastrophic effect on k_{cat}, which falls by a factor of 10^5. More dramatically, the mutant enzyme converts only one in six substrate molecules to product; the other five are degraded because the enzyme cannot hold onto the highly unstable enediol intermediate. The enediol diffuses into solution, where phosphate anion elimination is actually the thermodynamically favored reaction (Figure 10.23). In fact, an essential role of TIM is to prevent this elimination; to this end, the enzyme tenaciously binds the enediol intermediate, thereby ensuring that its only two avenues of escape are as either DHAP or GAP. Besides preventing the unstable enediol intermediate's escape, the lid also completely excludes water from the enzyme-active site, further maximizing the effectiveness of electrostatic interactions by lowering the dielectric.

Catalytic residues Isomerization involves transferring a hydrogen from C-3 of DHAP to C-2 of GAP together with transferring protons between the C-3 and C-2 oxygens. Considerable understanding of the general mechanisms of such tautomerisms suggests that the process will be initiated by abstraction of a proton by a base. In TIM, Glu 165 serves as the base and is ideally situated just above the substrate, where it can receive a proton during either the forward or reverse process. Because glutamate is bidentate, each of its oxygens can be positioned close to one substrate carbon; actual distances are 2.8 Å or 3.4 Å from C-2 or C-3, respectively. Moreover, the positioning of the glutamate oxygens creates an ideal geometry because abstraction of the proton perpendicular to the plane of the enediol provides the most favorable stereoelectronic orientations for forming the enediolate intermediate. The importance of precise positioning of the carboxylate anion is apparent in the reduction of k_{cat} by a factor of 10^3 in the Glu165Asp mutant, although the carboxylate moves only 1 Å. Moreover, with this mutant, the overall identical tautomerism between DHAP and GAP follows quite a different detailed mechanism. As pointed out in the previous discussion of subtilisin variants such as Ser221Ala (Section 10.3c), an enzyme may respond to structural insults by adopting a different pathway to accomplish the same overall conversion.

Even with the ideal placement of the catalytic Glu 165 carboxylate, proton abstraction from substrate involves considerable further assistance from other features of the active site. The pK_a of free glutamic acid is near 4.5, that of Glu 165 is 6–7. But the likely pK_a for the substrate proton that must be removed in the first catalytic step is around 15. The enzyme must provide for the bound substrate an environment that increases the proton's acidity, thereby assisting abstraction. Thus, a nearby acidic group withdraws electrons from the carbonyl oxygen and weakens the β C-H bond. This function is performed by His 95, which is located essentially equidistant from the two substrate oxygens (2.8 Å and 2.9 Å from the oxygens at C-2 and C-3, respectively, to N_ε of His 95). The effect of this histidine can be measured by the change in the C-O stretching frequency when DHAP binds to TIM; the binding lowers stretching frequency by 19 cm^{-1}, indicating a weakening of the C-O bond of the carbonyl group. In mutants where this histidine is replaced (e.g., His95Gln and His95Asn), enzymatic activity falls by factors of 10^2 and 10^4, respectively, and the C-O bond polarization is lost.

Histidine is usually protonated under physiological conditions, but TIM's crystal structure suggests this may not be true for His 95. N^{15} NMR (nuclear magnetic resonance) provides a powerful measure of the state of protonation of the imadazole ring of histidine. To observe His 95 alone, researchers removed two other histidines from TIM; there were no resulting changes in either the structure or activity. Expression of this variant in histidine auxotrophs grown in the presence of ^{15}N-histidine allowed the unique introduction of a label into this single histidine. Subsequent NMR observations confirmed a much reduced pK_a for His 95 in the folded active protein (pK_a less than 4.5), in contrast to a pK_a of 6.5 in the unfolded protein; thus, the functional enzyme has a neutral His 95. Why the enzyme uses a neutral imadazole rather than a more acidic imidazolum cation remains unclear, but the presence of helix dipoles in folded TIM likely accounts for the altered pK_a of His 95. His 95 interacts with the positive end of a short α-helix that points directly toward it in the folded protein.

As in the case of the serine proteases, a complex interaction among many elements in TIM creates a biological catalyst. These include the participation of specific functional groups, the modulation of the properties of functional groups by secondary structure, the active embrace of substrate to isolate it from water, and the creation of an environment with a very different dielectric from that of solvent. Acting in concert, these features, together with others not discussed or not yet recognized, create a perfect enzyme.

g. Evolutionary Improvement of a Damaged Enzyme

To study the evolution of an enzyme like TIM, which catalyzes the reaction at the maximum achievable, diffusion-controlled rate, researchers have used the defective Glu165Asp mutant as a starting point to see how easily more active variants can be generated by mutations elsewhere in the molecule. Recall that the activity of the Glu165Asp mutant is reduced by 10^3 over that of the wild type.

Because the codons for Glu (*GA G/A*) differ from those for Asp (*GA C/T*) by only a single base change, reversion of the Glu165Asp mutant to wild type can readily occur. To direct random mutations uniformly to other regions of the enzyme, researchers used ten synthetic oligonucleotides containing known frequencies of base variations at each position ("spiked" oligos) as primers for *in vitro* mutagenesis. This led to plasmids containing structural genes for TIM that, on average, contain a single second mutation per plasmid in addition to the Glu165Asp mutation. Transformation of *E. coli* followed by screening to identify colonies that manifested higher TIM activity than that of the Glu165Asp mutant led to the isolation of six pseudorevertants (Figure 10.24). In all cases, the changes were localized to the region of the active site. The αβ-barrel of TIM forms a relatively rigid scaffold such that only changes near the active site are likely to produce significantly beneficial effects on activity. The observed second site changes only increased the activity of the Glu165Asp mutant; when introduced into wild-type enzyme, they either had no or a deleterious effect. (After all, one cannot improve perfection.)

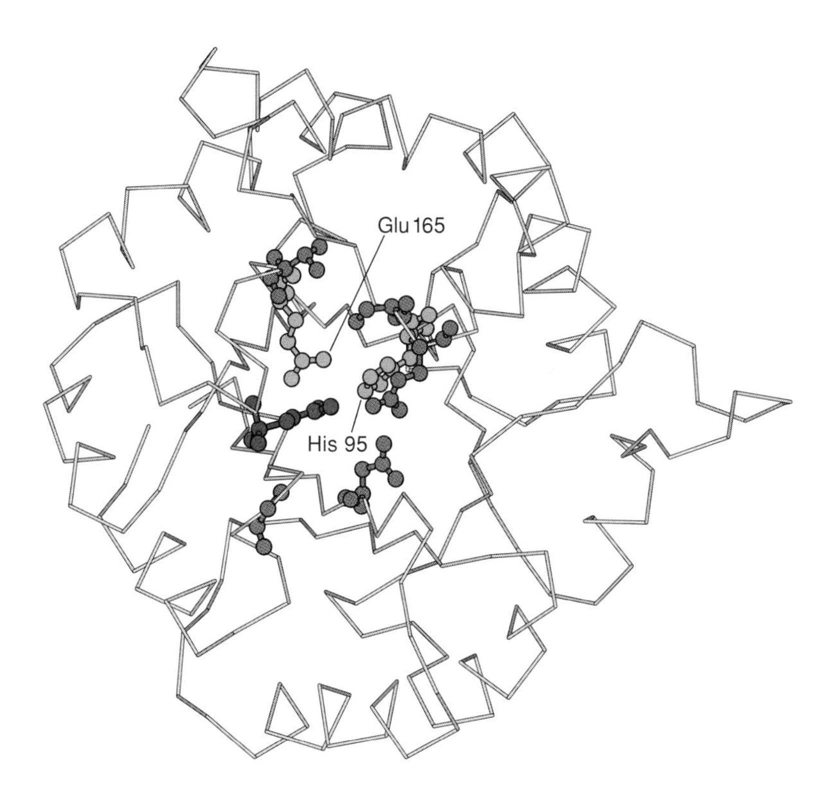

Figure 10.24
Amino acid changes that increase the efficiency of the Glu165Asp TIM mutant are shown in color; all are clustered near the active site residues, His 95 and Glu 165. The bound substrate is shown in darker color. (Adapted from J. D. Hermes et al. 1990. Searching Sequence Space by Definably Random Mutagenesis: Improving the Catalytic Potency of an Enzyme. *Proc. Natl. Acad. Sci. USA* 87 699.)

Against the Glu165Asp background, combinations of beneficial second mutations in different domains resulted in additive improvements; combinations of second site mutations in the same domain apparently interfered with each other and did not cooperatively improve activity. However, none of the single mutations or combinations restored the full activity of wild-type TIM. If that had occurred, it would have demonstrated that even within the constraint of the TIM structural skeleton, an ideal catalyst can be constructed in various ways. To achieve such a stunning result remains a challenge; the evolutionary conservation of the TIM catalytic apparatus suggests it will be very difficult if even possible.

Nonetheless, the ability to use mutagenic approaches to improve the activity of the defective Glu165Asp mutant lends hope to the goal of improving the activities of proteins that have not achieved evolutionary perfection. More importantly, using these techniques together with our ever-increasing insights into the origins of protein function will allow the creation of proteins with useful capabilities not presently available among the enormous range of known proteins.

h. Designing Enzymes with Novel Functions—Peptide Ligases

Modern techniques allow the creation of any protein composed of the 20 natural amino acids. Using mischarged tRNAs, investigators can synthesize only small amounts of proteins containing unnatural amino acids incorporated at specific sites. Overcoming this limitation would enormously expand the vocabulary of possible polypeptide structures. In principle, chemical synthesis could be used. However, conventional stepwise solid phase protein synthesis is a major challenge for peptides over about 30 amino acids. Although polypeptides containing as many as 120 amino acids have been prepared, purified, and folded into fully functional proteins, each such success represents a major achievement and requires great effort. In the analogous syntheses of long DNA chains, DNA ligases are used to join short pieces of chemically synthesized DNA. Such ligases occur naturally and function in the repair of damaged DNA and in DNA replication.

No analogous peptide ligases are known to occur naturally. Damaged proteins are not repaired but rather destroyed. Moreover, protein biosynthesis is a strictly stepwise process not involving the linkage of preformed peptides. However, the availability of peptide ligases would allow the construction of protein variants containing a limitless range of variations. Short peptides could be chemically synthesized and then joined by ligation under mild conditions.

Making a protease run backwards Many proteases exist. To alter them to catalyze the reverse reaction and act as efficient peptide ligases, the acyl enzyme intermediate should prefer attack by the N terminus of a polypeptide fragment with resultant formation of a new amide bond, rather than attack by water with resultant hydrolysis (Figure 10.25). Even with wild-type proteases, an exogenous amine can compete with water for the acyl enzyme, albeit with low efficiency. Suitable modification of a protease might reverse this preference and create a peptide ligase. For general synthetic applicability, the ligase should be able to operate on a relatively broad range of substrates. Accordingly, a protease such as subtilisin, with broad substrate specificity, provides a promising progenitor for the design of a ligase.

Thioesters are inherently more kinetically labile to aminolysis than to hydrolysis. Accordingly, a Ser221Cys change in subtilisin results in just over a 5×10^3 shift from preference for hydrolysis to attack on the thioacyl enzyme by an amine; the ratio of aminolysis to hydrolysis is 5.6×10^{-2} with native subtilisin and 3×10^2 with the Ser221Cys mutant. Unfortunately, the Ser221Cys mutation also drastically lowers the overall catalytic efficiency of the enzyme, probably because of steric overcrowding in the active site by the large sulfur. Previous

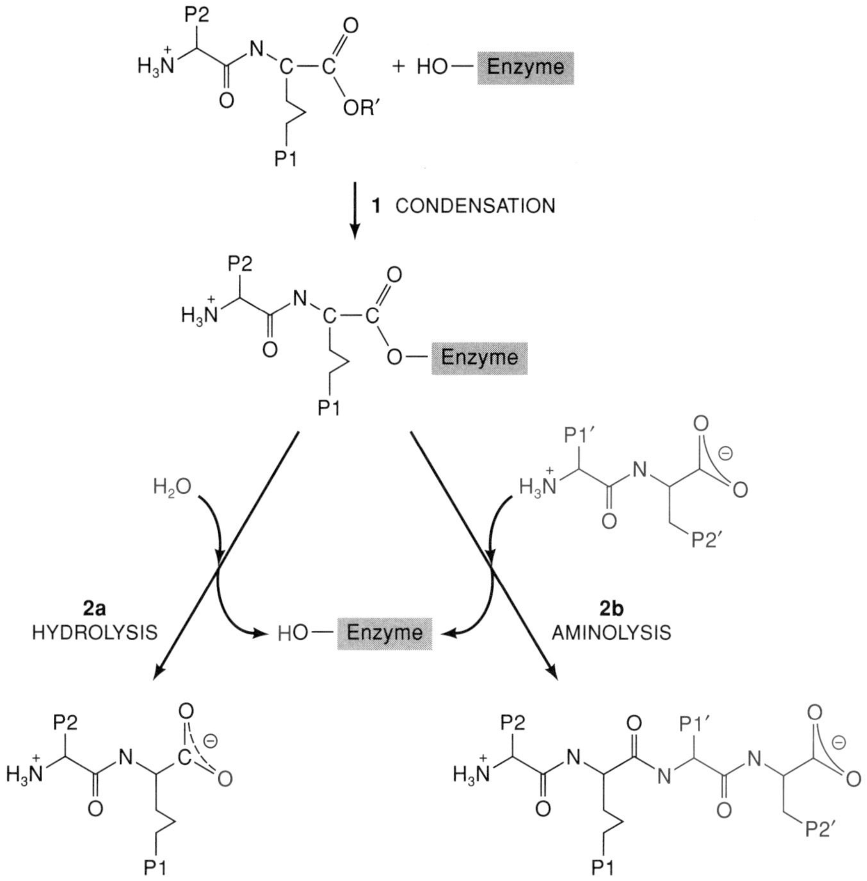

Figure 10.25
Serine protease as protein ligase. The first step in using a protein ligase is to form an acyl-enzyme intermediate with the N terminal fragment of the new protein covalently attached to subtilisin (1, condensation). The covalent intermediate can be cleaved unproductively by water (2a, hydrolysis), as in normal proteases, or by an attacking amino group from the peptide destined to become the C terminal fragment of a new protein (2b, aminolysis).

studies showed that a Pro225Ala mutation moves the helix containing the active site nucleophile away from the oxyanion hole and the catalytic His 64. Indeed, the double mutant Ser221Cys/Pro225Ala is ten times more efficient as a catalyst than the single Ser221Cys mutant and still retains 10 percent of the preference for aminolysis over hydrolysis of the thioacyl intermediate (a 500-fold increase in this ratio over that for the wild-type enzyme). Moreover, the greater enzymatic activity of the Ser221Cys/Pro225Ala mutant allows use of a less highly activated ester derivative of the peptide that will become the N terminal region of the newly formed amide bond; this significantly reduces problems associated with the inherent hydrolytic instability of more highly activated esters.

Coopting specificity mutants Knowledge about subtilisin suggests additional approaches for improving peptide ligase function. Up to seven residues spanning the scissile amide bond bind to native subtilisin in an extended β-sheet conformation, residues P_4 P_3 P_2 and P_1 preceding the scissile bond and $P_{1'}$, $P_{2'}$, and $P_{3'}$ following the scissile bond. The enzyme's preference for certain amino acids at P_1 largely determines the register in which an extended peptide binds. Residues

$P_{1'}$ through $P_{3'}$ also appreciably enhance catalytic efficiency. Thus, extending the ester substrate (which becomes the N terminal fragment of the new polypeptide) from peptide glycolate amide to peptide glycolate-Phe-Gly-NH_2 leads to a tenfold increase in efficiency (k_{cat}/K_M), primarily because of more favorable substrate binding (a decrease in K_M).

Although wild-type subtilisin has a relatively broad specificity, it nevertheless manifests certain preferences for the P_1 side chains. Variants that affect this P_1 specificity can be combined with the Ser221Cys/Pro225Ala double mutation to increase the range of N terminal fragments that can be ligated by subtilisin mutants. The wild-type enzyme works best when the P_1 site of the ester peptide is Tyr, Phe, Met, or Leu. The Gly166Ile mutant increases ligase activity for peptide esters with smaller side chains such as Ala in the P_1 site. The Gly166Asp mutant promotes ligations with lysine or arginine at P_1, and the Glu156Gln/Gly166Lys mutation in the pocket gives an enzyme that is effective with both Glu and Phe in P_1.

The efficiencies with which new peptide bonds can be created by coupling of peptide fragments catalyzed by a particular peptide ligase can vary greatly, depending on the amino acid residues on both sides of the bond to be formed. Understanding these preferences increases enormously the ability to design maximally efficient synthetic strategies. The sequence preferences of subtiligase for the first three residues of the C terminal peptide (the $P_{1'}$, $P_{2'}$, and $P_{3'}$ residues) have been explored by ligation of a single labelled reporter N terminal fragment either to a mixed panel of synthetic tripeptides or to a pool of substrate proteins. The products are then analyzed to identify those C terminal fragments whose couplings have been favored. The amino acid at $P_{1'}$ showed the greatest influence on ligation efficiency; the enzyme favored coupling of peptides containing Met, Phe, Lys, Leu, Arg, Ala, Ser, and His. At $P_{2'}$, the enzyme tolerated most amino acids, though Cys, Pro, or Gly were generally poor ligation substrates.

Secondary structure in either of the ligation partners can also greatly affect coupling efficiencies. For example, human growth hormone lacking its eight N terminal residues (Δ8hGH) has a favorable N terminal $P_{1'}$–$P_{3'}$ sequence (Leu-Phe-Asp) but nevertheless does not ligate efficiently. In hGH, helix 1 begins at residue 6; thus, a likely reason for the poor ligation is that the N terminal region of Δ8hGH exists as an α-helix and does not easily adopt the extended β-sheet conformation necessary to bind effectively in the cleft of the enzyme. To disrupt secondary structure of the substrate, one can add chaotropic agents such as guanidinium hydrochloride. But such agents also partially denature the enzyme itself; indeed, ligation efficiency of subtiligase operating on a good substrate drops by 50 percent in 1M guanidinium hydrochloride. To circumvent this problem, one can use an enzyme that has been altered to have enhanced structural stability as well as to be a good ligase. Mutations that protect subtilisin from denaturation have been created by random mutagenesis. Combination of five of these mutations (Met50Phe/Asn76Asp/Asn109Ser/Lys213Arg/Asn218Ser) into subtiligase creates "stabiligase," which retains more than 50 percent activity in 3M guanidinium hydrochloride. In fact, stabiligase is three times more effective than subtiligase in ligating a test sequence onto Δ8hGH in the presence of another chaotropic agent, 0.1 percent sodium dodecyl sulfate (SDS).

As a demonstration of the power of the peptide ligase presently available—and improvements are certain to follow—stabiligase has been employed in the chemical-biochemical synthesis of the 124 amino acid protein ribonuclease A from six synthetic peptide fragments. Because this approach does not restrict one to the 20 natural amino acids, RNase A variants containing 4-fluorohistidine at either or both histidine sites putatively involved directly in catalysis (His 12 and His 119) could be prepared and their rates of catalysis determined as a function of pH. The results indicated clearly that the state of protonation of these two histidines, as distinct from their inherent basicities, crucially influence the two catalytic steps in hydrolysis of RNA catalyzed by RNase A.

The creation of designer enzymes for specific novel and useful synthetic tasks is still in its infancy, but, as demonstrated in the development of peptide ligases, the approach holds enormous promise. Extensive trial-and-error attempts together with predictable, rational alterations in the structure of a native enzyme will be needed to generate new activities. A bootstrap strategy in which mutant enzymes composed of natural amino acids are used to create enzymes that in turn allow the synthesis of more effective enzymes containing regions of nonnatural subunits can be envisioned. The latter may be more effective than the former for the synthesis of themselves as well as to generate further variants. These approaches can lead to the creation of a wide range of complex molecules, perhaps not entirely polyamide in composition, that act as very specific molecular machines. Only imagination and cleverness limit the opportunities.

Further Reading

10.1

R. Wetzel. 1986. What is Protein Engineering? *Protein Eng.* 1 3–6.

R. J. Leatherbarrow and A. R. Fersht. 1986. Protein Engineering. *Protein Eng.* 1 7–16.

K. A. Dill. 1987. Protein Surgery. *Protein Eng.* 1 369–372.

S. J. Anthony-Cahill, M. C. Griffith, C. J. Noren, D. J. Suich, and P. G. Schultz. 1989. Site-Specific Mutagenesis with Unnatural Amino Acids. *Trends Brochem. Sci* 14 400–403.

K. A. Johnson and S. J. Benkovic. 1990. Analysis of Protein Function by Mutagenesis. *The Enzymes* 19 159–211.

10.2

T. Alber, S. Dao-Pin, K. Wilson, J. A. Wozniak, S. P. Cook, and B. W. Matthews. 1987. Contributions of Hydrogen Bonds of Thr 157 to the Thermodynamic Stability of Phage T4 Lysozyme. *Nature* 330 41–46.

B. W. Matthews, H. Nicholson, and W. J. Becktel. 1987. Enhanced Protein Thermostability from Site-Directed Mutations That Decrease the Entropy of Unfolding. *Proc. Natl. Acad. Sci. USA* 84 6663–6667.

K. R. Shoemaker, P. S. Kim, E. J. York, J. M. Stewart, and R. L. Baldwin. 1987. Tests of the Helix Dipole Model for Stabilization of α-Helices. *Nature* 326 563–567.

M. Matsumura, W. J. Becktel, and B. W. Matthews. 1988. Hydrophobic Stabilization in T4 Lysozyme Determined Directly by Multiple Substitutions of Ile 3. *Nature* 334 406–410.

T. Alber. 1989. Mutational Effects on Protein Stability. *Annu. Rev. Biochem.* 58 765–798.

M. Karpusas, W. A. Baase, M. Matsumura, and B. W. Matthews. 1989. Hydrophobic Packing in T4 Lysozyme Probed by Cavity-Filling Mutants. *Proc. Natl. Acad. Sci. USA* 86 8237–8241.

W. A. Lim and R. T. Sauer. 1989. Alternative Packing Arrangements in the Hydrophobic Core of λ Repressor. *Nature* 339 31–36.

M. Matsumura, W. J. Becktel, M. Levittt, and B. W. Matthews. 1989. Stabilization of Phage T4 Lysozyme by Engineered Disulfide Bonds. *Proc. Natl. Acad. Sci. USA* 86 6562–6566.

M. Matsumura and B. W. Matthews. 1989. Control of Enzyme Activity by an Engineered Disulfide Bond. *Science* 243 792–794.

M. Matsumura, G. Signor, and B. W. Matthews. 1989. Substantial Increase of Protein Stability by Multiple Disulphide Bonds. *Nature* 342 291–293.

C. Mitchinson and J. A. Wells. 1989. Protein Engineering of Disulfide Bonds in Subtilisin BPN′. *Biochem.* 28 4807–4815.

L. Serrano and A. R. Fersht. 1989. Capping and α-Helix Stability. *Nature* 342 296–299.

J. A. Bell, S. Dao-Pin, R. Faber, R. Jacobson, M. Karpusas, M. Matsumura, H. Nicholson, P. E. Pjura, D. E. Tronrud, L. H. Weaver, K. P. Wilson, J. A. Wozniak, X.-J. Zhang, T. Alber, and B. W. Matthews. 1990. Approaches Toward the Design of Proteins of Enhanced Thermostability. In W. G. Laver and G. M. Air (eds.), Use of X-Ray Crystallography in the Design of Antiviral Agents, pp. 233–245. Academic Press, San Diego.

J. U. Bowie, J. F. Reidhaar-Olson, W. A. Lim, and R. T. Sauer. 1990. Deciphering the Message in Protein Sequences: Tolerance to Amino Acid Substitutions. *Science* 247 1306–1310.

Z. Grabarek, R.-Y. Tan, J. Wang, T. Tao, and J. Gergely. 1990. Inhibition of Mutant Troponin C Activity by an Intra-Domain Disulphide Bond. *Nature* 345 132–135.

B. Katz and A. A, Kossiakoff. 1990. Crystal Structures of Subtilisin BPN′. Variants Containing Disulfide Bonds and Cavities: Concerted Structural Rearrangements Induced by Mutagenesis. *Proteins* 7 343–357.

T. Kiefhaber, H.-P. Grunert, U. Hahn, and F. X. Schmid. 1990. Replacement of a *Cis* Proline Simplifies the Mechanism of Ribonuclease T1 Folding. *Biochem.* 29 6475–6480.

S. Dao-Pin, U. Sauer, H. Nicholson, and B. W. Matthews. 1991. Contributions of Engineered Surface Salt Bridges to the Stability of T4 Lysozyme Determined by Directed Mutagenesis. *Biochem.* 30 7142–7153.

A. R. Fersht, M. Bycroft, A. Horovitz, J. T. Kellis Jr., A. Matouschek, and L. Serrano. 1991. Pathway and Stability of Protein Folding. *Phil. Trans. R. Soc. Lond.* B 332 171–176.

A. Mitraki, B. Fane, C. Haase-Pettingell, J. Sturtevant, and J. King. 1991. Global Suppression of Protein Folding Defects and Inclusion Body Formation. *Science* 253 54–58.

H. Nicholson, D. E. Anderson, S. Dao-Pin, and B. W. Matthews. 1991. Analysis of the Interaction Between Charged Side Chains and the α-Helix Dipole Using Designed Thermostable Mutants of Phage T4 Lysozyme. *Biochem.* 30 9816–9828.

J. A. Ellman, D. Mendel, and P. G. Schultz. 1992. Site Specific Incorporation of Novel Backbone Structures into Proteins. *Science* 255 197–199.

T. Kiefhaber, H.-P. Grunert, U. Hahn, and F. X. Schmid. 1992. Folding of RNase T1 Is Decelerated by a Specific Teriary Contact in a Folding Intermediate. *Proteins* 12 171–179.

B. A. Shirley, P. Stanssens, U. Hahn, and C. N. Pace. 1992. Contribution of Hydrogen Bonding to the Conformational Stability of Ribonuclease T1. *Biochem.* 31 725–732.

10.3

R. M. Stroud. 1974. A Family of Protein-Cutting Proteins. *Scientific American* July 74–88.

P. Bryan, M. W. Pantoliano, S. G. Quill, H.-Y. Hsiao, and T. Poulos. 1986. Site-Directed Mutagenesis and the Role of the Oxyanion Hole in Subtilisin. *Proc. Natl. Acad. Sci. USA* 83 3743–3745.

D. A. Estell, T. P. Graycar, J. V. Miller, D. B. Powers, J. P. Burnier, P. G. Ng, and J. A. Wells. 1986. Probing Steric and Hydrophobic Effects on Enzyme-Substrate Interactions by Protein Engineering. *Science* 233 659–663.

T. C. Alber, R. C. Davenport Jr., D. A. Giammona, E. Lolis, G. A. Petsko, and D. Ringe. 1987. Crystalography and Site-Directed Mutagenesis of Yeast Triosephosphate Isomerase: What Can We Learn About Catalysis from a "Simple" Enzyme? *Cold Spring Harbor Symp. Quant. Biol.* 52 603–613.

P. Carter and J. A. Wells. 1987. Engineering Enzyme Specificity by "Substrate-Assisted Catalysis." *Science* 237 394–399.

L. Gráf, C. S. Craik. A. Patthy, S. Roczniak, R. J. Fletterick, and W. J. Rutter. 1987. Selective Alteration of Substrate Specificity by Replacement of Aspartic Acid-189 with Lysine in the Binding Pocket of Trypsin. *Biochem.* 26 2616–2623.

J. R. Knowles. 1987. Tinkering with Enzymes: What Are We Learning? *Science* 236 1252–1258.

J. A. Wells, B. C. Cunningham, T. P. Graycar, and D. A. Estell. 1987. Recruitment of Substrate-Specificity Properties from One Enzyme into a Related One by Protein Engineering. *Proc. Natl. Acad. Sci. USA* 84 5167–5171.

J. A. Wells, D. B. Powers, R. R. Bott, T. P. Graycar, and D. A. Estell. 1987. Designing Substrate Specificity by Protein Engineering of Electrostatic Interactions. *Proc. Natl. Acad. Sci. USA* 84 1219–1223.

P. Carter and J. A. Wells. 1988. Dissecting the Catalytic Triad of a Serine Protease. *Nature* 332 564–568.

S. R. Sprang, R. J. Fletterick, L. Gráf, W. J. Rutter, and C. S. Craik. 1988. Studies of Specificity and Catalysis in Trypsin by Structural Analysis of Site-Directed Mutants. *CRC Crit. Rev. Biochem.* 8 225–236.

J. A. Wells and D. A. Estell. 1988. Subtilisin—An Enzyme Designed To Be Engineered. *Trends Biochem. Sci.* 13 291–297.

P. Carter and J. A. Wells. 1990. Functional Interaction Among Residues in Subtilisin BPN′. *Proteins* 7 335–342.

L. B. Evnin, J. R. Vásquez, and C. S. Craik. 1990. Substrate Specificity of Trypsin Investigated by Using a Genetic Selection. *Proc. Natl. Acad. Sci. USA* 87 6659–6663.

J. D. Hermes, S. C. Blacklow, and J. R. Knowles. 1990. Searching Sequence Space by Definably Random Mutagenesis: Improving the Catalytic Potency of an Enzyme. *Proc. Natl. Acad. Sci. USA* 87 696–700.

L. Abrahmsén, J. Tom, J. Burnier, K. A. Butcher, A. Kossiakoff, and J. A. Wells. 1991. Engineering Subtilisin and Its Substrates for Efficient Ligation of Peptide Bonds in Aqueous Solution. *Biochem.* 30 4151–4159.

J. R. Knowles. 1991. Enzyme Catalysis: Not Different, Just Better. *Nature* 350 121–124.

J. R. Knowles. 1991. To Build an Enzyme. *Phil. Trans. R. Soc. Lond.* B 332 115–121.

L. Hedstrom, L. Szilagyi, and W. J. Rutter. 1992. Converting Trypsin to Chymotrypsin: The Role of Surface Loops. *Science* 255 1249–1253.

T. K. Chang, D. Y. Jackson, J. P. Burnier, and J. A. Wells. 1994. Subtiligase: A Tool for Semisynthesis of Proteins. *Proc. Natl. Acad. Sci.* USA 91 12544–12548.

D. Y. Jackson, J. Burnier, C. Quan, M. Stanley, J. Tom, and J. A. Wells. 1994. A Designed Peptide Ligase for Total Synthesis of Ribonuclease A with Unnatural Catalytic Residues. *Science* 266 243–247.

S. Steinbacher, R. Seckler, S. Miller, B. Steipe, R. Huber, and P. Reinemer. 1994. Crystal Structure of P22 Tailspike Protein: Interdigitated Subunits in a Thermostable Trimer. *Science* 265 383–386.

J. J. Perona and C. S. Craik. 1995. Structural Basis of Substrate Specificity in the Serine Proteases. *Protein Science* 4 337–360.

CHAPTER 11

Genetic Modification of Animals

TIM STEWART

11.1. Genetically Modified Rodents as an Experimental Tool

a. History

The initial structural studies on cloned DNA and RNA (as cDNA) provided the impetus and resources for research into how eukaryotic genes are regulated. The development of the calcium phosphate precipitation method and, subsequently, other methods for the introduction of purified DNA into cells grown in tissue culture flasks (*Genes and Genomes,* section 5.7a) provided ways to assess the function of cloned DNAs. These experiments indicated that genes could be manipulated *in vitro* and introduced into mammalian cells, where they would be transcribed and the mRNA produced would be translated. These advances stimulated an exploration of the feasibility of introducing new genes not into cells in culture, but into the germ line of mice. If possible, this would allow an analysis of gene regulation during development and would also permit the study of how the misexpression of certain proteins affects a whole animal.

Two developments opened the way to the introduction of new genes into the germ line of an experimental mammal: first, DNA that was microinjected into the nuclei of eukaryotic cells grown in culture was shown to be transcribed; second, RNA and proteins could be injected into the cytoplasm of fertilized mouse eggs, where they were functional. Further encouragement was provided by the finding that exogenous viral sequences could be integrated into the mouse germ line by exposing preimplantation embryos to either purified viral DNA or replication-competent retroviruses. In 1980 and 1981, five groups independently succeeded in injecting recombinant DNA into the pronuclei of newly fertilized mouse eggs and generating **transgenic** mice, that is, mice whose entire complement of cells contained the newly introduced gene at the same chromosomal location.

The ability to create transgenic mice in which a particular cell type expresses a specific, newly introduced gene has greatly increased the understanding of mammalian biology. However, as was well appreciated by geneticists studying prokaryotes and yeast, this approach needs to be complemented by the analysis of mice in which either a particular cell type or the whole animal is unable to express a specific gene. The ability to produce such animals depends on two important technologies: first, the ability to grow totipotent stem cell lines in laboratory culture and, second, the ability of transfected DNA to integrate into a mammalian cell chromosome by homologous recombination. In this way, transgenic mice in which a variety of specific genes is structurally disrupted can be generated. The study of transgenic mice with newly acquired transgenes and of mice with modifications of specific endogenous genes (so-called **knockout mice**) has provided a number of intriguing insights into whole animal biology.

b. Techniques

Untargeted transgenic mice The method by which new genetic material is introduced into the germ line of mice, that is, the production of transgenic mice, was described in *Genes and Genomes,* figure IV.15, and is outlined in Figure 11.1. Briefly, fertilized eggs are collected from a female donor approximately 12 hours after mating. At this stage, the two pronuclei (a maternal pronucleus from the egg and a paternal pronucleus from the incoming sperm) have begun to enlarge and are easily visible using differential interference optics. The fertilized eggs are positioned and held by a blunt holding pipette using gentle back pressure, and a solution of the appropriate DNA (usually at 1 ng/μl) is injected into one of the two pronuclei. The eggs that survive the injection are then transferred into the infundibulum of a female mouse that had been mated the previous night to a vasectomized male. The mice that result from the injected eggs can be screened for the presence of the transgene by DNA blotting techniques (*Genes and Genomes,* section 6.1b) using DNA extracted from a biopsy of the tail.

Figure 11.1 (*facing page*)
Generation of transgenic mice. Fertilized eggs are obtained from donor females approximately 12 hours after fertilization. At this stage, the pronuclei are visible and can be directly microinjected with a DNA solution. The injected eggs are transferred to a pseudopregnant foster female where they develop to term. At 4 weeks of age the pups are weaned and DNA is extracted from a tail biopsy. Analysis of this DNA by Southern blot hybridization reveals the presence of the transgene.

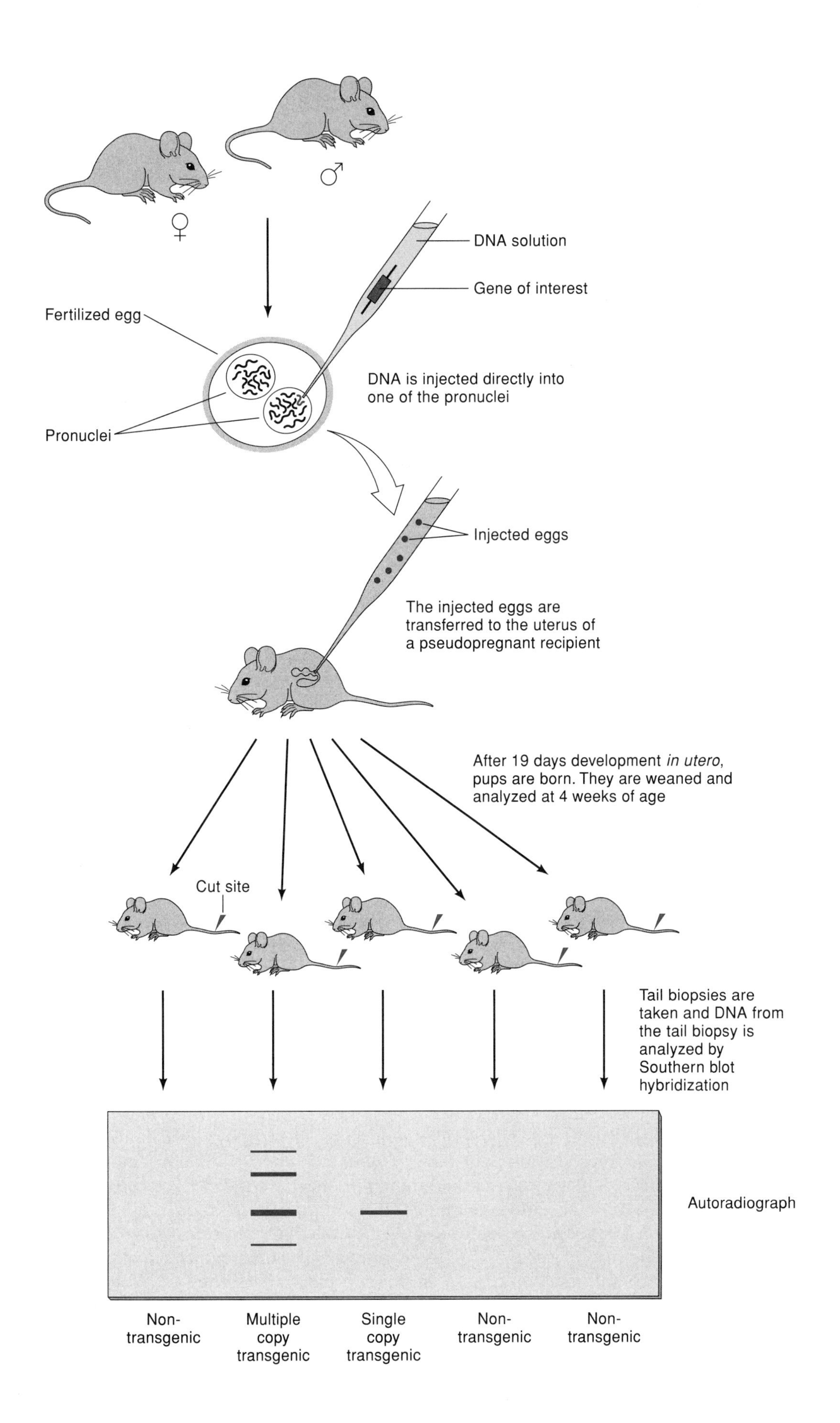

♀
♂
DNA solution
Gene of interest
Fertilized egg
Pronuclei
DNA is injected directly into one of the pronuclei
Injected eggs
The injected eggs are transferred to the uterus of a pseudopregnant recipient
After 19 days development *in utero*, pups are born. They are weaned and analyzed at 4 weeks of age
Cut site
Tail biopsies are taken and DNA from the tail biopsy is analyzed by Southern blot hybridization
Autoradiograph
Non-transgenic
Multiple copy transgenic
Single copy transgenic
Non-transgenic
Non-transgenic

The transgenic mice that develop from the injected zygote are called "founders" because they are used as the foundation stock for new lines of mice. They have one or more copies of the injected DNA integrated into the mouse genome usually linked in a tandem array. The chromosomal site of integration is presumed to be random, although there is little direct evidence to support this assumption; more to the point, the integrated DNA can appear at a very large number of different chromosomal sites. Although the DNA is injected into the one-cell zygote, the developmental stage at which integration into the mouse genome occurs must be later. As a consequence, the founder transgenic mouse can be a mosaic of transgenic and nontransgenic cells. This mosaicism is usually revealed by the founder having less than the expected 50 percent transgenic offspring. The first generation mice that do inherit the transgene are hemizygous for the transgenic locus, and, with few exceptions, this will be transmitted as a normal mendelian locus.

Expression of the transgene will depend on the sequence of the transgene and, to some extent, on where the transgene integrates into the mouse genome. The consequences for the animal will depend on the protein encoded and the level and site of expression. The ability to alter one or all of these aspects provides the opportunity to analyze gene regulation and protein function in a whole animal.

Targeted transgenic mice The ability to produce mice in which a particular endogenous gene is modified relies on the availability of totipotent stem cell lines that can grow in laboratory culture and can populate an embryo's germ line. Moreover, it is important to be able to identify those cells in which transfected DNA has integrated into a chromosome by homologous recombination.

The original totipotent stem cell lines were teratocarcinoma cells derived from spontaneous or induced teratomas. When grown as a solid tumor, these undifferentiated cell lines are able to differentiate into a wide variety of cell types. More important, if the cells are injected into a preimplantation embryo, the teratocarcinoma cells are able to contribute extensively to the developing fetus. In such experiments, the mice born subsequent to the injection of the teratocarcinoma cells into the blastocyst are normal. Furthermore, they are mosaic (i.e., composed of two different types of cells: one type derived from the recipient blastocyst and the other type derived from the injected teratocarcinoma cells). Not only are the teratocarcinoma cells able to contribute to all the somatic tissues; they are also able to differentiate into normal germ cells. In this way, genes present in the laboratory-grown teratocarcinoma cells are able to be incorporated into, and transmitted through, the germ line. Although the teratocarcinoma cells are able to differentiate into normal germ cells once incorporated into the developing embryo, it is difficult to maintain for prolonged periods this totipotent characteristic in teratocarcinoma cells grown in laboratory culture. This instability of the teratocarcinoma cells frustrated attempts to create mutant animals by mutating the cells in specific ways prior to injection into a blastocyst. Subsequently, totipotent cell lines were derived from normal preimplantation embryos. These cells, termed embryonic stem (ES) cells, colonize the germ line more reproducibly when introduced into preimplantation embryos and retain this ability even after undergoing prolonged periods in laboratory culture.

An important development that led to the generation of mice with modifications of specific chromosomal genes was the finding that DNA transfected into mammalian cells in culture can undergo homologous recombination with a chromosomal homologue. In this process, a transfected DNA segment interacts with its endogenous homologue, presumably by base pairing, resulting in either replacement of the endogenous DNA sequences by those present in the transfected DNA or integration of the transfected DNA at that site. If the transfected DNA contains a base change or a deletion compared to the endogenous gene, this mutation will become a permanent part of the ES cell's genome (Figure 11.2).

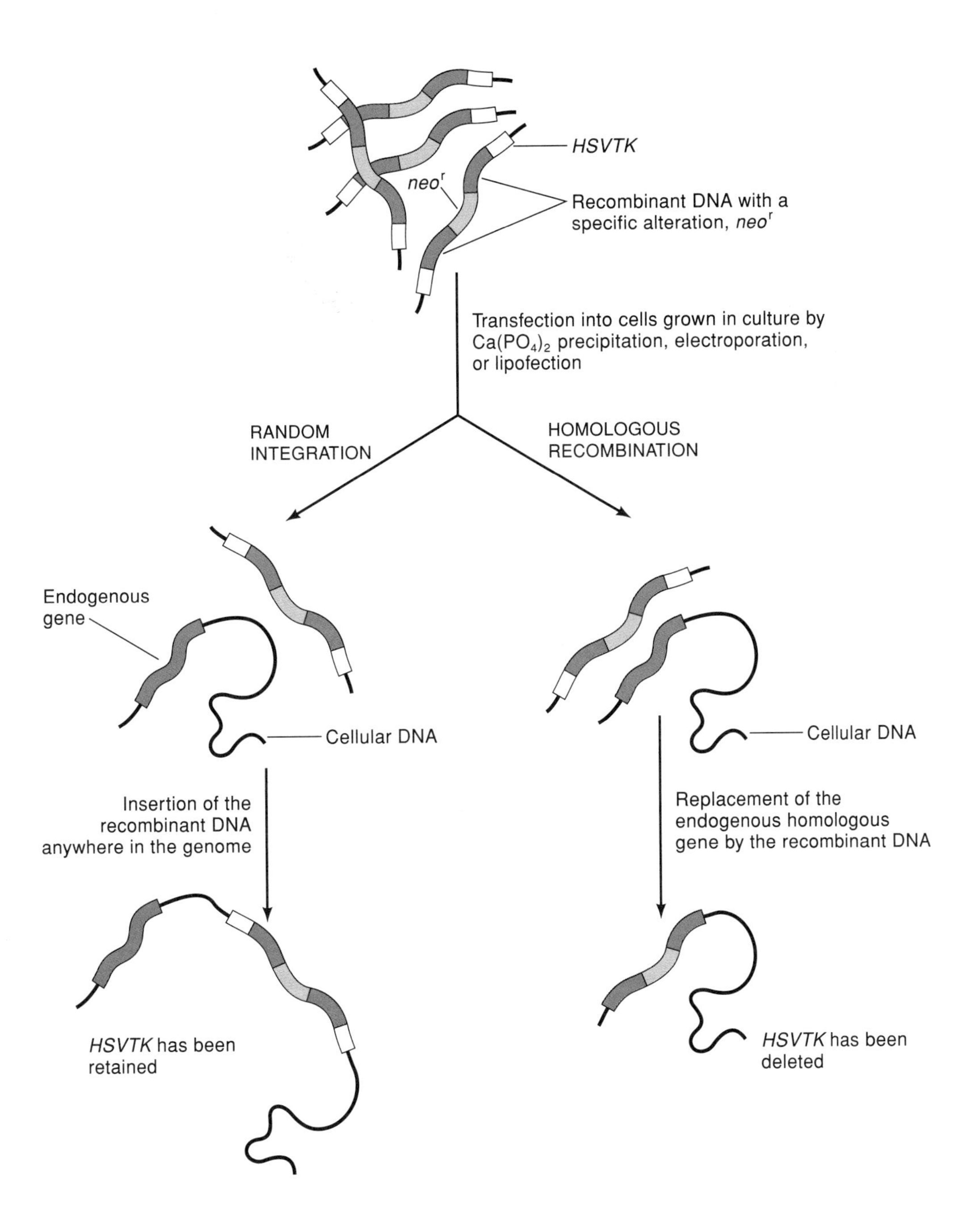

Figure 11.2
Homologous recombination of DNA transfected into cells. DNA that is transfected into cells grown in culture can integrate randomly into a chromosome or it can replace the endogenous sequence with those sequences present in the transfected DNA (homologous recombination). When homologous recombination occurs, regions of nonhomology at the ends of the DNA molecule (for example, the *HSVTK*) are removed. However, if random insertion occurs, the entire DNA molecule is retained.

Although the frequency of homologous recombination is low compared to the considerably higher frequency of nonhomologous recombinations, it is high enough to detect both the replacement and integration events. Thus, a cloned gene can be modified by recombinant DNA technology *in vitro* and introduced into cultured ES cells. Homologous recombination between the transfected mutant gene and its cellular homologue produces cells with a mutation in one of the chromosomal alleles. Cell clones with the specific desired mutation are isolated by selection for a marker associated with the transfected gene, and individual colonies are screened for alteration of the target gene by DNA blotting.

The gene constuct transfected into the ES cell line is termed a **targeting vector**. One widely used targeting vector contains the gene of interest with a copy of the bacterial gene (*neo*) that confers resistance to the antibiotic G418 inserted into an exon. This insertion of the *neo* gene serves two purposes: first, it creates an inactivating mutation in the gene of interest, and, second, it provides a selection for those cells that have incorporated the gene following transfection. The clones recovered following transfection and selection in G418 are of two types: the majority has the targeting vector integrated randomly into one or more mouse chromosomes, and the others (usually a variable minority) have the targeting vector integrated into a mouse chromosome by homologous recombination, that is, one normal endogenous allele has been replaced by the mutated gene present in the targeting vector.

Clones with targeted recombinations can be enriched by a process of negative selection. This procedure uses a targeting vector that includes the herpes simplex virus thymidine kinase gene (*HSV TK*) at one (or both) of its ends. The viral thymidine kinase (in contrast to the mammalian homologue) can utilize 1-(2-deoxy-2-fluoro-β-δ-arabinofuranosyl)-iodouracil (FIAU) as a substrate, converting it to a toxic metabolite that is incorporated into DNA and leads to the death of the cell. In most cases, a random insertion of the targeting vector will leave the *HSV TK* gene intact; consequently, in the presence of FIAU, the cells die. By contrast, homologous recombination results in the loss of the *HSV TK* genes from the ends of the targeting vector, and, if the cell has not also undergone a nonhomologous integration, the cells will survive (Figure 11.3). Clones that survive the double (G418 and FIAU) selection are putative homologous recombinants, but verification is generally obtained by examining the structure of the two alleles of the targeted gene either by DNA blotting analysis or by the use of the polymerase chain reaction (Figure 11.4). ES cells containing the desired mutation are then injected into a blastocyst, after which they differentiate and contribute significantly to the developing mouse embryo and adult. The value of this technique is that the mutations introduced *in vitro* can enter the germ line of the mosiac mice and thereby allow the breeding of future generations of transgenic animals.

The contribution of the injected ES cells to the mouse derived from an injected blastocyst can be followed by appropriate coat color markers (Figure 11.5). The ES cells are usually derived from a line of agouti mice (ones having a light brown coat color) and the blastocyst into which the ES cells are injected is usually derived from a mouse with an all black color. Mosaic mice, that is, those expressing genes derived from both the ES cells and the blastocyst, appear mottled or striped: black from the recipient blastocyst and agouti from the injected ES cells. In male mosaic mice, the spermatozoa will also be derived from either the recipient blastocyst cells or from the injected ES cells. To determine if the ES cells contributed to the mosaic animals' germ cells, the mosaic male mice are mated to the donor of the blastocyst, a black female. Because the agouti coat color is dominant, agouti colored offspring must have arisen from spermatozoa derived from the ES cells; by contrast, all black offspring must result from spermatozoa derived from cells contributed by the "black" blastocyst.

Generally, only one of the two alleles of the targeted gene in the ES cells is mutated; therefore, only half of the agouti offspring will have inherited a mu-

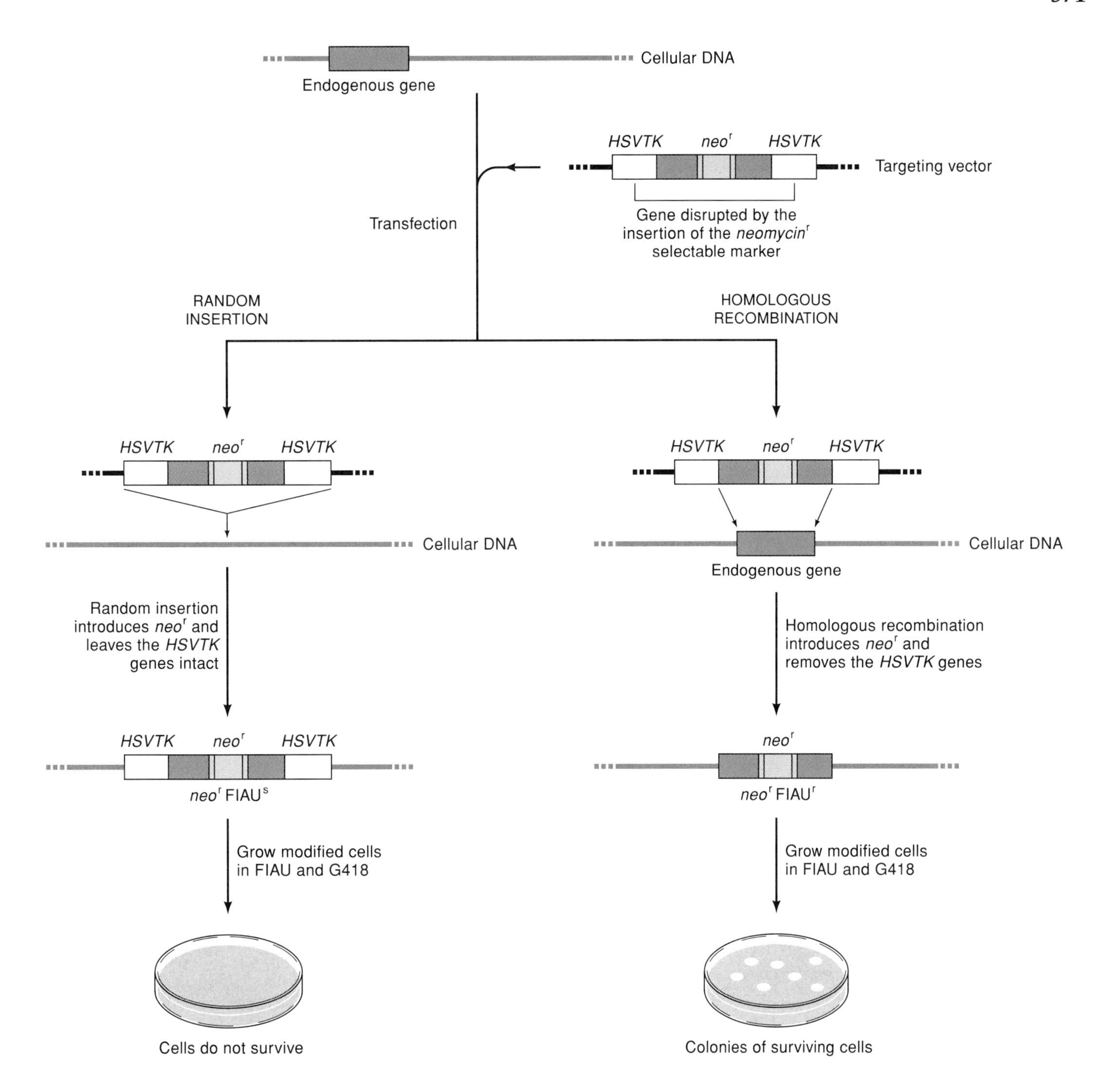

Figure 11.3
Positive/negative selection is used to enrich for ES cells that have undergone homologous recombination. The targeting vector contains a gene that has been disrupted by the insertion of the bacterial gene (*neo*^{r}) conferring resistance to the antibiotic G418, and has, at one or both ends, the thymidine kinase gene derived from the herpes simplex virus (*HSVTK*). In the presence of the drug FIAU, cells that express the *HSVTK* gene die. Cells that have undergone random integration retain the *HSVTK* gene and die in the presence of G418 and FIAU. In contrast, cells that have undergone homologous recombination retain the *neo*^{r} gene but lose the *HSVTK* gene. These cells survive in the presence of G418 and FIAU.

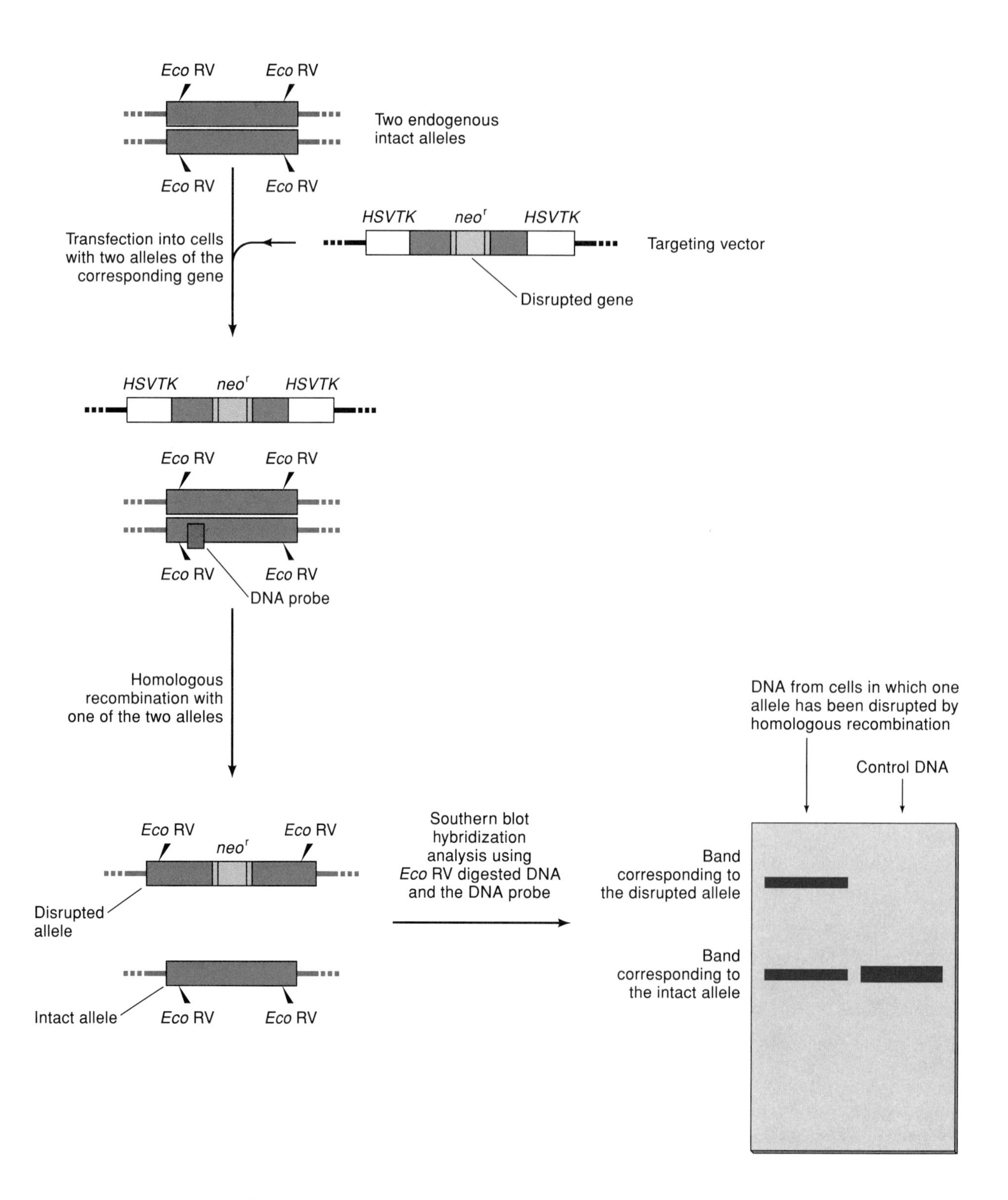

Figure 11.4
Identification of ES cells in which a homologous recombination event has occurred. Because homologous recombination introduces the *neo*^{r} gene into the targeted allele, the relationship between sequences recognized by restriction enzymes is different in the targeted allele as compared to the allele that is not disrupted. In the example shown, the distance between the *Eco* RV sites is increased. Southern blot hybridization analysis of the DNA extracted from the clone of ES cells in which homologous recombination has occurred reveals the two different alleles—the larger disrupted allele and the normal sized allele.

tant allele from their ES cell progenitor. Those animals carrying the modified allele can be distinguished from those with the unaltered allele by examining the structure of the targeted gene using Southern hybridization analysis. To obtain mice that have both mutant alleles, the heterozygous male and female transgenic mice are mated, and the offspring are monitored to identify the homozygotes. This approach is feasible provided that mice carrying the two null alleles are viable during development and postnatally.

This procedure extends the utility of the earlier transgenic technology by providing mice with specific types of mutations in endogenous genes. The physiological effects of such mutations can then be analyzed without concern for the effects of foreign neighboring sequences that often obfuscate the effects of randomly integrated transgenes.

11.2 Transcription

An early application of transgenic mice was to complement data on the regulation of transcription obtained in cell transfection experiments. Analysis of the sequences responsible for regulating gene transcription was hampered in cell experiments by three factors: first, in many cases, cell lines appropriate for a particular gene were not available; second, where the cells were available, it was not always possible using the calcium phosphate transformation procedure to introduce the new genes into the cells; and third, it was considered possible that, for a gene to be appropriately expressed in a differentiated cell, the gene might have had to undergo appropriate modification (e.g., methylation) during the preceding differentiation steps. Transgenic mice offered solutions to all these problems: the transgene being studied would be present not only in the appropriate cell types, but in the full spectrum of cells in which expression would not be appropriate, thus allowing a complete analysis of cell specificity. Furthermore, the transgene would be present throughout development, permitting any required epigenetic modifications. Finally, the most convincing rationale for transcriptional analysis in transgenic mice was that they are *in vivo* experiments. Normal developmental switches in transcription can be analyzed, and questions concerning whether cell lines maintained in culture are normal or not are irrelevant.

Many experiments have demonstrated cell-specific expression of transgenes. I outline examples to illustrate the potential of and problems associated with transgenic mice in the study of gene regulation.

a. Developmental Regulation: The Globin Locus

One of the first genes to be transferred into the mouse germ line was the human β-globin gene. These experiments all depended on untargeted transgenic mice. The globin genes in general, and the β-globin locus in particular, are attractive model systems for studying gene regulation (Chapter 6). Because the globin genes are transcribed only in erythroid cells, it is possible to investigate the mechanism by which transcription is limited to a certain cell type. Transcription of the various globin genes is developmentally regulated, that is, transcription switches from the embryonic to adult forms in mice and from embryonic to fetal to adult forms in humans. Thus, globin genes were seen to provide a way to analyze the events or factors controlling the developmental regulation of gene expression. Because the study of hematopoietic cell biology and globin molecular biology was well advanced, there was substantial scientific basis for being able to undertake and analyze the experiments.

The earliest experiments were designed to answer two very basic questions. Would a human globin gene, present as a normal nuclear gene in a mouse chro-

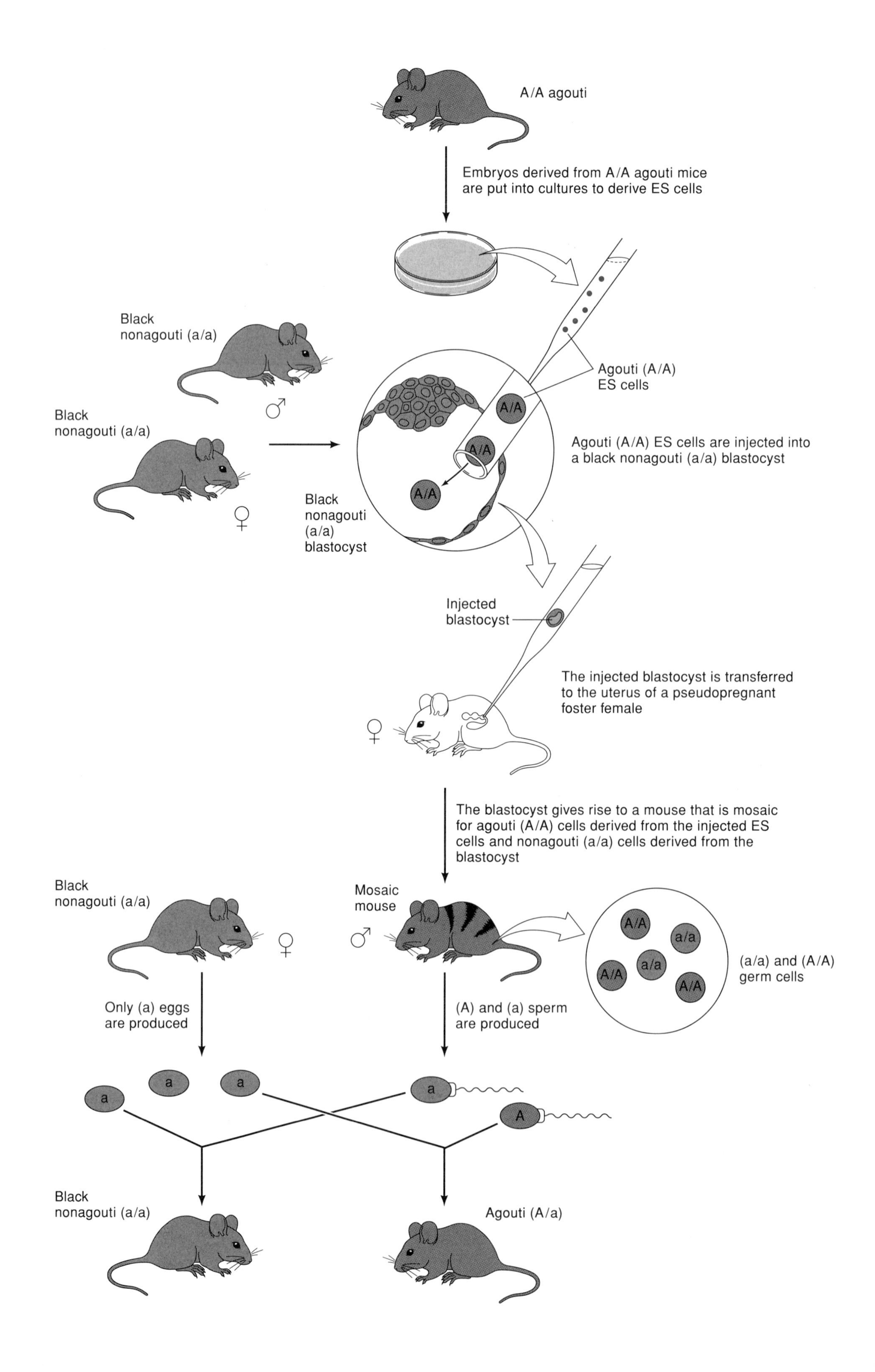
A/A agouti
Embryos derived from A/A agouti mice are put into cultures to derive ES cells
Black nonagouti (a/a)
Black nonagouti (a/a)
Agouti (A/A) ES cells
A/A
A/A
A/A
Agouti (A/A) ES cells are injected into a black nonagouti (a/a) blastocyst
Black nonagouti (a/a) blastocyst
Injected blastocyst
The injected blastocyst is transferred to the uterus of a pseudopregnant foster female
The blastocyst gives rise to a mouse that is mosaic for agouti (A/A) cells derived from the injected ES cells and nonagouti (a/a) cells derived from the blastocyst
Black nonagouti (a/a)
Mosaic mouse
A/A
a/a
a/a
A/A
A/A
(a/a) and (A/A) germ cells
Only (a) eggs are produced
(A) and (a) sperm are produced
a
a
a
a
A
Black nonagouti (a/a)
Agouti (A/a)

Figure 11.5 (*facing page*)
Identification of mice derived from the ES cells. The ES cells most commonly used are derived from a strain of mice that has an agouti (A/A) coat color. They are usually injected into blastocysts derived from a strain of black mice (a/a). The injected blastocysts are then transferred to the uterus of a pseudopregnant foster female. The mouse that develops from the black (a/a) blastocyst injected with agouti (A/A) ES cells is a mosaic with agouti and black stripes. The cells that give rise to the germ line of this animal can also be a mosaic of cells derived from the recipient blastocyst and the injected ES cells. Thus, each sperm that develops from a mosaic male mouse has a complement of chromosomes derived exclusively either from the ES cells (including a wild A type agouti (A) allele) or from the injected blastocyst (including a mutant agouti [a] allele). The offspring of the mosaic male mouse has a set of chromosomes derived from the mother and a set of chromosomes derived either from the ES cells or from the injected blastocyst. Because the wild-type agouti allele is dominant over the mutant agouti allele, the two types of offspring can be identified by coat color. Thus, if the mother is black (a/a), the offspring that has the set of chromosomes from the injected blastocyst is also black; the offspring that has the set of chromosomes from the ES cells is a/A, agouti.

---

mosome, be expressed at all? Would the gene be expressed exclusively, or at least primarily, in erythroid cells? The initial experiments were not encouraging; neither the rabbit β-globin gene nor the human β-globin gene was reproducibly expressed in erythroid cells of transgenic mice.

Following these early reports, there were attempts to improve the frequency and specificity of transcription of heterologous globin genes in transgenic mice. The first variant examined was a hybrid gene having mouse sequences at the 5′ end and human sequences at the 3′ end. In contrast to the earlier reports, this gene was expressed in erythroid tissues in at least some of the transgenic mice. It was not possible, however, to compare these results with the earlier experiments because there were differences in addition to replacing the 5′ half of the gene with mouse sequences. In the earlier experiments, the DNA introduced into the mouse germ line included λ-phage DNA that had been used as the vector in cloning the globin genes, whereas the hybrid gene was introduced substantially free of vector sequences, and there were indications that the vector sequences could inhibit transcription of the globin genes. Although this hybrid globin transgene was transcribed significantly less efficiently than the endogenous globin genes, there was both tissue-specific transcriptional regulation and appropriate developmental regulation.

In the developing mouse, prior to day 12, the predominant globin isoforms are the embryonic globin proteins synthesized by primitive nucleated erythroblasts. Erythropoiesis begins in the fetal liver around day 12 and then in the spleen and bone marrow around day 16. The predominant β-globin isoform synthesized by mouse erythroblasts in these latter two organs is the adult β-globin, and there is no detectable synthesis of the embryonic β-globin ε (Figure 11.6). In the transgenic mice, expression of the hybrid gene was detected in the fetal liver and adult marrow but not in blood taken from 11–12 day old fetuses. This is the result that would be expected if the hybrid globin gene is appropriately regulated in the mouse with transcription occurring in "adult" tissues (fetal liver and adult marrow) but not in embryonic tissues (blood islands of the 11.5 day old fetus). Although the level of expression was substantially less than that of the endogenous β-globin gene and the ratio of expression between fetal liver and adult bone marrow was different from the normal, these results encouraged

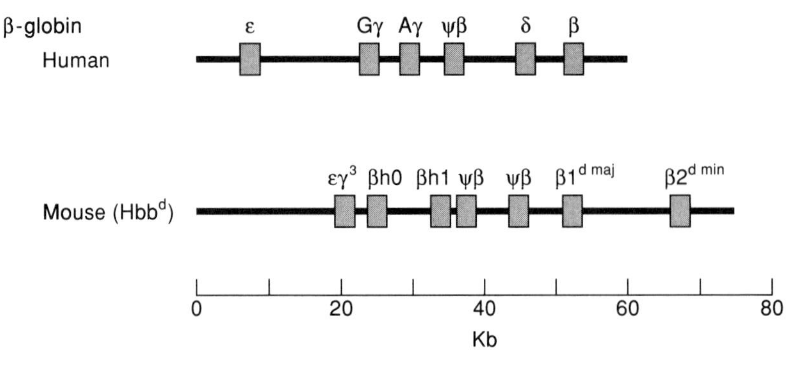

Figure 11.6
Chromosomal clustering of the mouse and human β-globin genes. The mouse contains a cluster of three embryonic β-globin genes ($\varepsilon\gamma^3$, β^{h0} and β^{h1}), two adult β-globin genes ($\beta 1^{dmaj}$ and $\beta 2^{dmin}$), and no fetal β-globin genes. In contrast, in humans there are two adult (δand β), two fetal ($^G\gamma$and $^A\gamma$), and one embryonic (ε) globin genes.

a search for the *cis*-acting sequences that regulate the expression of the globin gene.

This was carried out by generating a number of transgenic mice, each of which contained a human β-globin gene that had been truncated at a different position 5′ to the point at which transcription is initiated. The results indicated that erythroid-specific transcription was unaffected until all but the final 48 base pairs of 5′ sequence were removed. This report revealed a major logistical difficulty with this type of analysis, namely, that transcription of the transgene was influenced by its particular chromosomal position. Thus, the expression of the transgene can be influenced by sequences present at the transgene's integration site.

Although progress was being made in characterizing sequences partly responsible for cell- and stage-specific expression, the level of expression was still substantially below that of the endogenous genes, and the transgenes were particularly sensitive to chromosome position effects. This problem was resolved by the demonstration that most, if not all, of the chromosome position effects could be suppressed if the transgene included large pieces of DNA from the β-globin locus. Because these sequences are normally found at some distance (up to 40 kb) from the coding sequences, it seemed possible that they provide a "master switch" for allowing appropriate levels of β-globin gene transcription (see Chapter 6).

Another fundamental question was whether a human fetal β-globin gene would be transcribed in mouse erythroid tissues and, if so, would it function as a fetal gene, as it does in humans, or as an embryonic gene (mice have embryonic but no true fetal β-globin genes)? The experiments designed to answer these questions showed that a human fetal globin gene was expressed in transgenic mice in the same hematopoietic tissues as the mouse embryonic genes. This result suggests the possibility that the human fetal globin gene has sequences recognized by regulatory factors in mouse embryonic erythroid cells. In humans, these factors would be present in human fetal erythroid cells.

b. Differences in Expression of Genes in Transgenic Animals and Transfected Cell Cultures

Transfection of recombinant genes into cell lines identified several *cis*-acting DNA sequences that appeared to be important for gene regulation. A concern raised by this approach was the possibility that the defined sequences did not fully account for *in vivo* regulation. This concern prompted direct comparisons between the expression of recombinant genes in cell culture experiments and the expression of the same genes in transgenic mice.

MMTV Because the long terminal repeat (LTR) from the mouse mammary tumor virus (MMTV) permits heterologous genes to which the LTR is joined to be induced by glucocorticoids, it has been an important tool in defining the sequences responsible for transcriptional activation by glucocorticoids (*Genes and Genomes,* section 8.3f). The localization of these sequences to a region within 200 bp of the 3′ end of the LTR has been confirmed in transgenic mice. However, these transgenic mice experiments revealed a function for the LTR that had been inferred from the biology of the virus but never demonstrated in tissue culture experiments: the virus has target cell preferences in an animal. The cell transfection studies gave no indication that this tropism was mediated by the LTR, and, furthermore, it appeared that, with the exception of the 100 bp region responsible for the glucocorticoid regulation, all LTR sequences could be removed without altering transcription. In contrast, when the MMTV LTR was used as part of a fusion gene in transgenic mice, it was evident that the rest of the LTR is important. Thus, a transgene containing the 3′ end of the minimal MMTV LTR promoter fused to a reporter gene was expressed at significant levels only in the same cells (secretory epithelial cells) where MMTV virus genes are usually expressed. As would be expected from a promoter that lacked the glucocorticoid response elements, these particular fusion genes were not inducible by glucocorticoids.

Elastase Detailed analysis of the control regions flanking the rat elastase gene was carried out using both untargeted transgenic mice and transfections into cell lines. The elastase gene is normally expressed in the exocrine cells of the pancreas, and a relatively short DNA fragment (200 bp) upstream of the coding sequence is both necessary and sufficient to direct expression of heterologous genes in the appropriate cells. Within this 200 bp region, there are three subregions, each of 20–30 bp, that are important for regulating the level of transcription. To a large extent, there is concordance between the results obtained using the two different experimental systems (i.e., transgenic mice and cell transfections). However, one major difference was noted. In transfected cells, all three subregions are required; modification or deletion of any one reduces transcription at least 100-fold. In transgenic mice, however, the same modification reduces transcription by at most 10-fold. This increased sensitivity of the transfected genes to modifications of the control regions is independent of the cell line into which the gene is transfected, the reporter gene used, or whether stable or transient cell systems are used.

Comments These analyses of gene transcription in transgenic mice illustrate some of the strengths and weaknesses of this approach. In transgenic mice, the developmental regulation and tissue and cell specificity can be fully examined. In contrast, experiments relying on transfection of tissue culture cells are necessarily limited to a restricted number of cell types, and the differentiation of these cells is limited. Furthermore, in transgenic mice, gene regulation can be studied in cells for which there is no tissue culture counterpart (e.g., embryonic or fetal erythroid cells). The major limitations to this approach are the size, complexity, cost, and length of the experiment. To overcome the problems associated with chromosome position effects, investigators need to examine a large number of individual transgenic mice carrying each gene or construction to ensure that any observed effect is reproducible.

c. Methylation and Parental Imprinting

In some lines of transgenic mice, transgene expression depends on whether the transgene is inherited through the paternal or maternal germ line. This phenomenon, termed **parental imprinting**, can be associated with parent-specific methylations of the particular gene. The extent and pattern of methylation also appears to depend on both the transgene's sequence and the chromosomal locus into which it is inserted. For example, in a study of several lines of transgenic mice carrying a quail troponin I gene, there was an increased level of methylated cytosine residues in the transgenes and consequently less expression when these sequences were inherited from a female parent. Because this occurred in five of six independent lines examined, the transgene sequences themselves appear to be involved in the process. In contrast, parental imprinting of a fused immunoglobin-chloramphenicol acetyl transferase transgene was observed in only one of six lines of transgenic mice carrying this particular transgene, suggesting that, in this case, the methylation is probably more influenced by the chromosomal location. In another experiment, a transgene containing a region of the mouse immunoglobulin locus fused to the mouse *c-myc* gene was found to be hypermethylated and not transcribed when the transgene was maternally inherited. Conversely, the fusion gene was undermethylated and transcribed efficiently (only in the heart) when the transgene was paternally inherited.

11.3 Growth and Development

a. Mammalian Hox *Genes*

Genes that include a region encoding the 60 amino acid long homeo domain are critical for *Drosophila* morphological development (Chapter 9 and *Genes and Genomes,* Section 8.6a). The discovery of human and mouse homologues of these genes suggested that homeobox genes might also be critical in mammalian development. A correlation between the pattern of expression of these genes and the formation of discrete parts of the organism's structure was consistent with this suggestion. However, the key to a clear demonstration of the importance of the homeobox-containing genes in mammalian pattern formation was to study mice with homeobox-containing transgenes and gene knockouts produced by homologous recombination. Thus, untargeted transgenic mice that contained extra copies of the *Hox 1.1* gene were generated. The mice expressed this transgene in the same embryological structures as the endogenous *Hox 1.1* gene but at higher levels. There was, in addition, a low level of expression in areas in which the endogenous gene appeared not to be expressed. The transgenic mice expressing this transgene died during late embryogenesis as a result of the occurrence of posterior cervical somites.

Experiments in which specific homeobox genes were knocked out by recombination confirmed their importance in skeletal development. Thus, homozygous mice in which either the *Hox 1.5* or the *Hox 1.6* gene are knocked out have profound developmental abnormalities. In the case of homozygous mice lacking a functional *Hox 1.6* gene, rostral neurogenic structures fail to form (Figure 11.7). Similarly, mice lacking a functional *Hox 1.5* gene fail to form the rostral mesenchymal structures (thyroid, thymus, parotid, bone).

In *Drosophila,* inactivation of the single *engrailed* gene, which encodes a protein containing a homeobox domain, leads to a morphological transformation, the formation of an extra pair of wings. In mice, however, eliminating one of the two engrailed homologues (*en*2) has minimal consequences. Perhaps the remaining *engrailed* homologue is sufficient for normal development.

Several conclusions can be drawn from these experiments. First, for some

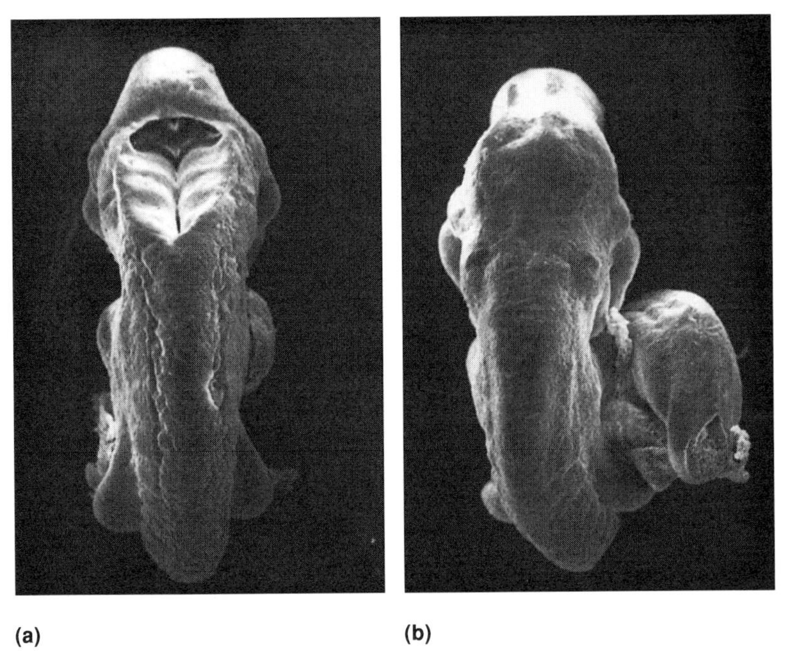

Figure 11.7
Phenotypic changes caused by the lack of *Hox 1.6*. (a) Embryos that have a mutation in the gene encoding *Hox-1.6* do not form rostral neural structures. (b) The normal embryo with rostral neural structures. (From T. Lufkin et al. 1991. Disruption of the Hox-1.6 Homeobox Gene Results in Defects in a Region Corresponding to its Rostral Domain of Expression. *Cell* 66 1105–1119. Reprinted with permission.)

homeobox genes (e.g., *en*2), there appears to be functional redundancy—the protein encoded by *en*1 may compensate for the absence of *en*2. Second, there are phenotypic consequences to knocking out some homeobox-containing genes (e.g., *Hox 1.5* or *Hox 1.6*), although the effects are seen in only a subset of cells in which the gene is expressed. For example, although *Hox 1.5* is expressed in the developing stomach, spleen, and kidney, these structures do not appear to be affected in the mice lacking this protein. Third, whereas deletion of homeobox-containing genes in *Drosophila* leads to transformation of segments to a form resembling a more anterior segment, this relationship does not appear to be true in mice. Fourth, both *Hox 1.5* and *Hox 1.6* are expressed in embryonic regions that, to a large extent, overlap; but deletions of the two genes lead to quite different phenotypes.

b. Hormonal Influences on Development

Both transgenic mice and mice with gene knockouts have been used to examine the hormonal control of peri- and postnatal development. For example, overexpression of human growth hormone (hGH) in transgenic mice leads to high systemic concentrations of hGH and accelerated growth as well as hepatomegaly, splenomegaly, reproductive dysfunctions, and insulin resistance. These experiments confirm the relationship between GH and linear growth deduced from human development. However, the hGH transgenic mice differ from human acromegalics in that the animals develop a fatal glomerularsclerosis, a condition not usually associated with acromegaly. Although hGH may have direct cellular activities, at least some of the actions of hGH are probably mediated through secondary effectors such as insulinlike growth factor I (IGF-I).

The role of IGF-I has been examined in both transgenic mice and mice in which the gene encoding this protein was disrupted. Although transgenic mice that overexpress IGF-I are larger than normal, this effect is not as dramatic as the result from the overexpression of hGH. However, mice lacking a functional IGF-I gene are markedly retarded in their development and smaller than normal. The most significant histological effect of the IGF-I deficiency is the lack of perinatal muscle development. A difficulty in interpreting these experiments is that each of the hormones examined is capable of binding to and activating several different receptors. For example, IGF-I will bind both the Type I IGF receptor and the insulin receptor, although the affinity for the insulin receptor is significantly less than that of insulin.

11.4 Immunology

a. Humoral Immune Response

The initiation and regulation of a humoral immunological response to a foreign antigen occurs at several steps (Chapter 7). These include the rearrangement and transcription of the immunoglobulin genes in the appropriate (B) lymphocyte, the synthesis by individual B cells of antibodies with only one specificity (**allelic exclusion**) and functional characteristic (**isotype exclusion**), **somatic mutation** that permits synthesis of antibodies with increased affinity for particular antigenic determinants, and the postulated role of any one antibody in regulating the synthesis of related antibodies (**idiotypic regulation**) (chapter 7). In addition, each mature B cell is capable of changing the type of the antibody without changing the antigenic specificity (**isotype switching**). These processes have been extensively investigated *in vivo* and in tissue culture and, more recently, by introducing already rearranged immunoglobulin genes into the mouse germ line. Because every somatic cell in the transgenic animals has a functionally rearranged antibody gene, these mice provide an *in vivo* model for investigating humoral immune response regulation.

Allelic exclusion Transgenic mice carrying a rearranged light chain (κ) immunoglobulin gene express this gene predominantly, but perhaps not exclusively, in B cells. The antibody-forming B cells of these mice can be immortalized by fusing them with a stably transformed partner, a myeloma cell. Such hybrid cells are referred to as **hybridomas**. Hybridomas derived from the previoiusly mentioned transgenic mice fall into a limited number of categories. Significantly, in those hybridomas that secrete an immunoglobulin composed of an endogenous heavy chain and the transgenic κ-light chain, there was no rearrangement of the endogenous κ-genes. Because rearrangement and expression of the heavy chain genes precedes the corresponding steps for the light chain genes, these results suggest that the existence of either an already expressed light chain or a functional secreted antibody protein can inhibit the rearrangement of the endogenous light chain genes. Thus, allelic exclusion of at least the light chain genes may be achieved by the shutting down of the rearrangement machinery by a light chain or a functional antibody.

As mentioned previously, the heavy chain genes are rearranged and transcribed in pre-B cells prior to the rearrangement and transcription of the light chain genes. The first type of heavy chain (μ) expressed in pre-B cells is made in two forms, either membrane bound or secreted. Thus, transgenic mice expressing a rearranged heavy chain transgene provide a useful experimental system to analyze the utilization of the endogenous light and heavy chain genes in pre-B cells that are expressing an already rearranged heavy chain transgene. Pre-B cells that have been immortalized by infection with a murine retrovirus, the Abelson murine leukemia virus (AbMuLV), permit the analysis of events subsequent to heavy chain expression. Approximately 60 percent of the chromosomes in the pre-B cells derived from these transgenic mice have an unrearranged heavy chain locus; by contrast, rearrangement of the heavy chain genes had occurred in all pre-B cells derived from nontransgenic mice. The implication is that expression of the heavy chain gene inhibits further heavy chain gene rearrangement. The transgene used in these studies was capable of making both the membrane-bound and the soluble form of the immunoglobulin (IgM). These two forms are derived from the same heavy chain gene by alternately spliced mRNAs. In subsequent experiments, the heavy chain transgene was engineered so that it could direct the synthesis of only the membrane-bound IgM protein or a transgene that encoded the γ-heavy chain isotype (IgG). In both experiments, the rearrangement and expression of the endogenous heavy chain genes were suppressed.

This allelic exclusion mediated by the transgenic immunoglobulin genes is not absolute. Different strains of mice have small variations in their heavy and light chain constant regions, which permits the chains to be distinguished by allele-specific antibodies. Mice in which the heavy chain transgene comes from a different mouse strain can be used to analyze whether the antibodies on the surface of individual mature B cells are from the transgene or from an endogenous gene. Two antibodies, one specific for the immunoglobulins derived from the endogenous heavy chain gene and one for the immunoglobulin derived from the heavy chain transgene, each labelled with a different fluorescent marker, can be used in conjunction with fluorescence-activated cell sorting (FACS) (*Genes and Genomes,* section 6.4) to determine the type of surface immunoglobulin. In transgenic mice with a rearranged immunoglobulin μ transgene, individual cells express surface IgM molecules with both endogenous and transgenic heavy chains, although the transgene's product predominates. Thus, in these transgenic mice, the expression of a rearranged heavy chain does not completely block further heavy chain gene rearrangements. The dimeric IgM molecules could contain one protein derived from the transgene and one protein from the endogenous gene; this fact obscures the estimate of the relative amount of the two kinds of heavy chains being made.

Somatic mutation Subsequent to stimulation of mature B cells by antigen, the V regions of the various IgG subclasses of immunoglobulin mutate at a frequency several orders of magnitude higher than nonimmunoglobulin genes. This hypermutation rate allows selection for immunoglobulins with increased affinity for the antigen. Neither the trigger nor the mechanism of the hypermutability is understood. Analysis of the antibodies synthesized by hybridomas derived from a mouse carrying a rearranged κ-transgene demonstrated that the process of rearrangement was not essential for somatic mutation. Nevertheless, the observed frequency of such mutations in the transgene was significantly less than that found with endogenous immunoglobulin genes. Inasmuch as the transgenes involved were not linked to the immunoglobulin light chain locus, it seems likely that there is not an absolute requirement for the target of the mutational process to be at a specific chromosomal location.

Lymphokines Although the interaction between a cell surface immunoglobulin and the appropriate antigen is essential for a normal humoral immune response, this is, in general, not sufficient. There is also a need for certain proteins called **lymphokines** or cytokines, produced by T cells and macrophages. The role of these proteins *in vivo* has been studied by knocking out the appropriate genes from mice. Although the immune systems of mice in which the genes encoding IL-2, IL-4, or interferon γ have been functionally eliminated are not grossly distorted, there are effects that become significant either when the animal is stressed or when the mutation is crossed into different mouse strains. One example of this is that mice that are unable to produce IL-4 do not make IgE in response to infection with nematodes. Another example is that mice lacking interferon γ are particularly sensitive to infection with intracellular bacteria.

b. Cellular Immune Responses

There are a number of significant differences between antigen recognition by the T cell receptor (TCR) and the immunoglobulins that serve as the B cell antigen receptor. First, the TCR recognizes linear peptide fragments as antigen rather than the complex surface shape recognized by an antibody. Second, the TCR usually responds only when the antigen is presented to the TCR in association with either Class I or Class II major histocompatibility complex (MHC) molecules.

The Class I MHC molecules have one major 44 kDa chain associated with a smaller 18 kDa β_2 microglobulin; the Class II MHC molecules are heterodimers of 28 kDa and 34 kDa proteins. In general, the MHC Class I molecule presents peptides to cytotoxic T cells (T_C) that are responsible for cell-mediated, antigen-specific killing of either infected or otherwise altered cells. The MHC Class II molecule presents peptides to helper T cells (T_H) that provide signals required to activate B cells to produce antibodies (Figure 11.8). This association between Class I presentation and T_C and Class II presentation and T_H is not absolute, but there is a strong association between the Class I and Class II presentations and the presence of accessory molecules on the respective T cells. Thus, T cells recognizing a Class I–associated antigen have the CD8 protein on their surface, and T cells recognizing Class II–associated antigen have the CD4 protein on their surface. Moreover, Class I and Class II MHCs tend to associate with antigens of different structures.

The maturation of T cells is a complex process by which immature T cells migrate from the bone marrow to the thymus, where selection takes place. This selection has two important aspects. First, functional T cells that recognize self-antigens with high affinity are eliminated and do not leave the thymus for peripheral lymphoid organs. Because immune recognition of self-antigens leads to destructive autoimmune disease, the achievement of tolerance to self-antigens is critical for the organism's survival. Second, mature T cells recognize foreign antigens only when antigen-presenting cells from the same organism present them to the T cells; this positive selection is also known as **MHC restriction**. Studies of mice carrying organ grafts (thymus and/or bone marrow) have shown that the thymus is the principal site responsible for phenomena.

These observations have given rise to several important questions. How, when, and where does positive selection (MHC restriction) occur? What is the nature of the bias of CD4 positive T cells for Class II recognition and CD8 positive T cells for Class I recognition? How is tolerance imposed? Mice that carry and express MHC Class II antigen transgenes and functionally rearranged TCR transgenes have contributed extensively to an understanding of developmental immunology, both by corroborating earlier experiments carried out using classical techniques and by providing novel insights.

Mice were produced that expressed either a TCR transgene encoding a receptor that recognizes an antigen presented by MHC Class I or a TCR transgene whose receptor recognizes a different antigen presented by MHC Class II. The majority of the T cells in the periphery of the first type of animal expressed the transgenic TCR and were CD8 positive. The mature T cells expressing the second type of transgenic TCR were CD4 positive. Thus, the bias for MHC Class II by CD4 positive T cells and for MHC Class I by CD8 positive cells appears to be restricted by the nature of the TCR/MHC plus antigen interaction.

The thymic cell type responsible for negative selection—the deletion of self-reactive T cells—has also been examined using transgenic mice. T cells bearing a

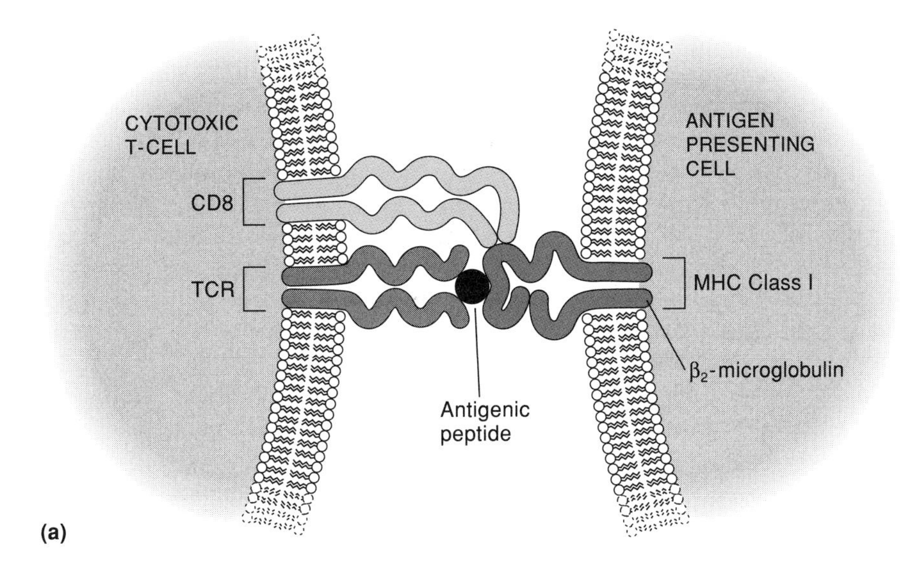

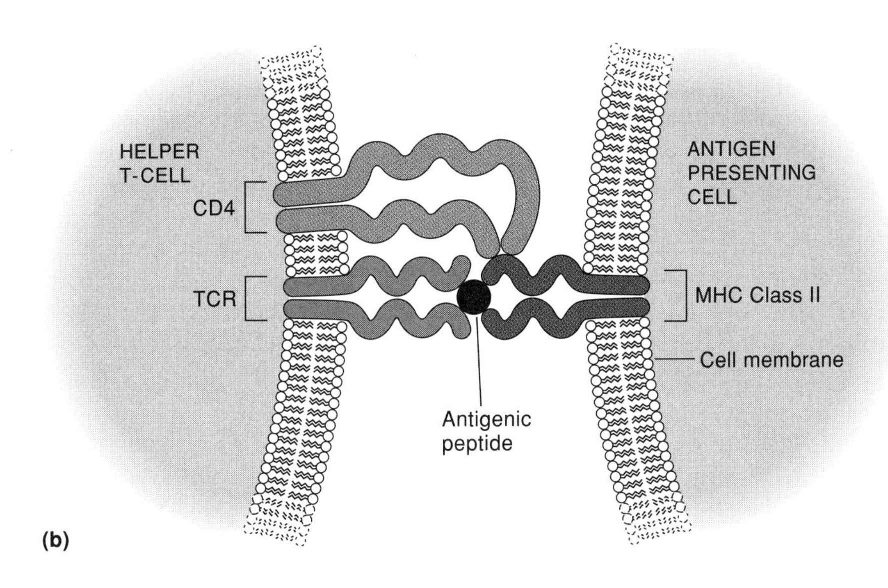

Figure 11.8
Interaction of the T cell receptor (TCR). (a) Antigen presented by MHC Class I molecule. (b) Antigen presented by MHC Class II molecule. In addition to the TCR-MHC/antigen interaction there is usually a requirement for contact between CD4 and MHC Class II antigen or CD8 and MHC Class I antigen.

TCR utilizing a particular variable region, V_β-5, have a high affinity for the type of MHC Class II antigen called I-E. In mice that express I-E, T cells expressing the V_β-5 TCR are deleted. Thus, the cells responsible for inducing this selection can be identified by using an I-E negative strain (C57Bl/6) to make transgenic mice that express the I-E molecule only on defined subsets of thymic cells. The results of these experiments support the view that while bone marrow-derived cells are primarily responsible for negative selection, epithelial cells in the thymic cortex may also paticipate.

Mice that no longer express specific cell surface proteins have been generated by the gene-targeting techniques described earlier. These animals have been particularly useful for assessing the role of these proteins in lymphocyte development and in cellular immune responses. For example, the expression of β_2-microglobulin is essential for the appearance of the MHC Class I antigens at the cell surface. Therefore, it was of interest to determine how transgenic mice with a knockout mutation in the β_2-microglobulin gene are affected, particularly with regard to the ability of MHC Class I molecules to present antigens. As anticipated, mutant mice that fail to express β_2-microglobulin are unable to develop CD8 positive T cells, although there are appropriate precursors in the thymus. This result suggests that intact MHC Class I molecules are critical for the development of the cell-mediated immune response.

Somewhat surprisingly, the mice lacking β_2-microglobulin are no more sensitive to viral infection than normal mice, despite their lack of MHC Class I expression and absence of CD8 cytotoxic T cells. Although they are fully capable of resisting a viral challenge, these mice are significantly more sensitive to a trypanosome infection.

c. Tolerance

As described in Section 11.4a, mice with either transgenic immunoglobulin genes or transgenic TCR genes express the relevant receptors on a significant proportion of the mature lymphocytes. In several experiments, the antigen these receptors would normally recognize was one expressed by the mouse. Thus, transgenic mice that carried and expressed a rearranged TCR specific for the male-restricted H-Y antigen were produced. Female transgenic mice expressed this receptor on a high proportion of their T cells, and these cells responded to the H-Y target antigen. By contrast, the T cells from male transgenic mice expressed this receptor on a smaller proportion of their peripheral T cells; those cells that expressed this receptor had lower levels of the coreceptor CD8 and did not respond to the H-Y antigen. Tolerance can apparently be induced by deleting cells with self-reactive TCRs and also by modulating the levels of accessory molecules such as CD8.

Comparable experiments have been carried out to examine B cell tolerance. In one example, two lines of transgenic mice were made. One line expressed high levels of a rearranged immunoglobulin specific for hen egg lysozyme (HEL), and the other line expressed the lysozyme itself, as if it were a normal self-protein. When these mice were interbred, offspring that had one or the other or both of the transgenes were obtained. Those mice that had just the HEL transgene had a normal immune system. Those that had just the rearranged anti-HEL immunoglobulin transgene had a high proportion of mature B cells that recognized HEL. The interesting results were obtained by mating animals producing HEL with ones producing anti-HEL to yield transgenic animals expressing both genes. Surprisingly, these double transgenic mice do not develop an autoimmune response. This is not achieved by deleting the B cells with the anti-self-antibodies, but rather by interfering with the signaling mechanism that leads to the formation of immunoglobulin-producing plasma cells.

Because tolerance appears to be primarily imposed by bone marrow–derived cells in the thymus, we can envisage how tolerance to proteins that traffic through the thymus develops. Less obvious is how tolerance develops to proteins restricted to nonthymic locations. This question has been addressed by generating transgenic mice that express specific antigens in defined nonthymic locations. In one experiment, transgenic mice that express MHC Class II antigens either on pancreatic endocrine cells or on pancreatic exocrine cells were made. In both experiments, the mice are tolerant to these antigens, but tolerance is achieved not by preventing the maturation of the self-reactive cells within the thymus (clonal deletion) but by some as yet poorly understood peripheral mechanism.

11.5 Oncogenesis

The link between specific genes and tumorigenesis was deduced from the discovery that certain cell-derived genes in retroviruses are responsible for tumor induction after infection of experimental animals by these viruses (Chapters 3 and 4). This association was extended to the cellular genes themselves by the demonstration that mutations affecting these genes' protein-coding regions or their transcriptional regulation could lead to oncogenic transformation of previously "normal" cells.

a. Tumorigenesis

In the intact animal, tumorigenesis is a complex process that, in addition to cellular transformation, requires alterations in the animal's immune surveillance and acquisition by incipient tumors of a blood supply. Transgenic mice carrying activated oncogenes have shed some light on these aspects of oncogenesis.

Several important conclusions can be drawn from these studies. First, different cell types display marked differences in the susceptibility to oncogenic transformation. Thus, mice containing a transgene in which the mouse mammary tumor virus LTR promoter is fused to the *c-myc* gene (MTV-*c-myc*) develop adenocarcinomas in the mammary gland but rarely in their salivary glands, although the expression of the MTV-*myc* gene in the salivary gland is comparable to that of the mammary gland. Similarly, although an MTV-*ras* transgene can transform a wider range of tissues than MTV-*myc,* it does not lead to seminal vesicle abnormalities despite a high level of Ras expression in this organ. The failure of these cellular oncogenes to transform specific cell types may not hold for their viral counterparts. Although there are unexplained tropisms with respect to their oncogenic potential, the expression of the cellular oncogenes is clearly necessary but insufficient for producing tumors.

Second, depending on the oncogene, the level of expression, and perhaps the cell type, additional events are required prior to tumor formation. Thus, in the mice carrying the MTV-c-*myc* transgene, most of the mammary epithelium appears normal, with transformation occurring in an apparently stochastic manner over the lifetime of the animal, usually in only one mammary gland. This failure to transform the entire mammary epithelium suggests that only a subset of the cells gives rise to tumors. That the tumors are in fact clonal is seen more definitively in lymphatic tumors arising in transgenic mice carrying a c-*myc* gene under the control of an immunoglobulin promoter and enhancer. This transgene is expressed at high levels specifically in lymphocytes, and it was therefore possible to analyze immunoglobulin rearrangements in the tumors and directly demonstrate their clonal origin. An exception to this was found in mice that overexpressed a *ras* transgene in the exocrine pancreas. The transcriptional regulatory region responsible for this organ-specific expression was derived from the elastase gene. The majority of these mice died perinatally as a consequence of widespread hyperplasia of the exocrine pancreas. This hyperplasia of all cells expressing the oncogene is not seen in transgenic mice expressing a *ras* oncogene in other tissues; for example, the tumors found in the MTV-*ras* transgenic mice were sporadic and did not involve the entire organ in which they were found. One possibility is that the pancreatic masses induced by the elastase-*ras* transgene are not fully transformed. Despite these differences, the organwide hyperplasia was striking and suggests that the timing of oncogene expression with respect to normal growth could be a critical event.

Third, different oncogenes can cooperate in producing tumors. For example, mice that contain both activated *ras* and *myc* transgenes driven by the same promoter (MTV) develop mammary tumors significantly faster than mice carrying either gene alone (Figure 11.9).

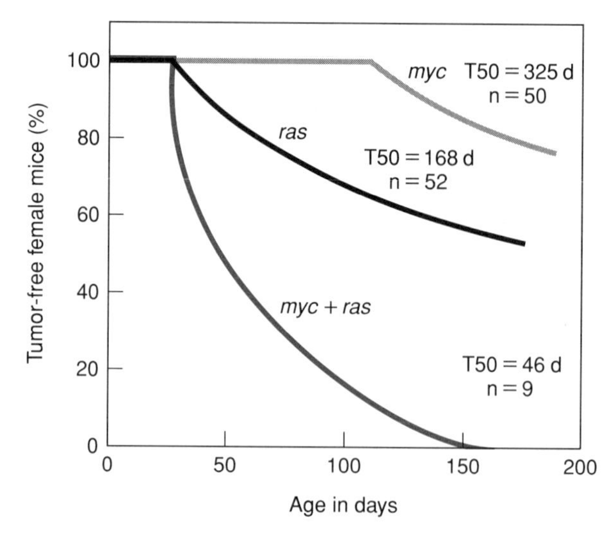

Figure 11.9
Enhanced rate of tumor formation in transgenic mice carrying both *ras* and *myc* oncogenes. The time for 50% of the mice to develop tumors (T50) is significantly less for transgenic mice that express both *myc* and *ras* transgenes as compared to transgenic mice that express either *myc* or *ras*. (From E. Sinn et al. 1987. Coexpression of MMTV/v-Ha-*ras* and MMTV/c-*myc* genes in Transgenic Mice. *Cell* 49 389–398. Adapted with permission.)

All the previously described experiments used relatively well characterized oncogenes. In contrast, the mechanism by which HTLV-I causes adult leukemia is not clear. This is due to the long latency between viral infection and tumor development, the lack of viral proteins associated with the leukemic cells, and the failure to demonstrate specific chromosomal integration sites for the provirus. As a test for *in vivo* function of one of the viral genes, transgenic mice that expressed the *tax* gene from the HTLV-I LTR were generated. These mice developed tumors, but, surprisingly, not the T cell leukemia associated with the virus in humans, but neurofibromas derived from the neural sheaths of peripheral nerves. The reason for the difference in specificity of cell transformation between mice and humans is not obvious, but the virus has been associated with the human neurological disorder tropical spastic paraparesis.

b. The Normal Function of Oncogenes

The ectopic expression of cellular oncogenes that have been activated by mutations in either the coding or the transcriptional regulatory region can lead to transformation and tumorigenesis. However, these experiments do not address the normal function of these proteins. This question has been addressed by creating mice in which these genes are knocked out; some of the results in such cases have been quite surprising. Notably, despite the widespread expression of the cellular form of the *src* gene, c-*src*, mice lacking a functional c-*src* gene as a result of a knockout are normal with one exception. They have an increased bone matrix deposition resulting from a failure adequately to form or activate osteoclasts, the cells responsible for bone resorption.

Genetic analysis of families whose members have a high incidence of tumors led to the observation that their tumor cells have, in many cases, specific chromosomal deletions. This suggested that the development of these tumors is associated with the loss of genes, as opposed to the activation of oncogenes. These findings implied that these genes are involved in preventing transformation, so the gene(s) (initially unidentified) were termed **tumor suppressor** genes. Subsequent cell and molecular biological analyses led to the identification of several candidate genes. One of these encodes a protein called p53 (its molecular weight is approximately 53 kDa). As was the case for the c-*src* mice lacking a functional gene, mice without a functional p53 gene are grossly normal in appearance and behavior, although they have a significantly higher incidence of spontaneous tumor formation.

11.6 Mutations

a. Insertional Mutagenesis

Although the mechanism for integrating a transgene at any particular chromosomal site has not been fully resolved, the end result is a disruption of the DNA sequence at the insertion point. Such disruption, termed **insertional mutagenesis**, can lead to phenotypically obvious mutations. The effects of the insertional mutations must be distinguished from the effects of the transgene's expression.

Insertional mutagenesis has been used extensively in the genetic analysis of, for example, flies and plants and was first demonstrated in mice by infection of preimplantation embryos with endogenous or exogenous retroviruses. In one example, the retrovirus insertion produced a recessive mutation that, when present in a homozygous form, resulted in prenatal lethality. The interrupted gene was identified by cloning the insertion site using the viral DNA as a hybridization probe. Then the uninterrupted DNA was cloned using the cellular DNA associated with the viral DNA as the probe. The gene's sequence revealed that it encodes α-1-collagen, an essential product during early mouse embryogenesis. In a second example, the mutation is also recessive, and homozygous mutant mice display a dilution of the coat color. The affected gene encodes a novel myosin heavy chain.

The insertions caused by integration of transgenes following the direct microinjection of DNA into the pronucleus of a fertilized egg cell can also be associated with phenotypically obvious consequences. Several examples in which the homozygotes fail to implant have been described. A detailed biological analysis of these is difficult because of problems associated with obtaining enough material from the affected embryos to document which of the embryos were affected homozygotes. Prenatal lethalities have also been induced in transgenic mice, apparently as a consequence of chromosomal translocation leading to a lethal chromosomal imbalance, although it is possible that the transgene also interrupts a required gene.

In addition, several nonlethal mutations associated with transgene integrations have been described. In one, the transgene was transmitted from a female parent to both male and female offspring, but the male transgenic mice rarely transmitted the transgene to their progeny. One interpretation is that the transgene caused a mutation that is revealed only in the postmeiotic spermatids. Several independent, nonallelic insertions that affect normal limb and digit formation have also been described (*Genes and Genomes*, perspective IV). This apparent preponderance of defects in limb development may be more apparent than real; gross defects in morphology or behavior are obvious and much easier to detect than changes in nonvisible properties, for example, metabolism.

b. Dominant Negative Mutations

Although transgenic mice may contain and express a new (dominant) gene, obtaining transgenic mice whose phenotypes resemble individuals with a recessive "loss of function" mutation is also possible. One way to achieve this is to generate a transgenic mouse in which a negative regulatory factor is secreted in close proximity to a target organ. Thus, a mutant whose phenotype resembled a loss of function of the growth hormone–secreting cells within the anterior pituitary (somatotrophs) was produced, not by directly altering growth hormone expression or the viability of these cells, but by producing human growth hormone locally. Such mice resulted from the expression of a human growth hormone transgene exclusively within the cerebral cortex. Although the level of expression was insufficient to influence distal tissues such as muscle, bone, and cartilage, the level of human growth hormone present in the hypothalamus was high enough to down-regulate endogenous mouse growth hormone synthesis by the mouse pituitary. As a consequence, growth of the mice was retarded.

Targeting an organ system A complete organ system has been destroyed by introducing a transgene that encodes a toxin and whose expression is limited to the target organ. For example, the diphtheria toxin has been expressed in the exocrine pancreas by using the elastase promoter to control transcription. Cells in which diphtheria toxin is made are killed, producing a phenotype in which the newborn pups have only a rudimentary pancreas. The limitation with this model is that such mice do not survive more than a few days after birth. However, tissue-, cell type–, or stage-specific promoters that express lethal gene products can be used to ablate specific tissues and cells at specific stages of development. Similarly, the timing of transgene activation can be controlled by linking it to an inducible promoter.

A potentially more useful variant of this approach has been termed the thymidine kinase knockout (TKO) (Figure 11.10). Transgenic mice that express the herpes simplex virus thymidine kinase in selected groups of cells (e.g., somatotrophs) were produced. Unlike the mammalian TK protein, the HSV TK uses (2-deoxy-2-fluoro β-D-arabino-furanosyl) FIAU as a substrate, converting it to a toxic metabolite. Thus, if mice expressing the *HSV TK* gene in their somatotrophs are injected with FIAU, these cells are selectively killed. A limitation of this approach is that only cells that are synthesizing DNA and dividing are susceptible to the induced death.

Antisense RNA The previous examples describe alterations in the function or viability of specific cells. More useful is the ability to block the expression of a specific gene. One potentially powerful method for this utilizes antisense RNA. mRNA normally exists and functions as a single-stranded template for protein assembly. However, mRNAs can, like DNAs, form a duplex with a complementary RNA (or DNA) strand; such duplexes do not provide functional mRNA. Thus, it seemed plausible that the function of a specific cellular mRNA could be blocked by synthesizing RNA complementary to the target mRNA in the particular cell. Using this approach, investigators have inhibited or even completely blocked the synthesis of specific proteins by cultured cells. The inhibition is generally believed to result from the formation of RNA duplexes at sequences essential for proper initiation of translation or at later stages of polypeptide chain assembly.

Transgenes have been used to generate complementary RNA targeted for specific mRNAs in mice. Unfortunately, the ratio of unsuccessful (unreported) to successful (reported) experiments appears to be high. The principal difficulty is most likely an insufficiently high level of the complementary RNA to inhibit the translation of the target mRNA. This shortcoming is illustrated by experiments attempting to reduce the synthesis of myelin basic protein. Inhibition of myelin basic protein synthesis was achieved with a transgene that expressed an RNA complementary to the myelin basic protein mRNA, but phenotypic consequences were seen only in a genetic background that was heterozygous for a null mutation in the mouse's myelin basic protein gene. In a second, more successful case,

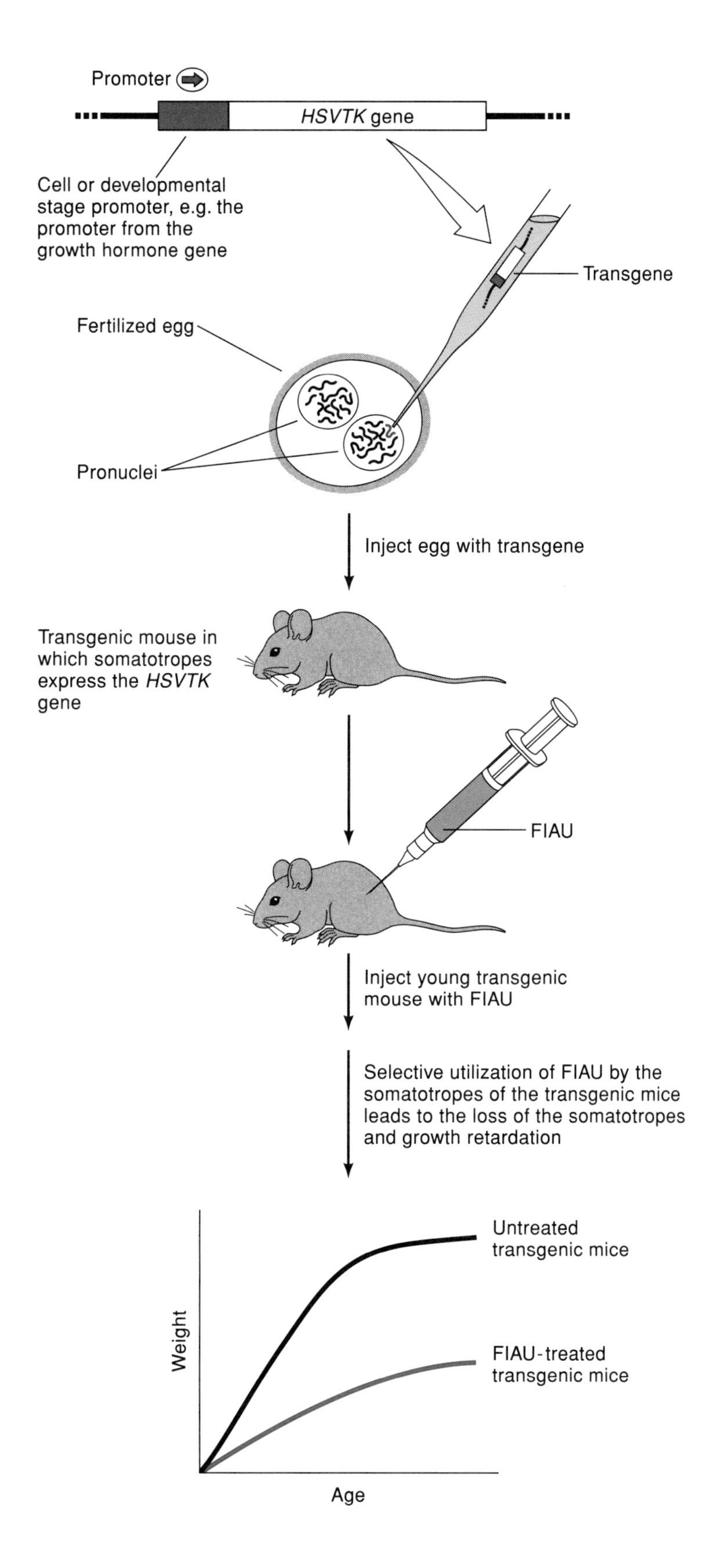

Figure 11.10
Ablation of specific cell types using the TKO protocol. Transgenic mice are generated in which the cells in the pituitary that express growth hormone (somatotrophs) also express the herpes simplex virus thymidine kinase (*HSVTK*) gene. These transgenic mice are then injected with FIAU. This metabolite is converted by the *HSVTK* to a toxic intermediate that kills the cells expressing the *HSVTK*. In this example the loss of the somatotrophs causes growth retardation.

the amount of glucocorticoid receptor present in specific cell types was lowered by expressing a transgene producing an antisense RNA to the glucocorticoid receptor mRNA. As a consequence of the transgene's action, there was an increase in circulating adrenocorticoid trophic hormone (ACTH) levels and a severe body wasting syndrome.

Inhibition of normal protein-protein interactions Many proteins function as part of multisubunit complexes. A mutant protein may still be competent to participate in complex formation but yield an inactive complex. Presumably, if the mutant protein is expressed at a sufficiently high level, most of the complexes will be inactivated, and a mutant phenotype will be produced. This approach was tested in a transgenic mouse expressing a mutant collagen. Such mice have the characteristics of the human disease osteogenesis imperfecta, a known consequence of a deficiency of normal collagen complexes. The presumption of this experiment is that the presence of the mutant collagen impairs the assembly of functional collagen complexes even though the endogenous collagen gene is expressing normal levels of authentic collagen. This approach can be exploited in other multisubunit systems.

11.7 Models of Human Diseases

As illustrated by the mice that model osteogenesis imperfecta, transgenic animals are being used to obtain an improved understanding of the pathogenesis of human diseases. This is especially so in those situations where the disease is likely to involve an interaction between genetic susceptibility and the environment, between multiple organ or cellular systems, or between multiple genetic factors. In situations such as these, the disease cannot be reproduced in a cell or organ culture, and the analysis of a whole animal is the only appropriate approach.

a. Type I Diabetes

One example of a disease that falls into this category is type I (insulin-dependent) diabetes, in which there is an autoimmunologic destruction of pancreatic β-islet cells and an ensuing hypoinsulinemia and hyperglycemia. A clear genetic susceptibility to this abnormality is conferred by specific alleles at the HLA locus. However, analysis of identical twins has shown that this genetic susceptibility is not sufficient; an environmental factor must also be involved. Three general hypotheses have been advanced to explain the pathogenesis of type I diabetes (Figure 11.11). One is the induction of the MHC proteins on the pancreatic β-cells. Because the MHC proteins are normally expressed at only very low levels on β-cells, an increased expression of these surface proteins may allow for efficient presentation of either endogenous or exogenous peptides to appropriate T cells, resulting in an autoimmune response. A second hypothesis is molecular mimicry. This hypothesis is based on the premise that the immune response to an invading pathogen's antigens is associated with an immune response that can cross-react with pancreatic β-cell antigens, thereby causing their immunologic destruction. The third hypothesis involves expression by the β-cells of one or more cytokines that stimulate a destructive inflammatory response. These three hypotheses have been tested using appropriate transgenic mice with the following results.

Increased expression of appropriate Class II MHC antigens on the β-cells leads to diabetes, but the pathology does not resemble type I diabetes in that there is no immune component responsible for the destruction of the β-cells. In

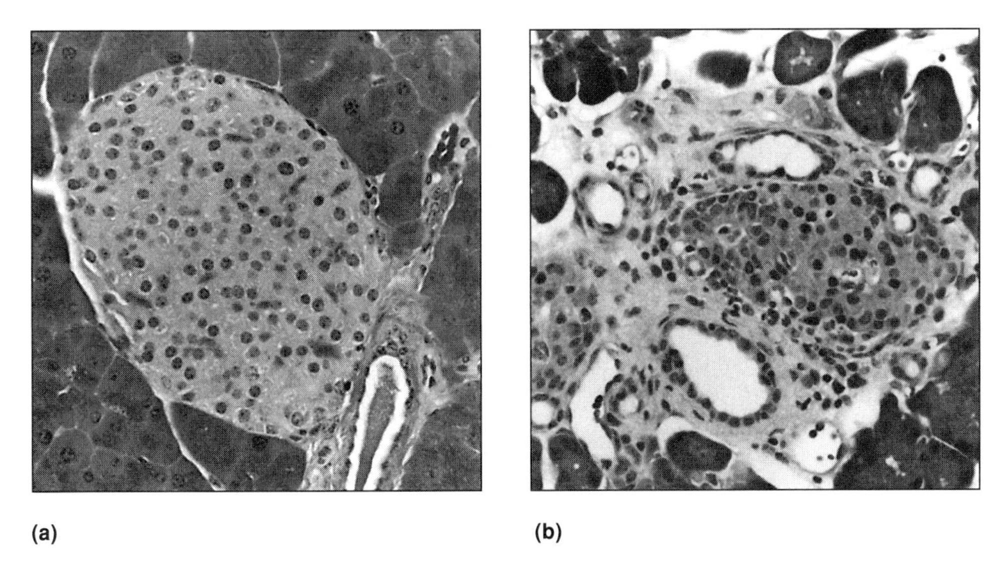

Figure 11.11
The insulitis characteristic of type I diabetes can be triggered by the expression of cytokines by the pancreatic β cells. (a) Approximately 80% of the cells in a normal islet are insulin-producing β cells. (b) In the islets of transgenic mice in which the β cells express interferon-α there is an invasion of leukocytes that coincides with the loss of the β cells and the development of diabetes. (From T. A. Stewart, B. Hultgren, X. Huang, S. Pitts-Meek, J. Hully, and N. J. Maclachlan. Induction of Type I Diabetes by Interferon α in Transgenic Mice. *Science* 260 1942–1946. Reproduced with permission.)

fact, when the MHC antigens expressed on the β-cells are different from those expressed in the rest of the animal, the immune system becomes tolerant to these "foreign" antigens. The concept of molecular mimicry has been tested by expressing a foreign surface antigen (non-MHC) in the β-cells and then stimulating the immune system to recognize this protein. When this "self-recognition" is induced by infection of the mice with a virus that infects cells within the islets and expresses the same antigen, the β-cells are destroyed, and diabetes results. Challenge to the immune system of a comparable transgenic mouse with soluble forms of the "foreign" protein results in the generation of autoantibodies and a minor lymphocytic infiltration around the islet, but this does not have pathological consequences. There are several experiments in which mice β-cells have been engineered to express cytokines from transgenes. For example, the expression of interferon α in β-cells produced both inflammation and selective loss of the β-cells and the consequent diabetes. Although these experiments have not led to a complete model for the development of type I diabetes, they have greatly increased our understanding of how tolerance is maintained and the conditions under which this form of self-protection may break down.

b. Hypertension

Another complex disease process that is beginning to be explored using transgenic rodents is hypertension. Genetic analysis of human populations has suggested that many genes contribute to a susceptibility to hypertension, and it is clear that many organ systems are involved. In addition, many environmental factors (smoking, diet, etc.) complicate the analysis. Hypertensive mice and rats have been produced by altering the renin/angiotensin system. Renin is the protease responsible for the conversion of angiotensinogen to angiotensin I, and angiotensinogen converting enzyme (ACE) produces angiotensin II from angiotensin I. Angiotensin II is a vasoconstrictor, and increased levels result in an increase in blood pressure. Transgenic rats that express a mouse renin gene were produced with the expectation that there would be an increase in circulating renin levels and consequently an increase in circulating angiotensin I and II, leading to hypertension. Such transgenic rats do indeed have a significantly increased blood pressure, but, surprisingly, there is no increase in circulating levels of angiotensin I or II or in renin (there is an increase in pro-renin levels). The vasculature was also found to express the renin transgene, and there is an increased release of angiotensin I and II from the transgenic blood vessels. These results suggest that local production of these products, rather than increased circulating levels, is responsible for the hypertension.

That this may not be the entire story is suggested by the results obtained from study of two lines of transgenic mice expressing angiotensinogen transgenes, albeit from different chromosomal integration sites. Both transgenic mouse lines have comparably increased levels of circulating angiotensin II, but only one developed an increased blood pressure. These lines differ with respect to the sites where the transgene is expressed, notably in specific brain regions, suggesting that the cellular sites of angiotensin II expression influence the onset of the hypertensive response.

c. Other Diseases

Transgenic mice have also been created to test hypotheses concerning the etiology of, for example, Down's syndrome, acromegaly, virally induced demyelinating diseases, Paget's disease, hepatitis, and familial amyloidotic polyneuropathy. Although such experiments could have a major impact on our understanding of these disorders, they may be inadequate if these diseases involve environmental influences or transient events (e.g., a rapidly cleared infection). For example, transgenic mice expressing the hepatitis surface antigen have been of considerable value in characterizing the chronic hepatitis B surface antigen carrier state. However, because such transgenic mice regard the virus's surface antigen encoded by the transgene as a self-protein, they fail to mount the significant immune response that follows a typical hepatitis B infection. If the immune response is an important part of the disease process, it is lacking in this model. Despite these problems, the use of transgenic mice to dissect and analyze the pathogenesis of human diseases remains an extremely powerful experimental tool.

Further Reading

11.1

T. P. Lin. 1967. Micropipetting Cytoplasm from the Mouse-Egg. *Nature* 216 162–163.

T. P. Lin. 1969. Microsurgery of Inner Cell Mass of Mouse Blastocysts. *Nature* 222 480–481.

R. L. Brinster. 1975. Can Teratocarcinoma Cells Colonize the Mouse Embryo? In M. I. Sherman and D. Solter (eds.), Teratomas and Differentiation, pp. 51–58. Academic Press, New York.

B. Mintz and K. Illmensee. 1975. Normal Genetically Mosaic Mice Produced from Malignant Teratocarcinoma Cells. *Proc. Natl. Acad. Sci. USA* 72 3585–3589.

R. L. Brinster. 1976. Participation of Teratocarcinoma Cells in Mouse Embryo Development. *Cancer Res.* 36 3412–3414.

A. J. Van-der-Eb, C. Mulder, F. L. Graham, and A. Houweling. 1977. Transformation with Specific Fragments of Adenovirus DNAs. I. Isolation of Specific Fragments with Transforming Activity of Adenovirus 2 and 5 DNA. *Gene* 2 115–132.

M. Wigler, S. Silverstein, L. S. Lee, A. Pellicer, Yc. Cheng, and R. Axel. 1977. Transfer of Purified Herpes Virus Thymidine Kinase Gene to Cultured Mouse Cells. *Cell* 11 223–232.

C. Mueller, A. Graessmann, and M. Graessmann. 1978. Mapping of Early SV40-Specific Functions by Microinjection of Different Early Viral DNA Fragments. *Cell* 15 579–585.

C. Shih, B. Z. Shilo, M. P. Goldfarb, A. Dannenberg, and R. A. Weinberg. 1979. Passage of Phenotypes of Chemically Transformed Cells via Transfection of DNA and Chromatin. *Proc. Natl. Acad. Sci. USA* 76 5714–5718.

J. W. Gordon, Q. A. Scangos, D. J. Plotkin, J. A. Barbosa, and F. H. Ruddle. 1980. Genetic Transformation of Mouse Embryos by Microinjection of Purified DNA. *Proc. Natl. Acad. Sci. USA* 77 7380–7384.

R. L. Brinster, H. Y. Chen, M. Trumbauer, A. W. Senear, R. Warren, and R. D. Palmiter. 1981. Somatic Expression of Herpes Thymidine Kinase in Mice Following Injection of a Fusion Gene into Eggs. *Cell* 27 223–231.

F. Costantini and E. Lacy. 1981. Introduction of a Rabbit Beta-Globin Gene into the Mouse Germ Line. *Nature* 294 92–94.

T. E. Wagner, P. C. Hoppe, J. D. Jollick, D. R. Scholl, R. L. Hodinka, and J. B. Gault. 1981. Microinjection of a Rabbit Beta-Globin Gene into Zygotes and Its Subsequent Expression in Adult Mice and Their Offspring. *Proc. Natl. Acad. Sci. USA* 78 6376–6380.

E. F. Wagner, T. A. Stewart, and B. Mintz, B. 1981. The Human β-Globin Gene and a Functional Viral Thymidine Kinase Gene in Developing Mice. *Proc. Natl. Acad. Sci. USA* 78 5016–5020.

A. Bradley, M. Evans, M. H. Kaufman, and E. Robertson. 1984. Formation of Germ-Line Chimæras from Embryo-Derived Teratocarcinoma Cell Lines. *Nature* 309 255–256.

M. R. Kuehn, A. Bradley, E. J. Robertson, and M. J. Evans. 1987. A Potential Model for Lesch-Nyhan Syndrome Through Introduction of HPRT Mutations into Mice. *Nature* 326 295–298.

M. Hooper, K. Hardy, A. Handyside, S. Hunter, and M. Monk. 1987. HPRT-Deficient (Lesch-Nyhan) Mouse Embryos Derived from Germline Colonization by Cultured Cells. *Nature* 326 292–295.

B. H. Koller, L. J. Hagemann, T. Doetschman, J. R. Hagaman, S. Huang, P. J. Williams, N. L. First, N. Maeda, and O. Smithies. 1989. Germ-Line Transmission of a Planned Alteration Made in a Hypoxanthine Phosphoribosyltransferase Gene by Homologous Recombinaiton in Embryonic Stem Cells. *Proc. Natl. Acad. Sci. USA* 86 8927–8931.

11.2

E. Lacy, S. Roberts, E. P. Evans, M. D. Burtenshaw, and F. D. Costantini. 1983. A Foreign Beta-Globin Gene in Transgenic Mice: Integration at Abnormal Chromosomal Positions and Expression in Inappropriate Tissues. *Cell* 34 343–358.

E. F. Wagner, L. Covarrubias, T. A. Stewart, and B. Mintz. 1983. Prenatal Lethalities in Mice Homozygous for Human Growth Hormone Gene Sequences Integrated in the Germ Line. *Cell* 35 647–655.

K. Chada, J. Magram, K. Raphael, G. Radice, E. Lacy, and F. Costantini. 1985. Specific Expression of a Foreign Beta-Globin Gene in Erythroid Cells of Transgenic Mice. *Nature* 314 377–380.

J. Magram, K. Chada, and F. Costantini. 1985. Developmental Regulation of a Cloned Adult Beta-Globin Gene in Transgenic Mice. *Nature* 315 338–340.

T. M. Townes, H. Y. Chen, J. B. Lingrel, R. D. Palmiter, and R. L. Brinster. 1985. Expression of Human Beta-Globin Genes in Transgenic Mice: Effects of a Flanking Metallothionein-Human Growth Hormone Fusion Gene. *Mol. Cell. Biol.* 5 1977–1983.

T. M. Townes, J. B. Lingrel, H. Y. Chen, R. L. Brinster, and R. D. Palmiter. 1985. Erythroid-Specific Expression of Human Beta-Globin Genes in Transgenic Mice. *EMBO J.* 4 1715–1723.

R. P. Woychik, T. A. Stewart, L. G. Davis, P. D'Eustachio, and P. Leder. 1985. An Inherited Limb Deformity Created by Insertional Mutagenesis in a Transgenic Mouse. *Nature* 318 36–40.

K. Chada, J. Magram, and F. Costantini. 1986. An Embryonic Pattern of Expression of a Human Fetal Globin Gene in Transgenic Mice. *Nature* 319 685–689.

G. Kollias, N. Wrighton, J. Hurst, and F. Grosveld. 1986. Regulated Expression of Human A Gamma-, Beta-, and Hybrid Gamma Beta-Globin Genes in Transgenic Mice: Manipulation of the Developmental Expression Patterns. *Cell* 46 89–94.

R. J. MacDonald, R. E. Hammer, G. H. Swift, B. P. Davis, and R. L. Brinster. 1986. Transgenic Progeny Inherit Tissue-Specific Expression of Rat Elastase I Genes. *DNA* 5 393–401.

R. R. Behringer, R. E. Hammer, R. L. Brinster, R. D. Palmiter, and T. M. Townes. 1987. Two 3′ Sequences Direct Adult Erythroid-Specific Expression of Human Beta-Globulin Genes in Transgenic Mice. *Proc. Natl. Acad. Sci. USA* 84 7056–7060.

R. E. Hammer, G. H. Swift, D. M. Ornitz, C. J. Quaife, R. D. Palmiter, R. L. Brinster, and R. J. MacDonald. 1987. The Rat Elastase I Regulatory Element Is an Enhancer That Directs Correct Cell Specificity and Developmental Onset of Expression in Transgenic Mice. *Mol. Cell. Biol.* 7 2956–2967.

G. Killias, J. Hurst, E. de Boer, and F. Grosveld. 1987. The Human Beta-Globin Gene Contains a Downstream Developmental Specific Enhancer. *Nucleic Acids Res.* 15 5739–5747.

W. Reik, A. Collick, M. L. Norris, S. C. Barton, and M. A. Surani. 1987. Genomic Imprinting Determines Methylation of Parental Alleles in Transgenic Mice. *Nature* 328 248–251.

C. Sapienza, A. C. Peterson, J. Rossant, and R. Balling. 1987. Degree of Methylation of Transgenes Is Dependent on Gamete of Origin. *Nature* 328 251–254.

J. L. Swain, T. A. Stewart, and P. Leder. 1987. Parental Legacy Determines Methylation and Expression of an Autosomal Transgene: A Molecular Mechanism for Parental Imprinting. *Cell* 50 719–727.

M. Trudel, J. Magram, L. Bruckner, and F. Costantini. 1987. Upstream G Gamma-Globin and Downstream Beta-Globin Sequences Required for Stage-Specific Expression in Transgenic Mice. *Mol. Cell. Biol.* 7 4024–4029.

B. P. Davis and R. J. MacDonald. 1988. Limited Transcription of Rat Elastase I Transgene Repeats in Transgenic Mice. *Genes & Development* 2 13–22.

T. A. Stewart, P. G. Hollingshead, and S. L. Pitts. 1988. Multiple Regulatory Domains in the Mouse Mammary Tumor Virus Long Terminal Repeat Revealed by Analysis of Fusion Genes in Transgenic Mice. *Mol. Cell. Biol.* 8 473–479.

G. H. Swift, F. Kruse, R. J. MacDonald, and R. E. Hammer. 1988. Differential Requirements for Cell-Specific Elastase I Enhancer Domains in Transfected Cells and Transgenic Mice. *Genes & Development* 3 687–696.

11.3

R. D. Palmiter, R. L. Brinster, R. E. Hammer, M. E. Trumbauer, M. G. Rosenfeld, N. C. Birnberg, and R. M. Evans. 1982. Dramatic Growth of Mice That Develop from Eggs Microinjected with Metallothionein-Growth Hormone Fusion Genes. *Nature* 300 611–615.

R. D. Palmiter, G. Norstedt, R. E. Gelinas, R. E. Hammer, and R. L. Brinster. 1983. Metallothionein-Human GH Fusion Genes Stimulate Growth of Mice. *Science* 222 809–814.

R. E. Hammer, R. L. Brinster, M. G. Rosenfeld, R. M. Evans, and K. E. Mayo. 1985. Expression of Human Growth Hormone-Releasing Factor in Transgenic Mice Results in Increased Somatic Growth. *Nature* 315 413–416.

L. S. Matthews, R. E. Hammer, R. R. Behringer, A. J. Dercole, G. I. Bell, R. L. Brinster, and R. D. Palmiter. 1988. Growth Enhancement of Transgenic Mice Expressing Human Insulin-Like Growth Factor I. *Endocrinology* 123 2827–2833.

K. E. Mayo, R. E. Hammer, L. W. Swanson, R. L. Brinster, M. G. Rosenfeld, and R. M. Evans. 1988. Dramatic Pituitary Hyperplasia in Transgenic Mice Expressing a Human Growth Hormone-Releasing Factor Gene. *Mol. Endocrinol.* 2 606–612.

R. Balling, G. Mutter, P. Gruss, and M. Kessel. 1989. Craniofacial Abnormalities Induced by Ectopic Expression of the Homeobox Gene Hox-1.1 in Transgenic Mice. *Cell* 58 337–347.

E. Borelli, R. A. Heyman, C. Arias, P. E. Sawchenko, and R. M. Evans. 1989. Transgenic Mice with Inducible Dwarfism. *Nature* 339 538–541.

P. G. Hollingshead, L. Martin, S. L. Pitts, and T. A. Stewart. 1989. A Dominant Phenocopy of Hypopituitarism in Transgenic Mice Resulting from Central Nervous System Synthesis of Human Growth Hormone. *Endocrinology* 125 1556–1564.

T. M. DeChiara, A. Efstratiadis, and E. J. Robertson. 1990. A Growth-Deficiency Phenotype in Heterozygous Mice Carrying an Insulin-Like Growth Factor II Gene Disrupted by Targeting. *Nature* 345 78–80.

M. Kessel, R. Balling, and P. Gruss. 1990. Variations of Cervical Vertebrae After Expression of a Hox-1.1 Transgene in Mice. *Cell* 61 301–308.

A. L. Joyner, K. Herrup, B. A. Auerbach, C. A. Davis, and J. Rossant. 1991. Subtle Cerebellar Phenotype in Mice Homozygous for a Targeted Deletion of the En-2 Homeobox. *Science* 251 1239–1243.

T. Lufkin, A. Dierich, M. LeMeur, M. Mark, and P. Chambon. 1991. Disruption of the Hox-1.6 Homeobox Gene Results in Defects in a Region Corresponding to Its Rostral Domain of Expression. *Cell* 66 1105–1119.

S. Kimura, J. J. Mullins, and B. Bunnemann. 1992. High Blood Pressure in Transgenic Mice Carrying the Rat Angiotensinogen Gene. *EMBO J.* 11 821–827.

11.4

R. L. Brinster, K. A. Ritchie, R. E. Hammer, R. L. O'Brien, B. Arp, and U. Storb. 1983. Expression of a Microinjected Immunoglobulin Gene in the Spleen of Transgenic Mice. *Nature* 306 332–336.

R. Grosschedl, D. Weaver, D. Baltimore, and F. Costantini. 1984. Introduction of a Mu Immunoglobulin Gene into the Mouse Germ Line: Specific Expression in Lymphoid Cells and Synthesis of Functional Antibody. *Cell* 38 647–658.

K. A. Ritchie, R. L. Brinster, and U. Storb. 1984. Allelic Exclusion and Control of Endogenous Immunoglobulin Gene Rearrangement in Kappa Transgenic Mice. *Nature* 312 517–520.

U. Storb, R. L. O'Brien, M. D. McMullen, K. Gollahon, and R. L. Brinster. 1984. High Expression of Cloned Immunoglobulin Kappa Gene in Transgenic Mice Is Restricted to B Lymphocytes. *Nature* 310 238–241.

S. Rusconi and G. Kohler. 1985. Transmission and Expression of a Specific Pair of Rearranged Immunoglobulin Mu and Kappa Genes in a Transgenic Mouse Line. *Nature* 314 330–334.

U. Storb, K. A. Denis, R. L. Brinster, and O. N. Witte. 1985. Pre-B Cells in Kappa-Transgenic Mice. *Nature* 316 356–358.

U. Storb, C. Pinkert, B. Arp, P. Engler, K. Gollahon, J. Manz, W. Brady, and R. L. Brinster. 1986. Transgenic Mice with Mu and Kappa Genes Encoding Antiphosphorylcholine Antibodies. *J. Exp. Med.* 164 627–641.

D. Weaver, M. H. Reis, C. Albanese, F. Costantini, D. Baltimore, and T. Imanishi-Kari. 1986. Altered Repertoire of Endogenous Immunoglobulin Gene Expression in Transgenic Mice Containing a Rearranged Mu Heavy Chain Gene. *Cell* 45 247–259.

K. Yamamura, A. Kudo, T. Ebihara, K. Kamino, K. Araki, Y. Kumahara, and T. Watanabe. 1986. Cell-Type-Specific and Regulated Expression of a Human Gamma 1 Heavy-Chain Immunoglobulin Gene in Transgenic Mice. *Proc. Natl. Acad. Sci. USA* 83 2152–2156.

A. Iglesias, M. Lamers, and G. Kohler. 1987. Expression of Immunoglobulin Delta Chain Causes Allelic Exclusion in Transgenic Mice. *Nature* 330 482–484.

F. Kievits, P. Ivanyi, P. Krimpenfort, A. Berns, and H. L. Ploegh. 1987. HLA-Restricted Recognition of Viral Antigens in HLA Transgenic Mice. *Nature* 329 447–449.

G. Widera, L. C. Burkly, C. A. Pinkert, E. C. Bottger, C. Cowing, R. D. Palmiter, R. L. Brinster, and R. A. Flavell. 1987. Transgenic Mice Selectively Lacking MHC Class II (I-E) Antigen Expression on B cells: An *in vivo* Approach to Investigate Ia Gene Function. *Cell* 51 175–187.

T. Yoshioka, C. Bieberich, G. Scangos, and G. Jay. 1987. A Transgenic Class I Antigen Is Recognized as Self and Functions as a Restriction Element. *J. Immunol.* 139 3861–3867.

H. Blüthmann, P. Kisielow, Y. Uematsu, M. Malissen, P. Krimpenfort, A. Berns, H. von Boehmer, and M. Steinmetz. 1988. T-Cell-Specific Deletion of T-Cell Receptor Transgenes Allows Functional Rearrangement of Endogenous α- and β-Genes. *Nature* 334 156–159.

C. C. Goodnow, J. Crosbie, S. Adelstein, T. B. Lavoie, S. J. Smith-Gill, R. A. Brink, H. Pritchard-Briscoe, L. S. Wotherspoon, R. H. Loblay, K. Raphael, R. J. Trent, and A. Basten. 1988. Altered Immunoglobulin Expression and Functional Silencing of Self-Reactive B Lymphocytes in Transgenic Mice. *Nature* 334 676–683.

P. Kisielow, H. Blüthmann, U. D. Staerz, M. Steinmetz, and H. von Boehmer. 1988. Tolerance in T-Cell-Receptor Transgenic Mice Involves Deletion of Nonmature $CD4^{+}8^{+}$ Thymocytes. *Nature* 333 742–746.

P. Kisielow, H. S. Teh, H. Blüthmann, and H. von Boehmer. 1988b. Positive Selection of Antigen-Specific T Cells in Thymus by Restricting MHC Molecules. *Nature* 335 730–733.

D. Lo, L. C. Burkly, G. Widera, C. Cowing, R. A. Flavell, R. D. Palmiter, and R. L. Brinster. 1988. Diabetes and Tolerance in Transgenic Mice Expressing Class II MHC Molecules in Pancreatic Beta Cells. *Cell* 53 159–168.

W. C. Sha, C. A. Nelson, R. D. Newberry, D. M. Kranz, J. H. Russell, and D. Y. Loh. 1988. Selective Expression of an Antigen Receptor on CD8-Bearing T Lymphocytes in Transgenic Mice. *Nature* 335 271–274.

H. S. Teh, P. Kisielow, B. Scott, H. Kishi, Y. Uematsu, H. Blüthmann, and H. von Boehmer. 1988. Thymic Major Histocompatibility Complex Antigens and the αβT-Cell Receptor Determine the CD4/CD8 Phenotype of T Cells. *Nature* 335 229–233.

Y. Uematsu, S. Ryser, Z. Dembic, P. Borgulya, P. Krimpenfort, A. Berns, H. von Boehmer, and M. Steinmetz. 1988. In Transgenic Mice the Introduced Functional T Cell Receptor Beta Gene Prevents Expression of Endogenous Beta Genes. *Cell* 52 831–841.

C. Benoist and D. Mathis. 1989. Positive Selection of the T Cell Repertoire: Where and When Does It Occur? *Cell* 58 1027–1033.

L. J. Berg, A. M. Pullen, B. F. de St. Groth, D. Mathis, C. Benoist, and M. M. Davis. 1989. Antigen/MHC-Specific T Cells Are Preferentially Exported from the Thymus in the Presence of Their MHC Ligand. *Cell* 58 1035–1046.

M. Zijlstra, E. Li, F. Sajjadi, S. Subramani, and R. Jaenisch. 1989. Germ-Line Transmission of a Disrupted Beta 2-Microglobulin Gene Produced by Homologous Recombination in Embryonic Stem Cells. *Nature* 342 435–438.

R. Kuhn, K. Rajewsky, and W. Muller. 1991. Generation and Analysis of Interleukin-4 Deficient Mice. *Science* 254 707–710.

H. Schorle, T. Holtschke, T. Hunig, A. Schimpl, and I. Horak. 1991. Development and Function of T Cells in Mice Rendered Interleukin-2 Deficient by Gene Targeting. *Nature* 352 621–624.

R. L. Tarleton, B. H. Koller, A. Latour, and M. Postan. 1992. Susceptibility of Beta-2 Microglobulin-Deficient Mice to *Trypanosoma Cruzi* Infection. *Nature* 356 338–340.

11.5

R. L. Brinster, H. Y. Chen, A. Messing, T. van Dyke, A. J. Levine, and R. D. Palmiter. 1984. Transgenic Mice Harboring SV40 T-Antigen Genes Develop Characteristic Brain Tumors. *Cell* 37 367–379.

T. A. Stewart, P. K. Pattengale, and P. Leder. 1984. Spontaneous Mammary Adenocarcinomas in Transgenic Mice That Carry and Express MTV/*myc* Fusion Genes. *Cell* 38 627–637.

R. D. Palmiter and R. L. Brinster. 1985. The c-*myc* Oncogene Driven by Immunoglobulin Enhancers Induces Lymphoid Malignancy in Transgenic Mice. *Nature* 318 533–538.

M. Nerenberg, S. H. Hinrichs, R. K. Reynolds, G. Khoury, and G. Jay. 1987. The TAT Gene of Human T-Lymphotropic Virus Type 1 Induces Mesenchymal Tumors in Transgenic Mice. *Science* 237 1324–1329.

D. M. Ornitz, R. E. Hammer, A. Messing, R. D. Palmiter, and R. L. Brinster. 1987. Pancreatic Neoplasia Induced by SV40 T-Antigen Expression in Acinar Cells of Transgenic Mice. *Science* 238 188–193.

E. Sinn, M. Muller, P. Pattengale, I. Tepler, R. Wallace, and P. Leder. 1987. Coexpression of MMTV/v-Ha-*ras* and MMTV/c-*myc* Genes in Transgenic Mice: Synergistic Action of Oncogenes *in vivo*. *Cell* 49 389–398.

P. Soriano, C. Montgomery, R. Geske, and A. Bradley. 1991. Targeted Disruption of the c-src Proto-Oncogene Leads to Osteopetrosis in Mice. *Cell* 64 693–702.

L. A. Donehower, M. Harvey, B. L. Slagle, M. J. McArthur, C. A. Montgomery Jr., J. S. Butel, and A. Bradley. 1992. Mice Deficient for p53 Are Developmentally Normal but Susceptible to Spontaneous Tumours. *Nature* 356 215–221.

11.6

R. D. Palmiter, T. M. Wilkie, H. Y. Chen, and R. L. Brinster. 1984. Transmission Distortion and Mosaicism in an Unusual Transgenic Mouse Pedigree. *Cell* 36 869–877.

J. W. Gordon. 1986. A Foreign Dihydrofolate Reductase Gene in Transgenic Mice Acts as a Dominant Mutation. *Mol. Cell. Biol.* 6 2158–2167.

L. M. Isola and J. W. Gordon. 1986. Systemic Resistance to Methotrexate in Transgenic Mice Carrying a Mutant Dihydrofolate Reductase Gene. *Proc. Natl. Acad. Sci. USA* 83 9621–9625.

M. L. Breitman, S. Clapoff, J. Rossant, L. C. Tsui, L. M. Glode, I. H. Maxwell, and A. Bernstein. 1987. Genetic Ablation: Targeted Expression of a Toxin Gene Causes Microphthalmia in Transgenic Mice. *Science* 238 1563–1565.

R. D. Palmiter, R. R. Behringer, C. J. Quaife, F. Maxwell, I. H. Maxwell, and R. L. Brinster. 1987. Cell Lineage Ablation in Transgenic Mice by Cell-Specific Expression of a Toxin Gene. *Cell* 50 435–443.

E. Borrelli, R. A. Heyman, C. Arias, P. E. Sawchenko, and R. M. Evans. 1988. Transgenic Mice with Inducible Dwarfism. *Nature* 339 538.

K. A. Mahon, P. A. Overbeek, and H. Westphal. 1988. Prenatal Lethality in a Transgenic Mouse Line Is the Result of a Chromosomal Translocation. *Proc. Natl. Acad. Sci. USA* 85 1165–1168.

A. Stacey, J. Bateman, T. Choi, T. Mascara, W. Cole, and R. Jaenisch. 1988. Perinatal Lethal Osteogenesis Imperfecta in Transgenic Mice Bearing an Engineered Mutant Pro-Alpha 1(I) Collagen Gene. *Nature* 332 131–136.

M. C. Pepin, F. Pothier, and N. Barden. 1992. Impaired Type II Glucocorticoid-Receptor Function in Mice Bearing Antisense RNA Transgene. *Nature* 355 725–728.

11.7

C. Babinet, H. Farza, D. Morello, M. Hadchouel, and C. Pourcel. 1985. Specific Expression of Hepatitis B Surface Antigen (HBsAg) in Transgenic Mice. *Science* 230 1160–1163.

F. V. Chisari, C. A. Pinkert, D. R. Milich, P. Filippi, A. McLachland, R. D. Palmiter, and R. L. Brinster. 1985. A Transgenic Mouse Model of the Chronic Hepatitis B Surface Antigen Carrier State. *Science* 230 1157–1160.

H. Sasaki, S. Tone, M. Nakazato, K. Yoshioka, H. Matsuo, Y. Kayo, and Y. Sakaki. 1986. Generation of Transgenic Mice Producing a Human Transthyretin Variant: A Possible Mouse Model for Familial Amyloidotic Polyneuropathy. *Biochem. Biophys. Res. Commun.* 139 794–799.

F. V. Chisari, P. Filippi, J. Buras, A. McLachland, H. Popper, C. A. Pinkert, R. D. Palmiter, and R. L. Brinster. 1987. Structural and Pathological Effects of Synthesis of Hepatitis B Virus Large Envelope Polypeptide in Transgenic Mice. *Proc. Natl. Acad. Sci. USA* 84 6909–6913.

S. H. Hinrichs, M. Nerenberg, R. K. Reynolds, G. Khoury, and G. Jay. 1987. A Transgenic Mouse Model for Human Neurofibromatosis. *Science* 237 1340–1343.

W. Reik, A. Collick, M. L. Norris, S. C. Barton, and M. A. Surani. 1987. Genomic Imprinting Determines Methylation of Parental Alleles in Transgenic Mice. *Nature* 328 248–251.

C. Sapienza, A. C. Peterson, J. Rossant, and R. Balling. 1987. Degree of Methylation of Transgenes Is Dependent on Gamete of Origin. *Nature* 328 251–254.

J. L. Swain, T. A. Stewart, and P. Leder. 1987. Parental Legacy Determines Methylation and Expression of an Autosomal Transgene: A Molecular Mechanism for Parental Imprinting. *Cell* 50 719–727.

N. Sarvetnick, D. Liggitt, S. L. Pitts, S. E. Hansen, and T. A. Stewart. 1988. Insulin-Dependent Diabetes Mellitus Induced in Transgenic Mice by Ectopic Expression of Class II MHC and Interferon-Gamma. *Cell* 52 773–782.

A. Stacey, J. Bateman, T. Choi, T. Mascara, W. Cole, and R. Jaenisch. 1988. Perinatal Lethal Osteogenesis Imperfecta in Transgenic Mice Bearing an Engineered Mutant Pro-Alpha 1(I) Collagen Gene. *Nature* 332 131–136.

J. J. Mullins, J. Peters, and D. Ganten. 1990. Fulminant Hypertension in Transgenic Rats Harbouring the Mouse Ren-2 Gene. *Nature* 344 541–544.

P. S. Ohashi, S. Oehen, K. Buerki, H. Pircher, C. T. Ohashi, and B. Odermatt. 1991. Ablation of "Tolerance" and Induction of Diabetes by Virus Infection in Viral Antigen Transgenic Mice. *Cell* 65 305–317.

M. B. Oldstone, M. Nerenberg, P. Southern, J. Price, and H. Lewicki. 1991. Virus Infection Triggers Insulin-Dependent Diabetes Mellitus in a Transgenic Model: Role of Anti-Self (Virus) Immune Response. *Cell* 65 319–331.

S. Kimura, J. J. Mullins, and B. Bunnemann, B. 1992. High Blood Pressure in Transgenic Mice Carrying the Rat Angiotensinogen Gene. *EMBO J.* 11 821–827.

CHAPTER 12

Genetic Manipulation of Plants

PATRICIA ZAMBRYSKI

This chapter covers two distinct topics, *Agrobacterium* plant cell transformation and flower development. These topics both give the reader a flavor for some of the techniques used to manipulate and study plants and present an area of intense study in plant biology.

The interaction of the soil bacterium *Agrobacterium tumefaciens* with plant cells has received considerable attention over the last century. This microorganism was shown to be the cause of crown gall, the name given to tumorous tissues usually formed on the stems of wounded plants. The initial interest was based on the hope that the study of plant neoplasia would provide some insight into animal tumors. During the last 20 years, the disease was shown to be due to the transfer of a particular segment of DNA from the bacterium to the plant cell and its expression following integration into the plant genome. With this result, interest expanded to the potential use of *Agrobacterium* as a vector for the genetic engineering of plant cells. This aspect of the *Agrobacterium*–plant cell interaction is perhaps most widely known.

The study of the interaction between *Agrobacterium* and plant cells has revealed a number of biologically interesting phenomena: (1) the complex series of events that results in the unidirectional transfer of DNA between two widely different organisms, the analysis of which has provided insight into other DNA mobile element phenomena; (2) the *Agrobacterium*–plant cell interaction provides a model system for the study of how bacterial–plant cell communication is initiated and maintained in the biologically and chemically complex soil environment; (3) because crown gall cells display either an undifferentiated or teratomatic growth phenotype, they provide insight into how growth and differentiation are controlled in plant cells, and perhaps of neoplasia in cells from other kingdoms; and (4) the DNA transfer process provides a natural vector system for the transformation of plant cells with any DNA of interest.

First I review the biological aspects of the *Agrobacterium* system. Then, I describe how this system has been modified for application to the genetic transformation of plant cells by specific DNA segments, mentioning how *Agrobacterium* can be used to probe other biological areas of interest. The last part of this chapter starts by providing a general review of plant development, highlighting the different strategies employed by plants and animals. Finally, I present the initiation and development of the reproductive pathway and flowering as an example of a complex plant developmental process. The understanding of these processes is a consequence, in part, of the use of the *Agrobacterium* to generate mutations in plants.

12.1. *Agrobacterium*–Plant Cell Interaction: General Concepts

The steps involved in the genetic transformation of plant cells by *Agrobacterium* are summarized in Figure 12.1. This process is highly regulated and is triggered only in the presence of susceptible wounded plant cells. In brief, (1) wounded plant cells produce low molecular weight phenolic compounds that serve as signal molecules to induce the expression of a particular set of *Agrobacterium* virulence (*vir*) genes and thereby (2) initiate the production of a DNA transfer intermediate that is (3) subsequently coated with proteins to form a transferable DNA-protein complex (T complex). Then (4) the T complex exits the bacterial cell, (5) enters the plant cell through the plant cell membrane, and (6) enters the plant cell nucleus. Finally (7) its DNA component is integrated stably into the plant genome, where its expression leads to tumorous growth.

A large (approximately 200 kb) plasmid within the bacterium contains the genes for plant transformation and tumor induction (Figure 12.2). A particular segment of this tumor-inducing (Ti) plasmid is transferred and integrated into the plant cell genome. The transferred segment is called the T-DNA. The T-DNA element contains genes that are selectively expressed in transformed plant cells and that specify the production of plant growth hormones (auxins and cytokinins) by novel enzymatic pathways. The overproduction of these plant growth hormones results in cellular dedifferentiation and tumorous growth.

A second region of the Ti-plasmid is essential for tumor formation. This virulence (*vir*) region acts as the master controller for whether or not T-DNA transfer occurs for two reasons. First, it encodes the protein products that constitute the functional machinery for mobilizing the T-DNA from *Agrobacterium* to the plant cell. Second, expression of this region occurs only when *Agrobacterium* is in the presence of susceptible plant cells. To *Agrobacterium*, a susceptible cell is a wounded plant cell, and such cells produce specific signal molecules for inducing *vir* expression.

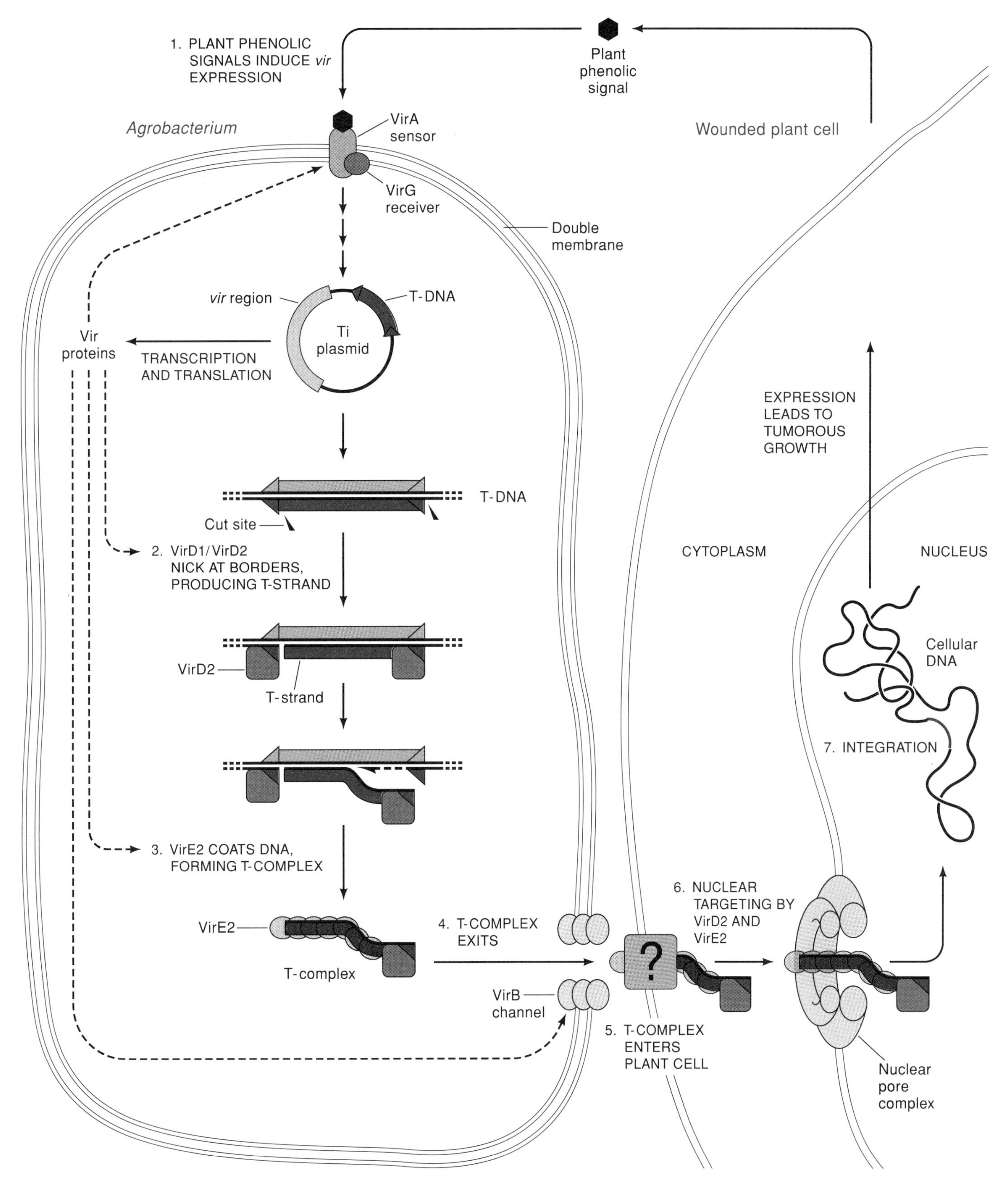

Figure 12.1
Outline of transformation of plant cells by *Agrobacterium.* The seven steps are briefly summarized: (1) induction of *vir* gene expression by plant wound phenolics, (2) production of a transferable T-DNA copy called the T-strand after VirD2-induced nicking at the T-DNA borders, (3) formation of a single-strand DNA protein complex (T-complex) after coating with VirE2, (4) exit of the T-complex from the bacterial cell, (5) entrance of the T-complex into the plant cell, (6) targeting and entrance of the T-complex into the nucleus, and (7) integration of the T-strand into plant cellular DNA.

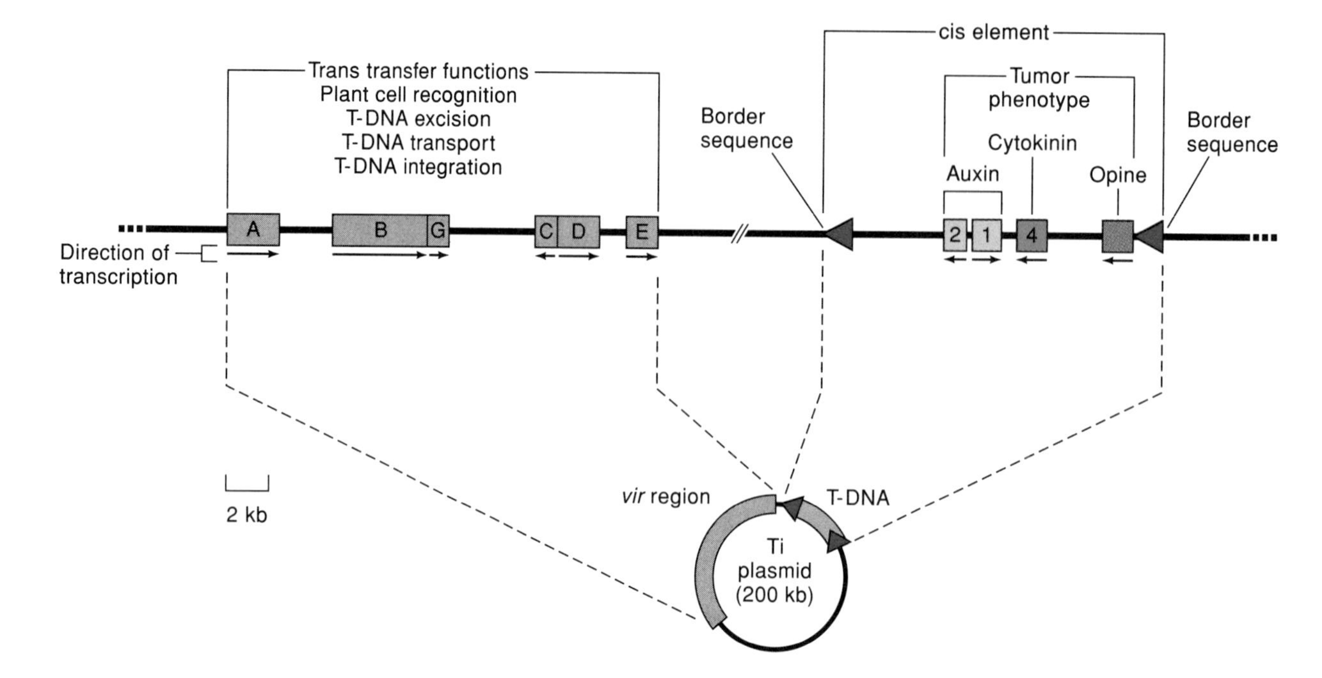

Figure 12.2
The Ti plasmid and the genetic components of virulence. The transcriptional and functional organization of the virulence (*vir*) region is as demonstrated first in Stachel and Nester, 1986. Arrows indicate the direction of transcription of the six *vir* genes (A, B, C, D, E, G). The T-DNA region is that of the nopaline Ti plasmid, which contains a single T-DNA segment. The polar 25 bp T-DNA border repeats are indicated by the black arrows at the end of the element. Three T-DNA genes, 1, 2, and 4 encode auxin and cytokinin synthesis; their overproduction results in the tumor phenotype. (Adapted from S. E. Stachel and E. W. Nester. 1986. The Genetic and Transcriptional Organization of the *vir* Region of the A6 Ti Plasmid of *Agrobacterium tumefaciens*. *EMBO J.* 5 1445–1454.)

Besides the two Ti-plasmid genetic elements (T-DNA and *vir* genes), several *Agrobacterium* chromosomal genes are essential for the transfer of T-DNA. Mutants at three loci, *chvA*, *chvB*, and *exoC*, affect the specific binding of *Agrobacterium* to plant cells during the infection process. These genes are involved in the synthesis or translocation of cyclic β-1,2 glucans and thus probably affect the composition of the bacterial glycoproteins that mediate the attachment to the plant cell. Another chromosomal locus, *chvE*, produces a periplasmic sugar-binding protein that enhances *vir* gene induction when the plant-derived *vir*-inducing factors are limiting. Because these chromosomal genes are expressed constitutively during bacterial growth, I do not discuss them further. I mention them to point out that T-DNA transfer also relies on common bacterial functions.

Another phenotypic characteristic of crown gall cells is the production of small sugar–amino acid derivatives called opines; the enzymes required for their synthesis are encoded by a portion of the T-DNA element. The production of opines is hypothesized to be the "biological rationale" for crown gall formation because *Agrobacterium* also carries genetic information for their catabolism. Thus, *Agrobacterium* transfers a DNA segment that results in the synthesis of opines by vigorously growing plant cells. Because the *Agrobacterium* can specifically use these compounds as carbon and nitrogen sources, they have a selective advantage over other soil organisms surrounding the crown gall. Two common opines are nopaline (the condensation product of arginine and α-ketoglutaric acid) and octopine (the condensation product of arginine and pyruvic acid).

Besides inducing expression of genes for opine catabolism, abundant opines in the soil microenvironment around a crown gall also induce *Agrobacteria* to express genes that promote conjugal transfer of the Ti-plasmid. Mating is initiated when opines induce a signaling cascade that results in the stimulation of bacterial production of a diffusible second messenger: a homoserine lactone derivative called conjugating factor. The conjugating factor in turn activates expression of the *Tra* conjugation genes.

Ti-plasmids are classified on the basis of their capacity to specify the machinery for opine synthesis and utilization. For example, some Ti-plasmids specify nopaline synthesis in crown gall tumor cells and encode catabolic enzymes for nopaline utilization in bacteria. Others are specialized for the same capabilities for octopine. The so-called nopaline and octopine types are the most commonly studied Ti-plasmids. Both are approximately 200 kb but are only 30 percent homologous. The homologous regions encode functions essential for plant cell transformation as well as plasmid replication and transfer.

12.2. Activation and Expression of Virulence Genes

The description of how *Agrobacterium* evolved to control and activate the T-DNA transfer process reveals both the logic and conservation of biological mechanisms. In nature, *Agrobacterium* infects only wounded plants. Researchers initially assumed these are the preferred targets because their cells present less of a physical barrier to bacterial penetration than the thick cell walls of unwounded plant cells. However, wounded but otherwise metabolically active plant cells excrete low molecular weight signal molecules that *Agrobacterium* recognize and that induce *vir* gene expression, thereby activating the T-DNA transfer process. The *vir*-inducing substances were purified from the culture media of wounded tobacco plant cells and identified as acetosyringone (AS) and hydroxy-acetosyringone (HO-AS)(Figure 12.3). These low molecular weight phenolic compounds alone, in chemically pure form, are capable of activating *vir* gene expression in the absence of plant cells. AS resembles metabolites of the major pathway (Figure 12.4) for the production of compounds such as lignin and flavonoids, which are important to plants subjected to stress or injury. Thus, the activation of *vir* genes makes sense; the production of phenolic compounds that act as *vir* inducers is stimulated by wounding and may provide a selective advantage by restricting *vir* gene expression to situations where the probability of successful infection is high. AS has also been shown to act as a chemical attractant for *Agrobacterium in vitro,* suggesting that its presence at wound sites in nature may serve a chemotactic role.

We now know that the soil microenvironment is a rich source of chemical signals for many bacteria and their plant hosts. For example, another well-studied system is the interaction between *Rhizobium* and plants leading to the production of nitrogen-fixing nodules. In this system, the production of specific flavonoid molecules produced by the host plant initiate the nodulation process by acting as inducers of bacterial *Nod* gene expression. Once *Nod* genes are

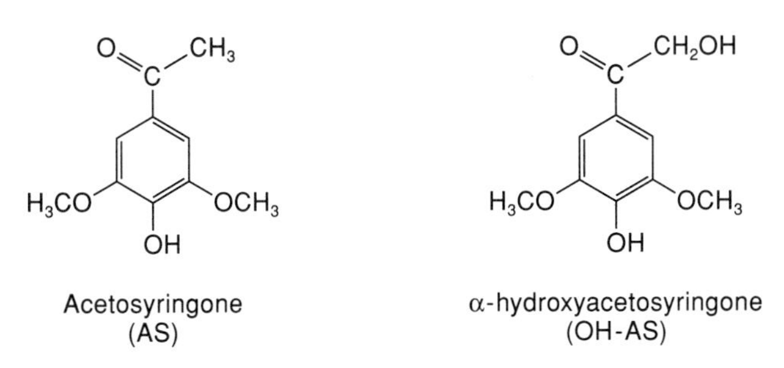

Figure 12.3
Chemical structure of plant signal molecules that induce *vir* gene expression.

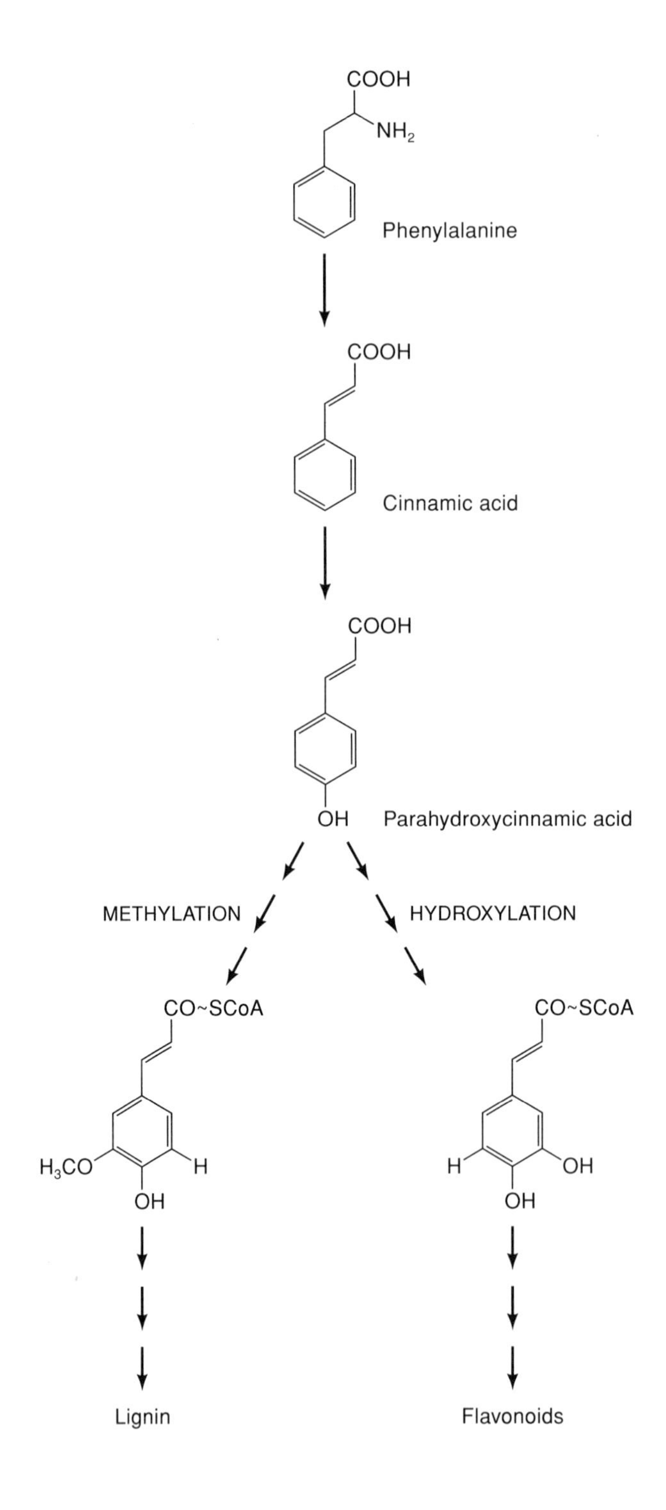

Figure 12.4
Outline of phenylpropanoid metabolism in plant cells.

turned on, the bacterium produces different lipooligosaccharides that act as signals to initiate changes in plant cell division patterns and morphology and ultimately lead to the formation of a nitrogen-fixing nodule. These Nod factors act as species-specific morphogens to their compatible host plant.

At least two components are necessary for *Agrobacterium* to link the recognition of plant phenolics to the expression of *vir* genes: extracellular recognition and intracellular response. These two different steps are mediated by the products of two of the *vir* genes, *vir*A and *vir*G. Current models on how VirA and VirG proteins function to activate *vir* gene transcription are based on the homology of these proteins to other pairs of bacterial proteins that act, respectively, as sensors or regulators of gene expression in response to environmental stimuli. Over 100 two-component bacterial sensor-regulator genes have been identified since 1986. The sensor component directly senses the environment, for example, changes in osmolarity or nutrients. The regulatory component acts as a positive transcriptional activator of genes whose expression represents the cell's response to the stimulus. Generally, the signal is transduced from the sensor to the regulator by phosphorylation, that is, sensors act as specific kinases of their regulator partners (Figure 12.5). Thus, *Agrobacterium* has evolved to utilize a well-conserved mechanism to activate *vir* gene expression.

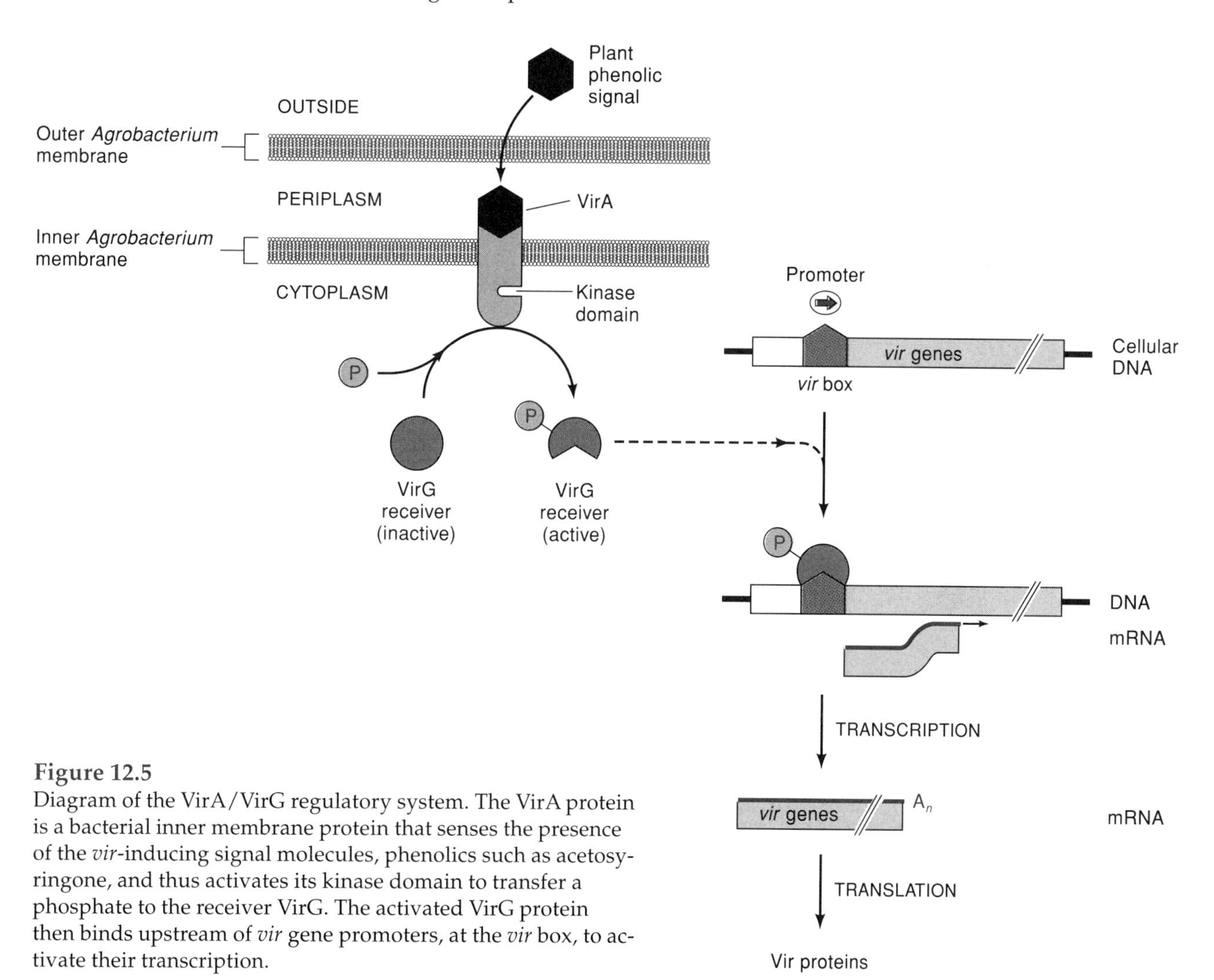

Figure 12.5
Diagram of the VirA/VirG regulatory system. The VirA protein is a bacterial inner membrane protein that senses the presence of the *vir*-inducing signal molecules, phenolics such as acetosyringone, and thus activates its kinase domain to transfer a phosphate to the receiver VirG. The activated VirG protein then binds upstream of *vir* gene promoters, at the *vir* box, to activate their transcription.

The VirA protein, the sensor, is located in the bacterial inner membrane and is oriented with its kinase domain in the cytoplasm. Recognition of plant signals activates VirA to transfer its phosphate to a specific site in VirG, thereby activating the regulatory component. Activated VirG binds upstream of the *vir* promoters to activate their transcription. Each of the *vir* promoters contains a conserved 12 bp sequence (TNCAATTGAAAPy), the *vir* box, which acts as a *cis*-acting regulatory sequence, binding the activated VirG transcriptional activator protein and thereby promoting *vir* gene expression.

Six *vir* genes have been shown to play a role in T-DNA transfer. *Vir*A and *vir*G, mentioned previously are the only monocistronic loci. The other *vir* genes produce polycistronic messengers that encode several proteins. The *vir*B and *vir*D operons are absolutely essential for T-DNA transfer; the *vir*C and *vir*E loci are essential on certain plant hosts, but not others. The *vir*C, *vir*D, and *vir*E loci are mainly involved in the synthesis and processing of the T-DNA transfer intermediate and encode 2, 4, and 2 proteins, respectively. The *vir*B locus is the largest, encoding 11 protein products, and is responsible for producing a membrane channel to allow exit of the T-DNA intermediate from the bacterial cell. Table 12.1 summarizes the molecular weights and likely functions of the Vir proteins.

Table 12.1 vir-*Specific Protein Products*

Locus	Size (kb)	General Role	Protein Products: Size (kDa)*	Amount	Location	Function
virA	2.0	Absolutely essential for virulence, regulatory system	92	+	M	AS sensor; sugar sensor; protein kinase
virG	1.0		30	++	C(?)	*vir* transcriptional regulator
virC	2.0	Attenuated virulence, increased host range	26, 23	+	?	Enhancement of T strand production
virE	2.0		7, 60.5	+++	C/M(?)	VirE2—single-stranded DNA-binding protein; unfolding the T strand
virD	4.5	Absolutely essential for virulence, T-DNA transfer machinery	16, 47, 21, 75	+	C(?)	VirD1—topoisomerase; helicase (?); VirD2–T-DNA border endonuclease; nuclear localization; helicase (?); primase (?); integrase (?)
virB	9.5		26, 12, 12, 87, 23, 32, 6, 26, 32, 41, 38	+/+++	M	T-DNA transfer structure (?); membrane channel (?); VirB11–ATPase; protein kinase

Note: Abbreviations: AS = acetosyringone, C = cytoplasm, M = membrane.
*Protein sizes are listed in order of arrangement relative to their respective promoters.
Source: Adapted from P. Zambryski, chronicles from the *Agrobacterium* - Plant Cell DNA Transfer Story. *Ann. Rev. Plant Physiol. Mol. Biol.* 43(1992), p. 465.

12.3 DNA Transfer

a. The T-DNA Element

The T-DNA element is defined as that segment of the Ti-plasmid that is homologous to sequences present in crown gall cells. The nopaline T-DNA is approximately 22 kb; the octopine Ti-plasmid contains two adjacent T-DNA regions, a left T-DNA (TL) element of 13 kb and a right T-DNA (TR) element of 7.8 kb. The TR element is unique to the octopine Ti-plasmid; however, most of TL is homologous to a portion of the nopaline T-DNA (Figure 12.6). The regions of homology determine the crown gall phenotype by encoding the synthesis of plant growth hormones (see Section 12.4a).

The structural limits of T-DNA were defined by comparing the nucleotide sequence at the ends of the T-DNA element following integration into the plant genome with the nucleotide sequence of the corresponding region of the Ti-plasmid. These analyses revealed little variation in the ends of the T-DNA from different tumor cell lines. In all cases, the homology between sequences present in the Ti-plasmid and those in the tumor DNA ends within or proximal to the 25 bp sequences that flank the T-DNA region of the Ti-plasmid as direct (albeit, imperfect) repeats (indicated as triangles in Figures 12.1, 12.2, and 12.6). These repeats are found to delimit all T-DNAs analyzed to date. Figure 12.7 shows a comparison between six terminal repeats, called **T-DNA borders**, two from nopaline, two from TL of octopine, and two from TR of octopine plasmids.

b. Production of a DNA Transfer Intermediate

T-DNA transfer is a complex process in that a specific segment of DNA is recognized and mobilized from the Ti-plasmid, transferred across the respective cell walls of the bacterium and plant cell, and integrated as a linear, nonpermuted segment into the plant's nuclear genome.

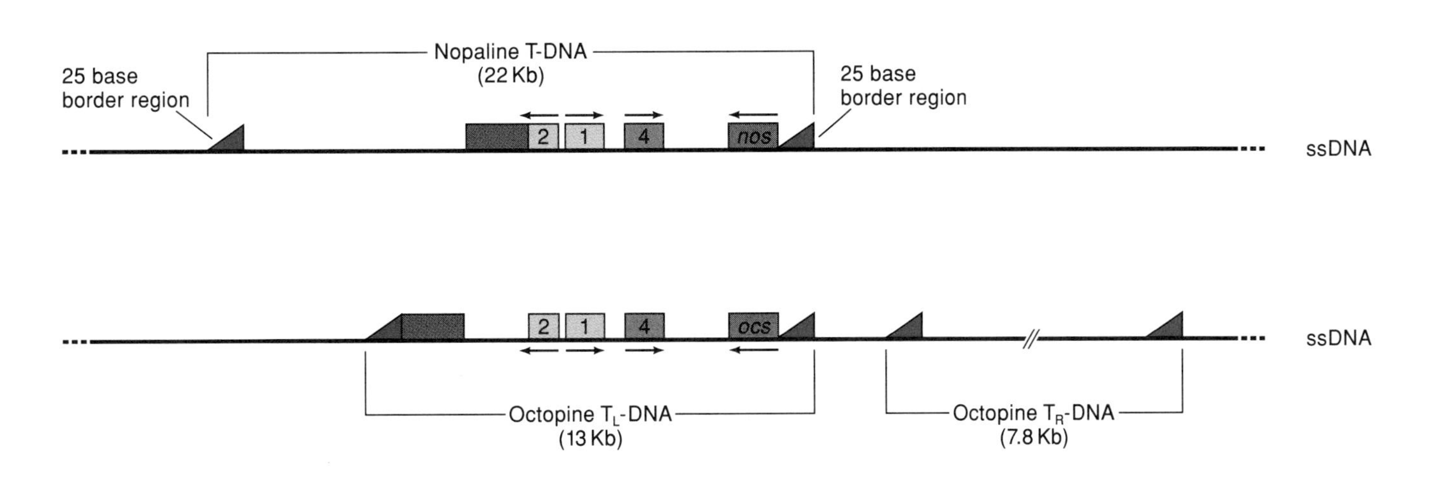

Figure 12.6
Comparison of the homologous regions of nopaline T-DNA and octopine TL-DNA. Similarly shaded regions indicate homology. The numbers 1, 2, and 4 refer to the genes encoding the synthesis of auxin and cytokinin (see also Figures 12.18, 12.19, 12.20, and accompanying text); arrows indicate the direction of transcription. *ocs* and *nos* refer to genes encoding octopine and nopaline synthase, respectively.

Nopaline L	GCTGGTGGCAGGATATATTGTGGTGTAAACAAATT
Nopaline R	GTGTTTGACAGGATATATTGGCGGGTAAACCTAAG
Octopine LL	AGCGGCGGCAGGATATATTCAATTGTAAATGGCTT
Octopine LR	CTGACTGGCAGGATATATACCGTTGTAATTTGAGC
Octopine RL	AAAGGTGGCAGGATATATCGAGGTGTAAAATATCA
Octopine RR	ACTGATGGCAGGATATATGCGGTTGTAATTCATTT
Consensus	TGGCAGGATATAT$^{TG}_{NC}$N$^{GG}_{NT}$TGTAAA$^{T}_{C}$

Figure 12.7
Comparison of the 25 bp border sequence flanking the T-DNAs of octopine and nopaline Ti plasmids. Nopaline L (left) and R (right) are the repeats flanking the T-DNA of nopaline-type Ti plasmids. Octopine-type Ti plasmids have two T-DNAs, termed left and right. Octopine LL, LR, RL, and RR refer to the left and right T-DNA borders from each of these T-DNA elements.

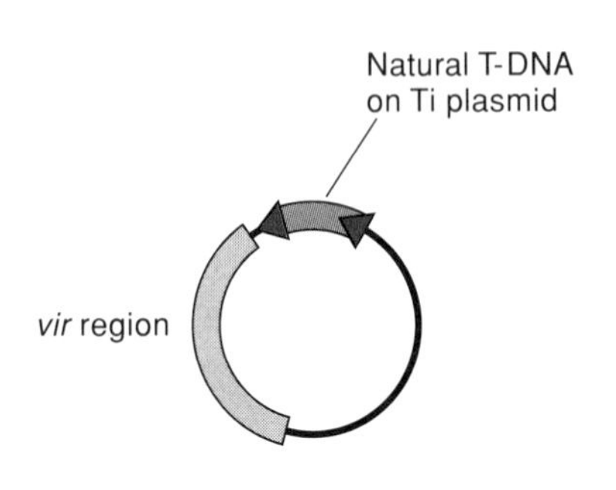

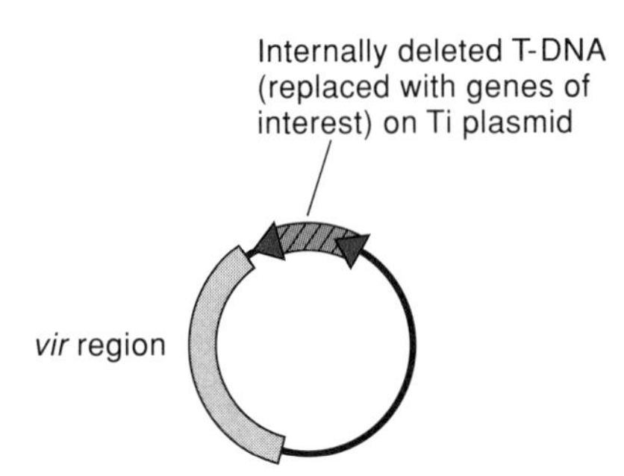

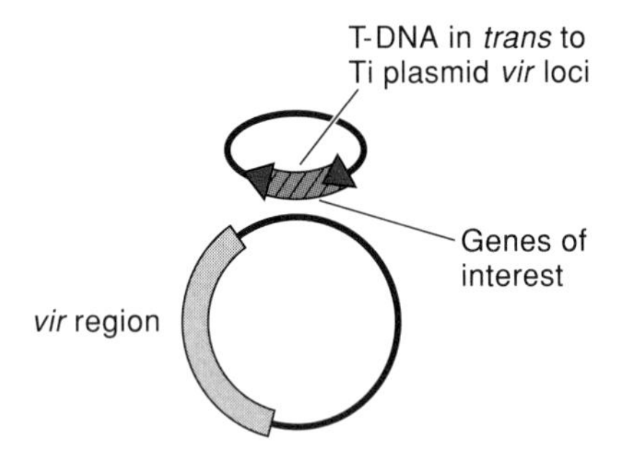

Figure 12.8
Different T-DNAs are effectively transferred to the plant cell. For effective transfer the only requirements are a complete *vir* region and the T-DNA border sequences (closed black arrows). The T-DNA border sequences can either reside in *cis* to the *vir* region or in *trans* on a separate replicon.

Essential structural and genetic characteristics for T-DNA element transfer During transfer, the T-DNA functions solely as a structural element; its internal portion can be fully deleted without affecting transfer. Thus, although the T-DNA is a mobile element, it is not like a classical transposon because its internal region does not specify products required for its movement. Another difference between T-DNA and transposons is that once integrated into plant DNA, T-DNA is stable and cannot move again.

The 25 bp direct repeats fully define the functional T-DNA; any DNA located between T-DNA borders will be efficiently transferred and integrated into the plant cell genome. In fact, the T-DNA need not be carried by the Ti-plasmid; if a DNA segment containing T-DNA borders is located on a separate plasmid or on the *Agrobacterium* chromosome in a cell that also carries a complete complement of the (*vir*) genes essential for transfer, it will be transferred to the plant cell during infection (Figure 12.8). This fact forms the basis of the design of modified and simplified T-DNA molecules useful as vectors for plant cell transformation by insertion of cloned DNA fragments of interest (see Section 12.5a). Moreover, these observations indicate that border sequences must be the structural substrates of the proteins that directly mediate the transfer process.

Genetic analyses of the 25 bp border sequences demonstrated that these sequences are polar in function (Figure 12.9). Thus, deletion of the left border re-

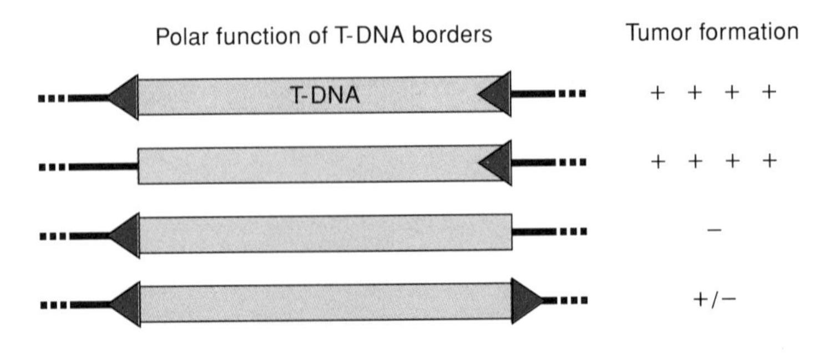

Figure 12.9
Functional polarity of 25 bp T-DNA borders. Deletion of the left T-DNA border does not affect T-DNA transfer, as measured by tumor formation. However, deletion of the right border abolishes transfer. Furthermore, the right border must be properly oriented for effective transfer to occur. Thus, T-DNA transfer is a polar process that initiates from the right side of the T-DNA.

peat has no significant effect on plant transformation, but deletion of the right repeat totally abolishes it. Furthermore, when the orientation of the right border is reversed with regard to its natural orientation, the efficient transfer and/or integration of the T-DNA sequences is greatly attenuated. These results suggest that T-DNA transfer might occur in a rightward to leftward fashion, determined by the orientation of the 25 bp border repeats. Thus, when the right border is present in its natural orientation, transfer includes the T-DNA–tumor–forming genes internal to the element; when only a left border is present, transfer occurs only away from the T-DNA element, and no tumor-forming genes are transferred to the plant cell.

A single-strand T-DNA intermediate Although genetic studies showed conclusively that the T-DNA borders are essential for T-DNA transfer and integration, they did not identify how the borders are recognized and utilized. Analysis of the structures homologous to T-DNA that are formed in *Agrobacterium* soon after induction of *vir* genes demonstrated that the first step is the introduction of nicks on the bottom strand of the 25 bp border sequence. Second, free, linear, single-stranded (ss) copies homologous to the bottom strand of the T-DNA accumulate (Figure 12.10). Presumably, the nicks in the 25 bp border sequences

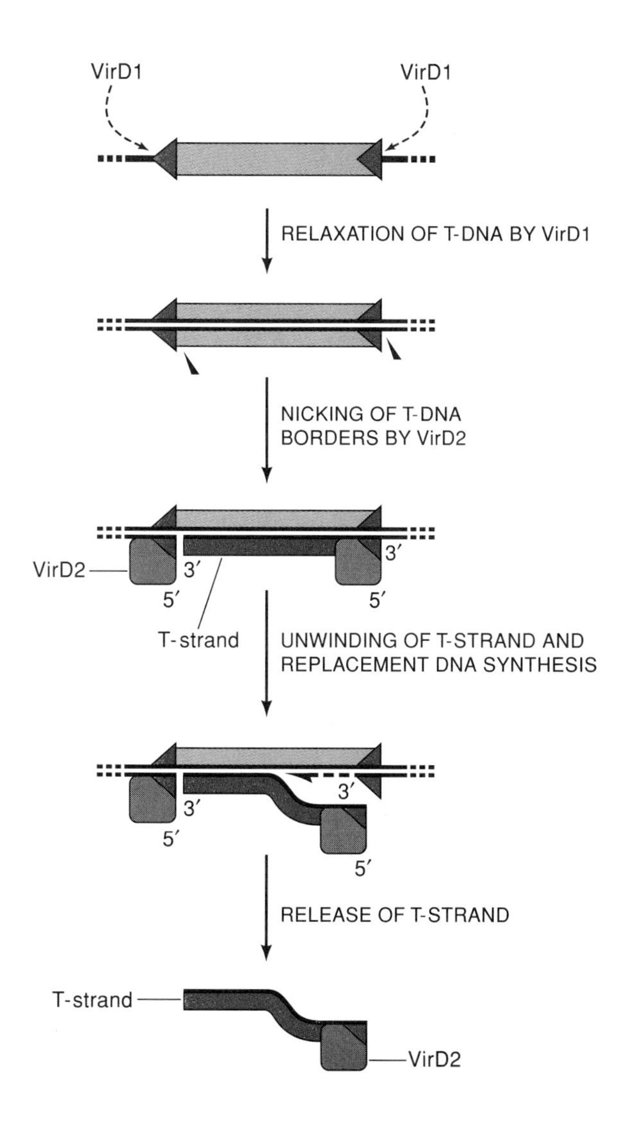

Figure 12.10
Synthesis of T-DNA transfer intermediate molecules (T-strands). The 25 bp T-DNA borders (black arrows) are used as sites of initiation and termination for the displacement of a copy of the lower strand of the T-DNA region, called the T-strand. Because genetic studies (see Figure 12.9) indicate that the right border likely acts to direct T-DNA transfer, it is assumed that displacement initiates at the right 25 bp repeat. The free 3′hydroxyl can then serve as a site for replacement strand synthesis by DNA replication machinery. Two Vir proteins, VirD1 and VirD2, form the endonuclease that recognizes and cleaves the lower strand of the T-DNA borders. The VirD1 protein may act to relax DNA in the vicinity of the borders, and VirD2 does the actual nicking. The VirD2 protein then remains covalently bound to the 5′end of the T-strand.

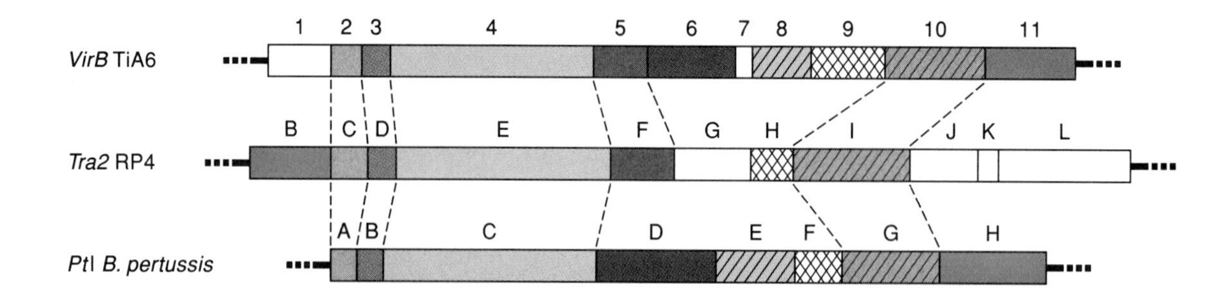

Figure 12.11
The conjugative transfer operon from the RP4 plasmid and the pertussis toxin (*Ptl*) secretion operon encode proteins with homology to the *virB* operon of *Agrobacterium* TiA6. Genes encoding related proteins are shown by identical shading. The eleven *virB* genes are numbered from 1 to 11, the eleven *Tra2* genes are lettered B through L, and the eight *Ptl* genes are lettered A through H. (Adapted from M. Lessl and E. Lanka. 1994. Common Mechanisms in Bacterial Conjugation and T-Mediated T-DNA Transfer to Plant Cells. *Cell* 77 321.)

provide sites for initiation and termination of the synthesis of the ss T-DNA copy, designated the **T strand**. T strand polarity corresponds to the genetically determined functional polarity of the 25 bp border sequences, that is, the 5′ end of the T strand is at the right T-DNA border, the proposed start of polar T-DNA transfer.

The mechanism of T-DNA transfer The discovery of a single-stranded T-DNA was the first clue to the possible mechanism of the transfer process. Thus, *Agrobacterium*-plant DNA transfer might be a modification of bacterial conjugation, except that the recipient is a plant cell. In conjugation, a newly synthesized single-stranded DNA is transferred between donor and recipient cells in a process requiring physical attachment between the two cells. *Agrobacterium* might have adapted an evolutionarily conserved mechanism to suit its ends.

Strong support for the model that T-DNA transfer is bacterial conjugation applied to plants has been accumulating since 1986. The first data came from a genetic study. The origin of transfer (oriT) from a conjugative *E. coli* plasmid, pRSF1010, was found to substitute for the T-DNA borders in directing DNA transfer to plant cells. This hybrid transfer system also required an intact *vir* region and a region of pRSF10101 encoding polypeptides involved in plasmid mobilization. Thus, the oriT of pRSF1010 and its cognate mobilization proteins presumably generate a conjugative DNA transfer intermediate that can be transferred to plant cells using the *Agrobacterium vir*-specific transfer machinery.

The second line of support for a conjugative model derives from a comparison of the physical and functional properties of the *Vir*B operon with two other bacterial operons (Figure 12.11). The *Tra*2 operon is required for conjugation by the broad host range plasmid RP4; six of the *Tra*2 gene products show significant homologies to six of the VirB proteins. The *Bordetella pertussis Ptl* operon is responsible for export of the pertussis toxin; eight Ptl proteins show homology to eight VirB proteins. Although only the VirB and Tra2 systems are involved in the DNA transport through bacterial membranes, the homology with the pertussis toxin system suggests that, in all three systems, similar proteins function during transport. In the case of the conjugative plasmid DNA or T strand, the proteins may provide the signals and recognition surfaces to carry DNA through the bacterial membranes.

Additionally, the DNA processing steps that generate transferable single-stranded DNA in plasmid DNA transfer between bacteria and in T-DNA transfer between *Agrobacterium* and plant cells are significantly similar. The RP4 nick region and the T-DNA border sequences share a 12 nucleotide consensus sequence: 5′-A C/T C/A T/C ATCCTG C/T C/A. The significance of the sequence similarities is underscored by striking structural and functional similarities in the enzymes acting at these sites.

Two *vir*-specific products, VirD1 and VirD2, are essential for T strand synthesis and are required to nick the T-DNA borders (Figure 12.10). VirD1 may be a site-specific topoisomerase that unwinds the DNA in the region of the T-DNA borders. VirD2 is an endonuclease specific for the T-DNA border; it cleaves the sequence on the lower strand. Although their exact functions are unknown, genetic studies indicate that the products of the *vir*C locus, VirC1 and VirC2, enhance T strand production.

No other *vir* products are critical for border nicking and T strand production. Other enzyme activities expected to be required, such as helicases and polymerases, might either be provided by constitutive bacterial enzymes or be part of VirD1 or VirD2 function. If these functions are bacterial, they are not specific to *Agrobacterium* because T strand production can occur in *E. coli* cells containing VirD1, VirD2, and a T-DNA substrate.

c. Formation of a Transferable DNA-Protein Complex

The generation of the T strand is really just the beginning of the complex journey that results in plant cell transformation. Following its formation, the T strand must pass through six barriers: bacterial inner membrane, cell wall and outer membrane, the plant cell wall, cell membrane, and finally the plant nuclear membrane. During this transit, the T strand must avoid degradation by nucleases. Thus, the T strand interacts with two Vir proteins, VirE2 and VirD2, to form a DNA-protein complex, the **T complex** (Figure 12.12). Besides protecting the T strand, the proteins provide signals to mediate its travels through the cellular and various envelope compartments.

The largest open reading frame of the *vir*E locus, VirE2, encodes a single-strand DNA-binding protein. *In vitro* experiments have shown that VirE2 binds any single-stranded DNA strongly, regardless of sequence; the binding is highly cooperative, suggesting that T strands are fully coated with adjacent VirE2 molecules. Indeed, the level of VirE2 allows its stoichiometric binding to the T strand. Because a single VirE2 molecule covers about 30 nucleotides, the 20 kb nopaline T strand requires 600 VirE2 molecules for complete coating and would produce a structure 3600 nm long (Figure 12.12). VirE2 renders single-stranded DNA com-

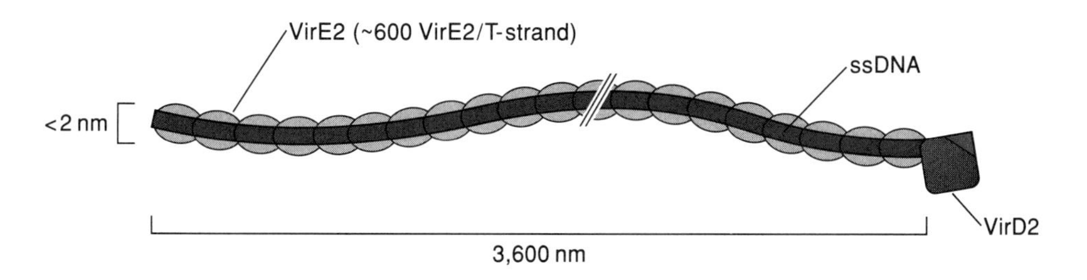

Figure 12.12
The nopaline-specific T-DNA transfer complex. The nopaline T-DNA is 22 kbp. The T-strand of this region would also be 22 kb long and is predicted to be coated along its length by the non-sequence-specific, single-strand binding protein, VirE2, encoded by the *virE* locus. The dimensions of VirE2 complexed with single strand DNA when viewed by electron microscopy suggest that the complex would be very thin, 2 nm wide, and very long, 3600 nm. A single molecule of VirD2 is covalently bound to the 5′(right) end of the T-strand.

pletely resistant to 3′ or 5′ exonucleases, as well as to endonucleases *in vitro*.

Protection of the T strand may be necessary but insufficient for efficient T strand transfer. Because free, single-stranded DNA strands form irregular, collapsed structures, T strand molecules may need to be unfolded prior to transport through bacterial and plant cell membranes. Binding of VirE2 to such molecules unfolds and extends them, forming exceptionally thin protein single-stranded DNA complexes measuring less than 2 nm in diameter (Figure 12.12). Thus, just as chaperonins mediate translocation of proteins through organelle membranes, VirE2 not only protects but also shapes T strands into an unfolded, transferable form.

T strand export probably occurs in a unidirectional fashion. All existing data suggest that synthesis and transfer initiate from the right T-DNA border, so the 5′ end of the T strand is probably the leading end in the transfer. A protein bound specifically to the 5′ end of the T strand could provide a piloting function. Indeed, the VirD2 protein becomes covalently linked to the 5′ end of the T strand, presumably when the T-DNA border is nicked.

d. Exit of the T Complex from the Bacterial Cell

The first passages the T complex makes are through the bacterial membranes and wall. Thus, some Vir proteins might be expected to alter the bacterial membrane, creating a transmembrane channel or pore. The products of the VirB locus are likely candidates to produce such a channel structure because most of the 11 VirB open reading frames encode transmembrane or membrane-associated proteins. Thus, proteins that traverse membranes usually require a signal sequence located at the amino terminus; the polypeptides predicted by the open reading frames VirB1, VirB5, VirB7, and VirB9 have such signal sequences. Second, many proteins anchored in membranes span the membrane with a contiguous stretch of about 20 mainly hydrophobic amino acids. VirB2, VirB3, VirB5, VirB6, VirB8, and VirB10 contain such membrane-spanning regions. Only VirB4 and VirB11 do not possess either of these two characteristic membrane-designating sequences. However, subcellular fractionation studies suggest that both VirB4 and VirB11, as well as the other VirB proteins, localize to the bacterial membrane. Thus, the *virB* products are the best candidates to form an exit channel for the T complex.

To date, there are only preliminary results relating to the possible structure of the VirB channel. These data suggest that many VirB proteins may span both the inner and outer membranes. The simplest model to explain these results hypothesizes that the VirB proteins interact tightly with one another, by analogy to interlocking puzzle pieces, thereby forming a channel that opens and closes in response to the transportable T complex substrate (Figure 12.13). The tight juxtaposition of the VirB proteins may bring the inner and outer membranes closer together, with the result that the distance for T complex travel is effectively shortened.

Future experiments will test this speculative model. Some insight may be gained by comparison with other complex transport systems, such as transport of conjugal DNA (Tra2 operon) or protein (pertussis toxin transport). Not only is *Agrobacterium*–plant cell conjugal DNA transfer intrinsically interesting, but elucidation of its detailed transfer mechanism may have application to the design of more effective strategies for plant transformation. Additionally, more widespread medical application is possible, for example, to the design of strategies to block plasmid DNA transfer between pathogenic organisms. While modern pharmaceuticals have revolutionized the treatment of disease, they have also brought on a new threat. Antibiotics have been used effectively in treating numerous bacterial diseases, but the bacteria have developed resistance to the drugs. The resistance genes are carried on small, promiscuous plasmids. Thus, conjugal transfer of such plasmids can account for the transfer of resistance

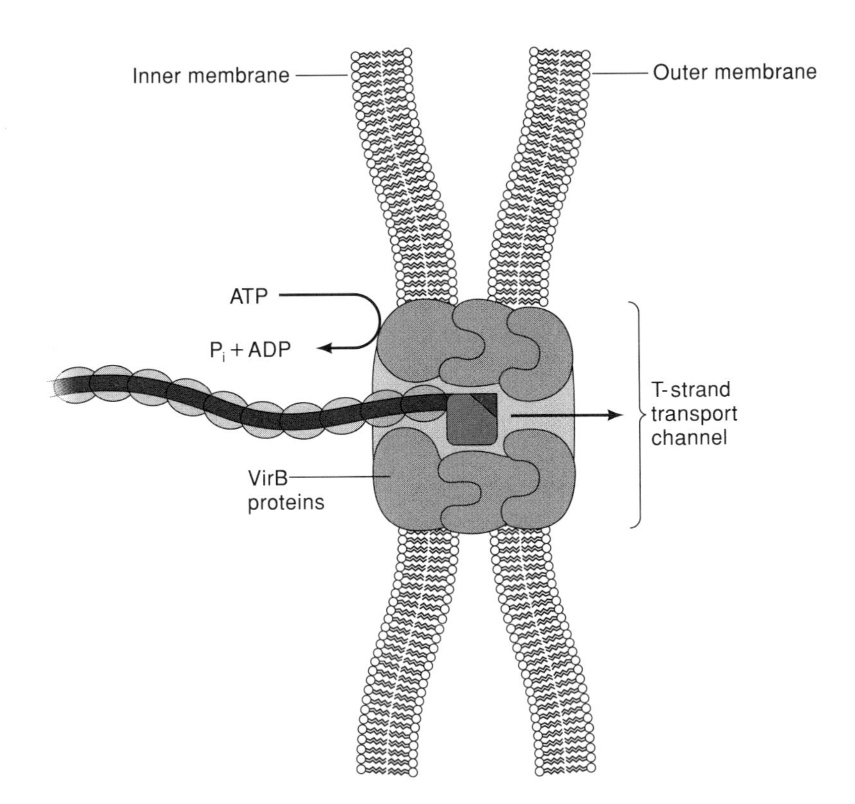

Figure 12.13
Hypothetical T-DNA transport channel composed of VirB proteins. Cell fractionation studies suggest that several VirB proteins localize to both the inner and outer bacterial membranes. Thus, the T-DNA transfer channel may bring these two membranes into close juxtaposition. The individual VirB proteins may interact with each other in a fashion analogous to interlocking puzzle pieces. The three types of puzzle pieces shown represent three types of VirB proteins, those located mainly in the outer or inner membranes, and those that interconnect them. Each puzzle piece may be composed of more than one protein. The T complex, composed of the T-strand with VirD2 at its 5′ end and cooperatively bound VirE2 along its length may move through this hypothetical channel in an ATP-dependent fashion. (Adapted from Y. R. Thorstenson et al. 1993. Subcellular Localization of Seven VirB Proteins of *Agrobacterium tumefaciens*. *J. Bacteriol.* 175 5233–5241.)

genes to other bacteria. Perhaps greater understanding of the mechanisms of conjugal DNA transfer will permit us to block this process and improve antibiotic effectiveness.

e. T Complex Passage Through the Plant Cell Membrane

Passage of the T complex through the cell membrane is the least understood step of the T-DNA transfer process, hence the big question mark in the black box of Figure 12.1. The existing information is very general, suggesting the involvement of cellulose fibrils and plant vitronectin analogues in the attachment of *Agrobacterium* to plant cells. Microscopic examinations reveal that the bacteria attach primarily at their poles. *Agrobacterium*–plant cell interaction may provide a useful tool to examine the architecture and function of plant cell surface proteins. Thus, the VirB proteins that localize to the outer membrane or surface of the bacterial cell may provide probes for isolation and characterization of receptors or general membrane proteins that facilitate import of the T complex into the plant cell.

Although the *Agrobacterium* T-DNA transfer system may share some features with other transfer systems, it is clearly unique in being the only known example of directed transfer and integration of DNA between widely different organisms. *Vir* gene activation and T strand synthesis and export can all be based on conserved bacterial mechanisms, but once the T strand enters the plant cell, it needs to rely on eukaryotic mechanisms for intracellular transport, nuclear uptake, and integration into the genome. Thus, there is a duality to the functions of some of the Vir proteins: the ability to function in prokaryotic and eukaryotic modes.

f. Targeting the T Complex to the Plant Cell Nucleus

Although the VirD2 and VirE2 proteins of the T complex play a role in T strand movement, this movement is unlikely to be passive and likely to involve recognition and targeting to bacterial and plant subcellular locations. Because nucleic acids generally do not specify signal sequences, the protein components of the T complex might be expected to provide this role. Some of the better defined signal sequences involved in membrane transport are those that specify uptake of proteins into the eukaryotic nucleus. Specific cytoplasmic receptor proteins recognize nuclear localizing sequences (NLS) and physically mediate nuclear uptake. Thus, the most direct means for T complex nuclear uptake would be via its associated VirD2 and VirE2 proteins.

The role of VirD2 Because the VirD2 protein is presumed to pilot the T strand from the bacterial cell into the plant cell nucleus, it was studied first for its nuclear localizing activity. Interestingly, *Agrobacterium* strains carrying a deletion of the C terminus of VirD2 are unable to elicit tumors on infected plants; however, only the N terminal 50 percent of VirD2 is required to produce T strands. This observation suggests that the C terminus of VirD2 plays a role after T strand formation. Conceivably, the tight binding of VirD2 to the 5′ end of the T strand provides signal sequences to facilitate T strand transport.

A comparison of the amino acid sequences of VirD2 proteins from three *Agrobacterium* strains indicates that their N terminal halves are well conserved, with 85 percent identity. In contrast, their C termini are highly diverged, with only 25 percent identity. Furthermore, the identity is clustered in a small region at the very terminus of the various VirD2 proteins. Within the last 30 amino acids, all the VirD2 proteins contain two short stretches of 4–5 amino acids, each with striking homology to the NLS of the SV40 large-T protein. Furthermore, that there are two basic regions suggests that VirD2 NLS may better resemble a more common type of NLS, the bipartite signal, first identified in nucleoplasmin, a *Xenopus* nuclear protein, and subsequently in several other nonplant karyophilic proteins. This latter signal consists of two basic regions separated by a variable (but not less than 4) amino acid spacer (Figure 12.14). Site-specific deletion of only these basic residues at the C terminus of VirD2 abolishes tumorigenicity.

When VirD2 is fused to a plant cell reporter gene and transfected into plant cells, reporter gene activity is confined to the plant cell nucleus. Furthermore,

Figure 12.14
Amino acid sequence homology between bipartite nuclear localization signals of *Agrobacterium* VirD2 protein (pTiC58) and VirE2 proteins from nopaline (pTiC58) and octopine (pTiA6) strains and nucleoplasmin from *Xenopus*. Basic residues are shaded and represent the two domains of bipartite nuclear localization signals.

VirE2 (pTiC58) NSE 1	K L R P E D R Y I Q T E – K Y G R R
VirE2 (pTiA6) NSE 1	K L R P E D R Y V Q T E – K Y G R R
VirE2 (pTiC58) NSE 2	K T K Y G S D T E I – – – K L K S K
VirE2 (pTiA6) NSE 2	K R R Y G G E T E I – – – K L K S K
VirD2 (pTiC58)	K R P R E D D D G E P S E R K R E R
Nucleoplasmin	K R P A A T K K A G G A – K K K K L

the VirD2 NLS alone are capable of directing nuclear uptake of the reporter. The reporter gene for these studies is β-glucuronidase (GUS), an enzyme that cleaves a variety of commercially available β-glucuronides, many of which produce spectrophotometric or fluorometric products (Figure 12.15).

The role of VirE2 The T complex is a very large structure containing a 20 kb T strand complexed with about 600 molecules of VirE2. Such a complex has a predicted length of 3600 nm and a mass of about 50 × 10^6 Da. The nuclear pore is about 60 nm wide. Thus, the T complex is 60 times longer than the dimensions of the nuclear pore, and its mass is about 1000 times greater than that of a single VirD2 molecule (50 kDa). Can such a large complex be transported into the nucleus by a single VirD2 molecule? Perhaps VirE2 also facilitates nuclear uptake. Indeed, the predicted amino acid sequence of VirE2 contains two stretches of amino acids with homology to the bipartite type NLS of nucleoplasmin and VirD2 (Figure 12.14). VirE2 nuclear localizing activity was confirmed by fusing it to the GUS reporter gene and monitoring nuclear uptake. Although the entire VirE2 protein was capable of efficient nuclear uptake equivalent to that observed with VirD2, deletion of either bipartite signal reduced nuclear uptake. These results suggest that each individual VirE2 bipartite NLS is intrinsically weaker than the single VirD2 bipartite signal.

Both VirD2 and VirE2 proteins are expected to be recognized by nuclear-binding proteins (NBP) to bring the T complex to the nuclear pore. Because the VirD2 NLS is a stronger signal, it may react more strongly with NBP, thereby targeting the T complex to the nuclear pore in a directed fashion. Thus, VirD2 may be critical for orienting the T complex into the nuclear pore. The weaker NLS of VirE2 along the length of the T complex may ensure uninterrupted nuclear uptake.

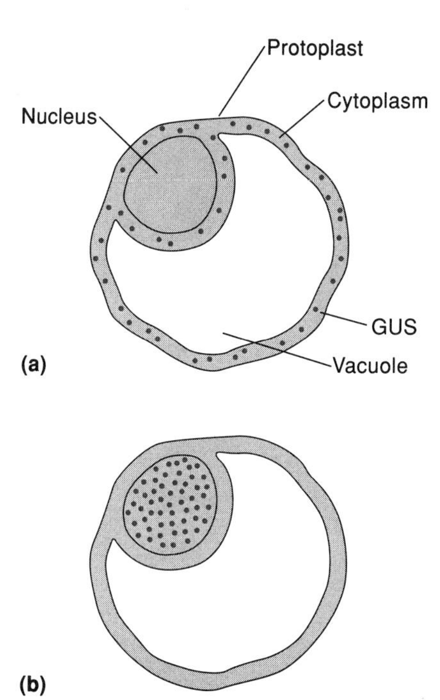

Figure 12.15
Localization of GUS activity in transiently transformed tobacco protoplasts. Two plant cell protoplasts are drawn. (a) GUS expression is localized to the cytoplasm. (b) The nuclear localization signals of VirD2 are fused to GUS, resulting in the nuclear localization of the GUS reporter.

A model for T complex uptake into the plant cell nucleus Figure 12.16 shows a model for T complex uptake into the plant cell nucleus. The evidence for VirD2- or VirE2-mediated nuclear uptake implies that the proteins themselves target to the nucleus. However, the direct demonstration that these proteins can mediate uptake of the T strand into the nucleus requires a more sophisticated approach. Because it is impossible to follow the movement of a single T strand from the bacterium into the plant, a strategy was employed to generate T complex *in vitro* and then follow the movement of a fluorescently labelled complex in a single plant cell by microscopy. To accomplish this, a DNA fragment was amplified by the polymerase chain reaction (PCR) using fluorescein-labelled nucleotides. The product was denatured to produce single-stranded DNA and VirE2 protein was added to coat the fluorescent single-stranded DNA. Within minutes of microinjecting the complex into the cytoplasm of single plant cells, the VirE2-complexed, single-stranded DNA accumulated in the plant cell nucleus (Figure 12.17). Fluorescent naked DNA without VirE2 coating did not accumulate in the nucleus. These results dramatically illustrate VirE2-dependent nuclear uptake of single-stranded DNA in living cells and provide strong support for the model of T complex nuclear import.

The movement of the T complex into the plant cell nucleus may represent a generalized process whereby unfolded, protein-coated DNA or RNA molecules move within the cell. Two obvious and possibly related processes are the nuclear import of enveloped and nonenveloped viruses and the nuclear transport of small nuclear ribonuclear protein complexes (snRNPs). Each of these processes has unique characteristics. However, all three systems involve specific localization (NLS) signals, gating of the nuclear pore, an unfolded structure, and directed transport. Even more generally, cellular analogues of VirE2 may serve as nucleic acid–specific molecular chaperones, coating, unfolding, and targeting them to and through the nuclear pore.

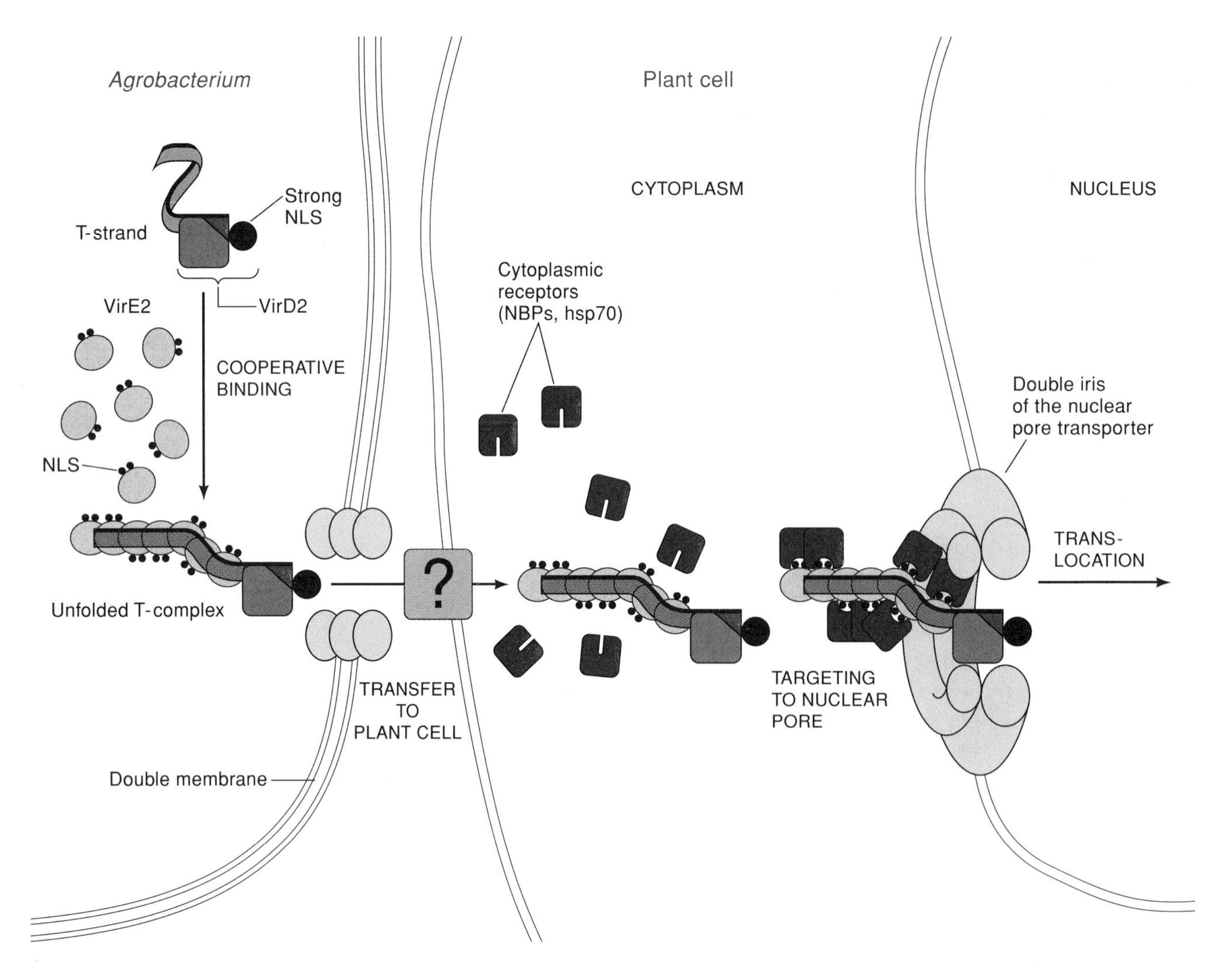

Figure 12.16
A model for nuclear transport of *Agrobacterium* T-strands. Nuclear localization signals are indicated as black balls, VirE2 (gray oval protein) has two bipartite nuclear localization signals, so two balls are indicated. VirD2 (dark gray square protein) has one strong nuclear localization signal and, hence, one black ball. Crescent-shaped proteins in the plant cell represent cytoplasmic receptor molecules that recognize the nuclear localization signals and translocate them to the nuclear pore. (Adapted from V. Citovsky and P. Zambryski. 1993. Transport of Nucleic Acids Through Membrane Channels. *Ann. Rev. Microbiol.* 47 167.)

Figure 12.17 (*facing page*)
VirE2 mediates nuclear uptake of single-strand DNA. (a) Fluorescently labeled DNA is generated by PCR in the presence of fluorescein-12-dUTP. The double-strand DNA is then rendered single-strand by melting and is either coated with VirE2 protein or allowed to reanneal. DNA alone or DNA coated with VirE2 is microinjected into single plant cells. The DNA alone sample (left) remains cytoplasmic; the DNA coated with VirE2 sample enters the plant cell nucleus within minutes of microinjection. (b) A single-stamen hair cell from *Tradescantia virginiana* under bright field optics (left panel) and under fluorescent optics injected with fluorescent DNA alone (middle panel) or after coating with VirE2 protein (right panel). (Adapted from J. Zupan, V. Citovsky, and P. Zambryski. 1996. *Agrobacterium* VirE2 protein mediates nuclear uptake of single-stranded DNA in plant cells. *Proc. Natl. Acad. Sci. USA* 93 2392.)

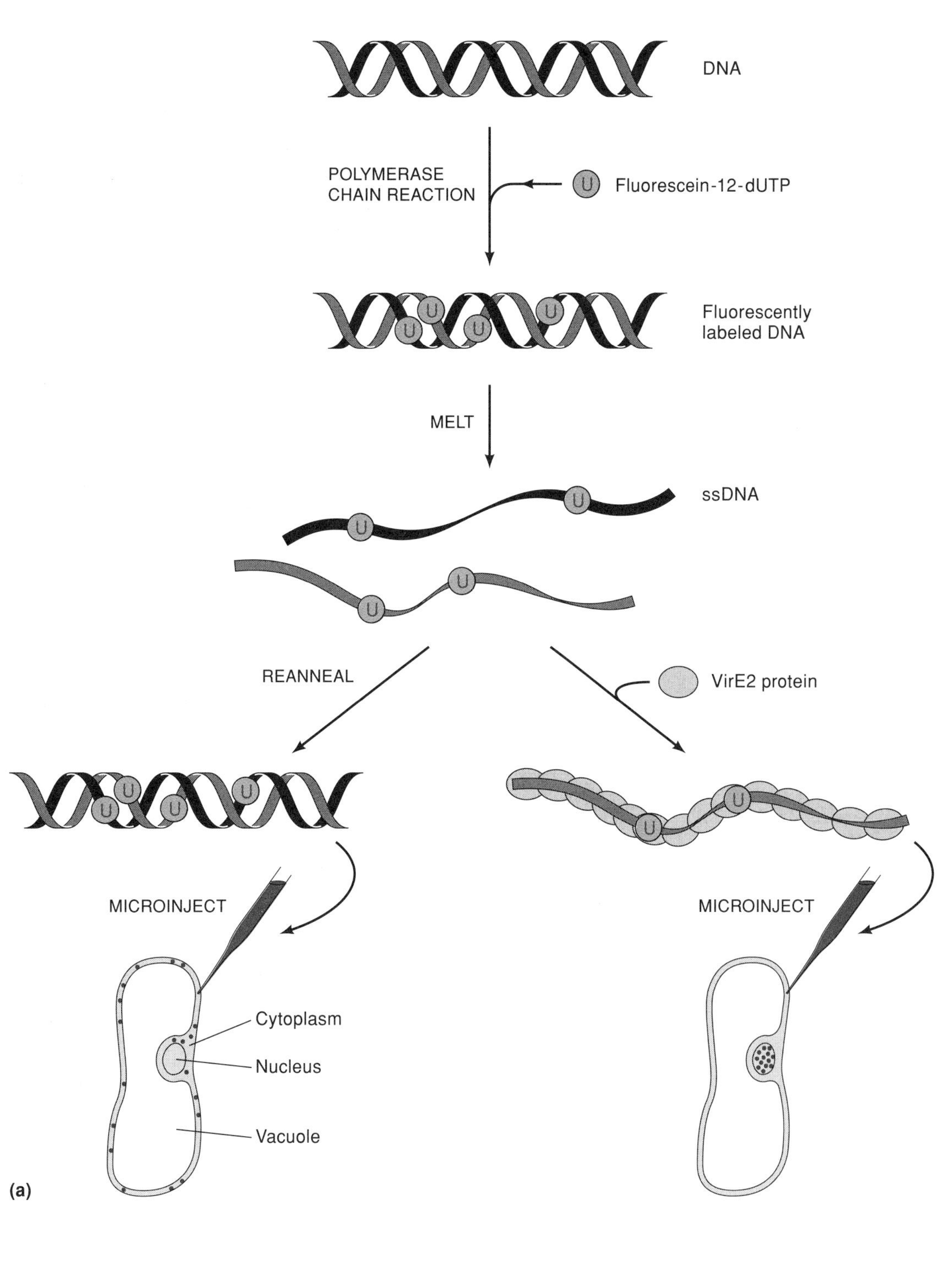
DNA
POLYMERASE
CHAIN REACTION
Fluorescein-12-dUTP
Fluorescently
labeled DNA
MELT
ssDNA
REANNEAL
VirE2 protein
MICROINJECT
MICROINJECT
Cytoplasm
Nucleus
Vacuole
(a)

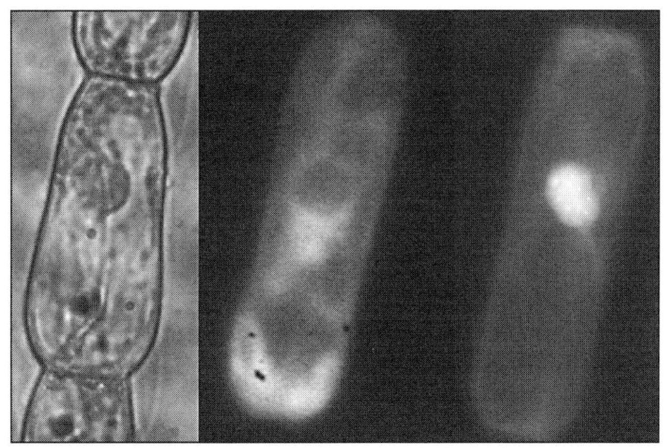
(b)

g. Integration of the T-DNA into the Plant Genome

With regard to T-DNA integration, structural analysis of the T-DNA–plant DNA junctions from independent integration events in different plant cell tumor lines demonstrated that the integration process is relatively precise with respect to the T-DNA border sequences. Thus, in the cases analyzed, the integration event, with respect to the T-DNA, occurred directly within or, in a few instances, close to the 25 bp sequence. Most frequently, the homology to T-DNA ends after the third base of the 25 bp repeat of the right T-DNA border and includes all but the first 3 bp of the 25 bp repeat of the left T-DNA border. These analyses confirm molecular studies that T strand production is initiated following *vir* gene activation by specific nicking at the T-DNA borders between the third and fourth base of the 25 bp repeat. Sequence analyses of integrated T-DNA copies also indicate that there is little or no specificity for the plant sequences present at the integration site and that deletions and/or rearrangements of plant sequences are occasionally associated with T-DNA integration. Generally, one to three copies of the T-DNA are found, but tandem arrays of stable T-DNA inserts have also been observed.

The molecular mechanism by which the T strand component of the T complex integrates into the plant genome is not yet known. Analyses of cloned integration sites reveals that the right side of the T-DNA copy, corresponding to the 5′ end of the T strand, is more precise than the left T-DNA end. Because there is no homology between T-DNA ends and genomic integration sites, most models posit illegitimate recombination as the route for T-DNA insertion. Illegitimate recombination is basically a two-step process: DNA ends are first generated and then they are joined indiscriminately. The single-stranded T strand provides one end, and this is presumed to join to breaks in the chromosomal DNA, possibly aided by the VirD2 protein. Free ends of plant DNA are probably generated during replication and repair processes, and the T-DNA may integrate into transcriptionally active regions. DNA undergoing replication or transcription offers better accessibility to invading DNA because nucleosomal complexes unravel during these processes.

T strand integration is very different from the integration of retrovirus or transposon DNA. These latter elements integrate in a relatively precise and undisturbing manner, in which a few bases at the target site are duplicated at each end of the integrated element. Also, retroviruses and transposons encode their own enzymes for integration. Besides the possibility that VirD2 functions to initiate T strand joining to chromosomal DNA breaks, the major steps in T strand integration most likely rely on host enzymes that are normally used for DNA synthesis, repair, and recombination.

12.4 Crown Gall Tumor

The T-DNA transfer process ultimately results in the phenotypic transformation of the infected plant cell into a crown gall tumor cell. This alteration of the properties of the transformed plant cell by the T-DNA element has been studied extensively and provides an interesting example of the types of responses that occur in plants in nature as a result of interactions with soil microorganisms.

a. Crown Gall Compared to Animal Tumors

Both crown gall tumor cells and virally transformed animal tumor cells result from the integration into a cellular genome of foreign DNA carrying genes whose subsequent expression leads to a neoplastic and often dedifferentiated phenotype. The closest parallels are the animal DNA tumor viruses. (See Chapter 1.) In these systems, the transforming genes encode functions related to cellular growth and/or differentiation, although the types of functions and their evolutionary origins are different. (See Chapter 3.) By contrast, retroviruses carry transforming genes that are typically derived from cellular prototypes that encode a variety of functions (e.g., protein tyrosine kinases, adenylate cyclase regulators, growth factor analogues) in various cellular locations (e.g., plasma membrane, Golgi, cytoplasm, nucleus). (See Chapter 4.)

Transformation of plant cells by T-DNA is a consequence of T-DNA genes that encode novel enzymes for the biosynthesis of the **phytohormones auxin** and **cytokinin**. The elevated levels of these molecules in crown gall cells are responsible for the neoplastic and hormone-independent growth. However, the T-DNA–transforming genes may have an evolutionary origin independent of that of their host (plant) analogues because the plant-encoded pathways for auxin biosynthesis involve chemical reactions different from those specified by the T-DNA genes.

The elucidation of the T-DNA-encoded phytohormone synthetic pathways provided further insight into the metabolism of auxin and cytokinin in both normal and transformed plant tissues. For example, plant cells normally control their auxin levels by converting auxin to conjugates that have only low growth factor activity. The T-DNA-directed auxin pathway circumvents this control pathway and actually capitalizes on it by utilizing these auxin conjugates as substrates for auxin synthesis.

b. The Molecular Basis of the Crown Gall Phenotype

In contrast to normal untransformed plant cells, T-DNA-transformed cells can grow and divide in tissue culture without addition of the plant growth hormones auxin and cytokinin because of several genes carried by the T-DNA segment. Thus, mutations in one region of the T-DNA result in crown gall tumors that produce numerous shoots, and mutations in another region cause crown gall tumors that produce numerous roots. Plant cells usually produce shoots when the cytokinin to auxin ratio is high; conversely, when the auxin to cytokinin ratio is high, roots are formed. In tumor cells, the levels of plant growth hormones are regulated by the T-DNA genes *1*, *2*, and *4* (Figure 12.18). Genes *1* and *2* together control auxin, and gene *4* controls cytokinin. Thus, a T-DNA that is $1^-2^-4^+$ results in tumors overproducing shoots, and a T-DNA that is $1^+2^+4^-$ results in a rooting tumor (Figure 12.18).

These data are easy to understand if one remembers that plant differentiation is controlled by a balance of these two classes of hormones; in the absence of one, the effects of the other take over. Furthermore, when the levels of both auxin and cytokinin are high, an undifferentiated phenotype results. This latter phenotype can be produced by growing normal untransformed cells in the presence of high concentrations of auxin and cytokinin. The crown gall tumor cell, however, can grow in tissue culture without added hormone because its cells already produce these compounds.

The genetic analysis of these hormone-producing T-DNA genes did not reveal whether the control of hormone levels was direct, by T-DNA–encoded genes specifying hormone biosynthesis, or indirect, by T-DNA–encoded regulatory element(s) that modified the expression of endogenous plant genes. This question was resolved by biochemical experiments; T-DNA genes *1* and *2* encode two auxin biosynthetic enzymes, and gene *4* encodes a cytokinin biosynthetic enzyme.

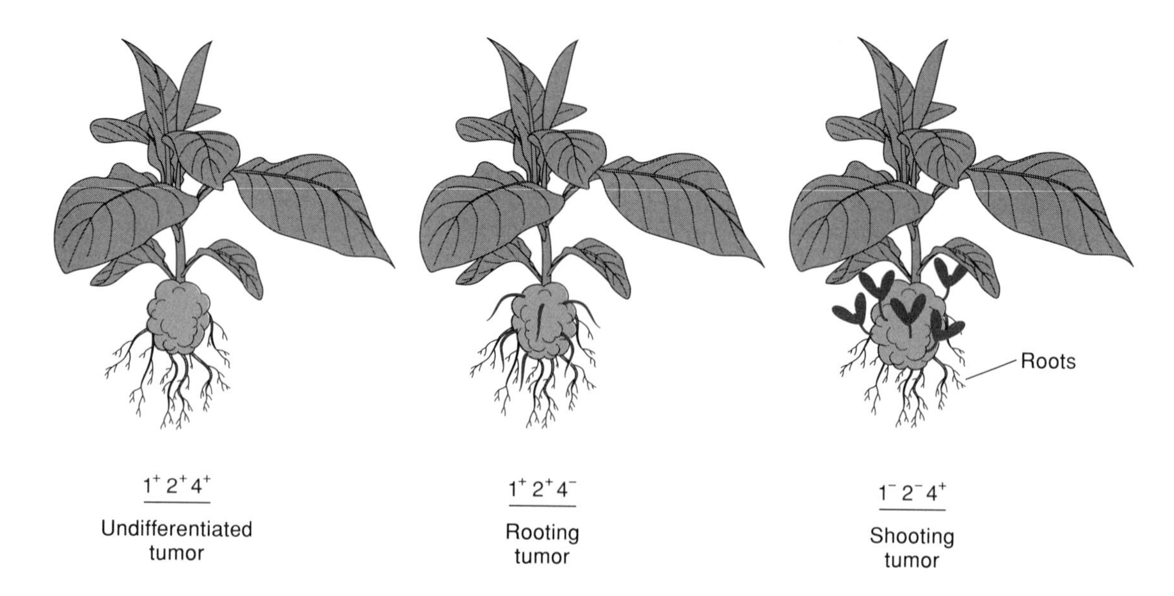

Figure 12.18
The T-DNA genes that control the crown gall phenotype. Mutations in T-DNA genes *1*, *2*, and *4* produce tumors with altered phenotypes. Because mutations in gene *4* produce rooty tumors and mutations in genes *1* and *2* produce shooty tumors, their wild-type gene products must control the synthesis of the plant growth regulators cytokinin and auxin, respectively.

Auxin synthesis Figure 12.19 summarizes the known pathways for synthesis of auxin: **indole-3-acetic acid (IAA)**. The amino acid tryptophan is modified by several enzymatic steps to produce IAA. The pathway shown in the center is encoded by the T-DNA of *Agrobacterium*: the first reaction is carried out by the gene *1* product, tryptophan monooxygenase, and the second reaction is carried out by the gene *2* product, amidohydrolase. Tumor cells containing only gene *1* in the their T-DNA accumulate up to 1000-fold more indole-3-acetamide than normal cells. Tumor cells containing only gene *2* can convert exogenously added indole-3-acetamide into active auxin; amidohydrolase can also convert indole -3-acetonitrile as well as storage forms of IAA (conjugates of IAA with glucose and myo-inositol) into active IAA.

The other reactions shown in Figure 12.19 are those known to occur in plant cells themselves. It is remarkable that the T-DNA element evolved to encode completely novel enzymes for auxin biosynthesis. *Agrobacterium* is not unique among bacteria in carrying genetic information for phytohormone biosynthesis. In fact, the experiments that established that T-DNA genes *1* and *2* encode enzymes involved in auxin biosynthesis were inspired by a comparison with another gram-negative soil pathogen, *Pseudomonas savastanoi.* This organism produces tumorous growths on olive and oleander plants. However, in contrast to tumors induced by *Agrobacterium,* the *Pseudomonas*-induced tissues are not genetically transformed but depend on the continuous production of auxin by bacteria at the infected site. *Pseudomonas* contain plasmids that carry two genes for IAA synthesis. These genes encode proteins that synthesize IAA by the same pathway as the *Agrobacterium* T-DNA genes *1* and *2* (i.e., tryptophan is converted to indole-3-acetamide by a tryptophan monooxygenase, and the indole-3-acetamide is converted to IAA by an amidohydrolase). Although the enzymes from these two bacteria are functionally equivalent, their structures are not iden-

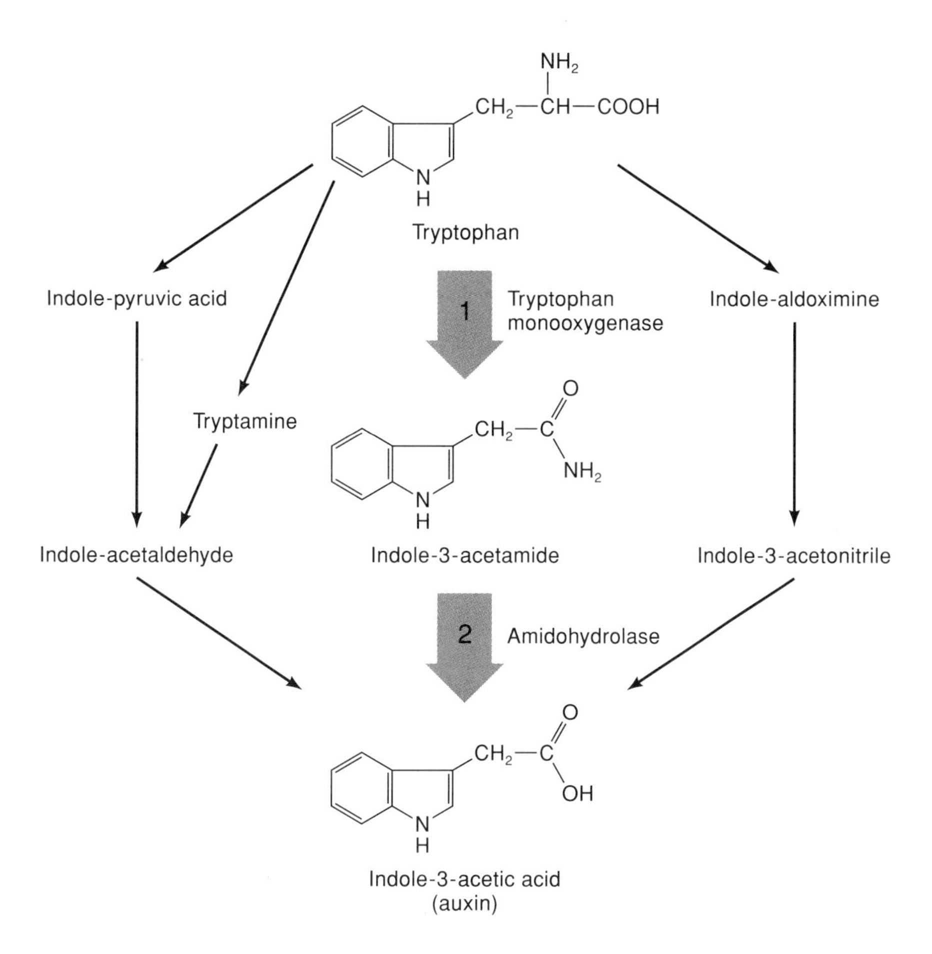

Figure 12.19
Biochemical pathways for auxin biosynthesis. Common plant pathways are indicated left and right. The center indicates the novel pathway to the production of indole-3-acetic acid that occurs in the presence of the gene *1* and gene *2* products of the T-DNA (indicated by the large, numbered arrows).

tical. There is 50 percent amino acid sequence homology for the gene *1*-type enzyme and only 27 percent for the gene *2*-type enzyme.

Cytokinin synthesis Figure 12.20 summarizes the essential steps in cytokinin biosynthesis in plants. The rate-limiting step is normally the production of isopentenyladenylic acid. However, crown gall cells contain elevated isopentenyltransferase activity. This activity derives from the expression of the T-DNA's gene *4* product. Extracts prepared from *E. coli* carrying gene *4* on expression vector plasmids contain high levels of isopentenyltransferase activity and produce authentic isopentenyladenylic acid. In crown gall cells, the newly synthesized isopentenyl derivative is rapidly converted to the active cytokinin transzeatin. As already described for the enzymes involved in auxin biosynthesis, *Pseudomonas* also contains a functionally related and structurally similar (50 percent amino acid homology) isopentenyltransferase.

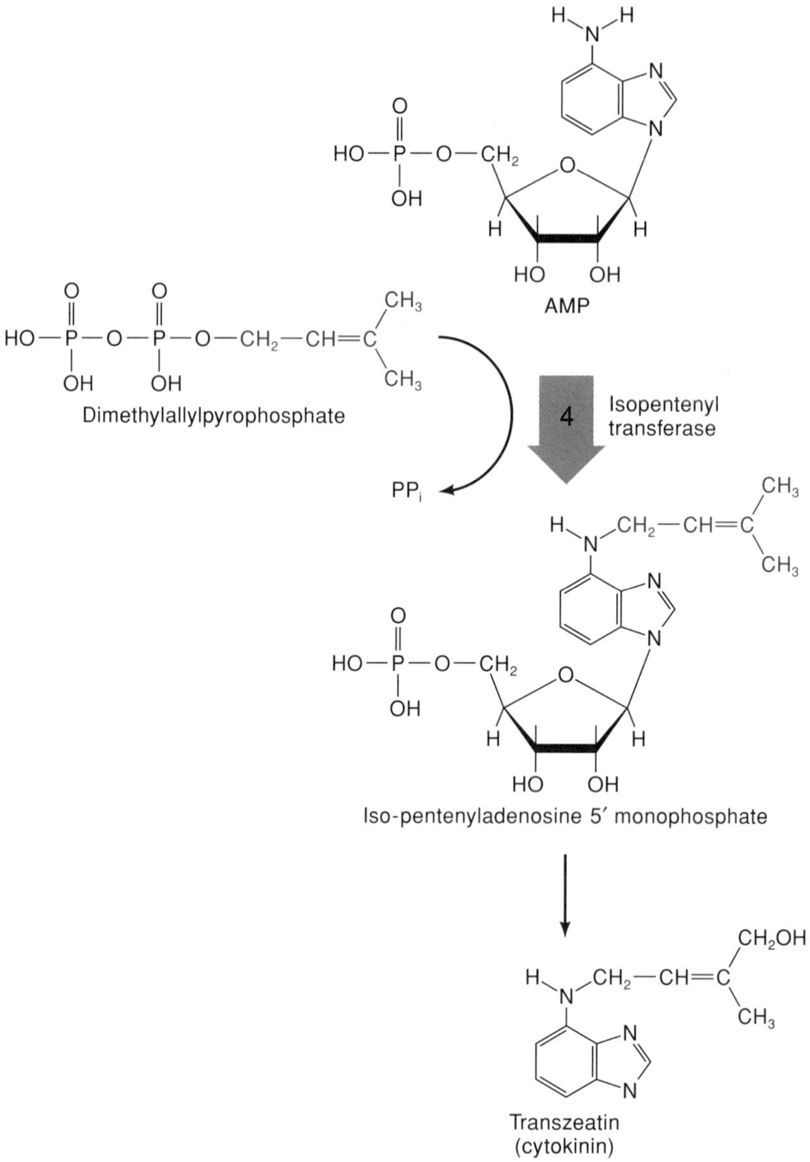

Figure 12.20
Cytokinin biosynthesis. Steps in the production of the active cytokinin, transzeatin, are outlined. Normally, the rate-limiting step is the production of isopentenyladenosine 5′monophosphate. However, the T-DNA gene *4* product (large arrow, numbered 4) produces an active isopentenyltransferase that leads to the elevated accumulation of this cytokinin precursor.

c. Origin of T-DNA-Transforming Genes

It is easy to envisage an exchange of genetic information between various prokaryotic organisms living in the soil. Such events could explain the presence of similar but evolutionary diverged genes in different bacteria such as *Agrobacterium* and *Pseudomonas*. Presumably, bacteria that live in close contact with plant cells benefit from promoting such interactions, thus explaining why they evolved to synthesize products of potential benefit to plant growth and differentiation. For example, in the *Pseudomonads*, such products are synthesized under the control of DNA sequences 5′ upstream of the coding regions, as is typical of prokaryotic genes. The *Agrobacterium* system is clearly more complex. Although the coding sequences for phytohormone biosynthesis are likely to be of bacterial origin (they do not hybridize with plant genomic DNA), the regulatory sequences controlling their transcription function in the eukaryotic plant cell. The existence of eukaryotic regulatory sequences in a prokaryotic cell hints at the possibility of DNA exchange between bacteria and plant cells in the soil microenvironment. *Agrobacterium* is certainly capable of promoting transfer to plant cells. Perhaps infections at the wound sites of broken or damaged plant cells provide DNA for uptake by *Agrobacterium tumefaciens.*

In summary, the study of the *Agrobacterium*–plant cell interaction provides fundamental insight into a variety of biologically relevant and interesting processes. However, *Agrobacterium* was actively used as a vector for plant gene transfer experiments in plants even before many of the molecular events underlying the T-DNA transformation process were understood.

12.5 Ti-Plasmid as a Vector for Gene Transfer to Plants

Once scientists firmly established that the interaction between *Agrobacterium* and wounded plant cells results in the transfer of the T-DNA segment from the Ti-plasmid into the plant genome, the question arose as to whether other DNA sequences, experimentally inserted within the T-DNA, would also be cotransferred. This could be tested with Ti-plasmid mutants that contained bacterial sequences (the 15 kbp Tn7 element) within the 22 kb nopaline T-DNA. Infection of plants with *Agrobacteria* carrying this modified T-DNA led to integration of a 37 kbp segment containing both the T-DNA and Tn7. So far, modified T-DNAs up to 50 kbp in size have been stably introduced into the genome of plant cells. The ability to transfer any DNA contained between the T-DNA borders to plant cells strongly supported *Agrobacterium* as a generalized vector for plant genetic engineering. The possibility of introducing DNA at will provides an efficient means both to study fundamental features of plant gene structure and regulation and to alter plants genetically. Although the ability to assay tumor cells for the expression of transforming DNA is useful, measuring expression in the differentiated tissues and cells of whole plants is even more advantageous.

The potential of *Agrobacterium* as a vector would not have been realized without plants' natural ability to regenerate themselves from single cells in culture (see Section 12.5e). This totipotency makes plant cells the optimal recipient for genetic engineering experiments because the introduced DNA can be studied in all plant tissues. Furthermore, regenerated plants are fertile, so the transmission of this DNA into progeny plants can be analyzed. However, for the Ti-plasmid to be a useful vector for transferring DNA into totipotent plant cells, it must first be altered to maintain its DNA transfer ability without causing tumors.

a. Nononcogenic Derivatives of the Ti-Plasmid Useful as Generalized Vectors

To be a useful vector, a Ti-plasmid should have five essential components: (1) T-DNA borders, (2) the entire *vir* region, (3) no internal T-DNA sequences encoding the tumorous crown gall phenotype, (4) a marker gene internal to the T-DNA that is expressed in transformed plant cells to allow their detection, and (5) an internal T-DNA sequence, usually one or more restriction sites, that allows the insertion of any DNA of interest.

The first Ti-plasmid vector was systematically designed to these specifications. Knowing the location of the T-DNA borders and the functions controlling plant cell dedifferentiation allowed the construction of a derivative of Ti-plas-

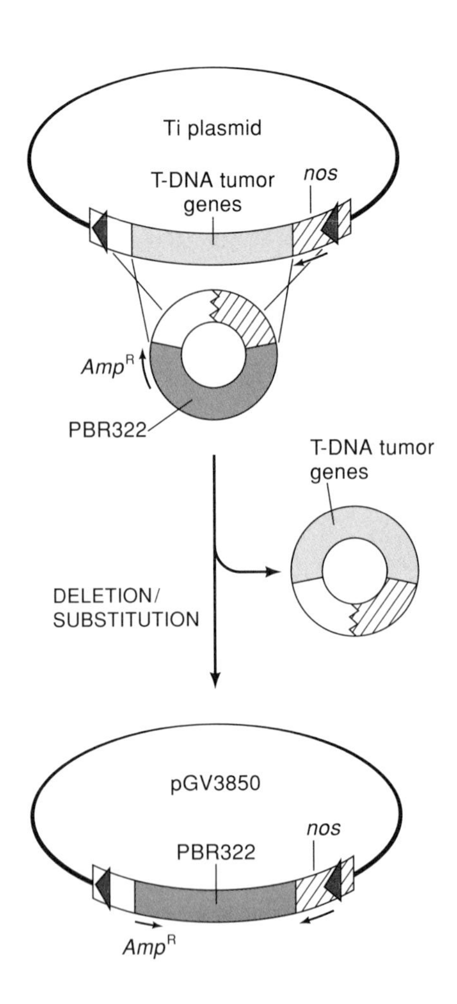

Figure 12.21
Construction of nononcogenic Ti plasmid vector pGV3850 by reverse genetics. A plasmid (pBR322-derived) carrying fragments with homology to the left and right borders of the T-DNA is used to select for a double recombination event that results in the deletion of the internal region of the T-DNA (light-colored shading) and its tumor (auxin- and cytokinin-producing) genes. The ampicillin resistance gene (*Amp*^{R}) of the plasmid serves as a selectable marker for the recombination events. pGV3850 contains functional left and right T-DNA borders (arrows) and the nopaline synthase gene (*nos*) bracketing the bacterial plasmid sequences (dark-colored shading).

mid (pGV3850) containing both a deletion and a substitution (Figure 12.21). Thus, most of the internal sequences of the T-DNA were deleted and replaced by sequences from the *E. coli* plasmid pBR322. The T-DNA region of this Ti-plasmid derivative also contained the nopaline synthase gene (*nos*), which, when transferred and expressed in plant cells, results in the production of nopaline. A test for nopaline identifies plant tissues transformed by the modified Ti-plasmid. Because genes *1, 2,* and *4* were deleted, pGV3850 can transform plant cells without causing oncogenic growth properties.

Cis *vectors* The pGV3850 nononcogenic Ti-plasmid derivative is an example of one type of Ti-vector, designated a ***cis* vector**. In this case, the DNA segment to be transferred to plant cells is inserted into, that is *cis* to, Ti plasmid sequences by homologous recombination. In contrast, ***trans* vectors** (see next subsection) contain modified T-DNA regions on small autonomously replicating plasmids that occur independently of the *Agrobacterium* Ti-plasmid (e.g., in *trans* to the Ti-plasmid).

The versatility of pGV3850 as an acceptor for foreign DNA sequences is illustrated in Figure 12.22. Any DNA sequence cloned into pBR322 or its derivatives can easily be inserted in *cis* into the pGV3850 T-DNA by a single homologous recombination between the pBR322 present in pGV3850 and the homologous sequences in the cloning vehicle. The foreign DNA is cloned and manipulated in *E. coli* and then mobilized to *Agrobacterium* using transmissible helper plasmids. Because the origin of pBR322 replication is not functional in *Agrobacterium,* the only way for the introduced pBR plasmid to be maintained in *Agrobacterium* is by cointegration with the Ti-plasmid. The cointegrates are selected for and maintained in *Agrobacterium* by including a drug resistance

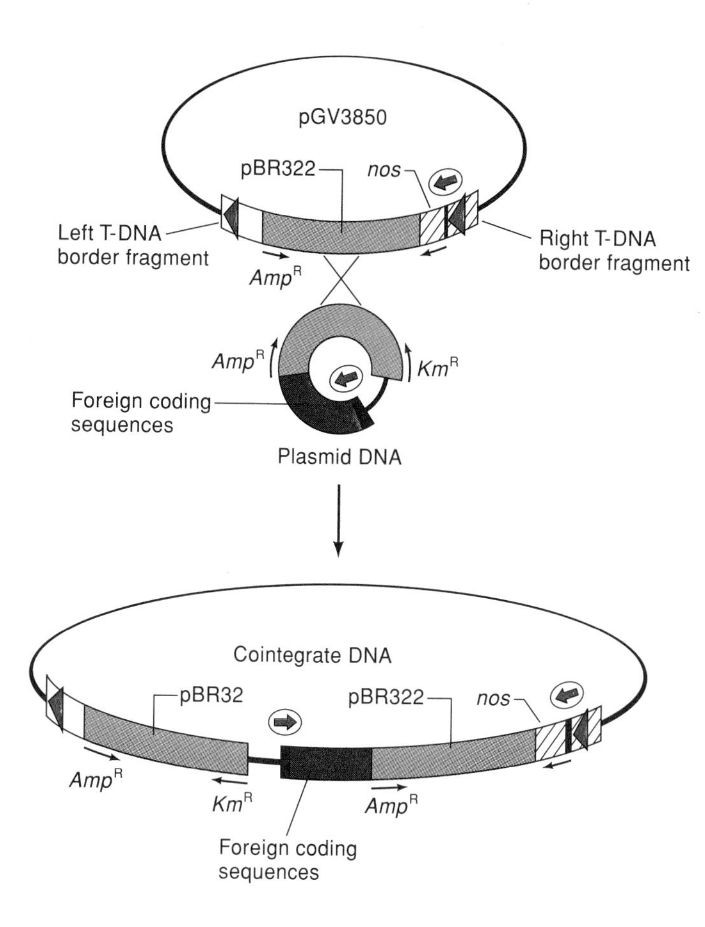

Figure 12.22
Use of pGV3850 as *cis* vector to accept foreign DNA of interest to transfer to the plant cell. The bacterial plasmid sequences (shaded) in the T-DNA indicate regions of homology that can be used to form cointegrates with other bacterial plasmids carrying genes of interest with foreign coding sequences. The only requirement is that the plasmids carrying genes of interest also carry an additional selectable marker gene, such as kanamycin resistance (Km^R), which serves as a selection to maintain the cointegrates. If *Agrobacterium* is grown in the presence of kanamycin, only cointegrates can grow because the bacterial plasmid origin of replication does not function as an independent replicon in *Agrobacterium.*

marker (e.g., kanamycin resistance) in addition to the ampicillin resistance gene in the pBR plasmid.

Trans vectors *Trans* (also called binary) vectors are based on the fact that the T-DNA element does not encode functions essential to its transfer (see Figure 12.8). Instead, the T-DNA is the structural substrate of the proteins encoded by the Ti *vir* region. Thus, the T-DNA can be located anywhere within *Agrobacterium* as long as it coexists with a full complement of the essential *vir* loci.

Trans T-DNA vectors should contain (1) the 25 bp T-DNA border sequences without the internal T-DNA transforming genes, (2) a marker for the detection of transformed plant cells, (3) an appropriately located set of restriction enzyme sites for insertion of foreign DNA of interest, (4) a bacterial resistance marker to select for its maintenance in *Agrobacterium,* and (5) a wide host range origin of replication functional in both *E. coli* (for cloning and manipulating foreign inserts) and *Agrobacterium* (for stable maintenance). *Trans* T-DNA vectors are used in combination with vir^+ T-DNA$^-$ or with vir^+ T-DNA$^+$ Ti-plasmids. In the latter case, the Ti T-DNA and *trans* T-DNA are chosen to produce different phenotypes in the transformed plant cell so that independent transformation events resulting from transfer of the individual T-DNAs can be distinguished.

Comparison of cis *and* trans *vectors* In general, *cis* and *trans* vectors are equally efficient in transferring DNA to plant cells. Both require the presence of Ti-plasmid sequences that overlap the T-DNA borders by at least a few hundred nucleotides. Vectors carrying only the 25 bp sequences without their normal contiguous sequences have a reduced efficiency (approximately 30 percent) of T-DNA transfer; presumably, the natural sequence context of these sequences promotes their most efficient utilization. Either *cis* or *trans* vectors can be used when transferring a single gene of interest. However, for "shotgun" cloning and transferring a population of DNAs, the *trans* vector is superior. With a *trans* vector, the efficiency of transfer of a population of plasmids depends only on the efficiency of plasmid mobilization from *E. coli* to *Agrobacterium.* However, a *cis* vector will produce a smaller population of cointegrate Ti-plasmids containing DNAs of interest because the frequency of their formation depends on both their mobilization from *E. coli* and their recombination with the acceptor Ti-plasmid.

b. Marker Genes for Plant Cells

The efficient detection of transformed plant cells using nononcogenic Ti-derived vectors requires that the transferred T-DNA segment also carry a selectable marker gene. Such a gene, when expressed, allows the selection of transformed from untransformed plant cells. Selectable markers are more advantageous than screenable markers such as nopaline because the latter do not endow the transformants with a readily distinguishable phenotype. As with bacteria and animal cells, antibiotic resistance is the most frequently used selectable marker. The coding sequences for resistance are taken from bacterial sources; the signals to allow their expression must be derived from sequences known to promote transcription in plant cells.

Construction of selectable marker genes To construct a versatile selectable marker gene to be expressed under a variety of plant growth conditions requires that the resistance sequence be placed under the control of constitutively active sequences. Genes isolated directly from plant cells are not a good source for transcriptional signals because they are usually highly regulated and expressed at particular developmental stages, in specific tissues, or under particular environmental conditions. Indeed, some of the best and most frequently used signals for constitutive gene expression are those normally found in the T-DNA. The signals from the genes encoding nopaline or octopine synthase have been extensively used. These genes, when present, are expressed in crown gall callus tissue as well as in all differentiated plant tissues regenerated from nontumorous calli.

The nucleotide sequences of the transcriptional signals for the nopaline and octopine synthase genes, as well as other T-DNA genes, bear close resemblance to the consensus sequences for the start and stop of transcription in other eukaryotic genes. Thus, the 5′ region flanking the T-DNA genes contains sequences homologous to the TATA-box often found 20 to 30 bases upstream of the start of transcription, and sequences related to the CAAT-box occur 60 to 80 nucleotides upstream of the transcriptional start site. In addition, sequences between 20 and 100 bases 5′ of the polyadenylation site in T-DNA transcripts strongly resemble the AATAAA consensus sequence similarly placed in animal genes.

Based on this information, scientists have constructed a series of chimeric genes using the promotor and terminator signals for transcription of the nopaline synthase (*nos*) gene. They introduced restriction endonuclease sites between these signals to allow the insertion of desired DNA sequences. They inserted the coding sequences from three different bacterial genes conferring resistance to kanamycin (Km), chloramphenicol (Cm), or methotrexate (Mtx) (from the Tn5 and Tn9 transposons and the R67 plasmid, respectively) between the *nos* control signals. Investigators then transferred all three constructs to *Agrobacterium,* cointegrated them with the nononcogenic Ti vector plasmid pGV3850 (Figure 12.22), and tested them for their ability to confer antibiotic resistance to plant cells.

Phenotypic expression of antibiotic resistance traits in plant cells Although *Agrobacterium* can be used to inoculate whole plants, the transformed tissues derived

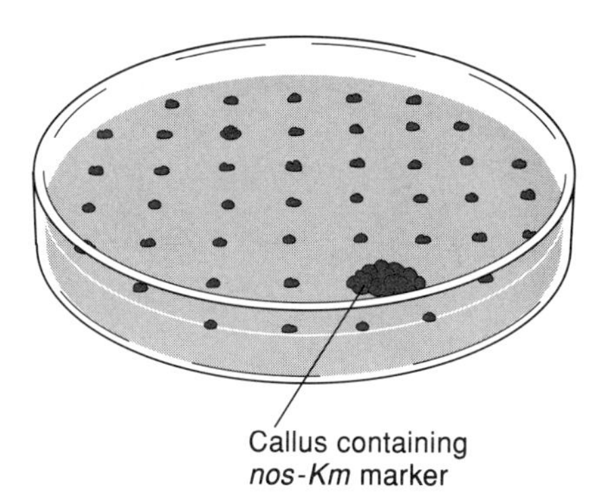

Figure 12.23
The nopaline promoter linked to the coding sequences for kanamycin resistance is useful as a selectable marker for plant cell transformation. Clonally derived successfully transformed calli are able to grow on medium supplemented with kanamycin. Calli that do not contain the selectable marker gene do not grow; they remain small and die on this medium.

from its infection represent a heterogenous population of transformation events. To study the transfer and expression of defined DNA segments, scientists need clonal tissue derived from the infection of single plant cells. To this end, single-cell protoplasts can be prepared from leaf tissues following treatment with enzymes to degrade their cell walls. The protoplasts are then cocultivated with *Agrobacterium* to produce independent transformants. If the *Agrobacterium* strain used for the infection carries a selectable marker gene, the transformed cells harboring this DNA can be detected following growth under selective conditions.

Cells derived from tobacco plants are sensitive to the antibiotics kanamycin, chloramphenicol, and methotrexate. However, tobacco cells transformed with *Agrobacterium* carrying the chimeric genes, *nos*-Km, *nos*-Cm, or *nos*-Mtx, are capable of growth in medium containing the respective antibiotic. These types of experiments demonstrated for the first time that foreign DNA sequences can be systematically transferred to and expressed in plant cells. The selective expression of these resistance traits is absolutely dependent on the transcriptional signals from the *nos* gene; constructs containing bacterial signals or lacking the *nos* promoter are unable to promote selective growth of plant cells in the presence of antibiotics.

The resistance traits are useful to select for transformed cells growing in tissue culture (Figure 12.23). However, whole plants containing these genes can also grow selectively in their presence. This property works especially well for the Km marker. For example, the transmission of the Km resistance gene from transformed plants to progeny can be easily scored by germinating seeds directly on Km-containing medium. Seedlings that contain the Km marker germinate normally and form plantlets in the presence of Km; untransformed seeds germinate into small **etiolated** (white, from the French word *étioler,* to grow pale or weak) plantlets that die after a few days. Figure 12.24 shows, for example, the growth on kanamycin of the F1 progeny from a transformed plant carrying one copy of the *nos*-Km marker in its genome; the resistance trait is expressed in a three (green) to one (white) ratio.

Other markers Besides selectable markers, certain nonselectable markers also provide useful tools to monitor gene expression in plant cells. Common markers of this type are the firefly or bacterial luciferase (*lux*) genes and the bacterial β-glucuronidase (*gus*) gene. Both systems utilize a variety of commercially available substrates that produce colored or fluorescent products. Because *lux* and *gus* are enzymes, even low activities give measurable products. Thus, when *lux*

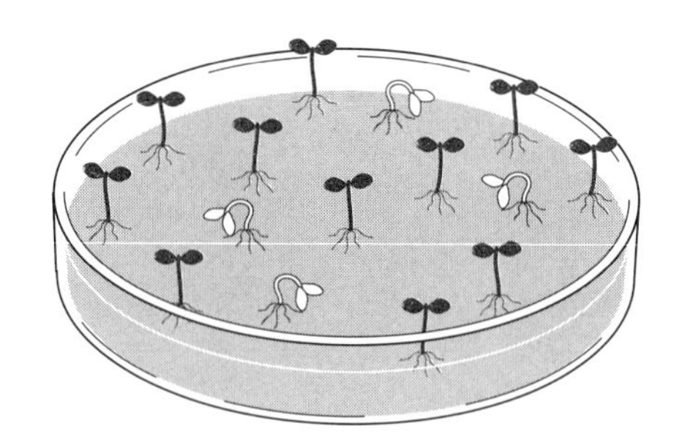

Figure 12.24
Germination of seedlings from the F1 progeny of a transformed kanamycin-resistant plant. Seeds from a self-fertilized kanamycin-resistant plant were planted in agar medium containing kanamycin. The resistant plants germinated and produced progeny plants within 3 weeks. The sensitive seeds also germinated, but after 1 week they did not grow further; after 2 more weeks they became etiolated (white) and died. The ratio of kanamycin-resistant to sensitive plants was 3 : 1, reflecting the Mendelian segregation of a heterozygous kanamycin-resistant F1 plant, where kanamycin resistance is a dominant trait. (From M. De Block, L. Herrera-Estrella, M. Van. Montagu, J. Schell, and P. Zambryski. 1984. *EMBO J.* 3 1681.)

or *gus* genes are fused to plant genes of interest, they provide a highly sensitive assay for gene expression. *lux* and *gus* products are also highly useful as reporters in histological assays for cell- and tissue-specific patterns of gene expression (see Figure 12.15).

Chimeric genes as probes for regulatory sequences controlling plant gene expression - Model experiments using the chimeric *nos* gene constructions suggested that a similar approach could be used to determine the sequences important for the regulated expression of endogenous plant genes. One of the most interesting aspects of plants is the regulation of their growth and development by light. The genes encoding the major proteins for photosynthesis are obvious candidates to study. Two gene families have been extensively studied: those encoding the small subunit of ribulose-1, 5-bisphosphate carboxylase (*rbc*S) and the chlorophyll a/b-binding protein (*cab*). The *rbc*S gene product participates in CO_2 fixation and its conversion into carbohydrates using the energy obtained during photosynthesis. The *cab* protein acts in a complex with chlorophyll to convert light energy into useful photosynthetic energy.

The *rbs*S and *cab* proteins function in the chloroplast; however, these as well as most chloroplast polypeptides are encoded by nuclear genes and are the products of cytoplasmic protein synthesis. Following synthesis, the proteins are imported into the chloroplast by highly specific translocation processes. Because *Agrobacterium* transfers DNA into the nuclear genome, it can be used as a tool to study the regulated synthesis of important nuclear-encoded chloroplast proteins.

Light, in the presence of photoreceptive pigments such as phytochrome, induces the expression of the *rbs*S and *cab* gene families by increasing their rates of transcription into functional mRNAs. To determine whether light-regulated gene expression is controlled by sequences 5′ to the transcriptional initiation sites, researchers constructed a chimeric gene using about 900 bp of the promoter region from an *rbc*S gene fused to CAT-coding sequences and the 3′ end of the *nos* gene. They transferred this construct to plant cells using *Agrobacterium,* grew the transformed tissues either in the light or dark, and assayed for CAT activity. The results are rather dramatic: the small subunit promoter directs the expression of CAT activity only when the tissue containing the chimeric gene is grown in the light. Control *nos*-CAT constructs produce identical CAT activity in light- and dark-grown tissues. The researchers further analyzed the regulatory fragment of the *rbc*S promoter region to show that it contains an enhancerlike element(s) that facilitates light-regulated transcription. Thus, these plant-specific transcriptional regulatory sequences may resemble the enhancers typical of animal systems.

Chimeric genes as probes for sequences controlling intracellular transport of plant proteins The protein products of the nuclear genes *cab* and *rbc*S must be transported following their synthesis in the cytoplasm to their sites of function in the chloroplast. Transport is accomplished, as in animal cells, by specific transit peptides (TP), which promote translocation across chloroplast membranes. This result was elegantly shown following transfer of a particular chimeric gene construct to plant cells using *Agrobacterium.* The chimeric gene TP*rbc*S-Km contained 1 kb of the 5′ region of the *rbc*S gene, including the promoter, downstream nontranslated sequences, and translated sequences encoding the first 57 amino acids of the *rbc*S polypeptide, fused to the coding sequences of neomycin phosphotransferase (NPT).

This chimeric gene promoted the phenotypic expression of Km resistance, and such tissues produced significant levels of NPT activity in crude extracts. Following subcellular fractionation, researchers specifically localized the NPT activity to the chloroplasts. Moreover, they found NPT activity to be localized to the stromal (soluble) fraction rather than the membrane fraction of the chloroplast. Because the *rbc*S protein is normally found in the chloroplast stroma, these results suggest that the first 57 amino acids of the *rbc*S protein, which were present in the chimeric gene construct, are sufficient to specify the exact intracellular location of even a foreign polypeptide (NPT). Control chimeric constructs, *nos*-Km or *rbc*S-Km (lacking the TP sequences of the *rbc*S protein), are not capable of directing NPT transport into the chloroplast. Thus, *Agrobacterium* can be used to transfer DNA to plant cells to ask a variety of questions important to basic plant cell metabolism.

c. Transgenic Plants

Although *Agrobacterium* infects a variety of dicotyledonous plants in nature, it is most often used in the laboratory to infect plant species that are easily grown and maintained under *in vitro* tissue culture conditions. Tobacco is the most widely used plant species. Historically, plant genes were isolated from several sources, depending on the interest of particular laboratories. For example, the first isolations of the *rbc*S and *cab* genes were from peas. Although pea cell culture is not advanced, the pea genes could be analyzed following their transfer to tobacco cells using the *Agrobacterium* system. Such "transgenic" plants simplify analyses of introduced genes because their size and sequence composition differ from those of related endogenous genes. Further, universal and evolutionarily conserved sequences that promote expression in different plant species can be defined.

Light regulation of genes introduced into transgenic plants Analysis of different transgenic tobacco plants containing chimeric gene fusions of either the pea *rbc*S promoter or pea *cab* promoter and a reporter for NPT activity revealed that the chimeric genes are expressed only in tissue containing mature chloroplasts (leaves, stems, sepals, and stigmas), and not in other tissues, such as petals and roots. However, when a larger selection of transgenic plants is sampled, the introduced genes are not always expressed in the expected places. These results presumably reflect the influence of the insertion site on expression of transferred genes. The T-DNA containing foreign DNA sequences would probably occasionally integrate close to (or at least within range of) a regulatory sequence that is expressed only at a particular developmental stage.

Such qualitative differences in the expression of introduced genes are relatively rare. However, quantitative differences in the rates of expression of the introduced genes are much more common. For example, an analysis of ten different transgenic plants for RNA transcribed from an introduced *rbc*S gene revealed that all ten plants show the appropriate qualitative differences in expression when grown in the light or dark; however, the quantitative levels of expression in all ten plants were dramatically different.

The expression of light-regulated chimeric genes in transgenic plants depends not only on light, but also on the presence of the appropriate structural and biochemical cofactors. For example, light-regulated gene expression depends on the presence of differentiated chloroplasts, and the addition of herbicides that arrest chloroplast development blocks expression of light-regulated reporter genes. Also, light-regulated expression of endogenous plant genes is mediated by the photoreceptive pigment, phytochrome. Whether or not phytochrome is a cofactor for light-regulated gene expression can be judged by measuring the effect of light of different wavelengths on gene expression; phytochrome activity is stimulated by red light and repressed by far red light, and the expression of the chimeric genes is similarly regulated.

Other chimeric genes It is impossible to mention all possible chimeric constructs that have been or might be designed to be useful for plant gene transfer experiments. A few examples will have to suffice. Thus, although the *nos* promoter directs a low level of constitutive gene expression in transformed plant cells, it is sometimes useful to promote high levels of constitutive expression. In this latter case, the strong promoters of the plant virus CAMV (cauliflower mosaic virus) are particularly useful. Or promoters from more diverged organisms can be tested to illustrate evolutionary relationships or constraints. For example, a 457 bp fragment containing the "heat shock" promoter from *Drosophila* directs heat-regulated expression of a reporter gene in tobacco. Further, the heat shock promoter can be used specifically to turn on ectopic gene expression, for example, to test the effects of regulatory genes during different developmental stages. That plants have upstream regulatory regions and coding sequences homologous to their better studied animal counterparts is not surprising in hindsight, although for quite some time, researchers assumed that plant cells might use distinctive regulatory and structural paradigms for developmental control. Thus, plant molecular biology can take advantage of the wealth of information as well as experimental strategies that have been developed using animal cells. The only limitations to these types of experiments are the availability of cloned sequences and the imagination to design appropriate experiments.

d. DNA Transfer as a Tool To Identify Plant Genes

The molecular analysis of plant genes is dependent on the number of available cloned plant genes. Methods to tag plant genes greatly facilitate their identification. For example, sequences essential to maintain normal plant morphology can be isolated if insertion of foreign DNA inactivates them and the aberrant phenotype can be visualized. A hybridization probe homologous to the inserted DNA can be used to clone the affected gene. The ultimate proof that the DNA at the insertion site encodes a particular phenotype then depends on the restoration of the wild-type phenotype by the introduction of the cloned wild-type DNA sequence.

The T-DNA element as an insertion mutagen An example of the utility of the T-DNA transfer system is its use to generate mutations in *Arabidopsis thaliana.* A

T-DNA element containing the Km resistance gene has been used to generate greater than 10,000 independent transformed lines of *Arabidopsis*. The procedure utilized for the initial transformation is critical to the success of the method. Basically, *Arabidopsis* seeds are germinated in the presence of *Agrobacteria* carrying a T-DNA element bearing the Km resistance marker gene. The resulting plants produce seeds that are then germinated on plates carrying Km.

The primary transformation (T1) does not involve any tissue culture, so the resulting plants do not exhibit any unstable phenotypic traits (somaclonal variation, see Section 12. 5e) such as can arise during a period of culture. The method does not require that the primary transformation be efficient (i.e., that the majority of seeds be Km resistant) because thousands of seeds can be germinated on a single petri dish. Km-sensitive seedlings are white and die shortly after germination (Figure 12.24). Km-resistant seedlings are green, healthy, and easily removed from the dish and grown into mature plants. Because Km resistance is a dominant marker and the first generation of T-DNA tagged lines is heterozygous, the T1 plants generally do not exhibit any special phenotypic traits. Thus, the visualization of recessive mutations leading to altered phenotypes is usually scored in the next generation, T2, in which 25 percent of the progeny will be homozygous for the Km resistance marker. This elegantly simple and reliable method has provided the scientific community with a wealth of mutant phenotypes in all aspects of plant growth and development, and their interrupted genes can be cloned directly by utilizing a T-DNA sequence as a probe.

Plant transposons as insertion mutagens Transposable elements are one of the most widely recognized mediators of genetic diversity. In fact, the numerous variegations generated by these elements in the kernels of the maize plant led to their discovery by Barbara McClintock in the first half of this century. Although the *Ac-Ds* system of maize is the most widely known plant transposon, the abundant presence of transposons in many plants in our surroundings is readily seen in the diverse and variegated color patterns of flowers.

Following the molecular cloning and characterization of the *Ac* element, researchers quickly realized that this transposon might serve to generate genetic diversity in other plant species. A model series of experiments demonstrated that *Ac* could move in tobacco following its introduction into the tobacco genome by *Agrobacterium* transformation. Since that time, the *Ac* element has been used as a tool to generate tagged lines in tomato and *Arabidopsis*. The *Ac* system has a unique property that highlights its utility compared to the *Agrobacterium* T-DNA-tagging system. This property is based on classical transposition.

There are two essential components to the the *Ac* system: (1) the ends of the element, which provide the sites for action of (2) the transposase, encoded by the internal portion of the element. Vectors for T-DNA transformation carrying one or the other of these components can be constructed. Transformed plants carrying an active transposase (without the ends of *Ac*) can be crossed to plants carrying an inactive *Ac* element containing the ends of *Ac* but no transposase. Following such a cross, the *Ac* end element is able to jump to additional sites. Because an *Ac* element generally jumps to linked sites, this strategy provides a way to produce a linked series of *Ac*-tagged genes. By including selectable and screenable markers linked to either the *Ac* ends or the transposase, researchers can cross out the active transposase, resulting in stable *Ac*-tagged lines.

e. Plant Culture Methods During Gene Transfer Experiments

The genetic transformation of plant cells using *Agrobacterium* can easily be accomplished in the laboratory. Whole plants and their tissues require simple growth media, moderate temperature, usually 22–24°C, and a light source for 16 hours each day. There are several commonly used strategies for the transfor-

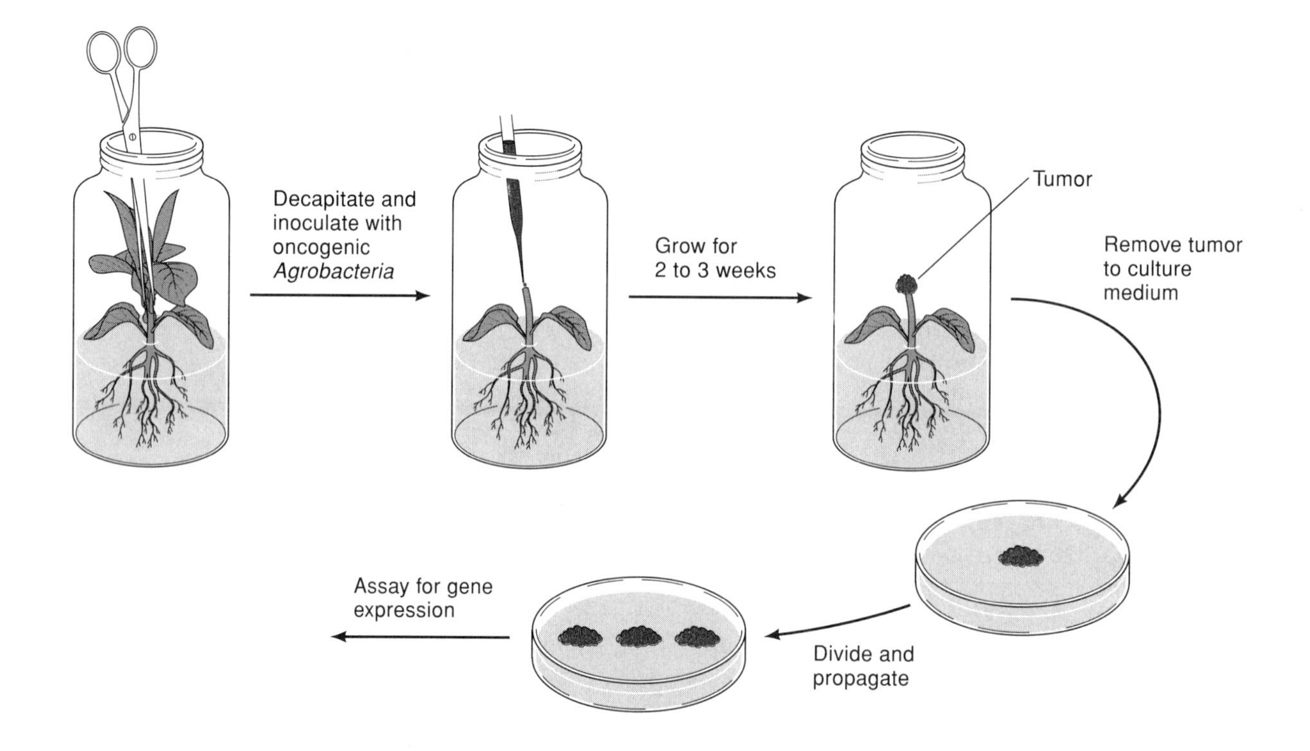

Figure 12.25
Generation of tumor tissue containing genes of interest. Decapitated tobacco plants are inoculated with tumorigenic *Agrobacterium* carrying genes of interest. Tumors are removed from plants and propagated axenically in tissue culture. The tissue is then assayed for gene expression by molecular and biochemical methods.

mation of plant cells using *Agrobacterium*; these methods have been developed mainly using cells and tissues derived from tobacco, petunia, or *Arabidopsis* plants.

Gene expression in transformed tumorous calli To establish quickly whether a particular DNA segment contains either regulatory or coding sequences, *Agrobacterium* carrying a DNA segment cloned (or recombined) into an oncogenic T-DNA element is allowed to infect wounded plantlets growing in sterile culture "pots" (Figure 12.25). Tumors usually form within two to three weeks. The tumorous tissue is then removed and placed in agar plant growth medium without the phytohormones (i.e., auxin or cytokinin), but with antibiotics to block the growth of *Agrobacterium* still present in the tumor tissue. The tumor tissue grows rapidly, and portions are used to score for expression of the inserted DNA either by direct tests for RNA or protein products or by growth under selective conditions. This approach was used in the first experiments demonstrating that the 5′ promoter regions of the *rbc*S and *cab* genes contain regulatory sequences sensitive to the presence of light.

Expression of introduced chimeric genes in whole plants clonally derived from single transformed cells To determine whether introduced genes are expressed in different plant tissues or to measure gene expression in independently transformed plants accurately requires that the individual plants analyzed be derived from transformation events occurring in single cells. Single plant cell protoplasts can be produced from leaf tissue after treatment with enzymes to digest plant cell walls (Figure 12.26). Following a two- to three-day recovery period, the proto-

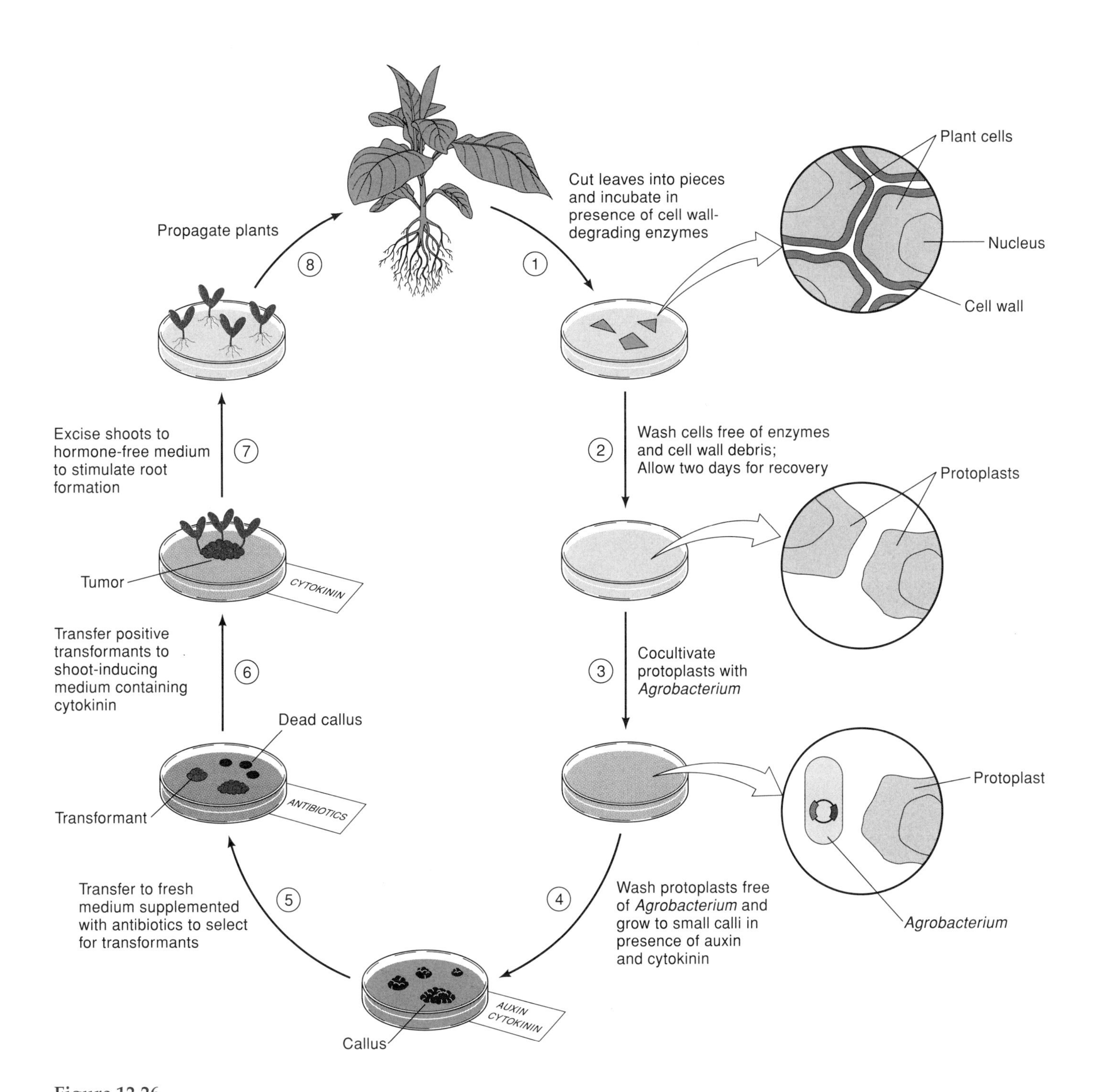

Figure 12.26
Cocultivation of *Agrobacterium* with single plant cell protoplasts. Single tobacco cells are prepared by treating leaf pieces with enzymes to digest plant cell walls. The cells are then washed free of enzymes and allowed to recover for 2 days. Single plant cells are then cocultivated with *Agrobacterium.* After incubation with *Agrobacterium* for 1 or 2 days, the cells are washed free of bacteria and allowed to grow into small calli in the presence of auxin and cytokinin. After a reasonably sized callus has formed, it is transferred to selective medium. Calli that retain their ability to grow under antibiotic selection are transferred to media containing cytokinin (without auxin) to promote shoot formation. Once shoots form, they can be transferred to media without growth hormones because the shoots will produce auxin themselves and stimulate root formation.

plasts are cocultivated with *Agrobacterium* carrying the DNA of interest linked to a selectable marker gene. Individual transformation events are detected following growth under selective conditions. Transformed calli are propagated on growth media containing both phytohormones, auxin and cytokinin. Once the tissue has reached a sufficient size, it is transferred to media containing only cytokinin to promote shoot formation. The shoots are excised from the callus tissue and placed on growth medium without phytohormones. The differentiated shoot material generally produces enough auxin to stimulate root formation. The transferred plantlets are maintained in the laboratory under sterile conditions (top shoots are removed to fresh pots for propagation) or in a greenhouse for vigorous growth and seed formation.

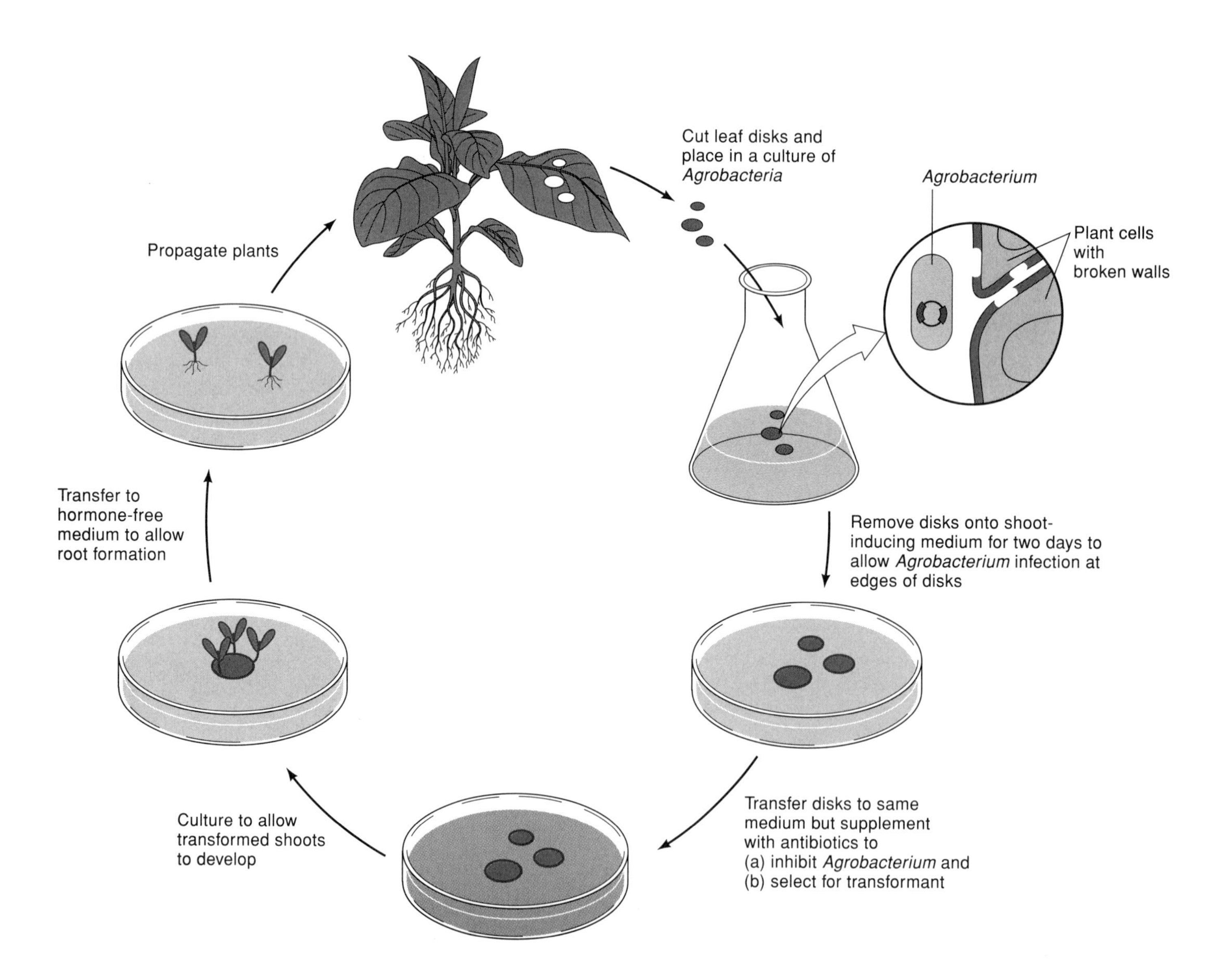

Figure 12.27
Leaf disk method for gene transfer. Disks are cut from tobacco leaves and placed into a culture of *Agrobacterium*. The cut surface breaks plant cell walls and allows access of *Agrobacterium*. The disks are washed and immediately placed on selective medium that also promotes shoot formation, then onto medium that selects against *Agrobacterium*, while retaining selection for transformants and shoots. Shoots that form are transformed and can be placed on medium without cytokinin to promote roots.

Except for the step where *Agrobacterium* is cultured in the presence of protoplasts for two days, the entire procedure is the same as that used by plant tissue culture specialists to produce **somaclonal** variants, (i.e., individual plants derived from somatic cells that display different phenotypic traits). This approach has been used to generate variants of several crop plants. Plants exhibiting improved growth or resistance to environmental stress can be propagated vegetatively by cuttings. Somaclonal variation is thought to result from chromosomal aberrations (duplication, deletions, etc.) in the original somatic tissue or induced during the culture period. The entire regeneration process, from single cell to plantlet, requires approximately three months of careful culture. However, several approaches dramatically shorten the production of transformed plants.

The leaf disk method for transferring genes to plants In the leaf disk method of transformation (Figure 12.27), disks are produced from leaves with a paper punch or cork borer. The disks are submerged in a culture of *Agrobacterium;* after gentle shaking to ensure that the cut leaf edges are infected, the disks are blotted dry and transferred to agar plates containing plant growth medium supplemented with cytokinin (to stimulate shoot regeneration). After two to three days to allow for *Agrobacterium* infection and T-DNA transfer, the disks are placed in medium containing antibiotics to block *Agrobacterium* growth and select for transformed cells.

After two to four weeks, shoots can be excised from the disks and transferred to medium to induce root formation (i.e., without cytokinin). The transformed plantlets are not clonal but likely represent the transformation of a few original cells because shoot development involves the formation of a shoot **meristem** (Figure 12.28) in which a limited number of cells participates in shoot formation and differentiation.

The leaf disk method combines gene transfer, plant regeneration, and effective selection of transformants into a simple and rapid process. In addition, the method allows use of *Agrobacterium* in gene transfer experiments with a wide

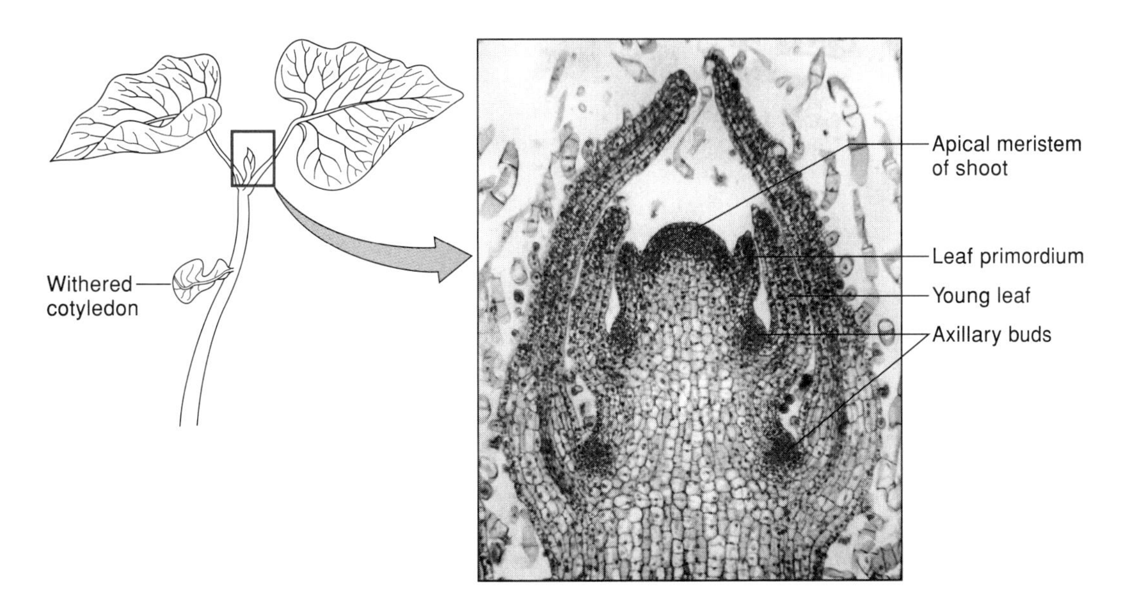

Figure 12.28
Shoot meristem. The undifferentiated tissue of the plant from which new organs arise is called the meristem. The shoot meristem is at the top (apex) of the plant. New organs, such as leaves, form off the flanks of the apical meristem. The cotyledon is the primary leaf formed just after germination of the seed. A very young organ is called a primordium.

range of those plant species susceptible to its infection because more plants are capable of regeneration from leaf explants than from the more drastic manipulations required for protoplast regeneration.

Agrobacterium *infiltration of flower buds* *Arabidopsis thaliana* has become a major model plant system. Therefore, I highlight a relatively new, elegantly simple, highly efficient method that dramatically improves the ease and efficiency of *Arabidopsis* transformation; this procedure involves no tissue culture. Transformation of seeds by *Agrobactrium* has been used to generate thousands of T-DNA–tagged lines, thus demonstrating that transformation can occur without elaborate wounding and tissue culture methods. In the new method, floral stems from mature *Arabidopsis* plants are immersed in media containing *Agrobacterium*. The plant is then put under vacuum to promote infiltration and infection by *Agrobacterium*. The plants are removed and allowed to grow and set seed; seeds are produced within a few weeks after infiltration. The seed is then scored for transformation, for example, by a selectable marker gene. The infiltration step dramatically improves transformation efficiency. Equally important is the fact that within two weeks infiltrated immature floral buds directly produce transformed seeds that can be scored directly for introduced traits. For example, this transformation method makes gene rescue experiments (where a cloned candidate gene is assayed for its ability to rescue a mutant phenotype) extremely rapid.

f. Genetic Engineering of Plants

The *Agrobacterium* system has only just begun to be exploited for the genetic engineering of plants containing desired traits. For example, it would be extremely valuable to alter plants to exhibit resistance to disease or environmental stress, to grow in the presence of herbicides designed to be toxic to nonagronomically important plants, or to improve their nutritive value. Before such traits can be engineered into plants of choice, the basic biology of each trait must be well understood.

To design plants resistant to a particular pathogen requires a detailed understanding of the disease itself. Genetic elements of the pathogen and host must be studied in detail, and potential regulatory genes must be cloned and altered to test for their ability to block disease development. One example illustrates the complexity. Many plants are subject to attack by insects, and insects in turn are susceptible to toxic compounds produced by fungal or bacterial pathogens. Thus, it is interesting to design a genetically engineered plant that synthesizes a product toxic to the insect and thereby avoids being the victim of the insect's hunger. Basic biological processes in three different organisms (bacterium, insect, and plant) must be understood to accomplish such an aim.

Even engineering endogenous plant genes requires a detailed understanding of the cellular and developmental processes that regulate their expression. For example, the seeds of many crop plants are their major consumable product. Seed storage proteins could be altered by *in vitro* genetic recombination techniques to contain higher proportions of essential amino acids, thereby increasing their nutritive value. A prerequisite for such engineering is an understanding of the developmental processes underlying the specific expression of storage protein genes in seed tissues.

Ongoing research certainly will decipher various biological events unique to plant growth and development, and *Agrobacterium* will be a tool in these investigations. However, there remains a major block to applying this knowledge to the genetic modification of the most agronomically important crops, grains such as rice, wheat, corn, barley, and oats, all of which are **monocotyledonous** flowering plants. Although *Agrobacterium* can effectively infect most plant species of the dicotyledonous subclass, monocotyledonous plants are resistant to its infection.

The molecular basis for the inability of *Agrobacterium* to infect monocots is not yet understood. Given the complexity of the T-DNA transfer process, there may be many required steps that are less efficient or unavailable in monocotyledonous cells. To date, we know that the block is not at the earliest step because monocots do produce the phenolic signal molecules that induce *vir* gene expression and initiate T-DNA transfer in *Agrobacterium*. We also know that the nuclear localization signals of the VirD2 and VirE2 molecules function in monocots, so the block is not at this step. However, there are two steps that might provide the explanation. First, properties of monocot cell walls and cell membranes compared to those of dicots may be sufficiently different and preclude *Agrobacterium* attachment and infection. Second, the mechanisms and machinery for illegitimate recombination in monocots may be incompatible with the requirements for T strand integration. Ongoing research aimed at deciphering further details of the transformation of dicotyledonous cells with *Agrobacterium* may eventually illuminate the reason for the block to infection in monocotyledonous cells.

In the meantime, other more generalized methods for DNA transfer to plant cells have been developed. Various direct gene transfer techniques have been adapted from similar methods used in animal cells. Among these are PEG (polyethylene glycol)-induced DNA uptake, calcium phosphate–mediated DNA precipitation and uptake, microinjection, liposome fusion, electroporation, and microprojectile bombardment. Another barrier to obtaining efficient transformation of monocots, their intractability to tissue culture methods, has been alleviated in recent years by major breakthroughs in techniques to regenerate important cereal crops. Methods for efficient transformation of monocots should soon be available.

Table 12.2 summarizes the classification of most of the common edible plants. Clearly, the dicots form a large part of our diet and agricultural industry. Protoplast culture and regeneration cannot yet be applied to all these dicot crop plants. However, all are susceptible to *Agrobacterium* infection, and by using simplified transformation procedures, such as infiltration, genetic engineering of any chosen dicot crop plant should be possible.

12.6 *Arabidopsis thaliana*, a Model Plant System

Although studying crop plants for their eventual improvement is a necessary and worthy endeavor, these plants are not easy to manipulate by classical and molecular genetic techniques. Rigorous genetics requires the ability to raise many successive generations, and typical crop plants have generative times of several months and require fields of space for their propagation. The ease of application of molecular genetic techniques depends on the size of the genome under investigation, and most crop or convenient laboratory plants have large genome sizes, similar to those found in animal cells. *Arabidopsis thaliana* is a plant species without these disadvantages and has been developed as a model plant system.

The major advantage of *Arabidopsis* is its small genome size: 7×10^7 bp (about five times the yeast genome and one-half the *Drosophila* genome). Table 12.3 compares the genome sizes of *Arabidopsis* and several representative plant species and the number of λ-clones required to obtain a complete genomic library of each. The small genome size of *Arabidopsis* reflects the relative absence of repeated DNA sequences compared to other plant species. In fact, most of the repeated DNA sequences in *Arabidopsis* are the numerous copies of the chloroplast genome. Nuclear DNA is mainly arranged as long blocks of unique DNA sequences, on average 120 kbp. This minimal genomic complexity simplifies an-

Table 12.2 *Supermarket Taxonomy*

Dicots

Family	Common Name	Scientific Name
Rosaceae	Apple	*Pyrus malus* L. (and other species)
	Blackberry	*Rubus canadensis* L. (and other species)
	Cherry	*Prunus avium* L.
	Peach	*Prunus persica*
	Pear	*Pyrus communis* L. (and other species)
	Plum	*Prunus domestica* L.
	Rose	*Rosa* spp
	Strawberry	*Fragaria virginiana* (and other species)
Cruciferae	Broccoli	*Brassica oleracea* L.
	Brussels sprouts	*Brassica oleracea* L.
	Cabbage	*Brassica oleracea* L.
	Cauliflower	*Brassica oleracea* L.
	Mustard	*Brassica alba*
	Radish	*Raphanus sativus* L.
	Turnip	*Brassica rapa* L.
	Watercress	*Nasturtium officinale*
Cucurbitaceae	Cantaloupe	*Cucumis melo* L.
	Cucumber and gourd	*Cucumis sativus* L.
	Pumpkin	*Cucurbita pepo* L. (and other species)
	Squash	*Cucurbita pepo* L.
	Watermelon	*Citrullus vulgaris*
Chenopodiaceae	Beet	*Beta vulgaris* L.
	Lamb's quarters	*Chenopodium album* L.
	Russian thistle	*Salsola kali* L.
	Spinach	*Spinaceae oleracea* L.
	Sugar beet	*Beta vulgaris* L.
Compositae	Artichoke	*Cynara scolymus* L.
	Dandelion	*Taraxacum officinale*
	Lettuce	*Lactuca sativa* L.
	Sunflower	*Helianthus annuus* L.
	Wormwood	*Artemisia absinthium* L.
Leguminosae	Alfalfa	*Medicago sativa* L.
	Bush bean	*Phaseolus vulgaris* L.
	Green pea	*Pisum sativum* L.
	Kidney bean	*Phaseolus vulgaris* L.
	Lima bean	*Phaseolus lunatus* L.
	Navy bean	*Phaseolus vulgaris* L.
	Pinto bean	*Phaseolus vulgaris* L.
	Soybean	*Glycine max*
Solanaceae	Eggplant	*Solanum melongena* L.
	Nightshade	*Solanum nigrum* L.
	Peppers	*Capsicum annuum* (and other species)
	Potato	*Solanum tuberosum* L.
	Tobacco	*Nicotiana tabacum* L.
	Tomato	*Lycopersicon esculentum*
Umbelliferae	Carrot	*Daucus carota* L.

Monocots

Family	Common Name	Scientific Name
Gramineae	Corn	*Zea mays* L.
	Crab grass	*Digitaria sanguinalis*
	Oats	*Avena sativa* L.
	Sorghum	*Sorghum bicolor*
	Sugar cane	*Saccharum officinarum* L.
	Wheat	*Triticum aestivum* L.
Bromeliaceae	Pineapple	*Ananas coinosus*
	Spanish moss	*Tillandsia usneoides* L.
Liliaceae	Asparagus	*Asparagus officinalis* L.
	Easter lily	*Lilium longiflorum*
	Garlic	*Allium sativum* L.
	Onion	*Allium cepa* L.
Orchidaceae	Lady slipper orchid	*Cypridpedium* spp
	Vanilla orchid	*Vanilla planifolia*

Source: W. M. Laetsch, *Basic Concepts in Botany.* Little, Brown, Boston, 1979, p. 465-471.

Table 12.3 *Comparison of Genomes Between* **Arabidopsis** *and Representative Plant Species*

Plant	Haploid Genome Size (in kilobase pairs)	Number of λ Clones in Complete Library	Average Size in Predominant Class of Single-Copy Sequences (in kilobase pairs)	Amount of Repetitive DNA in Haploid Genome (in kilobase pairs)
Arabidopsis	70,000	16,000	120	18,000
Mung bean	470,000	110,000	>6.7	160,000
Cotton	780,000	180,000	1.8	310,000
Tobacco	1,600,000	370,000	1.4	1,200,000
Soybean	1,800,000	440,000	<3	1,100,000
Pea	4,500,000	1,000,000	0.3	3,800,000
Wheat	5,900,000	1,400,000	1	4,400,000

Source: E. M. Meyerowitz and R. E. Pruitt, *Arabidopsis thaliana* and Plant Molecular Genetics. *Science* 229 (1985), p. 1214.

alysis by hybridization or cloning techniques; for example, "chromosomal walking" is a convenient approach to gene isolation.

Arabidopsis is a harmless weed of no particular value (Figure 12.29), a member of the mustard family (the Cruciferae, Table 12.2). Its generation time is only five to six weeks, and each plant can produce 10,000 seeds. Its small (20 cm) and delicate size allows dozens of plants to be grown in a small flowerpot. There are only five chromosomes, and many mutants with defects in biochemical and developmental pathways are available and useful and have visible and chromosomal genetic markers.

Initial work on cloning individual genes from *Arabidopsis* led to two generalizations that emphasize its utility for studying plant gene organization: (1) genes

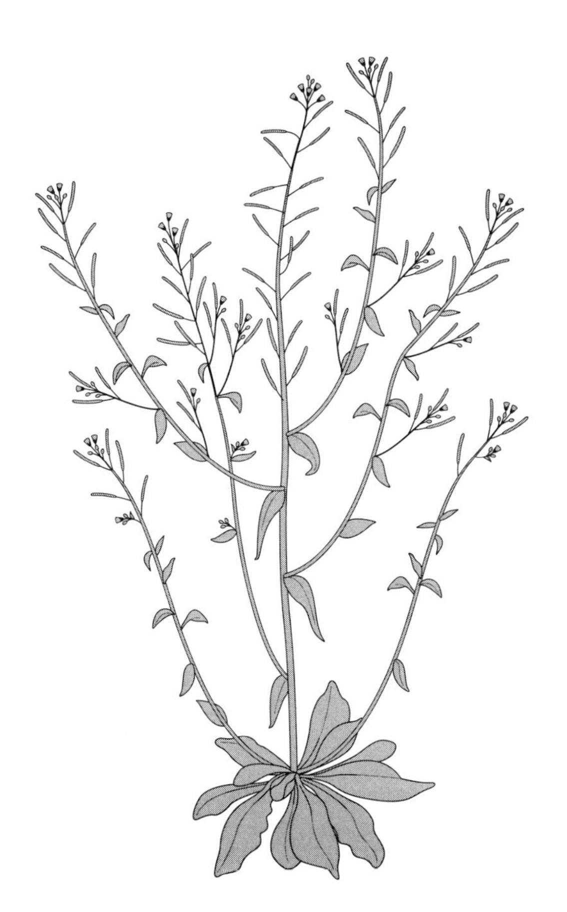

Figure 12.29
The model plant system, *Arabidopsis thaliana.* The aerial portion of the plant including leaves, inflorescence shoots, and flowers is illustrated. (Drawing adapted from original by Thi Pham.)

isolated from *Arabidopsis* cross-hybridize with homologous genes in a wide variety of plant species, and (2) proteins that are encoded in many plants by multiple genes or families of genes are encoded by single or few genes in *Arabidopsis.* The entire *Arabidopsis* genome has been cloned into cosmids and yeast artificial chromosomes (YACs), and these are rapidly being mapped to specific chromosomal locations. Already, the genome is richly dotted with RFLP (restriction fragment length polymorphism) markers, and recombinant inbred lines are available for quick mapping of newly cloned DNA segments, and the Arabidopsis Genome Project will undoubtedly provide a wealth of useful information.

a. Reproduction, Development, and Differentiation in Plants

Most readers will be familiar with well-studied examples of the complex genetic hierarchies that control developmental programs in, for example, *Drosophila* (Chapter 9). Recent investigations have provided us with an extremely useful paradigm for complex developmental processes in plants: flower development in the model plant system *Arabidopsis.* However, before embarking on a description of how the *Agrobacterium* T-DNA transfer system has contributed to our understanding of the plant developmental processes, I'll review how plants grow and develop. These programs are dramatically different from those followed in animal systems.

A comparison of plant and animal developmental strategies The animal and plant kingdoms utilize distinctly different strategies for development of multicellular organisms (Table 12.4). The most fundamental difference is that animals as well as animal cells can move, but plants and their cellular components cannot; this difference dictates how the organism responds to the environment, nourishes itself, and reproduces. For example, an animal can respond to an environmental change by both a behavioral response, such as running away, and a physiological response, such as changes in hormonal regulators that elicit a particular cellular response. Plants are limited to a physiological response.

Another major difference between plant and animal development is in the strategies employed to produce the whole organism. Animal development generally involves a series of complex pathways that lead from undifferentiated

Table 12.4 ***Comparison of Animal and Plant Developmental Strategies***

Developmental Strategy	Animal	Plant
Movement		
Of organism	+	−
Of cells	+	−
General adult developmental potential	− (Highly differentiated)	+ (Totipotent)
Sexual reproduction		
Time	Early, during embryogenesis	Late, in adult
Gametophyte generation	−	+
Fertilization	Single	Double
Development post fertilization	Continuous	Dormant stage in seed

cells to more and more lineages of differentiated cell types. Thus, animal development tends to restrict the developmental potential of individual cells. Once the whole organism is formed, an animal generally cannot reproduce its parts. In contrast, although plant cells also follow complex developmental pathways to form the organs of the mature plant, many plant cells remain **totipotent**. Totipotency has been exploited to develop plant tissue culture methods whereby a single plant cell can regenerate an entire plant. This property is especially useful for genetic engineering strategies described earlier in this chapter.

Moreover, major plant organ systems, the shoot and the root, retain the ability to differentiate new organs in the fully mature plant. Undifferentiated regions, termed **meristems** (Figure 12.28), reside at the root or shoot tips. The tip of the shoot is called the **apical meristem**; the cells at the apex of the meristem divide and maintain the meristem while cells along the flanks of the apical meristem give rise to peripheral meristems, forming branches that in turn produce organs such as leaves. Meristematic regions provide the mature plant with the ability continually to produce new roots, shoots, and leaves.

A comparison of plant and animal reproduction In animal development, formation of gametes and reproductive organs is determined early during embryonic development, and germ line cells do not contribute to the somatic cells of the animal. In contrast, a plant grows and matures before producing its reproductive organs. In mature plants, new organs grow vegetatively from the shoot and root meristematic regions. At a particular time, in response to developmental and environmental signals, the plant switches at the shoot apex from an undeterministic, vegetative mode to a determined mode of differentiation. This switch results in the production of the flower and gametes. Thus, plant reproductive development occurs late compared to animal development, and the cells that give rise to the plant germ line are derived from meristematic regions that previously gave rise to the vegetative (somatic) parts. If an animal were to form its reproductive organs analogously to plants, male and female reproductive organs would form at the extremities of the mature animal.

The early setting aside of the germ line in animal development has an important consequence: the reproductive cells are protected both from developmental signals that subsequently come into action and determine the production of the major animal organ systems and from mutagenic environmental influences. The plant reproductive system derives from cells of the mature plant that have been through many cell divisions as well as exposure to dramatic environmental changes throughout the life of the plant. However, plants have evolved an efficient mechanism to select against somatic mutations that may have occurred prior to gametogenesis; this mechanism is the basis of the gametophyte generation. After meiosis in either pollen (male) or ovule (female) progenitor cells, a series of mitotic divisions occurs to generate essential accessory cells; these divisions essentially provide a **haplosufficiency** test for critical cellular functions.

There are two other major differences between plants and animals in their reproductive cycles. First, fertilization in animals results from a single event, whereas in plants, two fertilizations occur. The pollen tube carries two haploid nuclei. At fertilization, one nucleus fertilizes the egg cell, and the other fertilizes the binucleate central cell of the ovary to form the progenitor of the triploid **endosperm**, the nutritive tissue that surrounds the embryo. The second difference lies in what occurs immediately after fertilization. In animals, the fertilized egg generally starts to develop immediately and never stops until the adult is formed. In contrast, in plants, there is generally a dormant stage following early embryonic development. That is, following the initiation of major organ system formation, the root and shoot, the embryo dessicates, and the surrounding ovule tissue hardens into a seed coat. Seed production provides plants with a mecha-

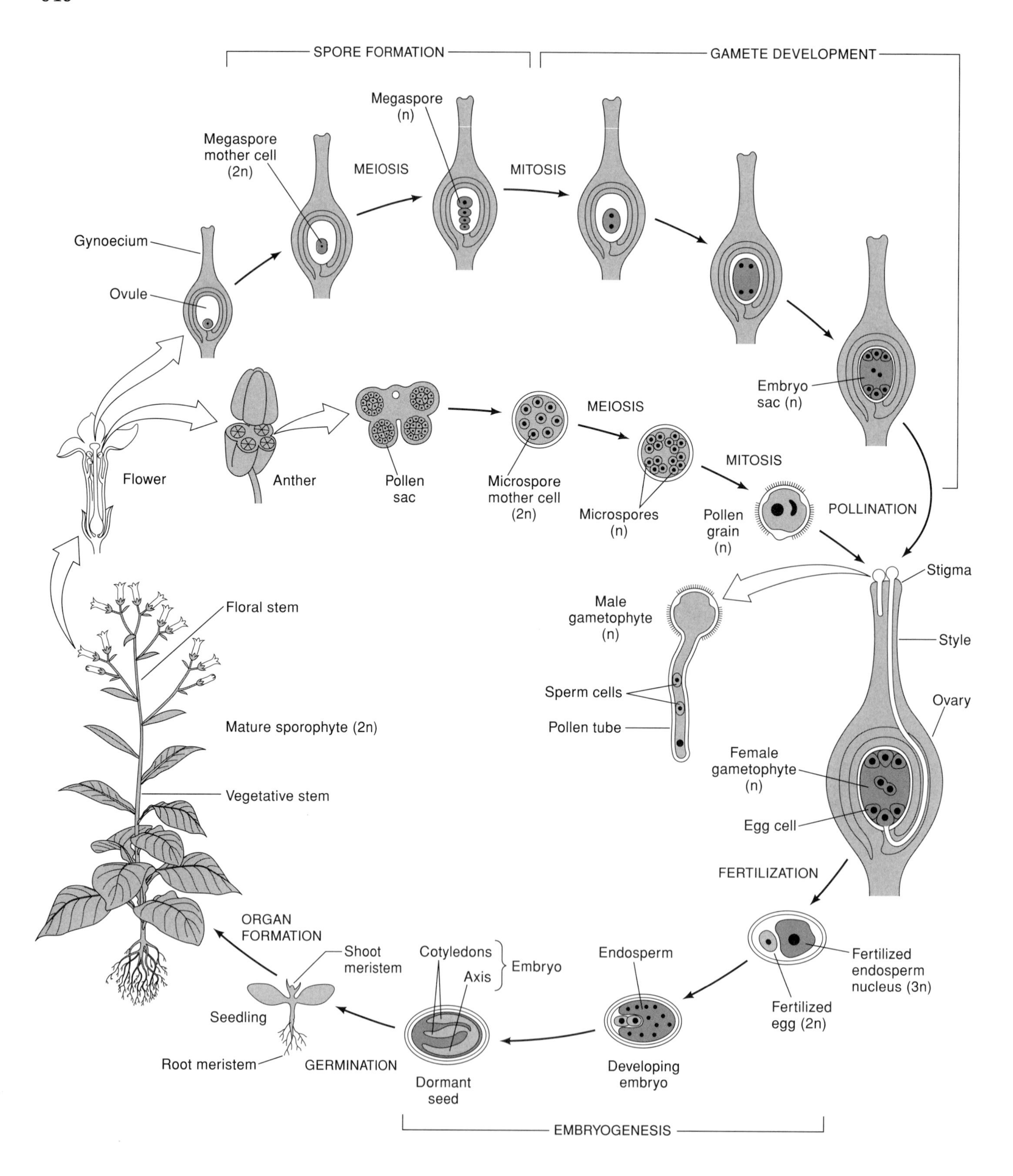

Figure 12.30
The life cycle of a flowering plant. (Adapted from R. Goldberg. 1988. Plants: Novel developmental processes. *Science* 240 1460.)

nism as well as a time to disperse and thus propagate over wide areas. Seed production is the plant's ultimate response to its natural immobility. Figure 12.30 diagrams the life cycle of a flowering plant.

b. Flowering

Although novel and fascinating developmental programs underlie leaf and root differentiation, the part of the plant that has captivated human beings down through the ages is the flower. Flowers have been studied and cultivated mainly for their beauty and enormous variety. They range from a few millimeters to over a meter in breadth, and their shape and symmetry stimulate even the casual observer to question how such geometrical patterns could possibly arise. In the age of molecular and cell biology, the flower presents itself in a new light, as perhaps the ultimate in plant developmental complexity. Flowers have an amazing diversity of form, but only a few flowering plants have been studied extensively in the laboratory: *Arabidopsis*, maize, petunia, snapdragon, and tobacco.

Interest in the mechanism of flower development is enhanced by the realization that it represents a dramatic alteration in the developmental program of the entire plant. When a plant is induced to flower, the meristematic region, which was previously undetermined and repetitively producing shoots, switches to a deterministic mode of development. The flower, once formed, cannot form new structures.

The mature *Arabidopsis* flower, for example, consists of four different organ types organized in a whorled arrangement (Figure 12.31). The outermost whorl

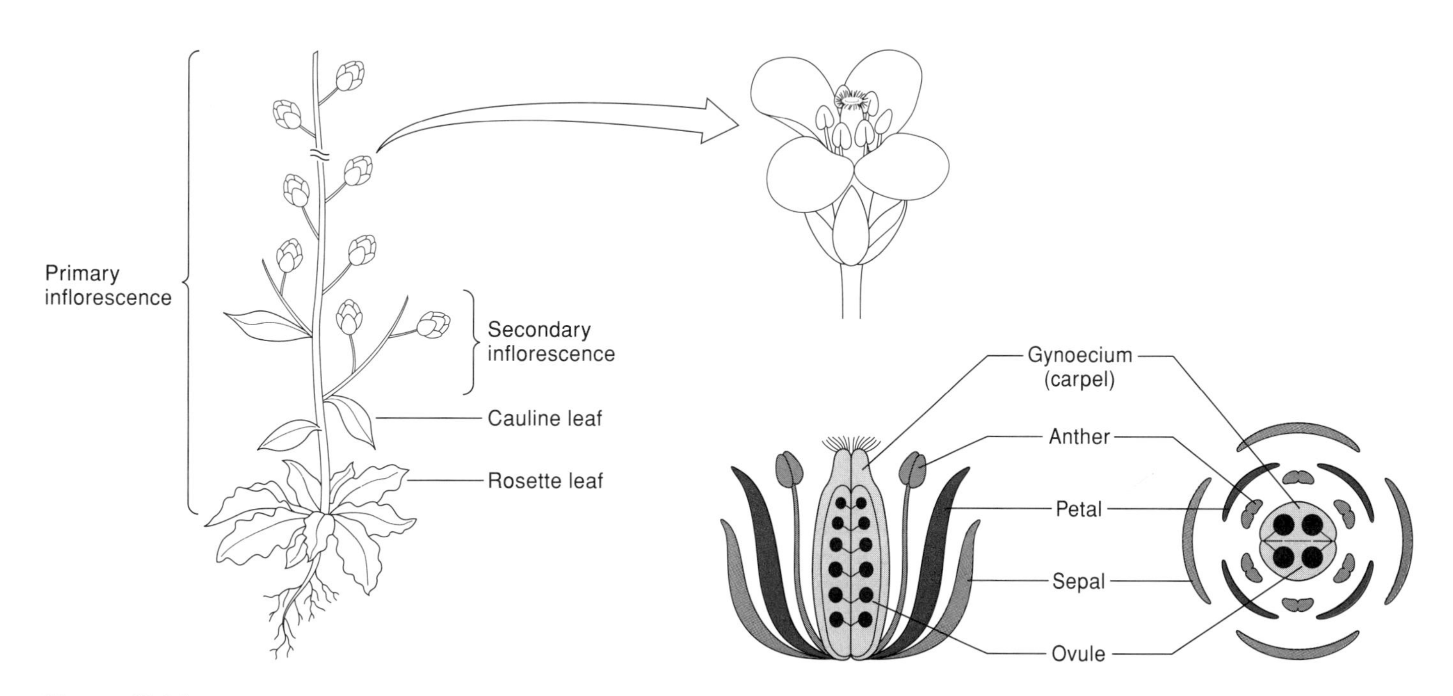

Figure 12.31
Diagrammatic representations of cross-sections and longitudinal sections of an *Arabidopsis* flower. On the left is shown a whole plant with the rosette leaves produced prior to flowering, the cauline leaves produced on the primary inflorescence stem, and flowers. On the right, to illustrate the different floral organs, are views of a flower. The outermost whorl contains four sepals, followed by the second whorl of four petals, the third whorl of six stamen consisting of a long filament and the pollen-bearing anthers, and the fourth innermost whorl containing the central gynoecium that houses the ovules.

has four **sepals** equally spaced in a cruciform arrangement; sepals are green and resemble leaves in their overall shape and cellular composition. The next whorl has four **petals** arranged in cruciform exactly 45° off from the sepal whorl (i.e., essentially in each of the spaces between the sepals). The third whorl has six **stamen** arranged either as pairs in the medial plane or singly in the lateral plane; the stamen are the male reproductive organs and are composed of a long filament that is topped by the pollen-bearing **anthers**. The central whorl has only one organ, the **gynoecium**, or female reproductive organ. The gynoecium is composed of three main regions. The top is the **stigmatic** surface containing elongated **papillar** cells that promote pollen grain germination. Just below the stigma is the **style**, the only solid part of the gynoecium; in the center of the style is the pollen-transmitting tract, which promotes directional pollen tube growth toward the ovary. The major structural part of the gynoecium is the **ovary**. The ovary is composed of two chambers, called **carpels**, each of which contains double rows of ovules attached to the ovary wall; the attachment region is the **placenta**. Following pollen germination, a pollen tube carrying the two sperm nuclei grows through the style and into the central inner wall (transmitting track) of the ovary. The tubes grow through the stem of the ovules (**funiculus**) and into the ovules themselves, where fertilization occurs.

The induction of flowering Most plants use environmental signals such as photoperiod, temperature, and water availability to regulate the transition from vegetative to reproductive growth. An endogenous transmissible signal may be transported to the apical meristem after the plant perceives an appropriate environmental stimulus. There is abundant controversy over the nature of the chemical control of flowering, and theories propose either a specific "florigen" or more general physiological signals resulting from altered light, nutrient supplies, or plant growth factors.

Arabidopsis floral induction sets off two new developmental programs: the production of the flower and the production of its accompanying branch or stem structure (Figure 12.31). Although many plants have an elongated upright stem and associated leaf-bearing branches, the vegetative, nonreproductive *Arabidopsis* plant is not upright. Instead, it is compact and grows by forming a series of leaves, called a "rosette," in a spiral arrangement close to the ground. Following the initiation of flowering, a stem bearing flowers and containing the apical meristem develops from the center of the rosette leaves. This flower-bearing stem has been called the **inflorescence**. The first inflorescence to form is called the primary inflorescence, and it produces flowers near the top as well as along its length. Additional secondary inflorescences bearing flowers along their lengths also arise as branches off the primary inflorescence.

Researchers had assumed that the inflorescence stem is the first structure developed during flowering because its appearance is the first obvious change. However, close ultrastructural examination of the shoot apical meristem soon after the induction of flowering revealed that the meristem first develops flowers and only secondarily develops the stem and its branches. Flowers arise in a spiral around the center of the meristem (Figure 12.32a), each at an angle between 130° and 150° from the previously initiated flower. The direction of new bud for-

---

Figure 12.32 (*facing page*)
Scanning election micrographs of *Arabidopsis* flowers. (a) Three panels showing the main flowering apex. (b) Six panels showing individual floral buds at stages 5 to 8 with sepals dissected away to reveal inner organs. (c) Three panels showing lateral views of developing floral buds from stages 9 to 12, again with outer organs dissected away to reveal inner organs. P = petal; S = sepal; A = anther; G = gynoecium. (From Smyth et al. 1990. Early flower development in Arabidopsis. *Plant Cell* 2 755.)

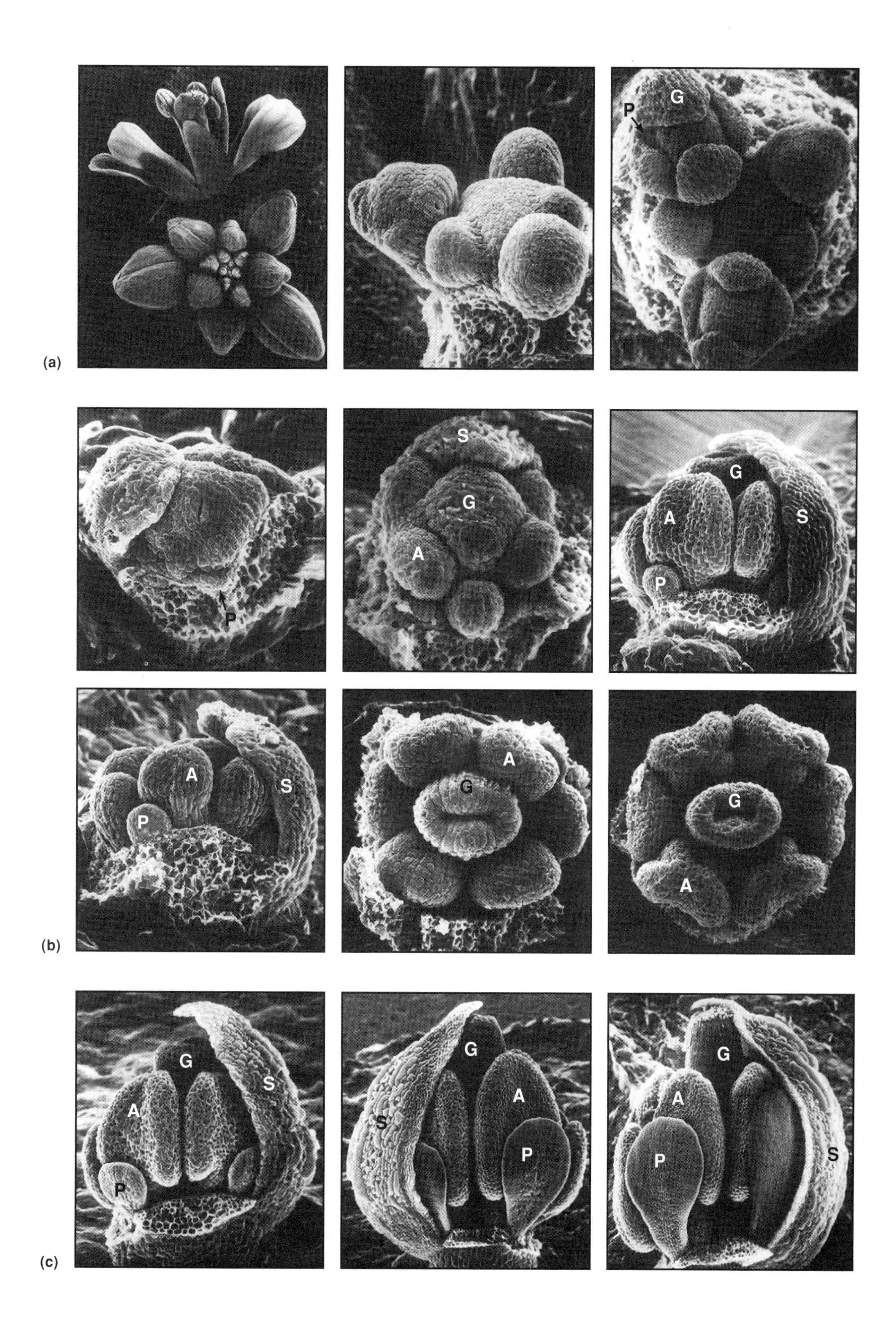
(a)
G
P
S
G
A
G
A
S
P
P
A
S
P
A
G
G
A
(b)
G
S
A
P
G
A
S
P
G
A
P
S
(c)

mation may be either clockwise or counterclockwise, but once a direction is established, all subsequent flowers arise in the same direction, forming either a lefthanded or righthanded helix.

Flower development is **acropetal**, that is, the oldest flowers are on the bottom and the youngest flowers on the top of the inflorescence stem. However, the organs within the flower develop in a **basipetal** direction, that is, the oldest part of the organ is at the top, and the youngest is at the bottom; this follows from the fact that organs arise as bulges off the meristem, and the older parts grow outward and upward.

The formation of the Arabidopsis *floral organs* Although flowers arise in a helical pattern, the organs of the flower arise in a whorled pattern. Thus, successive organs arise in concentric rings, where the older organs are on the outermost ring, and the younger organs arise as inner rings. For *Arabidopsis,* this means that the outer whorl of sepals develops first, followed by the petal whorl, then the stamen whorl, and finally the central whorl containing the gynoecium. The sepal, petal, stamen, and gynoecium are referred to as whorls 1, 2, 3, and 4, respectively, reflecting their order of initiation.

Each organ of the individual whorls initiates as a primordium that appears as a bump off the flanks of the meristem. The position of the primordial bumps marks the position of the mature organs. This spatial pattern is distinct and highly recognizable. The size of the primordia is also distinctive for the organs they will specify. Thus, the sepal, stamen, and gynoecium primordia are very large, but the petal primordia are tiny.

Figure 12.32a, b, and c shows the stages of early flower development. Landmark events divide early floral development into 12 recognizable stages (Table 12.5). Stage 1 begins with the initiation of a floral bump on the flanks of the meristem. Stage 2 starts when the flower primordium rises and separates from the meristem. Sepal primordia arise in stage 3 and then grow to overlie the primordium in stage 4. Petal and stamen primordia arise in stage 5 and are soon completely enclosed by the sepals in stage 6. During stage 6, petal primordia grow slowly while stamen primordia grow rapidly. Stage 7 is marked by the appearance of stalks on the stamen in the medial plane. The stamen then develops **locules** (that will enclose the pollen) at stage 8. Stage 9 is longer, and during this

Table 12.5 ***Summary of Stages of Flower Development in* A. thaliana**

Stage	Landmark Event at Beginning of Stage	Duration* (hr)	Age of Flower at End of Stage (days)
1	Flower buttress arises	24	1
2	Flower primordium forms	30	2.25
3	Sepal primordia arise	18	3
4	Sepals overlie flower meristem	18	3.75
5	Petal and stamen primordia arise	6	4
6	Sepals enclose bud	30	5.25
7	Long stamen primordia stalked at base	24	6.25
8	Locules appear in long stamens	24	7.25
9	Petal primordia stalked at base	60	9.75
10	Petals level with short stamens	12	10.25
11	Stigmatic papillae appear	30	11.5
12	Petals level with long stamens	42	13.25

*Estimated to nearest 6 hr.
Source: D. R. Smyth, J. L. Bowman, and E. M. Meyerowitz, Early Flower Development in *Arabidopsis. The Plant Cell* 2 (1990), p. 755.

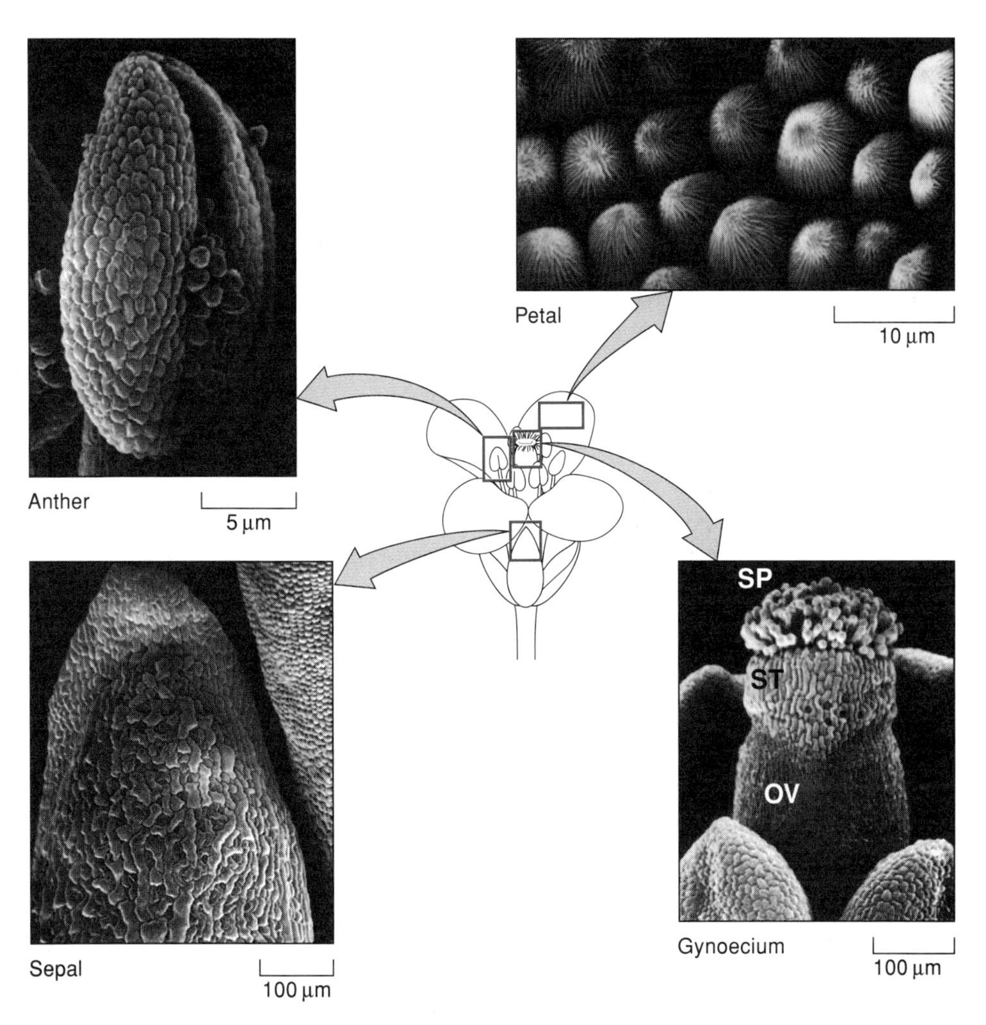

Figure 12.33
Surface morphology of mature and developing floral organs. The epidermal cells of the floral organs have dramatically different morphologies—from conical (petal) to elongated (sepal, style [ST], stigmatic papillae [SP]) to irregular (anthers). (From Smyth et al. 1990. Early flower development in Arabidopsis. *Plant Cell* 2 755.)

time, the petals become stalked. All organs elongate during this stage. Gynoecium development commences at stage 6 with the formation of an open-ended tube. During stage 10, the petals reach the length of the lateral stamens. Stigmatic papillae appear at the tip of the gynoecium in stage 11. At stage 12, the petals have reached the length of the median stamens. This final stage ends when the bud opens. The duration of each of the stages varies between 6 (stage 5) and 60 (stage 9) hours, with a total requirement of 13.25 days from initiation of the floral primordium to opening of the flower.

As the organs develop from primordia, their intrinsic cell types become highly distinctive (Figure 12.33). The outer epidermal cells, for example, can be used to discriminate among the organs. Sepal epidermal cells are large, irregular, and elongated and are thus the most leaflike of the floral organs. Petal epidermal cells are conical. Stamens have elongated cells along the filament and irregular, almost puzzle-shaped cells on the surface of the anther. The most complex organ, the gynoecium, has characteristic cells in all three major regions, from very regular, elongated cells on the surface of the ovary to papillar-type cells at the top of the stigma.

c. Genetic Dissection of Flower Development

Recently, mutations that alter either the induction or development of *Arabidopsis* flowers were isolated. These mutations provide a means to dissect this fundamental process at the molecular level. There are two main categories of mutations: those affecting the specification of the floral meristem and those required for floral organogenesis. I first review what is known about the genes specifying the identity of floral organs because there is more information about their function, and the mechanism of floral organogenesis provides a framework for predicting what genes specifying the floral meristem might be required to do to initiate the developmental program.

Specification of floral organs Mutations that affect floral organogenesis result in reduced, altered, or misplaced floral organs. Mutations in four genes, *APETAL2* (*AP2*), *AGAMOUS* (*AG*), *PISTILLATA* (*PI*), and *APETALA3* (*AP3*), cause misspecification of organ types. Generally, each of these genes affects development in two adjacent whorls of the flower. Thus, *ap2* mutants affect development in the first and second whorls such that carpels replace sepals, and stamens replace petals. The *ag* mutants affect flower development in the third and fourth whorls; stamens are replaced by petals, and the gynoecium is replaced by a second mutant flower so that a pattern of sepals, petals, petals is reiterated several times. The indeterminate development of *ag* flowers produces a roselike appearance. The generally opposite phenotypes of *ag* and *ap2* mutants suggested that their gene products are antagonistic or somehow regulate each other's expression. The *pi* or *ap3* mutants have similar phenotypes affecting the second and third whorls of the flower; sepals replace petals and carpellike organs replace stamens. By analogy to genes that affect *Drosophila* development and, when mutated, cause transformation of organs to abnormal positions, the *AG, AP2, PI,* and *AP3* genes are classified as **homeotic** (see Chapter 9).

Each of these genes affects development in two adjacent whorls: the first and second (*AP2*), the second and third (*PI/AP3*), or the third and fourth (*AG*). Thus, the flower primordium in its simplest form can be viewed as three concentric fields of gene activity (Figure 12.34a). A simple model was proposed whereby combinations of genes acting in overlapping domains are responsible for the specification of the four organ types (Figure 12.34b). The three developmental fields, A (whorls 1 and 2) , B (whorls 2 and 3), and C (whorls 3 and 4), represent the activity domains of *AP2, PI/AP3,* and *AG,* respectively, in the early wild-type flower primordium. Thus, each of these genes controls the identity of two adjacent whorls of organs. The model makes two main assumptions: (1) that the combinations of homeotic gene products present in each of the four flower whorls is responsible for specifying the developmental fate of the organs that will arise at that position and (2) that the products of AP2 and AG act antagonistically. Thus, *AP2* expression in the first whorl is responsible for sepal production, *AP2* expression in combination with *PI/AP3* expression in whorl 2 is responsible for petal production, and *PI/AP3* in combination with *AG* in whorl 3 is responsible for stamen specification. *AG* expression in whorl 4 is responsible for gynoecium development.

From this simple model, one can predict the phenotypes consequent to mutation expression in each of these four genes (Figure 12.34b). Thus, in an *ap2* mutant plant, the domain of expression of *AG* extends to whorls 1 and 2, resulting in the production of carpels and stamens in these domains. In an *ag* mutant, the domain of expression of *AP2* now extends to whorls 3 and 4, resulting in the production of petals and sepals in these regions. In the absence of *PI/AP3* function, there is no combinatorial gene activity with either *AP2* or *AG* in the second and third whorls. Thus, in a *pi* or *ap3* mutant, *AP2* expression in the second whorl results in sepal production just as in the first whorl, and *AG* expression in the third whorl results in carpel production. As further illustrated in Figure 12.34b, the

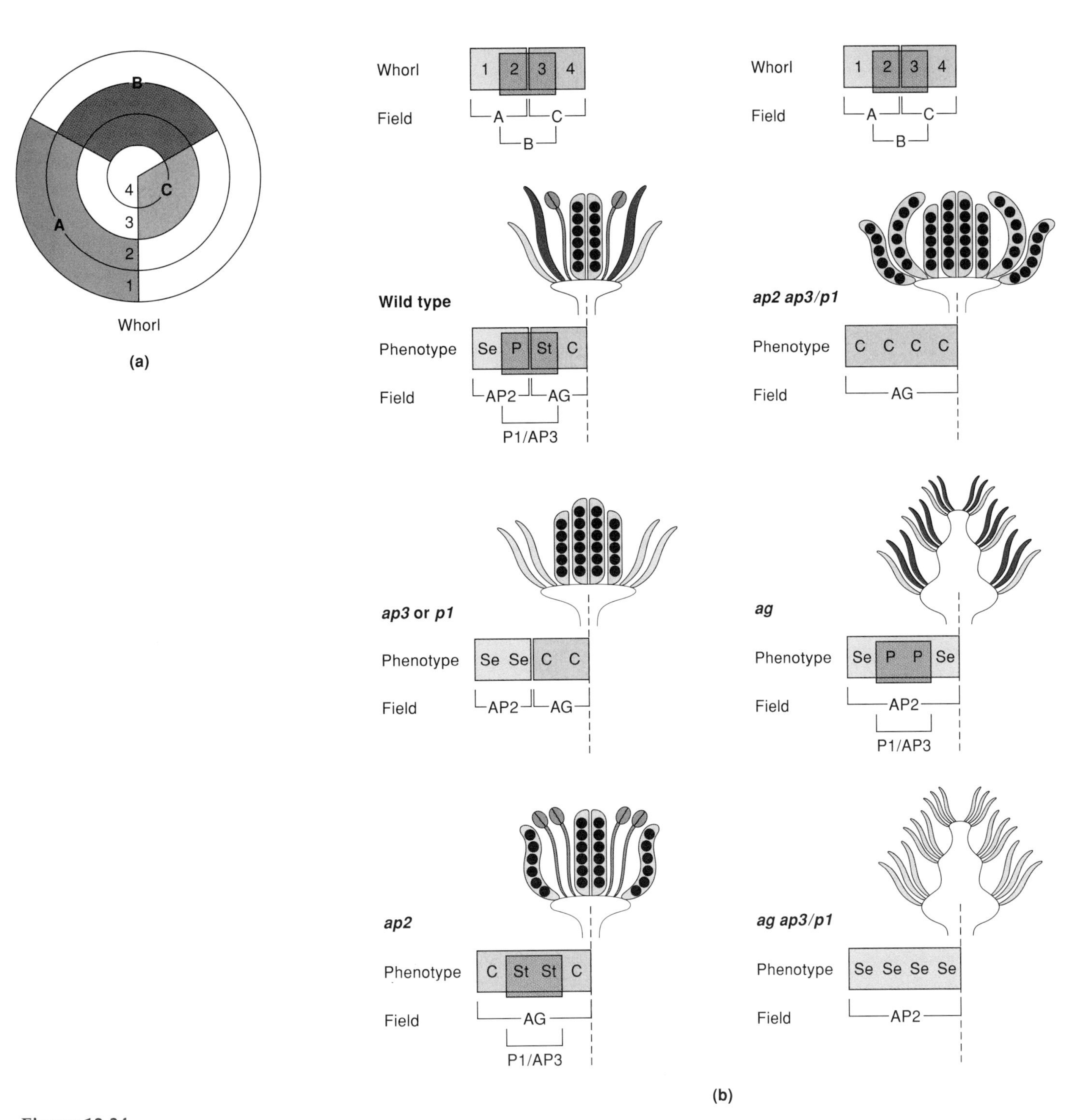

Figure 12.34
Schematic representation of a model depicting how three classes of floral genes could specify the identity of the four whorls of organs. (a) Diagram of the four whorls of a flower primordium (viewed from above) and the three regions, A, B, and C, that are the domains of action of the three classes of organ identity genes. A gene activity in the outer (1) whorl gives rise to sepals. A gene activity in combination with B gene activity in the second (2) whorl gives rise to petals. B gene activity in combination with C gene activity in the third (3) whorl gives rise to stamen. C gene activity in the fourth (4) whorl gives rise to the central gynoecium. (b) Genetic analysis that leads to the above model. A section through one-half of a floral primordium is represented as a set of boxes, with the regions representing each whorl shown at the top of each column. The genotype under consideration is listed at the left with the predicted distribution of gene products present in each genotype indicated by the upper case letters below the boxes. The predicted phenotypes of the organs in each whorl are shown in the boxes. Se = sepal; P = petal; St = stamen; C = carpel. A schematic drawing of a longitudinal section of each phenotype is depicted on the right, above the boxes. Each organ type is color coded. (Adapted from J. L. Bowman et al. 1991. Genetic Interactions Among Floral Homeotic Genes of *Arabidopsis. Development* 112 1.)

model makes clear predictions for the patterns of gene activity that govern the resulting phenotypes of double and triple mutants.

Expression of AP2, AG, PI, *and* AP3 The previous genetic model has been borne out for the most part by the molecular analysis of the homeotic gene products. *In situ* hybridization using probes homologous to each of these genes has been used to monitor the timing and pattern of expression of their homologous mRNAs throughout flower development, from primordia through to the morphological differentiation of the mature organs. As predicted, *AG* is expressed in the center of the flower meristem and continues to be expressed in this central region through to the formation of mature stamens and the gynoecium. As predicted, *AP3* is expressed in between the inner and outer regions of the floral meristem and is expressed in this region in the sepal and stamen primordia as well as in the mature organs. *PI* expression deviates from the model in its early mode of expression; it is expressed in whorls 2, 3, and 4 in very early flowers, before any morphological determination of petal, stamen, or gynoecium occurs. However, once morphological differentiation occurs, *PI* expression follows that of *AP3* in whorls 2 and 3.

The antagonism between *AG* and *AP2* is supported by the observation that *AG* expression extends to all four whorls in an *ap2* mutant. In addition, ectopic expression of *AG* under a constitutive (CaMV) promoter results in *ap2*-like flowers. However, the model is not directly supported by the expression patterns of *AP2* because it is detected in all four whorls throughout flower development. This last result implies that *AG* does not suppress *AP2* expression in whorls 3 and 4 and that *AP2* expression in these whorls does not suppress *AG* gene expression as it does in whorls 1 and 2. The data can be reconciled with the model if *AP2* supresses *AG* by acting in conjunction with another gene product whose expression is limited to whorls 1 and 2.

Gene products of AP2, AG, PI, *and* AP3 The molecular cloning of these four genes and knowledge of their predicted protein products allows predictions about the mechanism whereby they control flower organogenesis. The proteins encoded by *AG, PI,* and *AP3* are thought to belong to an evolutionarily conserved family of transcription factors. Each of their predicted gene products contains a highly conserved DNA-binding domain called the MADS-box (*M* from the yeast MCM1 factor, *A* from AGAMOUS, *D* from the snapdragon DEFICIENS protein, and *S* from the mammalian serum response factor). The proteins are believed to control the expression of a complex array of genes that ultimately specify the floral organs. The predicted product of *AP2* does not have homology to any known genes; however, it has a serine-rich acidic domain that may function in DNA binding.

The genetic and molecular analyses of four genes, *AP2, PI, AP3,* and *AG,* highlight some basic principles for the control of flower development. First, genes encoding transcriptional factors likely control a complex array of target genes that ultimately specify the fates of the flower primordia and their differentiation into mature organs. Second, these regulatory genes act in different combinations to control developmental fates. Third, the products of some of these identity genes, such as *AP2* and *AG,* control each other's activity. This highly regulative mode of development through control of gene activity is not unexpected; it has been used successfully throughout evolution in unicellular and multicellular organisms. Nor is it unexpected that the dramatic alteration of developmental pathways brought about by these homeotic genes reflects their predominant roles as trancriptional regulators. The simple ABC model for combinatorial gene action to specify the floral organ types is highly consistent with the organ identity genes encoding transcriptional factors. As has been shown often enough in other biological systems, different combinations of transcriptional

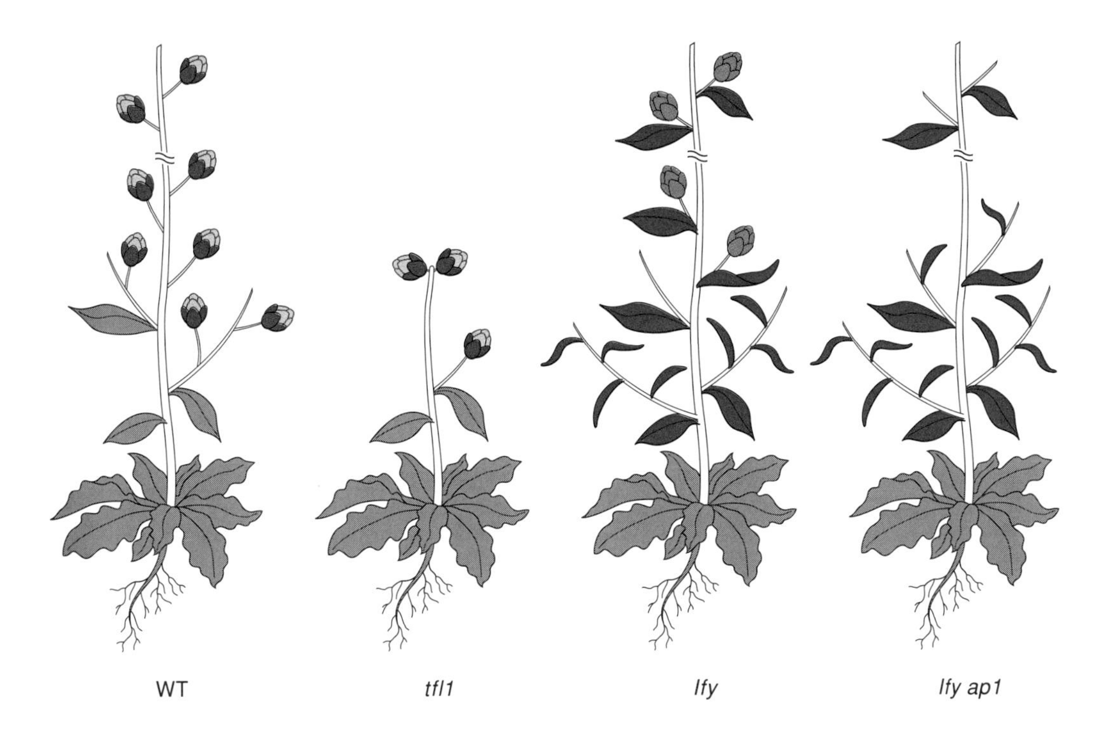

Figure 12.35
Genetic control of floral meristem production by *TERMINAL FLOWER, LEAFY,* and *APETALA1. WT, tfl1, lfy,* and *lfy ap1* refer to schematic representations of wild-type or mutant plants. *tfl1* mutants are characterized by early flowering and the conversion of the inflorescence meristem into a terminal flower. Severe *lfy* mutants are characterized by a partial block or delay in floral meristem production by the inflorescence meristem; flowers are replaced by shoots or flowerlike shoots (shown as shaded flowers). *lfy ap1* double mutants are characterized by a strong block in floral meristem production; flowers are replaced by lateral shoots. (Adapted from J. K. Okamuro et al. 1993. Regulation of *Arabidopsis* Flower Development. *Plant Cell* 5 1183.)

factors can interact in different ways at different promoters to specify novel developmental programs.

Floral meristem specification The genes that specify floral meristem induction can be best understood as affecting the overall floral program (see Figure 12.35). As mentioned , there is a major switch in development from a less determined mode in the vegetative plant to a more directed and deterministic mode in the reproductive plant. To date, two genes have been shown to affect dramatically the determinacy of the floral meristem: *TERMINAL FLOWER* (*TFL*) and *LEAFY* (*LFY*). The *tfl* mutants flower early, and the primary inflorescence branch produces a terminal flower but no lateral branches. There is a dramatic reduction in the number of flowers produced. The *TERMINAL FLOWER* gene may be responsible for maintaining some level of indeterminacy once flowering has been induced, to allow the formation of multiple flowering branches; in its absence, the plant has a short determined growth period, resulting in the production of one or few flowers. The *lfy* mutant plants produce greater than wild-type numbers of floral branches, and the "flowers" that form are green and leafy in appearance, consisting of whorls of sepallike and carpellike organs. The generally opposite phenotypes of *tfl* (more deterministic) and *lfy* (less deterministic) suggest that their gene products may act antagonistically.

Three other genes also affect the determinacy of the floral meristem: *APETALA1* (*AP1*), *APETELA2* (*AP2*), and *CAULIFLOWER* (*CAL*). For example, *ap1* or *ap2* mutations enhance the *lfy* phenotype because, in *lfy ap1* or *lfy ap2* double mutants, the lateral structures produced are more branched, containing mostly leaflike organs. The *cal* mutants have no visible phenotype, except when combined with *ap1*. However, *cal ap1* double mutants overproduce floral meristems, resulting in striking cauliflowerlike flowers. *AP1, LFY, AP2,* and *CAL* have distinct functions that also overlap at several stages of early flower development. Their products may mutually enhance each other's activity. This mutual stimulation might be explained by the fact that all four genes have been proposed to encode transcription factors.

Two mutations affect the determinacy of the flower itself. As previously mentioned, mutations in the *AG* gene produce flowers that do not terminate. Instead, *ag* mutant flowers have nested sets of sepals and petals that reiterate their pattern from the center of the flower. With *ap1* mutants, secondary flowers are produced in the axils of the primary flower's sepals; these secondary flowers can, in turn, produce additional secondary flowers.

Six genes mentioned thus far, *TFL, LFY, AP1, AP2, CAL,* and *AG,* have critical functions in transforming the development of the plant from a nonreproductive to reproductive program. *TFL* and *LFY* likely act earliest, followed by *AP1, AP2,* and *CAL.* The *AG* gene affects development in the center of the floral meristem, which is developmentally the last region to commit to the floral program. Because *ag* flowers do not terminate with the central gynoecium, the *AG* gene may be essential to terminate flower development.

Molecular studies indicate that *AP1* and *CAL* are analogous to *AG, PI,* and *AP3* in encoding proteins with a MADS-box and are thus likely to control transcription of genes important in floral meristem determinacy. *LFY* encodes a protein with proline-rich and acidic domains and may represent a novel type of transcriptional factor. Only *TFL* has not been cloned to date. The importance of *LFY* and *AP1* in specifying the switch to floral development is clearly demonstrated by their mRNA expression patterns. Although *AG, PI,* and *AP3* are expressed only in subregions of the early flower corresponding to their roles in organ development in particular whorls, *AP1* and *LFY* are expressed earlier and throughout the entire floral meristem prior to the development of the organ primordia. *LFY* is also distinguished from *AP1* by its expression at an even earlier developmental time in the central meristem that gives rise to the individual flower meristems.

Revised ABC model for floral development The ABC model presented in Figure 12.34b is attractive in its simplicity and ability to explain some of the initial observations made in genetic studies of the organ identity genes, *AP2, AG, PI,* and *AP3.* However, it needs to be revised to account for recent observations on genetic interactions among the organ and meristem identity genes, as well as gene expression patterns of all genes previously summarized. For example, mutations in both *ap2* and *ap1* affect the phenotypes of the first and second whorl organs; thus, *AP1* expression must also play a role in specification of organ identity in these two whorls. Further, although *AP1* has been proposed to affect early meristem activity, its more branched and less deterministic phenotype is enhanced by combination with *ap2*; therefore, *ap2* likely has an early role in specifying the floral meristem. Also, *AP1* is expressed in whorls 3 and 4 in *ag* mutants, suggesting that *AG* negatively regulates *AP1* expression in these whorls in the wild-type flower. A revised model accounts for these observations in Figure 12.36.

The revised model also proposes two scenarios to account for the fact that *AP2* is expressed in all whorls. Because *AP2* function is required to repress *AG* expression in whorls 1 and 2, either another factor, A1, acts with *AP2* in these whorls to repress *AG,* or there is a repressor, C1, of *AP2* function (but not its ex-

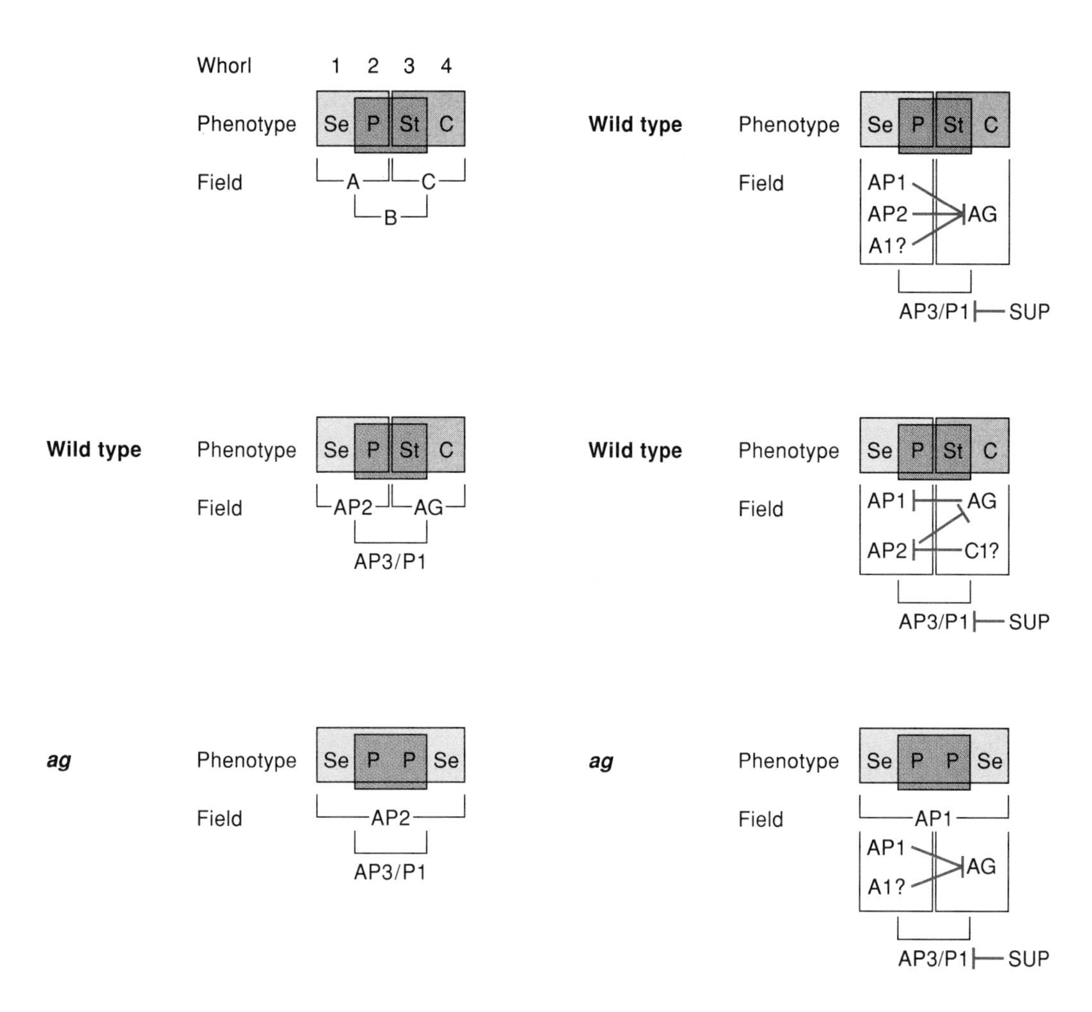

Figure 12.36
A revision of the ABC model for specification of floral organ identity. See text for details. (Adapted from H. Ma. 1994. The Unfolding Drama of Flower Development. *Genes Dev.* 8 745.)

pression) in whorls 3 and 4. Also, because *AP1* is an essential component of the A function (in whorls 1 and 2), it must function together with *AP2* in repressing *AG*. Furthermore, this implies that *AG* is essential for inhibiting *AP1* expression in whorls 3 and 4, an idea that is supported by the ectopic expression patterns of *AP1* in *ag* mutants. The antagonism between *AG* and *AP1* also accounts for the phenotype of plants ectopically expressing *AG*; that is, the phenotype resembles *ap2* because *AG* now inhibits *AP1* expression in the outer whorls.

Another aspect of the revised model concerns regulation of the B genes, *PI* and *AP3*. Analyses of the expression patterns in plants carrying mutations in either gene reveal that this pair of genes autoregulates its own expression. An additional gene, *SUPERMAN* (*SUP*), is required for proper expression of *PI* and *AP3* in the third whorl. *Sup* mutant plants extend the pattern of expression of *PI* and *AP3* to the center of the floral meristem, with the result that stamen production extends into the fourth whorl (Figure 12.37).

The revised model undoubtedly will be changed and updated as new information becomes available. However, it is useful as a framework upon which to hang new ideas and formulate new experiments to test its validity.

Additional types of mutations affecting floral development Nine genes that play a role in flower development have been described. It is not surprising that genes encoding transcriptional regulatory factors are important. However, many more genes important for floral development remain to be uncovered. Some of them can be discovered by assuming that important regulatory genes contain conserved domains. Thus, using *AG* gene sequences as probe, researchers identified six additional *AG*-like genes. Interestingly, five of them are expressed in the flower. The identification of additional functions that might be expected to play roles in flower development can be predicted from the functions already described in animal cell development. For example, membrane or intracellular receptors and protein kinases may signal developmental steps, and other proteins may control cell cycle times and planes of cell division.

Because plant cells do not move and are surrounded by cell walls, there might be novel, plant-specific, regulatory genes that influence cell shape. Thus, factors that determine the structure of the cell wall and extracellular matrix must play unique and critical roles in determining plant cell and ultimately plant tissue and organ architecture. Plant cell shape and morphology may also rely on the architecture of the cytoskeleton. Again due to immobility, control of plant cell destiny must rely largely on position relative to the surrounding cells and tissues. Intercellular communication between plant cells occurs through connections called **plasmodesmata** that must span the intervening cell wall material, another mechanism that appears to have evolved specifically in plants.

Besides the two categories of genes controlling meristem and organ development summarized previously, additional types of floral regulatory genes are hinted at by the novel phenotypes manifested by recently observed mutant alleles (Figure 12.37). Many of these genes were tagged by the T-DNA element of *Agrobacterium* using the method described in Section 12.5d. A few examples will give an idea of their effects. Mutations at *UNUSUAL FLORAL ORGANS (UFO)* suggest the gene has 2 roles, global floral meristem development at early stages, and regulating organ development. Different *ufo* mutations illustrate these different roles. The particular allele shown in Figure 12.37 produces a normal flower that is lacking only petals, reflecting the proposed later role of *UFO* in setting up a developmental boundary in the second whorl. A mutation at the *ETTIN (ETT)* locus is also shown in Figure 12.37. *ett* mutants predominantly affect development of the gynoecium, causing misspecifications of position along both the longitudinal and transverse axes. For example, stigmatic cells normally found at the top of the gynoecium appear along its length, and cells normally forming the central transmitting tract for pollen germination appear on the outside. The final example shown in Figure 12.37 is a mutation at the *TOUSLED* locus. *tsl* mutations have a global effect on floral development, causing a random loss of all floral organ types. *TSL* is expressed in all organs of the vegetative and reproductive plant but most abundantly in young floral buds, suggesting a general role in plant development as well as a critical role in the switch of the apical meristem from vegetative to reproductive development. *TSL* encodes a novel protein kinase whose function may depend on the availability of different relevant targets in different cell types.

Many mutants are affected in the timing of flowering. These mutations are particularly sensitive to light quality and duration, as well as changes in hormone balance. These time to flowering mutants illustrate that, besides floral specific genes, general plant metabolism is critical to establishing and maintaining the floral development program. One particularly interesting, time to flowering mutant, is embryonic flower (*emf*) (not shown). *emf* homozygous plants produce a single flower upon seed germination, bypassing vegetative development. Perhaps the wild type gene (*EMF*) product inhibits reproductive growth; if true, this result would imply the flower is a default developmental program. The determination of the function of these and many yet to be discovered genes will provide answers to numerous questions regarding this intricate and important pathway.

Figure 12.37
Photographs of recently identified mutants affecting floral development in *Arabidopsis.* Upper left shows *superman;* upper right shows petalless allele of *unusual floral organs;* lower left shows *ettin;* lower right shows *tousled.* (Courtesy of Judith Roe and Allen Sessions.)

12.7 Evolutionary Considerations and Floral Development

Although only *Arabidopsis* floral development has been presented here, studies to define genes for flower development in other tractable systems such as *Antirrhinummajus* (snapdragon) have revealed that nearly identical gene products are involved. Thus, comparing *Arabidopsis* to snapdragon, the *AP1* gene resembles *SQUAMOSA* (*SQA*), the *LFY* gene resembles *FLORICULA* (*FLO*), the *AP3* gene resembles *DEFICEINS* (*DEF*), the *PI* gene resembles *GLOBOSA* (*GLO*), and the *AG* gene resembes *PLENA* (*PLE*). Both the mutant phenotypes and the encoded gene products are highly conserved in these two different plant families, although the morphologies of *Arabidopsis* and snapdragon flowers are highly diverged. The analysis of the genetic and molecular hierarchies controlled by these few genes may imply that floral development across all flowering plants utilizes highly conserved regulatory genes. Presumably, the basic functional roles of the gene products, for example, to control transcription of target genes that specify floral morphogenesis, and their basic layout of structural domains are highly conserved. Thus, modifications of their exact expression patterns as well as alterations in their interactive partners may explain how such similar genes can control the enormous diversity in flower size and morphology found in nature.

Acknowledgments

I am indebted to colleagues in my own lab as well as numerous others throughout the community. Although the fact that the first version of this chapter was written in 1986, making this chapter double the trouble to execute in its final form, this aspect did give me a firsthand and sincere appreciation of the amount of progress in plant molecular biology that has been made in this relatively short period of time. The *Agrobacterium* research in my lab is supported by National Science Foundation Grant DCB8915613 and Arabidopsis research by Department of Energy Grant 88ER13882.

Further Reading

12.1

P. Zambryski, J. Tempe, and J. Schell. 1989. Transfer and Function of T-DNA Genes from *Agrobacterium* Ti and Ri Plasmids in Plants. *Cell* 56 193–201.

P. Zambryski. 1992. Chronicles from the *Agrobacterium*-Plant Cell DNA Transfer Story. *Ann. Rev. Plant Physiol. Mol. Biol.* 43 465–490.

12.2

S. E. Stachel, E. Messens, M. Van Montagu, and P. Zambryski. 1985. Identification of the Signal Molecules Produced by Wounded Plant Cells That Activate T-DNA Transfer in *Agrobacterium tumefaciens. Nature* 318 624–629.

S. E. Stachel and E. W. Nester. 1986. The Genetic and Transcriptional Organization of the *vir* Region of the A6 Ti Plasmid of *Agrobacterium tumefaciens. EMBO J.* 5 1445–1454.

S. E. Stachel and P. Zambryski. 1986. *virA* and *virG* Control the Plant Induced Activation of the T-DNA Transfer Process of *Agrobacterium tumefaciens. Cell* 46 325–333.

12.3

S. E. Stachel, B. Timmerman, and P. Zambryski. 1986. Generation of Single Stranded T-DNA Molecules During the Initial Stages of T-DNA Transfer from *Agrobacterium tumefaciens* to Plant Cells. *Nature* 322 706–712.

S. E. Stachel and P. Zambryski. 1986. *Agrobacterium tumefaciens* and the Susceptible Plant Cell: A Novel Adaptation of Extracellular Recognition and DNA Conjugation. *Cell* 47 155–157.

V. Citovsky, M. L. Wong, and P. Zambryski. 1989. Cooperative Interaction of *Agrobacterium* VirE2 Protein with Single Stranded DNA: Implications for the T-DNA Transfer Process. *Proc. Natl. Acad. Sci. USA* 86 1193–1197.

V. Citovsky and E. A. Howard. 1990. The Emerging Structure of the *Agrobacterium* T-DNA Transfer Complex. *Bioessays* 12 103–108.

G. Gheysen, R. Villarroel, and M. Van Montagu. 1991. Illegitimate Recombination in Plants: A Model for T-DNA Integration. *Genes and Devel.* 5 287–297.

V. Citovsky, J. R. Zupan, D. Warnick, and P. Zambryski. 1992. Nuclear Localization of *Agrobacterium* VirE2 Protein in Plant Cells. *Science* 256 1802–1805.

E. A. Howard, J. R. Zupan, V. Citovsky, and P. Zambryski. 1992. The VirD2 Protein of *A. tumefaciens* Contains a C-Terminal Bipartite Nuclear Localization Signal: Implication for Nuclear Uptake of DNA in Plant Cells. *Cell* 68 109–118.

V. Citovsky, and P. Zambryski. 1993. Transport of Nucleic Acids Through Membrane Channels: Snaking Through Small Holes. *Ann. Rev. Microbiol.* 47 167–197.

Y. R. Thorstenson, G. Kuldau, and P. Zambryski. 1993. Subcellular Localization of Seven VirB Proteins of *Agrobacterium tumefaciens*: Implication for the Formation of a T-DNA Transport Structure. *J. Bacteriol.* 175 5233–5241.

B. B. Berger and P. Christie. 1994. Genetic Complementation Analysis of the *Agrobacterium tumefaciens virB* Operon: *virB2* Through *virB11* Are Essential Virulence Genes. *J. Bacteriol.* 176 3646–3660.

M. Lessl and E. Lanka. 1994. Common Mechanisms in Bacterial Conjugation and T-Mediated T-DNA Transfer to Plant Cells. *Cell* 77 321–324.

12.4

A. C. Braun. 1982. A History of the Crown Gall Problem. In G. Kahl and J. Schell (eds.), Molecular Biology of Plant Tumors, pp. 155–208. Academic Press, New York.

R. O. Morris. 1986. Genes Specifying Auxin and Cytokinin Biosynthesis in Phytopathogens. *Ann Rev. Plant Physiol.* 37 509–538.

12.5

P. Zambryski, L. Herrera-Estrella, M. De Block, M. Van Montagu, and J. Schell. 1984. The Use of the Ti Plasmid of *Agrobacterium* To Study the Transfer and Expression of Foreign DNA in Plant Cells: New Vectors and Methods. In J. Setlow and A. Hollaender (eds.), Genetic Engineering, Principles and Methods, vol. 6, pp. 253–278. Plenum Press, New York.

R. Deblaere, A. Reynaerts, H. Hofte, J. P. Hernalsteens, J. Leemans, and M. Van Montagu. 1987. Vectors for Cloning in Plant Cells. In R. Wu and L. Grossman (eds.), Recombinant DNA, part D, Methods in Enzymology, vol. 153, pp. 277–292. Academic Press, New York.

K. Feldmann and M. D. Marks. 1987. *Agrobacterium*-Mediated Transformation of Germinating Seeds of *Arabidopsis thaliana*: A Non Tissue Culture Approach. *Mol. Gen. Genet.* 208 1–9.

R. A. Jefferson, T. A. Kavanagh, and M. W. Bevan. 1987. GUS Fusions: β-Glururonidase as a Sensitive and Versatile Gene Fusion Marker in Higher Plants. *EMBO J.* 6 3901–3907.

G. Gheysen, G. Angenon, and M. Van Montagu. 1992. Transgenic Plants: *Agrobacterium tumefaciens* Mediated Transformation and Its Use for Crop Improvement. In J. A. H. Murray (ed.), Transgenesis, pp. 187–232. Wiley, New York.

12.6

E. M. Meyerowitz. 1987. *Arabidopsis thaliana. Ann. Rev. Genet.* 21 93–111.

J. L. Bowman, D. R. Smyth, and E. M. Meyerowitz. 1991. Genetic Interactions Among Floral Homeotic Genes of *Arabidopsis. Development* 112 1–20.

E. S. Coen and E. M. Meyerowitz. 1991. The War of the Whorls: Genetic Interactions Controlling Flower Development. *Nature* 353 31–37.

R. Chassan and V. Walbot. 1993. Mechanisms of Plant Reproduction: Questions and Approaches. *The Plant Cell* 5 1139–1146.

J. K. Okamuro, B. G. W. den Boer, and K. D. Jofuku. 1993. Regulation of *Arabidopsis* Flower Development. *The Plant Cell* 5 1183–1193.

H. Ma. 1994. The Unfolding Drama of Flower Development: Recent Results from Genetic and Molecular Analyses. *Genes and Devel.* 8 745–756.

J. L. Roe, C. J. Rivin, R. A. Sessions, K. A. Feldmann, and P. Zambryski. 1993. The *TOUSLED* gene in *Arabidopsis thaliana* encodes a protein kinase homologue that is required for leaf and flower development. *Cell* 75 939–950.

R. A. Sessions and P. Zambryski. 1995. *Arabidopsis* gynoecium structure in the Wild Type, and in ettin mutants. *Development* 121 1519–1532.

Index

D

E

H

S

T

W

X

Y

Z